# PRINCIPLES OF ELECTRONIC MATERIALS AND DEVICES

**THIRD EDITION**

**S. O. Kasap**
*University of Saskatchewan*
*Canada*

McGraw-Hill Higher Education

Boston Burr Ridge, IL Dubuque, IA Madison, WI New York San Francisco St. Louis
Bangkok Bogotá Caracas Kuala Lumpur Lisbon London Madrid Mexico City
Milan Montreal New Delhi Santiago Seoul Singapore Sydney Taipei Toronto

The McGraw·Hill Companies

Higher Education

PRINCIPLES OF ELECTRONIC MATERIALS AND DEVICES,
THIRD EDITION

Published by McGraw-Hill, a business unit of The McGraw-Hill Companies, Inc., 1221 Avenue of the Americas, New York, NY 10020. 

This book is printed on acid-free paper.

3 4 5 6 7 8 9 0 DOC/DOC 0 9 8 7

ISBN-13: 978–0–07–295791–4
ISBN-10: 0–07–295791–3

Publisher: *Suzanne Jeans*
Senior Sponsoring Editor: *Michael S. Hackett*
Senior Developmental Editor: *Michelle L. Flomenhoft*
Executive Marketing Manager: *Michael Weitz*
Project Coordinator: *Melissa M. Leick*
Senior Production Supervisor: *Kara Kudronowicz*
Lead Media Project Manager: *Stacy A. Patch*
Media Technology Producer: *Eric A. Weber*
Designer: *Laurie B. Janssen*
Cover Designer: *Lisa Gravunder*
Cover Photos: *Credits on pages 77, 80, 112, 225, 582*
Lead Photo Research Coordinator: *Carrie K. Burger*
Compositor: *The GTS Companies/Los Angeles, CA Campus*
Typeface: *10/12 Times Roman*
Printer: *R. R. Donnelley Crawfordsville, IN*

**Library of Congress Cataloging-in-Publication Data**

Kasap, S. O. (Safa O.)
Principles of electronic materials and devices / S.O. Kasap. —3rd ed.
p. cm.
Includes index.
ISBN 0–07–295791–3 (hard copy : alk. paper)
1. Electric engineering—Materials. 2. Electric apparatus and appliances. I. Title.

TK453.K26 2006
621.382—dc22 2005000842
CIP

www.mhhe.com

# BRIEF CONTENTS

Arnold Johannes Wilhelm Sommerfeld (1868–1951) was responsible for the quantum mechanical free electron theory of metals covered in Chapter 4. Sommerfeld was the Director of Institute of Theoretical Physics, specially established for him, at Munich University.

SOURCE: AIP Emilio Segrè Visual Archives, Physics Today Collection.

Felix Bloch (left) and Lothar Wolfgang Nordheim (right). Nordheim (1899–1988) received his PhD from the University of Göttingen.

SOURCE: AIP Emilio Segrè Visual Archives, Uhlenbeck Collection.

# CONTENTS

GaAs ingots and wafers.
| SOURCE: Courtesy of Sumitomo Electric Industries, Ltd.

# PREFACE

## THIRD EDITION

The textbook represents a first course in electronic materials and devices for undergraduate students. With the additional topics in the accompanying CD, the text can also be used in a graduate introductory course in electronic materials for electrical engineers and material scientists. The third edition is an extensively revised and extended version of the second edition based on reviewer comments, with many new and expanded topics and numerous new worked examples and homework problems. While some of the changes appear to be minor, they have been, nonetheless, quite important in improving the text. For example, the intrinsic concentration $n_i$ in Si is now taken as $1 \times 10^{10}$ cm$^{-3}$, instead of the usual value of $1.45 \times 10^{10}$ cm$^{-3}$ found in many other textbooks; this change makes a significant difference in device-related calculations. A large number of new homework problems have been added, and more solved problems have been provided that put the concepts into applications. Bragg's diffraction law that is mentioned in several chapters is now explained in Appendix A for those readers who are unfamiliar with it.

The third edition is one of the few books on the market that has a broad coverage of electronic materials that today's scientists and engineers need. I believe that the revisions have improved the rigor without sacrificing the original semiquantitative approach that both the students and instructors liked. Some of the new and extended topics are as follows:

| | |
|---|---|
| Chapter 1 | Thermal expansion; atomic diffusion |
| Chapter 2 | Conduction in thin films; interconnects in microelectronics; electromigration |
| Chapter 3 | Planck's and Stefan's laws; atomic magnetic moment; Stern–Gerlach experiment |
| Chapter 4 | Field emission from carbon nanotubes; Grüneisen's thermal expansion |
| Chapter 5 | Piezoresistivity; amorphous semiconductors |
| Chapter 6 | LEDs; solar cells; semiconductor lasers |
| Chapter 7 | Debye relaxation; local field in dielectrics; ionic polarizability; Langevin dipolar polarization; dielectric mixtures |
| Chapter 8 | Pauli spin paramagnetism; band model of ferromagnetism; giant magnetoresistance (GMR); magnetic storage |
| Chapter 9 | Sellmeier and Cauchy dispersion relations; Reststrahlen or lattice absorption; luminescence and white LEDs |
| Appendices | Bragg's diffraction law and X-ray diffraction; luminous flux and brightness of radiation |

## ORGANIZATION AND FEATURES

In preparing the text, I tried to keep the general treatment and various proofs at a semiquantitative level without going into detailed physics. Many of the problems have been set to satisfy engineering accreditation requirements. Some chapters in the text have additional topics to allow a more detailed treatment, usually including quantum mechanics or more mathematics. Cross referencing has been avoided as much as possible without too much repetition and to allow various sections and chapters to be skipped as desired by the reader. The text has been written to be easily usable in one-semester courses by allowing such flexibility.

Some important features are

- The principles are developed with the minimum of mathematics and with the emphasis on physical ideas. Quantum mechanics is part of the course but without its difficult mathematical formalism.
- There are more than 170 worked examples or solved problems, most of which have a practical significance. Students learn by way of examples, however simple, and to that end nearly 250 problems have been provided.
- Even simple concepts have examples to aid learning.
- Most students would like to have clear diagrams to help them visualize the explanations and understand concepts. The text includes over 530 illustrations that have been professionally prepared to reflect the concepts and aid the explanations in the text.
- The end-of-chapter questions and problems are graded so that they start with easy concepts and eventually lead to more sophisticated concepts. Difficult problems are identified with an asterisk (*). Many practical applications with diagrams have been included. There is a regularly updated online extended *Solutions Manual* for all instructors; simply locate the McGraw-Hill website for this textbook.
- There is a glossary, *Defining Terms*, at the end of each chapter that defines some of the concepts and terms used, not only within the text but also in the problems.
- The end of each chapter includes a section *Additional Topics* to further develop important concepts, to introduce interesting applications, or to prove a theorem. These topics are intended for the keen student and can be used as part of the text for a two-semester course.
- The end of each chapter also includes a table *CD Selected Topics and Solved Problems* to enhance not only the subject coverage, but also the range of worked examples and applications. For example, the selected topic *Essential Mechanical Properties* can be used with Chapter 1 to obtain a broader coverage of elementary materials science. The selected topic *Thermoelectric Effects in Semiconductors* can be used with Chapters 5 and 6 to understand the origin of the Seebeck effect in semiconductors, and the reasons behind voltage drift in many semiconductor devices. There are numerous such selected topics and solved problems in the CD.
- The text is supported by McGraw-Hill's textbook website that contains resources, such as solved problems, for both students and instructors. Updates to various articles on the CD will be posted on this website.

## CD-ROM ELECTRONIC MATERIALS AND DEVICES: THIRD EDITION

The book has a CD-ROM that contains all the figures as large *color diagrams* in *PowerPoint* for the instructor, and class-ready notes for the students who do not have to draw the diagrams during the lectures. In addition, there are numerous *Selected Topics* and *Solved Problems* to extend the present coverage. These are listed in each chapter, and also at the end of the text. I strongly urge students to print out the CD's *Illustrated Dictionary of Electronic Materials and Devices: Third Student Edition,* to look up new terms and use the dictionary to refresh various concepts. This is probably the best feature of the CD.

## ACKNOWLEDGMENTS

My gratitude goes to my past and present graduate students and postdoctoral research fellows, who have kept me on my toes and read various sections of this book. I have been fortunate to have a colleague and friend like Charbel Tannous who, as usual, made many sharply critical but helpful comments, especially on Chapter 8. A number of reviewers, at various times, read various portions of the manuscript and provided extensive comments. A number of instructors also wrote to me with their own comments. I incorporated the majority of the suggestions, which I believe made this a better book. No textbook is perfect, and I'm sure that there will be more suggestions for the next edition. I'd like to personally thank them all for their invaluable critiques, some of whom include (alphabetically):

Çetin Aktik    University of Sherbrooke

Emily Allen    San Jose State University

Vasantha Amarakoon    New York State College of Ceramics at Alfred University

David Bahr    Washington State University

David Cahill    University of Illinois

David Cann    Iowa State University

Mark De Guire    Case Western Reserve University

Joel Dubow    University of Utah

Alwyn Eades    Lehigh University

Stacy Gleixner    San Jose State University

Mehmet Günes    Izmir Institute of Technology

Robert Johanson    University of Saskatchewan

Karen Kavanagh    Simon Fraser University

Furrukh Khan    Ohio State University

Michael Kozicki    Arizona State University

Eric Kvam    Purdue University

Hilary Lackritz    Purdue University

Long C. Lee    San Diego State University

Allen Meitzler    University of Michigan, Dearborn

Peter D. Moran    Michigan Technological University

Pierre Pecheur    University of Nancy, France

Aaron Peled Holon    Academic Institute of Technology, Israel

John Sanchez    University of Michigan, Ann Arbor

Christoph Steinbruchel    Rensselaer Polytechnic Institute

Charbel Tannous    Brest University, France

Linda Vanasupa    California Polytechnic State University

Steven M. Yalisove    University of Michigan, Ann Arbor

**Safa Kasap**
http://ElectronicMaterials.Usask.Ca

> "The important thing in science is not so much to obtain new facts as to discover new ways of thinking about them."
>
> *Sir William Lawrence Bragg*

***To Nicolette***

Silicon crystal ingots grown by the Czochralski crystal drawers in the background.
| SOURCE: Courtesy of MEMC, Electronic Materials, Inc.

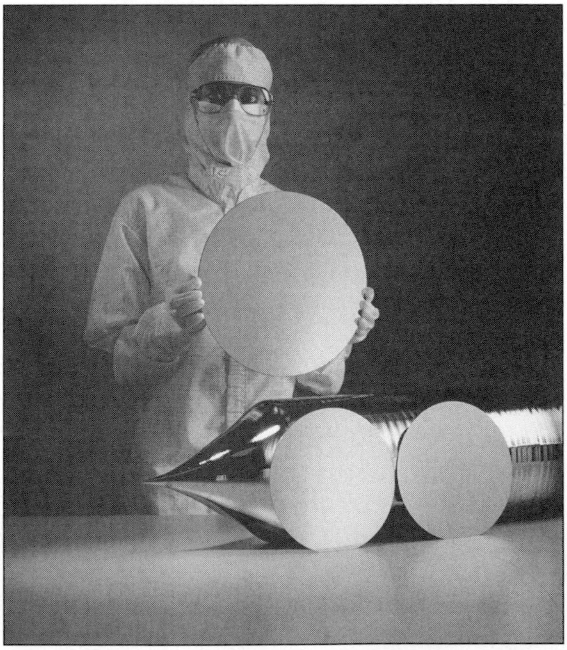

200 mm and 300 mm Si wafers.
| SOURCE: Courtesy of MEMC, Electronic Materials, Inc.

CHAPTER

# 1

# Elementary Materials Science Concepts[1]

Understanding the basic building blocks of matter has been one of the most intriguing endeavors of humankind. Our understanding of interatomic interactions has now reached a point where we can quite comfortably explain the macroscopic properties of matter, based on quantum mechanics and electrostatic interactions between electrons and ionic nuclei in the material. There are many properties of materials that can be explained by a classical treatment of the subject. In this chapter, as well as Chapter 2, we treat the interactions in a material from a classical perspective and introduce a number of elementary concepts. These concepts do not invoke any quantum mechanics, which is a subject of modern physics and is introduced in Chapter 3. Although many useful engineering properties of materials can be treated with hardly any quantum mechanics, it is impossible to develop the science of electronic materials and devices without modern physics.

## 1.1 ATOMIC STRUCTURE AND ATOMIC NUMBER

The model of the atom that we must use to understand the atom's general behavior involves quantum mechanics, a topic we will study in detail in Chapter 3. For the present, we will simply accept the following facts about a simplified, but intuitively satisfactory, atomic model called the **shell model,** based on the **Bohr model** (1913).

The mass of the atom is concentrated at the nucleus, which contains protons and neutrons. Protons are positively charged particles, whereas neutrons are neutral particles, and both have about the same mass. Although there is a Coulombic repulsion between the protons, all the protons and neutrons are held together in the nucleus by the

[1] This chapter may be skipped by readers who have already been exposed to an elementary course in materials science.

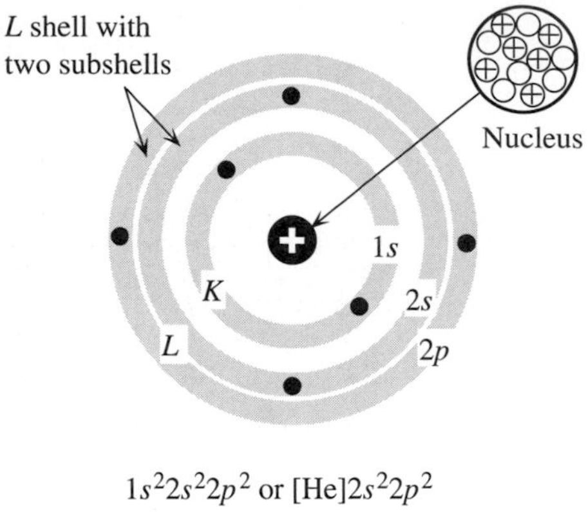


**Figure 1.1** The shell model of the carbon atom, in which the electrons are confined to certain shells and subshells within shells.

**strong force,** which is a powerful, fundamental, natural force between particles. This force has a very short range of influence, typically less than $10^{-15}$ m. When the protons and neutrons are brought together very closely, the strong force overcomes the electrostatic repulsion between the protons and keeps the nucleus intact. The number of protons in the nucleus is the **atomic number** $Z$ of the element.

The electrons are assumed to be orbiting the nucleus at very large distances compared to the size of the nucleus. There are as many orbiting electrons as there are protons in the nucleus. An important assumption in the Bohr model is that only certain orbits with fixed radii are stable around the nucleus. For example, the closest orbit of the electron in the hydrogen atom can only have a radius of 0.053 nm. Since the electron is constantly moving around an orbit with a given radius, over a long time period (perhaps $\sim 10^{-12}$ seconds on the atomic time scale), the electron would appear as a spherical negative-charge cloud around the nucleus and not as a single dot representing a finite particle. We can therefore view the electron as a charge contained within a spherical **shell** of a given radius.

Due to the requirement of stable orbits, the electrons therefore do not randomly occupy the whole region around the nucleus. Instead, they occupy various well-defined spherical regions. They are distributed in various shells and **subshells** within the shells, obeying certain occupation (or seating) rules.[2] The example for the carbon atom is shown in Figure 1.1.

The shells and subshells that define the whereabouts of the electrons are labeled using two sets of integers, $n$ and $\ell$. These integers are called the **principal** and **orbital angular momentum quantum numbers,** respectively. (The meanings of these names are not critical at this point.) The integers $n$ and $\ell$ have the values $n = 1, 2, 3, \ldots$, and $\ell = 0, 1, 2, \ldots, n-1$, and $\ell < n$. For each choice of $n$, there are $n$ values of $\ell$, so higher-order shells contain more subshells. The shells corresponding to $n = 1, 2, 3, 4, \ldots$

[2] In Chapter 3, in which we discuss the quantum mechanical model of the atom, we will see that these shells and subshells are spatial regions around the nucleus where the electrons are most likely to be found.

**Table 1.1** Maximum possible number of electrons in the shells and subshells of an atom

| | | Subshell | | | |
|---|---|---|---|---|---|
| | | $\ell = 0$ | 1 | 2 | 3 |
| $n$ | Shell | $s$ | $p$ | $d$ | $f$ |
| 1 | $K$ | 2 | | | |
| 2 | $L$ | 2 | 6 | | |
| 3 | $M$ | 2 | 6 | 10 | |
| 4 | $N$ | 2 | 6 | 10 | 14 |

are labeled by the capital letters $K, L, M, N, \ldots$, and the subshells denoted by $\ell = 0, 1, 2, 3, \ldots$ are labeled $s, p, d, f \ldots$. The subshell with $\ell = 1$ in the $n = 2$ shell is thus labeled the $2p$ subshell, based on the standard notation $n\ell$.

There is a definite rule to filling up the subshells with electrons; we cannot simply put all the electrons in one subshell. The number of electrons a given subshell can take is fixed by nature to be[3] $2(2\ell + 1)$. For the $s$ subshell ($\ell = 0$), there are two electrons, whereas for the $p$ subshell, there are six electrons, and so on. Table 1.1 summarizes the most number of electrons that can be put into various subshells and shells of an atom. Obviously, the larger the shell, the more electrons it can take, simply because it contains more subshells. The shells and subshells are filled starting with those closest to the nucleus as explained next.

The number of electrons in a subshell is indicated by a superscript on the subshell symbol, so the electronic structure, or configuration, of the carbon atom (atomic number 6) shown in Figure 1.1 becomes $1s^2 2s^2 2p^2$. The $K$ shell has only one subshell, which is full with two electrons. This is the structure of the inert element He. We can therefore write the electronic configuration more simply as $[\text{He}]2s^2 2p^2$. The general rule is put the nearest previous inert element, in this case He, in square brackets and write the subshells thereafter.

The electrons occupying the outer subshells are the farthest away from the nucleus and have the most important role in atomic interactions, as in chemical reactions, because these electrons are the first to interact with outer electrons on neighboring atoms. The outermost electrons are called **valence electrons** and they determine the **valency** of the atom. Figure 1.1 shows that carbon has four valence electrons in the $L$ shell.

When a subshell is full of electrons, it cannot accept any more electrons and it is said to have acquired a stable configuration. This is the case with the inert elements at the right-hand side of the Periodic Table, all of which have completely filled subshells and are rarely involved in chemical reactions. The majority of such elements are gases inasmuch as the atoms do not bond together easily to form a

[3] We will actually show this in Chapter 3 using quantum mechanics.

liquid or solid. They are sometimes used to provide an inert atmosphere instead of air for certain reactive materials.

In an atom such as the Li atom, there are two electrons in the $1s$ subshell and one electron in the $2s$ subshell. The atomic structure of Li is $1s^2 2s^1$. The third electron is in the $2s$ subshell, rather than any other subshell, because this is the arrangement of the electrons that results in the lowest overall energy for the whole atom. It requires energy (work) to take the third electron from the $2s$ to the $2p$ or higher subshells as will be shown in Chapter 3. Normally the zero energy reference corresponds to the electron being at infinity, that is, isolated from the atom. When the electron is inside the atom, its energy is negative, which is due to the attraction of the positive nucleus. An electron that is closer to the nucleus has a lower energy. The electrons nearer the nucleus are more closely bound and have higher binding energies. The $1s^2 2s^1$ configuration of electrons corresponds to the lowest energy structure for Li and, at the same time, obeys the occupation rules for the subshells. If the $2s$ electron is somehow excited to another outer subshell, the energy of the atom increases, and the atom is said to be **excited.**

The smallest energy required to remove a single electron from a neutral atom and thereby create a positive ion (*cation*) and an isolated electron is defined as the **ionization energy** of the atom. The Na atom has only a single valence electron in its outer shell, which is the easiest to remove. The energy required to remove this electron is 5.1 eV, which is the Na atom's ionization energy. The **electron affinity** represents the energy that is needed, or released, when we add an electron to a neutral atom to create a negative ion (*anion*). Notice that the ionization term implies the generation of a positive ion, whereas the electron affinity implies that we have created a negative ion. Certain atoms, notably the halogens (such as F, Cl, Br, I), can actually attract an electron to form a negative ion. Their electron affinities are negative. When we place an electron into a Cl atom, we find that an energy of 3.6 eV is *released*. The $Cl^-$ ion has a lower energy than the Cl atom, which means that it is energetically favorable to form a $Cl^-$ ion by introducing an electron into the Cl atom.

There is a very useful theorem in physics, called the **Virial theorem,** that allows us to relate the average kinetic energy $\overline{KE}$, average potential energy $\overline{PE}$, and average total or overall energy $\overline{E}$ of an electron in an atom, or electrons and nuclei in a molecule, through remarkably simple relationships,[4]

*Virial theorem*

$$\overline{E} = \overline{KE} + \overline{PE} \qquad \text{and} \qquad \overline{KE} = -\tfrac{1}{2}\overline{PE} \qquad [1.1]$$

For example, if we define zero energy for the H atom as the $H^+$ ion and the electron infinitely separated, then the energy of the electron in the H atom is $-13.6$ electron volts (eV). It takes 13.6 eV to ionize the H atom. The average $\overline{PE}$ of the electron, due to its Coulombic interaction with the positive nucleus, is $-27.2$ eV. Its average $\overline{KE}$ turns out to be 13.6 eV. Example 1.1 uses the Virial theorem to calculate the radius of the hydrogen atom, the velocity of the electron, and its frequency of rotation.

[4] While the final result stated in Equation 1.1 is elegantly simple, the actual proof is quite involved and certainly not trivial. As stated here, the Virial theorem applies to a system of charges that interact through electrostatic forces only.

**EXAMPLE 1.1**

**VIRIAL THEOREM AND THE BOHR ATOM** Consider the hydrogen atom in Figure 1.2 in which the electron is in the stable 1*s* orbit with a radius $r_o$. The ionization energy of the hydrogen atom is 13.6 eV.

*a.* It takes 13.6 eV to ionize the hydrogen atom, i.e., to remove the electron to infinity. If the condition when the electron is far removed from the hydrogen nucleus defines the zero reference of energy, then the total energy of the electron within the H atom is −13.6 eV. Calculate the average *PE* and average *KE* of the electron.

*b.* Assume that the electron is in a stable orbit of radius $r_o$ around the positive nucleus. What is the Coulombic *PE* of the electron? Hence, what is the radius $r_o$ of the electron orbit?

*c.* What is the velocity of the electron?

*d.* What is the frequency of rotation (oscillation) of the electron around the nucleus?

**SOLUTION**

*a.* From Equation 1.1 we obtain

$$\overline{E} = \overline{PE} + \overline{KE} = \tfrac{1}{2}\overline{PE}$$

or

$$\overline{PE} = 2\overline{E} = 2 \times (-13.6 \text{ eV}) = -27.2 \text{ eV}$$

The average kinetic energy is

$$\overline{KE} = -\tfrac{1}{2}\overline{PE} = 13.6 \text{ eV}$$

*b.* The Coulombic *PE* of interaction between two charges $Q_1$ and $Q_2$ separated by a distance $r_o$, from elementary electrostatics, is given by

$$PE = \frac{Q_1 Q_2}{4\pi\varepsilon_o r_o} = \frac{(-e)(+e)}{4\pi\varepsilon_o r_o} = -\frac{e^2}{4\pi\varepsilon_o r_o}$$

where we substituted $Q_1 = -e$ (electron's charge), and $Q_2 = +e$ (charge of nucleus). Thus the radius $r_o$ is

$$r_o = -\frac{(1.6 \times 10^{-19}\text{ C})^2}{4\pi(8.85 \times 10^{-12}\text{ F m}^{-1})(-27.2\text{ eV} \times 1.6 \times 10^{-19}\text{ J/eV})}$$

$$= 5.29 \times 10^{-11}\text{ m} \qquad \text{or} \qquad 0.0529 \text{ nm}$$

which is called the **Bohr radius** (also denoted $a_o$).

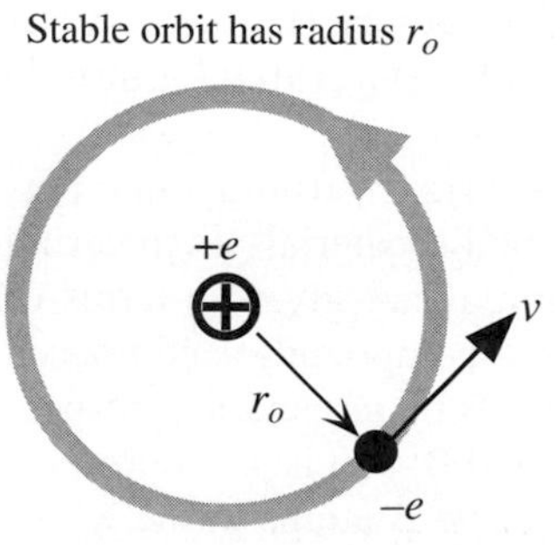


**Figure 1.2** The planetary model of the hydrogen atom in which the negatively charged electron orbits the positively charged nucleus.

*c.* Since $KE = \frac{1}{2}m_e v^2$, the average velocity is

$$v = \sqrt{\frac{KE}{\frac{1}{2}m_e}} = \sqrt{\frac{13.6 \text{ eV} \times 1.6 \times 10^{-19} \text{ J/eV}}{\frac{1}{2}(9.1 \times 10^{-31} \text{ kg})}} = 2.19 \times 10^6 \text{ m s}^{-1}$$

*d.* The period of orbital rotation $T$ is

$$T = \frac{2\pi r_o}{v} = \frac{2\pi(0.0529 \times 10^{-9} \text{ m})}{2.19 \times 10^6 \text{ m s}^{-1}} = 1.52 \times 10^{-16} \text{ seconds}$$

The orbital frequency $\nu = 1/T = 6.59 \times 10^{15}$ $\text{s}^{-1}$ (Hz).

## 1.2 ATOMIC MASS AND MOLE

We had defined the atomic number $Z$ as the number of protons in the nucleus of an atom. The **atomic mass number** $A$ is simply the total number of protons and neutrons in the nucleus. It may be thought that we can use the atomic mass number $A$ of an atom to gauge its atomic mass, but this is done slightly differently to account for the existence of different isotopes of an element; isotopes are atoms of a given element that have the same number of protons but a different number of neutrons in the nucleus. The **atomic mass unit** (amu) $u$ is a convenient atomic mass unit that is equal to $\frac{1}{12}$ of the mass of a neutral carbon atom which has a mass number $A = 12$ (6 protons and 6 neutrons). It has been found that $u = 1.66054 \times 10^{-27}$ kg.

The **atomic mass** or **relative atomic mass** or simply **atomic weight** $M_{\text{at}}$ of an element is the average atomic mass, in atomic mass units, of all the naturally occurring isotopes of the element. Atomic masses are listed in the Periodic Table. **Avogadro's number** $N_A$ is the number of atoms in exactly 12 grams of carbon-12, which is $6.022 \times 10^{23}$ to three decimal places. Since the atomic mass $M_{\text{at}}$ is defined as $\frac{1}{12}$ of the mass of the carbon-12 atom, it is straightforward to show that $N_A$ number of atoms of any substance has a mass equal to the atomic mass $M_{\text{at}}$ in grams.

A **mole** of a substance is that amount of the substance which contains exactly Avogadro's number $N_A$ of atoms or molecules that make up the substance. One mole of a substance has a mass as much as its atomic (molecular) mass in grams. For example, 1 mole of copper contains $6.022 \times 10^{23}$ number of copper atoms and has a mass of 63.55 grams. Thus, an amount of an element which has $6.022 \times 10^{23}$ atoms has a mass in grams equal to the atomic mass. This means we can express the atomic mass as grams per unit mole (g $\text{mol}^{-1}$). The atomic mass of Au is 196.97 amu or g $\text{mol}^{-1}$. Thus, a 10 gram bar of gold has (10 g) / (196.97 g $\text{mol}^{-1}$) or 0.0507 moles.

Frequently we have to convert the composition of a substance from atomic percentage to weight percentage, and vice versa. Compositions in materials engineering generally use weight percentages, whereas chemical formulas are given in terms of atomic composition. Suppose that a substance (an alloy or a compound) is composed of two elements, A and B. Let the *weight fractions* of A and B be $w_A$ and $w_B$, respectively. Let $n_A$ and $n_B$ be the *atomic* or *molar fractions* of A and B; that is, $n_A$ represents the fraction of type A atoms, $n_B$ represents the fraction of type B atoms in the whole substance, and $n_A + n_B = 1$. Suppose that the atomic masses of A and B are $M_A$ and $M_B$. Then $n_A$ and $n_B$ are given by

$$n_A = \frac{w_A/M_A}{w_A/M_A + w_B/M_B} \quad \text{and} \quad n_B = 1 - n_A \qquad [1.2]$$

*Weight to atomic percentage*

where $w_A + w_B = 1$. Equation 1.2 can be readily rearranged to obtain $w_A$ and $w_B$ in terms of $n_A$ and $n_B$.

---

**EXAMPLE 1.2**

**COMPOSITIONS IN ATOMIC AND WEIGHT PERCENTAGES** Consider a Pb–Sn solder that is 38.1 wt.% Pb and 61.9 wt.% Sn (this is the eutectic composition with the lowest melting point). What are the atomic fractions of Pb and Sn in this solder?

**SOLUTION**

For Pb, the weight fraction and atomic mass are respectively $w_A = 0.381$ and $M_A = 207.2$ g mol$^{-1}$ and for Sn, $w_B = 0.619$ and $M_B = 118.71$ g mol$^{-1}$. Thus, Equation 1.2 gives

$$n_A = \frac{w_A/M_A}{w_A/M_A + w_B/M_B} = \frac{(0.381)/(207.2)}{0.381/207.2 + 0.619/118.71}$$

$$= 0.261 \quad \text{or} \quad 26.1 \text{ at.\%}$$

and

$$n_B = \frac{w_B/M_B}{w_A/M_A + w_B/M_B} = \frac{(0.619)/(118.71)}{0.381/207.2 + 0.619/118.71}$$

$$= 0.739 \quad \text{or} \quad 73.9 \text{ at.\%}$$

Thus the alloy is 26.1 at.% Pb and 73.9 at.% Sn which can be written as $Pb_{0.261}Sn_{0.739}$.

---

# 1.3 BONDING AND TYPES OF SOLIDS

## 1.3.1 MOLECULES AND GENERAL BONDING PRINCIPLES

When two atoms are brought together, the valence electrons interact with each other and with the neighbor's positively charged nucleus. The result of this interaction is often the formation of a bond between the two atoms, producing a molecule. The formation of a bond means that the energy of the system of two atoms together must be less than that of the two atoms separated, so that the molecule formation is energetically favorable, that is, more stable. The general principle of molecule formation is illustrated in Figure 1.3a, showing two atoms brought together from infinity. As the two atoms approach each other, the atoms exert attractive and repulsive forces on each other as a result of mutual electrostatic interactions. Initially, the attractive force $F_A$ dominates over the repulsive force $F_R$. The net force $F_N$ is the sum of the two,

$$F_N = F_A + F_R$$

*Net force*

and this is initially attractive, as indicated in Figure 1.3a.

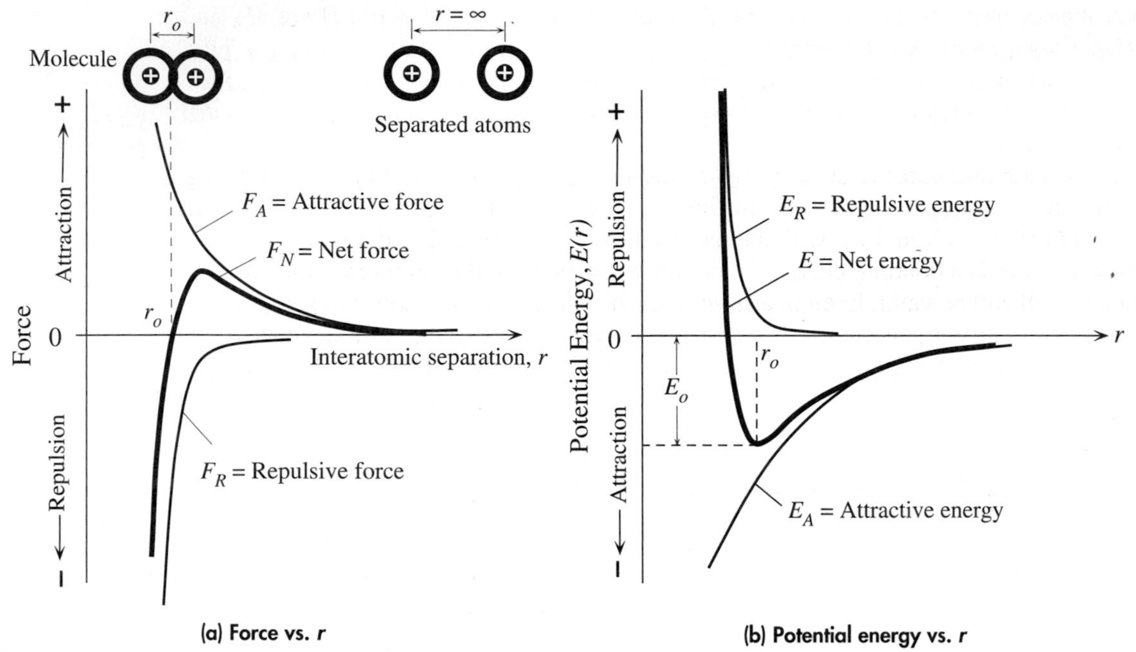


**Figure 1.3** (a) Force versus interatomic separation and (b) potential energy versus interatomic separation.

*Net force and potential energy*

The potential energy $E(r)$ of the two atoms can be found from[5]

$$F_N = \frac{dE}{dr}$$

by integrating the net force $F_N$. Figure 1.3a and b shows the variation of the net force $F_N(r)$ and the overall potential energy $E(r)$ with the interatomic separation $r$ as the two atoms are brought together from infinity. The lowering of energy corresponds to an attractive interaction between the two atoms.

The variations of $F_A$ and $F_R$ with distance are different. Force $F_A$ varies slowly, whereas $F_R$ varies strongly with separation and is strongest when the two atoms are very close. When the atoms are so close that the individual electron shells overlap, there is a very strong electron-to-electron shell repulsion and $F_R$ dominates. An equilibrium will be reached when the attractive force just balances the repulsive force and the net force is zero, or

*Net force in bonding between atoms*

$$F_N = F_A + F_R = 0 \quad [1.3]$$

In this state of equilibrium, the atoms are separated by a certain distance $r_o$, as shown in Figure 1.3. This distance is called the equilibrium separation and is effectively the **bond length.** On the energy diagram, $F_N = 0$ means $dE/dr = 0$, which means that the equilibrium of two atoms corresponds to the potential energy of the

[5] Remember that the change $dE$ in the $PE$ is the work done against the force, $dE = F_N\,dr$.

system acquiring its minimum value. Consequently, the molecule will only be formed if the energy of the two atoms as they approach each other can attain a minimum. This minimum energy also defines the bond energy of the molecule, as depicted in Figure 1.3b. An energy of $E_o$ is required to separate the two atoms, and this represents the **bond energy.**

Although we considered only two atoms, similar arguments also apply to bonding between many atoms, or between millions of atoms as in a typical solid. Although the actual details of $F_A$ and $F_R$ will change from material to material, the general principle that there is a bonding energy $E_o$ per atom and an equilibrium interatomic separation $r_o$ will still be valid. Even in a solid in the presence of many interacting atoms, we can still identify a general potential energy curve $E(r)$ per atom similar to the type shown in Figure 1.3b. We can also use the curve to understand the properties of the solid, such as the thermal expansion coefficient and elastic and bulk moduli.

### 1.3.2 COVALENTLY BONDED SOLIDS: DIAMOND

Two atoms can form a bond with each other by sharing some or all of their valence electrons and thereby reducing the overall potential energy of the combination. The covalent bond results from the sharing of valence electrons to complete the subshells of each atom. Figure 1.4 shows the formation of a covalent bond between two hydrogen atoms as they come together to form the $H_2$ molecule. When the $1s$ subshells overlap, the electrons are shared by both atoms and each atom now has a complete subshell. As illustrated in Figure 1.4, electrons 1 and 2 must now orbit both atoms; they therefore cross the overlap region more frequently, indeed twice as often. Thus, electron sharing,

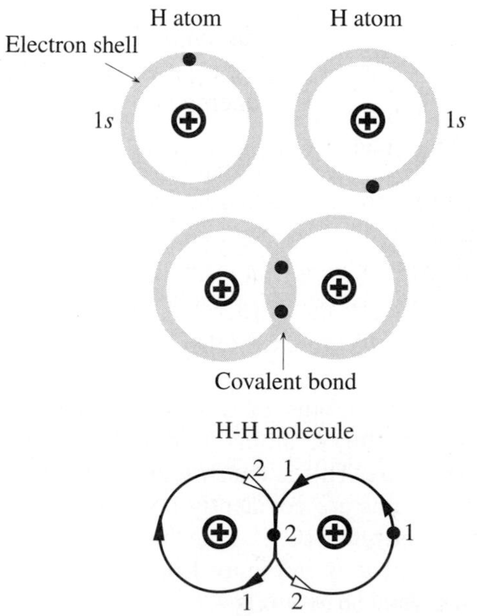


**Figure 1.4** Formation of a covalent bond between two H atoms, leading to the $H_2$ molecule. Electrons spend a majority of their time between the two nuclei, which results in a net attraction between the electrons and the two nuclei, which is the origin of the covalent bond.

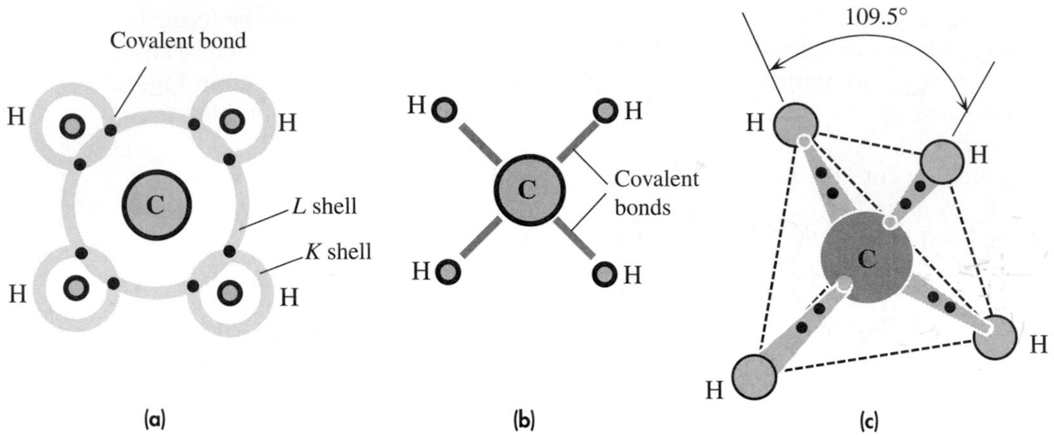


**Figure 1.5**
(a) Covalent bonding in methane, $CH_4$, which involves four hydrogen atoms sharing electrons with one carbon atom.
Each covalent bond has two shared electrons. The four bonds are identical and repel each other.
(b) Schematic sketch of $CH_4$ on paper.
(c) In three dimensions, due to symmetry, the bonds are directed toward the corners of a tetrahedron.

on average, results in a greater concentration of negative charge in the region between the two nuclei, which keeps the two nuclei bonded to each other. Furthermore, by synchronizing their motions, electrons 1 and 2 can avoid crossing the overlap region at the same time. For example, when electron 1 is at the far right (or left), electron 2 is in the overlap region; later, the situation is reversed.

The electronic structure of the carbon atom is $[He]2s^2 2p^2$ with four empty seats in the $2p$ subshell. The $2s$ and $2p$ subshells, however, are quite close. When other atoms are in the vicinity, as a result of interatomic interactions, the two subshells become indistinguishable and we can consider only the shell itself, which is the $L$ shell with a capacity of eight electrons. It is clear that the C atom with four vacancies in the $L$ shell can readily share electrons with four H atoms, as depicted in Figure 1.5, whereby the C atom and each of the H atoms attain complete shells. This is the $CH_4$ molecule, which is the gas methane. The repulsion between the electrons in one bond and the electrons in a neighboring bond causes the bonds to spread as far out from each other as possible, so that in three dimensions, the H atoms occupy the corners of an imaginary tetrahedron and the CH bonds are at an angle of 109.5° to each other, as sketched in Figure 1.5.

The C atom can also share electrons with other C atoms, as shown in Figure 1.6. Each neighboring C atom can share electrons with other C atoms, leading to a three-dimensional network of a covalently bonded structure. This is the structure of the precious diamond crystal, in which all the carbon atoms are covalently bonded to each other, as depicted in the figure. The **coordination number (CN)** is the number of nearest neighbors for a given atom in the solid. As is apparent in Figure 1.6, the coordination number for a carbon atom in the diamond crystal structure is 4.

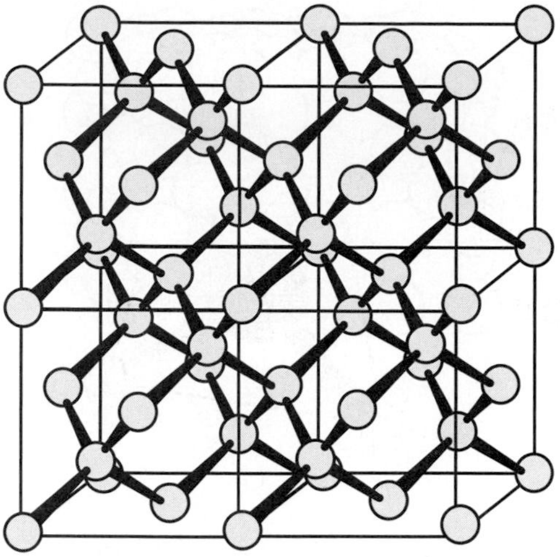

**Figure 1.6** The diamond crystal is a covalently bonded network of carbon atoms. Each carbon atom is bonded covalently to four neighbors, forming a regular three-dimensional pattern of atoms that constitutes the diamond crystal.

Due to the strong Coulombic attraction between the shared electrons and the positive nuclei, the covalent bond energy is usually the highest for all bond types, leading to very high melting temperatures and very hard solids: diamond is one of the hardest known materials.

Covalently bonded solids are also insoluble in nearly all solvents. The directional nature and strength of the covalent bond also make these materials nonductile (or nonmalleable). Under a strong force, they exhibit brittle fracture. Further, since all the valence electrons are locked in the bonds between the atoms, these electrons are not free to drift in the crystal when an electric field is applied. Consequently, the electrical conductivity of such materials is very poor.

### 1.3.3 METALLIC BONDING: COPPER

Metal atoms have only a few valence electrons, which are not very difficult to remove. When many metal atoms are brought together to form a solid, these valence electrons are lost from individual atoms and become collectively shared by all the ions. The valence electrons therefore become **delocalized** and form an **electron gas** or **electron cloud,** permeating the space between the ions, as depicted in Figure 1.7. The attraction between the negative charge of this electron gas and the metal ions more than compensates for the energy initially required to remove the valence electrons from the individual atoms. Thus, the bonding in a metal is essentially due to the attraction between the stationary metal ions and the freely wandering electrons between the ions.

The bond is a **collective sharing** of electrons and is therefore nondirectional. Consequently, the metal ions try to get as close as possible, which leads to **close-packed crystal** structures with high coordination numbers, compared to covalently bonded solids. In the particular example shown in Figure 1.7, $Cu^+$ ions are packed as closely as possible by the gluing effect of the electrons between the ions, forming a crystal

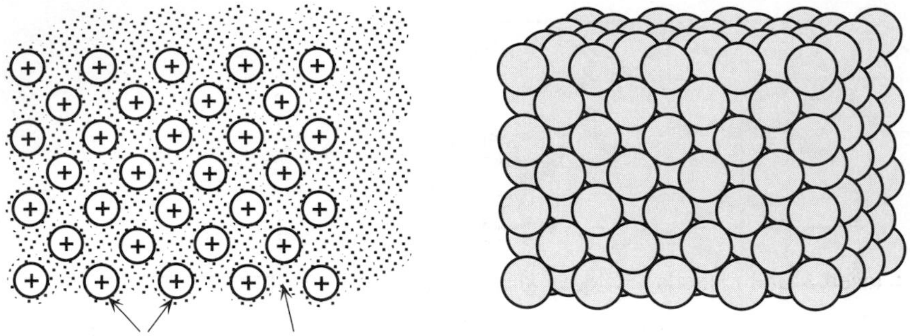


**Figure 1.7** In metallic bonding, the valence electrons from the metal atoms form a "cloud of electrons," which fills the space between the metal ions and "glues" the ions together through Coulombic attraction between the electron gas and the positive metal ions.

structure called the **face-centered cubic (FCC).** The FCC crystal structure, as explained later in Section 1.8, has $Cu^+$ ions at the corners of a cube and a $Cu^+$ at the center of each cube-face. (See Figure 1.31.)

The results of this type of bonding are dramatic. First, the nondirectional nature of the bond means that under an applied force, metal ions are able to move with respect to each other, especially in the presence of certain crystal defects (such as dislocations). Thus, metals tend to be ductile. Most importantly, however, the "free" valence electrons in the electron gas can respond readily to an applied electric field and drift along the force of the field, which is the reason for the high electrical conductivity of metals. Furthermore, if there is a temperature gradient along a metal bar, the free electrons can also contribute to the energy transfer from the hot to the cold regions, since they frequently collide with the metal ions and thereby transfer energy. Metals therefore, typically, also have good thermal conductivities; that is, they easily conduct heat. This is why when you touch your finger to a metal it feels cold because it conducts heat "away" from the finger to the ambient (making the fingertip "feel" cold).

## 1.3.4 IONICALLY BONDED SOLIDS: SALT

Common table salt, NaCl, is a classic example of a solid in which the atoms are held together by ionic bonding. Ionic bonding is frequently found in materials that normally have a metal and a nonmetal as the constituent elements. Sodium (Na) is an alkaline metal with only one valence electron that can easily be removed to form an $Na^+$ ion with complete subshells. The ion $Na^+$ looks like the inert element Ne, but with a positive charge. Chlorine has five electrons in its $3p$ subshell and can readily accept

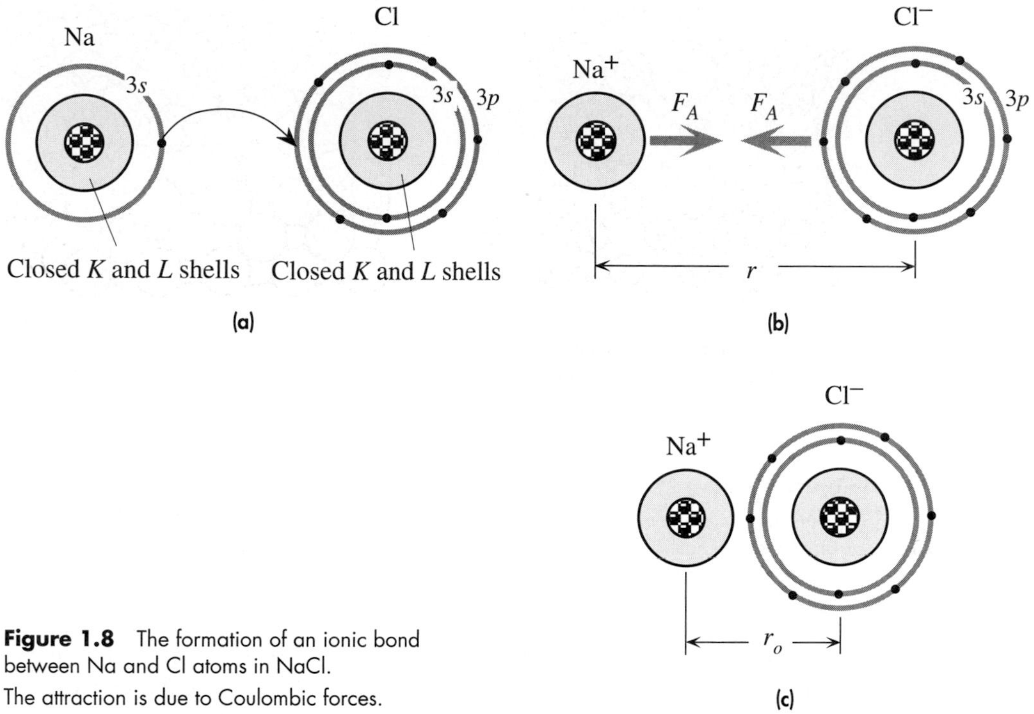


**Figure 1.8** The formation of an ionic bond between Na and Cl atoms in NaCl. The attraction is due to Coulombic forces.

one more electron to close this subshell. By taking the electron given up by the Na atom, the Cl atom becomes negatively charged and looks like the inert element Ar with a net negative charge. Transferring the valence electron of Na to Cl thus results in two oppositely charged ions, $Na^+$ and $Cl^-$, which are called the **cation** and **anion,** respectively, as shown in Figure 1.8. As a result of the Coulombic force, the two ions pull each other until the attractive force is just balanced by the repulsive force between the closed electron shells. Initially, energy is needed to remove the electron from the Na atom; this is the **energy of ionization.** However, this is more than compensated for by the energy of Coulombic attraction between the two resulting oppositely charged ions, and the net effect is a lowering of the potential energy of the $Na^+$ and $Cl^-$ ion pair.

When many Na and Cl atoms are ionized and brought together, the resulting collection of ions is held together by the Coulombic attraction between the $Na^+$ and $Cl^-$ ions. The solid thus consists of $Na^+$ cations and $Cl^-$ anions holding each other through the Coulombic force, as depicted in Figure 1.9. The Coulombic force around a charge is nondirectional; also, it can be attractive or repulsive, depending on the polarity of the interacting ions. There are also repulsive Coulombic forces between the $Na^+$ ions themselves and between the $Cl^-$ ions themselves. For the solid to be stable, each $Na^+$ ion must therefore have $Cl^-$ ions as nearest neighbors and vice versa so that like-ions are not close to each other.

The ions are in equilibrium and the solid is stable when the net potential energy is minimum, or $dE/dr = 0$. Figure 1.10 illustrates the variation of the net potential

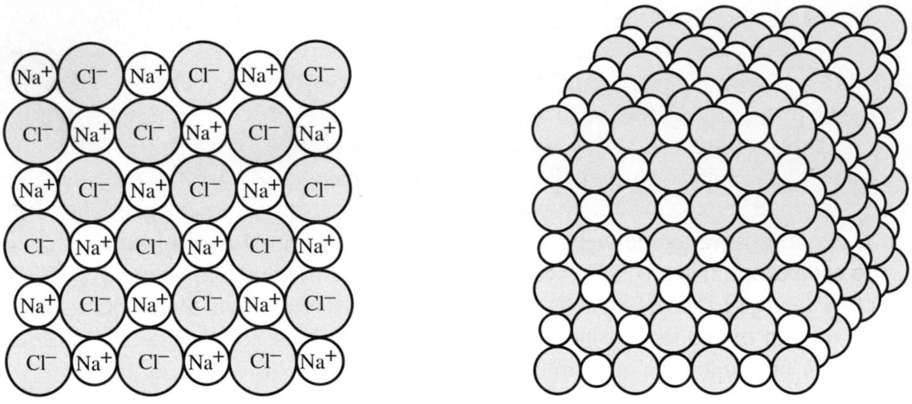


(a) (b)

**Figure 1.9**

(a) A schematic illustration of a cross section from solid NaCl. Solid NaCl is made of $Cl^-$ and $Na^+$ ions arranged alternatingly, so the oppositely charged ions are closest to each other and attract each other. There are also repulsive forces between the like-ions. In equilibrium, the net force acting on any ion is zero.

(b) Solid NaCl.

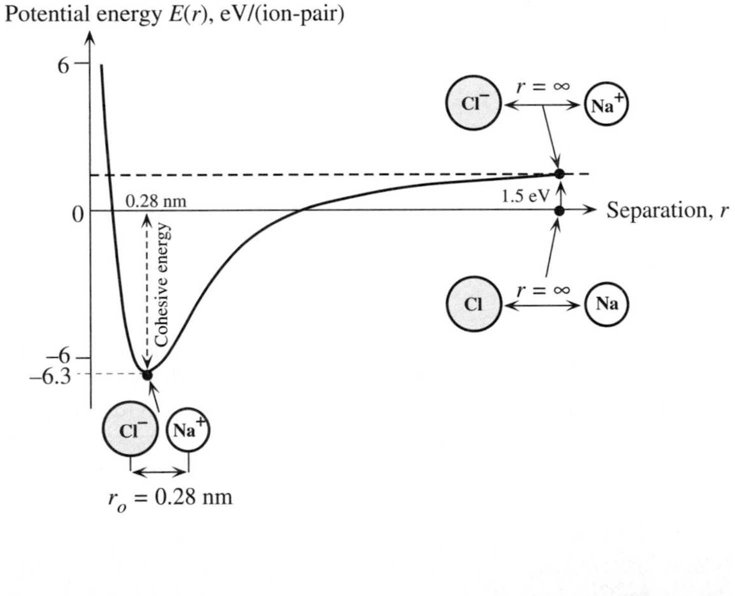


**Figure 1.10** Sketch of the potential energy per ion pair in solid NaCl. Zero energy corresponds to neutral Na and Cl atoms infinitely separated.

energy for a pair of ions as the interatomic distance $r$ is reduced from infinity to less than the equilibrium separation, that is, as the ions are brought together from infinity. Zero energy corresponds to separated Na and Cl atoms. Initially, about 1.5 eV is required to transfer the electron from the Na to Cl atom and thereby form $Na^+$ and $Cl^-$ ions. Then, as the ions come together, the energy is lowered, until it reaches a minimum at about 6.3 eV below the energy of the separated Na and Cl atoms. When $r = 0.28$ nm, the energy is minimum and the ions are in equilibrium. The bonding energy per ion in solid NaCl is thus 6.3/2 or 3.15 eV, as is apparent in Figure 1.10. The energy required to take solid NaCl apart into individual Na and Cl atoms is the **atomic cohesive energy** of the solid, which is 3.15 eV per atom.

In solid NaCl, the $Na^+$ and $Cl^-$ ions are thus arranged with each one having oppositely charged ions as its neighbors, to attain a minimum of potential energy. Since there is a size difference between the ions and since we must avoid like-ions getting close to each other, if we want to achieve a stable structure, each ion can have only six oppositely charged ions as nearest neighbors. Figure 1.9b shows the packing of $Na^+$ and $Cl^-$ ions in the solid. The number of nearest neighbors, that is, the **coordination number,** for both cations and anions in the NaCl crystal is 6.

A number of solids consisting of metal–nonmetal elements follow the NaCl example and have ionic bonding. They are called **ionic crystals** and, by virtue of their ionic bonding characteristics, share many physical properties. For example, LiF, MgO (magnesia), CsCl, and ZnS are all ionic crystals. They are strong, brittle materials with high melting temperatures compared to metals. Most become soluble in polar liquids such as water. Since all the electrons are within the rigidly positioned ions, there are no free or loose electrons to wander around in the crystal as in metals. Therefore, ionic solids are typically electrical insulators. Compared to metals and covalently bonded solids, ionically bonded solids have lower thermal conductivity since ions cannot readily pass vibrational kinetic energy to their neighbors.

---

**EXAMPLE 1.3**

**IONIC BONDING AND LATTICE ENERGY** The potential energy $E$ per $Na^+$–$Cl^-$ pair within the NaCl crystal depends on the interionic separation $r$ as

*Energy per ion pair in an ionic crystal*

$$E(r) = -\frac{e^2 M}{4\pi\varepsilon_o r} + \frac{B}{r^m} \qquad [1.4]$$

where the first term is the *attractive* and the second term is the *repulsive* potential energy, and $M$, $B$, and $m$ are constants explained in the following. If we were to consider the potential energy *PE* of one ion pair in isolation from all others, the first term would be a simple Coulombic interaction energy for the $Na^+$–$Cl^-$ pair, and $M$ would be 1. Within the NaCl crystal, however, a given ion, such as $Na^+$, interacts not only with its nearest six $Cl^-$ neighbors (Figure 1.9b), but also with its twelve second neighbors ($Na^+$), eight third neighbors ($Cl^-$), and so on, so the total or effective *PE* has a factor $M$, called the *Madelung constant,* that takes into account all these different Coulombic interactions. $M$ depends only on the geometrical arrangement of ions in the crystal, and hence on the particular crystal structure; for the FCC crystal structure, $M = 1.748$. The $Na^+$–$Cl^-$ ion pair also have a repulsive *PE* that is due to the repulsion between the electrons in filled electronic subshells of the ions. If the ions are pushed toward each other, the filled subshells begin to overlap, which results in a strong repulsion. The repulsive *PE* decays rapidly with distance and can be modeled by a short-range *PE* of the form $B/r^m$ as in the second term in Equation 1.4 where for $Na^+$–$Cl^-$, $m = 8$ and $B = 6.972 \times 10^{-96}$ J m$^8$. Find the equilibrium separation ($r_o$) of the ions in the crystal and the ionic bonding energy, defined as $-E(r_o)$. Given the *ionization energy* of Na (the energy to remove an electron) is 5.14 eV and the *electron affinity* of Cl (energy released when an electron is added) is 3.61 eV, calculate the *atomic cohesive energy* of the NaCl crystal as joules per mole.

**SOLUTION**

Bonding occurs when the potential energy $E(r)$ is a minimum at $r = r_o$ corresponding to the equilibrium separation between the $Na^+$ and $Cl^-$ ions. We differentiate $E(r)$ and set it to zero at $r = r_o$,

$$\frac{dE(r)}{dr} = \frac{e^2M}{4\pi\varepsilon_o r^2} - \frac{mB}{r^{m+1}} = 0 \qquad \text{at } r = r_o$$

Solving for $r_o$,

*Equilibrium ionic separation*

$$r_o = \left[\frac{4\pi\varepsilon_o Bm}{e^2M}\right]^{1/(m-1)} \tag{1.5}$$

Thus,

$$r_o = \left[\frac{4\pi(8.85 \times 10^{-12}\ \text{F m}^{-1})(6.972 \times 10^{-96}\ \text{J m}^8)(8)}{(1.6 \times 10^{-19}\text{C})^2(1.748)}\right]^{1/(8-1)}$$

$$= 0.281 \times 10^{-9}\text{m} \qquad \text{or} \qquad 0.28\ \text{nm}$$

The minimum energy $E_{\min}$ per ion pair is $E(r_o)$ and can be simplified further by substituting for $B$ in terms of $r_o$:

*Minimum PE at bonding*

$$E_{\min} = -\frac{e^2M}{4\pi\varepsilon_o r_o} + \frac{B}{r_o^m} = -\frac{e^2M}{4\pi\varepsilon_o r_o}\left(1 - \frac{1}{m}\right) \tag{1.6}$$

Thus,

$$E_{\min} = -\frac{(1.6 \times 10^{-19}\ \text{C})^2(1.748)}{4\pi(8.85 \times 10^{-12}\ \text{Fm}^{-1})(2.81 \times 10^{-10}\ \text{m})}\left(1 - \frac{1}{8}\right)$$

$$= -1.256 \times 10^{-18}\ \text{J} \qquad \text{or} \qquad -7.84\ \text{eV}$$

This is the energy with respect to two isolated $Na^+$ and $Cl^-$ ions. We need 7.84 eV to break up a $Na^+$–$Cl^-$ pair into isolated $Na^+$ and $Cl^-$ ions, which represents the *ionic cohesive energy.* Some authors call this ionic cohesive energy simply the **lattice energy.** To take the crystal apart into its neutral atoms, we have to transfer the electron from the $Cl^-$ ion to the $Na^+$ ion to obtain neutral Na and Cl atoms. It takes 3.61 eV to remove the electron from the $Cl^-$ ion, but 5.14 eV is released when it is put into the $Na^+$ ion. Thus, we need 7.84 eV + 3.61 eV but get back 5.14 eV.

Bond energy per Na–Cl pair = 7.84 eV + 3.61 eV − 5.14 eV = 6.31 eV

The *atomic cohesive energy* in terms of joules per mole is

$$E_{\text{cohesive}} = (6.31\ \text{eV})(1.6022 \times 10^{-19}\ \text{J/eV})(6.022 \times 10^{23}\text{mol}^{-1}) = 608\ \text{kJ mol}^{-1}$$

## 1.3.5 SECONDARY BONDING

Covalent, ionic, and metallic bonds between atoms are known as **primary bonds.** It may be thought that there should be no such bonding between the atoms of the inert elements as they have full shells and therefore cannot accept or lose any electrons, nor share any electrons. However, the fact that a solid phase of argon exists at low temperatures, below −189 °C, means that there must be some bonding mechanism between the Ar atoms. The magnitude of this bond cannot be strong because above −189 °C solid argon melts. Although each water molecule $H_2O$ is neutral overall, these molecules nonetheless attract each other to form the liquid state below 100 °C and the solid state below 0 °C. Between all atoms and molecules, there exists a weak type of attraction, the

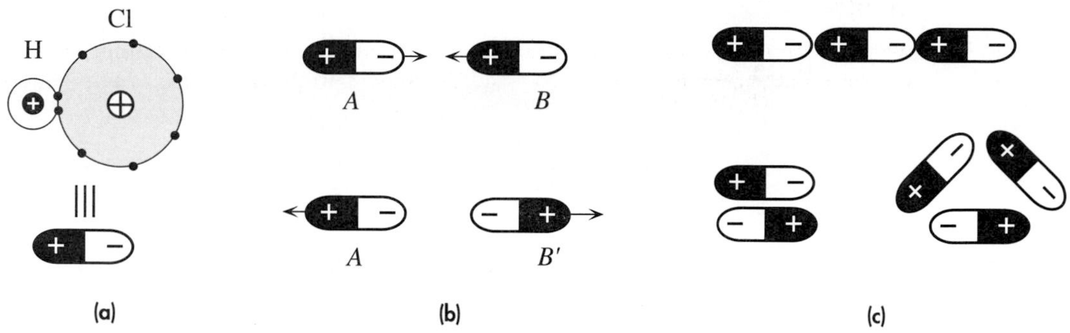


**Figure 1.11**
(a) A permanently polarized molecule is called an electric dipole moment.
(b) Dipoles can attract or repel each other depending on their relative orientations.
(c) Suitably oriented dipoles attract each other to form van der Waals bonds.

so-called van der Waals–London force, which is due to a net electrostatic attraction between the electron distribution of one atom and the positive nucleus of the other.

In many molecules the concentrations of negative and positive charges do not coincide. As apparent in the HCl molecule in Figure 1.11a, the electrons spend most of their time around the Cl nucleus, so the positive nucleus of the H atom is exposed (H has effectively donated its electron to the Cl atom) and the Cl-region acquires more negative charge than the H-region. An **electric dipole moment** occurs whenever a negative and a positive charge of equal magnitude are separated by a distance as in the $H^+$–$Cl^-$ molecule in Figure 1.11a. Such molecules are **polar,** and depending on their relative orientations, they can attract or repel each other as depicted in Figure 1.11b. Two dipoles arranged head to tail attract each other because the closest separation between charges on $A$ and $B$ is between the negative charge on $A$ and the positive charge on $B$, and the *net* result is an electrostatic attraction. The magnitude of the *net force* between two dipoles $A$ and $B$, however, does not depend on their separation $r$ as $1/r^2$ because there are both attractions and repulsions between the charges on $A$ and charges on $B$ and the net force is only *weakly* attractive. (In fact, the net force depends on $1/r^4$.) If the dipoles are arranged head to head or tail to tail, then, by similar arguments, the two dipoles repel each other. Suitably arranged dipoles can attract each other and form **van der Waals bonds** as illustrated in Figure 1.11c. The energies of such dipole arrangements as in Figure 1.11c are less than that of totally isolated dipoles and therefore encourage "bonding." Such bonds are weaker than primary bonds and are called **secondary bonds.**

The water molecule $H_2O$ is also polar and has a net dipole moment as shown in Figure 1.12a. The attractions between the positive charges on one molecule and the negative charges on a neighboring molecule lead to van der Waals bonding between the $H_2O$ molecules in water as illustrated in Figure 1.12b. When the positive charge of a dipole as in $H_2O$ arises from an exposed H nucleus, van der Waals bonding is referred to as **hydrogen bonding.** In ice, the $H_2O$ molecules, again attracted by van der Waals forces, bond to form a regular pattern and hence a crystal structure.

Van der Waals attraction also occurs between neutral atoms and nonpolar molecules. Consider the bonding between Ne atoms at low temperatures. Each has closed (or full) electron shells. The center of mass of the electrons in the closed shells, when

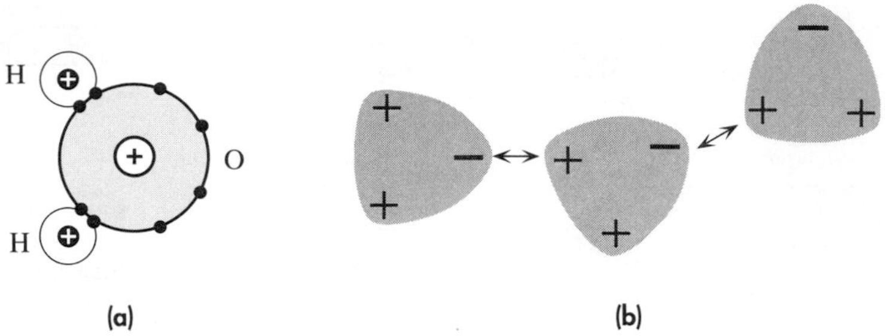


**Figure 1.12** The origin of van der Waals bonding between water molecules.
(a) The $H_2O$ molecule is polar and has a net permanent dipole moment.
(b) Attractions between the various dipole moments in water give rise to van der Waals bonding.

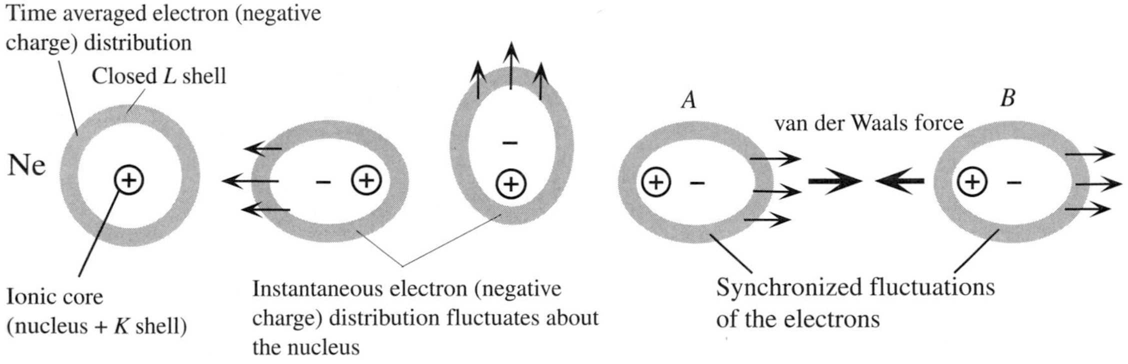


**Figure 1.13** Induced-dipole–induced-dipole interaction and the resulting van der Waals force.

averaged over time, coincides with the location of the positive nucleus. At any one instant, however, the center of mass is displaced from the nucleus due to various motions of the individual electrons around the nucleus as depicted in Figure 1.13. In fact, the center of mass of all the electrons fluctuates with time about the nucleus. Consequently, the electron charge distribution is not static around the nucleus but fluctuates asymmetrically, giving rise to an instantaneous dipole moment.

When two Ne atoms, $A$ and $B$, approach each other, the rapidly fluctuating negative charge distribution on one affects the motion of the negative charge distribution on the other. A lower energy configuration (*i.e.*, attraction) is produced when the fluctuations are synchronized so that the negative charge distribution on $A$ gets closer to the nucleus of the other, $B$, while the negative distribution on $B$ at that instant stays away from that on $A$ as shown in Figure 1.13. The strongest electrostatic interaction arises from the closest charges which are the displaced electrons in $A$ and the nucleus in $B$. This means that there will be a *net* attraction between the two atoms and hence a lowering of the net energy which in turn leads to bonding.

This type of attraction between two atoms is due to induced synchronization of the electronic motions around the nuclei, and we refer to this as *induced-dipole–induced-*

**Table 1.2** Comparison of bond types and typical properties (general trends)

| Bond Type | Typical Solids | Bond Energy (eV/atom) | Melt. Temp. (°C) | Elastic Modulus (GPa) | Density (g cm$^{-3}$) | Typical Properties |
|---|---|---|---|---|---|---|
| **Ionic** | NaCl (rock salt) | 3.2 | 801 | 40 | 2.17 | Generally electrical insulators. May become conductive at high temperatures. |
| | MgO (magnesia) | 10 | 2852 | 250 | 3.58 | High elastic modulus. Hard and brittle but cleavable. Thermal conductivity less than metals. |
| **Metallic** | Cu | 3.1 | 1083 | 120 | 8.96 | Electrical conductor. |
| | Mg | 1.1 | 650 | 44 | 1.74 | Good thermal conduction. High elastic modulus. Generally ductile. Can be shaped. |
| **Covalent** | Si | 4 | 1410 | 190 | 2.33 | Large elastic modulus. Hard and brittle. |
| | C (diamond) | 7.4 | 3550 | 827 | 3.52 | Diamond is the hardest material. Good electrical insulator. Moderate thermal conduction, though diamond has exceptionally high thermal conductivity. |
| **van der Waals: hydrogen bonding** | PVC (polymer) | | 212 | 4 | 1.3 | Low elastic modulus. Some ductility. |
| | $H_2O$ (ice) | 0.52 | 0 | 9.1 | 0.917 | Electrical insulator. Poor thermal conductivity. Large thermal expansion coefficient. |
| **van der Waals: induced dipole** | Crystalline argon | 0.09 | −189 | 8 | 1.8 | Low elastic modulus. Electrical insulator. Poor thermal conductivity. Large thermal expansion coefficient. |

*dipole.* It is weaker than permanent dipole interactions and at least an order of magnitude less than primary bonding. This is the reason why the inert elements Ne and Ar solidify at temperatures below 25 K (−248 °C) and 84 K (−189 °C). Induced dipole–induced dipole interactions also occur between nonpolar molecules such as $H_2$, $I_2$, $CH_4$, etc. Methane gas ($CH_4$) can be solidified at very low temperatures. Solids in which constituent molecules (or atoms) have been bonded by van der Waals forces are known as **molecular solids;** ice, solidified $CO_2$ (dry ice), $O_2$, $H_2$, $CH_4$, and solid inert gases, are typical examples.

Van der Waals bonding is responsible for holding the carbon chains together in polymers. Although the C-to-C bond in a C-chain is due to covalent bonding, the interaction between the C-chains arises from van der Waals forces and the interchain bonding is therefore of secondary nature. These bonds are weak and can be easily stretched or broken. Polymers therefore have substantially lower elastic moduli and melting temperatures than metals and ceramics.

Table 1.2 compares the energies involved in the five types of bonding found in materials. It also lists some important properties of these materials to show the correlation with the bond type and its energy. The greater is the bond energy, for example, the higher is the melting temperature. Similarly, strong bond energies lead to greater elastic moduli and smaller thermal expansion coefficients. Metals generally have the greatest electrical conductivity since only this type of bonding allows a very large number of free charges (conduction electrons) to wander in the solid and thereby contribute to electrical conduction. Electrical conduction in other types of solid may involve the motion of ions or charged defects from one fixed location to another.

### 1.3.6 MIXED BONDING

In many solids, the bonding between atoms is generally not just of one type; rather, it is a mixture of bond types. We know that bonding in the silicon crystal is totally covalent, because the shared electrons in the bonds are equally attracted by the neighboring positive ion cores and are therefore equally shared. When there is a covalent-type bond between two different atoms, the electrons become unequally shared, because the two neighboring ion cores are different and hence have different electron-attracting abilities. The bond is no longer purely covalent; it has some ionic character, because the shared electrons spend more time close to one of the ion cores. Covalent bonds that have an ionic character, due to an unequal sharing of electrons, are generally called **polar bonds.** Many technologically important semiconductor materials, such as III–V compounds (*e.g.*, GaAs), have polar covalent bonds. In GaAs, for example, the electrons in a covalent bond spend slightly more time around the $As^{5+}$ ion core than the $Ga^{+3}$ ion core.

**Electronegativity** is a relative measure of the ability of an atom to attract the electrons in a bond it forms with another atom. The *Pauling scale of electronegativity* assigns an electronegativity value $X$, a pure number, to various elements, the highest being 4 for F, and the lowest values being for the alkali metal atoms, for which $X$ are less than 1. In this scheme, the difference $X_A - X_B$ in the electronegativities of two atoms $A$ and $B$ is a measure of the polar or ionic character of the bond $A$–$B$ between $A$ and $B$. There is obviously no electronegativity difference for a covalent bond. While it is possible to calculate the fractional ionicity of a single bond between two different atoms using $X_A - X_B$, inside the crystal the overall ionic character can be substantially higher because ions can interact with distant ions further away than just the nearest neighbors, as we have found out in NaCl. Many technologically important semiconductor materials, such as III-V compounds (*e.g.*, GaAs) have polar covalent bonds. In GaAs, for example, the bond in the crystal is about 30 percent ionic in character ($X_{\text{As}} - X_{\text{Ga}} = 2.18 - 1.81 = 0.37$). In the ZnSe crystal, an important II-VI semiconductor, the bond is 63 percent ionic ($X_{\text{Se}} - X_{\text{Zn}} = 2.55 - 1.65 = 0.85$).[6]

**Ceramic** materials are compounds that generally contain metallic and nonmetallic elements. They are well known for their brittle mechanical properties, hardness, high

[6] Chemists use "Ionicity $= 1 - \exp[0.24(X_A - X_B)]$" to calculate the *ionicity* of the bond between $A$ and $B$. While this is undoubtedly useful in identifying the trend, it substantially underestimates the actual ionicity of bonding within the crystal itself. (It is left as an exercise to show this fact from the above $X_A$ and $X_B$ values.) The quoted ionicity percentages are from J. C. Phillips' book *Bonds and Bands in Semiconductors*, New York: Academic Press, 1973. By the way, the units of $X$ are sometimes quoted as Pauling units, after its originator Linus Pauling.

melting temperatures, and electrical insulating properties. The type of bonding in a ceramic material may be covalent, ionic, or a mixture of the two, in which the bond between the atoms involves some electron sharing and, to some extent, the partial formation of cations and anions; the shared electrons spend more time with one type of atom, which then becomes a partial anion while the other becomes a partial cation. Silicon nitride ($Si_3N_4$), magnesia (MgO), and alumina ($Al_2O_3$) are all ceramics, but they have different types of bonding: $Si_3N_4$ has covalent, MgO has ionic, and $Al_2O_3$ has a mixture of ionic and covalent bonding. All three are brittle, have high melting temperatures, and are electrical insulators.

**EXAMPLE 1.4**

**ENERGY OF SECONDARY BONDING** Consider the van der Waals bonding in solid argon. The potential energy as a function of interatomic separation can generally be modeled by the **Lennard–Jones 6–12 potential energy curve,** that is,

$$E(r) = -Ar^{-6} + Br^{-12}$$

where $A$ and $B$ are constants. Given that $A = 8.0 \times 10^{-77}$ J m$^6$ and $B = 1.12 \times 10^{-133}$ J m$^{12}$, calculate the bond length and bond energy (in eV) for solid argon.

**SOLUTION**

Bonding occurs when the potential energy is at a minimum. We therefore differentiate the Lennard–Jones potential $E(r)$ and set it to zero at $r = r_o$, the interatomic equilibrium separation or

$$\frac{dE}{dr} = 6Ar^{-7} - 12Br^{-13} = 0 \qquad \text{at } r = r_o$$

that is,

$$r_o^6 = \frac{2B}{A}$$

or

$$r_o = \left[\frac{2B}{A}\right]^{1/6}$$

Substituting $A = 8.0 \times 10^{-77}$ and $B = 1.12 \times 10^{-133}$ and solving for $r_o$, we find

$$r_o = 3.75 \times 10^{-10} \text{ m} \qquad \text{or} \qquad 0.375 \text{ nm}$$

When $r = r_o = 3.75 \times 10^{-10}$ m, the potential energy is at a minimum and corresponds to $-E_{\text{bond}}$, so

$$E_{\text{bond}} = \left|-Ar_o^{-6} + Br_o^{-12}\right| = \left|-\frac{8.0 \times 10^{-77}}{(3.75 \times 10^{-10})^6} + \frac{1.12 \times 10^{-133}}{(3.75 \times 10^{-10})^{12}}\right|$$

that is,

$$E_{\text{bond}} = 1.43 \times 10^{-20} \text{ J} \qquad \text{or} \qquad 0.089 \text{ eV}$$

Notice how small this energy is compared to primary bonding.

**EXAMPLE 1.5** **ELASTIC MODULUS** The elastic modulus, or Young's modulus $Y$, of a solid indicates its ability to deform elastically. The greater is the elastic modulus, the more effort is required for the same amount of elastic deformation given a constant sample geometry. When a solid is subjected to tensile forces $F$ acting on two opposite faces, as in Figure 1.14a, it experiences a **stress** $\sigma$ defined as the *force per unit area* $F/A$, where $A$ is the area on which $F$ acts. If the original length of the specimen is $L_o$, then the applied stress $\sigma$ stretches the solid by an amount $\delta L$. The **strain** $\varepsilon$ is the fractional increase in the length of the solid $\delta L/L_o$. As long as the applied force displaces the atoms in the solid by a small amount from their equilibrium positions, the deformation is elastic and recoverable when the forces are removed. The applied stress $\sigma$ and the resulting elastic strain $\varepsilon$ are related by the **elastic modulus** $Y$ by

*Definition of elastic modulus*

$$\sigma = Y\varepsilon \qquad [1.7]$$

The applied stress causes two neighboring atoms along the direction of force to be further separated. Their displacement $\delta r(= r - r_o)$ results in a net attractive force $\delta F_N$ between two neighboring atoms as indicated in Figure 1.14b (which is the same as Figure 1.3a) where $F_N$ is the net interatomic force. $\delta F_N$ attempts to restore the separation to equilibrium. This force $\delta F_N$, however, is balanced by a portion of the applied force acting on these atoms as in Figure 1.14a. If we were to proportion the area $A$ in Figure 1.14a among all the atoms on this area, each atom would have an area roughly $r_o^2$. (If there are $N$ atoms on $A$, $Nr_o^2 = A$.) The force $\delta F_N$ is therefore $\sigma r_o^2$. The strain $\varepsilon$ is $\delta r/r_o$. Thus, Equation 1.7 gives

$$\frac{\delta F_N}{r_o^2} = \sigma = Y\frac{\delta r}{r_o}$$

Clearly, $Y$ depends on the gradient of the $F_N$ versus $r$ curve at $r_o$, or the curvature of the minimum of $E$ versus $r$ at $r_o$,

*Elastic modulus and bonding*

$$Y = \frac{1}{r_o}\left[\frac{dF_N}{dr}\right]_{r=r_o} = \frac{1}{r_o}\left[\frac{d^2E}{dr^2}\right]_{r=r_o} \qquad [1.8]$$

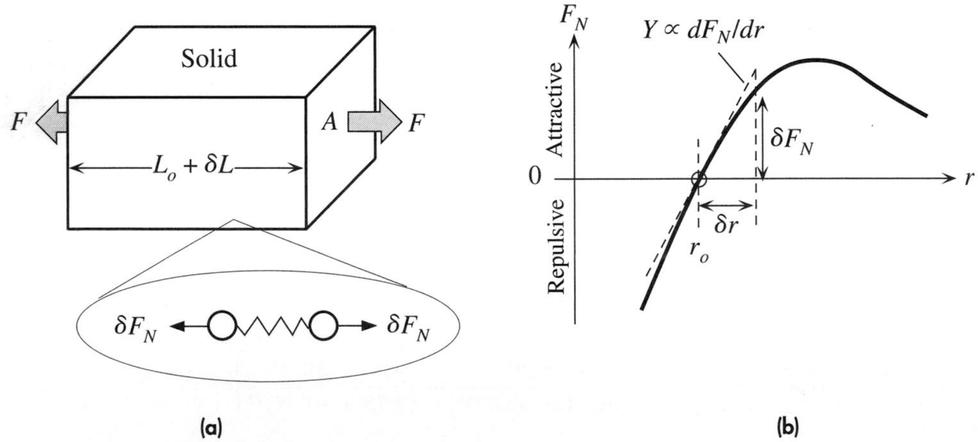


**Figure 1.14**

(a) Applied forces $F$ stretch the solid elastically from $L_o$ to $L_o + \delta L$. The force is divided among chains of atoms that make the solid. Each chain carries a force $\delta F_N$.

(b) In equilibrium, the applied force is balanced by the net force $\delta F_N$ between the atoms as a result of their increased separation.

The bonding energy $E_{\text{bond}}$ is the minimum of $E$ versus $r$ at $r_o$ (Figure 1.3b) and can be related to the curvature of $E$ versus $r$ which leads to[7]

*Elastic modulus and bond energy*

$$Y \approx f \frac{E_{\text{bond}}}{r_o^3} \tag{1.9}$$

where $f$ is a numerical factor (constant) that depends on the crystal structure and the type of bond (of the order of unity). The well-known Hooke's law for a spring expresses the magnitude of the net force $\delta F_N$ in terms of the displacement $\delta r$ by $\delta F_N = \beta|\delta r|$ where $\beta$ is the spring constant. Thus $Y = \beta/r_o$.

Solids with higher bond energies therefore tend to have higher elastic moduli as apparent in Table 1.2. Secondary bonding has both a smaller $E_{\text{bond}}$ and a larger $r_o$ than primary bonding and $Y$ is much smaller. For NaCl, from Figure 1.10, $E_{\text{bond}} = 6.3$ eV, $r_o = 0.28$ nm, and $Y$ is of the order of $\sim$50 GPa using Equation 1.9 and $f \approx 1$; and not far out from the value in Table 1.2.

## 1.4 KINETIC MOLECULAR THEORY

### 1.4.1 MEAN KINETIC ENERGY AND TEMPERATURE

The kinetic molecular theory of matter is a classical theory that can explain such seemingly diverse topics as the pressure of a gas, the heat capacity of metals, the average speed of electrons in a semiconductor, and electrical noise in resistors, among many interesting phenomena. We start with the kinetic molecular theory of gases, which considers a collection of gas molecules in a container and applies the classical equations of motion from elementary mechanics to these molecules. We assume that the collisions between the gas molecules and the walls of the container result in the gas pressure $P$. Newton's second law, $dp/dt$ = force, where $p = mv$ is the momentum, is used to relate the pressure $P$ (force per unit area) to the mean square velocity $\overline{v^2}$, and the number of molecules per unit volume $N/V$. The result can be stated simply as

*Kinetic molecular theory for gases*

$$PV = \tfrac{1}{3}Nm\overline{v^2} \tag{1.10}$$

where $m$ is the mass of the gas molecule. Comparing this theoretical derivation with the experimental observation that

$$PV = \left(\frac{N}{N_A}\right)RT$$

where $N_A$ is **Avogadro's number** and $R$ is the gas constant, we can relate the mean kinetic energy of the molecules to the temperature. Our objective is to derive Equation 1.10; to do so, we make the following assumptions:

1. The molecules are in constant random motion. Since we are considering a large number of molecules, perhaps $10^{20}$ m$^{-3}$, there are as many molecules traveling in one direction as in any other direction, so the center of mass of the gas is at rest.

[7] The mathematics and a more rigorous description may be found in the textbook's CD.

2. The range of intermolecular forces is short compared to the average separation of the gas molecules. Consequently,
   *a.* Intermolecular forces are negligible, except during a collision.
   *b.* The volume of the gas molecules (all together) is negligible compared to the volume occupied by the gas (that is, the container).
3. The duration of a collision is negligible compared to the time spent in free motion between collisions.
4. Each molecule moves with uniform velocity between collisions, and the acceleration due to the gravitational force or other external forces is neglected.
5. On average, the collisions of the molecules with one another and with the walls of the container are perfectly elastic. Collisions between molecules result in exchanges of kinetic energy.
6. Newtonian mechanics can be applied to describe the motion of the molecules.

We consider a collection of $N$ gas molecules within a cubic container of side $a$. We focus our attention on one of the molecules moving toward one of the walls. The velocity can be decomposed into two components, one directly toward the wall $v_x$, and the other parallel to the wall $v_y$, as shown in Figure 1.15. Clearly, the collision of the molecule, which is perfectly elastic, does not change the component $v_y$ along the wall, but reverses the perpendicular component $v_x$. The change in the momentum of the molecule following its collision with the wall is

$$\Delta p = 2mv_x$$

where $m$ is the mass of the molecule. Following its collision, the molecule travels back across the box, collides with the opposite face $B$, and returns to hit face $A$ again. The time interval $\Delta t$ is the time to traverse twice the length of the box, or $\Delta t = 2a/v_x$. Thus, every $\Delta t$ seconds, the molecule collides with face $A$ and changes its momentum by $2mv_x$. To find the force $F$ exerted by this molecule on face $A$, we need the rate of

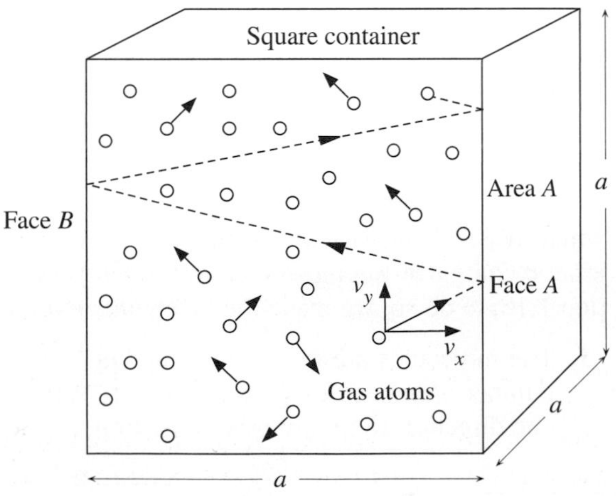


**Figure 1.15** The gas molecules in the container are in random motion.

change of momentum, or

$$F = \frac{\Delta p}{\Delta t} = \frac{2mv_x}{(2a/v_x)} = \frac{mv_x^2}{a}$$

The total pressure $P$ exerted by $N$ molecules on face $A$, of area $a^2$, is due to the sum of all individual forces $F$, or

$$P = \frac{\text{Total force}}{a^2} = \frac{mv_{x1}^2 + mv_{x2}^2 + \cdots + mv_{xN}^2}{a^3}$$
$$= \frac{m}{a^3}(v_{x1}^2 + v_{x2}^2 + \cdots + v_{xN}^2)$$

that is,

$$P = \frac{mN\overline{v_x^2}}{V}$$

where $\overline{v_x^2}$ is the average of $v_x^2$ for all the molecules and is called the *mean square velocity,* and $V$ is the volume $a^3$.

Since the molecules are in random motion and collide randomly with each other, thereby exchanging kinetic energy, the mean square velocity in the $x$ direction is the same as those in the $y$ and $z$ directions, or

$$\overline{v_x^2} = \overline{v_y^2} = \overline{v_z^2}$$

For any molecule, the velocity $v$ is given by

$$\overline{v^2} = \overline{v_x^2} + \overline{v_y^2} + \overline{v_z^2} = 3\overline{v_x^2}$$

The relationship between the pressure $P$ and the mean square velocity of the molecules is therefore

*Gas pressure in the kinetic theory*

$$P = \frac{Nm\overline{v^2}}{3V} = \frac{1}{3}\rho\overline{v^2} \qquad [1.11]$$

where $\rho$ is the density of the gas, or $Nm/V$. By using elementary mechanical concepts, we have now related the pressure exerted by the gas to the number of molecules per unit volume and to the mean square of the molecular velocity.

Equation 1.11 can be written explicitly to show the dependence of $PV$ on the mean kinetic energy of the molecules. Rearranging Equation 1.11, we obtain

$$PV = \frac{2}{3}N\left(\frac{1}{2}m\overline{v^2}\right)$$

where $\frac{1}{2}m\overline{v^2}$ is the average kinetic energy $\overline{KE}$ per molecule. If we consider one mole of gas, then $N$ is simply $N_A$, Avogadro's number.

Experiments on gases lead to the empirical gas equation

$$PV = \left(\frac{N}{N_A}\right)RT$$

where $R$ is the universal gas constant. Comparing this equation with the kinetic theory equation shows that the average kinetic energy per molecule must be proportional to the temperature.

*Mean kinetic energy per atom*

$$\overline{KE} = \frac{1}{2}m\overline{v^2} = \frac{3}{2}kT \qquad [1.12]$$

where $k = R/N_A$ is called the **Boltzmann constant.** Thus, the mean square velocity is proportional to the absolute temperature. This is a major conclusion from the kinetic theory, and we will use it frequently.

When heat is added to a gas, its internal energy and, by virtue of Equation 1.12, its temperature both increase. The rise in the internal energy per unit temperature is called the **heat capacity.** If we consider 1 mole of gas, then the heat capacity is called the **molar heat capacity** $C_m$. The total internal energy $U$ of 1 mole of monatomic gas (i.e., a gas with only one atom in each molecule) is

$$U = N_A\left(\frac{1}{2}m\overline{v^2}\right) = \frac{3}{2}N_AkT$$

so, from the definition of $C_m$, at constant volume, we have

*Molar heat capacity at constant volume*

$$C_m = \frac{dU}{dT} = \frac{3}{2}N_Ak = \frac{3}{2}R \qquad [1.13]$$

Thus, the heat capacity per mole of a monatomic gas at constant volume is simply $\frac{3}{2}R$. By comparison, we will see later that the heat capacity of metals is twice this amount. The reason for considering constant volume is that the heat added to the system then increases the internal energy without doing mechanical work by expanding the volume.

There is a useful theorem called **Maxwell's principle of equipartition of energy,** which assigns an average of $\frac{1}{2}kT$ to each independent energy term in the expression for the total energy of a system. A monatomic molecule can only have translational kinetic energy, which is the sum of kinetic energies in the $x$, $y$, and $z$ directions. The total energy is therefore

$$E = \frac{1}{2}mv_x^2 + \frac{1}{2}mv_y^2 + \frac{1}{2}mv_z^2$$

Each of these terms represents an independent way in which the molecule can be made to absorb energy. Each method by which a system can absorb energy is called a **degree of freedom.** A monatomic molecule has only three degrees of freedom. According to Maxwell's principle, for a collection of molecules in thermal equilibrium, each degree of freedom has an average energy of $\frac{1}{2}kT$, so the average kinetic energy of the monatomic molecule is $3(\frac{1}{2}kT)$.

A rigid diatomic molecule (such as an $O_2$ molecule) can acquire energy as translational motion and rotational motion, as depicted in Figure 1.16. Assuming the moment of inertia $I_x$ about the molecular axis (along $x$) is negligible, the energy of the molecule is

$$E = \frac{1}{2}mv_x^2 + \frac{1}{2}mv_y^2 + \frac{1}{2}mv_z^2 + \frac{1}{2}I_y\omega_y^2 + \frac{1}{2}I_z\omega_z^2$$

where $I_y$ and $I_z$ are moments of inertia about the $y$ and $z$ axes and $\omega_y$ and $\omega_z$ are angular velocities about the $y$ and $z$ axes (Figure 1.16).

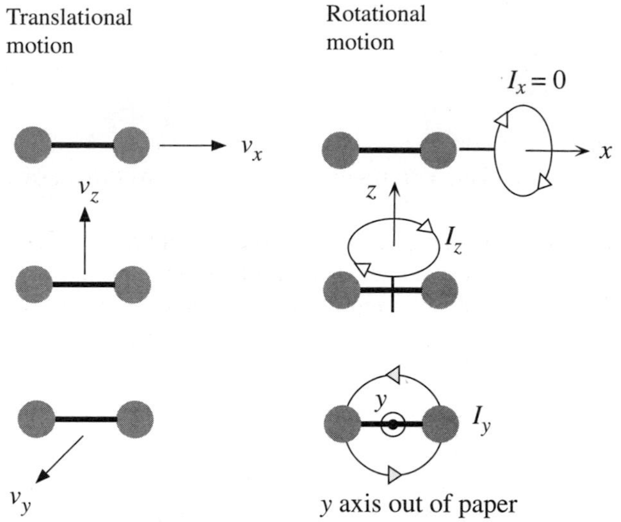


**Figure 1.16** Possible translational and rotational motions of a diatomic molecule. Vibrational motions are neglected.

This molecule has five degrees of freedom and hence an average energy of $5(\frac{1}{2}kT)$. Its molar heat capacity is therefore $\frac{5}{2}R$.

The atoms in the molecule will also vibrate by stretching or bending the bond, which behaves like a "spring." At room temperature, the addition of heat only results in the translational and rotational motions becoming more energetic (excited), whereas the molecular vibrations remain the same and therefore do not absorb energy. This occurs because the vibrational energy of the molecule can only change in finite steps; in other words, the vibrational energy is quantized. For many molecules, the energy required to excite a more energetic vibration is much more than the energy possessed by the majority of molecules. Therefore, energy exchanges via molecular collisions cannot readily excite more energetic vibrations; consequently, the contribution of molecular vibrations to the heat capacity is negligible.

In a solid, the atoms are bonded to each other and can only move by vibrating about their equilibrium positions. In the simplest view, a typical atom in a solid is joined to its neighbors by "springs" that represent the bonds, as depicted in Figure 1.17. If we consider a given atom, its potential energy as a function of displacement from the equilibrium position is such that if it is displaced slightly in any direction, it will experience a restoring force proportional to the displacement. Thus, this atom can acquire energy by vibrations in three directions. The energy associated with the $x$ direction, for example, is the kinetic energy of vibration plus the potential energy of the "spring," or $\frac{1}{2}mv_x^2 + \frac{1}{2}K_x x^2$, where $v_x$ is the velocity, $x$ is the extension of the spring, and $K_x$ is the spring constant, all along the $x$ direction. Clearly, there are similar energy terms in the $y$ and $z$ directions, so there are six energy terms in the total energy equation:

$$E = \frac{1}{2}mv_x^2 + \frac{1}{2}mv_y^2 + \frac{1}{2}mv_z^2 + \frac{1}{2}K_x x^2 + \frac{1}{2}K_y y^2 + \frac{1}{2}K_z z^2$$

We know that for simple harmonic motion, the average *KE* is equal to the average *PE*. Since, by virtue of the equipartition of energy principle, each average *KE* term has

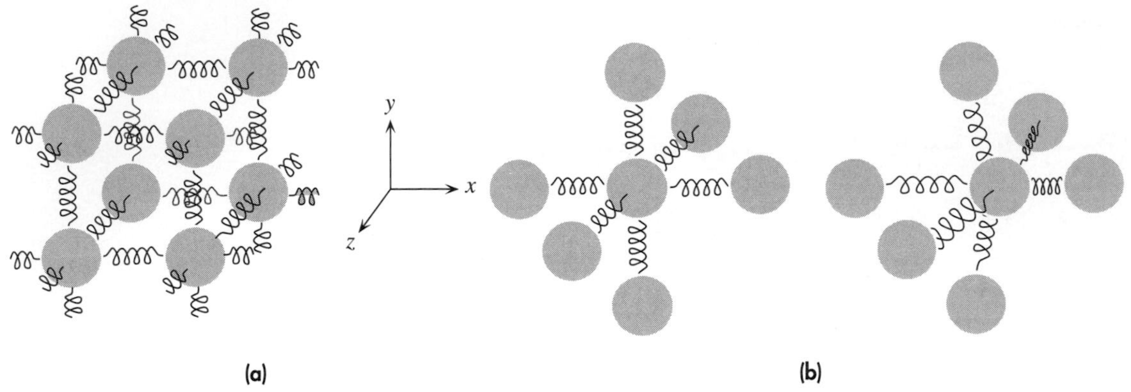


**Figure 1.17**
(a) The ball-and-spring model of solids, in which the springs represent the interatomic bonds. Each ball (atom) is linked to its neighbors by springs. Atomic vibrations in a solid involve three dimensions.
(b) An atom vibrating about its equilibrium position. The atom stretches and compresses its springs to its neighbors and has both kinetic and potential energy.

an energy of $\frac{1}{2}kT$, the average total energy per atom is $6(\frac{1}{2}kT)$. The internal energy $U$ per mole is

$$U = N_A 6\left(\frac{1}{2}kT\right) = 3RT$$

The molar heat capacity then becomes

*Dulong–Petit rule*

$$C_m = \frac{dU}{dT} = 3R = 25 \text{ J K}^{-1} \text{ mol}^{-1}$$

This is the **Dulong–Petit rule.**

The kinetic molecular theory of matter is one of the successes of classical physics, with a beautiful simplicity in its equations and predictions. Its failures, however, are numerous. For example, the theory fails to predict that, at low temperatures, the heat capacity increases as $T^3$ and that the resistivity of a metal increases linearly with the absolute temperature. We will explain the origins of these phenomena in Chapter 4.

**EXAMPLE 1.6**

**SPEED OF SOUND IN AIR** Calculate the root mean square (rms) velocity of nitrogen molecules in atmospheric air at 27 °C. Also calculate the root mean square velocity in one direction ($v_{\text{rms},x}$). Compare the speed of propagation of sound waves in air, 350 m s$^{-1}$, with $v_{\text{rms},x}$ and explain the difference.

**SOLUTION**

From the kinetic theory

$$\frac{1}{2}mv_{\text{rms}}^2 = \frac{3}{2}kT$$

so that

$$v_{\text{rms}} = \sqrt{\frac{3kT}{m}}$$

where $m$ is the mass of the nitrogen molecule $N_2$. The atomic mass of nitrogen is $M_{\text{at}} = 14\text{g mol}^{-1}$, so that in kilograms

$$m = \frac{2M_{\text{at}}(10^{-3})}{N_A}$$

Thus

$$v_{\text{rms}} = \left[\frac{3kN_AT}{2M_{\text{at}}(10^{-3})}\right]^{1/2} = \left[\frac{3RT}{2M_{\text{at}}(10^{-3})}\right]^{1/2}$$

$$= \left[\frac{3(8.314\text{ J mol}^{-1}\text{ K}^{-1})(300\text{ K})}{2(14\times10^{-3}\text{ kg mol}^{-1})}\right]^{1/2} = 517\text{ m s}^{-1}$$

Consider an rms velocity in one direction. Then

$$v_{\text{rms},x} = \sqrt{\overline{v_x^2}} = \sqrt{\frac{1}{3}\overline{v^2}} = \frac{1}{\sqrt{3}}v_{\text{rms}} = 298\text{ m s}^{-1}$$

which is slightly less than the velocity of sound in air (350 m $\text{s}^{-1}$). The difference is due to the fact that the propagation of a sound wave involves rapid compressions and rarefactions of air, and the result is that the propagation is not isothermal. Note that accounting for oxygen in air lowers $v_{\text{rms},x}$. (Why?)

---

**SPECIFIC HEAT CAPACITY** Estimate the heat capacity of copper per unit gram, given that its atomic mass is 63.6. **EXAMPLE 1.7**

**SOLUTION**

From the Dulong–Petit rule, $C_m = 3R$ for $N_A$ atoms. But $N_A$ atoms have a mass of $M_{\text{at}}$ grams, so the heat capacity per gram, the specific heat capacity $c_s$, is

$$c_s = \frac{3R}{M_{\text{at}}} = \frac{25\text{ J mol}^{-1}\text{ K}^{-1}}{63.6\text{ g mol}^{-1}}$$

$$\approx 0.39\text{ J g}^{-1}\text{ K}^{-1} \qquad \text{(The experimental value is 0.38 J g}^{-1}\text{ K}^{-1}.)$$

---

## 1.4.2 THERMAL EXPANSION

Nearly all materials expand as the temperature increases. This phenomenon is due to the asymmetric nature of the interatomic forces and the increase in the amplitude of atomic vibrations with temperature as expected from the kinetic molecular theory.

The potential energy curve $U(r)$ for two atoms separated by a distance $r$ is shown in Figure 1.18. In equilibrium the *PE* is a minimum at $U_{\text{min}} = -U_o$ and the bonding energy is simply $U_o$. The atoms are separated by the equilibrium separation $r_o$. However,

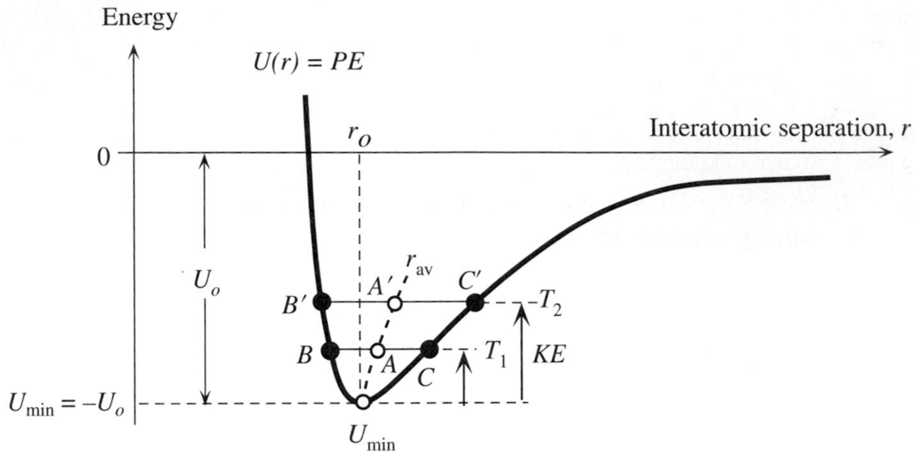


**Figure 1.18** The potential energy $PE$ curve has a minimum when the atoms in the solid attain the interatomic separation at $r = r_o$.
Because of thermal energy, the atoms will be vibrating and will have vibrational kinetic energy. At $T = T_1$, the atoms will be vibrating in such a way that the bond will be stretched and compressed by an amount corresponding to the $KE$ of the atoms. A pair of atoms will be vibrating between $B$ and $C$. Their average separation will be at $A$ and greater than $r_o$.

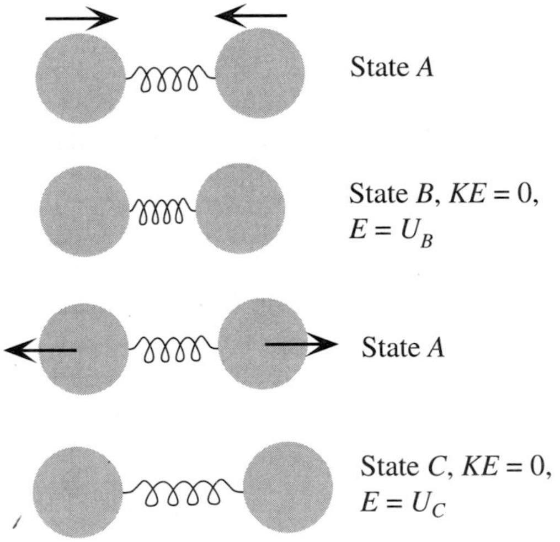


**Figure 1.19** Vibrations of atoms in the solid. We consider for simplicity a pair of atoms. Total energy is $E = PE + KE$, and this is constant for a pair of vibrating atoms executing simple harmonic motion. At $B$ and $C$, $KE$ is zero (atoms are stationary and about to reverse direction of oscillation) and $PE$ is maximum.

according to the kinetic molecular theory, atoms are vibrating about their equilibrium positions with a mean vibrational kinetic energy that increases with the temperature as $\frac{3}{2}kT$. At any instant the total energy $E$ of the pair of atoms is $U + KE$, and this is constant inasmuch as no external forces are being applied. The atoms will be vibrating about their equilibrium positions, stretching and compressing the bond, as depicted in Figure 1.19. At positions $B$ and $C$, $U$ is maximum and the $KE$ is zero; the atoms are stationary and about to reverse their direction of oscillation. Thus at $B$ and $C$ the total energy $E = U_B = U_C$ and the *PE* has increased from its minimum value $U_{\text{min}}$ by an amount equal to *KE*. The line *BC* corresponds to the total energy *E*. The atoms are confined to vibrate between *B* and *C*, executing simple harmonic motion and hence maintaining $E = U + KE =$ constant.

But the *PE* curve $U(r)$ is *asymmetric*. $U(r)$ is broader in the $r > r_o$ region. Thus, the atoms spend more time in the $r > r_o$ region, that is, more time stretching the bond than compressing the bond (with respect to the equilibrium length $r_o$). The average separation corresponds to point *A*,

$$r_{\text{av}} = \tfrac{1}{2}(r_B + r_C)$$

which is clearly greater than $r_o$. As the temperature increases, *KE* increases, the total energy *E* increases, and the atoms vibrate between wider extremes of the $U(r)$ curve, between $B'$ and $C'$. The new average separation at $A'$ is now greater than that at $A$: $r_{A'} > r_A$. Thus as the temperature increases, the average separation between the atoms also increases, which leads to the phenomenon of **thermal expansion.** If the *PE* curve were symmetric, then there would be no thermal expansion as the atoms would spend equal times in the $r < r_o$ and $r > r_o$ regions.

When the temperature increases by a small amount $\delta T$, the energy per atom increases by $C_{\text{atom}}\,\delta T$ where $C_{\text{atom}}$ is the heat capacity per atom (molar heat capacity divided by $N_A$). If $C_{\text{atom}}\,\delta T$ is large, then the line $B'C'$ in Figure 1.18 will be higher up on the energy curve and the average separation $A'$ will therefore be larger. Thus, the increase $\delta r_{\text{av}}$ in the average separation is proportional to $\delta T$. If the total length $L_o$ is made up of $N$ atoms, $L_o = N r_{\text{av}}$, then the change $\delta L$ in $L_o$ is proportional to $N\,\delta T$ or $L_o\,\delta T$. The proportionality constant is the **thermal coefficient of linear expansion,** or simply, **thermal expansion coefficient** $\lambda$, which is defined as the fractional change in length per unit temperature,

*Definition of thermal expansion coefficient*

$$\lambda = \frac{1}{L_o} \cdot \frac{\delta L}{\delta T} \qquad [1.14]$$

If $L_o$ is the original length at temperature $T_o$, then the length $L$ at temperature $T$, from Equation 1.14, is

*Thermal expansion*

$$L = L_o[1 + \lambda(T - T_o)] \qquad [1.15]$$

We note that $\lambda$ is a material property that depends on the nature of the bond. The variation of $r_{\text{av}}$ with $T$ in Figure 1.18 depends on the shape of the *PE* curve $U(r)$. Typically, $\lambda$ is larger for metallic bonding than for covalent bonding.

We can use a mathematical procedure (known as a Taylor expansion) to describe the $U(r)$ versus $r$ curve in terms of its minimum value $U_{\text{min}}$, plus correction terms that depend on the powers of the *displacement* $(r - r_o)$ from $r_o$,

*Potential energy of an atom*

$$U(r) = U_{\text{min}} + a_2(r - r_o)^2 + a_3(r - r_o)^3 + \cdots \qquad [1.16]$$

where $a_2$ and $a_3$ are coefficients that are related to the second and third derivatives of $U$ at $r_o$. The term $a_1(r - r_o)$ is missing because $dU/dr = 0$ at $r = r_o$ where $U = U_{\text{min}}$. The $U_{\text{min}}$ and $a_2(r - r_o)^2$ terms in Equation 1.16 give a parabola about $U_{\text{min}}$ which is a symmetric curve around $r_o$ and therefore does not lead to thermal expansion. The average location at any energy on a symmetric curve at $r_o$ is always at $r_o$. It is the $a_3$ term that gives the expansion because it leads to asymmetry. Thus, $\lambda$ depends on the amount of asymmetry, that is, $a_3/a_2$. The asymmetric *PE* curve in Figure 1.18 which has a finite cubic $a_3$ term as in Equation 1.16 does not lead to a perfect simple harmonic (sinusoidal) vibration about $r_o$ because the restoring force is not proportional to the displacement alone. Such oscillations are **unharmonic,** and the *PE* curve is said to possess an **unharmonicity** (terms such as $a_3$). Thermal expansion is an **unharmonic effect.**

The thermal expansion coefficient normally depends on the temperature, $\lambda = \lambda(T)$, and typically increases with increasing temperature, except at the lowest temperatures. We can always expand $\lambda(T)$ about some useful temperature such as $T_o$ to obtain a polynomial series in temperature terms up to the most significant term, usually the $T^2$ containing term. Thus, Equation 1.14 becomes

*Thermal expansion coefficient and temperature*

$$\frac{dL}{L_o\,dT} = \lambda(T) = A + B(T - T_o) + C(T - T_o)^2 + \cdots \qquad [1.17]$$

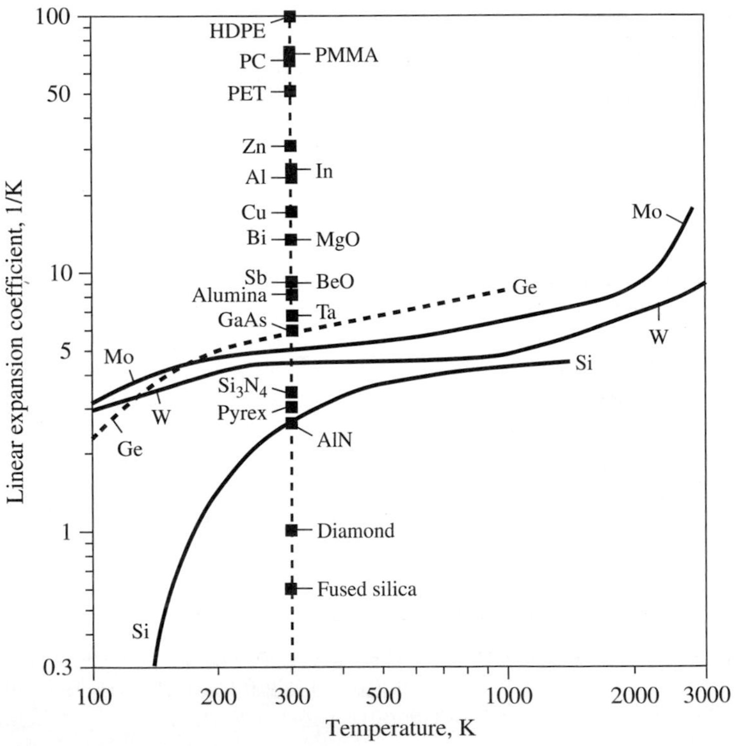


**Figure 1.20** Dependence of the linear thermal expansion coefficient $\lambda$ ($K^{-1}$) on temperature $T$ (K) on a log-log plot.

HDPE, high-density polyethylene; PMMA, polymethylmethacrylate (acrylic); PC, polycarbonate; PET, polyethylene terephthalate (polyester); fused silica, $SiO_2$; alumina, $Al_2O_3$.

SOURCE: Data extracted from various sources including G. A. Slack and S. F. Bartram, *J. Appl. Phys.*, **46**, 89, 1975.

where $A$, $B$, and $C$ are temperature-independent constants, and the expansion is about $T_o$. To find the total fractional change in the length $\Delta L/L_o$ from $T_o$ to $T$, we have to integrate $\lambda(T)$ with respect to temperature from $T_o$ to $T$. We can still employ Equation 1.15 provided that we use a properly defined mean value for the expansion coefficient from $T_o$ to $T$,

*Thermal expansion*

$$L = L_o[1 + \overline{\lambda}(T - T_o)] \quad [1.18]$$

*Mean thermal expansion coefficient*

where $$\overline{\lambda} = \frac{1}{(T - T_o)}\int_{T_o}^{T} \lambda(T)\, dT \quad [1.19]$$

Figure 1.20 shows the temperature dependence of $\lambda$ for various materials. In very general terms, except at very low (typically below 100 K) and very high temperatures (near the melting temperature), for most metals $\lambda$ does not depend strongly on the temperature; many engineers take $\lambda$ for a metal to be approximately temperature independent. There is a simple relationship between the linear expansion coefficient and the heat capacity of a material, which is discussed in Chapter 4.

**EXAMPLE 1.8**

**VOLUME EXPANSION COEFFICIENT** Suppose that the volume of a solid body at temperature $T_o$ is $V_o$. The volume expansion coefficient $\alpha_V$ of a solid body characterizes the change in its volume from $V_o$ to $V$ due to a temperature change from $T_o$ to $T$ by

*Volume expansion*

$$V = V_o[1 + \alpha_V(T - T_o)] \quad [1.20]$$

Show that $\alpha_V$ is given by

*Volume expansion coefficient*

$$\alpha_V = 3\lambda \quad [1.21]$$

Aluminum has a density of 2.70 g cm$^{-3}$ at 25 °C. Its thermal expansion coefficient is $24 \times 10^{-6}$ °C$^{-1}$. Calculate the density of Al at 350 °C.

**SOLUTION**

Consider the solid body in the form of a rectangular parallelepiped with sides $x_o$, $y_o$, and $z_o$. Then at $T_o$,

$$V_o = x_o y_o z_o$$

and at $T$,

$$V = [x_o(1 + \lambda\,\Delta T)][y_o(1 + \lambda\,\Delta T)][z_o(1 + \lambda\,\Delta T)]$$
$$= x_o y_o z_o(1 + \lambda\,\Delta T)^3$$

that is

$$V = x_o y_o z_o[1 + 3\lambda\,\Delta T + 3\lambda^2(\Delta T)^2 + \lambda^3(\Delta T)^3]$$

We can now substitute for $V$ from Equation 1.20, use $V_o = x_o y_o z_o$, and neglect the $\lambda^2(\Delta T)^2$ and $\lambda^3(\Delta T)^3$ terms compared with the $\lambda\,\Delta T$ term ($\lambda \ll 1$) to obtain,

$$V = V_o[1 + 3\lambda(T - T_o)] = V_o[1 + \alpha_V(T - T_o)]$$

Since density $\rho$ is mass/volume, volume expansion leads to a density reduction. Thus,

$$\rho = \frac{\rho_o}{1 + \alpha_V(T - T_o)} \approx \rho_o[1 - \alpha_V(T - T_o)]$$

For Al, the density at 350 °C is

$$\rho = 2.70[1 - 3(24 \times 10^{-6})(350 - 25)] = 2.637 \text{ g cm}^{-3}$$

**EXAMPLE 1.9**

*Thermal expansion coefficient of Si*

**EXPANSION OF Si** The expansion coefficient of silicon over the temperature range 120–1500 K is given by Okada and Tokumaru (1984) as

$$\lambda = 3.725 \times 10^{-6}[1 - e^{-5.88\times10^{-3}(T-124)}] + 5.548 \times 10^{-10}T \quad \textbf{[1.22]}$$

where $\lambda$ is in $K^{-1}$ (or $°C^{-1}$) and $T$ is in kelvins. At a room temperature of 20 °C, the above gives $\lambda = 2.51 \times 10^{-6}\ K^{-1}$. Calculate the fractional change $\Delta L/L_o$ in the length $L_o$ of the Si crystal from 20 to 320 °C, by (*a*) assuming a constant $\lambda$ equal to the room temperature value and (*b*) assuming the above temperature dependence. Calculate the mean $\overline{\lambda}$ for this temperature range.

**SOLUTION**

Assuming a constant we have

$$\frac{\Delta L}{L_o} = \lambda(T - T_0) = (2.51 \times 10^{-6}\ °C^{-1})(320 - 20) = 0.753 \times 10^{-3} \quad \text{or} \quad 0.075\%$$

With a temperature-dependent $\lambda(T)$,

$$\frac{\Delta L}{L_o} = \int_{T_o}^{T} \lambda(T)\, dT$$

$$= \int_{20+273}^{320+273} \{3.725 \times 10^{-6}[1 - e^{-5.88\times10^{-3}(T-124)}] + 5.548 \times 10^{-10}T\}\, dT$$

The integration can either be done numerically or analytically (both left as an exercise) with the result that

$$\frac{\Delta L}{L_o} = 1.00 \times 10^{-3} \quad \text{or} \quad 0.1\%$$

which is substantially more than when using a constant $\lambda$. The mean $\overline{\lambda}$ over this temperature range can be found from

$$\frac{\Delta L}{L_o} = \overline{\lambda}(T - T_o) \quad \text{or} \quad 1.00 \times 10^{-3} = \overline{\lambda}(320 - 20)$$

which gives $\overline{\lambda} = 3.33 \times 10^{-6}\ °C^{-1}$. A 0.1 percent change in length means that a 1 mm chip would expand by 1 micron.

## 1.5 MOLECULAR VELOCITY AND ENERGY DISTRIBUTION

Although the kinetic theory allows us to determine the root mean square velocity of the gas molecules, it says nothing about the distribution of velocities. Due to random collisions between the molecules and the walls of the container and between the molecules themselves, the molecules do not all have the same velocity. The velocity distribution of molecules can be determined experimentally by the simple scheme illustrated in Figure 1.21. Gas molecules are allowed to escape from a small aperture of a hot oven in which the substance is vaporized. Two blocking slits allow only those molecules that are moving along the line through the two slits to pass through, which results in a **collimated beam.** This beam is directed toward two rotating disks, which have slightly displaced

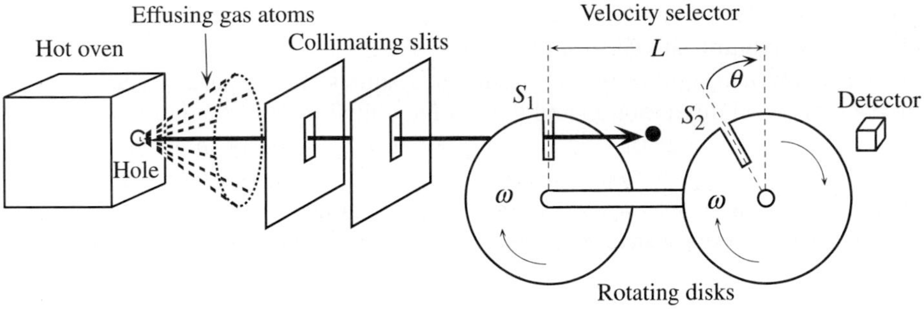


**Figure 1.21** Schematic diagram of a Stern-type experiment for determining the distribution of molecular speeds.

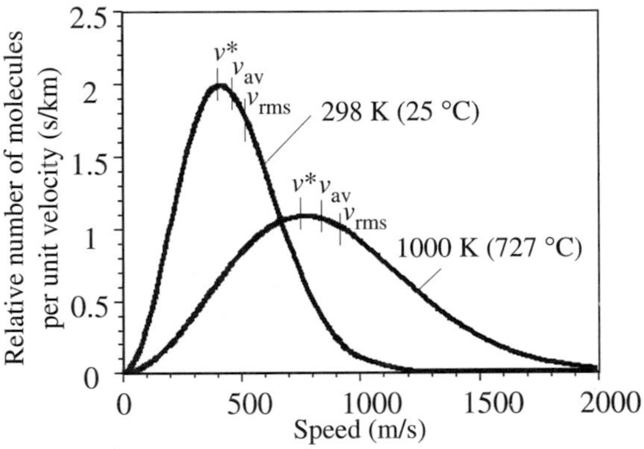


**Figure 1.22** Maxwell–Boltzmann distribution of molecular speeds in nitrogen gas at two temperatures. The ordinate is $dN/(N\,dv)$, the fractional number of molecules per unit speed interval in $(km/s)^{-1}$.

slits. The molecules that pass through the first slit can only pass through the second if they have a certain speed; that is, the exact speed at which the second slit lines up with the first slit. Thus, the two disks act as a speed selector. The speed of rotation of the disks determines which molecular speeds are allowed to go through. The experiment therefore measures the number of molecules $\Delta N$ with speeds in the range $v$ to $(v + \Delta v)$.

It is generally convenient to describe the number of molecules $dN$ with speeds in a certain range $v$ to $(v + dv)$ by defining a **velocity density function** $n_v$ as follows:

$$dN = n_v\, dv$$

where $n_v$ is the number of molecules per unit velocity that have velocities in the range $v$ to $(v + dv)$. This number represents the velocity distribution among the molecules and is a function of the molecular velocity $n_v = n_v(v)$. From the experiment, we can easily obtain $n_v$ by $n_v = \Delta N/\Delta v$ at various velocities. Figure 1.22 shows the velocity density function $n_v$ of nitrogen gas at two temperatures. The average ($v_{av}$), most probable ($v^*$), and rms ($v_{rms}$) speeds are marked to show their relative positions. As expected, these speeds all increase with increasing temperature. From various experiments of the type shown in Figure 1.21, the velocity distribution function $n_v$ has been widely studied and found to obey the following equation:

*Maxwell–Boltzmann distribution for molecular speeds*

$$n_v = 4\pi N\left(\frac{m}{2\pi kT}\right)^{3/2} v^2 \exp\left(-\frac{mv^2}{2kT}\right) \qquad [1.23]$$

where $N$ is the total number of molecules and $m$ is the molecular mass. This is the **Maxwell–Boltzmann distribution function,** which describes the statistics of particle velocities in thermal equilibrium. The function assumes that the particles do not interact with each other while in motion and that all the collisions are elastic in the sense that they involve an exchange of kinetic energy. Figure 1.22 clearly shows that molecules move around randomly, with a variety of velocities ranging from nearly zero to almost infinity. The kinetic theory speaks of their rms value only.

What is the energy distribution of molecules in a gas? In the case of a monatomic gas, the total energy $E$ is purely translational kinetic energy, so we can use $E = \frac{1}{2}mv^2$. To relate an energy range $dE$ to a velocity range $dv$, we have $dE = mv\,dv$. Suppose that $n_E$ is the number of atoms per unit volume per unit energy at an energy $E$. Then $n_E\,dE$ is the number of atoms with energies in the range $E$ to $(E + dE)$. These are also the atoms with velocities in the range $v$ to $(v + dv)$, because an atom with a velocity $v$ has an energy $E$. Thus,

$$n_E dE = n_v\,dv$$

i.e.,

$$n_E = n_v\left(\frac{dv}{dE}\right)$$

*Maxwell–Boltzmann distribution for translational kinetic energies*

If we substitute for $n_v$ and $(dv/dE)$, we obtain the expression for $n_E$ as a function of $E$:

$$n_E = \frac{2}{\pi^{1/2}}N\left(\frac{1}{kT}\right)^{3/2}E^{1/2}\exp\left(-\frac{E}{kT}\right) \qquad [1.24]$$

Thus, the total internal energy is distributed among the atoms according to the Maxwell–Boltzmann distribution in Equation 1.24. The exponential factor $\exp(-E/kT)$ is called the **Boltzmann factor.** Atoms have widely differing kinetic energies, but a mean energy of $\frac{3}{2}kT$. Figure 1.23 shows the Maxwell–Boltzmann energy distribution among the gas atoms in a tank at two temperatures. As the temperature increases, the distribution extends to higher energies. The area under the curve is the total number of molecules, which remains the same for a closed container.

Equation 1.24 represents the energy distribution among the $N$ gas atoms at any time. Since the atoms are continually colliding and exchanging energies, the energy of one

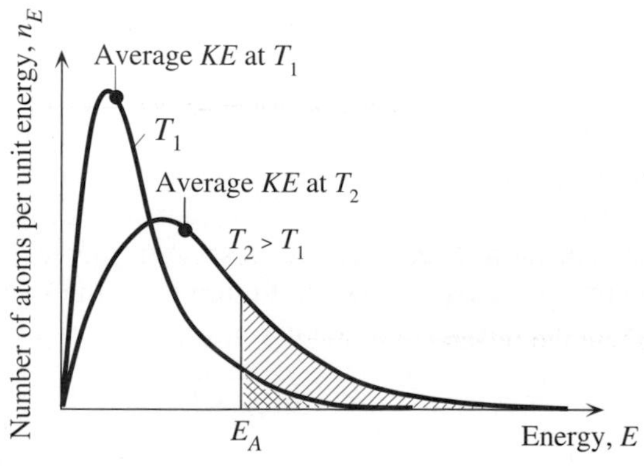


**Figure 1.23** Energy distribution of gas molecules at two different temperatures. The shaded area shows the number of molecules that have energies greater than $E_A$. This area depends strongly on the temperature as $\exp(-E_A/kT)$.

atom will sometimes be small and sometimes be large, but averaged over a long time, this energy will be $\frac{3}{2}kT$ as long as all the gas atoms are in thermal equilibrium (*i.e.*, the temperature is the same everywhere in the gas). Thus, we can also use Equation 1.24 to represent all possible energies an atom can acquire over a long period. There are a total of $N$ atoms, and $n_E\, dE$ of them have energies in the range $E$ to $(E + dE)$. Thus,

$$\text{Probability of energy being in } E \text{ to } (E + dE) = \frac{n_E\, dE}{N} \qquad [1.25]$$

When the probability in Equation 1.25 is integrated (*i.e.*, summed) for all energies ($E = 0$ to $\infty$), the result is unity, because the atom must have an energy somewhere in the range of zero to infinity.

What happens to the Maxwell–Boltzmann energy distribution law in Equation 1.24 when the total energy is not simply translational kinetic energy? What happens when we do not have a monatomic gas? Suppose that the total energy of a molecule (which may simply be an atom) in a system of $N$ molecules has vibrational and rotational kinetic energy contributions, as well as potential energy due to intermolecular interactions. In all cases, the number of molecules per unit energy $n_E$ turns out to contain the Boltzmann factor, and the energy distribution obeys what is called the **Boltzmann energy distribution:**

*Boltzmann energy distribution*

$$\frac{n_E}{N} = C\exp\left(-\frac{E}{kT}\right) \qquad [1.26]$$

where $E$ is the total energy ($KE + PE$), $N$ is the total number of molecules in the system, and $C$ is a constant that relates to the specific system (*e.g.*, a monatomic gas or a liquid). The constant $C$ may depend on the energy $E$, as in Equation 1.24, but not as strongly as the exponential term. Equation 1.26 is the **probability per unit energy** that a molecule in a given system has an energy $E$. Put differently, $(n_E\, dE)/N$ is the fraction of molecules in a small energy range $E$ to $E + dE$.

---

**EXAMPLE 1.10**

**MEAN AND RMS SPEEDS OF MOLECULES** Given the Maxwell–Boltzmann distribution law for the velocities of molecules in a gas, derive expressions for the mean speed ($v_{av}$), most probable speed ($v^*$), and rms velocity ($v_{rms}$) of the molecules and calculate the corresponding values for a gas of noninteracting electrons.

**SOLUTION**

The number of molecules with speeds in the range $v$ to $(v + dv)$ is

$$dN = n_v\, dv = 4\pi N\left(\frac{m}{2\pi kT}\right)^{3/2} v^2 \exp\left(-\frac{mv^2}{2kT}\right) dv$$

By definition, then, the mean speed is given by

*Mean speed*

$$v_{av} = \frac{\int v\, dN}{\int dN} = \frac{\int v n_v\, dv}{\int n_v\, dv} = \sqrt{\frac{8kT}{\pi m}}$$

where the integration is over all speeds ($v = 0$ to $\infty$). The mean square velocity is given by

$$\overline{v^2} = \frac{\int v^2\, dN}{\int dN} = \frac{\int v^2 n_v\, dv}{\int n_v\, dv} = \frac{3kT}{m}$$

*Root mean square velocity*

so the rms velocity is

$$v_{\text{rms}} = \sqrt{\frac{3kT}{m}}$$

where $n_v/N$ is the probability per unit speed that a molecule has a speed in the range $v$ to $(v + dv)$. Differentiating $n_v$ with respect to $v$ and setting this to zero, $dn_v/dv = 0$, gives the position of the peak of $n_v$ versus $v$, and thus the most probable speed $v^*$,

*Most probable speed*

$$v^* = \left[\frac{2kT}{m}\right]^{1/2}$$

Substituting $m = 9.1 \times 10^{-31}$ kg for electrons and using $T = 300$ K, we find $v^* = 95.3$ km s$^{-1}$, $v_{\text{av}} = 108$ km s$^{-1}$, and $v_{\text{rms}} = 117$ km s$^{-1}$, all of which are close in value. We often use the term **thermal velocity** to describe the mean speed of particles due to their thermal random motion. Also, the integrations shown are not trivial and they involve substitution and integration by parts.

## 1.6 HEAT, THERMAL FLUCTUATIONS, AND NOISE

Generally, thermal equilibrium between two objects implies that they have the same temperature, where temperature (from the kinetic theory) is a measure of the mean kinetic energy of the molecules. Consider a solid in a monatomic gas atmosphere such as He gas, as depicted in Figure 1.24. Both the gas and the solid are at the same temperature. The gas molecules move around randomly, with a mean kinetic energy given by $\frac{1}{2}m\overline{v^2} = \frac{3}{2}kT$, where $m$ is the mass of the gas molecule. We also know that the atoms in the solid vibrate with a mean kinetic energy given by $\frac{1}{2}M\overline{V^2} = \frac{3}{2}kT$, where $M$ is the mass of the solid atom and $V$ is the velocity of vibration. The gas molecules will collide with the atoms on the surface of the solid and will thus exchange energy with those solid atoms. Since both are at the same temperature, the solid atoms and gas molecules have the same mean kinetic energy, which means that over a long time, there will be no net transfer of energy from one to the other. This is basically what we mean by **thermal equilibrium.**

If, on the other hand, the solid is hotter than the gas, $T_{\text{solid}} > T_{\text{gas}}$, and thus $\frac{1}{2}M\overline{V^2} > \frac{1}{2}m\overline{v^2}$, then when an average gas molecule and an average solid atom collide,

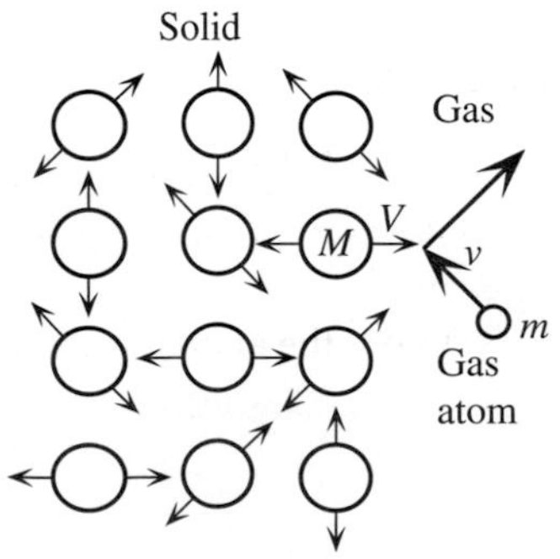


**Figure 1.24** Solid in equilibrium in air.
During collisions between the gas and solid atoms, kinetic energy is exchanged.

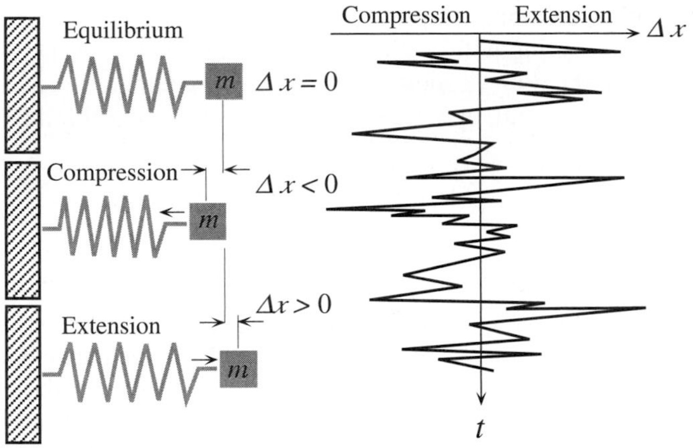


**Figure 1.25** Fluctuations of a mass attached to a spring, due to random bombardment by air molecules.

energy will be transferred from the solid atom to the gas molecule. As many more gas molecules collide with solid atoms, more and more energy will be transferred, until the mean kinetic energy of atoms in each substance is the same and they reach the same temperature: the bodies have **equilibrated.** The amount of energy transferred from the kinetic energy of the atoms in the hot solid to the kinetic energy of the gas molecules is called **heat.** Heat represents the energy transfer from the hot body to the cold body by virtue of the *random* motions and collisions of the atoms and molecules.

Although, over a long time, the energy transferred between two systems in thermal equilibrium is certainly zero, this does not preclude a net energy transfer from one to the other at one instant. For example, at any one instant, an average solid atom may be hit by a fast gas molecule with a speed at the far end of the Maxwell–Boltzmann distribution. There will then be a transfer of energy from the gas molecule to the solid atom. At another instant, a slow gas molecule hits the solid, and the reverse is true. Thus, although the mean energy transferred from one atom to the other is zero, the instantaneous value of this energy is not zero and varies randomly about zero.

As an example, consider a small mass attached to a spring, as illustrated in Figure 1.25. The gas or air molecules will bombard and exchange energy with the solid atoms. Some air molecules will be fast and some will be slow, which means that there will be an instantaneous exchange of energy. Consequently, the spring will be compressed when the bombarding air molecules are fast (more energetic) and extended when they are less energetic. This leads to a mechanical fluctuation of the mass about its equilibrium position, as depicted in Figure 1.25. These fluctuations make the measurement of the exact position of the mass uncertain, and it is futile to try to measure the position more accurately than these fluctuations permit.

If the mass $m$ compresses the spring by $\Delta x$, then at time $t$, the energy stored as potential energy in the spring is

$$PE(t) = \frac{1}{2}K(\Delta x)^2 \qquad [1.27]$$

where $K$ is the spring constant. At a later instant, this energy will be returned to the gas by the spring. The spring will continue to fluctuate because of the fluctuations in the velocity of the bombarding air molecules. Over a long period, the average value of $PE$ will be the same as $KE$ and, by virtue of the Maxwell equipartition of energy theorem, it will be given by

$$\overline{\frac{1}{2}K(\Delta x)^2} = \frac{1}{2}kT \qquad [1.28]$$

*Root mean square fluctuations of a body attached to a spring of stiffness K*

Thus, the rms value of the fluctuations of the mass about its equilibrium position is

$$(\Delta x)_{\text{rms}} = \sqrt{\frac{kT}{K}} \qquad [1.29]$$

To understand the origin of electrical noise, for example, we consider the thermal fluctuations in the instantaneous local electron concentration in a conductor, such as that shown in Figure 1.26. Because of fluctuations in the electron concentration at any one instant, end A of the conductor can become more negative with respect to end B, which will give rise to a voltage across the conductor. This fluctuation in the electron concentration is due to more electrons at that instant moving toward end A than toward B. At a later instant, the situation reverses and more electrons move toward B than toward A, resulting in end B becoming more negative and leading to a reversal of the voltage between A and B. Clearly, there will therefore be voltage fluctuations across the conductor, even though the mean voltage across it over a long period is always zero. If the conductor is connected to an amplifier, these voltage fluctuations will be amplified and recorded as noise at the output. This noise corrupts the actual signal at the amplifier input and is obviously undesirable. As engineers, we have to know how to calculate the magnitude of this noise. Although the mean voltage due to thermal fluctuations is zero, the rms value is not. The average voltage from a power outlet is zero, but the rms value is 120 V. We use the rms value to calculate the amount of average power available.

Consider a conductor of resistance $R$. To derive the noise voltage generated by $R$ we place a capacitor $C$ across this conductor, as in Figure 1.27, and we assume that both are at the same temperature; they are in thermal equilibrium. The capacitor is placed as a *convenient device* to obtain or derive the noise voltage generated by $R$. It should be emphasized that $C$ itself does not contribute to the source of the fluctuations (it generates no noise) but is inserted into the circuit to impose a finite *bandwidth* over which we will calculate the noise voltage. The reason is that all practical electric circuits have some kind of bandwidth, and the noise voltage we will derive depends on this bandwidth. Even if we remove the capacitor, there will still be stray capacitances; and if we short the conductor, the shorting wires will have some inductance that will also impose a bandwidth. As we mentioned previously, thermal fluctuations in the conductor give rise to voltage fluctuations across $R$. There is only so much average energy available in these thermal fluctuations, and this is the energy that is used to charge and discharge the external capacitor $C$. The voltage across the capacitor depends on how much energy that can be stored on it, which in turn depends on the thermal fluctuations in the conductor. Charging a capacitor to a voltage $v$ implies that an energy $E = \frac{1}{2}Cv^2$ is stored on the capacitor. The mean stored energy $\overline{E}$ in a thermal equilibrium system can only be

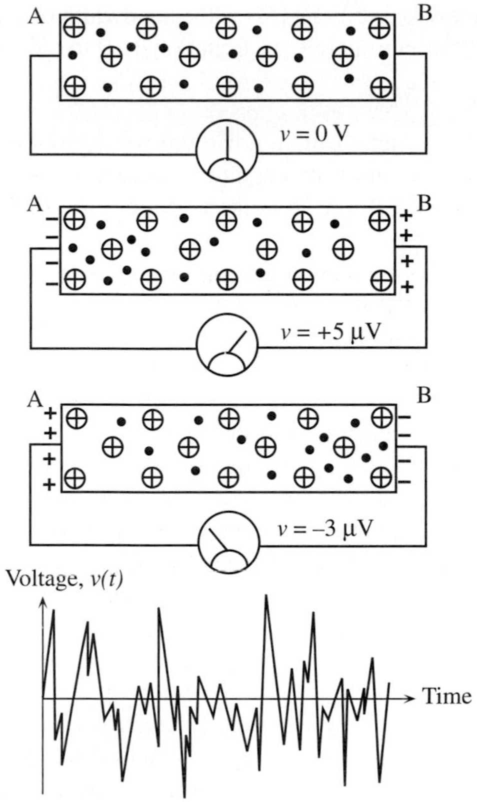


**Figure 1.26** Random motion of conduction electrons in a conductor, resulting in electrical noise.

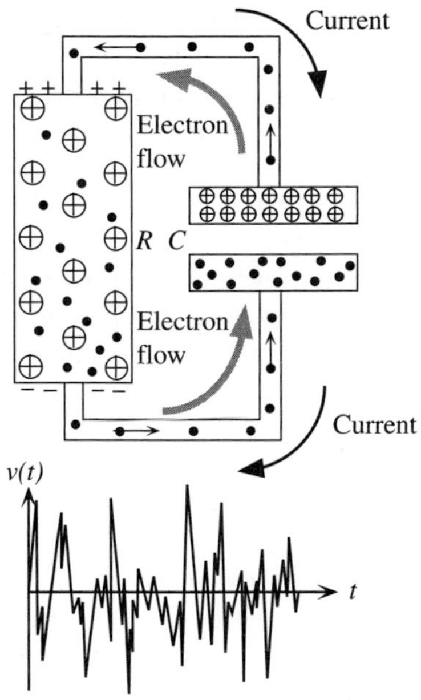


**Figure 1.27** Charging and discharging of a capacitor by a conductor, due to the random thermal motions of the conduction electrons.

$\frac{1}{2}kT$, according to the Maxwell energy equipartition theorem. Thus $\overline{E(t)}$, the mean energy stored on $C$ due to thermal fluctuations, is given by

$$\overline{E(t)} = \tfrac{1}{2}C\overline{v(t)^2} = \tfrac{1}{2}kT$$

We see that the mean square voltage across the capacitor is given by

$$\overline{v(t)^2} = \frac{kT}{C} \qquad [1.30]$$

Interestingly, the rms noise voltage across an $RC$ network seems to be independent of the resistance. However, the origin of the noise voltage arises from the electron fluctuations in the conductor and we must somehow reexpress Equation 1.30 to reflect this fact; that is, we must relate the electrical fluctuations to $R$.

The voltage fluctuations across the network will have many sinusoidal components, but only those below the cutoff frequency of the $RC$ network will contribute to the mean square voltage (that is, we effectively have a low-pass filter). If $B$ is the bandwidth of the $RC$ network,[8] then $B = 1/(2\pi RC)$ and we can eliminate $C$ in Equation 1.30 to obtain

$$\overline{v(t)^2} = 2\pi kTRB$$

This is the key equation for calculating the mean square noise voltage from a resistor over a bandwidth $B$. A more rigorous derivation makes the numerical factor 4 rather than $2\pi$. For a network with a bandwidth $B$, the **rms noise voltage** is therefore

*Root mean square noise voltage across a resistance*

$$v_{\text{rms}} = \sqrt{4kTRB} \qquad [1.31]$$

Equation 1.31 is known as the **Johnson resistor noise equation,** and it sets the lower limit of the magnitude of small signals that can be amplified. Note that Equation 1.31 basically tells us the rms value of the voltage fluctuations within a given bandwidth ($B$) and not the origin and spectrum (noise voltage vs. frequency) of the noise. The origin of noise is attributed to the random motions of electrons in the conductor (resistor), and Equation 1.31 is the fundamental description of electrical fluctuations; that is, the fluctuations in the conductor's instantaneous local electron concentration that charges and discharges the capacitor. To determine the rms noise voltage across a network with an impedance $Z(j\omega)$, all we have to do is find the real part of $Z$, which represents the resistive part, and use this for $R$ in Equation 1.31.

**EXAMPLE 1.11**

**NOISE IN AN *RLC* CIRCUIT** Most radio receivers have a tuned parallel-resonant circuit, which consists of an inductor $L$, capacitor $C$, and resistance $R$ in parallel. Suppose $L$ is 100 μH; $C$ is 100 pF; and $R$, the equivalent resistance due to the input resistance of the amplifier and to the loss in the coil (coil resistance plus ferrite losses), is about 200 kΩ. What is the minimum rms radio signal that can be detected?

**SOLUTION**

Consider the bandwidth of this tuned $RLC$ circuit, which can be found in any electrical engineering textbook:

$$B = \frac{f_o}{Q}$$

where $f_o = 1/[2\pi\sqrt{LC}]$ is the resonant frequency and $Q = 2\pi f_o CR$ is the quality factor. Substituting for $L$, $C$, and $R$, we get, $f_o = 10^7/2\pi = 1.6 \times 10^6$ Hz and $Q = 200$, which gives $B = 10^7/[2\pi(200)]$ Hz, or 8 kHz. The rms noise voltage is

$$v_{\text{rms}} = [4kTRB]^{1/2} = [4(1.38 \times 10^{-23}\ \text{J K}^{-1})(300\ \text{K})(200 \times 10^3\ \Omega)(8 \times 10^3\ \text{Hz})]^{1/2}$$
$$= 5.1 \times 10^{-6}\ \text{V} \qquad \text{or} \qquad 5.1\ \mu\text{V}$$

This rms voltage is within a bandwidth of 8 kHz centered at 1.6 MHz. This last information is totally absent in Equation 1.31. If we attempt to use

$$v_{\text{rms}} = \left[\frac{kT}{C}\right]^{1/2}$$

[8] A low-pass filter allows all signal frequencies up to the cutoff frequency $B$ to pass. $B$ is $1/(2\pi RC)$.

we get

$$v_{\text{rms}} = \left[\frac{(1.38 \times 10^{-23}\ \text{J K}^{-1})(300\ \text{K})}{100 \times 10^{-12}\ \text{F}}\right]^{1/2} = 6.4\,\mu\text{V}$$

However, Equation 1.30 was derived using the $RC$ circuit in Figure 1.27, whereas we now have an $LCR$ circuit. The correct approach uses Equation 1.31, which is generally valid, and the appropriate bandwidth $B$.

# 1.7 THERMALLY ACTIVATED PROCESSES

## 1.7.1 ARRHENIUS RATE EQUATION

Many physical and chemical processes strongly depend on temperature and exhibit what is called an **Arrhenius type behavior,** in which the rate of change is proportional to $\exp(-E_A/kT)$, where $E_A$ is a characteristic energy parameter applicable to the particular process. For example, when we store food in the refrigerator, we are effectively using the Arrhenius rate equation: cooling the food diminishes the rate of decay. Processes that exhibit an Arrhenius type temperature dependence are referred to as **thermally activated.**

For an intuitive understanding of a thermally activated process, consider a vertical filing cabinet that stands in equilibrium, with its center of mass at A, as sketched in Figure 1.28. Tilting the cabinet left or right increases the potential energy *PE* and requires external work. If we could supply this energy, we could move the cabinet over its edge and lay it flat, where its *PE* would be lower than at A. Clearly, since the *PE* at B is lower, this is a more stable position than A. Further, in going from A to B, we had to overcome a **potential energy barrier** of amount $E_A$, which corresponds to the cabinet standing on its edge with the center of mass at the highest point at A*. To topple the cabinet, we must first provide energy[9] equal to $E_A$ to take the center of mass to A*, from which point the cabinet, with the slightest encouragement, will fall spontaneously to B to attain the lowest *PE*. At the end of the whole tilting process, the internal energy change for the cabinet, $\Delta U$, is due to the change in the *PE* $(=mgh)$ from A to B, which is negative; B has lower *PE* than A.

Suppose, for example, a person with an average energy less than $E_A$ tries to topple the cabinet. Like everyone else, that person experiences energy fluctuations as a result of interactions with the environment (e.g., what type of day the person had). During one of those high-energy periods, he can topple the cabinet, even though most of the time he cannot do so because his average energy is less then $E_A$. The rate at which the cabinet is toppled depends on the number of times (frequency) the person tries and the probability that he possesses energy greater than $E_A$.

As an example of a thermally activated process, consider the diffusion of impurity atoms in a solid, one of which is depicted in Figure 1.29. In this example, the impurity atom is at an interatomic void A in the crystal, called an **interstitial site.** For the impurity atom to move from A to a neighboring void B, the atom must push the host neighbors apart as it moves across. This requires energy in much the same way

[9] According to the conservation of energy principle, the increase in the *PE* from A to A* must come from the external work.

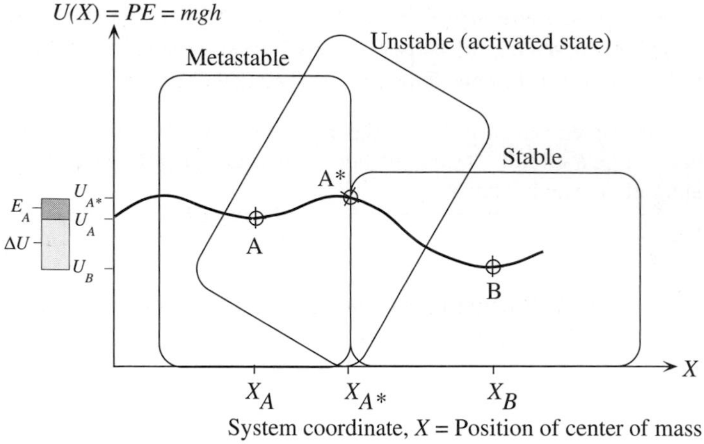


**Figure 1.28** Tilting a filing cabinet from state A to its edge in state A* requires an energy $E_A$.

After reaching A*, the cabinet spontaneously drops to the stable position B. The *PE* of state B is lower than A, and therefore state B is more stable than A.

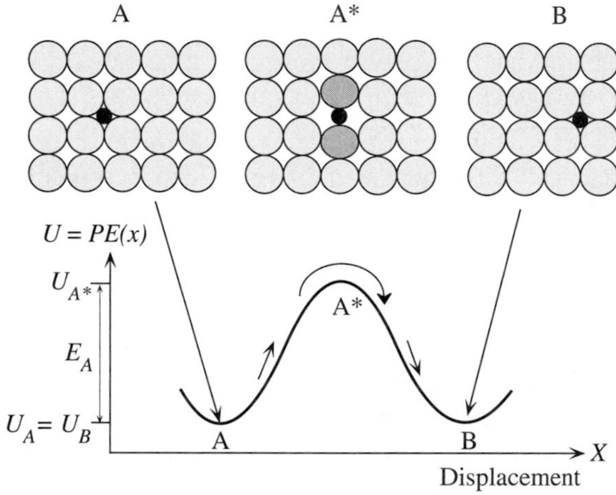


**Figure 1.29** Diffusion of an interstitial impurity atom in a crystal from one void to a neighboring void.

The impurity atom at position A must possess an energy $E_A$ to push the host atoms away and move into the neighboring void at B.

as does toppling the filing cabinet. There is a potential energy barrier $E_A$ to the motion of this atom from A to B.

Both the host and the impurity atoms in the solid vibrate about their equilibrium positions, with a distribution of energies, and they also continually exchange energies, which leads to energy fluctuations. In thermal equilibrium, at any instant, we can expect the energy distribution of the atoms to obey the Boltzmann distribution law (see Equation 1.26). The average kinetic energy per atom is vibrational and is $\frac{3}{2}kT$, which will not allow the impurity simply to overcome the *PE* barrier $E_A$, because typically $E_A \gg \frac{3}{2}kT$.

The rate of jump, called the **diffusion,** of the impurity from A to B depends on two factors. The first is the number of times the atom tries to go over the potential barrier, which is the vibrational frequency $v_o$, in the AB direction. The second factor is the probability that the atom has sufficient energy to overcome the *PE* barrier. Only during those times when the atom has an energy greater than the potential energy barrier $E_A = U_{A^*} - U_A$ will it jump across from A to B. During this diffusion process, the atom attains an **activated state,** labeled A* in Figure 1.29, with an energy $E_A$ above $U_A$, so the crystal internal energy is higher than $U_A$. $E_A$ is called the **activation energy.**

Suppose there are $N$ impurity atoms. At any instant, according to the Boltzmann distribution, $n_E\,dE$ of these will have kinetic energies in the range $E$ to $(E + dE)$, so the probability that an impurity atom has an energy $E$ greater than $E_A$ is

$$\text{Probability } (E > E_A) = \frac{\text{Number of impurities with } E > E_A}{\text{Total number of impurities}}$$

$$= \frac{\int_{E_A}^{\infty} n_E\,dE}{N} = A\exp\left(-\frac{E_A}{kT}\right)$$

where $A$ is a dimensionless constant that has only a weak temperature dependence. The rate of jumps, jumps per seconds, or simply the **frequency of jumps** $\vartheta$ from void to void is

*Rate for a thermally activated process*

$$\vartheta = (\text{Frequency of attempt along AB})(\text{Probability of } E > E_A)$$

$$= A v_o \exp\left(-\frac{E_A}{kT}\right) \qquad E_A = U_{A*} - U_A \qquad [1.32]$$

Equation 1.32 describes the rate of a thermally activated process, for which increasing the temperature causes more atoms to be energetic and hence results in more jumps over the potential barrier. Equation 1.32 is the well-known **Arrhenius rate equation** and is generally valid for a vast number of transformations, both chemical and physical.

### 1.7.2 ATOMIC DIFFUSION AND THE DIFFUSION COEFFICIENT

Consider the motion of the impurity atom in Figure 1.29. For simplicity, assume a two-dimensional crystal in the plane of the paper, as in Figure 1.30. The impurity atom has four neighboring voids into which it can jump. If $\theta$ is the angle with respect to the $x$ axis, then these voids are at directions $\theta = 0°$, $90°$, $180°$, and $270°$, as depicted in Figure 1.30. Each jump is in a random direction along one of these four angles. As the impurity atom jumps from void to void, it leaves its original location at O, and after $N$ jumps, after time $t$, it has been displaced from O to O′.

Let $a$ be the closest void-to-void separation. Each jump results in a displacement along $x$ which is equal to $a\cos\theta$, with $\theta = 0°, 90°, 180°$, or $270°$. Thus, each jump

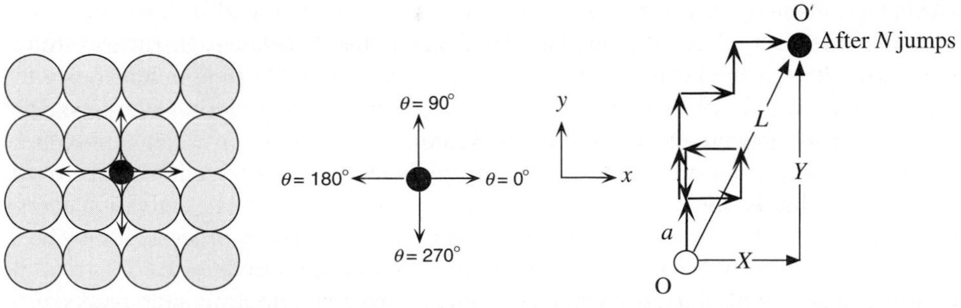


**Figure 1.30** An impurity atom has four site choices for diffusion to a neighboring interstitial vacancy. After $N$ jumps, the impurity atom would have been displaced from the original position at O.

results in a displacement along $x$ which can be $a$, 0, $-a$, or 0, corresponding to the four possibilities. After $N$ jumps, the mean displacement along $x$ will be close to zero, just as the mean voltage of the ac voltage from a power outlet is zero, even though it has an rms value of 120 V. We therefore consider the square of the displacements. The total square displacement, denoted $X^2$, is

$$X^2 = a^2\cos^2\theta_1 + a^2\cos^2\theta_2 + \cdots + a^2\cos^2\theta_N$$

Clearly, $\theta = 90°$ and $270°$ give $\cos^2\theta = 0$. Of all $N$ jumps, $\frac{1}{2}N$ are $\theta = 0$ and $180°$, each of which gives $\cos^2\theta = 1$. Thus,

$$X^2 = \frac{1}{2}a^2 N$$

There will be a similar expression for $Y^2$, which means that after $N$ jumps, the total square distance $L^2$ from O to O′ in Figure 1.30 is

$$L^2 = X^2 + Y^2 = a^2 N$$

The rate of jumping (frequency of jumps) is given by Equation 1.32

$$\vartheta = v_o A \exp\left(-\frac{E_A}{kT}\right)$$

so the time per jump is $1/\vartheta$. Time $t$ for $N$ jumps is $N/\vartheta$. Thus, $N = \vartheta t$ and

*Mean square displacement*

$$L^2 = a^2\vartheta t = 2Dt \qquad [1.33]$$

where, by definition, $D = \frac{1}{2}a^2\vartheta$, which is a constant that depends on the diffusion process, as well as the temperature, by virtue of $\vartheta$. This constant is generally called the **diffusion coefficient.** Substituting for $\vartheta$, we find

$$D = \frac{1}{2}a^2 v_o A \exp\left(-\frac{E_A}{kT}\right)$$

or

*Diffusion coefficient*

$$D = D_o \exp\left(-\frac{E_A}{kT}\right) \qquad [1.34]$$

where $D_o$ is a constant. The root square displacement $L$ in time $t$, from Equation 1.33, is given by $L = [2Dt]^{1/2}$. Since $L^2$ is evaluated from $X^2$ and $Y^2$, $L$ is known as the **root mean square (rms) displacement.**

The preceding specific example considered the diffusion of an impurity in a void between atoms in a crystal; this is a simple way to visualize the diffusion process. An impurity, indeed any atom, at a regular atomic site in the crystal can also diffuse around by various other mechanisms. For example, such an impurity can simultaneously exchange places with a neighbor. But, more significantly, if a neighboring atomic site has a *vacancy* that has been left by a missing host atom, then the impurity can simply jump into this vacancy. (Vacancies in crystals are explained in detail in Section 1.9.1; for the present, they simply correspond to missing atoms in the crystal.) The activation energy $E_A$ in Equation 1.34 is a measure of the difficulty of the diffusion process. It may be as simple as the energy (or work) required for an impurity atom to deform (or strain) the crystal around it as it jumps from one interstitial site to a neighboring interstitial site, as in Figure 1.29; or it may be more complicated, for example, involving vacancy creation.

Various Si semiconductor devices are fabricated by doping a single Si crystal with impurities (dopants) at high temperatures. For example, doping the Si crystal with phosphorus (P) gives the crystal a higher electrical conductivity. The P atoms substitute directly for Si atoms in the crystal. These dopants migrate from high to low dopant concentration regions in the crystal by diffusion, which occurs efficiently only at sufficiently high temperatures.

**EXAMPLE 1.12**

**DIFFUSION OF DOPANTS IN SILICON** The diffusion coefficient of P atoms in the Si crystal follows Equation 1.34 with $D_o = 10.5$ cm$^2$ s$^{-1}$ and $E_A = 3.69$ eV. What is the diffusion coefficient at a temperature of 1100 °C at which dopants such as P are diffused into Si to fabricate various devices? What is the rms distance diffused by P atoms in 5 minutes? Estimate, as an order of magnitude, how many jumps the P atom makes in 1 second if you take the jump distance to be roughly the mean interatomic separation, $\sim$0.27 nm.

**SOLUTION**

From Equation 1.34,

$$D = D_o \exp\left(-\frac{E_A}{kT}\right) = (10.5\ \text{cm}^2\ \text{s}^{-1}) \exp\left[-\frac{(3.69\ \text{eV})(1.602 \times 10^{-19}\ \text{J eV}^{-1})}{(1.381 \times 10^{-23}\ \text{J K}^{-1})(1100 + 273\ \text{K})}\right]$$

$$= 3.0 \times 10^{-13}\ \text{cm}^2\ \text{s}^{-1}$$

The rms distance $L$ diffused in a time $t = 5$ min $= 5 \times 60$ seconds is

$$L = \sqrt{2Dt} = [2(3.0 \times 10^{-13}\ \text{cm}^2\ \text{s}^{-1})(5 \times 60\ \text{s})]^{1/2} = 1.3 \times 10^{-5}\ \text{cm} \quad \text{or} \quad 13\ \mu\text{m}$$

Equation 1.33 was derived for a two-dimensional crystal as in Figure 1.30, and for an impurity diffusion. Nonetheless, we can still use it to estimate how many jumps a P atom makes in 1 second. From Equation 1.33, $\vartheta \approx 2D/a^2 \approx 2(3.0 \times 10^{-17}\ \text{m}^2\ \text{s}^{-1})/(0.27 \times 10^{-9}\ \text{m})^2 = 823$ jumps per second. It takes roughly 1 ms to make one jump. It is left as an exercise to show that at room temperature it will take a P atom $10^{46}$ years to make a jump! (Scientists and engineers know how to use thermally activated processes.)

## 1.8 THE CRYSTALLINE STATE

### 1.8.1 TYPES OF CRYSTALS

A **crystalline solid** is a solid in which the atoms bond with each other in a regular pattern to form a periodic collection (or array) of atoms, as shown for the copper crystal in Figure 1.31. The most important property of a crystal is **periodicity,** which leads to what is termed **long-range order.** In a crystal, the local bonding geometry is repeated many times at regular intervals, to produce a periodic array of atoms that constitutes the crystal structure. The location of each atom is well known by virtue of periodicity. There is therefore a long-range order, since we can always predict the atomic arrangement anywhere in the crystal. Nearly all metals, many ceramics and semiconductors, and various polymers are crystalline solids in the sense that the atoms or molecules are positioned on a **periodic array of points in space.**

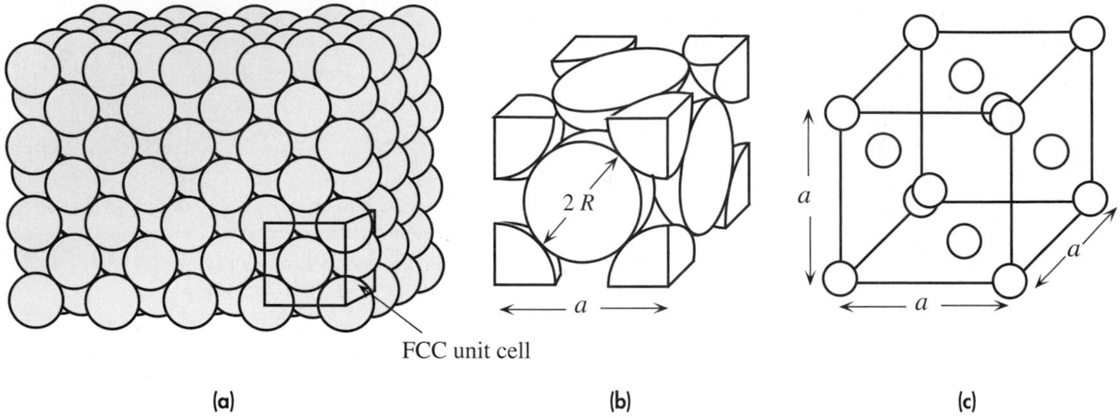


**Figure 1.31**

(a) The crystal structure of copper which is face-centered cubic (FCC). The atoms are positioned at well-defined sites arranged periodically, and there is a long-range order in the crystal.

(b) An FCC unit cell with close-packed spheres.

(c) Reduced-sphere representation of the FCC unit cell.

Examples: Ag, Al, Au, Ca, Cu, $\gamma$-Fe (>912 °C), Ni, Pd, Pt, Rh.

All crystals can be described in terms of a lattice and a basis.[10] A **lattice** is an infinite *periodic* array of geometric points in space, without any atoms. When we place an identical group of atoms (or molecules), called a **basis,** at each lattice point, we obtain the actual **crystal structure.** The crystal is thus a lattice plus a basis at each lattice point. In the copper crystal in Figure 1.31a, each lattice point has one Cu atom and the basis is a single Cu atom. As apparent from Figure 1.31a, the lattice of the copper crystal has **cubic symmetry** and is one of many possible lattices.

Since the crystal is essentially a periodic repetition of a small volume (or cell) of atoms in three dimensions, it is useful to identify the repeating unit so that the crystal properties can be described through this unit. The **unit cell** is the most convenient small cell in the crystal structure that carries the properties of the crystal. The repetition of the unit cell in three dimensions generates the whole crystal structure, as is apparent in Figure 1.31a for the copper crystal.

The unit cell of the copper crystal is cubic with Cu atoms at its corners and one Cu atom at the center of each face, as indicated in Figure 1.31b. The unit cell of Cu is thus said to have a **face-centered cubic (FCC)** structure. The Cu atoms are shared with neighboring unit cells. Effectively, then, only one-eighth of a corner atom is in the unit cell and one-half of the face-centered atom belongs to the unit cell, as shown in Figure 1.31b. This means there are effectively four atoms in the unit cell. The length of the cubic unit cell is termed the **lattice parameter** $a$ of the crystal structure. For Cu, for example, $a$ is 0.362 nm, whereas the radius $R$ of the Cu atom in the crystal is 0.128 nm.

[10] Lattice is a purely imaginary geometric concept whose only requirement is that the infinite array of points has periodicity. In many informal discussions, the term lattice or crystal lattice is used to mean the crystal structure itself. These concepts are further developed in Section 1.13 under Additional Topics.

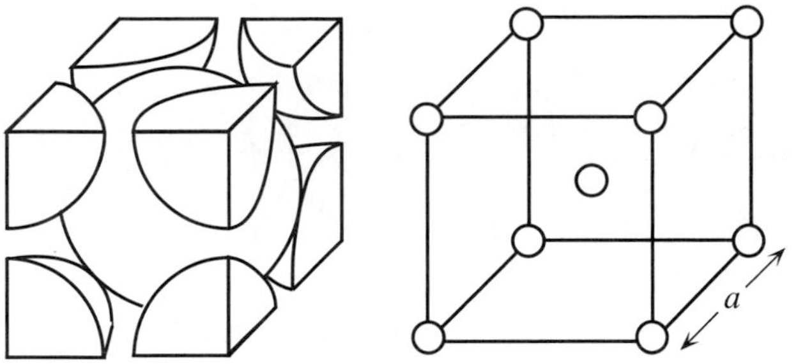


**Figure 1.32** Body-centered cubic (BCC) crystal structure.
(a) A BCC unit cell with close-packed hard spheres representing the Fe atoms.
(b) A reduced-sphere unit cell.

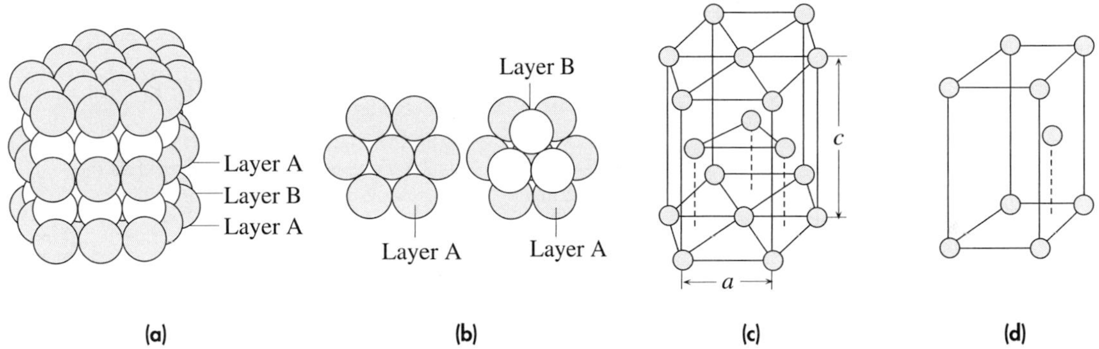


**Figure 1.33** The hexagonal close-packed (HCP) crystal structure.
(a) The hexagonal close-packed (HCP) structure. A collection of many Zn atoms. Color difference distinguishes layers (stacks).
(b) The stacking sequence of closely packed layers is ABAB.
(c) A unit cell with reduced spheres.
(d) The smallest unit cell with reduced spheres.

Assuming the Cu atoms are spheres that touch each other, we can geometrically relate $a$ and $R$. For clarity, it is often more convenient to draw the unit cell with the spheres reduced, as in Figure 1.31c.

The FCC crystal structure of Cu is known as a **close-packed crystal structure** because the Cu atoms are packed as closely as possible, as is apparent in Figure 1.31a and b. The volume of the FCC unit cell is 74 percent full of atoms, which is the maximum packing possible with identical spheres. By comparison, iron has a **body-centered cubic (BCC)** crystal structure and its unit cell is shown in Figure 1.32. The BCC unit cell has Fe atoms at its corners and one Fe atom at the center of the cell. The volume of the BCC unit cell is 68 percent full of atoms, which is lower than the maximum possible packing.

The FCC crystal structure is only one way to pack the atoms as closely as possible. For example, in zinc, the atoms are arranged as closely as possible in a hexagonal symmetry, to form the **hexagonal close-packed (HCP) structure** shown in Figure 1.33a. This structure corresponds to packing spheres as closely as possible first as one layer A, as shown in Figure 1.33b. You can visualize this by arranging six pennies as closely as possible on a table top. On top of layer A we can place an identical layer B, with the

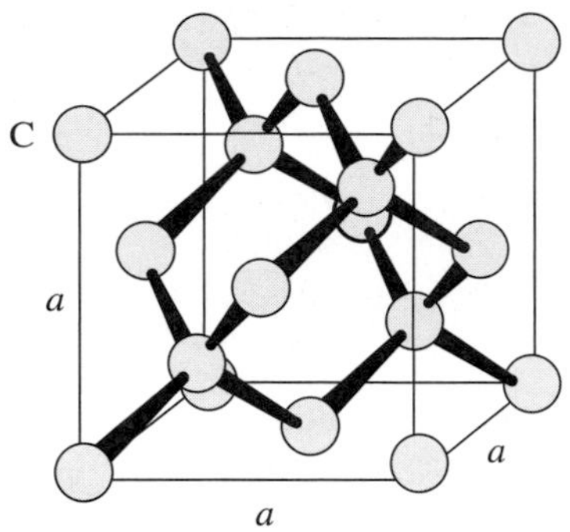


**Figure 1.34** The diamond unit cell which is cubic. The cell has eight atoms.

Gray Sn ($\alpha$-Sn) and the elemental semiconductors Ge and Si have this crystal structure.

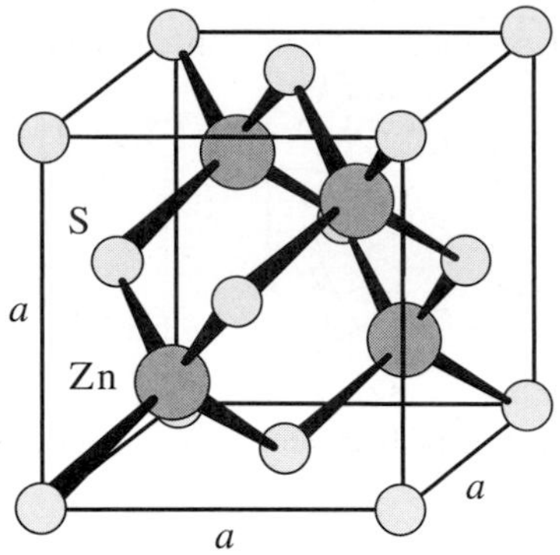


**Figure 1.35** The zinc blende (ZnS) cubic crystal structure.

Many important compound crystals have the zinc blende structure. Examples: AlAs, GaAs, GaP, GaSb, InAs, InP, InSb, ZnS, ZnTe.

spheres taking up the voids on layer A, as depicted in Figure 1.33b. The third layer can be placed on top of B and lined up with layer A. The stacking sequence is therefore ABAB. . . . A unit cell for the HCP structure is shown in Figure 1.33c, which shows that this is not a cubic structure. The unit cell shown, although convenient, is not the smallest unit cell. The smallest unit cell for the HCP structure is shown in Figure 1.33d and is called the **hexagonal unit cell.** The repetition of this unit cell will generate the whole HCP structure. The atomic packing density in the HCP crystal structure is 74 percent, which is the same as that in the FCC structure.

Covalently bonded solids, such as silicon and germanium, have a diamond crystal structure brought about by the directional nature of the covalent bond, as shown in Figure 1.34 (see also Figure 1.6). The rigid local bonding geometry of four Si–Si bonds in the tetrahedral configuration forces the atoms to form what is called the **diamond cubic crystal structure.** The unit cell in this case can be identified with the cubic structure. Although there are atoms at each corner and at the center of each face, indicating an FCC-like structure, there are four atoms within the cell as well. Thus, there are eight atoms in the unit cell. The diamond unit cell can actually be described in terms of an FCC lattice (a geometric arrangement of points) with each lattice point having a basis of two Si atoms. If we place the two Si atoms at each site appropriately, for example, one right at the lattice point, and the other displaced from it by a quarter lattice distance $a/4$ along the cube edges, we can easily generate the diamond unit cell. In the copper crystal, each FCC lattice point has one Cu atom, whereas in the Si crystal each lattice point has two Si atoms; thus there are $4 \times 2 = 8$ atoms in the diamond unit cell.

In the GaAs crystal, as in the silicon crystal, each atom forms four directional bonds with its neighbors. The unit cell looks like a diamond cubic, as indicated in Figure 1.35 but with the Ga and As atoms alternating positions. This unit cell is termed

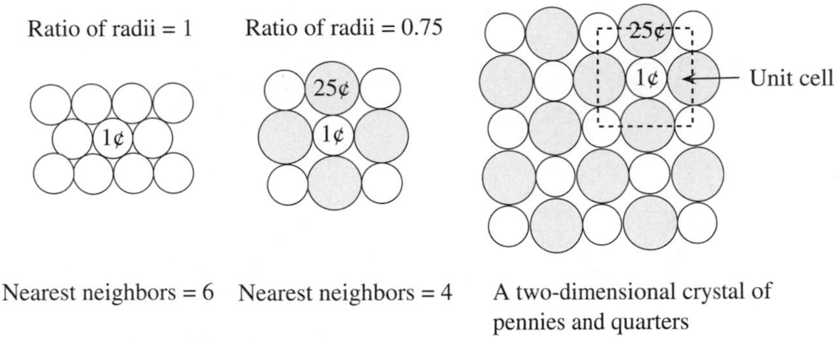


**Figure 1.36** Packing of coins on a table top to build a two-dimensional crystal.

the **zinc blende** structure after ZnS, which has this type of unit cell. Many important compound semiconductors have this crystal structure, GaAs being the most commonly known. The zinc blende unit cell can also be described in terms of a fundamental FCC lattice and a basis that has two atoms, Zn and S (or Ga and As). For example, we can place one Zn at each lattice point and one S atom displaced from the Zn by $a/4$ along the cube edges.

In ionic solids, the cations (*e.g.*, $Na^+$) and the anions ($Cl^-$) attract each other nondirectionally. The crystal structure depends on how closely the opposite ions can be brought together and how the same ions can best avoid each other while maintaining long-range order, or maintaining symmetry. These depend on the relative charge and relative size per ion.

To demonstrate the importance of the size effect in two dimensions, consider identical coins, say pennies (1-cent coins). At most, we can make six pennies touch one penny, as shown in Figure 1.36. On the other hand, if we use quarters[11] (25-cent coins) to touch one penny, at most only five quarters can do so. However, this arrangement cannot be extended to the construction of a two-dimensional crystal with periodicity. To fulfill the long-range symmetry requirement for crystals, we can only use four quarters to touch the penny and thereby build a two-dimensional "penny–quarter" crystal, which is shown in the figure. In the two-dimensional crystal, a penny has four quarters as nearest neighbors; similarly, a quarter has four pennies as nearest neighbors. A convenient unit cell is a square cell with one-quarter of a penny at each corner and a full penny at the center (as shown in the figure).

The three-dimensional equivalent of the unit cell of the penny–quarter crystal is the **NaCl unit cell** shown in Figure 1.37. The $Na^+$ ion is about half the size of the $Cl^-$ ion, which permits six nearest neighbors while maintaining long-range order. The repetition of this unit cell in three dimensions generates the whole NaCl crystal, which was depicted in Figure 1.9b.

A similar unit cell with $Na^+$ and $Cl^-$ interchanged is also possible and equally convenient. We can therefore describe the whole crystal with two interpenetrating FCC

[11] Although many are familiar with the United States coinage, any two coins with a size ratio of about 0.75 would work out the same.

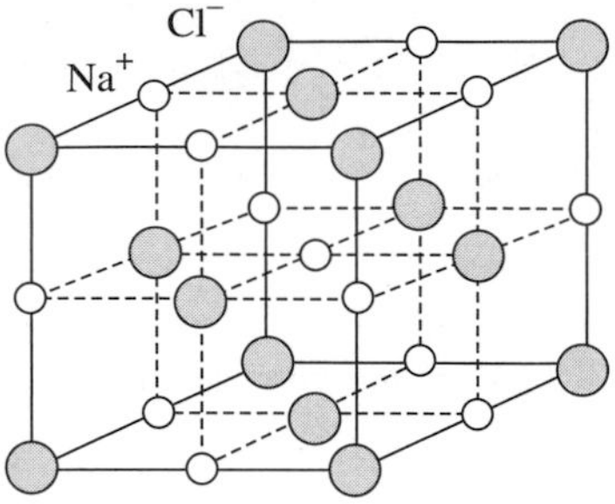


**Figure 1.37** A possible reduced-sphere unit cell for the NaCl (rock salt) crystal.

An alternative unit cell may have $Na^+$ and $Cl^-$ interchanged. Examples: AgCl, CaO, CsF, LiF, LiCl, NaF, NaCl, KF, KCl, MgO.

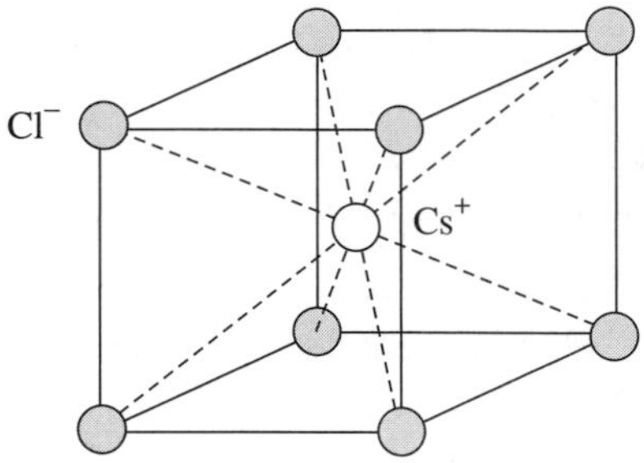


**Figure 1.38** A possible reduced-sphere unit cell for the CsCl crystal.

An alternative unit cell may have $Cs^+$ and $Cl^-$ interchanged. Examples: CsCl, CsBr, CsI, TlCl, TlBr, TlI.

unit cells, each having oppositely charged ions at the corners and face centers. Many ionic solids have the rock salt (NaCl) crystal structure.

When the cation and anions have equal charges and are about the same size, as in the CsCl crystal, the unit cell is called the **CsCl structure,** which is shown in Figure 1.38. Each cation is surrounded by eight anions (and vice versa), which are at

**Table 1.3** Properties of some important crystal structures

| Crystal Structure | *a* and *R* (*R* is the Radius of the Atom) | Coordination Number (CN) | Number of Atoms per Unit Cell | Atomic Packing Factor | Examples |
|---|---|---|---|---|---|
| Simple cubic | $a = 2R$ | 6 | 1 | 0.52 | No metals (Except Po) |
| BCC | $a = \frac{4R}{\sqrt{3}}$ | 8 | 2 | 0.68 | Many metals: $\alpha$–Fe, Cr, Mo, W |
| FCC | $a = \frac{4R}{\sqrt{2}}$ | 12 | 4 | 0.74 | Many metals: Ag, Au, Cu, Pt |
| HCP | $a = 2R$ <br> $c = 1.633a$ | 12 | 2 | 0.74 | Many metals: Co, Mg, Ti, Zn |
| Diamond | $a = \frac{8R}{\sqrt{3}}$ | 4 | 8 | 0.34 | Covalent solids: Diamond, Ge, Si, $\alpha$-Sn |
| Zinc blende | | 4 | 8 | 0.34 | Many covalent and ionic solids. Many compound semiconductors. ZnS, GaAs, GaSb, InAs, InSb |
| NaCl | | 6 | 4 cations | 0.67 | Ionic solids such as NaCl, AgCl, LiF, MgO, CaO |
| | | | 4 anions | (NaCl) | Ionic packing factor depends on relative sizes of ions. |
| CsCl | | 8 | 1 cation <br> 1 anion | | Ionic solids such as CsCl, CsBr, CsI |

the corners of a cube. This is not a true BCC unit cell because the atoms at various BCC lattice points are different. (As discussed in Section 1.13, CsCl has a simple cubic lattice with a basis that has one $Cl^-$ ion and one $Cs^+$ ion.)

Table 1.3 summarizes some of the important properties of the main crystal structures considered in this section.

---

**EXAMPLE 1.13**

**THE COPPER (FCC) CRYSTAL** Consider the FCC unit cell of the copper crystal shown in Figure 1.39.

a. How many atoms are there per unit cell?

*b.* If $R$ is the radius of the Cu atom, show that the lattice parameter $a$ is given by $a = R2\sqrt{2}$.

*c.* Calculate the **atomic packing factor** (APF) defined by

$$\text{APF} = \frac{\text{Volume of atoms in unit cell}}{\text{Volume of unit cell}}$$

*d.* Calculate the **atomic concentration** (number of atoms per unit volume) in Cu and the density of the crystal given that the atomic mass of Cu is 63.55 g $\text{mol}^{-1}$ and the radius of the Cu atom is 0.128 nm.

**SOLUTION**

*a.* There are four atoms per unit cell. The Cu atom at each corner is shared with eight other adjoining unit cells. Each Cu atom at the face center is shared with the neighboring unit cell. Thus, the number of atoms in the unit cell = 8 corners ($\frac{1}{8}$ atom) + 6 faces ($\frac{1}{2}$ atoms) = 4 atoms.

*b.* Consider the unit cell shown in Figure 1.39 and one of the cubic faces. The face is a square of side $a$ and the diagonal is $\sqrt{a^2 + a^2}$ or $a\sqrt{2}$. The diagonal has one atom at the center of diameter $2R$, which touches two atoms centered at the corners. The diagonal, going from corner to corner, is therefore $R + 2R + R$. Thus, $4R = a\sqrt{2}$ and $a = 4R/\sqrt{2} = R2\sqrt{2}$. Therefore, $a = 0.3620$ nm.

*c.*
$$\text{APF} = \frac{(\text{Number of atoms in unit cell}) \times (\text{Volume of atom})}{\text{Volume of unit cell}}$$

$$= \frac{4 \times \dfrac{4}{3}\pi R^3}{a^3} = \frac{\dfrac{4^2}{3}\pi R^3}{(R2\sqrt{2})^3} = \frac{4^2\pi}{3(2\sqrt{2})^3} = 0.74$$

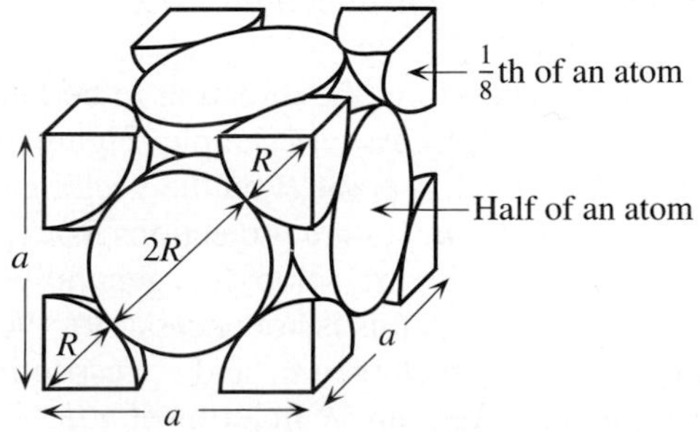


**Figure 1.39** The FCC unit cell. The atomic radius is $R$ and the lattice parameter is $a$.

*d.* In general, if there are $x$ atoms in the unit cell, the atomic concentration is

$$n_{at} = \frac{\text{Number of atoms in unit cell}}{\text{Volume of unit cell}} = \frac{x}{a^3}$$

Thus, for Cu

$$n_{at} = \frac{4}{(0.3620 \times 10^{-7}\ \text{cm})^3} = 8.43 \times 10^{22}\ \text{cm}^{-3}$$

There are $x$ atoms in the unit cell, and each atom has a mass of $M_{at}/N_A$ grams. The density $\rho$ is

$$\rho = \frac{\text{Mass of all atoms in unit cell}}{\text{Volume of unit cell}} = \frac{x\left(\dfrac{M_{at}}{N_A}\right)}{a^3}$$

that is, $$\rho = \frac{n_{at}M_{at}}{N_A} = \frac{(8.43 \times 10^{22}\ \text{cm}^{-3})(63.55\ \text{g mol}^{-1})}{6.022 \times 10^{23}\ \text{mol}^{-1}} = 8.9\ \text{g cm}^{-3}$$

The expression $\rho = (n_{at}M_{at})/N_A$ is independent of the crystal structure.

## 1.8.2 CRYSTAL DIRECTIONS AND PLANES

There can be a number of possibilities for choosing a unit cell for a given crystal structure, as is apparent in Figure 1.33c and d for the HCP crystal. As a convention, we generally represent the **geometry of the unit cell** as a parallelepiped with sides $a$, $b$, and $c$ and angles $\alpha$, $\beta$, and $\gamma$, as depicted in Figure 1.40a. The sides $a$, $b$, and $c$ and angles $\alpha$, $\beta$, and $\gamma$ are referred to as the **lattice parameters.** To establish a reference frame and to apply three-dimensional geometry, we insert an $xyz$ coordinate system. The $x$, $y$, and $z$ axes follow the edges of the parallelepiped and the origin is at the lower-left rear corner of the cell. The unit cell extends along the $x$ axis from 0 to $a$, along $y$ from 0 to $b$, and along $z$ from 0 to $c$.

For Cu and Fe, the unit-cell geometry has $a = b = c, \alpha = \beta = \gamma = 90°$, and cubic symmetry. For Zn, the unit cell has hexagonal geometry, with $a = b \neq c$, $\alpha = \beta = 90°$, and $\gamma = 120°$, as shown in Figure 1.33d.

In explaining crystal properties, we must frequently specify a direction in a crystal, or a particular plane of atoms. Many properties, for example, the elastic modulus, electrical resistivity, magnetic susceptibility, etc., are directional within the crystal. We use the convention described here for labeling crystal directions based on three-dimensional geometry.

All parallel vectors have the same indices. Therefore, the direction to be labeled can be moved to pass through the origin of the unit cell. As an example, Figure 1.40b shows a direction whose indices are to be determined. A point $P$ on the vector can be expressed by the coordinates $x_o$, $y_o$, $z_o$, where $x_o$, $y_o$, and $z_o$ are projections from point $P$ onto the $x$, $y$, and $z$ axes, respectively, as shown in Figure 1.40b. It is generally convenient to place $P$ where the line cuts a surface (though this is not necessary). We can express these coordinates in terms of the lattice parameters $a$, $b$, and $c$, respectively. We then have three coordinates, say $x_1$, $y_1$, and $z_1$, for point $P$ in terms of $a$, $b$, and $c$.

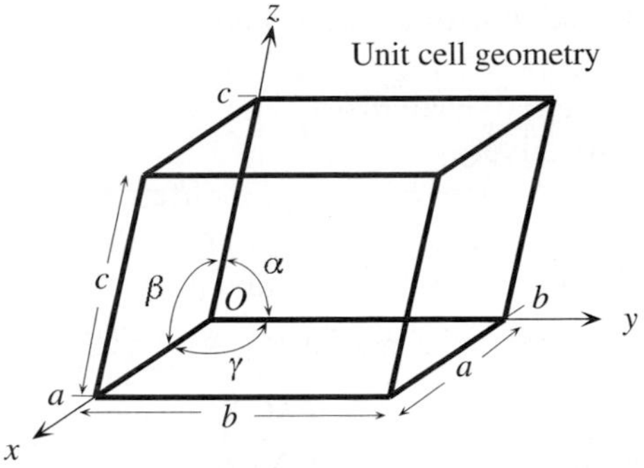


(a) A parallelepiped is chosen to describe the *geometry* of a unit cell. We line the *x*, *y*, and *z* axes with the edges of the parallelepiped taking the lower-left rear corner as the origin.

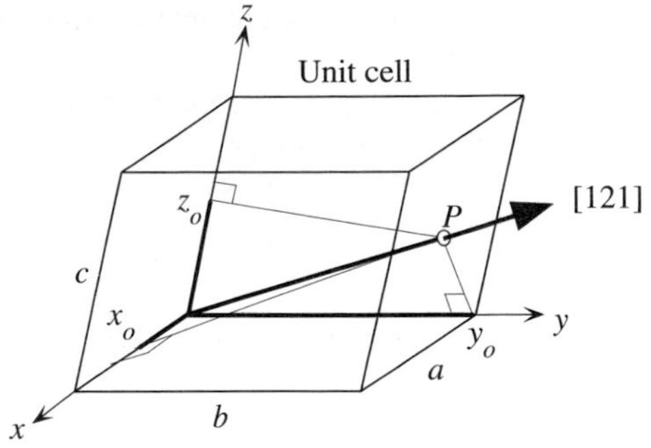


(b) Identification of a direction in a crystal.

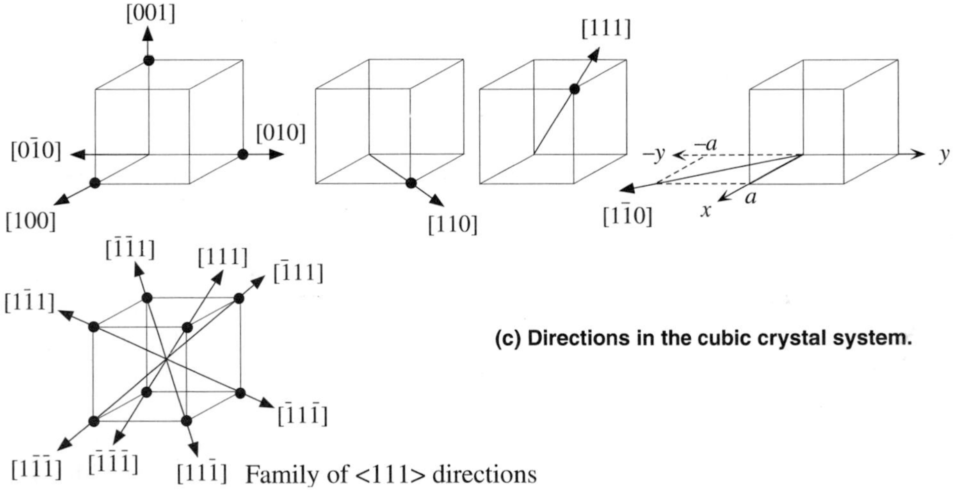


(c) Directions in the cubic crystal system.

**Figure 1.40**

For example, if

$$x_o,\ y_o,\ z_o \qquad \text{are} \qquad \frac{1}{2}a,\ b,\ \frac{1}{2}c$$

then $P$ is at

$$x_1,\ y_1,\ z_1 \qquad \textit{i.e.}, \qquad \frac{1}{2},\ 1,\ \frac{1}{2}$$

We then multiply or divide these numbers until we have the smallest integers (which may include 0). If we call these integers $u$, $v$, and $w$, then the direction is written in square brackets without commas as $[uvw]$. If any integer is a negative number, we use a bar on top of that integer. For the particular direction in Figure 1.40b, we therefore have [121].

Some of the important directions in a cubic lattice are shown in Figure 1. 40c. For example, the $x$, $y$, and $z$ directions in the cube are [100], [010], and [001], as shown. Reversing a direction simply changes the sign of each index. The negative $x$, $y$, and $z$ directions are $[\bar{1}00]$, $[0\bar{1}0]$, and $[00\bar{1}]$, respectively.

Certain directions in the crystal are equivalent because the differences between them are based only on our arbitrary decision for labeling $x$, $y$, and $z$ directions. For example, [100] and [010] are different simply because of the way in which we labeled the $x$ and $y$ axes. Indeed, directional properties of a material (*e.g.*, elastic modulus, and dielectric susceptibility) along the edge of the cube [100] are invariably the same as along the other edges, for example, [010] and [001]. All of these directions along the edges of the cube constitute a **family of directions,** which is any set of directions considered to be equivalent. We label a family of directions, for example, [100], [010], [001], . . . , by using a common notation, triangular brackets. Thus, $\langle 100 \rangle$ represents the family of six directions, [100], [010], [001], $[\bar{1}00]$, $[0\bar{1}0]$, and $[00\bar{1}]$ in a cubic crystal. Similarly, the family of diagonal directions in the cube, shown in Figure 1.40c, is denoted $\langle 111 \rangle$.

We also frequently need to describe a particular plane in a crystal. Figure 1.41 shows a general unit cell with a plane to be labeled. We use the following convention, called the **Miller indices of a plane,** for this purpose.

We take the intercepts $x_o$, $y_o$, and $z_o$ of the plane on the $x$, $y$, and $z$ axes, respectively. If the plane passes through the origin, we can use another convenient parallel plane, or simply shift the origin to another point. All planes that have been shifted by a lattice parameter have identical Miller indices.

We express the intercepts $x_o$, $y_o$, and $z_o$ in terms of the lattice parameters $a$, $b$, and $c$, respectively, to obtain $x_1$, $y_1$, and $z_1$. We then invert these numbers. Taking the reciprocals, we obtain

$$\frac{1}{x_1}, \frac{1}{y_1}, \frac{1}{z_1}$$

We then clear all fractions, without reducing to lowest integers, to obtain a set of integers, say $h$, $k$, and $\ell$. We then put these integers into parentheses, without commas, that is, $(hk\ell)$. For the plane in Figure 1.41a, we have

Intercepts $x_o$, $y_o$, and $z_o$ are $\frac{1}{2}a$, $1b$, and $\infty$ $c$.

Intercepts $x_1$, $y_1$, and $z_1$, in terms of $a$, $b$, and $c$, are $\frac{1}{2}$, 1, and $\infty$.

Reciprocals $1/x_1$, $1/y_1$, and $1/z_1$ are $1/\frac{1}{2}, 1/1, 1/\infty = 2, 1, 0$.

This set of numbers does not have fractions, so it is not necessary to clear fractions. Hence, the Miller indices $(hk\ell)$ are (210).

If there is a negative integer due to a negative intercept, a bar is placed across the top of the integer. Also, if parallel planes differ only by a shift that involves a multiple number of lattice parameters, then these planes may be assigned the same Miller indices. For example, the plane $(0\bar{1}0)$ is the $xz$ plane that cuts the $y$ axis at $-b$. If we shift the plane along $y$ by two lattice parameters ($2b$), it will cut the $y$ axis at $b$ and the Miller indices will become (010). In terms of the unit cell, the $(0\bar{1}0)$ plane is the same as the (010) plane, as shown in Figure 1.41b. Note that not all parallel planes are

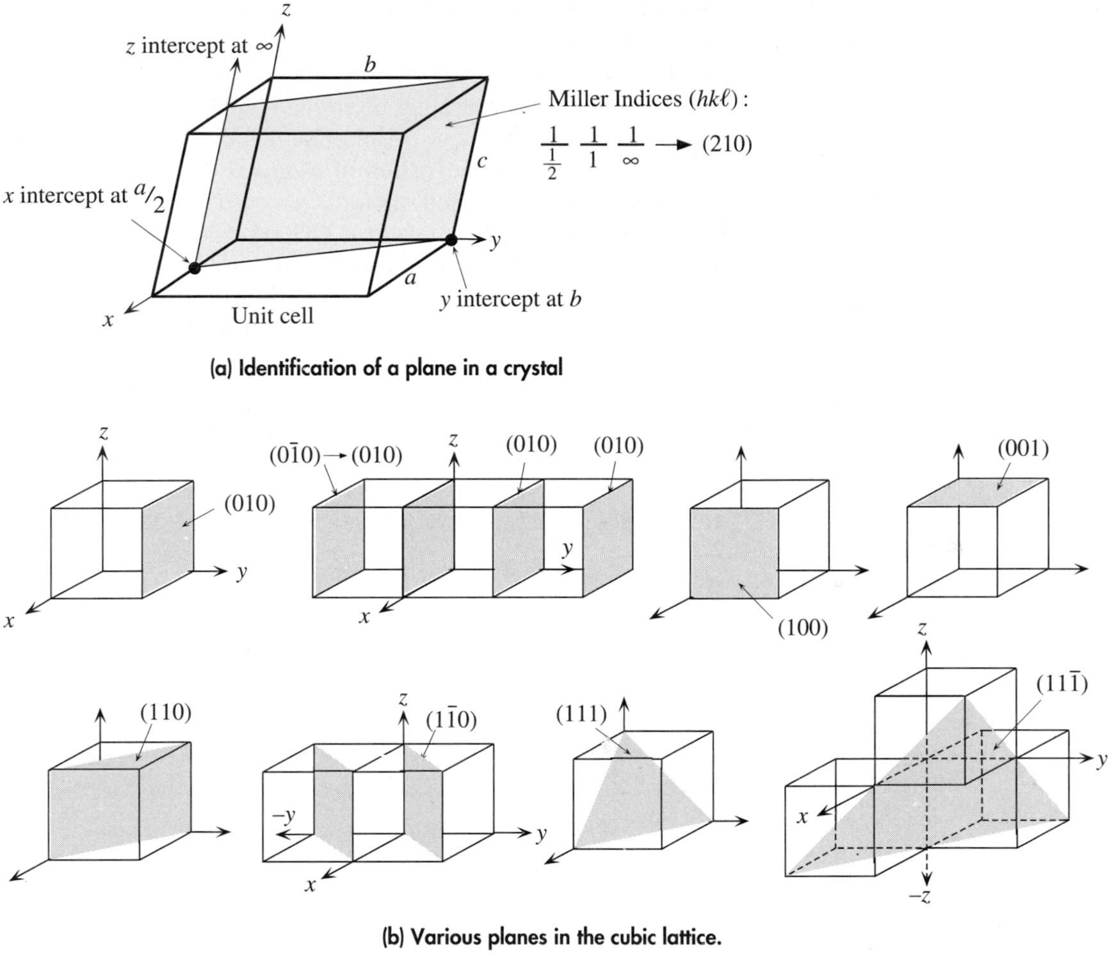


**Figure 1.41** Labeling of crystal planes and typical examples in the cubic lattice.

identical. Planes can have the same Miller indices *only* if they are separated by a multiple of the lattice parameter. For example, the (010) plane is not identical to the (020) plane, even though they are geometrically parallel. In terms of the unit cell, plane (010) is a face of the unit cell cutting the $y$ axis at $b$, whereas (020) is a plane that is halfway inside the unit cell, cutting the $y$ axis at $\frac{1}{2}b$. The planes contain different numbers of atoms. The (020) plane cannot be shifted by the lattice parameter $b$ to coincide with plane (010).

It is apparent from Figure 1.41b that in the case of the cubic crystal, the $[hk\ell]$ direction is always perpendicular to the $(hk\ell)$ plane.

Certain planes in the crystal belong to a **family of planes** because their indices differ only as a consequence of the arbitrary choice of axis labels. For example, the indices of the (100) plane become (010) if we switch the $x$ and $y$ axes. All the (100),

(010), and (001) planes, and hence the parallel $(\bar{1}00)$, $(0\bar{1}0)$, $(00\bar{1})$ planes, form a family of planes, conveniently denoted by curly brackets as {100}.

Frequently we need to know the number of atoms per unit area on a given plane $(hk\ell)$. For example, if the surface concentration of atoms is high on one plane, then that plane may encourage oxide growth more rapidly than another plane where there are less atoms per unit area. **Planar concentration of atoms** is the number of atoms per unit area, that is, the surface concentration of atoms, on a given plane in the crystal. Among the {100}, {110}, and {111}, planes in FCC crystals, the most densely packed planes, those with the highest planar concentration, are {111} planes and the least densely packed are {110}.

**EXAMPLE 1.14**

**MILLER INDICES AND PLANAR CONCENTRATION** Consider the plane shown in Figure 1.42a, which passes through one side of a face and the center of an opposite face in the FCC lattice. The plane passes through the origin at the lower-left rear corner. We therefore shift the origin to say point $O'$ at the lower-right rear corner of the unit cell. In terms of $a$, the plane cuts the $x$, $y$, and $z$ axes at $\infty$, $-1$, $\frac{1}{2}$, respectively. We take the reciprocals to obtain, 0, $-1$, 2. Therefore, the Miller indices are $(0\bar{1}2)$.

To calculate the planar concentration $n_{(hk\ell)}$ on a given $(hk\ell)$ plane, we consider a bound area $A$ of the $(hk\ell)$ plane within the unit cell as in Figure 1.42b. Only atoms whose centers lie on $A$ are involved in $n_{(hk\ell)}$. For each atom, we then evaluate what portion of the atomic cross section (a circle in two dimensions) cut by the plane $(hk\ell)$ is contained within $A$. Consider the Cu FCC crystal with $a = 0.3620$ nm.

The (100) plane corresponds to a cube face and has an area $A = a^2$. There is one full atom at the center; that is, the (100) plane cuts through one full atom, one full circle in two dimensions, at the face center as in Figure 1.42b. However, not all corner atoms are within $A$. Only a quarter of a circle is within the bound area $A$ in Figure 1.42b.

$$\text{Number of atoms in } A = (4 \text{ corners}) \times \left(\tfrac{1}{4} \text{ atom}\right) + 1 \text{ atom at face center} = 2$$

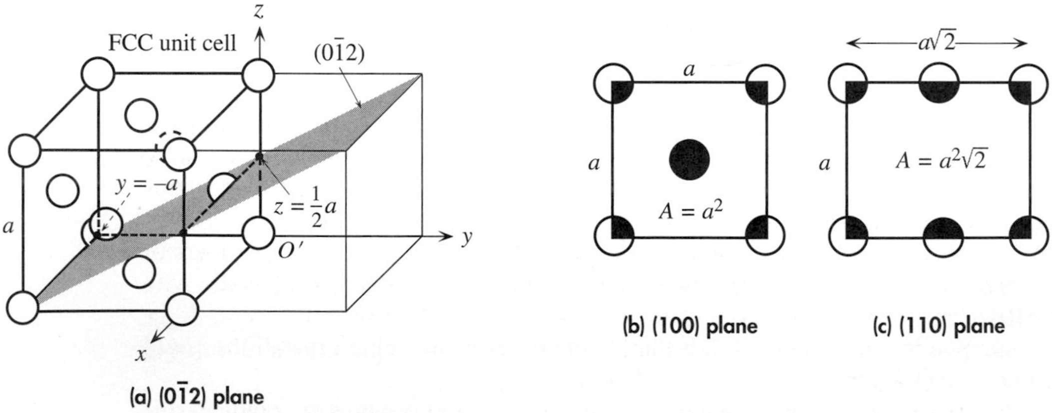


**Figure 1.42** The $(0\bar{1}2)$ plane and planar concentrations in an FCC crystal.

Planar concentration $n_{(100)}$ of (100) is

$$n_{(100)} = \frac{4(\frac{1}{4}) + 1}{a^2} = \frac{2}{a^2} = \frac{2}{(0.3620 \times 10^{-9}\ \text{m})^2} = 15.3 \text{ atoms nm}^{-2}$$

Consider the (110) plane as in Figure 1.42c. The number of atoms in the area $A = (a)(a\sqrt{2})$ defined by two face diagonals and two cube sides is

$$(4 \text{ corners}) \times \left(\tfrac{1}{4} \text{ atom}\right) + (2 \text{ face diagonals}) \times \left(\tfrac{1}{2} \text{ atom at diagonal center}\right) = 2$$

Planar concentration on (110) is

$$n_{(110)} = \frac{4(\frac{1}{4}) + 2(\frac{1}{2})}{(a)(a\sqrt{2})} = \frac{2}{a^2\sqrt{2}} = 10.8 \text{ atoms nm}^{-2}$$

Similar for the (111) plane, $n_{(111)}$ is 17.0 atoms nm$^{-2}$. Clearly the (111) planes are the most and (110) planes are the least densely packed among the (100), (110), and (111) planes.

---

## 1.8.3 ALLOTROPY AND CARBON

Certain substances can have more than one crystal structure, iron being one of the best-known examples. This characteristic is termed **polymorphism** or **allotropy.** Below 912 °C, iron has the BCC structure and is called $\alpha$-Fe. Between 912 °C and 1400 °C, iron has the FCC structure and is called $\gamma$-Fe. Above 1400 °C, iron again has the BCC structure and is called $\delta$-Fe. Since iron has more than one crystal structure, it is called **polymorphic.** Each iron crystal structure is an allotrope or a polymorph.

The allotropes of iron are all metals. Furthermore, one allotrope changes to another at a well-defined temperature called a **transition temperature,** which in this case is 912 °C.

Many substances have allotropes that exhibit widely different properties. Moreover, for some polymorphic substances, the transformation from one allotrope to another cannot be achieved by a change of temperature, but requires the application of pressure, as in the transformation of graphite to diamond.

Carbon has three important crystalline allotropes: diamond, graphite, and the newly discovered **buckminsterfullerene.** These crystal structures are shown in Figure 1.43a, b and c, respectively, and their properties are summarized in Table 1.4. Graphite is the carbon form that is stable at room temperature. Diamond is the stable form at very high pressures. Once formed, diamond continues to exist at atmospheric pressures and below about 900 °C, because the transformation rate of diamond to graphite is virtually zero under these conditions. Graphite and diamond have widely differing properties, which lead to diverse applications. For example, graphite is an electrical conductor, whereas diamond is an insulator. Diamond is the hardest substance known. On the other hand, the carbon layers in graphite can readily slide over each other under shear stresses, because the layers are only held together by weak secondary bonds (van der Waals bonds). This is the reason for graphite's lubricating properties.

Buckminsterfullerene is another polymorph of carbon. In the buckminsterfullerene molecule (called the "buckyball"), 60 carbon atoms bond with each other to form a

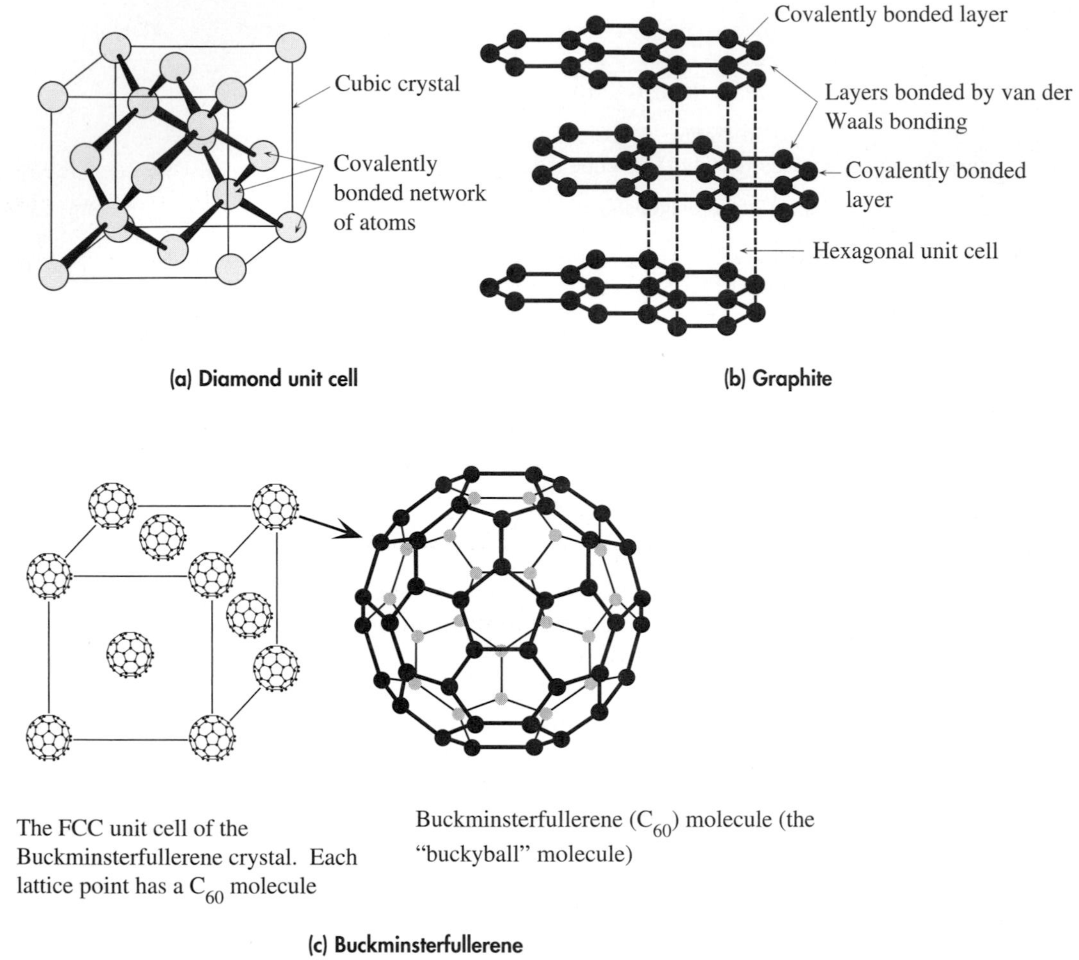


**Figure 1.43** The three allotropes of carbon.

perfect soccer ball–type molecule. The $C_{60}$ molecule has 12 pentagons and 20 hexagons joined together to form a spherical molecule, with each C atom at a corner, as depicted in Figure 1.43c. The molecules are produced in the laboratory by a carbon arc in a partial atmosphere of an inert gas (He); they are also found in the soot of partial combustion. The crystal form of buckminsterfullerene has the FCC structure, with each $C_{60}$ molecule occupying a lattice point and being held together by van der Waals forces, as shown in Figure 1.43c. The Buckminsterfullerene crystal is a semiconductor, and its compounds with alkali metals, such as $K_3C_{60}$, exhibit superconductivity at low temperatures (below 18 K). Mechanically, it is a soft material.

Diamond, graphite, and the fullerene crystals are not the only crystalline allotropes of carbon, and neither are they the only structural forms of carbon. For example, **lonsdaleite,** which is another crystalline allotrope, is *hexagonal diamond*

**Table 1.4** Crystalline allotropes of carbon ($\rho$ is the density and $Y$ is the elastic modulus or Young's modulus)

| | Graphite | Diamond | Buckminsterfullerene Crystal |
|---|---|---|---|
| **Structure** | Covalent bonding within layers. Van der Waals bonding between layers. Hexagonal unit cell. | Covalently bonded network. Diamond crystal structure. | Covalently bonded $C_{60}$ spheroidal molecules held in an FCC crystal structure by van der Waals bonding. |
| **Electrical and thermal properties** | Good electrical conductor. Thermal conductivity comparable to metals. | Very good electrical insulator. Excellent thermal conductor, about five times more than silver or copper. | Semiconductor. Compounds with alkali metals (*e.g.*, $K_3C_{60}$) exhibit superconductivity. |
| **Mechanical properties** | Lubricating agent. Machinable. Bulk graphite: $Y \approx 27$ GPa $\rho = 2.25$ g cm$^{-3}$ | The hardest material. $Y = 827$ GPa $\rho = 3.25$ g cm$^{-3}$ | Mechanically soft. $Y \approx 18$ GPa $\rho = 1.65$ g cm$^{-3}$ |
| **Comment** | Stable allotrope at atmospheric pressure | High-pressure allotrope. | Laboratory synthesized. Occurs in the soot of partial combustion. |
| **Uses, potential uses** | Metallurgical crucibles, welding electrodes, heating elements, electrical contacts, refractory applications. | Cutting tool applications. Diamond anvils. Diamond film coated drills, blades, bearings, etc. Jewelry. Heat conductor for ICs. Possible thin-film semiconductor devices, as the charge carrier mobilities are large. | Possible future semiconductor or superconductivity applications. |

in which each C atom covalently bonds to four neighbors, as in diamond, but the crystal structure has hexagonal symmetry. (It forms from graphite on meteors when the meteors impact the Earth; currently it is only found in Arizona.) **Amorphous carbon** has no crystal structure (no long-range order), so it is not a crystalline allotrope, but many scientists define it as a form or phase of carbon, or as a structural "allotrope." The recently discovered **carbon nanotubes** are thin and long carbon tubes, perhaps 10 to 100 microns long but only several nanometers in diameter, hence the name *nanotube*. They are tubes made from rolling a graphite sheet into a tube and then capping the ends with hemispherical buckyballs. The carbon tube is really a single macromolecule rather than a crystal in its traditional sense[12]; it is a structural form of carbon. Carbon nanotubes have many interesting and remarkable properties and offer much potential for various applications in electronics; the most topical currently being carbon nanotube field emission devices. (Chapter 4 has an example.)

[12] It is possible to define a unit cell on the surface of a carbon nanotube and apply various crystalline concepts, as some scientists have done. To date, however, there seems to be no single crystal of carbon nanotubes in the same way that there is a fullerene crystal in which the $C_{60}$ molecules are bonded to form an FCC structure.

## 1.9 CRYSTALLINE DEFECTS AND THEIR SIGNIFICANCE

By bringing all the atoms together to try to form a perfect crystal, we lower the total potential energy of the atoms as much as possible for that particular structure. What happens when the crystal is grown from a liquid or vapor; do you always get a perfect crystal? What happens when the temperature is raised? What happens when impurities are added to the solid?

There is no such thing as a perfect crystal. We must therefore understand the types of defects that can exist in a given crystal structure. Quite often, key mechanical and electrical properties are controlled by these defects.

### 1.9.1 POINT DEFECTS: VACANCIES AND IMPURITIES

Above the absolute zero temperature, all crystals have atomic vacancies or atoms missing from lattice sites in the crystal structure. The vacancies exist as a requirement of thermal equilibrium and are called **thermodynamic defects.** Vacancies introduce disorder into the crystal by upsetting the perfect periodicity of atomic arrangements.

We know from the kinetic molecular theory that all the atoms in a crystal vibrate about their equilibrium positions with a distribution of energies, a distribution that closely resembles the Boltzmann distribution. At some instant, there may be one atom with sufficient energy to break its bonds and jump to an adjoining site on the surface, as depicted in Figure 1.44. This leaves a vacancy behind, just below the surface. This vacancy can then diffuse into the bulk of the crystal, because a neighboring atom can diffuse into it.

This latter process of vacancy creation has been shown to be a sequence of events in Figure 1.44. Suppose that $E_v$ is the average energy required to create such a vacancy. Then only a fraction, $\exp(-E_v/kT)$, of all the atoms in the crystal can have

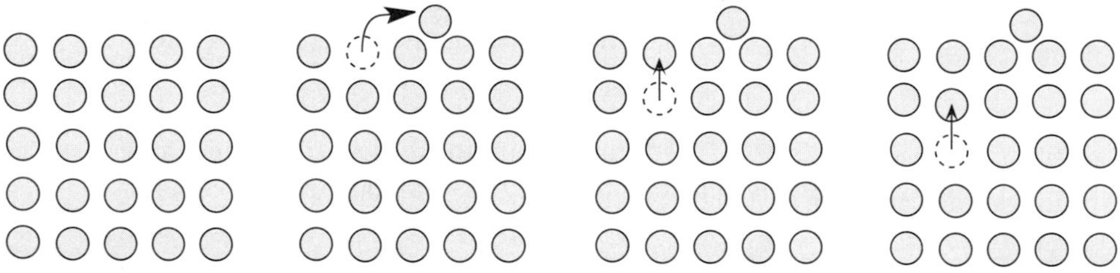

**(a) Perfect crystal without vacancies**

**(b) An energetic atom at the surface breaks bonds and jumps on to a new adjoining position on the surface. This leaves behind a vacancy.**

**(c) An atom in the bulk diffuses to fill the vacancy thereby displacing the vacancy toward the bulk.**

**(d) Atomic diffusions cause the vacancy to diffuse into the bulk.**

**Figure 1.44** Generation of a vacancy by the diffusion of an atom to the surface and the subsequent diffusion of the vacancy into the bulk.

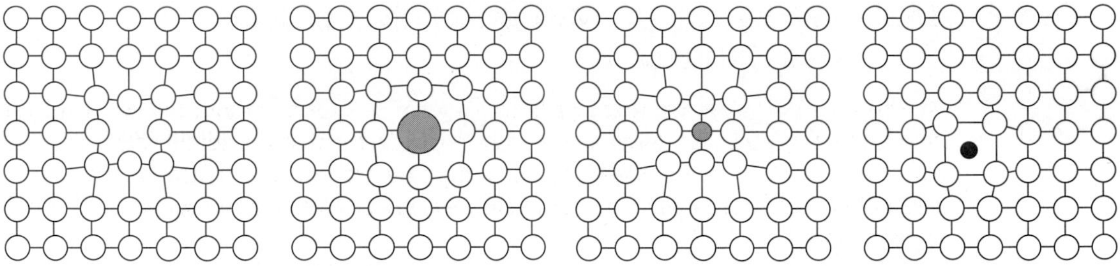

(a) A vacancy in the crystal.

(b) A substitutional impurity in the crystal. The impurity atom is larger than the host atom.

(c) A substitutional impurity in the crystal. The impurity atom is smaller than the host atom.

(d) An interstitial impurity in the crystal. The impurity occupies an empty space between host atoms.

**Figure 1.45** Point defects in the crystal structure.
The regions around the point defect become distorted; the lattice becomes strained.

sufficient energy to create vacancies. If the number of atoms per unit volume in the crystal is $N$, then the vacancy concentration $n_v$ is given by[13]

*Equilibrium concentration of vacancies*

$$n_v = N \exp\left(-\frac{E_v}{kT}\right) \qquad [1.35]$$

At all temperatures above absolute zero, there will always be an equilibrium concentration of vacancies, as dictated by Equation 1.35. Although we considered only one possible vacancy creation process in Figure 1.44 there are other processes that also create vacancies. Furthermore, we have shown the vacancy to be the same size in the lattice as the missing atom, which is not entirely true. The neighboring atoms around a vacancy close in to take up some of the slack, as shown in Figure 1.45a. This means that the crystal lattice around the vacancy is distorted from the perfect arrangement over a few atomic dimensions. The vacancy volume is therefore smaller than the volume of the missing atom.

Vacancies are only one type of **point defect** in a crystal structure. Point defects generally involve lattice changes or distortions of a few atomic distances, as depicted in Figure 1.45. The crystal structure may contain impurities, either naturally or as a consequence of intentional addition, as in the case of silicon crystals grown for microelectronics. If the impurity atom substitutes directly for the host atom, the result is called a **substitutional impurity** and the resulting crystal structure is that of a **substitutional solid solution,** as shown in Figure 1.45b and c. When a Si crystal is "doped" with small amounts of arsenic (As) atoms, the As atoms substitute directly for the Si atoms in the Si crystal; that is, the arsenic atoms are substitutional impurities. The impurity atom can also place itself in an interstitial site, that is, in a void between the host atoms, as

[13] The proper derivation of the vacancy concentration involves considering thermodynamics and equilibrium concepts. In the actual thermodynamic expression, the pre-exponential term in Equation 1.35 is not unity but a factor that depends on the change in the *entropy* of the crystal upon vacancy creation. For nearly all practical purposes Equation 1.35 is sufficient.

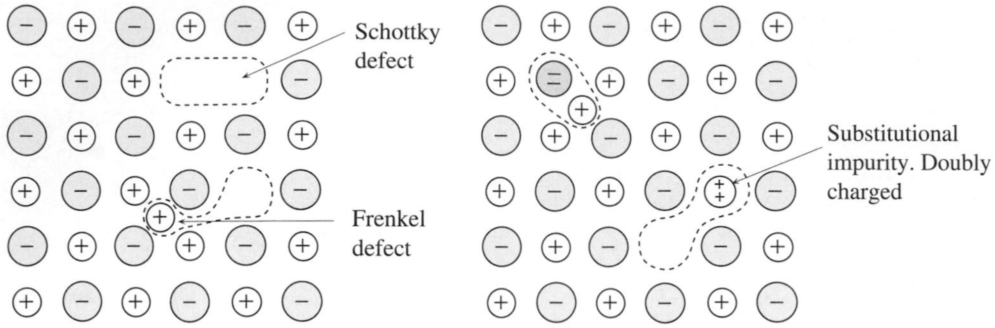


**(a) Schottky and Frenkel defects in an ionic crystal.**

**(b) Two possible imperfections caused by ionized substitutional impurity atoms in an ionic crystal.**

**Figure 1.46** Point defects in ionic crystals.

carbon does in the BCC iron crystal. In that case, the impurity is called an **interstitial impurity,** as shown in Figure 1.45d.

In general, the impurity atom will have both a different valency and a different size. It will therefore distort the lattice around it. For example, if a substitutional impurity atom is larger than the host atom, the neighboring host atoms will be pushed away, as in Figure 1.45b. The crystal region around an impurity is therefore distorted from the perfect periodicity and the lattice is said to be **strained around a point defect.** A smaller substitutional impurity atom will pull in the neighboring atoms, as in Figure 1.45c. Typically, interstitial impurities tend to be small atoms compared to the host atoms, a typical example being the small carbon atom in the BCC iron crystal.

In an ionic crystal, such as NaCl, which consists of anions ($Cl^-$) and cations ($Na^+$), one common type of defect is called a **Schottky defect.** This involves a missing cation–anion pair (which may have migrated to the surface), so the neutrality is maintained, as indicated in Figure 1.46a. These Schottky defects are responsible for the major optical and electrical properties of alkali halide crystals. Another type of defect in the ionic crystal is the **Frenkel defect,** which occurs when a host ion is displaced into an interstitial position, leaving a vacancy at its original site. The interstitial ion and the vacancy pair constitute the Frenkel defect, as identified in Figure 1.46a. For the AgCl crystal, which has predominantly Frenkel defects, an $Ag^+$ is in an interstitial position. The concentration of such Frenkel defects is given by Equation 1.35, with an appropriate defect creation energy $E_{\text{defect}}$ instead of $E_v$.

Ionic crystals can also have substitutional and interstitial impurities that become ionized in the lattice. Overall, the ionic crystal must be neutral. Suppose that an $Mg^{2+}$ ion substitutes for an $Na^+$ ion in the NaCl crystal, as depicted in Figure 1.46b. Since the overall crystal must be neutral, either one $Na^+$ ion is missing somewhere in the crystal, or an additional $Cl^-$ ion exists in the crystal. Similarly, when a doubly charged negative ion, such as $O^{2-}$, substitutes for $Cl^-$, there must either be an additional cation (usually in an interstitial site) or a missing $Cl^-$ somewhere in order to maintain charge neutrality in the crystal. The most likely type of defect depends on the composition of the ionic solid and the relative sizes and charges of the ions.

**EXAMPLE 1.15**

**VACANCY CONCENTRATION IN A METAL** The energy of formation of a vacancy in the aluminum crystal is about 0.70 eV. Calculate the fractional concentration of vacancies in Al at room temperature, 300 K, and very close to its melting temperature 660 °C. What is the vacancy concentration at 660 °C given that the atomic concentration in Al is about $6.0 \times 10^{22}$ cm$^{-3}$?

**SOLUTION**

Using Equation 1.35, the fractional concentration of vacancies are as follows:
At 300 K,

$$\frac{n_v}{N} = \exp\left(-\frac{E_v}{kT}\right) = \exp\left[-\frac{(0.70\text{ eV})(1.6 \times 10^{-19}\text{ J eV}^{-1})}{(1.38 \times 10^{-23}\text{ J K}^{-1})(300\text{ K})}\right]$$

$$= 1.7 \times 10^{-12}$$

At 660 °C or 933 K,

$$\frac{n_v}{N} = \exp\left(-\frac{E_v}{kT}\right) = \exp\left[-\frac{(0.70\text{ eV})(1.6 \times 10^{-19}\text{ J eV}^{-1})}{(1.38 \times 10^{-23}\text{ J K}^{-1})(933\text{ K})}\right]$$

$$= 1.7 \times 10^{-4}$$

That is, almost 1 in 6000 atomic sites is a vacancy. The atomic concentration $N$ in Al is about $6.0 \times 10^{22}$ cm$^{-3}$, which means that the vacancy concentration $n_v$ at 660 °C is

$$n_v = (6.0 \times 10^{22}\text{ cm}^{-3})(1.7 \times 10^{-4}) = 1.0 \times 10^{19}\text{ cm}^{-3}$$

The mean vacancy separation (on the order of $n_v^{-1/3}$) at 660 °C is therefore roughly 5 nm. The mean atomic separation in Al is ~0.3 nm ($\sim N^{-1/3}$), so the mean separation between vacancies is only about 20 atomic separations! (A more accurate version of Equation 1.35, with an entropy term, shows that the vacancy concentration is even higher than the estimate in this example.) The increase in the linear thermal expansion coefficient of a metal with temperature near its melting temperature, as shown for Mo in Figure 1.20, has been attributed to the generation of vacancies in the crystal.

**EXAMPLE 1.16**

**VACANCY CONCENTRATION IN A SEMICONDUCTOR** The energy of vacancy formation in the Ge crystal is about 2.2 eV. Calculate the fractional concentration of vacancies in Ge at 938 °C, just below its melting temperature. What is the vacancy concentration given that the atomic mass $M_{at}$ and density $\rho$ of Ge are 72.64 g mol$^{-1}$ and 5.32 g cm$^{-3}$, respectively? Neglect the change in the density with temperature which is small compared with other approximations in Equation 1.35.

**SOLUTION**

Using Equation 1.34, the fractional concentration of vacancies at 938 °C or 1211 K is

$$\frac{n_v}{N} = \exp\left(-\frac{E_v}{kT}\right) = \exp\left[-\frac{(2.2\text{ eV})(1.6 \times 10^{-19}\text{ J eV}^{-1})}{(1.38 \times 10^{-23}\text{ J K}^{-1})(1211\text{ K})}\right] = 7.0 \times 10^{-10}$$

which is orders of magnitude less than that for Al at its melting temperature in Example 1.15; vacancies in covalent crystals cost much more energy than those in metals.

The number of Ge atoms per unit volume is

$$N = \frac{\rho N_A}{M_{at}} = \frac{(5.32 \text{ g cm}^{-3})(6.022 \times 10^{23} \text{ g mol}^{-1})}{72.64 \text{ g mol}^{-1}} = 4.41 \times 10^{22} \text{ cm}^{-3}$$

so that at 938 °C,

$$n_v = (4.4 \times 10^{22} \text{ cm}^{-3})(7.0 \times 10^{-10}) = 3.1 \times 10^{13} \text{ cm}^{-3}$$

Only 1 in $10^9$ atoms is a vacancy.

## 1.9.2 LINE DEFECTS: EDGE AND SCREW DISLOCATIONS

A line defect is formed in a crystal when an atomic plane terminates within the crystal instead of passing all the way to the end of the crystal, as depicted in Figure 1.47a. The edge of this short plane of atoms is therefore like a line running inside the crystal. The planes neighboring (*i.e.,* above) this short plane are dislocated (displaced) with respect to those below the line. We therefore call this type of defect an **edge dislocation** and use an inverted ⊤ symbol. The vertical line corresponds to the half-plane of atoms in the crystal, as illustrated in Figure 1.47a. It is clear that the atoms around the dislocation line have been effectively displaced from their perfect-crystal equilibrium positions, which results in atoms being out of registry above and below the dislocation. The atoms above the dislocation line are pushed together, whereas those below it are pulled apart, so there are regions of compression and tension above and below the dislocation line, respectively, as depicted by the shaded region around the dislocation line in Figure 1.47b. Therefore, around a dislocation line, we have a **strain field** due to the stretching or compressing of bonds.

The energy required to create a dislocation is typically in the order of 100 eV per nm of dislocation line. On the other hand, it takes only a few eV to form a point defect,

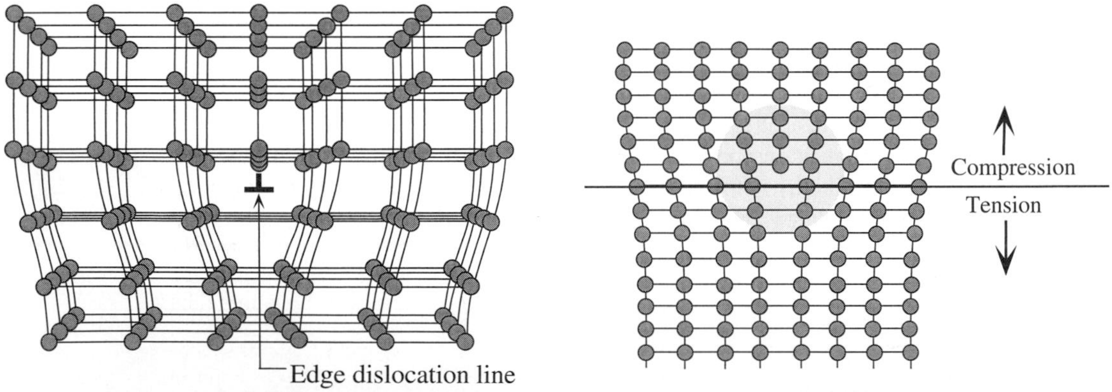


**(a) Dislocation is a line defect. The dislocation shown runs into the paper.**

**(b) Around the dislocation there is a strain field as the atomic bonds have been compressed above and stretched below the dislocation line.**

**Figure 1.47** Dislocation in a crystal. This is a line defect, which is accompanied by lattice distortion and hence a lattice strain around it.

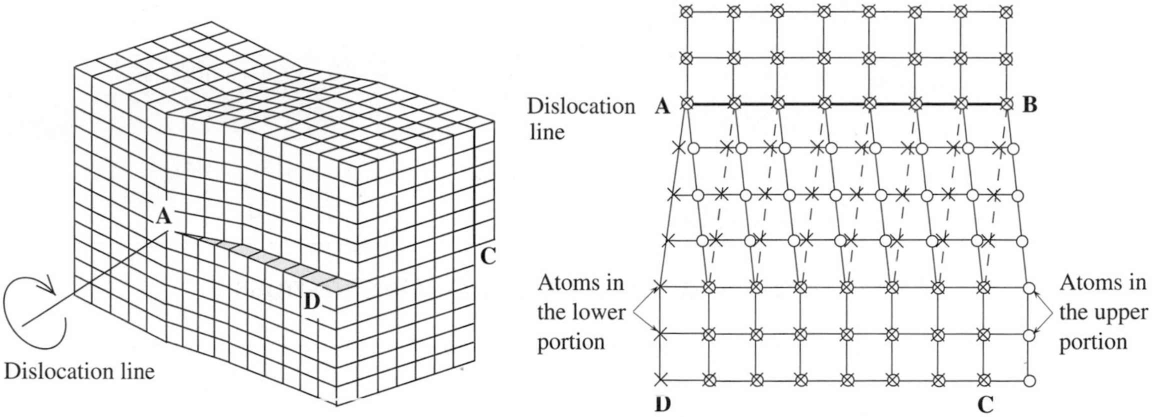


(a) A screw dislocation in a crystal

(b) The screw dislocation in (a) as viewed from above

**Figure 1.48** A screw dislocation, which involves shearing one portion of a perfect crystal with respect to another, on one side of a line (AB).

which is a few nanometers in dimension. In other words, forming a number of point defects is energetically more favorable than forming a dislocation. Dislocations are not equilibrium defects. They normally arise when the crystal is deformed by stress, or when the crystal is actually being grown.

Another type of dislocation is the **screw dislocation,** which is essentially a shearing of one portion of the crystal with respect to another, by one atomic distance, as illustrated in Figure 1.48a. The displacement occurs on either side of the **screw dislocation line.** The circular arrow around the line symbolizes the screw dislocation. As we move away from the dislocation line, the atoms in the upper portion become more out of registry with those below; at the edge of the crystal, this displacement is one atomic distance, as illustrated in Figure 1.48b.

Both edge and screw dislocations are generally created by stresses resulting from thermal and mechanical processing. A line defect is not necessarily either a pure edge or a pure screw dislocation; it can be a mixture, as depicted in Figure 1.49. Screw dislocations frequently occur during crystal growth, which involves atomic stacking on the surface of a crystal. Such dislocations aid crystallization by providing an additional "edge" to which the incoming atoms can attach, as illustrated in Figure 1.50. To explain, if an atom arrives at the surface of a perfect crystal, it can only attach to one atom in the plane below. However, if there is a screw dislocation, the incoming atom can attach to an edge and thereby form more bonds; hence, it can lower its potential energy more than anywhere else on the surface. With incoming atoms attaching to the edges, the growth occurs spirally around the screw dislocation, and the final crystal surface reflects this spiral growth geometry.

The phenomenon of **plastic** or **permanent deformation** of a metal depends totally on the presence and motions of dislocations, as discussed in elementary books on the mechanical properties of materials. In the case of electrical properties of metals,

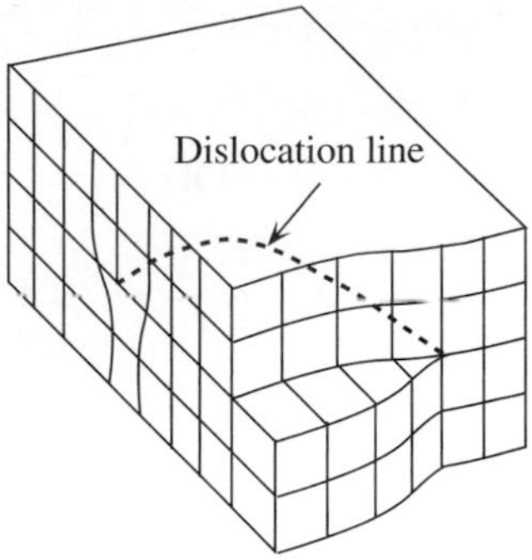


**Figure 1.49** A mixed dislocation.

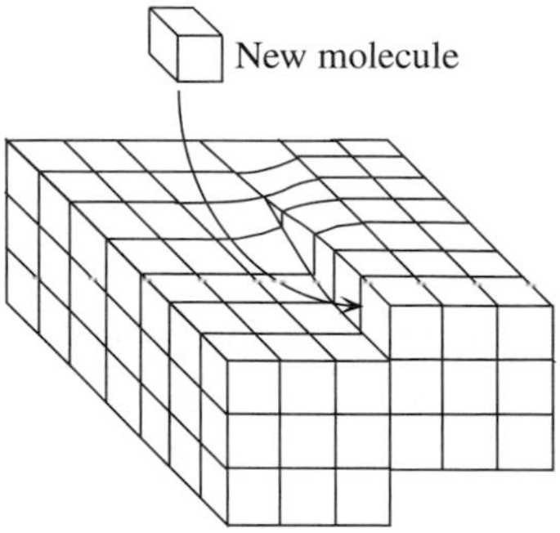


**Figure 1.50** Screw dislocation aids crystal growth because the newly arriving atom can attach to two or three atoms instead of one atom and thereby form more bonds.

Growth spiral on the surface of a polypropylene crystal due to screw dislocation aided crystal growth.

SOURCE: Photo by Phillip Geil, Courtesy of Case Western Reserve University.

we will see in Chapter 2 that dislocations increase the resistivity of materials, cause significant leakage current in a *pn* junction, and give rise to unwanted noise in various semiconductor devices. Fortunately, the occurrence of dislocations in semiconductor crystals can be controlled and nearly eliminated. In a metal interconnection line on a chip, there may be an average of $10^4$–$10^5$ dislocation lines per $mm^2$ of crystal, whereas a silicon crystal wafer that is carefully grown may typically have only 1 dislocation line per $mm^2$ of crystal.

### 1.9.3 Planar Defects: Grain Boundaries

Many materials are polycrystalline; that is, they are composed of many small crystals oriented in different directions. In fact, the growth of a flawless single crystal from what is called the **melt** (liquid) requires special skills, in addition to scientific knowledge. When a liquid is cooled to below its freezing temperature, solidification does not occur at every point; rather, it occurs at certain sites called **nuclei,** which are small crystal-like structures containing perhaps 50 to 100 atoms. Figure 1.51a to c depicts a typical solidification process from the melt. The liquid atoms adjacent to a nucleus diffuse into the nucleus, thereby causing it to grow in size to become a small crystal, or a crystallite, called a **grain.** Since the nuclei are randomly oriented when they are formed, the grains have random crystallographic orientations during crystallite growth. As the liquid between the grains is consumed, some grains meet and obstruct each other. At the end of solidification, therefore, the whole structure has grains with irregular shapes and orientations, as shown in Figure 1.51c.

It is apparent from Figure 1.51c that in contrast to a single crystal, a polycrystalline material has grain boundaries where differently oriented crystals meet. As indicated in Figure 1.52, the atoms at the grain boundaries obviously cannot follow their

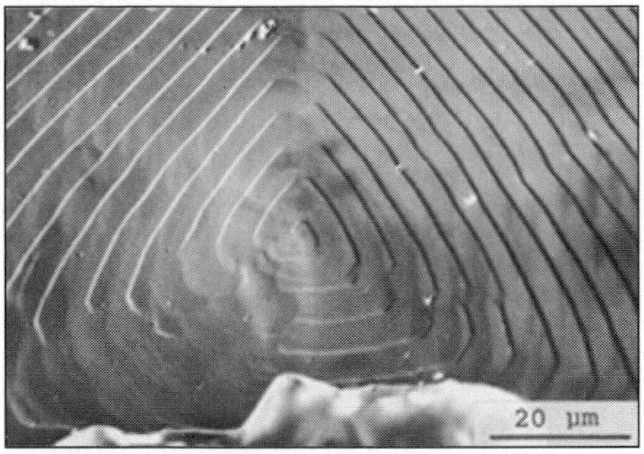


The photograph of the surface of a synthetic diamond grown on the (111) surface of natural diamond from sodium carbonate solvent at 5.5 GPa and 1600 °C.

SOURCE: Courtesy of Dr. Hisao Kanda, National Institute for Materials Science, Ibaraki, Japan.

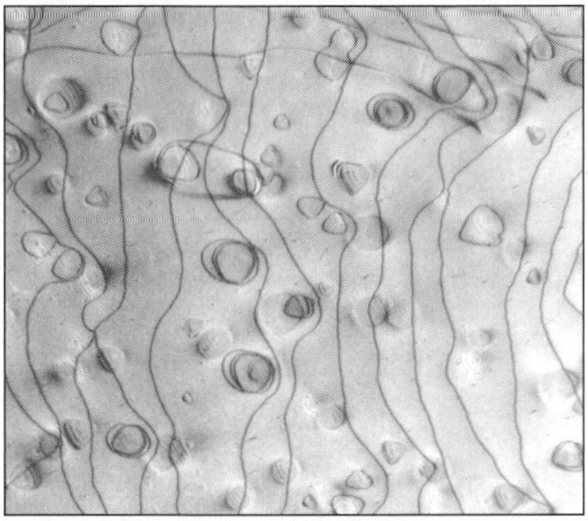

Dislocations can be seen by examining a thin slice of the sample under a transmission electron microscope (TEM). They appear as dark lines and loops as shown here in a Ni–Si alloy single crystal. The loop dislocations are around $Ni_3Si$ particles inside the crystal. The sample had been mechanically deformed, which generates dislocations.

SOURCE: Courtesy of Professor John Humphreys, UMIST, England. (J. Humphreys and V. Ramaswamy in *High Voltage Electron Microscopy*, ed. P. R. Swann. C. J. Humphreys and M. J. Goringe, New York: Academic Press, 1974, p. 26.)

Left: A polycrystalline diamond film on the (100) surface of a single crystal silicon wafer. The film thickness is 6 microns and the SEM magnification is 6000.

Right: A 6-micron-thick CVD diamond film grown on a single crystal silicon wafer. SEM magnification is 8000.

SOURCE: Courtesy of Dr. Paul May, The School of Chemistry, University of Bristol, England.

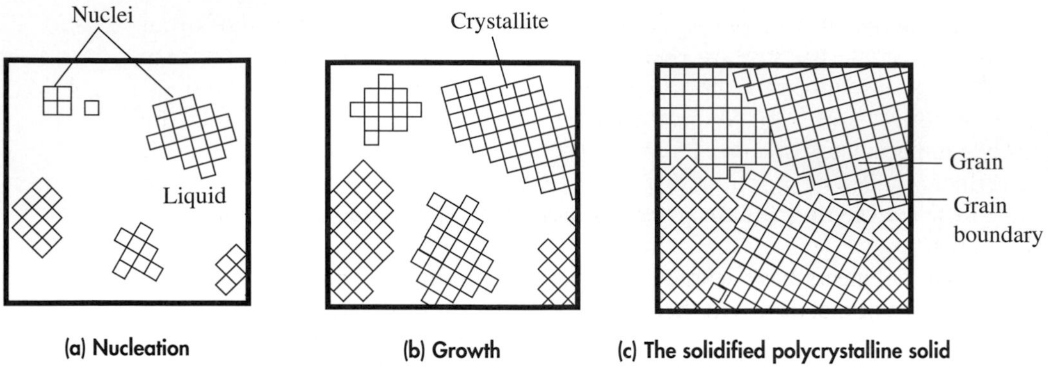


**Figure 1.51** Solidification of a polycrystalline solid from the melt. For simplicity, cubes represent atoms.

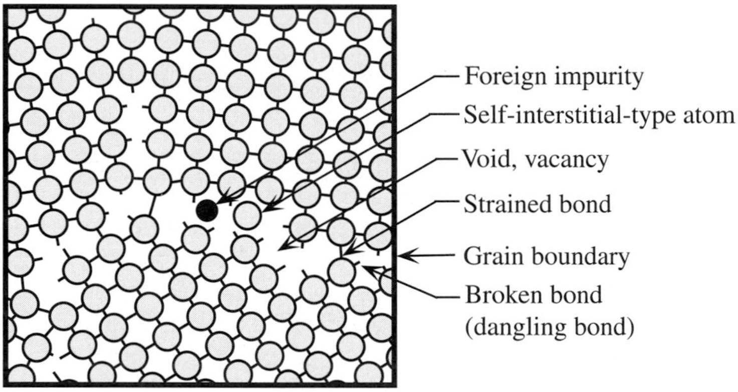


**Figure 1.52** The grain boundaries have broken bonds, voids, vacancies, strained bonds, and interstitial-type atoms.

The structure of the grain boundary is disordered, and the atoms in the grain boundaries have higher energies than those within the grains.

natural bonding habits, because the crystal orientation suddenly changes across the boundary. Therefore, there are both voids at the grain boundary and stretched and broken bonds. In addition, in this region, there are misplaced atoms that do not follow the crystalline pattern on either side of the boundary. Consequently, the grain boundary represents a high-energy region per atom with respect to the energy per atom within the bulk of the grains themselves. The atoms can diffuse more easily along a grain boundary because (a) less bonds need to be broken due to the presence of voids and (b) the bonds are strained and easily broken anyway. In many polycrystalline materials, impurities therefore tend to congregate in the grain boundary region. We generally refer to the atomic arrangement in the grain boundary region as being **disordered** due to the presence of the voids and misplaced atoms.

Since the energy of an atom at the grain boundary is greater than that of an atom within the grain, these grain boundaries are nonequilibrium defects; consequently, they try to reduce in size to give the whole structure a lower potential energy. At or around room temperature, the atomic diffusion process is slow; thus, the reduction in the grain boundary is insignificant. At elevated temperatures, however, atomic diffusion allows big grains to grow, at the expense of small grains, which leads to **grain coarsening (grain growth)** and hence to a reduction in the grain boundary area.

Mechanical engineers have learned to control the grain size, and hence the mechanical properties of metals to suit their needs, through various thermal treatment cycles. For electrical engineers, the grain boundaries become important when designing electronic devices based on polysilicon or any polycrystalline semiconductor. For example, in highly polycrystalline materials, particularly thin-film semiconductors (*e.g.*, polysilicon), the resistivity is invariably determined by polycrystallinity, or grain size, of the material, as discussed in Chapter 2.

### 1.9.4 CRYSTAL SURFACES AND SURFACE PROPERTIES

In describing crystal structures, we assume that the periodicity extends to infinity which means that the regular array of atoms is not interrupted anywhere by the presence of real surfaces of the material. In practice, we know that all substances have real surfaces. When the crystal lattice is abruptly terminated by a surface, the atoms at the surface cannot fulfill their bonding requirements as illustrated in Figure 1.53. For simplicity, the figure shows a Si crystal schematically sketched in two dimensions where each atom in the bulk of the crystal has four covalent bonds, each covalent bond

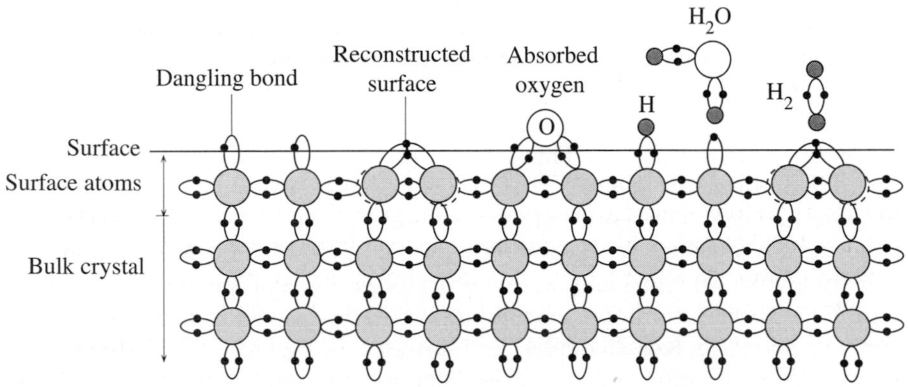


**Figure 1.53** At the surface of a hypothetical two-dimensional crystal, the atoms cannot fulfill their bonding requirements and therefore have broken, or dangling, bonds.
Some of the surface atoms bond with each other; the surface becomes reconstructed. The surface can have physisorbed and chemisorbed atoms.

having two electrons.[14] The atoms at the surface are left with **dangling bonds,** bonds that are half full, only having one electron. These dangling bonds are looking for atoms to which they can bond. Two neighboring surface atoms can share each other's dangling bond electrons, that is, form a surface bond with each other. This bonding between surface atoms causes a slight displacement of the surface atoms and leads to a surface that has been **reconstructed.**

Atoms from the environment can also bond with the atoms on the crystal surface. For example, a hydrogen atom can be captured by a dangling bond at the surface to form a chemical bond as a result of which hydrogen becomes **absorbed.** Primary bonding of foreign atoms to a crystal surface is called **chemisorption.** The H atom in Figure 1.53 forms a covalent bond with a Si atom and hence becomes **chemisorbed.** However, the $H_2O$ molecule cannot form a covalent bond, but, because of hydrogen bonding, it can form a secondary bond with a surface Si atom and become **adsorbed.** Secondary bonding of foreign atoms or molecules to a crystal surface is called **physisorption** (*physical adsorption*). Water molecules in the air can readily become adsorbed at the surface of a crystal. Although the figure also shows a physisorbed $H_2$ molecule as an example, this normally occurs at very low temperatures where crystal vibrations are too weak to quickly dislodge the $H_2$ molecule. It should be remarked that in many cases, atoms or molecules from the environment become adsorbed at the surface for only a certain period of time; they have a certain sticking or dwell time. For example, at room temperature, inert gases stick to a metal surface only for a duration of the order of microseconds, which is extremely long compared with the vibrational period of the crystal atoms ($\sim 10^{-12}$ seconds). A dangling bond can capture a free electron from the environment if one is available in its vicinity. The same idea applies to a dangling bond at a grain boundary as in Figure 1.52.

At sufficiently high temperatures, some of the absorbed foreign surface atoms can diffuse into the crystal volume to become bulk impurities. Many substances have a natural oxide layer on the surface that starts with the chemical bonding of oxygen atoms to the surface atoms and the subsequent growth of the oxide layer. For example, aluminum surfaces always have a thin aluminum oxide layer. In addition, the surface of the oxide often has adsorbed organic species of atoms usually from machining and handling. The surface condition of a Si crystal wafer in microelectronics is normally controlled by first etching the surface and then oxidizing it at a high temperature to form a $SiO_2$ **passivating layer** on the crystal surface. This oxide layer is an excellent barrier against the diffusion of impurity atoms into the crystal. (It is also an excellent electrical insulator.)

Figure 1.53 shows only some of the possibilities at the surface of a crystal. Generally the surface structure depends greatly on the mode of surface formation, which invariably involves thermal and mechanical processing, and previous environmental history. One visualization of a crystal surface is based on the **terrace-ledge-kink model,** the so-called **Kossel model,** as illustrated in Figure 1.54. The surface has ledges, kinks, and various imperfections such as holes and dislocations, as well as impurities which can diffuse to and from the surface. The dimensions of the various imperfections (*e.g.,* the step size) depend on the process that generated the surface.

---

[14] Not all possibilities shown in Figure 1.53 occur in practice; their occurrences depend on the preparation method of the crystal.

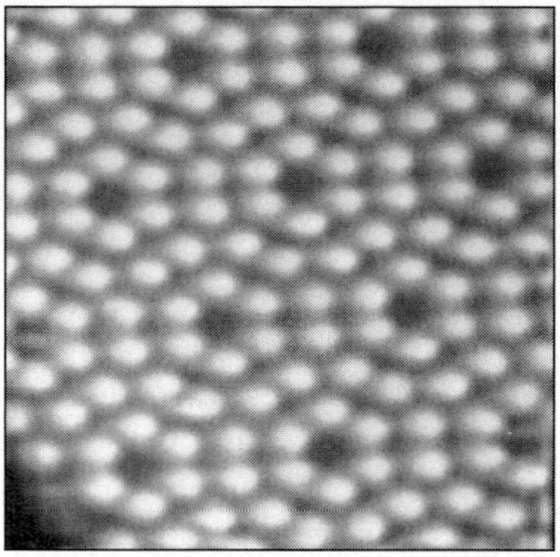

Atomic arrangements on a reconstructed (111) surface of a Si crystal as seen by a surface tunneling microscope.

SOURCE: Courtesy of Burleigh Instruments, Inc.

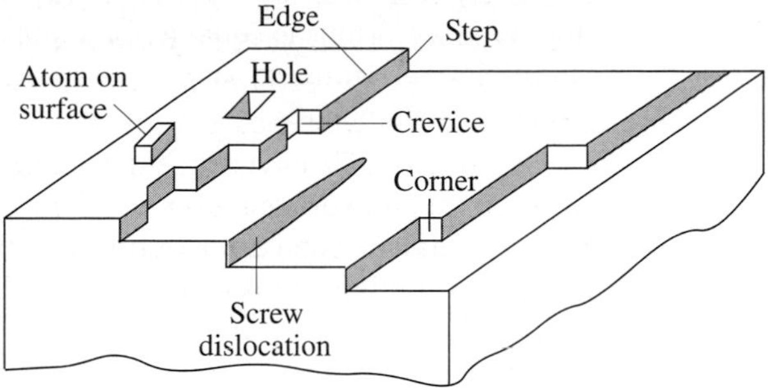


**Figure 1.54** Typically, a crystal surface has many types of imperfections, such as steps, ledges, kinks, crevices, holes, and dislocations.

## 1.9.5 STOICHIOMETRY, NONSTOICHIOMETRY, AND DEFECT STRUCTURES

**Stoichiometric compounds** are those that have an integer ratio of atoms, for example, as in $CaF_2$ where two F atoms bond with one Ca atom. Similarly, in the compound ZnO, if there is one O atom for every Zn atom, the compound is stoichiometric, as schematically illustrated in Figure 1.55a. Since there are equal numbers of $O^{2-}$ anions and $Zn^{2+}$ cations, the crystal overall is neutral. It is also possible to have a nonstoichiometric ZnO in which there is excess zinc. This may result if, for example, there is insufficient oxygen during the preparation of the compound. The $Zn^{2+}$ ion has a radius of 0.074 nm, which is about 1.9 times smaller than the $O^{2-}$ anion (radius of 0.14 nm), so it is much easier for a $Zn^{2+}$ ion to enter an interstitial site than the $O^{2-}$ ion or the Zn atom itself, which has a radius of 0.133 nm. Excess Zn atoms therefore occupy interstitial sites as $Zn^{2+}$ cations. Even though the excess zinc atoms are still ionized within the crystal, their lost electrons cannot be taken by oxygen atoms, which are all

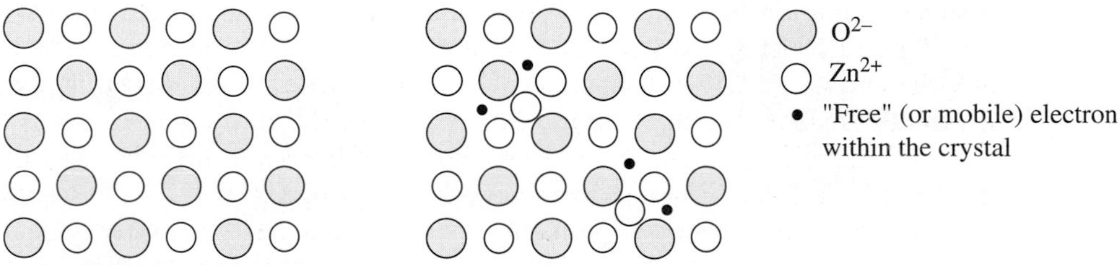


**(a) Stoichiometric ZnO crystal with equal number of anions and cations and no free electrons**

**(b) Nonstoichiometric ZnO crystal with excess Zn in interstitial sites as $Zn^{2+}$ cations**

**Figure 1.55** Stoichiometry and nonstoichiometry and the resulting defect structure.

$O^{2-}$ anions, as indicated in Figure 1.55b. Thus, the nonstoichiometric ZnO with excess Zn has $Zn^{2+}$ cations in interstitial sites and mobile electrons within the crystal, which can contribute to the conduction of electricity. Overall, the crystal is neutral, as the number of $Zn^{2+}$ ions is equal to the number of $O^{2-}$ ions plus two electrons from each excess Zn. The structure shown in Figure 1.55b is a defect structure, since it deviates from the stoichiometry.

## 1.10 SINGLE-CRYSTAL CZOCHRALSKI GROWTH

The fabrication of discrete and integrated circuit (IC) solid-state devices requires semiconductor crystals with impurity concentrations as low as possible and crystals that contain very few imperfections. A number of laboratory techniques are available for growing high-purity semiconductor crystals. Generally, they involve either solidification from the melt or condensation of atoms from the vapor phase. The initial process in IC fabrication requires large single-crystal wafers that are typically 15 cm in diameter and 0.6 mm thick. These wafers are cut from a long, cylindrical single Si crystal (typically, 1–2 m in length).

Large, single Si crystals for IC fabrication are often grown by the **Czochralski method,** which involves growing a single-crystal ingot from the melt, using solidification on a seed crystal, as schematically illustrated in Figure 1.56a. Molten Si is held in a quartz (crystalline $SiO_2$) crucible in a graphite susceptor, which is either heated by

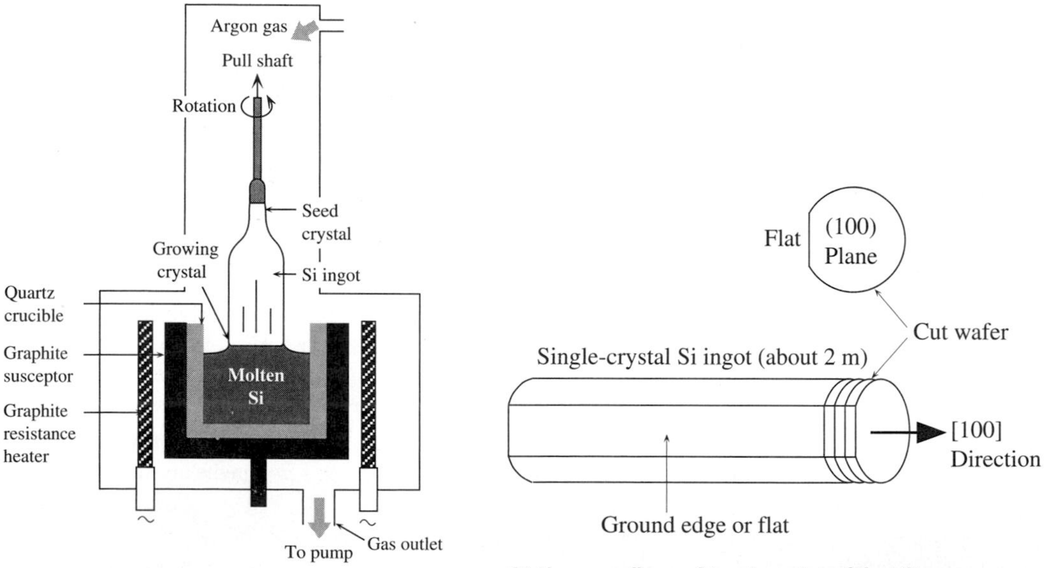


**(a) Schematic illustration of the growth of a single-crystal Si ingot by the Czochralski technique.**

**(b) The crystallographic orientation of the silicon ingot is marked by grounding a flat. The ingot can be as long as 2 m. Wafers are cut using a rotating annula diamond saw. Typical wafer thickness is 0.6–0.7 mm.**

**Figure 1.56**

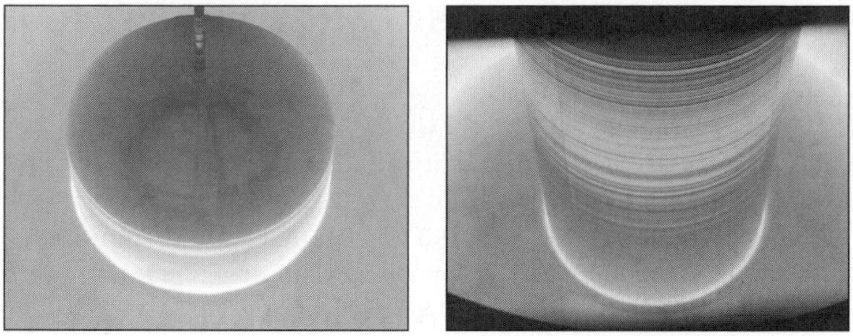

Silicon ingot being pulled from the melt in a Czochralski crystal drawer.
SOURCE: Courtesy of MEMC Electronic Materials, Inc.

a graphite resistance heater or by a radio frequency induction coil (a process called **RF heating**).[15] A small dislocation-free crystal, called a **seed,** is lowered to touch the melt and then slowly pulled out of the melt; a crystal grows by solidifying on the seed crystal. The seed is rotated during the pulling stage, to obtain a cylindrical ingot. To suppress evaporation from the melt and prevent oxidation, argon gas is passed through the system.

Initially, as the crystal is withdrawn, its cross-sectional area increases; it then reaches a constant value determined by the temperature gradients, heat losses, and the rate of pull. As the melt solidifies on the crystal, heat of fusion is released and must be conducted away; otherwise, it will raise the temperature of the crystal and remelt it. The area of the melt–crystal interface determines the rate at which this heat can be conducted away through the crystal, whereas the rate of pull determines the rate at which latent heat is released. Although the analysis is not a simple one, it is clear that to obtain an ingot with a large cross-sectional area, the pull speed must be slow. Typical growth rates are a few millimeters per minute.

The sizes and diameters of crystals grown by the Czochralski method are obviously limited by the equipment, though crystals 20–30 cm in diameter and 1–2 m in length are routinely grown for the IC fabrication industry. Also, the crystal orientation of the seed and its flatness with melt surface are important engineering requirements. For example, for very large scale integration (VLSI), the seed is placed with its (100) plane flat to the melt, so that the axis of the cylindrical ingot is along the [100] direction.

Following growth, the Si ingot is usually ground to a specified diameter. Using X-ray diffraction, the crystal orientation is identified and either a flat or an edge is ground along the ingot, as shown in Figure 1.56b. Subsequently, the ingot is cut into thin wafers by a rotating annular diamond saw. To remove any damage to the wafer surfaces caused by sawing and obtain flat, parallel surfaces, the wafers are lapped (ground flat with alumina powder and glycerine), chemically etched, and then polished. The wafers are then used in IC fabrication, usually as a substrate for the growth of a thin layer of crystal from the vapor phase.

The Czochralski technique is also used for growing Ge, GaAs, and InP single crystals, though each case has its own particular requirements. The main drawback of the Czochralski technique is that the final Si crystal inevitably contains oxygen impurities dissolved from the quartz crucible.

---

15 The induced eddy currents in the graphite give rise to $I^2R$ heating of the graphite susceptor.

# 1.11 GLASSES AND AMORPHOUS SEMICONDUCTORS

## 1.11.1 GLASSES AND AMORPHOUS SOLIDS

A characteristic property of the crystal structure is its periodicity and degree of symmetry. For each atom, the number of neighbors and their exact orientations are well defined; otherwise, the periodicity would be lost. There is therefore a **long-range order** resulting from strict adherence to a well-defined bond length and **relative bond angle** (or exact orientation of neighbors). Figure 1.57a schematically illustrates the presence of a clear, long-range order in a hypothetical two-dimensional crystal. Taking an arbitrary origin, we can predict the position of each atom anywhere in the crystal. We can perhaps use this to represent crystalline $SiO_2$ (silicon dioxide), for example, in two dimensions. In reality, a Si atom bonds with four oxygen atoms to form a tetrahedron, and the tetrahedra are linked at the corners to create a three-dimensional crystal structure.

Not all solids exhibit crystallinity. Many substances exist in a noncrystalline or amorphous form, due to their method of formation. For example, $SiO_2$ can have an amorphous structure, as illustrated schematically in two dimensions in Figure 1.57b. In the amorphous phase, $SiO_2$ is called **vitreous silica,** a form of glass, which has wide engineering applications, including optical fibers. The structure shown in the figure for vitreous silica is essentially that of a frozen liquid, or a **supercooled liquid.** Vitreous silica is indeed readily obtained by cooling the melt.

Many amorphous solids are formed by rapidly cooling or quenching the liquid to temperatures where the atomic motions are so sluggish that crystallization is virtually halted. (The cooling rate is measured relative to the crystallization rate, which depends on atomic diffusion.) We refer to these solids as **glasses.** In the liquid state, the atoms

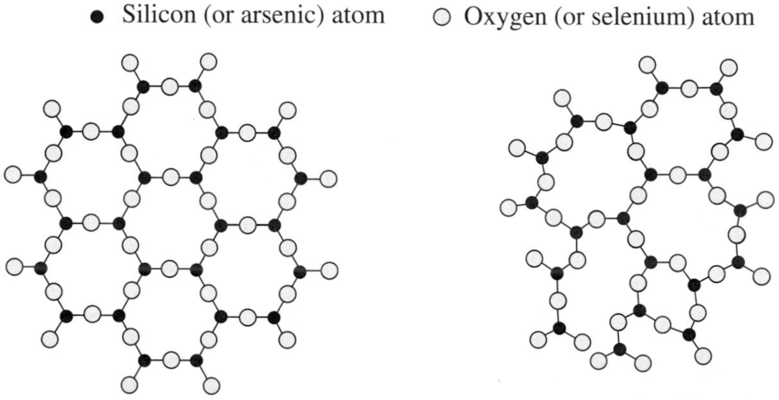


(a) A crystalline solid reminiscent of crystalline $SiO_2$ (density = 2.6 g cm$^{-3}$)

(b) An amorphous solid reminiscent of vitreous silica ($SiO_2$) cooled from the melt (density = 2.27 g cm$^{-3}$)

**Figure 1.57** Crystalline and amorphous structures illustrated schematically in two dimensions.

have sufficient kinetic energy to break and make bonds frequently and to bend and twist their bonds. There are bond angle variations, as well as rotations of various atoms around bonds **(bond twisting).** Thus, the bonding geometry around each atom is not necessarily identical to that of other atoms, which leads to the loss of long-range order and the formation of an amorphous structure, as illustrated in Figure 1.57b for the same material in Figure 1.57a. We may view Figure 1.57b as a snapshot of the structure of a liquid. As we move away from a reference atom, after the first and perhaps the second neighbors, random bending and twisting of the bonds is sufficient to destroy long-range order. The amorphous structure therefore lacks the long-range order of the crystalline state.

To reach the glassy state, the temperature is rapidly dropped well below the melting temperature where the atomic diffusion processes needed for arranging the atoms into a crystalline structure are infinitely slow on the time scale of the observation. The liquid structure thus becomes frozen. Figure 1.57b shows that for an amorphous structure, the coordination of each atom is well defined, because each atom must satisfy its chemical bonding requirement, but the whole structure lacks long-range order. Therefore, there is only a **short-range order** in an amorphous solid. The structure is a **continuous random network** of atoms (often called a CRN model of an amorphous solid). As a consequence of the lack of long-range order, amorphous materials do not possess such crystalline imperfections as grain boundaries and dislocations, which is a distinct advantage in certain engineering applications.

Whether a liquid forms a glass or a crystal structure on cooling depends on a combination of factors, such as the nature of the chemical bond between the atoms or molecules, the viscosity of the liquid (which determines how easily the atoms move), the rate of cooling, and the temperature relative to the melting temperature. For example, the oxides $SiO_2$, $B_2O_3$, $GeO_2$, and $P_2O_5$ have directional bonds that are a mixture of covalent and ionic bonds and the liquid is highly viscous. These oxides readily form glasses on cooling from the melt. On the other hand, it is virtually impossible to quench a pure metal, such as copper, from the melt, bypass crystallization, and form a glass. The metallic bonding is due to an electron gas permeating the space between the copper ions, and that bonding is nondirectional, which means that on cooling, copper ions are readily (and hence, quickly) shifted with respect to each other to form the crystal. There are, however, a number of metal–metal ($Cu_{66}Zr_{33}$) and metal–metalloid alloys ($Fe_{80}B_{20}$, $Pd_{80}Si_{20}$) that form glasses if quenched at ultrahigh cooling rates of $10^6$–$10^8$ °C s$^{-1}$. In practice, such cooling rates are achieved by squirting a thin jet of the molten metal against a fast-rotating, cooled copper cylinder. On impact, the melt is frozen within a few milliseconds, producing a long ribbon of metallic glass. The process is known as **melt spinning** and is depicted in Figure 1.58.

Many solids used in various applications have an amorphous structure. The ordinary window glass $(SiO_2)_{0.8}(Na_2O)_{0.2}$ and the majority of glassware are common examples. Vitreous silica ($SiO_2$) mixed with germania ($GeO_2$) is used extensively in optical fibers. The insulating oxide layer grown on the Si wafer during IC fabrication is the amorphous form of $SiO_2$. Some intermetallic alloys, such as $Fe_{0.8}B_{0.2}$, can be rapidly quenched from the liquid (as shown in Figure 1.58) to obtain a glassy metal used in low-loss transformer cores. Arsenic triselenide, $As_2Se_3$, has a crystal structure that resembles the two-dimensional sketch in Figure 1.57a, where an As atom (valency III) bonds with three Se atoms, and a Se atom (valency VI) bonds with two As atoms. In the amorphous

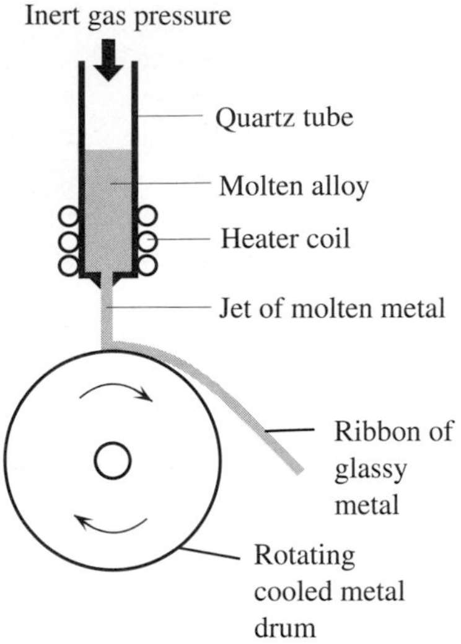


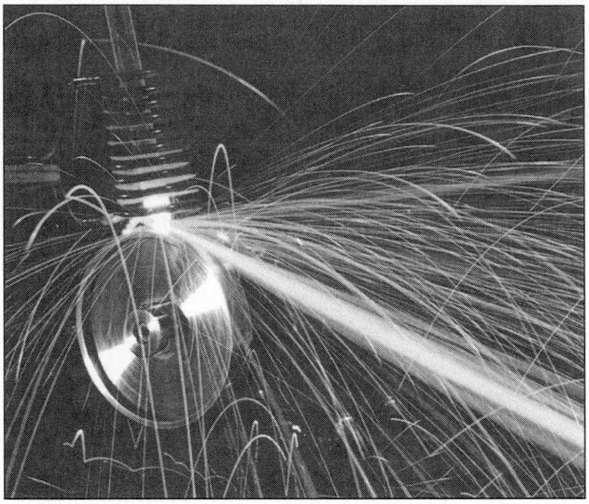

**Figure 1.58** It is possible to rapidly quench a molten metallic alloy, thereby bypassing crystallization, and forming a glassy metal commonly called a metallic glass.
The process is called *melt spinning*.

Melt spinning involves squirting a jet of molten metal onto a rotating cool metal drum. The molten jet is instantly solidified into a glassy metal ribbon which is a few microns in thickness. The process produces roughly 1 to 2 kilometers of ribbon per minute.
| SOURCE: Photo courtesy of the Estate of Fritz Goro.

phase, this crystal structure looks like the sketch in Figure 1.57b, in which the bonding requirements are only locally satisfied. The crystal can be prepared by condensation from the vapor phase, or by cooling the melt. The vapor-grown films of amorphous $As_2Se_3$ are used in some photoconductor drums in the photocopying industry.

## 1.11.2 CRYSTALLINE AND AMORPHOUS SILICON

A silicon atom in the silicon crystal forms four tetrahedrally oriented, covalent bonds with four neighbors, and the repetition of this exact bonding geometry with a well-defined bond length and angle leads to the diamond structure shown in Figure 1.6. A simplified two-dimensional sketch of the Si crystal is shown in Figure 1.59. The crystal has a clear long-range order. Single crystals of Si are commercially grown by the Czochralski crystal pulling technique.

It is also possible to grow amorphous silicon, denoted by a-Si, by the condensation of Si vapor onto a solid surface, called a substrate. For example, an electron beam is used to vaporize a silicon target in a vacuum; the Si vapor then condenses on a metallic substrate to form a thin layer of solid noncrystalline silicon. The technique, which is schematically depicted in Figure 1.60, is referred to as **electron beam deposition.** The structure of amorphous Si (a-Si) lacks the long-range order of crystalline Si (c-Si), even though each Si atom in a-Si, on average, prefers to bond

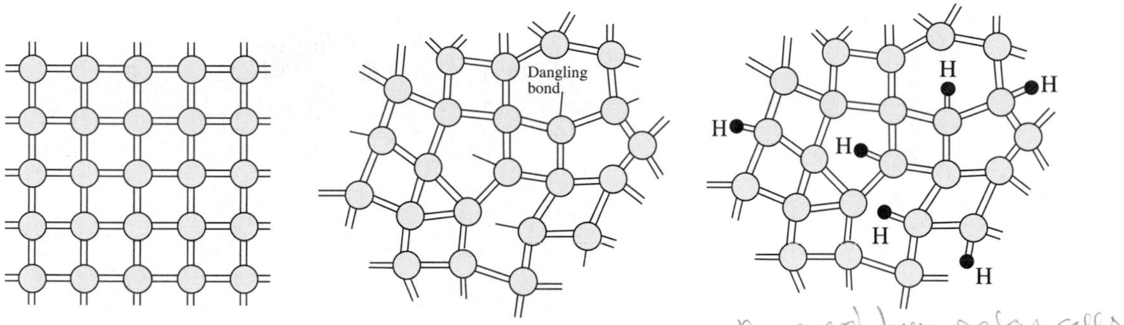


**(a) Two-dimensional schematic representation of a silicon crystal.**

**(b) Two-dimensional schematic representation of the structure of amorphous silicon. The structure has voids and dangling bonds and there is no long-range order.**

**(c) Two-dimensional schematic representation of the structure of hydrogenated amorphous silicon. The number of hydrogen atoms shown is exaggerated.**

**Figure 1.59** Silicon can be grown as a semiconductor crystal or as an amorphous semiconductor film. Each line represents an electron in a bond. A full covalent bond has two lines, and a broken bond has one line.

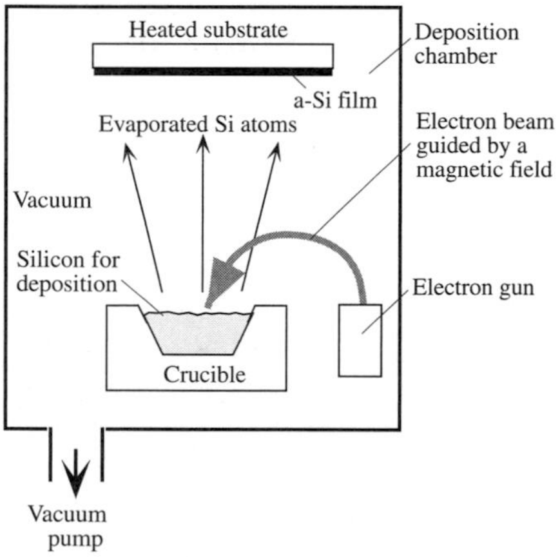


**Figure 1.60** Amorphous silicon, a-Si, can be prepared by an electron beam evaporation of silicon.

Silicon has a high melting temperature, so an energetic electron beam is used to melt the crystal in the crucible locally and thereby vaporize Si atoms. Si atoms condense on a substrate placed above the crucible, to form a film of a-Si.

with four neighbors. The difference is that the relative angles between the Si–Si bonds in a-Si deviate considerably from those in the crystal, which obey a strict geometry. Therefore, as we move away from a reference atom in a-Si, eventually the periodicity for generating the crystalline structure is totally lost, as illustrated schematically in Figure 1.59. Furthermore, because the Si–Si bonds do not follow the equilibrium geometry, the bonds are strained and some are even missing, simply because the formation of a bond causes substantial bond bending. Consequently, the a-Si structure has many voids and incomplete bonds, or **dangling bonds,** as schematically depicted in Figure 1.59.

One way to reduce the density of dangling bonds is simply to terminate a dangling bond using hydrogen. Since hydrogen only has one electron, it can attach itself to a

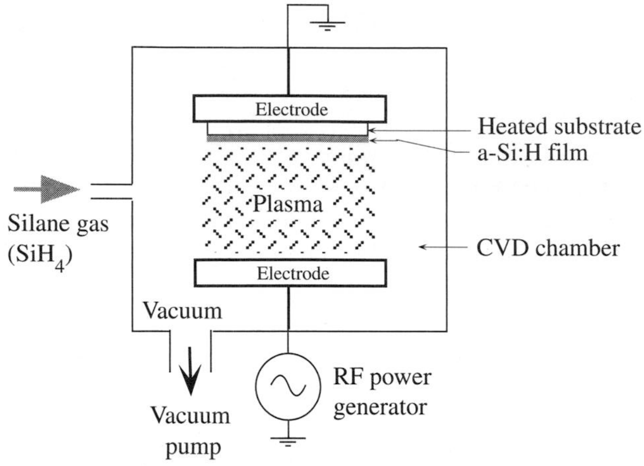


**Figure 1.61** Hydrogenated amorphous silicon, a-Si:H, is generally prepared by the decomposition of silane molecules in a radio frequency (RF) plasma discharge.
Si and H atoms condense on a substrate to form a film of a-Si:H.

dangling bond, that is, passivate the dangling bond. The structure resulting from hydrogen in amorphous silicon is called **hydrogenated amorphous Si (a-Si:H).**

Many electronic devices, such as a-Si:H solar cells, are based on a-Si being deposited with H to obtain a-Si:H, in which the hydrogen concentration is typically 10 at.% (atomic %). The process involves the decomposition of silane gas, $SiH_4$, in an electrical plasma in a vacuum chamber. Called **plasma-enhanced chemical vapor deposition (PECVD),** the process is illustrated schematically in Figure 1.61. The silane gas molecules are dissociated in the plasma, and the Si and H atoms then condense onto a substrate to form a film of a-Si:H. If the substrate temperature is too hot, the atoms on the substrate surface will have sufficient kinetic energy, and hence the atomic mobility, to orient themselves to form a polycrystalline structure. Typically, the substrate temperature is ~250 °C. The advantage of a-Si:H is that it can be grown on large areas, for such applications as photovoltaic cells, flat panel thin-film transistor (TFT) displays, and the photoconductor drums used in some photocopying machines. Table 1.5 summarizes the properties of crystalline and amorphous silicon, in terms of structure and applications.

**Table 1.5** Crystalline and amorphous silicon

| | Crystalline Si (c-Si) | Amorphous Si (a-Si) | Hydrogenated a-Si (a-Si:H) |
|---|---|---|---|
| Structure | Diamond cubic. | Short-range order only. On average, each Si covalently bonds with four Si atoms. Has microvoids and dangling bonds. | Short-range order only. Structure typically contains 10% H. Hydrogen atoms passivate dangling bonds and relieve strain from bonds. |
| Typical preparation | Czochralski technique. | Electron beam evaporation of Si. | Chemical vapor deposition of silane gas by RF plasma. |
| Density (g $cm^{-3}$) | 2.33 | About 3–10% less dense. | About 1–3% less dense. |
| Electronic applications | Discrete and integrated electronic devices. | None | Large-area electronic devices such as solar cells, flat panel displays, and some photoconductor drums used in photocopying. |

## 1.12 SOLID SOLUTIONS AND TWO-PHASE SOLIDS

### 1.12.1 ISOMORPHOUS SOLID SOLUTIONS: ISOMORPHOUS ALLOYS

A **phase** of a material has the same composition, structure, and properties everywhere, so it is a homogeneous portion of the chemical system under consideration. In a given chemical system, one phase may be in contact with another phase. For example, at 0 °C, iced water will have solid and liquid phases in contact. Each phase, ice and water, has a distinct structure.

A bartender knows that alcohol and water are totally miscible; she can dilute whisky with as much water as she likes. When the two liquids are mixed, the molecules are randomly mixed with each other and the whole liquid is a homogenous mixture of the molecules. The liquid therefore has one phase; the properties of the liquid are the same everywhere. The same is not true when we try to mix water and oil. The mixture consists of two distinctly separate phases, oil and water, in contact. Each phase has a different composition, even though both are liquids.

Many solids are a homogeneous mixture of two types of separate atoms. For example, when nickel atoms are added to copper, Ni atoms substitute directly for the Cu atoms, and the resulting solid is a **solid solution,** as depicted in Figure 1.62a. The structure remains an FCC crystal whatever the amount of Ni we add, from 100% Cu to 100% Ni. The solid is a homogenous mixture of Cu and Ni atoms, with the same structure everywhere in the solid solution, which is called an **isomorphous solid solution.** The atoms in the majority make up the **solvent,** whereas the atoms in the minority are the **solute,** which is dissolved in the solvent. For a Cu–Ni alloy with a Ni content of less than 50 at.%, copper is the solvent and nickel is the solute.

The substitution of solute atoms for solvent atoms at various lattice sites of the solvent can be either random (disordered) or ordered. The two cases are schematically illustrated in Figure 1.62a and b, respectively. In many solid solutions, the substitution is random, but for certain compositions, the substitution becomes ordered. There is a

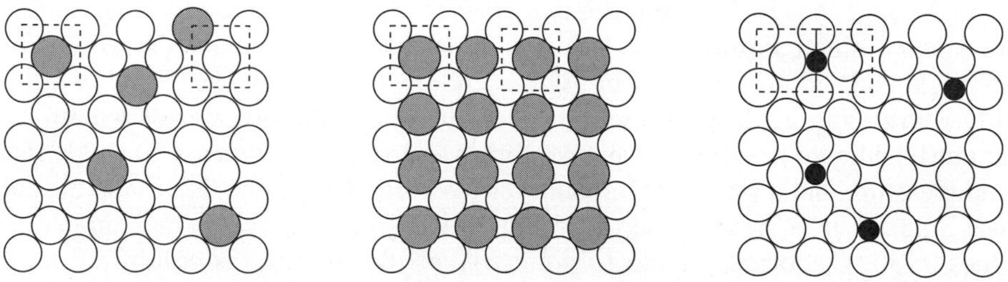

**(a) Disordered substitutional solid solution. Example: Cu–Ni alloys ({100} planes)**

**(b) Ordered substitutional solid solution. Example: Cu–Zn alloy of composition 50% Cu–50% Zn. ({110} planes)**

**(c) Interstitial solid solution. Example: Small number of C atoms in FCC Fe (austenite). ({100} planes)**

**Figure 1.62** Solid solutions can be disordered substitutional, ordered substitutional, and interstitial substitutional.

Only one phase within the alloy has the same composition, structure, and properties everywhere.

distinct ordering of atoms around each solute atom such that the crystal structure resembles that of a compound. For example, $\beta'$ brass has the composition 50 at.% Cu–50 at.% Zn. Each Zn atom is surrounded by eight Cu atoms and vice versa, as depicted in two dimensions in Figure 1.62b. The structure is that of a metallic compound between Cu and Zn.

Another type of solid solution is the **interstitial solid solution,** in which solute atoms occupy interstitial sites, or voids between atoms, in the crystal. Figure 1.62c shows an example in which a small number of carbon atoms have been dissolved in a $\gamma$-iron crystal (FCC) at high temperatures.

## 1.12.2 PHASE DIAGRAMS: Cu–Ni AND OTHER ISOMORPHOUS ALLOYS

The Cu–Ni alloy is isomorphous. Unlike pure copper or pure nickel, when a Cu–Ni alloy melts, its melting temperature is not well defined. The alloy melts over a range of temperatures in which both the liquid and the solid coexist as a heterogeneous mixture. It is therefore instructive to know the phases that exist in a chemical system at various temperatures as a function of composition, and this need leads to the use of phase diagrams.

Suppose we take a crucible of molten copper and allow it to cool. Above its melting temperature (1083 °C), there is only the liquid phase. The temperature drops with time, as shown in Figure 1.63a, until at the melting or fusion temperature at point $L_0$ when copper crystals begin to **nucleate** (solidify) in the crucible. During solidification, the temperature remains constant. As long as we have both the liquid and solid phases coexisting, the temperature remains constant at 1083 °C. During this time, heat is given off as the Cu atoms in the melt attach themselves to the Cu crystals. This heat is called the **heat of fusion.** Once all the liquid has solidified (point $S_0$), the temperature begins to drop as the solid cools. There is therefore a sharp melting temperature for copper, at 1083 °C.

If we were to cool pure nickel from its melt, we would observe a behavior similar to that of pure copper, with a well-defined melting temperature at 1453 °C.

Now suppose we cool the melt of a Cu–Ni alloy with a composition[16] of 80 wt.% Cu and 20 wt.% Ni. In the melt, the two species of atoms are totally miscible, and there is only a single liquid phase. As the cooling proceeds, we reach the temperature 1195 °C, identified as point $L_{20}$ in Figure 1.63a, where the first crystals of Cu–Ni alloy begin to appear. In this case, however, the temperature does not remain constant until the liquid is solidified, but continues to drop. Thus, there is no single melting temperature, but a range of temperatures over which both the liquid and the solid phases coexist in a heterogeneous mixture. We find that when the temperature reaches 1130 °C, corresponding to point $S_{20}$, all the liquid has solidified. Below 1130 °C, we have a single-phase solid that is an isomorphous solid solution of Cu and Ni. If we repeat these experiments for other compositions, we find a similar behavior; that is, freezing occurs over a transition temperature range. The beginning and end

[16] In materials science, we generally prefer to give alloy composition in wt.%, which henceforth will simply be %.

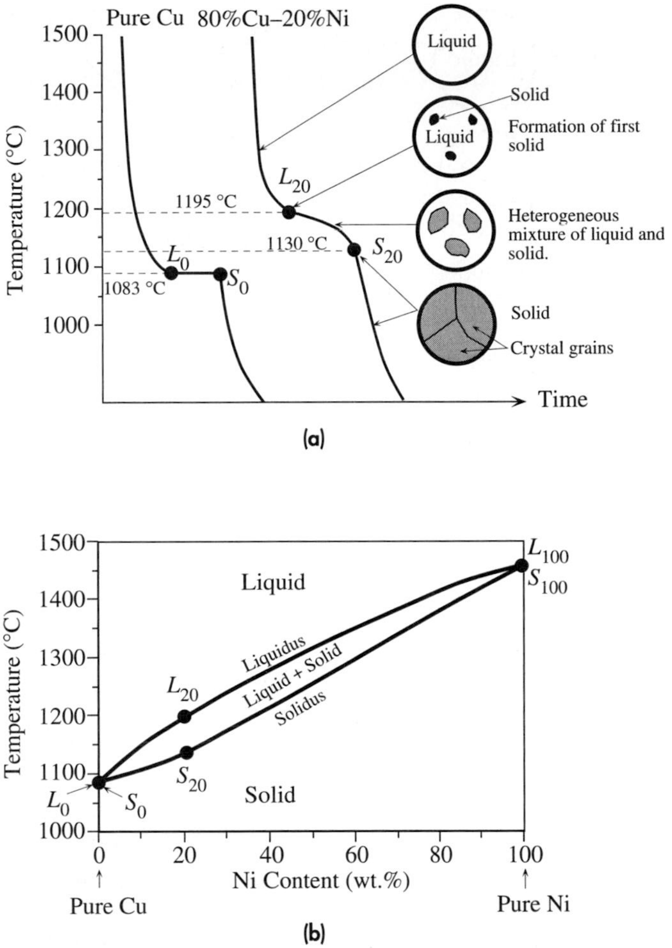


**Figure 1.63** Solidification of an isomorphous alloy such as Cu–Ni.
(a) Typical cooling curves.
(b) The phase diagram marking the regions of existence for the phases.

of solidification, at points $L$ and $S$, respectively, depend on the specific composition of the alloy.

To characterize the freezing or melting behavior of other compositions of Cu–Ni alloys, we can plot the temperatures for the beginning and end of solidification versus the composition and identify those temperature regions where various phases exist, as shown in Figure 1.63b. When we join all the points corresponding to the beginning of freezing, that is, all the $L$ points, we obtain what is called the **liquidus curve.** For any given composition, only the liquid phase can exist above the liquidus curve. If we join all the points where the liquid has totally solidified, that is, all the $S$ points, we have a curve called the **solidus curve.** At any temperature and composition below the solidus curve, we can only have the solid phase. The region between

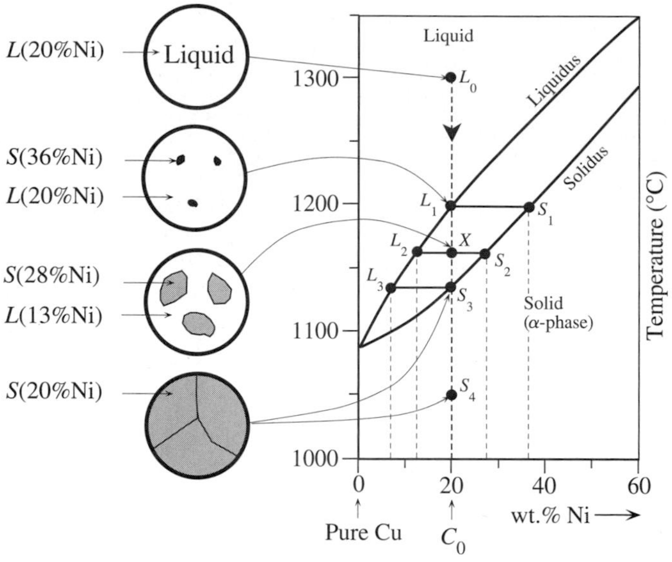


**Figure 1.64** Cooling of a 80% Cu–20% Ni alloy from the melt to the solid state.

the liquidus and solidus curves marks where a heterogeneous mixture of liquid and solid phases exists.

Let's follow the cooling behavior of the 80% Cu–20% Ni alloy from the melt at 1300 °C down to the solid state at 1000 °C, as shown in Figure 1.64. The vertical dashed line at 20% Ni represents the overall composition of the alloy (the whole chemical system) and the cooling process corresponds to movement down this dashed line, starting from the liquid phase at $L_0$.

When the Cu–Ni alloy begins to solidify at 1195 °C, at point $L_1$, the first solid that forms is richer in Ni content. The only solid that can exist at this temperature has a composition $S_1$, which has a greater Ni content than the liquid, as shown in Figure 1.64. Intuitively, we can see this by noting that Cu, the component with the lower melting temperature, prefers to remain in the liquid, whereas Ni, which has a higher melting temperature, prefers to remain in the solid. When the temperature drops further, say to 1160 °C (indicated by $X$ in the figure), the alloy is a heterogeneous mixture of liquid and solid. At this temperature, the only solid that can coexist with the liquid has a composition $S_2$. The liquid has the composition $L_2$. Since the liquid has lost some of its Ni atoms, the liquid composition is less than that at $L_1$. The liquidus and solidus curves therefore give the compositions of the liquid and solid phases coexisting in the heterogeneous mixture during melting.

At 1160 °C, the overall composition of the alloy (the whole chemical system) is still 20% Ni and is represented by point $X$ in the phase diagram. When the temperature reaches 1130 °C, nearly all the liquid has been solidified. The solid has the composition $S_3$, which is 20% Ni, as we expect since the whole alloy is almost all solid. The last drops of the liquid in the alloy have the composition $L_3$, since at this temperature,

**Table 1.6** Phase in the 80% Cu–20% Ni isomorphous alloy

| Temperature, °C | Phases | Composition | Amount |
|---|---|---|---|
| 1300 | Liquid only | $L_0 = 20\%$ Ni | 100% |
| 1195 | Liquid and solid | $L_1 = 20\%$ Ni | 100% |
| | | $S_1 = 36\%$ Ni | First solid appears |
| 1160 | Liquid and solid | $L_2 = 13\%$ Ni | 53.3% |
| | | $S_2 = 28\%$ Ni | 46.7% |
| 1130 | Liquid and solid | $L_3 = 7\%$ Ni | The last liquid drop |
| | | $S_3 = 20\%$ Ni | 100% |
| 1050 | Solid only | $S_4 = 20\%$ Ni | 100% |

only the liquid with this composition can coexist with the solid at $S_3$. Table 1.6 summarizes the phases and their compositions, as observed during the cooling process depicted in Figure 1.64. By convention, all solid phases that can exist are labeled by different Greek letters. Since we can only have one solid phase, this is labeled the $\alpha$-phase.

During the solidification process depicted in Figure 1.64, the solid composition changes from $S_1$ to $S_2$ to $S_3$. We tacitly assume that the cooling is sufficiently slow to allow time for atomic diffusion to change the composition of the whole solid. Therefore, the phase diagram in Figure 1.63b, which assumes near equilibrium conditions during cooling, is termed an **equilibrium phase diagram.** If the cooling is fast, there will be limited time for atomic diffusion in the solid phase, and the resulting solid will have a composition variation. The inner core will correspond to the solidification at $S_1$ and will be Ni rich. Since the solidification occurs quickly, the Ni atoms do not have time to diffuse out from the inner core to allow the composition in the solid to change from $S_1$ to $S_2$ to $S_3$. Thus, the outer region, the final solidification, will be Ni deficient (or Cu rich); its composition is not $S_3$ but less, because $S_3$ is the average composition in the whole solid. The solid structure will be **cored,** as depicted in Figure 1.65. The cooling process is then said to have occurred under nonequilibrium conditions, which leads to a segregation of the elements in the grains. Under nonequilibrium cooling conditions we cannot quantitatively use the equilibrium phase diagram in Figure 1.63b. The diagram can only serve as a qualitative guide.

The amounts of liquid and solid in the mixture can be determined from the phase diagram using the **lever rule,** which is based on the fact that the total mass of the alloy

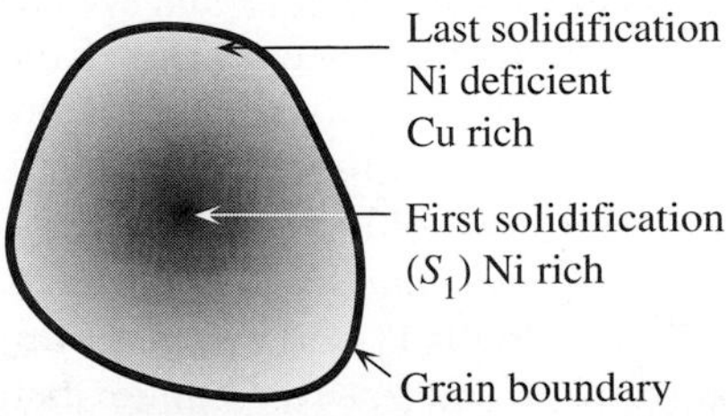


**Figure 1.65** Segregation in a grain due to rapid cooling (nonequilibrium cooling).

remains the same throughout the entire cooling process. Let $W_L$ and $W_S$ be the **weight** (or **mass**) **fraction** of the liquid and solid phases in the alloy mixture. The compositions of the liquid and solid are denoted as $C_L$ and $C_S$, respectively. The overall composition of the alloy is denoted $C_O$, which is the overall weight fraction of Ni in the alloy.

If we take the alloy to have a weight of unity, then the conservation of mass means that

$$W_L + W_S = 1$$

Further, the weight fraction of Ni in both the liquid and solid must add up to the composition $C_O$ of Ni in the whole alloy, or

$$C_L W_L + C_S W_S = C_O$$

We can substitute for $W_S$ in the above equation to find the weight fraction of the liquid and then that of the solid phase, as follows:

*Lever rules*

$$W_L = \frac{C_S - C_O}{C_S - C_L} \quad \text{and} \quad W_S = \frac{C_O - C_L}{C_S - C_L} \qquad [1.36]$$

To apply Equation 1.36, we first draw a line (called a **tie line**) from $L_2$ to $S_2$ corresponding to $C_L$ and $C_S$, as shown in Figure 1.64. The line represents a "horizontal lever" and point $X$ at $C_O$ at this temperature is the lever's fulcrum. The lengths of the lever arms from the fulcrum to the liquidus and solidus curves are $(C_O - C_L)$ and $(C_S - C_O)$, respectively. The lever must be balanced by the weights $W_L$ and $W_S$ attached to the ends. The total length of the lever is $(C_S - C_L)$. At 1160 °C, $C_L = 0.13$ (13% Ni) and $C_S = 0.28$ (28% Ni), so the weight fraction of the liquid phase is

$$W_L = \frac{C_S - C_O}{C_S - C_L} = \frac{0.28 - 0.20}{0.28 - 0.13} = 0.533 \quad \text{or} \quad 53.3\%$$

Similarly, the weight fraction of the solid phase is $1 - 0.533$ or 0.467.

### 1.12.3 ZONE REFINING AND PURE SILICON CRYSTALS

**Zone refining** is used for the production of high-purity crystals. Silicon, for example, has a high melting temperature, so any impurities present in the crystal decrease the melting temperature. This is similar to the depression of the melting temperature of pure Ni by the addition of Cu, as shown by the right-hand side of Figure 1.63b. We can represent the phase diagram of Si with small impurities as shown in Figure 1.66. Consider what happens if we have a rod of the solid and we melt only the left end by applying heat locally (using RF heating, for example). At the same time, we move the melted zone toward the right by moving the heater. We therefore melt the solid at $A$ and refreeze it at $B$, as shown in Figure 1.67a.

The solid has an impurity concentration of $C_O$; when it melts at $A$, the melt initially also has the same concentration $C_L = C_O$. However, at temperature $T_B$, the melt

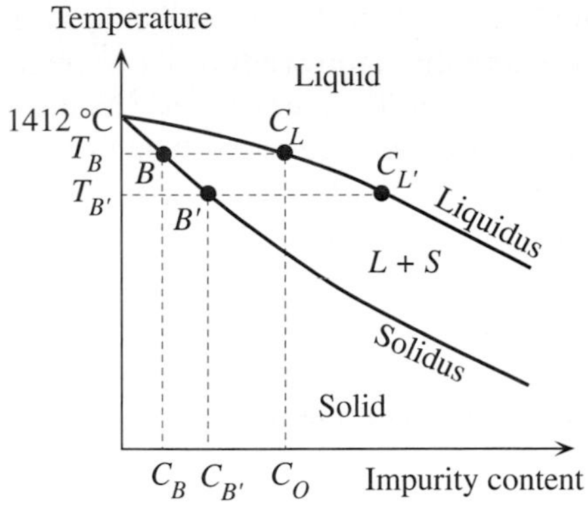


**Figure 1.66** The phase diagram of Si with impurities near the low-concentration region.

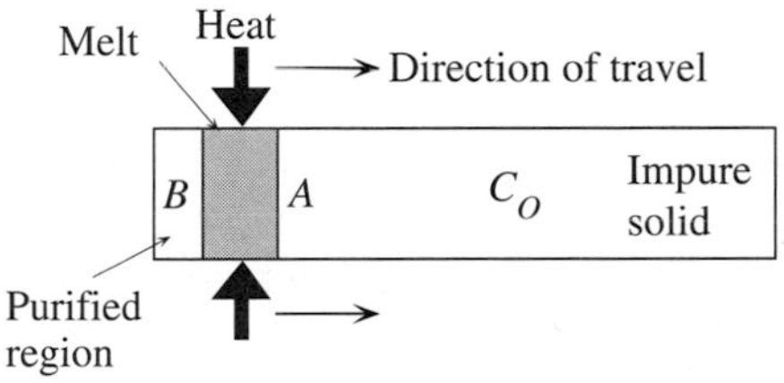


**(a) Heat is applied locally starting at one end. The impurity concentration in the refrozen solid at *B* is $C_B < C_O$. The impurity concentration in the melt is $C_{L'} > C_O$.**

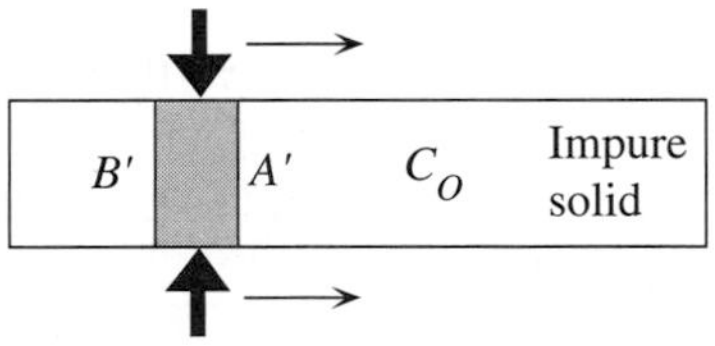


**(b) As the torch travels toward the right, the refrozen solid at *B'* has $C_{B'}$ where $C_B < C_{B'} < C_O$. The impurity concentration in the melt is now even greater than $C_{L'}$.**

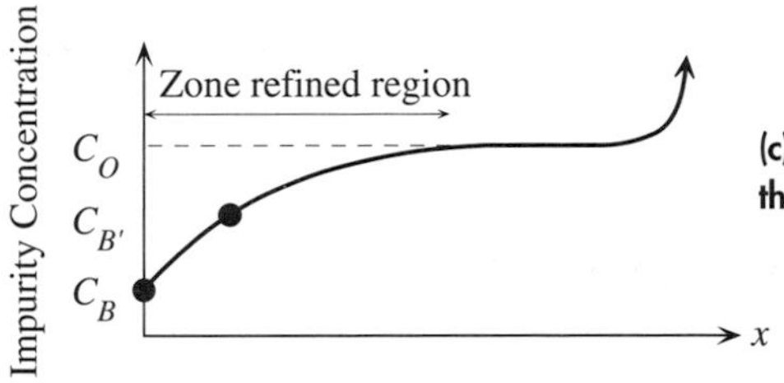


**(c) The impurity concentration profile in the refrozen solid after one pass.**

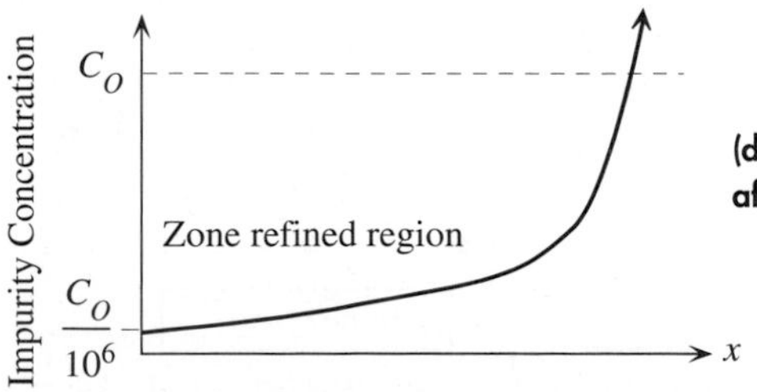


**(d) Typical impurity concentration profile after many passes.**

**Figure 1.67** The principle of zone refining.

begins to solidify. At the start of solidification the solid that freezes has a composition $C_B$, which is considerably less than $C_O$, as is apparent in Figure 1.66. The cooling at $B$ occurs rapidly, so the concentration $C_B$ cannot adjust to the equilibrium value at the end of freezing. Thus, the solid that freezes at $B$ has a lower concentration of impurities. The impurities have been pushed out of the solid at $B$ and into the melt, whose impurity concentration increases from $C_L$ to $C_{L'}$.

Next, refreezing at $B'$, shown in Figure 1.67b, occurs at a lower temperature $T_{B'}$, because the melt concentration $C_{L'}$ is now greater than $C_O$. The solid that freezes at $B'$ has the concentration $C_{B'}$, shown in Figure 1.66, which is greater than $C_B$ but less than $C_O$. As the melted zone is floated toward the right, the melt that is solidified at $B$, $B'$, etc., has a higher and higher impurity concentration, until its impurity content reaches that of the impure solid, at which point the concentration remains at $C_O$. When the melted zone approaches the far right where the freezing is halted, the impurities in the final melt appear in the last frozen region at the far right. The resulting impurity concentration profile is schematically depicted in Figure 1.67c. The region of impurity concentration below $C_O$ is the **zone refined** section of the rod. The zone refining procedure can be repeated again, starting from the left toward the right, to reduce the impurity concentration even lower. The impurity concentration profile after many passes is sketched in Figure 1.67d. Although the profile is nonuniform, due to the segregation effect, the impurity concentrations in the zone refined section may be as low as a factor of $10^{-6}$.

## 1.12.4 BINARY EUTECTIC PHASE DIAGRAMS AND Pb–Sn SOLDERS

When we dissolve salt in water, we obtain a brine solution. If we continue to add more salt, we eventually reach the solubility limit of salt in the solution, and the excess salt remains as a solid at the bottom of the container. We then have two coexisting phases: brine (liquid solution) and salt (solid), as shown in Figure 1.68. The solubility limit of one component in another in a mixture is represented by a **solvus curve** shown schematically in Figure 1.68 for salt in brine. In the solid state, there are many elements that can only be dissolved in small amounts in another solid.

Lead in the solid phase has an FCC crystal structure, and tin has a BCT (body-centered tetragonal) structure. Although the two elements are totally miscible in any

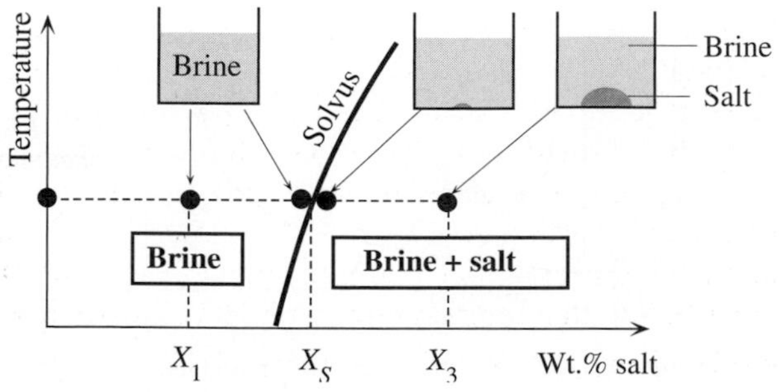


**Figure 1.68** We can only dissolve so much salt in brine (solution of salt in water).
Eventually we reach the solubility limit at $X_S$, which depends on the temperature. If we add more salt, then the excess salt does not dissolve and coexists with the brine. Past $X_S$ we have two phases, brine (solution) and salt (solid).

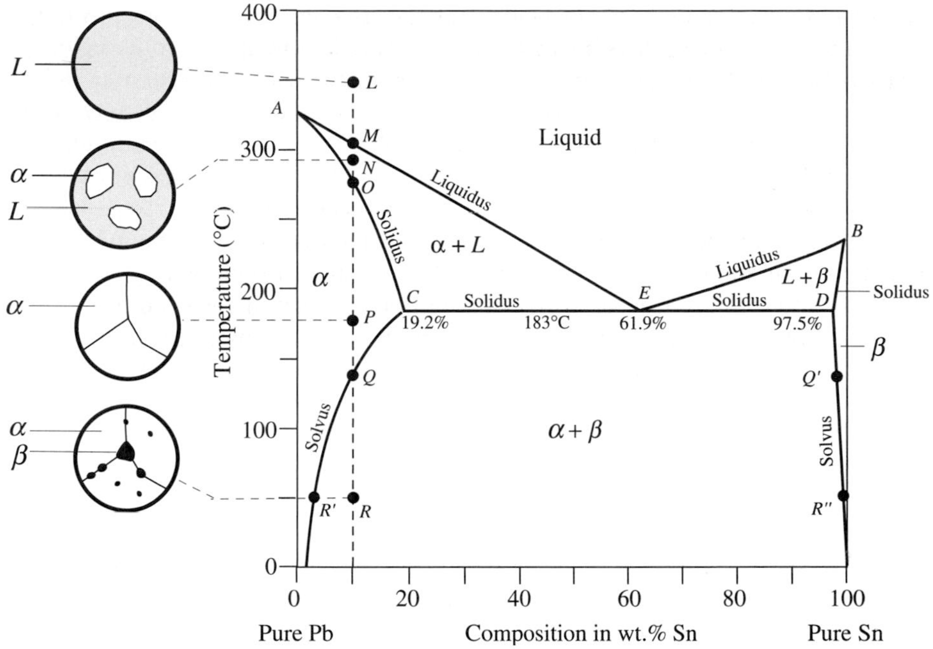


**Figure 1.69** The equilibrium phase diagram of the Pb–Sn alloy.
The microstructures on the left show the observations at various points during the cooling of a 90% Pb–10% Sn from the melt along the dashed line (the *overall alloy composition* remains constant at 10% Sn).

proportion when melted, this is not so in the solid state. We can only dissolve so much Sn in solid Pb, and vice versa. We quickly reach the solubility limit, and the resulting solid is a mixture of two distinctly different solid phases. One solid phase, labeled $\alpha$, is Pb rich and has the FCC structure with some Sn atoms dissolved in the crystal. The amount of Sn dissolved in $\alpha$ is given by the solvus curve of Sn in $\alpha$ at that temperature. The other phase, labeled $\beta$, is Sn rich and has the BCT structure with some Pb atoms dissolved in it. The amount of Pb dissolved in $\beta$ is given by the solvus curve of Pb in $\beta$ at that temperature.

The existence of various phases and their compositions as a function of temperature are given by the equilibrium phase diagram for the Pb–Sn alloy, shown in Figure 1.69. This is called an equilibrium **eutectic phase diagram.** The liquidus and solidus curves, as usual, mark the borders for the liquid and solid phases. Between the liquidus and solidus curves, we have a heterogeneous mixture of melt and solid. Unlike the Cu–Ni case, the melting temperature of both elements here is depressed with alloying. The liquidus and solidus curves thus decrease from both ends, starting at $A$ and $B$. They meet at a point $E$, called the **eutectic point,** at 61.9% Sn and 183 °C. This point has a special significance: No liquid can exist below this temperature, so 183 °C is the lowest melting temperature of the alloy.

In addition, we must insert the solvus curves at both the Pb and Sn ends to mark the extent of solid-state solubility and hence identify the two-phase solid region. The solvus curve for the solubility limit of Sn in Pb meets the solidus curve at point $C$, 19.2% Sn. Similarly, the solubility limit of Pb in Sn meets the solvus curve at $D$. A characteristic feature of this phase diagram is that $CD$ is a straight line through $E$ at 183 °C. Below 183 °C, between the two solvus curves, we have a solid with two phases, $\alpha$ and $\beta$. This is identified as $\alpha + \beta$ in the diagram.

The usefulness of such a phase diagram is best understood by examining the phase transformations and microstructures during the cooling of a melt of a given composition alloy. Consider a 90% Pb–10% Sn alloy being cooled from the melt at 350 °C (point $L$) where there is only one phase, the liquid phase. At point $M$, 315 °C, few nuclei of the $\alpha$-phase appear in the liquid. The composition of the $\alpha$-phase is given by the solidus curve at 315 °C and is about 5% Sn. At point $N$, 290 °C, there is more $\alpha$-phase in the mixture. The compositions of the liquid and $\alpha$-phases are given respectively by the liquidus and solidus curves at 290 °C. At point $O$, 275 °C, all liquid has been solidified into the $\alpha$-phase, which then has the composition 10% Sn.

Between $M$ and $O$, the alloy is a coexistent mixture of the liquid phase (melt) and the solid $\alpha$-phase. At point $P$, 175 °C, we still have only the $\alpha$-phase. When we reach the solvus curve at point $Q$, 140 °C, we can no longer keep all the Sn dissolved in the $\alpha$-phase, as we have reached the solubility limit of Sn in $\alpha$. Some of the Sn atoms must diffuse out from the $\alpha$-phase; they do so by forming a second solid phase, which is the $\beta$-phase. The $\beta$-phase nucleates within the $\alpha$-phase (usually at the grain boundaries, where atomic diffusion occurs readily). The $\beta$-phase will contain as much dissolved Pb as is allowed by the solubility of Pb in the $\beta$-phase, which is given by the solvus curve on the Sn side and marked as point $Q'$, about 98% Sn. Thus, the microstructure is now a mixture of the $\alpha$ and $\beta$ phases.

As cooling proceeds, the two phases continue to coexist, but their relative proportions change. At $R$, 50 °C, the alloy is a mixture of the $\alpha$-phase given by $R'$(4% Sn) and the $\beta$-phase given by $R''$(99% Sn). The relative amounts of $\alpha$ and $\beta$ phases are given by the lever rule. Figure 1.69 illustrates the microstructure of the 90% Pb–10% Sn alloy as it is cooled.

An interesting phenomenon can be observed when we cool an alloy of the eutectic composition 38.1% Pb–61.9% Sn from the melt. The cooling process and the observed microstructures are illustrated in Figure 1.70; the microstructures are on the right. The temperature–time profile is also depicted in Figure 1.70. At point $L$, 350 °C, the alloy is all liquid; as it cools, its temperature drops until point $E$ at 183 °C. At $E$, the temperature remains constant and a solid phase nucleates within the melt. With time, the amount of solid grows until all the liquid is solidified and the temperature begins to drop again. This behavior is much like that of a pure element, for which melting occurs at a well-defined temperature. This behavior only occurs for the eutectic composition (61.9% Sn), because this is the composition at which the liquidus and solidus curves meet at one temperature. Generally, the liquid with the eutectic composition will solidify through the **eutectic transformation** at the eutectic temperature, or

*Eutectic transformation*

$$L_{61.9\%\ \text{Sn}} \rightarrow \alpha_{19.2\%\ \text{Sn}} + \beta_{97.5\%\ \text{Sn}} \quad (183\ ^\circ\text{C}) \qquad [1.37]$$

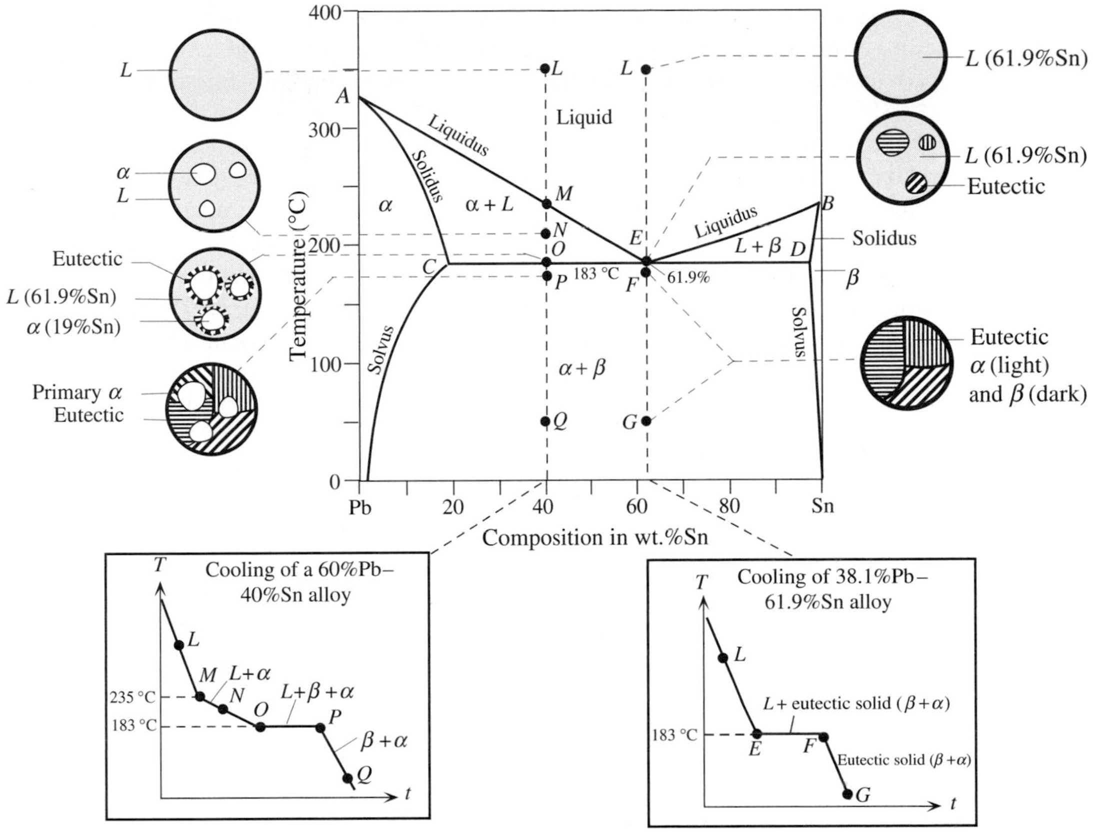


**Figure 1.70** The alloy with the eutectic composition cools like a pure element, exhibiting a single solidification temperature at 183 °C.
The solid has the special eutectic structure. The alloy with the composition 60% Pb–40% Sn when solidified is a mixture of primary $\alpha$ and eutectic solid.

The solid that forms from the eutectic solidification has a special microstructure, consisting of alternating plates, or **lamellae,** of $\alpha$ and $\beta$ phases, as shown in Figure 1.70. This is called the **eutectic microstructure** (or **eutectic solid**). The formation of a Pb-rich $\alpha$-phase and an Sn-rich $\beta$-phase from the 61.9% Sn liquid requires the redistribution of the two types of atoms by atomic diffusion. Atomic diffusions are easier in the liquid than in the solid. The formation of a solid with alternating $\alpha$ and $\beta$ layers allows the Pb and Sn atoms to diffuse in the liquid without having to move over long distances. The eutectic structure is not a phase itself, but a mixture of the two phases, $\alpha$ and $\beta$.

When cooled from the melt, an alloy with a composition between 19.2% Sn and 61.9% Sn solidifies into a mixture of $\alpha$-phase and a eutectic solid (a mixture of $\alpha$ and $\beta$ phases). Consider the cooling of an alloy with a composition of 40% Sn, starting from the liquid phase $L$ at 350 °C, as shown in Figure 1.70. At point $M$ (235 °C), the first solid, the $\alpha$-phase, nucleates. Its composition is about 15% Sn. At $N$, 210 °C, the alloy is a mixture of liquid, composition 50% Sn, and $\alpha$-phase, composition 18% Sn. The composition of the liquid thus moves along the liquidus line from $M$ toward $E$. At 183 °C, the liquid has the composition 61.9% Sn, or the eutectic composition, and therefore undergoes the eutectic transformation indicated in Equation 1.37. There is still $\alpha$-phase in the alloy, but its composition is now 19.2% Sn; it does not take part in the eutectic transformation of the liquid. During the eutectic transformation, the temperature remains constant. When all the liquid has been solidified, we have a mixture of the preexisting $\alpha$-phase, called **primary** $\alpha$ (or **proeutectic** $\alpha$), and the newly formed eutectic solid. The final microstructure is shown in Figure 1.70 and consists of a primary $\alpha$ and a eutectic solid; therefore, two solid phases, $\alpha$ and $\beta$, coexist.

During cooling between points $M$ and $O$, the alloy 60% Pb–40% Sn is a mixture of melt and $\alpha$-phase, and it exhibits plastic-like characteristics while solidifying. Further, the temperature range for the solidification is about 183 °C to 235 °C, or about 50 °C. Such an alloy is preferable for such uses as soldering wiped joints to join pipes together, giving the plumber sufficient play for adjusting and wiping the joint. On the other hand, a solder with the eutectic composition (commercially, this is 40% Pb–60% Sn solder, which is close to the eutectic) has the lowest melting temperature and solidifies quickly. The liquid also has good wetting properties. Therefore, 40% Pb–60% Sn is widely used for soldering semiconductor devices, where good wetting and minimal exposure to high temperature are required.

**EXAMPLE 1.17** **THE 60% Pb–40% Sn ALLOY** Consider the solidification of the 60% Pb–40% Sn alloy. What are the phases, compositions, and weight fractions of various phases existing in the alloy at 250 °C, 210 °C, 183. 5 °C (just above 183 °C), and 182.5 °C (just below 183 °C)?

**SOLUTION**

We again refer to the phase diagram in Figure 1.70 to identify which phases exist at what temperatures. At 250 °C, we only have the liquid phase. At 210 °C, point $N$, the liquid and the $\alpha$-phase are in equilibrium. The composition of the $\alpha$-phase is given by the solidus line; at 210 °C, $C_\alpha = 18\%$ Sn. The composition of the liquid is given by the liquidus line; at 210 °C, $C_L = 50\%$ Sn. To find the weight fraction of $\alpha$ the alloy, we use the lever rule,

$$W_\alpha = \frac{C_L - C_O}{C_L - C_\alpha} = \frac{50 - 40}{50 - 18} = 0.313$$

From $W_\alpha + W_L = 1$, we obtain the weight fraction of the liquid phase, $W_L = 1 - 0.313 = 0.687$.

At 183.5 °C, point $O$, the composition of the $\alpha$-phase is 19.2% Sn corresponding to $C$ and that of the liquid is 61.9% Sn corresponding to $E$. The liquid therefore has the eutectic composition. The weight fractions are

$$W_\alpha = \frac{C_L - C_O}{C_L - C_\alpha} = \frac{61.9 - 40}{61.9 - 19.2} = 0.513$$

$$W_L = 1 - 0.513 = 0.487$$

As expected, the amount of $\alpha$-phase increases during solidification; at the same time, its composition changes along the solidus curve. Just above 183 °C, about half the alloy is the solid $\alpha$-phase and the other half is liquid with the eutectic composition. Thus, on solidification, the liquid

**Table 1.7** The 60% Pb–40% Sn alloy

| Temperature (°C) | Phases | Composition | Mass (g) | Microstructure and Comment |
|---|---|---|---|---|
| 250 | $L$ | 40% Sn | 100 | |
| 235 | $L$ | 40% Sn | 100 | The first solid ($\alpha$-phase) nucleates in the liquid. |
| | $\alpha$ | 15% Sn | 0 | |
| 210 | $L$ | 50% Sn | 68.7 | Mixture of liquid and $\alpha$ phases. More solid forms. Compositions change. |
| | $\alpha$ | 18% Sn | 31.3 | |
| 183.5 | $L$ | 61.9% Sn | 48.7 | Liquid has the eutectic composition. |
| | $\alpha$ | 19.2% Sn | 51.3 | |
| 182.5 | $\alpha$ | 19.2% Sn | 73.4 | Eutectic ($\alpha$ and $\beta$ phases) and primary $\alpha$-phase. |
| | $\beta$ | 97.5% Sn | 26.6 | |

I Assume mass of the alloy is 100 g.

undergoes the eutectic transformation and forms the eutectic solid. Just below 183 °C, therefore, the microstructure is the primary $\alpha$-phase and the eutectic solid. Stated differently, below 183 °C, the $\alpha$ and $\beta$ phases coexist, and $\beta$ is in the eutectic structure. The weight fraction of the eutectic phase is the same as that of the liquid just above 183 °C, from which it was formed. The weight fractions of $\alpha$ and $\beta$ in the whole alloy are given by the lever rule applied at point $P$, or

$$W_\alpha = \frac{C_\beta - C_O}{C_\beta - C_\alpha} = \frac{97.5 - 40}{97.5 - 19.2} = 0.734$$

$$W_\beta = \frac{C_O - C_\alpha}{C_\beta - C_\alpha} = \frac{40 - 19.2}{97.5 - 19.2} = 0.266$$

The microstructure at room temperature will be much like that just below 183 °C, at which the alloy is a two phase solid because atomic diffusions in the solid will not be sufficiently fast to allow the compositions to change. Table 1.7 summarizes the phases that exist in this alloy at various temperatures.

# ADDITIONAL TOPICS

## 1.13 BRAVAIS LATTICES

An infinite periodic array of geometric points in space defines a **space lattice** or simply a **lattice.** Strictly, a lattice does not contain any atoms or molecules because it is simply an imaginary array of geometric points. A two-dimensional *simple square* lattice is shown in Figure 1.71a. In three dimensions, Figure 1.71a would correspond to the simple cubic (SC) lattice. The actual crystal is obtained from the lattice by placing an identical group of atoms (or molecules) at each lattice point. The identical group of atoms is called the **basis** of the crystal structure. Thus, conceptually, as illustrated in Figure 1.71a to c,

$$\textbf{Crystal} = \textbf{Lattice} + \textbf{Basis}$$

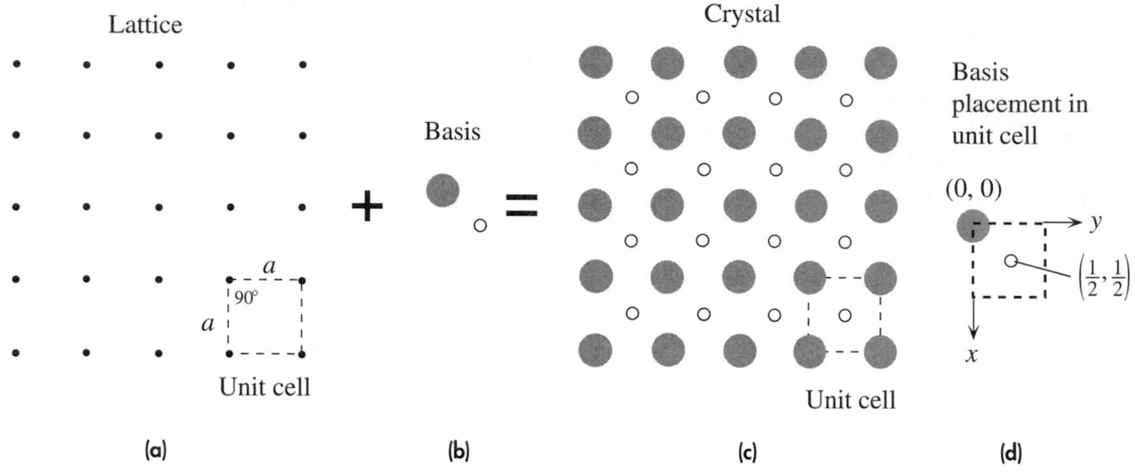


**Figure 1.71**
(a) A simple square lattice. The unit cell is a square with a side $a$.
(b) Basis has two atoms.
(c) Crystal = Lattice + Basis. The unit cell is a simple square with two atoms.
(d) Placement of basis atoms in the crystal unit cell.

The unit cell of the two-dimensional lattice in Figure 1.71a is a square which is characterized by the length $a$ of one of the sides; $a$ is called a **lattice parameter.** A given lattice can generate different *patterns* of atoms depending on the basis. The lattice in Figure 1.71a with the two-atom basis in Figure 1.71b produces the crystal in Figure 1.71c. Although the latter crystal appears as a body-centered square (similar to BCC in three dimensions), it is nonetheless a *simple square lattice* with two atoms comprising the basis. Suppose that the basis had only one atom; then the crystal would appear as the simple square lattice in Figure 1.71a (with each point now being an atom). The *patterns* in Figure 1.71a and c are different but the underlying lattice is the same. Because they have the same lattice, the two crystals would have certain identical symmetries. For example, for both crystals, a rotation by 90° about a lattice point would produce the same crystal structure.

To fully characterize the crystal, we also have to specify the locations of the basis atoms in the unit cell as in Figure 1.71d. By convention, we place a Cartesian coordinate system at the rear-left corner of the unit cell with the $x$ and $y$ axes along the square edges. We indicate the coordinates $(x_i, y_i)$ of each $i$th atom in terms of the lattice parameters along $x$ and $y$. Thus, the atoms in the unit cell in Figure 1.71d are at (0, 0) and at $(\frac{1}{2}, \frac{1}{2})$. The CsCl unit cell in Figure 1.38 appears as BCC, but it can be described by a SC lattice and a basis that has one $Cl^-$ ion and one $Cs^+$ ion. The ions in the SC unit cell are located at (0, 0, 0) and at the cell center at $(\frac{1}{2}, \frac{1}{2}, \frac{1}{2})$. Similarly, the NaCl crystal in Figure 1.37 is an FCC lattice with a basis of $Na^+$ and $Cl^-$ ions.

The diamond unit cell of silicon is an FCC lattice with two Si atoms constituting the basis. The two Si atoms are placed at (0, 0, 0) and $(\frac{1}{4}, \frac{1}{4}, \frac{1}{4})$. Most of the important

III–V compound semiconductors such as GaAs, AlAs, InAs, InP, etc., which are widely used in numerous optoelectronic devices, have the zinc blende (ZnS) unit cell. The zinc blende unit cell consists of an FCC lattice and a basis that has the Zn and S atoms placed at (0, 0, 0) and $(\frac{1}{4}, \frac{1}{4}, \frac{1}{4})$, respectively.

We generally represent the *geometry* of the unit cell of a lattice as a parallelepiped with sides $a, b, c$ and angles $\alpha, \beta, \gamma$ as depicted in Figure 1.40*a*. In the case of copper and iron, the geometry of the unit cell has $a = b = c, \alpha = \beta = \gamma = 90°$, and cubic

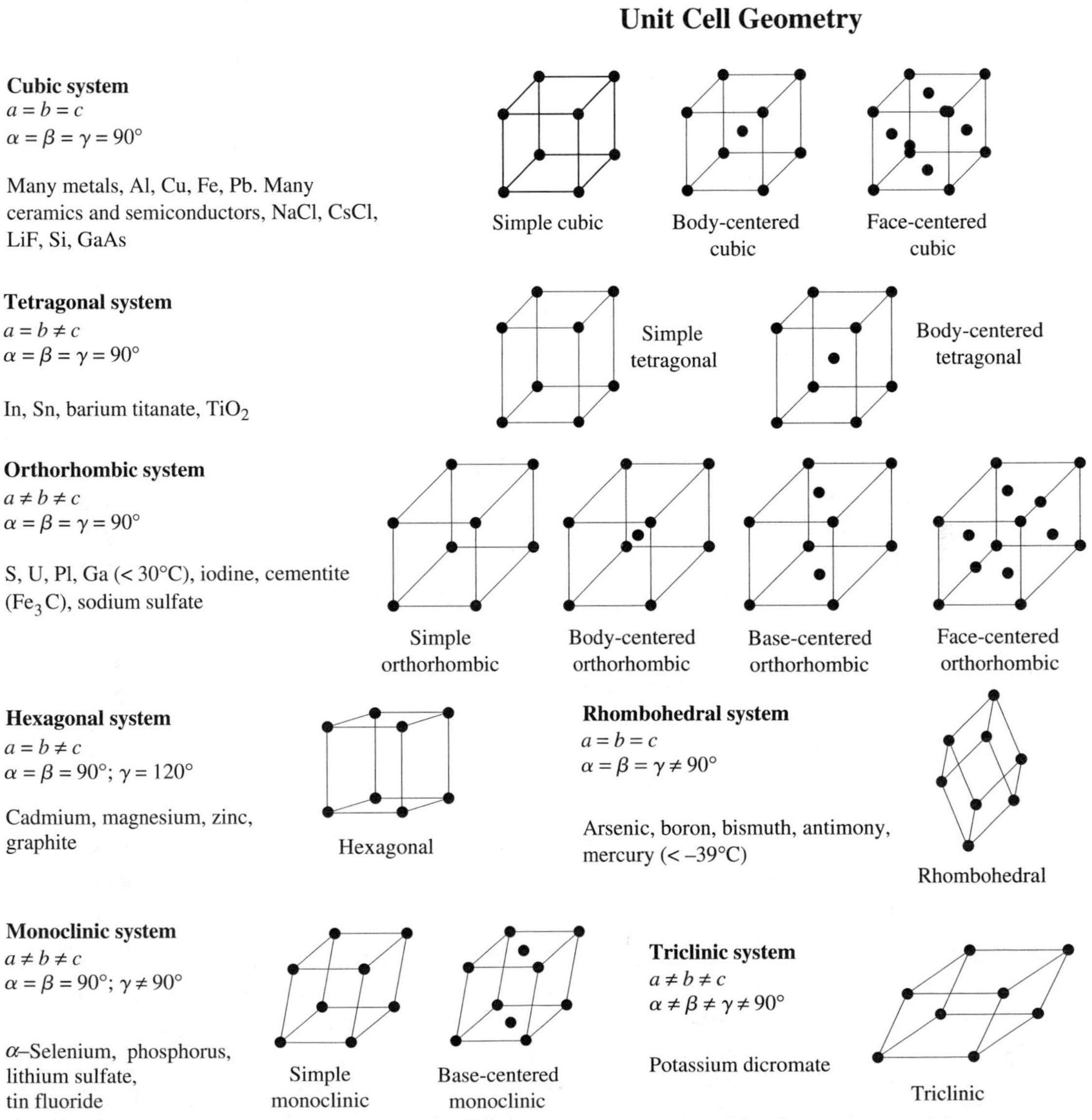


**Figure 1.72** The seven crystal systems (unit-cell geometries) and fourteen Bravais lattices.

symmetry. For Zn, the unit cell has hexagonal geometry with $a = b \neq c$, $\alpha = \beta = 90°$, and $\gamma = 120°$ as shown in Figure 1.33d. Based on different lattice parameters, there are *seven* possible distinct unit-cell geometries, which we call **crystal systems** each with a particular distinct symmetry. The seven crystal systems are depicted in Figure 1.72 with typical examples. We are already familiar with the cubic and hexagonal systems. The seven crystal systems only categorize the unit cells based on the geometry of the unit cell and not in terms of the symmetry and periodicity of the lattice points. (One should not confuse the unit-cell geometry with the lattice, which is a periodic array of points.) In the cubic system, for example, there are three possible distinct lattices corresponding to SC, BCC, and FCC which are shown in Figure 1.72. All three have the same cubic geometry: $a = b = c$ and $\alpha = \beta = \gamma = 90°$.

Many distinctly different lattices, or distinct patterns of points, exist in three dimensions. There are 14 distinct lattices whose unit cells have one of the seven geometries as indicated in Figure 1.72. Each of these is called a **Bravais lattice.** The copper crystal, for example, has the FCC Bravais lattice, but arsenic, antimony, and bismuth crystals have the rhombohedral Bravais lattice. Tin's unit cell belongs to the **tetragonal** crystal system, and its crystal lattice is a **body-centered tetragonal (BCT).**

## CD Selected Topics and Solved Problems

**Selected Topics**

Units and Conversions
Bonding: Bond Energies and Elastic Moduli
Secondary Bonding
Cohesive Energy: Ionic Bonding and Madelung Constant
Elementary Crystals
X-Ray Diffraction and Crystal Structures
Essential Mechanical Properties
Diffusion
Diffusion and Oxidation
Thermal Expansion
Surface Tension of Crystals

**Solved Problems**

van der Waals Bonding: Secondary Bonding and Bulk Modulus
Elementary Concepts in Material Science: Mean Atomic Separation, Bulk and Surface Atomic Concentrations, and Density
Elementary Crystals
Ionic Crystals

## DEFINING TERMS

**Activated state** is the state that occurs temporarily during a transformation or reaction when the reactant atoms or molecules come together to form a particular arrangement (intermediate between reactants and products) that has a higher potential energy than the reactants. The potential energy barrier between the activated state and the reactants is the activation energy.

**Activation energy** is the potential energy barrier against the formation of a product. In other words, it is the minimum energy that the reactant atom or molecule must have to be able to reach the activated state and hence form a product.

**Amorphous solid** is a solid that exhibits no crystalline structure or long-range order. It only possesses a short-range order in the sense that the nearest neighbors of an atom are well defined by virtue of chemical bonding requirements.

**Anion** is an atom that has gained negative charge by virtue of accepting one or more electrons. Usually, atoms of nonmetallic elements can gain electrons easily to become anions. Anions become attracted to the anode (positive terminal) in ionic conduction. Typical anions are the halogen ions $F^-$, $Cl^-$, $Br^-$, and $I^-$.

**Atomic mass** (or **relative atomic mass** or **atomic weight**) $M_{at}$ of an element is the average atomic mass, in atomic mass units (amu), of all the naturally occurring isotopes of the element. Atomic masses are listed in the Periodic Table. The amount of an element that has $6.022 \times 10^{23}$ atoms (the Avogadro number of atoms) has a mass in grams equal to the atomic mass.

**Atomic mass unit (amu)** is a convenient mass measurement equal to one-twelfth of the mass of a neutral carbon atom that has a mass number of $A = 12$ (6 protons and 6 neutrons). It has been found that amu = $1.66054 \times 10^{27}$ kg, which is equivalent to $10^{-3}/N_A$, where $N_A$ is Avogadro's number.

**Atomic packing factor (APF)** is the fraction of volume actually occupied by atoms in a crystal.

**Avogadro's number** ($N_A$) is the number of atoms in exactly 12 g of carbon-12. It is $6.022 \times 10^{23}$. Since atomic mass is defined as one-twelfth of the mass of the carbon-12 atom, the $N_A$ number of atoms of any substance has a mass equal to the atomic mass $M_{at}$, in grams.

**Basis** represents an atom, a molecule, or a collection of atoms, that is placed at each lattice point to generate the true crystal structure of a substance. All crystals are thought of as a lattice with each point occupied by a basis.

**Bond energy** or **binding energy** is the work (or energy) needed to separate two atoms infinitely from their equilibrium separation in the molecule or solid.

**Bulk modulus** $K$ is volume stress (pressure) needed per unit elastic volume strain and is defined by $p = -K\Delta$, where $p$ is the applied volume stress (pressure) and $\Delta$ is the volume strain. $K$ indicates the extent to which a body can be reversibly (and hence elastically) deformed in volume by an applied pressure.

**Cation** is an atom that has gained positive charge by virtue of losing one or more electrons. Usually, metal atoms can lose electrons easily to become cations. Cations become attracted to the cathode (negative terminal) in ionic conduction, as in gaseous discharge. The alkali metals, Li, Na, K, . . . , easily lose their valence electron to become cations, $Li^+$, $Na^+$, $K^+$, . . .

**Coordination number** is the number of nearest neighbors around a given atom in the crystal.

**Covalent bond** is the sharing of a pair of valence electrons between two atoms. For example, in $H_2$, the two hydrogen atoms share their electrons, so that each has a closed shell.

**Crystal** is a three-dimensional periodic arrangement of atoms, molecules, or ions. A characteristic property of the crystal structure is its periodicity and a degree of symmetry. For each atom, the number of neighbors and their exact orientations are well defined; otherwise the periodicity will be lost. Therefore, a long-range order results from strict adherence to a well-defined bond length and relative bond angle (that is, exact orientation of neighbors).

**Crystallization** is a process by which crystals of a substance are formed from another phase of that substance. Examples are solidification just below the fusion temperature from the melt, or condensation of the molecules from the vapor phase onto a substrate. The crystallization process initially requires the formation of small crystal nuclei, which contain a limited number (perhaps $10^3$–$10^4$) of atoms or molecules of the substance. Following nucleation, the nuclei grow by atomic diffusion from the melt or vapor.

**Diffusion** is the migration of atoms by virtue of their random thermal motions.

**Diffusion coefficient** is a measure of the rate at which atoms diffuse. The rate depends on the nature of the diffusion process and is typically temperature dependent. The diffusion coefficient is defined as the magnitude of diffusion flux per unit concentration gradient.

**Dislocation** is a line imperfection within a crystal that extends over many atomic distances.

**Edge dislocation** is a line imperfection within a crystal that occurs when an additional, short plane of atoms does not extend as far as its neighbors. The edge of this short plane constitutes a line of atoms where the bonding is irregular, that is, a line of imperfection called an edge dislocation.

**Elastic modulus** or **Young's modulus** ($Y$) is a measure of the ease with which a solid can be elastically deformed. The greater $Y$ is, the more difficult it is to deform the solid elastically. When a solid of length $\ell$ is subjected to a tensile stress $\sigma$ (force per unit area), the solid will extend elastically by an amount $\delta\ell$ where $\delta\ell/\ell$ is the strain $\varepsilon$. Stress and strain are related by $\sigma = Y\varepsilon$, so $Y$ is the stress needed per unit elastic strain.

**Electric dipole moment** is formed when a positive charge $+Q$ is separated from a negative charge $-Q$ of equal magnitude. Even though the net charge is zero, there is nonetheless an electric dipole moment formed by the two charges $-Q$ and $+Q$ being separated by a finite distance. Just as two charges exert a Coulombic force on each other, two dipoles also exert an electrostatic force on each other that depends on the separation of dipoles and their relative orientation.

**Electron affinity** represents the energy that is needed to add an electron to a neutral atom to create a negative ion (*anion*). When an electron is added to Cl to form $Cl^-$, energy is actually released.

**Electronegativity** is a relative measure of the ability of an atom to attract the electrons in a bond it forms with another atom. The *Pauling scale of electronegativity* assigns an electronegativity value (a pure number) $X$ to various elements, the highest being 4 for F, and the lowest values being for the alkali metal atoms, for which $X$ are less than 1. The difference $X_A - X_B$ in the electronegativities of two atoms $A$ and $B$ is a measure of the polar or ionic character of the bond $A$–$B$ between $A$ and $B$. A molecule $A$–$B$ would be polar, that is, possess a dipole moment, if $X_A$ and $X_B$ are different.

**Equilibrium** between two systems requires mechanical, thermal, and chemical equilibrium. Mechanical equilibrium means that the pressure should be the same in the two systems, so that one does not expand at the expense of the other. Thermal equilibrium implies that both have the same temperature. Equilibrium within a single-phase substance (*e.g.*, steam only or hydrogen gas only) implies uniform pressure and temperature within the system.

**Equilibrium state** of a system is the state in which the pressure and temperature in the system are uniform throughout. We say that the system possesses mechanical and thermal equilibrium.

**Eutectic composition** is an alloy composition of two elements that results in the lowest melting temperature compared to any other composition. A eutectic solid has a structure that is a mixture of two phases. The eutectic structure is usually special, such as alternating lamellae.

**Face-centered cubic (FCC) lattice** is a cubic lattice that has one lattice point at each corner of a cube and one at the center of each face. If there is a chemical species (atom or a molecule) at each lattice point, then the structure is an FCC crystal structure.

**Frenkel defect** is an ionic crystal imperfection that occurs when an ion moves into an interstitial site, thereby creating a vacancy in its original site. The imperfection is therefore a pair of point defects.

**Grain** is an individual crystal within a polycrystalline material. Within a grain, the crystal structure and orientation are the same everywhere and the crystal is oriented in one direction only.

**Grain boundary** is a surface region between differently oriented, adjacent grain crystals. The grain boundary contains a lattice mismatch between adjacent grains.

**Heat** is the amount of energy transferred from one system to another (or between the system and its surroundings) as a result of a temperature difference. Heat is not a new form of energy, but rather the transfer of energy from one body to another by virtue of the random motions of their molecules. When a hot body is in contact with a cold body, energy is transferred from the hot body to the cold one. The energy that is transferred is the excess mean kinetic energy of the molecules in the hot body. Molecules in the hot body have a higher mean kinetic energy and vibrate more violently. As a result of the collisions between the molecules, there is a net transfer of energy (heat) from the hot body to the cold one, until the molecules in both bodies have the same mean kinetic energy, that is, until their temperatures become equal.

**Heat capacity** at constant volume is the increase in the total energy $E$ of the system per degree increase in the temperature of the system with the volume remaining constant: $C = (\partial E/\partial T)_V$. Thus, the heat added to the system does no mechanical work due to a volume change but increases the internal energy. **Molar heat capacity** is the heat capacity for 1 mole of a substance. **Specific heat capacity** is the heat capacity per unit mass.

**Interstitial site (interstice)** is an unoccupied space between the atoms (or ions, or molecules) in a crystal.

**Ionization energy** is the energy required to remove an electron from a neutral atom; normally the most outer electron that has the least binding energy to the nucleus is removed to ionize an atom.

**Isomorphous** describes a structure that is the same everywhere (from *iso,* uniform, and *morphology,* structure).

**Isotropic substance** is a material that has the same property in all directions.

**Kinetic molecular theory** assumes that the atoms and molecules of all substances (gases, liquids, and solids) above absolute zero of temperature are in constant motion. Monatomic molecules (*e.g.,* He, Ne) in a gas exhibit constant and random translational motion, whereas the atoms in a solid exhibit constant vibrational motion.

**Lattice** is a regular array of points in space with a discernible periodicity. There are 14 distinct lattices possible in three-dimensional space. When an atom or molecule is placed at each lattice point, the resulting regular structure is a crystal structure.

**Lattice parameters** are (a) the lengths of the sides of the unit cell, and (b) the angles between the sides.

**Mechanical work** is qualitatively defined as the energy expended in displacing a constant force through a distance. When a force **F** is moved a distance $d\mathbf{x}$, work done $dW = \mathbf{F} \cdot d\mathbf{x}$. When we lift a body such as an apple of mass $m$ (100 g) by a distance $h$ (1 m), we do work by an amount $F\ \Delta x = mgh$ (1 J), which is then stored as the gravitational potential energy of the body. We have transferred energy from ourselves to the potential energy of the body by exchanging energy with it in the form of work. Further, in lifting the apple, the molecules have been displaced in orderly fashion, all upwards. Work therefore involves an orderly displacement of atoms and molecules of a substance in complete contrast to heat. When the volume $V$ of a substance changes by $dV$ when the pressure is $P$, the mechanical work involved is $P\ dV$ and is called the ***PV* work.**

**Metallic bonding** is the binding of metal atoms in a crystal through the attraction between the positive metal ions and the mobile valence electrons in the crystal. The valence electrons permeate the space between the ions.

**Miller indices** $(hk\ell)$ are indices that conveniently identify parallel planes in a crystal. Consider a plane with the intercepts, $x_1$, $y_1$, and $z_1$, in terms of lattice parameters $a$, $b$, and $c$. (For a plane passing through the origin, we shift the origin or use a parallel plane.) Then, $(hk\ell)$ are obtained by taking the reciprocals of $x_1$, $y_1$, and $z_1$ and clearing all fractions.

**Miscibility** of two substances is a measure of the mutual solubility of those two substances when they are in the same phase, such as liquid.

**Mole** of a substance is that amount of the substance that contains $N_A$ number of atoms (or molecules), where $N_A$ is Avogadro's number ($6.023 \times 10^{23}$). One mole of a substance has a mass equal to its atomic (molecular) mass, in grams. For example, 1 mole of copper contains $6.023 \times 10^{23}$ atoms and has a mass of 63.55 g.

**Phase** of a system is a homogeneous portion of the chemical system that has the same composition, structure, and properties everywhere. In a given chemical system, one phase may be in contact with another phase of the system. For example, iced water at 0 °C will have solid and liquid phases in contact. Each phase, solid ice and liquid water, has a distinct structure.

**Phase diagram** is a temperature versus composition diagram in which the existence and coexistence of various phases are identified by regions and lines. Between the liquidus and solidus lines, for example, the material is a heterogeneous mixture of the liquid and solid phases.

**Planar concentration of atoms** is the number of atoms per unit area on a given $(hk\ell)$ plane in a crystal.

**Polarization** is the separation of positive and negative charges in a system, which results in a net electric dipole moment.

**Polymorphism** or **allotropy** is a material attribute that allows the material to possess more than one crystal structure. Each possible crystal structure is called a polymorph. Generally, the structure of the polymorph depends on the temperature and pressure, as well as on the method of preparation of the solid. (For example, diamond can be prepared from graphite by the application of very high pressures.)

**Primary bond** is a strong interatomic bond, typically greater then 1 eV/atom, that involves ionic, covalent, or metallic bonding.

**Property** is a system characteristic or an attribute that we can measure. Pressure, volume, temperature, mass, energy, electrical resistivity, magnetization, polarization, and color are all properties of matter. Properties such as pressure, volume, and temperature can only be attributed to a system of many particles (which we treat as a continuum). Note that heat and work are not properties of a substance; instead, they represent energy transfers involved in producing changes in the properties.

**Saturated solution** is a solution that has the maximum possible amount of solute dissolved in a given amount of solvent at a specified temperature and pressure.

**Schottky defect** is an ionic crystal imperfection that occurs when a pair of ions is missing, that is, when there is a cation and anion pair vacancy.

**Screw dislocation** is a crystal defect that occurs when one portion of a perfect crystal is twisted or skewed with respect to another portion on only one side of a line.

**Secondary bond** is a weak bond, typically less than 0.1 eV/atom, which is due to dipole–dipole interactions between the atoms or molecules.

**Solid solution** is a homogeneous crystalline phase that contains two or more chemical components.

**Solute** is the minor chemical component of a solution; the component that is usually added in small amounts to a solvent to form a solution.

**Solvent** is the major chemical component of a solution.

**Stoichiometric compounds** are compounds with an integer ratio of atoms, as in $CaF_2$, in which two fluorine atoms bond with one calcium atom.

**Strain** is a relative measure of the deformation a material exhibits under an applied stress. Under an applied tensile (or compressive) stress, strain $\varepsilon$ is the change in the length per unit original length. When a shear stress is applied, the deformation involves a shear angle. **Shear strain** is the tangent of the shear angle that is developed by the application of the shearing stress. **Volume strain** $\Delta$ is the change in the volume per unit original volume.

**Stress** is force per unit area. When the applied force $F$ is perpendicular to the area $A$, stress $\sigma = F/A$ is either tensile or compressive. If the applied force is tangential to the area, then stress is **shear stress,** $\tau = F/A$.

**Thermal expansion** is the change in the length or volume of a substance due to a change in the temperature. **Linear coefficient of thermal expansion** $\lambda$ is the fractional change in the length per unit temperature change or $\Delta L/L_o = \lambda\, \Delta T$. Volume coefficient of expansion $\alpha_V$ is the fractional change in the volume per unit temperature change; $\alpha_V \approx 3\lambda$.

**Unit cell** is the most convenient small cell in a crystal structure that carries the characteristics of the crystal. The repetition of the unit cell in three dimensions generates the whole crystal structure.

**Vacancy** is a point defect in a crystal, where a normally occupied lattice site is missing an atom.

**Valence electrons** are the electrons in the outer shell of an atom. Since they are the farthest away from the nucleus, they are the first electrons involved in atom-to-atom interactions.

**Young's modulus** see **elastic modulus.**

# QUESTIONS AND PROBLEMS

**1.1** **Virial theorem** The Li atom has a nucleus with a $+3e$ positive charge, which is surrounded by a full $1s$ shell with two electrons, and a single valence electron in the outer $2s$ subshell. The atomic radius of the Li atom is about 0.17 nm. Using the Virial theorem, and assuming that the valence electron sees the nuclear $+3e$ shielded by the two $1s$ electrons, that is, a net charge of $+e$, estimate the ionization energy of Li (the energy required to free the 2*s* electron). Compare this value with the experimental value of 5.39 eV. Suppose that the actual nuclear charge seen by the valence electron is not $+e$ but a little higher, say $+1.25e$, due to the imperfect shielding provided by the closed 1*s* shell. What would be the new ionization energy? What is your conclusion?

1.2 **Atomic mass and molar fractions**

*a.* Consider a multicomponent alloy containing *N* elements. If $w_1, w_2, \ldots, w_N$ are the weight fractions of components $1, 2, \ldots, N$ in the alloy and $M_1, M_2, \ldots, M_N$ are the respective atomic masses of the elements, show that the atomic fraction of the *i*th component is given by

$$n_i = \frac{w_i/M_i}{\dfrac{w_1}{M_1} + \dfrac{w_2}{M_2} + \cdots + \dfrac{w_N}{M_N}}$$

*Weight to atomic percentage*

*b.* Suppose that a substance (compound or an alloy) is composed of *N* elements, $A, B, C, \ldots$ and that we know their atomic (or molar) fractions $n_A, n_B, n_C, \ldots$. Show that the weight fractions $w_A, w_B, w_C, \ldots$ are given by

*Atomic to weight percentage*

$$w_A = \frac{n_A M_A}{n_A M_A + n_B M_B + n_C M_C + \cdots}$$

$$w_B = \frac{n_B M_B}{n_A M_A + n_B M_B + n_C M_C + \cdots}$$

*c.* Consider the semiconducting II–VI compound cadmium selenide, CdSe. Given the atomic masses of Cd and Se, find the weight fractions of Cd and Se in the compound and grams of Cd and Se needed to make 100 grams of CdSe.

*d.* A Se–Te–P glass alloy has the composition 77 wt.% Se, 20 wt.% Te, and 3 wt.% P. Given their atomic masses, what are the atomic fractions of these constituents?

1.3 **The covalent bond** Consider the $H_2$ molecule in a simple way as two touching H atoms, as depicted in Figure 1.73. Does this arrangement have a lower energy than two separated H atoms? Suppose that electrons totally correlate their motions so that they move to avoid each other as in the snapshot in Figure 1.73. The radius $r_o$ of the hydrogen atom is 0.0529 nm. The electrostatic potential energy of two charges $Q_1$ and $Q_2$ separated by a distance $r$ is given by $Q_1 Q_2/(4\pi\varepsilon_o r)$. Using the virial theorem as in Example 1.1 consider the following:

*a.* Calculate the total electrostatic potential energy *PE* of all the charges when they are arranged as shown in Figure 1.73. In evaluating the *PE* of the whole collection of charges you must consider all pairs of charges and, at the same time, avoid double counting of interactions between the same pair of charges. The total *PE* is the sum of the following: electron 1 interacting with the proton at a distance $r_o$ on the left, proton at $r_o$ on the right, and electron 2 at a distance $2r_o$ + electron 2 interacting with a proton at $r_o$ and another proton at $3r_o$ + two protons, separated by $2r_o$, interacting with each other. Is this configuration energetically favorable?

*b.* Given that in the isolated H atom the *PE* is $2 \times (-13.6\ \text{eV})$, calculate the change in *PE* in going from two isolated H atoms to the $H_2$ molecule. Using the virial theorem, find the change in the total energy and hence the covalent bond energy. How does this compare with the experimental value of 4.51 eV?

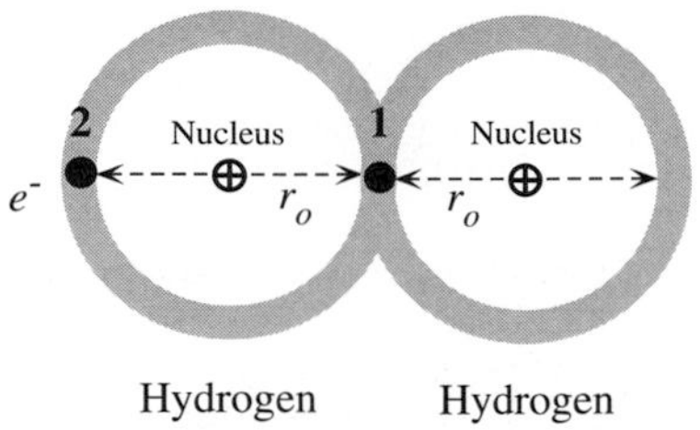


**Figure 1.73** A simplified view of the covalent bond in $H_2$. A snapshot at one instant.

1.4 **Ionic bonding and CsCl** The potential energy $E$ per $Cs^+$–$Cl^-$ pair within the CsCl crystal depends on the interionic separation $r$ in the same fashion as in the NaCl crystal,

*Energy per ion pair in ionic crystals*

$$E(r) = -\frac{e^2 M}{4\pi\varepsilon_o r} + \frac{B}{r^m} \qquad [1.38]$$

where for CsCl, $M = 1.763$, $B = 1.192 \times 10^{-104}$ J m$^9$ or $7.442 \times 10^{-5}$ eV (nm)$^9$, and $m = 9$. Find the equilibrium separation ($r_o$) of the ions in the crystal and the ionic bonding energy, that is, the ionic cohesive energy, and compare the latter value to the experimental value of 657 kJ mol$^{-1}$. Given that the *ionization energy* of Cs is 3.89 eV and the *electron affinity* of Cl (energy released when an electron is added) is 3.61 eV, calculate the atomic cohesive energy of the CsCl crystal as joules per mole.

1.5 **Madelung constant** If we were to examine the NaCl crystal in three dimensions, we would find that each $Na^+$ ion has

6 $Cl^-$ ions as *nearest* neighbors at a distance $r$

12 $Na^+$ ions as *second* nearest neighbors at a distance $r\sqrt{2}$

8 $Cl^-$ ions as *third* nearest neighbors at a distance $r\sqrt{3}$

and so on. Show that the electrostatic potential energy of the $Na^+$ atom can be written as

*Madelung constant M for NaCl*

$$E(r) = -\frac{e^2}{4\pi\varepsilon_o r}\left[6 - \frac{12}{\sqrt{2}} + \frac{8}{\sqrt{3}} - \cdots\right] = -\frac{e^2 M}{4\pi\varepsilon_o r}$$

where $M$, called the **Madelung constant,** is given by the summation in the square brackets for this particular ionic crystal structure (NaCl). Calculate $M$ for the first three terms and compare it with $M = 1.7476$, its value had we included the higher terms. What is your conclusion?

*1.6 **Bonding and bulk modulus** In general, the potential energy $E$ per atom, or per ion pair, in a crystal as a function of interatomic (interionic) separation $r$ can be written as the sum of an attractive *PE* and a repulsive *PE*,

*General PE curve for bonding*

$$E(r) = -\frac{A}{r^n} + \frac{B}{r^m} \qquad [1.39]$$

where $A$ and $n$ are constants characterizing the attractive *PE* and $B$ and $m$ are constants characterizing the repulsive *PE*. This energy is minimum when the crystal is in equilibrium. The magnitude of the minimum energy and its location $r_o$ define the bonding energy and the equilibrium interatomic (or interionic) separation, respectively.

When a pressure $P$ is applied to a solid, its original volume $V_o$ shrinks to $V$ by an amount $\Delta V = V - V_o$. The bulk modulus $K$ relates the volume strain $\Delta V/V$ to the applied pressure $P$ by

*Bulk modulus definition*

$$P = -K\frac{\Delta V}{V_o} \qquad [1.40]$$

The bulk modulus $K$ is related to the energy curve. In its simplest form (assuming a simple cubic unit cell) $K$ can be estimated from Equation 1.39 by

*Bulk modulus*

$$K = \frac{1}{9cr_o}\left[\frac{d^2E}{dr^2}\right]_{r=r_o} \qquad [1.41]$$

where $c$ is a numerical factor, of the order of unity, given by $b/p$ where $p$ is the number of atoms or ion pairs in the unit cell and $b$ is a numerical factor that relates the cubic unit cell lattice parameter $a_o$ to the equilibrium interatomic (interionic) separation $r_o$ by $b = a_o^3/r_o^3$.

*a.* Show that the bond energy and equilibrium separation are given by

$$E_{\text{bond}} = \frac{A}{r_o^n}\left(1 - \frac{n}{m}\right) \quad \text{and} \quad r_o = \left(\frac{Bm}{An}\right)^{1/(m-n)}$$

*b.* Show that the bulk modulus is given by

$$K = \frac{An}{9cr_o^{n+3}}(m-n) \qquad \text{or} \qquad K = \frac{mnE_{\text{bond}}}{9cr_o^3}$$

*c.* For a NaCl-type crystal, $Na^+$ and $Cl^-$ ions touch along the cube edge so that $r_o = (a_o/2)$. Thus, $a^3 = 2^3 r_o^3$ and $b = 2^3 = 8$. There are four ion pairs in the unit cell, $p = 4$. Thus, $c = b/p = 8/4 = 2$. Using the values from Example 1.3, calculate the bulk modulus of NaCl.

***1.7 Van der Waals bonding** Below 24.5 K, Ne is a crystalline solid with an FCC structure. The interatomic interaction energy per atom can be written as

$$E(r) = -2\varepsilon\left[14.45\left(\frac{\sigma}{r}\right)^6 - 12.13\left(\frac{\sigma}{r}\right)^{12}\right] \quad \text{(eV/atom)}$$

where $\varepsilon$ and $\sigma$ are constants that depend on the polarizability, the mean dipole moment, and the extent of overlap of core electrons. For crystalline Ne, $\varepsilon = 3.121 \times 10^{-3}$ eV and $\sigma = 0.274$ nm.

*a.* Show that the equilibrium separation between the atoms in an inert gas crystal is given by $r_o = (1.090)\sigma$. What is the equilibrium interatomic separation in the Ne crystal?

*b.* Find the bonding energy per atom in solid Ne.

*c.* Calculate the density of solid Ne (atomic mass = 20.18).

**1.8 Kinetic molecular theory**

*a.* In a particular Ar-ion laser tube the gas pressure due to Ar atoms is about 0.1 torr at 25 °C when the laser is off. What is the concentration of Ar atoms per $cm^3$ at 25 °C in this laser? (760 torr = 1 atm = $1.013 \times 10^5$ Pa.)

*b.* In the He–Ne laser tube He and Ne gases are mixed and sealed. The total pressure $P$ in the gas is given by contributions arising from He and Ne atoms:

$$P = P_{\text{He}} + P_{\text{Ne}}$$

where $P_{\text{He}}$ and $P_{\text{Ne}}$ are the *partial pressures* of He and Ne in the gas mixture, that is, pressures due to He and Ne gases alone,

$$P_{\text{He}} = \frac{N_{\text{He}}}{N_A}\left(\frac{RT}{V}\right) \qquad \text{and} \qquad P_{\text{Ne}} = \frac{N_{\text{Ne}}}{N_A}\left(\frac{RT}{V}\right)$$

In a particular He–Ne laser tube the ratio of He and Ne atoms is 7:1, and the total pressure is about 1 torr at 22 °C. Calculate the concentrations of He and Ne atoms in the gas at 22 °C. What is the pressure at an operating temperature of 130 °C?

**1.9 Kinetic molecular theory** Calculate the effective (rms) speeds of the He and Ne atoms in the He–Ne gas laser tube at room temperature (300 K).

***1.10 Kinetic molecular theory and the Ar-ion laser** An argon-ion laser has a laser tube that contains Ar atoms that produce the laser emission when properly excited by an electrical discharge. Suppose that the gas temperature inside the tube is 1300 °C (very hot).

*a.* Calculate the mean speed ($v_{\text{av}}$), rms velocity ($v_{\text{rms}} = \sqrt{\overline{v^2}}$), and the rms speed ($v_{\text{rms},x} = \sqrt{\overline{v_x^2}}$) in one particular direction of the Ar atoms in the laser tube, assuming 1300 °C . (See Example 1.10.)

*b.* Consider a light source that is emitting waves and is moving toward an observer, somewhat like a whistling train moving toward a passenger. If $f_o$ is the frequency of the light waves emitted at the source, then, due to the *Doppler effect*, the observer measures a higher frequency $f$ that depends on the velocity $v_{\text{Ar}}$ of the source moving toward the observer and the speed $c$ of light,

$$f = f_o\left(1 + \frac{v_{\text{Ar}}}{c}\right)$$

It is the Ar ions that emit the laser output light in the Ar-ion laser. The emission wavelength $\lambda_o = c/f_o$ is 514.5 nm. Calculate the wavelength $\lambda$ registered by an observer for those atoms that are moving with a mean speed $v_{av}$ toward the observer. Those atoms that are moving away from the observer will result in a lower observed frequency because $v_{Ar}$ will be negative. Estimate the width of the wavelengths (the difference between the longest and shortest wavelengths) emitted by the Ar-ion laser.

***1.11** **Vacuum deposition** Consider air as composed of nitrogen molecules $N_2$.

*a.* What is the concentration $n$ (number of molecules per unit volume) of $N_2$ molecules at 1 atm and 27 °C?

*b.* Estimate the mean separation between the $N_2$ molecules.

*c.* Assume each molecule has a finite size that can be represented by a sphere of radius $r$. Also assume that $\ell$ is the **mean free path,** defined as the mean distance a molecule travels before colliding with another molecule, as illustrated in Figure 1.74a. If we consider the motion of one $N_2$ molecule, with all the others stationary, it is apparent that if the path of the traveling molecule crosses the cross-sectional area $S = \pi(2r)^2$, there will be a collision. Since $\ell$ is the mean distance between collisions, there must be at least one stationary molecule within the volume $S\ell$,

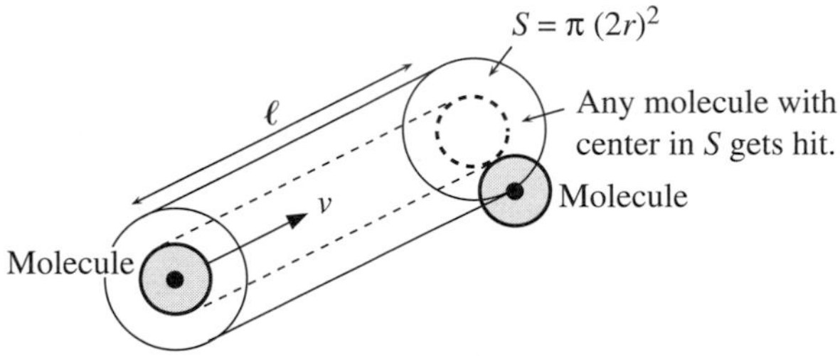


**(a) A molecule moving with a velocity $v$ travels a mean distance $\ell$ between collisions. Since the collision cross-sectional area is $S$, in the volume $S\ell$ there must be at least one molecule. Consequently, $n(S\ell) = 1$.**

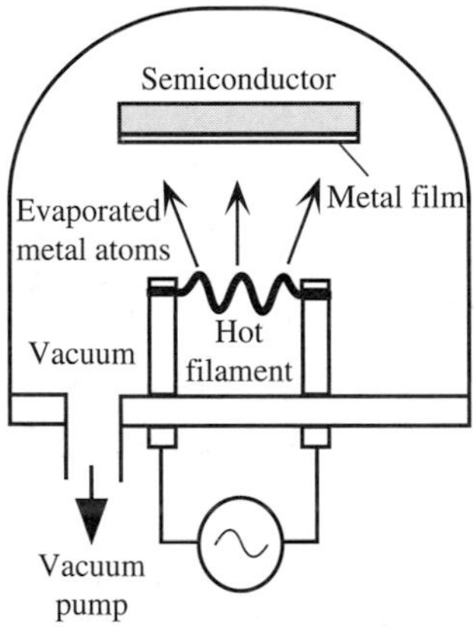


**(b) Vacuum deposition of metal electrodes by thermal evaporation.**

**Figure 1.74**

Walter Houser Brattain (1902–1987), experimenting with metal contacts on copper oxide (1935) at Bell Telephone Labs. A vacuum evaporation chamber is used to deposit the metal electrode.

| SOURCE: Bell Telephone Laboratories, courtesy AIP Emilio Segrè Visual Archives.

as shown in Figure 1.74a. Since $n$ is the concentration, we must have $n(S\ell) = 1$ or $\ell = 1/(\pi 4r^2 n)$. However, this must be corrected for the fact that all the molecules are in motion, which only introduces a numerical factor, so that

$$\ell = \frac{1}{2^{1/2}4\pi r^2 n}$$

Assuming a radius $r$ of 0.1 nm, calculate the mean free path of $N_2$ molecules between collisions at 27 °C and 1 atm.

*d.* Assume that an Au film is to be deposited onto the surface of a Si chip to form metallic interconnections between various devices. The deposition process is generally carried out in a vacuum chamber and involves the condensation of Au atoms from the vapor phase onto the chip surface. In one procedure, a gold wire is wrapped around a tungsten filament, which is heated by passing a large current through the filament (analogous to the heating of the filament in a light bulb) as depicted in Figure 1.74b. The Au wire melts and wets the filament, but as the temperature of the filament increases, the gold evaporates to form a vapor. Au atoms from this vapor then condense onto the chip surface, to solidify and form the metallic connections. Suppose that the source (filament)-to-substrate (chip) distance $L$ is 10 cm. Unless the mean free path of air molecules is much longer than $L$, collisions between the metal atoms and air molecules will prevent the deposition of the Au onto the chip surface. Taking the mean free path $\ell$ to be $100L$, what should be the pressure inside the vacuum system? (Assume the same $r$ for Au atoms.)

**1.12 Heat capacity**

*a.* Calculate the heat capacity per mole and per gram of $N_2$ gas, neglecting the vibrations of the molecule. How does this compare with the experimental value of 0.743 J g$^{-1}$ K$^{-1}$?

*b.* Calculate the heat capacity per mole and per gram of $CO_2$ gas, neglecting the vibrations of the molecule. How does this compare with the experimental value of 0.648 J K$^{-1}$ g$^{-1}$? Assume that the $CO_2$ molecule is linear (O−C−O) so that it has two rotational degrees of freedom.

*c.* Based on the Dulong–Petit rule, calculate the heat capacity per mole and per gram of solid silver. How does this compare with the experimental value of 0.235 J K$^{-1}$ g$^{-1}$?

*d.* Based on the Dulong–Petit rule, calculate the heat capacity per mole and per gram of the silicon crystal. How does this compare with the experimental value of 0.71 J K$^{-1}$ g$^{-1}$?

**1.13 Dulong–Petit atomic heat capacity** Express the Dulong–Petit rule for the molar heat capacity as heat capacity per atom and in the units of eV K$^{-1}$ per atom, called the **atomic heat capacity.** CsI is an ionic crystal used in optical applications that require excellent infrared transmission at very long wavelengths (up to 55 μm). It has the CsCl crystal structure with one $Cs^+$ and one $I^-$ ion in the unit cell. Calculate the specific heat capacity of CsI and compare it with the experimental value of 0.20 J K$^{-1}$ g$^{-1}$. What is your conclusion?

**1.14 Dulong–Petit specific heat capacity of alloys and compounds**

*a.* Consider an alloy $AB$, such as solder, or a compound material such as MgO, composed of $n_A$, atomic fractions of $A$, and $n_B$, atomic fractions of $B$. (The atomic fraction of $A$ is the same as its molar fraction.) Let $M_A$ and $M_B$ be the atomic weights of $A$ and $B$, in g mol$^{-1}$. The mean atomic weight per atom in the alloy or compound is then

$$\overline{M} = n_A M_A + n_B M_B$$

*Average atomic weight*

Show that the *Dulong–Petit rule* for the specific heat capacity $c_s$ leads to

$$c_s = \frac{C_m}{\overline{M}} = \frac{25}{n_A M_A + n_B M_B} \quad \text{J K}^{-1}\,\text{g}^{-1}$$

*Specific heat capacity*

*b.* Calculate the specific heat capacity of Pb–Sn solder assuming that its composition is 38 wt.% Pb and 62 wt.% Sn.

*c.* Calculate the specific heat capacities of Pb and Sn individually as $c_{sA}$ and $c_{sB}$, respectively, and then calculate the $c_s$ for the alloy using

*Alloy specific heat capacity*

$$c_s = c_{sA}w_A + c_{sB}w_B$$

where $w_A$ and $w_B$ are the weight fractions of $A$ (Pb) and $B$ (Sn) in the alloy (solder). Compare your result with part (*b*). What is your conclusion?

*d.* ZnSe is an important optical material (used in infrared windows and lenses and high-power $CO_2$ laser optics) and also an important II–VI semiconductor that can be used to fabricate blue-green laser diodes. Calculate the specific heat capacity of ZnSe, and compare the calculation to the experimental value of 0.345 J $K^{-1}$ $g^{-1}$.

**1.15 Thermal expansion**

*a.* If $\lambda$ is the thermal expansion coefficient, show that the thermal expansion coefficient for an area is $2\lambda$. Consider an aluminum square sheet of area 1 $cm^2$. If the thermal expansion coefficient of Al at room temperature (25 °C) is about $24 \times 10^{-6}$ $K^{-1}$, at what temperature is the percentage change in the area +1%?

*b.* A particular incandescent light bulb (100 W, 120 V) has a tungsten (W) filament of length 57.9 cm and a diameter of 63.5 μm. Calculate the length of the filament at 2300 °C, the approximate operating temperature of the filament inside the bulb. The linear expansion coefficient $\lambda$ of W is approximately $4.50 \times 10^{-6}$ $K^{-1}$ at 300 K. How would you improve your calculation?

**1.16 Thermal expansion of Si** The expansion coefficient of silicon over the temperature range 120–1500 K is given by Okada and Tokumaru (1984) as

*Silicon linear expansion coefficient*

$$\lambda = 3.725 \times 10^{-6}[1 - e^{-5.88\times10^{-3}(T-124)}] + 5.548 \times 10^{-10}T$$

where $\lambda$ is in $K^{-1}$ (or $°C^{-1}$) and $T$ is in kelvins.

*a.* By expanding the above function around 20 °C (293 K) show that,

*Silicon linear expansion coefficient*

$$\lambda = 2.5086 \times 10^{-6} + (8.663 \times 10^{-9})(T - 293) - (2.3839 \times 10^{-11})(T - 293)^2$$

*b.* The change $\delta\rho$ in the density due to a change $\delta T$ in the temperature, from Example 1.8, is given by

$$\delta\rho = -\rho_o\alpha_V\ \delta T = -3\rho_o\lambda\ \delta T$$

Given the density of Si as 2.329 g $cm^{-3}$ at 20 °C, calculate the density at 1000 °C by using the full expression and by using the polynomials expansion of $\lambda$. What is your conclusion?

**1.17 Thermal expansion of GaP and GaAs**

*a.* GaP has the zinc blende structure. The linear expansion coefficient in GaP has been measured as follows: $\lambda = 4.65 \times 10^{-6}$ $K^{-1}$ at 300 K; $5.27 \times 10^{-6}$ $K^{-1}$ at 500 K; $5.97 \times 10^{-6}$ $K^{-1}$ at 800 K. Calculate the coefficients, $A$, $B$, and $C$ in

$$\frac{dL}{L_o\,dT} = \lambda(T) = A + B(T - T_o) + C(T - T_o)^2 + \cdots$$

where $T_o = 300$ K. The lattice constant of GaP, $a$, at 27 °C is 0.5451 nm. Calculate the lattice constant at 300 °C.

*b.* The linear expansion coefficient of GaAs over 200–1000 K is given by

*GaAs linear expansion coefficient*

$$\lambda = 4.25 \times 10^{-6} + (5.82 \times 10^{-9})T - (2.82 \times 10^{-12})T^2$$

where $T$ is in kelvins. The lattice constant $a$ at 300 K is 0.56533 nm. Calculate the lattice constant and the density at −40°C.

**1.18 Electrical noise** Consider an amplifier with a bandwidth $B$ of 5 kHz, corresponding to a typical speech bandwidth. Assume the input resistance of the amplifier is 1 MΩ. What is the rms noise voltage at the input? What will happen if the bandwidth is doubled to 10 kHz? What is your conclusion?

**1.19** **Thermal activation** A certain chemical oxidation process (*e.g.*, $SiO_2$) has an activation energy of 2 eV atom$^{-1}$.

*a.* Consider the material exposed to pure oxygen gas at a pressure of 1 atm at 27 °C. Estimate how many oxygen molecules per unit volume will have energies in excess of 2 eV? (Consider the numerical integration of Equation 1.24.)

*b.* If the temperature is 900 °C, estimate the number of oxygen molecules with energies more than 2 eV. What happens to this concentration if the pressure is doubled?

**1.20** **Diffusion in Si** The diffusion coefficient of boron (B) atoms in a single crystal of Si has been measured to be $1.5 \times 10^{-18}$ m$^2$ s$^{-1}$ at 1000 °C and $1.1 \times 10^{-16}$ m$^2$ s$^{-1}$ at 1200 °C.

*a.* What is the activation energy for the diffusion of B, in eV/atom?

*b.* What is the preexponential constant $D_o$?

*c.* What is the rms distance (in micrometers) diffused in 1 hour by the B atom in the Si crystal at 1200 °C and 1000 °C?

*d.* The diffusion coefficient of B in polycrystalline Si has an activation energy of 2.4–2.5 eV/atom and $D_o = (1.5 - 6) \times 10^{-7}$ m$^2$ s$^{-1}$. What constitutes the diffusion difference between the single crystal sample and the polycrystalline sample?

**1.21** **Diffusion in $SiO_2$** The diffusion coefficient of P atoms in $SiO_2$ has an activation energy of 2.30 eV/atom and $D_o = 5.73 \times 10^{-9}$ m$^2$ s$^{-1}$. What is the rms distance diffused in 1 hour by P atoms in $SiO_2$ at 1200 °C?

**1.22** **BCC and FCC crystals**

*a.* Molybdenum has the BCC crystal structure, a density of 10.22 g cm$^{-3}$, and an atomic mass of 95.94 g mol$^{-1}$. What is the atomic concentration, lattice parameter $a$, and atomic radius of molybdenum?

*b.* Gold has the FCC crystal structure, a density of 19.3 g cm$^{-3}$, and an atomic mass of 196.97 g mol$^{-1}$. What is the atomic concentration, lattice parameter $a$, and atomic radius of gold?

**1.23** **BCC and FCC crystals**

*a.* Tungsten (W) has the BCC crystal structure. The radius of the W atom is 0.1371 nm. The atomic mass of W is 183.8 amu (g mol$^{-1}$). Calculate the number of W atoms per unit volume and density of W.

*b.* Platinum (Pt) has the FCC crystal structure. The radius of the Pt atom is 0.1386 nm. The atomic mass of Pt is 195.09 amu (g mol$^{-1}$). Calculate the number of Pt atoms per unit volume and density of Pt.

**1.24** **Planar and surface concentrations** Niobium (Nb) has the BCC crystal with a lattice parameter $a = 0.3294$ nm. Find the planar concentrations as the number of atoms per nm$^2$ of the (100), (110), and (111) planes. Which plane has the most concentration of atoms per unit area? Sometimes the number of atoms per unit area $n_{\text{surface}}$ on the surface of a crystal is estimated by using the relation $n_{\text{surface}} = n_{\text{bulk}}^{2/3}$, where $n_{\text{bulk}}$ is the concentration of atoms in the bulk. Compare $n_{\text{surface}}$ values with the planar concentrations that you calculated and comment on the difference. [*Note:* The BCC (111) plane does not cut through the center atom and the (111) has one-sixth of an atom at each corner.]

**1.25** **Diamond and zinc blende** Si has the diamond and GaAs has the zinc blende crystal structure. Given the lattice parameters of Si and GaAs, $a = 0.543$ nm and $a = 0.565$ nm, respectively, and the atomic masses of Si, Ga, and As as 28.08, 69.73, and 74.92, respectively, calculate the density of Si and GaAs. What is the atomic concentration (atoms per unit volume) in each crystal?

**1.26** **Zinc blende, NaCl, and CsCl**

*a.* InAs is a III-V semiconductor that has the zinc blende structure with a lattice parameter of 0.606 nm. Given the atomic masses of In (114.82 g mol$^{-1}$) and As (74.92 g mol$^{-1}$), find the density.

*b.* CdO has the NaCl crystal structure with a lattice parameter of 0.4695 nm. Given the atomic masses of Cd (112.41 g mol$^{-1}$) and O (16.00 g mol$^{-1}$), find the density.

*c.* KCl has the same crystal structure as NaCl. The lattice parameter $a$ of KCl is 0.629 nm. The atomic masses of K and Cl are 39.10 g $mol^{-1}$ and 35.45 g $mol^{-1}$, respectively. Calculate the density of KCl.

**1.27 Crystallographic directions and planes** Consider the cubic crystal system.

*a.* Show that the line $[hk\ell]$ is perpendicular to the $(hk\ell)$ plane.

*b.* Show that the spacing between adjacent $(hk\ell)$ planes is given by

$$d = \frac{a}{\sqrt{h^2 + k^2 + \ell^2}}$$

**1.28 Si and $SiO_2$**

*a.* Given the Si lattice parameter $a = 0.543$ nm, calculate the number of Si atoms per unit volume, in $nm^{-3}$.

*b.* Calculate the number of atoms per $m^2$ and per $nm^2$ on the (100), (110), and (111) planes in the Si crystal as shown in Figure 1.75. Which plane has the most number of atoms per unit area?

*c.* The density of $SiO_2$ is 2.27 g $cm^{-3}$. Given that its structure is amorphous, calculate the number of molecules per unit volume, in $nm^{-3}$. Compare your result with (a) and comment on what happens when the surface of an Si crystal oxidizes. The atomic masses of Si and O are 28.09 and 16, respectively.

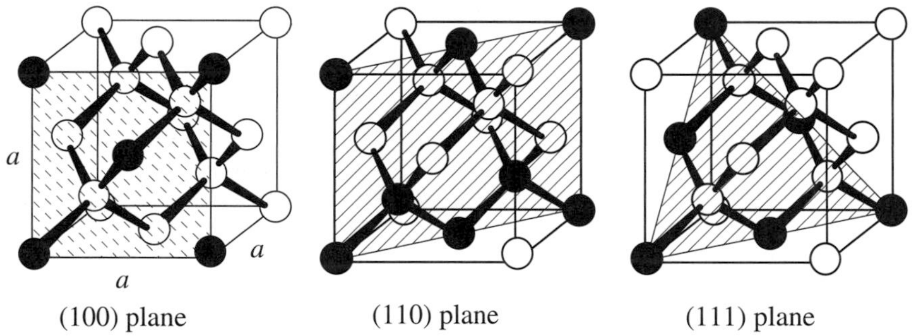


**Figure 1.75** Diamond cubic crystal structure and planes.
Determine what portion of a black-colored atom belongs to the plane that is hatched.

**1.29 Vacancies in metals**

*a.* The energy of formation of a vacancy in the copper crystal is about 1 eV. Calculate the concentration of vacancies at room temperature (300 K) and just below the melting temperature, 1084 °C. Neglect the change in the density which is small.

*b.* The following table shows the energies of vacancy formation in various metals with *close-packed* crystal structures and the melting temperature $T_m$. Plot $E_v$ in eV versus $T_m$ in kelvins, and explore if there is a correlation between $E_v$ and $T_m$. Some materials engineers take $E_v$ to be very roughly $10kT_m$. Do you think that they are correct? (Justify.)

| | Metal | | | | | | | | |
|---|---|---|---|---|---|---|---|---|---|
| | **Al** | **Ag** | **Au** | **Cu** | **Mg** | **Pt** | **Pb** | **Ni** | **Pd** |
| Crystal | FCC | FCC | FCC | FCC | HCP | FCC | FCC | FCC | FCC |
| $E_v$ (eV) | 0.70–0.76 | 1.0–1.1 | 0.90–0.98 | 1–1.28 | 0.89 | 1.3–1.5 | 0.50 | 1.63–1.79 | 1.54–1.85 |
| $T_m$ (°C) | 660 | 962 | 1064 | 1085 | 650 | 1768 | 328 | 1455 | 1555 |

**1.30** **Vacancies in silicon** In device fabrication, Si is frequently doped by the diffusion of impurities (dopants) at high temperatures, typically 950–1200°C. The energy of vacancy formation in the Si crystal is about 3.6 eV. What is the equilibrium concentration of vacancies in a Si crystal at 1000 °C? Neglect the change in the density with temperature which is less than 1 percent in this case.

**1.31** **Pb–Sn solder** Consider the soldering of two copper components. When the solder melts, it wets both metal surfaces. If the surfaces are not clean or have an oxide layer, the molten solder cannot wet the surfaces and the soldering fails. Assume that soldering takes place at 250 °C, and consider the diffusion of Sn atoms into the copper (the Sn atom is smaller than the Pb atom and hence diffuses more easily).

*a.* The diffusion coefficient of Sn in Cu at two temperatures is $D = 1.69 \times 10^{-9}$ cm$^2$ hr$^{-1}$ at 400 °C and $D = 2.48 \times 10^{-7}$ cm$^2$ hr$^{-1}$ at 650 °C. Calculate the rms distance diffused by an Sn atom into the copper, assuming the cooling process takes 10 seconds.

*b.* What should be the composition of the solder if it is to begin freezing at 250 °C?

*c.* What are the components (phases) in this alloy at 200 °C? What are the compositions of the phases and their relative weights in the alloy?

*d.* What is the microstructure of this alloy at 25 °C? What are weight fractions of the $\alpha$ and $\beta$ phases assuming near equilibrium cooling?

**1.32** **Pb–Sn solder** Consider 50% Pb–50% Sn solder.

*a.* Sketch the temperature-time profile and the microstructure of the alloy at various stages as it is cooled from the melt.

*b.* At what temperature does the solid melt?

*c.* What is the temperature range over which the alloy is a mixture of melt and solid? What is the structure of the solid?

*d.* Consider the solder at room temperature following cooling from 182 °C. Assume that the rate of cooling from 182 °C to room temperature is faster than the atomic diffusion rates needed to change the compositions of the $\alpha$ and $\beta$ phases in the solid. Assuming the alloy is 1 kg, calculate the masses of the following components in the solid:
1. The primary $\alpha$.
2. $\alpha$ in the whole alloy.
3. $\alpha$ in the eutectic solid.
4. $\beta$ in the alloy. (Where is the $\beta$-phase?)

*e.* Calculate the specific heat of the solder given the atomic masses of Pb (207.2) and Sn (118.71).

Walter Houser Brattain (1902–1987), one of the inventors of the transistor, looking at a vacuum evaporator used for depositing metal film electrodes on semiconductors (1937).

| SOURCE: AIP Emilio Segrè Visual Archives, Brattain Collection.

Highly magnified scanning electron microscope (SEM) view of IBM's six-level copper interconnect technology in an integrated circuit chip. The aluminum in transistor interconnections in a silicon chip has been replaced by copper that has a higher conductivity (by nearly 40%) and also a better ability to carry higher current densities without electromigration. Lower copper interconnect resistance means higher speeds and lower *RC* constants (1997).

| SOURCE: Courtesy of IBM Corporation.

SEM view of three levels of copper interconnect metallization in IBM's new faster CMOS integrated circuits (1997).

| SOURCE: Courtesy of IBM Corporation.

CHAPTER

# 2

# Electrical and Thermal Conduction in Solids

Electrical conduction involves the motion of charges in a material under the influence of an applied electric field. A material can generally be classified as a conductor if it contains a large number of "free" or mobile charge carriers. In metals, due to the nature of metallic bonding, the valence electrons from the atoms form a sea of electrons that are free to move within the metal and are therefore called conduction electrons. In this chapter, we will treat the conduction electrons in metal as "free charges" that can be accelerated by an applied electric field. In the presence of an electric field, the conduction electrons attain an average velocity, called the drift velocity, that depends on the field. By applying Newton's second law to electron motion and using such concepts as mean free time between electron collisions with lattice vibrations, crystal defects, impurities, etc., we will derive the fundamental equations that govern electrical conduction in solids. A key concept will be the drift mobility, which is a measure of the ease with which charge carriers in the solid drift under the influence of an external electric field.

Good electrical conductors, such as metals, are also known to be good thermal conductors. The conduction of thermal energy from higher to lower temperature regions in a metal involves the conduction electrons carrying the energy. Consequently, there is an innate relationship between the electrical and thermal conductivities, which is supported by theory and experiments.

# 2.1 CLASSICAL THEORY: THE DRUDE MODEL

## 2.1.1 METALS AND CONDUCTION BY ELECTRONS

The electric current density $J$ is defined as the net amount of charge flowing across a unit area per unit time, that is,

*Current density definition*

$$J = \frac{\Delta q}{A\,\Delta t}$$

where $\Delta q$ is the net quantity of charge flowing through an area $A$ in time $\Delta t$. Figure 2.1 shows the net flow of electrons in a conductor section of cross-sectional area $A$ in the presence of an applied field $\mathcal{E}_x$. Notice that the direction of electron motion is opposite to that of the electric field $\mathcal{E}_x$ and of conventional current, because the electrons experience a Coulombic force $e\mathcal{E}_x$ in the $x$ direction, due to their negative charge.

We know that the conduction electrons are actually moving around randomly[1] in the metal, but we will assume that as a result of the application of the electric field $\mathcal{E}_x$, they all acquire a net velocity in the $x$ direction. Otherwise, there would be no net flow of charge through area $A$.

The average velocity of the electrons in the $x$ direction at time $t$ is denoted $v_{dx}(t)$. This is called the **drift velocity,** which is the instantaneous velocity $v_x$ in the $x$ direction averaged over many electrons (perhaps, $\sim 10^{28}$ m$^{-3}$); that is

*Definition of drift velocity*

$$v_{dx} = \frac{1}{N}[v_{x1} + v_{x2} + v_{x3} + \cdots + v_{xN}] \qquad \textbf{[2.1]}$$

where $v_{xi}$ is the $x$ direction velocity of the $i$th electron, and $N$ is the number of conduction electrons in the metal. Suppose that $n$ is the number of electrons per unit volume in the conductor ($n = N/V$). In time $\Delta t$, electrons move a distance $\Delta x = v_{dx}\,\Delta t$, so the total charge $\Delta q$ crossing the area $A$ is $enA\,\Delta x$. This is valid because all the electrons within distance $\Delta x$ pass through $A$; thus, $n(A\,\Delta x)$ is the total number of electrons crossing $A$ in time $\Delta t$.

The current density in the $x$ direction is

$$J_x = \frac{\Delta q}{A\,\Delta t} = \frac{enAv_{dx}\,\Delta t}{A\,\Delta t} = env_{dx}$$

This general equation relates $J_x$ to the average velocity $v_{dx}$ of the electrons. It must be appreciated that the average velocity at one time may not be the same as at another time, because the applied field, for example, may be changing: $\mathcal{E}_x = \mathcal{E}_x(t)$. We therefore allow for a time-dependent current by writing

*Current density and drift velocity*

$$J_x(t) = env_{dx}(t) \qquad \textbf{[2.2]}$$

To relate the current density $J_x$ to the electric field $\mathcal{E}_x$, we must examine the effect of the electric field on the motion of the electrons in the conductor. To do so, we will consider the copper crystal.

[1] All the conduction electrons are "free" within the metal and move around randomly, being scattered from vibrating metal ions, as we discuss in this chapter.

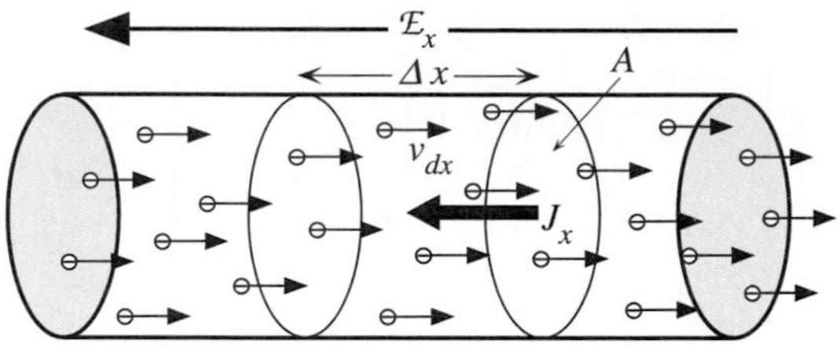


**Figure 2.1** Drift of electrons in a conductor in the presence of an applied electric field.
Electrons drift with an average velocity $v_{dx}$ in the $x$ direction.

The copper atom has a single valence electron in its 4*s* subshell, and this electron is loosely bound. The solid metal consists of positive ion cores, $Cu^+$, at regular sites, in the face-centered cubic (FCC) crystal structure. The valence electrons detach themselves from their parents and wander around freely in the solid, forming a kind of electron cloud or gas. These mobile electrons are free to respond to an applied field, creating a current density $J_x$. The valence electrons in the electron gas are therefore **conduction electrons.**

The attractive forces between the negative electron cloud and the $Cu^+$ ions are responsible for metallic bonding and the existence of the solid metal. (This simplistic view of metal was depicted in Figure 1.7 for copper.) The electrostatic attraction between the conduction electrons and the positive metal ions, like the electrostatic attraction between the electron and the proton in the hydrogen atom, results in the conduction electron having both potential energy *PE* and kinetic energy *KE*. The conduction electrons move about the crystal lattice in the same way that gas atoms move randomly in a cylinder. Although the average *KE* for gas atoms is $\frac{3}{2}kT$, this is not the case for electrons in a metal, because these electrons strongly interact with the metal ions and with each other as a result of electrostatic interactions.

The mean *KE* of the conduction electrons in a metal is primarily determined by the electrostatic interaction of these electrons with the positive metal ions and also with each other. For most practical purposes, we will therefore neglect the temperature dependence of the mean *KE* compared with other factors that control the behavior of the conduction electrons in the metal crystal. We can speculate from Example 1.1, that the magnitude of mean *KE* must be comparable to the magnitude of the mean *PE* of electrostatic interaction[2] or, stated differently, to the metal bond energy which is several electron volts per atom. If $u$ is the **mean speed** of the conduction electrons, then, from electrostatic interactions alone, we expect $\frac{1}{2}m_e u^2$ to be several electron volts which means that $u$ is typically $\sim 10^6$ m s$^{-1}$. This purely classical and intuitive reasoning is not sufficient, however, to show that the mean speed $u$ is relatively temperature insensitive and much greater than that expected from kinetic molecular theory. The true reasons are quantum mechanical and are discussed in Chapter 4. (They arise from what is called the Pauli exclusion principle.)

[2] There is a theorem in classical mechanics called the **virial theorem,** which states that for a collection of particles, the mean *KE* has half the magnitude of the mean *PE* if the only forces acting on the particles are such that they follow an inverse square law dependence on the particle–particle separation (as in Coulombic and gravitational forces).

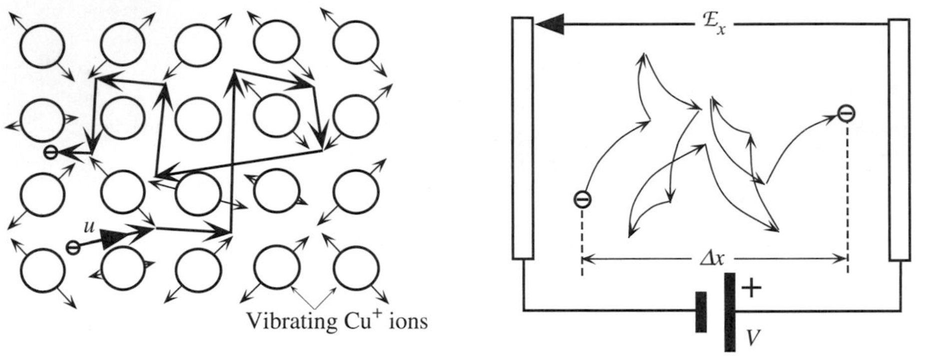


(a) A conduction electron in the electron gas moves about randomly in a metal (with a mean speed $u$) being frequently and randomly scattered by thermal vibrations of the atoms. In the absence of an applied field there is no net drift in any direction.

(b) In the presence of an applied field, $\mathcal{E}_x$, there is a net drift along the $x$ direction. This net drift along the force of the field is superimposed on the random motion of the electron. After many scattering events the electron has been displaced by a net distance, $\Delta x$, from its initial position toward the positive terminal.

**Figure 2.2** Motion of a conduction electron in a metal.

In general, the copper crystal will not be perfect and the atoms will not be stationary. There will be crystal defects, vacancies, dislocations, impurities, etc., which will scatter the conduction electrons. More importantly, due to their thermal energy, the atoms will vibrate about their lattice sites (equilibrium positions), as depicted in Figure 2.2a. An electron will not be able to avoid collisions with vibrating atoms; consequently, it will be "scattered" from one atom to another. In the absence of an applied field, the path of an electron may be visualized as illustrated in Figure 2.2a, where scattering from lattice vibrations causes the electron to move randomly in the lattice. On those occasions when the electron reaches a crystal surface, it becomes "deflected" (or "bounced") back into the crystal. Therefore, in the absence of a field, after some duration of time, the electron crosses its initial $x$ plane position again. Over a long time, the electrons therefore show no net displacement in any one direction.

When the conductor is connected to a battery and an electric field is applied to the crystal, as shown in Figure 2.2b, the electron experiences an acceleration in the $x$ direction in addition to its random motion, so after some time, it will drift a finite distance in the $x$ direction. The electron accelerates along the $x$ direction under the action of the force $e\mathcal{E}_x$, and then it suddenly collides with a vibrating atom and loses the gained velocity. Therefore, there is an average velocity in the $x$ direction, which, if calculated, determines the current via Equation 2.2. Note that since the electron experiences an acceleration in the $x$ direction, its trajectory between collisions is a parabola, like the trajectory of a golf ball experiencing acceleration due to gravity.

To calculate the drift velocity $v_{dx}$ of the electrons due to applied field $\mathcal{E}_x$, we first consider the velocity $v_{xi}$ of the $i$th electron in the $x$ direction at time $t$. Suppose its last collision was at time $t_i$; therefore, for time $(t - t_i)$, it accelerated *free of collisions,* as

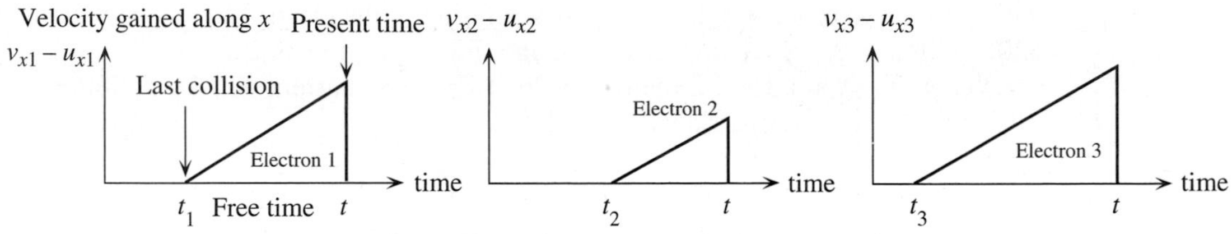


**Figure 2.3** Velocity gained in the x direction at time $t$ from the electric field ($\mathcal{E}_x$) for three electrons. There will be $N$ electrons to consider in the metal.

indicated in Figure 2.3. Let $u_{xi}$ be the velocity of electron $i$ in the $x$ direction just after the collision. We will call this the initial velocity. Since $e\mathcal{E}_x/m_e$ is the acceleration of the electron, the velocity $v_{xi}$ in the $x$ direction at time $t$ will be

$$v_{xi} = u_{xi} + \frac{e\mathcal{E}_x}{m_e}(t - t_i)$$

However, this is only for the $i$th electron. We need the average velocity $v_{dx}$ for all such electrons along $x$. We average the expression for $i = 1$ to $N$ electrons, as in Equation 2.1. We assume that immediately after a collision with a vibrating ion, the electron may move in any random direction; that is, it can just as likely move along the negative or positive $x$, so that $u_{xi}$ averaged over many electrons is zero. Thus,

*Drift velocity*

$$v_{dx} = \frac{1}{N}[v_{x1} + v_{x2} + \cdots + v_{xN}] = \frac{e\mathcal{E}_x}{m_e}\overline{(t - t_i)}$$

where $\overline{(t - t_i)}$ is the **average free time** for $N$ electrons between collisions.

Suppose that $\tau$ is the mean free time, or the **mean time between collisions** (also known as the **mean scattering time**). For some electrons, $(t - t_i)$ will be greater than $\tau$, and for others, it will be shorter, as shown in Figure 2.3. Averaging $(t - t_i)$ for $N$ electrons will be the same as $\tau$. Thus, we can substitute $\tau$ for $\overline{(t - t_i)}$ in the previous expression to obtain

$$v_{dx} = \frac{e\tau}{m_e}\mathcal{E}_x \qquad [2.3]$$

Equation 2.3 shows that the drift velocity increases linearly with the applied field. The constant of proportionality $e\tau/m_e$ has been given a special name and symbol. It is called the **drift mobility** $\mu_d$, which is defined as

*Definition of drift mobility*

$$v_{dx} = \mu_d\mathcal{E}_x \qquad [2.4]$$

where

*Drift mobility and mean free time*

$$\mu_d = \frac{e\tau}{m_e} \qquad [2.5]$$

Equation 2.5 relates the drift mobility of the electrons to their mean scattering time $\tau$. To reiterate, $\tau$, which is also called the **relaxation time,** is directly related to the microscopic processes that cause the scattering of the electrons in the metal; that is, lattice vibrations, crystal imperfections, and impurities, to name a few.

From the expression for the drift velocity $v_{dx}$, the current density $J_x$ follows immediately by substituting Equation 2.4 into 2.2, that is,

*Ohm's law*

$$J_x = en\mu_d \mathcal{E}_x \tag{2.6}$$

Therefore, the current density is proportional to the electric field and the conductivity $\sigma$ is the term multiplying $\mathcal{E}_x$, that is,

*Unipolar conductivity*

$$\sigma = en\mu_d \tag{2.7}$$

It is gratifying that by treating the electron as a particle and applying classical mechanics ($F = ma$), we are able to derive Ohm's law. We should note, however, that we assumed $\tau$ to be independent of the field.

Drift mobility is important because it is a widely used electronic parameter in semiconductor device physics. The drift mobility gauges how fast electrons will drift when driven by an applied field. If the electron is not highly scattered, then the mean free time between collisions will be long, $\tau$ will be large, and by Equation 2.5, the drift mobility will also be large; the electrons will therefore be highly mobile and be able to "respond" to the field. However, a large drift mobility does not necessarily imply high conductivity, because $\sigma$ also depends on the concentration of conduction electrons $n$.

The mean time between collisions $\tau$ has further significance. Its reciprocal $1/\tau$ represents the **mean frequency of collisions** or **scattering events;** that is, $1/\tau$ is the mean probability per unit time that the electron will be scattered (see Example 2.1). Therefore, during a small time interval $\delta t$, the probability of scattering will be $\delta t/\tau$. The probability of scattering per unit time $1/\tau$ is time independent and depends only on the nature of the electron scattering mechanism.

There is one important assumption in the derivation of the drift velocity $v_{dx}$ in Equation 2.3. We obtained $v_{dx}$ by averaging the velocities $v_{xi}$ of $N$ electrons along $x$ at one instant, as defined in Equation 2.1. The drift velocity therefore represents the average velocity of *all* the electrons along $x$ at one instant; that is, $v_{dx}$ is a number average at one instant. Figure 2.2b shows that after many collisions, after a time interval $\Delta t \gg \tau$, an electron would have been displaced by a net distance $\Delta x$ along $x$. The term $\Delta x/\Delta t$ represents the effective velocity with which the electron drifts along $x$. It is an average velocity for one electron over many collisions, that is, over a long time (hence, $\Delta t \gg \tau$), so $\Delta x/\Delta t$ is a time average. Provided that $\Delta t$ contains many collisions, it is reasonable to expect that the drift velocity $\Delta x/\Delta t$ from the time average for one electron is the same as the drift velocity $v_{dx}$ per electron from averaging for all electrons at one instant, as in Equation 2.1, or

*Drift velocity*

$$\frac{\Delta x}{\Delta t} = v_{dx}$$

The two velocities are the same only under steady-state conditions ($\Delta t \gg \tau$). The proof may be found in more advanced texts.

EXAMPLE 2.1

**PROBABILITY OF SCATTERING PER UNIT TIME AND THE MEAN FREE TIME** If $1/\tau$ is defined as the mean probability per unit time that an electron is scattered, show that the mean time between collisions is $\tau$.

**SOLUTION**

Consider an infinitesimally small time interval $dt$ at time $t$. Let $N$ be the number of unscattered electrons at time $t$. The probability of scattering during $dt$ is $(1/\tau)\,dt$, and the number of scattered electrons during $dt$ is $N(1/\tau)\,dt$. The change $dN$ in $N$ is thus

$$dN = -N\left(\frac{1}{\tau}\right)dt$$

The negative sign indicates a reduction in $N$ because, as electrons become scattered, $N$ decreases. Integrating this equation, we can find $N$ at any time $t$, given that at time $t = 0$, $N_0$ is the total number of unscattered electrons. Therefore,

*Unscattered electron concentration*

$$N = N_0 \exp\left(-\frac{t}{\tau}\right)$$

This equation represents the number of unscattered electrons at time $t$. It reflects an exponential decay law for the number of unscattered electrons. The **mean free time** $\bar{t}$ can be calculated from the mathematical definition of $\bar{t}$,

*Mean free time*

$$\bar{t} = \frac{\int_0^\infty tN\,dt}{\int_0^\infty N\,dt} = \tau$$

where we have used $N = N_0 \exp(-t/\tau)$. Clearly, $1/\tau$ is the mean probability of scattering per unit time.

EXAMPLE 2.2

**ELECTRON DRIFT MOBILITY IN METALS** Calculate the drift mobility and the mean scattering time of conduction electrons in copper at room temperature, given that the conductivity of copper is $5.9 \times 10^5\ \Omega^{-1}\ \text{cm}^{-1}$. The density of copper is $8.96\ \text{g cm}^{-3}$ and its atomic mass is $63.5\ \text{g mol}^{-1}$.

**SOLUTION**

We can calculate $\mu_d$ from $\sigma = en\mu_d$ because we already know the conductivity $\sigma$. The number of free electrons $n$ per unit volume can be taken as equal to the number of Cu atoms per unit volume, if we assume that each Cu atom donates one electron to the conduction electron gas in the metal. One mole of copper has $N_A$ ($6.02 \times 10^{23}$) atoms and a mass of 63.5 g. Therefore, the number of copper atoms per unit volume is

$$n = \frac{dN_A}{M_{\text{at}}}$$

where $d$ = density = $8.96\ \text{g cm}^{-3}$, and $M_{\text{at}}$ = atomic mass = 63.5 ($\text{g mol}^{-1}$). Substituting for $d$, $N_A$, and $M_{\text{at}}$, we find $n = 8.5 \times 10^{22}$ electrons $\text{cm}^{-3}$.

The electron drift mobility is therefore

$$\mu_d = \frac{\sigma}{en} = \frac{5.9 \times 10^5\ \Omega^{-1}\ \text{cm}^{-1}}{[(1.6 \times 10^{-19}\ \text{C})(8.5 \times 10^{22}\ \text{cm}^{-3})]}$$

$$= 43.4\ \text{cm}^2\ \text{V}^{-1}\ \text{s}^{-1}$$

From the drift mobility we can calculate the mean free time $\tau$ between collisions by using Equation 2.5,

$$\tau = \frac{\mu_d m_e}{e} = \frac{(43.4 \times 10^{-4}\ \text{m}^2\ \text{V}^{-1}\ \text{s}^{-1})(9.1 \times 10^{-31}\ \text{kg})}{1.6 \times 10^{-19}\ \text{C}} = 2.5 \times 10^{-14}\ \text{s}$$

Note that the mean speed $u$ of the conduction electrons is about $1.5 \times 10^6$ m s$^{-1}$, so that their mean free path is about 37 nm.

**EXAMPLE 2.3**

**DRIFT VELOCITY AND MEAN SPEED** What is the applied electric field that will impose a drift velocity equal to 0.1 percent of the mean speed $u(\sim 10^6$ m s$^{-1})$ of conduction electrons in copper? What is the corresponding current density and current through a Cu wire of diameter 1 mm?

**SOLUTION**

The drift velocity of the conduction electrons is $v_{dx} = \mu_d \mathcal{E}_x$, where $\mu_d$ is the drift mobility, which for copper is 43.4 cm$^2$ V$^{-1}$ s$^{-1}$ (see Example 2.2). With $v_{dx} = 0.001u = 10^3$ m s$^{-1}$, we have

$$\mathcal{E}_x = \frac{v_{dx}}{\mu_d} = \frac{10^3\ \text{m s}^{-1}}{43.4 \times 10^{-4}\ \text{m}^2\ \text{V}^{-1}\ \text{s}^{-1}} = 2.3 \times 10^5\ \text{V m}^{-1} \qquad \text{or} \qquad 230\ \text{kV m}^{-1}$$

This is an unattainably large electric field in a metal. Given the conductivity $\sigma$ of copper, the equivalent current density is

$$J_x = \sigma \mathcal{E}_x = (5.9 \times 10^7\ \Omega^{-1}\ \text{m}^{-1})(2.3 \times 10^5\ \text{V m}^{-1})$$
$$= 1.4 \times 10^{13}\ \text{A m}^{-2} \qquad \text{or} \qquad 1.4 \times 10^7\ \text{A mm}^{-2}$$

This means a current of $1.1 \times 10^7$ A through a 1 mm diameter wire! It is clear from this example that for all practical purposes, even under the highest working currents and voltages, the drift velocity is much smaller than the mean speed of the electrons. Consequently, when an electric field is applied to a conductor, for all practical purposes, the mean speed is unaffected.

**EXAMPLE 2.4**

**DRIFT VELOCITY IN A FIELD: A CLOSER LOOK** There is another way to explain the observed dependence of the drift velocity on the field, and Equation 2.3. Consider the path of a conduction electron in an applied field $\mathcal{E}$ as shown in Figure 2.4. Suppose that at time $t = 0$ the electron has just been scattered from a lattice vibration. Let $u_{x1}$ be the initial velocity in the $x$ direction just after this initial collision (to which we assign a collision number of zero). We will assume that immediately after a collision, the velocity of the electron is in a random *direction*. Suppose that the first collision occurs at time $t_1$. Since $e\mathcal{E}_x/m_e$ is the acceleration, the distance $s_1$ covered in the $x$ direction during the free time $t_1$ will be

*Distance traversed along x before collision*

$$s_1 = u_{x1}t_1 + \frac{1}{2}\left(\frac{e\mathcal{E}_x}{m_e}\right)t_1^2$$

At time $t_1$, the electron collides with a lattice vibration (its first collision), and the velocity is randomized again to become $u_{x2}$. The whole process is then repeated during the next interval which lasts for a free time $t_2$, and the electron traverses a distance $s_2$ along $x$, and so on. To find the overall distance traversed by the electron after $p$ such scattering events, we sum all the

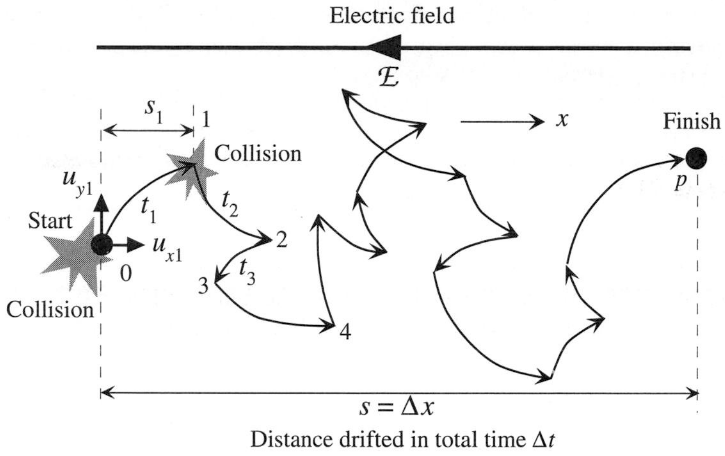


**Figure 2.4** The motion of a single electron in the presence of an electric field $\mathcal{E}$. During a time interval $t_i$, the electron traverses a distance $s_i$ along $x$. After $p$ collisions, it has drifted a distance $s = \Delta x$.

above distances $s_1, s_2, \ldots$ for $p$ free time intervals,

$$s = s_1 + s_2 + \cdots + s_p = [u_{x1}t_1 + u_{x2}t_2 + \cdots + u_{xp}t_p] + \frac{1}{2}\left(\frac{e\mathcal{E}_x}{m_e}\right)[t_1^2 + t_1^2 + \cdots + t_p^2] \quad \textbf{[2.8]}$$

Since after a collision the "initial" velocity $u_x$ is always random, the first term has $u_x$ values that are randomly negative and positive, so for many collisions (large $p$) the first term on the right-hand side of Equation 2.8 is nearly zero and can certainly be neglected compared with the second term. Thus, after many collisions, the net distance $s = \Delta x$ traversed in the $x$ direction is given by the second term in Equation 2.8, which is the electric field induced displacement term. If $\overline{t^2}$ is the **mean square free time,** then

*Distance drifted after p scattering events*

$$s = \frac{1}{2}\left(\frac{e\mathcal{E}_x}{m_e}\right)p\overline{t^2}$$

where $$\overline{t^2} = \frac{1}{p}[t_1^2 + t_1^2 + \cdots + t_p^2]$$

*Mean square free time definition*

Suppose that $\tau$ is the **mean free time between collisions,** where $\tau = (t_1 + t_2 + \cdots + t_p)/p$. Then from straightforward elementary statistics it can be shown that $\overline{t^2} = 2(\bar{t})^2 = 2\tau^2$. So in terms of the mean free time $\tau$ between collisions, the overall distance $s = \Delta x$ drifted in the $x$ direction after $p$ collisions is

$$s = \frac{e\mathcal{E}_x}{m_e}(p\tau^2)$$

Further, since the total time $\Delta t$ taken for these $p$ scattering events is simply $p\tau$, the drift velocity $v_{dx}$ is given by $\Delta x/\Delta t$ or $s/(p\tau)$, that is,

*Drift velocity and mean free time*

$$v_{dx} = \frac{e\tau}{m_e}\mathcal{E}_x \quad \textbf{[2.9]}$$

This is the same expression as Equation 2.3, except that $\tau$ is defined here as the average free time for a single electron over a long time, that is, over many collisions, whereas previously it was the mean free time averaged over many electrons. Further, in Equation 2.9 $v_{dx}$ is an average drift for an electron over a long time, over many collisions. In Equation 2.1 $v_{dx}$ is the average velocity averaged over all electrons at one instant. For all practical purposes, the two are equivalent. (The equivalence breaks down when we are interested in events over a time scale that is comparable to one scattering, $\sim 10^{-14}$ second.)

The drift mobility $\mu_d$ from Equation 2.9 is identical to that of Equation 2.5, $\mu_d = e\tau/m_e$. Suppose that the mean speed of the electrons (not the drift velocity) is $u$. Then an electron moves a distance $\ell = u\tau$ in mean free time $\tau$, which is called the **mean free path.** The drift mobility and conductivity become,

*Drift mobility and conductivity and mean free path*

$$\mu_d = \frac{e\ell}{m_e u} \qquad \text{and} \qquad \sigma = en\mu_d = \frac{e^2 n\ell}{m_e u} \tag{2.10}$$

Equations 2.3 and 2.10 both assume that after each collision the velocity is randomized. The scattering process, lattice scattering, is able to randomize the velocity in one single scattering. In general not all electron scattering processes can randomize the velocity in one scattering process. If it takes more than one collision to randomize the velocity, then the electron is able to carry with it some velocity gained from a previous collision and hence possesses a higher drift mobility. In such cases one needs to consider the effective mean free path a carrier has to move to eventually randomize the velocity gained; this is a point considered in Chapter 4 when we calculate the resistivity at low temperatures.

## 2.2 TEMPERATURE DEPENDENCE OF RESISTIVITY: IDEAL PURE METALS

When the conduction electrons are only scattered by thermal vibrations of the metal ions, then $\tau$ in the mobility expression $\mu_d = e\tau/m_e$ refers to the mean time between scattering events by this process. The resulting conductivity and resistivity are denoted by $\sigma_T$ and $\rho_T$, where the subscript $T$ represents "thermal vibration scattering."

To find the temperature dependence of $\sigma$, we first consider the temperature dependence of the mean free time $\tau$, since this determines the drift mobility. An electron moving with a mean speed $u$ is scattered when its path crosses the cross-sectional area $S$ of a scattering center, as depicted in Figure 2.5. The scattering center

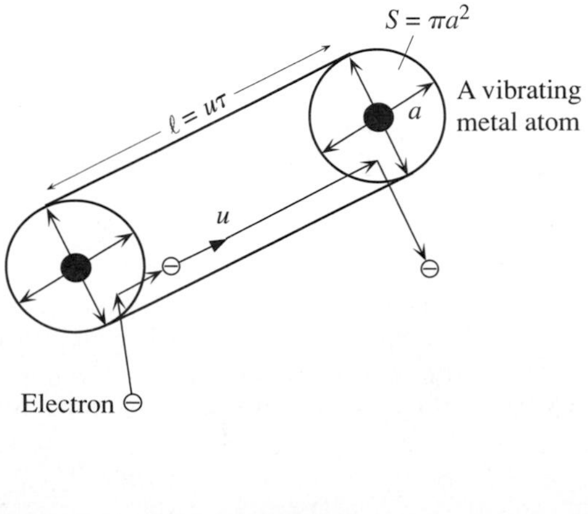


**Figure 2.5** Scattering of an electron from the thermal vibrations of the atoms.
The electron travels a mean distance $\ell = u\tau$ between collisions. Since the scattering cross-sectional area is $S$, in the volume $S\ell$ there must be at least one scatterer, $N_s(Su\tau) = 1$.

may be a vibrating atom, impurity, vacancy, or some other crystal defect. Since $\tau$ is the mean time taken for one scattering process, the **mean free path** $\ell$ of the electron between scattering processes is $u\tau$. If $N_s$ is the concentration of scattering centers, then in the volume $S\ell$, there is one scattering center, that is, $(Su\tau)N_s = 1$. Thus, the mean free time is given by

*Mean free time between collisions*

$$\tau = \frac{1}{SuN_s} \qquad [2.11]$$

The mean speed $u$ of conduction electrons in a metal can be shown to be only slightly temperature dependent.[3] In fact, electrons wander randomly around in the metal crystal with an almost constant mean speed that depends largely on their concentration and hence on the crystal material. Taking the number of scattering centers per unit volume to be the atomic concentration, the temperature dependence of $\tau$ then arises essentially from that of the cross-sectional area $S$. Consider what a free electron "sees" as it approaches a vibrating crystal atom as in Figure 2.5. Because the atomic vibrations are random, the atom covers a cross-sectional area $\pi a^2$, where $a$ is the amplitude of the vibrations. If the electron's path crosses $\pi a^2$, it gets scattered. Therefore, the mean time between scattering events $\tau$ is inversely proportional to the area $\pi a^2$ that scatters the electron, that is, $\tau \propto 1/\pi a^2$.

The thermal vibrations of the atom can be considered to be simple harmonic motion, much the same way as that of a mass $M$ attached to a spring. The average kinetic energy of the oscillations is $\frac{1}{4}Ma^2\omega^2$, where $\omega$ is the oscillation frequency. From the kinetic theory of matter, this average kinetic energy must be on the order of $\frac{1}{2}kT$. Therefore,

$$\tfrac{1}{4}Ma^2\omega^2 \approx \tfrac{1}{2}kT$$

so $a^2 \propto T$. Intuitively, this is correct because raising the temperature increases the amplitude of the atomic vibrations. Thus,

$$\tau \propto \frac{1}{\pi a^2} \propto \frac{1}{T} \qquad \text{or} \qquad \tau = \frac{C}{T}$$

where $C$ is a temperature-independent constant. Substituting for $\tau$ in $\mu_d = e\tau/m_e$, we obtain

$$\mu_d = \frac{eC}{m_e T}$$

So, the resistivity of a metal is

$$\rho_T = \frac{1}{\sigma_T} = \frac{1}{en\mu_d} = \frac{m_e T}{e^2 nC}$$

[3] The fact that the mean speed of electrons in a metal is only weakly temperature dependent can be proved from what it called the Fermi–Dirac statistics for the collection of electrons in a metal (see Chapter 4). This result contrasts sharply with the kinetic molecular theory of gases (Chapter 1), which predicts that the mean speed of molecules is proportional to $\sqrt{T}$. For the time being, we simply use a constant mean speed $u$ for the conduction electrons in a metal.

*Pure metal resistivity due to thermal vibrations of the crystal*

that is,

$$\rho_T = AT \qquad \textbf{[2.12]}$$

where $A$ is a temperature-independent constant. This shows that the resistivity of a pure metal wire increases linearly with the temperature, and that the resistivity is due simply to the scattering of conduction electrons by the thermal vibrations of the atoms. We term this conductivity **lattice-scattering-limited conductivity.**

**EXAMPLE 2.5**

**TEMPERATURE DEPENDENCE OF RESISTIVITY** What is the percentage change in the resistance of a pure metal wire from Saskatchewan's summer to winter, neglecting the changes in the dimensions of the wire?

**SOLUTION**

Assuming 20 °C for the summer and perhaps −30 °C for the winter, from $R \propto \rho = AT$, we have

$$\frac{R_{\text{summer}} - R_{\text{winter}}}{R_{\text{summer}}} = \frac{T_{\text{summer}} - T_{\text{winter}}}{T_{\text{summer}}} = \frac{(20 + 273) - (-30 + 273)}{(20 + 273)}$$

$$= 0.171 \qquad \text{or} \qquad 17\%$$

Notice that we have used the absolute temperature for $T$. How will the outdoor cable power losses be affected?

**EXAMPLE 2.6**

**DRIFT MOBILITY AND RESISTIVITY DUE TO LATTICE VIBRATIONS** Given that the mean speed of conduction electrons in copper is $1.5 \times 10^6$ m s$^{-1}$ and the frequency of vibration of the copper atoms at room temperature is about $4 \times 10^{12}$ s$^{-1}$, estimate the drift mobility of electrons and the conductivity of copper. The density $d$ of copper is 8.96 g cm$^{-3}$ and the atomic mass $M_{\text{at}}$ is 63.56 g mol$^{-1}$.

**SOLUTION**

The method for calculating the drift mobility and hence the conductivity is based on evaluating the mean free time $\tau$ via Equation 2.11, that is, $\tau = 1/SuN_s$. Since $\tau$ is due to scattering from atomic vibrations, $N_s$ is the atomic concentration,

$$N_s = \frac{dN_A}{M_{\text{at}}} = \frac{(8.96 \times 10^3 \text{ kg m}^{-3})(6.02 \times 10^{23} \text{ mol}^{-1})}{63.56 \times 10^{-3} \text{ kg mol}^{-1}}$$

$$= 8.5 \times 10^{28} \text{ m}^{-3}$$

The cross-sectional area $S = \pi a^2$ depends on the amplitude $a$ of the thermal vibrations as shown in Figure 2.5. The average kinetic energy $KE_{\text{av}}$ associated with a vibrating mass $M$ attached to a spring is given by $KE_{\text{av}} = \frac{1}{4}Ma^2\omega^2$, where $\omega$ is the angular frequency of the vibration ($\omega = 2\pi 4 \times 10^{12}$ rad s$^{-1}$). Applying this equation to the vibrating atom and equating the average kinetic energy $KE_{\text{av}}$ to $\frac{1}{2}kT$, by virtue of equipartition of energy theorem, we have $a^2 = 2kT/M\omega^2$ and thus

$$S = \pi a^2 = \frac{2\pi kT}{M\omega^2} = \frac{2\pi(1.38 \times 10^{-23} \text{ J K}^{-1})(300 \text{ K})}{\left(\dfrac{63.56 \times 10^{-3} \text{ kg mol}^{-1}}{6.022 \times 10^{23} \text{ mol}^{-1}}\right)(2\pi \times 4 \times 10^{12} \text{ rad s}^{-1})^2}$$

$$= 3.9 \times 10^{-22} \text{ m}^2$$

Therefore,

$$\tau = \frac{1}{SuN_s} = \frac{1}{(3.9 \times 10^{-22}\text{ m}^2)(1.5 \times 10^6\text{ m s}^{-1})(8.5 \times 10^{28}\text{ m}^{-3})}$$
$$= 2.0 \times 10^{-14}\text{ s}$$

The drift mobility is

$$\mu_d = \frac{e\tau}{m_e} = \frac{(1.6 \times 10^{-19}\text{ C})(2.0 \times 10^{-14}\text{ s})}{(9.1 \times 10^{-31}\text{ kg})}$$
$$= 3.5 \times 10^{-3}\text{ m}^2\text{ V}^{-1}\text{ s}^{-1} = 35\text{ cm}^2\text{ V}^{-1}\text{ s}^{-1}$$

The conductivity is then

$$\sigma = en\mu_d = (1.6 \times 10^{-19}\text{ C})(8.5 \times 10^{22}\text{ cm}^{-3})(35\text{ cm}^2\text{ V}^{-1}\text{ s}^{-1})$$
$$= 4.8 \times 10^5\ \Omega^{-1}\text{ cm}^{-1}$$

The experimentally measured value for the conductivity is $5.9 \times 10^5\ \Omega^{-1}\text{ cm}^{-1}$, so our crude calculation based on Equation 2.11 is actually only 18 percent lower, which is not bad for an estimate. (As we might have surmised, the agreement is brought about by using reasonable values for the mean speed $u$ and the atomic vibrational frequency $\omega$. These values were taken from quantum mechanical calculations, so our evaluation for $\tau$ was not truly based on classical concepts.)

## 2.3 MATTHIESSEN'S AND NORDHEIM'S RULES

### 2.3.1 MATTHIESSEN'S RULE AND THE TEMPERATURE COEFFICIENT OF RESISTIVITY ($\alpha$)

The theory of conduction that considers scattering from lattice vibrations only works well with pure metals; unfortunately, it fails for metallic alloys. Their resistivities are only weakly temperature dependent. We must therefore search for a different type of scattering mechanism.

Consider a metal alloy that has randomly distributed impurity atoms. An electron can now be scattered by the impurity atoms because they are not identical to the host atoms, as illustrated in Figure 2.6. The impurity atom need not be larger than the host atom; it can be smaller. As long as the impurity atom results in a local distortion of the crystal lattice, it will be effective in scattering. One way of looking at the scattering process from an impurity is to consider the scattering cross section. What actually scatters the electron is a local, unexpected change in the potential energy $PE$ of the electron as it approaches the impurity, because the force experienced by the electron is given by

$$F = -\frac{d(PE)}{dx}$$

For example, when an impurity atom of a different size compared to the host atom is placed into the crystal lattice, the impurity atom distorts the region around it, either by

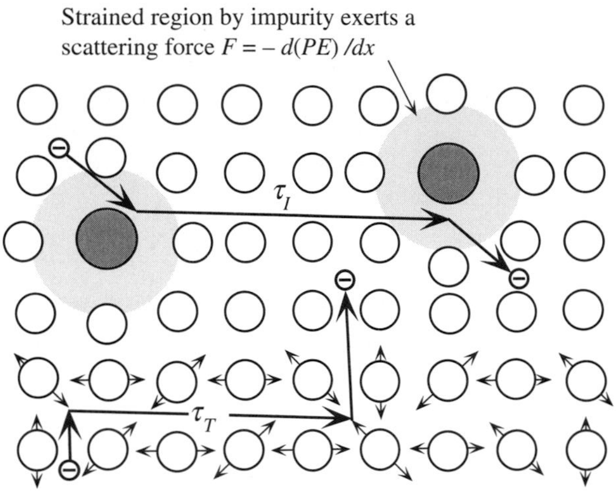


**Figure 2.6** Two different types of scattering processes involving scattering from impurities alone and from thermal vibrations alone.

pushing the host atoms farther away, or by pulling them in, as depicted in Figure 2.6. The cross section that scatters the electron is the lattice region that has been elastically distorted by the impurity (the impurity atom itself and its neighboring host atoms), so that in this zone, the electron suddenly experiences a force $F = -d(PE)/dx$ due to a sudden change in the $PE$. This region has a large scattering cross section, since the distortion induced by the impurity may extend a number of atomic distances. These impurity atoms will therefore hinder the motion of the electrons, thereby increasing the resistance.

We now effectively have two types of mean free times between collisions: one, $\tau_T$, for scattering from thermal vibrations only, and the other, $\tau_I$, for scattering from impurities only. We define $\tau_T$ as the mean time between scattering events arising from thermal vibrations alone and $\tau_I$ as the mean time between scattering events arising from collisions with impurities alone. Both are illustrated in Figure 2.6.

In general, an electron may be scattered by both processes, so the effective mean free time $\tau$ between any two scattering events will be less than the individual scattering times $\tau_T$ and $\tau_I$. The electron will therefore be scattered when it collides with either an atomic vibration or an impurity atom. Since in unit time, $1/\tau$ is the net probability of scattering, $1/\tau_T$ is the probability of scattering from lattice vibrations alone, and $1/\tau_I$ is the probability of scattering from impurities alone, then within the realm of elementary probability theory for independent events, we have

*Overall frequency of scattering*

$$\frac{1}{\tau} = \frac{1}{\tau_T} + \frac{1}{\tau_I} \quad [2.13]$$

In writing Equation 2.13 for the various probabilities, we make the reasonable assumption that, to a greater extent, the two scattering mechanisms are essentially independent. Here, the effective mean scattering time $\tau$ is clearly smaller than both $\tau_T$ and $\tau_I$. We can also interpret Equation 2.13 as follows: In unit time, the overall number of collisions $(1/\tau)$ is the sum of the number of collisions with thermal vibrations alone $(1/\tau_T)$ and the number of collisions with impurities alone $(1/\tau_I)$.

The drift mobility $\mu_d$ depends on the effective scattering time $\tau$ via $\mu_d = e\tau/m_e$, so Equation 2.13 can also be written in terms of the drift mobilities determined by the various scattering mechanisms. In other words,

$$\frac{1}{\mu_d} = \frac{1}{\mu_L} + \frac{1}{\mu_I} \qquad [2.14]$$

*Effective drift mobility*

where $\mu_L$ is the **lattice-scattering-limited drift mobility,** and $\mu_I$ is the **impurity-scattering-limited drift mobility.** By definition, $\mu_L = e\tau_T/m_e$ and $\mu_I = e\tau_I/m_e$. The effective (or overall) resistivity $\rho$ of the material is simply $1/en\mu_d$, or

$$\rho = \frac{1}{en\mu_d} = \frac{1}{en\mu_L} + \frac{1}{en\mu_I}$$

which can be written

$$\rho = \rho_T + \rho_I \qquad [2.15]$$

*Matthiessen's rule*

where $1/en\mu_L$ is defined as the resistivity due to scattering from thermal vibrations, and $1/en\mu_I$ is the resistivity due to scattering from impurities, or

$$\rho_T = \frac{1}{en\mu_L} \qquad \text{and} \qquad \rho_I = \frac{1}{en\mu_I}$$

*Resistivities due to lattice and impurity scattering*

The final result in Equation 2.15 simply states that the effective resistivity $\rho$ is the sum of two contributions. First, $\rho_T = 1/en\mu_L$ is the resistivity due to scattering by thermal vibrations of the host atoms. For those near-perfect pure metal crystals, this is the dominating contribution. As soon as we add impurities, however, there is an additional resistivity, $\rho_I = 1/en\mu_I$, which arises from the scattering of the electrons from the impurities. The first term is temperature dependent because $\tau_T \propto T^{-1}$ (see Section 2.2), but the second term is not.

The mean time $\tau_I$ between scattering events involving electron collisions with impurity atoms depends on the separation between the impurity atoms and therefore on the concentration of those atoms (see Figure 2.6). If $\ell_I$ is the mean separation between the impurities, then the mean free time between collisions with impurities alone will be $\ell_I/u$, which is temperature independent because $\ell_I$ is determined by the impurity concentration $N_I$ (*i.e.*, $\ell_I = N_I^{-1/3}$), and the mean speed of the electrons $u$ is nearly constant in a metal. In the absence of impurities, $\tau_I$ is infinitely long, and thus $\rho_I = 0$. The summation rule of resistivities from different scattering mechanisms, as shown by Equation 2.15, is called **Matthiessen's rule.**

There may also be electrons scattering from dislocations and other crystal defects, as well as from grain boundaries. All of these scattering processes add to the resistivity of a metal, just as the scattering process from impurities. We can therefore write the effective resistivity of a metal as

$$\rho = \rho_T + \rho_R \qquad [2.16]$$

*Matthiessen's rule*

where $\rho_R$ is called the **residual resistivity** and is due to the scattering of electrons by impurities, dislocations, interstitial atoms, vacancies, grain boundaries, etc. (which means that $\rho_R$ also includes $\rho_I$). The residual resistivity shows very little temperature dependence, whereas $\rho_T = AT$, so the effective resistivity $\rho$ is given by

$$\rho \approx AT + B \qquad [2.17]$$

where $A$ and $B$ are temperature-independent constants.

Equation 2.17 indicates that the resistivity of a metal varies almost linearly with the temperature, with $A$ and $B$ depending on the material. Instead of listing $A$ and $B$ in resistivity tables, we prefer to use a temperature coefficient that refers to small, normalized changes around a reference temperature. The **temperature coefficient of resistivity (TCR)** $\alpha_0$ is defined as the fractional change in the resistivity per unit temperature increase at the reference temperature $T_0$, that is,

*Definition of temperature coefficient of resistivity*

$$\alpha_0 = \frac{1}{\rho_0}\left[\frac{\delta\rho}{\delta T}\right]_{T=T_0} \qquad [2.18]$$

where $\rho_0$ is the resistivity at the reference temperature $T_0$, usually 273 K (0 °C) or 293 K (20 °C), and $\delta\rho = \rho - \rho_0$ is the change in the resistivity due to a small increase in temperature, $\delta T = T - T_0$.

When the resistivity follows the behavior $\rho \approx AT + B$ in Equation 2.17, then according to Equation 2.18, $\alpha_0$ is constant over a temperature range $T_0$ to $T$, and Equation 2.18 leads to the well-known equation,

*Temperature dependence of resistivity*

$$\rho = \rho_0[1 + \alpha_0(T - T_0)] \qquad [2.19]$$

Equation 2.19 is actually only valid when $\alpha_0$ is constant over the temperature range of interest, which requires Equation 2.17 to hold. Over a limited temperature range, this will usually be the case. Although it is not obvious from Equation 2.19, we should note that $\alpha_0$ depends on the reference temperature $T_0$, by virtue of $\rho_0$ depending on $T_0$.

The equation $\rho = AT$, which we used for pure-metal crystals to find the change in the resistance with temperature, is only approximate; nonetheless, for pure metals, it is useful to recall in the absence of tabulated data. To determine how good the formula $\rho = AT$ is, put it in Equation 2.19, which leads to $\alpha_0 = T_0^{-1}$. If we take the reference temperature $T_0$ as 273 K (0 °C), then $\alpha_0$ is simply 1/273 K; stated differently, Equation 2.19 is then equivalent to $\rho = AT$.

Table 2.1 shows that $\rho \propto T$ is not a bad approximation for some of the familiar pure metals used as conductors (Cu, Al, Au, etc.), but it fails badly for others, such as indium, antimony, and, in particular, the magnetic metals, iron and nickel.

The temperature dependence of the resistivity of various metals is shown in Figure 2.7, where it is apparent that except for the magnetic materials, such as iron and nickel, the linear relationship $\rho \propto T$ seems to be approximately obeyed almost all the way to the melting temperature for many pure metals. It should also be noted that for the alloys, such as nichrome (Ni–Cr), the resistivity is essentially dominated by the residual resistivity, so the resistivity is relatively temperature insensitive, with a very small TCR.

**Table 2.1** Resistivity, thermal coefficient of resistivity $\alpha_0$ at 273 K (0 °C) for various metals. The resistivity index $n$ in $\rho \propto T^n$ for some of the metals is also shown.

| Metal | $\rho_0$ (n$\Omega$ m) | $\alpha_0 \left(\frac{1}{\text{K}}\right)$ | $n$ | Comment |
|---|---|---|---|---|
| Aluminum, Al | 25.0 | $\frac{1}{233}$ | 1.20 | |
| Antimony, Sb | 38 | $\frac{1}{196}$ | 1.40 | |
| Copper, Cu | 15.7 | $\frac{1}{232}$ | 1.15 | |
| Gold, Au | 22.8 | $\frac{1}{251}$ | 1.11 | |
| Indium, In | 78.0 | $\frac{1}{196}$ | 1.40 | |
| Platinum, Pt | 98 | $\frac{1}{255}$ | 0.94 | |
| Silver, Ag | 14.6 | $\frac{1}{244}$ | 1.11 | |
| Tantalum, Ta | 117 | $\frac{1}{294}$ | 0.93 | |
| Tin, Sn | 110 | $\frac{1}{217}$ | 1.11 | |
| Tungsten, W | 50 | $\frac{1}{220}$ | 1.20 | |
| Iron, Fe | 84.0 | $\frac{1}{152}$ | 1.80 | Magnetic metal; $273 < T < 1043$ K |
| Nickel, Ni | 59.0 | $\frac{1}{125}$ | 1.72 | Magnetic metal; $273 < T < 627$ K |

SOURCE: Data were extracted and combined from several sources. Typical values.

Frequently, the resistivity versus temperature behavior of pure metals can be empirically represented by a power law of the form

*Resistivity of pure metals*

$$\rho = \rho_0 \left[\frac{T}{T_0}\right]^n \qquad [2.20]$$

where $\rho_0$ is the resistivity at the reference temperature $T_0$, and $n$ is a characteristic index that best fits the data. Table 2.1 lists some typical $n$ values for various pure metals above 0 °C. It is apparent that for the nonmagnetic metals, $n$ is close to unity, whereas it is closer to 2 than 1 for the magnetic metals Fe and Ni. In iron, for example, the conduction electron is not scattered simply by atomic vibrations, as in copper, but is affected by its magnetic interaction with the Fe ions in the lattice. This leads to a complicated temperature dependence.

Although our oversimplified theoretical analysis predicts a linear $\rho = AT + B$ behavior for the resistivity down to the lowest temperatures, this is not true in reality,

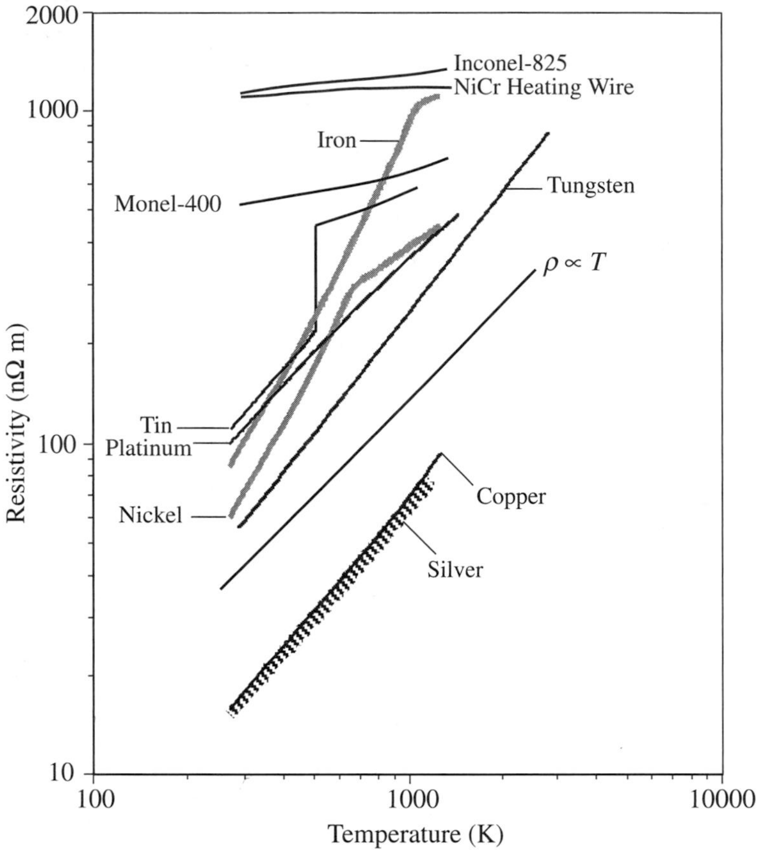


**Figure 2.7** The resistivity of various metals as a function of temperature above 0 °C.

Tin melts at 505 K, whereas nickel and iron go through a magnetic-to-nonmagnetic (Curie) transformation at about 627 K and 1043 K, respectively. The theoretical behavior ($\rho \sim T$) is shown for reference.

SOURCE: Data selectively extracted from various sources, including sections in *Metals Handbook,* 10th ed., 2 and 3. Metals Park, Ohio: ASM, 1991.

as depicted for copper in Figure 2.8. As the temperature decreases, typically below ~100 K for many metals, our simple and gross assumption that all the atoms are vibrating with a constant frequency fails. Indeed, the number of atoms that are vibrating with sufficient energy to scatter the conduction electrons starts to decrease rapidly with decreasing temperature, so the resistivity due to scattering from thermal vibrations becomes more strongly temperature dependent. The mean free time $\tau = 1/SuN_s$ becomes longer and strongly temperature dependent, leading to a smaller resistivity than the $\rho \propto T$ behavior. A full theoretical analysis, which is beyond the scope of this chapter, shows that $\rho \propto T^5$. Thus, at the lowest temperature, from Matthiessen's rule, the resistivity becomes $\rho = DT^5 + \rho_R$, where $D$ is a constant. Since the slope of $\rho$ versus $T$ is $d\rho/dT = 5DT^4$, which tends to zero as $T$ becomes small, we have $\rho$ curving

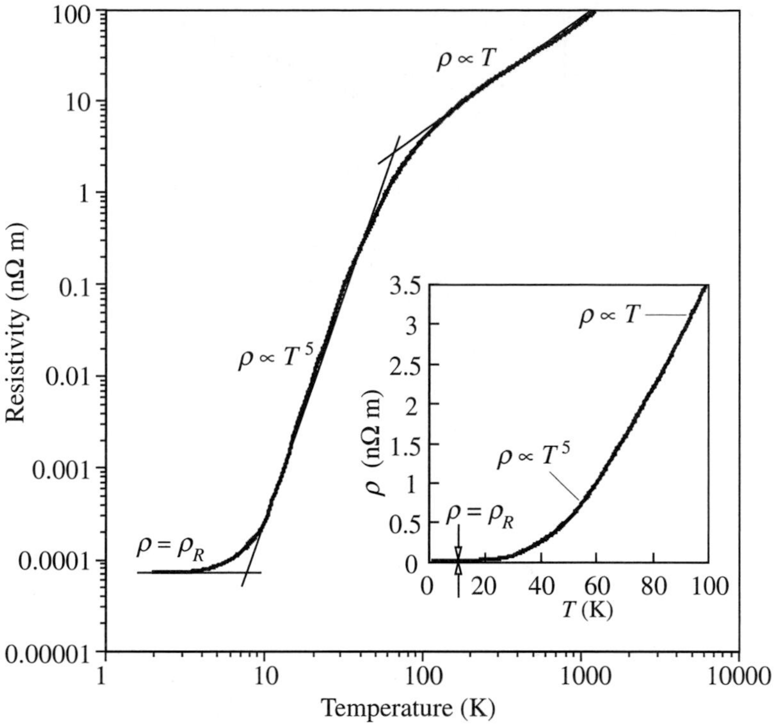


**Figure 2.8** The resistivity of copper from lowest to highest temperatures (near melting temperature, 1358 K) on a log-log plot.
Above about 100 K, $\rho \propto T$, whereas at low temperatures, $\rho \propto T^5$, and at the lowest temperatures $\rho$ approaches the residual resistivity $\rho_R$. The inset shows the $\rho$ vs. $T$ behavior below 100 K on a linear plot. ($\rho_R$ is too small on this scale.)

toward $\rho_R$ as $T$ decreases toward 0 K. This is borne out by experiments, as shown in Figure 2.8 for copper. Therefore, at the lowest temperatures of interest, the resistivity is limited by scattering from impurities and crystal defects.[4]

**EXAMPLE 2.7**

**MATTHIESSEN'S RULE** Explain the typical resistivity versus temperature behavior of annealed and cold-worked (deformed) copper containing various amounts of Ni as shown in Figure 2.9.

**SOLUTION**

When small amounts of nickel are added to copper, the resistivity increases by virtue of Matthiessen's rule, $\rho = \rho_T + \rho_R + \rho_I$, where $\rho_T$ is the resistivity due to scattering from thermal vibrations; $\rho_R$ is the residual resistivity of the copper crystal due to scattering from crystal defects, dislocations, trace impurities, etc.; and $\rho_I$ is the resistivity arising from Ni addition

[4] At sufficiently low temperatures (typically, below 10–20 K for many metals and below ~135 K for certain ceramics) certain materials exhibit superconductivity in which the resistivity vanishes ($\rho = 0$), even in the presence of impurities and crystal defects. Superconductivity and its quantum mechanical origin will be explained in Chapter 8.

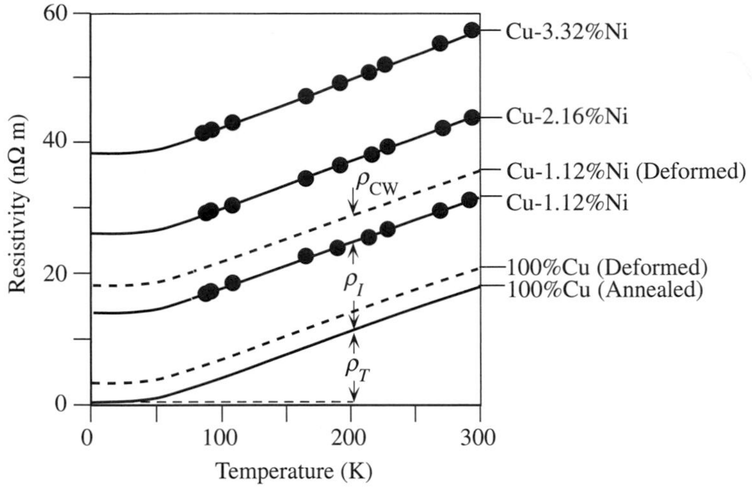


**Figure 2.9** Typical temperature dependence of the resistivity of annealed and cold-worked (deformed) copper containing various amounts of Ni in atomic percentage.

SOURCE: Data adapted from J.O. Linde, *Ann Pkysik*, 5, 219 (Germany, 1932).

alone (scattering from Ni impurity regions). Since $\rho_I$ is temperature independent, for small amounts of Ni addition, $\rho_I$ will simply shift up the $\rho$ versus $T$ curve for copper, by an amount proportional to the Ni content, $\rho_I \propto N_{\text{Ni}}$, where $N_{\text{Ni}}$ is the Ni impurity concentration. This is apparent in Figure 2.9, where the resistivity of Cu–2.16% Ni is almost twice that of Cu–1.12% Ni. Cold working (CW) or deforming a metal results in a higher concentration of dislocations and therefore increases the residual resistivity $\rho_R$ by $\rho_{\text{CW}}$. Thus, cold-worked samples have a resistivity curve that is shifted up by an additional amount $\rho_{\text{CW}}$ that depends on the extent of cold working.

**EXAMPLE 2.8**

**TEMPERATURE COEFFICIENT OF RESISTIVITY α AND RESISTIVITY INDEX *n*** If $\alpha_0$ is the temperature coefficient of resistivity (TCR) at temperature $T_0$ and the resistivity obeys the equation

$$\rho = \rho_0\left[\frac{T}{T_0}\right]^n$$

show that

$$\alpha_0 = \frac{n}{T_0}\left[\frac{T}{T_0}\right]^{n-1}$$

What is your conclusion?

Experiments indicate that $n = 1.2$ for W. What is its $\alpha_0$ at 20 °C? Given that, experimentally, $\alpha_0 = 0.00393$ for Cu at 20 °C, what is $n$?

**SOLUTION**

Since the resistivity obeys $\rho = \rho_0(T/T_0)^n$, we substitute this equation into the definition of TCR,

$$\alpha_0 = \frac{1}{\rho_0}\left[\frac{d\rho}{dT}\right] = \frac{n}{T_0}\left[\frac{T}{T_0}\right]^{n-1}$$

It is clear that, in general, $\alpha_0$ depends on the temperature $T$, as well as on the reference temperature $T_0$. The TCR is only independent of $T$ when $n = 1$.

At $T = T_0$, we have

$$\frac{\alpha_0 T_0}{n} = 1 \qquad \text{or} \qquad n = \alpha_0 T_0$$

For W, $n = 1.2$, so at $T = T_0 = 293$ K, we have $\alpha_{293\,\text{K}} = 0.0041$, which agrees reasonably well with $\alpha_{293\,\text{K}} = 0.0045$, frequently found in data books.

For Cu, $\alpha_{293\,\text{K}} = 0.00393$, so that $n = 1.15$, which agrees with the experimental value of $n$.

**EXAMPLE 2.9**

**TCR AT DIFFERENT REFERENCE TEMPERATURES** If $\alpha_1$ is the temperature coefficient of resistivity (TCR) at temperature $T_1$ and $\alpha_0$ is the TCR at $T_0$, show that

$$\alpha_1 = \frac{\alpha_0}{1 + \alpha_0(T_1 - T_0)}$$

**SOLUTION**

Consider the resistivity at temperature $T$ in terms of $\alpha_0$ and $\alpha_1$:

$$\rho = \rho_0[1 + \alpha_0(T - T_0)] \qquad \text{and} \qquad \rho = \rho_1[1 + \alpha_1(T - T_1)]$$

These equations are expected to hold at any temperature $T$, so the first and second equations at $T_1$ and $T_0$, respectively, give

$$\rho_1 = \rho_0[1 + \alpha_0(T_1 - T_0)] \qquad \text{and} \qquad \rho_0 = \rho_1[1 + \alpha_1(T_0 - T_1)]$$

These two equations can be readily solved to eliminate $\rho_0$ and $\rho_1$ to obtain

$$\alpha_1 = \frac{\alpha_0}{1 + \alpha_0(T_1 - T_0)}$$

**EXAMPLE 2.10**

**TEMPERATURE OF THE FILAMENT OF A LIGHT BULB**

*a.* Consider a 40 W, 120 V incandescent light bulb. The tungsten filament is 0.381 m long and has a diameter of 33 $\mu$m. Its resistivity at room temperature is $5.51 \times 10^{-8}$ $\Omega$ m. Given that the resistivity of the tungsten filament varies at $T^{1.2}$, estimate the temperature of the bulb when it is operated at the rated voltage, that is, when it is lit directly from a power outlet, as shown schematically in Figure 2.10. Note that the bulb dissipates 40 W at 120 V.

*b.* Assume that the electrical power dissipated in the tungsten wire is radiated from the surface of the filament. The radiated electromagnetic power at the absolute temperature $T$ can

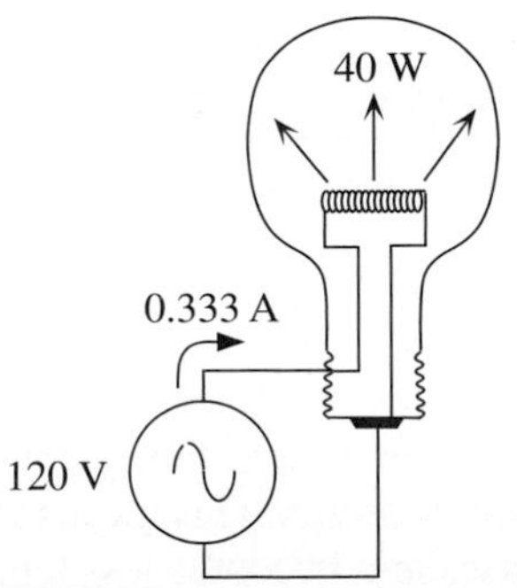


**Figure 2.10** Power radiated from a light bulb is equal to the electrical power dissipated in the filament.

be described by **Stefan's law,** as follows:

$$P_{\text{radiated}} = \epsilon\sigma_S A\left(T^4 - T_0^4\right)$$

where $\sigma_S$ is Stefan's constant ($5.67 \times 10^{-8}$ W m$^{-2}$ K$^{-4}$), $\epsilon$ is the emissivity of the surface (0.35 for tungsten), $A$ is the surface area of the tungsten filament, and $T_0$ is the room temperature (293 K). For $T^4 \gg T_0^4$, the equation becomes

$$P_{\text{radiated}} = \epsilon\sigma_S A T^4$$

Assuming that all the electrical power is radiated as electromagnetic waves from the surface, estimate the temperature of the filament and compare it with your answer in part (a).

**SOLUTION**

*a.* When the bulb is operating at 120 V, it is dissipating 40 W, which means that the current is

$$I = \frac{P}{V} = \frac{40\text{ W}}{120\text{ V}} = 0.333\text{ A}$$

The resistance of the filament at the operating temperature $T$ must be

$$R = \frac{V}{I} = \frac{120}{0.333} = 360\ \Omega$$

Since $R = \rho L/A$, the resistivity of tungsten at the operating temperature $T$ must be

$$\rho(T) = \frac{R(\pi D^2/4)}{L} = \frac{360\ \Omega\ \pi(33 \times 10^{-6}\text{ m})^2}{4(0.381\text{ m})} = 8.08 \times 10^{-7}\ \Omega\text{ m}$$

But, $\rho(T) = \rho_0(T/T_0)^{1.2}$, so that

$$T = T_0\left(\frac{80.8 \times 10^{-8}}{5.51 \times 10^{-8}}\right)^{1/1.2}$$

$$= 2746\text{ K} \quad \text{or} \quad 2473\ ^\circ\text{C} \quad \text{(melting temperature of W is about 3680, K)}$$

*b.* To calculate $T$ from the radiation law, we note that $T = [P_{\text{radiated}}/\epsilon\sigma_S A]^{1/4}$.
The surface area is

$$A = L(\pi D) = (0.381)(\pi 33 \times 10^{-6}) = 3.95 \times 10^{-5}\text{ m}^2$$

Then,

$$T = \left[\frac{P_{\text{radiated}}}{\epsilon\sigma_S A}\right]^{1/4} = \left[\frac{40\text{ W}}{(0.35)(5.67 \times 10^{-8}\text{ W m}^{-2}\text{ K}^{-4})(3.95 \times 10^{-5}\text{ m}^2)}\right]^{1/4}$$

$$= [5.103 \times 10^{13}]^{1/4} = 2673\text{ K} \quad \text{or} \quad 2400\ ^\circ\text{C}$$

The difference between the two methods is less than 3 percent.

## 2.3.2 SOLID SOLUTIONS AND NORDHEIM'S RULE

In an isomorphous alloy of two metals, that is, a binary alloy that forms a solid solution, we would expect Equation 2.15 to apply, with the temperature-independent impurity contribution $\rho_I$ increasing with the concentration of solute atoms. This means that as the alloy concentration increases, the resistivity $\rho$ increases and becomes less temperature

**Table 2.2** The effect of alloying on the resistivity

| Material | Resistivity at 20 °C (nΩ m) | $\alpha$ at 20 °C (1/K) |
|---|---|---|
| Nickel | 69 | 0.006 |
| Chrome | 129 | 0.003 |
| Nichrome | 1120 | 0.0003 |

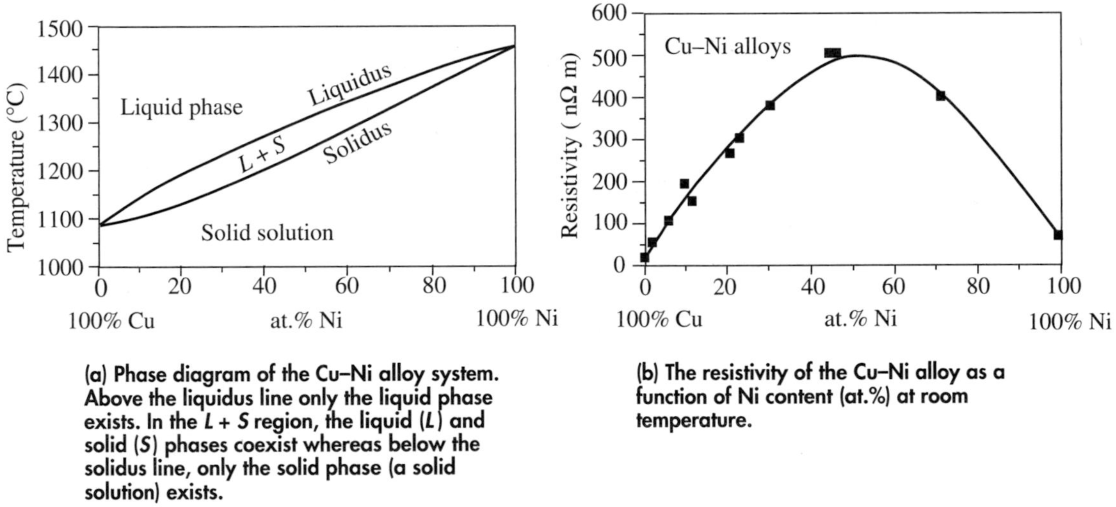


**(a) Phase diagram of the Cu–Ni alloy system. Above the liquidus line only the liquid phase exists. In the *L* + *S* region, the liquid (*L*) and solid (*S*) phases coexist whereas below the solidus line, only the solid phase (a solid solution) exists.**

**(b) The resistivity of the Cu–Ni alloy as a function of Ni content (at.%) at room temperature.**

**Figure 2.11** The Cu–Ni alloy system.
SOURCE: Data extracted from *Metals Handbook*, 10th ed., 2 and 3, Metals Park, Ohio: ASM, 1991, and M. Hansen and K. Anderko, *Constitution of Binary Alloys*, New York: McGraw-Hill, 1958.

dependent as $\rho_I$ overwhelms $\rho_T$, leading to $\alpha \ll 1/273$. This is the advantage of alloys in resistive components. Table 2.2 shows that when 80% nickel is alloyed with 20% chromium, the resistivity of Ni increases almost 16 times. In fact, the alloy is called **nichrome** and is widely used as a heater wire in household appliances and industrial furnaces.

As a further example of the resistivity of a solid solution, consider the copper– nickel alloy. The phase diagram for this alloy system is shown in Figure 2.11a. It is clear that the alloy forms a one-phase solid solution for all compositions. Both Cu and Ni have the same FCC crystal structure, and since the Cu atom is only slightly larger than the Ni atom by about ~3 percent (easily checked on the Periodic Table), the Cu–Ni alloy will therefore still be FCC, but with Cu and Ni atoms randomly mixed, resulting in a solid solution. When Ni is added to copper, the impurity resistivity $\rho_I$ in Equation 2.15 will increase with the Ni concentration. Experimental results for this alloy system are shown in Figure 2.11b. It should be apparent that when we reach 100% Ni, we again have a pure metal whose resistivity must be small. Therefore, $\rho$ versus Ni concentration must pass through a maximum, which for the Cu–Ni alloy seems to be at around ~50% Ni.

There are other binary solid solutions that reflect similar behavior to that depicted in Figure 2.11, such as Cu–Au, Ag–Au, Pt–Pd, Cu–Pd, to name a few. Quite often, the use of an alloy for a particular application is necessitated by the mechanical properties, rather than the desired electrical resistivity alone. For example, brass, which is 70% Cu–30% Zn in solid solution, has a higher strength compared to pure copper; as such, it is a suitable metal for the prongs of an electrical plug.

An important semiempirical equation that can be used to predict the resistivity of an alloy is **Nordheim's rule** which relates the impurity resistivity $\rho_I$ to the atomic fraction $X$ of solute atoms in a solid solution, as follows:

*Nordheim's rule for solid solutions*

$$\rho_I = CX(1 - X) \qquad [2.21]$$

where $C$ is the constant termed the **Nordheim coefficient,** which represents the effectiveness of the solute atom in increasing the resistivity. Nordheim's rule assumes that the solid solution has the solute atoms randomly distributed in the lattice, and these random distributions of impurities cause the electrons to become scattered as they whiz around the crystal. For sufficiently small amounts of impurity, experiments show that the increase in the resistivity $\rho_I$ is nearly always simply proportional to the impurity concentration $X$, that is, $\rho_I \propto X$, which explains the initial approximately equal increments of rise in the resistivity of copper with 1.11% Ni and 2.16% Ni additions as shown in Figure 2.9. For dilute solutions, Nordheim's rule predicts the same linear behavior, that is, $\rho_I = CX$ for $X \ll 1$.

Table 2.3 lists some typical Nordheim coefficients for various additions to copper and gold. The value of the Nordheim coefficient depends on the type of solute and the solvent. A solute atom that is drastically different in size to the solvent atom will result in a bigger increase in $\rho_I$ and will therefore lead to a larger $C$. An important assumption

**Table 2.3** Nordheim coefficient $C$ (at 20 °C) for dilute alloys obtained from $\rho_I = CX$ and $X < 1$ at.%*

| Solute in Solvent (element in matrix) | $C$ (nΩ m) | Maximum Solubility at 25 °C (at.%) |
|---|---|---|
| Au in Cu matrix | 5500 | 100 |
| Mn in Cu matrix | 2900 | 24 |
| Ni in Cu matrix | 1200 | 100 |
| Sn in Cu matrix | 2900 | 0.6 |
| Zn in Cu matrix | 300 | 30 |
| Cu in Au matrix | 450 | 100 |
| Mn in Au matrix | 2410 | 25 |
| Ni in Au matrix | 790 | 100 |
| Sn in Au matrix | 3360 | 5 |
| Zn in Au matrix | 950 | 15 |

*NOTE: For many isomorphous alloys $C$ may be different at higher concentrations; that is, it may depend on the composition of the alloy.

SOURCES: D.G. Fink and D. Christiansen, eds., *Electronics Engineers' Handbook,* 2nd ed., New York, McGraw-Hill, 1982. J. K. Stanley, *Electrical and Magnetic Properties of Metals,* Metals Park, OH, American Society for Metals, 1963. Solubility data from M. Hansen and K. Anderko, *Constitution of Binary Alloys,* 2nd ed., New York, McGraw-Hill, 1985.

in Nordheim's rule in Equation 2.21 is that the alloying does not significantly vary the number of conduction electrons per atom in the alloy. Although this will be true for alloys with the same valency, that is, from the same column in the Periodic Table (*e.g.*, Cu–Au, Ag–Au), it will not be true for alloys of different valency, such as Cu and Zn. In pure copper, there is just one conduction electron per atom, whereas each Zn atom can donate two conduction electrons. As the Zn content in brass is increased, more conduction electrons become available per atom. Consequently, the resistivity predicted by Equation 2.21 at high Zn contents is greater than the actual value because $C$ refers to dilute alloys. To get the correct resistivity from Equation 2.21 we have to lower $C$, which is equivalent to using an effective Nordheim coefficient $C_{\text{eff}}$ that decreases as the Zn content increases. In other cases, for example, in Cu–Ni alloys, we have to increase $C$ at high Ni concentrations to account for additional electron scattering mechanisms that develop with Ni addition. Nonetheless, the Nordheim rule is still useful for predicting the resistivities of dilute alloys, particularly in the low-concentration region.

With Nordheim's rule in Equation 2.21, the resistivity of an alloy of composition $X$ is

*Combined Matthiessen and Nordheim rules*

$$\rho = \rho_{\text{matrix}} + CX(1 - X) \qquad [2.22]$$

where $\rho_{\text{matrix}} = \rho_T + \rho_R$ is the resistivity of the matrix due to scattering from thermal vibrations and from other defects, in the absence of alloying elements. To reiterate, the value of $C$ depends on the alloying element and the matrix. For example, $C$ for gold in copper would be different than $C$ for copper in gold, as shown in Table 2.3.

In solid solutions, at some concentrations of certain binary alloys, such as 75% Cu–25% Au and 50% Cu–50% Au, the annealed solid has an orderly structure; that is, the Cu and Au atoms are not randomly mixed, but occupy regular sites. In fact, these compositions can be viewed as pure compound—like the solids $Cu_3Au$ and CuAu. The resistivities of $Cu_3Au$ and CuAu will therefore be less than the same composition random alloy that has been quenched from the melt. As a consequence, the resistivity $\rho$ versus composition $X$ curve does not follow the dashed parabolic curve throughout; rather, it exhibits sharp falls at these special compositions, as illustrated in Figure 2.12.

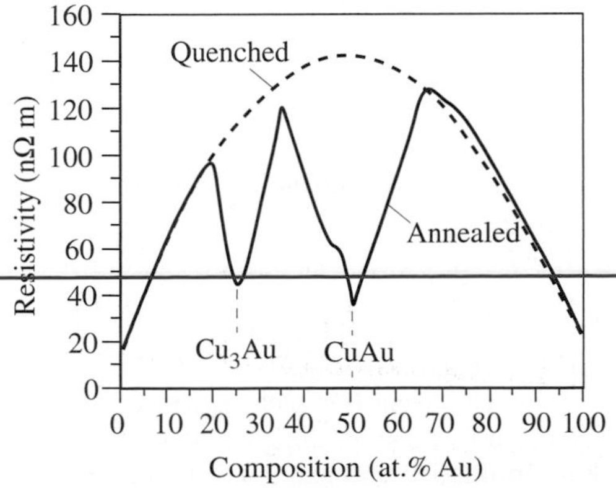


**Figure 2.12** Electrical resistivity vs. composition at room temperature in Cu–Au alloys.

The quenched sample (dashed curve) is obtained by quenching the liquid, and the Cu and Au atoms are randomly mixed. The resistivity obeys the Nordheim rule. When the quenched sample is annealed or the liquid is slowly cooled (solid curve), certain compositions ($Cu_3Au$ and CuAu) result in an ordered crystalline structure in which the Cu and Au atoms are positioned in an ordered fashion in the crystal and the scattering effect is reduced.

**EXAMPLE 2.11**

**NORDHEIM'S RULE** The alloy 90 wt.% Au–10 wt.% Cu is sometimes used in low-voltage dc electrical contacts, because pure gold is mechanically soft and the addition of copper increases the hardness of the metal without sacrificing the corrosion resistance. Predict the resistivity of the alloy and compare it with the experimental value of 108 nΩ m.

**SOLUTION**

We apply Equation 2.22, $\rho(X) = \rho_{Au} + CX(1 - X)$ but with 10 wt.% Cu converted to the atomic fraction for $X$. If $w$ is the weight fraction of Cu, $w = 0.1$, and if $M_{Au}$ and $M_{Cu}$ are the atomic masses of Au and Cu, then the atomic fraction $X$ of Cu is given by (see Example 1.2),

$$X = \frac{w/M_{Cu}}{w/M_{Cu} + (1-w)/M_{Au}} = \frac{0.1/63.55}{(0.1/63.55) + (0.90/197)} = 0.256$$

Given that $\rho_{Au} = 22.8$ nΩ m and $C = 450$ nΩ m,

$$\rho = \rho_{Au} + CX(1 - X) = (22.8\ \text{n}\Omega\ \text{m}) + (450\ \text{n}\Omega\ \text{m})(0.256)(1 - 0.256)$$
$$= 108.5\ \text{n}\Omega\ \text{m}$$

This value is only 0.5% different from the experimental value.

**EXAMPLE 2.12**

**RESISTIVITY DUE TO IMPURITIES** The mean speed of conduction electrons in copper is about $1.5 \times 10^6$ m s$^{-1}$. Its room temperature resistivity is 17 nΩ m, and the atomic concentration $N_{at}$ in the crystal is $8.5 \times 10^{22}$ cm$^{-3}$. Suppose that we add 1 at.% Au to form a solid solution. What is the resistivity of the alloy, the effective mean free path, and the mean free path due to collisions with Au atoms only?

**SOLUTION**

According to Table 2.3, the Nordheim coefficient $C$ of Au in Cu is 5500 nΩ m. With $X = 0.01$ (1 at.%), the overall resistivity from Equation 2.22 is

$$\rho = \rho_{matrix} + CX(1 - X) = 17\ \text{n}\Omega\ \text{m} + (5500\ \text{n}\Omega\ \text{m})(0.01)(1 - 0.01)$$
$$= 17\ \text{n}\Omega\ \text{m} + 54.45\ \text{n}\Omega\ \text{m} = 71.45\ \text{n}\Omega\ \text{m}$$

Suppose that $\ell$ is the overall or effective mean free path and $\tau$ is the effective mean free time between scattering events (includes both scattering from lattice vibrations and impurities). Since $\ell = u\tau$, and the effective drift mobility $\mu_d = e\tau/m_e$, the expression for the conductivity becomes

*Conductivity and mean free path*

$$\sigma = en\mu_d = \frac{e^2 n\tau}{m_e} = \frac{e^2 n\ell}{m_e u}$$

We can now calculate the effective mean free path $\ell$ in the alloy given that copper has a valency of *I* and the electron concentration $n = N_{at}$,

$$\frac{1}{71.5 \times 10^{-9}\ \Omega\ \text{m}} = \frac{(1.6 \times 10^{-19}\text{C})^2(8.5 \times 10^{28}\ \text{m}^{-3})\ell}{(9.1 \times 10^{-31}\text{kg})(1.5 \times 10^6\ \text{m s}^{-1})}$$

which gives $\ell = 8.8$ nm. We can repeat the calculation for pure copper using $\sigma = 1/\rho_{matrix} = 1/(17 \times 10^{-9}\ \Omega\ \text{m})$ to find $\ell_{Cu} = 37$ nm. The mean free path is reduced approximately by 4 times by adding only 1 at.% Au. The mean free path $\ell_I$ due to scattering from impurities only can be found from Equation 2.13 multiplied through by $1/u$, or by using

Matthiessen's rule in Equation 2.14:

$$\frac{1}{\ell} = \frac{1}{\ell_{Cu}} + \frac{1}{\ell_I}$$

Substituting $\ell_{Cu} = 37$ nm and $\ell = 8.8$ nm, we find $\ell_I = 11.5$ nm.

We can take these calculations one step further. If $N_I$ is the impurity concentration in the alloy, then $N_I = 0.01N_{at} = 0.01(8.5 \times 10^{28}\ \text{m}^{-3}) = 8.5 \times 10^{26}\ \text{m}^{-3}$. The mean separation $d_I$ between the impurities can be estimated roughly from $d_I \approx 1/N_I^{1/3}$, which gives $d_I \approx 1.0$ nm. It is clear that not all Au atoms can be involved in scattering the electrons since $\ell_I$ is much longer than $d_I$. (Another way to look at it is to say that it takes more than just one collision with an impurity to randomize the velocity of the electron.)

# 2.4 RESISTIVITY OF MIXTURES AND POROUS MATERIALS

## 2.4.1 HETEROGENEOUS MIXTURES

Nordheim's rule only applies to solid solutions that are single-phase solids. In other words, it is valid for homogeneous mixtures in which the atoms are mixed at the atomic level throughout the solid, as in the Cu–Ni alloy. The classic problem of determining the effective resistivity of a multiphase solid is closely related to the evaluation of the effective dielectric constant, effective thermal conductivity, effective elastic modulus, effective Poisson's ratio, etc., for a variety of mixtures, including such composite materials as fiberglass. Indeed, many of the mixture rules are identical.

Consider a material with two distinct phases $\alpha$ and $\beta$, which are stacked in layers as illustrated in Figure 2.13a. Let us evaluate the effective resistivity for current flow

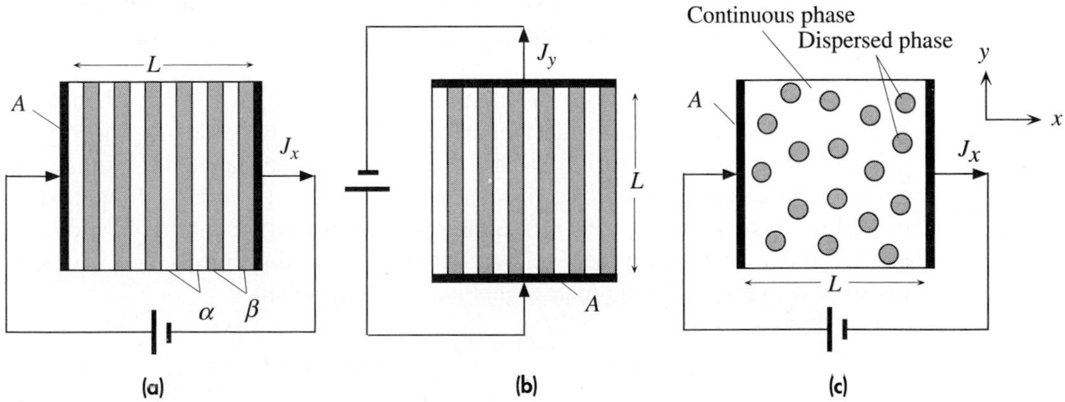


**Figure 2.13** The effective resistivity of a material with a layered structure.
(a) Along a direction perpendicular to the layers.
(b) Along a direction parallel to the plane of the layers.
(c) Materials with a dispersed phase in a continuous matrix.

in the $x$ direction. Since the layers are in series, the effective resistance $R_{\text{eff}}$ for the whole material is

*Effective resistance*

$$R_{\text{eff}} = \frac{L_\alpha \rho_\alpha}{A} + \frac{L_\beta \rho_\beta}{A} \quad [2.23]$$

where $L_\alpha$ is the total length (thickness) of the $\alpha$-phase layers, and $L_\beta$ is the total length of the $\beta$-phase layers, $L_\alpha + L_\beta = L$ is the length of the sample, and $A$ is the cross-sectional area. Let $\chi_\alpha$ and $\chi_\beta$ be the volume fractions of the $\alpha$ and $\beta$ phases. The effective resistance is defined by

$$R_{\text{eff}} = \frac{L\rho_{\text{eff}}}{A}$$

where $\rho_{\text{eff}}$ is the **effective resistivity.** Using $\chi_\alpha = L_\alpha/L$ and $\chi_\beta = L_\beta/L$ in Equation 2.23, we find

*Resistivity–mixture rule*

$$\rho_{\text{eff}} = \chi_\alpha \rho_\alpha + \chi_\beta \rho_\beta \quad [2.24]$$

which is called the **resistivity–mixture rule** (or the **series rule of mixtures**).

If we are interested in the effective resistivity in the $y$ direction, as shown in Figure 2.13b, obviously the $\alpha$ and $\beta$ layers are in parallel, so an effective conductivity could be calculated in the same way as we did for the series case to find the **parallel rule of mixtures,** that is,

*Conductivity–mixture rule*

$$\sigma_{\text{eff}} = \chi_\alpha \sigma_\alpha + \chi_\beta \sigma_\beta \quad [2.25]$$

where $\sigma$ is the electrical conductivity of those phases identified by the subscript. Notice that the parallel rule uses the conductivity, and the series rule uses the resistivity. Equation 2.25 is often referred to as the **conductivity–mixture rule.**

Although these two rules refer to special cases, in general, for a random mixture of phase $\alpha$ and phase $\beta$, we would not expect either equation to apply rigorously. When the resistivities of two randomly mixed phases are not markedly different, the series mixture rule can be applied at least approximately, as we will show in Example 2.13.

However, if the resistivity of one phase is appreciably different than the other, there are two semiempirical rules that are quite useful in materials engineering.[5] Consider a heterogeneous material that has a dispersed phase (labeled $d$), in the form of particles, in a continuous phase (labeled $c$) that acts as a matrix, as depicted in Figure 2.13c. Assume that $\rho_c$ and $\rho_d$ are the resistivities of the continuous and dispersed phases, and $\chi_c$ and $\chi_d$ are their volume fractions. If the dispersed phase is much more resistive with respect to the matrix, that is, $\rho_d > 10\rho_c$, then

*Mixture rule*

$$\rho_{\text{eff}} = \rho_c \frac{\left(1 + \frac{1}{2}\chi_d\right)}{(1 - \chi_d)} \qquad (\rho_d > 10\rho_c) \quad [2.26]$$

[5] Over the years, the task of predicting the resistivity of a mixture has challenged many theorists and experimentalists, including Lord Rayleigh who, in 1892, published an excellent exposition on the subject in the Philosophical Magazine. An extensive treatment of mixtures can be found in a paper by J. A. Reynolds and J. M. Hough published in 1957 (Proceedings of the Physical Society, 70, no. 769, London), which contains nearly all the mixture rules for the resistivity.

On the other hand, if $\rho_d < (\rho_c/10)$, then

$$\rho_{\text{eff}} = \rho_c \frac{(1-\chi_d)}{(1+2\chi_d)} \qquad (\rho_d < 0.1\rho_c) \tag{2.27}$$

*Mixture rule*

We therefore have at least four mixture rules at our disposal, the uses of which depend on the mixture geometry and the resistivities of the various phases. The problem is identifying which one to use for a given material, which in turn requires a knowledge of the microstructure and properties of the constituents. It should be emphasized that, at best, Equations 2.24 to 2.27 provide only a reasonable estimate of the effective resistivity of the mixture.[6]

Equations 2.26 and 2.27 are simplified special cases of a more general mixture rule due to Reynolds and Hough (1957). Consider a mixture that consists of a continuous conducting phase with a conductivity $\sigma_c$ that has dispersed spheres of another phase of conductivity $\sigma_d$ and of volume fraction $\chi$, similar to Figure 2.13c. The effective conductivity of the mixture is given by

*Reynolds and Hough rule for mixture of dispersed phases*

$$\frac{\sigma - \sigma_c}{\sigma + 2\sigma_c} = \chi \frac{\sigma_d - \sigma_c}{\sigma_d + 2\sigma_c} \tag{2.28}$$

It is assumed that the spheres are randomly dispersed in the material. It is left as an exercise to show that if $\sigma_d \ll \sigma_c$, then Equation 2.28 reduces to Equation 2.26. A good application would be the calculation of the effective resistivity of porous carbon electrodes, which can be 50–100 percent higher than the resistivity of bulk polycrystalline carbon (graphite). If, on the other hand, $\sigma_d \gg \sigma_c$, the dispersed phase is very conducting, for example, silver particles mixed into a graphite paste to increase the conductivity of the paste, then Equation 2.28 reduces to Equation 2.27. The usefulness of Equation 2.28 cannot be underestimated inasmuch as there are many types of materials in engineering that are mixtures of one type or another.

**EXAMPLE 2.13**

**THE RESISTIVITY-MIXTURE RULE** Consider a two-phase alloy consisting of phase $\alpha$ and phase $\beta$ randomly mixed as shown in Figure 2.14a. The solid consists of a random mixture of two types of resistivities, $\rho_\alpha$ of $\alpha$ and $\rho_\beta$ of $\beta$. We can divide the solid into a bundle of $N$ parallel fibers of length $L$ and cross-sectional area $A/N$, as shown in Figure 2.14b. In this fiber (infinitesimally thin), the $\alpha$ and $\beta$ phases are in series, so if $\chi_\alpha = V_\alpha/V$ is the volume fraction of phase $\alpha$ and $\chi_\beta$ is that of $\beta$, then the total length of all $\alpha$ regions present in the fiber is $\chi_\alpha L$, and the total length of $\beta$ regions is $\chi_\beta L$. The two resistances are in series, so the fiber resistance is

$$R_{\text{fiber}} = \frac{\rho_\alpha(\chi_\alpha L)}{(A/N)} + \frac{\rho_\beta(\chi_\beta L)}{(A/N)}$$

But the resistance of the solid is made up of $N$ such fibers in parallel, that is,

$$R_{\text{solid}} = \frac{R_{\text{fiber}}}{N} = \frac{\rho_\alpha \chi_\alpha L}{A} + \frac{\rho_\beta \chi_\beta L}{A}$$

[6] More accurate mixture rules have been established for various types of mixtures with components possessing widely different properties, which the keen reader can find in P. L. Rossiter, *The Electrical Resistivity of Metals and Alloys* (Cambridge University Press, Cambridge, 1987).

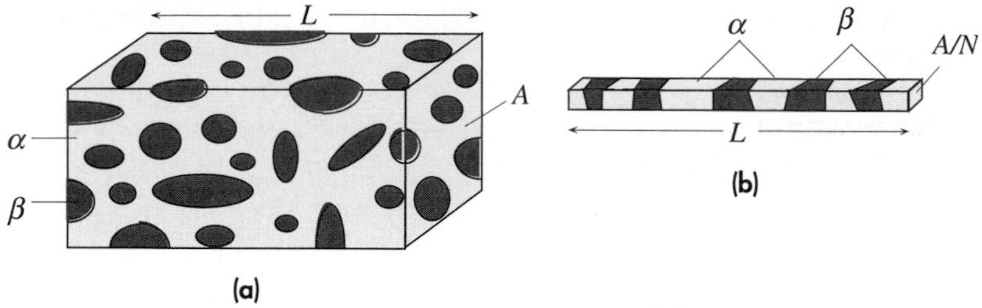


**Figure 2.14**
(a) A two-phase solid.
(b) A thin fiber cut out from the solid.

By definition, $R_{\text{solid}} = \rho_{\text{eff}} L/A$, where $\rho_{\text{eff}}$ is the effective resistivity of the material, so

$$\frac{\rho_{\text{eff}} L}{A} = \frac{\rho_\alpha \chi_\alpha L}{A} + \frac{\rho_\beta \chi_\beta L}{A}$$

Thus, for a two-phase solid, the effective resistivity will be

*Resistivity mixture rule*

$$\rho_{\text{eff}} = \chi_\alpha \rho_\alpha + \chi_\beta \rho_\beta$$

If the densities of the two phases are not too different, we can use weight fractions instead of volume fractions. The series rule fails when the resistivities of the phases are vastly different. A major (and critical) tacit assumption here is that the current flow lines are all parallel, so that no current crosses from one fiber to another. Only then can we say that the effective resistance is $R_{\text{fiber}}/N$.

**EXAMPLE 2.14**

**A COMPONENT WITH DISPERSED AIR PORES** What is the effective resistivity of 95/5 (95% Cu–5% Sn) bronze, which is made from powdered metal containing dispersed pores at 15% (volume percent, vol.%). The resistivity of 95/5 bronze is $1 \times 10^{-7}$ Ω m.

**SOLUTION**

Pores are infinitely more resistive ($\rho_d = \infty$) than the bronze matrix, so we use Equation 2.26,

$$\rho_{\text{eff}} = \rho_c \frac{1 + \frac{1}{2}\chi_d}{1 - \chi_d} = (1 \times 10^{-7}\ \Omega\ \text{m}) \frac{1 + \frac{1}{2}(0.15)}{1 - 0.15} = 1.27 \times 10^{-7}\ \Omega\ \text{m}$$

**EXAMPLE 2.15**

**COMBINED NORDHEIM AND MIXTURE RULES** Brass is an alloy composed of Cu and Zn. The alloy is a solid solution for Zn content less than 30 wt.%. Consider a brass component made from sintering 90 at.% Cu and 10 at.% Zn brass powder. The component contains dispersed air pores at 15% (vol.%). The Nordheim coefficient $C$ of Zn in Cu is 300 nΩ m, under very dilute conditions. Each Zn atom donates two, whereas each Cu atom of the matrix donates one conduction electron, so that the Cu–Zn alloy has a higher electron concentration than in the Cu crystal itself. Predict the effective resistivity of this brass component.

**SOLUTION**

We first calculate the resistivity of the alloy without the pores, which forms the continuous phase in the powdered material. The simple Nordheim's rule predicts that

$$\rho_{\text{brass}} = \rho_{\text{copper}} + CX(1 - X) = 17\ \text{n}\Omega\ \text{m} + 300(0.1)(1 - 0.1) = 44\ \text{n}\Omega\ \text{m}$$

The experimental value, about 40 nΩ m, is actually less because Zn has a valency of 2, and when a Zn atom replaces a host Cu atom, it donates two electrons instead of one. We can very roughly adjust the calculated resistivity by noting that a 10 at.% Zn addition increases the conduction electron concentration by 10% and hence reduces the resistivity $\rho_{\text{brass}}$ by 10% to 40 nΩ m.

The powdered metal has $\chi_d = 0.15$, which is the volume fraction of the dispersed phase, that is, the air pores, and $\rho_c = \rho_{\text{brass}} = 40$ nΩ m is the resistivity of the continuous matrix. The effective resistivity of the powdered metal is given by

$$\rho_{\text{eff}} = \rho_c \frac{1 + \frac{1}{2}\chi_d}{1 - \chi_d} = (40 \text{ n}\Omega \text{ m})\frac{1 + \frac{1}{2}(0.15)}{1 - (0.15)} = 50.6 \text{ n}\Omega \text{ m}$$

If we use the simple conductivity mixture rule, $\rho_{\text{eff}}$ is 47.1 nΩ m, and it is underestimated.

The *effective* Nordheim coefficient $C_{\text{eff}}$ at the composition of interest is about 255 nΩ m, which would give $\rho_{\text{brass}} = \rho_o + C_{\text{eff}}X(1 - X) = 40$ nΩ m. It is left as an exercise to show that the effective number of conduction electrons per atom in the alloy is $1 + X$ so that we must divide the $\rho_{\text{brass}}$ calculated above by $(1 + X)$ to obtain the correct resistivity of brass if we use the listed value of $C$ under dilute conditions. (See Question 2.8.)

## 2.4.2 TWO-PHASE ALLOY (Ag–Ni) RESISTIVITY AND ELECTRICAL CONTACTS

Certain binary alloys, such as Pb–Sn and Cu–Ag, only exhibit a single-phase alloy structure over very small composition ranges. For most compositions, these alloys form a two-phase heterogeneous mixture of phases $\alpha$ and $\beta$. A typical phase diagram for such a eutectic binary alloy system is shown in Figure 2.15a, which could be a

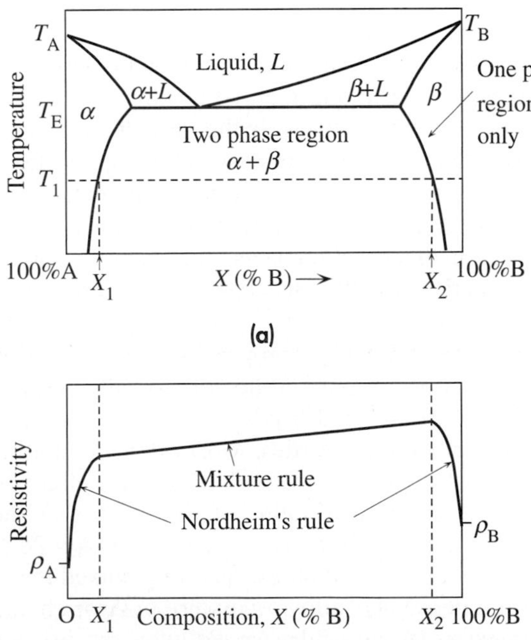


(a)

(b)

**Figure 2.15** Eutectic-forming alloys, *e.g.*, Cu–Ag.
(a) The phase diagram for a binary, eutectic-forming alloy.
(b) The resistivity versus composition for the binary alloy.

schematic scheme for the Cu–Ag system or the Pb–Sn system. The phase diagram identifies the phases existing in the alloy at a given temperature and composition. If the overall composition $X$ is less than $X_1$, then at $T_1$, the alloy will consist of phase $\alpha$ only. This phase is Cu rich. When the composition $X$ is between $X_1$ and $X_2$, then the alloy will consist of the two phases $\alpha$ and $\beta$ randomly mixed. The phase $\alpha$ is Cu rich (that is, it has composition $X_1$) and the phase $\beta$ is Ag rich (composition $X_2$). The relative amounts of each phase are determined by the well-known **lever rule,** which means that we can determine the volume fractions of $\alpha$ and $\beta$, $\chi_\alpha$ and $\chi_\beta$, as the alloy composition is changed from $X_1$ to $X_2$.

For this alloy system, the dependence of the resistivity on the alloy composition is shown in Figure 2.15b. Between O and $X_1$ (% Ag), the solid is one phase (isomorphous); therefore, in this region, $\rho$ increases with the concentration of Ag by virtue of Nordheim's rule. At $X_1$, the solubility limit of Ag in Cu is reached, and after $X_1$, a second phase, which is $\beta$ rich, is formed. Thus, in the composition range $X_1$ to $X_2$, we have a mixture of $\alpha$ and $\beta$ phases, so $\rho$ is given by Equation 2.24 for mixtures and is therefore less than that for a single-phase alloy of the same composition. Similarly, at the Ag end ($X_2 < X < 100\%$), as Cu is added to Ag, between 100% Ag and the solubility limit at $X_2$, the resistivity is determined by Nordheim's rule. The expected behavior of the resistivity of an eutectic binary alloy over the whole composition range is therefore as depicted in Figure 2.15b.

Electrical, thermal, and other physical properties make copper the most widely used metallic conductor. For many electrical applications, high-conductivity copper, having extremely low oxygen and other impurity contents, is produced. Although aluminum has a conductivity of only about half that of copper, it is also frequently used as an electrical conductor. On the other hand, silver has a higher conductivity than copper, but its cost prevents its use, except in specialized applications. Switches often have silver contact specifications, though it is likely that the contact metal is actually a silver alloy. In fact, silver has the highest electrical and thermal conductivity and is consequently the natural choice for use in electrical contacts. In the form of alloys with various other metals, it is used extensively in make-and-break switching applications for currents of up to about 600 A. The precious metals, gold, platinum, and palladium, are extremely resistant to corrosion; consequently, in the form of various alloys, particularly with Ag, they are widely used in electrical contacts. For example, Ag–Ni alloys are common electrical contact materials for the switches in many household appliances.

It is frequently necessary to improve the mechanical properties of a metal alloy without significantly impairing its electrical conductivity. Solid-solution alloying improves mechanical strength, but at the expense of conductivity. A compromise must often be found between electrical and mechanical properties. Most often, strength is enhanced by introducing a second phase that does not have such an adverse effect on the conductivity. For example, Ag–Pd alloys form a solid solution such that the resistivity increases appreciably due to Nordheim's rule. The resistivity of Ag–Pd is mainly controlled by the scattering of electrons from Pd atoms randomly mixed in the Ag matrix. In contrast, Ag and Ni form a two-phase alloy, a mixture of Ag-rich and Ni-rich phases. The Ag–Ni alloy is almost as strong as the Ag–Pd alloy, but it has a lower resistivity because the mixture rule volume averages the two resistivities.

## 2.5 THE HALL EFFECT AND HALL DEVICES

An important phenomenon that we can comfortably explain using the "electron as a particle" concept is the Hall effect, which is illustrated in Figure 2.16. When we apply a magnetic field in a perpendicular direction to the applied field (which is driving the current), we find there is a transverse field in the sample that is perpendicular to the direction of both the applied field $\mathcal{E}_x$ and the magnetic field $B_z$, that is, in the $y$ direction. Putting a voltmeter across the sample, as in Figure 2.16, gives a voltage reading $V_H$. The applied field $\mathcal{E}_x$ drives a current $J_x$ in the sample. The electrons move in the $-x$ direction, with a drift velocity $v_{dx}$. Because of the magnetic field, there is a force (called the **Lorentz force**) acting on each electron and given by $F_y = -ev_{dx}B_z$. The direction of this Lorentz force is the $-y$ direction, which we can show by applying the corkscrew rule, because, in vector notation, the force **F** acting on a charge $q$ moving with a velocity **v** in a magnetic field **B** is given through the vector product

$$\mathbf{F} = q\mathbf{v} \times \mathbf{B} \qquad [2.29]$$

*Lorentz force*

All moving charges experience the Lorentz force in Equation 2.29 as shown schematically in Figure 2.17. In our example of a metal in Figure 2.16, this Lorentz force is the $-y$ direction, so it pushes the electrons downward, as a result of which there is a negative charge accumulation near the bottom of the sample and a positive charge near the top of the sample, due to exposed metal ions (*e.g.*, $Cu^+$).

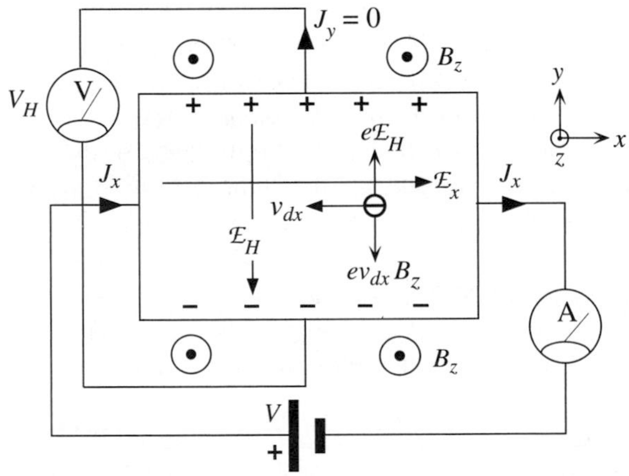


**Figure 2.16** Illustration of the Hall effect. The $z$ direction is out of the plane of the paper. The externally applied magnetic field is along the $z$ direction.

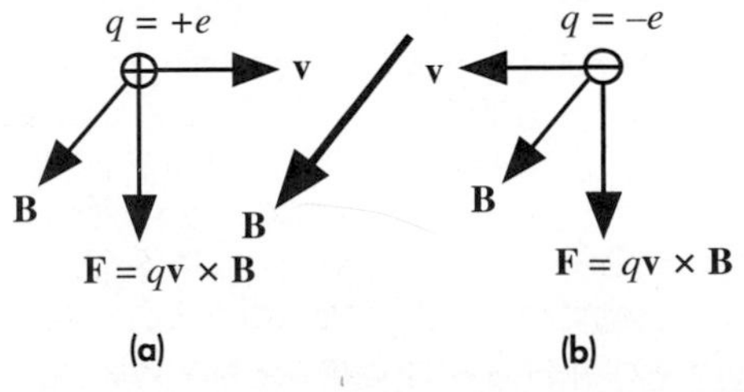


**Figure 2.17** A moving charge experiences a Lorentz force in a magnetic field.
(a) A positive charge moving in the $x$ direction experiences a force downward.
(b) A negative charge moving in the $-x$ direction also experiences a force downward.

The accumulation of electrons near the bottom results in an internal electric field $\mathcal{E}_H$ in the $-y$ direction. This is called the **Hall field** and gives rise to a Hall voltage $V_H$ between the top and bottom of the sample. Electron accumulation continues until the increase in $\mathcal{E}_H$ is sufficient to stop the further accumulation of electrons. When this happens, the magnetic-field force $ev_{dx}B_z$ that pushes the electrons down just balances the force $e\mathcal{E}_H$ that prevents further accumulation. Therefore, in the steady state,

$$e\mathcal{E}_H = ev_{dx}B_z$$

However, $J_x = env_{dx}$. Therefore, we can substitute for $v_{dx}$ to obtain $e\mathcal{E}_H = J_xB_z/n$ or

$$\mathcal{E}_H = \left(\frac{1}{en}\right)J_xB_z \qquad [2.30]$$

A useful parameter called the **Hall coefficient** $R_H$ is defined as

*Definition of Hall coefficient*

$$R_H = \frac{\mathcal{E}_y}{J_xB_z} \qquad [2.31]$$

The quantity $R_H$ measures the resulting Hall field, along $y$, per unit transverse applied current and magnetic field. The larger $R_H$, the greater $\mathcal{E}_y$ for a given $J_x$ and $B_z$. Therefore, $R_H$ is a gauge of the magnitude of the Hall effect. A comparison of Equations 2.30 and 2.31 shows that for metals,

*Hall coefficient for electron conduction*

$$R_H = -\frac{1}{en} \qquad [2.32]$$

The reason for the negative sign is that $\mathcal{E}_H = -\mathcal{E}_y$, which means that $\mathcal{E}_H$ is in the $-y$ direction.

Inasmuch as $R_H$ depends inversely on the free electron concentration, its value in metals is much less than that in semiconductors. In fact, Hall-effect devices (such as magnetometers) always employ a semiconductor material, simply because the $R_H$ is larger. Table 2.4 lists the Hall coefficients of various metals. Note that this is negative

**Table 2.4** Hall coefficient and Hall mobility ($\mu_H = |\sigma R_H|$) of selected metals

| Metal | $n$ [m$^{-3}$] ($\times 10^{28}$) | $R_H$ (Experimental) [m$^3$ A$^{-1}$ s$^{-1}$] ($\times 10^{-11}$) | $\mu_H = \lvert\sigma R_H\rvert$ [m$^2$ V$^{-1}$ s$^{-1}$] ($\times 10^{-4}$) |
|---|---|---|---|
| Ag | 5.85 | −9.0 | 57 |
| Al | 18.06 | −3.5 | 13 |
| Au | 5.90 | −7.2 | 31 |
| Be | 24.2 | +3.4 | ? |
| Cu | 8.45 | −5.5 | 32 |
| Ga | 15.3 | −6.3 | 3.6 |
| In | 11.49 | −2.4 | 2.9 |
| Mg | 8.60 | −9.4 | 22 |
| Na | 2.56 | −25 | 53 |

SOURCES: Data from various sources, including C. Nording and J. Osterman, *Physics Handbook*, Bromley, England: Chartwell-Bratt Ltd., 1982.

Magnetically operated Hall-effect position sensor as available from Micro Switch.

for most metals, although a few metals exhibit a positive Hall coefficient (see Be in Table 2.4). The reasons for the latter involve the band theory of solids, which we will discuss in Chapter 4.

Since the Hall voltage depends on the product of two quantities, the current density $J_x$ and the transverse applied magnetic field $B_z$, we see that the effect naturally multiplies two independently variable quantities. Therefore, it provides a means of carrying out a multiplication process. One obvious application is measuring the power dissipated in a load, where the load current and voltage are multiplied. There are many instances when it is necessary to measure magnetic fields, and the Hall effect is ideally suited to such applications. Commercial Hall-effect magnetometers can measure magnetic fields as low as 10 nT, which should be compared to the earth's magnetic field of ~50 μT. Depending on the application, manufacturers use different semiconductors to obtain the desired sensitivity. Hall-effect semiconductor devices are generally inexpensive, small, and reliable. Typical commercial, linear Hall-effect sensor devices are capable of providing a Hall voltage of ~10 mV per mT of applied magnetic field.

The Hall effect is also widely used in magnetically actuated electronic switches. The application of a magnetic field, say from a magnet, results in a Hall voltage that is amplified to trigger an electronic switch. The switches invariably use Si and are readily available from various companies. Hall-effect electronic switches are used as noncontacting keyboard and panel switches that last almost forever, as they have no mechanical contact assembly. Another advantage is that the electrical contact is "bounce" free. There are a variety of interesting applications for Hall-effect switches, ranging from ignition systems, to speed controls, position detectors, alignment controls, brushless dc motor commutators, etc.

**EXAMPLE 2.16**

**HALL-EFFECT WATTMETER** The Hall effect can be used to implement a wattmeter to measure electrical power dissipated in a load. The schematic sketch of the Hall-effect wattmeter is shown in Figure 2.18, where the Hall-effect sample is typically a semiconductor material (usually Si). The load current $I_L$ passes through two coils, which are called current coils and are shown as $C$ in Figure 2.18. These coils set up a magnetic field $B_z$ such that $B_z \propto I_L$. The Hall-effect sample is positioned in this field between the coils. The voltage $V_L$ across the load drives a current

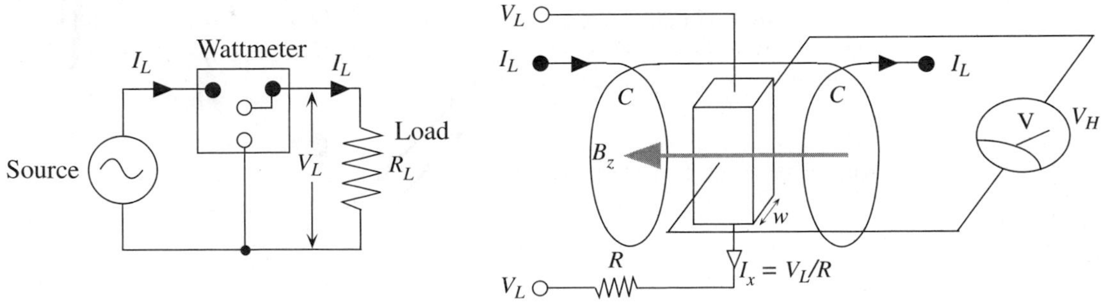


**Figure 2.18** Wattmeter based on the Hall effect.
Load voltage and load current have $L$ as subscript; $C$ denotes the current coils for setting up a magnetic field through the Hall-effect sample (semiconductor).

$I_x = V_L/R$ through the sample, where $R$ is a series resistance that is much larger than the resistance of the sample and that of the load. Normally, the current $I_x$ is very small and negligible compared to the load current. If $w$ is the width of the sample, then the measured Hall voltage is

$$V_H = w\mathcal{E}_H = wR_H J_x B_z \propto I_x B_z \propto V_L I_L$$

which is the electrical power dissipated in the load. The voltmeter that measures $V_H$ can now be calibrated to read directly the power dissipated in the load.

**EXAMPLE 2.17** **HALL MOBILITY** Show that if $R_H$ is the Hall coefficient and $\sigma$ is the conductivity of a metal, then the drift mobility of the conduction electrons is given by

$$\mu_d = |\sigma R_H| \qquad \textbf{[2.33]}$$

The Hall coefficient and conductivity of copper at 300 K have been measured to be $-0.55 \times 10^{-10}$ m$^3$ A$^{-1}$ s$^{-1}$ and $5.9 \times 10^7$ Ω$^{-1}$ m$^{-1}$, respectively. Calculate the drift mobility of electrons in copper.

**SOLUTION**

Consider the expression for

$$R_H = \frac{-1}{en}$$

Since the conductivity is given by $\sigma = en\mu_d$, we can substitute for $en$ to obtain

$$R_H = \frac{-\mu_d}{\sigma} \qquad \text{or} \qquad \mu_d = -R_H\sigma$$

which is Equation 2.33. The drift mobility can thus be determined from $R_H$ and $\sigma$.

The product of $\sigma$ and $R_H$ is called the **Hall mobility** $\mu_H$. Some values for the Hall mobility of electrons in various metals are listed in Table 2.4. From the expression in Equation 2.33, we get

$$\mu_d = -(-0.55 \times 10^{-10}\ \text{m}^3\ \text{A}^{-1}\ \text{s}^{-1})(5.9 \times 10^7\ \Omega^{-1}\ \text{m}^{-1}) = 3.2 \times 10^{-3}\ \text{m}^2\ \text{V}^{-1}\ \text{s}^{-1}$$

It should be mentioned that Equation 2.33 is an oversimplification.The actual relationship involves a numerical factor that multiplies the right term in Equation 2.33. The factor depends on the charge carrier scattering mechanism that controls the drift mobility.

**EXAMPLE 2.18** **CONDUCTION ELECTRON CONCENTRATION FROM THE HALL EFFECT** Using the electron drift mobility from Hall-effect measurements (Table 2.4), calculate the concentration of conduction electrons in copper, and then determine the average number of electrons contributed to the free electron gas per copper atom in the solid.

**SOLUTION**

The number of conduction electrons is given by $n = \sigma/e\mu_d$. The conductivity of copper is $5.9 \times 10^7$ Ω$^{-1}$ m$^{-1}$, whereas from Table 2.4, the electron drift mobility is $3.2 \times 10^{-3}$ m$^2$ V$^{-1}$ s$^{-1}$. So,

$$n = \frac{(5.9 \times 10^7\ \Omega^{-1}\ \text{m}^{-1})}{[(1.6 \times 10^{-19}\ \text{C})(3.2 \times 10^{-3}\ \text{m}^2\ \text{V}^{-1}\ \text{s}^{-1})]} = 1.15 \times 10^{29}\ \text{m}^{-3}$$

Since the concentration of copper atoms is $8.5 \times 10^{28}$ m$^{-3}$, the average number of electrons contributed per atom is $(1.15 \times 10^{29}\ \text{m}^{-3})/(8.5 \times 10^{28}\ \text{m}^{-3}) \approx 1.36$.

# 2.6 THERMAL CONDUCTION

## 2.6.1 THERMAL CONDUCTIVITY

Experience tells us that metals are both good electrical and good thermal conductors. We may therefore surmise that the free conduction electrons in a metal must also play a role in heat conduction. Our conjecture is correct for metals, but not for other materials. The transport of heat in a metal is accomplished by the electron gas (conduction electrons), whereas in nonmetals, the conduction is due to lattice vibrations.

When a metal piece is heated at one end, the amplitude of the atomic vibrations, and thus the average kinetic energy of the electrons, in this region increases, as depicted in Figure 2.19. Electrons gain energy from energetic atomic vibrations when the two collide. By virtue of their increased random motion, these energetic electrons then transfer the extra energy to the colder regions by colliding with the atomic vibrations there. Thus, electrons act as "energy carriers."

The thermal conductivity of a material, as its name implies, measures the ease with which heat, that is, thermal energy, can be transported through the medium. Consider the metal rod shown in Figure 2.20, which is heated at one end. Heat will flow from the hot end to the cold end. Experiments show that the rate of heat flow, $Q' = dQ/dt$, through a thin section of thickness $\delta x$ is proportional to the temperature gradient $\delta T/\delta x$ and the cross-sectional area $A$, so

$$Q' = -A\kappa\frac{\delta T}{\delta x} \quad [2.34]$$

*Fourier's law of thermal conduction*

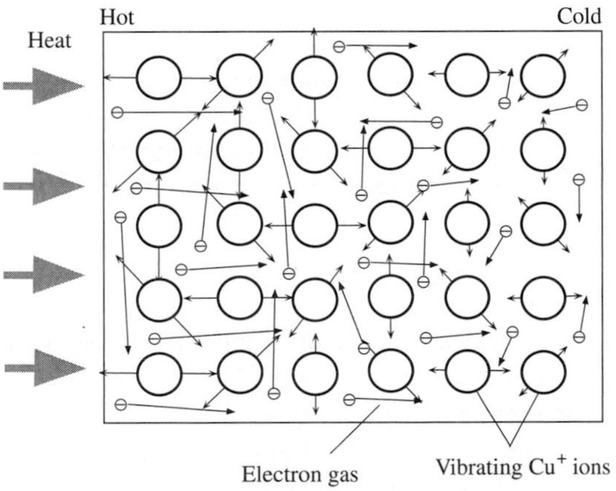


**Figure 2.19** Thermal conduction in a metal involves transferring energy from the hot region to the cold region by conduction electrons.

More energetic electrons (shown with longer velocity vectors) from the hotter regions arrive at cooler regions, collide with lattice vibrations, and transfer their energy. Lengths of arrowed lines on atoms represent the magnitudes of atomic vibrations.

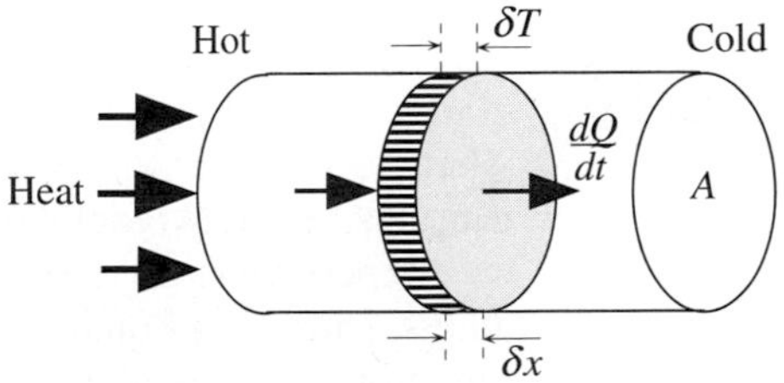


**Figure 2.20** Heat flow in a metal rod heated at one end.

Consider the rate of heat flow, $dQ/dt$, across a thin section $\delta x$ of the rod. The rate of heat flow is proportional to the temperature gradient $\delta T/\delta x$ and the cross-sectional area $A$.

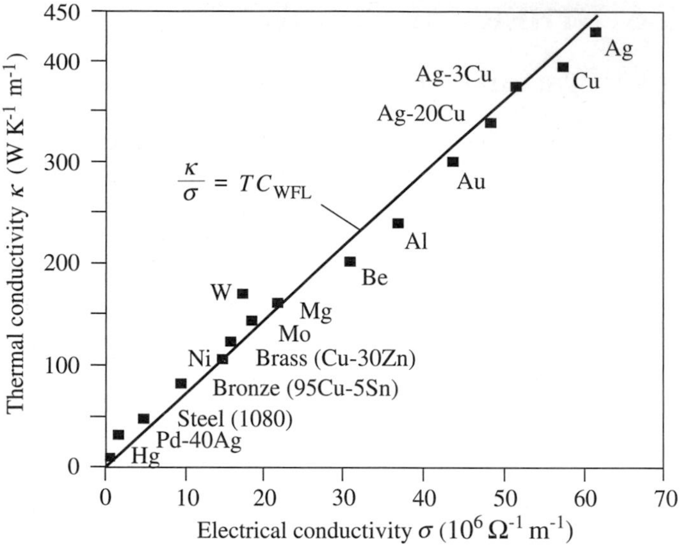


**Figure 2.21** Thermal conductivity $\kappa$ versus electrical conductivity $\sigma$ for various metals (elements and alloys) at 20 °C.
The solid line represents the WFL law with $C_{WFL} \approx 2.44 \times 10^8$ W Ω K$^{-2}$.

where $\kappa$ is a material-dependent **constant of proportionality** that we call the **thermal conductivity.** The negative sign indicates that the heat flow direction is that of decreasing temperature. Equation 2.34 is often referred to as **Fourier's law** of heat conduction and is a defining equation for $\kappa$. The driving force for the heat flow is the temperature gradient $\delta T/\delta x$. If we compare Equation 2.34 with Ohm's law for the electric current $I$, we see that

*Ohm's law of electrical conduction*

$$I = -A\sigma \frac{\delta V}{\delta x} \qquad [2.35]$$

which shows that in this case, the driving force is the potential gradient, that is, the electric field.[7] In metals, electrons participate in the processes of charge and heat transport, which are characterized by $\sigma$ and $\kappa$, respectively. Therefore, it is not surprising to find that the two coefficients are related by the **Wiedemann–Franz–Lorenz law,**[8] which is

*Wiedemann–Franz–Lorenz law*

$$\frac{\kappa}{\sigma T} = C_{WFL} \qquad [2.36]$$

where $C_{WFL} = \pi^2 k^2/3e^2 = 2.44 \times 10^{-8}$ W Ω K$^{-2}$ is a constant called the **Lorenz number** (or the Wiedemann–Franz–Lorenz coefficient).

Experiments on a wide variety of metals, ranging from pure metals to various alloys, show that Equation 2.36 is reasonably well obeyed at close to room temperature and above, as illustrated in Figure 2.21. Since the electrical conductivity of pure metals is inversely proportional to the temperature, we can immediately conclude that the thermal conductivity of these metals must be relatively temperature independent at room temperature and above.

[7] Recall that $J = \sigma\mathcal{E}$ which is equivalent to Equation 2.35.

[8] Historically, Wiedemann and Franz noted in 1853 that $\kappa/\sigma$ is the same for all metals at the same temperature. Lorenz in 1881 showed that $\kappa/\sigma$ is proportional to the temperature with a proportionality constant that is nearly the same for many metals. The law stated in Equation 2.36 reflects both observations. By the way, Lorenz, who was a Dane, should not be confused with Lorentz, who was Dutch.

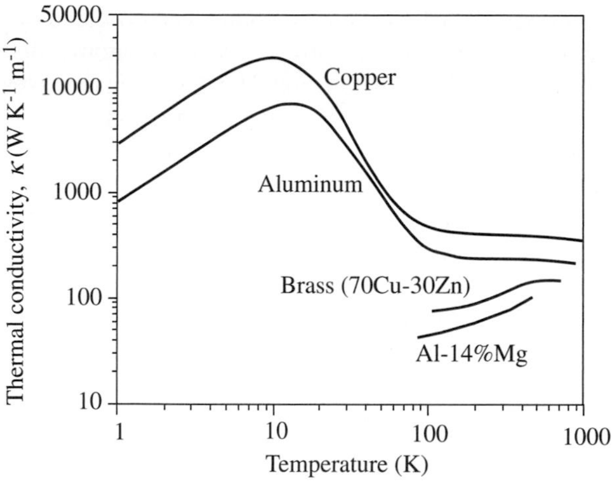


**Figure 2.22** Thermal conductivity versus temperature for two pure metals (Cu and Al) and two alloys (brass and Al–14% Mg).

SOURCE: Data extracted from Y. S. Touloukian, *et al.*, *Thermophysical Properties of Matter*, vol. 1: "Thermal Conductivity, Metallic Elements and Alloys," New York: Plenum, 1970.

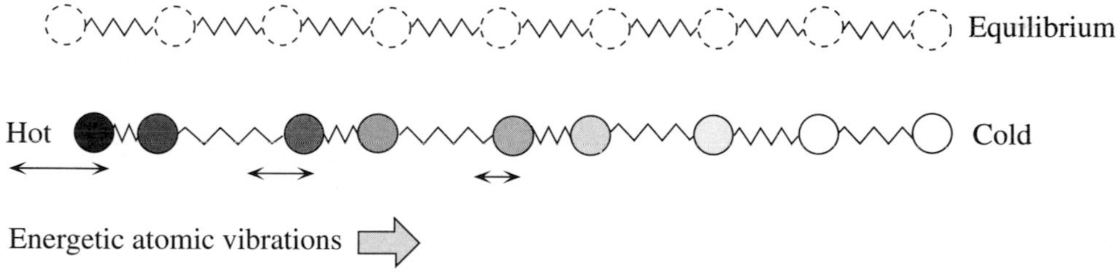


**Figure 2.23** Conduction of heat in insulators involves the generation and propagation of atomic vibrations through the bonds that couple the atoms (an intuitive figure).

Figure 2.22 shows the temperature dependence of $\kappa$ for copper and aluminum down to the lowest temperatures. It can be seen that for these two metals, above ~100 K, the thermal conductivity becomes temperature independent, in agreement with Equation 2.36. Qualitatively, above ~100 K, $\kappa$ is constant, because heat conduction depends essentially on the rate at which the electron transfers energy from one atomic vibration to another as it collides with them (Figure 2.19). This rate of energy transfer depends on the mean speed of the electron $u$, which increases only fractionally with the temperature. In fact, the fractionally small increase in $u$ is more than sufficient to carry the energy from one collision to another and thereby excite more energetic lattice vibrations in the colder regions.

Nonmetals do not have any free conduction electrons inside the crystal to transfer thermal energy from hot to cold regions of the material. In nonmetals, the energy transfer involves lattice vibrations, that is, atomic vibrations of the crystal. We know that we can view the atoms and bonds in a crystal as balls connected together through springs as shown for one chain of atoms in Figure 2.23. As we know from the kinetic molecular theory, all the atoms would be vibrating and the average vibrational kinetic energy would be proportional to the temperature. Intuitively, as depicted in Figure 2.23, when we heat one end of a crystal, we set up large-amplitude atomic vibrations at this hot end. The springs *couple* the vibrations to neighboring atoms and thus allow the large-amplitude vibrations to propagate, as a **vibrational wave,** to the cooler regions of the crystal. If we were to grab the left-end atom in Figure 2.23 and vibrate it violently, we would be sending vibrational waves down the ball-spring-ball chain. The efficiency of heat transfer depends not only on the efficiency of coupling between the atoms, and hence on the nature of interatomic bonding, but also on how the vibrational waves propogate in the crystal and how they are scattered by crystal imperfections and by their interactions with other vibrational waves; this topic is discussed in Chapter 4. The stronger the coupling, the greater will be the thermal conductivity, a trend that is intuitive but also borne out by experiments. Diamond has an exceptionally strong covalent bond and also has a very high thermal conductivity; $\kappa \approx 1000$ W m$^{-1}$ K$^{-1}$. On the other hand, polymers have weak secondary bonding between the polymer chains and their thermal conductivities are very poor; $\kappa < 1$ W m$^{-1}$ K$^{-1}$.

The thermal conductivity, in general, depends on the temperature. Different classes of materials exhibit different $\kappa$ values and also different $\kappa$ versus $T$ behavior. Table 2.5

**Table 2.5** Typical thermal conductivities of various classes of materials at 25 °C

| Material | $\kappa$ (W m$^{-1}$ K$^{-1}$) |
|---|---|
| Pure metal | |
| Nb | 52 |
| Fe | 80 |
| Zn | 113 |
| W | 178 |
| Al | 250 |
| Cu | 390 |
| Ag | 420 |
| Metal alloys | |
| Stainless steel | 12–16 |
| 55% Cu–45% Ni | 19.5 |
| 70% Ni–30% Cu | 25 |
| 1080 steel | 50 |
| Bronze (95% Cu–5% Sn) | 80 |
| Brass (63% Cu–37% Zn) | 125 |
| Dural (95% Al–4% Cu–1% Mg) | 147 |
| Ceramics and glasses | |
| Glass-borosilicate | 0.75 |
| Silica-fused ($SiO_2$) | 1.5 |
| $S_3N_4$ | 20 |
| Alumina ($Al_2O_3$) | 30 |
| Sapphire ($Al_2O_3$) | 37 |
| Beryllium (BeO) | 260 |
| Diamond | ~1000 |
| Polymers | |
| Polypropylene | 0.12 |
| PVC | 0.17 |
| Polycarbonate | 0.22 |
| Nylon 6,6 | 0.24 |
| Teflon | 0.25 |
| Polyethylene, low density | 0.3 |
| Polyethylene, high density | 0.5 |

summarizes $\kappa$ at room temperature for various classes of materials. Notice how ceramics have a very large range of $\kappa$ values.

**EXAMPLE 2.19**

**THERMAL CONDUCTIVITY** A 95/5 (95% Cu–5% Sn) bronze bearing made of powdered metal contains 15% (vol.%) porosity. Calculate its thermal conductivity at 300 K, given that the electrical conductivity of 95/5 bronze is $10^7\ \Omega^{-1}\ \text{m}^{-1}$.

**SOLUTION**

Recall that in Example 2.14, we found the electrical resistivity of the same bronze by using the mixture rule in Equation 2.26 in Section 2.4. We can use the same mixture rule again here, but we need the thermal conductivity of 95/5 bronze. From $\kappa/\sigma T = C_{\text{WFL}}$, we have

$$\kappa = \sigma T C_{\text{WFL}} = (1 \times 10^7)(300)(2.44 \times 10^{-8}) = 73.2\ \text{W m}^{-1}\ \text{K}^{-1}$$

Thus, the effective thermal conductivity is

$$\frac{1}{\kappa_{\text{eff}}} = \frac{1}{\kappa_c}\left[\frac{1+\frac{1}{2}\chi_d}{1-\chi_d}\right] = \frac{1}{(73.2\ \text{W m}^{-1}\ \text{K}^{-1})}\left[\frac{1+\frac{1}{2}(0.15)}{1-0.15}\right]$$

so that

$$\kappa_{\text{eff}} = 57.9\ \text{W m}^{-1}\ \text{K}^{-1}$$

## 2.6.2 Thermal Resistance

Consider a component of length $L$ that has a temperature difference $\Delta T$ between its ends as in Figure 2.24a. The temperature gradient is $\Delta T/L$. Thus, the rate of heat flow, or the **heat current,** is

$$Q' = A\kappa\frac{\Delta T}{L} = \frac{\Delta T}{(L/\kappa A)} \qquad [2.37]$$

*Fourier's law*

This should be compared with Ohm's law in electric circuits,

$$I = \frac{\Delta V}{R} = \frac{\Delta V}{(L/\sigma A)} \qquad [2.38]$$

*Ohm's law*

where $\Delta V$ is the voltage difference across a conductor of resistance $R$, and $I$ is the electric current.

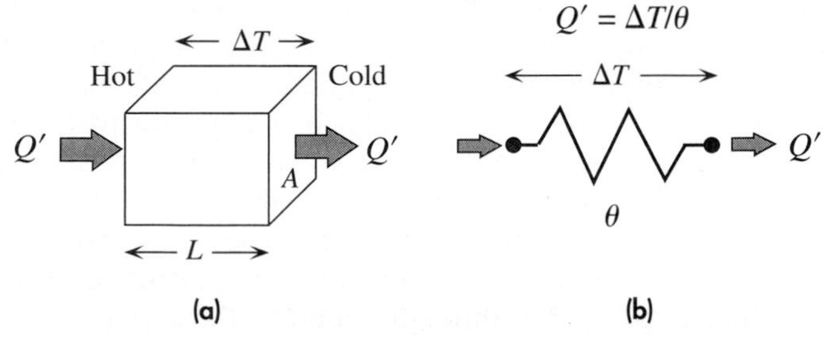


**Figure 2.24** Conduction of heat through a component in (a) can be modeled as a thermal resistance $\theta$ shown in (b) where $Q' = \Delta T/\theta$.

In analogy with electrical resistance, we may define **thermal resistance** $\theta$ by

*Definition of thermal resistance*

$$Q' = \frac{\Delta T}{\theta} \tag{2.39}$$

where, in terms of thermal conductivity,

*Thermal resistance*

$$\theta = \frac{L}{\kappa A} \tag{2.40}$$

The rate of heat flow $Q'$ and the temperature difference $\Delta T$ correspond to the electric current $I$ and potential difference $\Delta V$, respectively. Thermal resistance is the thermal analog of electrical resistance and its thermal circuit representation is shown in Figure 2.24b.

**EXAMPLE 2.20** **THERMAL RESISTANCE** A brass disk of electrical resistivity 50 nΩ m conducts heat from a heat source to a heat sink at a rate of 10 W. If its diameter is 20 mm and its thickness is 30 mm, what is the temperature drop across the disk, neglecting the heat losses from the surface?

**SOLUTION**

We first determine the thermal conductivity:

$$\kappa = \sigma T C_{\text{WFL}} = (5 \times 10^{-8}\ \Omega\ \text{m})^{-1}(300\ \text{K})(2.44 \times 10^{-8}\ \text{W}\ \Omega\ \text{K}^{-2})$$
$$= 146\ \text{W m}^{-1}\ \text{K}^{-1}$$

The thermal resistance is

$$\theta = \frac{L}{\kappa A} = \frac{(30 \times 10^{-3}\ \text{m})}{\pi(10 \times 10^{-3}\ \text{m})^2(146\ \text{W m}^{-1}\ \text{K}^{-1})} = 0.65\ \text{K W}^{-1}$$

Therefore, the temperature drop is

$$\Delta T = \theta Q' = (0.65\ \text{K W}^{-1})(10\ \text{W}) = 6.5\ \text{K or °C}$$

## 2.7 ELECTRICAL CONDUCTIVITY OF NONMETALS

All metals are good conductors because they have a very large number of conduction electrons free inside the metal. We should therefore expect solids that do not have metallic bonding to be very poor conductors, indeed insulators. Figure 2.25 shows the range of conductivities exhibited by a variety of solids. Based on typical values of the conductivity, it is possible to empirically classify various materials into conductors, semiconductors, and insulators as in Figure 2.25. It is apparent that nonmetals are not perfect insulators with zero conductivity. There is no well-defined sharp boundary between what we call insulators and semiconductors. Conductors are intimately identified with metals. It is more appropriate to view insulators as **high resistivity** (or **low conductivity**) **materials.** In general terms, current conduction is due to the drift of mobile charge carriers through a solid by the application of

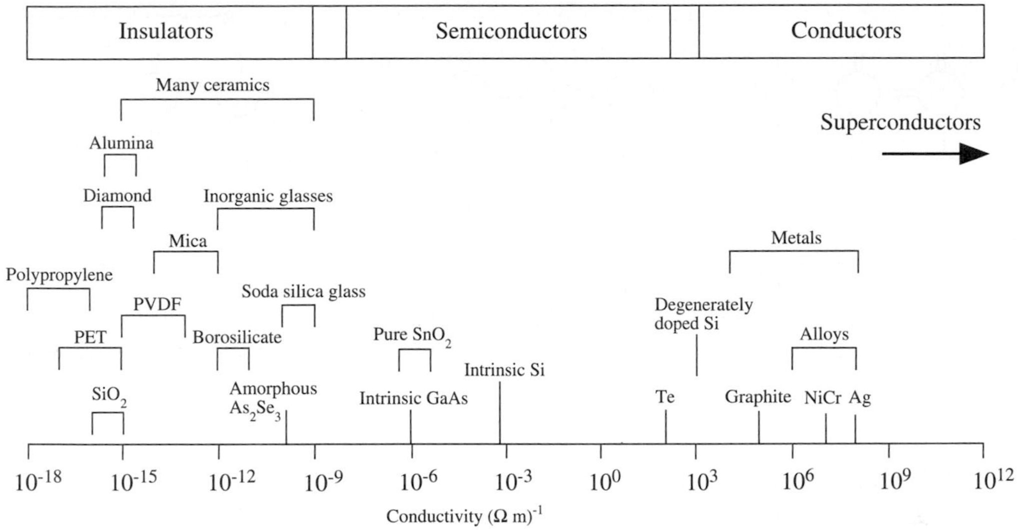


**Figure 2.25** Range of conductivities exhibited by various materials.

an electric field. Each of the drifting species of charge carriers contributes to the observed current. In metals, there are only free electrons. In nonmetals there are other types of charge carriers that can drift.

## 2.7.1 SEMICONDUCTORS

A perfect Si crystal has each Si atom bonded to four neighbors, and each covalent bond has two shared electrons as we had shown in Figure 1.59a. We know from classical physics (the kinetic molecular theory and Boltzmann distribution) that all the atoms in the crystal are executing vibrations with a distribution of energies. As the temperature increases, the distribution spreads to higher energies. Statistically some of the atomic vibrations will be sufficiently energetic to rupture a bond as indicated in Figure 2.26a. This releases an electron from the bond which is *free* to wander inside the crystal. The free electron can drift in the presence of an applied field; it is called a **conduction electron.** As an electron has been removed from a region of the crystal that is otherwise neutral, the broken-bond region has a *net positive charge*. This broken-bond region is called a **hole** ($h^+$). An electron in a neighboring bond can jump and repair this bond and thereby create a hole in its original site as shown in Figure 2.26b. Effectively, the hole has been displaced in the opposite direction to the electron jump by this *bond switching.* Holes can also wander in the crystal by the repetition of bond switching. When a field is applied, both holes and electrons contribute to electrical conduction as in Figure 2.26c. For all practical purposes, these holes behave as if they were *free* positively charged particles (independent of the original electrons) inside the crystal. In the presence of an applied field, holes drift along the field direction and contribute to conduction just as the free electrons

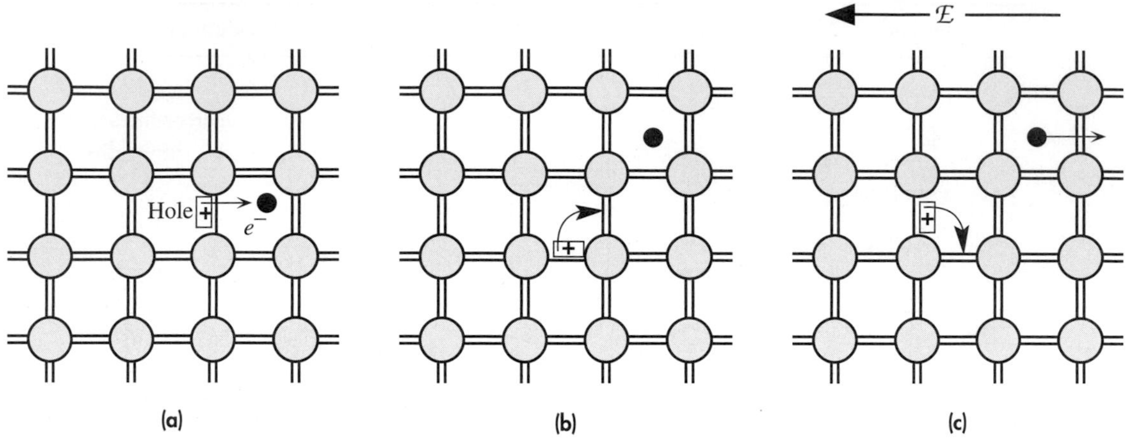


**Figure 2.26**

(a) Thermal vibrations of the atoms rupture a bond and release a free electron into the crystal. A hole is left in the broken bond, which has an effective positive charge.

(b) An electron in a neighboring bond can jump and repair this bond and thereby create a hole in its original site; the hole has been displaced.

(c) When a field is applied, both holes and electrons contribute to electrical conduction.

released from the broken bonds drift in the opposite direction and contribute to conduction.

It is also possible to create free electrons or holes by intentionally doping a semiconductor crystal, that is substituting impurity atoms for some of the Si atoms. Defects can also generate free carriers. The simplest example is nonstoichiometric ZnO that is shown in Figure 1.55b which has excess Zn. The electrons from the excess Zn are free to wander in the crystal and hence contribute to conduction.

Suppose that $n$ and $p$ are the concentrations of electrons and holes in a semiconductor crystal. If electrons and holes have drift mobilities of $\mu_e$ and $\mu_h$, respectively, then the overall conductivity of the crystal is given by

*Conductivity of a semiconductor*

$$\sigma = ep\mu_h + en\mu_e \qquad [2.41]$$

Unless a semiconductor has been heavily doped, the concentrations $n$ and $p$ are much smaller than the electron concentration in a metal. Even though carrier drift mobilities in most semiconductors are higher than electron drift mobilities in metals, semiconductors have much lower conductivities due to their lower concentration of free charge carriers.

**EXAMPLE 2.21** **HALL EFFECT IN SEMICONDUCTORS** The Hall effect in a sample where there are both negative and positive charge carriers, for example, electrons and holes in a semiconductor, involves not only the concentrations of electrons and holes, $n$ and $p$, respectively, but also the electron and hole drift mobilities, $\mu_e$ and $\mu_h$. We first have to reinterpret the relationship between the drift velocity and the electric field $\mathcal{E}$.

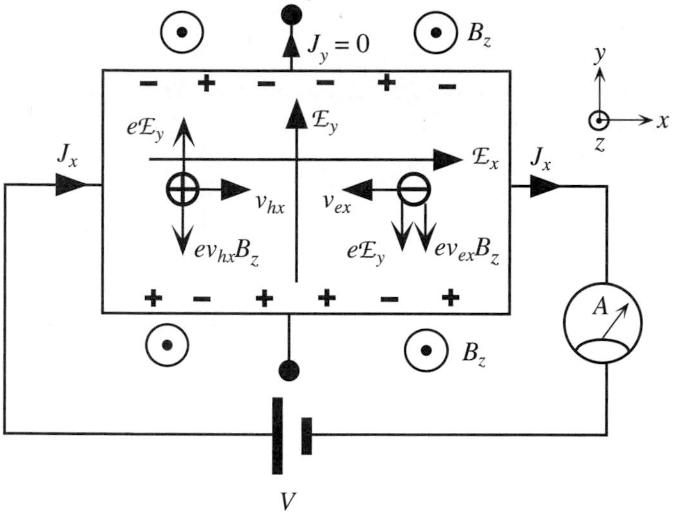


**Figure 2.27** Hall effect for ambipolar conduction as in a semiconductor where there are both electrons and holes.
The magnetic field $B_z$ is out from the plane of the paper. Both electrons and holes are deflected toward the bottom surface of the conductor and consequently the Hall voltage depends on the relative mobilities and concentrations of electrons and holes.

If $\mu_e$ is the drift mobility and $v_e$ is the drift velocity of the electrons, then we already know that $v_e = \mu_e\mathcal{E}$. This has been derived by considering the *net electrostatic force* $e\mathcal{E}$ acting on a single electron and the imparted acceleration $a = e\mathcal{E}/m_e$. The drift is therefore due to the net force $F_{\text{net}} = e\mathcal{E}$ experienced by a conduction electron. If we were to keep $e\mathcal{E}$ as the *net force* $F_{\text{net}}$ acting on a single electron, then we would have found

$$v_e = \frac{\mu_e}{e}F_{\text{net}} \qquad [2.42]$$

*Drift velocity and net force*

Equation 2.42 emphasizes the fact that drift is due to a net force $F_{\text{net}}$ acting on an electron. A similar expression would also apply to the drift of a hole in a semiconductor.

When both electrons and holes are present in a semiconductor sample, both charge carriers experience a Lorentz force in the same direction since they would be drifting in the opposite directions as illustrated in Figure 2.27. Thus, both holes and electrons tend to pile near the bottom surface. The magnitude of the Lorentz force, however, will be different since the drift mobilities and hence drift velocities will be different in general. Once equilibrium is reached, there should be no current flowing in the $y$ direction as we have an open circuit. Suppose that more holes have accumulated near the bottom surface so there is a built-in electric field $\mathcal{E}_y$ along $y$ as shown in Figure 2.27. Suppose that $v_{ey}$ and $v_{hy}$ are the *usual* electron and hole drift velocities in the $-y$ and $+y$ directions, respectively, as if the electric field $\mathcal{E}_y$ existed alone in the $+y$ direction. The net current along $y$ is zero, which means that

$$J_y = J_h + J_e = epv_{hy} + env_{ey} = 0 \qquad [2.43]$$

From Equation 2.43 we obtain

$$pv_{hy} = -nv_{ey} \qquad [2.44]$$

We note that either the electron or the hole drift velocity must be reversed from its usual direction; for example, holes drifting in the opposite directon to $\mathcal{E}_y$. The net force acting on the charge carriers cannot be zero. This is impossible when two types of carriers are involved and both carriers are drifting along $y$ to give a net current $J_y$ that is zero. This is what Equation 2.43 represents. We therefore conclude that, along $y$, both the electron and the hole must experience a driving force to drift them. The net force experienced by the carriers, as shown in Figure 2.27, is

$$F_{hy} = e\mathcal{E}_y - ev_{hx}B_z \qquad \text{and} \qquad -F_{ey} = e\mathcal{E}_y + ev_{ex}B_z \tag{2.45}$$

where $v_{hx}$ and $v_{ex}$ are the hole and electron drift velocities, respectively, along $x$. In general, the drift velocity is determined by the net force acting on a charge carrier; that is, from Equation 2.42

$$F_{hy} = \frac{ev_{hy}}{\mu_h} \qquad \text{and} \qquad -F_{ey} = \frac{ev_{ey}}{\mu_e}$$

so that Equation 2.45 becomes,

$$\frac{ev_{hy}}{\mu_h} = e\mathcal{E}_y - ev_{hx}B_z \qquad \text{and} \qquad \frac{ev_{ey}}{\mu_e} = e\mathcal{E}_y + ev_{ex}B_z$$

where $v_{hy}$ and $v_{ey}$ are the hole and electron drift velocities along $y$. Substituting $v_{hx} = \mu_h\mathcal{E}_x$ and $v_{ex} = \mu_e\mathcal{E}_x$, these become

$$\frac{v_{hy}}{\mu_h} = \mathcal{E}_y - \mu_h\mathcal{E}_x B_z \qquad \text{and} \qquad \frac{v_{ey}}{\mu_e} = \mathcal{E}_y + \mu_e\mathcal{E}_x B_z \tag{2.46}$$

From Equation 2.46 we can substitute for $v_{hy}$ and $v_{ey}$ in Equation 2.44 to obtain

$$p\mu_h\mathcal{E}_y - p\mu_h^2\mathcal{E}_x B_z = -n\mu_e\mathcal{E}_y - n\mu_e^2\mathcal{E}_x B_z$$

or

$$\mathcal{E}_y(p\mu_h + n\mu_e) = B_z\mathcal{E}_x\left(p\mu_h^2 - n\mu_e^2\right) \tag{2.47}$$

We now consider what happens along the $x$ direction. The total current density is finite and is given by the usual expression,

*Current density along x*

$$J_x = epv_{hx} + env_{ex} = (p\mu_h + n\mu_e)e\mathcal{E}_x \tag{2.48}$$

We can use Equation 2.48 to substitute for $\mathcal{E}_x$ in Equation 2.47, to obtain

$$e\mathcal{E}_y(n\mu_e + p\mu_h)^2 = B_z J_x\left(p\mu_h^2 - n\mu_e^2\right)$$

The Hall coefficient, by definition, is $R_H = \mathcal{E}_y/J_x B_z$, so

*Hall effect for ambipolar conduction*

$$R_H = \frac{p\mu_h^2 - n\mu_e^2}{e(p\mu_h + n\mu_e)^2} \tag{2.49}$$

or

*Hall effect for ambipolar conduction*

$$R_H = \frac{p - nb^2}{e(p + nb)^2} \tag{2.50}$$

where $b = \mu_e/\mu_h$. It is clear that the Hall coefficient depends on both the drift mobility ratio and the concentrations of holes and electrons. For $p > nb^2$, $R_H$ will be positive and for $p < nb^2$, it will be negative. We should note that when only one type of carrier is involved, for example, electrons only, the $J_y = 0$ requirement means that $J_y = env_{ey} = 0$, or $v_{ey} = 0$. The drift velocity along $y$ can only be zero, if the net driving force $F_{ey}$ along $y$ is zero. This occurs when $e\mathcal{E}_y - ev_{ex}B_z = 0$, that is, when the Lorentz force just balances the force due to the built-in field.

**EXAMPLE 2.22** **HALL COEFFICIENT OF INTRINSIC SILICON** At room temperature, a pure silicon crystal (called **intrinsic silicon**) has electron and hole concentrations $n = p = n_i = 1.0 \times 10^{10}$ cm$^{-3}$, and electron and hole drift mobilities $\mu_e = 1350$ cm$^2$ V$^{-1}$ s$^{-1}$ and $\mu_h = 450$ cm$^2$ V$^{-1}$ s$^{-1}$. Calculate the Hall coefficient and compare it with a typical metal.

**SOLUTION**

Given $n = p = n_i = 1.0 \times 10^{10}$ cm$^{-3}$, $\mu_e = 1350$ cm$^2$ V$^{-1}$ s$^{-1}$, and $\mu_h = 450$ cm$^2$ V$^{-1}$ s$^{-1}$, we have

$$b = \frac{\mu_e}{\mu_h} = \frac{1350}{450} = 3$$

Then from Equation 2.50,

$$R_H = \frac{(1.0 \times 10^{16}\ \text{m}^{-3}) - (1.0 \times 10^{16}\ \text{m}^{-3})(3)^2}{(1.6 \times 10^{-19}\ \text{C})[(1.0 \times 10^{16}\ \text{m}^{-3}) + (1.0 \times 10^{16}\ \text{m}^{-3})(3)]^2}$$
$$= -312\ \text{m}^3\ \text{A}^{-1}\ \text{s}^{-1}$$

which is orders of magnitude larger than that for a typical metal. All Hall-effect devices use a semiconductor rather than a metal sample.

## 2.7.2 IONIC CRYSTALS AND GLASSES

Figure 2.28a shows how crystal defects in an ionic crystal lead to mobile charges that can contribute to the conduction process. All crystalline solids possess vacancies and interstitial atoms as a requirement of thermal equilibrium. Many solids have interstitial impurities which are often ionized or charged. These interstitial ions can jump, *i.e.*, diffuse, from one interstitial site to another and hence drift by diffusion in the presence of a field. A positive ion at an interstitial site such as that shown in Figure 2.28a always prefers to jump into a neighboring interstitial site along the direction of the field because it experiences an effective force in this direction. When an ion with

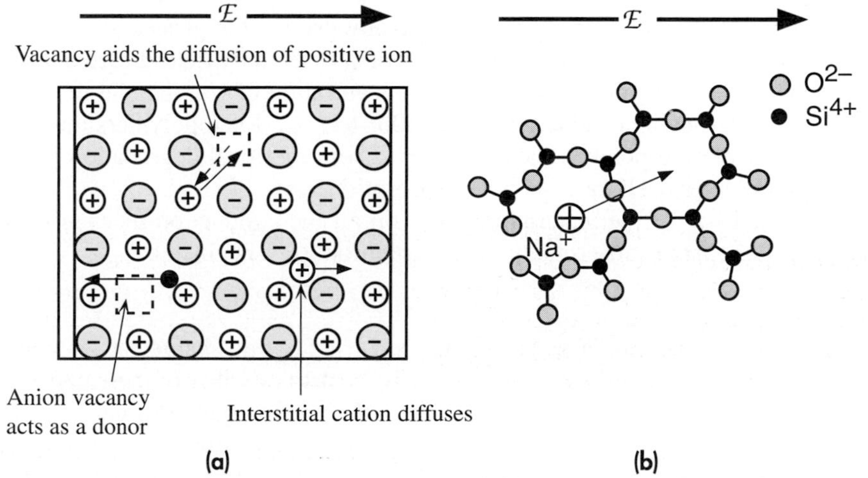


**Figure 2.28** Possible contributions to the conductivity of ceramic and glass insulators.
(a) Possible mobile charges in a ceramic.
(b) An $Na^+$ ion in the glass structure diffuses and therefore drifts in the direction of the field.

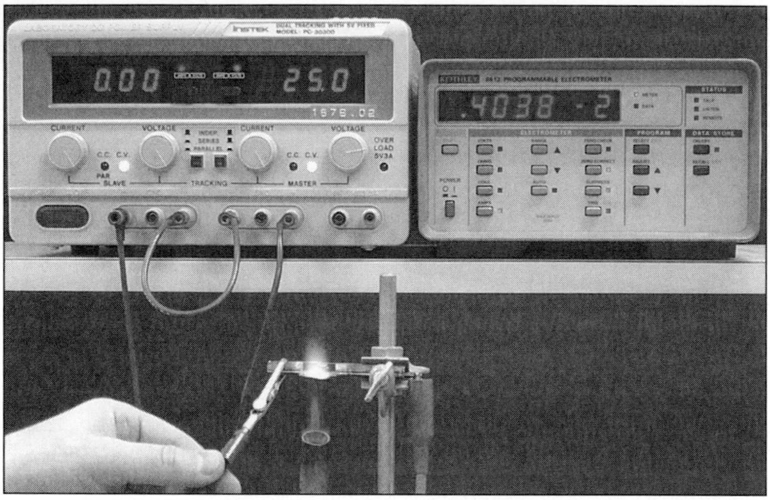


This soda glass rod when heated under a torch becomes electrically conducting. It passes 4 mA when the voltage is 50 V (2 × 25 V); a resistance of 12.5 kΩ! Ordinary soda glass at room temperature is an insulator but can be quite conducting at sufficiently high temperatures.

charge $q_{\text{ion}}$ jumps a distance $d$ along the field, its potential energy decreases by $q_{\text{ion}}\mathcal{E}d$. If it tries to jump in the opposite direction, it has to do work $q_{\text{ion}}\mathcal{E}d$ against the force of the field.

Deviations from stoichiometry in compound solids often lead to the generation of mobile electrons (or holes) and point defects such as vacancies. Therefore, there are electrons, holes, and various mobile ions available for conduction under an applied field as depicted in Figure 2.28a. Many glasses and polymers contain a certain concentration of mobile ions in the structure. An example of a $Na^+$ ion in silica glass is shown in Figure 2.28b. Aided by the field, the $Na^+$ can jump from one interstice to a neighboring interstice along the field and thereby drift in the glass and contribute to current conduction. The conduction process is then essentially field-directed diffusion. Ordinary window glass, in fact, has a high concentration of $Na^+$ ions in the structure and becomes reasonably conducting above 300–400 °C. Some polymers may contain ions derived from the polymerization process, from the local degradation (dissociation) of the polymer itself, or from water absorption.

Conductivity $\sigma$ of the material depends on all the conduction mechanisms with each species of charge carrier making a contribution, so it is given by

*General conductivity*

$$\sigma = \sum q_i n_i \mu_i \qquad [2.51]$$

where $n_i$ is the concentration, $q_i$ is the charge carried by the charge carrier species of type $i$ (for electrons and holes $q_i = e$), and $\mu_i$ is the drift mobility of these carriers. The dominant conduction mechanism in Equation 2.51 is often quite difficult to uniquely identify. Further, it may change with temperature, composition, and ambient conditions such as the air pressure as in some oxide ceramics. For many insulators, whether ceramic, glass, or polymer, it has been found that, in the majority of cases, the conductivity follows an exponential or Arrhenius-type temperature dependence so that $\sigma$ is

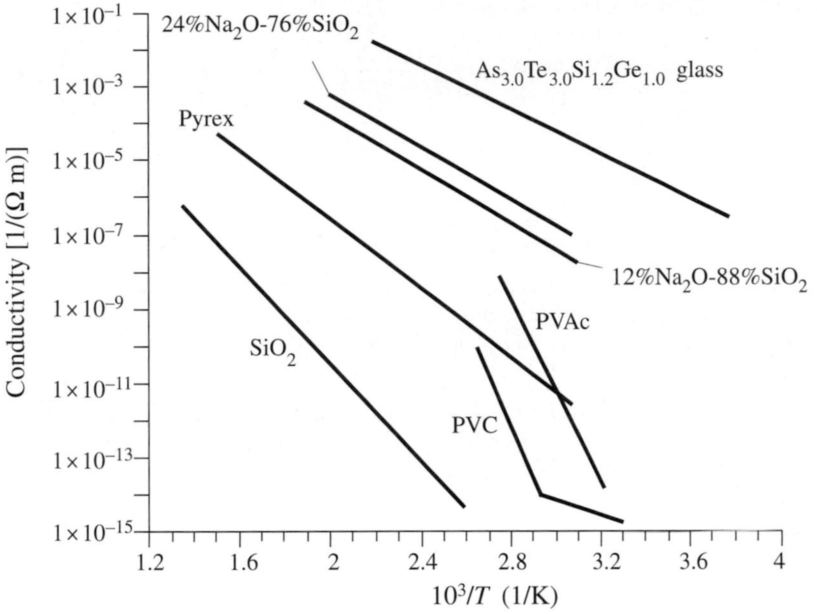


**Figure 2.29** Conductivity versus reciprocal temperature for various low-conductivity solids.

| SOURCE: Data selectively combined from numerous sources.

**thermally activated,**

$$\sigma = \sigma_o \exp\left(-\frac{E_\sigma}{kT}\right) \qquad [2.52]$$

*Temperature dependence of conductivity*

where $E_\sigma$ is the **activation energy for conductivity.**

Figure 2.29 shows examples of the temperature dependence of conductivity for various high-resistivity solids: oxide ceramics, glasses, and polymers. When Equation 2.51 is plotted as $\log(\sigma)$ versus $1/T$, the result is a straight line with a negative slope that indicates the activation energy $E_\sigma$. Equation 2.52 is useful in predicting the conductivity at different temperatures and evaluating the temperature stability of the insulator.

**EXAMPLE 2.23**

**CONDUCTIVITY OF A SODA–SILICATE GLASS** Figure 2.29 shows the temperature dependence of 12% $Na_2O$–88% $SiO_2$, soda–silicate glass which has 12 mol% $Na_2O$ and 88 mol% $SiO_2$. Calculate the activation energy of conductivity and compare this with the activation energy for the diffusion of $Na^+$ ions in the soda–silicate glass structure which is in the range 0.65–0.75 eV.

**SOLUTION**

According to Equation 2.52 when $\ln(\sigma)$ is plotted against $1/T$, the slope should be $-E_\sigma/k$. If the conductivity at temperatures $T_1$ and $T_2$ are $\sigma_1$ and $\sigma_2$, respectively, then the slope of the straight line for 12% $Na_2O$–88% $SiO_2$ in Figure 2.29 is

$$\text{Slope} = \frac{\ln(\sigma_2/\sigma_1)}{(1/T_2 - 1/T_1)} = -\frac{E_\sigma}{k}$$

Taking $\sigma_1 = 10^{-4}\ \Omega^{-1}\ \text{m}^{-1}$ and $\sigma_2 = 10^{-6}\ \Omega^{-1}\ \text{m}^{-1}$ in Figure 2.29, we find $1/T_1 = 0.00205$ and $1/T_2 = 0.00261$. Then, $E_\sigma$ as eV is

$$E_\sigma = -\frac{\ln(\sigma_2/\sigma_1)}{(1/T_2 - 1/T_1)}\frac{k}{e} = -\frac{\ln(10^{-6}/10^{-4})}{(0.00261 - 0.00205)}\frac{1.38 \times 10^{-23}}{1.602 \times 10^{-19}} = 0.71\ \text{eV}$$

A similar calculation for the 24% $Na_2O$–76% $SiO_2$ gives an activation energy of 0.69 eV.

Both of these activation energies are comparable with the activation energy for the diffusion of $Na^+$ ions in the structure. Thus, $Na^+$ diffusion is responsible for the conductivity.

**EXAMPLE 2.24**

**DRIFT MOBILITY DUE TO IONIC CONDUCTION** The soda–silicate glass of composition 20% $Na_2O$–80% $SiO_2$ and density of approximately 2.4 g cm$^{-3}$ has a conductivity of $8.25 \times 10^{-6}\ \Omega^{-1}\ \text{m}^{-1}$ at 150 °C. If conduction occurs by the diffusion of $Na^+$ ions, what is their drift mobility?

**SOLUTION**

We can calculate the drift mobility $\mu_i$ of the $Na^+$ ions from the conductivity expression $\sigma = q_i n_i \mu_i$ where $q_i$ is the charge of the ion $Na^+$, so that it is $+e$, and $n_i$ is the concentration of $Na^+$ ions in the structure. For simplicity we can take the glass to be made of $(Na_2O)_{0.2}(SiO_2)_{0.8}$ units. The atomic masses of Na, O, and Si are 23, 16, and 28.1, respectively. The atomic mass of $(Na_2O)_{0.2}(SiO_2)_{0.8}$ is

$$M_{\text{at}} = 0.2[2(23) + 1(16)] + 0.8[1(28.1) + 2(16)]$$
$$= 60.48\ \text{g mol}^{-1}\ \text{of}\ (Na_2O)_{0.2}(SiO_2)_{0.8}$$

The number of $(Na_2O)_{0.2}(SiO_2)_{0.8}$ units per unit volume can be found from the density $d$ by

$$n = \frac{dN_A}{M_{\text{at}}} = \frac{(2.4 \times 10^3\ \text{kg m}^{-3})(6.02 \times 10^{23}\ \text{mol}^{-1})}{(10^{-3}\ \text{kg/g})(60.48\ \text{g mol}^{-1})}$$
$$= 2.39 \times 10^{28}\ (Na_2O)_{0.2}(SiO_2)_{0.8}\ \text{units m}^{-3}$$

The concentration $n_i$ of $Na^+$ ions is the concentration of Na atoms as each would be ionized. Then $n_i$ can be expressed as $n_i = n_{\text{Na}} =$ [atomic fraction of Na in $(Na_2O)_{0.2}(SiO_2)_{0.8}$] $\times\ n$.

$$n_i = \left[\frac{0.2(2)}{0.2(2+1) + 0.8(1+2)}\right](2.39 \times 10^{28}\ \text{m}^{-3}) = 3.186 \times 10^{27}\ \text{m}^{-3}$$

and

$$\mu_i = \frac{\sigma}{en_i} = \frac{(8.25 \times 10^{-6}\ \Omega^{-1}\ \text{m}^{-1})}{(1.60 \times 10^{-19}\ \text{C})(3.186 \times 10^{27}\ \text{m}^{-3})} = 1.62 \times 10^{-14}\ \text{m}^2\ \text{V}^{-1}\ \text{s}^{-1}$$

This is an extremely small drift mobility, by orders of magnitude, compared with the typical electron drift mobility in metals and semiconductors. The reason is that the drift involves the $Na^+$ ion jumping from one site to another by a diffusion process. This diffusion requires overcoming a potential energy barrier, typically 0.5 to 1 eV, which limits drastically the rate of diffusion by virtue of the Boltzmann factor.

# ADDITIONAL TOPICS

## 2.8 SKIN EFFECT: HF RESISTANCE OF A CONDUCTOR

Consider the cylindrical conductor shown in Figure 2.30a, which is carrying a current $I$ into the paper ($\times$). The magnetic field $B$ of $I$ is clockwise. Consider two magnetic field values $B_1$ and $B_2$, which are shown in Figure 2.30a. $B_1$ is inside the core and $B_2$ is just outside the conductor.

Assume that the conductor is divided into two conductors. The hypothetical cut is taken just outside of $B_1$. The conductor in Figure 2.30a is now cut into a hollow cylinder and a smaller solid cylinder, as shown in Figure 2.30b and c, respectively. The currents $I_1$ and $I_2$ in the solid and hollow cylinders sum to $I$. We can arrange things and choose $B_1$ such that our cut gives $I_1 = I_2 = \frac{1}{2}I$. Obviously, $I_1$ flowing in the inner conductor is threaded (or linked) by both $B_1$ and $B_2$. (Remember that $B_1$ is just inside the conductor in Figure 2.30b, so it threads at least 99% of $I_1$.) On the other hand, the outer conductor is only threaded by $B_2$, simply because $I_2$ flows in the hollow cylinder and there is no current in the hollow, which means that $B_1$ is not threaded by $I_2$. Clearly, $I_1$ threads more magnetic field than $I_2$ and thus conductor (c) has a higher inductance than (b). Recall that **inductance** is defined as the *total magnetic flux threaded per unit current*. Consequently, an ac current will prefer paths near the surface where the inductivc impedance is smaller. As the frequency increases, the current is confined more and more to the surface region.

For a given conductor, we can assume that most of the current flows in a surface region of depth $\delta$, called the **skin depth,** as indicated in Figure 2.31. In the central region,

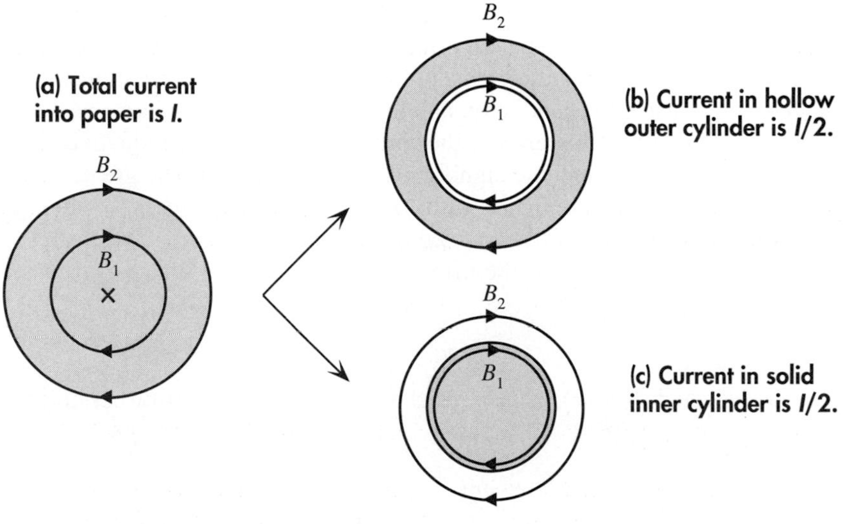


**Figure 2.30** Illustration of the skin effect.
A hypothetical cut produces a hollow outer cylinder and a solid inner cylinder. Cut is placed where it would give equal current in each section. The two sections are in parallel so that the currents in (b) and (c) sum to that in (a).

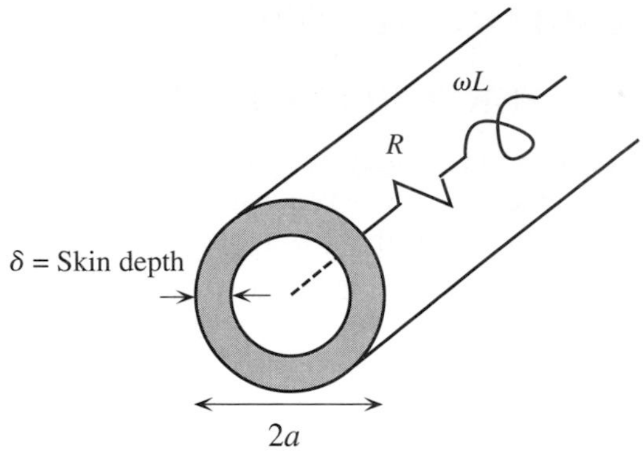


**Figure 2.31** At high frequencies, the core region exhibits more inductive impedance than the surface region, and the current flows in the surface region of a conductor defined approximately by the skin depth, $\delta$.

the current will be negligibly small. The skin depth will obviously depend on the frequency $\omega$. To find $\delta$, we must solve Maxwell's equations in a conductive medium, a tedious task that, fortunately, has been done by others. We can therefore simply take the result that the skin depth $\delta$ is given by

*Skin depth for conduction*

$$\delta = \frac{1}{\sqrt{\frac{1}{2}\omega\sigma\mu}} \tag{2.53}$$

where $\omega$ is the angular frequency of the current, $\sigma$ is the conductivity ($\sigma$ is constant from dc up to $\sim 10^{14}$ Hz in metals), and $\mu$ is the magnetic permeability of the medium, which is the product of the absolute (free space) permeability $\mu_o$ and the relative permeability $\mu_r$.

We can imagine the central conductor as a resistance $R$ in series with an inductance $L$. Intuitively, those factors that enhance the inductive impedance $\omega L$ over the resistance $R$ will also tend to emphasize the skin effect and will hence tend to decrease the skin depth. For example, the greater the permeability of the conducting medium, the stronger the magnetic field inside the conductor, and hence the larger the inductance of the central region. The higher the frequency of the current, the greater the inductive impedance $\omega L$ compared with $R$ and the more significant is the skin effect. The greater is the conductivity $\sigma$, the smaller is $R$ compared with $\omega L$ and hence the more important is the skin effect. All these dependences are accounted for in Equation 2.53.

With the skin depth known, the effective cross-sectional area is given approximately by

$$A = \pi a^2 - \pi(a-\delta)^2 \approx 2\pi a\delta$$

where $\delta^2$ is neglected ($\delta \ll a$). The ac resistance $r_{ac}$ of the conductor per unit length is therefore

*HF resistance per unit length due to skin effect*

$$r_{ac} = \frac{\rho}{A} \approx \frac{\rho}{2\pi a\delta} \tag{2.54}$$

where $\rho$ is the ac resistivity at the frequency of interest, which for all practical purposes is equal to the dc resistivity of the metal. Equation 2.54 clearly shows that as $\omega$ increases, $\delta$ decreases, by virtue of $\delta \propto \omega^{-1/2}$ and, as a result, $r_{ac}$ increases.

From this discussion, it is obvious that the skin effect arises because the magnetic field of the ac current in the conductor restricts the current flow to the surface region within a depth of $\delta < a$. Since the current can only flow in the surface region, there is an effective increase in the resistance due to a decrease in the cross-sectional area for current flow. Taking this effective area for current flow as $2\pi a\delta$ leads to Equation 2.54.

The skin effect plays an important role in electronic engineering because it limits the use of solid-core conductors in high-frequency applications. As the signal frequencies reach and surpass the gigahertz ($10^9$ Hz) range, the transmission of the signal over a long distance becomes almost impossible through an ordinary, solid-metal conductor. We must then resort to pipes (or waveguides).

**EXAMPLE 2.25**

**SKIN EFFECT FROM DIMENSIONAL ANALYSIS** Using dimensional analysis, obtain the general form of the equation for the skin depth $\delta$ in terms of the angular frequency of the current $\omega$, conductivity $\sigma$, and permeability $\mu$.

**SOLUTION**

The skin effect depends on the angular frequency $\omega$ of the current, the conductivity $\sigma$, and the magnetic permeability $\mu$ of the conducting medium. In the most general way, we can group these effects as

$$[\delta] = [\omega]^x [\sigma]^y [\mu]^z$$

where the indices $x$, $y$, and $z$ are to be determined. We then substitute the dimensions of each quantity in this expression. The dimensions of each, in terms of the fundamental units, are as follows:

| Quantity | Units | Fundamental Units | Comment |
|---|---|---|---|
| $\delta$ | m | m | |
| $\omega$ | $s^{-1}$ | $s^{-1}$ | |
| $\sigma$ | $\Omega^{-1}\ m^{-1}$ | $C^2\ s\ kg^{-1}\ m^{-3}$ | $\Omega = V\ A^{-1} = (J\ C^{-1})(C\ s^{-1})^{-1}$ $= N\ m\ s\ C^{-2} = (kg\ m\ s^{-2})(m\ s\ C^{-2})$ |
| $\mu$ | $Wb\ A^{-1}\ m^{-1}$ | $kg\ m\ C^{-2}$ | $Wb = T\ m^2 = (N\ A^{-1}\ m^{-1})(m^2)$ $= (kg\ m\ s^{-2})(C^{-1}\ s)(m)$ |

Therefore,

$$[m] = [s^{-1}]^x [C^2\ s\ kg^{-1}\ m^{-3}]^y [kg\ m\ C^{-2}]^z$$

Matching the dimensions of both sides, we see that $y = z$; otherwise C and kg do not cancel.

| | | | |
|---|---|---|---|
| For m | $1 = -3y + z$ | | |
| For s | $0 = -x + y$ | | |
| For C or kg | $0 = 2y - 2z$ | or | $0 = -y + z$ |

Clearly, $x = y = z = -\frac{1}{2}$ is the only possibility. Then, $\delta \propto [\omega\sigma\mu]^{-1/2}$. It should be reemphasized that the dimensional analysis is not a proof of the skin depth expression, but a consistency check that assures confidence in the equation.

**EXAMPLE 2.26** **SKIN EFFECT IN AN INDUCTOR** What is the change in the dc resistance of a copper wire of radius 1 mm for an ac signal at 10 MHz? What is the change in the dc resistance at 1 GHz? Copper has $\rho_{dc} = 1.70 \times 10^{-8}$ Ω m or $\sigma_{dc} = 5.9 \times 10^7$ Ω$^{-1}$ m$^{-1}$ and a relative permeability near unity.

**SOLUTION**

Per unit length, $r_{dc} = \rho_{dc}/\pi a^2$ and at high frequencies, from Equation 2.54, $r_{ac} = \rho_{dc}/2\pi a\delta$. Therefore, $r_{ac}/r_{dc} = a/2\delta$.

We need to find $\delta$. From Equation 2.53, at 10 MHz we have

$$\delta = \left[\tfrac{1}{2}\omega\sigma_{dc}\mu\right]^{-1/2} = \left[\tfrac{1}{2} \times 2\pi \times 10 \times 10^6 \times 5.9 \times 10^7 \times 1.257 \times 10^{-6}\right]^{-1/2}$$
$$= 2.07 \times 10^{-5} \text{ m} = 20.7\ \mu\text{m}$$

Thus

$$\frac{r_{ac}}{r_{dc}} = \frac{a}{2\delta} = \frac{(10^{-3} \text{ m})}{(2 \times 2.07 \times 10^{-5} \text{ m})} = 24.13$$

The resistance has increased by 24 times. At 1 GHz, the increase is 240 times. Furthermore, the current is confined to a surface region of about $\sim 2 \times 10^{-5}$(20 μm) at 10 MHz and $\sim 2 \times 10^{-6}$ m (2 μm) at 1 GHz, so most of the material is wasted. This is exactly the reason why solid conductors would not be used for high-frequency work. As very high frequencies, in the gigahertz range and above, are reached, the best bet would be to use pipes (waveguides).

One final comment is appropriate. An inductor wound from a copper wire would have a certain $Q$ (quality factor) value[9] that depends inversely on its resistance. At high frequencies, $Q$ would drop, because the current would be limited to the surface of the wire. One way to overcome this problem is to use a thick conductor that has a surface coating of higher-conductivity metal, such as silver. This is what the early radio engineers practiced. In fact, tank circuits of high-power radio transmitters often have coils made from copper tubes with a coolant flowing inside.

## 2.9 THIN METAL FILMS

### 2.9.1 CONDUCTION IN THIN METAL FILMS

The resistivity of a material, as listed in materials tables and in our analysis of conduction, refers to the resistivity of the material in bulk form; that is, any dimension of the specimen is much larger than the mean free path for electron scattering. In such cases resistivity is determined by scattering from lattice vibrations and, if significant, scattering from various impurities and defects in the crystal. In certain applications,

[9] The $Q$ value refers to the quality factor of an inductor, which is defined by $Q = \omega_o L/R$, where $\omega_o$ is the resonant frequency, $L$ is the inductance, and $R$ is the resistance due to the losses in the inductor.

notably microelectronics, metal films are widely used to provide electrical conduction paths to and from the semiconductor devices. Various methods are used to deposit thin films. In many applications, the metal film is simply deposited onto a substrate, such as a semiconductor or an insulator (*e.g.*, $SiO_2$), by **physical vapor deposition** (PVD), that is, by **vacuum deposition,** which typically involves either evaporation or sputtering. In **thermal evaporation,** the metal is evaporated from a heated source in a vacuum chamber as depicted in Figure 1.74. As the metal atoms, evaporated from the source, impinge and adhere to the semiconductor surface, they form a metal film which is often highly polycrystalline. Stated differently, the metal atoms in the vapor condense to form a metal film on a suitably placed substrate. In **electron beam deposition,** an energetic electron beam is used to melt and evaporate the metal. **Sputtering** is a vacuum deposition process that involves bombarding a metal target material with energetic Ar ions, which dislodges the metal atoms and then condenses them onto a substrate. The use of sputtering is quite common in microelectronic fabrication. Copper metal interconnect films used in microelectronics are usually grown by **electrodeposition,** that is, using electroplating, an electrochemical process, to deposit the metal film onto the required chip areas. In many applications, especially in microelectronics, we are interested in the resistivity of a metal film in which the thickness of the film or the average size of the grains is comparable to the mean distance between scattering events $\ell_{\text{bulk}}$ (the mean free path) in the bulk material. In such cases, the resistivity of the metal film is greater than the corresponding resistivity of the bulk crystal. A good example is the resistivity of interconnects and various metal films used in the "shrinking" world of microelectronics, in which more and more transistors are packed into a single Si crystal, and various device dimensions are scaled down.

### 2.9.2 RESISTIVITY OF THIN FILMS

***Polycrystalline Films and Grain Boundary Scattering*** In a highly polycrystalline sample the conduction electrons are more likely to be scattered by grain boundaries than by other processes as depicted in Figure 2.32a. Consider the resistivity due to scattering from grain boundaries alone as shown in Figure 2.32b. The conduction electron is free within a grain, but becomes scattered at the grain boundary. Its mean free path $\ell_{\text{grains}}$ is therefore roughly equal to the average grain size $d$. If $\lambda = \ell_{\text{crystal}}$ is

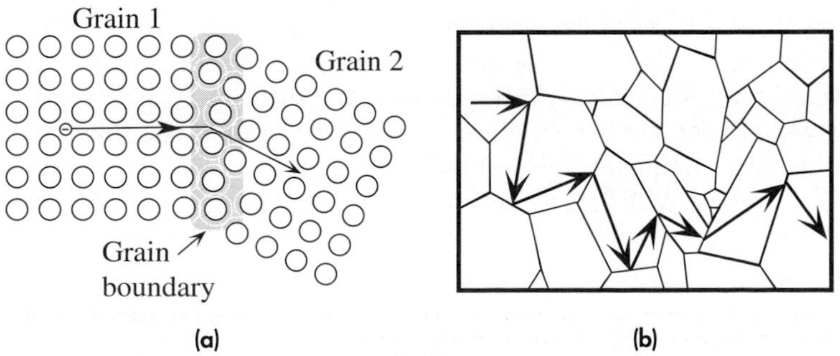


**Figure 2.32**
(a) Grain boundaries cause scattering of the electron and therefore add to the resistivity by Matthiessen's rule.
(b) For a very grainy solid, the electron is scattered from grain boundary to grain boundary and the mean free path is approximately equal to the mean grain diameter.

the mean free path of the conduction electrons in the *single crystal* (no grain boundaries), then

*Mean free path in polycrystalline sample*

$$\frac{1}{\ell} = \frac{1}{\ell_{\text{crystal}}} + \frac{1}{\ell_{\text{grains}}} = \frac{1}{\lambda} + \frac{1}{d} \quad [2.55]$$

The resistivity is inversely proportional to the mean free path which means that the resistivity of the bulk single crystal $\rho_{\text{crystal}} \propto 1/\lambda$ and the resistivity of the polycrystalline sample $\rho \propto 1/\ell$. Thus,

*Resistivity of a polycrystalline sample*

$$\frac{\rho}{\rho_{\text{crystal}}} = 1 + \left(\frac{\lambda}{d}\right) \quad [2.56]$$

Polycrystalline metal films with a smaller grain diameter $d$ (*i.e.*, more grainy films) will have a higher resistivity.

In a more rigorous theory we have to consider a number of effects. It may take more than one scattering at a grain boundary to totally randomize the velocity, so we need to calculate the effective mean free path that accounts for how many collisions are needed to randomize the velocity. There is a possibility that the electron may be totally reflected back at a grain boundary (bounce back). Suppose that the probability of reflection at a grain boundary is $R$. If $d$ is the average grain size (diameter), then the popular **Mayadas–Shatkez formula** is approximately given by[10]

*Resistivity due to grain boundary scattering*

$$\frac{\rho}{\rho_{\text{crystal}}} \approx 1 + 1.33\beta \quad [2.57a]$$

where

$$\beta = \frac{\lambda}{d}\left(\frac{R}{1-R}\right) \quad [2.57b]$$

Equation 2.57a is in the form of Matthiessen's rule and indicates that the grain boundary scattering contribution $\rho_{\text{grains}}$ to the overall resistivity is $(1.33\beta)\rho_{\text{crystal}}$. The approximate sign in Equation 2.57 implies that Matthiessen's rule is "approximately," though reasonably well, obeyed. For copper, typical $R$ values are 0.24 to 0.40, and $R$ is somewhat smaller for Al. Equation 2.57 for a Cu film with $R \approx 0.3$ predicts $\rho/\rho_{\text{crystal}} \approx 1.20$ for $d \approx 3\lambda$ or a grain size $d \approx 120$ nm since the bulk crystal $\lambda \approx 40$ nm.

***Surface Scattering*** Consider the scattering of electrons from the surfaces of a conducting film as in Figure 2.33. Take the film thickness as $D$. Assume that the scattering from the surface is *inelastic;* that is, the electron loses the gained velocity from the field. Put differently, the direction of the electron after the scattering process is *independent* of the direction before the scattering process. This type of scattering is called *nonspecular*. (If the electron is elastically reflected from the surface just like a rubber ball bouncing off a wall, then there is no increase in the resistivity.) It is unlikely that one surface scattering will completely randomize the electron's velocity. The mean free path $\ell_{\text{surf}}$ of the electron will depend on its direction right after the scattering

[10] This is obtained by expanding the original long expression about $\beta = 1$ to the first term. To two decimal places, the expansion is $1 + 1.33\beta$.

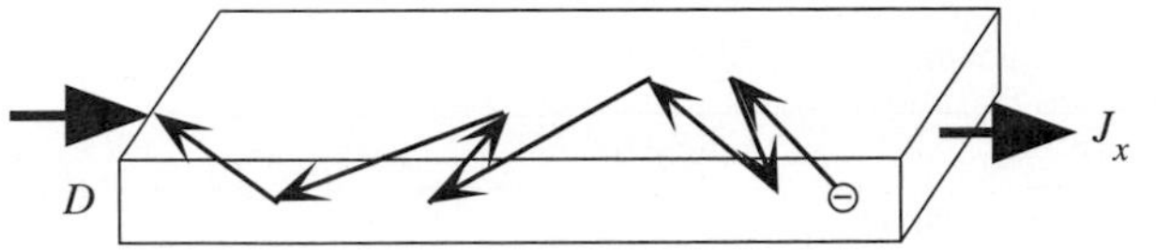


**Figure 2.33** Conduction in thin films may be controlled by scattering from the surfaces.

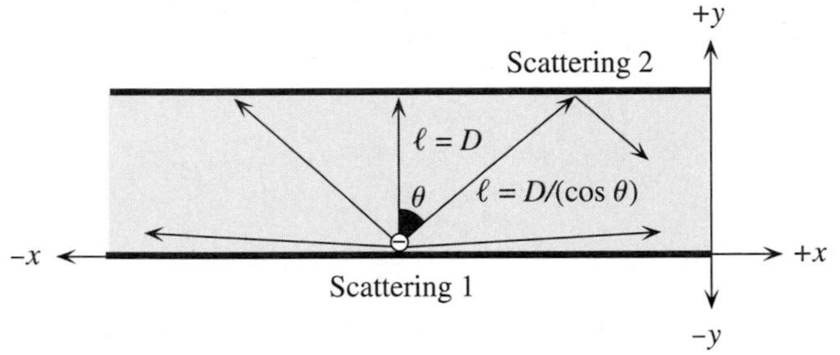


**Figure 2.34** The mean free path of the electron depends on the angle $\theta$ after scattering.

process as depicted in Figure 2.34. For example, if the angle $\theta$ after surface scattering is zero, (the electron moves transversely to the film length), then $\ell_{surf} = D$. In general, the mean free path $\ell_{surf}$ will be $D/(\cos\theta)$ as illustrated in Figure 2.34.

Consider the surface scattering example in Figure 2.34 where the electron is scattered from the bottom surface. If the scattering of the electron were truly random, then the probability of being scattered in a direction back into the film, that is, in the $+y$ direction, would be 0.5 on average. However, the electron's direction right after the surface scattering is not totally random because we know that the electron cannot leave the film; thus $\theta$ is between $-\pi/2$ and $+\pi/2$ and cannot be between $-\pi$ and $+\pi$. The electron's velocity after the first surface scattering must have a $y$ component along $+y$ and not along $-y$. The electron can only acquire a velocity component along $-y$ again after the second surface scattering as shown in Figure 2.34. It therefore takes two collisions to randomize the velocity, which means that the *effective mean free path* must be twice as long, that is $2D/\cos\theta$. To find the overall mean free path $\ell$ for calculating the resistivity we must use Matthiessen's rule. If $\lambda$ is the mean free path of the conduction electrons in the *bulk crystal* (no surface scattering), then

$$\frac{1}{\ell} = \frac{1}{\lambda} + \frac{1}{\ell_{surf}} = \frac{1}{\lambda} + \frac{\cos\theta}{2D} \qquad [2.58]$$

*Mean free path in a film*

We have to average for all possible $\theta$ values per scattering, that is, $\theta$ from $-\pi/2$ to $+\pi/2$. Once this is done we can relate $\ell$ to $\lambda$ as follows:

$$\frac{\lambda}{\ell} = 1 + \frac{\lambda}{\pi D}$$

*Averaged mean free path in a film*

The resistivity of the bulk crystal is $\rho_{bulk} \propto 1/\lambda$, and the resistivity of the film is $\rho \propto 1/\ell$. Thus,

$$\frac{\rho}{\rho_{bulk}} = 1 + \frac{1}{\pi}\left(\frac{\lambda}{D}\right) \qquad [2.59]$$

*Resistivity of a conducting thin film*

**Table 2.6** Resistivities of some thin Cu and Au films at room temperature

| Film | *D* (nm) | *d* (nm) | $\rho$ (n$\Omega$ m) | Comment |
|---|---|---|---|---|
| | | | **Cu films (Polycrystalline)** | |
| Cu on TiN, W, and TiW [1] | >250 | 186 | 21 | Chemical vapor deposition (CVD). |
| | | 45 | 32 | Substrate temperature 200 °C. $\rho$ depends on $d$ not $D$ = 250–900 nm. |
| Cu on 500 nm $SiO_2$ [2] | 20.5 | | 35 | Thermal evaporation. Substrate at RT. |
| | 37 | | 27 | |
| Cu on Si (100) [3] | 52 | | 38 | Sputtered Cu films. Annealing at 150 °C has no effect. $R \approx 0.40$ and $p \approx 0$. |
| | 100 | | 22 | |
| Cu on glass [4] | 40 | | 50 | As deposited |
| | 40 | | 29 | Annealed at 200 °C |
| | 40 | | 25 | Annealed at 250 °C |
| | | | | All thermal evaporated and PC. |
| | | | **Au films** | |
| Au epitaxial film on mica | 30 | | 25 | Single crystal on mica. $p \approx 0.8$. Specular scattering. |
| Au PC film on mica | 30 | | 54 | PC. Sputtered on mica. $p$ is small. |
| Au film on glass | 30 | | 70 | PC. Evaporated onto glass. $p$ is small. Nonspecular scattering. |
| Au on glass [5] | 40 | 8.5 | 92 | PC. Sputtered films. $R$ = 0.27–0.33. |
| | 40 | 3.8 | 189 | |

NOTE: PC-polycrystalline film, RT-room temperature, $D$ = film thickness, $d$ = average grain size. At RT for Cu, $\lambda$ = 38–40 nm, and for Au, $\lambda$ = 36–38 nm.

SOURCES: Data selectively combined from various sources, including [1] S. Riedel *et al., Microelec. Engin.* **33,** 165, 1997; [2] H. D. Liu *et al., Thin Solid Films,* **34,** 151, 2001; [3] J. W. Lim *et al., Appl. Surf. Sci.* **217,** 95, 2003. [4] R. Suri *et al., J. Appl. Phys.,* **46,** 2574, 1975; [5] R. H. Cornely and T. A. Ali, *J. Appl. Phys.,* **49,** 4094, 1978.

A more rigorous calculation modifies the numerical factor $1/\pi$ and also considers what fraction $p$ of surface collisions is specular and results in[11]

*Surface scattering resistivity*

$$\frac{\rho}{\rho_{\text{bulk}}} \approx 1 + \frac{3\lambda}{8D}(1-p) \qquad \frac{D}{\lambda} > 0.3 \qquad \textbf{[2.60]}$$

which is valid down to about $D \approx 0.3\lambda$. Equation 2.60 is in Matthiessen's rule format, which means that the second term is the fractional contribution of the surfaces to the resistivity. It can be seen that for elastic or specular scattering $p = 1$ and there is no change in the resistivity. For $p = 0$, Equation 2.60 predicts $\rho/\rho_{\text{bulk}} \approx 1.20$ for roughly $D \approx 1.9\lambda$ or a thickness $D \approx 75$ nm for Cu for which $\lambda \approx 40$ nm. The value of $p$ depends on the film preparation method (*e.g.*, sputtering, epitaxial growth) and the substrate on which the film has been deposited.

Equation 2.60 involves scattering from two surfaces, that is, from the two interfaces of the film. In general the two interfaces will not be identical and hence will have different $p$ coefficients; $p$ in Equation 2.60 is some mean $p$ value. Table 2.6

[11] This is known as the **Fuchs–Sondheimer equation** in a simplified form.

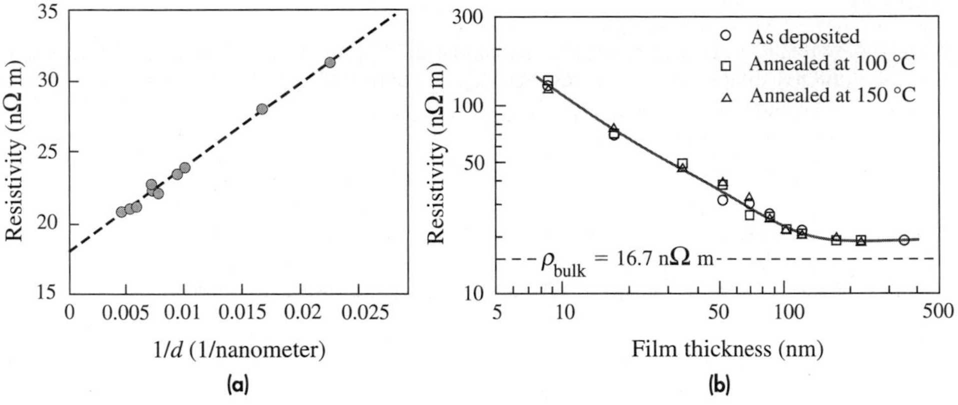


**Figure 2.35**

(a) $\rho_{\text{film}}$ of Cu polycrystalline films versus reciprocal mean grain size (diameter) $1/d$. Film thickness $D = 250$–900 nm does not affect the resistivity. The straight line is $\rho_{\text{film}} = 17.8$ nΩ m + (595 nΩ m nm)$(1/d)$.

(b) $\rho_{\text{film}}$ of thin Cu polycrystalline films versus film thickness $D$. In this case, annealing (heat treating) the films to reduce the polycrystallinity does not significantly affect the resistivity because $\rho_{\text{film}}$ is controlled mainly by surface scattering.

SOURCES: Data extracted from (a) S. Riedel *et al.*, *Microelec. Engin.* **33**, 165, 1997, and (b) W. Lim *et al.*, *Appl. Surf. Sci.*, **217**, 95, 2003.

summarizes the resistivity of thin Cu and Au films deposited by various preparation techniques. Notice the large difference between the Au films deposited on a noncrystalline glass substrate and on a crystalline mica substrate. Such differences between films are typically attributed to different values of $p$. The $p$ value can also change (increase) when the film is annealed. Obviously, the polycrystallinity of the film will also affect the resistivity as discussed previously. Typically, most epitaxial thin films, unless very thin ($D \ll \lambda$), deposited onto heated crystalline substrates exhibit highly specular scattering with $p = 0.9$–1.

It is generally very difficult to separate the effects of surface and grain boundary scattering in thin polycrystalline films; the contribution from grain boundary scattering is likely to exceed that from the surfaces. In any event, both contributions, by Matthiessen's general rule, increase the overall resistivity. Figure 2.35a shows an example in which the resistivity $\rho_{\text{film}}$ of thin Cu polycrystalline films is due to grain boundary scattering, and thickness has no effect ($D$ was 250–900 nm and much greater than $\lambda$). The resistivity $\rho_{\text{film}}$ is plotted against the reciprocal mean grain size $1/d$, which then follows the expected linear behavior in Equation 2.57a. On the other hand, Figure 2.35b shows the resistivity of Cu films as a function of film thickness $D$. In this case, annealing (heat treating) the films to reduce the polycrystallinity does not significantly affect the resistivity because $\rho_{\text{film}}$ is controlled primarily by surface scattering and is given by Equation 2.60.

**THIN-FILM RESISTIVITY** Consider the data presented in Figure 2.35a. What can you conclude from the plot given that the mean free path $\lambda \approx 40$ nm in Cu? **EXAMPLE 2.27**

**SOLUTION**

Consider the results in Figure 2.35a. It is stated that the film thickness $D = 250$–$900$ nm does not affect the resistivity, which implies that $\rho_{\text{film}}$ is controlled only by the grain size $d$. From Equation 2.57a and b we expect

$$\rho_{\text{film}} \approx \rho_{\text{crystal}}(1 + 1.33\beta) \approx \rho_{\text{crystal}} + 1.33\rho_{\text{crystal}}\left(\frac{R}{1-R}\right)\frac{\lambda}{d}$$

This equation represents the observed line when $\rho_{\text{film}}$ is plotted against $1/d$ as in Figure 2.35a. The $\rho_{\text{film}} - 1/d$ line has an intercept given by 17.8 nΩ m and a slope given by 595 (nΩ m) (nm). The intercept approximately matches the bulk resistivity $\rho_{\text{crystal}}$ of Cu. The slope is

$$\text{Slope} \approx 1.33\rho_{\text{crystal}}\left(\frac{R}{1-R}\right)\lambda$$

or

$$595(\text{n}\Omega\text{ m})(\text{nm}) \approx 1.33(17.8\text{ n}\Omega\text{ m})\left(\frac{1}{R^{-1}-1}\right)(40\text{ nm})$$

Solving this equation yields $R \approx 0.39$ for these copper films.

## 2.10 INTERCONNECTS IN MICROELECTRONICS

An integrated circuit (IC) is a single crystal of Si that contains millions of transistors that have been fabricated within this one crystal. **Interconnects** are simply metal conductors that are used to wire the devices together to implement the desired overall operation of the IC; see the photographs in Figure 2.36. Aluminum and Al alloys, or Al silicides, have been the workhouse of the interconnects, but today's fast chips rely on copper interconnects, which have three distinct advantages. First, copper has a resistivity that is about 40 percent lower than that of Al. In high-transistor-density chips in which various voltages are switched on and off, what limits the speed of operation is the $RC$ time constant, that is, the time constant that is involved in charging and discharging the capacitance between the interconnects, and the input capacitance of the transistor; usually the former dominates. The $RC$ is substantially reduced with Cu replacing Al so that the chip speed is faster. The second advantage is that a lower overall interconnect resistance leads to a lower power consumption, lower $I^2R$.

The third advantage is that copper has superior resistance to **electromigration,** a process in which metal atoms are forced to migrate by a large current density. Such electromigration can eventually lead to a failure of the interconnect. The current density in interconnects with a small cross-sectional area can be very high, and hence the electron drift velocities can also be very high. As these fast electrons collide with the metal ions there is a momentum transfer that slowly drifts the metal ions. Thus, the metal ions are forced to slowly migrate as a result of being bombarded by drifting electrons; the migration is in the direction of electron flow (not current flow). This atomic migration can deplete or accumulate material in certain local regions of the interconnect structure. The result is that electromigration can lead to voids (material depletion) or hillocks (material accumulation), and eventually there may be a break or a short between interconnects (an interconnect failure). The electromigration effects are reduced

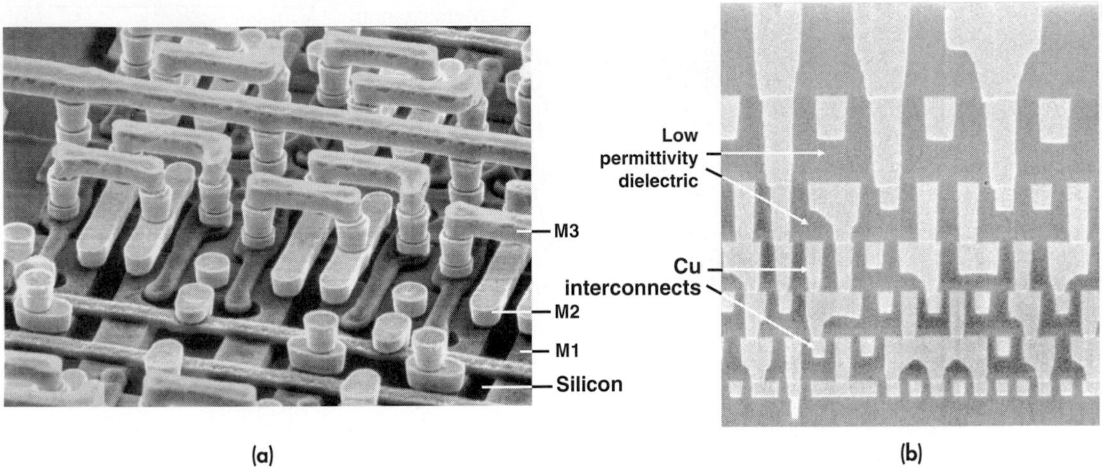


**Figure 2.36**

(a) Metal interconnects wiring devices on a silicon crystal. Three different metallization levels M1, M2, and M3 are used. The dielectric between the interconnects has been etched away to expose the interconnect structure.

(b) Cross section of a chip with seven levels of metallization, M1 to M7. The image is obtained with a scanning electron microscope (SEM).

| SOURCES: (a) Courtesy of IBM. (b) Courtesy of Mark Bohr, Intel.

in Cu interconnects because the Cu atoms are heavier and cannot be as easily migrated by an electric current as are Al atoms.

There is a relatively simple expression for estimating the ***RC*** **time constant** of multilevel interconnects that is useful in comparing various interconnect technologies and the effects of interconnect metal resistance $\rho$, the relative permittivity $\varepsilon_r$ of the interlevel dielectric (insulation) between the interconnects, and the geometry of the whole interconnect wiring. First consider a simple interconnect line, as in Figure 2.37a, whose thickness is $T$, width is $W$, and length is $L$. Its resistance $R$ is simply $\rho L/(TW)$. In the chip, this interconnect will have other interconnects around it as shown in a simplified way in Figure 2.37b. It will couple with all these different conductors around it and will have an overall (effective) capacitance $C_{\text{eff}}$. $RC_{\text{eff}}$ is what we know as the $RC$ time constant associated with the interconnect line in Figure 2.37b.

Suppose that the interconnect is an $M$th-level metallization. It will have a series of many "horizontal" neighbors along this $M$th level. Let $X$ be the nearest edge-to-edge *separation* and $P$ be the *pitch* of these horizontal neighbors at the $M$th level. The pitch $P$ refers to the separation from center to center, or the periodicity of interconnects; $P = W + X$. At a height $H$ above the interconnect there will be a line running at the $(M + 1)$ level. Similarly there will be an interconnect line at a distance $H$ below at the $(M - 1)$ level. We can identify two sets of capacitances. $C_V$ represents the capacitance in the vertical direction, between the interconnect and its upper or lower neighbor. $C_H$ is the lateral capacitance in the horizontal direction, between a neighbor on the right or left. Both are shown in Figure 2.37c. The interconnect therefore has two $C_V$ and two

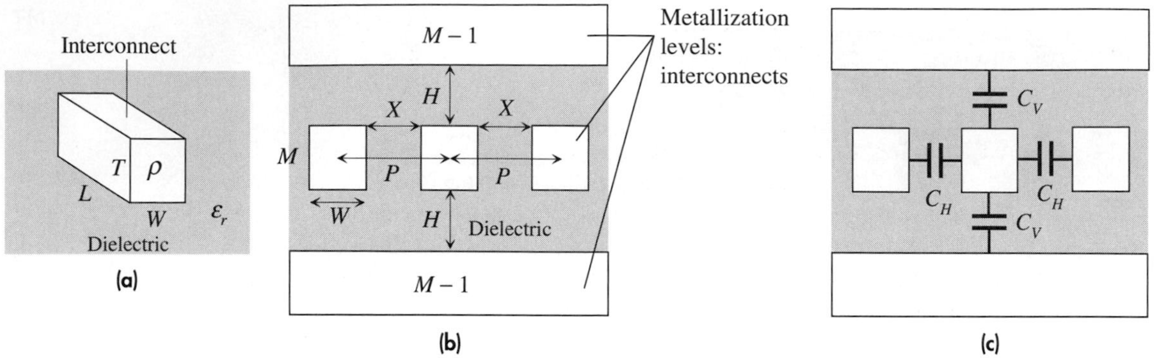


**Figure 2.37**

(a) A single line interconnect surrounded by dielectric insulation.

(b) Interconnects crisscross each other. There are three levels of interconnect: $M-1$, $M$, and $M+1$.

(c) An interconnect has vertical and horizontal capacitances $C_V$ and $C_H$.

$C_H$, four capacitances in total, and all are in parallel as shown in Figure 2.37c. From the simple parallel plate capacitance formula we can write

$$C_H = \frac{\varepsilon_o \varepsilon_r TL}{X} \qquad \text{and} \qquad C_V = \frac{\varepsilon_o \varepsilon_r WL}{H}$$

*Effective capacitance in multilevel interconnect structures*

Usually $C_H$ is greater than $C_V$. From Figure 2.37c, the effective capacitance $C_{\text{eff}} = 2(C_H + C_V)$,

$$C_{\text{eff}} = 2\varepsilon_o \varepsilon_r L\left(\frac{T}{X} + \frac{W}{H}\right) \qquad \textbf{[2.61]}$$

*RC time constant in multilevel interconnect structures*

which is the **effective multilevel interconnect capacitance.** We now multiply this with $R = \rho L/(TW)$ to obtain the $RC$ time constant,

$$RC = 2\varepsilon_o \varepsilon_r \rho\left(\frac{L^2}{TW}\right)\left(\frac{T}{X} + \frac{W}{H}\right) \qquad \textbf{[2.62]}$$

Equation 2.62 is only an approximate first-order calculation, but, nonetheless, it turns out to be quite a useful equation for roughly predicting the *RC* time constant and hence the speed of multilevel interconnect based high-transistor-density chips.[12] Most significantly, it highlights the importance of *three* influencing effects: the resistivity of the interconnect metal; relative permittivity $\varepsilon_r$ of the dielectric insulation between the conductors; and the geometry or "architecture" of the interconnects $L$, $T$, $W$, $X$, and $H$. Notice that $L$ appears as $L^2$ in Equation 2.62 and has

[12] A more rigorous theory would consider the interconnect system as having a distributed resistance and a distributed capacitance, similar to a transmission line; a topical research area. The treatment here is more than sufficient to obtain approximate results and understand the factors that control the interconnect delay time.

significant control on the overall *RC*. Equation 2.62 does not obviously include the time it takes to turn on and off the individual transistors connected to the interconnects. In a high-transistor-density chip, the latter is smaller than the interconnect *RC* time constant.

The reduction in the interconnect resistivity $\rho$ by the use of Cu instead of Al has been a commendable achievement, and cuts down *RC* significantly. Further reduction in $\rho$ is limited because Cu already has a very small resistivity; the smallest $\rho$ is for Ag which is only about 5 percent lower. Current research efforts for reducing *RC* further are concentrated on mainly two factors. First is the reduction of $\varepsilon_r$ as much as possible by using dielectrics such as *fluorinated* $SiO_2$ (known as FSG) for which $\varepsilon_r = 3.6$, or, more importantly, using what are called **low-*k* dielectric materials** ($k$ stands for $\varepsilon_r$) such as various polymers or porous dielectrics[13] that have a lower $\varepsilon_r$, typically 2–3, which is a substantial reduction from 3.6. The second is the development of optimized interconnect geometries that reduce $L^2$ in Equation 2.62. ($T$, $W$, $X$, and $H$ are all of comparable size, so $L^2$ is the most dominant geometric factor.)

The ratio of the thickness $T$ to width $W$ of an interconnect is called the **aspect ratio,** $A_r = T/W$. This ratio is typically between 1 to 2. Very roughly, in many cases, $X$ and $W$ are the same, $X \approx W$ and $X \approx P/2$ (see Figure 2.37b). Then Equation 2.62 simplifies further,

*RC time constant in multilevel interconnect structures*

$$RC \approx 2\varepsilon_o\varepsilon_r\rho L^2\left(\frac{4}{P^2} + \frac{1}{T^2}\right) \qquad [2.63]$$

The signal delays between the transistors on a chip arise from the interconnect *RC* time constant. Equations 2.62 and 2.63 are often also used to calculate the **multilevel interconnect delay time.** Suppose that we take some typical values, $L \approx 10$ mm, $T \approx 1$ $\mu$m, $P \approx 1$ $\mu$m, $\rho = 17$ n$\Omega$ m for a Cu interconnect, and $\varepsilon_r \approx 3.6$ for FSG; then $RC \approx 0.43$ ns, not a negligible value in today's speed hungry computing.

**EXAMPLE 2.28**

**MULTILEVEL INTERCONNECT *RC* TIME CONSTANT** In a particular high-transistor-density IC where copper is used as the interconnect, one level of the multilevel interconnects has the following characteristics: pitch $P = 0.45$ $\mu$m, $T = 0.36$ $\mu$m, $A_R = 1.6$, $H = X$, and $\varepsilon_r \approx 3.6$. Find the effective capacitance per millimeter of interconnect length, and the *RC* delay time per $L^2$ as ps/mm$^2$ (as normally used in industry).

**SOLUTION**

Since $A_R = T/W$, $W = T/A_R = 0.36/1.6 = 0.225$ $\mu$m. Further, from Figure 2.37b, $P = W + X$, so that $X = P - W = 0.45 - 0.225 = 0.225$ $\mu$m. $H = X = 0.225$ $\mu$m. Thus, Equation 2.61 for $L = 1$ mm $= 10^{-3}$ m gives

$$C_{\text{eff}} = 2\varepsilon_o\varepsilon_r L\left(\frac{T}{X} + \frac{W}{H}\right) = 2(8.85 \times 10^{-12})(3.6)(10^{-3})\left[\frac{0.36}{0.225} + \frac{0.225}{0.225}\right] = 0.17 \text{ pF}$$

[13] The mixture rules mentioned in this chapter turn up again in a different but recognizable form for predicting the overall relative permittivity of porous dielectrics.

which is about 0.2 pF per millimeter of interconnect. The $RC$ time constant per $L^2$ is

$$\frac{RC}{L^2} = 2\varepsilon_o\varepsilon_r\rho\left(\frac{1}{TW}\right)\left(\frac{T}{X}+\frac{W}{H}\right) = 2\varepsilon_o\varepsilon_r\rho\left(\frac{1}{WX}+\frac{1}{TH}\right)$$

$$= 2(8.85\times10^{-12})(3.6)(17\times10^{-9})$$

$$\left[\frac{1}{(0.225\times10^{-6})(0.225\times10^{-6})}+\frac{1}{(0.36\times10^{-6})(0.225\times10^{-6})}\right]$$

$$= 3.4\times10^{-5}\ \text{s m}^{-2} \qquad \text{or} \qquad 34\ \text{ps mm}^{-2}$$

## 2.11 ELECTROMIGRATION AND BLACK'S EQUATION

Interconnects have small cross-sectional dimensions, and consequently the current densities can be quite large. Figure 2.38a depicts how the continual bombardment of lattice atoms (metal ions) by many "fast" conduction electrons in high-current-density regions can transfer enough momentum to a host metal atom to migrate it, that is, diffuse it along a suitable path in the crystal. The bombarded metal atom has to jump to a suitable lattice location to migrate, which is usually easiest along grain boundaries or surfaces where there is sufficient space as depicted in Figure 2.38a and b. Grain boundaries that are parallel to the electron flow therefore can migrate atoms more efficiently than grain boundaries in other directions. Atomic diffusion can also occur along a surface of the interconnect, that is, along an interface between the interconnect metal and the neighboring material. The final result of atomic migration is usually either material depletion or accumulation as depicted in Figure 2.38c. The depletion of material

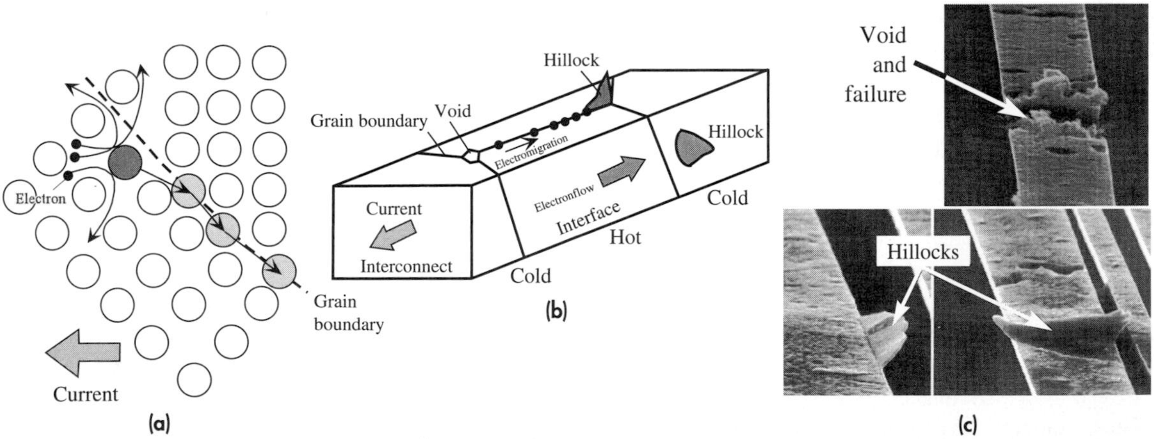


**Figure 2.38**

(a) Electrons bombard the metal ions and force them to slowly migrate.

(b) Formation of voids and hillocks in a polycrystalline metal interconnect by the electromigration of metal ions along grain boundaries and interfaces.

(c) Accelerated tests on a 3 μm chemical vapor deposited Cu line: $T = 200$ °C and $J = 6$ MA cm$^{-2}$. The photos show void formation and fatal failure (break), and hillock formation.

| SOURCE: Courtesy of L. Arnaud *et al.*, *Microelectronics Reliability*, **40**, 86, 2000.

leads to a **void** and a possible eventual break in the interconnect. The accumulation of material leads to a **hillock** and a short between lines. Interconnect failure by electromigration is measured by the **mean time to 50 percent failure** $t_{MTF}$. There are two factors that control the rate of electromigration $R_{EM}$. First is the activation energy $E_A$ involved in migrating (diffusing) the metal atom, and the second is the rate at which the atoms are bombarded with electrons, which depends on the current density $J$. Thus,

*Electromigration rate*

$$R_{EM} \propto J^n \exp\left(-\frac{E_A}{kT}\right)$$

in which the rate is proportional to $J^n$, instead of just $J$ because it is found experimentally that $n \geq 1$. From the electromigration rate we can find the average time $t_{MTF}$ it takes for 50 percent failure of interconnects because this time is inversely proportional to the electromigration rate just given:

*Black's electromigration failure equation*

$$t_{MTF} = A_B J^{-n} \exp\left(\frac{E_A}{kT}\right) \quad \textbf{[2.64]}$$

where $A_B$ is a constant. Equation 2.64 is known as **Black's equation,** and it is extremely useful in extrapolating high-temperature failure tests to normal operating temperatures. Electromigration-induced interconnect failures are typically examined at elevated temperatures where the failure times are over a measurable time scale in the laboratory (perhaps several hours or a few days). These experiments are called **accelerated failure** tests because they make use of the fact that at high temperatures the electromigration failure occurs more quickly. The results are then extrapolated to room temperature using Black's equation.

Typically electromigration occurs along grain boundaries or along various interfaces that the interconnect has with its surroundings, the semiconductor, dielectric material, etc. The diffusion coefficient has a lower activation energy $E_A$ for these migration paths than for diffusion within the volume of the crystal. The electromigration process therefore depends on the microstructure of the interconnect metal, and its interfaces. Usually another metal, called a **barrier,** is deposited to occupy the interface space between the interconnect and the semiconductor or the oxide. The barrier *passivates* the interface, rendering it relatively inactive in terms of providing an electromigration path. An interconnect can also have a temperature gradient along it. (The heat generated by $I^2R$ may be conducted away faster at the ends of the interconnect, leaving the central region hotter.) Electromigration would be faster in the hot region and very slow (almost stationary) in the cold region since it is a thermally activated process. Consequently a pileup of electromigrated atoms can occur as atoms are migrated from hot to cold regions along the interconnect, leading to a hillock.[14]

Pure Al suffers badly from electromigration problems and is usually alloyed with small amounts of Cu, called Al(Cu), to reduce electromigration to a tolerable level. But the resistivity increases. (Why?) In recent Cu interconnects, the most important diffusion path seems to be the interface between the Cu surface and the dielectric. Surface coating of these Cu interconnects provides control over electromigration failures.

[14] Somewhat like a traffic accident pileup in which speeding cars run into stationary cars ahead of them.

## CD Selected Topics and Solved Problems

| Selected Topics | Solved Problems |
|---|---|
| Conduction in Metals: Electrical and Thermal Conduction | Radiation Theory of the Fuse |
| Joule's Law | The Strain Gauge |
| Hall Effect | |
| Heat Transfer in Electrical Engineering: Fundamentals and Applications | |
| Thermal Conductivity | |

## DEFINING TERMS

**Alloy** is a metal that contains more than one element.

**Brass** is a copper-rich Cu–Zn alloy.

**Bronze** is a copper-rich Cu–Sn alloy.

**Drift mobility** is the drift velocity per unit applied field. If $\mu_d$ is the drift mobility, then the defining equation is $v_d = \mu_d \mathcal{E}$, where $v_d$ is the drift velocity and $\mathcal{E}$ is the field.

**Drift velocity** is the average electron velocity, over all the conduction electrons in the conductor, in the direction of an applied electrical force ($F = -e\mathcal{E}$ for electrons). In the absence of an applied field, all the electrons move around randomly, and the average velocity over all the electrons in any direction is zero. With an applied field $\mathcal{E}_x$, there is a net velocity per electron $v_{dx}$, in the direction opposite to the field, where $v_{dx}$ depends on $\mathcal{E}_x$ by virtue of $v_{dx} = \mu_d \mathcal{E}_x$, where $\mu_d$ is the drift mobility.

**Electrical conductivity** ($\sigma$) is a property of a material that quantifies the ease with which charges flow inside the material along an applied electric field or a voltage gradient. The conductivity is the inverse of electrical resistivity $\rho$. Since charge flow is caused by a voltage gradient, $\sigma$ is the rate of charge flow across a unit area per unit voltage gradient, $J = \sigma \mathcal{E}$.

**Electromigration** is current density–induced diffusion of host metal atoms due to their repeated bombardment by conduction electrons at high current densities; the metal atoms migrate in the direction of electron flow. **Black's equation** describes the mean time to failure of metal film interconnects due to electromigration failure.

**Fourier's law** states that the rate of heat flow $Q'$ through a sample, due to thermal conduction, is proportional to the temperature gradient $dT/dx$ and the cross-sectional area $A$, that is, $Q' = -\kappa A(dT/dx)$, where $\kappa$ is the thermal conductivity.

**Hall coefficient** ($R_H$) is a parameter that gauges the magnitude of the Hall effect. If $\mathcal{E}_y$ is the electric field in the $y$ direction, due to a current density $J_x$ along $x$ and a magnetic field $B_z$ along $z$, then $R_H = \mathcal{E}_y / J_x B_z$.

**Hall effect** is a phenomenon that occurs in a conductor carrying a current when the conductor is placed in a magnetic field perpendicular to the current. The charge carriers in the conductor are deflected by the magnetic field, giving rise to an electric field (Hall field) that is perpendicular to both the current and the magnetic field. If the current density $J_x$ is along $x$ and the magnetic field $B_z$ is along $z$, then the Hall field is along either $+y$ or $-y$, depending on the polarity of the charge carriers in the material.

**Heterogeneous mixture** is a mixture in which the individual components remain physically separate and possess different chemical and physical properties; that is, a mixture of different phases.

**Homogeneous mixture** is a mixture of two or more chemical species in which the chemical properties (*e.g.*, composition) and physical properties (*e.g.*, density, heat capacity) are uniform throughout. A homogeneous mixture is a solution.

**Interconnects** are various thin metal conductors in a Si integrated circuit that connect various devices to implement the required wiring of the devices. In modern ICs, these interconnects are primarily electrodeposited Cu films.

**Ionic conduction** is the migration of ions in the material as a result of field-directed diffusion. When a positive ion in an interstitial site jumps to a neighboring interstitial site in the direction of the field, it lowers its potential energy which is a favorable process. If it jumps in the opposite direction, then it has to do work against the force of the field which is undesirable. Thus the diffusion of the positive ion is directed along the field.

**Isomorphous phase diagram** is a phase diagram for an alloy that has unlimited solid solubility.

**Joule's law** relates the power dissipated per unit volume $P_{\text{vol}}$ by a current-carrying conductor to the applied field $\mathcal{E}$ and the current density $J$, such that $P_{\text{vol}} = J\mathcal{E} = \sigma\mathcal{E}^2$.

**Lorentz force** is the force experienced by a moving charge in a magnetic field. When a charge $q$ is moving with a velocity $\mathbf{v}$ in a magnetic field $\mathbf{B}$, the charge experiences a force $\mathbf{F}$ that is proportional to the magnitude of its charge $q$, its velocity $\mathbf{v}$, and the field $\mathbf{B}$, such that $\mathbf{F} = q\mathbf{v} \times \mathbf{B}$.

**Magnetic field, magnetic flux density, or magnetic induction (B)** is a vector field quantity that describes the magnitude and direction of the *magnetic force* exerted on a moving charge or a current-carrying conductor. The magnetic force is essentially the Lorentz force and excludes the electrostatic force $q\mathcal{E}$.

**Magnetic permeability** ($\mu$) or simply permeability is a property of the medium that characterizes the effectiveness of a medium in generating as much magnetic field as possible for given external currents. It is the product of the permeability of free space (vacuum) or absolute permeability ($\mu_o$) and relative permeability of the medium ($\mu_r$), *i.e.*, $\mu = \mu_o\mu_r$.

**Magnetometer** is an instrument for measuring the magnitude of a magnetic field.

**Matthiessen's rule** gives the overall resistivity of a metal as the sum of individual resistivities due to scattering from thermal vibrations, impurities, and crystal defects. If the resistivity due to scattering from thermal vibrations is denoted $\rho_T$ and the resistivities due to scattering from crystal defects and impurities can be lumped into a single resistivity term called the residual resistivity $\rho_R$, then $\rho = \rho_T + \rho_R$.

**Mean free path** is the mean distance traversed by an electron between scattering events. If $\tau$ is the mean free time between scattering events and $u$ is the mean speed of the electron, then the mean free path is $\ell = u\tau$.

**Mean free time** is the average time it takes to scatter a conduction electron. If $t_i$ is the free time between collisions (between scattering events) for an electron labeled $i$, then $\tau = \overline{t_i}$ averaged over all the electrons. The drift mobility is related to the mean free time by $\mu_d = e\tau/m_e$. The reciprocal of the mean free time is the mean probability per unit time that a conduction electron will be scattered; in other words, the mean frequency of scattering events.

**Nordheim's rule** states that the resistivity of a solid solution (an isomorphous alloy) due to impurities $\rho_I$ is proportional to the concentrations of the solute $X$ and the solvent $(1 - X)$.

**Phase** (in materials science) is a physically homogeneous portion of a materials system that has uniform physical and chemical characteristics.

**Relaxation time** is an equivalent term for the mean free time between scattering events.

**Residual resistivity** ($\rho_R$) is the contribution to the resistivity arising from scattering processes other than thermal vibrations of the lattice, for example, impurities, grain boundaries, dislocations, point defects.

**Skin effect** is an electromagnetic phenomenon that, at high frequencies, restricts ac current flow to near the surface of a conductor to reduce the energy stored in the magnetic field.

**Solid solution** is a crystalline material that is a homogeneous mixture of two or more chemical species. The mixing occurs at the atomic scale, as in mixing alcohol and water. Solid solutions can be substitutional (as in Cu–Ni) or interstitial (for example, C in Fe).

**Stefan's law** is a phenomenological description of the energy radiated (as electromagnetic waves) from a surface per second. When a surface is heated to a temperature $T$, it radiates net energy at a rate given by $P_{\text{radiated}} = \epsilon\sigma_S A(T^4 - T_0^4)$, where $\sigma_S$ is Stefan's constant ($5.67 \times 10^{-8}$ W m$^{-2}$ K$^{-4}$), $\epsilon$ is the emissivity of the surface, $A$ is the surface area, and $T_0$ is the ambient temperature.

**Temperature coefficient of resistivity (TCR)** ($\alpha_0$) is defined as the fractional change in the electrical resistivity of a material per unit increase in the temperature with respect to some reference temperature $T_0$.

**Thermal conductivity** ($\kappa$) is a property of a material that quantifies the ease with which heat flows along the material from higher to lower temperature regions. Since heat flow is due to a temperature gradient, $\kappa$ is the rate of heat flow across a unit area per unit temperature gradient.

**Thermal resistance** ($\theta$) is a measure of the difficulty with which heat conduction takes place along a material sample. The thermal resistance is defined as the temperature drop per unit heat flow, $\theta = \Delta T/Q'$. It depends on both the material and its geometry. If the heat losses from the surfaces are negligible, then $\theta = L/\kappa A$, where $L$ is the length of the sample (along heat flow) and $A$ is the cross-sectional area.

**Thermally activated conductivity** means that the conductivity increases in an exponential fashion with temperature as in $\sigma = \sigma_o \exp(-E_\sigma/kT)$ where $E_\sigma$ is the activation energy.

**Thin film** is a conductor whose thickness is typically less than ~1 micron; the thickness is also much less than the width and length of the conductor. Typically thin films have a higher resistivity than the corresponding bulk material due to the grain boundary and surface scattering.

# QUESTIONS AND PROBLEMS

**2.1** **Electrical conduction** Na is a monovalent metal (BCC) with a density of 0.9712 g cm$^{-3}$. Its atomic mass is 22.99 g mol$^{-1}$. The drift mobility of electrons in Na is 53 cm$^2$ V$^{-1}$ s$^{-1}$.

*a.* Consider the collection of conduction electrons in the solid. If each Na atom donates one electron to the electron sea, *estimate* the mean separation between the electrons. (Note: If $n$ is the concentration of particles, then the particles' mean separation $d = 1/n^{1/3}$.)

*b.* Estimate the mean separation between an electron ($e^-$) and a metal ion ($Na^+$), assuming that most of the time the electron prefers to be between two neighboring $Na^+$ ions. What is the approximate Coulombic interaction energy (in eV) between an electron and an $Na^+$ ion?

*c.* How does this electron/metal-ion interaction energy compare with the average thermal energy per particle, according to the kinetic molecular theory of matter? Do you expect the kinetic molecular theory to be applicable to the conduction electrons in Na? If the mean electron/metal-ion interaction energy is of the same order of magnitude as the mean *KE* of the electrons, what is the mean speed of electrons in Na? Why should the mean kinetic energy be comparable to the mean electron/metal-ion interaction energy?

*d.* Calculate the electrical conductivity of Na and compare this with the experimental value of $2.1 \times 10^7$ $\Omega^{-1}$ m$^{-1}$ and comment on the difference.

**2.2** **Electrical conduction** The resistivity of aluminum at 25 °C has been measured to be $2.72 \times 10^{-8}$ $\Omega$ m. The thermal coefficient of resistivity of aluminum at 0 °C is $4.29 \times 10^{-3}$ K$^{-1}$. Aluminum has a valency of 3, a density of 2.70 g cm$^{-3}$, and an atomic mass of 27.

*a.* Calculate the resistivity of aluminum at −40 °C.

*b.* What is the thermal coefficient of resistivity at −40 °C?

*c.* Estimate the mean free time between collisions for the conduction electrons in aluminum at 25 °C, and hence estimate their drift mobility.

*d.* If the mean speed of the conduction electrons is about $2.0 \times 10^6$ m s$^{-1}$, calculate the mean free path and compare this with the interatomic separation in Al (Al is FCC). What should be the thickness of an Al film that is deposited on an IC chip such that its resistivity is the same as that of bulk Al?

*e.* What is the percentage change in the power loss due to Joule heating of the aluminum wire when the temperature drops from 25 °C to −40 °C?

**2.3** **Conduction in gold** Gold is in the same group as Cu and Ag. Assuming that each Au atom donates one conduction electron, calculate the drift mobility of the electrons in gold at 22 °C. What is the mean free path of the conduction electrons if their mean speed is $1.4 \times 10^6$ m s$^{-1}$? (Use $\rho_o$ and $\alpha_o$ in Table 2.1.)

**2.4** **Effective number of conduction electrons per atom**

*a.* Electron drift mobility in tin (Sn) is 3.9 cm$^2$ V$^{-1}$ s$^{-1}$. The room temperature (20 °C) resistivity of Sn is about 110 nΩ m. Atomic mass $M_{at}$ and density of Sn are 118.69 g mol$^{-1}$ and 7.30 g cm$^{-3}$, respectively. How many "free" electrons are donated by each Sn atom in the crystal? How does this compare with the position of Sn in Group IVB of the Periodic Table?

*b.* Consider the resistivity of few selected metals from Groups I to IV in the Periodic Table in Table 2.7. Calculate the number of conduction electrons contributed per atom and compare this with the location of the element in the Periodic Table. What is your conclusion?

**Table 2.7** Selection of metals from Groups I to IV in the Periodic Table

| Metal | Periodic Group | Valency | Density (g cm$^{-3}$) | Resistivity (nΩ m) | Mobility (cm$^2$ V$^{-1}$s$^{-1}$) |
|---|---|---|---|---|---|
| Na | IA | 1 | 0.97 | 42.0 | 53 |
| Mg | IIA | 2 | 1.74 | 44.5 | 17 |
| Ag | IB | 1 | 10.5 | 15.9 | 56 |
| Zn | IIB | 2 | 7.14 | 59.2 | 8 |
| Al | IIIB | 3 | 2.7 | 26.5 | 12 |
| Sn | IVB | 4 | 7.30 | 110 | 3.9 |
| Pb | IVB | 4 | 11.4 | 206 | 2.3 |

NOTE: Mobility from Hall-effect measurements.

**2.5** **TCR and Matthiessen's rule** Determine the temperature coefficient of resistivity of pure iron and of electrotechnical steel (Fe with 4% C), which are used in various electrical machinery, at two temperatures: 0 °C and 500 °C. Comment on the similarities and differences in the resistivity versus temperature behavior shown in Figure 2.39 for the two materials.

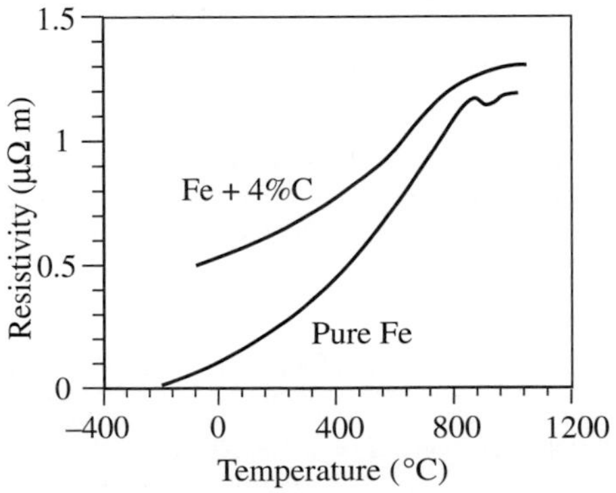


**Figure 2.39** Resistivity versus temperature for pure iron and 4% C steel.

***2.6 TCR of isomorphous alloys**

*a.* Show that for an isomorphous alloy $A\%–B\%$ ($B\%$ solute in $A\%$ solvent), the temperature coefficient of resistivity $\alpha_{AB}$ is given by

$$\alpha_{AB} \approx \frac{\alpha_A \rho_A}{\rho_{AB}}$$

where $\rho_{AB}$ is the resistivity of the alloy ($AB$) and $\rho_A$ and $\alpha_A$ are the resistivity and TCR of pure $A$. What are the assumptions behind this equation?

*b.* Determine the composition of the Cu–Ni alloy that will have a TCR of $4 \times 10^{-4}$ K$^{-1}$, that is, a TCR that is an order of magnitude less than that of Cu. Over the composition range of interest, the resistivity of the Cu–Ni alloy can be calculated from $\rho_{\text{CuNi}} \approx \rho_{\text{Cu}} + C_{\text{eff}} X(1 - X)$, where $C_{\text{eff}}$, the effective Nordheim coefficient, is about 1310 nΩ m.

**2.7 Resistivity of isomorphous alloys and Nordheim's rule** What are the maximum atomic and weight percentages of Cu that can be added to Au without exceeding a resistivity that is twice that of pure gold? What are the maximum atomic and weight percentages of Au that can be added to pure Cu without exceeding twice the resistivity of pure copper? (Alloys are normally prepared by mixing the elements in weight.)

**2.8 Nordheim's rule and brass** Brass is a Cu–Zn alloy. Table 2.8 shows some typical resistivity values for various Cu–Zn compositions in which the alloy is a solid solution (up to 30% Zn).

*a.* Plot $\rho$ versus $X(1 - X)$. From the slope of the best-fit line find the mean (effective) Nordheim coefficient $\overline{C}$ for Zn dissolved in Cu over this compositional range.

*b.* Since $X$ is the atomic fraction of Zn in brass, for each atom in the alloy, there are $X$ Zn atoms and $(1 - X)$ Cu atoms. The conduction electrons consist of each Zn donating two electrons and each copper donating one electron. Thus, there are $2(X) + 1(1 - X) = 1 + X$ conduction electrons per atom. Since the conductivity is proportional to the electron concentration, the combined Nordheim–Matthiessens rule must be scaled up by $(1 + X)$,

$$\rho_{\text{brass}} = \frac{\rho_o + CX(1 - X)}{(1 + X)}$$

Plot the data in Table 2.8 as $\rho(1 + X)$ versus $X(1 - X)$. From the best-fit line find $C$ and $\rho_o$. What is your conclusion? (Compare the correlation coefficients of the best-fit lines in your two plots.[15])

**Table 2.8** Cu–Zn brass alloys

| Zn at.% in Cu–Zn | 0 | 0.34 | 0.5 | 0.93 | 3.06 | 4.65 | 9.66 | 15.6 | 19.59 | 29.39 |
|---|---|---|---|---|---|---|---|---|---|---|
| Resistivity nΩ m | 17 | 18.1 | 18.84 | 20.7 | 26.8 | 29.9 | 39.1 | 49.0 | 54.8 | 63.5 |

SOURCE: H. A. Fairbank, *Phys. Rev.*, **66,** 274, 1944.

**2.9 Resistivity of solid solution metal alloys: testing Nordheim's rule** Nordheim's rule accounts for the increase in the resistivity resulting from the scattering of electrons from the random distribution of impurity (solute) atoms in the host (solvent) crystal. It can nonetheless be quite useful in approximately

[15] More rigorously, $\rho_{\text{brass}} = \rho_{\text{matrix}} + C_{\text{eff}} X(1-X)$, in which $\rho_{\text{matrix}}$ is the resistivity of the perfect matrix. Accounting for the extra electrons, $\rho_{\text{matrix}} \approx \rho_o/(1+X)$, where $\rho_o$ is the pure metal matrix resistivity and $C_{\text{eff}}$ is the Nordheim coefficient at the composition of interest, given by $C_{\text{eff}} \approx C/(1+X)^{2/3}$. (It is assumed that the atomic concentration does not change significantly.) As always, there are also other theories; part *b* is more than sufficient for most practical purposes.

predicting the resistivity at one composition of a solid solution metal alloy, given the value at another composition. Table 2.9 lists some solid solution metal alloys and gives the resistivity $\rho$ at one composition $X$ and asks for a prediction $\rho'$ based on Nordheim's rule at another composition $X'$. Fill in the table for $\rho'$ and compare the predicted values with the experimental values, and comment.

**Table 2.9** Resistivities of some solid solution metal alloys

| | Alloy | | | | | | | |
|---|---|---|---|---|---|---|---|---|
| | Ag–Au | Au–Ag | Cu–Pd | Ag–Pd | Au–Pd | Pd–Pt | Pt–Pd | Cu–Ni |
| $X$ (at.%) | 8.8% Au | 8.77% Ag | 6.2% Pd | 10.1% Pd | 8.88% Pd | 7.66% Pt | 7.1% Pd | 2.16% Ni |
| $\rho_o$ (nΩ m) | 16.2 | 22.7 | 17 | 16.2 | 22.7 | 108 | 105.8 | 17 |
| $\rho$ at $X$ (nΩ m) | 44.2 | 54.1 | 70.8 | 59.8 | 54.1 | 188.2 | 146.8 | 50 |
| $C_{eff}$ | | | | | | | | |
| $X'$ | 15.4% Au | 24.4% Ag | 13% Pd | 15.2% Pd | 17.1% Pd | 15.5% Pt | 13.8% Pd | 23.4% Ni |
| $\rho'$ at $X'$ (nΩ m) | | | | | | | | |
| $\rho'$ at $X'$ (nΩ m) Experimental | 66.3 | 107.2 | 121.6 | 83.8 | 82.2 | 244 | 181 | 300 |

NOTE: First symbol (*e.g.*, Ag in AgAu) is the matrix (solvent) and the second (Au) is the added solute. $X$ is in at.%, converted from traditional weight percentages reported with alloys. $C_{eff}$ is the effective Nordheim coefficient in $\rho = \rho_o + C_{eff} X(1-X)$.

***2.10** **TCR and alloy resistivity** Table 2.10 shows the resistivity and TCR ($\alpha$) of Cu–Ni alloys. Plot TCR versus $1/\rho$, and obtain the best-fit line. What is your conclusion? Consider the Matthiessen rule, and explain why the plot should be a straight line. What is the relationship between $\rho_{Cu}$, $\alpha_{Cu}$, $\rho_{CuNi}$, and $\alpha_{CuNi}$? Can this be generalized?

**Table 2.10** Cu–Ni alloys, resistivity, and TCR

| | Ni wt.% in Cu–Ni | | | | |
|---|---|---|---|---|---|
| | 0 | 2 | 6 | 11 | 20 |
| Resistivity (nΩ m) | 17 | 50 | 100 | 150 | 300 |
| TCR (ppm °C$^{-1}$) | 4270 | 1350 | 550 | 430 | 160 |

NOTE: ppm-parts per million, *i.e.*, $10^{-6}$.

**2.11** **Electrical and thermal conductivity of In** Electron drift mobility in indium has been measured to be 6 cm$^2$ V$^{-1}$ s$^{-1}$. The room temperature (27 °C) resistivity of In is $8.37 \times 10^{-8}$ Ω m, and its atomic mass and density are 114.82 amu or g mol$^{-1}$ and 7.31 g cm$^{-3}$, respectively.

*a.* Based on the resistivity value, determine how many free electrons are donated by each In atom in the crystal. How does this compare with the position of In in the Periodic Table (Group IIIB)?

*b.* If the mean speed of conduction electrons in In is $1.74 \times 10^8$ cm s$^{-1}$, what is the mean free path?

*c.* Calculate the thermal conductivity of In. How does this compare with the experimental value of 81.6 W m$^{-1}$ K$^{-1}$?

**2.12** **Electrical and thermal conductivity of Ag** The electron drift mobility in silver has been measured to be 56 cm$^2$ V$^{-1}$ s$^{-1}$ at 27 °C. The atomic mass and density of Ag are given as 107.87 amu or g mol$^{-1}$ and 10.50 g cm$^{-3}$, respectively.

*a.* Assuming that each Ag atom contributes one conduction electron, calculate the resistivity of Ag at 27 °C. Compare this value with the measured value of $1.6 \times 10^{-8}$ Ωm at the same temperature and suggest reasons for the difference.

*b.* Calculate the thermal conductivity of silver at 27 °C and at 0 °C.

**2.13 Mixture rules** A 70% Cu–30% Zn brass electrical component has been made of powdered metal and contains 15 vol.% porosity. Assume that the pores are dispersed randomly. Given that the resistivity of 70% Cu–30% Zn brass is 62 nΩ m, calculate the effective resistivity of the brass component using the simple conductivity mixture rule, Equation 2.26, and the Reynolds and Hough rule.

**2.14 Mixture rules**

*a.* A certain carbon electrode used in electrical arcing applications is 47 percent porous. Given that the resistivity of graphite (in polycrystalline form) at room temperature is about 9.1 $\mu\Omega$ m, estimate the effective resistivity of the carbon electrode using the appropriate Reynolds and Hough rule and the simple conductivity mixture rule. Compare your estimates with the measured value of 18 $\mu\Omega$ m and comment on the differences.

*b.* Silver particles are dispersed in a graphite paste to increase the effective conductivity of the paste. If the volume fraction of dispersed silver is 30 percent, what is the effective conductivity of this paste?

**2.15 Ag–Ni alloys (contact materials) and the mixture rules** Silver alloys, particularly Ag alloys with the precious metals Pt, Pd, Ni, and Au, are extensively used as contact materials in various switches. Alloying Ag with other metals generally increases the hardness, wear resistance, and corrosion resistance at the expense of electrical and thermal conductivity. For example, Ag–Ni alloys are widely used as contact materials in switches in domestic appliances, control and selector switches, circuit breakers, and automotive switches up to several hundred amperes of current. Table 2.11 shows the resistivities of four Ag–Ni alloys used in make-and-break as well as disconnect contacts with current ratings up to ~100 A.

*a.* Ag–Ni is a two-phase alloy, a mixture of Ag-rich and Ni-rich phases. Using an appropriate mixture rule, predict the resistivity of the alloy and compare with the measured values in Table 2.11. Explain the difference between the predicted and experimental values.

*b.* Compare the resistivity of Ag–10% Ni with that of Ag–10% Pd in Table 2.9. The resistivity of the Ag–Pd alloy is almost a factor of 3 greater. Ag–Pd is an isomorphous solid solution, whereas Ag–Ni is a two-phase mixture. Explain the difference in the resistivities of Ag–Ni and Ag–Pd.

**Table 2.11** Resistivity of Ag–Ni contact alloys for switches

| | Ni % in Ag–Ni | | | | | |
|---|---|---|---|---|---|---|
| | 0 | 10 | 15 | 20 | 30 | 100 |
| $\rho$(nΩ m) | 16.9 | 20.9 | 23.6 | 25 | 31.1 | 71.4 |
| $d$(g cm$^{-3}$) | 10.5 | 10.3 | 9.76 | 9.4 | 9.47 | 8.9 |
| Hardness VHN | 30 | 50 | 55 | 60 | 65 | 80 |

NOTE: Compositions are in wt.%. Ag–10% Ni means 90% Ag–10% Ni. Vickers hardness number (VHN) is a measure of the hardness or strength of the alloy, and *d* is density.

**2.16 Ag–W alloys (contact materials) and the mixture rule** Silver–tungsten alloys are frequently used in heavy-duty switching applications (*e.g.*, current-carrying contacts and oil circuit breakers) and in arcing tips. Ag–W is a two-phase alloy, a mixture of Ag-rich and W-rich phases. The measured resistivity and density for various Ag–W compositions are summarized in Table 2.12.

*a.* Plot the resistivity and density of the Ag–W alloy against the W content (wt.%)

*b.* Show that the density of the mixture, $d$, is given by

$$d^{-1} = w_\alpha d_\alpha^{-1} + w_\beta d_\beta^{-1}$$

where $w_\alpha$ is the weight fraction of phase $\alpha$, $w_\beta$ is the weight fraction of phase $\beta$, $d_\alpha$ is the density of phase $\alpha$, and $d_\beta$ is the density of phase $\beta$.

*c.* Show that the resistivity mixture rule is

*Mixture rule and weight fractions*

$$\rho = \rho_\alpha \frac{d w_\alpha}{d_\alpha} + \rho_\beta \frac{d w_\beta}{d_\beta}$$

where $\rho$ is the resistivity of the alloy (mixture), $d$ is the density of the alloy (mixture), and subscripts $\alpha$ and $\beta$ refer to phases $\alpha$ and $\beta$, respectively.

*d.* Calculate the density $d$ and the resistivity $\rho$ of the mixture for various values of W content (in wt.%) and plot the calculated values in the same graph as the experimental values. What is your conclusion?

**Table 2.12** Dependence of resistivity in Ag–W alloy on composition as a function of wt.% W

| | W(wt.%) | | | | | | | | | | | | |
|---|---|---|---|---|---|---|---|---|---|---|---|---|---|
| | **0** | **10** | **15** | **20** | **30** | **40** | **65** | **70** | **75** | **80** | **85** | **90** | **100** |
| $\rho$ (n$\Omega$ m) | 16.2 | 18.6 | 19.7 | 20.9 | 22.7 | 27.6 | 35.5 | 38.3 | 40 | 46 | 47.9 | 53.9 | 55.6 |
| $d$ (g cm$^{-3}$) | 10.5 | 10.75 | 10.95 | 11.3 | 12 | 12.35 | 14.485 | 15.02 | 15.325 | 16.18 | 16.6 | 17.25 | 19.1 |

NOTE: $\rho$ = resistivity and $d$ = density.

**2.17** **Thermal conduction** Consider brass alloys with an $X$ atomic fraction of Zn. Since Zn addition increases the number of conduction electrons, we have to scale the final alloy resistivity calculated from the simple Matthiessen–Nordheim rule in Equation 2.22 down by a factor $(1 + X)$ (see Question 2.8) so that the resistivity of the alloy is $\rho \approx [\rho_o + CX(1 - X)]/(1 + X)$ in which $C = 300$ n$\Omega$ m and $\rho_o = \rho_{Cu} = 17$ n$\Omega$ m.

*a.* An 80 at.% Cu–20 at.% Zn brass disk of 40 mm diameter and 5 mm thickness is used to conduct heat from a heat source to a heat sink.

(1) Calculate the thermal resistance of the brass disk.

(2) If the disk is conducting heat at a rate of 100 W, calculate the temperature drop along the disk.

*b.* What should be the composition of brass if the temperature drop across the disk is to be halved?

**2.18** **Thermal resistance** Consider a thin insulating disk made of mica to electrically insulate a semiconductor device from a conducting heat sink. Mica has $\kappa = 0.75$ W m$^{-1}$ K$^{-1}$. The disk thickness is 0.1 mm, and the diameter is 10 mm. What is the thermal resistance of the disk? What is the temperature drop across the disk if the heat current through it is 25 W?

***2.19** **Thermal resistance** Consider a coaxial cable operating under steady-state conditions when the current flow through the inner conductor generates Joule heat at a rate $P = I^2R$. The heat generated per second by the core conductor flows through the dielectric; $Q' = I^2R$. The inner conductor reaches a temperature $T_i$, whereas the outer conductor is at $T_o$. Show that the thermal resistance $\theta$ of the hollow cylindrical insulation for heat flow in the radial direction is

*Thermal resistance of hollow cylinder*

$$\theta = \frac{(T_i - T_o)}{Q'} = \frac{\ln(b/a)}{2\pi\kappa L} \qquad [2.65]$$

where $a$ is the inside (core conductor) radius, $b$ is the outside radius (outer conductor), $\kappa$ is the thermal conductivity of the insulation, and $L$ is the cable length. Consider a coaxial cable that has a copper core conductor and polyethylene (PE) dielectric with the following properties: Core conductor resistivity $\rho = 19$ n$\Omega$ m, core radius $a = 4$ mm, dielectric thickness $b - a = 3.5$ mm, dielectric thermal conductivity $\kappa = 0.3$ W m$^{-1}$ K$^{-1}$. The outside temperature $T_o$ is 25 °C. The cable is carrying a current of 500 A. What is the temperature of the inner conductor?

**2.20** **The Hall effect** Consider a rectangular sample, a metal or an $n$-type semiconductor, with a length $L$, width $W$, and thickness $D$. A current $I$ is passed along $L$, perpendicular to the cross-sectional area $WD$. The face $W \times L$ is exposed to a magnetic field density $B$. A voltmeter is connected across the width, as shown in Figure 2.40, to read the Hall voltage $V_H$.

*a.* Show that the Hall voltage recorded by the voltmeter is

*Hall voltage*

$$V_H = \frac{IB}{Den}$$

*b.* Consider a 1-micron-thick strip of gold layer on an insulating substrate that is a candidate for a Hall probe sensor. If the current through the film is maintained at constant 100 mA, what is the magnetic field that can be recorded per $\mu$V of Hall voltage?

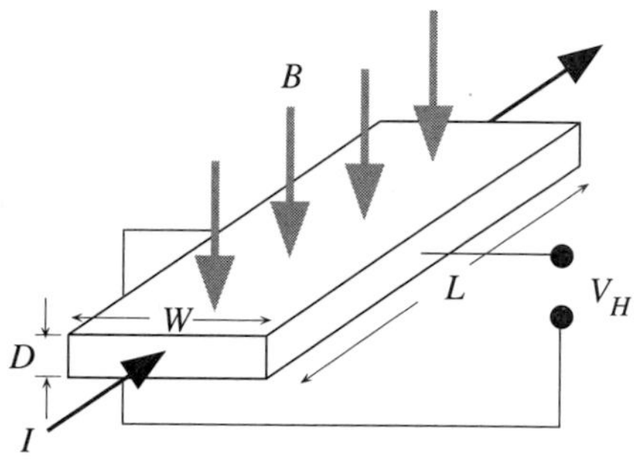


**Figure 2.40** Hall effect in a rectangular material with length *L*, width *W*, and thickness *D*. The voltmeter is across the width *W*.

**2.21** **The strain gauge** A **strain gauge** is a transducer attached to a body to measure its fractional elongation $\Delta L/L$ under an applied load (force) $F$. The gauge is a grid of many folded runs of a thin, resistive wire glued to a flexible backing, as depicted in Figure 2.41. The gauge is attached to the body under test such that the resistive wire length is parallel to the strain.

*a.* Assume that the elongation does not change the resistivity and show that the change in the resistance $\Delta R$ is related to the strain $\varepsilon = \Delta L/L$ by

*Strain gauge equation*

$$\Delta R \approx R(1 + 2\nu)\varepsilon \qquad [2.66]$$

where $\nu$ is the **Poisson ratio,** which is defined by

*Poisson ratio*

$$\nu = -\frac{\text{Transverse strain}}{\text{Longitudinal strain}} = -\frac{\varepsilon_t}{\varepsilon_l} \qquad [2.67]$$

where $\varepsilon_l$ is the strain along the applied load, that is, $\varepsilon_l = \Delta L/L = \varepsilon$, and $\varepsilon_t$ is the strain in the transverse direction, that is, $\varepsilon_t = \Delta D/D$, where $D$ is the diameter (thickness) of the wire.

*b.* Explain why a nichrome wire would be a better choice than copper for the strain gauge (consider the TCR).

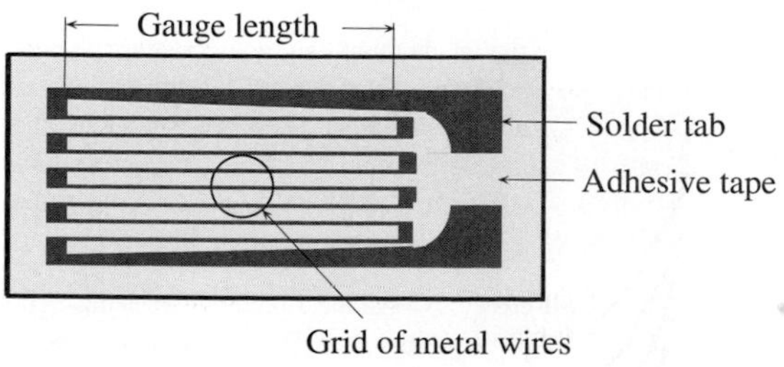


**Figure 2.41** The strain gauge consists of a long, thin wire folded several times along its length to form a grid as shown and embedded in a self-adhesive tape. The ends of the wire are attached to terminals (solder pads) for external connections. The tape is stuck on the component for which the strain is to be measured.

*c.* How do temperature changes affect the response of the gauge? Consider the effect of temperature on $\rho$. Also consider the differential expansion of the specimen with respect to the gauge wire such that even if there is no applied load, there is still strain, which is determined by the differential expansion coefficient, $\lambda_{\text{specimen}} - \lambda_{\text{gauge}}$, where $\lambda$ is the thermal coefficient of linear expansion: $L = L_0[1 + \lambda(T - T_0)]$, where $T_0$ is the reference temperature.

*d.* The gauge factor for a transducer is defined as the fractional change in the measured property $\Delta R/R$ per unit input signal ($\varepsilon$). What is the gauge factor for a metal-wire strain gauge, given that for most metals, $\nu \approx \frac{1}{3}$?

*e.* Consider a strain gauge that consists of a nichrome wire of resistivity 1 μΩ m, a total length of 1 m, and a diameter of 25 μm. What is $\Delta R$ for a strain of $10^{-3}$? Assume that $\nu \approx \frac{1}{3}$.

*f.* What will $\Delta R$ be if constantan wire with a resistivity of 500 nΩ m is used?

**2.22 Thermal coefficients of expansion and resistivity**

*a.* Consider a thin metal wire of length $L$ and diameter $D$. Its resistance is $R = \rho L/A$, where $A = \pi D^2/4$. By considering the temperature dependence of $L$, $A$, and $\rho$ individually, show that

*Change in R with temperature*

$$\frac{1}{R}\frac{dR}{dT} = \alpha_0 - \lambda_0$$

where $\alpha_0$ is the temperature coefficient of resistivity (TCR), and $\lambda_0$ is the temperature coefficient of linear expansion (thermal expansion coefficient or expansivity), that is,

$$\lambda_0 = L_0^{-1}\left(\frac{dL}{dT}\right)_{T=T_0} \quad \text{or} \quad \lambda_0 = D_0^{-1}\left(\frac{dD}{dT}\right)_{T=T_0}$$

*Note:* Consider differentiating $R = \rho L/[(\pi D^2)/4]$ with respect to $T$ with each parameter, $\rho$, $L$, and $D$, having a temperature dependence.

Given that typically, for most pure metals, $\alpha_0 \approx 1/273\ \text{K}^{-1}$ and $\lambda_0 \approx 2 \times 10^{-5}\ \text{K}^{-1}$, confirm that the temperature dependence of $\rho$ controls $R$, rather than the temperature dependence of the geometry. Is it necessary to modify the given equation for a wire with a noncircular cross section?

*b.* Is it possible to design a resistor from a suitable alloy such that its temperature dependence is almost nil? Consider the TCR of an alloy of two metals $A$ and $B$, for which $\alpha_{AB} \approx \alpha_A \rho_A/\rho_{AB}$.

**2.23 Temperature of a light bulb filament**

*a.* Consider a 100 W, 120 V incandescent bulb (lamp). The tungsten filament has a length of 0.579 m and a diameter of 63.5 μm. Its resistivity at room temperature is 56 nΩ m. Given that the resistivity of the filament can be represented as

$$\rho = \rho_0\left[\frac{T}{T_0}\right]^n \quad \textbf{[2.68]}$$

*Resistivity of W*

where $T$ is the temperature in K, $\rho_0$ is the resistance of the filament at $T_0$ K, and $n = 1.2$, estimate the temperature of the bulb when it is operated at the rated voltage, that is, directly from the main outlet. Note that the bulb dissipates 100 W at 120 V.

*b.* Suppose that the electrical power dissipated in the tungsten wire is totally radiated from the surface of the filament. The radiated power at the absolute temperature $T$ can be described by Stefan's law

$$P_{\text{radiated}} = \epsilon\sigma_S A\left(T^4 - T_0^4\right) \quad \textbf{[2.69]}$$

*Radiated power*

where $\sigma_S$ is Stefan's constant ($5.67 \times 10^{-8}\ \text{W m}^{-2}\ \text{K}^{-4}$), $\epsilon$ is the emissivity of the surface (0.35 for tungsten), $A$ is the surface area of the tungsten filament, and $T_0$ is room temperature (293 K). Obviously, for $T > T_0$, $P_{\text{radiated}} = \epsilon\sigma_S A T^4$.

Assuming that all the electrical power is radiated from the surface, estimate the temperature of the filament and compare it with your answer in part (a).

*c.* If the melting temperature of W is 3407 °C, what is the voltage that guarantees that the light bulb will blow?

**2.24 Einstein relation and ionic conductivity** In the case of ionic conduction, ions have to jump from one interstice to the neighboring one. This process involves overcoming a potential energy barrier just like atomic diffusion, and drift and diffusion are related. The drift mobility $\mu$ of ions is proportional to the diffusion coefficient $D$ because drift is limited by the atomic diffusion process. The **Einstein relation** relates the two by

*Einstein relation*

$$\frac{D}{\mu} = \frac{kT}{e} \qquad [2.70]$$

Diffusion coefficient of the $Na^+$ ion in sodium silicate ($Na_2O$–$SiO_2$) glasses at 400 °C is $3.4 \times 10^{-9}$ cm$^2$ s$^{-1}$. The density of such glasses is approximately 2.4 g cm$^{-3}$. Calculate the ionic conductivity and resistivity of (17.5 mol% $Na_2O$)(82.5 mol% $SiO_2$) sodium silicate glass at 400 °C and compare your result with the experimental values of the order of $10^4$ Ω cm for the resistivity.

**2.25 Skin effect**

*a.* What is the skin depth for a copper wire carrying a current at 60 Hz? The resistivity of copper at 27 °C is 17 nΩ m. Its relative permeability $\mu_r \approx 1$. Is there any sense in using a conductor for power transmission which has a diameter more than 2 cm?

*b.* What is the skin depth for an iron wire carrying a current at 60 Hz? The resistivity of iron at 27 °C is 97 nΩ m. Assume that its relative permeability $\mu_r \approx 700$. How does this compare with the copper wire? Discuss why copper is preferred over iron for power transmission even though iron is nearly 100 times cheaper than copper.

**2.26 Thin films**

*a.* Consider a polycrystalline copper film that has $R = 0.40$. What is the approximate mean grain size $d$ in terms of the mean free path $\lambda$ in the bulk that would lead to the polycrystalline Cu film having a resistivity that is $1.5\rho_{\text{bulk}}$. If the mean free path in the crystal is about 40 nm at room temperature, what is $d$?

*b.* What is the thickness $D$ of a copper film in terms of $\lambda$ in which surface scattering increases the film resistivity to $1.2\rho_{\text{bulk}}$ if the specular scattering fraction $p$ is 0.5?

*c.* Consider the data of Lim *et al.* (2003) presented in Table 2.13. Show that the excess resistivity, *i.e.* resistivity above that of bulk Cu, is roughly proportional to the reciprocal film thickness.

**Table 2.13** Resistivity $\rho_{\text{film}}$ of a copper film as a function of thickness $D$.

| $D$ (nm) | 8.61 | 17.2 | 34.4 | 51.9 | 69 | 85.8 | 102.6 | 120.3 | 173.2 | 224.3 |
|---|---|---|---|---|---|---|---|---|---|---|
| $\rho_{\text{film}}$ (nΩ m) | 121.8 | 75.3 | 46.1 | 38.5 | 32.1 | 25.2 | 22.0 | 20.5 | 19.9 | 18.8 |

NOTE: Film annealed at 150 °C.
SOURCE: Data extracted from J. W. Lim *et al.*, *Appl. Surf. Sci.* **217,** 95, 2003.

**2.27 Interconnects** Consider a high-transistor-density CMOS chip in which the interconnects are copper with a pitch $P$ of 500 nm, interconnect thickness $T$ of 400 nm, aspect ratio 1.4, and $H = X$. The dielectric is FSG with $\varepsilon_r = 3.6$. Consider two cases, $L = 1$mm and $L = 10$ mm, and calculate the overall effective interconnect capacitance $C_{\text{eff}}$ and the $RC$ delay time. Suppose that Al, which is normally Al with about 4 wt.% Cu in the microelectronics industry with a resistivity 31 nΩ m, is used as the interconnect. What is the corresponding $RC$delay time?

***2.28 Thin 50 nm interconnects** Equation 2.60 is for conduction in a thin film of thickness $D$ and assumes scattering from two surfaces, which yields an additional resistivity $\rho_2 = \rho_{\text{bulk}}\frac{3}{8}(\lambda/D)(1-p)$. An interconnect line in an IC is not quite a thin film and has four surfaces (interfaces), because the thickness $T$ of the conductor is comparable to the width $W$. If we assume $T = W$, we can very roughly take $\rho_4 \approx \rho_2 + \rho_2 \approx \rho_{\text{bulk}}\frac{3}{4}(\lambda/D)(1-p)$ in which $D = T$. (The exact expression is more complicated, but the latter will suffice for this problem.) In addition there will be a contribution from grain boundary scattering, (Equation 2.57a). For simplicity assume $T \approx W \approx X \approx H \approx 60$ nm, $\lambda = 40$ nm, $p = 0.5$ and $\varepsilon_r = 3.6$. If the mean grain size $d$ is roughly 40 nm and $R = 0.4$, estimate the resistivity of the interconnect and hence the $RC$ delay for a 1 mm interconnect.

**2.29** **TCR of thin films** Consider Matthiessen's rule applied to a thin film. Show that, very approximately, the product of the thermal coefficient of resistivity (TCR) $\alpha_{\text{film}}$ and the resistivity $\rho_{\text{film}}$ is equivalent to the product of the bulk TCR and resistivity:

$$\alpha_{\text{film}}\rho_{\text{film}} \approx \alpha_{\text{bulk}}\rho_{\text{bulk}}$$

**2.30** **Electromigration** Although electromigration-induced failure in Cu metallization is less severe than in Al metallization, it can still lead to interconnect failure depending on current densities and the operating temperature. In a set of experiments carried out on electroplated Cu metallization lines, failure of the Cu interconnects have been examined under accelerated tests (at elevated temperatures). The mean lifetime $t_{50}$ (time for 50 percent of the lines to break) have been measured as a function of current density $J$ and temperature $T$ at a given current density. The results are summarized in Table 2.14.

*a.* Plot semilogarithmically $t_{50}$ versus $1/T$ ($T$ in Kelvins) for the first three interconnects. Al(Cu) and Cu ($1.3 \times 0.7\mu\text{m}^2$) have single activation energies $E_A$. Calculate $E_A$ for these interconnects. Cu ($1.3 \times 0.7\mu\text{m}^2$) exhibits different activation energies for the high-and low-temperature regions. Estimate these $E_A$.

*b.* Plot on a log-log plot $t_{50}$ versus $J$ at 370 °C. Show that at low $J$, $n \approx 1.1$ and at high $J$, $n \approx 1.8$.

**Table 2.14** Results of electromigration failure experiments on various Al and Cu interconnects

| Al(Cu) [$J = 25$ mA/μm$^2$, $A = 0.35 \times 0.2$ $(\mu\text{m})^2$] | | Cu [$J = 25$ mA/μm$^2$, $A = 0.24 \times 0.28$ $(\mu\text{m})^2$] | | Cu [$J = 25$ mA/μm$^2$, $A = 1.3 \times 0.7$ $(\mu\text{m})^2$] | | Cu ($T = 370$ °C) | |
|---|---|---|---|---|---|---|---|
| **$T$ (°C)** | **$t_{50}$ (hr)** | **$T$ (°C)** | **$t_{50}$ (hr)** | **$T$ (°C)** | **$t_{50}$ (hr)** | **J mA μm$^{-2}$** | **$t_{50}$ (hr)** |
| 365 | 0.11 | 397 | 2.87 | 395 | 40.3 | 3.54 | 131.5 |
| 300 | 0.98 | 354 | 12.8 | 360 | 196 | 11.7 | 25.2 |
| 259 | 5.73 | 315 | 70.53 | 314 | 825 | 24.8 | 14.9 |
| 233 | 15.7 | 269 | 180 | 285 | 2098 | 49.2 | 4.28 |
| | | 232 | 899 | | | 74.1 | 2.29 |
| | | | | | | 140 | 0.69 |

NOTE: $A$ = width × height in micron$^2$.
SOURCE: Data extracted from R. Rosenberg *et al.*, (IBM, T. J. Watson Research Center, *Annu. Rev. Mater. Sci.*, **30,** 229, 2000, figures 29 and 31, and subject to small extraction errors.)

Gordon Teal (Left) and Morgan Sparks fabricated the first grown-junction Ge transistor in 1950–1951 at Bell Labs. Gordon Teal started at Bell Labs but later moved to Texas Instruments where he lead the development of the first commercial Si transistor; the first Si transistor was made at Bell Labs by Morris Tanenbaum.
SOURCE: Courtesy of Bell Laboratories, Lucent Technologies.

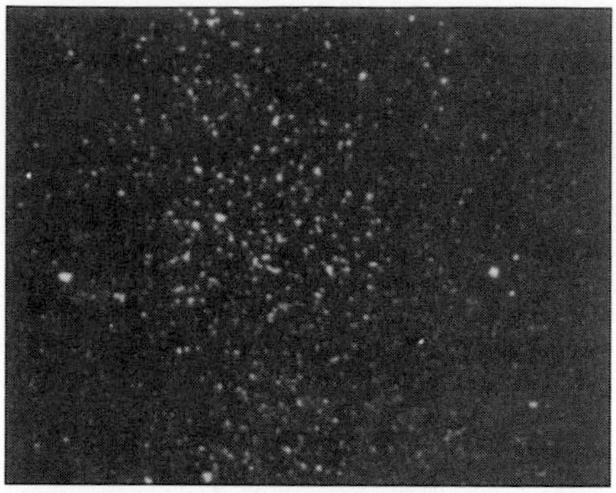

$3 \times 10^3$ photons

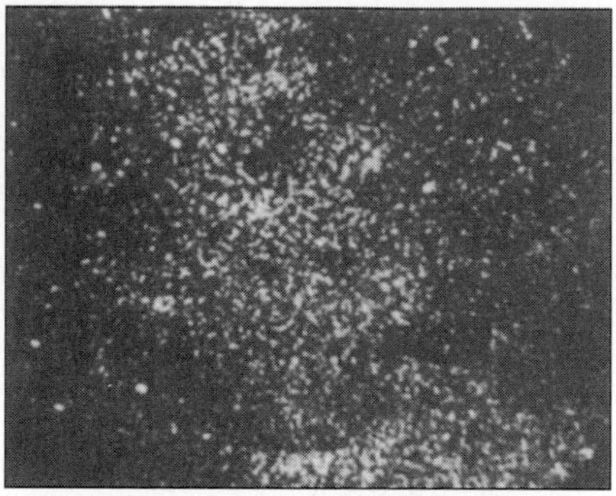

$1.2 \times 10^4$ photons

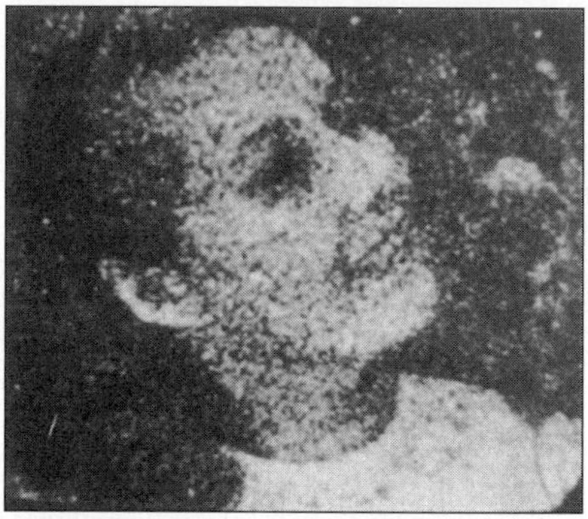

$9.3 \times 10^4$ photons

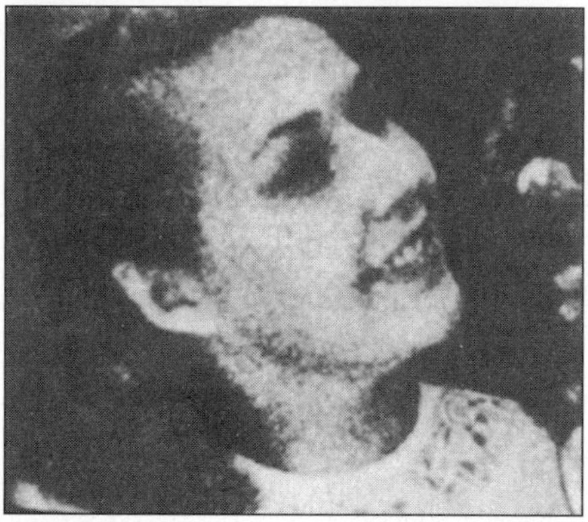

$7.6 \times 10^5$ photons

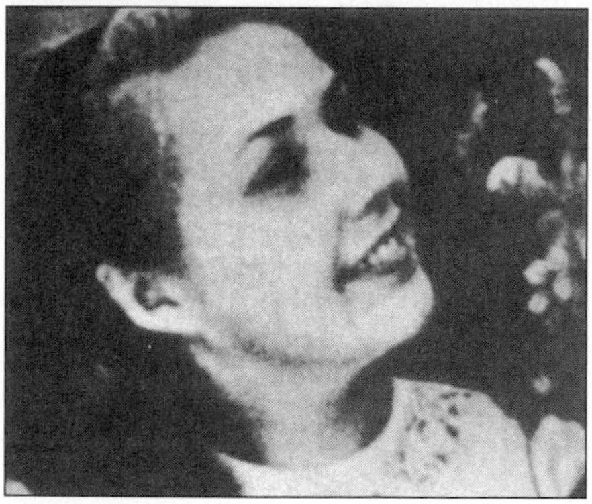

$3.6 \times 10^6$ photons

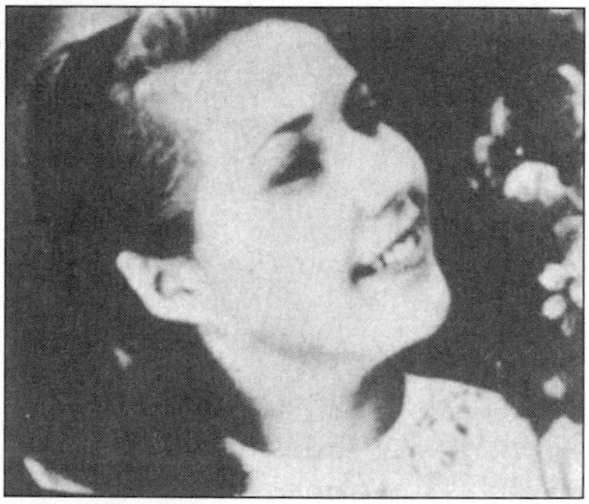

$2.8 \times 10^7$ photons

These electronic images were made with the number of photons indicated. The discrete nature of photons means that a large number of photons are needed to constitute an image with satisfactorily discernable details.

SOURCE: A. Rose, "Quantum and noise limitations of the visual process" *J. Opt. Soc. of America,* vol. 43, 715, 1953. (Courtesy of OSA.)

CHAPTER

# 3

# Elementary Quantum Physics

The triumph of modern physics is the triumph of quantum mechanics. Even the simplest experimental observation that the resistivity of a metal depends linearly on the temperature can only be explained by quantum physics, simply because we must take the mean speed of the conduction electrons to be nearly independent of temperature. The modern definitions of voltage and ohm, adopted in January 1990 and now part of the IEEE standards, are based on Josephson and quantum Hall effects, both of which are quantum mechanical phenomena.

One of the most important discoveries in physics has been the wave–particle duality of nature. The electron, which we have so far considered to be a particle and hence to be obeying Newton's second law ($F = ma$), can also exhibit wave-like properties quite contrary to our intuition. An electron beam can give rise to diffraction patterns and interference fringes, just like a light wave. Interference and diffraction phenomena displayed by light can only be explained by treating light as an electromagnetic wave. But light can also exhibit particle-like properties in which it behaves as if it were a stream of discrete entities ("photons"), each carrying a linear momentum and each interacting discretely with electrons in matter (just like a particle colliding with another particle).

## 3.1 PHOTONS

### 3.1.1 LIGHT AS A WAVE

In introductory physics courses, light is considered to be a wave. Indeed, such phenomena as interference, diffraction, refraction, and reflection can all be explained by the theory of waves. In all these phenomena, a ray of light is considered to be an **electromagnetic (EM) wave** with a given frequency, as depicted in Figure 3.1. The electric and magnetic fields, $\mathcal{E}_y$ and $B_z$, of this wave are perpendicular to each other and to the direction of propagation $x$. The electric field $\mathcal{E}_y$ at position $x$ at time $t$ may be

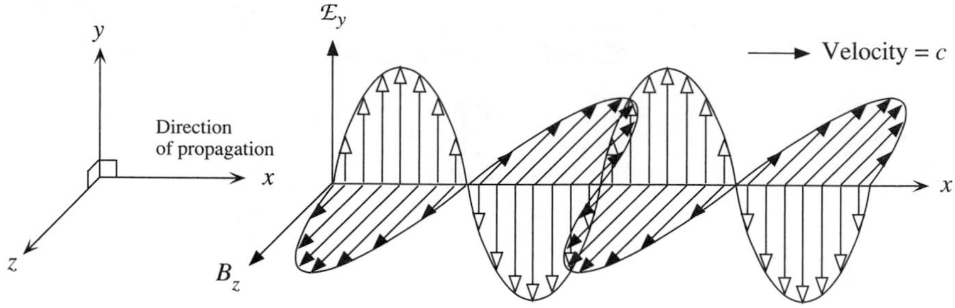


**Figure 3.1** The classical view of light as an electromagnetic wave.
An electromagnetic wave is a traveling wave with time-varying electric and magnetic fields that are perpendicular to each other and to the direction of propagation.

described by

*Traveling wave*

$$\mathcal{E}_y(x, t) = \mathcal{E}_o \sin(kx - \omega t) \qquad [3.1]$$

where $k$ is the wavenumber (propagation constant) related to the wavelength $\lambda$ by $k = 2\pi/\lambda$, and $\omega$ is the angular frequency of the wave (or $2\pi\nu$, where $\nu$ is the frequency). A similar equation describes the variation of the magnetic field $B_z$ (directed along $z$) with $x$ at any time $t$. Equation 3.1 represents a traveling wave in the $x$ direction, which, in the present example, is a sinusoidally varying function (Figure 3.1). The velocity of the wave (strictly the phase velocity) is

$$c = \frac{\omega}{k} = \nu\lambda$$

where $\nu$ is the frequency. The intensity $I$, that is, the energy flowing per unit area per second, of the wave represented by Equation 3.1 is given by

*Intensity of light wave*

$$I = \frac{1}{2}c\varepsilon_o\mathcal{E}_o^2 \qquad [3.2]$$

where $\varepsilon_o$ is the absolute permittivity.

Understanding the wave nature of light is fundamental to understanding interference and diffraction, two phenomena that we experience with sound waves almost on a daily basis. Figure 3.2 illustrates how the interference of secondary waves from the two slits $S_1$ and $S_2$ gives rise to the dark and bright fringes (called **Young's fringes**) on a screen placed at some distance from the slits. At point $P$ on the screen, the waves emanating from $S_1$ and $S_2$ interfere constructively, if they are in phase. This is the case if the path difference between the two rays is an integer multiple of the wavelength $\lambda$, or

$$S_1P - S_2P = n\lambda$$

where $n$ is an integer. If the two waves are out of phase by a path difference of $\lambda/2$, or

$$S_1P - S_2P = \left(n + \frac{1}{2}\right)\lambda$$

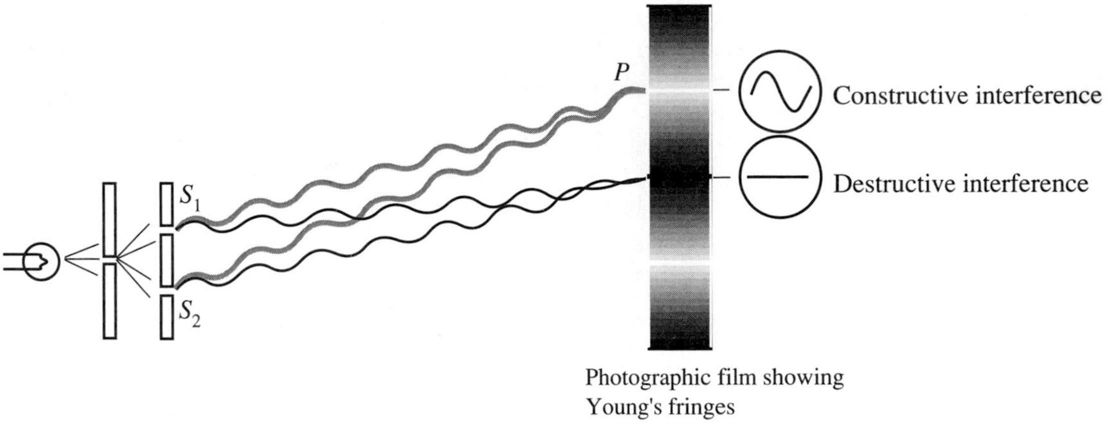


**Figure 3.2** Schematic illustration of Young's double-slit experiment.

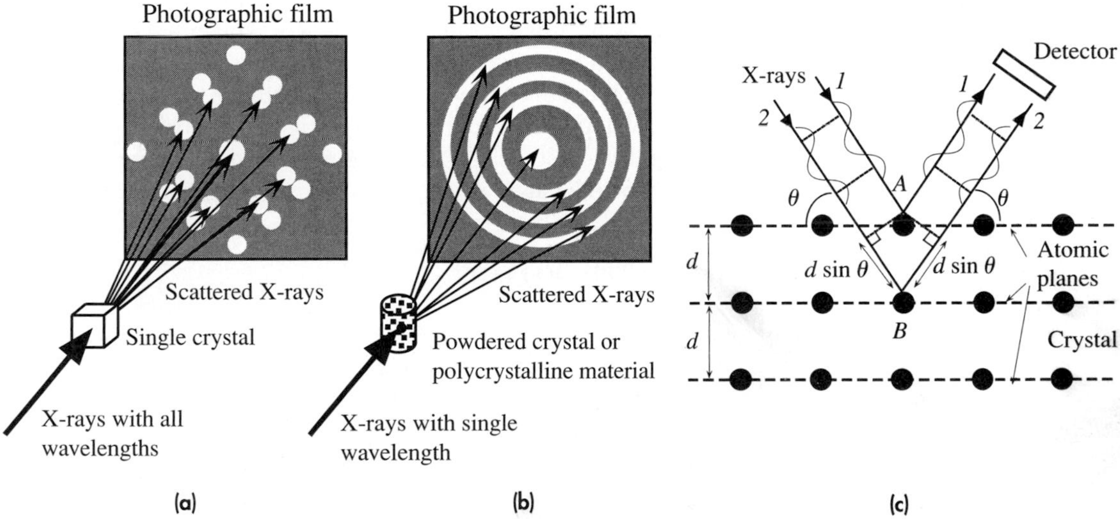


**Figure 3.3** Diffraction patterns obtained by passing X-rays through crystals can only be explained by using ideas based on the interference of waves.

(a) Diffraction of X-rays from a single crystal gives a diffraction pattern of bright spots on a photographic film.

(b) Diffraction of X-rays from a powdered crystalline material or a polycrystalline material gives a diffraction pattern of bright rings on a photographic film.

(c) X-ray diffraction involves the constructive interference of waves being "reflected" by various atomic planes in the crystal.

then the waves interfere destructively and the intensity at point $P$ vanishes. Thus, in the $y$ direction, the observer sees a pattern of bright and dark fringes.

When X-rays are incident on a crystalline material, they give rise to typical diffraction patterns on a photographic plate, as shown in Figure 3.3a and b, which can only be explained by using wave concepts. For simplicity, consider two waves, *1* and *2*, in an X-ray beam. The waves are initially in phase, as shown in Figure 3.3c. Suppose that wave *1* is "reflected" from the first plane of atoms in the crystal, whereas wave *2* is "reflected" from the second plane.[1] After reflection, wave *2* has traveled an additional distance equivalent to $2d \sin\theta$ before reaching wave *1*. The path difference between the two waves is $2d \sin\theta$, where $d$ is the separation of the atomic planes. For constructive interference, this must be $n\lambda$, where $n$ is an integer. Otherwise, waves *1* and *2* will interfere destructively and will cancel each other. Waves reflected from adjacent atomic planes interfere constructively to constitute a diffracted beam *only* when the path difference between the waves is an integer multiple of the wavelength, and this will only be the case for certain directions. Therefore the *condition* for the existence of a diffracted beam is

*Bragg diffraction condition*

$$2d \sin\theta = n\lambda \qquad n = 1, 2, 3, \ldots \qquad [3.3]$$

The condition expressed in Equation 3.3, for observing a diffracted beam, forms the whole basis for identifying and studying various crystal structures (the science of crystallography). The equation is referred to as **Bragg's law,** and arises from the constructive interference of waves.

Aside from exhibiting wave-like properties, light can behave like a stream of "particles" of zero rest-mass. As it turns out, the only way to explain a vast number of experiments is to view light as a stream of discrete entities or energy packets called **photons,** each carrying a quantum of energy $h\nu$, and momentum $h/\lambda$, where $h$ is a universal constant that can be determined experimentally, and $\nu$ is the frequency of light. This photonic view of light is drastically different than the simple wave picture and must be examined closely to understand its origin.

### 3.1.2 THE PHOTOELECTRIC EFFECT

Consider a quartz glass vacuum tube with two metal electrodes, a photocathode and an anode, which are connected externally to a voltage supply $V$ (variable and reversible) via an ammeter, as schematically illustrated in Figure 3.4. When the cathode is illuminated with light, if the frequency $\nu$ of the light is greater than a certain critical value $\nu_0$, the ammeter registers a current $I$, even when the anode voltage is zero (*i.e.*, the supply is bypassed). When light strikes the cathode, electrons are emitted with sufficient kinetic energy to reach the opposite electrode. Applying a positive voltage to the anode helps to collect more of the electrons and thus increases the current, until it saturates because all the photoemitted electrons have been collected. The current, then, is limited by the rate of supply of photoemitted electrons. If, on the other hand, we apply a negative voltage to the anode, we can "push" back the photoemitted electrons and hence reduce the current $I$. Figure 3.5a shows the dependence of the photocurrent on the anode voltage, for one particular frequency of light.

Recall that when an electron traverses a voltage difference $V$, its potential energy changes by $eV$ (potential difference is defined as work done per unit charge). When a negative voltage is applied to the anode, the electron has to do work to get to this electrode, and this work comes from its kinetic energy just after photoemission. When the negative anode voltage $V$ is equal to $V_0$, which just "extinguishes" the current $I$, we

[1] Strictly, one must consider the scattering of waves from the electrons in individual atoms (*e.g.*, atoms $A$ and $B$ in Figure 3.3c) and examine the constructive interference of these scattered waves, which leads to the same condition as that derived in Equation 3.3.

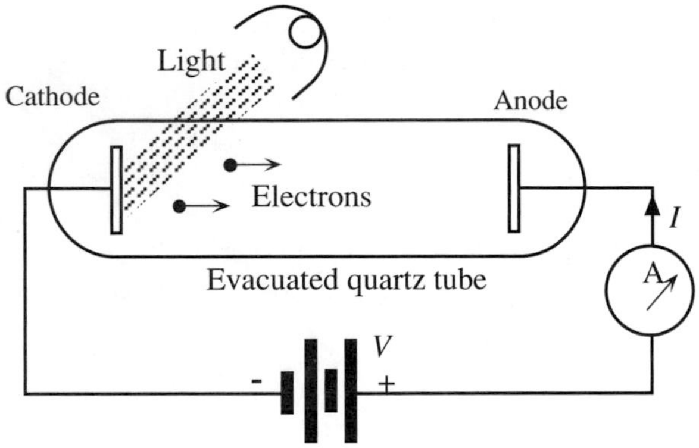


**Figure 3.4** The photoelectric effect.

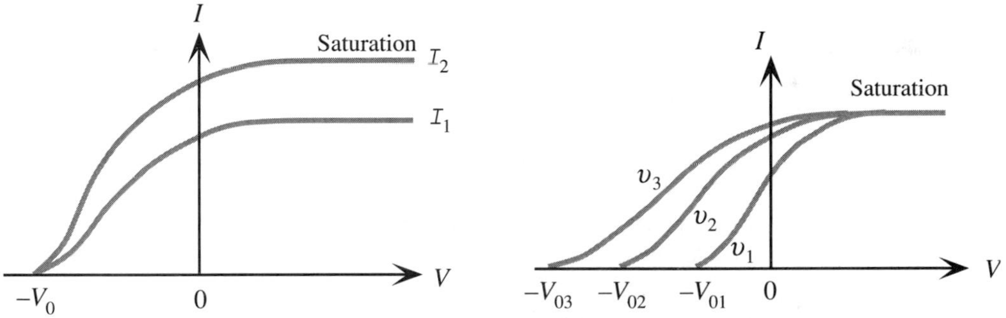


**(a) Photoelectric current versus voltage when the cathode is illuminated with light of identical wavelength but different intensities ($I$). The saturation current is proportional to the light intensity.**

**(b) The stopping voltage and therefore the maximum kinetic energy of the emitted electron increases with the frequency of light, $\upsilon$. (The light intensity is not the same; it is adjusted to keep the saturation current the same.)**

**Figure 3.5** Results from the photoelectric experiment.

know that the potential energy "gained" by the electron is just the kinetic energy lost by the electron, or

$$eV_0 = \frac{1}{2}m_e v^2 = KE_m$$

where $v$ is the velocity and $KE_m$ is the kinetic energy of the electron just after photoemission. Therefore, we can conveniently measure the maximum kinetic energy $KE_m$ of the emitted electrons.

For a given frequency of light, increasing the intensity of light $I$ requires the *same* voltage $V_0$ to extinguish the current; that is, the $KE_m$ of emitted electrons is independent of the light intensity $I$. This is quite surprising. However, increasing the intensity does increase the saturation current. Both of these effects are noted in the *I–V* results shown in Figure 3.5a.

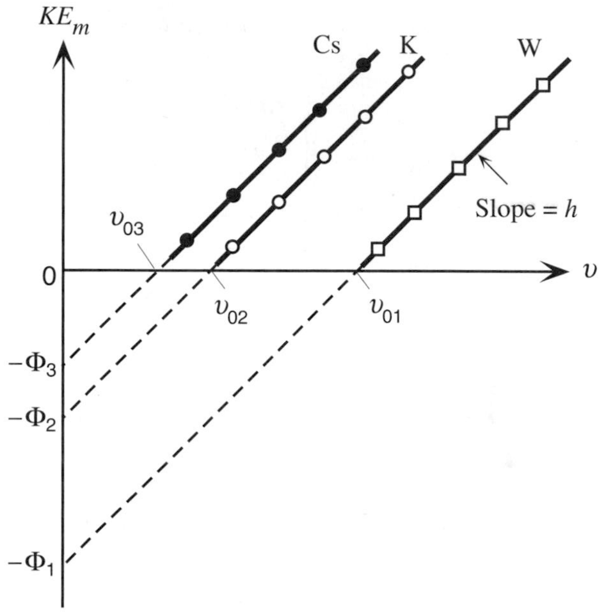


**Figure 3.6** The effect of varying the frequency of light and the cathode material in the photoelectric experiment. The lines for the different materials have the same slope $h$ but different intercepts.

Since the magnitude of the saturation photocurrent depends on the light intensity $I$, whereas the *KE* of the emitted electron is independent of $I$, we are forced to conclude that only the *number* of electrons ejected depends on the light intensity. Furthermore, if we plot $KE_m$ (from the $V_0$ value) against the light frequency $\nu$ for different electrode metals for the cathode, we find the typical behavior shown in Figure 3.6. This shows that the *KE* of the emitted electron depends on the frequency of light. The experimental results shown in Figure 3.6 can be summarized by a statement that relates the $KE_m$ of the electron to the frequency of light and the electrode metal, as follows:

*Photoemitted electron maximum KE*

$$KE_m = h\nu - h\nu_0 \qquad [3.4]$$

where $h$ is the slope of the straight line and is independent of the type of metal, whereas $\nu_0$ depends on the electrode material for the photocathode (*e.g.*, $\nu_{01}$, $\nu_{02}$, etc.). Equation 3.4 is essentially a succinct statement of the experimental observations of the photoelectric effect as exhibited in Figure 3.6. The constant $h$ is called **Planck's constant,** which, from the slope of the straight lines in Figure 3.6, can be shown to be about $6.6 \times 10^{-34}$ J s. This was beautifully demonstrated by Millikan in 1915, in an excellent series of photoelectric experiments using different photocathode materials.

The successful interpretation of the photoelectric effect was first given in 1905 by Einstein, who proposed that light consists of "energy packets," each of which has the magnitude $h\nu$. We can call these energy quanta **photons.** When one photon strikes an electron, its energy is transferred to the electron. The whole photon becomes absorbed by the electron. Yet, an electron in a metal is in a lower state of potential energy (*PE*) than in vacuum, by an amount $\Phi$, which we call the **work function** of the metal, as illustrated in Figure 3.7. The lower *PE* is what keeps the electron in the metal; otherwise, it would "drop out."

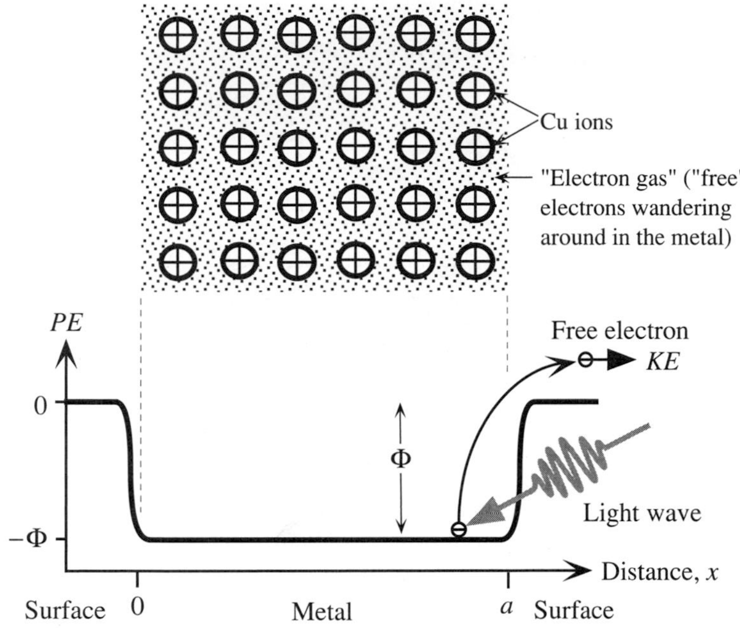


**Figure 3.7** The *PE* of an electron inside the metal is lower than outside by an energy called the workfunction of the metal.
Work must be done to remove the electron from the metal.

This lower *PE* is a result of the Coulombic attraction interaction between the electron and the positive metal ions. Some of the photon energy $h\nu$ therefore goes toward overcoming this *PE* barrier. The energy that is left $(h\nu - \Phi)$ gives the electron its *KE*. The work function $\Phi$ changes from one metal to another. Photoemission only occurs when $h\nu$ is greater than $\Phi$. This is clearly borne out by experiment, since a critical frequency $\nu_0$ is needed to register a photocurrent. When $\nu$ is less than $\nu_0$, even if we use an extremely intense light, no current exists because no photoemission occurs, as demonstrated by the experimental results in Figure 3.6. Inasmuch as $\Phi$ depends on the metal, so does $\nu_0$. Therefore, in Einstein's interpretation $h\nu_0 = \Phi$. In fact, the measurement of $\nu_0$ constitutes one method of determining the work function of the metal.

This explanation for the photoelectric effect is further supported by the fact that the work function $\Phi$ from $h\nu_0$ is in good agreement with that from thermionic emission experiments. There is an apparent similarity between the *I–V* characteristics of the phototube and that of the vacuum tube used in early radios. The only difference is that in the vacuum tube, the emission of electrons from the cathode is achieved by heating the cathode. Thermal energy ejects some electrons over the *PE* barrier $\Phi$. The measurement of $\Phi$ by this thermionic emission process agrees with that from photoemission experiments.

In the photonic interpretation of light, we still have to resolve the meaning of the intensity of light, because the classical intensity in Equation 3.2

$$I = \frac{1}{2} c\varepsilon_o \mathcal{E}_o^2$$

*Classical light intensity*

is obviously not acceptable. Increasing the intensity of illumination in the photoelectric experiment increases the saturation current, which means that more electrons are

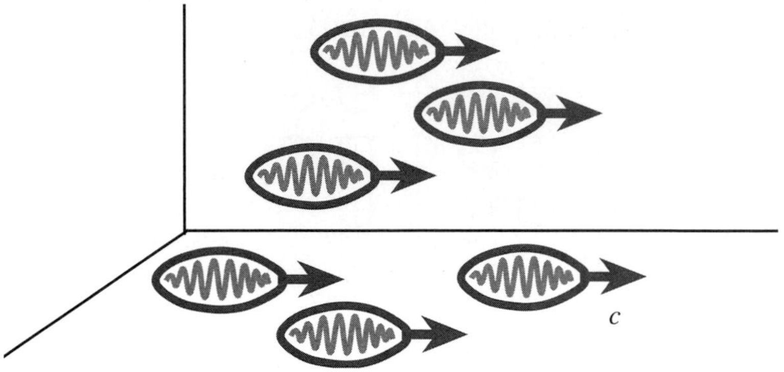


**Figure 3.8** Intuitive visualization of light consisting of a stream of photons (not to be taken too literally).
SOURCE: R. Serway, C. J. Moses, and C. A. Moyer, *Modern Physics*, Saunders College Publishing, 1989, p. 56, figure 2.16(b).

emitted per unit time. We therefore infer that the cathode must be receiving more photons per unit time at higher intensities. By definition, "intensity" refers to the amount of energy flowing through a unit area per unit time. If the number of photons crossing a unit area per unit time is the **photon flux,** denoted by $\Gamma_{ph}$, then the flow of energy through a unit area per unit time, the **light intensity,** is the product of this photon flux and the energy per photon, that is,

*Light intensity*

$$I = \Gamma_{ph} h\nu \quad [3.5]$$

where

*Photon flux*

$$\Gamma_{ph} = \frac{\Delta N_{ph}}{A \Delta t} \quad [3.6]$$

in which $\Delta N_{ph}$ is the net number of photons crossing an area $A$ in time $\Delta t$. With the energy of a photon given as $h\nu$ and the intensity of light defined as $\Gamma_{ph} h\nu$, the explanation for the photoelectric effect becomes self-consistent. The interpretation of light as a stream of photons can perhaps be intuitively imagined as depicted in Figure 3.8.

**EXAMPLE 3.1** **ENERGY OF A BLUE PHOTON** What is the energy of a blue photon that has a wavelength of 450 nm?

**SOLUTION**

The energy of the photon is given by

$$E_{ph} = h\nu = \frac{hc}{\lambda} = \frac{(6.6 \times 10^{-34}\ \text{J s})(3 \times 10^{8}\ \text{m s}^{-1})}{450 \times 10^{-9}\ \text{m}} = 4.4 \times 10^{-19}\ \text{J}$$

Generally, with such small energy values, we prefer electron–volts (eV), so the energy of the photon is

$$\frac{4.4 \times 10^{-19}\ \text{J}}{1.6 \times 10^{-19}\ \text{J/eV}} = 2.75\ \text{eV}$$

**EXAMPLE 3.2**

**THE PHOTOELECTRIC EXPERIMENT** In the photoelectric experiment, green light, with a wavelength of 522 nm, is the longest-wavelength radiation that can cause the photoemission of electrons from a clean sodium surface.

*a.* What is the work function of sodium, in electron–volts?

*b.* If UV (ultraviolet) radiation of wavelength 250 nm is incident to the sodium surface, what will be the kinetic energy of the photoemitted electrons, in electron–volts?

*c.* Suppose that the UV light of wavelength 250 nm has an intensity of 20 mW $cm^{-2}$. If the emitted electrons are collected by applying a positive bias to the opposite electrode, what will be the photoelectric current density?

**SOLUTION**

*a.* At threshold, the photon energy just causes photoemissions; that is, the electron just overcomes the potential barrier $\Phi$. Thus, $hc/\lambda_0 = e\Phi$, where $\Phi$ is the work function in eV, and $\lambda_0$ is the longest wavelength.

$$\Phi = \frac{hc}{e\lambda_0} = \frac{(6.626 \times 10^{-34}\text{ J s})(3 \times 10^8\text{ m s}^{-1})}{(1.6 \times 10^{-19}\text{ J/eV})(522 \times 10^{-9}\text{ m})} = 2.38\text{ eV}$$

*b.* The energy of the incoming photon $E_{ph}$ is $(hc/\lambda)$, so the excess energy over $e\Phi$ goes to the kinetic energy of the electron. Thus,

$$KE = \frac{hc}{e\lambda} - \Phi = \frac{(6.626 \times 10^{-34}\text{ J s})(3 \times 10^8\text{ m s}^{-1})}{(1.6 \times 10^{-19}\text{ J/eV})(250 \times 10^{-9}\text{ m})} - 2.38\text{ eV} = 2.58\text{ eV}$$

*c.* The light intensity (defined as energy flux) is given by $I = \Gamma_{ph}(hc/\lambda)$, where $\Gamma_{ph}$ is the number of photons arriving per unit area per unit time; that is, photon flux and $(hc/\lambda)$ is the energy per photon. Thus, if each photon releases one electron, the electron flux will be equal to the photon flux, and the current density, which is the charge flux, will be

$$J = e\Gamma_{ph} = \frac{eI\lambda}{hc} = \frac{(1.6 \times 10^{-19}\text{ C})(20 \times 10^{-3} \times 10^4\text{ J s}^{-1}\text{ m}^{-2})(250 \times 10^{-9}\text{ m})}{(6.626 \times 10^{-34}\text{ J s})(3 \times 10^8\text{ m s}^{-1})}$$

$$= 40.3\text{ A m}^{-2} \quad \text{or} \quad 4.0\text{ mA cm}^{-2}$$

## 3.1.3 Compton Scattering

When an X-ray strikes an electron, it is deflected, or "scattered." In addition, the electron moves away after the interaction, as depicted in Figure 3.9. The wavelength of the incoming and scattered X-rays can readily be measured. The frequency $\nu'$ of the scattered X-ray is less than the frequency $\nu$ of the incoming X-ray. When the *KE* of the electron is determined, we find that

$$KE = h\nu - h\nu'$$

Since the electron now also has a momentum $p_e$, then from the conservation of linear momentum law, we are forced to accept that the X-ray also has a momentum. The Compton effect experiments showed that the momentum of the photon is related to its wavelength by

$$p = \frac{h}{\lambda} \qquad [3.7]$$

*Momentum of a photon*

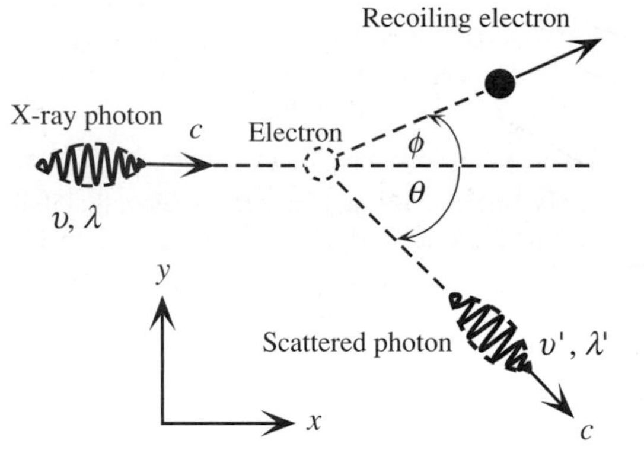


**Figure 3.9** Scattering of an X-ray photon by a "free" electron in a conductor.

We see that a photon not only has an energy $h\nu$, but also a momentum $p$, and it interacts as if it were a discrete entity like a particle. Therefore, when discussing the properties of a photon, we must consider its energy and momentum as if it were a particle.

We should mention that the description of the Compton effect shown in Figure 3.9 is, in fact, the inference from a more practical experiment involving the scattering of X-rays from a metal target. A collimated monochromatic beam of X-rays of wavelength $\lambda_0$ strikes a conducting target, such as graphite, as illustrated in Figure 3.10a. A conducting target contains a large number of nearly "free" electrons (conduction electrons), which can scatter the X-rays. The scattered X-rays are detected at various angles $\theta$ with respect to the original direction, and their wavelength $\lambda'$ is measured. The result of the experiment is therefore the scattered wavelength $\lambda'$ measured at various scattering angles $\theta$, as shown in Figure 3.10b. It turns out that the $\lambda'$ versus $\theta$ results agree with the conservation of linear momentum law applied to an X-ray photon colliding with an electron with the momentum of the photon given precisely by Equation 3.7.

The photoelectric experiment and the Compton effect are just two convincing experiments in modern physics that force us to accept that light can have particle-like properties. We already know that it can also exhibit wave-like properties, in such experiments as Young's interference fringes. We are then faced with what is known as the wave–particle dilemma. How do we know whether light is going to behave like a wave or a particle? The properties exhibited by light depend very much on the nature of the experiment. Some experiments will require the wave model, whereas others may use the particulate interpretation of light. We should perhaps view the two interpretations as two complementary ways of modeling the behavior of light when it interacts with matter, accepting the fact that light has a dual nature. Both models are needed for a full description of the behavior of light.

The expressions for the energy and momentum of the photon, $E = h\nu$ and $p = h/\lambda$, can also be written in terms of the angular frequency $\omega(= 2\pi\nu)$ and the wave number $k$, defined as $k = 2\pi/\lambda$. If we define $\hbar = h/2\pi$, then

*Photon energy and momentum*

$$E = h\nu = \hbar\omega \qquad \text{and} \qquad p = \frac{h}{\lambda} = \hbar k \tag{3.8}$$

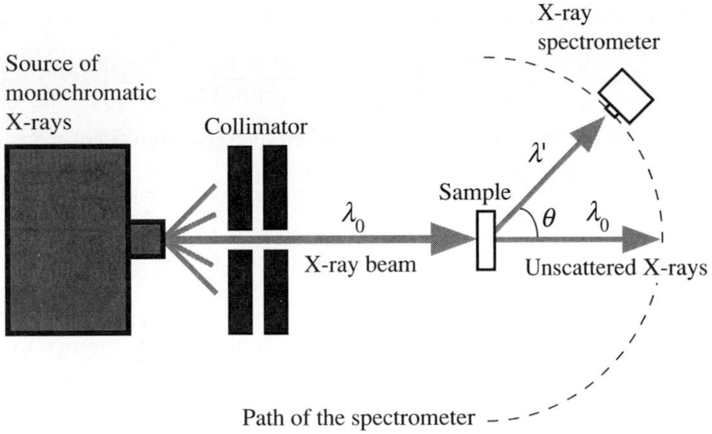


(a) A schematic diagram of the Compton experiment

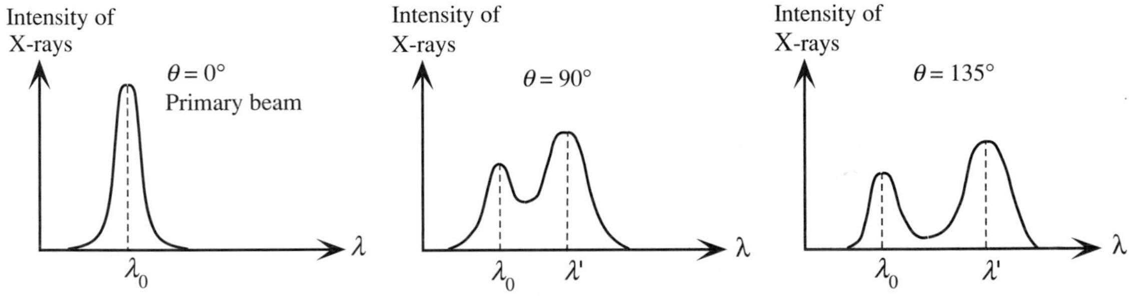


(b) Results from the Compton experiment

**Figure 3.10** The Compton experiment and its results.

---

**EXAMPLE 3.3**

**X-RAY PHOTON ENERGY AND MOMENTUM** X-rays are photons with very short wavelengths that can penetrate or pass through objects, hence their use in medical imaging, security scans at airports, and many other applications including X-ray diffraction studies of crystal structures. Typical X-rays used in mammography (medical imaging of breasts) have a wavelength of about 0.6 angstrom (1 Å $= 10^{-10}$ m). Calculate the energy and momentum of an X-ray photon with this wavelength, and the velocity of a *corresponding* electron that has the same momentum.

**SOLUTION**

The photon energy $E_{ph}$ is given by

$$E_{ph} = h\nu = \frac{hc}{\lambda} = \frac{(6.6 \times 10^{-34}\ \text{J s})(3 \times 10^{8}\ \text{m s}^{-1})}{0.6 \times 10^{-10}\ \text{m}} \times \frac{\text{eV J}^{-1}}{1.6 \times 10^{-19}}$$

$$= 2.06 \times 10^{4}\ \text{eV} \quad \text{or} \quad 20.6\ \text{keV}$$

The momentum $p$ of this X-ray photon is

$$p = \frac{h}{\lambda} = \frac{6.6 \times 10^{-34}\ \text{J s}}{0.6 \times 10^{-10}\ \text{m}} = 1.1 \times 10^{-23}\ \text{kg m s}^{-1}$$

A corresponding electron with the same momentum, $m_e v_{\text{electron}} = p$, would have a velocity

$$v_{\text{electron}} = \frac{p}{m_e} = \frac{1.1 \times 10^{-23}\ \text{kg m s}^{-1}}{9.1 \times 10^{-31}\ \text{kg}} = 1.2 \times 10^7\ \text{m s}^{-1}$$

This is much greater than the average speed of conduction (free) electrons whizzing around inside a metal, which is $\sim 10^6$ m s$^{-1}$.

## 3.1.4 BLACK BODY RADIATION

Experiments indicate that all objects emit and absorb energy in the form of radiation, and the intensity of this radiation depends on the radiation wavelength and temperature of the object. This radiation is frequently termed **thermal radiation.** When the object is in thermal equilibrium with its surroundings, that is, at the same temperature, the object absorbs as much radiation energy as it emits. On the other hand, when the temperature of the object is above the temperature of its surroundings, there is a net emission of radiation energy. The maximum amount of radiation energy that can be emitted by an object is called the **black body radiation.** Although, in general, the intensity of the radiated energy depends on the material's surface, the radiation emitted from a cavity with a small aperture is independent of the material of the cavity and corresponds very closely to black body radiation.

The intensity of the emitted radiation has the spectrum (*i.e.*, intensity vs. wavelength characteristic), and the temperature dependence illustrated in Figure 3.11. It is useful to define a **spectral irradiance** $I_\lambda$ as the emitted radiation intensity (power per unit area) per unit wavelength, so that $I_\lambda\, \delta\lambda$ is the intensity in a small range of wavelengths $\delta\lambda$. Figure 3.11 shows the typical $I_\lambda$ versus $\lambda$ behavior of black body radiation at two temperatures. We assume that the characteristics of the radiation emerging from the aperture represent those of the radiation within the cavity.

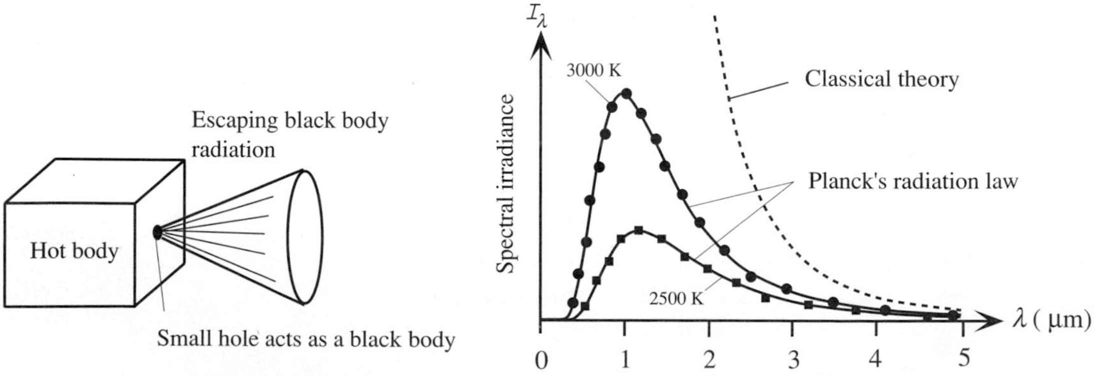


**Figure 3.11** Schematic illustration of black body radiation and its characteristics.
Spectral irradiance versus wavelength at two temperatures (3000 K is about the temperature of the incandescent tungsten filament in a light bulb).

Classical physics predicts that the acceleration and deceleration of the charges due to various thermal vibrations, oscillations, or motions of the atoms in the surface region of the cavity material result in electromagnetic waves of the emissions. These waves then interfere with each other, giving rise to many types of standing electromagnetic waves with different wavelengths in the cavity. Each wave contributes an energy $kT$ to the emitted intensity. If we calculate the number of standing waves within a small range of wavelength, the classical prediction leads to the **Rayleigh–Jeans law** in which $I_\lambda \propto 1/\lambda^4$ and $I_\lambda \propto T$, which are not in agreement with the experiment, especially in the short-wavelength range (see Figure 3.11).

Max Planck (1900) was able to show that the experimental results can be explained if we assume that the radiation within the cavity involves the emission and absorption of discrete amounts of light energy by the oscillation of the molecules of the cavity material. He assumed that oscillating molecules emit and absorb a quantity of energy that is an integer multiple of a discrete energy quantum that is determined by the frequency $\nu$ of the radiation and given by $h\nu$. This is what we now call a photon. He then considered the energy distribution (the statistics) in the molecular oscillations and took the probability of an oscillator possessing an energy $nh\nu$ (where $n$ is an integer) to be proportional to the Boltzmann factor, $\exp(-nh\nu/kT)$. He eventually derived the mathematical form of the black body radiation characteristics in Figure 3.11. Planck's black body radiation formula for $I_\lambda$ is generally expressed as

$$I_\lambda = \frac{2\pi hc^2}{\lambda^5\left[\exp\left(\dfrac{hc}{\lambda kT}\right) - 1\right]} \qquad [3.9]$$

*Planck's radiation law*

where $k$ is the Boltzmann constant. Planck's radiation law based on the emission and absorption of photons is in excellent agreement with all observed black body radiation characteristics as depicted in Figure 3.11.

Planck's radiation law is undoubtedly one of the major successes of modern physics. We can take Equation 3.9 one step further and derive *Stefan's black body radiation law* that was used in Chapter 2 to calculate the rate of radiation energy emitted from the hot filament of a light bulb. If we integrate $I_\lambda$ over all wavelengths,[2] we will obtain the total radiative power $P_S$ emitted by a black body per unit surface area at a temperature $T$,

$$P_S = \int_0^\infty I_\lambda \, d\lambda = \left(\frac{2\pi^5 k^4}{15c^2h^3}\right)T^4 = \sigma_S T^4 \qquad [3.10]$$

*Stefan's black body radiation law*

where

$$\sigma_S = \frac{2\pi^5 k^4}{15c^2h^3} = 5.670 \times 10^{-8} \text{ W m}^{-2} \text{ K}^{-4} \qquad [3.11]$$

*Stefan's constant*

[2] The integration of Equation 3.9 can be done by looking up definite integral tables in math handbooks—we only need the result of the mathematics, which is Equation 3.10. The $P_S$ in Equation 3.10 is sometimes called the *radiant emittance*. Stefan's law is also known as the Stefan-Boltzmann law.

Equation 3.10 in which $P_S = \sigma_S T^4$ is **Stefan's law** for black body radiation, and the $\sigma_S$ in Equation 3.11 is the **Stefan constant** with a value of approximately $5.67 \times 10^{-8}$ W m$^{-2}$ K$^{-4}$. Stefan's law was known before Planck used quantum physics to derive his black body radiation law embedded in $I_\lambda$. A complete explanation of Stefan's law and the value for $\sigma_S$ however had to wait for Planck's law. The $h$ in Equation 3.10 or 3.11 is a clear pointer that the origin of Stefan's law lies in quantum physics.

**EXAMPLE 3.4**

**STEFAN'S LAW AND THE LIGHT BULB** Stefan's law as stated in Equation 3.10 applies to a perfect black body that is emitting radiation into its environment which is at absolute zero. If the environment or the surroundings of the black body is at a finite temperature $T_o$, than the surroundings would also be emitting radiation. The same black body will then also absorb radiation from its environment. By definition, a black body is not only a perfect emitter of radiation but also a perfect absorber of radiation. The rate of radiation absorbed from the environment per unit surface is again given by Equation 3.10 but with $T_o$ instead of $T$ since it is the surroundings that are emitting the radiation. Thus, $\sigma_S T_o^4$ is the absorbed radiation rate from the surroundings, so

$$\text{Net rate of radiative power emission per unit surface area} = \sigma_S T^4 - \sigma_S T_o^4$$

Further, not all surfaces are perfect black bodies. Black body emission is the maximum possible emission from a surface at a given temperature. A real surface emits less than a black body. **Emissivity** $\varepsilon$ of a surface measures the efficiency of a surface in terms of a black body emitter; it is the ratio of the emitted radiation from a real surface to that emitted from a black body at a given temperature and over the same wavelength range. The *total* net rate of radiative power emission becomes

*Stefan's law for a real surface*

$$P_{\text{radiation}} = S\varepsilon\sigma_S\left(T^4 - T_o^4\right) \qquad [3.12]$$

where $S$ is the surface area that is emitting the radiation. Consider the tungsten filament of a 100 W light bulb in a lamp. When we switch the lamp on, the current through the filament generates heat which quickly heats up the filament to an operating temperature $T_f$. At this temperature, the electric energy that is input into the bulb is radiated away from the filament as radiation energy. A typical 100 W bulb filament has a length of 57.9 cm and a diameter of 63.5 μm. Its surface area is then

$$S = \pi(63.5 \times 10^{-6}\text{ m})(0.579\text{ m}) = 1.155 \times 10^{-4}\text{ m}^2$$

The emissivity $\varepsilon$ of tungsten is about 0.35. Assuming that under steady-state operation all the electric power that is input into the bulb's filament is radiated away,

$$\begin{aligned}100\text{ W} = P_{\text{radiation}} &= S\varepsilon\sigma_S\left(T_f^4 - T_o^4\right)\\ &= (1.155 \times 10^{-4}\text{ m}^2)(0.35)(5.67 \times 10^{-8}\text{ W m}^{-2}\text{ K}^{-4})\left(T_f^4 - 300^4\right)\end{aligned}$$

Solving we find,

$$T_f = 2570\text{ K} \qquad \text{or} \qquad 2297\text{ °C}$$

which is well below the melting temperature of tungsten which is 3422 °C. The second term that has $T_o^4$ has very little effect on the calculation as radiation absorption from the environment is practically nil compared with the emitted radiation at $T_f$.

The shift in the spectral intensity emitted from a black body with temperature is of particular interest to many photoinstrumention engineers. The peak spectral intensity in Figure 3.11 occurs at a wavelength $\lambda_{\max}$, which, by virtue of Equation 3.9, depends on the temperature of the black body. By substituting a new variable $x = hc/(kT\lambda)$ into Equation 3.9 and differentiating it, or plotting it against $x$, we can show that the peak occurs when

*Wien's displacement law*

$$\lambda_{max} T \approx 2.89 \times 10^{-3} \text{ m K}$$

which is known as **Wien's displacement law.** The peak emission shifts to lower wavelengths as the temperature increases. We can calculate the wavelength $\lambda_{max}$ corresponding to the peak in the spectral distribution of emitted radiation from our 100 W lamp: $\lambda_{max} = (2.89 \times 10^{-3} \text{ m K})/(2570 \text{ K}) = 1.13 \text{ µm}$ (in the infrared).

# 3.2 THE ELECTRON AS A WAVE

## 3.2.1 DE BROGLIE RELATIONSHIP

It is apparent from the photoelectric and Compton effects that light, which we thought was a wave, can behave as if it were a stream of particulate-like entities called photons. Can electrons exhibit wave-like properties? Again, this depends on the experiment and on the energy of the electrons.

When the interference and diffraction experiments in Figures 3.2 and 3.3 are repeated with an electron beam, very similar results are found to those obtainable with light and X-rays. When we use an electron beam in Young's double-slit experiment, we observe high- and low-intensity regions (*i.e.*, Young's fringes), as illustrated in Figure 3.12. The interference pattern is viewed on a fluorescent TV screen. When an energetic electron beam hits an Al polycrystalline sample, it produces diffraction rings on a fluorescent screen (Figure 3.13), just like X-rays do on a photographic

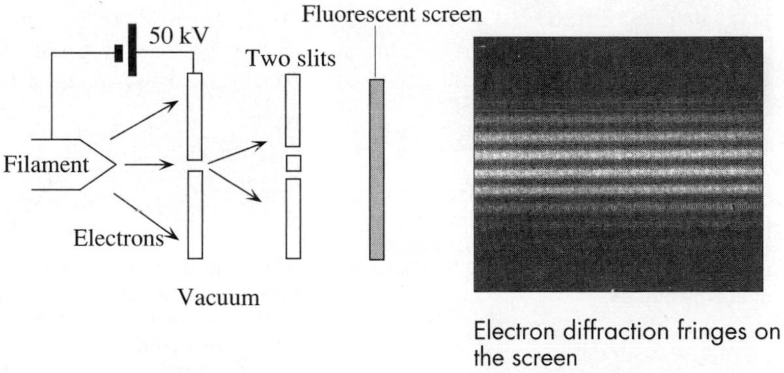


**Figure 3.12** Young's double-slit experiment with electrons involves an electron gun and two slits in a cathode ray tube (CRT) (hence, in vacuum).

Electrons from the filament are accelerated by a 50 kV anode voltage to produce a beam that is made to pass through the slits. The electrons then produce a visible pattern when they strike a fluorescent screen (*e.g.*, a TV screen), and the resulting visual pattern is photographed.

SOURCE: Pattern from C. Jönsson, D. Brandt, and S. Hirschi, *Am. J. Physics*, **42,** 1974, p. 9, figure 8. Used with permission.

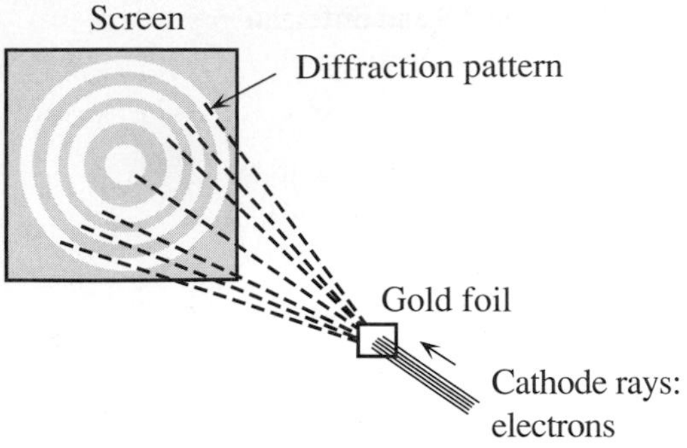


(a) Thomson diffracted electrons by using a thin gold foil and produced a diffraction pattern on the screen of his apparatus in (b). The foil was polycrystalline, so the diffraction pattern was circular rings.

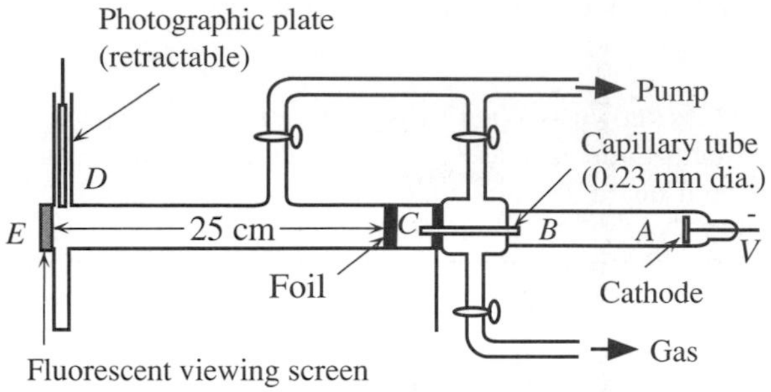


(b) In Thomson's electron diffraction apparatus a beam of electrons is generated in tube *A*, passed through collimating tube *B*, and made to impinge on a thin gold foil *C*. The transmitted electrons impinge on the fluorescent screen *E*, or a photographic plate *D*, which could be lowered into the path. The entire apparatus was evacuated during the experiment.

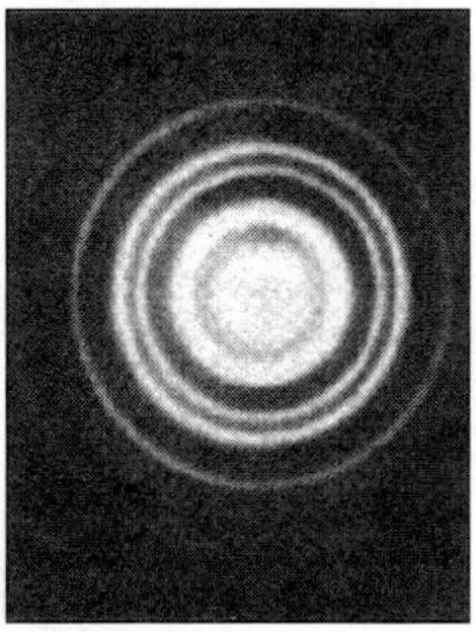

(c) Electron diffraction pattern obtained by G. P. Thomson using a gold foil target.

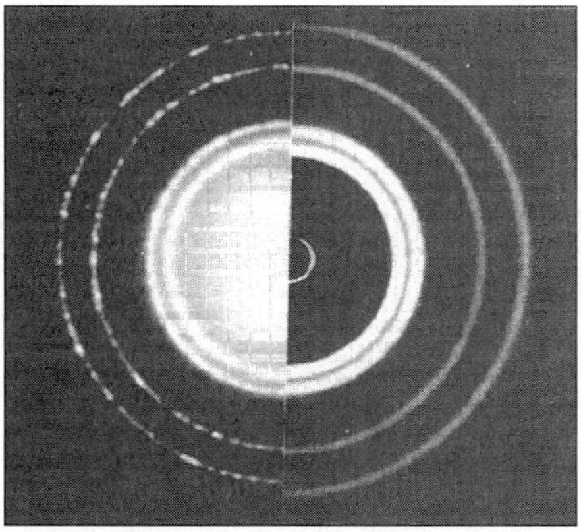

(d) Composite photograph showing diffraction patterns produced with an aluminum foil by X-rays and electrons of similar wavelength. Left: X-rays of $\lambda = 0.071$ nm. Right: Electrons of energy 600 eV.

(e) Diffraction pattern produced by 40 keV electrons passing through zinc oxide powder. The distortion of the pattern was produced by a small magnet placed between the sample and the photographic plate. An X-ray diffraction pattern would not be affected by a magnetic field.

**Figure 3.13** The diffraction of electrons by crystals gives typical diffraction patterns that would be expected if waves were being diffracted, as in X-ray diffraction with crystals.

SOURCE: (b) from G. P. Thomson, *Proceedings of the Royal Society,* A117, no. 600, 1928; (c) and (d) from A. P. French and F. Taylor, *An Introduction to Quantum Mechanics*, Norton, New York, 1978, p. 75; (e) from R. B. Leighton, *Principles of Modern Physics*, New York: McGraw-Hill, 1959, p. 84.

plate. The diffraction pattern obtained with an electron beam (Figure 3.13) means that the electrons are obeying the Bragg diffraction condition $2d \sin\theta = n\lambda$ just as much as the X-ray waves.

Since we know the interatomic spacing $d$ and we can measure the angle of diffraction $2\theta$, we can readily evaluate the wavelength $\lambda$ associated with the wave-like behavior of the electrons. Furthermore, from the accelerating voltage $V$ in the electron tube, we can also determine the momentum of the electrons, because the kinetic energy gained by the electrons, $(p^2/2m_e)$, is equal to $eV$. Simply by adjusting the accelerating voltage $V$, we can therefore study how the wavelength of the electron depends on the momentum.

As a result of such studies and other similar experiments, it has been found that an electron traveling with a momentum $p$ behaves like a wave of wavelength $\lambda$ given by

$$\lambda = \frac{h}{p} \qquad [3.13]$$

*Wavelength of the electron*

This is just the reverse of the equation for the momentum of a photon given its wavelength. The same equation therefore relates wave-like and particle-like properties to and from each other. Thus,

$$\lambda = \frac{h}{p} \quad \text{or} \quad p = \frac{h}{\lambda}$$

*De Broglie relations*

is an equation that exposes the wave–particle duality of nature. It was first hypothesized by De Broglie in 1924. As an example, we can calculate the wavelengths of a number of particle-like objects:

*a.* A 50 gram golf ball traveling at a velocity of 20 m s$^{-1}$.

The wavelength is

$$\lambda = \frac{h}{mv} = \frac{6.63 \times 10^{-34}\ \text{J s}}{(50 \times 10^{-3}\ \text{kg})(20\ \text{m s}^{-1})} = 6.63 \times 10^{-34}\ \text{m}$$

The wavelength is so small that this golf ball will not exhibit any wave effects. Firing a stream of golf balls at a wall will not result in "diffraction rings" of golf balls.

*b.* A proton traveling at 2200 m s$^{-1}$.

Using $m_p = 1.67 \times 10^{-27}$ kg, we have $\lambda = (h/mv) \approx 0.18$ nm. This is only slightly smaller than the interatomic distance in crystals, so firing protons at a crystal can result in diffraction. (Recall that to get a diffraction peak, we must satisfy the Bragg condition, $2d \sin\theta = n\lambda$.) Protons, however, are charged, so they can penetrate only a small distance into the crystal. Hence, they are not used in crystal diffraction studies.

*c.* Electron accelerated by 100 V.

This voltage accelerates the electron to a $KE$ equal to $eV$. From $KE = p^2/2m_e = eV$, we can calculate $p$ and hence $\lambda = h/p$. The result is $\lambda = 0.123$ nm. Since this is comparable to typical interatomic distances in solids, we would see a diffraction pattern when an electron beam strikes a crystal. The actual pattern is determined by the Bragg diffraction condition.

### 3.2.2 TIME-INDEPENDENT SCHRÖDINGER EQUATION

The experiments in which electrons exhibit interference and diffraction phenomena show quite clearly that, under certain conditions, the electron can behave as a wave; in other words, it can exhibit wave-like properties. There is a general equation that describes this wave-like behavior and, with the appropriate potential energy and boundary conditions, will predict the results of the experiments. The equation is called the **Schrödinger equation** and it forms the foundations of quantum theory. Its fundamental nature is analogous to the classical physics assertion of Newton's second law, $F = ma$, which of course cannot be proved. As a fundamental equation, Schrödinger's has been found to successfully predict every observable physical phenomenon at the atomic scale. Without this equation, we will not be able to understand the properties of electronic materials and the principles of operation of many semiconductor devices. We introduce the equation through an analogy.

A traveling electromagnetic wave resulting from sinusoidal current oscillations, or the traveling voltage wave on a long transmission line, can generally be described by a traveling-wave equation of the form

$$\mathcal{E}(x, t) = \mathcal{E}_o \exp j(kx - \omega t) = \mathcal{E}(x) \exp(-j\omega t) \quad \textbf{[3.14]}$$

where $\mathcal{E}(x) = \mathcal{E}_0 \exp(jkx)$ represents the spatial dependence, which is separate from the time variation. We assume that no transients exist to upset this perfect sinusoidal propagation. We note that the time dependence is harmonic and therefore predictable. For this reason, in ac circuits we put aside the $\exp(-j\omega t)$ term until we need the instantaneous magnitude of the voltage.

The average intensity $I_{\text{av}} = \frac{1}{2}c\varepsilon_o\mathcal{E}_o^2$ depends on the square of the amplitude. In Young's double-slit experiment, the intensity varies along the *y* direction, which means that $\mathcal{E}_o^2$ for the resultant wave depends on *y*. In the electron version of this experiment in Figure 3.12, what changes in the *y* direction is the probability of observing electrons; that is, there are peaks and troughs in the probability of finding electrons along *y*, just like the $\mathcal{E}_o^2$ variation along *y*. We should therefore attach some probability interpretation to the wave description of the electron.

In 1926, Max Born suggested a probability wave interpretation for the wave-like behavior of the electron.

$$\mathcal{E}(x, t) = \mathcal{E}_o \sin(kx - \omega t)$$

is a plane traveling **wavefunction** for an electric field; experimentally, we measure and interpret the *intensity* of a wave, namely $|\mathcal{E}(x, t)|^2$. There may be a similar wave function for the electron, which we can represent by a function $\Psi(x, t)$. According to Born, the significance of $\Psi(x, t)$ is that its amplitude squared represents the probability of finding the electron per unit distance. Thus, in three dimensions, if $\Psi(x, y, z, t)$ represents the wave property of the electron, it must have one of the following interpretations:

$|\Psi(x, y, z, t)|^2$ is the probability of finding the electron per unit volume at $x, y, z$ at time *t*.

$|\Psi(x, y, z, t)|^2\, dx\, dy\, dz$ is the probability of finding the electron in a small elemental volume $dx\, dy\, dz$ at $x, y, z$ at time *t*.

If we are just considering one dimension, then the wavefunction is $\Psi(x, t)$, and $|\Psi(x, t)|^2\, dx$ is the probability of finding the electron between $x$ and $(x + dx)$ at time $t$.

We should note that since only $|\Psi|^2$ has meaning, not $\Psi$, the latter function need not be real; it can be a complex function with real and imaginary parts. For this reason, we tend to use $\Psi^* \Psi$, where $\Psi^*$ is the complex conjugate of $\Psi$, instead of $|\Psi|^2$, to represent the probability per unit volume.

To obtain the wavefunction $\Psi(x, t)$ for the electron, we need to know how the electron interacts with its environment. This is embodied in its potential energy function $V = V(x, t)$, because the net force the electron experiences is given by

$$F = -dV/dx.$$

For example, if the electron is attracted by a positive charge (*e.g.*, the proton in a hydrogen atom), then it clearly has an electrostatic potential energy given by

$$V(r) = -\frac{e^2}{4\pi\varepsilon_o r}$$

where $r = \sqrt{x^2 + y^2 + z^2}$ is the distance between the electron and the proton.

If the *PE* of the electron is time independent, which means that $V = V(x)$ in one dimension, then the spatial and time dependences of $\Psi(x, t)$ can be separated, just as in Equation 3.14, and the **total wavefunction** $\Psi(x, t)$ of the electron can be written as

*Steady-state total wave function*

$$\Psi(x, t) = \psi(x) \exp\left(-\frac{jEt}{\hbar}\right) \quad [3.15]$$

where $\psi(x)$ is the electron wavefunction that describes only the spatial behavior, and $E$ is the energy of the electron. The temporal behavior is simply harmonic, by virtue of $\exp(-jEt/\hbar)$, which corresponds to $\exp(-j\omega t)$ with an angular frequency $\omega = E/\hbar$. The fundamental equation that describes the electron's behavior by determining $\psi(x)$ is called the **time-independent Schrödinger equation.** It is given by the famous equation

*Schrödinger's equation for one dimension*

$$\frac{d^2\psi}{dx^2} + \frac{2m}{\hbar^2}(E - V)\psi = 0 \quad [3.16a]$$

where $m$ is the mass of the electron.

This is a second-order differential equation. It should be reemphasized that the potential energy $V$ in Equation 3.16a depends only on $x$. If the potential energy of the electron depends on time as well, that is, if $V = V(x, t)$, then in general $\Psi(x, t)$ cannot be written as $\psi(x)\exp(-jEt/\hbar)$. Instead, we must use the full version of the Schrödinger equation, which is discussed in more advanced textbooks.

In three dimensions, there will be derivatives of $\psi$ with respect to $x$, $y$, and $z$. We use the calculus notation $(\partial\psi/\partial x)$, differentiating $\psi(x, y, z)$ with respect to $x$ but keeping $y$ and $z$ constant. Similar notations $\partial\psi/\partial y$ and $\partial\psi/\partial z$ are used for derivatives with respect to $y$ alone and with respect to $z$ alone, respectively. In three dimensions, Equation 3.16a becomes

*Schrödinger's equation for three dimensions*

$$\frac{\partial^2\psi}{\partial x^2} + \frac{\partial^2\psi}{\partial y^2} + \frac{\partial^2\psi}{\partial z^2} + \frac{2m}{\hbar^2}(E - V)\psi = 0 \quad [3.16b]$$

where $V = V(x, y, z)$ and $\psi = \psi(x, y, z)$.

Equation 3.16b is a fundamental equation, called the time-independent Schrödinger equation, the solution of which gives the steady-state behavior of the electron in a time-independent potential energy environment described by $V = V(x, y, z)$. By solving Equation 3.16b, we will know the probability distribution and the energy of the electron. Once $\psi(x, y, z)$ has been determined, the total wavefunction for the electron is given by Equation 3.15 so that

$$|\Psi(x, y, z, t)|^2 = |\psi(x, y, z)|^2$$

which means that the steady-state probability distribution of the electron is simply $|\psi(x, y, z)|^2$.

The time-independent Schrödinger equation can be viewed as a "mathematical crank." We input the potential energy of the electron and the boundary conditions, turn the crank, and get the probability distribution and the energy of the electron under steady-state conditions.

Two important boundary conditions are often used to solve the Schrödinger equation. First, as an analogy, when we stretch a string between two fixed points and put it into a steady-state vibration, there are no discontinuities or kinks along the string. We can therefore intelligently guess that because $\psi(x)$ represents wave-like behavior, it must be a smooth function without any discontinuities.

The first boundary condition is that $\Psi$ must be continuous, and the second is that $d\Psi/dx$ must be continuous. In the steady state, these two conditions translate directly to $\psi$ and $d\psi/dx$ being continuous. Since the probability of finding the electron is represented by $|\psi|^2$, this function must be single-valued and smooth, without any discontinuities, as illustrated in Figure 3.14. The enforcement of these boundary conditions results in strict requirements on the wavefunction $\psi(x)$, as a result of which only certain wavefunctions are acceptable. These wavefunctions are called the **eigenfunctions** (characteristic functions) of the system, and they determine the behavior and energy of the electron under steady-state conditions. The eigenfunctions $\psi(x)$ are also called **stationary states,** inasmuch as we are only considering steady-state behavior.

It is important to note that the Schrödinger equation is generally applicable to all matter, not just the electron. For example, the equation can also be used to describe the behavior of a proton, if the appropriate potential energy $V(x, y, z)$ and mass ($m_{\text{proton}}$) are used. Wavefunctions associated with particles are frequently called **matter waves.**

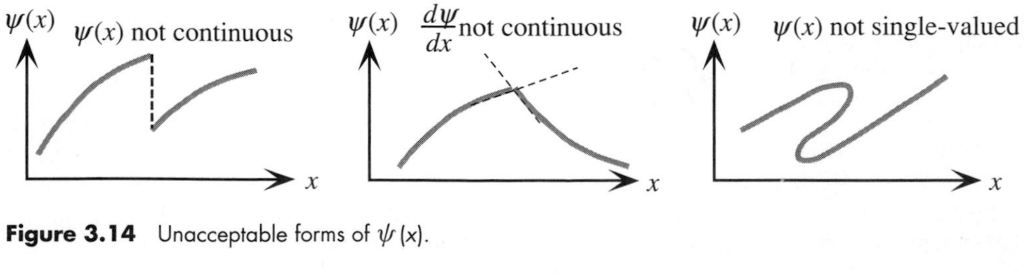


**Figure 3.14** Unacceptable forms of $\psi(x)$.

**EXAMPLE 3.5**

**THE FREE ELECTRON** Solve the Schrödinger equation for a free electron whose energy is $E$. What is the uncertainty in the position of the electron and the uncertainty in the momentum of the electron?

**SOLUTION**

Since the electron is free, its potential energy is zero, $V = 0$. In the Schrödinger equation, this leads to

$$\frac{d^2\psi}{dx^2} + \frac{2m}{\hbar^2}E\psi = 0$$

We can write this as

$$\frac{d^2\psi}{dx^2} + k^2\psi = 0$$

where we defined $k^2 = (2m/\hbar^2)E$. Solving the differential equation, we get

$$\psi(x) = A\exp(jkx) \qquad \text{or} \qquad B\exp(-jkx)$$

The total wavefunction is obtained by multiplying $\psi(x)$ by $\exp(-jEt/\hbar)$. We can define a fictitious frequency for the electron by $\omega = E/\hbar$ and multiply $\psi(x)$ by $\exp(-j\omega t)$:

$$\Psi(x, t) = A\exp j(kx - \omega t) \qquad \text{or} \qquad B\exp j(-kx - \omega t)$$

Each of these is a traveling wave. The first solution is a traveling wave in the $+x$ direction, and the second one is in the $-x$ direction. Thus, the free electron has a traveling wave solution with a wavenumber $k = 2\pi/\lambda$, that can have any value. The energy $E$ of the electron is simply $KE$, so

$$KE = E = \frac{(\hbar k)^2}{2m}$$

When we compare this with the classical physics expression $KE = (p^2/2m)$, we see that the momentum is given by

$$p = \hbar k \qquad \text{or} \qquad p = \frac{h}{\lambda}$$

This is the de Broglie relationship. The latter therefore results naturally from the Schrödinger equation for a free electron.

The probability distribution for the electron is

$$|\psi(x)|^2 = |A\exp j(kx)|^2 = A^2$$

which is constant over the entire space. Thus, the electron can be anywhere between $x = -\infty$ and $x = +\infty$. The uncertainty $\Delta x$ in its position is infinite. Since the electron has a well-defined wavenumber $k$, its momentum $p$ is also well-defined by virtue of $p = \hbar k$. The uncertainty $\Delta p$ in its momentum is thus zero.

**EXAMPLE 3.6**

**WAVELENGTH OF AN ELECTRON BEAM** Electrons are accelerated through a 100 V potential difference to strike a polycrystalline aluminum sample. The diffraction pattern obtained indicates that the highest intensity and smallest angle diffraction, corresponding to diffraction from the (111) planes, has a diffraction angle of 30.4°. From X-ray studies, the separation of the (111) planes is 0.234 nm. What is the wavelength of the electron and how does it compare with that from the de Broglie relationship?

**SOLUTION**

Since we know the angle of diffraction $2\theta\,(=30.4^\circ)$ and the interplanar separation $d\,(=0.234$ nm), we can readily calculate the wavelength of the electron from the Bragg condition for diffraction, $2d\sin\theta = n\lambda$. With $n = 1$,

$$\lambda = 2d\sin\theta = 2(0.234\text{ nm})\sin(15.2^\circ) = 0.1227\text{ nm}$$

This is the wavelength of the electron.

When an electron is accelerated through a voltage $V$, it gains $KE$ equal to $eV$, so $p^2/2m = eV$ and $p = (2meV)^{1/2}$. This is the momentum imparted by the potential difference $V$. From the de Broglie relationship, the wavelength should be

$$\lambda = \frac{h}{p} = \frac{h}{(2meV)^{1/2}}$$

or

$$\lambda = \left(\frac{h^2}{2meV}\right)^{1/2}$$

Substituting for $e$, $h$, and $m$, we obtain

$$\lambda = \frac{1.226\text{ nm}}{V^{1/2}}$$

The experiment uses 100 V, so the de Broglie wavelength is

$$\lambda = \frac{1.226\text{ nm}}{V^{1/2}} = \frac{1.226\text{ nm}}{100^{1/2}} = 0.1226\text{ nm}$$

which is in excellent agreement with that determined from the Bragg condition.

## 3.3 INFINITE POTENTIAL WELL: A CONFINED ELECTRON

Consider the behavior of the electron when it is confined to a certain region, $0 < x < a$. Its $PE$ is zero inside that region and infinite outside, as shown in Figure 3.15. The electron cannot escape, because it would need an infinite $PE$. Clearly the probability $|\psi|^2$ of finding the electron per unit volume is zero outside $0 < x < a$. Thus, $\psi = 0$ when $x \le 0$ and $x \ge a$, and $\psi$ is determined by the Schrödinger equation in $0 < x < a$ with $V = 0$. Therefore, in the region $0 < x < a$

$$\frac{d^2\psi}{dx^2} + \frac{2m}{\hbar^2}E\psi = 0 \qquad [3.17]$$

This is a second-order linear differential equation. As a general solution, we can take

$$\psi(x) = A\exp(jkx) + B\exp(-jkx)$$

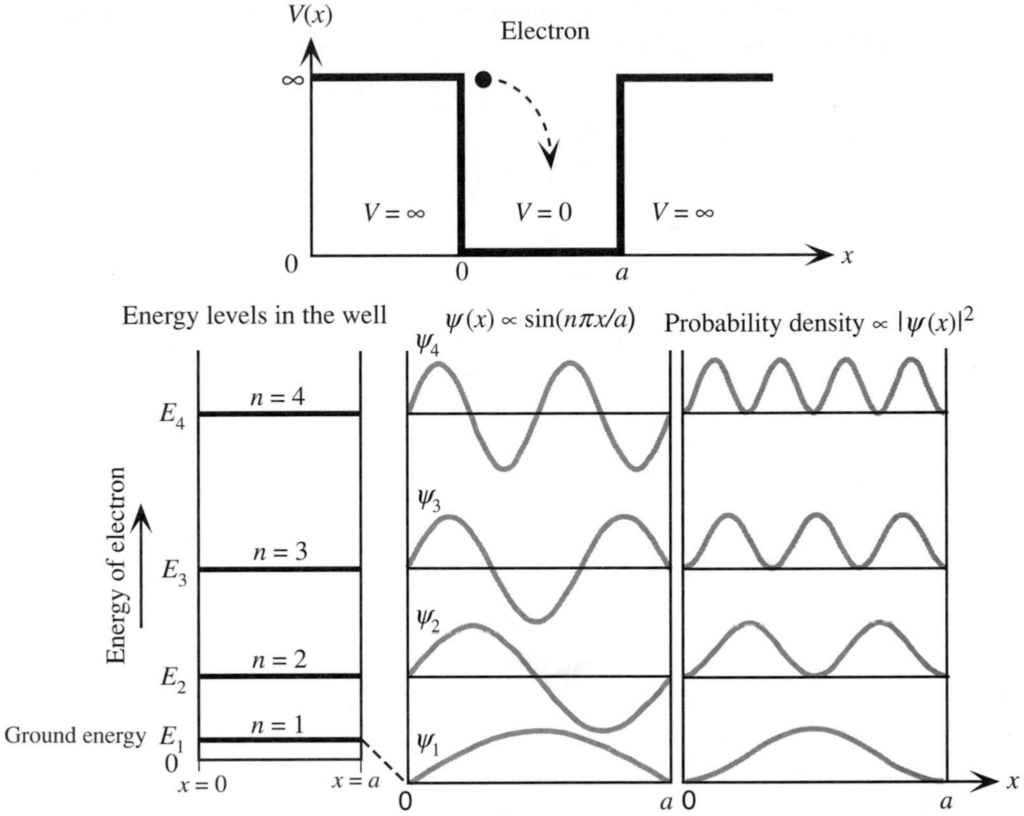


**Figure 3.15** Electron in a one-dimensional infinite *PE* well.
The energy of the electron is quantized. Possible wavefunctions and the probability distributions for the electron are shown.

where $k$ is some constant (to be determined) and substitute this in Equation 3.17 to find $k$. We first note that $\psi(0) = 0$; therefore, $B = -A$, so that

$$\psi(x) = A[\exp(jkx) - \exp(-jkx)] = 2Aj \sin kx \qquad [3.18]$$

We now substitute this into the Schrödinger Equation 3.17 to relate the energy $E$ to $k$. Thus, Equation 3.17 becomes

$$-2Ajk^2(\sin kx) + \left(\frac{2m}{\hbar^2}\right)E(2Aj \sin kx) = 0$$

which can be rearranged to obtain the energy of the electron:

$$E = \frac{\hbar^2 k^2}{2m} \qquad [3.19]$$

Since the electron has no $PE$ within the well, its total energy $E$ is kinetic energy $KE$, and we can write

$$E = KE = \frac{p_x^2}{2m}$$

where $p_x$ is its momentum. Comparing this with Equation 3.19, we see that the momentum of the electron must be

$$p_x = \pm\hbar k \qquad [3.20]$$

The momentum $p_x$ may be in the $+x$ direction or the $-x$ direction (which is the reason for $\pm$), so the **average momentum** is actually zero, $p_{av} = 0$.

We have already seen this relationship, when we defined $k$ as $2\pi/\lambda$ (wavenumber) for a free traveling wave. So the constant $k$ here is a wavenumber-type quantity even though there is no distinct traveling wave. Its value is determined by the boundary condition at $x = a$ where $\psi = 0$, or

$$\psi(a) = 2Aj \sin ka = 0$$

The solution to $\sin ka = 0$ is simply $ka = n\pi$, where $n = 1, 2, 3, \ldots$ is an integer. We exclude $n = 0$ because it will result in $\psi = 0$ everywhere (no electron at all).

We notice immediately that $k$, and therefore the energy of the electron, can only have certain values; they are **quantized** by virtue of $n$ being an integer. Here, $n$ is called a **quantum number.** For each $n$, there is a special wavefunction

*Wavefunction in infinite PE well*

$$\psi_n(x) = 2Aj \sin\left(\frac{n\pi x}{a}\right) \qquad [3.21]$$

which is called an eigenfunction.[3] All $\psi_n$ for $n = 1, 2, 3 \ldots$ constitute the eigenfunctions of the system. Each eigenfunction identifies a possible state for the electron. For each $n$, there is one special $k$ value, $k_n = n\pi/a$, and hence a special energy value $E_n$, since

$$E_n = \frac{\hbar^2 k_n^2}{2m}$$

that is,

*Electron energy in infinite PE well*

$$E_n = \frac{\hbar^2(\pi n)^2}{2ma^2} = \frac{h^2 n^2}{8ma^2} \qquad [3.22]$$

The energies $E_n$ defined by Equation 3.22 with $n = 1, 2, 3 \ldots$ are called **eigenenergies** of the system.

We still have not completely solved the problem, because $A$ has yet to be determined. To find $A$, we use what is called the **normalization condition.** The total probability of finding the electron in the whole region $0 < x < a$ is unity, because we know the electron is somewhere in this region. Thus, $|\psi|^2\, dx$ summed between $x = 0$ and

[3] From the German meaning "characteristic function."

$x = a$ must be unity, or

$$\int_{x=0}^{x=a} |\psi(x)|^2\,dx = \int_{x=0}^{x=a} \left|2Aj\sin\left(\frac{n\pi x}{a}\right)\right|^2 dx = 1$$

*Normalization condition*

Carrying out the simple integration, we find

$$A = \left(\frac{1}{2a}\right)^{1/2}$$

The resulting wavefunction for the electron is thus

$$\psi_n(x) = j\left(\frac{2}{a}\right)^{1/2} \sin\left(\frac{n\pi x}{a}\right) \qquad [3.23]$$

We can now summarize the behavior of an electron in a one-dimensional *PE* well. Its wavefunction and energy, shown in Figure 3.15, are given by Equations 3.23 and 3.22, respectively. Both depend on the quantum number $n$. The energy of the electron increases with $n^2$, so the minimum energy of the electron corresponds to $n = 1$. This is called the **ground state,** and the energy of the ground state is the lowest energy the electron can possess. Note also that the energy of the electron in this potential well cannot be zero, even though the *PE* is zero. Thus, the electron always has *KE*, even when it is in the ground state.

The **node** of a wavefunction is defined as the point where $\psi = 0$ inside the well. It is apparent from Figure 3.15 that the ground wavefunction $\psi_1$ with the lowest energy has no nodes, $\psi_2$ has one node, $\psi_3$ has two nodes, and so on. Thus, the energy increases as the number of nodes increases in a wavefunction.

It may seem surprising that the energy of the electron is quantized; that is, that it can only have finite values, given by Equation 3.22. The electron cannot be made to take on any value of energy, as in the classical case. If the electron behaved like a particle, then an applied force $F$ could impart any value of energy to it, because $F = dp/dt$ (Newton's second law), or $p = \int F dt$. By applying a force $F$ for a time $t$, we can give the electron a *KE* of

$$E = \frac{p^2}{2m} = \left(\frac{1}{2}m\right)\left[\int F\,dt\right]^2$$

However, Equation 3.22 tells us that, in the microscopic world, the energy can only have quantized values. The two conflicting views can be reconciled if we consider the energy difference between two consecutive energy levels, as follows:

$$\Delta E = E_{n+1} - E_n = \frac{h^2(2n+1)}{8ma^2}$$

*Energy separation in infinite PE well*

As $a$ increases to macroscopic dimensions, $a \to \infty$, the electron is completely free and $\Delta E \to 0$. Since $\Delta E = 0$, the energy of a completely free electron ($a = \infty$) is continuous. The energy of a confined electron, however, is quantized, and $\Delta E$ depends on the dimension (or size) of the potential well confining the electron.

In general, an electron will be "contained" in a spatial region of three dimensions, within which the *PE* will be lower (hence the confinement). We must then solve the

Schrödinger equation in three dimensions. The result is three quantum numbers that characterize the behavior of the electron.

Examination of the wavefunctions $\psi_n$ in Figure 3.15 shows that these are either *symmetric* or *antisymmetric* with respect to the center of the well at $x = \frac{1}{2}a$. The symmetry of a wavefunction is called its **parity.** Whenever the potential energy function $V(x)$ exhibits symmetry about a certain point $C$, for example, about $x = \frac{1}{2}a$ in Figure 3.15, then the wavefunctions have either **even parity** (such as $\psi_1, \psi_3, \ldots$ that are symmetric) or have **odd parity** (such as $\psi_2, \psi_4, \ldots$ that are antisymmetric).

**EXAMPLE 3.7**

**ELECTRON CONFINED WITHIN ATOMIC DIMENSIONS** Consider an electron in an infinite potential well of size 0.1 nm (typical size of an atom). What is the ground energy of the electron? What is the energy required to put the electron at the third energy level? How can this energy be provided?

**SOLUTION**

The electron is confined in an infinite potential well, so its energy is given by

$$E_n = \frac{h^2 n^2}{8ma^2}$$

We use $n = 1$ for the ground level and $a = 0.1$ nm. Therefore,

$$E_1 = \frac{(6.6 \times 10^{-34}\ \text{J s})^2(1)^2}{8(9.1 \times 10^{-31}\ \text{kg})(0.1 \times 10^{-9}\ \text{m})^2} = 6.025 \times 10^{-18}\ \text{J} \quad \text{or} \quad 37.6\ \text{eV}$$

The frequency of the electron associated with this energy is

$$\omega = \frac{E}{\hbar} = \frac{6.025 \times 10^{-18}\ \text{J}}{1.055 \times 10^{-34}\ \text{J s}} = 5.71 \times 10^{16}\ \text{rad s}^{-1} \quad \text{or} \quad \nu = 9.092 \times 10^{15}\ \text{s}^{-1}$$

The third energy level $E_3$ is

$$E_3 = E_1 n^2 = (37.6\ \text{eV})(3)^2 = 338.4\ \text{eV}$$

The energy required to take the electron from 37.6 eV to 338.4 eV is 300.8 eV. This can be provided by a photon of exactly that energy; no less, and no more. Since the photon energy is $E = h\nu = hc/\lambda$, or

$$\lambda = \frac{hc}{E} = \frac{(6.6 \times 10^{-34}\ \text{J s})(3 \times 10^{8}\ \text{m s}^{-1})}{300.8\ \text{eV} \times 1.6 \times 10^{-19}\ \text{C}}$$

$$= 4.12\ \text{nm}$$

which is an X-ray photon.

**EXAMPLE 3.8**

**ENERGY OF AN APPLE IN A CRATE** Consider a macroscopic object of mass 100 grams (say, an apple) confined to move between two rigid walls separated by 1 m (say, a typical size of a large apple crate). What is the minimum speed of the object? What should the quantum number $n$ be if the object is moving with a speed 1 m s$^{-1}$? What is the separation of the energy levels of the object moving with that speed?

**SOLUTION**

Since the object is within rigid walls, we take the *PE* outside the walls as infinite and use

$$E_n = \frac{h^2 n^2}{8ma^2}$$

to find the ground-level energy. With $n = 1$, $a = 1$ m, $m = 0.1$ kg, we have

$$E_1 = \frac{(6.6 \times 10^{-34}\ \text{J s})^2 (1)^2}{8(0.1\ \text{kg})(1\ \text{m})^2} = 5.45 \times 10^{-67}\ \text{J} = 3.4 \times 10^{-48}\ \text{eV}$$

Since this is kinetic energy, $\frac{1}{2} m v_1^2 = E_1$, so the minimum speed is

$$v_1 = \sqrt{\frac{2E_1}{m}} = \sqrt{\frac{2(5.45 \times 10^{-67}\ \text{J})}{0.1\ \text{kg}}} = 3.3 \times 10^{-33}\ \text{m s}^{-1}$$

This speed cannot be measured by any instrument; therefore, for all practical purposes, the apple is at rest in the crate (a relief for the fruit grocer). The time required for the object to move a distance of 1 mm is $3 \times 10^{29}$ s or $10^{21}$ years, which is more than the present age of the universe!

When the object is moving with a speed 1 m s$^{-1}$,

$$KE = \frac{1}{2} m v^2 = \frac{1}{2}(0.1\ \text{kg})(1\ \text{m s}^{-1})^2 = 0.05\ \text{J}$$

This must be equal to $E_n = h^2 n^2 / 8ma^2$ for some value of $n$

$$n = \left(\frac{8ma^2 E_n}{h^2}\right)^{1/2} = \left[\frac{8(0.1\ \text{kg})(1\ \text{m})^2 (0.05\ \text{J})}{(6.6 \times 10^{-34}\ \text{J s})^2}\right]^{1/2} = 3.03 \times 10^{32}$$

which is an enormous number. The separation between two energy levels corresponds to a change in $n$ from $3.03 \times 10^{32}$ to $3.03 \times 10^{32} + 1$. This is such a negligibly small change in $n$ that for all practical purposes, the energy levels form a continuum. Thus,

$$\Delta E = E_{n+1} - E_n = \frac{h^2(2n+1)}{8ma^2}$$

$$= \frac{[(6.6 \times 10^{-34}\ \text{J s})^2 (2 \times 3.03 \times 10^{32} + 1)]}{[8(0.1\ \text{kg})(1\ \text{m})^2]}$$

$$= 3.30 \times 10^{-34}\ \text{J} \quad \text{or} \quad 2.06 \times 10^{-15}\ \text{eV}$$

This energy separation is not detectable by any instrument. So for all practical purposes, the energy of the object changes continuously.

We see from this example that in the limit of large quantum numbers, quantum predictions agree with the classical results. This is the essence of **Bohr's correspondence principle.**

## 3.4 HEISENBERG'S UNCERTAINTY PRINCIPLE

The wavefunction of a free electron corresponds to a traveling wave with a single wavelength $\lambda$, as shown in Example 3.5. The traveling wave extends over all space, along all $x$, with the same amplitude, so the probability distribution function is uniform

Werner Heisenberg (1901–1976) received the Nobel prize in physics in 1932 for the uncertainty principle. This photo was apparently taken in 1936, while he was lecturing on quantum mechanics. "An expert is someone who knows some of the worst mistakes that can be made in his subject, and how to avoid them." W. Heisenberg.
SOURCE: AIP Emilio Segrè Visual Archives.

throughout the whole of space. The uncertainty $\Delta x$ in the position of the electron is therefore infinite. Yet, the uncertainty $\Delta p_x$ in the momentum of the electron is zero, because $\lambda$ is well-defined, which means that we know $p_x$ exactly from the de Broglie relationship, $p_x = h/\lambda$.

For an electron trapped in a one-dimensional infinite *PE* well, the wavefunction extends from $x = 0$ to $x = a$, so the uncertainty in the position of the electron is $a$. We know that the electron is within the well, but we cannot pinpoint with certainty exactly where it is. The momentum of the electron is either $p_x = \hbar k$ in the $+x$ direction or $-\hbar k$ in the $-x$ direction. The uncertainty $\Delta p_x$ in the momentum is therefore $2\hbar k$; that is, $\Delta p_x = 2\hbar k$. For the ground-state wavefunction, which corresponds to $n = 1$, we have $ka = \pi$. Thus, $\Delta p_x = 2\hbar\pi/a$. Taking the product of the uncertainties in $x$ and $p$, we get

$$(\Delta x)(\Delta p_x) = (a)\left(\frac{2\hbar\pi}{a}\right) = h$$

In other words, the product of the position and momentum uncertainties is simply $h$. This relationship is fundamental; and it constitutes a limit to our knowledge of the behavior of a system. *We cannot exactly and simultaneously know both the position and momentum of a particle along a given coordinate.* In general, if $\Delta x$ and $\Delta p_x$ are the respective uncertainties in the simultaneous measurement of the position and momentum of a particle along a particular coordinate (such as $x$), the **Heisenberg uncertainty principle** states that[4]

*Heisenberg uncertainty principle for position and momentum*

$$\Delta x\, \Delta p_x \gtrsim \hbar \qquad \textbf{[3.24]}$$

We are therefore forced to conclude that as previously stated, because of the wave nature of quantum mechanics, we are unable to determine exactly and simultaneously the position and momentum of a particle along a given coordinate. There will be an uncertainty $\Delta x$ in the position and an uncertainty $\Delta p_x$ in the momentum of the particle

[4] The Heisenberg uncertainty principle is normally written in terms of $\hbar$ rather than $h$. Further, in some physics texts, $\hbar$ in Equation 3.24 has a factor $\frac{1}{2}$ multiplying it.

and these uncertainties will be related by Heisenberg's uncertainty relationship in Equation 3.24.

These uncertainties are not in any way a consequence of the accuracy of a measurement or the precision of an instrument. Rather, they are the theoretical limits to what we can determine about a system. They are part of the quantum nature of the universe. In other words, even if we build the most perfectly engineered instrument to measure the position and momentum of a particle at one instant, we will still be faced with position and momentum uncertainties $\Delta x$ and $\Delta p_x$ such that $\Delta x\,\Delta p_x > \hbar$.

There is a similar uncertainty relationship between the uncertainty $\Delta E$ in the energy $E$ (or angular frequency $\omega$) of the particle and the time duration $\Delta t$ during which it possesses the energy (or during which its energy is measured). We know that the $kx$ part of the wave leads to the uncertainty relation $\Delta x\,\Delta p_x > \hbar$ or $\Delta x\,\Delta k \geq 1$. By analogy we should expect a similar relationship for the $\omega t$ part, or $\Delta\omega\,\Delta t \geq 1$. This hypothesis is true, and since $E = \hbar\omega$, we have the uncertainty relation for the particle energy and time:

*Heisenberg uncertainty principle for energy and time*

$$\Delta E\,\Delta t \gtrsim \hbar \qquad [3.25]$$

Note that the uncertainty relationships in Equations 3.24 and 3.25 have been written in terms of $\hbar$, rather than $h$, as implied by the electron in an infinite potential energy well ($\Delta x\,\Delta p_x \geq h$). In general there is also a numerical factor of $\frac{1}{2}$ multiplying $\hbar$ in Equations 3.24 and 3.25 which comes about when we consider a Gaussian spread for all possible position and momentum values. The proof is not presented here, but can be found in advanced quantum mechanics books.

It is important to note that the uncertainty relationship applies only when the position and momentum are measured in the same direction (such as the $x$ direction). On the other hand, the exact momentum, along, say, the $y$ direction and the exact position, along, say, the $x$ direction can be determined exactly, since $\Delta x\,\Delta p_y$ need not satisfy the Heisenberg uncertainty relationship (in other words, $\Delta x\,\Delta p_y$ can be zero).

### THE MEASUREMENT TIME AND THE FREQUENCY OF WAVES: AN ANALOGY WITH $\Delta E\,\Delta t \geq \hbar$ — EXAMPLE 3.9

Consider the measurement of the frequency of a sinusoidal wave of frequency 1000 Hz (or cycles/s). Suppose we can only measure the number of cycles to an accuracy of 1 cycle, because we need to receive a whole cycle to record it as one complete cycle. Then, in a time interval of $\Delta t = 1$ s, we will register $1000 \pm 1$ cycles. The uncertainty $\Delta f$ in the frequency is 1 cycle/1 s or 1 Hz. If $\Delta t$ is 2 s, we will measure $2000 \pm 1$ cycles, and the uncertainty $\Delta f$ will be 1 cycle/2 s or $\frac{1}{2}$ cycle/s or $\frac{1}{2}$ Hz. Thus, $\Delta f$ decreases with $\Delta t$.

Suppose that in a time interval $\Delta t$, we measure $N \pm 1$ cycles. Since the uncertainty is 1 cycle in a time interval $\Delta t$, the uncertainty in $f$ will be

$$\Delta f = \frac{(1\text{ cycle})}{\Delta t} = \frac{1}{\Delta t}\text{ Hz}$$

Since $\omega = 2\pi f$, we have

$$\Delta\omega\,\Delta t = 2\pi$$

In quantum mechanics, under steady-state conditions, an object has a time-oscillating wavefunction with a frequency $\omega$ which is related to its energy $E$ by $\omega = E/\hbar$ (see Equation 3.15).

Substituting this into the previous relationship gives

$$\Delta E \, \Delta t = h$$

The uncertainty in the energy of a quantum object is therefore related, in a fundamental way, to the time duration during which the energy is observed. Notice that we again have $h$, as for $\Delta x \, \Delta p_x = h$, though the quantum mechanical uncertainty relationship in Equation 3.25 has $\hbar$.

**EXAMPLE 3.10**

**THE UNCERTAINTY PRINCIPLE ON THE ATOMIC SCALE** Consider an electron confined to a region of size 0.1 nm, which is the typical dimension of an atom. What will be the uncertainty in its momentum and hence its kinetic energy?

**SOLUTION**

We apply the Heisenberg uncertainty relationship, $\Delta x \, \Delta p_x \approx \hbar$, or

$$\Delta p_x \approx \frac{\hbar}{\Delta x} = \frac{1.055 \times 10^{-34}\ \text{J s}}{0.1 \times 10^{-9}\ \text{m}} = 1.055 \times 10^{-24}\ \text{kg m s}^{-1}$$

The uncertainty in the velocity is therefore

$$\Delta v = \frac{\Delta p_x}{m_e} = \frac{1.055 \times 10^{-24}\ \text{kg m s}^{-1}}{9.1 \times 10^{-31}\ \text{kg}} = 1.16 \times 10^{6}\ \text{m s}^{-1}$$

We can take this uncertainty to represent the order of magnitude of the actual speed. The kinetic energy associated with this momentum is

$$KE = \frac{\Delta p_x^2}{2m_e} = \frac{(1.055 \times 10^{-24}\ \text{kg m s}^{-1})^2}{2(9.1 \times 10^{-31}\ \text{kg})}$$
$$= 6.11 \times 10^{-19}\ \text{J} \quad \text{or} \quad 3.82\ \text{eV}$$

**EXAMPLE 3.11**

**THE UNCERTAINTY PRINCIPLE WITH MACROSCOPIC OBJECTS** Estimate the minimum velocity of an apple of mass 100 g confined to a crate of size 1 m.

**SOLUTION**

Taking the uncertainty in the position of the apple as 1 m, the apple is somewhere in the crate,

$$\Delta p_x \approx \frac{\hbar}{\Delta x} = \frac{1.05 \times 10^{-34}\ \text{J s}}{1\ \text{m}} = 1.05 \times 10^{-34}\ \text{kg m s}^{-1}$$

So the minimum uncertainty in the velocity is

$$\Delta v_x = \frac{\Delta p_x}{m} = \frac{1.05 \times 10^{-34}\ \text{kg m s}^{-1}}{0.1\ \text{kg}} = 1.05 \times 10^{-33}\ \text{m s}^{-1}$$

The quantum nature of the universe implies that the apple in the crate is moving with a velocity on the order of $10^{-33}$ m s$^{-1}$. This cannot be measured by any instrument; indeed, it would take the apple $\sim 10^{19}$ years to move an atomic distance of 0.1 nm.

## 3.5 TUNNELING PHENOMENON: QUANTUM LEAK

To understand the tunneling phenomenon, let us examine the thrilling events experienced by the roller coaster shown in Figure 3.16a. Consider what the roller coaster can do when released from rest at a height *A*. The conservation of energy means that the carriage can reach *B* and at most *C*, but certainly not beyond *C* and definitely not *D* and *E*. Classically, there is no possible way the carriage will reach *E* at the other side of the potential barrier *D*. An extra energy corresponding to the height difference, $D - A$, is needed. Anyone standing at *E* will be quite safe. Ignoring frictional losses, the roller coaster will go back and forth between *A* and *C*.

Now, consider an analogous event on an atomic scale. An electron moves with an energy *E* in a region $x < 0$ where the potential energy *PE* is zero; therefore, *E* is solely kinetic energy. The electron then encounters a potential barrier of "height" $V_o$, which

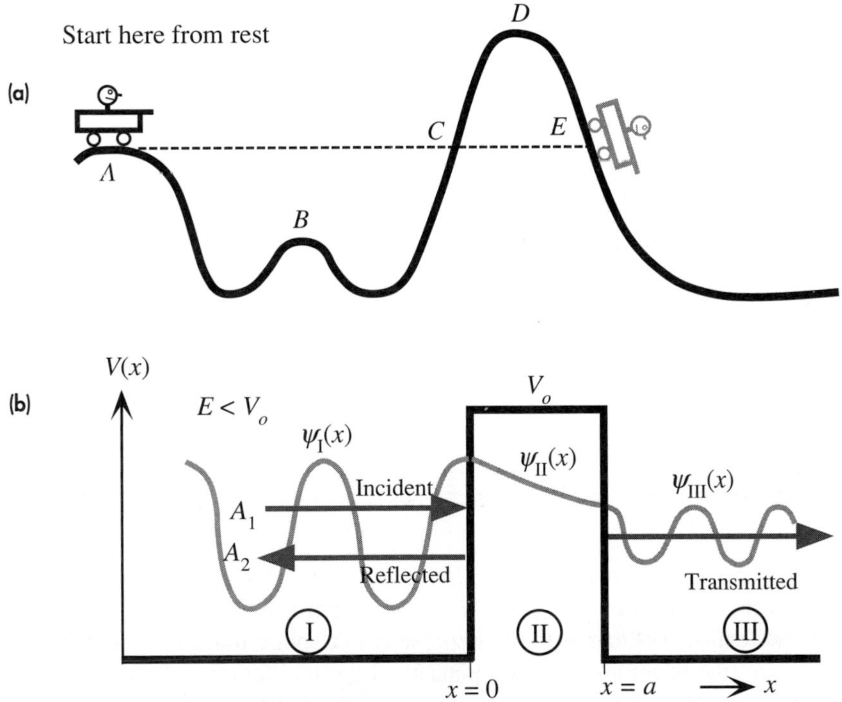


**Figure 3.16**

(a) The roller coaster released from *A* can at most make it to *C*, but not to *E*. Its *PE* at *A* is less than the *PE* at *D*. When the car is at the bottom, its energy is totally *KE*. *CD* is the energy barrier that prevents the car from making it to *E*. In quantum theory, on the other hand, there is a chance that the car could tunnel (leak) through the potential energy barrier between *C* and *E* and emerge on the other side of the hill at *E*.

(b) The wavefunction for the electron incident on a potential energy barrier ($V_o$). The incident and reflected waves interfere to give $\psi_I(x)$. There is no reflected wave in region III. In region II, the wavefunction decays with *x* because $E < V_o$.

is greater than $E$ at $x = 0$. The extent (width) of the potential barrier is $a$. On the other side of the potential barrier, $x > a$, the $PE$ is again zero. What will the electron do? Classically, just like the roller coaster, the electron should bounce back and thus be confined to the region $x < 0$, because its total energy $E$ is less than $V_o$. In the quantum world, however, there is a distinct possibility that the electron will "tunnel" through the potential barrier and appear on the other side; it will leak through.

To show this, we need to solve the Schrödinger equation for the present choice of $V(x)$. Remember that the only way the Schrödinger equation will have the solution $\psi(x) = 0$ is if the $PE$ is infinite, that is, $V = \infty$. Therefore, within any zero or finite $PE$ region, there will always be a solution $\psi(x)$ and there always will be some probability of finding the electron.

We can divide the electron's space into three regions, I, II, and III, as indicated in Figure 3.16b. We can then solve the Schrödinger equation for each region, to obtain three wavefunctions $\psi_{\text{I}}(x)$, $\psi_{\text{II}}(x)$, and $\psi_{\text{III}}(x)$. In regions I and III, $\psi(x)$ must be traveling waves, as there is no $PE$ (the electron is free and moving with a kinetic energy $E$). In zone II, however, $E - V_o$ is negative, so the general solution of the Schrödinger equation is the sum of an exponentially decaying function and an exponentially increasing function. In other words,

$$\psi_{\text{I}}(x) = A_1 \exp(jkx) + A_2 \exp(-jkx) \qquad \text{[3.26a]}$$

$$\psi_{\text{II}}(x) = B_1 \exp(\alpha x) + B_2 \exp(-\alpha x) \qquad \text{[3.26b]}$$

$$\psi_{\text{III}}(x) = C_1 \exp(jkx) + C_2 \exp(-jkx) \qquad \text{[3.26c]}$$

are the wavefunctions in which

$$k^2 = \frac{2mE}{\hbar^2} \qquad \text{[3.27]}$$

and

$$\alpha^2 = \frac{2m(V_o - E)}{\hbar^2} \qquad \text{[3.28]}$$

Both $k^2$ and $\alpha^2$, and hence $k$ and $\alpha$, in Equations 3.26a to c are positive numbers. This means that $\exp(jkx)$ and $\exp(-jkx)$ represent traveling waves in opposite directions, and $\exp(-\alpha x)$ and $\exp(\alpha x)$ represent an exponential decay and rise, respectively. We see that in region I, $\psi_{\text{I}}(x)$ consists of the incident wave $A_1 \exp(jkx)$ in the $+x$ direction, and a reflected wave $A_2 \exp(-jkx)$, in the $-x$ direction. Furthermore, because the electron is traveling toward the right in region III, there is no reflected wave, so $C_2 = 0$.

We must now apply the boundary conditions and the normalization condition to determine the various constants $A_1$, $A_2$, $B_1$, $B_2$, and $C_1$. In other words, we must match the three waveforms in Equations 3.26a to c at their boundaries ($x = 0$ and $x = a$) so that they form a continuous single-valued wavefunction. With the boundary conditions enforced onto the wavefunctions $\psi_{\text{I}}(x)$, $\psi_{\text{II}}(x)$, and $\psi_{\text{III}}(x)$, all the constants can be determined in terms of the amplitude $A_1$ of the incoming wave. The relative probability that the electron will tunnel from region I through to III is defined as the **transmission**

**coefficient $T$,** and this depends very strongly on both the relative *PE* barrier height $(V_o - E)$ and the width $a$ of the barrier. The final result that comes out from a tedious application of the boundary conditions is

$$T = \frac{|\psi_{\mathrm{III}}(x)|^2}{|\psi_{\mathrm{I}}(\text{incident})|^2} = \frac{C_1^2}{A_1^2} = \frac{1}{1 + D\sinh^2(\alpha a)} \quad [3.29]$$

*Probability of tunneling*

where

$$D = \frac{V_o^2}{4E(V_o - E)} \quad [3.30]$$

and $\alpha$ is the rate of decay of $\psi_{\mathrm{II}}(x)$ as expressed in Equation 3.28. For a wide or high barrier, using $\alpha a \gg 1$ in Equation 3.29 and $\sinh(\alpha a) \approx \frac{1}{2}\exp(\alpha a)$, we can deduce

*Probability of tunneling through*

$$T = T_o \exp(-2\alpha a) \quad [3.31]$$

where

$$T_o = \frac{16E(V_o - E)}{V_o^2} \quad [3.32]$$

By contrast, the relative probability of reflection is determined by the ratio of the square of the amplitude of the reflected wave to that of the incident wave. This quantity is the **reflection coefficient $R$,** which is given by

$$R = \frac{A_2^2}{A_1^2} = 1 - T \quad [3.33]$$

*Reflection coefficient*

We can now summarize the entire tunneling affair as follows. When an electron encounters a potential energy barrier of height $V_o$ greater than its energy $E$, there is a finite probability that it will leak through that barrier. This probability depends sensitively on the energy and width of the barrier. For a wide potential barrier, the probability of tunneling is proportional to $\exp(-2\alpha a)$, as in Equation 3.31. The wider or higher the potential barrier, the smaller the chance of the electron tunneling.

One of the most remarkable technological uses of the tunneling effect is in the scanning tunneling microscope (STM), which elegantly maps out the surfaces of solids. A conducting probe is brought so close to the surface of a solid that electrons can tunnel from the surface of the solid to the probe, as illustrated in Figure 3.17. When the probe is far removed, the wavefunction of an electron decays exponentially outside the material, by virtue of the potential energy barrier being finite (the work function is ~10 eV). When the probe is brought very close to the surface, the wavefunction penetrates into the probe and, as a result, the electron can tunnel from the material into the probe. Without an applied voltage, there will be as many electrons tunneling from the material to the probe as there are going in the opposite direction from the probe to the material, so the net current will be zero.

On the other hand, if a positive bias is applied to the probe with respect to the material, as shown in Figure 3.17, an electron tunneling from the material to the probe will see a lower potential barrier than one tunneling from the probe to the material. Consequently, there will be a net current from the probe to the material and this current

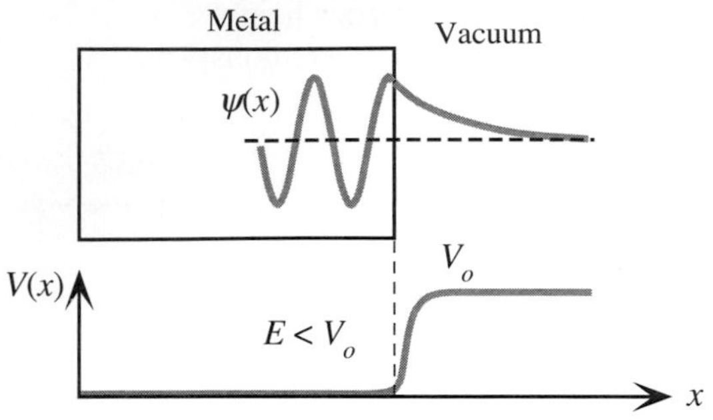


(a) The wavefunction decays exponentially as we move away from the surface because the *PE* outside the metal is $V_o$ and the energy of the electron, $E < V_o$.

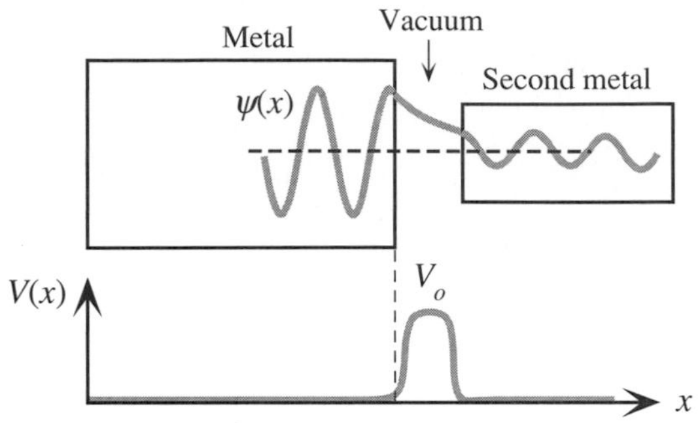


(b) If we bring a second metal close to the first metal, then the wavefunction can penetrate into the second metal. The electron can tunnel from the first metal to the second.

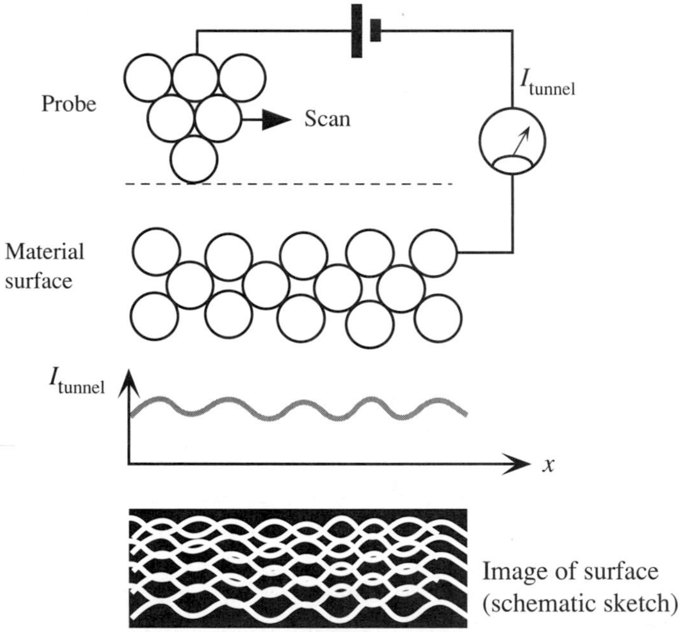


(c) The principle of the scanning tunneling microscope. The tunneling current depends on $\exp(-2\alpha a)$ where $a$ is the distance of the probe from the surface of the specimen and $\alpha$ is a constant.

**Figure 3.17**

will depend very sensitively on the separation $a$ of the probe from the surface, by virtue of Equation 3.31.

Because the tunneling current is extremely sensitive to the width of the potential barrier, the tunneling current is essentially dominated by electrons tunneling to the probe atom nearest to the surface. Thus, the probe tip has an atomic dimension. By scanning the surface of the material with the probe and recording the tunneling current

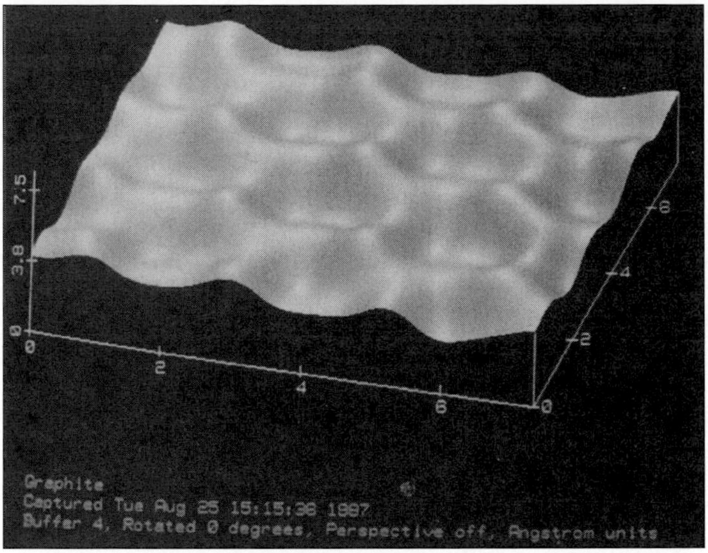


**Figure 3.18** Scanning tunneling microscope (STM) image of a graphite surface where contours represent electron concentrations within the surface, and carbon rings are clearly visible. The scale is in 2 Å.
| SOURCE: Courtesy of Veeco Instruments, Metrology Division, Santa Barbara, CA.

the user can map out the surface topology of the material with a resolution comparable to the atomic dimension. The probe motion along the surface, and also perpendicular to the surface, is controlled by piezoelectric transducers to provide sufficiently small and smooth displacements. Figure 3.18 shows an STM image of a graphite surface, on which the hexagonal carbon rings can be clearly seen. Notice that the scale is 0.2 nm (2 Å). The contours in the image actually represent electron concentrations within the surface since it is the electrons that tunnel from the graphite surface to the probe tip. The astute reader will notice that not all the carbon atoms in a hexagonal ring are at the same height; three are higher and three are lower. The reason is that the exact electron concentration on the surface is also influenced by the second layer of atoms underneath the top layer. The overall effect makes the electron concentration

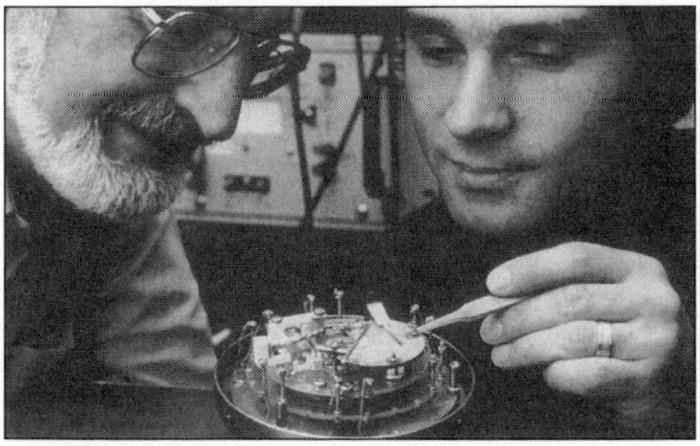

STM's inventors Gerd Binning (right) and Heinrich Rohrer (left), at IBM Zurich Research Laboratory with one of their early devices. They won the 1986 Nobel prize for the STM.
| SOURCE: Courtesy of IBM Zurich Research Laboratory.

An STM image of a Ni (110) surface.
| SOURCE: Courtesy of IBM.

change (alternate) from one atomic site to a neighboring site within the hexagonal rings. STM was invented by Gerd Binning and Heinrich Rohrer at the IBM Research Laboratory in Zurich, for which they were awarded the 1986 Nobel prize.[5]

**EXAMPLE 3.12** **TUNNELING CONDUCTION THROUGH METAL-TO-METAL CONTACTS** Consider two copper wires separated only by their surface oxide layer (CuO). Classically, since the oxide layer is an insulator, no current should be possible through the two copper wires. Suppose that for the conduction ("free") electrons in copper, the surface oxide layer looks like a square potential energy barrier of height 10 eV. Consider an oxide layer thickness of 5 nm and evaluate the transmission coefficient for conduction electrons in copper, which have a kinetic energy of about 7 eV. What will be the transmission coefficient if the oxide barrier is 1 nm?

**SOLUTION**

We can calculate $\alpha$ from

$$\alpha = \left[\frac{2m(V_o - E)}{\hbar^2}\right]^{1/2}$$
$$= \left[\frac{2(9.1 \times 10^{-31}\text{ kg})(10\text{ eV} - 7\text{ eV})(1.6 \times 10^{-19}\text{ J/eV})}{(1.05 \times 10^{-34}\text{ J s})^2}\right]^{1/2}$$
$$= 8.9 \times 10^9\text{ m}^{-1}$$

so that

$$\alpha a = (8.9 \times 10^9\text{ m}^{-1})(5 \times 10^{-9}\text{ m}) = 44.50$$

Since this is greater than unity, we use the wide-barrier transmission coefficient in Equation 3.31.

Now,

$$T_o = \frac{16E(V_o - E)}{V_o^2} = \frac{16(7\text{ eV})(10\text{ eV} - 7\text{ eV})}{(10\text{ eV})^2} = 3.36$$

Thus,

$$T = T_o \exp(-2\alpha a)$$
$$= 3.36 \exp[-2(8.9 \times 10^9\text{ m}^{-1})(5 \times 10^{-9}\text{ m})] = 3.36 \exp(-89)$$
$$\approx 7.4 \times 10^{-39}$$

an incredibly small number.

With $a = 1$ nm,

$$T = 3.36 \exp[-2(8.9 \times 10^9\text{ m}^{-1})(1 \times 10^{-9}\text{ m})]$$
$$= 3.36 \exp(-17.8) \approx 6.2 \times 10^{-8}$$

Notice that reducing the layer thickness by five times increases the transmission probability by $10^{31}$! Small changes in the barrier width lead to enormous changes in the transmission

[5] The IBM Research Laboratory in Zurich, Switzerland, received both the 1986 and the 1987 Nobel prizes. The first was for the scanning tunneling microscope by Gerd Binning and Heinrich Rohrer. The second was awarded to Georg Bednorz and Alex Müller for the discovery of high-temperature superconductors which we will examine in Chapter 8.

probability. We should note that when a voltage is applied across the two wires, the potential energy height is altered ($PE$ = charge × voltage), which results in a large increase in the transmission probability and hence results in a current.

**EXAMPLE 3.13**

**SIGNIFICANCE OF A SMALL $h$** Estimate the probability that a roller coaster carriage that weighs 100 kg released from point $A$ in Figure 3.16a from a height at 10 m can reach point $E$ over a hump that is 15 m high and 10 m wide. What will this probability be in a universe where $\hbar \approx 10$ kJ s?

**SOLUTION**

The total energy of the carriage at height $A$ is

$$E = PE = mg(\text{height}) = (100\text{ kg})(10\text{ m s}^{-2})(10\text{ m}) = 10^4\text{ J}$$

Suppose that as a first approximation, we can approximate the hump as a square hill of height 15 m and width 10 m. The $PE$ required to reach the peak would be

$$V_o = mg(\text{height}) = (100\text{ kg})(10\text{ m s}^{-2})(15\text{ m}) = 1.5 \times 10^4\text{ J}$$

Applying this, we have

$$\alpha^2 = \frac{2m(V_o - E)}{\hbar^2} = \frac{2(100\text{ kg})(1.5 \times 10^4\text{ J} - 10^4\text{ J})}{(1.05 \times 10^{-34}\text{ J s})^2} = 9.07 \times 10^{73}\text{ m}^{-2}$$

"Just like the good old ghost of the middle ages." In a world where $h$ is of the order of unity, one can expect tunneling surprises.

SOURCE: George Gamow, *Mr. Tompkins in Paperback*, Cambridge, England, University Press, 1965, p. 96. Used with permission.

and so

$$\alpha = 9.52 \times 10^{36}\ \mathrm{m}^{-1}$$

With $a = 10$ m, we have $\alpha a \gg 1$, so we can use the wide-barrier tunneling equation,

$$T = T_o \exp(-2\alpha a)$$

where

$$T_o = \frac{16[E(V_o - E)]}{V_o^2} = 3.56$$

Thus,

$$T = 3.56 \exp[-2(9.52 \times 10^{36}\ \mathrm{m}^{-1})(10\ \mathrm{m})] = 3.56 \exp(-1.9 \times 10^{38})$$

which is a fantastically small number, indicating that it is impossible for the carriage to tunnel through the hump.

Suppose that $\hbar \approx 10$ kJ s. Then

$$\alpha^2 = \frac{2m(V_o - E)}{\hbar^2} = \frac{2(100\ \mathrm{kg})(1.5 \times 10^4\ \mathrm{J} - 10^4\ \mathrm{J})}{(10^4\ \mathrm{J\ s})^2} = 0.01\ \mathrm{m}^{-2}$$

so that $\alpha = 0.1\ \mathrm{m}^{-1}$. Clearly, $\alpha a = 1$, so we must use

$$T = [1 + D \sinh^2(\alpha a)]^{-1}$$

where

$$D = \frac{V_o^2}{[4E(V_o - E)]} = 1.125$$

Thus,

$$T = [1 + 1.125 \sinh^2(1)]^{-1} = 0.39$$

Thus, after three goes, the carriage would tunnel to the other side (giving the person standing at $E$ the shock of his life).

## 3.6 POTENTIAL BOX: THREE QUANTUM NUMBERS

To examine the properties of a particle confined to a region of space, we take a three-dimensional space with a volume marked by $a, b, c$ along the $x, y, z$ axes. The *PE* is zero ($V = 0$) inside the space and is infinite on the outside, as illustrated in Figure 3.19. This is a three-dimensional potential energy well. The electron essentially lives in the "box." What will the behavior of the electron be in this box? In this case we need to solve the three-dimensional version of the Schrödinger equation,[6] which is

*Schrödinger equation in three dimensions*

$$\frac{\partial^2 \psi}{\partial x^2} + \frac{\partial^2 \psi}{\partial y^2} + \frac{\partial^2 \psi}{\partial z^2} + \frac{2m}{\hbar^2}(E - V)\psi = 0 \quad [3.34]$$

[6] The term $\partial\psi/\partial x$ simply means differentiating $\psi(x, y, z)$ with respect to $x$ while keeping $y$ and $z$ constant, just like $d\psi/dx$ in one dimension.

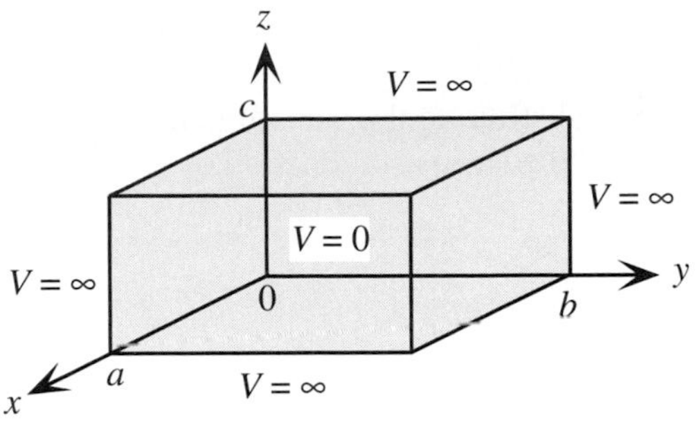


**Figure 3.19** Electron confined in three dimensions by a three-dimensional infinite *PE* box.
Everywhere inside the box, $V = 0$, but outside, $V = \infty$. The electron cannot escape from the box.

with $V = 0$ in $0 < x < a$, $0 < y < b$, and $0 < z < c$, and $V$ infinite outside. We can try to solve this by separating the variables via $\psi(x, y, z) = \psi_x(x)\,\psi_y(y)\,\psi_z(z)$. Substituting this back into Equation 3.34, we can obtain three ordinary differential equations, each just like the one for the one-dimensional potential well. Having found $\psi_x(x)$, $\psi_y(y)$, and $\psi_z(z)$ we know that the total wavefunction is simply the product,

$$\psi(x, y, z) = A\sin(k_x x)\sin(k_y y)\sin(k_z z) \qquad [3.35]$$

where $k_x$, $k_y$, $k_z$, and $A$ are constants to be determined. We can then apply the boundary conditions at $x = a$, $y = b$, and $z = c$ to determine the constants $k_x$, $k_y$, and $k_z$ in the same way we found $k$ for the one-dimensional potential well. If $\psi(x, y, z) = 0$ at $x = a$, then $k_x$ will be quantized via

$$k_x a = n_1 \pi$$

where $n_1$ is a quantum number, $n_1 = 1, 2, 3, \ldots$. Similarly, if $\psi(x, y, z) = 0$ at $y = b$ and $z = c$, then $k_y$ and $k_z$ will be quantized, so that, overall, we will have

$$k_x = \frac{n_1\pi}{a} \qquad k_y = \frac{n_2\pi}{b} \qquad k_z = \frac{n_3\pi}{c} \qquad [3.36]$$

where $n_1$, $n_2$, and $n_3$ are quantum numbers, each of which can be any integer except zero.

We notice immediately that in three dimensions, we have three quantum numbers $n_1$, $n_2$, and $n_3$ associated with $\psi_x(x)$, $\psi_y(y)$, and $\psi_z(z)$. The eigenfunctions of the electron, denoted by the quantum numbers $n_1$, $n_2$, and $n_3$, are now given by

*Electron wavefunction in infinite PE well*

$$\psi_{n_1 n_2 n_3}(x, y, z) = A\sin\left(\frac{n_1\pi x}{a}\right)\sin\left(\frac{n_2\pi y}{b}\right)\sin\left(\frac{n_3\pi z}{c}\right) \qquad [3.37]$$

Notice that these consist of the products of infinite one-dimensional *PE* well-type wavefunctions, one for each dimension, and each has its own quantum number $n$. Each possible eigenfunction can be labeled a **state** for the electron. Thus, $\psi_{111}$ and $\psi_{121}$ are two possible states.

To find the constant $A$ in Equation 3.37, we need to use the normalization condition that $|\psi_{n_1 n_2 n_3}(x, y, z)|^2$ integrated over the volume of the box must be unity, since the electron is somewhere in the box. The result for a square box is $A = (2/a)^{3/2}$.

We can find the energy of the electron by substituting the wavefunction in Equation 3.35 into the Schrödinger Equation 3.34. The energy as a function of $k_x$, $k_y$, $k_z$ is then found to be

$$E = E(k_x, k_y, k_z) = \frac{\hbar^2}{2m}\left(k_x^2 + k_y^2 + k_z^2\right)$$

which is quantized by virtue of $k_x$, $k_y$, and $k_z$ being quantized. We can write this energy in terms of $n_1^2$, $n_2^2$, and $n_3^2$ by using Equation 3.36, as follows:

$$E_{n_1 n_2 n_3} = \frac{h^2}{8m}\left(\frac{n_1^2}{a^2} + \frac{n_2^2}{b^2} + \frac{n_3^2}{c^2}\right)$$

For a square box for which $a = b = c$, the energy is

*Electron energy in infinite PE box*

$$E_{n_1 n_2 n_3} = \frac{h^2\left(n_1^2 + n_2^2 + n_3^2\right)}{8ma^2} = \frac{h^2 N^2}{8ma^2} \qquad [3.38]$$

where $N^2 = (n_1^2 + n_2^2 + n_3^2)$, which can only have certain integer values. It is apparent that the energy now depends on three quantum numbers. Our conclusion is that in three dimensions, we have three quantum numbers, each one arising from boundary conditions along one of the coordinates. They quantize the energy of the electron via Equation 3.38 and its momentum in a particular direction, such as, $p_x = \pm\hbar k_x = \pm(hn_1/2a)$, though the average momentum is zero.

The lowest energy for the electron is obviously equal to $E_{111}$, not zero. The next energy level corresponds to $E_{211}$, which is the same as $E_{121}$ and $E_{112}$, so there are three states (*i.e.*, $\psi_{211}$, $\psi_{121}$, $\psi_{112}$) for this energy. The number of states that have the same energy is termed the **degeneracy** of that energy level. The second energy level $E_{211}$ is thus **three-fold degenerate.**

**EXAMPLE 3.14** **NUMBER OF STATES WITH THE SAME ENERGY** How many states (eigenfunctions) are there at energy level $E_{443}$ for a square potential energy box?

**SOLUTION**

This energy level corresponds to $n_1 = 4$, $n_2 = 4$, and $n_3 = 3$, but the energy depends on

$$N^2 = n_1^2 + n_2^2 + n_3^2 = 4^2 + 4^2 + 3^2 = 41$$

via Equation 3.38. As long as $N^2 = 41$ for any choice of $(n_1, n_2, n_3)$, not just (4, 4, 3), the energy will be the same.

The value $N^2 = 41$ can be obtained from (4, 4, 3), (4, 3, 4), and (3, 4, 4) as well as (6, 2, 1), (6, 1, 2), (2, 6, 1), (2, 1, 6), (1, 6, 2), and (1, 2, 6). There are thus three states from (4, 4, 3) combinations and six from (6, 2, 1) combinations, giving nine possible states, each with a distinct wavefunction, $\psi_{n_1 n_2 n_3}$. However, all these $\psi_{n_1 n_2 n_3}$ for the electron have the same energy $E_{443}$.

# 3.7 HYDROGENIC ATOM

## 3.7.1 ELECTRON WAVEFUNCTIONS

Consider the behavior of the electron in a hydrogenic (hydrogen-like) atom, which has a nuclear charge of $+Ze$, as depicted in Figure 3.20. For the hydrogen atom, $Z = 1$, whereas for an ionized helium atom $He^+$, $Z = 2$. For a doubly ionized lithium atom $Li^{++}$, $Z = 3$, and so on. The electron is attracted by a positive nuclear charge and therefore has a Coulombic *PE*,

*Electron PE in hydrogenic atom*

$$V(r) = \frac{-Ze^2}{4\pi\varepsilon_o r} \quad [3.39]$$

Since force $F = -dV/dr$, Equation 3.39 is simply a statement of Coulomb's force between the positive charge $+Ze$ of the nucleus and the negative charge $-e$ of the electron. The task of finding $\psi(x, y, z)$ and the energy $E$ of the electron now involves putting $V(r)$ from Equation 3.39 into the Schrödinger equation with $r = \sqrt{x^2 + y^2 + z^2}$ and solving it.

Fortunately, the problem has a spherical symmetry, and we can solve the Schrödinger equation by transforming it into the $r, \theta, \phi$ coordinates shown in Figure 3.20. Even then, obtaining a solution is not easy. We must then ensure that the solution for $\psi(r, \theta, \phi)$ satisfies all the boundary conditions, as well as being single-valued and continuous with a continuous derivative. For example, when we go $2\pi$ around the $\phi$ coordinate, $\psi(r, \theta, \phi)$ should come back to its original value, or $\psi(r, \theta, \phi) = \psi(r, \theta, \phi + 2\pi)$, as is apparent from an examination of Figure 3.20. Along the radial

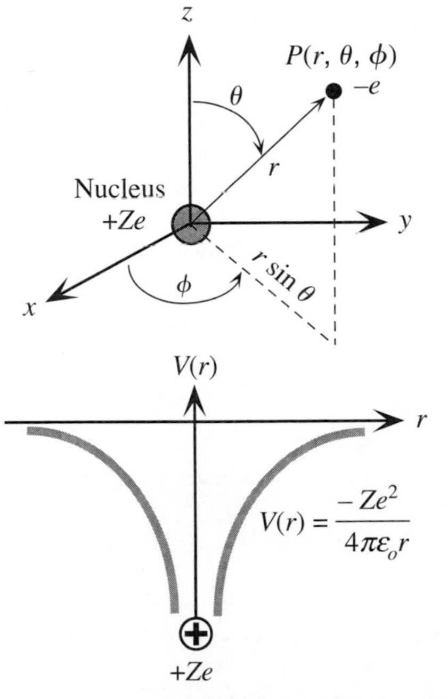


**Figure 3.20** The electron in the hydrogenic atom is attracted by a central force that is always directed toward the positive nucleus.
Spherical coordinates centered at the nucleus are used to describe the position of the electron. The *PE* of the electron depends only on *r*.

coordinate, we need $\psi(r, \theta, \phi) \to 0$ as $r \to \infty$; otherwise, the total probability will diverge when $|\psi(r, \theta, \phi)|^2$ is integrated over all space. In an analogy with the three-dimensional potential well, there should be three quantum numbers to characterize the wavefunction, energy, and momentum of the electron. The three quantum numbers are called the **principal, orbital angular momentum,** and **magnetic quantum numbers** and are respectively denoted by $n$, $\ell$, and $m_\ell$. Unlike the three-dimensional potential well, however, not all the quantum numbers run as independent positive integers.

The solution to the Schrödinger equation $\psi(r, \theta, \phi)$ depends on three variables, $r, \theta, \phi$. The wavefunction $\psi(r, \theta, \phi)$ can be written as the product of two functions

$$\psi(r, \theta, \phi) = R(r)\, Y(\theta, \phi)$$

where $R(r)$ is a radial function depending only on $r$, and $Y(\theta, \phi)$ is called the **spherical harmonic,** which expresses the angular dependence of the wavefunction. These functions are characterized by the quantum numbers $n, \ell, m_\ell$. The radial part $R(r)$ depends on $n$ and $\ell$, whereas the spherical harmonic depends on $\ell$ and $m_\ell$, so

$$\psi(r, \theta, \phi) = \psi_{n,\ell,m_\ell}(r, \theta, \phi) = R_{n,\ell}(r)\, Y_{\ell,m_\ell}(\theta, \phi) \quad [3.40]$$

By solving the Schrödinger equation, these functions have already been evaluated. It turns out that we can only assign certain values to the quantum numbers $n$, $\ell$, and $m_\ell$ to obtain acceptable solutions, that is, $\psi_{n,\ell,m_\ell}(r, \theta, \phi)$ that are well behaved: single-valued and with $\psi$ and the gradient of $\psi$ continuous. We can summarize the allowed values of $n, \ell, m_\ell$ as follows:

| | |
|---|---|
| Principal quantum number | $n = 1, 2, 3, \ldots$ |
| Orbital angular momentum quantum number | $\ell = 0, 1, 2, \ldots, (n-1) < n$ |
| Magnetic quantum number | $m_\ell = -\ell, -(\ell-1), \ldots, 0, \ldots, (\ell-1), \ell$ or $\lvert m_\ell\rvert \le \ell$ |

The $\ell$ values carry a special notation inherited from spectroscopic terms. The first four $\ell$ values are designated by the first letters of the terms *sharp, principal, diffuse,* and *fundamental,* whereas the higher $\ell$ values follow from *f* onwards, as $g, h, i,$ etc. For example, any state $\psi_{n,\ell,m_\ell}$ that has $\ell = 0$ is called an $s$ state, whereas that which has $\ell = 1$ is termed a $p$ state. We can also use $n$ as a prefix to $\ell$ to identify $n$. Thus $\psi_{n,\ell,m_\ell}$ with $n = 2$ and $\ell = 0$ corresponds to the $2s$ state. The notation for identifying the $\ell$ value and labeling a state is summarized in Table 3.1.

**Table 3.1** Labeling of various $n\ell$ possibilities

| | $\ell$ | | | | |
|---|---|---|---|---|---|
| $n$ | 0 | 1 | 2 | 3 | 4 |
| 1 | 1*s* | | | | |
| 2 | 2*s* | 2*p* | | | |
| 3 | 3*s* | 3*p* | 3*d* | | |
| 4 | 4*s* | 4*p* | 4*d* | 4*f* | |
| 5 | 5*s* | 5*p* | 5*d* | 5*f* | 5*g* |

**Table 3.2** The radial and spherical harmonic parts of the wavefunction in the hydrogen atom ($a_o = 0.0529$ nm)

| $n$ | $\ell$ | $R(r)$ | $m_\ell$ | $Y(\theta, \phi)$ | | |
|---|---|---|---|---|---|---|
| 1 | 0 | $\left(\frac{1}{a_o}\right)^{3/2} 2\exp\left(-\frac{r}{a_o}\right)$ | 0 | $\frac{1}{2\sqrt{\pi}}$ | | |
| 2 | 0 | $\left(\frac{1}{2a_o}\right)^{3/2}\left(2-\frac{r}{a_o}\right)\exp\left(-\frac{r}{2a_o}\right)$ | 0 | $\frac{1}{2\sqrt{\pi}}$ | | |
| 2 | 1 | $\left(\frac{1}{2a_o}\right)^{3/2}\left(\frac{r}{\sqrt{3}a_o}\right)\exp\left(-\frac{r}{2a_o}\right)$ | 0 | $\frac{1}{2}\sqrt{\frac{3}{\pi}}\cos\theta$ | | |
| | | | 1 | $\frac{1}{2}\sqrt{\frac{3}{2\pi}}\sin\theta e^{j\phi}$ | $\propto \sin\theta\cos\phi$ | Correspond to $m_\ell = -1$ and $+1$. |
| | | | −1 | $\frac{1}{2}\sqrt{\frac{3}{2\pi}}\sin\theta e^{-j\phi}$ | $\propto \sin\theta\sin\phi$ | |

Table 3.2 summarizes the functional forms of $R_{n,\ell}(r)$ and $Y_{\ell,m_\ell}(\theta, \phi)$. For $\ell = 0$ (the $s$ states), the angular dependence of $Y_{0,0}(\theta, \phi)$ is constant, which means that $\psi(r, \theta, \phi)$ is spherically symmetrical about the nucleus. For the $\ell = 1$ and higher states, there is a strong directionality to the wavefunctions with respect to each other. The radial part $R_{n,\ell}(r)$ is sketched in Figure 3.21a for two choices of $n$ and $\ell$. Notice that $R_{n,\ell}(r)$ is largest at $r = 0$, when $\ell = 0$. However, this does not mean that the electron will be mainly at $r = 0$, because the probability of finding the electron at a distance $r$ actually depends on $r^2|R_{n,\ell}(r)|^2$, which vanishes as $r \to 0$.

Let us examine the probability of finding the electron at a distance $r$ within a thin spherical shell of radius $r$ and thickness $\delta r$ (assumed to be very small). The directional dependence of the probability will be determined by the function $Y_{\ell,m_\ell}(\theta, \phi)$. We can average this over all directions (all angles $\theta$ and $\phi$) to obtain $\overline{Y_{\ell,m_\ell}(\theta, \phi)}$, which turns out to be simply $1/4\pi$. The volume of the spherical shell is $\delta V = 4\pi r^2 \delta r$. The probability of finding the electron in this shell is then

$$|(\overline{Y_{\ell,m_\ell}(\theta, \phi)})(R_{n,\ell}(r))|^2 \times (4\pi r^2 \delta r)$$

If $\delta P(r)$ represents the probability that the electron is in this spherical shell of thickness $\delta r$, then

$$\delta P(r) = |R_{n,\ell}(r)|^2 r^2 \delta r \quad [3.41]$$

The **radial probability density** $P_{n,\ell}(r)$ is defined as the probability per unit radial distance, that is, $dP/dr$ which from Equation 3.41 is $|R_{n,\ell}(r)|^2 r^2$. The latter vanishes at the nucleus and peaks at certain locations, as shown in Figure 3.21b. This behavior implies that the probability of finding the electron within a thin spherical shell close to the nucleus also disappears. For $n = 1$, and $\ell = 0$, for example, the maximum probability is at $r = a_o = 0.0529$ nm, which is called the **Bohr radius.** Therefore, if the electron is in the $1s$ state, it spends most of its time at a distance $a_o$. Notice that the probability distribution does not depend on $m_\ell$, but only on $n$ and $\ell$.

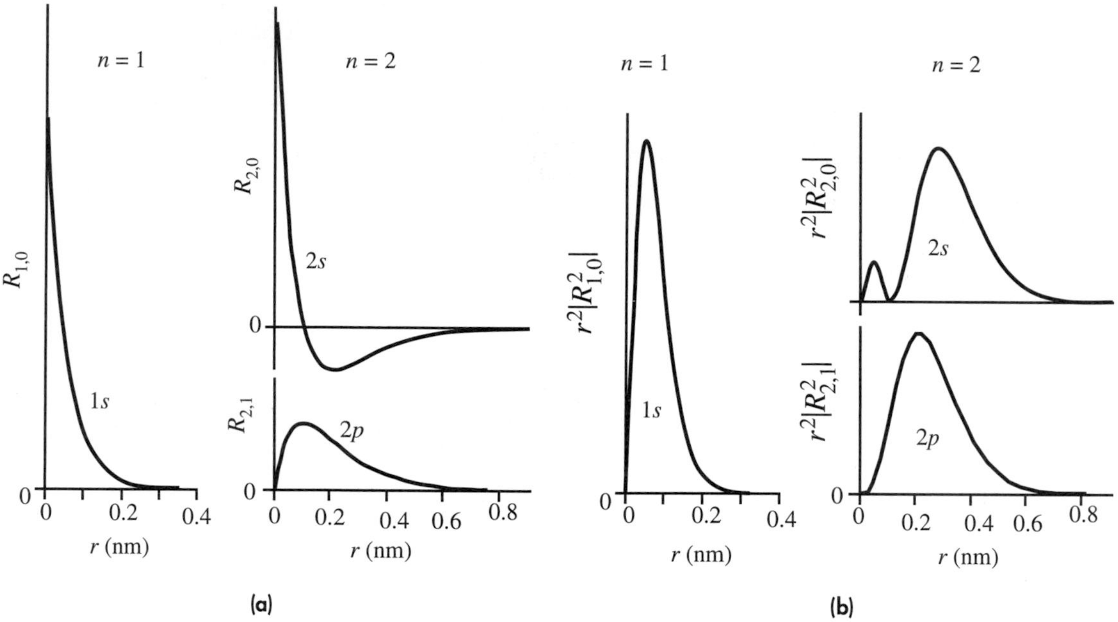


**Figure 3.21**

(a) Radial wavefunctions of the electron in a hydrogenic atom for various $n$ and $\ell$ values.
(b) $r^2\,|R^2_{n,\ell}|$ gives the radial probability density. Vertical axis scales are linear in arbitrary units.

Table 3.2 summarizes the nature of the functions $R_{n,\ell}(r)$ and $Y_{\ell,m_\ell}(\theta,\phi)$ for various $n, \ell, m_\ell$ values. Each possible wavefunction $\psi_{n,\ell,m_\ell}(r,\theta,\phi)$ with a particular choice of $n, \ell, m_\ell$ constitutes a **quantum state** for the electron. The function $\psi_{n,\ell,m_\ell}(r,\theta,\phi)$ basically describes the behavior of the electron in the atom in probabilistic terms, as distinct from a well-defined line orbit for the electron, as one might expect from classical mechanics. For this reason, $\psi_{n,\ell,m_\ell}(r,\theta,\phi)$ is often referred to as an **orbital,** in contrast to the classical theory, which assigns an orbit to the electron.

Figure 3.22a shows the polar plots of $Y_{\ell,m_\ell}(\theta,\phi)$ for $s$ and $p$ orbitals. The radial distance from the origin in the polar plot represents the magnitude of $Y_{\ell,m_\ell}(\theta,\phi)$, which depends on the angles $\theta$ and $\phi$. The polar plots of the probability distribution $|Y_{\ell,m_\ell}(\theta,\phi)|^2$ are shown in Figure 3.22b. Although for the $s$ states, $Y_{1,0}(\theta,\phi)$ is spherically symmetric, resulting in a spherically symmetrical probability distribution around the nucleus, this is not so for $\ell = 1$ and higher states.

For example, each of the $p$ states has a distinctly directional character, as illustrated in the polar plots in Figure 3.22. The angular dependence of $|\psi_{2,1,0}(r,\theta,\phi)|$, for which $m_\ell = 0$, is such that most of the probability is oriented along the $z$ axis. This wavefunction is referred to as the $2p_z$ orbital. The two wavefunctions for $m_\ell = \pm 1$ are often represented by $\psi_{2p_x}(r,\theta,\phi)$ and $\psi_{2p_y}(r,\theta,\phi)$, or more simply, $2p_x$ and $2p_y$ orbitals, which do not possess a specific $m_\ell$ individually, but together represent the two $m_\ell = \pm 1$ wavefunctions. The angular dependence of $2p_x$ and $2p_y$ are essentially along the $x$ and $y$ directions. Thus, the three orbitals for $m_\ell = 0, \pm 1$ are all oriented perpendicular to each other, as depicted in Figure 3.22.

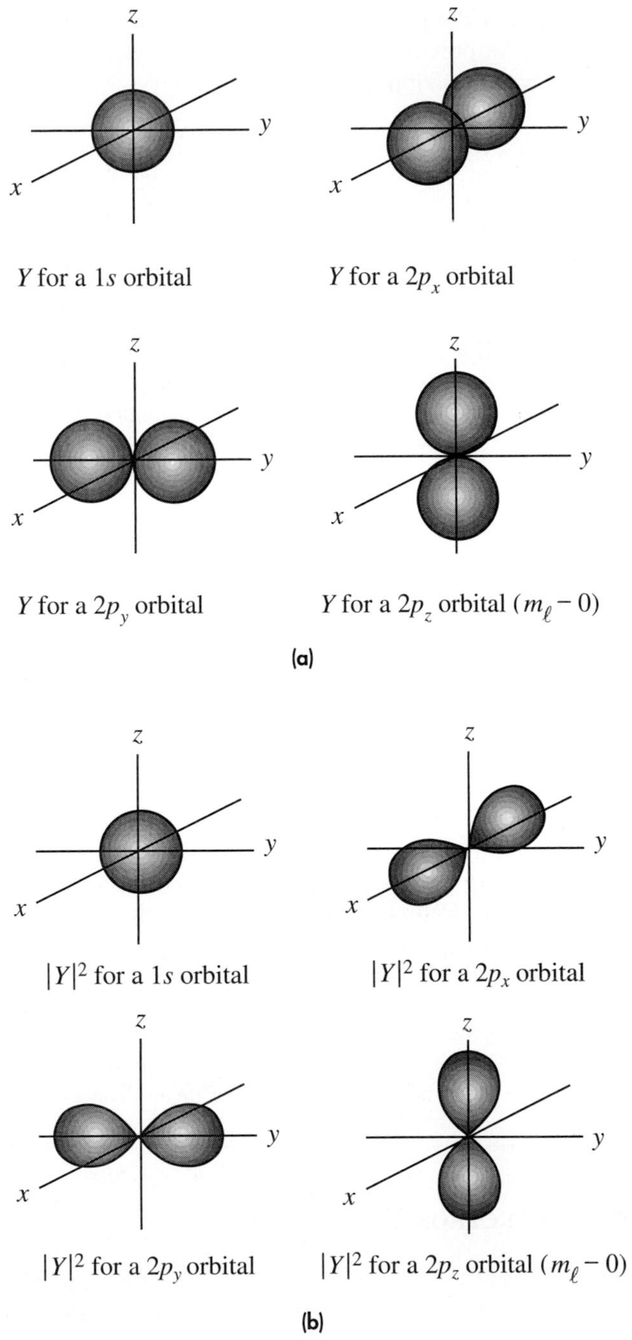


**Figure 3.22**
(a) The polar plots of $Y_{n,\ell}(\theta, \phi)$ for 1s and 2p states.
(b) The angular dependence of the probability distribution, which is proportional to $|Y_{n,\ell}(\theta, \phi)|^2$.

It should be noted that the probability distributions in Figures 3.21b and 3.22b do not depend on time. As previously mentioned, under steady-state conditions, the magnitude of the total wavefunction is

$$|\Psi(r,\theta,\phi,t)| = \left|\psi(r,\theta,\phi)\exp\left(-\frac{jEt}{\hbar}\right)\right| = |\psi(r,\theta,\phi)|$$

which is independent of time.

**EXAMPLE 3.15**

**PROBABILITY DENSITY FUNCTION** The quantity $|R_{n,\ell}(r)|^2 r^2$ in Equation 3.41 is called the **radial probability density function** and is simply written as $P_{n,\ell}(r)$. Thus, $dP(r) = P_{n,\ell}(r)\,dr$ is the probability of finding the electron between $r$ and $r + dr$. We can use $P_{n,\ell}(r)$ to conveniently calculate the probability of finding the electron within a certain region of the atom, or to find the mean distance of the electron from the nucleus, and so on. For example, the electron in the 1$s$ orbital has the wavefunction shown for $n = 1$, $\ell = 0$ in Table 3.2, which decays exponentially,

$$R_{n,\ell}(r) = 2a_o^{-3/2}\exp\left(-\frac{r}{a_o}\right)$$

The *total* probability of finding the electron inside the Bohr radius $a_o$ can be found by summing (integrating) $P_{n,\ell}\,dr$ from $r = 0$ to $r = a_o$,

$$P_{\text{total}}(r < a_o) = \int_0^{a_o} P_{n,\ell}(r)\,dr = \int_0^{a_o} |R_{n,\ell}(r)|^2 r^2\,dr$$

$$= \int_0^{a_o} 4a_o^{-3}\exp\left(-\frac{2r}{a_o}\right)r^2\,dr = 0.32 \qquad \text{or} \qquad 32\%$$

The integration is not trivial but can nonetheless be done as indicated by the result 0.32 above. Thirty-two percent of the time the electron is therefore closer to the nucleus than the Bohr radius.

*Average distance of electron from nucleus*

The mean distance $\bar{r}$ of the electron, from the definition of the mean, becomes

$$\bar{r} = \int_0^{\infty} r P_{n,\ell}(r)\,dr = \frac{a_o n^2}{Z}\left[\frac{3}{2} - \frac{\ell(\ell+1)}{2n^2}\right] \qquad \textbf{[3.42]}$$

where we have simply inserted the result of the integration for various orbitals. (Again we take the mathematics as granted.) For the 1$s$ orbital, in the hydrogen atom, $Z = 1$, $n = 1$, and $\ell = 0$, so $\bar{r} = \frac{3}{2}a_o$, further than the Bohr radius. Notice that the mean distance $\bar{r}$ of the electron increases as $n^2$.

## 3.7.2 QUANTIZED ELECTRON ENERGY

Once the wavefunctions $\psi_{n,\ell,m_\ell}(r,\theta,\phi)$ have been found, they can be substituted into the Schrödinger equation to find the possible energies of the electron. These turn out to depend only on the principal quantum number $n$. The energy is given by

*Electron energy in hydrogenic atom*

$$E_n = -\frac{me^4Z^2}{8\varepsilon_o^2h^2n^2} \qquad \textbf{[3.43a]}$$

or

$$E_n = -\frac{Z^2 E_I}{n^2} = -\frac{Z^2(13.6 \text{ eV})}{n^2} \qquad [3.43b]$$

where

*Ionization energy of hydrogen*

$$E_I = \frac{me^4}{8\varepsilon_o^2 h^2} = 2.18 \times 10^{-18} \text{ J} \quad \text{or} \quad 13.6 \text{ eV} \qquad [3.43c]$$

This corresponds to the energy required to remove the electron in the hydrogen atom ($Z = 1$) from the lowest energy level $E_1$ (at $n = 1$) to infinity; hence, it represents the **ionization energy.** The energy $E_n$ in Equation 3.43b is negative with respect to that for the electron completely isolated from the nucleus (at $r = \infty$, therefore $V = 0$). Thus, when the electron is in the vicinity of the nucleus, $+Ze$, it has a lower energy, which is a favorable situation (hence, formation of the hydrogenic atom is energetically favorable). In general, the energy required to remove an electron from the $n$th shell to $n = \infty$ (where the electron is free) is called the **ionization energy for the *n*th shell,** which from Equation 4.43b is simply $|E_n|$ or $(13.6 \text{ eV})Z^2/n^2$.

Since the energy is quantized, the lowest energy of the electron corresponds to $n = 1$, which is $-13.6$ eV. The next higher energy value it can have is $E_2 = -3.40$ eV when $n = 2$, and so on, as sketched in Figure 3.23. Normally, the electron will take up a state corresponding to $n = 1$, because this has the lowest energy, called the **ground energy.** Its wavefunction corresponds to $\psi_{100}(r, \theta, \phi)$, which has a probability peak at $r = a_o$ and no angular dependence, as indicated in Figures 3.21 and 3.22.

The electron can only become excited to the next energy level if it is supplied by the right amount of energy $E_2 - E_1$. A photon of energy $h\nu = E_2 - E_1$ can readily supply this energy when it strikes the electron. The electron then gets excited to the state with $n = 2$ by absorbing the photon, and its wavefunction changes to $\psi_{210}(r, \theta, \phi)$, which has the maximum probability at $r = 4a_o$. The electron thus spends most of its time in this excited state, at $r = 4a_o$. It can return from the excited state at $E_2$ to the ground state at $E_1$ by emitting a photon of energy $h\nu = E_2 - E_1$.

By virtue of the quantization of energy, we see that the emission of light from excited atoms can only have certain wavelengths: those corresponding to transitions from higher quantum-number states to lower ones. In fact, in spectroscopic analysis, these wavelengths can be used to identify the elements, since each element has its unique set of emission and absorption wavelengths arising from a unique set of energy levels. Figure 3.24 illustrates the origin of the emission and absorption spectra of atoms, which are a direct consequence of the quantization of the energy.

The electrons in atoms can also be excited by other means, for example, by collisions with other atoms as a result of heating a gas. Figure 3.25 depicts how collisions with other atoms can excite an electron to higher energies. If an impinging atom has sufficient kinetic energy, it can impart just the right energy to excite the electron to a higher energy level. Since the total energy must be conserved, the incoming atom will lose some of its kinetic energy in the process. The excited electron can later return to

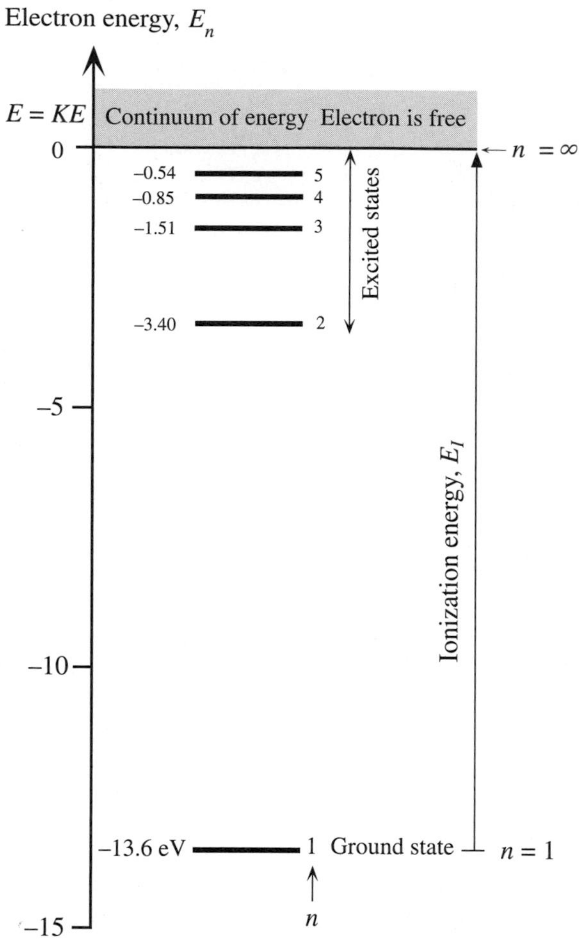


**Figure 3.23** The energy of the electron in the hydrogen atom ($Z = 1$).

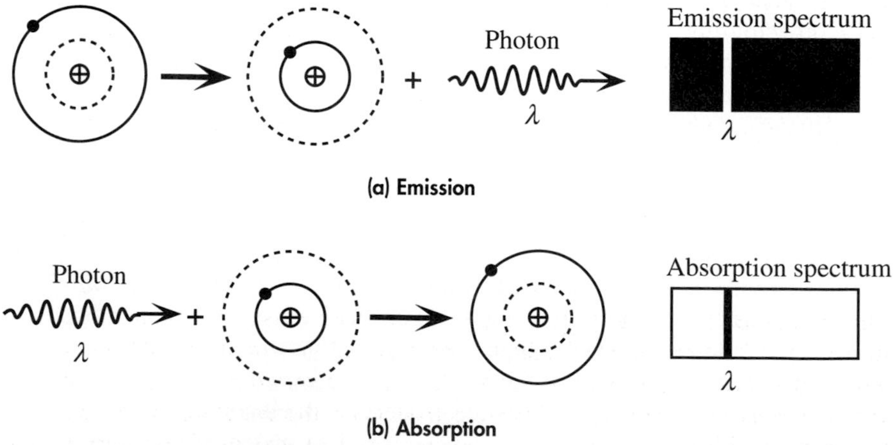


**Figure 3.24** The physical origin of spectra.

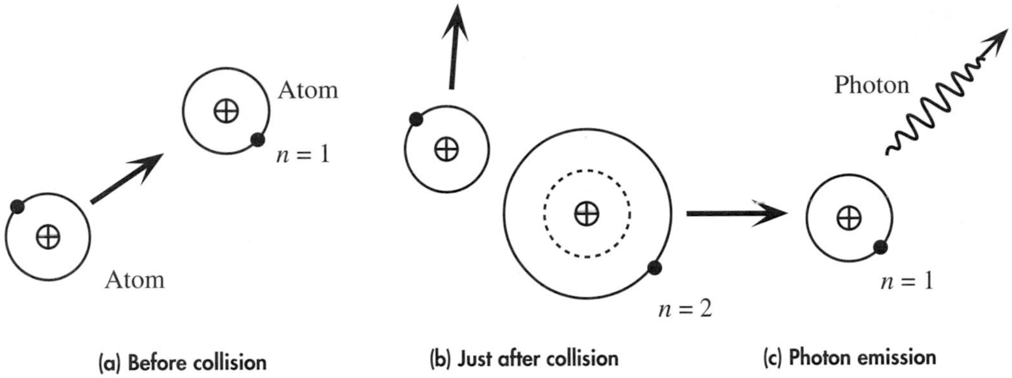


**Figure 3.25** An atom can become excited by a collision with another atom. When it returns to its ground energy state, the atom emits a photon.

its ground state by emitting a photon. Excitation by atomic collisions is the process by which we obtain light from an electrical discharge in gases, a quantum phenomenon we experience every day as we read a neon sign. Indeed, this is exactly how the Ne atoms in the common laboratory HeNe laser are excited, via atomic collisions between Ne and He atoms.

Since the principal quantum number determines the energy of the electron and also the position of maximum probability, as we noticed in Figure 3.21, various $n$ values define electron **shells,** within which we can most likely find the electron. These shells are customarily labeled $K, L, M, N, \ldots$, corresponding to $n = 1, 2, 3, \ldots$. For each $n$ value, there are a number of $\ell$ values that determine the spatial distribution of the electron. For a given $n$, each $\ell$ value constitutes a **subshell.** For example, we often talk about 3$s$, 3$p$, 3$d$ subshells within the $M$ shell. From the radial dependence of the electron's wavefunction $\psi_{n,\ell,m_\ell}(r, \theta, \phi)$, shown in Figure 3.21, we see that for higher values of $n$, which correspond to more energetic states, the mean distance of the electron from the nucleus increases. In fact, we observe from Figure 3.21 that an orbital with $\ell = n - 1$ (*e.g.*, 1$s$, 2$p$) exhibits a single maximum in its radial probability distribution, and this maximum rapidly moves farther away from the nucleus as $n$ increases. By examining the electron wavefunctions, we can show that the location of the maxima for these $\ell = n - 1$ states are at

*Maximum probability for $\ell = n - 1$*

$$r_{\max} = \frac{n^2 a_o}{Z} \qquad \text{for} \qquad \ell = n - 1 \qquad \textbf{[3.44]}$$

where $a_o$ is the radius of the ground state (0.0529 nm). The maximum probability radius $r_{\max}$ in Equation 3.44 is the Bohr radius. Note that $r_{\max}$ in Equation 3.44 is for $\ell = n - 1$ states only. For other $\ell$ values, there are multiple maxima, and we must think in terms of the average position of the electron from the nucleus. When we evaluate the average position from $\psi_{n,\ell,m_\ell}(r, \theta, \phi)$, we see that it depends on both $n$ and $\ell$; strongly on $n$ and weakly on $\ell$.

**EXAMPLE 3.16**

**THE IONIZATION ENERGY OF $He^+$** What is the energy required to further ionize $He^+$ ions to $He^{++}$?

**SOLUTION**

$He^+$ is a hydrogenic atom with one electron attracted by a nucleus with a $+2e$ charge. Thus $Z = 2$. The energy of the electron in a hydrogenic atom (in eV) is given by

$$E_n(\text{eV}) = -\frac{Z^2\,13.6}{n^2}$$

Since $Z = 2$, the energy required to ionize $He^+$ further is

$$|E_1| = |-(2^2)13.6| = 54.4\,\text{eV}$$

**EXAMPLE 3.17**

**IONIZATION ENERGY AND EFFECTIVE $Z$** The Li atom has a nucleus with a $+3e$ positive charge, which is surrounded by a full $1s$ orbital with two electrons, and a single valence electron in the outer $2s$ orbital as shown in Figure 3.26a. Intuitively we expect the valence electron to see the nuclear $+3e$ charge shielded by the two $1s$ electrons, that is, a net charge of $+1e$. It seems that we should be able to predict the ionization energy of the $2s$ electron by using the hydrogenic atom model and by taking $Z = 1$ and $n = 2$ as indicated in Figure 3.26b. However, according to quantum mechanics, the $2s$ electron has a probability distribution that has two peaks as shown in Figure 3.21; a major peak outside the $1s$ orbital, and a small peak around the $1s$ orbital. Thus, although the $2s$ electron spends a substantial time outside the $1s$ orbital, it does nonetheless penetrate the $1s$ shell and get close to the nucleus. Instead of experiencing a net $+1e$ of nuclear charge, it now experiences an effective nuclear charge that is greater than $+1e$, which we can represent as $+Z_{\text{effective}}e$, where we have used an *effective* $Z$. Thus, the ionization energy from Equation 3.43 is

*Ionization and effective nuclear charge*

$$E_{I,n} = \frac{Z^2_{\text{effective}}(13.6\text{ eV})}{n^2} \qquad [3.45]$$

The experimental ionization energy of Li is 5.39 eV which corresponds to creating a $Li^+$ ion and an isolated electron. Calculate the effective nuclear charge seen by the $2s$ electron.

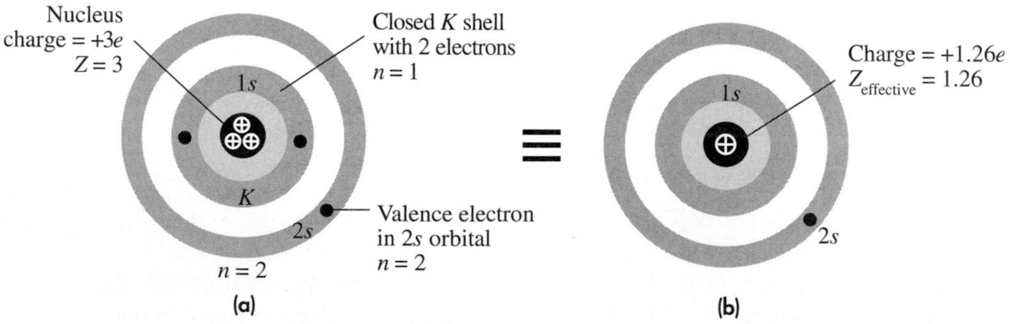


**Figure 3.26**
(a) The Li atom has a nucleus with charge $+3e$; two electrons in the $K$ shell, which is closed; and one electron in the $2s$ orbital.
(b) A simple view of (a) would be one electron in the $2s$ orbital that sees a single positive charge, $Z = 1$.

**SOLUTION**

The most outer electron in the Li atom is in the 2$s$ orbital, which is the electron that is removed in the ionization process. For this 2$s$ electron, $n = 2$, and hence from Equation 3.45

$$5.39 \text{ eV} = \frac{Z^2_{\text{effective}}(13.6 \text{ eV})}{(2)^2}$$

Solving, we find $Z_{\text{effective}} = 1.26$. If we simply use $Z = 1$ in Equation 3.45, we would find $E_{I,n} = 3.4$ eV, too small compared with the experimental value because, according to its probability distribution, the electron spends some time close to the nucleus, and hence increases its binding energy (stronger attraction). Variables $Z$ and $Z_{\text{effective}}$ should not be confused. $Z$ is the integer number of protons in the nucleus of the simple hydrogenic atom that are attracting the electron, as in H, $He^+$, or $Li^{++}$. $Z_{\text{effective}}$ is a convenient way of describing what the outer electron experiences in an atom because we would like to continue to use the simple expression for $E_{I,n}$, Equation 3.45, which was originally derived for a hydrogenic atom.

## 3.7.3 Orbital Angular Momentum and Space Quantization

The electron in the atom has an orbital angular momentum $L$. The electron is attracted to the nucleus by a central force, just like the Earth is attracted by the central gravitational force of the sun and thus possesses an orbital angular momentum. It is well known that in classical mechanics, under the action of a central force, both the total energy ($KE + PE$) and the orbital angular momentum ($L$) of an orbiting object are conserved. In quantum mechanics, the orbital angular momentum of the electron, like its energy, is also quantized, but by the quantum number $\ell$. The magnitude of $L$ is given by

*Orbital angular momentum*

$$L = \hbar[\ell(\ell + 1)]^{1/2} \qquad [3.46]$$

where $\ell = 0, 1, 2, \ldots < n$. Thus, for an electron in the ground state, $n = 1$ and $\ell = 0$, the angular momentum is zero, which is surprising since we always think of the electron as orbiting the nucleus. In the ground state, the spherical harmonic is a constant, independent of the angles $\theta$ and $\phi$, so the electron has a spherically symmetrical probability distribution that depends only on $r$.

The quantum numbers $n$ and $\ell$ quantize the energy and the magnitude of the orbital angular momentum. What is the significance of $m_\ell$? In the presence of an external magnetic field $B_z$, taken arbitrarily in the $z$ direction, the component of the angular momentum along the $z$ axis, $L_z$, is also quantized and is given by

*Orbital angular momentum along $B_z$*

$$L_z = m_\ell \hbar \qquad [3.47]$$

Therefore, the quantum number $m_\ell$ quantizes the component of the angular momentum along the direction of an external magnetic field $B_z$, which for reference purposes is taken along $z$, as illustrated in Figure 3.27. Therefore, $m_\ell$, is appropriately called the **magnetic quantum number.** For any given $\ell$, quantum mechanics requires that $m_\ell$ must have values in the range $-\ell, -(\ell - 1), \ldots, -1, 0, 1, \ldots, (\ell - 1), \ell$. We see that $|m_\ell| \leq \ell$. Moreover, $m_\ell$ can be negative, since $L_z$ can be negative or positive, depending on the orientation of the angular momentum vector $\mathbf{L}$. Since $|m_\ell| \leq \ell$, $\mathbf{L}$ can never align with the magnetic field along $z$; instead, it makes an angle with $B_z$, an

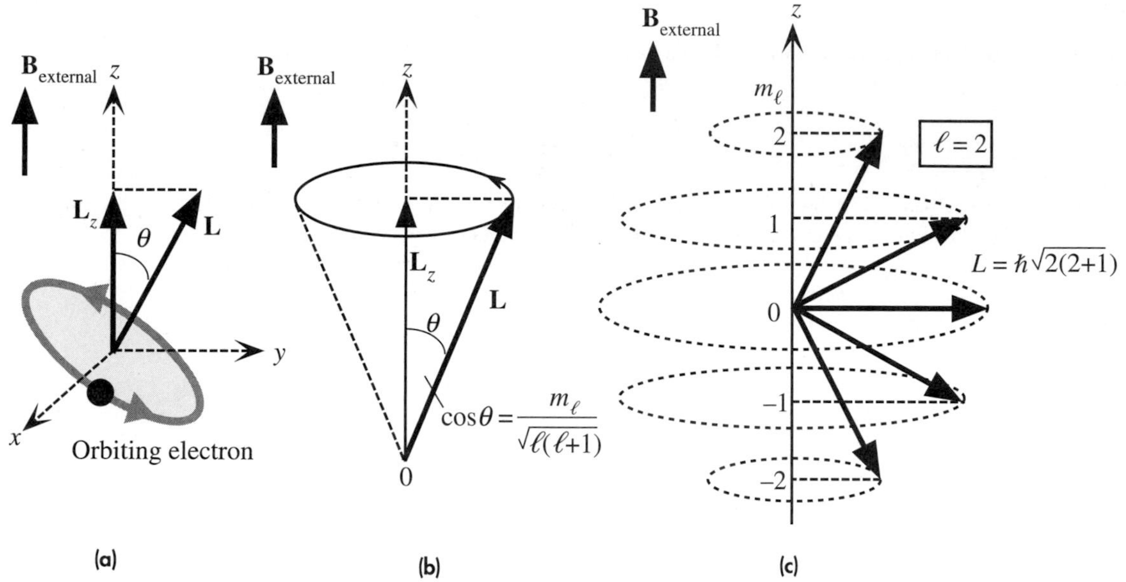


**Figure 3.27**

(a) The electron has an orbital angular momentum, which has a quantized component $L$ along an external magnetic field $B_{\text{external}}$.

(b) The orbital angular momentum vector **L** rotates about the $z$ axis. Its component $L_z$ is quantized; therefore, the **L** orientation, which is the angle $\theta$, is also quantized. **L** traces out a cone.

(c) According to quantum mechanics, only certain orientations ($\theta$) for **L** are allowed, as determined by $\ell$ and $m_\ell$.

angle that is determined by $\ell$ and $m_\ell$. We say that **L** is **space quantized.** Space quantization is illustrated in Figure 3.27 for $\ell = 2$.

Since the energy of the electron does not depend on either $\ell$ or $m_\ell$ we can have a number of possible states for a given energy. For example, when the energy is $E_2$, then $n = 2$, which means that $\ell = 0$ or 1. For $\ell = 1$, we have $m_\ell = -1, 0, 1$, so there are a total of three different orbitals for the electron.

Since the electron has a quantized orbital angular momentum, when an electron interacts with a photon, the electron must obey the law of the conservation of angular momentum, much as an ice skater does sudden fast spins by pulling in her arms. All experiments indicate that the photon has an intrinsic angular momentum with a constant magnitude given by $\hbar$. Therefore, when a photon of energy $h\nu = E_2 - E_1$ is absorbed, the angular momentum of the electron must change. This means that following photon absorption or emission, both the principal quantum number $n$ and the orbital angular momentum quantum number $\ell$ must change.

The rules that govern which transitions are allowed from one state to another as a consequence of photon absorption or emission are called **selection rules.** As a result of photon absorption or emission, we must have

*Selection rules for EM radiation*

$$\Delta\ell = \pm 1 \qquad \text{and} \qquad \Delta m_\ell = 0, \pm 1 \tag{3.48}$$

As an example, consider the excitation of the electron in the hydrogen atom from the ground energy $E_1$ to a higher energy level $E_2$. The photon energy $h\nu$ must be

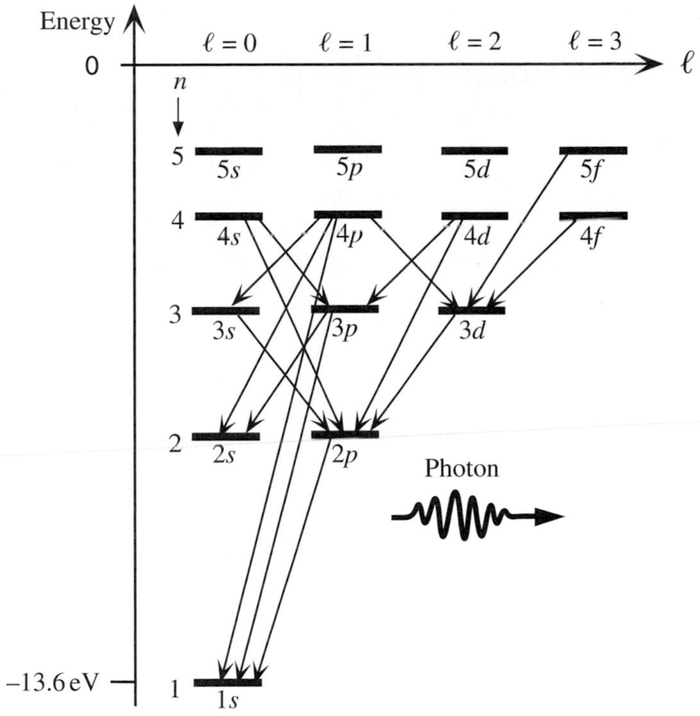


**Figure 3.28** An illustration of the allowed photon emission processes.
Photon emission involves $\Delta\ell = \pm 1$.

exactly $E_2 - E_1$. The wavefunction of the $1s$ ground state is $\psi_{1,0,0}$, whereas there are four wavefunctions at $E_2$: one $2s$ state, $\psi_{2,0,0}$; and three $2p$ states, $\psi_{2,1,-1}$, $\psi_{2,1,0}$, and $\psi_{2,1,1}$. The excited electron cannot jump into the $2s$ state, because $\Delta\ell$ must be $\pm 1$, so it enters a $2p$ state corresponding to one of the orbitals $\psi_{2,1,-1}$, $\psi_{2,1,0}$, or $\psi_{2,1,1}$. Various allowed transitions for photon emission in the hydrogen atom are indicated in Figure 3.28.

**EXAMPLE 3.18**

**EXCITATION BY ELECTRON–ATOM COLLISIONS IN A GAS DISCHARGE TUBE** A projectile electron with a velocity $2.1 \times 10^6$ m s$^{-1}$ collides with a hydrogen atom in a gas discharge tube. Find the $n$th energy level to which the electron in the hydrogen atom gets excited. Calculate the possible wavelengths of radiation that will be emitted from the excited H atom as the electron returns to its ground state.

**SOLUTION**

The energy of the electron in the hydrogen atom is given by $E_n$ (eV) $= -13.6/n^2$. The electron must be excited from its ground state $E_1 = -13.6$ eV to a quantized energy level $-(13.6/n^2)$ eV. The change in the energy is $\Delta E = (-13.6/n^2) - (-13.6)$ eV. This must be supplied by the incoming projectile electron, which has an energy of

$$E = \frac{1}{2}mv^2 = \frac{1}{2}(9.1 \times 10^{-31}\ \text{kg})(2.1 \times 10^6\ \text{m s}^{-1})^2$$

$$= 2.01 \times 10^{-18}\ \text{J} \quad \text{or} \quad 12.5\ \text{eV}$$

Therefore,

$$12.5\,\text{eV} = 13.6\,\text{eV} - \left[\frac{(13.6\,\text{eV})}{n^2}\right]$$

Solving this for $n$, we find

$$n^2 = \frac{13.6}{(13.6 - 12.5)} = 12.36$$

so $n = 3.51$. But $n$ can only be an integer; thus, the electron gets excited to the level $n = 3$ where its energy is $E_3 = -13.6/3^2 = -1.51$ eV.

The energy of the incoming electron after the collision is less by

$$(E_3 - E_1) = 13.6 - 1.51 = 12.09\,\text{eV}$$

Since the initial energy of the incoming electron was 12.5 eV, it leaves the collision with a kinetic energy of $12.5 - 12.09 = 0.41$ eV. From the $E_3$ level, the electron can undergo a transition from $n = 3$ to $n = 1$,

$$\Delta E_{31} = -1.51\,\text{eV} - (-13.6\,\text{eV}) = 12.09\,\text{eV}$$

The emitted radiation will have a wavelength $\lambda$ given by $hc/\lambda = \Delta E$, so that

$$\lambda_{31} = \frac{hc}{\Delta E_{31}} = \frac{(6.626 \times 10^{-34}\,\text{J s})(3 \times 10^8\,\text{m s}^{-1})}{12.09 \times 1.6 \times 10^{-19}\,\text{J}}$$

$$= 1.026 \times 10^{-7}\,\text{m} \qquad \text{or} \qquad 102.6\,\text{nm} \qquad \text{(in the ultraviolet region)}$$

Another possibility is the transition from $n = 3$ to $n = 2$, for which

$$\Delta E_{32} = -1.51\,\text{eV} - (-3.40\,\text{eV}) = 1.89\,\text{eV}$$

This will give a wavelength

$$\lambda_{32} = \frac{hc}{\Delta E_{32}} = 656\,\text{nm}$$

which is in the red region of the visible spectrum. For the transition from $n = 2$ to $n = 1$,

$$\Delta E_{21} = -3.40\,\text{eV} - (-13.6\,\text{eV}) = 10.2\,\text{eV}$$

which results in the emission of a photon of wavelength $\lambda_{21} = hc/\Delta E_{21} = 121.5$ nm. Note that each transition obeys $\Delta\ell = \pm 1$.

**EXAMPLE 3.19** **THE FRAUNHOFER LINES IN THE SUN'S SPECTRUM** The light from the sun includes extremely sharp "dark lines" at certain wavelengths, superimposed on a bright continuum at all other wavelengths, as discovered by Josef von Fraunhofer in 1829. One of these dark lines occurs in the orange range and another in the blue. Fraunhofer measured their wavelengths to be 6563 Å and 4861 Å, respectively. With the aid of Figure 3.23, show that these are spectral lines from the hydrogen atom spectrum. (They are called the $H_\alpha$ and $H_\beta$ Fraunhofer lines. Such lines provided us with the first clues to the chemical composition of the sun.)

**SOLUTION**

The energy of the electron in a hydrogenic atom is

$$E_n = -\frac{Z^2 E_I}{n^2}$$

where $E_I = me^4/(8\varepsilon_0^2 h^2)$. Photon emission resulting from a transition from quantum number $n_2$ to $n_1$ has an energy

$$\Delta E = E_{n_2} - E_{n_1} = -Z^2 E_I \left( \frac{1}{n_2^2} - \frac{1}{n_1^2} \right)$$

From $h\nu = hc/\lambda = \Delta E$, we have

*Emitted wavelengths for transitions in hydrogenic atom*

$$\frac{1}{\lambda} = \left( \frac{E_I}{hc} \right) Z^2 \left( \frac{1}{n_1^2} - \frac{1}{n_2^2} \right) = R_\infty Z^2 \left( \frac{1}{n_1^2} - \frac{1}{n_2^2} \right)$$

where $R_\infty = E_I/hc = 1.0974 \times 10^7 \text{ m}^{-1}$. The equation for $\lambda$ is called the **Balmer–Rydberg formula,** and $R_\infty$ is called the **Rydberg constant.** We apply the Balmer–Rydberg formula with $n_1 = 2$ and $n_2 = 3$ to obtain

$$\frac{1}{\lambda} = (1.0974 \times 10^7 \text{ m}^{-1})(1^2)\left( \frac{1}{2^2} - \frac{1}{3^2} \right) = 1.524 \times 10^6 \text{ m}^{-1}$$

to get $\lambda = 6561$ Å. We can also apply the Balmer–Rydberg formula with $n_1 = 2$ and $n_2 = 4$ to get $\lambda = 4860$ Å.

---

**EXAMPLE 3.20**

**GIANT ATOMS IN SPACE** Radiotelescopic studies by B. Höglund and P. G. Mezger (*Science* vol. 150, p. 339, 1965) detected a 5009 MHz electromagnetic radiation in space. Show that this radiation comes from excited hydrogen atoms as they undergo transitions from $n = 110$ to 109. What is the size of such an excited hydrogen atom?

**SOLUTION**

Since the energy of the electron is $E_n = -(Z^2 E_I/n^2)$, the energy of the emitted photon in the transition from $n_2$ to $n_1$ is

$$h\nu = E_{n_2} - E_{n_1} = Z^2 E_I \left(n_1^{-2} - n_2^{-2}\right)$$

With $n_2 = 110, n_1 = 109$, and $Z = 1$, the frequency is

$$\nu = \frac{Z^2 E_I \left(n_1^{-2} - n_2^{-2}\right)}{h}$$

$$= \frac{[(1.6 \times 10^{-19} \times 13.6)][(109^{-2} - 110^{-2})]}{(6.626 \times 10^{-34})}$$

$$= 5 \times 10^9 \text{ s}^{-1} \quad \text{or} \quad 5000 \text{ MHz}$$

The size of the atom from Equation 3.44 is on the order of

$$2r_{\text{max}} = 2n^2 a_o = 2(110^2)(52.918 \times 10^{-12} \text{ m}) = 1.28 \times 10^{-6} \text{ m} \quad \text{or} \quad 1.28\ \mu\text{m}$$

A giant atom!

---

## 3.7.4 Electron Spin and Intrinsic Angular Momentum S

One aspect of electron behavior does not come from the simple Schrödinger equation. That is the spin of the electron about its own axis, which is analogous to the 24-hour

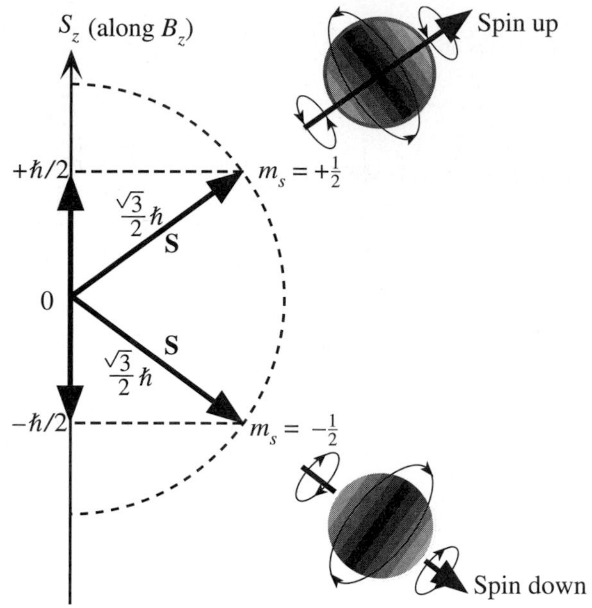


**Figure 3.29** Spin angular momentum exhibits space quantization. Its magnitude along $z$ is quantized, so the angle of **S** to the $z$ axis is also quantized.

spin of Earth around its axis.[7] Earth has an orbital angular momentum due to its motion around the sun, and an intrinsic or spin angular momentum due to its rotation about its own axis. Similarly, the electron has a **spin** or **intrinsic angular momentum,** denoted by **S**. In classical mechanics, in the absence of external torques, spin angular momentum is conserved. In quantum mechanics, this spin angular momentum is quantized, in a manner similar to that of orbital angular momentum. The magnitude of the spin has been found to be constant, with a quantized component $S_z$ in the $z$ direction along a magnetic field:

*Electron spin*

$$S = \hbar[s(s+1)]^{1/2} \qquad s = \frac{1}{2} \tag{3.49}$$

*Spin along magnetic field*

$$S_z = m_s\hbar \qquad m_s = \pm\frac{1}{2} \tag{3.50}$$

where, in an analogy with $\ell$ and $m_\ell$, we use the quantum numbers $s$ and $m_s$, which are called the **spin** and **spin magnetic quantum numbers.** Contrary to our past experience with quantum numbers, $s$ and $m_s$ are not integers, but are $\frac{1}{2}$ and $\pm\frac{1}{2}$, respectively. The existence of electron spin was put forward by Goudsmit and Uhlenbeck in 1925 and derived by Dirac from relativistic quantum theory, which is beyond the scope of this book. Figure 3.29 illustrates the spin angular momentum of the electron and the two possibilities for $S_z$. When $S_z = +\frac{1}{2}\hbar$, using classical orbital motion as an analogy, we

[7] Do not take the meaning of "spin" too literally, as in classical mechanics. Remember that the electron is assumed to have wave-like properties, which can have no classical spin.

**Table 3.3** The four quantum numbers for the hydrogenic atom

| | | | |
|---|---|---|---|
| $n$ | Principal quantum number | $n = 1, 2, 3, \ldots$ | Quantizes the electron energy |
| $\ell$ | Orbital angular momentum quantum number | $\ell = 0, 1, 2, \ldots (n-1)$ | Quantizes the magnitude of orbital angular momentum $L$ |
| $m_\ell$ | Magnetic quantum number | $m_\ell = 0, \pm 1, \pm 2, \ldots, \pm \ell$ | Quantizes the orbital angular momentum component along a magnetic field $B_z$ |
| $m_s$ | Spin magnetic quantum number | $m_s = \pm \dfrac{1}{2}$ | Quantizes the spin angular momentum component along a magnetic field $B_z$ |

can label the spin of the electron as being in the clockwise direction, so $S_z = -\frac{1}{2}\hbar$ can be labeled as a counterclockwise spin. However, no such true clockwise or counterclockwise spinning of the electron can in reality[8] be identified. When $S_z = +\frac{1}{2}\hbar$, we could just as easily label the electron spin as "up," and call it "down" when $S_z = -\frac{1}{2}\hbar$. This terminology is used henceforth in this book.

Since the magnitude of the electron spin is constant, which is a remarkable fact, and is determined by $s = \frac{1}{2}$, we need not mention it further. It can simply be regarded as a fundamental property of the electron, in much the same way as its mass and charge. We do, however, need to specify whether $m_s = +\frac{1}{2}$ or $-\frac{1}{2}$, since each of these selections gives the electron a different behavior. We therefore need four quantum numbers to specify what the electron is doing. Each state of the electron needs the spin magnetic quantum number $m_s$, in addition to $n$, $\ell$, and $m_\ell$. For each orbital $\psi_{n,\ell,m_\ell}(r, \theta, \phi)$, we therefore have two possibilities: $m_s = \pm\frac{1}{2}$. The quantum numbers $n$, $\ell$, and $m_\ell$ determine the spatial extent of the electron by specifying the form of $\psi_{n,\ell,m_\ell}(r, \theta, \phi)$, whereas $m_s$ determines the "direction" of the electron's spin. A full description of the behavior of the electron must therefore include all four quantum numbers $n$, $\ell$, $m_\ell$, and $m_s$.

An **electronic state** is a wavefunction that defines both the spatial ($\psi_{n,\ell,m_\ell}$) and spin ($m_s$) properties of an electron. Frequently, an electronic state is simply denoted $\psi_{n,\ell,m_\ell,m_s}$, which adds the spin quantum number to the orbital wavefunction.

The quantum numbers are extremely important, because they quantize the various properties of the electron: its total energy, orbital angular momentum, and the orbital and spin angular momenta along a magnetic field. Their significance is summarized in Table 3.3.

The spin angular momentum **S**, like the orbital angular momentum, is **space quantized.** $S_z = \pm(\frac{1}{2}\hbar)$ is smaller than $S = \hbar\sqrt{3}/2$, which means that $S$ can never line up with $z$, or a magnetic field, and the angle $\theta$ between **S** and the $z$ axis can only have two values corresponding to $m_\ell = +\frac{1}{2}$ and $-\frac{1}{2}$, which means that $\cos\theta = S_z/S = \pm 1/\sqrt{3}$. Classically, $S_z$ of a spinning object, or the orientation of **S** to the $z$-axis, can be any value inasmuch as classical spin has no space quantization.

[8] The explanation in terms of spin and its two possible orientational directions ("clockwise" and "counterclockwise") serve as mental aids in visualizing a quantum mechanical phenomenon. One question, however, is, "If the electron is a wave, what is spinning?"

### 3.7.5 MAGNETIC DIPOLE MOMENT OF THE ELECTRON

Consider the electron orbiting the nucleus with an angular frequency $\omega$ as illustrated in Figure 3.30a. The orbiting electron is equivalent to a current loop. The equivalent current $I$ due to the orbital motion of the electron is given by the charge flowing per unit time, $I = \text{charge/period} = -e(\omega/2\pi)$. The negative sign indicates that current $I$ flows in the opposite direction to the electron motion. The magnetic field around the current loop is similar to that of a permanent magnet as depicted in Figure 3.30a. The magnetic moment is defined as $\mu = IA$, the product of the current and the area enclosed by the current loop. It is a vector normal to the surface $A$ in a direction determined by the corkscrew rule applied to the circulation of the current $I$. If $r$ is the radius of the orbit (current loop), then the magnetic moment is

$$\mu = IA = \left(-\frac{e\omega}{2\pi}\right)(\pi r^2) = -\frac{e\omega r^2}{2}$$

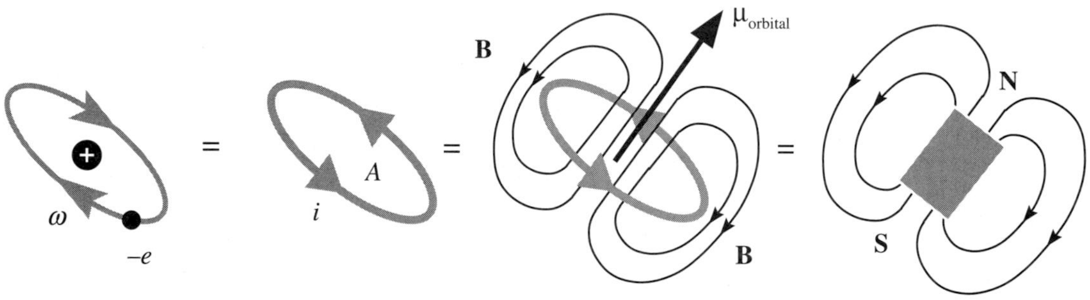


**(a) The orbiting electron is equivalent to a current loop that behaves like a bar magnet.**

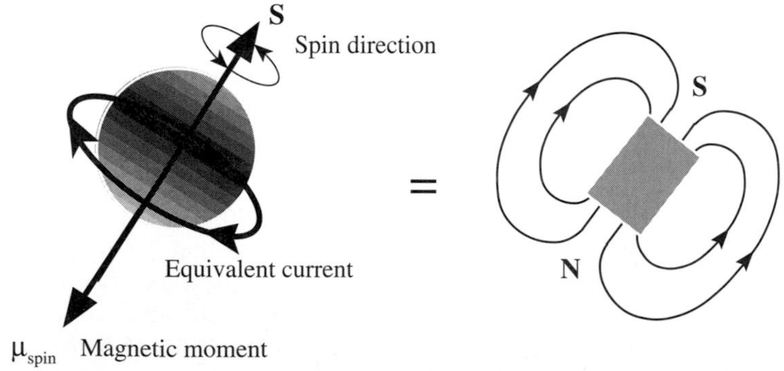


**(b) The spinning electron can be imagined to be equivalent to a current loop as shown. This current loop behaves like a bar magnet, just as in the orbital case.**

**Figure 3.30**

Consider now the orbital angular momentum $L$, which is the linear momentum $p$ multiplied by the radius $r$, or

$$L = pr = m_e vr = m_e \omega r^2$$

Using this, we can substitute for $\omega r^2$ in $\mu = -e\omega r^2/2$ to obtain

$$\mu = -\frac{e}{2m_e}L$$

In vector notation, using the subscript "orbital" to identify the origin of the magnetic moment,

*Orbital magnetic moment*

$$\boldsymbol{\mu}_{\text{orbital}} = -\frac{e}{2m_e}\mathbf{L} \qquad [3.51]$$

This means that the orbital magnetic moment $\boldsymbol{\mu}_{\text{orbital}}$ is in the opposite direction to that of the orbital angular momentum **L** and is related to it by a constant $(e/2m_e)$.

Similarly, the spin angular momentum of the electron **S** leads to a **spin magnetic moment** $\boldsymbol{\mu}_{\text{spin}}$, which is in the opposite direction to **S** and given by

*Spin magnetic moment*

$$\boldsymbol{\mu}_{\text{spin}} = -\frac{e}{m_e}\mathbf{S} \qquad [3.52]$$

which is shown in Figure 3.30b. Notice that there is no factor of 2 in the denominator. We see that, as a consequence of the orbital motion and also of spin, the electron has two distinct magnetic moments. These moments act on each other, just like two magnets interact with each other. The result is a coupling of the orbital and the spin angular momenta **L** and **S** and their precession about the total angular momentum $\mathbf{J} = \mathbf{L} + \mathbf{S}$, which is discussed in Section 3.7.6.

Since both **L** and **S** are quantized, so are the orbital and spin magnetic moments $\boldsymbol{\mu}_{\text{orbital}}$ and $\boldsymbol{\mu}_{\text{spin}}$. In the presence of an external magnetic field **B**, the electron has an additional energy term that arises from the interaction of these magnetic moments with **B**. We know from electromagnetism that a magnetic dipole (equivalent to a magnet) placed in a magnetic field **B** will have a potential energy $PE$. (A free magnet will rotate to align with the magnetic field, as in a compass, and thereby reduce the $PE$.) The *potential energy* $E_{BL}$ due to $\mu_{\text{orbital}}$ and $B$ interacting is given by

*Potential energy of a magnetic moment*

$$E_{BL} = -\mu_{\text{orbital}} B \cos\theta$$

where $\theta$ is the angle between $\mu_{\text{orbital}}$ and $B$. The potential energy $E_{BL}$ is minimum when $\mu_{\text{orbital}}$ (the magnet) and $B$ are parallel, $\theta = 0$. We know that, by definition, the $z$ axis is always along an external field **B**, and $L_z$ is the component of $L$ along $z$ (along $B$), and is quantized, so that $L_z = L\cos\theta = m_\ell \hbar$. We can substitute for $\mu_{\text{orbital}}$ to find

*Potential energy of orbital angular momentum in B*

$$E_{BL} = \left(\frac{e}{2m_e}\right)LB\cos\theta = \left(\frac{e}{2m_e}\right)L_z B = \left(\frac{e\hbar}{2m_e}\right)m_\ell B$$

which depends on $m_\ell$, and it is minimum for the smallest $m_\ell$. Since $m_\ell = -\ell, \ldots, 0, \ldots, +\ell$, negative and positive values through zero, the electron's energy splits into a number of levels determined by $m_\ell$. Similarly, the spin magnetic moment $\boldsymbol{\mu}_{\text{spin}}$ and

*Potential energy of orbital angular momentum in B*

**B** interact to give the electron a potential energy $E_{SL}$,

$$E_{SL} = \left(\frac{e\hbar}{m_e}\right) m_s B$$

which depends on $m_s$. Since $m_s = \pm\frac{1}{2}$, $E_{SL}$ has only two values, positive ($m_s = +\frac{1}{2}$) and negative ($m_s = -\frac{1}{2}$), which add and subtract from the electron's energy depending on whether the spin is up or down. Thus, in an external magnetic field, the electron's spin splits the energy level into two levels. The separation $\Delta E_{SL}$ of the split levels is $(e\hbar/m_e)B$, which is 0.12 meV $T^{-1}$, very small compared with the energy $E_n$ in the absence of the field. It should also be apparent that a single wavelength emission $\lambda_o$ corresponding to a particular transition from $E_{n'}$ to $E_n$ will now be split into a number of closely spaced wavelengths around $\lambda_o$. Although the separation $\Delta E_{SL}$ is small, it is still more than sufficient even at moderate fields to be easily detected and used in various applications. As it turns out, spin splitting of the energy in a field can be fruitfully used to study the electronic structures of not only atoms and molecules, but also various defects in semiconductors in what is called *electron spin resonance*.

**EXAMPLE 3.21**

**STERN–GERLACH EXPERIMENT AND SPIN** The Stern–Gerlach experiment is quite famous for demonstrating the spin of the electron and its space quantization. A neutral silver atom has one outer valence electron in a 4*s* orbital and looks much like the hydrogenic atom. (We can simply ignore the inner filled subshells in the Ag atom.). The 4*s* electron has no orbital angular momentum. Because of the *spin* of this one outer 4*s* electron, the whole Ag atom has a spin magnetic moment $\boldsymbol{\mu}_{\text{spin}}$. When Otto Stern and Walther Gerlach (1921–1922) passed a beam of Ag atoms through a nonuniform magnetic field, they found that the narrow beam split into two distinct beams as depicted in Figure 3.31a. The interpretation of the experiment was that the Ag atom's magnetic moment along the field direction can have only two values, hence the split beam. This observation agrees with the quantum mechanical fact that in a field along $z$, $\mu_{\text{spin},z} = -(e/m_e)m_s\hbar$ where $m_s = +\frac{1}{2}$ or $-\frac{1}{2}$; that is, the electron's spin can have only two values parallel to the field, or in other words, the electron spin is *space quantized.*

In the Stern–Gerlach experiment, the nonuniform magnetic field is generated by using a big magnet with shaped poles as in Figure 3.31a. The N-pole is sharp and the S-pole is wide, so the magnetic field lines get closer toward the N-pole and hence the magnetic field increases towards the N-pole. (This is much like a sharp point having a large electric field.) Whenever a magnetic moment, which we take to be a simple bar magnet, is in a nonuniform field, its poles experience different forces, say $\mathbf{F}_{\text{large}}$ and $\mathbf{F}_{\text{small}}$, and hence the magnet, overall, experiences a net force. The direction of the net force depends on the orientation of the magnet with respect to the $z$ axis as illustrated in Figure 3.31b for two differently oriented magnets representing magnetic moments labeled as 1 and 2. The S-pole of magnet 1 is in the high field region and experiences a bigger pull ($\mathbf{F}_{\text{large}}$) from the big magnet's N-pole than the small force ($\mathbf{F}_{\text{small}}$) pulling the N-pole of 1 to the big magnet's S-pole. Hence magnet 1 is pulled toward the N-pole and is deflected up. The overall force on a magnetic moment is the difference between $\mathbf{F}_{\text{large}}$ and $\mathbf{F}_{\text{small}}$, and its direction here is determined by the force on whichever pole is in the high field region. Magnet 2 on the other hand has its N-pole in the high field region, and hence is pushed away from the big magnet's N-pole and is deflected down. If the magnet is at right angles to the $z$ axis ($\theta = \pi/2$), it would experience no net force as both of its poles would be in the same field. This magnetic moment would pass through undeflected.

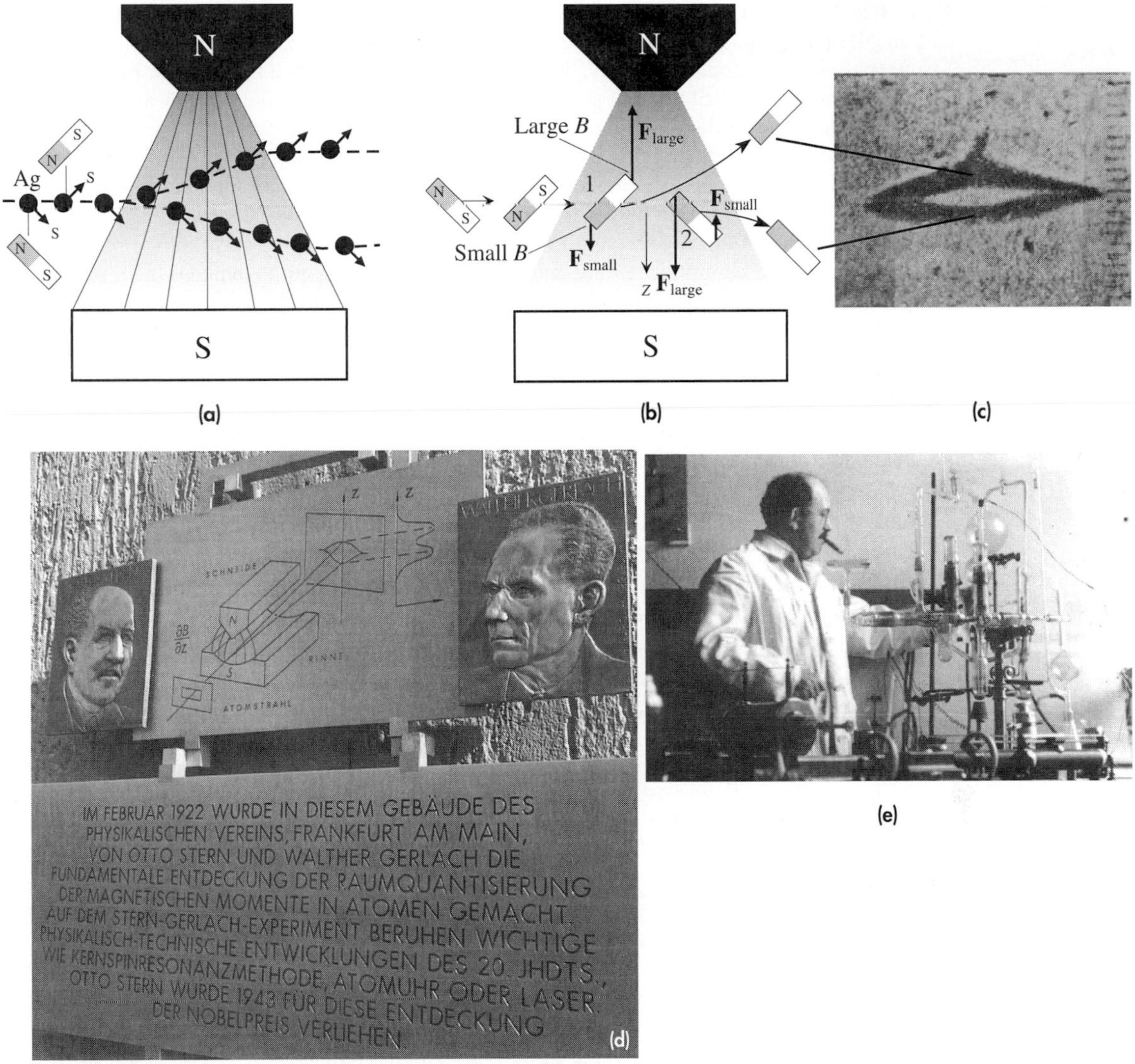


**Figure 3.31**

(a) Schematic illustration of the Stern–Gerlach experiment. A stream of Ag atoms passing through a nonuniform magnetic field splits into two.

(b) Explanation of the Stern–Gerlach experiment.

(c) Actual experimental result recorded on a photographic plate by Stern and Gerlach (O. Stern and W. Gerlach, *Zeitschr. fur. Physik*, **9**, 349, 1922.) When the field is turned off, there is only a single line on the photographic plate. Their experiment is somewhat different than the simple sketches in (a) and (b) as shown in (d).

(d) Stern–Gerlach memorial plaque at the University of Frankfurt. The drawing shows the original Stern–Gerlach experiment in which the Ag atom beam is passed along the long-length of the external magnet to increase the time spent in the nonuniform field, and hence increase the splitting.

(e) The photo on the lower right is Otto Stern (1888–1969), standing and enjoying a cigar while carrying out an experiment. Otto Stern won the Nobel prize in 1943 for development of the molecular beam technique.

SOURCES: (d) Courtesy of Horst Schmidt-Böcking from B. Friedrich and D. Herschbach, "Stern and Gerlach: How a Bad Cigar Helped Reorient Atomic Physics," *Physics Today*, December 2003, pp. 53–59. (e) AIP Emilio Segrè Visual Archives, Segrè Collection.

When we pass a stream of classical magnetic moments through a nonuniform field, there will be all possible orientations of the magnetic moment, from $-\pi$ to $+\pi$, with the field because there is no space quantization. Classically, the Ag atoms passing through a nonuniform field would be deflected through a distribution of angles and would not split into two distinct beams. The actual result of Stern and Gerlach's experiment is shown in Figure 3.31c, which is their photographic recording of a flat line-beam of Ag atoms passing through a long nonuniform field. In the absence of the field, the image is a simple horizontal line, the cross section of the beam. With the field turned on, the line splits into two. The edges of the line do not experience splitting because the field is very weak in the edge region. In the actual experiment, as shown in Figure 3.31c, an Ag atomic beam is passed along the long-length of the external magnet to increase the time spent in the nonuniform field, and hence increase the splitting. The physics remains the same.

### 3.7.6 TOTAL ANGULAR MOMENTUM J

The orbital angular momentum **L** and the spin angular momentum **S** add to give the electron a total angular momentum **J** = **L** + **S**, as illustrated in Figure 3.32. There are a number of possibilities for the total angular momentum **J**, based on the relative orientations of **L** and **S**. For example, for a given **L**, we can add **S** either in parallel or antiparallel, as depicted in Figure 3.32a and b, respectively.

Since in classical physics the total angular momentum of a body (not experiencing an external torque) must be conserved, we can expect $J$ (the magnitude of **J**) to be quantized. This turns out to be true. The magnitude of **J** and its $z$ component along an external magnetic field are quantized via

*Total angular momentum*

$$J = \hbar[j(j+1)]^{1/2} \qquad [3.53]$$

$$J_z = m_j\hbar \qquad [3.54]$$

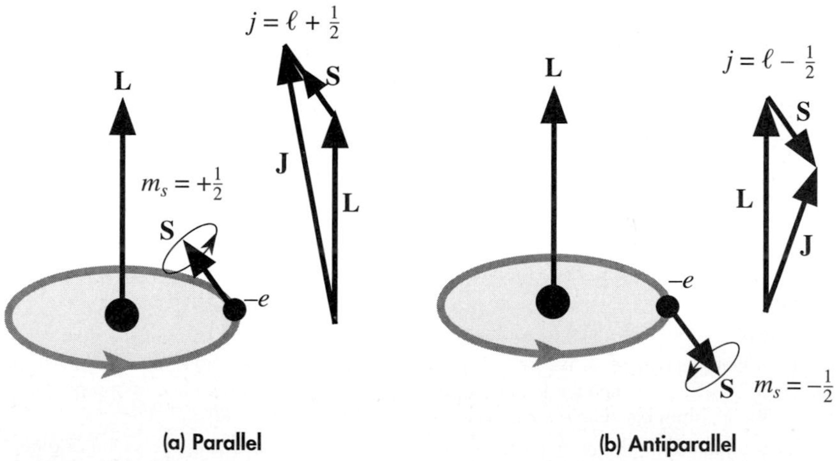


**Figure 3.32** Orbital angular momentum vector **L** and spin angular momentum vector **S** can add either in parallel as in (a) or antiparallel, as in (b).
The total angular momentum vector **J** = **L** + **S**, has a magnitude $J = \sqrt{[j(j+1)]}$, where in (a) $j = \ell + \frac{1}{2}$ and in (b) $j = \ell - \frac{1}{2}$.

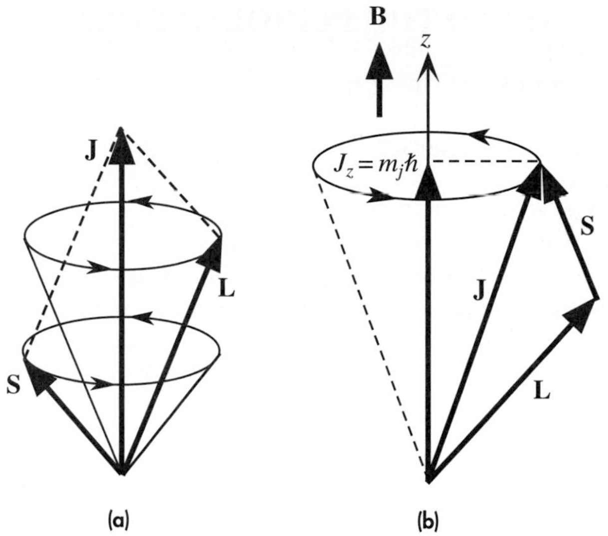


**Figure 3.33**
(a) The angular momentum vectors **L** and **S** precess around their resultant total angular momentum vector **J**.
(b) The total angular momentum vector is space quantized. Vector **J** precesses about the *z* axis, along which its component must be $m_j\hbar$.

where both $j$ and $m_j$ are quantum numbers[9] like $\ell$ and $m_\ell$, but $j$ and $m_j$ can have fractional values. A rigorous theory of quantum mechanics shows that when $\ell > s$, the quantum numbers for the total angular momentum are given by $j = \ell + s$ and $\ell - s$ and $m_j = \pm j, \pm(j-1)$. For example, for an electron in a $p$ orbital, where $\ell = 1$, we have $j = \frac{3}{2}$ and $\frac{1}{2}$, and $m_j = \frac{3}{2}, \frac{1}{2}, -\frac{1}{2}$, and $-\frac{3}{2}$. However, when $\ell = 0$ (as for all $s$ orbitals), we have $j = s = \frac{1}{2}$ and $m_j = m_s = \pm\frac{1}{2}$, which are the only possibilities. We note from Equations 3.53 and 3.54 that $|J_z| < J$ and both are quantized, which means that **J** is space quantized; its orientation (or angle) with respect to the $z$ axis is determined by $j$ and $m_j$.

The spinning electron actually experiences a magnetic field $\mathbf{B}_{\text{int}}$ due to its orbital motion around the nucleus. If we were sitting on the electron, then in our reference frame, the positively charged nucleus would be orbiting around us, which would be equivalent to a current loop. At the center of this current loop, there would be an "internal" magnetic field $\mathbf{B}_{\text{int}}$, which would act on the magnetic moment of the spinning electron to produce a torque. Since **L** and **S** add to give **J**, and since the latter quantity is space quantized (or conserved), then as a result of the internal torque on the electron, we must have **L** and **S** synchronously precessing about **J**, as illustrated in Figure 3.33a. If there is an external magnetic field **B** taken to be along $z$, this torque will act on the net magnetic moment due to **J** to cause this quantity to precess about **B**, as depicted in Figure 3.33b. Remember that the component along the $z$ axis must be quantized and equal to $m_j\hbar$, so the torque can only cause precession. To understand the precession of the electron's angular momentum about the magnetic field **B**, think of a spinning top that precesses about the gravitational field of Earth.

---

[9] The quantum number $j$ as used here should not be confused with $j$ for $\sqrt{-1}$.

# 3.8 THE HELIUM ATOM AND THE PERIODIC TABLE

## 3.8.1 He ATOM AND PAULI EXCLUSION PRINCIPLE

In the He atom, there are two electrons in the presence of a nucleus of charge $+2e$, as depicted in Figure 3.34. (Obviously, in higher-atomic-number elements, there will be $Z$ electrons around a nucleus of charge $+Ze$.) The *PE* of an electron in the He atom consists of two interactions. The first is due to the Coulombic attraction between itself and the positive nucleus; the second is due to the mutual repulsion between the two electrons. The *PE* function $V$ of any one of the electrons, for example, that labeled as 1, therefore depends on both its distance from the nucleus $r_1$ and the separation of the two electrons $r_{12}$. The *PE* of electron 1 thus depends on the locations of both the electrons, or

*PE of one electron in He atom*

$$V(r_1, r_{12}) = -\frac{2e^2}{4\pi\varepsilon_o r_1} + \frac{e^2}{4\pi\varepsilon_o r_{12}} \quad [3.55]$$

When we use this *PE* in the Schrödinger equation for a single electron, we find the wavefunction and energy of one of the electrons in the He atom. We thus obtain the **one-electron wavefunction** and the **energy of one electron** within a many-electron atom.

One immediate and obvious result is that the energy of an electron now depends not only on $n$ but also on $\ell$, because the electron–electron potential energy term (the second term in Equation 3.55, which contains $r_{12}$) depends on the relative orientations of the electron orbitals, which change $r_{12}$. We therefore denote the electron energy by $E_{n,\ell}$. The dependence on $\ell$ is weaker than on $n$, as shown in Figure 3.35. As $n$ and $\ell$ increase, $E_{n,\ell}$ also increases. Notice, however, that the energy of a 4$s$ state is lower than that of a 3$d$ state, and the same pattern also occurs at 4$s$ and 5$s$.

One of the most important theorems in quantum physics is the **Pauli exclusion principle,** which is based on experimental observations. This principle states that *no two electrons within a given system (e.g., an atom) may have all four identical quantum numbers, $n$, $\ell$, $m_\ell$, and $m_s$*. Each set of values for $n$, $\ell$, $m_\ell$, and $m_s$ represents a possible electronic state, that is, a wavefunction denoted by $\psi_{n,\ell,m_\ell,m_s}$, that the electron may (or may not) acquire. For example, an electron with the quantum numbers given by 2, 1, 1, $\frac{1}{2}$ will have a definite wavefunction $\psi_{n,\ell,m_\ell,m_s} = \psi_{2,1,1,1/2}$, and it is said to be

**Figure 3.34** A helium-like atom.
The nucleus has a charge of $+Ze$, where $Z = 2$ for He. If one electron is removed, we have the $He^+$ ion, which is equivalent to the hydrogenic atom with $Z = 2$.

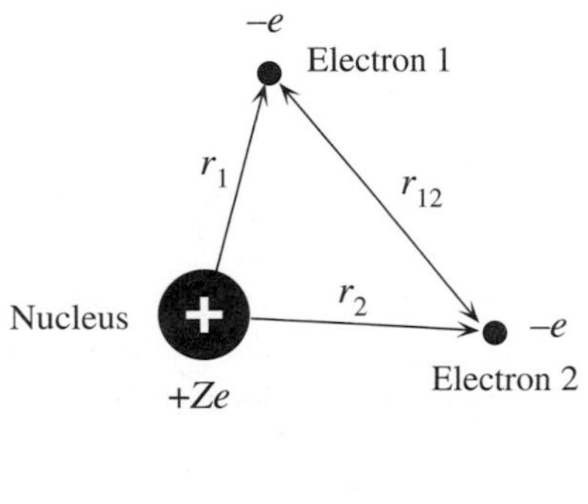

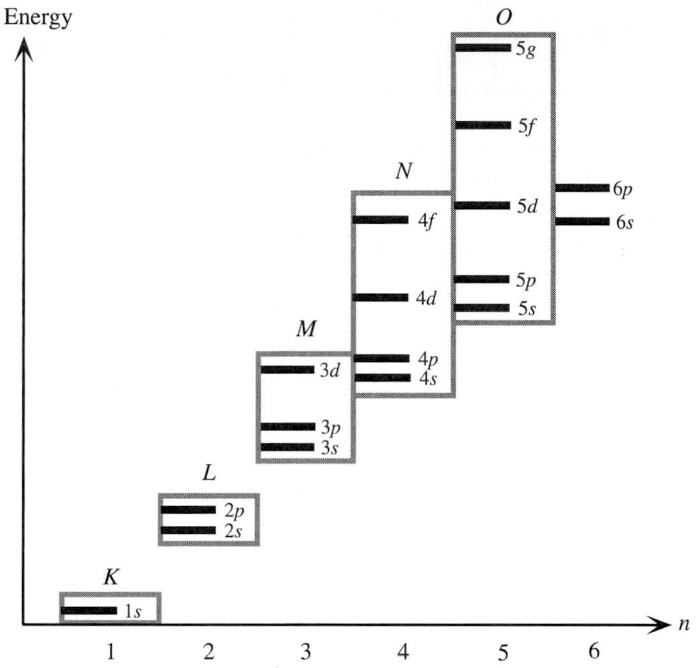


**Figure 3.35** Energy of various one-electron states. The energy depends on both $n$ and $\ell$.

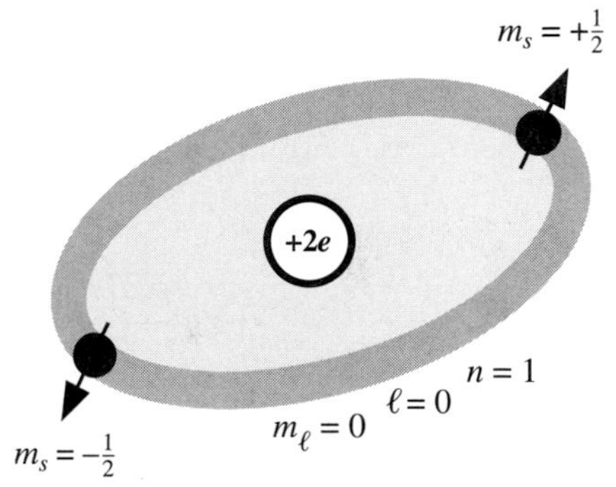


**Figure 3.36** Paired spins in an orbital.

in the state $2p$, $m_\ell = 1$ and spin up. Its energy will be $E_{2p}$. The Pauli exclusion principle requires that no other electron be in this same state.

The orbital motion of an electron is determined by $n$, $\ell$, and $m_\ell$, whereas $m_s$ determines the spin direction (up or down). Suppose two electrons are in the same orbital state, with identical $n$, $\ell$, $m_\ell$. By the Pauli exclusion principle, they would have to spin in opposite directions, as shown in Figure 3.36. One would have to spin "up" and the other "down." In this case we say that the electrons are **spin paired.** Two electrons can thus have the same orbitals (occupy the same region of space) if they pair their spins. However, the Pauli exclusion principle prevents a third electron from entering this orbital, since $m_s$ can only have two values.

Using the Pauli exclusion principle, we can determine the electronic structure of many-electron atoms. For simplicity, we will use a box to represent an orbital state defined by a set of $n$, $\ell$, $m_\ell$ values. Each box can take two electrons at most, with their spins paired. When we put an electron into a box, we are essentially assigning a wavefunction to that electron; that is, we are defining its orbital $n$, $\ell$, $m_\ell$. We use an arrow to show whether the electron is spinning up or down. As depicted in Figure 3.37, we arrange all the boxes to correspond to the electronic subshells. As an example, consider boron, which has five electrons. The first electron enters the $1s$ orbital at the lowest energy. The second also enters this orbital by spinning in the opposite direction. The third goes into the $n = 2$ orbital. The lowest energy there is in the $s$ orbitals corresponding to $\ell = 0$ and $m_\ell = 0$. The fourth electron can also enter the $2s$

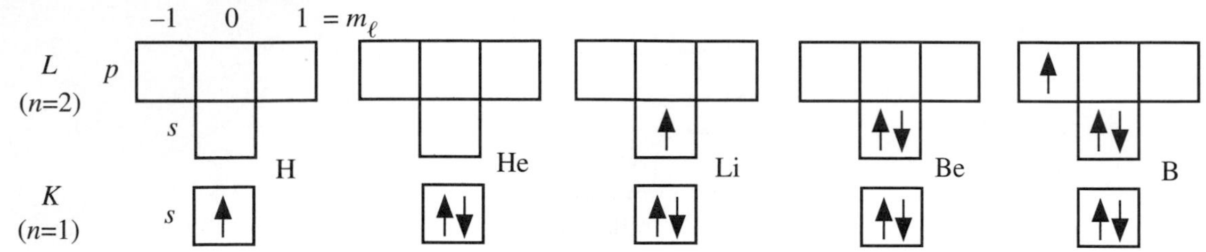


**Figure 3.37** Electronic configurations for the first five elements. Each box represents an orbital $\psi(n, \ell, m_\ell)$.

orbital, provided that it spins in the opposite direction. Similarly, the fifth must go into another orbital, and the next nearest low-energy orbitals are those having $\ell = 1$ ($p$ states) and $m_\ell = -1, 0, +1$. The final electronic structure of the B atom is shown in Figure 3.37.

We see that because the electron energy depends on $n$ and $\ell$, there are a number of states for a given energy $E_{n,\ell}$. Each of these states corresponds to different sets of $m_\ell$ and $m_s$. For example, the energy $E_{2,1}$ (or $E_{2p}$) corresponding to $n = 2$, $\ell = 1$ has six possible states, arising from $m_\ell = -1, 0, 1$ and $m_s = +\frac{1}{2}, -\frac{1}{2}$. Each $m_\ell$ state can have an electron spinning up or down, $m_s = +\frac{1}{2}$ or $m_s = -\frac{1}{2}$, respectively.

**EXAMPLE 3.22** **THE NUMBER OF STATES AT AN ENERGY LEVEL** Enumerate and identify the states corresponding to the energy level $E_{3d}$, or $n = 3$, $\ell = 2$.

**SOLUTION**

When $n = 3$ and $\ell = 2$, $m_\ell$ and $m_s$ can have these following values: $m_\ell = -2, -1, 0, 1, 2$, and $m_s = +\frac{1}{2}, -\frac{1}{2}$. This means there are 10 combinations. The possible wavefunctions (electron states) are

- $\psi_{3,2,2,1/2}$; $\psi_{3,2,1,1/2}$; $\psi_{3,2,0,1/2}$; $\psi_{3,2,-1,1/2}$; $\psi_{3,2,-2,1/2}$, all of which have spins up ($m_s = +\frac{1}{2}$)
- $\psi_{3,2,2,-1/2}$; $\psi_{3,2,1,-1/2}$; $\psi_{3,2,0,-1/2}$; $\psi_{3,2,-1,-1/2}$; $\psi_{3,2,-2,-1/2}$, all of which have spins down ($m_s = -\frac{1}{2}$)

## 3.8.2 HUND'S RULE

In the many-electron atom, the electrons take up the lowest-energy orbitals and obey the Pauli exclusion principle. However, the Pauli exclusion principle does not determine how any two electrons distribute themselves among the many states of a given $n$ and $\ell$. For example, there are six $2p$ states corresponding to $m_\ell = -1, 0, +1$, with each $m_\ell$ having $m_s = \pm\frac{1}{2}$. The two electrons could pair their spins and enter a given $m_\ell$ state, or they could align their spins (same $m_s$) and enter different $m_\ell$ states. An experimental fact deducted from spectroscopic studies shows that *electrons in the same* $n$, $\ell$ *orbitals prefer their spins to be parallel* (same $m_s$). This is known as **Hund's rule.**

The origin of Hund's rule can be readily understood. If electrons enter the same $m_\ell$ state by pairing their spins (different $m_s$), their quantum numbers $n$, $\ell$, $m_\ell$ will be the same and they will both occupy the same region of space (same $\psi_{n,\ell,m_\ell}$ orbital). They will then experience a large Coulombic repulsion and will have a large Coulombic potential energy. On the other hand, if they parallel their spins (same $m_s$), they will each have a different $m_\ell$ and will therefore occupy different regions of space (different $\psi_{n,\ell,m_\ell}$ orbitals), thereby reducing their Coulombic repulsion.

The oxygen atom has eight electrons and its electronic structure is shown in Figure 3.38. The first two electrons enter the $1s$ box (orbital). The next two enter the $2s$ box. But $p$ states can accommodate six electrons, so the remaining four electrons have a choice. Hund's rule forces three of the four electrons to enter the boxes corresponding to $m_\ell = -1, 0, +1$, all with their spins parallel. The last electron can go into any of the $2p$ boxes, but it has no choice for spin. It must pair its spin with the electron already in the box. Thus, the oxygen atom has two unpaired electrons in half-occupied orbitals, as indicated in Figure 3.38. Since these two unpaired electrons spin in the same direction, they give the O atom a net angular momentum. An angular momentum due to charge rotation (*i.e.*, spin) gives rise to a magnetic moment $\boldsymbol{\mu}$. If there is an external magnetic field present, then $\boldsymbol{\mu}$ experiences a force given by $\boldsymbol{\mu} \cdot d\mathbf{B}/dx$. Oxygen atoms will therefore be deflected by a nonuniform magnetic field, as experimentally observed.

Following the Pauli exclusion principle and Hund's rule, it is not difficult to build the electronic structure of various elements in the Periodic Table. There are only a few instances of unusual behavior in the energy levels of the electronic states. The $4s$ state happens to be energetically lower than the $3d$ states, so the $4s$ state fills up first. Similarly, the $5s$ state is at a lower energy than the $4d$ states. These features are summarized in the energy diagram of Figure 3.35. There is a neat shorthand way of writing the electronic structure of any atom. To each $n\ell$ state, we attach a superscript to represent the number of electrons in those $n\ell$ states. For example, for oxygen, we write $1s^2 2s^2 2p^4$, or simply $[\mathrm{He}]2s^2 2p^4$, since $1s^2$ is a full (closed) shell corresponding to He.

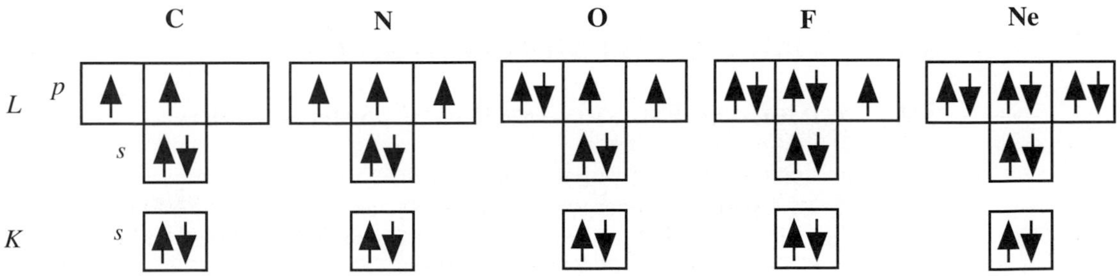


**Figure 3.38** Electronic configurations for C, N, O, F, and Ne atoms.
Notice that in C, N, and O, Hund's rule forces electrons to align their spins. For the Ne atom, all the $K$ and $L$ orbitals are full.

**EXAMPLE 3.23** **HUND'S RULE** The Fe atom has the electronic structure [Ar]$3d^6 4s^2$. Show that the Fe atom has four unpaired electrons and therefore a net angular momentum and a magnetic moment due to spin.

**SOLUTION**

In a closed subshell, for example, $2p$ subshell with six states given by $m_\ell = -1, 0, +1$ and $m_s = \pm\frac{1}{2}$, all $m_\ell$ and $m_s$ values have been taken up by electrons, so each $m_\ell$ orbital is occupied and has paired electrons. Each positive $m_\ell$ (or $m_s$) value assigned to an electron is canceled by the negative $m_\ell$ (or $m_s$) value assigned to another electron in the subshell. Therefore, *there is no net angular momentum from a closed subshell.* Only unfilled subshells contribute to the overall angular momentum. Thus, only the six electrons in the $3d$ subshell need be considered.

There are five $d$ orbitals, corresponding to $m_\ell = -2, -1, 0, 1, 2$. Five of the six electrons obey Hund's rule and align their spins, with each taking one of the $m_\ell$ values.

| $m_\ell = -2$ | $-1$ | $0$ | $1$ | $2$ |
|---|---|---|---|---|
| ↑ | ↑ | ↑ | ↑ | ↑ |
| ↓ | | | | |

The sixth must take the same $m_\ell$ as another electron. This is only possible if they pair their spins. Consequently, there are four electrons with unpaired spins in the Fe atom, which gives the Fe atom a net angular momentum. The Fe atom therefore possesses a magnetic moment as a result of four electrons having their charges spinning in the same direction.

Many *isolated* atoms possess unpaired spins and hence also possess a magnetic moment. For example, the isolated Ag atom has one outer $5s$ electron with an unpaired spin and hence it is magnetic; it can be deflected in a magnetic field. The silver crystal, however, is nonmagnetic. In the crystal, the $5s$ electrons become detached to form the electron gas (metallic bonding) where they pair their spins, and the silver crystal has no net magnetic moment. The iron crystal is magnetic because the constituent Fe atoms retain at least two of the unpaired electron spins which then all align in the same direction to give the crystal an overall magnetic moment; iron is a magnetic metal.[10]

# 3.9 STIMULATED EMISSION AND LASERS

## 3.9.1 STIMULATED EMISSION AND PHOTON AMPLIFICATION

An electron can be excited from an energy level $E_1$ to a higher energy level $E_2$ by the absorption of a photon of energy $h\nu = E_2 - E_1$, as show in Figure 3.39a. When an electron at a higher energy level transits down in energy to an unoccupied energy level,

[10] This qualitative explanation is discussed in Chapter 8.

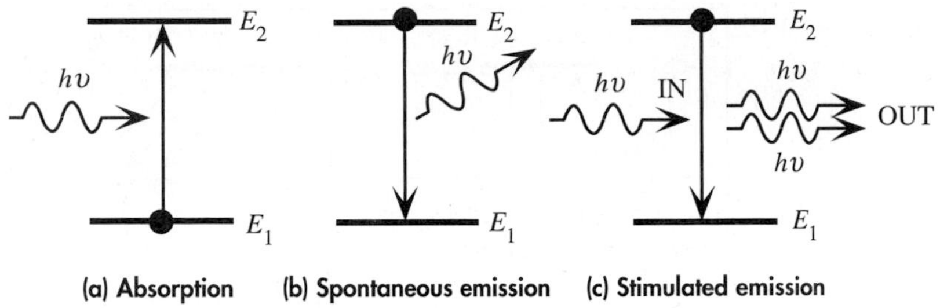


**Figure 3.39** Absorption, spontaneous emission, and stimulated emission.

it emits a photon. There are essentially two possibilities for the emission process. The electron can spontaneously undergo the downward transition by itself, or it can be induced to do so by another photon.

In **spontaneous emission,** the electron falls in energy from level $E_2$ to $E_1$ and emits a photon of energy $h\nu = E_2 - E_1$, as indicated in Figure 3.39b. The transition is only spontaneous if the state with energy $E_1$ is not already occupied by another electron. In classical physics, when a charge accelerates and decelerates, as in an oscillatory motion, with a frequency $\nu$, it emits an electromagnetic radiation also of frequency $\nu$. The emission process during the transition of the electron from $E_2$ to $E_1$ appears as if the electron is oscillating with a frequency $\nu$.

In **stimulated emission,** an incoming photon of energy $h\nu = E_2 - E_1$ stimulates the emission process by inducing the electron at $E_2$ to transit down to $E_1$. The emitted photon is in phase with the incoming photon, it is going in the same direction, and it has the same frequency, since it must also have the energy $E_2 - E_1$, as shown in Figure 3.39c. To get a feel for what is happening during stimulated emission, imagine the electric field of the incoming photon coupling to the electron and thereby driving it with the same frequency as the photon. The forced oscillation of the electron at a frequency $\nu = (E_2 - E_1)/h$ causes the electron to emit electromagnetic radiation, for which the electric field is totally in phase with that of the stimulating photon. When the incoming photon leaves the site, the electron can return to $E_1$, because it has emitted a photon of energy $h\nu = E_2 - E_1$.

Stimulated emission is the basis for photon amplification, since one incoming photon results in two outgoing photons, which are in phase. It is possible to achieve a practical light amplifying device based on this phenomenon. From Figure 3.39c, we see that to obtain stimulated emission, the incoming photon should not be absorbed by another electron at $E_1$. When we are considering using a collection of atoms to amplify light, we must therefore require that the majority of the atoms be at the energy level $E_2$. If this were not the case, the incoming photons would be absorbed by the atoms at $E_1$. When there are more atoms at $E_2$ than at $E_1$, we have what is called a **population inversion.** It should be apparent that with two energy levels, we can never achieve a population at $E_2$ greater than that at $E_1$, because, in the steady state, the incoming photon flux will cause as many upward excitations as downward stimulated emissions.

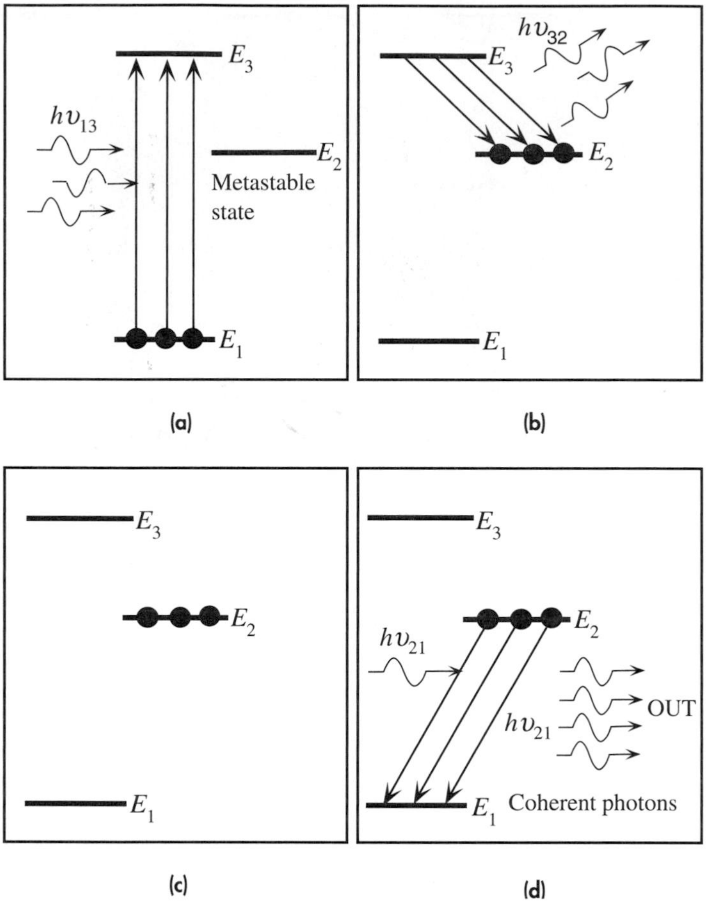


**Figure 3.40** The principle of the LASER.
(a) Atoms in the ground state are pumped up to energy level $E_3$ by incoming photons of energy $h\nu_{13} = E_3 - E_1$.
(b) Atoms at $E_3$ rapidly decay to the metastable state at energy level $E_2$ by emitting photons or emitting lattice vibrations: $h\nu_{32} = E_3 - E_2$.
(c) Since the states at $E_2$ are metastable, they quickly become populated, and there is a population inversion between $E_2$ and $E_1$.
(d) A random photon of energy $h\nu_{21} = E_2 - E_1$ can initiate stimulated emission. Photons from this stimulated emission can themselves further stimulate emissions, leading to an avalanche of stimulated emissions and coherent photons being emitted.

Let us consider the three-energy-level system shown in Figure 3.40. Suppose an external excitation causes the atoms[11] in this system to become excited to energy level $E_3$. This is called the **pump energy level,** and the process of exciting the atoms to $E_3$ is called **pumping.** In the present case, **optical pumping** is used, although this is not the only means of taking the atoms to $E_3$. Suppose further that the atoms in $E_3$ decay rapidly to energy level $E_2$, which happens to correspond to a state that does not rapidly and spontaneously decay to a lower energy state. In other words, the state at $E_2$ is a **long-lived state.**[12] Quite often, the long-lived states are referred to as **metastable states.** Since the atoms cannot decay rapidly from $E_2$ to $E_1$, they accumulate at this energy level, causing a population inversion between $E_2$ and $E_1$ as pumping takes more and more atoms to $E_3$ and hence to $E_2$.

[11] An atom is in an excited state when one (or more) of its electrons is excited from the ground energy to a higher energy level. The ground state of an atom has all the electrons in their lowest energy states consistent with the Pauli exclusion principle and Hund's rule.

[12] We will not examine what causes certain states to be long lived; we will simply accept that these states do not decay rapidly and spontaneously to lower energy states.

Arthur L. Schawlow in 1961 with a ruby laser built by his Stanford group. The solid state laser was a dark ruby crystal containing $Cr^{3+}$ ions. Lasing is obtained by stimulated emission from the $Cr^{3+}$ ions. Arthur Schawlow won the Nobel prize in Physics in 1981 for his contribution to the development of laser spectroscopy.

SOURCE: Stanford University, courtesy of AIP Emilio Segrè Visual Archives.

When one atom at $E_2$ decays spontaneously, it emits a photon, which can go on to a neighboring atom and cause that to execute stimulated emission. The photons from the latter can then go on to the next atom at $E_2$ and cause that atom to emit by stimulated emission, and so on. The result is an avalanche effect of stimulated emission processes with all the photons in phase, so the light output is a large collection of coherent photons. This is the principle of the ruby laser in which the energy levels $E_1$, $E_2$, and $E_3$ are those of the $Cr^{+3}$ ion in the $Al_2O_3$ crystal. At the end of the avalanche of stimulated emission processes, the atoms at $E_2$ will have returned to $E_1$ and can be pumped again to repeat the stimulated emission cycle again. The emission from $E_2$ to $E_1$ is called the **lasing emission.**

The system we have just described for photon amplification is a **LASER,** an acronym for light amplification by stimulated emission of radiation. In the ruby laser, pumping is achieved by using a xenon flashlight. The lasing atoms are chromium ions ($Cr^{3+}$) in a crystal of alumina $Al_2O_3$ (sapphire). The ends of the ruby crystal are silvered to reflect the stimulated radiation back and forth so that its intensity builds up, in much the same way we build up voltage oscillations in an electric oscillator circuit. One of the mirrors is partially silvered to allow some of this radiation to be tapped out. What comes out is a highly coherent radiation with a high intensity. The coherency and the well-defined wavelength of this radiation are what make it distinctly different from a random stream of different-wavelength photons emitted from a tungsten bulb.

## 3.9.2 HELIUM–NEON LASER

With the helium–neon (HeNe) laser, the actual operation is not simple, since we need to know such things as the energy states of the whole atom. We will therefore only consider the lasing emission at 632.8 nm, which gives the well-known red color to the laser light. The actual stimulated emission occurs from the Ne atoms; He atoms are used to excite the Ne atoms by atomic collisions.

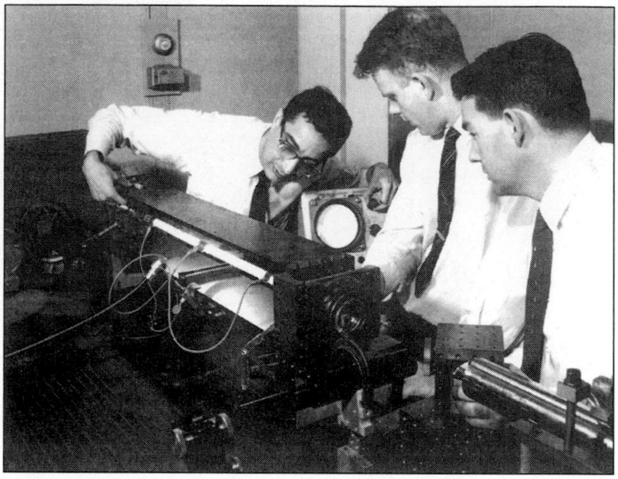

Ali Javan and his associates William Bennett Jr. and Donald Herriott at Bell Labs were first to successfully demonstrate a continuous wave (cw) helium–neon laser operation (1960).
| SOURCE: Courtesy of Bell Labs, Lucent Technologies.

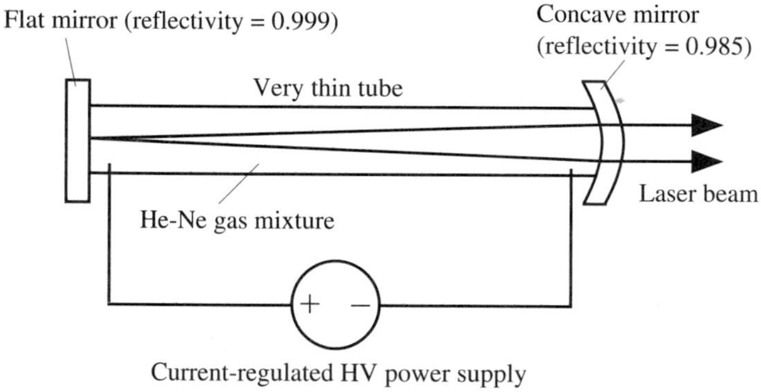


**Figure 3.41** Schematic illustration of the HeNe laser.

A modern stabilized HeNe laser.
| SOURCE: Courtesy of Melles Griot.

Ne is an inert gas with a ground state $(1s^2 2s^2 2p^6)$, which is represented as $(2p^6)$ when the inner closed $1s$ and $2s$ subshells are ignored. If one of the electrons from the $2p$ orbital is excited to a $5s$ orbital, the excited configuration $(2p^5 5s^1)$ is a state of the Ne atom that has higher energy. Similarly, He is an inert gas with the ground-state configuration of $(1s^2)$. The state of He when one electron is excited to a $2s$ orbital can be represented as $(1s^1 2s^1)$, which has higher energy.

The HeNe laser consists of a gaseous mixture of He and Ne atoms in a gas discharge tube, as shown schematically in Figure 3.41. The ends of the tube are mirrored to reflect the stimulated radiation and to build up the intensity within the cavity. If sufficient dc high voltage is used, electrical discharge is obtained within the tube, causing the He atoms to become excited by collisions with the drifting electrons. Thus,

$$\text{He} + e^- \rightarrow \text{He}^* + e^-$$

where He* is an excited He atom.

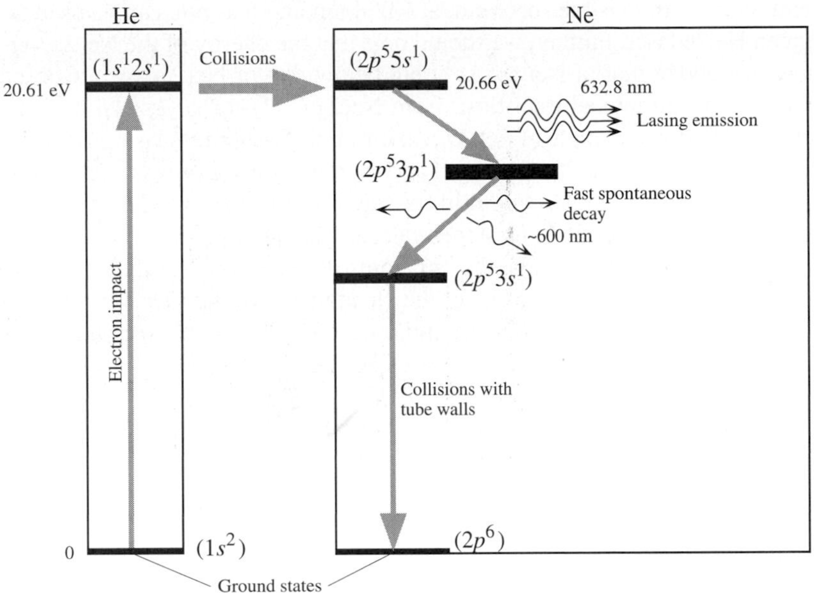


**Figure 3.42** The principle of operation of the HeNe laser. Important HeNe laser energy levels (for 632.8 nm emission).

The excitation of the He atom by an electron collision puts the second electron in He into a $2s$ state, so the excited He atom, He$^*$, has the configuration $(1s^12s^1)$. This atom is metastable (long lasting) with respect to the $(1s^2)$ state, as shown schematically in Figure 3.42. He$^*$ cannot spontaneously emit a photon and decay down to the $(1s^2)$ ground state because $\Delta\ell$ must be $\pm 1$. Thus, a large number of He$^*$ atoms build up during the electrical discharge.

When an excited He atom collides with a Ne atom, it transfers its energy to the Ne atom by resonance energy exchange. This happens because, by good fortune, Ne has an empty energy level, corresponding to the $(2p^55s^1)$ configuration, which matches that of $(1s^12s^1)$ of He$^*$. The collision process excites the Ne atom and de-excites He$^*$ down to its ground energy, that is,

$$\text{He}^* + \text{Ne} \rightarrow \text{He} + \text{Ne}^*$$

With many He$^*$–Ne collisions in the gaseous discharge, we end up with a large number of Ne$^*$ atoms and a population inversion between the $(2p^55s^1)$ and $(2p^53p^1)$ states of the Ne atom, as indicated in Figure 3.42. The spontaneous emission of a photon from one Ne$^*$ atom falling from $5s$ to $3p$ gives rise to an avalanche of stimulated emission processes, which leads to a lasing emission with a wavelength of 632.8 nm, in the red.

There are a few interesting facts about the HeNe laser, some of which are quite subtle. First, the $(2p^55s^1)$ and $(2p^53p^1)$ electronic configurations of the Ne atom actually have a spread of energies. For example for Ne$(2p^55s^1)$, there are four closely spaced energy levels. Similarly, for Ne$(2p^53p^1)$, there are 10 closely separated energies. We can therefore achieve population inversion with respect to a number of energy levels. As a result, the lasing emissions from the HeNe laser contain a variety of wavelengths. The two lasing emissions in the visible spectrum, at 632.8 nm and 543 nm, can be used to build a red or green HeNe laser. Further, we should note that the energy of the Ne($2p^54p^1$) state (not shown) is above that of Ne($2p^53p^1$) but below that of Ne($2p^55s^1$). Consequently, there will also be stimulated transitions from Ne($2p^55s^1$) to Ne($2p^54p^1$), and hence a lasing emission at a wavelength of ~3.39 $\mu$m infrared. To suppress lasing emissions at the unwanted wavelengths (*e.g.*, the infrared) and to obtain lasing only at the wavelength of interest, we can make the reflecting mirrors wavelength selective. This way the optical cavity builds up optical oscillations at the selected wavelength.

From ($2p^53p^1$) energy levels, the Ne atoms decay rapidly to the ($2p^53s^1$) energy levels by spontaneous emission. Most of the Ne atoms with the ($2p^53s^1$) configuration, however, cannot simply return to the ground state $2p^6$, because the return of the electron in $3s$ requires that its spin be flipped to close the $2p$ subshell. An electromagnetic radiation cannot change the electron spin. Thus, the Ne($2p^53s^1$) energy levels are metastable. The only possible means of returning to the ground state (and for the next repumping act) is collisions with the walls of the laser tube. Therefore, we cannot increase the power obtainable from a HeNe laser simply by increasing the laser tube diameter, because that will accumulate more Ne atoms at the metastable ($2p^53s^1$) states.

A typical HeNe laser, illustrated in Figure 3.41, consists of a narrow glass tube that contains the He and Ne gas mixture (typically, the He to Ne ratio is 10:1). The lasing emission intensity increases with tube length, since more Ne atoms are then used in stimulated emission. The intensity decreases with increasing tube diameter, since Ne atoms in the ($2p^53s^1$) states can only return to the ground state by collisions with the walls of the tube. The ends of the tube are generally sealed with a flat mirror (99.9 percent reflecting) at one end and, for easy alignment, a concave mirror (98.5 percent reflecting) at the other end, to obtain an optical cavity within the tube. The outer surface of the concave mirror is ground to behave like a convergent lens, to compensate for the divergence in the beam arising from reflections from the concave mirror. The output radiation from the tube is typically a beam of diameter 0.5–2 mm and a divergence of 1 milliradians at a power of a few milliwatts. In high-power HeNe lasers, the mirrors are external to the tube. In addition, Brewster windows are fused at the ends of the laser tube, to allow only polarized light to be transmitted and amplified within the cavity, so that the output radiation is polarized (that is, has electric field oscillations in one plane).

**EXAMPLE 3.24** **EFFICIENCY OF THE HeNe LASER** A typical low-power 2.5 mW HeNe laser tube operates at a dc voltage of 2 kV and carries a current of 5 mA. What is the efficiency of the laser?

**SOLUTION**

From the definition of efficiency,

$$\text{Efficiency} = \frac{\text{Output power}}{\text{Input power}}$$

$$= \frac{(2.5 \times 10^{-3}\text{ W})}{(5 \times 10^{-3}\text{ A})(2000\text{ V})} = 0.00025 \quad \text{or} \quad 0.025\%$$

### 3.9.3 LASER OUTPUT SPECTRUM

The output radiation from a laser is not actually at one single well-defined wavelength corresponding to the lasing transition. Instead, the output covers a spectrum of wavelengths with a central peak. This is not a simple consequence of the Heisenberg uncertainty principle (which does broaden the output). Predominantly, it is a result of the broadening of the emitted spectrum by the **Doppler effect.** We recall from the kinetic molecular theory that gas atoms are in random motion, with an average translational kinetic energy of $\frac{3}{2}kT$. Suppose that these gas atoms emit radiation of frequency $\nu_0$ which we label as the source frequency. Then, due to the Doppler effect, when a gas atom moves toward an observer, the latter detects a higher frequency $\nu_2$, given by

$$\nu_2 = \nu_0\left(1 + \frac{v_x}{c}\right)$$

*Doppler effect*

where $v_x$ is the relative velocity of the atom with respect to the observer and $c$ is the speed of light. When the atom moves away, the observer detects a smaller frequency, which corresponds to

$$\nu_1 = \nu_0\left(1 - \frac{v_x}{c}\right)$$

*Doppler effect*

Since the atoms are in random motion, the observer will detect a range of frequencies, due to this Doppler effect. As a result, the frequency or wavelength of the output radiation from a gas laser will have a "linewidth" of $\Delta\nu = \nu_2 - \nu_1$, called a Doppler-broadened **linewidth** of a laser radiation. Other mechanisms also broaden the output spectrum, but we will ignore these at present.

The reflections from the laser end mirrors give rise to traveling waves in opposite directions within the cavity. Since the waves are in phase, they interfere constructively, to set up a standing wave—in other words, stationary oscillations. Some of the energy in this wave is tapped by the 99 percent reflecting mirror to get an output, in much the same way that we tap the energy from an oscillating field in an $LC$ circuit by attaching an antenna to it.

Only standing waves with certain wavelengths can be maintained within the optical cavity, just as only certain acoustic wavelengths can be obtained from musical instruments. Any standing wave in the cavity must have a half-wavelength $\lambda/2$ that fits into the cavity length $L$, or

$$n\left(\frac{\lambda}{2}\right) = L \qquad [3.56]$$

*Laser cavity modes*

where $n$ is an integer called the **mode number** of the standing wave. Each possible standing wave within the laser tube (cavity) satisfying Equation 3.56 is called a **cavity mode.** The laser output thus has a broad spectrum with peaks at certain wavelengths corresponding to various cavity modes existing within the Doppler-broadened emission curve. Figure 3.43 shows the expected output from a typical gas laser. At wavelengths satisfying Equation 3.56, that is, representing certain cavity modes, we have intensity spikes in the output. The net envelope of the output

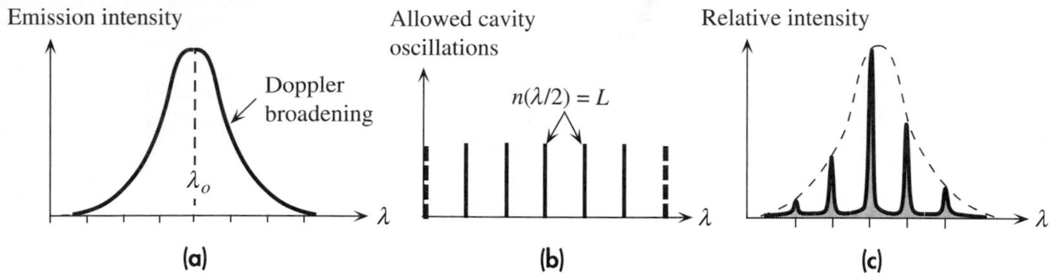


**Figure 3.43**
(a) Doppler-broadened emission versus wavelength characteristics of the lasing medium.
(b) Allowed oscillations and their wavelengths within the optical cavity.
(c) The output spectrum is determined by satisfying (a) and (b) simultaneously.

radiation is a Gaussian distribution, which is essentially due to the Doppler-broadened linewidth.

Even though we can try to get as parallel a beam as possible by lining the mirrors up perfectly, we will still be faced with diffraction effects at the output. When the output laser beam hits the end of the laser tube, it becomes diffracted, so the emerging beam is necessarily divergent. Simple diffraction theory can readily predict the divergence angle.

**EXAMPLE 3.25** **DOPPLER-BROADENED LINEWIDTH** Calculate the Doppler-broadened linewidths $\Delta\nu$ and $\Delta\lambda$ for the HeNe laser transition $\lambda = 632.8$ nm, if the gas discharge temperature is about 127 °C. The atomic mass of Ne is 20.2 g mol$^{-1}$.

**SOLUTION**

Due to the Doppler effect, the laser radiation from gas lasers is broadened around a central frequency $\nu_o$, which corresponds to the source frequency. Higher frequencies detected will be due to radiations emitted from atoms moving toward the observer, and lower frequencies detected will be the result of emissions from atoms moving away from the observer. Therefore, the width of the observed frequencies will be approximately

*Doppler-broadened frequency width*

$$\Delta\nu = \nu_o\left(1 + \frac{v_x}{c}\right) - \nu_o\left(1 - \frac{v_x}{c}\right) = \frac{2\nu_o v_x}{c}$$

From $\lambda = c/\nu$, we obtain the following by differentiation:

$$\frac{d\lambda}{d\nu} = -\frac{c}{\nu^2} = -\frac{\lambda}{\nu} = -\frac{\lambda^2}{c}$$

We need to know $v_x$, which is given by kinetic theory as $v_x^2 = kT/m$. For the HeNe laser, the Ne atoms lase, so

$$m = \frac{20.2 \times 10^{-3}\ \text{kg mol}^{-1}}{6.023 \times 10^{23}\ \text{mol}^{-1}} = 3.35 \times 10^{-26}\ \text{kg}$$

Thus

$$v_x = \left[\frac{(1.38 \times 10^{-23}\,\text{J K}^{-1})(127 + 273\ \text{K})}{(3.35 \times 10^{-26}\ \text{kg})}\right]^{1/2} = 406\ \text{m s}^{-1}$$

The central frequency is

$$\nu_o = \frac{c}{\lambda_o} = \frac{3 \times 10^8 \text{ m s}^{-1}}{632.8 \times 10^{-9} \text{ m}} = 4.74 \times 10^{14} \text{ s}^{-1}$$

The frequency linewidth is

$$\Delta\nu = \frac{(2\nu_o v_x)}{c} = \frac{2(4.74 \times 10^{14} \text{ s}^{-1})(406 \text{ m s}^{-1})}{3 \times 10^8 \text{ m s}^{-1}} = 1.283 \text{ GHz}$$

To get $\Delta\lambda$, we use $d\lambda/d\nu = -\lambda/\nu$, so that

$$\Delta\lambda = \Delta\nu \left| -\frac{\lambda_o}{\nu_o} \right| = \frac{(1.283 \times 10^9 \text{ Hz})(632.8 \times 10^{-9} \text{ m})}{4.74 \times 10^{14} \text{ s}^{-1}}$$

$$= 1.71 \times 10^{-12} \text{ m} \quad \text{or} \quad 0.0017 \text{ nm}$$

# ADDITIONAL TOPICS

## 3.10 OPTICAL FIBER AMPLIFIERS

A light signal that is traveling along an optical fiber communications link over a long distance suffers marked attenuation. It becomes necessary to regenerate the light signal at certain intervals for long-haul communications over several thousand kilometers. Instead of regenerating the optical signal by photodetection, conversion to an electrical signal, amplification, and then conversion back from electrical to light energy by a laser diode, it becomes practical to amplify the signal directly by using an optical amplifier. The photons in an optical signal have a wavelength of 1550 nm, and optical amplifiers have to amplify signal photons at this wavelength.

One practical **optical amplifier** is based on the **erbium ($Er^{3+}$ ion) doped fiber amplifier (EDFA).**[13] The core region of an optical fiber is doped with $Er^{3+}$ ions. The host fiber core material is a glass based on $SiO_3$–$GeO_2$ and perhaps some other glass-forming oxides such as $Al_2O_3$. It is easily fused to a long-distance optical fiber by a technique called splicing.

When the $Er^{3+}$ ion is implanted in the host glass material, it has the energy levels indicated in Figure 3.44 where $E_1$ corresponds to the lowest energy possible consistent with the Pauli exclusion principle and Hund's rule. One of the convenient energy levels for optically pumping the $Er^{3+}$ ion is at $E_3$, approximately 1.27 eV above the ground energy level. The $Er^{3+}$ ions are optically pumped, usually from a laser diode, to excite them to $E_3$. The wavelength for this pumping is about 980 nm. The $Er^{3+}$ ions decay rapidly from $E_3$ to a **long-lived** energy level at $E_2$ which has a long lifetime of ~10 ms (very long on the atomic scale). The decay $E_3$ to $E_2$ involves energy losses by radiationless transitions (generation of crystal vibrations) and are very rapid. Thus, more and more $Er^{3+}$ ions accumulate at $E_2$ which is 0.80 eV above the ground energy. The accumulation of $Er^{3+}$ ions at $E_2$ leads to a population inversion between $E_2$ and $E_1$. Signal photons at 1550 nm have an energy of 0.80 eV, or $E_2 - E_1$, and give rise to *stimulated transitions* of $Er^{3+}$ ions from $E_2$ to $E_1$. Any $Er^{3+}$ ions left at $E_1$, however, will

[13] EDFA was first reported in 1987 by E. Desurvire, J. R. Simpson, and P. C. Becker and, within a short period, AT&T began deploying EDFA repeaters in long-haul fiber communications in 1994.

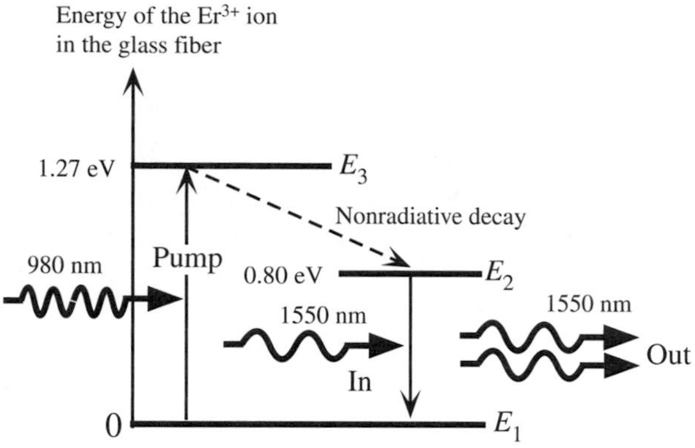


**Figure 3.44** Energy diagram for the $Er^{3+}$ ion in the glass fiber medium and light amplification by stimulated emission from $E_2$ to $E_1$.
Dashed arrows indicate radiationless transitions (energy emission by lattice vibrations).

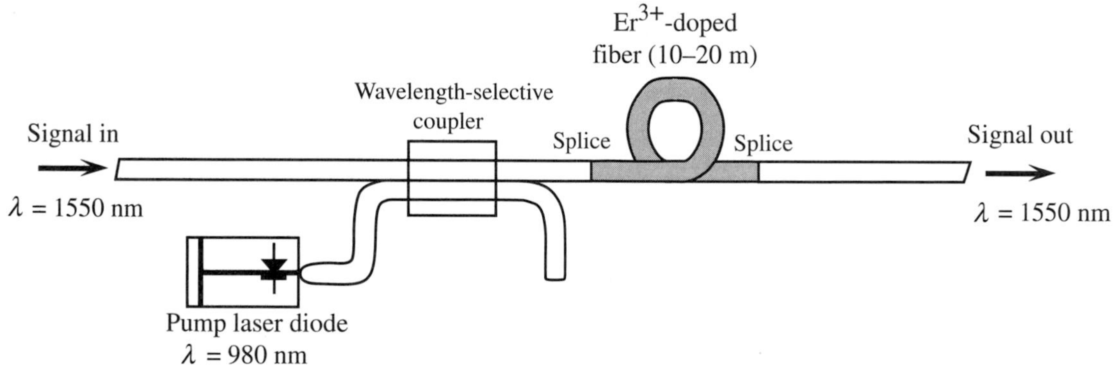


**Figure 3.45** A simplified schematic illustration of an EDFA (optical amplifier).
The erbium-ion doped fiber is pumped by feeding the light from a laser pump diode, through a coupler, into the erbium-ion doped fiber.

*absorb* the incoming 1550 nm photons to reach $E_2$. To achieve light amplification we must therefore have stimulated emission exceeding absorption. This is only possible if there are more $Er^{3+}$ ions at the $E_2$ level than at the $E_1$ level, that is, if we have population inversion. With sufficient optical pumping, population inversion is readily achieved.

In practice the erbium-doped fiber is inserted into the fiber communications line by splicing as shown in the simplified schematic diagram in Figure 3.45 and it is pumped from a laser diode through a coupling fiber arrangement which allows only the pumping wavelength to be coupled.

## CD Selected Topics and Solved Problems

| Selected Topics | Solved Problems |
|---|---|
| Compton Scattering | Modern Physics: Photoelectric Experiment, Ionization Energy |
| Stimulated Emission and Laser Principles | He–Ne Laser Problem |
| Stimulated Emission and Optical Amplifiers | |
| Time-Dependent Schrödinger Equation | |

# DEFINING TERMS

**Angular momentum L** about a point $O$ is defined as $\mathbf{L} = \mathbf{p} \times \mathbf{r}$, where **p** is the linear momentum and **r** is the position vector of the body from $O$. For a circular orbit around $O$, the angular momentum is orbital and $L = pr = mvr$.

**Bragg diffraction law** describes the diffraction of an X-ray beam by a crystal in which the interplanar separation $d$ of a given set of atomic planes causing the X-ray diffraction is related to the diffraction angle $2\theta$ and the wavelength $\lambda$ of the X-rays through $2d \sin\theta = n\lambda$ where $n$ is an integer, usually unity.

**Complementarity principle** suggests that the wave model and the particle model are complementary models in that one model alone cannot be used to explain all the observations in nature. For example, the electron diffraction phenomenon is best explained by the wave model, whereas in the Compton experiment, the electron is treated as a particle; that is, it is deflected by an impinging photon that imparts an additional momentum to the electron.

**Compton effect** is the scattering of a high-energy photon by a "free" electron. The effect is experimentally observed when an X-ray beam is scattered from a target that contains many conduction ("free") electrons, such as a metal or graphite.

**De Broglie** relationship relates the wave-like properties (*e.g.*, wavelength $\lambda$) of matter to its particle-like properties (*e.g.*, momentum $p$) via $\lambda = h/p$.

**Diffraction** is the bending of waves as a result of the interaction of the waves with an object of size comparable to the wavelength. If the object has a regular pattern, periodicity, an incident beam of waves can be bent (diffracted) in certain well-defined directions that depend on the periodicity, which is used in the X-ray diffraction study of crystals.

**Doppler effect** is the change in the measured frequency of a wave due to the motion of the source relative to the observer. In the case of electromagnetic radiation, if $v$ is the relative velocity of the source object toward the observer and $\nu_o$ is the source frequency, then the measured electromagnetic wave frequency is $\nu = \nu_o[1 + (v/c)]$ for $(v/c) \ll 1$.

**Energy density** $\rho_E$ is the amount of energy per unit volume. In a region where the electric field is $\mathcal{E}$, the energy stored per unit volume is $\frac{1}{2}\varepsilon_0\mathcal{E}^2$.

**Flux** is a term used to describe the rate of flow through a unit area. If $\Delta N$ is the number of particles flowing through an area $A$ in time $\Delta t$, then particle flux $\Gamma$ is defined as $\Gamma = \Delta N/(A\Delta t)$. If an amount of energy $\Delta E$ flows through an area $A$ in time $\Delta t$, energy flux is $\Gamma_E = \Delta E/(A\Delta t)$, which defines the intensity $(\mathcal{I})$ of an electromagnetic wave.

**Flux in radiometry** is the flow of radiation (electromagnetic wave) energy per unit time in watts. It is simply the radiation power that is flowing. In contrast, the photon or particle flux refers to the number of photons or particles flowing per unit time per unit area. **Radiant flux emitted** by a source refers to the radiation power in watts that is emitted. Flux in radiometry normally has either *radiant* or *luminous* as an adjective, *e.g.*, radiant flux, luminous flux.

**Ground state** is the state of the electron with the lowest energy.

**Heisenberg's uncertainty principle** states that the uncertainty $\Delta x$ in the position of a particle and the uncertainty $\Delta p_x$ in its momentum in the $x$ direction obey $(\Delta x)(\Delta p_x) \gtrsim \hbar$. This is a consequence of the wave nature of matter and has nothing to do with the precision of measurement. If $\Delta E$ is the uncertainty in the energy of a particle during a time $\Delta t$, then according to the uncertainty principle, $(\Delta E)(\Delta t) \gtrsim \hbar$. To measure the energy of a particle without any uncertainty means that we would need an infinitely long time $\Delta t \to \infty$.

**Hund's rule** states that electrons in a given subshell $n\ell$ try to occupy separate orbitals (different $m_\ell$) and keep their spins parallel (same $m_s$). In doing so, they achieve a lower energy than pairing their spins (different $m_s$) and occupying the same orbital (same $m_\ell$).

**Intensity** $(\mathcal{I})$ is the flow of energy per unit area per unit time. It is equal to an energy flux.

**LASER (light amplification by stimulated emission of radiation)** is a device within which photon multiplication by stimulated emission produces an output radiation that is nearly monochromatic and coherent (vis-à-vis an incoherent stream of photons from a tungsten light bulb). Furthermore, the output beam has very little divergence.

**Luminous flux or power** $\Phi_v$ is a measure of flow of "visual energy" per unit time that takes into account the wavelength dependence of the efficiency of the human eye, that is, whether the energy that is flowing is perceptible to the human eye. It is a measure of "brightness." One lumen of luminous flux is obtained from a 1.58 mW light source emitting a single wavelength of 555 nm (green).

**Magnetic quantum number** $m_\ell$ specifies the component of the orbital angular momentum $L_z$ in the direction of a magnetic field along $z$ so that $L_z = \pm\hbar m_\ell$, where $m_\ell$ can be a negative or positive integer from $-\ell$ to $+\ell$ including 0, that is, $-\ell, -(\ell-1), \ldots, 0, \ldots, (\ell-1), \ell$. The orbital $\psi$ of the electron depends on $m_\ell$, as well as on $n$ and $\ell$. The $m_\ell$, however, generally determines the angular variation of $\psi$.

**Orbital** is a region of space in an atom or molecule where an electron with a given energy may be found. Two electrons with opposite spins can occupy the same orbital. An orbit is a well-defined path for an electron, but it cannot be used to describe the whereabouts of the electron in an atom or molecule, because the electron has a probability distribution. The wavefunction $\psi_{n\ell m_\ell}(r, \theta, \phi)$ is often referred to as an orbital that represents the spatial distribution of the electron, since $|\psi_{n\ell m_\ell}(r, \theta, \phi)|^2$ is the probability of finding the electron per unit volume at $(r, \theta, \phi)$.

**Orbital (angular momentum) quantum number** specifies the magnitude of the orbital angular momentum of the electron via $L = \hbar\sqrt{[\ell(\ell+1)]}$, where $\ell$ is the orbital quantum number with values $0, 1, 2, 3, \ldots, n-1$. The $\ell$ values 0, 1, 2, 3 are labeled the $s, p, d, f$ states.

**Orbital wavefunction** describes the spatial dependence of the electron, not its spin. It is $\psi(r, \theta, \phi)$, which depends on $n$, $\ell$, and $m_\ell$, with the spin dependence $m_s$ excluded. Generally, $\psi(r, \theta, \phi)$ is simply called an orbital.

**Pauli exclusion principle** requires that no two electrons in a given system may have the same set of quantum numbers, $n, \ell, m_\ell, m_s$. In other words, no two electrons can occupy a given state $\psi(n, \ell, m_\ell, m_s)$. Equivalently, up to two electrons with opposite spins can occupy a given orbital $\psi(n, \ell, m_\ell)$.

**Photoelectric effect** is the emission of electrons from a metal upon illumination with a frequency of light above a critical value which depends on the material. The kinetic energy of the emitted electron is independent of the light intensity and dependent on the light frequency $\nu$, via $KE = h\nu - \Phi$ where $h$ is Planck's constant and $\Phi$ is a material-related constant called the **work function.**

**Photon** is a quantum of energy $h\nu$ (where $h$ is Planck's constant and $\nu$ is the frequency) associated with electromagnetic radiation. A photon has a zero rest mass and a momentum $p$ given by the de Broglie relationship $p = h/\lambda$, where $\lambda$ the wavelength. A photon does have a "moving mass" of $h\nu/c^2$, so it experiences gravitational attraction from other masses. For example, light from a star gets deflected as it passes by the sun.

**Population inversion** is the phenomenon of having more atoms occupy an excited energy level $E_2$, higher than a lower energy level, $E_1$, which means that the normal equilibrium distribution is reversed; that is, $N(E_2) > N(E_1)$. Population inversion occurs temporarily as a result of the excitation of a medium (pumping). If left on its own, the medium will eventually return to its equilibrium population distribution, with more atoms at $E_1$ than at $E_2$. For gas atoms, this means $N(E_2)/N(E_1) \approx \exp[-(E_2 - E_1)/kT]$.

**Principal quantum number** $n$ is an integer quantum number with values $1, 2, 3, \ldots$ that characterizes the total energy of an electron in an atom. The energy increases with $n$. With the other quantum numbers $\ell$ and $m_\ell$, $n$ determines the orbital of the electron in an atom, or $\psi_{n\ell m_\ell}(r, \theta, \phi)$. The values $n = 1, 2, 3, 4, \ldots$ are labeled the $K, L, M, N, \ldots$ shells, within each of which there may be subshells based on $\ell = 0, 1, 2, \ldots (n-1)$ and corresponding to the $s, p, d, \ldots$ states.

**Pumping** means exciting atoms from their ground states to higher energy states.

**Radiant** is a common adjective used to imply the involvement of radiation, that is, electromagnetic waves, in the noun that it qualifies; *e.g., radiant energy* is the energy transmitted by radiation.

**Radiant power** is radiation energy flowing, or emitted from a source, per unit time, which is also known as **optical power** even if the wavelength is not within the visible spectrum. **Radiant flux** signifies radiant power flow in radiometry, measured in watts.

**Radiation** normally signifies a traveling electromagnetic wave that is carrying energy. Due to the particle-like behavior of waves, radiation can also mean a *stream of photons*.

**Schrödinger equation** is a fundamental equation in nature, the solution of which describes the wave-like behavior of a particle. The equation cannot be derived from a more fundamental law. Its validity is based on its ability to predict any known physical phenomena. The solution requires as input the potential energy function $V(x, y, z, t)$ of the particle and the boundary and initial conditions. The *PE* function $V(x, y, z, t)$ describes the interaction of the particle with its environment. The time-independent Schrödinger equation describes the wave behavior of a particle under steady-state conditions, that is, when the *PE* is time-independent $V(x, y, z)$. If $E$ is the total energy and $\nabla^2 = (\partial^2/\partial x^2 + \partial^2/\partial y^2 + \partial^2/\partial z^2)$, then

$$\nabla^2\psi + \left(\frac{2m}{\hbar^2}\right)[E - V(x, y, z)]\psi = 0$$

The solution of the time-independent Schrödinger equation gives the wavefunction $\psi(x, y, z)$ of the electron and its energy $E$. The interpretation of the wavefunction $\psi(x, y, z)$ is that $|\psi(x, y, z)|^2$ is the probability of finding the electron per unit volume at point $x, y, z$.

**Selection rules** determine what values of $\ell$ and $m_\ell$ are allowed for an electron transition involving the emission and absorption of electromagnetic radiation, that is, a photon. In summary, $\Delta\ell = \pm 1$ and $\Delta m_\ell = 0, \pm 1$. The spin number $m_s$ of the electron remains unchanged. Within an atom, the transition of the electron from one state $\psi(n, \ell, m_\ell, m_s)$ to another $\psi(n', \ell', m'_\ell, m'_s)$, due to collisions with other atoms or electrons, does not necessarily obey the selection rules.

**Spin of an electron S** is its intrinsic angular momentum (analogous to the spin of Earth around its own axis), which is space quantized to have two possibilities. The magnitude of the electron's spin is a constant, $\hbar\sqrt{3}/2$, but its component along a magnetic field in the $z$ direction is $m_s\hbar$, where $m_s$ is the **spin magnetic quantum number,** which is $+\frac{1}{2}$ or $-\frac{1}{2}$.

**Spontaneous emission** is the phenomenon in which a photon is emitted when an electron in a high energy state $\psi(n, \ell, m_\ell, m_s)$ with energy $E_2$ spontaneously falls down to a lower, unoccupied energy state $\psi(n', \ell', m'_\ell, m'_s)$ with energy $E_1$. The photon energy is $h\upsilon = (E_2 - E_1)$. Since the emitted photon has an angular momentum, the orbital quantum number $\ell$ of the electron must change, that is $\Delta\ell = \ell' - \ell = \pm 1$.

**State** is a possible wavefunction for the electron that defines its spatial (orbital) and spin properties. For example, $\psi(n, \ell, m_\ell, m_s)$ is a state of the electron. From the Schrödinger equation, each state corresponds to a certain electron energy $E$. We use the terms state of energy $E$, or *energy state.* There is generally more than one state $\psi$ with the same energy $E$.

**Stimulated emission** is the phenomenon in which an incoming photon of energy $h\upsilon = E_2 - E_1$ interacts with an electron in a high-energy state $\psi(n, \ell, m_\ell, m_s)$ at $E_2$, and induces that electron to oscillate down to a lower, unoccupied energy state, $\psi(n', \ell', m'_\ell, m'_s)$ at $E_1$. The photon emitted by stimulation has the same energy and phase as the incoming photon, and it moves in the same direction. Consequently, stimulated emission results in two coherent photons, with the same energy, traveling in the same direction. The stimulated emission process must obey the selection rule $\Delta\ell = \ell' - \ell = \pm 1$, just as spontaneous emission must.

**Tunneling** is the penetration of an electron through a potential energy barrier by virtue of the electron's wave-like behavior. In classical mechanics, if the energy $E$ of the electron is less than the *PE* barrier $V_o$, the electron cannot cross the barrier. In quantum mechanics, there is a distinct probability that the electron will "tunnel" through the barrier to appear on the other side. The probability of tunneling depends very strongly on the height and width of the *PE* barrier.

**Wave** is a periodically occurring disturbance, such as the displacement of atoms from their equilibrium positions in a solid carrying sound waves, or a periodic variation in a measurable quantity, such as the electric field $\mathcal{E}(x, t)$ in a medium or space. In a traveling wave, energy is transferred from one location to another by the oscillations. For example, $\mathcal{E}_y(x, t) = \mathcal{E}_o \sin(kx - \omega t)$ is a traveling wave in the $x$ direction, where $k = 2\pi/\lambda$ and $\omega = 2\pi\nu$. The electric field in the $y$ direction varies periodically along $x$, with a period $\lambda$ called the wavelength. The field also varies with time, with a period $1/\nu$, where $\nu$ is the frequency. The wave propagates along the $x$ direction with a velocity of propagation $c$. Electromagnetic waves are transverse waves in which the electric and magnetic fields $\mathcal{E}_y(x, t)$ and $B_z(x, t)$ are at right angles to each other, as well as to the direction of propagation $x$. A traveling wave in the electric field must be accompanied by a similar traveling wave in the associated magnetic field $B_z(x, t) = B_{zo} \sin(kx - \omega t)$. Typical wave-like properties are interference and diffraction.

**Wave equation** is a general partial differential equation in classical physics, of the form

$$v^2 \frac{\partial^2 u}{\partial x^2} - \frac{\partial^2 u}{\partial t^2} = 0$$

the solution of which describes the space and time dependence of the displacement $u(x, t)$ from equilibrium or zero, given the boundary conditions. The parameter $v$ in the wave equation is the propagation velocity of the wave. In the case of electromagnetic waves in a vacuum, the wave equation describes the variation of the electric (or magnetic) field $\mathcal{E}(x, t)$ with space and time, $(c^2\partial^2\mathcal{E}/\partial x^2) - (\partial^2\mathcal{E}/\partial t^2) = 0$, where $c$ is the speed of light.

**Wavefunction** $\Psi(x, y, z, t)$ is a probability-based function used to describe the wave-like properties of a particle. It is obtained by solving the Schrödinger equation, which in turn requires a knowledge of the *PE* of the particle and the boundary and initial conditions. The term $|\Psi(x, y, z, t)|^2$ is the probability per unit volume of finding the electron at $(x, y, z)$ at time $t$. In other words, $|\Psi(x, y, z, t)|^2\, dx\, dy\, dz$ is the probability of finding the electron in the small volume $dx\, dy\, dz$ at $(x, y, z)$ at time $t$. Under steady-state conditions, the wavefunction can be separated into a space-dependent component and a time-dependent component, *i.e.*, $\Psi(x, y, z, t) = \psi(x, y, z) \exp(-jEt/\hbar)$, where $E$ is the energy of the particle and $\hbar = h/2\pi$. The spatial component $\psi(x, y, z)$ satisfies the time-independent Schrödinger equation.

**Wavenumber** (or **wavevector**) **k** is the number of waves per $2\pi$ of length, that is, $k = 2\pi/\lambda$.

**Work function** is the minimum energy required to remove an electron from inside a metal to vacuum.

**X-rays** are electromagnetic waves of wavelength typically in the range 10 pm–1 nm, which is shorter than ultraviolet light wavelengths. X-rays can be diffracted by crystals due to their wave-like properties.

# QUESTIONS AND PROBLEMS

**3.1 Photons and photon flux**

*a.* Consider a 1 kW AM radio transmitter at 700 kHz. Calculate the number of photons emitted from the antenna per second.

*b.* The average intensity of sunlight on Earth's surface is about 1 kW $m^{-2}$. The maximum intensity is at a wavelength around 800 nm. Assuming that all the photons have an 800 nm wavelength, calculate the number of photons arriving on Earth's surface per unit time per unit area. What is the magnitude of the electric field in the sunlight?

*c.* Suppose that a solar cell device can convert each sunlight photon into an electron, which can then give rise to an external current. What is the maximum current that can be supplied per unit area ($m^2$) of this solar cell device?

**3.2 Yellow, cyan, magenta, and white** *Three primary colors*, red, green, and blue (RGB), can be added together in various proportions to generate any color on various displays and light emitting devices in what is known as the *additive theory of color*. For example, yellow can be generated from adding red and green, cyan from blue and green, and magenta from red and blue.

*a.* A device engineer wants to use three light emitting diodes (LEDs) to generate various colors in an LED-based color display that is still in the research stage. His three LEDs have wavelengths of 660 nm for red, 563 nm for green, and 450 nm for blue. He simply wishes to generate the yellow and cyan by mixing equal optical powers from these LEDs; *optical power*, or *radiant power*, is defined as the radiation energy emitted per unit time. What are the numbers of red and blue photons needed (to the nearest integer) to generate yellow and cyan, respectively, for every 100 green photons?

*b.* An equi-energy white light is generated by mixing red, green, and blue light in equal optical powers. Suppose that the wavelengths are 700 nm for red, 546 nm for green, and 436 nm for blue (which is one set of possible standard primary colors). Suppose that the optical power in each primary color is 0.1 W. Calculate the *total photon flux* (photons per second) needed from each primary color.

*c.* There are bright white LEDs on the market that generate the white light by mixing yellow (a combination of red and green) with blue emissions. The inexpensive types use a single blue LED to generate a strong blue radiation, some of which is absorbed by a phosphor in front of the LED which then emits yellow light. The yellow and the blue passing through the phosphor mix and make up the white light. In one type of white LED, the blue and yellow wavelengths are 450 nm and 564 nm, respectively. White light can be generated by setting the optical (radiative) power ratio of yellow to blue light emerging from the LED to be about 1.74. What is the ratio of the number of blue to yellow photons needed? (Sometimes the mix is not perfect and the white LED light tends to have a noticeable slight blue tint.) If the total optical power output from the white LED is 100 mW, calculate the blue and yellow total photon fluxes (photons per second).

**3.3** **Brightness of laser pointers** The brightness of a light source depends not only on the radiation (optical) power emitted by the source but also on its wavelength because the human eye perceives each wavelength with a different efficiency. The visual "brightness" of a source as observed by an average daylight-adapted eye is proportional to the radiation power emitted, called the *radiant flux* $\Phi_e$, and the efficiency of the eye to detect the spectrum of the emitted radiation. While the eye can see a red color source, it cannot see an infrared source and the brightness of the infrared source would be zero. The **luminous flux** $\Phi_v$ is a measure of *brightness*, in lumens (lm), and is defined by

$$\Phi_v = \Phi_e \times (633 \text{ lm W}^{-1}) \times \eta_{\text{eye}} \tag{3.57}$$

*Luminous flux, brightness*

where $\Phi_e$ is the radiant flux or the radiation power emitted (in watts) and $\eta_{\text{eye}} = \eta_{\text{eye}}(\lambda)$ is the *relative luminous efficiency* (or the relative sensitivity) of an average light-adapted eye which depends on the wavelength; $\eta_{\text{eye}}$ is a Gaussian looking function with a peak of unity at 555 nm. (See Figure 3.46 for $\eta_{\text{eye}}$ vs. $\lambda$.) One lumen of luminous flux, or brightness, is obtained from a 1.58 mW light source emitting at a single wavelength of 555 nm (green). A typical 60 W incandescent lamp provides roughly 900 lm. When we buy a light bulb, we are buying lumens. Consider one 5 mW red 650 nm laser pointer, and another weaker 2 mW green 532 nm laser: $\eta_{\text{eye}}(650 \text{ nm}) = 0.11$ and $\eta_{\text{eye}}(532 \text{ nm}) = 0.86$. Find the luminous flux (brightness) of each laser pointer. Which is brighter? Calculate the number of photons emitted per unit time, the total *photon flux*, by each laser.

**3.4** **Human eye** Photons passing through the pupil are focused by the lens onto the retina of the eye and are detected by two types of photosensitive cells, called *rods* and *cones*, as visualized in Figure 3.46. Rods are highly sensitive photoreceptors with a peak response at a wavelength of about 507 nm (green-cyan). They do not register color and are responsible for our vision under dimmed light conditions, termed **scotopic vision.** Cones are responsible for our color perception and daytime vision, called **photopic vision.** These three types of cone photoreceptors are sensitive to blue, green, and red at wavelengths, respectively, of 430 nm, 535 nm, and 575 nm. All three cones have an overall peak response at 555 nm (green), which represents the peak response of an average daylight-adapted eye or in our photopic vision.

*a.* Calculate the photon energy (in eV) for the peak responsivity for each of the photoreceptors in the eye (one rod and three cones).

*b.* Various experiments (the most well known being by Hecht et al., *J. Opt. Soc. America*, **38**, 196, 1942) have tested the threshold sensitivity of the dark-adapted eye and have estimated that visual

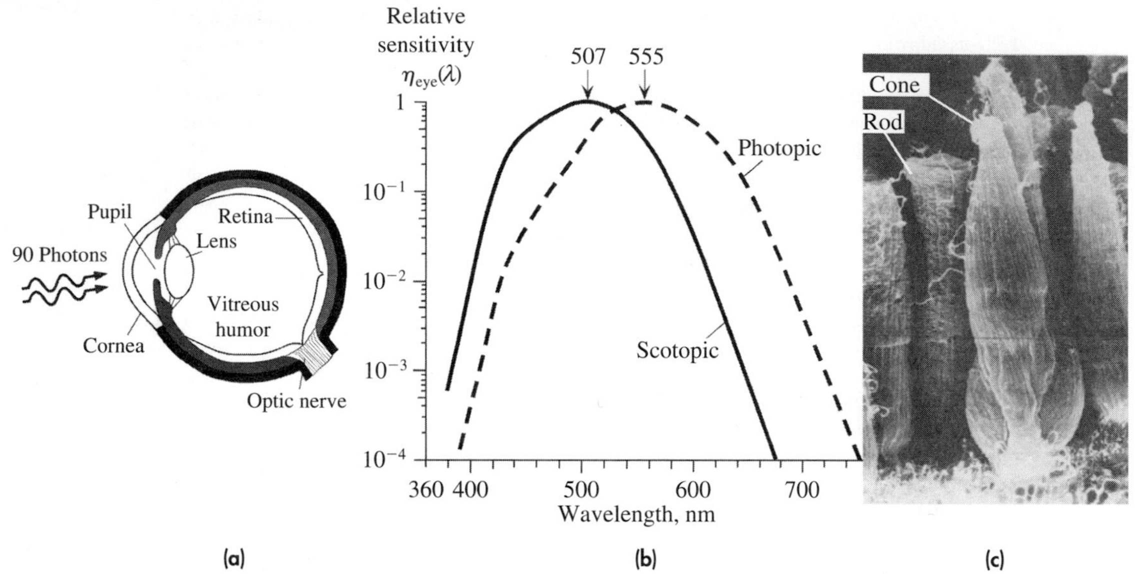


**Figure 3.46**

(a) The retina in the eye has photoreceptors that can sense the incident photons on them and hence provide necessary visual perception signals. It has been estimated that for minimum visual perception there must be roughly 90 photons falling on the cornea of the eye.

(b) The wavelength dependence of the relative efficiency $\eta_{eye}(\lambda)$ of the eye is different for daylight vision, or *photopic* vision (involves mainly cones), and for vision under dimmed light, or *scotopic* vision, which represents the dark-adapted eye, and involves rods.

(c) SEM photo of rods and cones in the retina.

SOURCE: Dr. Frank Werblin, University of California, Berkeley.

perception requires a minimum of roughly 90 photons to be incident onto the cornea in front of the eye's pupil and within 1/10 second. Taking 90 incident photons every 100 ms as the threshold sensitivity, calculate the total photon flux (photons per second), total energy in eV (within 100 ms), and the optical power that is needed for threshold visual perception.

*c.* Not all photons incident on the eye make it to the actual photoreceptors in the retina. It has been estimated that only 1 in 10 photons arriving at the eye's cornea actually make it to rod photoreceptors, due to various reflections and absorptions in the eye and other loss mechanisms. Thus, only nine photons make it to photoreceptors on the retina.[14] It is estimated that the nine test photons fall randomly onto a circular area of about 0.0025 $mm^2$. What is the estimated threshold intensity for visual perception? If there are 150,000 rods $mm^{-2}$ in this area of the eye, estimate the number of rods in this test spot. If there are a large number of rods, more than 100 in this spot, then it is likely that no single rod receives more than one photon since the nine photons arrive randomly. Thus, a rod must be able to *sense* a single photon, but it takes nine excited rods, somehow summed up by the visual system, to generate the visual sensation. Do you agree with the latter conclusion?

*d.* It is estimated that at least 200,000 photons per second must be incident on the eye to generate a color sensation by exciting the cones. Assuming that this occurs at the peak sensitivity at 555 nm,

[14] Sometimes one comes across a statement that the eye can detect a single photon. While a rod photoreceptor can indeed sense a single photon (or, put differently, a photon can activate a single rod), the overall human visual perception needs roughly nine photons at around 507 nm to consciously register a visual sensation.

and that as in part (b) only about 10 percent of the photons make it to the retina, estimate the threshold optical power stimulating the cones in the retina.

3.5 **X-ray photons** In *chest radiology*, a patient's chest is exposed to X-rays, and the X-rays passing through the patient are recorded on a photographic film to generate an X-ray image of the chest for medical diagnosis. The average wavelength of X-rays in chest radiology is about 0.2 Å (0.02 nm). Numerous measurements indicate that the patient, on average, is exposed to a total radiation energy per unit area of roughly 0.1 μJ $cm^{-2}$ for one chest X-ray image. Find the photon energy used in chest radiology, and the average number of photons incident on the patient per unit area (per $cm^2$).

*3.6 **X-rays, exposure, and roentgens** X-rays are widely used in many applications such as medical imaging, security scans, X-ray diffraction studies of crystals, and for examining defects such as cracks in objects and structures. X-rays are highly energetic photons that can easily penetrate and pass through various objects. Different materials attenuate X-rays differently, so when X-rays are passed through an object, the emerging X-rays can be recorded on a photographic film, or be captured by a modern flat panel X-ray image detector, to generate an X-ray image of the interior of the object; this is called **radiography.** X-rays also cause ionization in a medium and hence are known as *ionization radiation*. The amount of exposure (denoted by $X$) to X-rays, ionizing radiation, is measured in terms of the ionizing effects of the X-ray photons. One **roentgen** (1 R) is defined as an X-ray exposure that ionizes 1 $cm^3$ of air to generate 0.33 nC of charge in this volume at standard temperature and pressure (STP). When a body is exposed to X-rays, it will receive a certain amount of radiation energy per unit area, called **energy fluence** $\Psi_E$, that is, so many joules per $cm^2$, that depends on the exposure $X$. If $X$ in roentgens is the exposure, then the energy fluence is given by

$$\Psi_E = \left[\frac{8.73 \times 10^{-6}}{\mu_{\text{en,air}}/\rho_{\text{air}}}\right] X \quad \text{J cm}^{-2} \qquad [3.58]$$

*Fluence and roentgens*

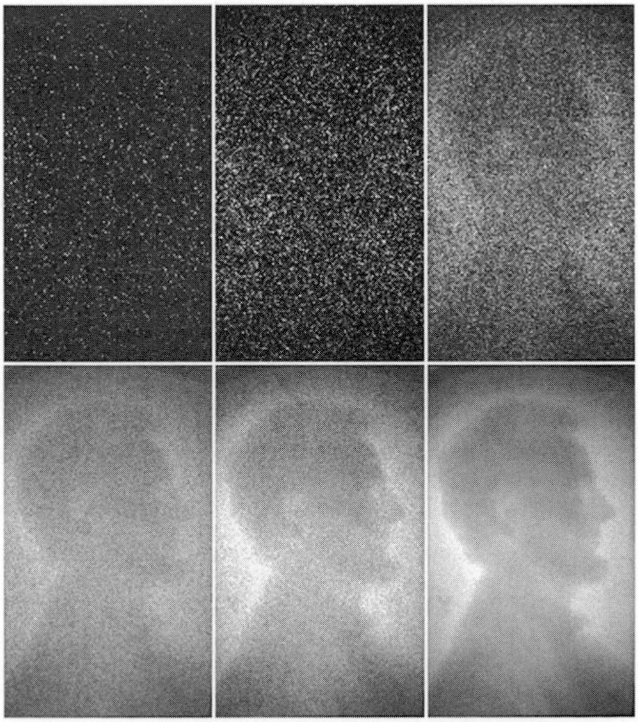

X-ray image of an American one-cent coin captured using an X-ray a-Se HARP camera. The first image at the top left is obtained under extremely low exposure, and the subsequent images are obtained with increasing exposure of approximately one order of magnitude between each image. The slight attenuation of the X-ray photons by Lincoln provides the image. The image sequence clearly shows the discrete nature of X-rays, and hence their description in terms of photons.

SOURCE: Courtesy of Dylan Hunt and John Rowlands, Sunnybrook Hospital, University of Toronto.

where $\Psi_E$ is in J cm$^{-2}$, and $\mu_{en,air}/\rho_{air}$ is the *mass energy absorption coefficient* of air in cm$^2$ g$^{-1}$ at the photon energy $E_{ph}$ of interest; the $\mu_{en,air}/\rho_{air}$ values are listed in radiological tables. For example, for 1 R of exposure, $X = 1$, $E_{ph} = 20$ keV, and $\mu_{en,air}/\rho_{air} = 0.539$ cm$^2$ g$^{-1}$. Equation 3.58 gives $\Psi_E = 1.62 \times 10^{-5}$ J cm$^{-2}$ incident on the object.

*a.* In mammography (X-ray imaging of the breasts for breast cancer), the average photon energy is about 20 keV, and the X-ray mean exposure is 12 mR. At $E_{ph} = 20$ keV, $\mu_{en,air}/\rho_{air} = 0.539$ cm$^2$ g$^{-1}$. Find the mean energy incident per unit area in μJ cm$^{-2}$, and the mean number of X-ray photons incident per unit area (photons cm$^{-2}$), called **photon fluence** $\Phi$.

*b.* In chest radiography, the average photon energy is about 60 keV, and the X-ray mean exposure is 300 μR. At $E_{ph} = 60$ keV, $\mu_{en,air}/\rho_{air} = 0.0304$ cm$^2$ g$^{-1}$. Find the mean energy incident per unit area in μJ cm$^{-2}$, and the mean number of X-ray photons incident per unit area.

*c.* A modern flat panel *X-ray image detector* is a large area image sensor that has numerous arrays of tiny pixels (millions) all tiled together to make one large continuous image sensor. Each pixel is an independent X-ray detector and converts the X-rays it receives to an electrical signal. Each tiny detector is responsible for capturing a small pixel of the whole image. (Typically, the image resolution is determined by the detector pixel size.) Each pixel in a particular experimental chest radiology X-ray sensor is 150 μm × 150 μm. If the mean exposure is 300 μR, what is the number of photons received by each pixel detector? If each pixel is required to have at least 10 photons for an acceptable signal-to-noise ratio, what is the minimum exposure required in μR?

**3.7** **Photoelectric effect** A photoelectric experiment indicates that violet light of wavelength 420 nm is the longest wavelength radiation that can cause the photoemission of electrons from a particular multi-alkali photocathode surface.

*a.* What is the work function of the photocathode surface, in eV?

*b.* If a UV radiation of wavelength 300 nm is incident upon the photocathode surface, what will be the maximum kinetic energy of the photoemitted electrons, in eV?

*c.* Given that the UV light of wavelength 300 nm has an intensity of 20 mW cm$^{-2}$, if the emitted electrons are collected by applying a positive bias to the opposite electrode, what will be the photoelectric current density in mA cm$^{-2}$?

**3.8** **Photoelectric effect and quantum efficiency** Cesium metal is to be used as the photocathode material in a photoemissive electron tube because electrons are relatively easily removed from a cesium surface. The work function of a clean cesium surface is 1.9 eV.

*a.* What is the longest wavelength of radiation which can result in photoemission?

*b.* If blue radiation of wavelength 450 nm is incident onto the Cs photocathode, what will be the kinetic energy of the photoemitted electrons in eV? What should be the voltage required on the opposite electrode to extinguish the external photocurrent?

*c.* **Quantum efficiency** (*QE*) of a photocathode is defined by,

*Quantum efficiency definition*

$$\text{Quantum efficiency} = \frac{\text{Number of photoemitted electrons}}{\text{Number of incident photons}} \tag{3.59}$$

*QE* is 100 percent if each incident photon ejects one electron. Suppose that blue light of wavelength 450 nm with an intensity of 30 mW cm$^{-2}$ is incident on a Cs photocathode that is a circular disk of diameter 6 mm. If the emitted electrons are collected by applying a positive bias voltage to the anode, and the photocathode has a *QE* of 25 percent, what will be the photoelectric current?

**3.9** **Photoelectric effect** A multi-alkali metal alloy is to be used as the photocathode material in a photoemissive electron tube. The work function of the metal is 1.6 eV, and the photocathode area is 0.5 cm$^2$. Suppose that blue light of wavelength 420 nm with an intensity of 50 mW cm$^{-2}$ is incident on the photocathode.

*a.* If the photoemitted electrons are collected by applying a positive bias to the anode, what will be the photoelectric current density assuming that the quantum efficiency $\eta$ is 15 percent? *Quantum efficiency* as a percentage is the number of photoemitted electrons per 100 absorbed photons and is defined in Equation 3.59. What is the kinetic energy of a photoemitted electron at 420 nm?

*b.* What should be the voltage and its polarity to extinguish the current?

*c.* What should be the intensity of an incident red light beam of wavelength 600 nm that would give the same photocurrent if the quantum efficiency is 5 percent at this wavelength? (Normally the quantum efficiency depends on the wavelength.)

***3.10 Planck's law and photon energy distribution of radiation** Planck's law, stated in Equation 3.9, provides the spectral distribution of the black body radiation intensity in terms of wavelength through $I_\lambda$, intensity per unit wavelength. Suppose that we wish to find the distribution in terms of frequency $\nu$ or photon energy $h\nu$. Frequency $\nu = c/\lambda$ and the wavelength range $\lambda$ to $\lambda + d\lambda$ corresponds to a frequency range $\nu$ to $\nu + d\nu$. ($d\lambda$ and $d\nu$ have opposite signs since $\nu$ increases as $\lambda$ decreases.) The intensity $I_\lambda\, d\lambda$ in $\lambda$ to $\lambda + d\lambda$ must be the same as the intensity in $\nu$ to $\nu + d\nu$, which we can write as $I_\nu\, d\nu$ where $I_\nu$ is the radiation intensity per unit frequency. Thus,

$$I_\nu = I_\lambda \left| \frac{d\lambda}{d\nu} \right|$$

The magnitude sign is needed because $\lambda = c/\nu$ results in a negative $d\lambda/d\nu$, and $I_\nu$ must be positive by definition. We can simply substitute $\lambda = c/\nu$ for $\lambda$ in $I_\lambda$ and obtain $I_\lambda$ as a function of $\nu$, and then find $|d\lambda/d\nu|$ to find $I_\nu$ from the preceding expression.

*a.* Show that

$$I_\nu = \frac{2\pi(h\nu)^3}{c^2h^2[\exp(h\nu/kT) - 1]} \qquad [3.60]$$

*Black body photon energy distribution*

Equation 3.60 is written to highlight that it is a function of the *photon energy* $h\nu$, which is in joules in Equation 3.60 but can be converted to eV by dividing by $1.6 \times 10^{-19}$ J eV$^{-1}$.

*b.* If we integrate $I_\nu$ over all photon energies (numerically on a calculator or a computer from 0 to say 6 eV), we would obtain the total intensity at a temperature $T$. Find the total intensity $I_T$ emitted at $T = 2600$ K (a typical incandescent light bulb filament temperature) and at 6000 K (roughly representing the sun's spectrum). Plot $y = I_\nu/I_T$ versus the photon energy in eV. What are the photon energies for the peaks in the distributions? Calculate the corresponding wavelength for each using $\lambda = c/\nu$ and then compare these wavelengths with those predicted by Wien's law, $\lambda_{max}T \approx 2.89 \times 10^{-3}$ m K.

**3.11 Wien's law** The maximum in the intensity distribution of black body radiation depends on the temperature. Substitute $x = hc/(\lambda kT)$ in Planck's law and plot $I_\lambda$ versus $x$ and find $\lambda_{max}$ which corresponds to the peak of the distribution, and hence derive Wien's law. Find the peak intensity wavelength $\lambda_{max}$ for a 40 W light bulb given that its filament operates at roughly 2400 °C.

**3.12 Diffraction by X-rays and an electron beam** Diffraction studies on a polycrystalline Al sample using X-rays gives the smallest diffraction angle ($2\theta$) of 29.5° corresponding to diffraction from the (111) planes. The lattice parameter $a$ of Al (FCC), is 0.405 nm. If we wish to obtain the same diffraction pattern (same angle) using an electron beam, what should be the voltage needed to accelerate the electron beam? Note that the interplanar separation $d$ for planes $(h, k, \ell)$ and the lattice parameter $a$ for cubic crystals are related by $d = a/(h^2 + k^2 + \ell^2)^{1/2}$.

**3.13 Heisenberg's uncertainty principle** Show that if the uncertainty in the position of a particle is on the order of its de Broglie wavelength, then the uncertainty in its momentum is about the same as the momentum value itself.

**3.14 Heisenberg's uncertainty principle** An excited electron in an Na atom emits radiation at a wavelength 589 nm and returns to the ground state. If the mean time for the transition is about 20 ns, calculate the inherent width in the emission line. What is the length of the photon emitted?

**3.15 Tunneling**

*a.* Consider the phenomenon of tunneling through a potential energy barrier of height $V_o$ and width $a$, as shown in Figure 3.16. What is the probability that the electron will be reflected? Given the transmission coefficient $T$, can you find the reflection coefficient $R$? What happens to $R$ as $a$ or $V_o$ or both become very large?

*b.* For a wide barrier ($\alpha a \gg 1$), show that $T_o$ can at most be 4 and that $T_o = 4$ when $E = \frac{1}{2}V_o$.

**3.16 Electron impact excitation**

*a.* A projectile electron of kinetic energy 12.2 eV collides with a hydrogen atom in a gas discharge tube. Find the $n$th energy level to which the electron in the hydrogen atom gets excited.

*b.* Calculate the possible wavelengths of radiation (in nm) that will be emitted from the excited H atom in part (a) as the electron returns to its ground state. Which one of these wavelengths will be in the visible spectrum?

*c.* In neon street lighting tubes, gaseous discharge in the Ne tube involves electrons accelerated by the electric field impacting Ne atoms and exciting some of them to the $2p^5 3p^1$ states, as shown in Figure 3.42. What is the wavelength of emission? Can the Ne atom fall from the $2p^5 3p^1$ state to the ground state by spontaneous emission?

**3.17 Line spectra of hydrogenic atoms** Spectra of hydrogen-like atoms are classified in terms of electron transitions to a common lower energy level.

*a.* All transitions from energy levels $n = 2, 3, \ldots$ to $n = 1$ (the $K$ shell) are labeled $K$ lines and constitute the **Lyman series.** The spectral line corresponding to the smallest energy difference ($n = 2$ to $n = 1$) is labeled the $K_\alpha$ line, next is labeled $K_\beta$, and so on. The transition from $n = \infty$ to $n = 1$ has the largest energy difference and defines the greatest photon energy (shortest wavelength) in the $K$ series; hence it is called the absorption edge $K_{\alpha e}$. What is the range of wavelengths for the $K$ lines? What is $K_{\alpha e}$? Where are these lines with respect to the visible spectrum?

*b.* All transitions from energy levels $n = 3, 4, \ldots$ to $n = 2$ ($L$ shell) are labeled $L$ lines and constitute the **Balmer series.** What is the range of wavelengths for the $L$ lines (*i.e.*, $L_\alpha$ and $L_{\alpha e}$)? Are these in the visible range?

*c.* All transitions from energy levels $n = 4, 5, \ldots$ to $n = 3$ ($M$ shell) are labeled $M$ lines and constitute the **Paschen series.** What is the range of wavelengths for the $M$ lines? Are these in the visible range?

*d.* How would you expect the spectral lines to depend on the atomic number $Z$?

**3.18 Ionization energy and effective Z**

*a.* Consider the singly ionized Li ion, $Li^+$, which has lost its $2s$ electron. If the energy required to ionize one of the $1s$ electrons in $Li^+$ is 18.9 eV, calculate the effective nuclear charge seen by a $1s$ electron, that is, $Z_{\text{effective}}$ in the hydrogenic atom ionization energy expression in Equation 3.45; $E_{I,n} = (Z_{\text{effective}}/n)^2(13.6\text{ eV})$.

*b.* The B atom has a total of five electrons, two in the $1s$ orbital, two in the $2s$, and one in the $2p$. The experimental ionization energy of B is 8.30 eV. Calculate $Z_{\text{effective}}$.

*c.* The experimental ionization energy of Na is 3.49 eV. Calculate the effective nuclear charge seen by the $3s$ valence electron.

*d.* The chemical tables typically list the first, second, and third ionization energies $E_1$, $E_2$, $E_3$, respectively, and so on. Consider Al. $E_1$ represents the energy required to remove the first electron from neutral Al; $E_2$, the second electron from $Al^+$; $E_3$, the third electron from $Al^{2+}$ to generate $Al^{3+}$. For Al, experimentally, $E_1 = 6.0$ eV, $E_2 = 18.8$ eV, and $E_3 = 28.4$ eV. For each case find the $Z_{\text{effective}}$ seen by the electron that is removed.

**3.19 Atomic and ionic radii** The maximum in the radial probability distribution of an electron in a hydrogen-like atom is given by Equation 3.38, that is, $r_{\max} = (n^2 a_o)/Z$, for $\ell = n - 1$. The average distance $\bar{r}$ of an electron from the nucleus can be calculated by using the definition of an average and the probability distribution function $P_{n,\ell}(r)$, that is,

$$\bar{r} = \int_0^\infty r P_{n,\ell}(r)\, dr = \frac{a_o n^2}{Z}\left[\frac{3}{2} - \frac{\ell(\ell+1)}{2n^2}\right]$$

*Average distance of electron from nucleus*

in which the right-hand side represents the result of the integration (which has been done by physicists). Calculate $r_{max}$ and $\bar{r}$ for the $2p$ valence electron in the B atom. Which value is closer to the radius of the B atom, 0.085 nm, given in the Period Table? Consider only the outermost electrons, and calculate $r_{average}$ for Li, $Li^+$, $Be^{2+}$, and B, and compare with the experimental values of the atomic or ionic sizes: 0.15 nm for Li, 0.070 nm for $Li^+$, 0.035 nm for $Be^{2+}$, and 0.085 nm for B.

***3.20 X-rays and the Moseley relation** X-rays are photons with wavelengths in the range 0.01–10 nm, with typical energies in the range 100 eV to 100 keV. When an electron transition occurs in an atom from the $L$ to the $K$ shell, the emitted radiation is generally in the X-ray spectrum. For all atoms with atomic number $Z > 2$, the $K$ shell is full. Suppose that one of the electrons in the $K$ shell has been knocked out by an energetic projectile electron impacting the atom (the projectile electron would have been accelerated by a large voltage difference). The resulting vacancy in the $K$ shell can then be filled by an electron in the $L$ shell transiting down and emitting a photon. The emission resulting from the $L$ to $K$ shell transition is labeled the $K_\alpha$ line. The table shows the $K_\alpha$ line data obtained for various materials.

| | Material | | | | | | | | |
|---|---|---|---|---|---|---|---|---|---|
| | **Mg** | **Al** | **S** | **Ca** | **Cr** | **Fe** | **Cu** | **Rb** | **W** |
| Z | 12 | 13 | 16 | 20 | 24 | 26 | 29 | 37 | 74 |
| $K_\alpha$ line (nm) | 0.987 | 0.834 | 0.537 | 0.335 | 0.229 | 0.194 | 0.154 | 0.093 | 0.021 |

*a.* If $\nu$ is the frequency of emission, plot $\nu^{1/2}$ against the atomic number $Z$ of the element.

*b.* H. G. Moseley, while still a graduate student of E. Rutherford in 1913, found the empirical relationship

$$\nu^{1/2} = B(Z - C)$$

*Moseley relation*

where $B$ and $C$ are constants. What are $B$ and $C$ from the plot? Can you give a simple explanation as to why $K_\alpha$ absorption should follow this relationship?

Henry G. J. Moseley (1887–1915), around 1910, carrying out experiments at Balliol-Trinity Laboratory at Oxford.

SOURCE: University of Oxford Museum of Science, courtesy AIP Emilio Segre Visual Archives.

**3.21** **The He atom** Suppose that for the He atom, zero energy is taken to be the two electrons stationary at infinity (and infinitely apart) from the nucleus ($He^{++}$). Estimate the energy (in eV) of the electrons in the He atom by neglecting the electron–electron repulsion, that is, neglecting the potential energy due to the mutual Coulombic repulsion between the electrons. How does this compare with the experimental value of −79 eV? How strong is the electron–electron repulsion energy?

**3.22** **Excitation energy of He** In the HeNe laser, an energetic electron is accelerated by the applied field impacts and excites the He from its ground state, $1s^2$, to an excited state He*, $1s^12s^1$, which has one of the electrons in the $2s$ orbital. The ground energy of the He atom is −79 eV with respect to both electrons isolated at infinity, which defines the zero energy. Consider the $1s^12s^1$ state. If we neglect the electron–electron interactions, we can calculate the energy of the $1s$ and $2s$ electrons using the energy for a hydrogenic atom, $E_n = -(Z^2/n^2)(13.6\text{ eV})$. We can then add the electron–electron interaction energy by assuming that the $1s$ and $2s$ electrons are effectively separated by $3a_o$, which is the difference, $4a_o - 1a_o$, between the $1s$ and $2s$ Bohr radii. Calculate the overall energy of He* and hence the excitation energy from He to He*. The experimental value is about 20.6 eV.

**3.23** **Electron affinity** The fluorine atom has the electronic configuration $[He]2s^2p^5$. The F atom can actually capture an electron to become a $F^-$ ion, and release energy, which is listed as its *electron affinity,* 328 kJ $mol^{-1}$. We will assume that the two $1s$ electrons in the closed $K$ shell (very close to the nucleus) and the two electrons in the $2s$ orbitals will shield four positive charges and thereby expose $+9e - 4e = +5e$ for the $2p$ orbital. Suppose that we try to calculate the energy of the $F^-$ ion by simply assuming that the additional electron is attracted by an effective positive charge, $+e(5 - Z_{2p})$ or $+eZ_{\text{effective}}$, where $Z_{2p}$ is the overall shielding effect of the five electrons in the $2p$ orbital, so that the tenth electron we have added sees an effective charge of $+eZ_{\text{effective}}$. Calculate $Z_{2p}$ and $Z_{\text{effective}}$. The F atom does not enjoy losing an electron. The ionization energy of the F atom is 1681 kJ $mol^{-1}$. What is the $Z_{\text{effective}}$ that is experienced by a $2p$ electron? (Note: 1 kJ $mol^{-1}$ = 0.01036 eV/atom.)

***3.24** **Electron spin resonance (ESR)** It is customary to write the spin magnetic moment of an electron as

*Spin magnetic moment*

$$\mu_{\text{spin}} = -\frac{ge}{2m_e}S$$

where $S$ is the spin angular momentum, and $g$ is a numerical factor, called the **g factor,** which is 2 for a free electron. Consider the interaction of an electron's spin with an external magnetic field. Show that the additional potential energy $E_{BS}$ is given by

*Electron spin in a magnetic field*

$$E_{BS} = \beta g m_s B$$

where $\beta = e\hbar/2m_e$ is called the **Bohr magneton.** Frequently **electron spin resonance** is used to examine various defects and impurities in semiconductors. A defect such as a dangling bond, for example, will have a single unpaired electron in an orbital and thus will possess a spin magnetic moment. A strong magnetic field is applied to the specimen to split the energy level $E_1$ of the unpaired spin to two levels $E_1 - E_{BS}$ and $E_1 + E_{BS}$, separated by $\Delta E_{BS}$. The electron occupies the lower level $E_1 - E_{BS}$. Electromagnetic waves (usually in the microwave range) of known frequency $\nu$, and hence of known photon energy $h\nu$, are passed through the specimen. The magnetic field $B$ is varied until the EM waves are absorbed by the specimen, which corresponds to the excitation of the electron at each defect from $E_1 - E_{BS}$ to $E_1 + E_{BS}$, that is, $h\nu = \Delta E_{BS}$ at a certain field $B$. This maximum absorption condition is called **electron spin resonance,** as the electron's spin is made to resonate with the EM wave. If $B = 2$ T, calculate the frequency of the EM waves needed for ESR, taking $g = 2$. Note: For many molecules, and impurities and defects in crystals, $g$ is not exactly 2, because the electron is in a different environment in each case. The experimentally measured value of $g$ can be used to characterize molecules, impurities, and defects.

**3.25** **Spin–orbit coupling** An electron in an atom will experience an internal magnetic field $B_{\text{int}}$ because, from the electron's reference frame, it is the positive nucleus that is orbiting the electron. The electron will "see" the nucleus, take as charge $+e$, circling around it, which is equivalent to a current $I = +ef$ where $f$ is the electron's frequency of rotation around the nucleus. The current $I$ generates the internal magnetic field $B_{int}$ at the electron. From electromagnetism texts, $B_{int}$ is given by

$$B_{int} = \frac{\mu_o I}{2r}$$

where $r$ is the radius of the electron's orbit and $\mu_o$ is the absolute permeability. Show that

*Internal magnetic field at an electron in an atom*

$$B_{int} = \frac{\mu_o e}{4\pi m_e r^3} L$$

Consider the hydrogen atom with $Z = 1$, $2p$ orbital, $n = 2$, $\ell = 1$, and take $r \approx n^2 a_o$. Calculate $B_{int}$.

The electron's spin magnetic moment $\boldsymbol{\mu}_{spin}$ will couple with this internal field, which means that the electron will now possess a magnetic potential energy $E_{SL}$ that is due to the coupling of the *spin* with the *orbital motion,* called **spin-orbit coupling.** $E_{SL}$ will be either negative or positive, with only two values, depending on whether the electron's spin magnetic moment is along or opposite $\mathbf{B}_{int}$. Take $z$ along $B_{int}$ so that $E_{SL} = -B_{int}\mu_{spin,z}$, where $\mu_{spin,z}$ is $\mu_{spin}$ along $z$, and then show that the energy $E_2$ of the $2p$ orbital splits into two closely separated levels whose separation is

*Spin–orbit coupling potential energy*

$$\Delta E_{SL} = \left(\frac{e\hbar}{m_e}\right) B_{int}$$

Calculate $\Delta E_{SL}$ in eV and compare it with $E_2(n = 2)$and the separation $\Delta E = E_2 - E_1$. (The exact calculation of $E_{SL}$ is much more complicated, but the calculated value here is sufficiently close to be useful.) What is the effect of $E_{SL}$ on the observed emission spectrum from the H-atom transition from $2p$ to $1s$? What is the separation of the two wavelengths? The observation is called **fine structure splitting.**

**3.26** **Hund's rule** For each of the following atoms and ions, sketch the electronic structure, using a box for an orbital wavefunction, and an arrow (up or down for the spin) for an electron.

*a.* Aluminum, $[Ne]3s^2p^1$

*b.* Silicon, $[Ne]3s^2p^2$

*c.* Phosphorus, $[Ne]3s^2p^3$

*d.* Sulfur, $[Ne]3s^2p^4$

*e.* Chlorine, $[Ne]3s^2p^5$

*f.* Titanium, $[Ar]3d^24s^2$

*g.* Vanadium, $[Ar]3d^34s^2$

*h.* Manganese, $[Ar]3d^54s^2$

*i.* Cobalt, $[Ar]3d^74s^2$

*j.* $Cu^{2+}$, $[Ar]3d^94s^0$

**3.27** **Hund's rule** The carbon atom has the electronic structure $2s^22p^2$ in its ground state. The ground state and various possible excited states of C are shown in Figure 3.47. The following energies are known for the states $a$ to $e$ in Figure 3.47, not in any particular order: 0, 7.3 eV, 4.1 eV, 7.9 eV, and 1.2 eV. Using reasonable arguments match these energies to the states $a$ to $e$. Use Hund's rule to establish the ground state with 0 eV. If you have to flip a spin to go from the ground to another configuration, that would cost energy. If you have to move an electron from a lower $s$ to $p$ or from $p$ to a higher $s$, that would cost a lot of energy. Two electrons in the same orbital (obviously with paired electrons) would have Coulombic repulsion energy.

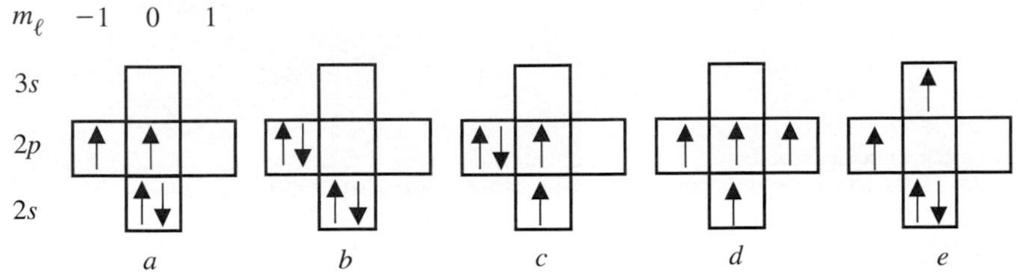


**Figure 3.47**
Some possible states of the carbon atom, not in any particular order.

**3.28** **The HeNe laser** A particular HeNe laser operating at 632.8 nm has a tube that is 40 cm long. The operating gas temperature is about 130 °C.

*a.* Calculate the Doppler-broadened linewidth $\Delta\lambda$ in the output spectrum.

*b.* What are the $n$ values that satisfy the resonant cavity condition? How many modes are therefore allowed?

*c.* Calculate the frequency separation and the wavelength separation of the laser modes. How do these change as the tube warms up during operation? Taking the linear expansion coefficient to be $10^{-6}$ $K^{-1}$, estimate the change in the mode frequency separation.

**3.29** **$Er^{3+}$-doped fiber amplifier** When the $Er^{3+}$ ion in the $Er^{3+}$-doped fiber amplifer (EDFA) is pumped with 980 nm of radiation, the $Er^{3+}$ ions absorb energy from the pump signal and become excited to $E_3$ (Figure 3.44). Later the $Er^{3+}$ ions at $E_2$ are stimulated to add energy (coherent photons) to the signal at 1550 nm. What is the wasted energy (in eV) from the pump to the signal at each photon amplification step? (This energy is lost as heat in the glass medium.) An Er-doped fiber amplifier is 10 m long, and the radius of the core is 5 μm. The Er concentration in the core is $10^{18}$ $cm^{-3}$. The nominal power gain of the amplifier is 100 (or 20 dB). The pump wavelength is 980 nm, and the signal wavelength is 1550 nm. If the output power from the amplifier is 100 mW and assuming the signal and pump are confined to the core, what is the minimum intensity of the pump signal? How much power is wasted in this EDFA? (The pump must provide enough photons to pump the $Er^{3+}$ ions needed to generate the additional output photons over that of input photons. The concentration of $Er^{3+}$ ions in the fiber is given for information only.)

Wolfgan Pauli (1900–1958) won the Nobel prize in 1945 for his contributions to quantum mechanics. His exclusion principle was announced in 1925. "I don't mind your thinking slowly; I mind your publishing faster than you think." (Translation from German. Attributed to Pauli by H. Coblaus. From A. L. Mackay, *A Dictionary of Scientific Quotations*, Bristol: IOP Publishing, 1991, p. 191.)

SOURCE: AIP Emilio Segrè Visual Archives, Goudsmit Collection.

Arthur Holly Compton (1892–1962) won the Nobel prize in physics in 1927 for his discovery of the Compton effect with C. T. R. Wilson in 1923.

SOURCE: King Features Syndicate, Inc., New York and Argonne National Laboratory, courtesy AIP Emilio Segrè Visual Archives.

Theodore Harold Maiman was born in 1927 in Los Angeles, son of an electrical engineer. He studied engineering physics at Colorado University, while repairing electrical appliances to pay for college, and then obtained a Ph.D. from Stanford. Theodore Maiman constructed this first laser in 1960 while working at Hughes Research Laboratories. There is a vertical chromium ion–doped ruby rod in the center of a helical xenon flash tube. The ruby rod has mirrored ends. The xenon flash provides optical pumping of the chromium ions in the ruby rod. The output is a pulse of red laser light.

SOURCE: Courtesy of HRL Laboratories, LLC, Malibu, California.

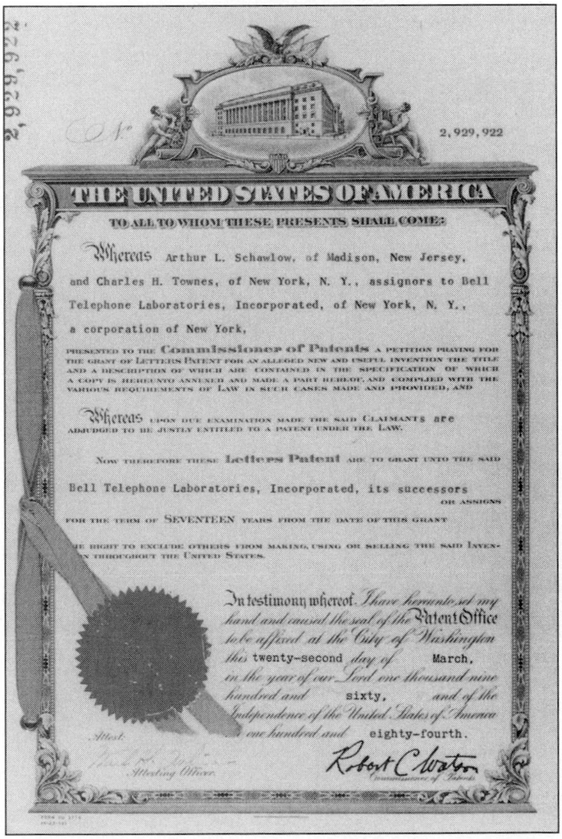

No 2,929,922

THE UNITED STATES OF AMERICA

TO ALL TO WHOM THESE PRESENTS SHALL COME:

Whereas Arthur L. Schawlow, of Madison, New Jersey, and Charles H. Townes, of New York, N. Y., assignors to Bell Telephone Laboratories, Incorporated, of New York, N. Y., a corporation of New York,

presented to the Commissioner of Patents a petition praying for the grant of Letters Patent for an alleged new and useful invention the title and a description of which are contained in the specification of which a copy is hereunto annexed and made a part hereof, and complied with the various requirements of Law in such cases made and provided, and

Whereas upon due examination made the said Claimants are adjudged to be justly entitled to a patent under the Law.

Now therefore these Letters Patent are to grant unto the said Bell Telephone Laboratories, Incorporated, its successors or assigns for the term of SEVENTEEN years from the date of this grant the right to exclude others from making, using or selling the said Invention throughout the United States.

In testimony whereof I have hereunto set my hand and caused the seal of the Patent Office to be affixed at the City of Washington this twenty-second day of March, in the year of our Lord one thousand nine hundred and sixty, and of the Independence of the United States of America one hundred and eighty-fourth.

Attest:

Attesting Officer.

Robert C. Watson
Commissioner of Patents.

The patent for the invention of the laser by Charles H. Townes and Arthur L. Schawlow in 1960 (Courtesy of Bell Laboratories). The laser patent was later bitterly disputed for almost three decades in "the patent wars" by Gordon Gould, an American physicist, and his designated agents. Gordon Gould eventually received the U.S. patent for optical pumping of the laser in 1977 since the original laser patent did not detail such a pumping procedure. In 1987 he also received a patent for the gas discharge laser, thereby winning his 30 year patent war. His original notebook even contained the word *laser*.

Motorola's prototype flat panel display based on the Fowler–Nordheim field emission principle. The display is 14 cm in diagonal and 3.5 mm thick with a viewing angle 160°. Each pixel (325 $\mu$m thick) uses field emission of electrons from microscopic sharp point sources (icebergs). Emitted electrons impinge on colored phosphors on a screen and cause light emission by cathodoluminescence. There are millions of these microscopic field emitters to constitute the image.

SOURCE: Courtesy of Dr. Babu Chalamala, Flat Panel Display Division, Motorola.

Left: A scanning electron microscope image of an array of electron field emitters (icebergs). Center: One iceberg. Right: A cross section of a field emitter. Each iceberg is a source of electron emission arising from Fowler–Nordheim field emission; for further information see B. Chalamala, et al., *IEEE Spectrum,* April 1998, pp. 42–51.

SOURCE: Courtesy of Dr. Babu Chalamala, Flat Panel Display Division, Motorola.

# CHAPTER 4

# Modern Theory of Solids

One of the great successes of modern physics has been the application of quantum mechanics or the Schrödinger equation to the behavior of molecules and solids. For example, quantum mechanics explains the nature of the bond between atoms, and its consequences. How can carbon bond with four other carbon atoms? What determines the direction and strength of a bond? An intuitively obvious outcome from quantum mechanics is that the energy of the electron is still quantized in the molecule. In addition, the application of quantum mechanics to many atoms, as in a solid, leads to energy bands within which the electron energy levels are almost continuous. The electron energy falls within possible values in a band of energies. It is nearly impossible to comprehend the principles of operation of modern solid-state electronic devices without a good grasp of the band theory of solids. Since we are dealing with a large number of electrons in the solid, we must consider a statistical way of describing their behavior, just as we use the Maxwell distribution of velocities to explain the behavior of gas atoms. An equally important question, therefore, is "What is the probability that an electron is in a state with energy $E$ within an energy band?"

## 4.1 HYDROGEN MOLECULE: MOLECULAR ORBITAL THEORY OF BONDING

Consider what happens when two hydrogen atoms approach each other to form the hydrogen molecule. This is the H–H (or $H_2$) system. Let us examine the energy levels of the H–H system as a function of the interatomic distance $R$. When the atoms are infinitely separated, each atom has its own set of energy levels, labeled $1s$, $2s$, $2p$, etc. The electron energy in each atom is $-13.6$ eV with respect to the "free" state (electron infinitely separated from the parent nucleus). The energy of the two isolated hydrogen atoms is twice $-13.6$ eV.

As the atoms approach closer, the electrons interact both with each other and with the other nuclei. To obtain the wavefunctions and the new energy of the electrons, we need to find the new potential energy function *PE* for the electrons in this new environment and then solve the Schrödinger equation with this new *PE* function. The new energy is actually *lower* than twice $-13.6$ eV, which means that the $H_2$ formation is energetically favorable.

The bond formation between two H atoms can be easily explained by describing the behavior of the electron within the molecule. We use a **molecular orbital** $\psi$, which depends on the interaction of individual atomic wavefunctions and is regarded as an electron wavefunction within the molecule.

In the $H_2$ molecule, we cannot have two sets of identical atomic $\psi_{1s}$ orbitals, for two reasons. First, this would violate the Pauli exclusion principle, which requires that, in a given system of electrons (those within the $H_2$ molecule), we cannot have two sets of identical quantum numbers. When the atoms were separated, we did not have this problem, because we had two isolated systems.

Second, as the two atoms approach each other, as shown in Figure 4.1, the atomic $\psi_{1s}$ wavefunctions overlap. This overlap produces two new wavefunctions with different energies and hence different quantum numbers. When the two atomic wavefunctions interfere, they can overlap either in phase (both positive or both negative) or out of phase

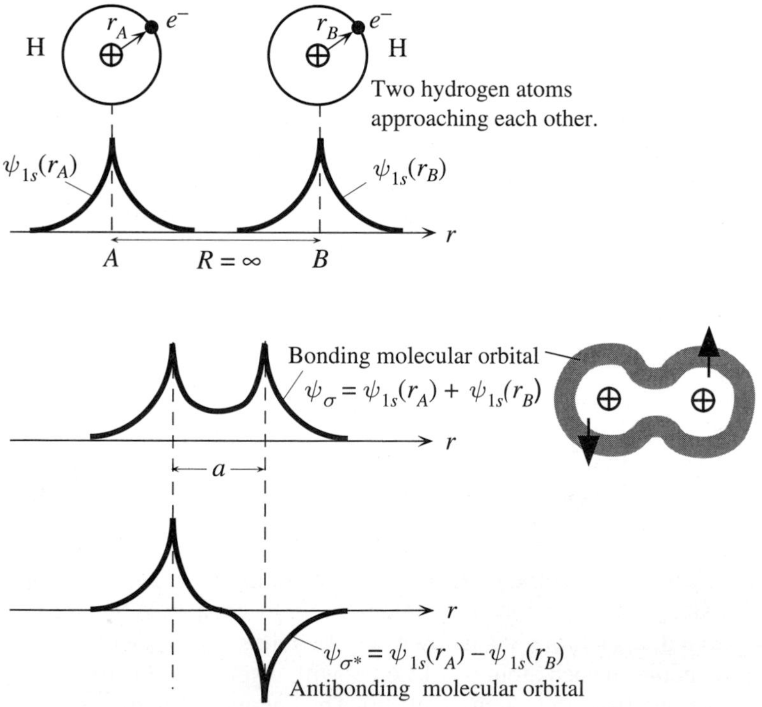


**Figure 4.1** Formation of molecular orbitals, bonding, and antibonding ($\psi_\sigma$ and $\psi_{\sigma*}$) when two H atoms approach each other.
The two electrons pair their spins and occupy the bonding orbital $\psi_\sigma$.

(one positive and the other negative), as a result of which two molecular orbitals are formed. These are conventionally labeled $\psi_\sigma$ and $\psi_{\sigma^*}$ as illustrated in Figure 4.1. Thus, two of the molecular orbitals in the H–H system are

$$\psi_\sigma = \psi_{1s}(r_A) + \psi_{1s}(r_B) \qquad [4.1]$$

$$\psi_{\sigma^*} = \psi_{1s}(r_A) - \psi_{1s}(r_B) \qquad [4.2]$$

where the two hydrogen atoms are labeled $A$ and $B$, and $r_A$ and $r_B$ are the respective distances of the electrons from their parent nucleus. In generating two separate molecular orbitals $\psi_\sigma$ and $\psi_{\sigma^*}$ from a linear combination of two identical atomic orbitals $\psi_{1s}$, we have used the **linear combination of atomic orbitals (LCAO)** method.

The first molecular orbital $\psi_\sigma$ is *symmetric* and has considerable magnitude between the nuclei, whereas the second $\psi_{\sigma^*}$, is *antisymmetric* and has a node between the nuclei. The resulting electron probability distributions $|\psi_\sigma|^2$ and $|\psi_{\sigma^*}|^2$ are shown in Figure 4.2.

In an analogy to hydrogenic wavefunctions, since $\psi_{\sigma^*}$ has a node, we would expect it to have a higher energy than the $\psi_\sigma$ orbital and therefore a different energy quantum number, which means that the Pauli exclusion principle is no longer violated. We can also expect that because $|\psi_\sigma|^2$ has an appreciable electron concentration between the two nuclei, the electrostatic *PE*, and hence the total energy for the wavefunction $\psi_\sigma$, will be lower than that for $\psi_{\sigma^*}$, as well as those for the individual atomic wavefunctions.

Of course, the true wavefunctions of the electrons in the $H_2$ system must be determined by solving the Schrödinger equation, but an intelligent guess is that these must look like $\psi_\sigma$ and $\psi_{\sigma^*}$. We can therefore use $\psi_\sigma$ and $\psi_{\sigma^*}$ in the Schrödinger equation, with the correct form of the *PE* term $V$, to evaluate the energies $E_\sigma$ and $E_{\sigma^*}$ of $\psi_\sigma$ and $\psi_{\sigma^*}$, respectively, as a function of $R$. The *PE* function $V$ in the H–H system has positive *PE* contributions arising from electron–electron repulsions and proton–proton

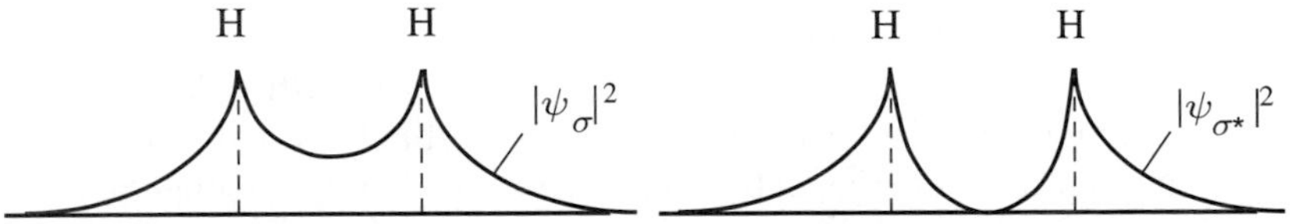


**(a)** Electron probability distributions for bonding and antibonding orbitals, $\psi_\sigma$ and $\psi_{\sigma^*}$.

**(b)** Lines representing contours of constant probability (darker lines represent greater relative probability).

**Figure 4.2**

(a) Energy of $\psi_\sigma$ and $\psi_{\sigma^*}$ vs. the interatomic separation $R$.

(b) Schematic diagram showing the changes in the electron energy as two isolated H atoms, far left and far right, come together to form a hydrogen molecule.

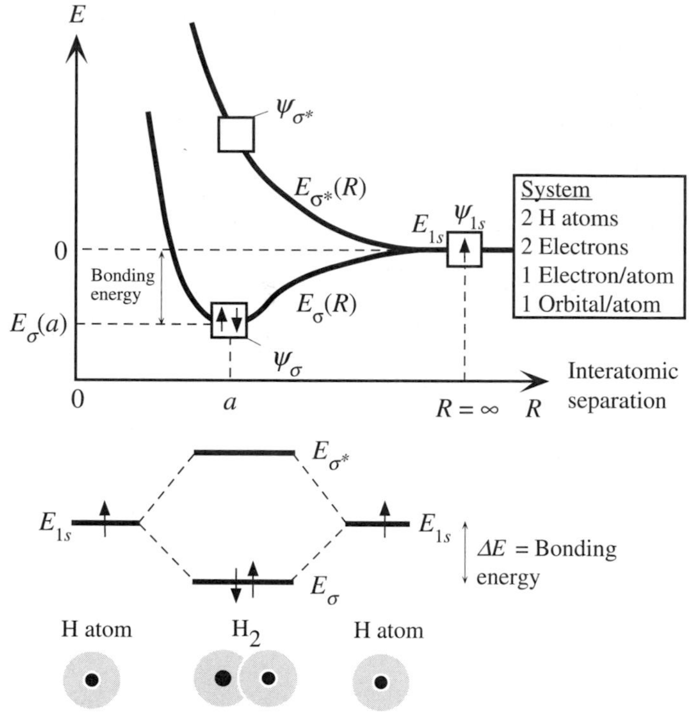


**Figure 4.3** Electron energy in the system comprising two hydrogen atoms.

repulsions, but negative *PE* contributions arising from the attractions of the two electrons to the two protons.

The two energies, $E_\sigma$ and $E_{\sigma^*}$, are widely different, with $E_\sigma$ below $E_{1s}$ and $E_{\sigma^*}$ above $E_{1s}$, as shown schematically in Figure 4.3a. As $R$ decreases and the two H atoms get closer, the energy of the $\psi_\sigma$ orbital state passes through a minimum at $R = a$. Each orbital state can hold two electrons with spins paired, and within the two hydrogen atoms, we have two electrons. If these enter the $\psi_\sigma$ orbital and pair their spins, then this new configuration is energetically more favorable than two isolated H atoms. It corresponds to the hydrogen molecule $H_2$. The energy difference between that of the two isolated H atoms and the $E_\sigma$ minimum energy at $R = a$ is the bonding energy, as illustrated in Figure 4.3a. When the two electrons in the $H_2$ molecule occupy the $\psi_\sigma$ orbital, their probability distribution (and hence, the negative charge distribution) is such that the negative *PE*, arising from the attractions of these two electrons to the two protons, is stronger in magnitude than the positive *PE*, arising from electron–electron repulsions and proton–proton repulsions and the kinetic energy of the two electrons. Therefore, the $H_2$ molecule is energetically stable.

The wavefunction $\psi_\sigma$ corresponding to the lowest electron energy is called the **bonding orbital,** and $\psi_{\sigma^*}$ is the **antibonding orbital.** When two atoms are brought together, the two identical atomic wavefunctions combine in two ways to generate two different molecular orbitals, each with a different energy. Effectively, then, an atomic

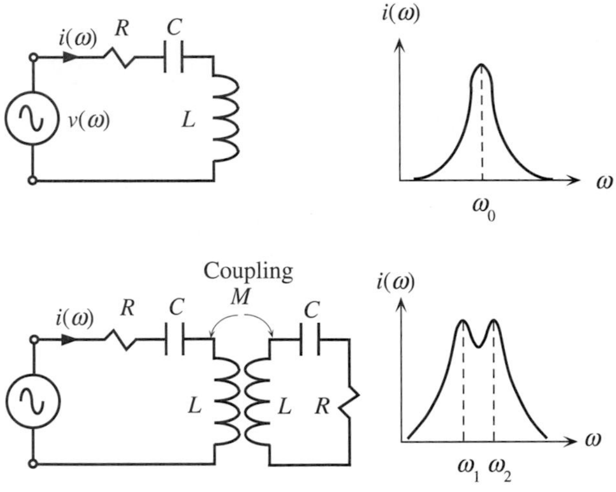


**(a) There is one resonant frequency, $\omega_0$, in an isolated *RLC* circuit.**

**(b) There are two resonant frequencies in two coupled *RLC* circuits: one below and the other above $\omega_0$.**

**Figure 4.4**

energy level, such as $E_{1s}$, splits into two, $E_\sigma$ and $E_{\sigma^*}$. The splitting is due to the interaction (or overlap) between the atomic orbitals. Figure 4.3b schematically illustrates the changes in the electron energy levels as two isolated H atoms are brought together to form the $H_2$ molecule.

The splitting of a one-atom energy level when a molecule is formed is analogous to the splitting of the resonant frequency in an *RLC* circuit when two such circuits are brought together and coupled. Consider the *RLC* circuit shown in Figure 4.4a. The circuit is excited by an ac voltage source. The current peaks at the resonant frequency $\omega_0$, as indicated in Figure 4.4a. When two such identical *RLC* circuits are coupled together and driven by an ac voltage source, the current develops two peaks, at frequencies $\omega_1$ and $\omega_2$, below and above $\omega_0$, as illustrated in Figure 4.4b. The two peaks at $\omega_1$ and $\omega_2$ are due to the mutual inductance that couples the two circuits, allowing them to interact. From this analogy, we can intuitively accept the energy splitting observed in Figure 4.3a.

Consider what happens when two He atoms come together. Recall that the $1s$ orbital has paired electrons and is full. The $1s$ atomic energy level will again split into two levels, $E_\sigma$ and $E_{\sigma^*}$, associated with the molecular orbitals $\psi_\sigma$ and $\psi_{\sigma^*}$, as illustrated in Figure 4.5. However, in the He–He system, there are four electrons, so two occupy the $\psi_\sigma$ orbital state and two go to the $\psi_{\sigma^*}$ orbital state. Consequently, the system energy is not lowered by bringing the two He atoms closer. Furthermore, quantum mechanical calculations show that the antibonding energy level $E_{\sigma^*}$ shifts higher than the bonding level $E_\sigma$ shifts lower. By the same token, although we could put an additional electron at $E_{\sigma^*}$ in $H_2$ to make $H_2^-$, we could not make $H_2^{2-}$ by placing two electrons at $E_{\sigma^*}$.

From the He–He example, we can conclude that, as a general rule, the overlap of full atomic orbital states does not lead to bonding. In fact, full orbitals repel each other, because any overlap results in an increase in the system energy. To form a bond

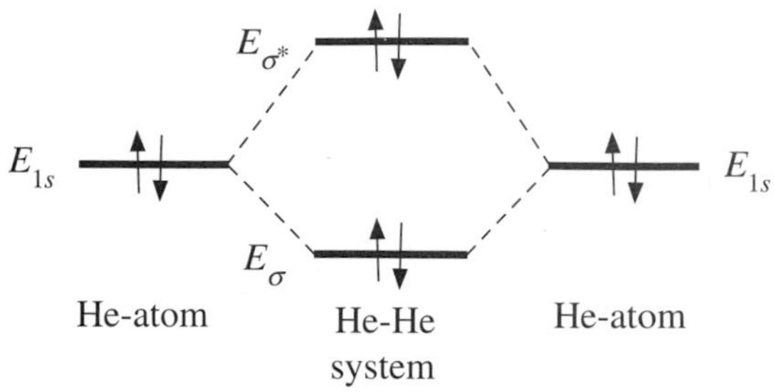


**Figure 4.5** Two He atoms have four electrons. When He atoms come together, two of the electrons enter the $E_\sigma$ level and two the $E_{\sigma^*}$ level, so the overall energy is greater than two isolated He atoms.

between two atoms, we essentially need an overlap of half-occupied orbitals, as in the $H_2$ molecule.

**EXAMPLE 4.1**

**HYDROGEN HALIDE MOLECULE (HF)** We already know that H has a half-occupied $1s$ orbital, which can take part in bonding. Since the F atom has the electronic structure $1s^2 2s^2 p^5$, two of the $p$ orbitals are full and one $p$ orbital, $p_x$, is half full. This means that only the $p_x$ orbital can participate in bonding. Figure 4.6 shows the electron orbitals in both H and F. When the H atom and the F atom approach each other to form an HF molecule, the $\psi_{1s}$ orbital of H overlaps the $p_x$ orbital of F. There are two possibilities for the overlap. First, $\psi_{1s}$ and $p_x$ can overlap in phase (both positive or both negative), to give a $\psi_\sigma$ orbital that does not have a node between H and F, as shown in Figure 4.6. Second, they can overlap out of phase (one positive and the other negative), so that the overlap orbital $\psi_{\sigma^*}$ has a node (similar to $\psi_{\sigma^*}$ in Figure 4.1). We know from hydrogen atomic wavefunctions in Chapter 3 that orbitals with more nodes have higher energies. The molecular orbital $\psi_\sigma$ therefore corresponds to a bonding orbital with a lower energy than the $\psi_{\sigma^*}$ orbital. The two electrons, one from $\psi_{1s}$ and the other from $p_x$, enter the $\psi_\sigma$ orbital with spins paired, thereby forming a bond between H and F.

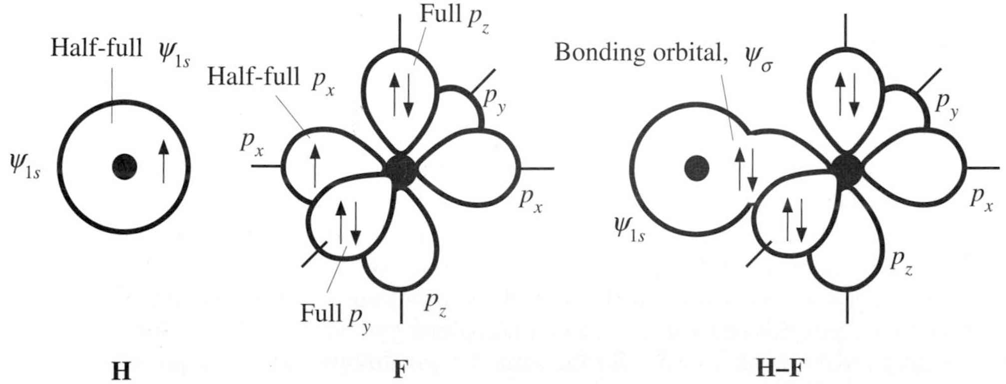


**Figure 4.6** H has one half-empty $\psi_{1s}$ orbital.
F has one half-empty $p_x$ orbital but full $p_y$ and $p_z$ orbitals. The overlap between $\psi_{1s}$ and $p_x$ produces a bonding orbital and an antibonding orbital. The two electrons fill the bonding orbital and thereby form a covalent bond between H and F.

# 4.2 BAND THEORY OF SOLIDS

## 4.2.1 ENERGY BAND FORMATION

When we bring three hydrogen atoms (labeled $A$, $B$, and $C$) together, we generate three separate molecular orbital states, $\psi_a$, $\psi_b$, and $\psi_c$, from three $\psi_{1s}$ atomic states. Again, this occurs in three different ways, as illustrated in Figure 4.7a. As in the case of the $H_2$ molecule, each molecular orbital must be either *symmetric* or *antisymmetric* with respect to center atom $B$.[1] The orbitals that satisfy even and odd requirements are

$$\psi_a = \psi_{1s}(A) + \psi_{1s}(B) + \psi_{1s}(C) \tag{4.3a}$$

$$\psi_b = \psi_{1s}(A) - \psi_{1s}(C) \tag{4.3b}$$

$$\psi_c = \psi_{1s}(A) - \psi_{1s}(B) + \psi_{1s}(C) \tag{4.3c}$$

where $\psi_{1s}(A)$, $\psi_{1s}(B)$, and $\psi_{1s}(C)$ are the $1s$ atomic wavefunctions centered around the atoms $A$, $B$, and $C$, respectively, as shown in Figure 4.7a. For example, the wavefunction $\psi_{1s}(A)$ represents $\psi_{1s}(r_A)$, which is centered around $A$ and has the form $\exp(-r_A/a_o)$, where $r_A$ is the distance from the nucleus of $A$, and $a_o$ is the Bohr radius. Notice that $\psi_{1s}(B)$ is missing in Equation 4.3b, so $\psi_b$ is antisymmetric.

The energies $E_a$, $E_b$, and $E_c$ of $\psi_a$, $\psi_b$, and $\psi_c$ can be calculated from the Schrödinger equation by using the *PE* function of this system (the *PE* also includes proton–proton repulsions). It is clear that since $\psi_a$, $\psi_b$, and $\psi_c$ are different, their energies $E_a$, $E_b$, and $E_c$ are also different. Consequently, the $1s$ energy level splits into three separate levels, corresponding to the energies of $\psi_a$, $\psi_b$, and $\psi_c$, as depicted by Figure 4.7b. By analogy with the electron wavefunctions in the hydrogen atom, we can argue that if the molecular wavefunction has more nodes, its energy is higher. Thus, $\psi_a$ has the lowest energy $E_a$, $\psi_b$ has the next higher energy $E_b$, and $\psi_c$ has the highest energy $E_c$, as shown in Figure 4.7b. There are three electrons in the three-hydrogen system. The first two pair their spins and enter orbital $\psi_a$ at energy $E_a$, and the third enters orbital $\psi_b$ at energy $E_b$. Comparing Figures 4.7 and 4.3, we notice that although $H_2$ and $H_3$ both have two electrons in the lowest energy level, $H_3$ also has an extra electron at the higher energy level ($E_b$), which tends to increase the net energy of the atom. Thus, the $H_3$ molecule is much less stable than the $H_2$ molecule.[2]

Now consider the formation of a solid. Take $N$ Li (lithium) atoms from infinity and bring them together to form the Li metal. Lithium has the electronic configuration $1s^2 2s^1$, which is somewhat like the hydrogen atom, since the $K$ shell is closed and the third electron is alone in the $2s$ orbital.

Based on our previous discussions, we assume that the atomic energy levels will split into $N$ separate energy levels. Since the $1s$ subshell is full and is close to the nucleus, it will not be affected much by the interatomic interactions; consequently, the energy of

[1] The reason is that the molecule *A–B–C*, when *A*, *B*, and *C* are identical atoms, is symmetric with respect to *B*. Thus each wavefunction must have odd or even parity (Chapter 3).

[2] See G. Pimentel and R. Spratley, *Understanding Chemistry*, San Francisco: Holden-Day, Inc., 1972, pp. 682–687 for an excellent discussion.

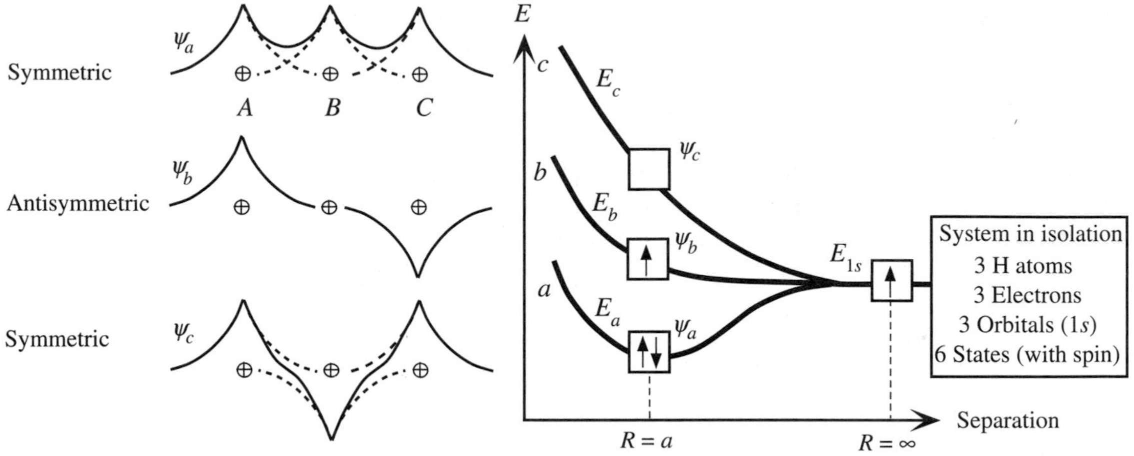


**(a) Three molecular orbitals from three $\psi_{1s}$ atomic orbitals overlapping in three different ways.**

**(b) The energies of the three molecular orbitals, labeled *a*, *b*, and *c*, in a system with three H atoms.**

**Figure 4.7**

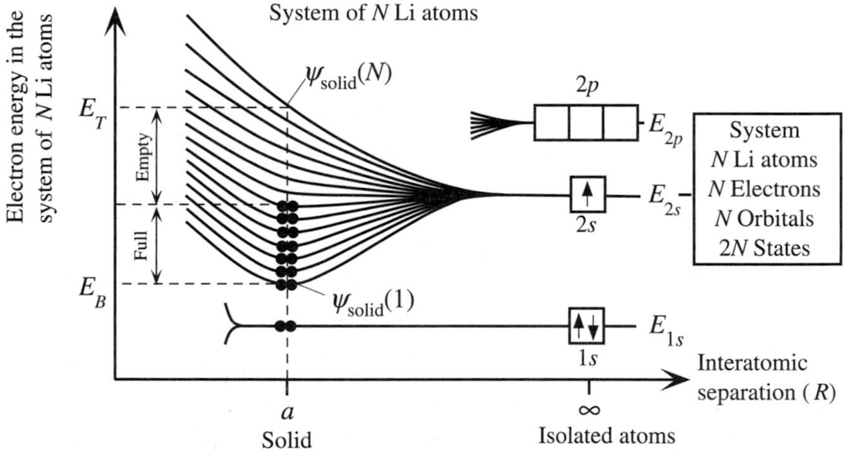


**Figure 4.8** The formation of a 2*s* energy band from the 2*s* orbitals when *N* Li atoms come together to form the Li solid.

There are *N* 2*s* electrons, but 2*N* states in the band. The 2*s* band is therefore only half full. The atomic 1*s* orbital is close to the Li nucleus and remains undisturbed in the solid. Thus, each Li atom has a closed *K* shell (full 1*s* orbital).

this state will experience only negligible splitting, if any. Since the $1s$ electrons will stay close to their parent nuclei, we will not consider them during formation of the solid.

In the system of $N$ isolated Li atoms, we have $N$ electrons in $N$ $\psi_{2s}$ orbitals at the energy $E_{2s}$, as illustrated in Figure 4.8 (at infinite interatomic separation). Let us assume that $N$ is large (typically, $\sim 10^{23}$). As $N$ atoms are brought together to form the solid, the energy level at $E_{2s}$ splits into $N$ finely separated energy levels. The maximum width of the energy splitting depends on the closest interatomic distance $a$ in the solid, as apparent in Figure 4.3a. The atoms separated by a distance greater than $R = a$ give rise to a lesser amount of energy splitting. The interatomic interactions between $N\psi_{2s}$ orbitals thus spread the $N$ energy levels between the bottom and top levels, $E_B$ and $E_T$, respectively, which are determined by the closest interatomic distance $a$. Put differently, $E_B$ and $E_T$ are determined by the distance between nearest neighbors. It is obvious that with $N$ very large, the energy separation between two consecutive energy levels is very small; indeed, it is almost infinitesimal and not as exaggerated as in Figure 4.8.

Remember that each energy level $E_i$ in the Li metal of Figure 4.8 is the energy of an electron wavefunction $\psi_{\text{solid}}(i)$ in the solid, where $\psi_{\text{solid}}(i)$ is one particular combination of the $N$ atomic wavefunctions $\psi_{2s}$. There are $N$ different ways to combine $N$ atomic wavefunctions $\psi_{2s}$, since each can be added in phase or out of phase, as is apparent in Equations 4.3a to c (see also Figure 4.7a and b). For example, when all $N$ $\psi_{2s}$ are summed in phase, the resulting wavefunction $\psi_{\text{solid}}(1)$ is like $\psi_a$ in Equation 4.3a, and it has the lowest energy. On the other hand, when $N$ $\psi_{2s}$ are summed with alternating phases, $+ - + \cdots$, the resulting wavefunction $\psi_{\text{solid}}(N)$ is like $\psi_c$, and it has the highest energy. Other combinations of $\psi_{2s}$ give rise to different energy values between $E_B$ and $E_T$.

The single $2s$ energy level $E_{2s}$ therefore splits into $N$ ($\sim 10^{23}$) finely separated energy levels, forming an **energy band,** as illustrated in Figure 4.8. Consequently, there are $N$ separate energy levels, each of which can take two electrons with opposite spins. The $N$ electrons fill all the levels up to and including the level at $N/2$. Therefore, the band is half full. We do not mean literally that the band is full to the half-energy point. The levels are not spread equally over the band from $E_B$ to $E_T$, which means that the band cannot be full to the half-energy point. Half filled simply means half the states in the band are filled from the bottom up.

We have generated a half-filled band from a half-filled isolated $2s$ energy level. The energy band resulting from the splitting of the atomic $2s$ energy level is loosely termed the **$2s$ band.** By the same token, the atomic $1s$ levels are full, so any $1s$ band that forms from these $1s$ states will also be full. We can get an idea of the separation of energy levels in the $2s$ band by noting that the maximum separation, $E_T - E_B$, between the top and bottom of the band is on the order of 10 eV, but there are some $10^{23}$ atoms, giving rise to $10^{23}$ energy levels between $E_B$ and $E_T$. Thus, the energy levels are finely separated, forming, for all practical purposes, a continuum of energy levels.

The $2p$ energy level, as well as the higher levels at $3s$ and so on, also split into finely separated energy levels, as shown in Figure 4.9. In fact, some of these energy levels overlap the $2s$ band; hence, they provide further energy levels and "extend" the $2s$ band into higher energy levels, as indicated in Figure 4.10, which shows how energy bands in metals are often represented. The vertical axis is the electron energy. The top of the $2s$ band, which is half full, overlaps the bottom of the $2p$ band, which itself

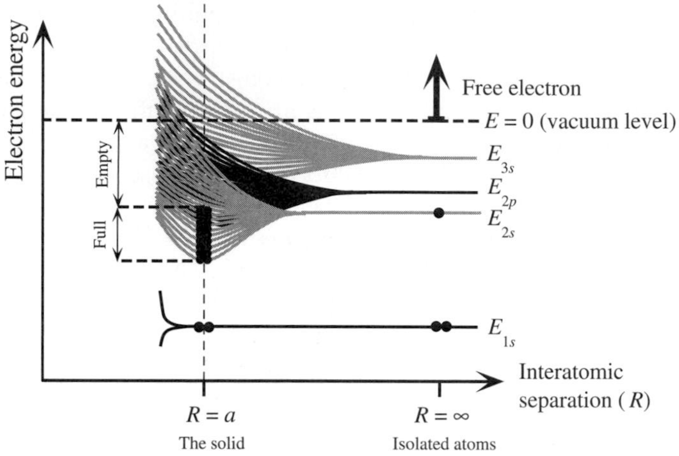


**Figure 4.9** As Li atoms are brought together from infinity, the atomic orbitals overlap and give rise to bands.
Outer orbitals overlap first. The $3s$ orbitals give rise to the $3s$ band, $2p$ orbitals to the $2p$ band, and so on. The various bands overlap to produce a single band in which the energy is nearly continuous.

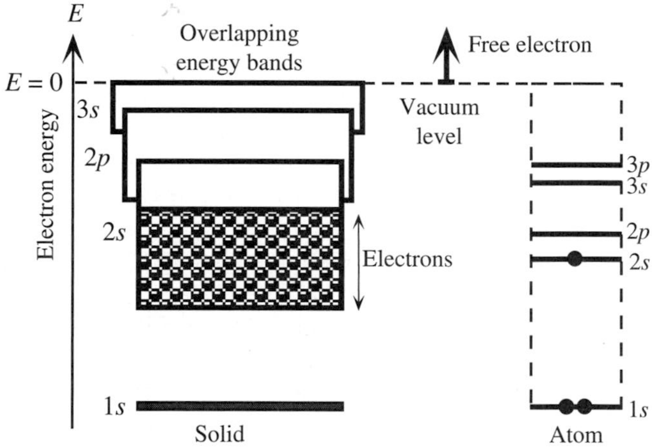


**Figure 4.10** In a metal, the various energy bands overlap to give a single energy band that is only partially full of electrons.
There are states with energies up to the vacuum level, where the electron is free.

is overlapped near the top by the $3s$ band. We therefore have a band of energies that stretches from the bottom of the $2s$ band all the way to the vacuum level, as depicted in Figure 4.11. The reader may wonder what happened to the $3d$, $4s$, etc., bands. In the solid, the energies of these bands (including the top portion of the $3s$ band) are above the vacuum level, and the electron is free and far from the solid before it can acquire those energies.

At a temperature of absolute zero, or nearly so, the thermal energy is insufficient to excite the electrons to higher energy levels, so all the electrons pair their spins and fill each energy level from $E_B$ up to an energy level $E_{FO}$ that we call the Fermi level at 0 K, as shown in Figure 4.11. The energy value for the Fermi level depends on where we take the reference energy. For example, if we take the vacuum level as the zero reference, then for the Li metal, $E_{FO}$ is at $-2.5$ eV. The Fermi level is normally measured with respect to the bottom of the band, in which case, it is simply termed the Fermi energy and denoted $E_{FO}$. For the Li metal, $E_{FO}$ is 4.7 eV, which is with respect to the bottom of the band. The Fermi level has considerable significance, as we will discover later in this chapter.

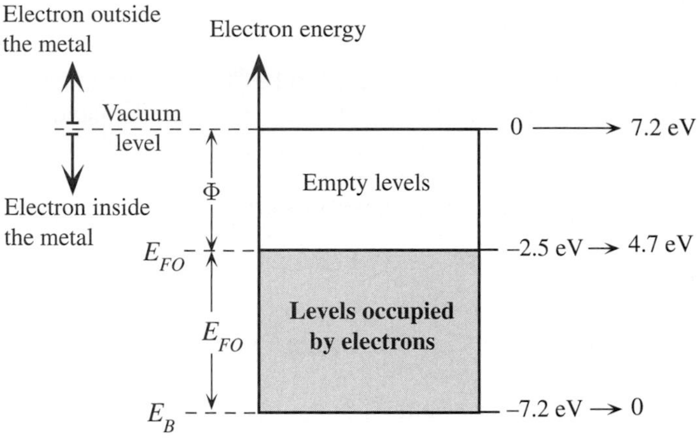


**Figure 4.11** Typical electron energy band diagram for a metal.
All the valence electrons are in an energy band, which they only partially fill. The top of the band is the vacuum level, where the electron is free from the solid ($PE = 0$).

At absolute zero, all the energy levels up to the Fermi level are full. The energy required to excite an electron from the Fermi level to the vacuum level, that is, to liberate the electron from the metal, is called the **work function** Φ of the metal. As the temperature increases, some of the electrons get excited to higher energy levels. To determine the probability of finding an electron at an energy level $E$, we must consider what is called "particle statistics," a topic that is key to understanding the behavior of electronic devices. Clearly, the probability of finding an electron at 0 K at some energy $E < E_{FO}$ is unity, and at $E > E_{FO}$, the probability is zero. Table 4.1 summarizes the Fermi energy and work function of a few selected metals.

The electrons in the energy band of a metal are loosely bound valence electrons which become free in the crystal and thereby form a kind of **electron gas.** It is this electron gas that holds the metal ions together in the crystal structure and constitutes the metallic bond. This intuitive interpretation is shown in Figure 4.9. When solid Li is formed from $N$ atoms, the $N$ electrons fill all the lower energy levels up to $N/2$. The energy of the system of $N$ Li atoms, according to Figure 4.9, is therefore much less than that of $N$ isolated Li atoms by virtue of the $N$ electrons taking up lower energy levels. It must be emphasized that the electrons within a band do not belong to any specific atom but to the whole solid. We cannot identify a given electron in the band with a certain Li atom. All the $2s$ electrons essentially form an electron gas and have energies that fall within the energy band. These electrons are constantly moving around in the metal which in terms of quantum mechanics means that their wavefunctions must be of the traveling wave type and not the type that localizes the electron around a given atom (*e.g.*, $\psi_{n,\ell,m_\ell}$ in the hydrogen atom). We can represent each electron with a wavevector $k$ so that its momentum $p$ is $\hbar k$.

**Table 4.1** Fermi energy and work function of selected metals

| | Metal | | | | | | | |
|---|---|---|---|---|---|---|---|---|
| | Ag | Al | Au | Cs | Cu | Li | Mg | Na |
| Φ (eV) | 4.5 | 4.28 | 5.0 | 2.14 | 4.65 | 2.3 | 3.7 | 2.75 |
| $E_{FO}$ (eV) | 5.5 | 11.7 | 5.5 | 1.58 | 7.0 | 4.7 | 7.1 | 3.2 |

### 4.2.2 PROPERTIES OF ELECTRONS IN A BAND

Since the electrons inside the metal crystal are considered to be "free," their energy is *KE*. These electrons occupy all the energy levels up to $E_{FO}$ as shown in the band diagram of Figure 4.12a. The energy $E$ of an electron in a metal increases with its momentum $p$ as $p^2/2m_e$. Figure 4.12b shows the energy versus momentum behavior of the electrons in a hypothetical one-dimensional crystal. The energy increases with momentum whether the electron is moving toward the left or right. Electrons take on all available momentum values until their energy reaches $E_{FO}$. For every electron that is moving right (such as $a$), there is another (such as $b$) with the same energy but moving left with the same magnitude of momentum. Thus, the average momentum is zero and there is no net current.

Consider what happens when an electric field $\mathcal{E}_x$ is applied in the $-x$ direction. The electron $a$ at the Fermi level and moving along in the $+x$ direction experiences a force $e\mathcal{E}_x$ along the same direction. It therefore accelerates and gains momentum and hence has the energy as shown in Figure 4.12c. (The actual energy gained from the field is very small compared with $E_{FO}$, so Figure 4.12c is highly exaggerated.) The electron $a$ at $E_{FO}$ can move to higher energy levels because these adjacent higher levels are empty. The momentum state vacated by $a$ is filled by the electron immediately below which now gains energy and moves up, and so on. An electron that is moving in the $-x$ direction, however, is decelerated (its momentum decreases) and hence loses energy as indicated by $b$ moving to $b'$ in Figure 4.12c. The electrons that are moving in the $+x$ direction gain energy, and those that are moving in the $-x$ direction, lose energy. The whole electron momentum distribution therefore shifts in the $+x$ direction as in Figure 4.12c. Eventually the electron $a$, now at $a'$, is scattered by a lattice vibration.

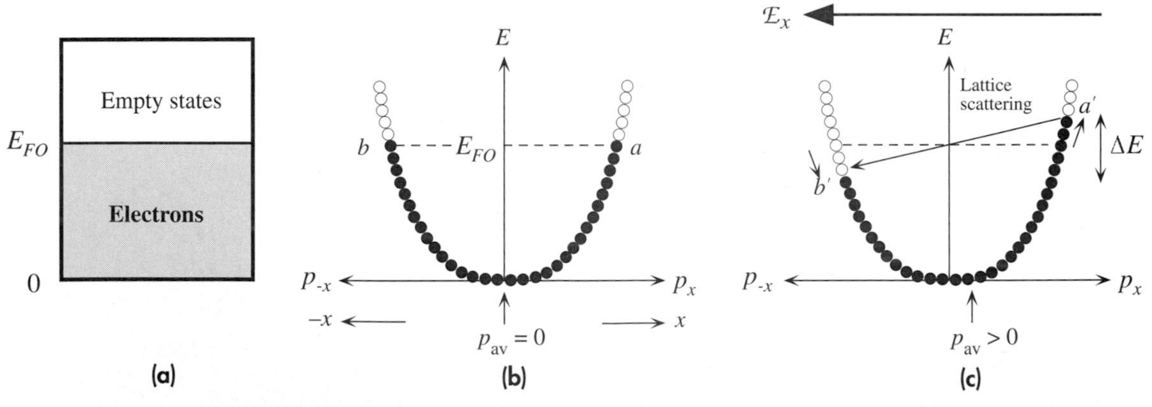


**Figure 4.12**

(a) Energy band diagram of a metal.

(b) In the absence of a field, there are as many electrons moving right as there are moving left. The motions of two electrons at each energy cancel each other as for *a* and *b*.

(c) In the presence of a field in the $-x$ direction, the electron *a* accelerates and gains energy to *a′* where it is scattered to an empty state near $E_{FO}$ but moving in the $-x$ direction. The average of all momenta values is along the $+x$ direction and results in a net electric current.

Typically lattice vibrations have small energies but substantial momentum. The scattered electron must find an *unoccupied* momentum state with roughly the same energy, and it must change its momentum substantially. The electron at $a'$ is therefore scattered to an empty state around $E_{FO}$ but with a momentum in the opposite direction. Its momentum is *flipped* as shown in Figure 4.12c. The average momentum of the electrons is no longer zero but finite in the $+x$ direction. Consequently there is a current flow in the $-x$ direction, along the field, as determined by this average momentum $p_{av}$. Notice that $a$ moves up to $a'$ and $b$ falls down to $b'$. Under steady-state conduction, lattice scattering simply replenishes the electrons at $b'$ from $a'$. Notice that for energies below $b'$, for every electron moving right there is another moving left with the same momentum magnitude that cancels it. Thus, electrons below the $b'$ energy level do *not* contribute to conduction and are excluded from further consideration. Notice that electrons above the $b'$ level are only moving right and their momenta are not canceled. Thus, the conductivity is determined by the electrons in the energy range $\Delta E$ from $b'$ to $a'$ about the Fermi level as shown in Figure 4.12c. Further, as the energy change from $a$ to $a'$ is orders of magnitude smaller than $E_{FO}$, we can summarize that conduction occurs by the drift of electrons at the Fermi level.[3] (If we were to calculate $\Delta E$ for a typical metal for typical currents, it would be $\sim 10^{-6}$ eV whereas $E_{FO}$ is 1–10 eV. The shift in the distribution in Figure 4.12c is very small indeed; $a'$ and $b'$, for all practical purposes, are at the Fermi level.)

Conduction can be explained very simply and intuitively in terms of a band diagram as shown in Figure 4.13. Notice that the application of the electric field bends the energy band, because the electrostatic *PE* of the electron is $-eV(x)$ where $V(x)$ is the voltage at position $x$. However, $V(x)$ changes linearly from 0 to $V$, by virtue of $dV/dx = -\mathcal{E}_x$. Since $E = -eV(x)$ adds to the energy of the electron, the energy band must bend to account for the additional electrostatic energy. Since only the electrons near $E_{FO}$ contribute to electrical conduction, we can represent this by drifting the electrons at $E_{FO}$ down the potential hill. Although these electrons possess a very high mean velocity ($\sim 10^6$ m s$^{-1}$), as determined by the Fermi energy, they drift very slowly ($10^{-2}$–$10^{-1}$ m s$^{-1}$) with a velocity that is drift mobility × field.

When a metal is illuminated, provided the wavelength of the radiation is correct, it will cause emission of electrons from the metal as in the photoelectric effect. Since $\Phi$ is the "minimum energy" required to excite an electron into the vacuum level (out from the metal), the longest wavelength radiation required is $hc/\lambda = \Phi$.

Addition of heat to a metal can excite some of the electrons in the band to higher energy levels. Thus heat can also be absorbed by the conduction electrons of a metal. We also know that the addition of heat increases the amplitude of atomic vibrations. We can therefore guess that the heat capacity of a metal has two terms which are due to energy absorption by the lattice vibrations and energy absorption by conduction electrons. It turns out that at room temperature the energy absorption by lattice vibrations dominates the heat capacity whereas at the lowest temperatures the electronic contribution is important.

---

[3] In some books (including the first edition of this textbook) it is stated that the electrons at $E_{FO}$ can gain energy from the field and contribute to conduction but not those deep in the band (below $b'$). This is a simplified statement of the fact that at a level below $E_{FO}$ there is one electron moving along in the $+x$ direction and gaining energy and another one at the same energy but moving along in the $-x$ direction and losing energy so that an average electron at this level does not gain energy.

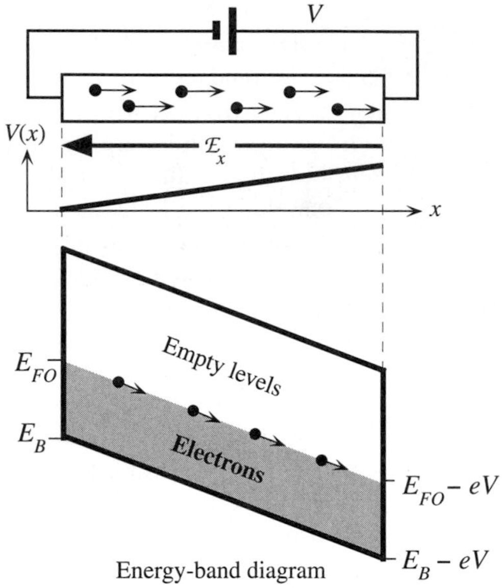


**Figure 4.13** Conduction in a metal is due to the drift of electrons around the Fermi level.

When a voltage is applied, the energy band is bent to be lower at the positive terminal so that the electron's potential energy decreases as it moves toward the positive terminal.

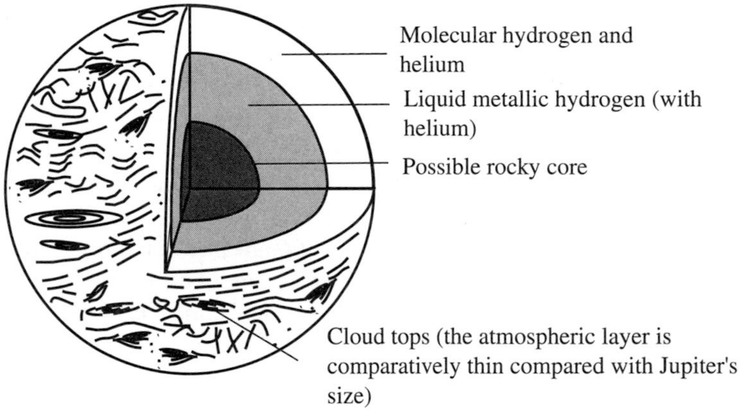


**Figure 4.14** The interior of Jupiter is believed to contain liquid hydrogen, which is metallic.

SOURCE: Drawing adapted from T. Hey and P. Walters, *The Quantum Universe*, Cambridge, MA: Cambridge University Press, 1988, p. 96, figure 7.1.

**EXAMPLE 4.2** **METALLIC LIQUID HYDROGEN IN JUPITER AND ITS MAGNETIC FIELD** The surface of Jupiter, as visualized schematically in Figure 4.14, mainly consists of a mixture of molecular hydrogen and He gases. Deep in the planet, however, the pressure is so tremendous that the hydrogen molecular bond breaks, leaving a dense ocean of hydrogen atoms. Hydrogen has only one electron in the 1*s* energy level. When atoms are densely packed, the 1*s* energy level forms an energy band, which is then only half filled. This is just like the Li metal, which means we can treat liquid hydrogen as a liquid metal, with electrical properties reminiscent of liquid mercury. Liquid hydrogen can sustain electric currents, which in turn can give rise to the magnetic fields on Jupiter. The origin of the electric currents are not known with certainty. We do know, however, that the core of the planet is hot and emanates heat, which causes convection currents. Temperature differences can readily give rise to electric currents, by virtue of thermoelectric effects, as discussed in Section 4.8.2.

**EXAMPLE 4.3**

**WHAT MAKES A METAL?** The Be atom has an electronic structure of $1s^2 2s^2$. Although the Be atom has a full $2s$ energy level, solid Be is a metal. Why?

**SOLUTION**

We will neglect the $K$ shell ($1s$ state), which is full and very close to the nucleus, and consider only the higher energy states. In the solid, the $2s$ energy level splits into $N$ levels, forming a $2s$ band. With $2N$ electrons, each level is occupied by spin-paired electrons. The $2s$ band is therefore full. However, the empty $2p$ band, from the empty $2p$ energy levels, overlaps the $2s$ band, thereby providing empty energy levels to these $2N$ electrons. Thus, the conduction electrons are in an energy band that is only partially filled; they can gain energy from the field to contribute to electrical conduction. Solid Be is therefore a metal.

**EXAMPLE 4.4**

**FERMI SPEED OF CONDUCTION ELECTRONS IN A METAL** In copper, the Fermi energy of conduction electrons is 7.0 eV. What is the speed of the conduction electrons around this energy?

**SOLUTION**

Since the conduction electrons are not bound to any one atom, their *PE* must be zero within the solid (but large outside), so all their energy is kinetic. For conduction electrons around the Fermi energy $E_{FO}$ with a speed $v_F$, we have

$$\frac{1}{2}mv_F^2 = E_{FO}$$

so that

$$v_F = \sqrt{\frac{2E_{FO}}{m_e}} = \sqrt{\frac{2(1.6 \times 10^{-19}\ \text{J/eV})(7.0\ \text{eV})}{(9.1 \times 10^{-31}\ \text{kg})}} = 1.6 \times 10^6\ \text{m s}^{-1}$$

Although the Fermi energy depends on the properties of the energy band, to a good approximation it is only weakly temperature dependent, so $v_F$ will be relatively temperature insensitive, as we will show later in Section 4.7.

## 4.3 SEMICONDUCTORS

The Si atom has 14 electrons, which distribute themselves in the various atomic energy levels as shown in Figure 4.15. The inner shells ($n = 1$ and $n = 2$) are full and therefore "closed." Since these shells are near the nucleus, when Si atoms come together to form the solid, they are not much affected and they stay around the parent Si atoms. They can therefore be excluded from further discussion. The $3s$ and $3p$ subshells are farther away from the nucleus. When two Si atoms approach, these electrons strongly interact with each other. Therefore, in studying the formation of bands in the Si solid, we will only consider the $3s$ and $3p$ levels.

The first task is to examine why Si actually bonds with four neighbors, since the $3s$ orbital is full and there are only two electrons in the $3p$ orbitals. The full $3s$ orbital should not overlap a neighbor and become involved in bonding. Since only two $3p$ orbitals are half full, bonds should be formed with two neighboring Si atoms. In reality,

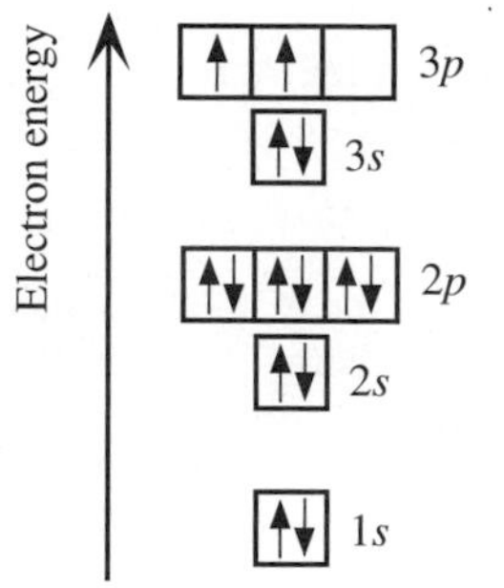


**Figure 4.15** The electronic structure of Si.

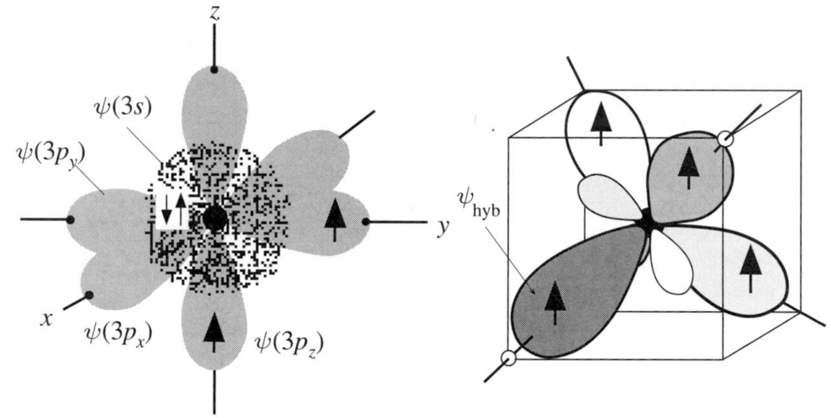


**(a) Isolated Si** **(b) Si just before bonding**

**Figure 4.16**

(a) Si is in Group IV in the Periodic Table. An isolated Si atom has two electrons in the $3s$ and two electrons in the $3p$ orbitals.

(b) When Si is about to bond, the one $3s$ orbital and the three $3p$ orbitals become perturbed and mixed to form four hybridized orbitals, $\psi_{\text{hyb}}$, called $sp^3$ orbitals, which are directed toward the corners of a tetrahedron. The $\psi_{\text{hyb}}$ orbital has a large major lobe and a small back lobe. Each $\psi_{\text{hyb}}$ orbital takes one of the four valence electrons.

the $3s$ and $3p$ energy levels are quite close, and when five Si atoms approach each other, the interaction results in the four orbitals $\psi(3s)$, $\psi(3p_x)$, $\psi(3p_y)$, and $\psi(3p_z)$ mixing together to form four new **hybrid orbitals,** which are directed in tetrahedral directions; that is, each one is aimed as far away from the others as possible, as illustrated in Figure 4.16. We call this process $sp^3$ **hybridization,** since one $s$ orbital and three $p$ orbitals are mixed. (The superscript 3 on $p$ has nothing to do with the number of electrons; it refers to the number of $p$ orbitals used in the hybridization.)

The four $sp^3$ hybrid orbitals, $\psi_{\text{hyb}}$, each have one electron, so they are half occupied. This means that four Si atoms can have their orbitals $\psi_{\text{hyb}}$ overlap to form bonds with one Si atom, which is what actually happens; thus, one Si atom bonds with four other Si atoms in tetrahedral directions.

In the same way, one Si atom bonds with four H atoms to form the important gas $SiH_4$, known as silane, which is widely used in the semiconductor technology to fabricate Si devices. In $SiH_4$, four hybridized orbitals of the Si atom overlap with the 1*s* orbitals of four H atoms. In exactly the same way, one carbon atom bonds with four hydrogen atoms to form methane, $CH_4$.

There are two ways in which the hybrid orbital $\psi_{hyb}$ can overlap with that of the neighboring Si atom to form two molecular orbitals. They can add in phase (both positive or both negative) or out of phase (one positive and the other negative) to produce a bonding or an antibonding molecular orbital $\psi_B$ and $\psi_A$, respectively, with energies $E_B$ and $E_A$. Each Si–Si bond thus corresponds to two paired electrons in a bonding molecular orbital $\psi_B$. In the solid, there are $N(\sim 5 \times 10^{22}\ \text{cm}^{-3})$ Si atoms, and there are nearly as many such $\psi_B$ bonds. The interactions between the $\psi_B$ orbitals (*i.e.*, the Si–Si bonds) lead to the splitting of the $E_B$ energy level to $N$ levels, thereby forming an energy band labeled the **valence band** (VB) by virtue of the valence electrons it contains. Since the energy level $E_B$ is full, so is the valence band. Figure 4.17 illustrates the formation of the VB from $E_B$.

In the solid, the interactions between the $N$ number of $\psi_A$ orbitals result in the splitting of the energy level $E_A$ to $N$ levels and the formation of an energy band that is

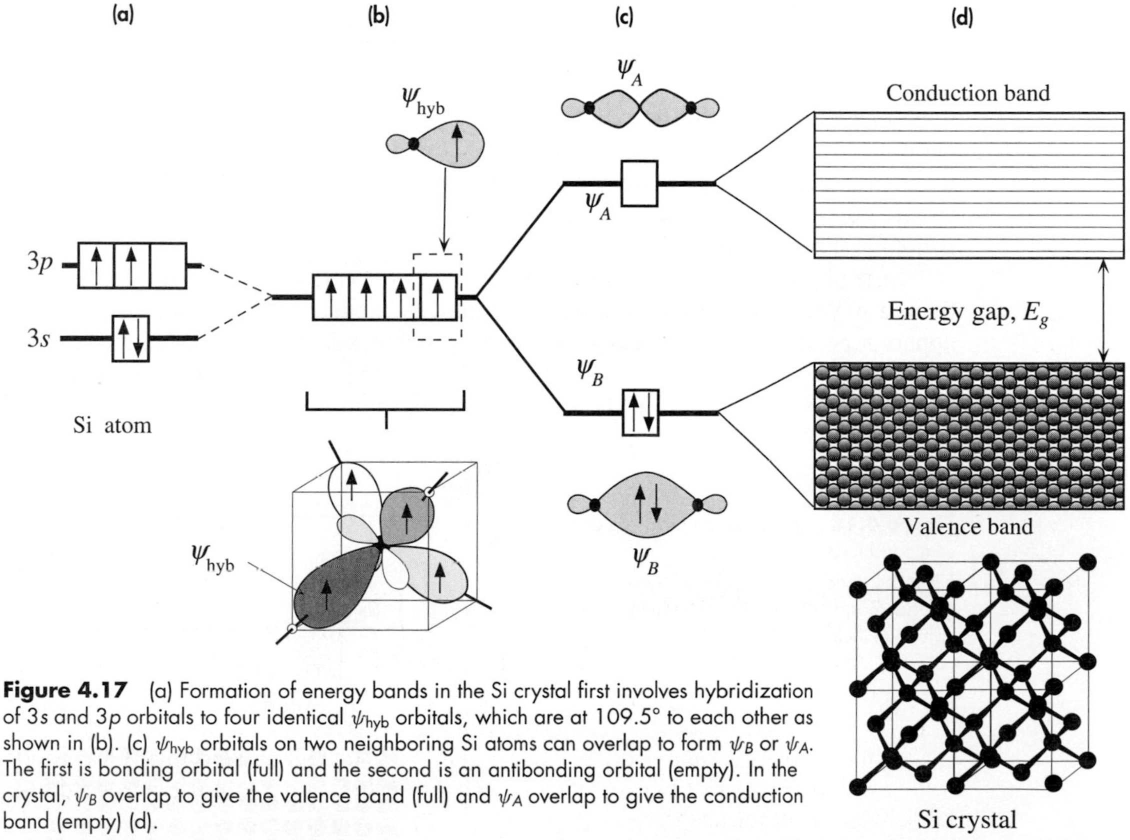


**Figure 4.17** (a) Formation of energy bands in the Si crystal first involves hybridization of 3*s* and 3*p* orbitals to four identical $\psi_{hyb}$ orbitals, which are at 109.5° to each other as shown in (b). (c) $\psi_{hyb}$ orbitals on two neighboring Si atoms can overlap to form $\psi_B$ or $\psi_A$. The first is bonding orbital (full) and the second is an antibonding orbital (empty). In the crystal, $\psi_B$ overlap to give the valence band (full) and $\psi_A$ overlap to give the conduction band (empty) (d).

completely empty and separated from the full valence band by a definite energy gap $E_g$. In this energy region, there are no states; therefore, the electron cannot have energy with a value within $E_g$. The energy band formed from $N\psi_A$ orbitals is a **conduction band** (CB), as also indicated in Figure 4.17.

The electronic states in the VB (and also in the CB) extend throughout the whole solid, because they result from $N\psi_B$ orbitals interfering and overlapping each other. As before $N\psi_B$, orbitals can overlap in $N$ different ways to produce $N$ distinct wavefunctions $\psi_{\text{vb}}$ that extend throughout the solid. We cannot relate a particular electron to a particular bond or site because the wavefunctions $\psi_{\text{vb}}$ corresponding to the VB energies are not concentrated at a single location. The electrical properties of solids are based on the fact that in solids, such as semiconductors and insulators, there are certain bands of allowed energies for the electrons, and these bands are separated by energy gaps, that is, bandgaps. The valence and conduction bands for the ideal Si crystal shown in Figure 4.17 are separated by an **energy gap,** or a **bandgap,** $E_g$, in which there are no allowed electron energy levels.

At temperatures above absolute zero, the atoms in a solid vibrate due to their thermal energy. Some of the atoms can acquire a sufficiently high energy from thermal fluctuations to strain and rupture their bonds. Physically, there is a possibility that the atomic vibration will impart sufficient energy to the electron for it to surmount the bonding energy and leave the bond. The electron must then enter a higher energy state. In the case of Si, this means entering a state in the CB, as shown in Figure 4.18. If there is an applied electric field $\mathcal{E}_x$ in the $+x$ direction, then the excited electron will be acted on by a force $-e\mathcal{E}_x$ and it will try to move in the $-x$ direction. For it to do so, there must be empty higher energy levels, so that as the electron accelerates and gains energy, it moves up in the band. When an electron collides with a lattice vibration, it loses the energy acquired from the field and drops down within the CB. Again, it should be emphasized that states in an energy band are extended; that is, the electron is not localized to any one atom.

Note also that the thermal generation of an electron from the VB to the CB leaves behind a VB state with a missing electron. This unoccupied electron state has an apparent positive charge, because this crystal region was neutral prior to the removal of the electron. The VB state with the missing electron is called a **hole** and is denoted $h^+$. The hole can "move" in the direction of the field by exchanging places with a

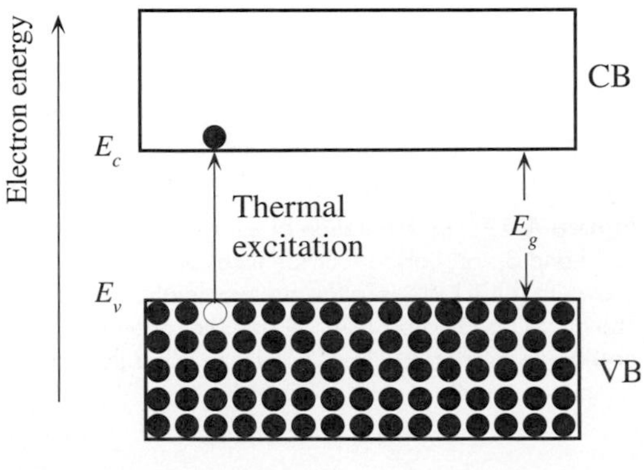


**Figure 4.18** Energy band diagram of a semiconductor.
CB is the conduction band and VB is the valence band. At 0 K, the VB is full with all the valence electrons.

neighboring valence electron hence it contributes to conduction, as will be discussed in Chapter 5.

**EXAMPLE 4.5**

**CUTOFF WAVELENGTH OF A Si PHOTODETECTOR** What wavelengths of light can be absorbed by a Si photodetector given $E_g = 1.1$ eV? Can such a photodetector be used in fiber-optic communications at light wavelengths of 1.31 $\mu$m and 1.55 $\mu$m?

**SOLUTION**

The energy bandgap $E_g$ of Si is 1.1 eV. A photon must have at least this much energy to excite an electron from the VB to the CB, where the electron can drift. Excitation corresponds to the breaking of a Si–Si bond. A photon of less energy does not get absorbed, because its energy will put the electron in the bandgap where there are no states. Thus, $hc/\lambda > E_g$ gives

$$\lambda < \frac{hc}{E_g} = \frac{(6.6 \times 10^{-34}\ \text{J s})(3 \times 10^{8}\ \text{m s}^{-1})}{(1.1\ \text{eV})(1.6 \times 10^{-19}\ \text{J/eV})}$$

$$= 1.13 \times 10^{-6}\ \text{m} \qquad \text{or} \qquad 1.1\ \mu\text{m}$$

Since optical communications networks use wavelengths of 1.3 and 1.55 $\mu$m, these light waves will not be absorbed by Si and thus cannot be detected by a Si photodetector.

## 4.4 ELECTRON EFFECTIVE MASS

When an electric field $\mathcal{E}_x$ is applied to a metal, an electron near the Fermi level can gain energy from the field and move to higher energy levels, as shown in Figure 4.12. The external force $F_{ext} = e\mathcal{E}_x$ is in the $x$ direction, and it drives the electron along $x$. The acceleration of the electron is still given by $a = F_{ext}/m_e$, where $m_e$ is the mass of the electron in vacuum.

The law $F_{ext} = m_e a$ cannot strictly be valid for the electron inside a solid, because the electron interacts with the host ions and experiences internal forces $F_{int}$ as it moves around, as depicted in Figure 4.19. The electron therefore has a $PE$ that varies with distance. Recall that we interpret mass as inertial resistance against acceleration per unit

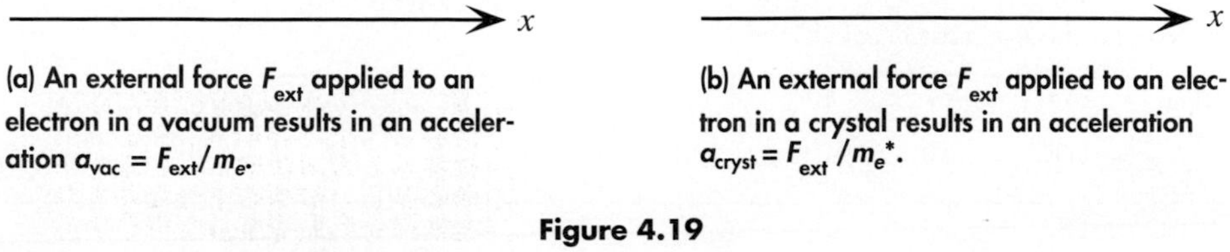


(a) An external force $F_{ext}$ applied to an electron in a vacuum results in an acceleration $a_{vac} = F_{ext}/m_e$.

(b) An external force $F_{ext}$ applied to an electron in a crystal results in an acceleration $a_{cryst} = F_{ext}/m_e^*$.

**Figure 4.19**

applied force. When an external force $F_{\text{ext}}$ is applied to an electron in the vacuum level, as in Figure 4.19a, the electron will accelerate by an amount

$$a_{\text{vac}} = \frac{F_{\text{ext}}}{m_e} \tag{4.4}$$

as determined by its mass $m_e$ in vacuum.

When the same force $F_{\text{ext}}$ is applied to the electron inside a crystal, the acceleration of the electron will be different, because it will also experience internal forces, as shown in Figure 4.19b. Its acceleration in the crystal will be

$$a_{\text{cryst}} = \frac{F_{\text{ext}} + F_{\text{int}}}{m_e} \tag{4.5}$$

where $F_{\text{int}}$ is the sum of all the internal forces acting on the electron, which is quite different than Equation 4.4. To the outside agent applying the force $F_{\text{ext}}$, the electron will appear to be exhibiting a different inertial mass, since its acceleration will be different. It would be most useful for the external agent if the effect of the internal forces in $F_{\text{int}}$ could be accounted for in a simple way, and if the acceleration could be calculated from the external force $F_{\text{ext}}$ alone, through something like Equation 4.4. This is indeed possible.

In a crystalline solid, the atoms are arranged periodically, and the variation of $F_{\text{int}}$, and hence the *PE*, or $V(x)$, of the electron with distance along $x$, is also periodic. In principle, then, the effect on the electron motion can be predicted and accounted for. When we solve the Schrödinger equation with the periodic *PE*, or $V(x)$, we essentially obtain the effect of these internal forces on the electron motion. It has been found that when the electron is in a band that is not full, we can still use Equation 4.4, but instead of the mass in vacuum $m_e$, we must use the effective mass $m_e^*$ of the electron in that particular crystal. The effective mass is a quantum mechanical quantity that behaves in the same way as the inertial mass in classical mechanics. The acceleration of the electron in the crystal is then simply

$$a_{\text{cryst}} = \frac{F_{\text{ext}}}{m_e^*} \tag{4.6}$$

The effects of all internal forces are incorporated into $m_e^*$. It should be emphasized that $m_e^*$, is obtained theoretically from the solution of the Schrödinger equation for the electron in a particular crystal, a task that is by no means trivial. However, the effective mass can be readily measured. For some of the familiar metals, $m_e^*$ is very close to $m_e$. For example, in copper, $m_e^* = m_e$ for all practical purposes, whereas in lithium $m_e^* = 1.28m_e$, as shown in Table 4.2. On the other hand, $m_e^*$ for many metals and

**Table 4.2** Effective mass $m_e^*$ of electrons in some metals

| Metal | Ag | Au | Bi | Cu | K | Li | Na | Ni | Pt | Zn |
|---|---|---|---|---|---|---|---|---|---|---|
| $\frac{m_e^*}{m_e}$ | 0.99 | 1.10 | 0.047 | 1.01 | 1.12 | 1.28 | 1.2 | 28 | 13 | 0.85 |

semiconductors is appreciably different than the electron mass in vacuum and can even be negative. ($m_e^*$ depends on the properties of the band that contains the electron. This is further discussed in Section 5.11.)

## 4.5 DENSITY OF STATES IN AN ENERGY BAND

Although we know there are many energy levels (perhaps $\sim 10^{23}$) in a given band, we have not yet considered how many states (or electron wavefunctions) there are per unit energy per unit volume in that band. Consider the following *intuitive* argument. The crystal will have $N$ atoms and there will be $N$ electron wavefunctions $\psi_1, \psi_2, \ldots, \psi_N$ that represent the electron within the whole crystal. These wavefunctions are constructed from $N$ different combinations of atomic wavefunctions, $\psi_A, \psi_B, \psi_C, \ldots$ as schematically illustrated in Figure 4.20a,[4] starting with

$$\psi_1 = \psi_A + \psi_B + \psi_C + \psi_D + \cdots$$

all the way to alternating signs

$$\psi_N = \psi_A - \psi_B + \psi_C - \psi_D + \cdots$$

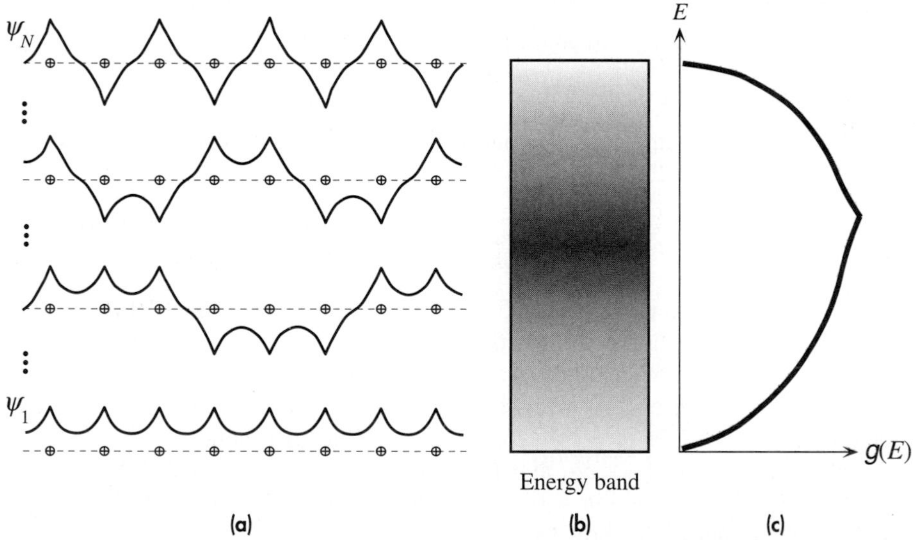


**Figure 4.20**

(a) In the solid there are $N$ atoms and $N$ extended electron wavefunctions from $\psi_1$ all the way to $\psi_N$. There are many wavefunctions, states, that have energies that fall in the central regions of the energy band.

(b) The distribution of states in the energy band; darker regions have a higher number of states.

(c) Schematic representation of the density of states $g(E)$ versus energy $E$.

[4] This intuitive argument, as schematically depicted in Figure 4.20a, is obviously highly simplified because the solid is three-dimensional (3-D) and we should combine the atomic wavefunctions not on a linear chain but on a 3-D lattice. In the 3-D case there are large numbers of wavefunctions with energies that fall in the central regions of the band.

and there are $N(\sim 10^{23})$ combinations. The lowest-energy wavefunction will be $\psi_1$ constructed by adding all atomic wavefunctions (all in phase), and the highest-energy wavefunction will be $\psi_N$ from alternating the signs of the atomic wavefunctions, which will have the highest number of nodes. Between these two extremes, especially around $N/2$, there will be many combinations that will have comparable energies and fall near the middle of the band. (By analogy, if we arrange $N = 10$ coins by heads and tails, there will be many combinations of coins in which there are 5 heads and 5 tails, and only one combination in which there are 10 heads or 10 tails.) We therefore expect the number of energy levels, each corresponding to an electron wavefunction in the crystal, in the central regions of the band to be very large as depicted in Figure 4.20b and c.

Figure 4.20c illustrates schematically how the energy and volume density of electronic states change across an energy band. We define the **density of states** $g(E)$ such that $g(E)\,dE$ is the number of states (*i.e.*, wavefunctions) in the energy interval $E$ to $(E + dE)$ per unit volume of the sample. Thus, the number of states per unit volume up to some energy $E'$ is

$$S_v(E') = \int_0^{E'} g(E)\,dE \qquad \textbf{[4.7]}$$

which is called the total number of states per unit volume with energies less than $E'$. This is denoted $S_v(E')$.

To determine the density of states function $g(E)$, we must first determine the number of states with energies less than $E'$ in a given band. This is tantamount to calculating $S_v(E')$ in Equation 4.7. Instead, we will improvise and use the energy levels for an electron in a three-dimensional potential well. Recall that the energy of an electron in a cubic *PE* well of size $L$ is given by

$$E = \frac{h^2}{8m_e L^2}\left(n_1^2 + n_2^2 + n_3^2\right) \qquad \textbf{[4.8]}$$

where $n_1$, $n_2$, and $n_3$ are integers 1, 2, 3, . . . . The spatial dimension $L$ of the well now refers to the size of the entire solid, as the electron is confined to be somewhere inside that solid. Thus, $L$ is very large compared to atomic dimensions, which means that the separation between the energy levels is very small. We will use Equation 4.8 to describe the energies of **free electrons** inside the solid (as in a metal).

Each combination of $n_1$, $n_2$, and $n_3$ is one electron orbital state. For example, $\psi_{n_1,n_2,n_3} = \psi_{1,1,2}$ is one possible orbital state. Suppose that in Equation 4.8 $E$ is given as $E'$. We need to determine how many combinations of $n_1, n_2, n_3$ (*i.e.*, how many $\psi$) have energies less than $E'$, as given by Equation 4.8. Assume that $(n_1^2 + n_2^2 + n_3^2) = n'^2$. The object is to enumerate all possible choices of integers for $n_1$, $n_2$, and $n_3$ that satisfy $n_1^2 + n_2^2 + n_3^2 \le n'^2$.

The two-dimensional case is easy to solve. Consider $n_1^2 + n_2^2 \le n'^2$ and the two-dimensional ***n*-space** where the axes are $n_1$ and $n_2$, as shown in Figure 4.21. The two-dimensional space is divided by lines drawn at $n_1 = 1, 2, 3, \ldots$ and $n_2 = 1, 2, 3, \ldots$ into infinitely many boxes (squares), each of which has a unit area and represents a possible state $\psi_{n_1,n_2}$. For example, the state $n_1 = 1, n_2 = 3$ is shaded, as is that for $n_1 = 2, n_2 = 2$.

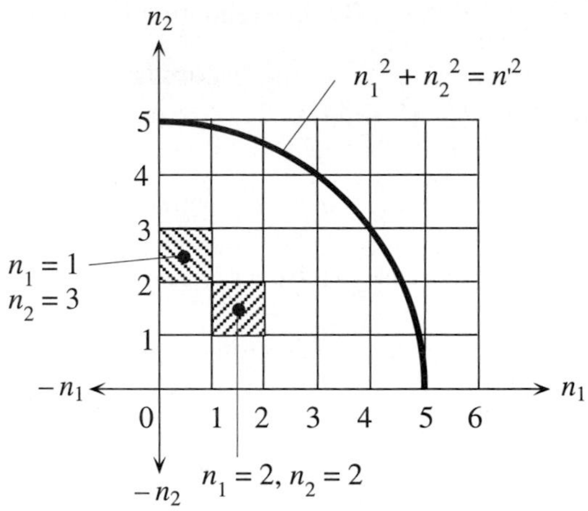


**Figure 4.21** Each state, or electron wavefunction in the crystal, can be represented by a box at $n_1$, $n_2$.

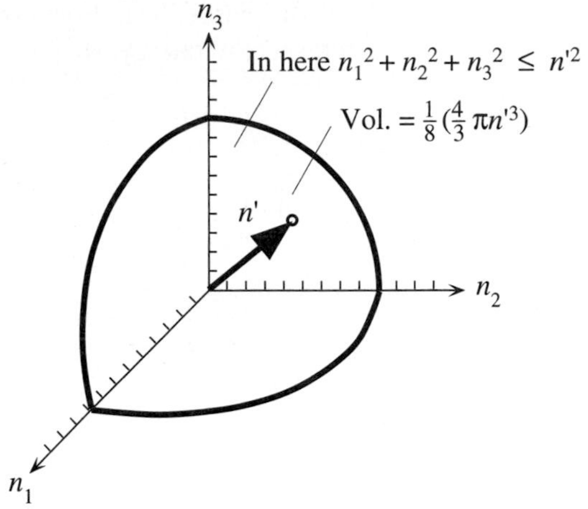


**Figure 4.22** In three dimensions, the volume defined by a sphere of radius $n'$ and the positive axes $n_1$, $n_2$, and $n_3$, contains all the possible combinations of positive $n_1$, $n_2$, and $n_3$ values that satisfy $n_1^2 + n_2^2 + n_3^2 \le n'^2$.

Clearly, the area contained by $n_1$, $n_2$ and the circle defined by $n'^2 = n_1^2 + n_2^2$ (just like $r^2 = x^2 + y^2$) is the number of states that satisfy $n_1^2 + n_2^2 \le n'^2$. This area is $\frac{1}{4}(\pi n'^2)$.

In the three-dimensional case, $n_1^2 + n_2^2 + n_3^2 \le n'^2$ is required, as indicated in Figure 4.22. This is the volume contained by the positive $n_1$, $n_2$, and $n_3$ axes and the surface of a sphere of radius $n'$. Each state has a unit volume, and within the sphere, $n_1^2 + n_2^2 + n_3^2 \le n'^2$ is satisfied. Therefore, the number of orbital states $S_{\text{orb}}(n')$ within this volume is given by

$$S_{\text{orb}}(n') = \frac{1}{8}\left(\frac{4}{3}\pi n'^3\right) = \frac{1}{6}\pi n'^3$$

Each orbital state can take two electrons with opposite spins, which means that the number of states, including spin, is given by

$$S(n') = 2S_{\text{orb}}(n') = \frac{1}{3}\pi n'^3$$

We need this expression in terms of energy. Substituting $n'^2 = 8m_e L^2 E'/h^2$ from Equation 4.8 in $S(n')$, we get

$$S(E') = \frac{\pi L^3 (8m_e E')^{3/2}}{3h^3}$$

Since $L^3$ is the physical volume of the solid, the number of states per unit volume $S_v(E')$ with energies $E \le E'$ is

$$S_v(E') = \frac{\pi (8m_e E')^{3/2}}{3h^3} \quad [4.9]$$

Furthermore, from Equation 4.7, $dS_v/dE = g(E)$. By differentiating Equation 4.9 with respect to energy, we get

*Density of states*

$$g(E) = (8\pi 2^{1/2})\left(\frac{m_e}{h^2}\right)^{3/2} E^{1/2} \quad [4.10]$$

Equation 4.10 shows that the density of states $g(E)$ increases with energy as $E^{1/2}$ from the bottom of the band. As we approach the top of the band, according to our understanding in Figure 4.20d, $g(E)$ should decrease with energy as $(E_{top} - E)^{1/2}$, where $E_{top}$ is the top of the band, so that as $E \rightarrow E_{top}$, $g(E) \rightarrow 0$. The electron mass $m_e$ in Equation 4.10 should be the *effective mass* $m_e^*$ as in Equation 4.6. Further, Equation 4.10 strictly applies only to *free electrons* in a crystal. However, we will frequently use it to approximate the true $g(E)$ versus $E$ behavior near the band edges for both metals and semiconductors.

Having found the distribution of the electron energy states, Equation 4.10, we now wish to determine the number of states that actually contain electrons; that is, the probability of finding an electron at an energy level $E$. This is given by the Fermi–Dirac statistics.

As an example, one convenient way of calculating the population of a city is to find the density of houses in that city (*i.e.*, the number of houses per unit area), multiply that by the probability of finding a human in a house, and finally, integrate the result over the area of the city. The problem is working out the chances of actually finding someone at home, using a mathematical formula. For those who like analogies, if $g(A)$ is the density of houses and $f(A)$ is the probability that a house is occupied, then the population of the city is

$$n = \int_{\text{City}} f(A)g(A)\, dA$$

where the integration is done over the entire area of the city. This equation can be used to find the number of electrons per unit volume within a band. If $E$ is the electron energy and $f(E)$ is the probability that a state with energy $E$ is occupied, then

$$n = \int_{\text{Band}} f(E)g(E)\, dE$$

where the integration is done over all the energies of the band.

**EXAMPLE 4.6** **X-RAY EMISSION AND THE DENSITY OF STATES IN A METAL** Consider what happens when a metal such as Al is bombarded with high-energy electrons. The inner atomic energy levels are not disturbed in the solid, so these inner levels remain as distinct single levels, each one localized to the parent atom. When an energetic electron hits an electron in one of the inner atomic energy levels, it knocks out this electron from the metal leaving behind a vacancy in the inner core as depicted in Figure 4.23a. An electron in the energy band of the solid can then fall down to occupy this empty state and emit a photon in the process. The energy difference between the energies in the band and the inner atomic level is in the X-ray range, so the emitted photon is an X-ray photon. Since electrons occupy the band from the bottom $E_B$ to the Fermi level $E_F$, the

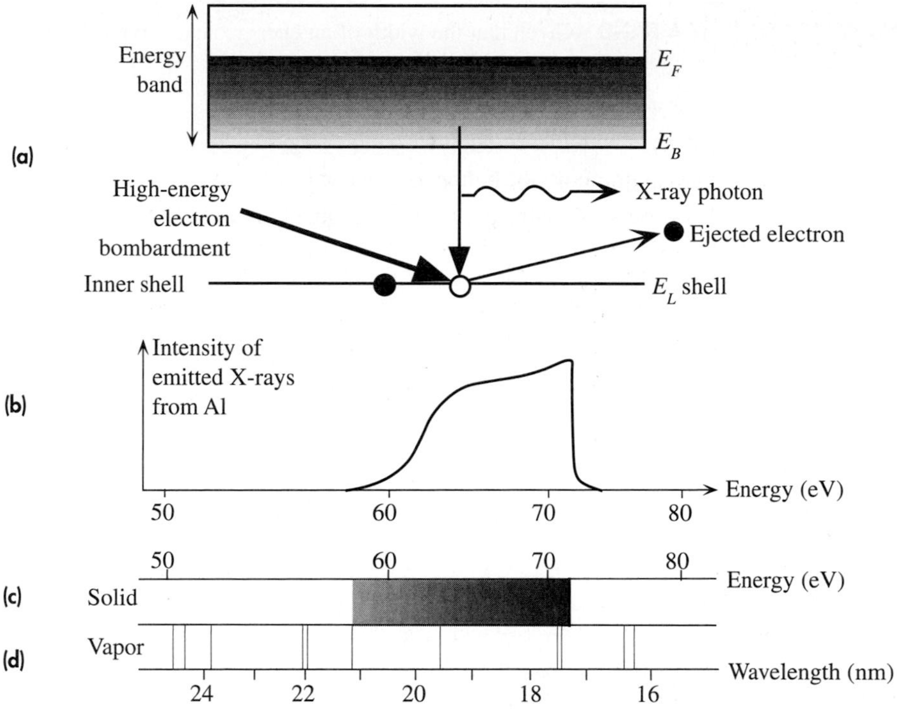


**Figure 4.23**

(a) High-energy electron bombardment knocks out an electron from the closed inner *L* shell leaving an empty state. An electron from the energy band of the metal drops into the *L* shell to fill the vacancy and emits a soft X-ray photon in the process.

(b) The spectrum (intensity versus photon energy) of soft X-ray emission from a metal involves a range of energies corresponding to transitions from the bottom of the band and from the Fermi level to the *L* shell. The intensity increases with energy until around $E_F$ where it drops sharply.

(c) and (d) contrast the emission spectra from a solid and vapor (isolated gas atoms).

emitted X-ray photons have a range of energies corresponding to transitions from $E_B$ and $E_F$ to the inner atomic level as shown in Figure 4.23b. These energies are in the soft X-ray spectrum. We assumed that the levels above $E_F$ are almost empty, though, undoubtedly, there is no sharp transition from full to empty levels at $E_F$. Further, since the density of states increases from $E_B$ toward $E_F$, there are more and more electrons that can fall down to the atomic level as we move from $E_B$ toward $E_F$. Therefore the intensity of the emitted X-ray radiation increases with energy until the energy reaches the Fermi level beyond which there are only a small number of electrons available for the transit. Figure 4.23c and d contrasts the emission spectra from an aluminum crystal (solid) and its vapor. The line spectra from a vapor become an emission band in the spectrum of the solid.

The X-ray intensity emitted from Al in Figure 4.23 starts to rise at around 60 eV and then sharply falls around 72 eV. Thus the energy range is 12 eV, which represents approximately the Fermi energy with respect to the bottom of the band, that is, $E_F \approx 72 - 60 = 12$ eV with respect to $E_B$.

**EXAMPLE 4.7** **DENSITY OF STATES IN A BAND** Given that the width of an energy band is typically ~10 eV, calculate the following, in per $cm^3$ and per eV units:

*a.* The density of states at the center of the band.

*b.* The number of states per unit volume within a small energy range $kT$ about the center.

*c.* The density of states at $kT$ above the bottom of the band.

*d.* The number of states per unit volume within a small energy range of $kT$ to $2kT$ from the bottom of the band.

**SOLUTION**

The density of states, or the number of states per unit energy range per unit volume $g(E)$, is given by

$$g(E) = (8\pi 2^{1/2})\left(\frac{m_e}{h^2}\right)^{3/2} E^{1/2}$$

which gives the number of states per cubic meter per Joule of energy. Substituting $E = 5$ eV, we have

$$g_{\text{center}} = (8\pi 2^{1/2})\left[\frac{9.1 \times 10^{-31}}{(6.626 \times 10^{-34})^2}\right]^{3/2} (5 \times 1.6 \times 10^{-19})^{1/2} = 9.50 \times 10^{46}\ \text{m}^{-3}\ \text{J}^{-1}$$

Converting to $cm^{-3}$ and $eV^{-1}$, we get

$$g_{\text{center}} = (9.50 \times 10^{46}\ \text{m}^{-3}\ \text{J}^{-1})(10^{-6}\ \text{m}^3\ \text{cm}^{-3})(1.6 \times 10^{-19}\ \text{J}\,\text{eV}^{-1})$$
$$= 1.52 \times 10^{22}\ \text{cm}^{-3}\ \text{eV}^{-1}$$

If $\delta E$ is a small energy range (such as $kT$), then, by definition, $g(E)\,\delta E$ is the number of states per unit volume in $\delta E$. To find the number of states per unit volume within $kT$ at the center of the band, we multiply $g_{\text{center}}$ by $kT$ or $(1.52 \times 10^{22}\ \text{cm}^{-3}\ \text{eV}^{-1})(0.026\ \text{eV})$ to get $3.9 \times 10^{20}\ \text{cm}^{-3}$. This is not a small number!

At $kT$ above the bottom of the band, at 300 K ($kT = 0.026$ eV), we have

$$g_{0.026} = (8\pi 2^{1/2})\left[\frac{9.1 \times 10^{-31}}{(6.626 \times 10^{-34})^2}\right]^{3/2} (0.026 \times 1.6 \times 10^{-19})^{1/2}$$
$$= 6.84 \times 10^{45}\ \text{m}^{-3}\ \text{J}^{-1}$$

Converting to $cm^{-3}$ and $eV^{-1}$ we get

$$g_{0.026} = (6.84 \times 10^{45}\ \text{m}^{-3}\ \text{J}^{-1})(10^{-6}\ \text{m}^3\ \text{cm}^{-3})(1.6 \times 10^{-19}\ \text{J}\,\text{eV}^{-1})$$
$$= 1.10 \times 10^{21}\ \text{cm}^{-3}\ \text{eV}^{-1}$$

Within $kT$, the volume density of states is

$$(1.10 \times 10^{21}\ \text{cm}^{-3}\ \text{eV}^{-1})(0.026\ \text{eV}) = 2.8 \times 10^{19}\ \text{cm}^{-3}$$

This is very close to the bottom of the band and is still very large.

### TOTAL NUMBER OF STATES IN A BAND

**EXAMPLE 4.8**

*a.* Based on the overlap of atomic orbitals to form the electron wavefunction in the crystal, how many states should there be in a band?

*b.* Consider the density of states function

$$g(E) = (8\pi 2^{1/2})\left(\frac{m_e}{h^2}\right)^{3/2} E^{1/2}$$

By integrating $g(E)$, estimate the total number of states in a band per unit volume, and compare this with the atomic concentration for silver. For silver, we have $E_{FO} = 5.5$ eV and $\Phi = 4.5$ eV. (Note that "state" means a distinct wavefunction, including spin.)

### SOLUTION

*a.* We know that when $N$ atoms come together to form a solid, $N$ atomic orbitals can overlap $N$ different ways to produce $N$ orbitals or $2N$ states in the crystal, since each orbital has two states, spin up and spin down. These states form the band.

*b.* For silver, $E_{FO} = 5.5$ eV and $\Phi = 4.5$ eV, so the width of the energy band is 10 eV. To estimate the total volume density of states, we assume that the density of states $g(E)$ reaches its maximum at the center of the band $E = E_{\text{center}} = 5$ eV. Integrating $g(E)$ from the bottom of the band, $E = 0$, to the center, $E = E_{\text{center}}$, yields the number of states per unit volume up to the center of the band. This is half the total number of states in the whole band, that is, $\frac{1}{2}S_{\text{band}}$, where $S_{\text{band}}$ is the number of states per unit volume in the band and is determined by

$$\frac{1}{2}S_{\text{band}} = \int_0^{E_{\text{center}}} g(E)\,dE = \frac{16\pi 2^{1/2}}{3}\left(\frac{m_e}{h^2}\right)^{3/2} E_{\text{center}}^{3/2}$$

or

$$\frac{1}{2}S_{\text{band}} = \frac{16\pi 2^{1/2}}{3}\left[\frac{9.1\times10^{-31}\text{ kg}}{(6.626\times10^{-34}\text{ J s})^2}\right]^{3/2}(5\text{ eV}\times1.6\times10^{-19}\text{ J/eV})^{3/2}$$

$$= 5.08\times10^{28}\text{ m}^{-3} = 5.08\times10^{22}\text{ cm}^{-3}$$

Thus

$$S_{\text{band}} = 10.16\times10^{22}\text{ states cm}^{-3}$$

We must now calculate the number of atoms per unit volume in silver. Given the density $d = 10.5$ g cm$^{-3}$ and the atomic mass $M_{\text{at}} = 107.9$ g mol$^{-1}$ of silver, the atomic concentration is

$$n_{\text{Ag}} = \frac{dN_A}{M_{\text{at}}} = 5.85\times10^{22}\text{ atoms cm}^{-3}$$

As expected, the density of states is almost twice the atomic concentration, even though we used a crude approximation to estimate the density of states.

# 4.6 STATISTICS: COLLECTIONS OF PARTICLES

## 4.6.1 BOLTZMANN CLASSICAL STATISTICS

Given a collection of particles in random motion and colliding with each other,[5] we need to determine the concentration of particles in the energy range $E$ to $(E + dE)$. Consider the process shown in Figure 4.24, in which two electrons with energies $E_1$ and $E_2$ interact and then move off in different directions, with energies $E_3$ and $E_4$. Let the probability of an electron having an energy $E$ be $P(E)$, where $P(E)$ is the fraction of electrons with an energy $E$. Assume there are no restrictions to the electron energies, that is, we can ignore the Pauli exclusion principle. The probability of this event is then $P(E_1)P(E_2)$. The probability of the reverse process, in which electrons with energies $E_3$ and $E_4$ interact, is $P(E_3)P(E_4)$. Since we have thermal equilibrium, that is, the system is in equilibrium, the forward process must be just as likely as the reverse process, so

$$P(E_1)P(E_2) = P(E_3)P(E_4) \tag{4.11}$$

Furthermore, the energy in this collision must be conserved, so we also need

$$E_1 + E_2 = E_3 + E_4 \tag{4.12}$$

We therefore need to find the $P(E)$ that satisfies both Equations 4.11 and 4.12. Based on our experience with the distribution of energies among gas molecules, we can guess that the solution for Equations 4.11 and 4.12 would be

*Boltzmann probability function*

$$P(E) = A\exp\left(-\frac{E}{kT}\right) \tag{4.13}$$

where $k$ is the Boltzmann constant, $T$ is the temperature, and $A$ is a constant. We can show that Equation 4.13 is a solution to Equations 4.11 and 4.12 by a simple substitution. Equation 4.13 is the **Boltzmann probability function** and is shown in Figure 4.25. The probability of finding a particle at an energy $E$ therefore decreases exponentially with energy. We assume, of course, that any number of particles may have a given energy $E$. In other words, there is no restriction such as permitting only one particle per state at an energy $E$, as in the Pauli exclusion principle. The term $kT$ appears in Equation 4.13 because the average energy as calculated by using $P(E)$ then agrees with experiments. (There is no $kT$ in Equations 4.11 and 4.12.)

Suppose that we have $N_1$ particles at energy level $E_1$ and $N_2$ particles at a higher energy $E_2$. Then, by Equation 4.13, we have

*Boltzmann statistics*

$$\frac{N_2}{N_1} = \exp\left(-\frac{E_2 - E_1}{kT}\right) \tag{4.14}$$

[5] From Chapter 1, we can associate this with the kinetic theory of gases. The energies of the gas molecules, which are moving around randomly, are distributed according to the Maxwell–Boltzmann statistics.

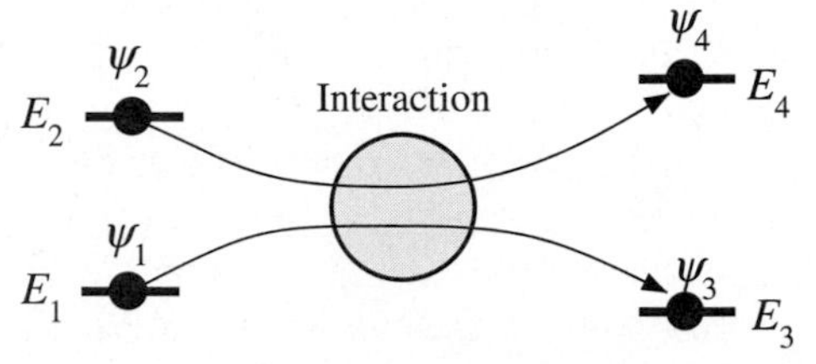


**Figure 4.24** Two electrons with initial wavefunctions $\psi_1$ and $\psi_2$ at $E_1$ and $E_2$ interact and end up at different energies $E_3$ and $E_4$. Their corresponding wavefunctions are $\psi_3$ and $\psi_4$.

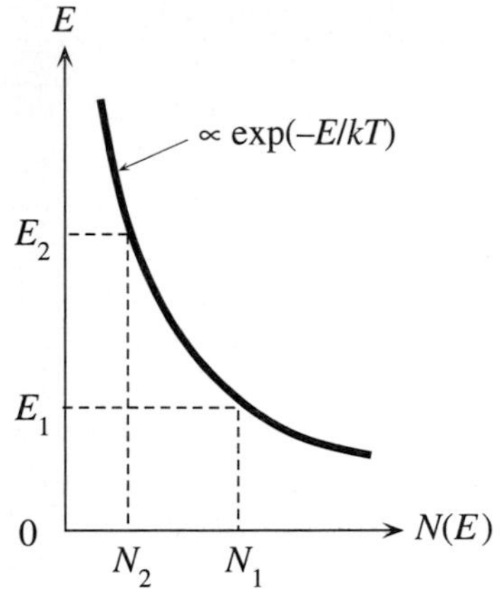


**Figure 4.25** The Boltzmann energy distribution describes the statistics of particles, such as electrons, when there are many more available states than the number of particles.

If $E_2 - E_1 \gg kT$, then $N_2$ can be orders of magnitude smaller than $N_1$. As the temperature increases, $N_2/N_1$ also increases. Therefore, increasing the temperature populates the higher energy levels.

Classical particles obey the Boltzmann statistics. Whenever there are many more states (by orders of magnitude) than the number of particles, the likelihood of two particles having the same set of quantum numbers is negligible and we do not have to worry about the Pauli exclusion principle. In these cases, we can use the Boltzmann statistics. An important example is the statistics of electrons in the conduction band of a semiconductor where, in general, there are many more states than electrons.

### 4.6.2 FERMI–DIRAC STATISTICS

Now consider the interaction for which no two electrons can be in the same quantum state, which is essentially obedience to the Pauli exclusion principle, as shown in Figure 4.24. We assume that we can have only one electron in a particular quantum state $\psi$ (including spin) associated with the energy value $E$. We therefore need those states that have energies $E_3$ and $E_4$ to be not occupied. Let $f(E)$ be the probability that an electron is in such a state, with energy $E$ in this new interaction environment. The probability of the forward event in Figure 4.24 is

$$f(E_1)f(E_2)[1 - f(E_3)][1 - f(E_4)]$$

The square brackets represent the probability that the states with energies $E_3$ and $E_4$ are empty. In thermal equilibrium, the reverse process, the electrons with $E_3$ and $E_4$ interacting to transfer to $E_1$ and $E_2$, has just as equal a likelihood as the forward process.

Paul Adrien Maurice Dirac (1902–1984) received the 1933 Nobel prize for physics with Erwin Schrödinger. His first degree was in electrical engineering from Bristol University. He obtained his PhD in 1926 from Cambridge University under Ralph Fowler.

| SOURCE: Courtesy of AIP Emilio Segrè Visual Archives.

Thus, $f(E)$ must satisfy the equation

$$f(E_1)f(E_2)[1 - f(E_3)][1 - f(E_4)] = f(E_3)f(E_4)[1 - f(E_1)][1 - f(E_2)] \quad \textbf{[4.15]}$$

In addition, for energy conservation, we must have

$$E_1 + E_2 = E_3 + E_4 \quad \textbf{[4.16]}$$

By an "intelligent guess," the solution to Equations 4.15 and 4.16 is

$$f(E) = \frac{1}{1 + A\exp\left(\dfrac{E}{kT}\right)} \quad \textbf{[4.17]}$$

where $A$ is a constant. You can check that this is a solution by substituting Equation 4.17 into 4.15 and using Equation 4.16. The reason for the term $kT$ in Equation 4.17 is not obvious from Equations 4.15 and 4.16. It appears in Equation 4.17 so that the mean properties of this system calculated by using $f(E)$ agree with experiments. Letting $A = \exp(-E_F/kT)$, we can write Equation 4.17 as

*Fermi–Dirac statistics*

$$f(E) = \frac{1}{1 + \exp\left(\dfrac{E - E_F}{kT}\right)} \quad \textbf{[4.18]}$$

where $E_F$ is a constant called the **Fermi energy.** The probability of finding an electron in a state with energy $E$ is given by Equation 4.18, which is called the **Fermi–Dirac function.**

The behavior of the Fermi–Dirac function is shown in Figure 4.26. Note the effect of temperature. As $T$ increases, $f(E)$ extends to higher energies. At energies of a few $kT$ (0.026 eV) above $E_F$, $f(E)$ behaves almost like the Boltzmann function

$$f(E) = \exp\left[-\frac{(E - E_F)}{kT}\right] \qquad (E - E_F) \gg kT \quad \textbf{[4.19]}$$

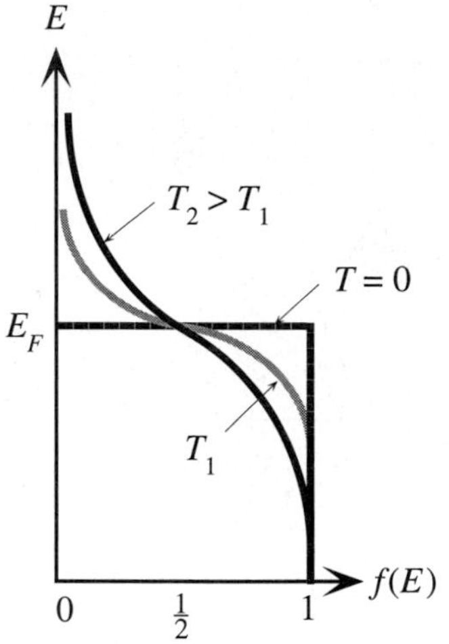


**Figure 4.26**
The Fermi–Dirac function $f(E)$ describes the statistics of electrons in a solid. The electrons interact with each other and the environment, obeying the Pauli exclusion principle.

Above absolute zero, at $E = E_F$, $f(E_F) = \frac{1}{2}$. We define the Fermi energy as that energy for which the probability of occupancy $f(E_F)$ equals $\frac{1}{2}$. The approximation to $f(E)$ in Equation 4.19 at high energies is often referred to as the **Boltzmann tail** to the Fermi–Dirac function.

# 4.7 QUANTUM THEORY OF METALS

## 4.7.1 FREE ELECTRON MODEL[6]

We know that the number of states $g(E)$ for an electron, per unit energy per unit volume, increases with energy as $g(E) \propto E^{1/2}$. We have also calculated that the probability of an electron being in a state with an energy $E$ is the Fermi–Dirac function $f(E)$. Consider the energy band diagram for a metal and the density of states $g(E)$ for that band, as shown in Figure 4.27a and b, respectively.

At absolute zero, all the energy levels up to $E_F$ are full. At 0 K, $f(E)$ has the step form at $E_F$ (Figure 4.26). This clarifies why $E_F$ in $f(E)$ is termed the Fermi energy. At 0 K, $f(E) = 1$ for $E < E_F$, and $f(E) = 0$ for $E > E_F$, so at 0 K, $E_F$ separates the empty and full energy levels. This explains why we restricted ourselves to 0 K or thereabouts when we introduced $E_F$ in the band theory of metals.

At some finite temperature, $f(E)$ is *not* zero beyond $E_F$, as indicated in Figure 4.27c. This means that some of the electrons are excited to, and thereby occupy, energy levels above $E_F$. If we multiply $g(E)$, by $f(E)$, we obtain the number of electrons per unit energy per unit volume, denoted $n_E$. The distribution of electrons in the energy levels is described by $n_E = g(E)\, f(E)$.

Since $f(E) = 1$ for $E \ll E_F$, the states near the bottom of the band are all occupied; thus, $n_E \propto E^{1/2}$ initially. As $E$ passes through $E_F$, $f(E)$ starts decreasing

[6] The free electron model of metals is also known as the Sommerfeld model.

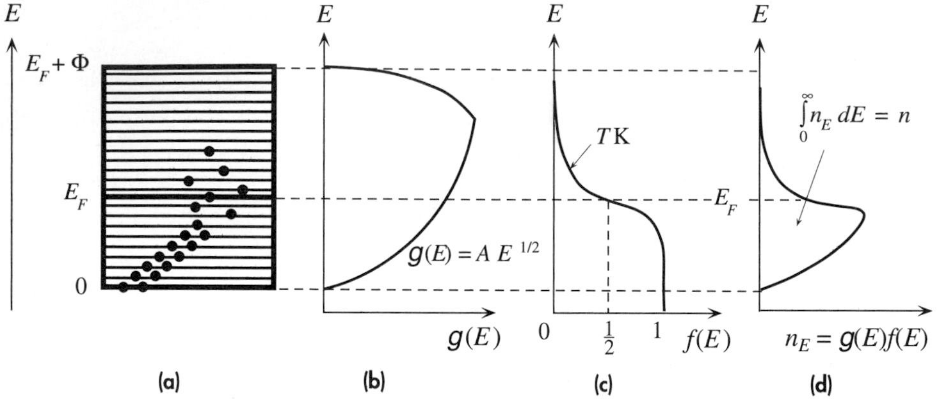


**Figure 4.27**

(a) Above 0 K, due to thermal excitation, some of the electrons are at energies above $E_F$.

(b) The density of states, $g(E)$ versus $E$ in the band.

(c) The probability of occupancy of a state at an energy $E$ is $f(E)$.

(d) The product $g(E)f(E)$ is the number of electrons per unit energy per unit volume, or the electron concentration per unit energy. The area under the curve on the energy axis is the concentration of electrons in the band.

sharply. As a result, $n_E$ takes a turn and begins to decrease sharply as well, as depicted in Figure 4.27d.

In the small energy range $E$ to $(E + dE)$, there are $n_E\, dE$ electrons per unit volume. When we sum all $n_E\, dE$ from the bottom to the top of the band ($E = 0$ to $E = E_F + \Phi$), we get the total number of valence electrons per unit volume, $n$, in the metal, as follows:

$$n = \int_0^{\text{Top of band}} n_E\, dE = \int_0^{\text{Top of band}} g(E) f(E)\, dE \qquad \textbf{[4.20]}$$

Since $f(E)$ falls very sharply when $E > E_F$, we can carry the integration to $E = \infty$, rather than to $(E_F + \Phi)$, because $f \to 0$ when $E \gg E_F$. Putting in the functional forms of $g(E)$ and $f(E)$ (e.g., from Equations 4.10 and 4.18), we obtain

$$n = \frac{8\pi 2^{1/2} m_e^{3/2}}{h^3} \int_0^\infty \frac{E^{1/2} dE}{1 + \exp\left(\dfrac{E - E_F}{kT}\right)} \qquad \textbf{[4.21]}$$

If we could integrate this, we would obtain an expression relating $n$ and $E_F$. At 0 K, however, $E_F = E_{FO}$ and the integrand exists only for $E < E_{FO}$. If we integrate at 0 K, Equation 4.21 yields

*Fermi energy at $T = 0$ K*

$$E_{FO} = \left(\frac{h^2}{8m_e}\right)\left(\frac{3n}{\pi}\right)^{2/3} \qquad \textbf{[4.22]}$$

It may be thought that $E_F$ is temperature independent, since it was sketched that way in Figure 4.26. However, in our derivation of the Fermi–Dirac statistics, there was no restriction that demanded this. Indeed, since the number of electrons in a band is fixed, $E_F$ at a temperature $T$ is implicitly determined by Equation 4.21, which can be solved to express $E_F$ in terms of $n$ and $T$. It turns out that at 0 K, $E_F$ is given by Equation 4.22, and it changes very little with temperature. In fact, by utilizing various mathematical approximations, it is not too difficult to integrate Equation 4.21 to obtain the **Fermi energy** at a temperature $T$, as follows:

$$E_F(T) = E_{FO}\left[1 - \frac{\pi^2}{12}\left(\frac{kT}{E_{FO}}\right)^2\right] \qquad [4.23]$$

*Fermi energy at T (K)*

which shows that $E_F(T)$ is only weakly temperature dependent, since $E_{FO} \gg kT$.

The Fermi energy has an important significance in terms of the average energy $E_{av}$ of the conduction electrons in a metal. In the energy range $E$ to $(E + dE)$, there are $n_E\,dE$ electrons with energy $E$. The average energy of an electron will therefore be

$$E_{av} = \frac{\int E n_E\,dE}{\int n_E\,dE} \qquad [4.24]$$

If we substitute $g(E)f(E)$ for $n_E$ and integrate, the result at 0 K is

$$E_{av}(0) = \frac{3}{5}E_{FO} \qquad [4.25]$$

*Average energy per electron at 0 K*

Above absolute zero, the **average energy** is approximately

$$E_{av}(T) = \frac{3}{5}E_{FO}\left[1 + \frac{5\pi^2}{12}\left(\frac{kT}{E_{FO}}\right)^2\right] \qquad [4.26]$$

*Average energy per electron at T (K)*

Since $E_{FO} \gg kT$, the second term in the square brackets is much smaller than unity, and $E_{av}(T)$ shows only a very weak temperature dependence. Furthermore, in our model of the metal, the electrons are free to move around within the metal, where their potential energy *PE* is zero, whereas outside the metal, it is $E_F + \Phi$ (Figure 4.11). Therefore, their energy is purely kinetic. Thus, Equation 4.26 gives the average *KE* of the electrons in a metal

$$\frac{1}{2}m_e v_e^2 = E_{av} \approx \frac{3}{5}E_{FO}$$

where $v_e$ is the root mean square (rms) speed of the electrons, which is simply called the **effective speed.** The effective speed $v_e$ depends on the Fermi energy $E_{FO}$ and is relatively insensitive to temperature. Compare this with the behavior of molecules in an ideal gas. In that case, the average $KE = \frac{3}{2}kT$, so $\frac{1}{2}mv^2 = \frac{3}{2}kT$. Clearly, the average speed of molecules in a gas increases with temperature.

The relationship $\frac{1}{2}mv_e^2 \approx \frac{3}{5}E_{FO}$ is an important conclusion that comes from the application of quantum mechanical concepts, ideas that lead to $g(E)$ and $f(E)$ and so on. It cannot be proved without invoking quantum mechanics. The fact that the average electronic speed is nearly constant is the only way to explain the observation that the resistivity of a metal is proportional to $T$ (and not $T^{3/2}$), as we saw in Chapter 2.

## 4.7.2 CONDUCTION IN METALS

We know from our energy band discussions that in metals only those electrons in a small range $\Delta E$ around the Fermi energy $E_F$ contribute to electrical conduction as shown in Figure 4.12c. The concentration $n_F$ of these electrons is approximately $g(E_F)\,\Delta E$ inasmuch as $\Delta E$ is very small. The electron $a$ moves to $a'$, as shown in Figure 4.12b and c, and then it is scattered to an empty state above $b'$. In steady conduction, all the electrons in the energy range $\Delta E$ that are moving to the right are not canceled by any moving to the left and hence contribute to the current. An electron at the bottom of the $\Delta E$ range gains energy $\Delta E$ to move $a'$ in a time interval $\Delta t$ that corresponds to the scattering time $\tau$. It gains a momentum $\Delta p_x$. Since $\Delta p_x/\Delta t =$ external force $= e\mathcal{E}_x$, we have $\Delta p_x = \tau e\mathcal{E}_x$. The electron $a$ has an energy $E = p_x^2/(2m_e^*)$ which we can differentiate to obtain $\Delta E$ when the momentum changes by $\Delta p_x$,

$$\Delta E = \frac{p_x}{m_e^*}\Delta p_x = \frac{(m_e^* v_F)}{m_e^*}(\tau e\mathcal{E}_x) = ev_F\tau\mathcal{E}_x$$

The current $J_x$ is due to all the electrons in the range $\Delta E$ which are moving toward the right in Figure 4.12c,

$$J_x = en_F v_F = e[g(E_F)\,\Delta E]v_F = e[g(E_F)ev_F\tau\mathcal{E}_x]v_F = e^2 v_F^2 \tau g(E_F)\mathcal{E}_x$$

The conductivity is therefore

$$\sigma = e^2 v_F^2 \tau g(E_F)$$

However, the numerical factor is wrong because Figure 4.12c considers only a hypothetical one-dimensional crystal. In a three-dimensional crystal, the conductivity is one-third of the conductivity value just determined:

*Conductivity of Fermi-level electrons*

$$\sigma = \frac{1}{3}e^2 v_F^2 \tau g(E_F) \qquad [4.27]$$

This conductivity expression is in sharp contrast with the classical expression in which all the electrons contribute to conduction. According to Equation 4.27, what is important is the density of states at the Fermi energy $g(E_F)$. For example, Cu and Mg are metals with valencies I and II. Classically, Cu and Mg atoms each contribute one and two conduction electrons, respectively, into the crystal. Thus, we would expect Mg to have higher conductivity. However, the Fermi level in Mg is where the top tail of the $3s$ band overlaps the bottom tail of the $3p$ band where the density of states is small. In Cu, on the other hand, $E_F$ is nearly in the middle of the $4s$ band where the density of states is high. Thus, Mg has a lower conductivity than Cu.

The scattering time $\tau$ in Equation 4.27 assumes that the scattered electrons at $E_F$ remain in the same energy band. In certain metals, there are two different energy bands that overlap at $E_F$. For example, in Ni (see Figure 4.61), $3d$ and $4s$ bands overlap at $E_F$. An electron can be scattered from the $4s$ to the $3d$ band, and vice versa. Electrons in the $3d$ band have very low drift mobilities and effectively do not contribute to conduction, so only $g(E_F)$ of the $4s$ band operates in Equation 4.27.

Since $4s$ to $3d$ band scattering is an additional scattering mechanism, by virtue of Matthiessen's rule, the scattering time $\tau$ for the $4s$ band electrons is shortened. Thus, Ni has poorer conductivity than Cu.

In deriving Equation 4.27 we did not assume a particular density of states model. If we now apply the *free electron model* for $g(E_F)$ as in Equation 4.10, and also relate $E_F$ to the total number of conduction electrons per unit volume $n$ as in Equation 4.22, we would find that the conductivity is the same as the **Drude model,** that is,

$$\sigma = \frac{e^2 n\tau}{m_e} \qquad [4.28]$$

*Drude model and free electrons*

**EXAMPLE 4.9**

**MEAN SPEED OF CONDUCTION ELECTRONS IN A METAL** Calculate the Fermi energy $E_{FO}$ at 0 K for copper and estimate the average speed of the conduction electrons in Cu. The density of Cu is 8.96 g cm$^{-3}$ and the relative atomic mass (atomic weight) is 63.5.

**SOLUTION**

Assuming each Cu atom donates one free electron, we can find the concentration of electrons from the density $d$, atomic mass $M_{at}$, and Avogadro's number $N_A$, as follows:

$$n = \frac{dN_A}{M_{at}} = \frac{8.96 \times 6.02 \times 10^{23}}{63.5}$$

$$= 8.5 \times 10^{22}\ \text{cm}^{-3} \quad \text{or} \quad 8.5 \times 10^{28}\ \text{m}^{-3}$$

The Fermi energy at 0 K is given by Equation 4.22:

$$E_{FO} = \left(\frac{h^2}{8m_e}\right)\left(\frac{3n}{\pi}\right)^{2/3}$$

Substituting $n = 8.5 \times 10^{28}$ m$^{-3}$ and the values for $h$ and $m_e$, we obtain

$$E_{FO} = 1.1 \times 10^{-18}\ \text{J} \quad \text{or} \quad 7\ \text{eV}$$

To estimate the mean speed of the electrons, we calculate the rms speed $v_e$ from $\frac{1}{2}m_e v_e^2 = \frac{3}{5}E_{FO}$. The mean speed will be close to the rms speed. Thus, $v_e = (6E_{FO}/5m_e)^{1/2}$. Substituting for $E_{FO}$ and $m_e$, we find $v_e = 1.2 \times 10^6$ m s$^{-1}$.

**EXAMPLE 4.10**

**CONDUCTION IN SILVER** Consider silver whose density of states $g(E)$ was calculated in Example 4.8, assuming a *free electron model* for $g(E)$ as in Equation 4.10. For silver, $E_F = 5.5$ eV, so from Equation 4.10, the density of states at $E_F$ is $g(E_F) = 1.60 \times 10^{28}$ m$^{-3}$ eV$^{-1}$. The velocity of Fermi electrons, $v_F = (2E_F/m_e)^{1/2} = 1.39 \times 10^6$ m s$^{-1}$. The conductivity $\sigma$ of Ag at room temperature is $62.5 \times 10^6\ \Omega^{-1}$ m$^{-1}$. Substituting for $\sigma$, $g(E_F)$, and $v_F$ in Equation 4.27,

$$\sigma = 62.5 \times 10^6 = \frac{1}{3}e^2 v_F^2 \tau g(E_F) = \frac{1}{3}(1.6 \times 10^{-19})^2(1.39 \times 10^6)^2\tau\left(\frac{1.60 \times 10^{28}}{1.6 \times 10^{-19}}\right)$$

we find $\tau = 3.79 \times 10^{-14}$ s. The *mean free path* $\ell = v_F\tau = 53$ nm. The *drift mobility* of $E_F$ electrons is $\mu = e\tau/m_e = 67$ cm$^2$ V$^{-1}$ s$^{-1}$.

From Example 4.8, since Ag has a valency of I, the concentration of conduction electrons is $n = n_{Ag} = 5.85 \times 10^{28}$ m$^{-3}$. Substituting for $n$ and $\sigma$ in Equation 4.28 gives

$$\sigma = 62.5 \times 10^6 = \frac{e^2 n \tau}{m_e} = \frac{(1.6 \times 10^{-19})^2 (5.85 \times 10^{28}) \tau}{(9.1 \times 10^{-31})}$$

we find $\tau = 3.79 \times 10^{-14}$ s as expected because we have used the free electron model.

## 4.8 FERMI ENERGY SIGNIFICANCE

### 4.8.1 METAL–METAL CONTACTS: CONTACT POTENTIAL

Suppose that two metals, platinum (Pt) with a work function 5.36 eV and molybdenum (Mo) with a work function 4.20 eV, are brought together, as shown in Figure 4.28a. We know that in metals, all the energy levels up to the Fermi level are full. Since the Fermi level is higher in Mo (due to a smaller $\Phi$), the electrons in Mo are more energetic. They therefore immediately go over to the Pt surface (by tunneling), where there are empty states at lower energies, which they can occupy. This electron transfer from Mo to the Pt surface reduces the total energy of the electrons in the Pt–Mo system, but at the same time, the Pt surface becomes negatively charged with respect to the Mo surface. Consequently, a contact voltage (or a potential difference) develops at the junction between Pt and Mo, with the Mo side being positive.

The electron transfer from Mo to Pt continues until the contact potential is large enough to prevent further electron transfer: the system reaches equilibrium. It should be apparent that the transfer of energetic electrons from Mo to Pt continues until the two Fermi levels are lined up, that is, until the Fermi level is uniform and the same in both metals, so that no part of the system has more (or less) energetic electrons, as

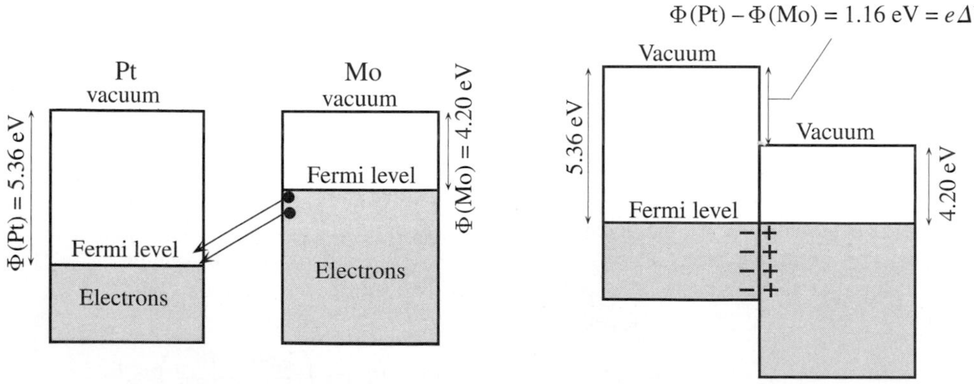


**(a) Electrons are more energetic in Mo, so they tunnel to the surface of Pt.**

**(b) Equilibrium is reached when the Fermi levels are lined up.**

**Figure 4.28** When two metals are brought together, there is a contact potential $\Delta V$.

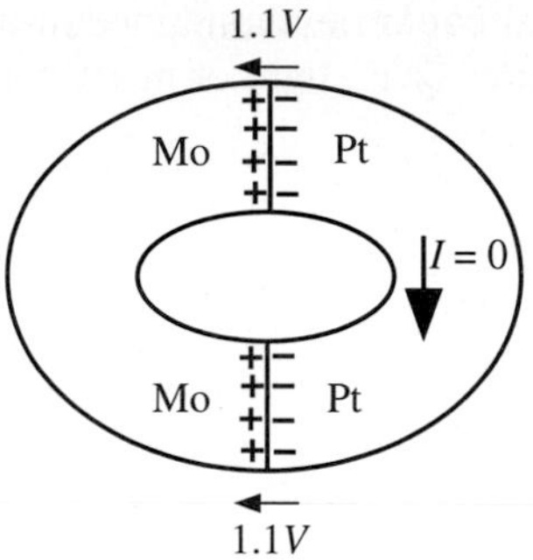


**Figure 4.29** There is no current when a closed circuit is formed by two different metals, even though there is a contact potential at each contact.

The contact potentials oppose each other.

illustrated in Figure 4.28b. Otherwise, the energetic electrons in one part of the system will flow toward a region with lower energy states. Under these conditions, the Pt–Mo system is in equilibrium. The contact voltage $\Delta V$ is determined by the difference in the work functions, that is,

$$e\,\Delta V = \Phi(\text{Pt}) - \Phi(\text{Mo}) = 5.36\,\text{eV} - 4.20\,\text{eV} = 1.16\,\text{eV}$$

We should note that away from the junction on the Mo side, we must still provide an energy of $\Phi = 4.20$ eV to free an electron, whereas away from the junction on the Pt side, we must provide $\Phi = 5.36$ eV to free an electron. This means that the vacuum energy level going from Mo to Pt has a step $\Delta\Phi$ at the junction. Since we must do work equivalent to $\Delta\Phi$ to get a free electron (*e.g.*, on the metal surface) from the Mo surface to the Pt surface, this represents a voltage of $\Delta\Phi/e$ or 1.16 V.

From the second law of thermodynamics,[7] this contact voltage cannot do work; that is, it cannot drive current in an external circuit. To see this, we can close the Pt metal–Mo metal circuit to form a ring, as depicted in Figure 4.29. As soon as we close the circuit, we create another junction with a contact voltage that is equal and opposite to that of the first junction. Consequently, going around the circuit, the net voltage is zero and the current is therefore zero.

There is a deep significance to the Fermi energy $E_F$, which should at least be mentioned. For a given metal the Fermi energy represents the free energy per electron called the **electrochemical potential** $\mu$. In other words, the Fermi energy is a measure of the potential of an electron to do electrical work ($e \times V$) or nonmechanical work, through chemical or physical processes.[8] In general, when two metals are brought into contact, the Fermi level (with respect to a vacuum) in each will be different. This difference means a difference in the chemical potential $\Delta\mu$, which in turn means that the system will do external work, which is obviously not possible. Instead, electrons are immediately transferred from one metal to the other, until the free energy per electron $\mu$ for the whole system is minimized and is uniform across the two metals, so that

[7] By the way, the second law of thermodynamics simply says that you cannot extract heat from a system in thermal equilibrium and do work (*i.e.*, charge × voltage).

[8] A change in any type of *PE* can, in principle, be used to do work, that is, $\Delta(PE)$ = work done. Chemical *PE* is the potential to do nonmechanical work (*e.g.*, electrical work) by virtue of physical or chemical processes. The chemical *PE* per electron is $E_F$ and $\Delta E_F$ = electrical work per electron.

$\Delta\mu = 0$. We can guess that if the Fermi level in one metal could be maintained at a higher level than the other, by using an external energy source (*e.g.*, light or heat), for example, then the difference could be used to do electrical work.

### 4.8.2 THE SEEBECK EFFECT AND THE THERMOCOUPLE

Consider a conductor such as an aluminum rod that is heated at one end and cooled at the other end as depicted in Figure 4.30. The electrons in the hot region are more energetic and therefore have greater velocities than those in the cold region.[9]

Consequently there is a net diffusion of electrons from the hot end toward the cold end which leaves behind exposed positive metal ions in the hot region and accumulates electrons in the cold region. This situation prevails until the electric field developed between the positive ions in the hot region and the excess electrons in the cold region prevents further electron motion from the hot to the cold end. A voltage therefore develops between the hot and cold ends, with the hot end at positive potential. The potential difference $\Delta V$ across a piece of metal due to a temperature difference $\Delta T$ is called the **Seebeck effect.**[10] To gauge the magnitude of this effect we introduce a special coefficient which is defined as the potential difference developed per unit temperature difference, or

*Thermo-electric power or Seebeck coefficient*

$$S = \frac{dV}{dT} \qquad [4.29]$$

By convention, the sign of $S$ represents the potential of the cold side with respect to the hot side. If electrons diffuse from the hot end to the cold end as in Figure 4.30, then the cold side is negative with respect to the hot side and the Seebeck coefficient is *negative* (as for aluminum).

In some metals, such as copper, this intuitive explanation fails to explain why electrons actually diffuse from the cold to the hot region, giving rise to *positive* Seebeck coefficients; the polarity of the voltage in Figure 4.30 is actually reversed for copper. The reason is that the net diffusion process depends on how the mean free path $\ell$ and the mean free time (due to scattering from lattice vibrations) change with the electron energy, which can be quite complicated. Typical Seebeck coefficients for various selected metals are listed in Table 4.3.

Consider two neighboring regions H (hot) and C (cold) with widths corresponding to the mean free paths $\ell$ and $\ell'$ in H and C as depicted in Figure 4.31a. Half the electrons in H would be moving in the $+x$ direction and the other half in the $-x$ direction. Half of the electrons in H therefore cross into C, and half in C cross into H. Suppose that, very roughly, the electron concentration $n$ in H and C is about the same. The number of electrons crossing from H to C is $\frac{1}{2}n\ell$, and the number crossing from C to H is $\frac{1}{2}n\ell'$. Then,

$$\text{Net diffusion from H to C} \propto \tfrac{1}{2}n(\ell - \ell') \qquad [4.30]$$

[9] The conduction electrons around the Fermi energy have a mean speed that has only a small temperature dependence. This small change in the mean speed with temperature is, nonetheless, intuitively significant in appreciating the thermoelectric effect. The actual effect, however, depends on the mean free path as discussed later.

[10] Thomas Seebeck observed the thermoelectric effect in 1821 using two different metals as in the thermocouple, which is the only way to observe the phenomenon. It was William Thomson (Lord Kelvin) who explained the observed effect.

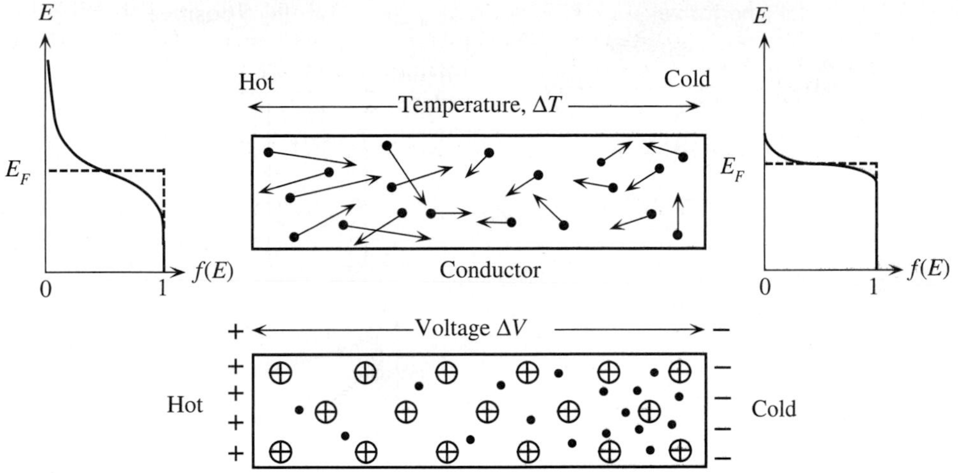


**Figure 4.30** The Seebeck effect.
A temperature gradient along a conductor gives rise to a potential difference.

Suppose that the scattering of electrons is such that $\ell$ increases strongly with the electron energy. Then electrons in H, which are more energetic, have a longer mean free path, that is, $\ell > \ell'$ as shown in Figure 4.31a. This means that the net migration is from H to C and $S$ is negative, as in aluminum. In those metals such as copper in which $\ell$ decreases strongly with the energy, electrons in the cold region have a longer mean free path, $\ell' > \ell$ as shown in Figure 4.31b. The net electron migration is then from C to H and $S$ is positive. Even this qualitative explanation is not quite correct because $n$ is not the same in H and C (diffusion changes $n$) and, further, we neglected the change in the mean scattering time with the electron energy.

The coefficient $S$ is widely referred to as the **thermoelectric power** even though this term is misleading, as it refers to a voltage difference rather than power. A more appropriate recent term is the **Seebeck coefficient.** $S$ is a material property that depends on temperature, $S = S(T)$, and is tabulated for many materials as a function of

**Table 4.3** Seebeck coefficients of selected metals (from various sources)

| Metal | $S$ at 0 °C ($\mu$V K$^{-1}$) | $S$ at 27 °C ($\mu$V K$^{-1}$) | $E_F$ (eV) | $x$ |
|---|---|---|---|---|
| Al | −1.6 | −1.8 | 11.6 | 2.78 |
| Au | +1.79 | +1.94 | 5.5 | −1.48 |
| Cu | +1.70 | +1.84 | 7.0 | −1.79 |
| K | | −12.5 | 2.0 | 3.8 |
| Li | +14 | | 4.7 | −9.7 |
| Mg | −1.3 | | 7.1 | 1.38 |
| Na | | −5 | 3.1 | 2.2 |
| Pd | −9.00 | −9.99 | | |
| Pt | −4.45 | −5.28 | | |

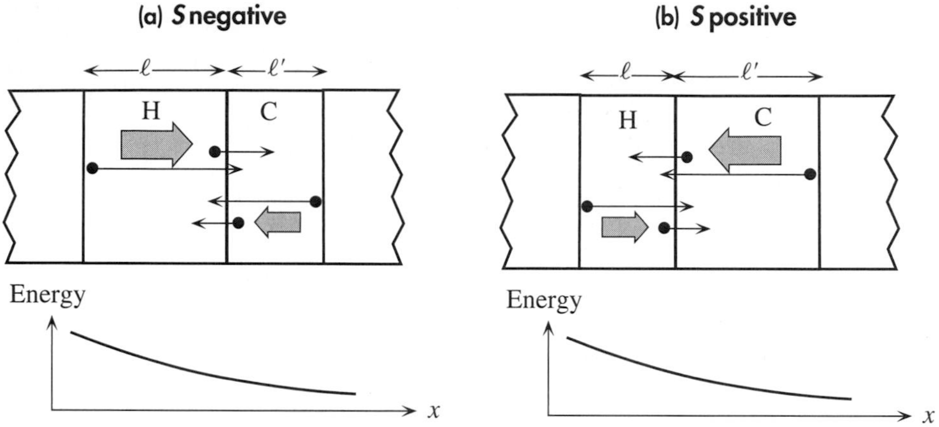


**Figure 4.31** Consider two neighboring regions H (hot) and C (cold) with widths corresponding to the mean free paths $\ell$ and $\ell'$ in H and C.
Half the electrons in H would be moving in the $+x$ direction and the other half in the $-x$ direction. Half of the electrons in H therefore cross into C, and half in C cross into H.

temperature. Given the Seebeck coefficient $S(T)$ for a material, Equation 4.29 yields the voltage difference between two points where temperatures are $T_o$ and $T$ as follows:

$$\Delta V = \int_{T_o}^{T} S\,dT \quad [4.31]$$

A proper explanation of the Seebeck effect has to consider how electrons around the Fermi energy $E_F$, which contribute to electrical conduction, are scattered by lattice vibrations, impurities, and crystal defects. This scattering process controls the mean free path and hence the Seebeck coefficient (Figure 4.31). The scattered electrons need empty states, which in turn requires that we consider how the density of states changes with the energy as well. Moreover, in certain metals such as Ni, there are overlapping partially filled bands and the Fermi electron can be scattered from one electronic band to another, for example from the 4*s* band to the 3*d* band, which must also be considered (see Question 4.25). The Seebeck coefficient for many metals is given by the **Mott and Jones equation,**

*Mott and Jones thermoelectric power*

$$S \approx -\frac{\pi^2 k^2 T}{3eE_{FO}} x \quad [4.32]$$

where $x$ is a numerical constant that takes into account how various charge transport parameters (such as $\ell$) depend on the electron energy. A few examples for $x$ are given in Table 4.3. The reason for the $kT/E_{FO}$ factor in Equation 4.32 is that only those electrons about a $kT$ around the Fermi level $E_{FO}$ are involved in the transport and scattering processes. Equation 4.32 does not apply directly to transition metals (Ni, Pd, Pt) that have overlapping bands. These metals have a negative Seebeck coefficient that is proportional to temperature as in Equation 4.32, but the exact expression depends on the band structure.

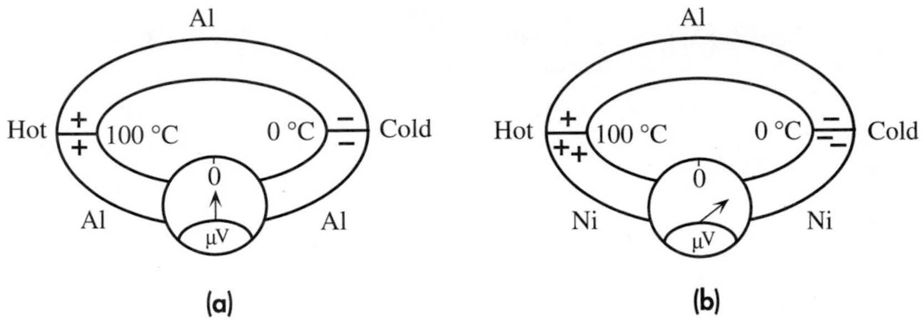


**Figure 4.32**

(a) If Al wires are used to measure the Seebeck voltage across the Al rod, then the net emf is zero.

(b) The Al and Ni have different Seebeck coefficients. There is therefore a net emf in the Al–Ni circuit between the hot and cold ends that can be measured.

Suppose that we try to measure the voltage difference $\Delta V$ across the aluminum rod by using aluminum connecting wires to a voltmeter as indicated in Figure 4.32a. The same temperature difference now also exists across the aluminum connecting wires; therefore an identical voltage also develops across the connecting wires, opposing that across the aluminum rod. Consequently no net voltage will be registered by the voltmeter. It is, however, possible to read a net voltage difference, if the connecting wires are of different material, that is, have a different Seebeck coefficient from that of aluminum. Then the thermoelectric voltage across this material is different than that across the aluminum rod, as in Figure 4.32b.

The Seebeck effect is fruitfully utilized in the thermocouple (TC), shown in Figure 4.32b, which uses two different metals with one junction maintained at a reference temperature $T_o$ and the other used to sense the temperature $T$. The voltage across each metal element depends on its Seebeck coefficient. The potential difference between the two wires will depend on $S_A - S_B$. By virtue of Equation 4.31, the electromotive force (emf) between the two wires, $V_{AB} = \Delta V_A - \Delta V_B$, is then given by

*Thermocouple emf between metals A and B*

$$V_{AB} = \int_{T_o}^{T} (S_A - S_B)\, dT = \int_{T_o}^{T} S_{AB}\, dT \qquad [4.33]$$

where $S_{AB} = S_A - S_B$ is defined as the thermoelectric power for the thermocouple pair $A$-$B$. For the chromel-alumel (K-type) TC, for example, $S_{AB} \approx 40\ \mu\text{V K}^{-1}$ at 300 K.

The output voltage from a TC pair obviously depends on the two metals used. Instead of tabulating the emf from all possible pairs of materials in the world, which would be a challenging task, engineers have tabulated the emfs available when a given material is used with a reference metal which is chosen to be platinum. The reference junction is kept at 0 °C (273.16 K) which corresponds to a mixture of ice and water. Some typical materials and their emfs are listed in Table 4.4.

Using the expression for the Seebeck coefficient, Equation 4.32, in Equation 4.33, and then integrating, leads to the familiar thermocouple equation,

*Thermocouple equation*

$$V_{AB} = a\, \Delta T + b(\Delta T)^2 \qquad [4.34]$$

**Table 4.4** Thermoelectric emf for metals at 100 and 200 °C with respect to Pt and the reference junction at 0 °C

| Material | emf (mV) At 100 °C | At 200 °C |
|---|---|---|
| Copper, Cu | 0.76 | 1.83 |
| Aluminum, Al | 0.42 | 1.06 |
| Nickel, Ni | −1.48 | −3.10 |
| Palladium, Pd | −0.57 | −1.23 |
| Platinum, Pt | 0 | 0 |
| Silver, Ag | 0.74 | 1.77 |
| Alumel | −1.29 | −2.17 |
| Chromel | 2.81 | 5.96 |
| Constantan | −3.51 | −7.45 |
| Iron, Fe | 1.89 | 3.54 |
| 90% Pt–10% Rh (platinum-rhodium) | 0.643 | 1.44 |

where $a$ and $b$ are the thermocouple coefficients and $\Delta T = T - T_o$ is the temperature with respect to the reference temperature $T_o$ (273.16 K). The inference from Equation 4.34 is that the emf output from the thermocouple wires does not depend linearly on the temperature difference $\Delta T$. Figure 4.33 shows the emf output versus temperature for various thermocouples. It should be immediately obvious that the voltages are small, typically a few tens of a microvolt per degree temperature difference. At

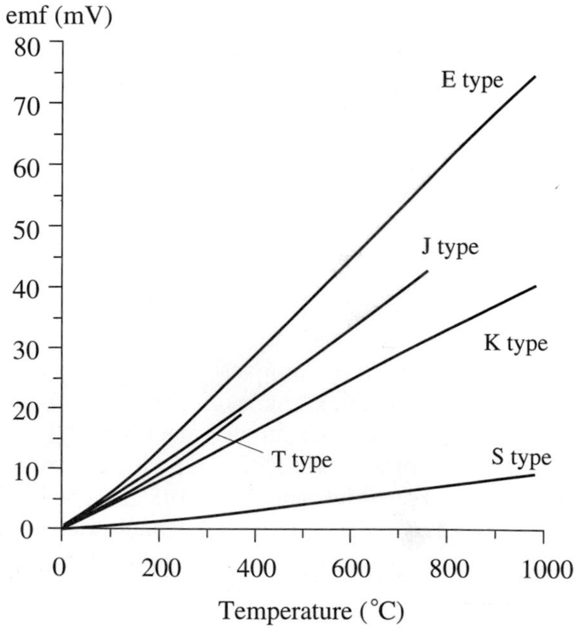


**Figure 4.33** Output emf versus temperature (°C) for various thermocouples between 0 to 1000 °C.

0 °C, by definition, the TC emf is zero. The K-type thermocouple, the chromel-alumel pair, is a widely employed general-purpose thermocouple sensor up to about 1200 °C.

**EXAMPLE 4.11**

**THE THERMOCOUPLE EMF** Consider a thermocouple pair from Al and Cu which have Fermi energies and $x$ as in Table 4.3. Estimate the emf available from this thermocouple if one junction is held at 0 °C and the other at 100 °C.

**SOLUTION**

We essentially have the arrangement shown in Figure 4.32b but with Cu replacing Ni and Cu having the cold end positive ($S$ is positive). For each metal there will be a voltage across it, given by integrating the Seebeck coefficient from $T_o$ (at the low temperature end) to $T$. From the Mott and Jones equation,

$$\Delta V = \int_{T_o}^{T} S\, dT = \int_{T_o}^{T} -\frac{x\pi^2 k^2 T}{3eE_{FO}}\, dT = -\frac{x\pi^2 k^2}{6eE_{FO}}\left(T^2 - T_o^2\right)$$

The available emf ($V_{AB}$) is the difference in $\Delta V$ for the two metals ($A$ and $B$), so

$$V_{AB} = \Delta V_A - \Delta V_B = -\frac{\pi^2 k^2}{6e}\left[\frac{x_A}{E_{FAO}} - \frac{x_B}{E_{FBO}}\right]\left(T^2 - T_o^2\right)$$

where in this example $T = 373$ K and $T_o = 273$ K.

For Al ($A$), $E_{FAO} = 11.6$ eV, $x_A = 2.78$, and for copper ($B$), $E_{FBO} = 7.0$ eV, $x_B = -1.79$. Thus,

$$V_{AB} = -189\ \mu\text{V} - (+201\ \mu\text{V}) = -390\ \mu\text{V}$$

Thermocouple emf calculations that closely represent experimental observations require thermocouple voltages for various metals listed against some reference metal. The reference is usually Pt with the reference junction at 0 °C. From Table 4.4 we can read Al–Pt and Cu–Pt emfs as $V_{\text{Al-Pt}} = 0.42$ mV and $V_{\text{Cu-Pt}} = 0.76$ mV at 100 °C with the experimental error being around ±0.01 mV, so that for the Al–Cu pair,

$$V_{\text{Al-Cu}} = V_{\text{Al-Pt}} - V_{\text{Cu-Pt}} = 0.42\ \text{mV} - 0.76\ \text{mV} = -0.34\ \text{mV}$$

There is a reasonable agreement with the calculation using the Mott and Jones equation.

**EXAMPLE 4.12**

**THE THERMOCOUPLE EQUATION** We know that we can only measure differences between thermoelectric powers of materials. When two different metals $A$ and $B$ are connected to make a thermocouple, as in Figure 4.32b, then the net emf is the voltage difference between the two elements. From Example 4.11,

$$\Delta V_{AB} = \Delta V_A - \Delta V_B = \int_{T_o}^{T} (S_A - S_B)\, dT = \int_{T_o}^{T} S_{AB}\, dT$$

$$= -\frac{\pi^2 k^2}{6e}\left[\frac{x_A}{E_{FAO}} - \frac{x_B}{E_{FBO}}\right]\left(T^2 - T_o^2\right)$$

$$= C\left(T^2 - T_o^2\right)$$

where $C$ is a constant that is independent of $T$ but dependent on the material properties ($x$, $E_{FO}$ for the metals).

We can now expand $V_{AB}$ about $T_o$ by using Taylor's expansion

$$F(T) \approx F(T_o) + \Delta T\ (dF/dT)_o + \tfrac{1}{2}(\Delta T)^2(d^2F/dT^2)_o$$

where the function $F = V_{AB}$ and $\Delta T = T - T_o$ and the derivatives are evaluated at $T_o$. The result is the thermocouple equation:

$$V_{AB}(T) = a(\Delta T) + b(\Delta T)^2$$

where the coefficients $a$ and $b$ are $2CT_o$ and $C$, respectively.

It is clear that the magnitude of the emf produced depends on $C$ or $S_A - S_B$, which we can label as $S_{AB}$. The greater the thermoelectric power difference $S_{AB}$ for the TC, the larger the emf produced. For the copper constantan TC, $S_{AB}$ is about 43 $\mu$V K$^{-1}$.

## 4.9 THERMIONIC EMISSION AND VACUUM TUBE DEVICES

### 4.9.1 Thermionic Emission: Richardson–Dushman Equation

Even though most of us view vacuum tubes as electrical antiques, their basic principle of operation (electrons emitted from a heated cathode) still finds application in cathode ray and X-ray tubes and various RF microwave vacuum tubes, such as triodes, tetrodes, klystrons, magnetrons, and traveling wave tubes and amplifiers. Therefore, it is useful to examine how electrons are emitted when a metal is heated.

When a metal is heated, the electrons become more energetic as the Fermi–Dirac function extends to higher temperatures. Some of the electrons have sufficiently large energies to leave the metal and become free. This situation is self-limiting because as the electrons accumulate outside the metal, they prevent more electrons from leaving the metal. (Put differently, emitted electrons leave a net positive charge behind, which pulls the electrons in.) Consequently, we need to replenish the "lost" electrons and collect the emitted ones, which is done most conveniently using the vacuum tube arrangement in a closed circuit, as shown in Figure 4.34a. The cathode, heated by a filament, emits electrons. A battery connected between the cathode and the anode replenishes

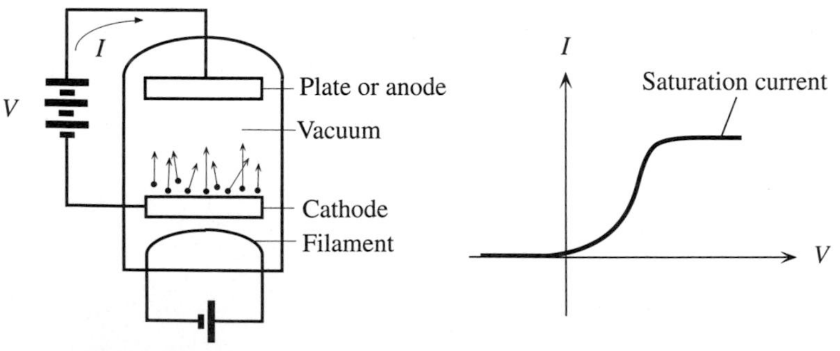


**(a) Thermionic electron emission in a vacuum tube.**

**(b) Current-voltage characteristics of a vacuum diode.**

**Figure 4.34**

the cathode electrons and provides a positive bias to the anode to collect the thermally emitted electrons from the cathode. The vacuum inside the tube ensures that the electrons do not collide with the air molecules and become dispersed, with some even being returned to the cathode by collisions. Therefore, the vacuum is essential. The current due to the flow of emitted electrons from the cathode to the anode depends on the anode voltage as indicated in Figure 4.34b. The current increases with the anode voltage until, at sufficiently high voltages, all the emitted electrons are collected by the anode and the current *saturates*. The **saturation current** of the vacuum diode depends on the rate of thermionic emission of electrons which we will derive below. The vacuum tube in Figure 4.34a acts as a **rectifier** because there is no current flow when the anode voltage becomes negative; the anode then repels the electrons.

We know that only those electrons with energies greater than $E_F + \Phi$ (Fermi energy + work function) which are moving toward the surface can leave the metal. Their number depends on the temperature, by virtue of the Fermi–Dirac statistics. Figure 4.35 shows how the concentration of conduction electrons with energies above $E_F + \Phi$ increases with temperature. We know that conduction electrons behave as if they are free within the metal. We can therefore take the *PE* to be zero within the metal, but $E_F + \Phi$ outside the metal. The energy $E$ of the electron within the metal is then purely kinetic, or

$$E = \frac{1}{2}m_e v_x^2 + \frac{1}{2}m_e v_y^2 + \frac{1}{2}m_e v_z^2 \tag{4.35}$$

Suppose that the surface of the metal is perpendicular to the direction of emission, say along $x$. For an electron to be emitted from the surface, its $KE = \frac{1}{2}mv_x^2$ along $x$ must be greater than the potential energy barrier $E_F + \Phi$, that is,

$$\frac{1}{2}mv_x^2 > E_F + \Phi \tag{4.36}$$

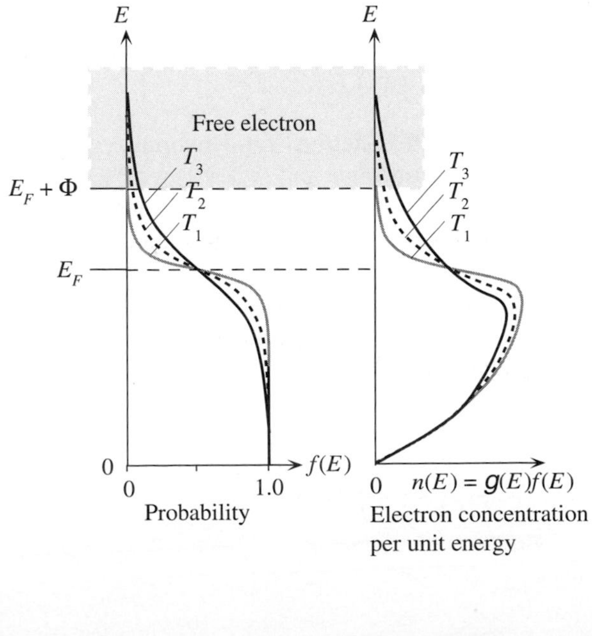


**Figure 4.35** Fermi–Dirac function $f(E)$ and the energy density of electrons $n(E)$ (electrons per unit energy and per unit volume) at three different temperatures.

The electron concentration extends more and more to higher energies as the temperature increases. Electrons with energies in excess of $E_F + \Phi$ can leave the metal (thermionic emission).

Left to right: Owen Williams Richardson, Robert Andrews Millikan, and Arthur Holly Compton at an international conference on nuclear physics, Rome, 1931. Richardson won the physics Nobel prize in 1928 for thermionic emission.

SOURCE: Amaldi Archives, Dipartimento di Fisica, Università La Sapienza, Rome; courtesy of AIP Emilio Segrè Visual Archives.

Let $dn(v_x)$ be the number of electrons moving along $x$ with velocities in the range $v_x$ to $(v_x + dv_x)$, with $v_x$ satisfying emission in Equation 4.36. These electrons will be emitted when they reach the surface. Their number $dn(v_x)$ can be determined from the density of states and the Fermi–Dirac statistics, since energy and velocity are related through Equation 4.35. Close to $E_F + \Phi$, the Fermi–Dirac function will approximate the Boltzmann distribution, $f(E) = \exp[-(E - E_F)/kT]$. The number $dn(v_x)$ is therefore at least proportional to this exponential energy factor.

The emission of $dn(v_x)$ electrons will give a thermionic current density $dJ_x = ev_x\, dn(v_x)$. This must be integrated (summed) for all velocities satisfying Equation 4.36 to obtain the total current density $J_x$, or simply $J$. Since $dn(v_x)$ includes an exponential energy function, the integration also leads to an exponential. The final result is

*Richardson–Dushman thermionic emission equation*

$$J = B_o T^2 \exp\left(-\frac{\Phi}{kT}\right) \qquad [4.37]$$

where $B_o = 4\pi e m_e k^2/h^3$. Equation 4.37 is called the **Richardson–Dushman equation,** and $B_o$ is the Richardson–Dushman constant, whose value is $1.20 \times 10^6$ A m$^{-2}$ K$^{-2}$. We see from Equation 4.37 that the emitted current from a heated cathode varies exponentially with temperature and is sensitive to the work function $\Phi$ of the cathode material. Both factors are apparent in Equation 4.37.

The wave nature of electrons means that when an electron approaches the surface, there is a probability that it may be reflected back into the metal, instead of being emitted over the potential barrier. As the potential energy barrier becomes very large, $\Phi \rightarrow \infty$, the electrons are totally reflected and there is no emission. Taking into account that waves can be reflected, the thermionic emission equation is appropriately modified to

*Thermionic emission*

$$J = B_e T^2 \exp\left(-\frac{\Phi}{kT}\right) \qquad [4.38]$$

where $B_e = (1 - R)B_o$ is the **emission constant** and $R$ is the reflection coefficient. The value of $R$ will depend on the material and the surface conditions. For most metals, $B_e$ is about half of $B_o$, whereas for some oxide coatings on Ni cathodes used in thermionic tubes, $B_e$ can be as low as $1 \times 10^2$ A m$^{-2}$ K$^{-2}$.

Equation 4.37 was derived by neglecting the effect of the applied field on the emission process. Since the anode is positively biased with respect to the cathode, the field will not only collect the emitted electrons (by drifting them to the anode), but will also enhance the process of thermal emission by lowering the potential energy barrier $\Phi$.

There are many thermionic emission–based vacuum tubes that find applications in which it is not possible or practical to use semiconductor devices, especially at high-power and high-frequency operation at the same time, such as in radio and TV broadcasting, radars, microwave communications; for example, a tetrode vacuum tube in radio broadcasting equipment has to handle hundreds of kilowatts of power. X-ray tubes operate on the thermionic emission principle in which electrons are thermally emitted, and then accelerated and impacted on a metal target to generate X-ray photons.

---

**EXAMPLE 4.13**

**VACUUM TUBES** It is clear from the Richardson–Dushman equation that to obtain an efficient thermionic cathode, we need high temperatures and low work functions. Metals such as tungsten (W) and tantalum (Ta) have high melting temperatures but high work functions. For example, for W, the melting temperature $T_m$ is 3680 °C and its work function is about 4.5 eV. Some metals have low work functions, but also low melting temperatures, a typical example being Cs with $\Phi = 1.8$ eV and $T_m = 28.5$ °C. If we use a thin film coating of a low $\Phi$ material, such as ThO or BaO, on a high-melting-temperature base metal such as W, we can maintain the high melting properties and obtain a lower $\Phi$. For example, Th on W has a $\Phi = 2.6$ eV and $T_m = 1845$ °C. Most vacuum tubes use indirectly heated cathodes that consist of the oxides of B, Sr, and Ca on a base metal of Ni. The operating temperatures for these cathodes are typically 800 °C.

A certain transmitter-type vacuum tube has a cylindrical Th-coated W (thoriated tungsten) cathode, which is 4 cm long and 2 mm in diameter. Estimate the saturation current if the tube is operated at a temperature of 1600 °C, given that the emission constant is $B_e = 3.0 \times 10^4$ A m$^{-2}$ K$^{-2}$ for Th on W.

**SOLUTION**

We apply the Richardson–Dushman equation with $\Phi = 2.6$ eV, $T = (1600 + 273)$ K $= 1873$ K, and $B_e = 3.0 \times 10^4$ A m$^{-2}$ K$^{-2}$, to find the maximum current density that can be obtained from the cathode at 1873 K, as follows:

$$J = (3.0 \times 10^4 \text{ A m}^{-2} \text{ K}^{-2})(1873 \text{ K})^2 \exp\left[-\frac{(2.6 \times 1.6 \times 10^{-19})}{(1.38 \times 10^{-23} \times 1873)}\right]$$

$$= 1.08 \times 10^4 \text{ A m}^{-2}$$

The emission surface area is

$$A = \pi(\text{diameter})(\text{length}) = \pi(2 \times 10^{-3})(4 \times 10^{-2}) = 2.5 \times 10^{-4} \text{ m}^2$$

so the saturation current, which is the maximum current obtainable (*i.e.*, the thermionic current), is

$$I = JA = (1.08 \times 10^4 \text{ A m}^{-2})(2.5 \times 10^{-4} \text{ m}^2) = 2.7 \text{ A}$$

---

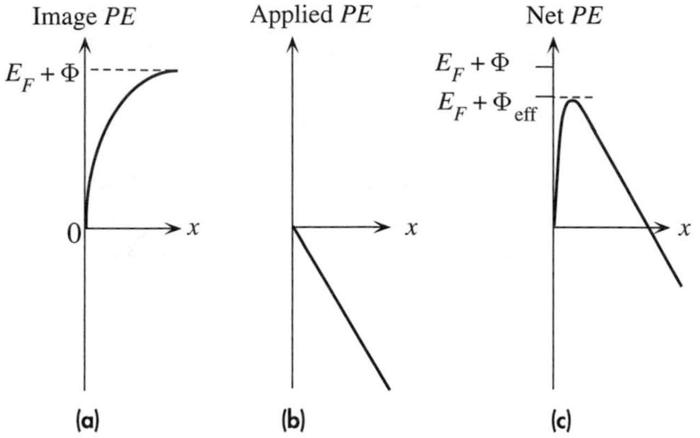


**Figure 4.36**
(a) *PE* of the electron near the surface of a conductor.
(b) Electron *PE* due to an applied field, that is, between cathode and anode.
(c) The overall *PE* is the sum.

## 4.9.2 SCHOTTKY EFFECT AND FIELD EMISSION

When a positive voltage is applied to the anode with respect to the cathode, the electric field at the cathode helps the thermionic emission process by lowering the *PE* barrier $\Phi$. This is called the **Schottky effect.** Consider the *PE* of the electron just outside the surface of the metal. The electron is pulled in by the effective positive charge left in the metal. To represent this attractive *PE* we use the **theorem of image charges** in electrostatics,[11] which says that an electron at a distance $x$ from the surface of a conductor possesses a potential energy that is

$$PE_{\text{image}}(x) = -\frac{e^2}{16\pi\varepsilon_o x} \qquad \textbf{[4.39]}$$

where $\varepsilon_o$ is the absolute permittivity.

This equation is valid for $x$ much greater than the atomic separation $a$; otherwise, we must consider the interaction of the electron with the individual ions. Further, Equation 4.39 has a reference level of zero *PE* at infinity ($x = \infty$), but we defined $PE = 0$ to be inside the metal. We must therefore modify Equation 4.39 to conform to our definition of zero *PE* as a reference. Figure 4.36a shows how this "image *PE*" varies with $x$ in this system. In the region $x < x_o$, we artificially bring $PE_{\text{image}}(x)$ to zero at $x = 0$, so our definition $PE = 0$ within the metal is maintained. Far away from the surface, the *PE* is expected to be $(E_F + \Phi)$ (and not zero, as in Equation 4.39), so we modify Equation 4.39 to read

$$PE_{\text{image}}(x) = (E_F + \Phi) - \frac{e^2}{16\pi\varepsilon_o x} \qquad \textbf{[4.40]}$$

The present model, which takes $PE_{\text{image}}(x)$ from 0 to $(E_F + \Phi)$ along Equation 4.40, is in agreement with the thermionic emission analysis, since the electron must still overcome a *PE* barrier of $E_F + \Phi$ to escape.

[11] An electron at a distance $x$ from the surface of a conductor experiences a force as if there were a positive charge of $+e$ at a distance $2x$ from it. The force is $e^2/[4\pi\varepsilon_o(2x)^2]$ or $e^2/[16\pi\varepsilon_o x^2]$. The result is called the image charge theorem. Integrating the force gives the potential energy in Equation 4.39.

From the definition of potential, which is potential energy per unit charge, when a voltage difference is applied between the anode and cathode, there is a *PE* gradient just outside the surface of the metal, given by $eV(x)$, or

$$PE_{\text{applied}}(x) = -ex\mathcal{E} \quad \textbf{[4.41]}$$

where $\mathcal{E}$ is the applied field and is assumed, for all practical purposes, to be uniform. The variation of $PE_{\text{applied}}(x)$ with $x$ is depicted in Figure 4.36b. The total $PE(x)$ of the electron outside the metal is the sum of Equations 4.40 and 4.41, as sketched in Figure 4.36c,

$$PE(x) = (E_F + \Phi) - \frac{e^2}{16\pi\varepsilon_o x} - ex\mathcal{E} \quad \textbf{[4.42]}$$

Note that the $PE(x)$ outside the metal no longer goes up to $(E_F + \Phi)$, and the *PE* barrier against thermal emission is effectively reduced to $(E_F + \Phi_{\text{eff}})$, where $\Phi_{\text{eff}}$ is a new effective work function that takes into account the effect of the applied field. The new barrier $(E_F + \Phi_{\text{eff}})$ can be found by locating the maximum of $PE(x)$, that is, by differentiating Equation 4.42 and setting it to zero. The **effective work function** in the presence of an applied field is therefore

$$\Phi_{\text{eff}} = \Phi - \left(\frac{e^3\mathcal{E}}{4\pi\varepsilon_o}\right)^{1/2} \quad \textbf{[4.43]}$$

This lowering of the work function by the applied field, as predicted by Equation 4.43, is the **Schottky effect.** The current density is given by the Richardson–Dushman equation, but with $\Phi_{\text{eff}}$ instead of $\Phi$,

*Field-assisted thermionic emission*

$$J = B_e T^2 \exp\left[-\frac{(\Phi - \beta_S \mathcal{E}^{1/2})}{kT}\right] \quad \textbf{[4.44]}$$

where $\beta_S = [e^3/4\pi\varepsilon_o]^{1/2}$ is the **Schottky coefficient,** whose value is $3.79 \times 10^{-5}$ $(\text{eV}/\sqrt{\text{V m}^{-1}})$.

When the field becomes very large, for example, $\mathcal{E} > 10^7$ V cm$^{-1}$, the $PE(x)$ outside the metal surface may bend sufficiently steeply to give rise to a narrow *PE* barrier. In this case, there is a distinct probability that an electron at an energy $E_F$ will tunnel through the barrier and escape into vacuum, as depicted in Figure 4.37. The likelihood of tunneling depends on the effective height $\Phi_{\text{eff}}$ of the *PE* barrier above $E_F$, as well as the width $x_F$ of the barrier at energy level $E_F$. Since tunneling is temperature independent, the emission process is termed **field emission.** The tunneling probability $P$ was calculated in Chapter 3, and depends on $\Phi_{\text{eff}}$ and $x_F$ through the equation[12]

$$P \approx \exp\left[\frac{-2(2m_e\Phi_{\text{eff}})^{1/2}x_F}{\hbar}\right]$$

We can easily find $x_F$ by noting that when $x = x_F$, $PE(x_F)$ is level with $E_F$, as shown in Figure 4.37. From Equation 4.42, when the field is very strong, then around

[12] In Chapter 3 we showed that the transmission probability $T = T_o \exp(-2\alpha a)$ where $\alpha^2 = 2m(V_o - E)/\hbar^2$ and $a$ is the barrier width. The pre-exponential constant $T_o$ can be taken to be ~1. Clearly $V_o - E = \Phi_{\text{eff}}$ since electrons with $E = E_F$ are tunneling and $a = x_F$.

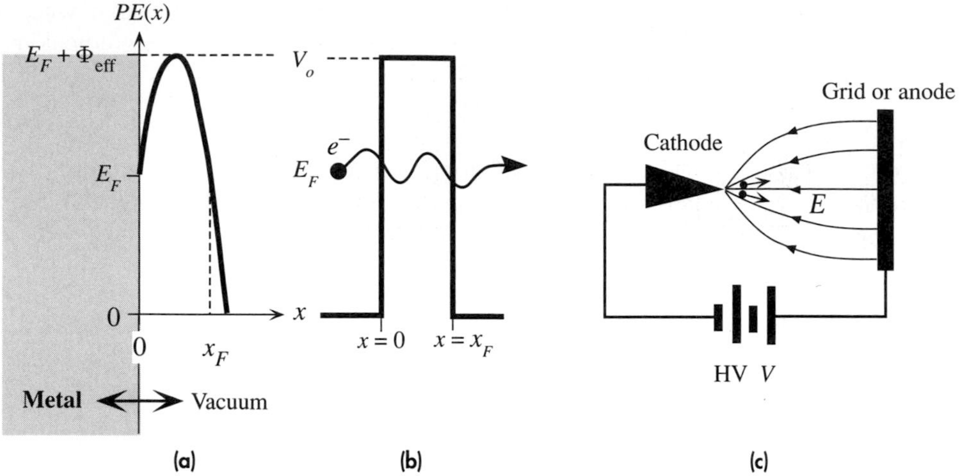


**Figure 4.37**

(a) Field emission is the tunneling of an electron at an energy $E_F$ through the narrow *PE* barrier induced by a large applied field.

(b) For simplicity, we take the barrier to be rectangular.

(c) A sharp point cathode has the maximum field at the tip where the field emission of electrons occurs.

$x \approx x_F$ the second term is negligible compared to the third, so putting $x = x_F$ and $PE(x_F) = E_F$ in Equation 4.42 yields $\Phi = e\mathcal{E}x_F$. Substituting $x_F = \Phi/e\mathcal{E}$ in Equation 4.45, we can obtain the tunneling probability $P$

*Field-assisted tunneling probability*

$$P \approx \exp\left[-\frac{2(2m_e\Phi_{\text{eff}})^{1/2}\Phi}{e\hbar\mathcal{E}}\right] \tag{4.45}$$

Equation 4.45 represents the probability $P$ that an electron in the metal at $E_F$ will tunnel out from the metal, as in Figure 4.37a and b, and become field-emitted. In a more rigorous analysis we have to consider that electrons not just at $E_F$ but at energies below $E_F$ can also tunnel out (though with lower probability) and we have to abandon the rough rectangular $PE(x)$ approximation in Figure 4.37b.

To calculate the current density $J$ we have to consider how many electrons are moving toward the surface per second and per unit area, the electron flux, and then multiply this flow by the probability that they will tunnel out. The final result of the calculations is the **Fowler–Nordheim equation,** which still has the exponential field dependence in Equation 4.45,

*Field-assisted tunneling: Fowler–Nordheim equation*

$$J_{\text{field-emission}} \approx CE^2 \exp\left(-\frac{\mathcal{E}_c}{\mathcal{E}}\right) \tag{4.46a}$$

in which $C$ and $\mathcal{E}_c$ are temperature-independent constants

$$C = \frac{e^3}{8\pi h\Phi} \quad \text{and} \quad \mathcal{E}_c = \frac{8\pi(2m_e\Phi^3)^{1/2}}{3eh} \tag{4.46b}$$

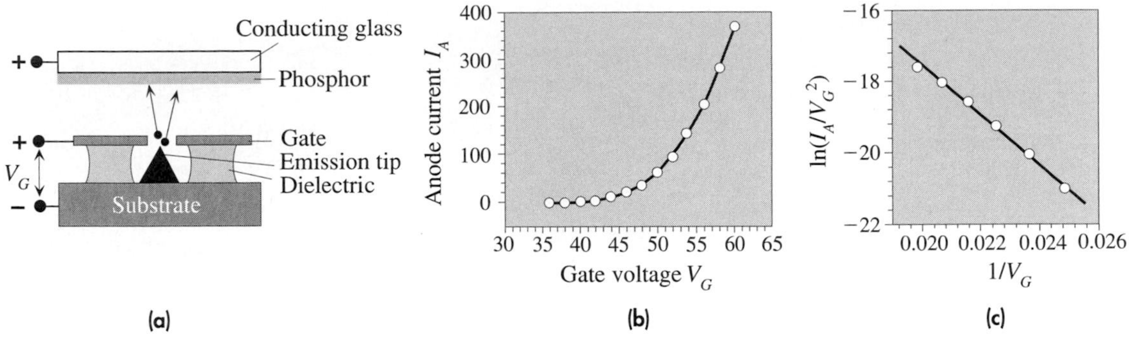


**Figure 4.38**
(a) Spindt-type cathode and the basic structure of one of the pixels in the FED.
(b) Emission (anode) current versus gate voltage.
(c) Fowler–Nordheim plot that confirms field emission.

that depend on the work function $\Phi$ of the metal. Equation 4.46a can also be used for field emission of electrons from a metal into an insulating material by using the electron *PE* barrier $\Phi_B$ from metal's $E_F$ into the insulator's conduction band (where the electron is free) instead of $\Phi$.

Notice that the field $\mathcal{E}$ in Equation 4.46a has taken over the role of temperature in thermionic emission in Equation 4.38. Since field-assisted emission depends exponentially on the field via Equation 4.46a, it can be enhanced by shaping the cathode into a cone with a sharp point where the field is maximum and the electron emission occurs from the tip as depicted in Figure 4.37c. The field $\mathcal{E}$ in Equation 4.46a is the *effective field* at the tip of the cathode that emits the electrons.

A popular field-emission tip design is based on the **Spindt tip cathode,** named after its originator. As shown in Figure 4.38a, the emission cathode is an iceberg-type sharp cone and there is a positively biased **gate** above it with a hole to extract the emitted electrons. A positively biased **anode** draws and accelerates the electrons passing through the gate toward it, which impinge on a phosphor screen to generate light by **cathodoluminescence,** a process in which light is emitted from a material when it is bombarded with electrons. Arrays of such electron field-emitters are used in field emission displays (FEDs) to generate bright images with vivid colors. Color is obtained by using red, green, and blue phosphors. The field at the tip is controlled by the potential difference between the gate and the cathode, the gate voltage $V_G$, which therefore controls field emission. Since $\mathcal{E} \propto V_G$, Equation 4.46a can be written to obtain the emission current or the anode current $I_A$ as

*Fowler–Nordheim anode current in a field emission device*

$$I_A = aV_G^2 \exp\left(-\frac{b}{V_G}\right) \quad \text{[4.47]}$$

where $a$ and $b$ are constants that depend on the particular field-emitting structure and cathode material. Figure 4.38b shows the dependence of $I_A$ on $V_G$. There is a very sharp increase with the voltage once the threshold voltages (around ~45 V in Figure 4.38b) are reached to start the electron emission. Once the emission is fully operating,

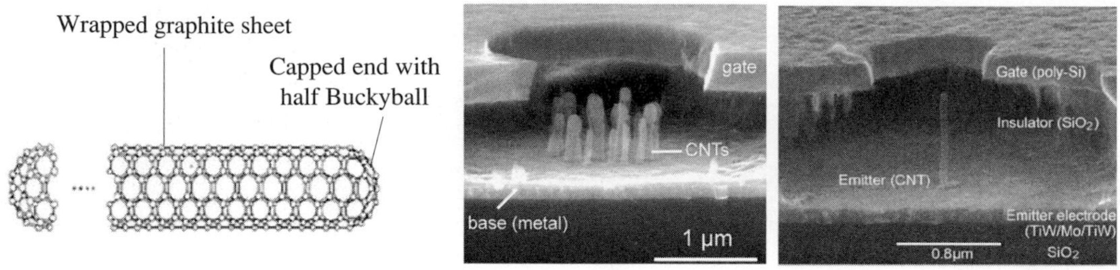


(a) (b) (c)

**Figure 4.39**

(a) A carbon nanotube (CNT) is a whisker-like, very thin and long carbon molecule with rounded ends, almost the perfect shape to be an electron field-emitter.

(b) Multiple CNTs as electron emitters.

(c) A single CNT as an emitter.

SOURCE: Courtesy of Professor W. I. Milne, University of Cambridge; G. Pirio *et al.*, *Nanotechnology*, **13**, 1, 2002.

$I_A$ versus $V_G$ follows the Fowler–Nordheim emission. A plot of $\ln(I_A/V_G^2)$ versus $1/V_G$ is a straight line as shown in Figure 4.38c.

Field emission has a number of distinct advantages. It is much more power efficient than thermionic emission which requires heating the cathode to high temperatures. In principle, field emission can be operated at high frequencies (fast switching times) by reducing various capacitances in the emission device or controlling the electron flow with a grid. Field emission has a number of important realized and potential applications: field emission microscopy, microwave amplifiers (high power and wide bandwidth), parallel electron beam microscopy, nanolithography, portable X-ray generators, and FEDs. For example, FEDs are thin flat displays ($\sim$2 mm thick), that have a low power consumption, quick start, and most significantly, a wide viewing angle of about 170°. Monochrome FEDs are already on the market, and color FEDs are expected to be commercialized soon, probably before the fourth edition of this text.

Typically molybdenum, tungsten, and hafnium have been used as the field-emission tip materials. Micromachining (microfabrication) has lead to the use of Si emission tips as well. Good electron emission characteristics have been also reported for diamond-like carbon films. Recently there has been a particular interest in using carbon nanotubes as emitters. A **carbon nanotube** (CNT) is a very thin filament-like carbon molecule whose diameter is in the nanometer range but whose length can be quite long, *e.g.*, 10–100 microns, depending on how it is grown or prepared. A CNT is made by rolling a graphite sheet into a tube and then capping the ends with hemispherical buckminsterfullerene molecules (a half Buckyball) as shown in Figure 4.39a. Depending on how the graphite sheet is rolled up, the CNT may be a metal or a semiconductor[13]. The high aspect ratio (length/diameter) of the CNT makes it an efficient

[13] Carbon nanotubes can be single-walled or multiwalled (when the graphite sheets are wrapped more than once) and can have quite complicated structures. There is no doubt that they possess some remarkable properties, so it is likely that CNTs will eventually be used in various engineering applications. See, for example, M. Baxendale, *J. Mater. Sci.: Mater Electron*, **14**, 657, 2003.

electron emitter. If one were to wonder what is the best shape for an efficient field emission tip, one might guess that it should be a sharp cone with some suitable apex angle. However, it turns out that the best emitter is actually a whisker-type thin filament with a rounded tip, much like a CNT. It is as if the CNT has been designed by nature to be the best field emitter. Figure 4.39b and c shows SEM photographs of two CNT Spindt-type emitters. Figure 4.39b has several CNTs, and Figure 4.39c just one CNT for electron emission. (Which is more efficient?)

**EXAMPLE 4.14**

**FIELD EMISSION** Field emission displays operate on the principle that electrons can be readily emitted from a microscopic sharp point source (*cathode*) that is biased negatively with respect to a neighboring electrode (*gate* or *grid*) as depicted in Figure 4.38a. Emitted electrons impinge on colored phosphors on a screen and cause light emission by cathodoluminescence. There are millions of these microscopic field emitters to constitute the image. A particular field emission cathode in a field-emission-type flat panel display gives a current of 61.0 μA when the voltage between the cathode and the grid is 50 V. The current is 279 μA when the voltage is 58.2 V. What is the current when the voltage is 56.2 V?

**SOLUTION**

Equation 4.47 related $I_A$ to $V_G$,

$$I_A = aV_G^2 \exp\left(-\frac{b}{V_G}\right)$$

where $a$ and $b$ are constants that can be determined from the two sets of data given. Thus,

$$61.0\ \mu\text{A} = a50^2 \exp\left(-\frac{b}{50}\right) \quad \text{and} \quad 279\ \mu\text{A} = a58.2^2 \exp\left(-\frac{b}{58.2}\right)$$

Dividing the first by the second gives

$$\frac{61.0}{279} = \frac{50^2}{58.2^2} \exp\left[-b\left(\frac{1}{50} - \frac{1}{58.2}\right)\right]$$

which can be solved to obtain $b = 431.75$ V and hence $a = 137.25\ \mu\text{A/V}^2$. At $V = 58.2$ V,

$$I = (137.25)(56.2)^2 \exp\left(-\frac{431.75}{56.2}\right) = 200\ \mu\text{A}$$

The experimental value for this device was 202 μA, which happens to be the device in Figure 4.37b (close).

# 4.10 PHONONS

## 4.10.1 HARMONIC OSCILLATOR AND LATTICE WAVES

***Quantum Harmonic Oscillator*** In the classical picture of a solid, the constituent atoms are held together by bonds which can be represented by springs. According to the kinetic molecular theory, the atoms in a solid are constantly vibrating about their equilibrium positions by stretching and compressing their springs. The oscillations are

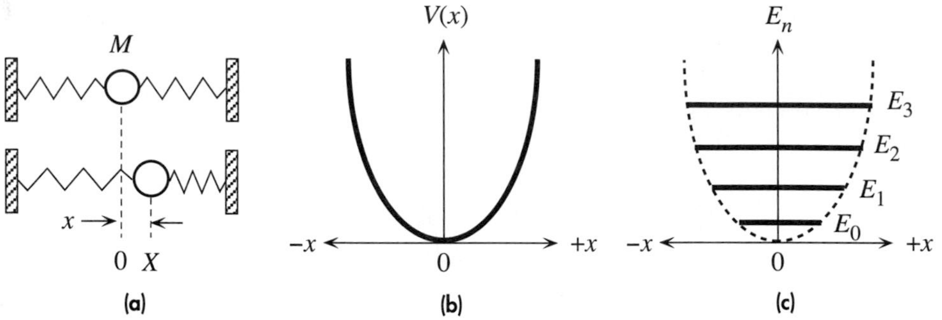


**Figure 4.40**

(a) Harmonic vibrations of an atom about its equilibrium position assuming its neighbors are fixed.
(b) The *PE* curve $V(x)$ versus displacement from equilibrium, $x$.
(c) The energy is quantized.

assumed to be simple harmonic so that the average kinetic and potential energies are the same. Figure 4.40a shows a one-dimensional independent simple harmonic oscillator that represents an atom of mass $M$ attached by springs to fixed neighbors. The potential energy $V(x)$ is a function of displacement $x$ from equilibrium. For small displacements, $V(x)$ is parabolic in $x$, as indicated in Figure 4.40b, that is,

*Harmonic potential energy*

$$V(x) = \frac{1}{2}\beta x^2 \tag{4.48}$$

where $\beta$ is a spring constant. The instantaneous energy, in principle, can be of any value. Equation 4.48 neglects the cubic term and is therefore symmetric about the equilibrium position at $x = 0$. It is called a **harmonic** approximation to the *PE* curve.

In modern physics, the energy of such a harmonic oscillator must be calculated using the *PE* in Equation 4.48 in the Schrödinger equation so that

*Schrödinger equation: harmonic oscillator*

$$\frac{d^2\psi}{dx^2} + \frac{2M}{\hbar^2}\left(E - \frac{1}{2}\beta x^2\right)\psi = 0 \tag{4.49}$$

The solution of Equation 4.49 shows that the energy $E_n$ of such a harmonic oscillator is quantized,

*Energy of a harmonic oscillator*

$$E_n = \left(n + \frac{1}{2}\right)\hbar\omega \tag{4.50}$$

where $\omega$ is the angular frequency of the vibrations[14] and $n$ is a quantum number 0, 1, 2, 3, .... The oscillation frequency is determined by the spring constant $\beta$ and the mass $M$ through $\omega = (\beta/M)^{1/2}$. Figure 4.40c shows the allowed energies of the quantum mechanical harmonic oscillator.

[14] Henceforth frequency will imply $\omega$.

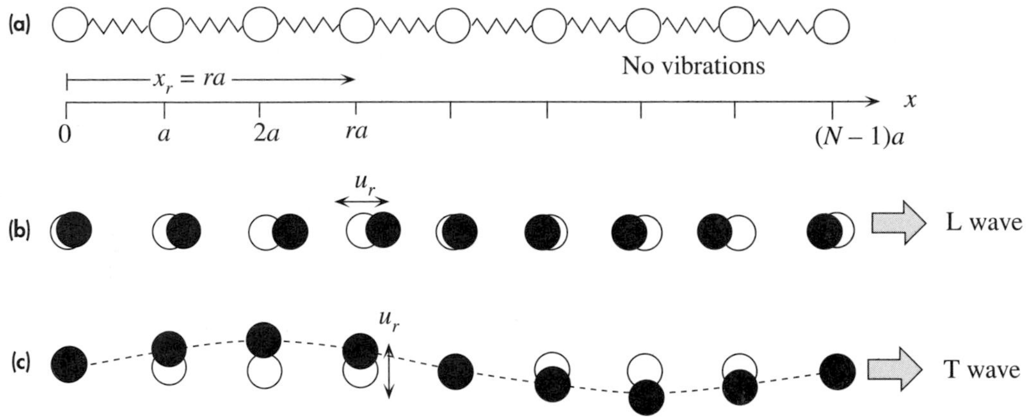


**Figure 4.41**
(a) A chain of *N* atoms through a crystal in the absence of vibrations.
(b) Coupled atomic vibrations generate a traveling longitudinal (L) wave along *x*. Atomic displacements ($u_r$) are parallel to *x*.
(c) A transverse (T) wave traveling along *x*. Atomic displacements ($u_r$) are perpendicular to the *x* axis. (b) and (c) are snapshots at one instant.

It is apparent that the minimum energy of the oscillator can never be zero but must be a finite value that is $E_0 = \frac{1}{2}\hbar\omega$. This energy is called the **zero-point energy.** As the temperature approaches 0 K, the harmonic oscillator would have an energy of $E_0$ and not zero. The energy levels are equally spaced by an amount $\hbar\omega$, which represents the amount of energy absorbed or emitted by the oscillator when it is excited and de-excited to a neighboring energy level. The vibrational energies of a molecule due to its atoms vibrating relative to each other, *e.g.*, the vibrations of the $Cl_2$ molecule in which the Cl–Cl bond is stretched and compressed, can also be described by Equation 4.50.

***Phonons*** Atoms in a solid are coupled to each other by bonds. Atomic vibrations are therefore also coupled. These coupled vibrations lead to waves that involve cooperative vibrations of many atoms and cannot be represented by independent vibrations of individual atoms. Figure 4.41a shows a chain of atoms in a crystal. As an atom vibrates it transfers its energy to neighboring vibrating atoms and the coupled vibrations produce traveling wave-trains in the crystal.[15] (Consider grabbing and strongly vibrating the first atom in the atomic chain in Figure 4.41a. Your vibrations will be coupled and transferred by the springs to neighboring atoms in the chain along *x*.) Two examples are shown in Figure 4.41b and c. In the first, the atomic vibrations are parallel to the direction of propagation *x* and the wave is a **longitudinal wave.** In the second, the vibrations are transverse to the direction of propagation and the corresponding wave is a **transverse wave.** Suppose that $x_r$ is the position of the *r*th atom in the absence of vibrations, that is, $x_r = ra$, where *r* is an integer from 0 to *N*, the number of atoms in the chain, as indicated in Figure 4.41a. By writing the mechanical equations (Newton's

[15] In the presence of coupling, the individual atoms do not execute simple harmonic motion.

second law) for the coupled atoms in Figure 4.41a, we can show that the displacement $u_r$ from equilibrium at a location $x_r$ is given by a **traveling-wave**-like behavior,[16]

*Traveling-wave-type lattice vibrations*

$$u_r = A\exp[j(Kx_r - \omega t)] \qquad [4.51]$$

where $A$ is the amplitude, $K$ is a wavevector, and $\omega$ is the angular frequency. Notice that the $Kx_r$ term is very much like the usual $kx$ phase term of a traveling wave propagating in a continuous medium; the only difference is that $Kx_r$ exists at discrete $x_r$ locations. The wave-train described by Equation 4.51 in the crystal is called a **lattice wave.** Along the $x$ direction it has a **wavelength** $\Lambda = 2\pi/K$ over which the longitudinal (or transverse) displacement $u_r$ repeats itself. The displacement $u_r$ repeats itself at one location over a time period $2\pi/\omega$. A wave traveling in the opposite direction to Equation 4.51 is of course also possible. Indeed, two oppositely traveling waves of the same frequency can interfere to set up a stationary wave which is also a lattice wave.

The lattice wave described by Equation 4.51 is a *harmonic oscillation* with a frequency $\omega$ that itself has no coupling to another lattice wave. The energy possessed by this lattice vibration is *quantized* in much the same way as the energy of the quantized harmonic oscillator in Equation 4.50. The energy of a lattice vibration therefore can only be multiples of $\hbar\omega$ above the zero-point energy, $\frac{1}{2}\hbar\omega$. The quantum of energy $\hbar\omega$ is therefore the smallest unit of lattice vibrational energy that can be added or subtracted from a lattice wave. The quantum of lattice vibration $\hbar\omega$ is called a **phonon** in analogy with the quantum of electromagnetic radiation, the photon. Whenever a lattice vibration interacts with another lattice vibration, an electron or a photon, in the crystal, it does so as if it had possessed a momentum of $\hbar K$. Thus,

*Phonon energy*

$$E_{\text{phonon}} = \hbar\omega = h\nu \qquad [4.52]$$

*Phonon momentum*

$$p_{\text{phonon}} = \hbar K \qquad [4.53]$$

The frequency of vibrations $\omega$ and the wavevector $K$ of a lattice wave are related. If we were to use Equation 4.51 in the mechanical equations that describe the coupled atomic vibrations, we would find that

*Dispersion relation*

$$\omega = 2\left(\frac{\beta}{M}\right)^{1/2}\left|\sin\left(\frac{1}{2}Ka\right)\right| \qquad [4.54]$$

which relates $\omega$ and $K$ and is called the **dispersion relation.** Figure 4.42 shows how the frequency $\omega$ of the lattice waves increases with increasing wavevector $K$, or decreasing wavelength $\Lambda$. From Equation 4.54, there can be no frequencies higher than $\omega_{\max} = 2(\beta/M)^{1/2}$, which is the **lattice cut-off frequency.** Both longitudinal and transverse waves exhibit this type of dispersion relationship shown in Figure 4.42a though their exact $\omega$–$K$ curves would be different depending on the nature of interatomic bonding and the crystal structure. The dispersion relation in Equation 4.54 is periodic in $K$ with a period $2\pi/a$. Only values of $K$ in the range $-\pi/a < K < \pi/a$ are physically meaningful. A point $A$ with $K_A$ is the same as a point $B$ with $K_B$ because we can shift $K$ by the period, $2\pi/a$ as shown in Figure 4.42a.

[16] The exponential notation for a wave is convenient, but we have to consider only the real part to actually represent the wave in the physical world.

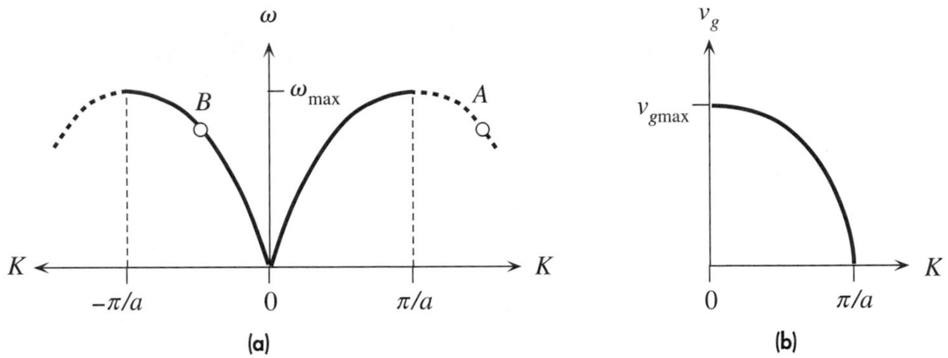


**Figure 4.42**
(a) Frequency $\omega$ versus wavevector $K$ relationship for lattice waves.
(b) Group velocity $v_g$ versus wavevector $K$.

The velocity at which traveling waves carry energy is called the **group velocity** $v_g$ of the wave.[17] It depends on the slope $d\omega/dK$ of the $\omega$–$K$ dispersion curve, so for lattice waves,

$$v_g = \frac{d\omega}{dK} = \left(\frac{\beta}{M}\right)^{1/2} a \cos\left(\frac{1}{2}Ka\right) \quad [4.55]$$

*Group velocity*

which is shown in Figure 4.42b. Points $A$ and $B$ in Figure 4.42a have the same group velocity and are equivalent.

The number of distinct or independent lattice waves, with different wavevectors, in a crystal is not infinite but depends on the number of atoms $N$. Consider a linear crystal as in Figure 4.43 with many atoms. We will take $N$ to be large and ignore the difference between $N$ and $N-2$. The lattice waves in this crystal would be standing waves represented by two oppositely traveling waves. The crystal length $L = Na$ can support multiples of the half-wavelength $\frac{1}{2}\Lambda$ as indicated in Figure 4.43,

$$q\frac{\Lambda}{2} = L = Na \qquad q = 1, 2, 3, \ldots \quad [4.56a]$$

*Vibrational modes*

or

$$K = \frac{q\pi}{L} = \frac{q\pi}{Na} \qquad q = 1, 2, 3, \ldots \quad [4.56b]$$

*Vibrational modes*

where $q$ is an integer. Each particular $K$ value $K_q$ represents one distinct lattice wave with a particular frequency as determined by the dispersion relation. Four examples are shown in Figure 4.43. Each of these $K_q$ values defines a **mode** or **state of lattice vibration.** Each mode is an independent lattice vibration. Its energy can be increased or decreased only by a quantum amount of $\hbar\omega$. Since $K_q$ values outside the range $-\pi/a < K < \pi/a$ are the same as those in that range ($A$ and $B$ are the same

[17] For those readers who are not familiar with the group velocity concept, this is discussed in Chapter 9 without prerequisite material.

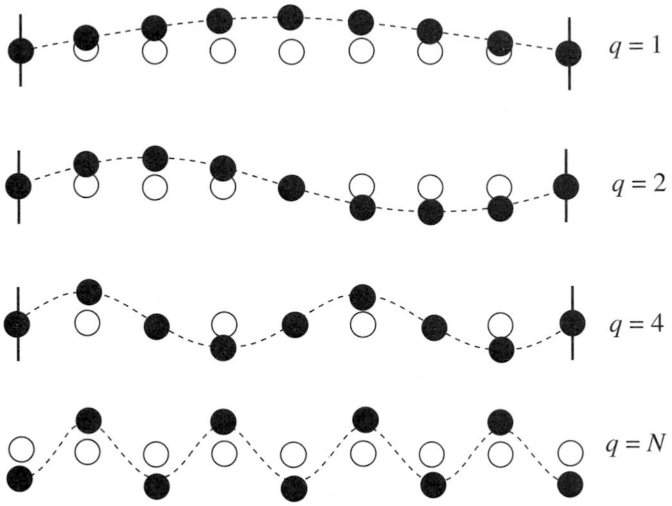


**Figure 4.43** Four examples of standing waves in a linear crystal corresponding to $q = 1, 2, 4$, and $N$.
$q$ is maximum when alternating atoms are vibrating in opposite directions. A portion from a very long crystal is shown.

in Figure 4.42a), it is apparent that the maximum value of $q$ is $N$ and thus the **number of modes** is also $N$. Notice that as $q$ increases, $\Lambda$ decreases. The smallest $\Lambda$ occurs when alternating atoms in the crystal are moving in opposite directions which corresponds to $\frac{1}{2}\Lambda = a$, that is, $q = N$, as shown in Figure 4.43. In terms of the wavevector, $K = 2\pi/\Lambda = \pi/a$. Smaller wavelengths or longer wavevectors are meaningless and correspond to shifting $K$ by a multiple of $2\pi/a$. Since $N$ is large, the $\omega$ versus $K$ curve in Figure 4.42a consists of very finely separated distinct points, each corresponding to a particular $q$, analogous to the energy levels in an energy band.

The above ideas for the linear chain of atoms can be readily extended to a three-dimensional crystal. If $L_x$, $L_y$, and $L_z$ are the sides of the solid along the $x$, $y$, and $z$ axes, with $N_x$, $N_y$, and $N_z$ number of atoms, respectively, then the wavevector components along $x$, $y$, and $z$ are

*Lattice vibrational modes in 3-D*

$$K_x = \frac{q_x \pi}{L_x} \qquad K_y = \frac{q_y \pi}{L_y} \qquad K_z = \frac{q_z \pi}{L_z} \tag{4.57}$$

where the integers $q_x$, $q_y$, and $q_z$ run from 1 to $N_x$, $N_y$, and $N_z$, respectively. The total number of permitted modes is $N_x N_y N_z$ or $N$, the total number of atoms in the solid. Vibrations however can be set up independently along the $x$, $y$, and $z$ directions so that the actual *number of independent modes* is $3N$.

### 4.10.2 DEBYE HEAT CAPACITY

The heat capacity of a solid represents the increase in the internal energy of the crystal per unit increase in the temperature. The increase in the internal energy is due to an increase in the energy of lattice vibrations. This is generally true for all the solids except metals at very low temperatures where the heat capacity is due to the electrons near the Fermi level becoming excited to higher energies. For most practical temperature ranges of interest, the heat capacity of solids is determined by the excitation of lattice vibrations. The **molar heat capacity** $C_m$ is the increase in the internal energy $U_m$ of a crystal of $N_A$ atoms per unit increase in the temperature at constant volume,[18] that is, $C_m = dU_m/dT$.

The simplest approach to calculating the average energy is first to assume that all the lattice vibrational modes have the same frequency $\omega$. (We will account for different modes having different frequencies later.) If $E_n$ is the energy of a harmonic oscillator such as a lattice vibration, then the average energy, by definition, is given by

$$\overline{E} = \frac{\sum_{n=0}^{\infty} E_n P(E_n)}{\sum_{n=0}^{\infty} P(E_n)} \quad [4.58]$$

*Average energy of oscillators*

where $P(E_n)$ is the probability that the vibration has the energy $E_n$ which is proportional to the Boltzmann factor. Thus we can use $P(E_n) \propto \exp(-E_n/kT)$ and $E_n = (n + \frac{1}{2})\hbar\omega$ in Equation 4.58. We can drop the zero-point energy as this does not affect the heat capacity (which deals with energy *changes*). The substitution and calculation of Equation 4.58 yields the vibrational mean energy at a frequency $\omega$,

$$\overline{E}(\omega) = \frac{\hbar\omega}{\exp\left(\frac{\hbar\omega}{kT}\right) - 1} \quad [4.59]$$

*Average energy of oscillators at ω*

This energy increases with temperature. Each phonon has an energy of $\hbar\omega$. Thus, the *phonon concentration in the crystal increases with temperature*; increasing the temperature creates more phonons.

To find the internal energy due to *all* the lattice vibrations we must also consider how many modes there are at various frequencies, that is, the distribution of the modes over the possible frequencies, the spectrum of the vibrations. Suppose that $g(\omega)$ is the number of modes per unit frequency, that is, $g(\omega)$ is the **density of vibrational states** or modes. Then $g(\omega)\, d\omega$ is the number of states in the range $d\omega$. The internal energy $U_m$ of all lattice vibrations for 1 mole of solid is

$$U_m = \int_0^{\omega_{max}} \overline{E}(\omega) g(\omega)\, d\omega \quad [4.60]$$

*Internal energy of all lattice vibrations*

The integration is up to a certain allowed maximum frequency $\omega_{max}$ (Figure 4.42a). The density of states $g(\omega)$ for the lattice vibrations can be found in a similar fashion to the density of states for electrons in an energy band, and we will simply quote the result,

$$g(\omega) \approx \frac{3V}{2\pi^2} \frac{\omega^2}{v^3} \quad [4.61]$$

*Density of states for lattice vibrations*

[18] Constant volume in the definition means that the heat added to the system increases the internal energy without doing mechanical work by changing the volume.

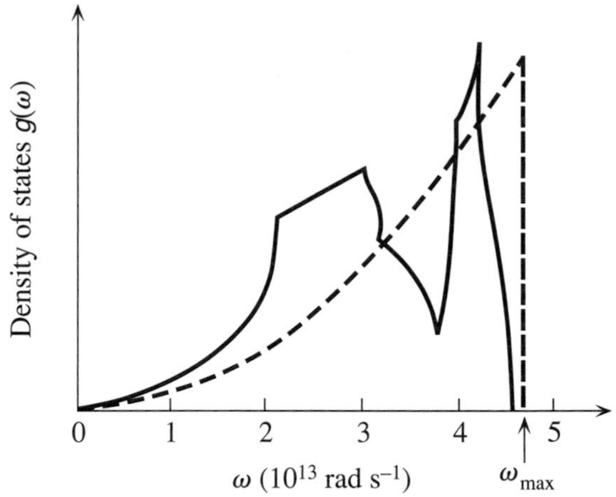


**Figure 4.44** Density of states for phonons in copper. The solid curve is deduced from experiments on neutron scattering. The broken curve is the three-dimensional Debye approximation, scaled so that the areas under the two curves are the same.

This requires that $\omega_{max} \approx 4.5 \times 10^{13}$ rad s$^{-1}$, or a Debye characteristic temperature $T_D = 344$ K.

where $v$ is the mean velocity of longitudinal and transverse waves in the solid and $V$ is the volume of the crystal. Figure 4.44 shows the spectrum $g(\omega)$ for a real crystal such as Cu and the expression in Equation 4.61. The maximum frequency is $\omega_{max}$ and is determined by the fact that the total number of modes up to $\omega_{max}$ must be $3N_A$. It is called the **Debye frequency.** Thus, integrating $g(\omega)$ up to $\omega_{max}$ we find,

*Debye frequency*

$$\omega_{max} \approx v(6\pi^2 N_A/V)^{1/3} \quad [4.62]$$

This maximum frequency $\omega_{max}$ corresponds to an energy $\hbar\omega_{max}$ and to a temperature $T_D$ defined by,

*Debye temperature*

$$T_D = \frac{\hbar\omega_{max}}{k} \quad [4.63]$$

and is called the **Debye temperature.** Qualitatively, it represents the temperature above which all vibrational frequencies are executed by the lattice waves.

Thus, by using Equations 4.59 to 4.63 in Equation 4.60 we can evaluate $U_m$ and hence differentiate $U_m$ with respect to temperature to obtain the molar heat capacity at constant volume,

*Heat capacity: lattice vibrations*

$$C_m = 9R\left(\frac{T}{T_D}\right)^3 \int_0^{T_D/T} \frac{x^4 e^x \, dx}{(e^x - 1)^2} \quad [4.64]$$

which is the Debye heat capacity expression.

Figure 4.45 represents the constant-volume molar heat capacity $C_m$ of nearly all crystals, Equation 4.64, as a function of temperature, normalized with respect to the Debye temperature. The **Dulong–Petit rule** of $C_m = 3R$ is only obeyed when $T > T_D$. Notice that $C_m$ at $T = 0.5T_D$ is $0.825(3R)$ whereas at $T = T_D$ it is $0.952(3R)$. For most practical purposes, $C_m$ is to within 6 percent of $3R$ when the temperature is at $0.9T_D$. For example, for copper $T_D = 315$ K and above

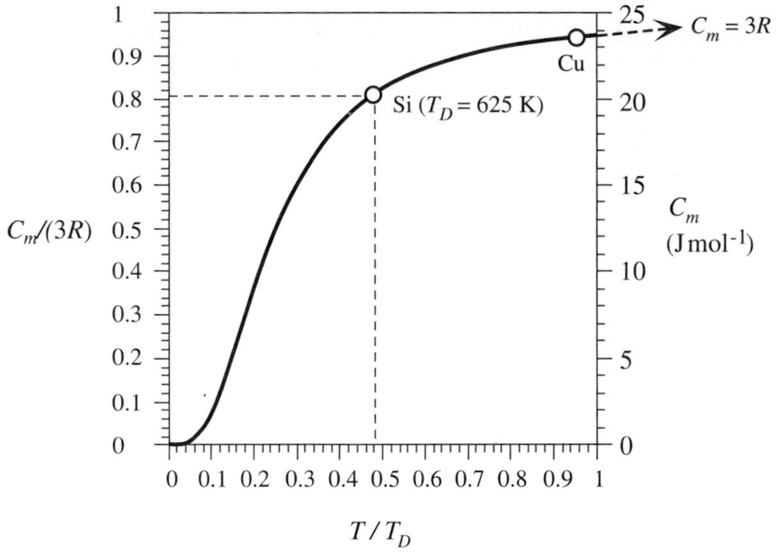


**Figure 4.45** Debye constant-volume molar heat capacity curve.
The dependence of the molar heat capacity $C_m$ on temperature with respect to the Debye temperature: $C_m$ versus $T/T_D$. For Si, $T_D = 625$ K, so at room temperature (300 K), $T/T_D = 0.48$ and $C_m$ is only 0.81 ($3R$).

about $0.9T_D$, that is, above 283 K (or 10 °C), $C_m \approx 3R$, as borne out by experiments.[19] Table 4.5 provides typical values for $T_D$, and heat capacities for a few selected elements. It is left as an exercise to check the accuracy of Equation 4.64 for predicting the heat capacity given the $T_D$ values. At the lowest temperatures when $T \ll T_D$, Equation 4.64 predicts that $C_m \propto T^3$, and this is indeed observed in low-temperature heat capacity experiments on a variety of crystals.[20]

It is useful to provide a physical picture of the Debye model inherent in Equation 4.64. As the temperature increases from near zero, the increase in the crystal's vibrational energy is due to *more* phonons being created and *higher* frequencies being excited. The phonon concentration increases as $T^3$, and the mean phonon energy increases as $T$. Thus, the internal energy increases as $T^4$. At temperatures above $T_D$, increasing the temperature creates *more* phonons but does not increase the mean phonon energy and does not excite higher frequencies. All frequencies up to $\omega_{max}$ have now been excited. The internal energy increases only due to more phonons being created. The phonon concentration and hence the internal energy increase as $T$; the heat capacity is constant as expected from Equation 4.64.

[19] Sometimes it is stated that the Debye temperature is a characteristics temperature for each material at which all the atoms are able to possess vibrational kinetic energies in accordance with the Maxwell equipartition of energy principle; that is, the average vibrational kinetic energy will be $\frac{3}{2}kT$ per atom and average potential energy will also be $\frac{3}{2}kT$. This means that the average energy per atom is $3kT$, and hence the heat capacity is $3kN_A$ or $3R$ per mole which is the *Dulong–Petit rule*.

[20] Well-known exceptions are glasses, noncrystalline solids, whose heat capacity is proportional to $a_1T + a_2T^3$, where $a_1$ and $a_2$ are constants.

**Table 4.5** Debye temperatures $T_D$, heat capacities, and thermal conductivities of selected elements

| | Crystal | | | | | | | |
|---|---|---|---|---|---|---|---|---|
| | Ag | Be | Cu | Diamond | Ge | Hg | Si | W |
| $T_D$ (K)* | 215 | 1000 | 315 | 1860 | 360 | 100 | 625 | 310 |
| $C_m$ (J K$^{-1}$ mol$^{-1}$)† | 25.6 | 16.46 | 24.5 | 6.48 | 23.38 | 27.68 | 19.74 | 24.45 |
| $c_s$ (J K$^{-1}$ g$^{-1}$)† | 0.237 | 1.825 | 0.385 | 0.540 | 0.322 | 0.138 | 0.703 | 0.133 |
| $\kappa$ (W m$^{-1}$ K$^{-1}$)† | 429 | 200 | 400 | 1000 | 60 | 8.65 | 148 | 173 |

* $T_D$ is obtained by fitting the Debye curve to the experimental molar heat capacity data at the point $C_m = \frac{1}{2}$ (3R).
† $C_m$, $c_s$, and $\kappa$ are at 25 °C.
SOURCE: $T_D$ data from J. De Launay, *Solid State Physics*, vol. 2, F. Seitz and D. Turnbull, eds., Academic Press, New York, 1956.

It is apparent that, above the Debye temperature, the increase in temperature leads to the creation of more phonons. In Chapters 1 and 2, using classical concepts only, we had mentioned that increasing the temperature increases the magnitude of atomic vibrations. This simple and intuitive classical concept in terms of modern physics corresponds to creating more phonons with temperature. We can use the photon analogy from Chapter 3. When we increase the intensity of light of a given frequency, classically we simply increase the electric field (magnitude of the vibrations), but in modern physics we have to increase the number of photons flowing per unit area.

**EXAMPLE 4.15** **SPECIFIC HEAT CAPACITY OF Si** Find the specific heat capacity $c_s$ of a silicon crystal at room temperature given $T_D = 625$ K for Si.

**SOLUTION**

At room temperature, $T = 300$ K, $(T/T_D) = 0.48$, and, from Figure 4.45, the molar heat capacity is

$$C_m = 0.81(3R) = 20.2 \text{ J K}^{-1} \text{ mol}^{-1}$$

The specific heat capacity $c_s$ from the Debye curve is

$$c_s = \frac{C_m}{M_{at}} \approx \frac{(0.81 \times 25 \text{ J K}^{-1} \text{ mol}^{-1})}{(28.09 \text{ g mol}^{-1})} = 0.72 \text{ J K}^{-1} \text{ g}^{-1}$$

The experimental value of 0.70 J K$^{-1}$ g$^{-1}$ is very close to the Debye value.

**EXAMPLE 4.16** **SPECIFIC HEAT CAPACITY OF GaAs** Example 4.15 applied Equation 4.64, the Debye molar heat capacity $C_m$, to the silicon crystal in which all atoms are of the same type. It was relatively simple to calculate the specific heat capacity $c_s$ (what is really used in engineering) from the molar heat capacity $C_m$ by using $c_s = C_m/M_{\text{at}}$ where $M_{\text{at}}$ is the atomic mass of the type of atom (only one) in the crystal. When the crystal has two types of atoms, we must modify the specific heat capacity derivation. We can still keep the symbol $C_m$ to represent the Debye molar heat capacity given in Equation 4.64. Consider a GaAs crystal that has $N_A$ units of GaAs, that is,

1 mole of GaAs. There will be 1 mole ($N_A$ atoms) of Ga and 1 mole of As atoms. To a reasonable approximation we can assume that each mole of Ga and As contributes a $C_m$ amount of heat capacity so that the total heat capacity of 1 mole GaAs will be $C_m + C_m$ or $2C_m$, a maximum of 50 J K$^{-1}$ mol$^{-1}$. The total mass of this 1 mole of GaAs is $M_{\text{Ga}} + M_{\text{As}}$. Thus, the specific heat capacity of GaAs is

*Specific heat capacity of GaAs*

$$c_s = \frac{C_{\text{total}}}{M_{\text{total}}} = \frac{C_m + C_m}{M_{\text{Ga}} + M_{\text{As}}} = \frac{2C_m}{M_{\text{Ga}} + M_{\text{As}}}$$

which can alternatively be written as

*Specific heat capacity of a polyatomic crystal*

$$c_s = \frac{C_m}{\frac{1}{2}(M_{\text{Ga}} + M_{\text{As}})} = \frac{C_m}{\overline{M}}$$

where $\overline{M} = (M_{\text{Ga}} + M_{\text{As}})/2$ is the average atomic mass of the constituent atoms. Although we derived $c_s$ for GaAs, it can also be applied to other compounds by suitably calculating an average atomic mass $\overline{M}$. GaAs has a Debye temperature $T_D = 344$ K, so that at a room temperature of 300 K, $T/T_D = 0.87$, and from Figure 4.45, $C_m/(3R) = 0.94$. Therefore,

$$c_s = \frac{C_m}{\overline{M}} = \frac{(0.94)(25 \text{ J K}^{-1} \text{ mol}^{-1})}{\frac{1}{2}(69.72 \text{ g mol}^{-1} + 74.92 \text{ g mol}^{-1})} = 0.325 \text{ J K}^{-1} \text{ g}^{-1}$$

At $-40$ °C, $T/T_D = 0.68$, and $C_m/(3R) = 0.90$, so the new $c_s = (0.90/0.94)(0.325) = 0.311$ J K$^{-1}$ g$^{-1}$, which is not a large change in $c_s$.

The heat capacity per unit volume $C_v$ can be found from $C_v = c_s\rho$, where $\rho$ is the density. Thus, at 300 K, $C_v = (0.325 \text{ J K}^{-1} \text{ g}^{-1})(5.32 \text{ g cm}^{-3}) = 1.73$ J K$^{-1}$ cm$^{-3}$. The calculated $c_s$ match the reported experimental values very closely.

---

**EXAMPLE 4.17**

**LATTICE WAVES AND SOUND VELOCITY** Consider *longitudinal* waves in a linear crystal and three atoms at $r - 1$, $r$, and $r + 1$ as in Figure 4.46. The displacement of each atom from equilibrium in the $+x$ direction is $u_{r-1}$, $u_r$, and $u_{r+1}$, respectively. Consider the $r$th atom. Its bond with the left neighbor stretches by $(u_r - u_{r-1})$. Its bond with the right neighbor stretches by $(u_{r+1} - u_r)$. The left spring exerts a force $\beta(u_r - u_{r-1})$, and the right spring exerts a force $\beta(u_{r+1} - u_r)$. The net force on the $r$th atom is mass × acceleration,

$$\text{Net force} = \beta(u_{r+1} - u_r) - \beta(u_r - u_{r-1}) = M\frac{d^2u_r}{dt^2}$$

*Wave equation*

so

$$M\frac{d^2u_r}{dt^2} = \beta(u_{r+1} - 2u_r + u_{r-1}) \qquad [4.65]$$

This is the **wave equation** that describes the coupled longitudinal vibrations of the atoms in the crystal. A similar expression can also be derived for transverse vibrations. We can substitute Equation 4.51 in Equation 4.65 to show that Equation 4.51 is indeed a solution of the wave

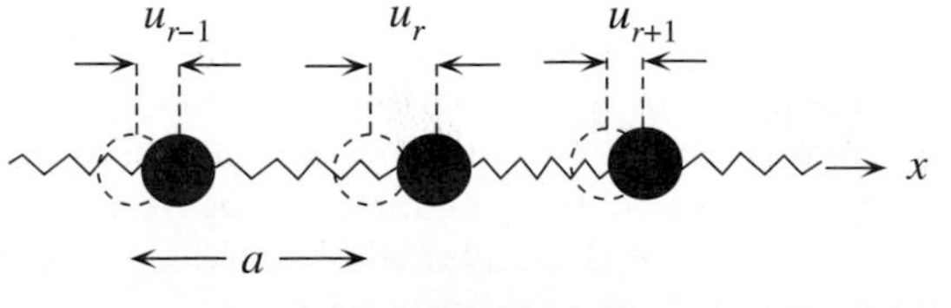


**Figure 4.46** Atoms executing longitudinal vibrations parallel to x.

equation. It is assumed that the crystal response is **linear,** that is, the net force is proportional to net displacement.

*Long-wavelength group velocity*

The **group velocity** of lattice waves is given by Equation 4.55. For sufficiently small $K$, or long wavelengths, such that $\frac{1}{2}Ka \ll 1$,

$$v_g = \left(\frac{\beta}{M}\right)^{1/2} a \cos\left(\frac{1}{2}Ka\right) \approx \left(\frac{\beta}{M}\right)^{1/2} a \qquad [4.66]$$

which is a constant. It is the slope of the straight-line region of $\omega$ versus $K$ curve for small $K$ values in Figure 4.42. Furthermore, the elastic modulus $Y$ depends on the slope of the net force versus displacement curve as derived in Example 1.5. From Equation 4.48 $F_N = dV/dx = \beta x$ and hence $Y = \beta/a$. Moreover, each atom occupies a volume of $a^3$, so the density $\rho$ is $M/a^3$. Substituting both of these results in Equation 4.66 yields

*Longitudinal elastic wave velocity*

$$v_g \approx \left(\frac{Y}{\rho}\right)^{1/2} \qquad [4.67]$$

The relationship has to be modified for an actual crystal incorporating a small numerical factor multiplying $Y$. Aluminum has a density of 2.7 g cm$^{-3}$ and $Y = 70$ GPa, so the long-wavelength longitudinal velocity from Equation 4.67 is 5092 m s$^{-1}$. The sound velocity in Al is 5100 m s$^{-1}$, which is very close.

## 4.10.3 Thermal Conductivity of Nonmetals

In nonmetals the heat transfer involves lattice vibrations, that is, phonons. The heat absorbed in the hot region increases the amplitudes of the lattice vibrations, which is the same as generating more phonons. These new phonons travel toward the cold regions and thereby transport the lattice energy from the hot to cold end. The **thermal conductivity** $\kappa$ measures the rate at which heat can be transported through a medium per unit area per unit temperature gradient. It is proportional to the rate at which a medium can absorb energy; that is, $\kappa$ is proportional to the heat capacity. $\kappa$ is also proportional to the rate at which phonons are transported which is determined by their mean velocity $v_{\text{ph}}$. In addition, of course, $\kappa$ is proportional to the *mean free path* $\ell_{\text{ph}}$ that a phonon has to travel before losing its momentum just as the electrical conductivity is proportional to the electron's mean free path. A rigorous classical treatment gives $\kappa$ as

*Thermal conductivity due to phonons*

$$\kappa = \tfrac{1}{3}C_v v_{\text{ph}} \ell_{\text{ph}} \qquad [4.68]$$

where $C_v$ is the heat capacity per unit volume. The mean free path $\ell_{\text{ph}}$ depends on various processes that can scatter the phonons and *hinder* their propagation along the direction of heat flow. Phonons collide with other phonons, crystal defects, impurities, and crystal surfaces.

The mean phonon velocity $v_{\text{ph}}$ is constant and approximately independent of temperature. At temperatures above the Debye temperature, $C_v$ is constant and, thus, $\kappa \propto \ell_{\text{ph}}$. The mean free path of phonons at these temperatures is determined by phonon–phonon collisions, that is, phonons interacting with other phonons as depicted in Figure 4.47. Since the phonon concentration $n_{\text{ph}}$ increases with temperature, $n_{\text{ph}} \propto T$, the mean free path decreases as $\ell_{\text{ph}} \propto 1/T$. Thus, $\kappa$ decreases with increasing temperature as observed for most crystals at sufficiently high temperatures.

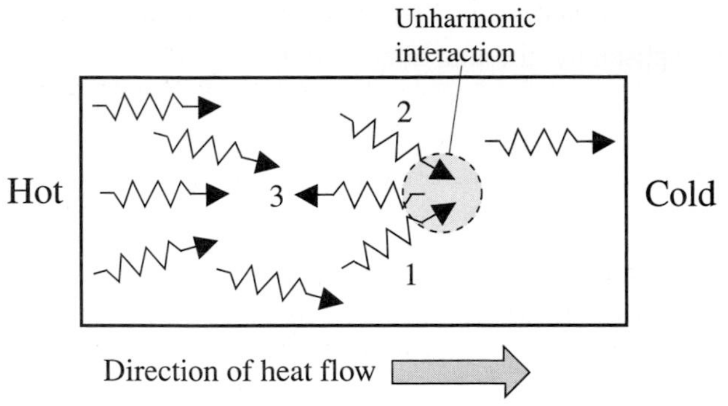


**Figure 4.47** Phonons generated in the hot region travel toward the cold region and thereby transport heat energy. Phonon–phonon unharmonic interaction generates a new phonon whose momentum is toward the hot region.

The phonon–phonon collisions that are responsible for limiting the thermal conductivity, that is, scattering the phonon momentum in the opposite direction to the heat flow, are due to the **unharmonicity (asymmetry)** of the interatomic potential energy curve. Stated differently, the net force $F$ acting on an atom is not simply $\beta x$ but also has an $x^2$ term; it is **nonlinear.** The greater the asymmetry or nonlinearity, the larger is the effect of such momentum flipping collisions. The same asymmetry that is responsible for thermal expansion of solids is also responsible for determining the thermal conductivity. When two phonons 1 and 2 interact in a crystal region as in Figure 4.47, the *nonlinear* behavior and the *periodicity* of the lattice cause a new phonon 3 to be generated. This new phonon 3 has the same energy as the sum of 1 and 2, but it is traveling in the wrong direction! (The frequency of 3 is the sum of the frequencies of 1 and 2.)

At low temperatures there are two factors. The phonon concentration is too low for phonon–phonon collisions to be significant. Instead, the mean free path $\ell_{ph}$ is determined by phonon collisions with crystal imperfections, most significantly, crystal surfaces and grain boundaries. Thus, $\ell_{ph}$ depends on the sample geometry and crystallinity. Further, as we expect from the Debye model, $C_v$ depends on $T^3$, so $\kappa$ has the same temperature dependence as $C_v$, that is, $\kappa \propto T^3$. Between the two temperature regimes $\kappa$ exhibits a peak as shown in Figure 4.48 for sapphire (crystalline $Al_2O_3$) and

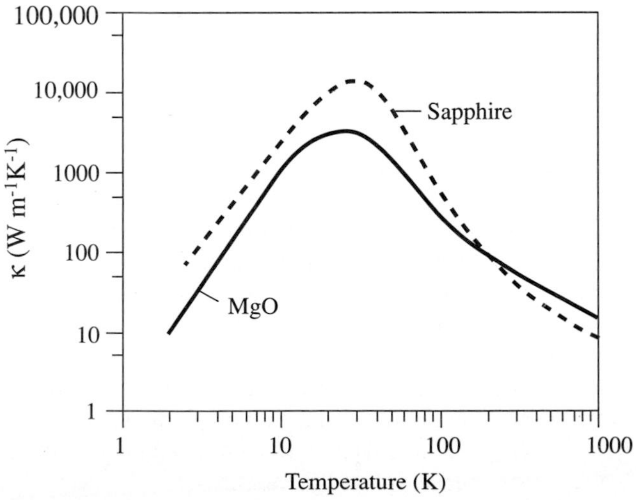


**Figure 4.48** Thermal conductivity of sapphire and MgO as a function of temperature.

MgO crystals. Even though there are no conduction electrons in these two example crystals, they nonetheless exhibit substantial thermal conductivity.

**EXAMPLE 4.18**

**PHONONS IN GaAs** Estimate the phonon mean free path in GaAs at room temperature 300 K and at 20 K from its $\kappa$, $C_v$, and $v_{\text{ph}}$, using Equation 4.68. At room temperature, semiconductor data handbooks list the following for GaAs: $\kappa = 45$ W m$^{-1}$ K$^{-1}$, elastic modulus $Y = 85$ GPa, density $\rho = 5.32$ g cm$^{-3}$, and specific heat capacity $c_s = 0.325$ J K$^{-1}$ g$^{-1}$. At 20 K, $\kappa = 4000$ W m$^{-1}$ K$^{-1}$ and $c_s = 0.0052$ J K$^{-1}$ g$^{-1}$. $Y$ and $\rho$ and hence $v_{\text{ph}}$ do not change significantly with temperature compared with the changes in $\kappa$ and $C_v$ with temperature.

**SOLUTION**

The phonon velocity $v_{\text{ph}}$ from Equation 4.67 is approximately

$$v_{\text{ph}} \approx \sqrt{\frac{Y}{\rho}} = \sqrt{\frac{85 \times 10^9 \text{ N m}^{-2}}{5.32 \times 10^3 \text{ kg m}^{-3}}} = 4000 \text{ m s}^{-1}$$

Heat capacity per unit volume $C_v = c_s\rho = (325 \text{ J K}^{-1} \text{ kg}^{-1})(5320 \text{ kg m}^{-3}) = 1.73 \times 10^6 \text{ J K}^{-1}$ m$^{-3}$. From Equation 4.68, $\kappa = \frac{1}{3}C_v v_{\text{ph}} \ell_{\text{ph}}$,

$$\ell_{\text{ph}} = \frac{3\kappa}{C_v v_{\text{ph}}} = \frac{(3)(45 \text{ W m}^{-1} \text{ K}^{-1})}{(1.73 \times 10^6 \text{ J K}^{-1} \text{ m}^{-3})(4000 \text{ m s}^{-1})} = 2.0 \times 10^{-8}\text{m} \quad \text{or} \quad 20 \text{ nm}$$

We can easily repeat the calculation at 20 K, given $\kappa \approx 4000$ W m$^{-1}$ K$^{-1}$ and $c_s = 5.2$ J K$^{-1}$ kg$^{-1}$, so $C_v = c_s\rho \approx (5.2 \text{ J K}^{-1} \text{ kg}^{-1})(5320 \text{ kg m}^{-3}) = 2.77 \times 10^4 \text{ J K}^{-1} \text{ m}^{-3}$. $Y$ and $\rho$ and hence $v_{\text{ph}}$ ($\approx 4000$ m s$^{-1}$), do not change significantly with temperature compared with $\kappa$ and $C_v$. Thus,

$$\ell_{\text{ph}} = \frac{3\kappa}{C_v v_{\text{ph}}} \approx \frac{(3)(4 \times 10^3 \text{ W m}^{-1} \text{ K}^{-1})}{(2.77 \times 10^4 \text{ J K}^{-1} \text{ m}^{-3})(4000 \text{ m s}^{-1})} = 1.1 \times 10^{-4}\text{m} \quad \text{or} \quad 0.011 \text{ cm}$$

For small specimens, the above phonon mean free path will be comparable to the sample size, which means that $\ell_{\text{ph}}$ will actually be limited by the sample size. Consequently $\kappa$ will depend on the sample dimensions, being smaller for smaller samples, similar to the dependence of the electrical conductivity of thin films on the film thickness.

### 4.10.4 ELECTRICAL CONDUCTIVITY

Except at low temperatures, the electrical conductivity of metals is primarily controlled by scattering of electrons around $E_F$ by lattice vibrations, that is, phonons. These electrons have a speed $v_F = (2E_F/m_e)^{1/2}$ and a momentum of magnitude $m_e v_F$. We know that the electrical conductivity $\sigma$ is proportional to the mean collision time $\tau$ of the electrons, that is, $\sigma \propto \tau$. This scattering time assumes that each scattering process is 100 percent efficient in randomizing the electron's momentum, that is, destroying the momentum gained from the field, which may not be the case. If it takes on average $N$ collisions to randomize the electron's momentum, and $\tau$ is the mean time between the scattering events, then the *effective* scattering time is simply $N\tau$ and $\sigma \propto N\tau$. ($1/N$ indicates the efficiency of each scattering process in randomizing the velocity.)

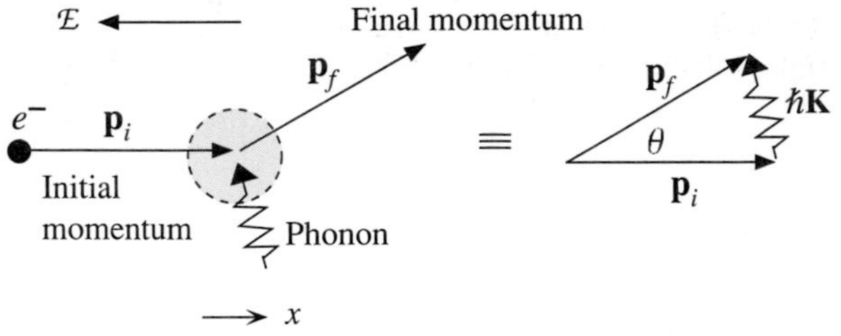


**Figure 4.49** Low-angle scattering of a conduction electron by a phonon.

Figure 4.49 shows an example in which an electron with an initial momentum $\mathbf{p}_i$ collides with a lattice vibration of momentum $\hbar\mathbf{K}$. The result of the interaction is that the electron's momentum is deflected through a small angle $\theta$ to $\mathbf{p}_f$ which still has a component along the original direction $x$. This is called a low-angle scattering process. It will take many such collisions to reverse the electron's momentum which corresponds to flipping the momentum along the $+x$ direction to the $-x$ direction. Recall that the momentum gained from the field is actually very small compared with the momentum of the electron which is $m_e v_F$. A scattered electron must have an energy close to $E_F$ because lower energy states are filled. Thus, $\mathbf{p}_i$ and $\mathbf{p}_f$ have approximately the same magnitude $p_i = p_f = m_e v_F$ as shown in Figure 4.49.

At temperatures above the Debye temperature, we can assume that most of the phonons are vibrating with the Debye frequency $\omega_{\text{max}}$ and the phonon concentration $n_{\text{ph}}$ increases as $T$. These phonons have sufficient energies and momenta to fully scatter the electron on impact. Thus,

$$\sigma \propto \tau \propto \frac{1}{n_{\text{ph}}} \propto \frac{1}{T} \tag{4.69a}$$

*Electrical conductivity* $T > T_D$

When $T < T_D$, the phonon concentration follows $n_{\text{ph}} \propto T^3$, and the mean phonon energy $\overline{E}_{\text{ph}} \propto T$, because, as the temperature is raised, higher frequencies are excited. However, these phonons have low energy and small momenta, thus they only cause small-angle scattering processes as in Figure 4.49. The average phonon momentum $\hbar K$ is also proportional to the temperature (recall that at low frequencies Figure 4.42a shows that $\hbar\omega \propto \hbar K$). It will take many such collisions, say $N$, to flip the electron's momentum by $2m_e v_F$ from $+m_e v_F$ to $-m_e v_F$. During each collision, a phonon of momentum $\hbar K$ is absorbed as shown in Figure 4.49. Thus, if all phonons deflected the electron in the same angular direction, the collisions would sequentially add to $\theta$ in Figure 4.49, and we will need $(2m_e v_F)/(\hbar K)$ number of steps to flip the electron's momentum. The actual collisions add $\theta$'s randomly and the process is similar to particle diffusion, random walk, in Example 1.12 ($L^2 = Na^2$, where $L$ = displaced distance after $N$ jumps and $a$ = jump step). Thus,

$$N = \frac{(2m_e v_F)^2}{(\hbar K)^2} \propto \frac{1}{T^2}$$

The conductivity is therefore given by

$$\sigma \propto N\tau \propto \frac{N}{n_{\text{ph}}} \propto \frac{1}{T^5} \tag{4.69b}$$

*Electrical conductivity* $T < T_D$

which is indeed observed for Cu in Figure 2.8 when $T < T_D$ over the range where impurity scattering is negligible.

# ADDITIONAL TOPICS

## 4.11 BAND THEORY OF METALS: ELECTRON DIFFRACTION IN CRYSTALS

A rigorous treatment of the band theory of solids involves extensive quantum mechanical analysis and is beyond the scope of this book. However, we can attain a satisfactory understanding through a semiquantitative treatment.

We know that the wavefunction of the electron moving freely along $x$ in space is a traveling wave of the spatial form $\psi_k(x) = \exp(jkx)$, where $k$ is the wavevector $k = 2\pi/\lambda$ of the electron and $\hbar k$ is its momentum. Here, $\psi_k(x)$ represents a traveling wave because it must be multiplied by $\exp(-j\omega t)$, where $\omega = E/\hbar$, to get the total wavefunction $\Psi(x, t) = \exp[j(kx - \omega t)]$.

We will assume that an electron moving freely within the crystal and within a given energy band should also have a traveling wave type of wavefunction,

$$\psi_k(x) = A\exp(jkx) \tag{4.70}$$

where $k$ is the electron wavevector in the crystal and $A$ is the amplitude. This is a reasonable expectation, since, to a first order, we can take the *PE* of the electron inside a solid as zero, $V = 0$. Yet, the *PE* must be large outside, so the electron is contained within the crystal. When the *PE* is zero, Equation 4.70 is a solution to the Schrödinger equation. The momentum of the electron described by the traveling wave Equation 4.70 is then $\hbar k$ and its energy is

$$E_k = \frac{(\hbar k)^2}{2m_e} \tag{4.71}$$

The electron, as a traveling wave, will freely propagate through the crystal. However, not all traveling waves, can propagate in the lattice. The electron cannot have any $k$ value in Equation 4.70 and still move through the crystal. Waves can be reflected and diffracted, whether they are electron waves, X-rays, or visible light. Diffraction occurs when reflected waves interfere constructively. Certain $k$ values will cause the electron wave to be diffracted, preventing the wave from propagating.

The simplest illustration that certain $k$ values will result in the electron wave being diffracted is shown in Figure 4.50 for a hypothetical linear lattice in which diffraction is simply a reflection (what we call diffraction becomes Bragg reflection). The electron is assumed to be propagating in the forward direction along $x$ with a traveling wave function of the type in Equation 4.70. At each atom, some of this wave will be reflected. At $A$, the reflected wave is $A'$ and has a magnitude $A'$. If the reflected waves $A'$, $B'$, and $C'$ will reinforce each other, a full reflected wave will be created, traveling in the backward direction. The reflected waves $A'$, $B'$, $C'$, ... will reinforce each other if the path difference between $A'$, $B'$, $C'$, ... is $n\lambda$, where $\lambda$ is the wavelength and $n = 1, 2, 3, \ldots$ is an integer. When wave $B'$ reaches $A'$, it has traveled an additional

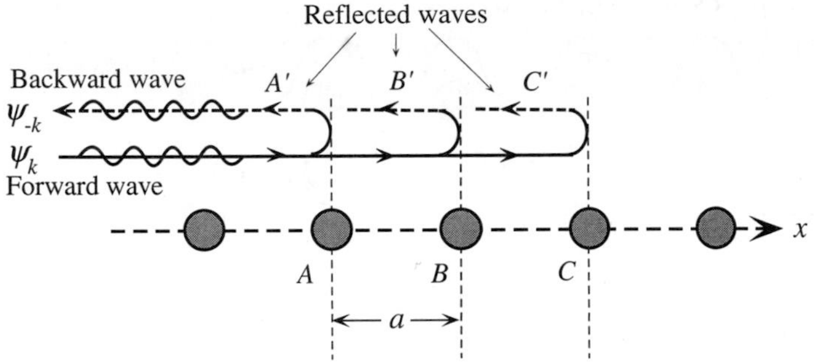


**Figure 4.50** An electron wave propagation through a linear lattice.
For certain $k$ values, the reflected waves at successive atomic planes reinforce each other, giving rise to a reflected wave traveling in the backward direction. The electron cannot then propagate through the crystal.

distance of $2a$. The path difference between $A'$ and $B'$ is therefore $2a$. For $A'$ and $B'$ to reinforce each other, we need

$$2a = n\lambda \qquad n = 1, 2, 3, \ldots$$

Substituting $\lambda = 2\pi/k$, we obtain the condition in terms of $k$

$$k = \frac{n\pi}{a} \qquad n = 1, 2, 3, \ldots \tag{4.72}$$

Thus, whenever $k$ is such that it satisfies the condition in Equation 4.72, all the reflected waves reinforce each other and produce a backward-traveling, reflected wave of the following form (with a negative $k$ value):

$$\psi_{-k}(x) = A\exp(-jkx) \tag{4.73}$$

This wave will also probably suffer a reflection, since its $k$ satisfies Equation 4.72, and the reflections will continue. The crystal will then contain waves traveling in the forward and backward directions. These waves will interfere to give **standing waves** inside the crystal. Hence, whenever the $k$ value satisfies Equation 4.72, traveling waves cannot propagate through the lattice. Instead, there can only be standing waves. For $k$ satisfying Equation 4.72, the electron wavefunction consists of waves $\psi_k$ and $\psi_{-k}$ interfering in two possible ways to give two possible standing waves:

$$\psi_c(x) = A\exp(jkx) + A\exp(-jkx) = A_c\cos\left(\frac{n\pi x}{a}\right) \tag{4.74}$$

$$\psi_s(x) = A\exp(jkx) - A\exp(-jkx) = A_s\sin\left(\frac{n\pi x}{a}\right) \tag{4.75}$$

The probability density distributions $|\psi_c(x)|^2$ and $|\psi_s(x)|^2$ for the two standing waves are shown in Figure 4.51. The first standing wave $\psi_c(x)$ is at a maximum on the ion cores, and the other $\psi_s(x)$ is at a maximum between the ion cores. Note also that both the standing waves $\psi_c(x)$ and $\psi_s(x)$ are solutions to the Schrödinger equation.

The closer the electron is to a positive nucleus, the lower is its electrostatic $PE$, by virtue of $-e^2/4\pi\varepsilon_o r$. The $PE$ of the electron distribution in $\psi_c(x)$ is lower than that in $\psi_s(x)$, because the maxima for $\psi_c(x)$ are nearer the positive ions. Therefore, the energy of the electron in $\psi_c(x)$ is lower than that of the electron in $\psi_s(x)$, or $E_c < E_s$.

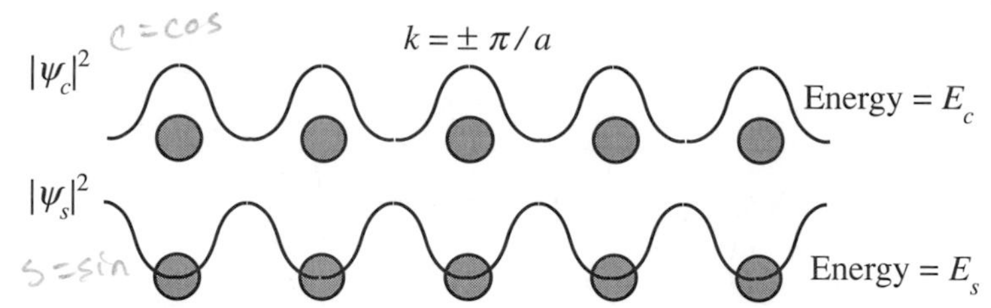


**Figure 4.51** Forward and backward waves in the crystal with $k = \pm\pi/a$ give rise to two possible standing waves $\psi_c$ and $\psi_s$. Their probability density distributions $|\psi_c|^2$ and $|\psi_s|^2$ have maxima either at the ions or between the ions, respectively.

It is not difficult to evaluate the energies $E_c$ and $E_s$. The kinetic energy of the electron is the same in both $\psi_c(x)$ and $\psi_s(x)$, because these wavefunctions have the same $k$ value and $KE$ is given by $(\hbar k)^2/2m_e$. However, there is an electrostatic $PE$ arising from the interaction of the electron with the ion cores, and this $PE$ is different for the two wavefunctions. Suppose that $V(x)$ is the electrostatic $PE$ of the electron at position $x$. We then must find the average, using the probability density distribution. Given that $|\psi_c(x)|^2\,dx$ is the probability of finding the electron at $x$ in $dx$, the potential energy $V_c$ of the electron is simply $V(x)$ averaged over the entire linear length $L$ of the crystal. Thus, the potential energy $V_c$ for $\psi_c(x)$ is

$$V_c = \frac{1}{L}\int_0^L V(x)|\psi_c(x)|^2\,dx = -V_n \qquad [4.76]$$

where $V_n$ is the numerical result of the integration, which depends on $k = n\pi/a$ or $n$, by virtue of Equation 4.74. The integration in Equation 4.76 is a negative number that depends on $n$. We do not need to evaluate the integral, as we only need its final numerical result.

Using $|\psi_s(x)|^2$, we can also find $V_s$, the $PE$ associated with $\psi_s(x)$. The result is that $V_s$ is a positive quantity given by $+V_n$, where $V_n$ is again the numerical result of the integration in Equation 4.76, which depends on $n$. The energies of the wavefunctions $\psi_c$ and $\psi_s$ whenever $k = n\pi/a$ are

$$E_c = \frac{(\hbar k)^2}{2m_e} - V_n \qquad k = \frac{n\pi}{a} \qquad [4.77]$$

$$E_s = \frac{(\hbar k)^2}{2m_e} + V_n \qquad k = \frac{n\pi}{a} \qquad [4.78]$$

Clearly, whenever $k$ has the critical values $n\pi/a$, there are only two possible values for the energies $E_c$ and $E_s$ as determined by Equations 4.77 and 4.78; no other energies are allowed in between. These two energies are separated by $2V_n$.

Away from the critical $k$ values determined by $k = n\pi/a$, the electron simply propagates as a traveling wave; the wave does not get reflected. The energy is then given by the free-running wave solution to the Schrödinger equation, that is, Equation 4.71,

$$E_k = \frac{(\hbar k)^2}{2m_e} \qquad \text{Away from } k = \frac{n\pi}{a} \qquad [4.79]$$

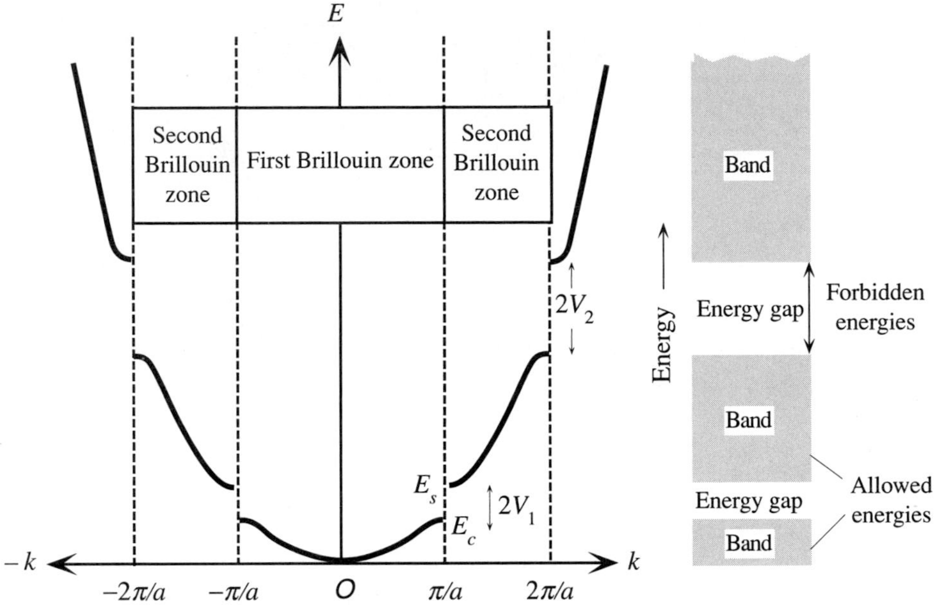


**Figure 4.52** The energy of the electron as a function of its wavevector $k$ inside a one-dimensional crystal.
There are discontinuities in the energy at $k = \pm n\pi/a$, where the waves suffer Bragg reflections in the crystal. For example, there can be no energy value for the electron between $E_c$ and $E_s$. Therefore, $E_s - E_c$ is an energy gap at $k = \pm\pi/a$. Away from the critical $k$ values, the $E$–$k$ behavior is like that of a free electron, with $E$ increasing with $k$ as $E = (\hbar k)^2/2m_e$. In a solid, these energies fall within an energy band.

It seems that the energy of the electron increases parabolically with $k$ along Equation 4.79 and then suddenly, at $k = n\pi/a$, it suffers a sharp discontinuity and increases parabolically again. Although the discontinuities at the critical points $k = n\pi/a$ are expected, by virtue of the Bragg reflection of waves, reflection effects will still be present to a certain extent, even within a small region around $k = n\pi/a$. The individual reflections shown in Figure 4.50 do not occur exactly at the origins of the atoms at $x = a, 2a, 3a, \ldots$. Rather, they occur over some distance, since the wave must interact with the electrons in the ion cores to be reflected. We therefore expect $E$–$k$ behavior to deviate from Equation 4.79 in the neighborhood of the critical points, even if $k$ is not exactly $n\pi/a$. Figure 4.52 shows the $E$–$k$ behavior we expect, based on these arguments.

In Figure 4.52, we notice that there are certain energy ranges occurring at $k = \pm(n\pi/a)$ in which there are no allowed energies for the electron. As we saw previously, the electron cannot possess an energy between $E_c$ and $E_s$ at $k = \pi/a$. These energy ranges form **energy gaps** at the critical points $k = \pm(n\pi/a)$.

The range of $k$ values from zero to the first energy gap at $k = \pm(\pi/a)$ defines a zone of $k$ values called the **first Brillouin zone.** The zone between the first and second energy gap defines the **second Brillouin zone,** and so on. The Brillouin zone boundaries therefore identify where the energy discontinuities, or gaps, occur along the $k$ axis.

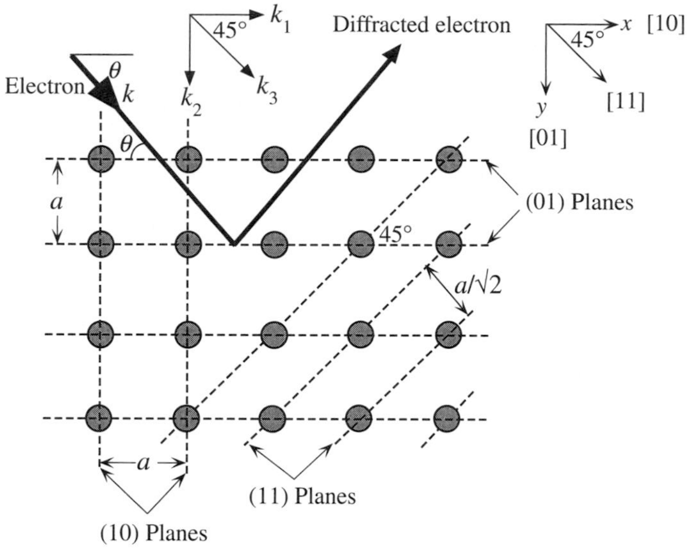


**Figure 4.53** Diffraction of the electron in a two-dimensional crystal.
Diffraction occurs whenever $k$ has a component satisfying $k_1 = \pm n\pi/a$, $k_2 = \pm n\pi/a$, or $k_3 = \pm n\pi\sqrt{2}/a$. In general terms, diffraction occurs when $k \sin\theta = n\pi/d$.

Electron motion in the three-dimensional crystal can be readily understood based on the concepts described here. For simplicity, we consider an electron propagating in a two-dimensional crystal, which is analogous, for example, to propagation in the $xy$ plane of a crystal, as depicted in Figure 4.53. For certain $k$ values and in certain directions, the electron will suffer diffraction and will be unable to propagate in the crystal.

Suppose that the electron's $k$ vector along $x$ is $k_1$. Whenever $k_1 = \pm n\pi/a$, the electron will be diffracted by the planes perpendicular to $x$, that is, the (10) planes.[21] Similarly, it will be diffracted by the (01) planes whenever its $k$ vector along $y$ is $k_2 = \pm n\pi/a$. The electron can also be diffracted by the (11) planes, whose separation is $a/\sqrt{2}$. If the component of $k$ perpendicular to the (11) plane is $k_3$, then whenever $k_3 = \pm n\pi(\sqrt{2}/a)$, the electron will experience diffraction. These diffraction conditions can all be expressed through the **Bragg diffraction condition** $2d \sin\theta = n\lambda$, or

*Bragg diffraction condition*

$$k \sin\theta = \frac{n\pi}{d} \qquad [4.80]$$

where $d$ is the interplanar separation and $n$ is an integer; $d = a$ for (10) planes, and $d = a/\sqrt{2}$ for (11) planes.

When we plot the energy of the electron as a function of $k$, we must consider the direction of $k$, since the diffraction behavior in Equation 4.80 depends on $\sin\theta$. Along $x$, at $\theta = 0$, the energy gap occurs at $k = \pm(n\pi/a)$. Along $\theta = 45°$, it is at $k = \pm n\pi(\sqrt{2}/a)$, which is farther away. The $E$–$k$ behavior for the electron in the two-dimensional lattice is shown in Figure 4.54 for the [10] and [11] directions. The figure shows that the first energy gap along $x$, in the [10] direction, is at $k = \pi/a$. Along the [11] direction, which is at 45° to the $x$ axis, the first gap is at $k = \pi\sqrt{2}/a$.

[21] We use Miller indices in two dimensions by dropping the third digit but keeping the same interpretation. The direction along $x$ is [10] and the plane perpendicular to $x$ is (10).

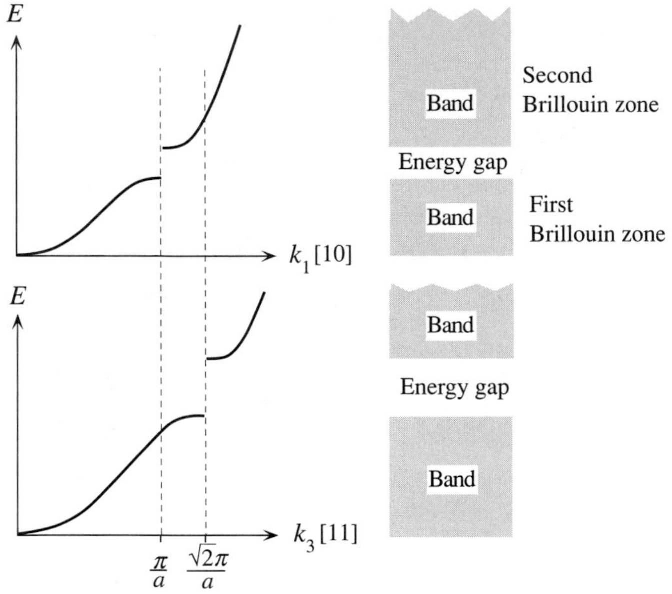


**Figure 4.54** The $E$–$k$ behavior for the electron along different directions in the two-dimensional crystal.
The energy gap along [10] is at $\pi/a$ whereas it is at $\sqrt{2}\pi/a$ along [11].

When we consider the overlap of the energy bands along [10] and [11], in the case of a metal, there is no apparent energy gap. The electron can always find any energy simply by changing its direction.

The effects of overlap between energy bands and of energy gaps in different directions are illustrated in Figure 4.55. In the case of a semiconductor, the energy gap along [10] overlaps that along [11], so there is an overall energy gap. The electron in the semiconductor cannot have an energy that falls into this energy gap.

The first and second Brillouin zones for the two-dimensional lattice of Figure 4.53 are shown in Figure 4.56. The zone boundaries mark the occurrences of energy gaps in $k$ space (space defined by $k$ axes along the $x$ and $y$ directions). When we look at the $E$–$k$ behavior, we must consider the crystal directions. This is most conveniently done by plotting energy contours in $k$ space, as in Figure 4.57. Each contour connects all those values of $k$ that possess the same energy. A point such as $P$ on an energy contour gives the value of $k$ for that energy along the direction $OP$. Initially, the energy contours are circles, as the energy follows $(\hbar k)^2/2m_e$ behavior, whatever the direction of $k$. However, near the critical values, that is, near the Brillouin zone boundaries, $E$ increases more slowly than the parabolic relationship, as is apparent in Figure 4.52. Therefore, the circles begin to bulge as critical $k$ values are approached. In Figure 4.57, the high-energy contours are concentrated in the corners of the zone, simply because the critical value is reached last along [11]. The energy contours do not continue smoothly across the zone boundary, because of the energy discontinuity in the $E$–$k$ relationship at the boundary. Indeed, Figure 4.54 shows that the lowest energy in the second Brillouin zone may be lower than the highest energy in the first Brillouin zone.

There are two cases of interest. In the first, there is no apparent energy gap, as in Figure 4.57a, which corresponds to Figure 4.55a. The electron can have any energy

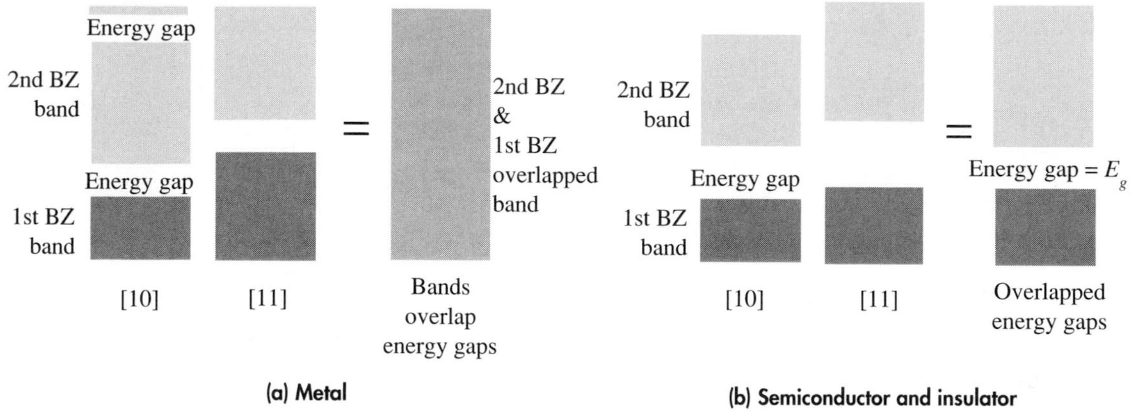


**Figure 4.55**

(a) For the electron in a metal, there is no apparent energy gap because the second BZ (Brillouin zone) along [10] overlaps the first BZ along [11]. Bands overlap the energy gaps. Thus, the electron can always find any energy by changing its direction.

(b) For the electron in a semiconductor, there is an energy gap arising from the overlap of the energy gaps along the [10] and [11] directions. The electron can never have an energy within this energy gap $E_g$.

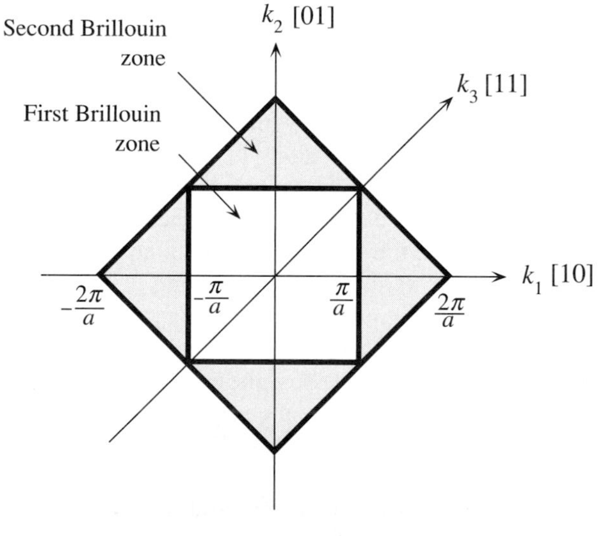


**Figure 4.56** The Brillouin zones in two dimensions for the cubic lattice.

The Brillouin zones identify the boundaries where there are discontinuities in the energy (energy gaps).

value. In the second case, there is a range of energies that are not allowed, as shown in Figure 4.57b, which corresponds to Figure 4.55b.

In three dimensions, the $E$–$k$ energy contour in Figure 4.57 becomes a surface in three-dimensional $k$ space. To understand the use of such $E$–$k$ contours or surfaces, consider that an $E$–$k$ contour (or a surface) is made of many finely separated individual points, each representing a possible electron wavefunction $\psi_k$ with a possible

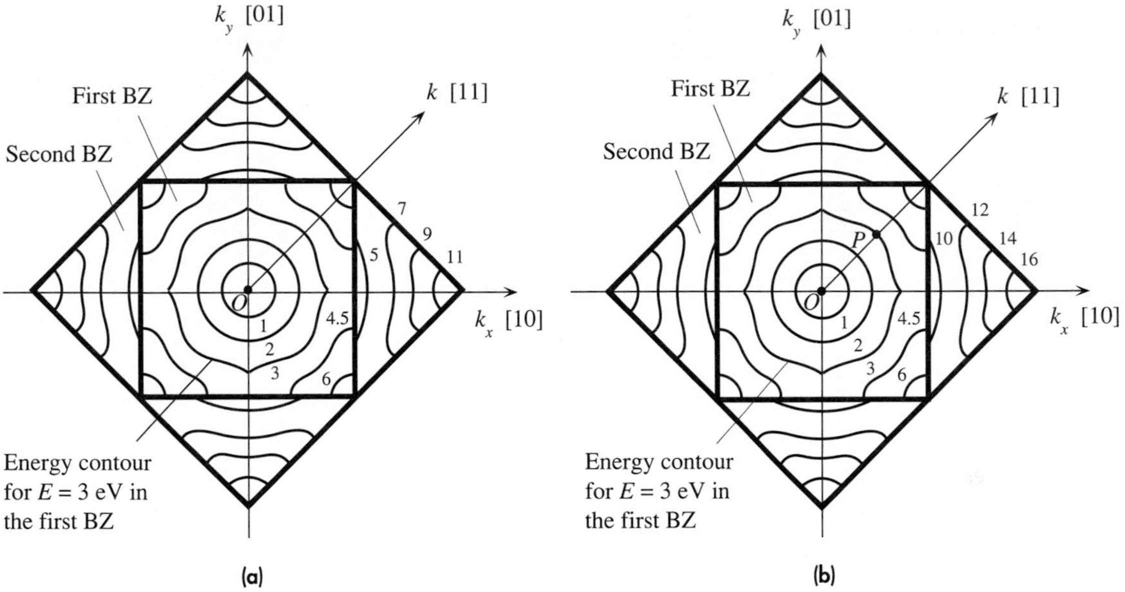


**Figure 4.57** Energy contours in $k$ space (space defined by $k_x$, $k_y$).
Each contour represents the same energy value. Any point $P$ on the contour gives the values of $k_x$ and $k_y$ for that energy in that direction from $O$. For point $P$, $E = 3$ eV and $OP$ along [11] is $k$.
(a) In a metal, the lowest energy in the second zone (5 eV) is lower than the highest energy (6 eV) in the first zone. There is an overlap of energies between the Brillouin zones.
(b) In a semiconductor or an insulator, there is an energy gap between the highest energy contour (6 eV) in the first zone and the lowest energy contour (10 eV) in the second zone.

energy $E$. At absolute zero, all the energies up to the Fermi energy are taken by the valence electrons. In $k$ space, the energy surface, corresponding to the Fermi energy is termed the **Fermi surface.** The shape of this Fermi surface provides a means of interpreting the electrical and magnetic properties of solids.

For example, Na has one 3$s$ electron per atom. In the solid, the 3$s$ band is half full. The electrons take energies up to $E_F$, which corresponds to a spherical Fermi surface within the first Brillouin zone, as indicated in Figure 4.58a. We can then say that all the valence electrons (or nearly all) in this alkali solid exhibit an $E = (\hbar k)^2/2m_e$ type of behavior, as if they were free. When an external force is applied, such as an electric or magnetic field, we can treat the electron behavior as if it were free inside the metal with a constant mass. This is a desirable simplification for studying such metals. We can illustrate this desirability with an example. The Hall coefficient $R_H$ derived in Chapter 2 was based on treating the electron as if it were a free particle inside the metal, or

$$R_H = -\frac{1}{en} \qquad \textbf{[4.81]}$$

For Na, the experimental value of $R_H$ is $-2.50 \times 10^{-10}$ m$^3$ C$^{-1}$. Using the density (0.97 g cm$^{-3}$) and atomic mass (23) of Na and one valence electron per atom, we can

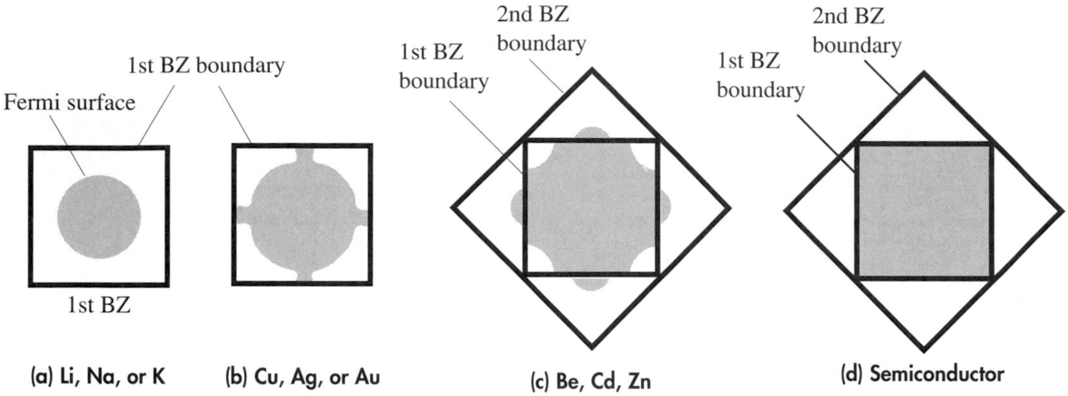


**Figure 4.58** Schematic sketches of Fermi surfaces in two dimensions, representing various materials qualitatively.
(a) Monovalent group IA metals.
(b) Group IB metals.
(c) Be (Group IIA), Zn, and Cd (Group IIB).
(d) A semiconductor.

calculate $n = 2.54 \times 10^{28}$ m$^{-3}$ and $R_H = -2.46 \times 10^{-10}$ m$^3$ C$^{-1}$, which is very close to the experimental value.

In the case of Cu, Ag, and Au (the IB metals in the Periodic Table), the Fermi surface is inside the first Brillouin zone, but it is not spherical as depicted in Figure 4.58b. Also, it touches the centers of the zone boundaries. Some of those electrons near the zone boundary behave quite differently than $E = (\hbar k)^2/2m_e$, although the majority of the electrons in the sphere do exhibit this type of behavior. To an extent, we can expect the free electron derivations to hold. The experimental value of $R_H$ for Cu is $-0.55 \times 10^{-10}$ m$^3$ C$^{-1}$, whereas the expected value, based on Equation 4.81 with one electron per atom, is $-0.73 \times 10^{-10}$ m$^3$ C$^{-1}$, which is noticeably greater than the experimental value.

The divalent metals Be, Mg, and Ca have closed outer $s$ subshells and should have a full $s$ band in the solid. Recall that electrons in a full band cannot respond to an applied field and drift. We also know that there should be an overlap between the $s$ and $p$ bands, forming one partially filled continuous energy band, so these metals are indeed conductors. In terms of Brillouin zones, their structure is based on Figure 4.55a, which has the second zone overlapping the first Brillouin zone. The Fermi surface extends into the second zone and the corners of the first zone are empty, as depicted in Figure 4.58c. Since there are empty energy levels next to the Fermi surface, the electrons can gain energy and drift in response to an applied field. But the surface is not spherical; indeed, near the corners of the first zone, it even has the wrong curvature. Therefore, it is no longer possible to describe these electrons on the Fermi surface as obeying $E = (\hbar k)^2/2m_e$. When a magnetic field is applied to a drifting electron to bend its trajectory, its total behavior is different than that expected when it is acting as a free particle. The external force changes the momentum $\hbar k$ and the corresponding change in the energy depends on the Fermi surface and can be quite complicated. To finish the example on the Hall coefficient, we note that based on two valence electrons per atom (Group IIA), the Hall coefficient for Be should be $-0.25 \times 10^{-10}$ m$^3$ C$^{-1}$, but the measured value is a positive coefficient of $+2.44 \times 10^{-10}$ m$^3$ C$^{-1}$. Equation 4.81 is therefore useless. It seems that the electrons moving at the Fermi surface of Be are equivalent to the motion of positive charges (like holes), so the Hall effect registers a positive coefficient.

The Fermi surface of a semiconductor is simply the boundary of the first Brillouin zone, because there is an energy gap between the first and the second Brillouin zones, as depicted in Figure 4.55b. In a semiconductor, all the energy levels up to the energy gap are taken up by the valence electrons. The first Brillouin zone forms the valence band and the second forms the conduction band.

## 4.12 GRÜNEISEN'S MODEL OF THERMAL EXPANSION

We considered thermal expansion in Section 1.4.2 where the principle is illustrated in Figure 1.18, which shows the potential energy curve $U(r)$ for two atoms separated by a distance $r$ in a crystal. At temperature $T_1$ we know that the atoms will be vibrating about their equilibrium positions between positions $B$ and $C$, compressing ($B$) and stretching ($C$) the bond between them. The line $BC$ corresponds to the total energy $E$ of the pair of atoms. The average separation at $T_1$ is at $A$, halfway between $B$ and $C$. We also know that the *PE* curve $U(r)$ is *asymmetric,* and it is this asymmetry that leads to the phenomenon of thermal expansion. When the temperature increases from $T_1$ to $T_2$, the atoms vibrate between $B'$ and $C'$ and the average separation between the atoms also increases, from $A$ to $A'$, which we identified as *thermal expansion*. If the *PE* curve were symmetric, then there would be no thermal expansion.

Since the linear expansion coefficient $\lambda$ is related to the shape of the *PE* curve, $U(r)$, it is also related to the elastic bulk modulus $K$ that measures how difficult it is to stretch or compress the bonds. $K$ depends on $U(r)$ in the same way that the elastic modulus $Y$ depends on $U(r)$ as explained in Example 1.5.[22] Further, $\lambda$ also depends on the amount of increase from $BC$ to $B'C'$ per degree of increase in the temperature. $\lambda$ must therefore also depend on the heat capacity. When the temperature increases by a small amount $\delta T$, the energy per atom increases by $(C_v\, \delta T)/N$ where $C_v$ is the heat capacity per unit volume and $N$ is the number of atoms per unit volume. If $C_v\, \delta T$ is large, then the line $B'C'$ in Figure 1.18 will be higher up on the energy curve and the average separation $A'$ will therefore be larger. Thus, the larger is the heat capacity, the greater is the interatomic separation, which means $\lambda \propto C_v$. Further, the average separation, point $A$, depends on how much the bonds are stretched and compressed. For large

[22] $K$ is a measure of the elastic change in the volume of a body in response to an applied pressure; large $K$ means a small change in volume for a given pressure. $Y$ is a measure of the elastic change in the length of the body in response to an applied stress; large $Y$ means a small change in length. Both involve stretching or compressing bonds.

amounts of displacement from equilibrium, the average $A$ will be greater as more asymmetry of the $PE$ curve is used. Thus, the smaller is the elastic modulus $K$, the greater is $\lambda$; we see that $\lambda \propto C_v/K$.

*Asymmetric potential energy curve*

If we were to expand $U(r)$ about its minimum value $U_{\min}$ at $r = r_o$, we would obtain the Taylor expansion,

$$U(r) = U_{\min} + a_2(r - r_o)^2 + a_3(r - r_o)^3 + \cdots$$

where $a_2$ and $a_3$ are coefficients related to the second and third derivatives of $U$ at $r_o$. The term $(r - r_o)$ is missing because we are expanding a series about $U_{\min}$ where $dU/dr = 0$. The $U_{\min}$ and the $a_2(r - r_o)^2$ term give a parabola about $U_{\min}$ which is a symmetric curve around $r_o$ and therefore does not lead to thermal expansion. It is the $a_3$ term that gives the expansion because it leads to asymmetry. Thus the amount of expansion $\lambda$ also depends on the amount of asymmetry with respect to symmetry, that is $a_3/a_2$. Thus,

*Linear expansion coefficient*

$$\lambda \propto \frac{a_3}{a_2}\frac{C_v}{K}$$

The ratio of $a_3$ and $a_2$ depends on the nature of the bond. A simplified analytical treatment (beyond the scope of this book) gives $\lambda$ as

*Grüneisen's law*

$$\lambda \approx 3\gamma\frac{C_v}{K} \qquad [4.82]$$

where $\gamma$ is a "constant" called the *Grüneisen parameter*. The Grüneisen constant $\gamma$ is approximately $-(r_oa_3)/(2a_2)$ where $r_o$ is the equilibrium atomic separation, and thus $\gamma$ represents the asymmetry of the energy curve. The approximate equality simply emphasizes the number of assumptions that are typically made in deriving Equation 4.82. The Grüneisen parameter $\gamma$ is of the order of unity for many materials; experimentally, $\gamma = 0.1 - 1$. We can also write the Grüneisen law in terms of the molar heat capacity $C_m$ (heat capacity per mole) or the specific heat capacity $c_s$ (heat capacity per unit mass). If $\rho$ is the density, and $M_{\text{at}}$ is the atomic mass of the constituent atoms of the crystal, then

*Grüneisen's law*

$$\lambda = 3\gamma\frac{\rho C_m}{M_{\text{at}}K} = 3\gamma\frac{\rho c_s}{K} \qquad [4.83]$$

We can calculate the Grüneisen parameter $\gamma$ for materials that possess different types of interatomic bonding and thereby obtain typical values for $\gamma$. This would also expose the extent of unharmonicity in the bonding. Given the experimental values for $\lambda$, $K$, $\rho$ and $c_s$, the Grüneisen parameters have been calculated from Equation 4.83 and are listed in Table 4.6. An interesting feature of the results is that the experimental $\gamma$ values, within a factor of 2–3, are about the same, at least to an order of magnitude. Equation 4.83 also indicates that the $\lambda$ versus $T$ behavior should resemble the $C_v$ versus $T$ dependence, which is approximately the case if one compares Figure 1.20 with Figure 4.45. ($K$ does not change much with temperature.) There is one notable difference. At very low temperatures $\lambda$ can change sign and become negative for certain crystals, whereas $C_v$ cannot.

**Table 4.6** The Grüneisen parameter for some selected materials with different types of interatomic bonding

| Material | $\rho$ (g cm$^{-3}$) | $\lambda$ ($\times 10^{-6}$ K$^{-1}$) | $K$ (GPa) | $c_s$ (J kg$^{-1}$ K$^{-1}$) | $\gamma$ |
|---|---|---|---|---|---|
| Iron (metallic, BCC) | 7.9 | 12.1 | 170 | 444 | 0.20 |
| Copper (metallic, FCC) | 8.96 | 17 | 140 | 380 | 0.23 |
| Germanium (covalent) | 5.32 | 6 | 77 | 322 | 0.09 |
| Glass (covalent-ionic) | 2.45 | 8 | 70 | 800 | 0.10 |
| NaCl (ionic) | 2.16 | 39.5 | 28 | 880 | 0.19 |
| Tellurium (mixed) | 6.24 | 18.2 | 40 | 202 | 0.19 |
| Polystyrene (van der Waals) | 1.05 | 100 | 3 | 1200 | 0.08 |

## CD Selected Topics and Solved Problems

**Selected Topics**

Hall Effect
Thermal Conductivity
Thermoelectric Effects in Metals: Thermocouples
Thermal Expansion (Grüneisen's Law)

**Solved Problems**

The Water Molecule

## DEFINING TERMS

**Average energy** $E_{av}$ of an electron in a metal is determined by the Fermi–Dirac statistics and the density of states. It increases with the Fermi energy and also with the temperature.

**Boltzmann statistics** describes the behavior of a collection of particles (*e.g.*, gas atoms) in terms of their energy distribution. It specifies the number of particles $N(E)$ with given energy, through $N(E) \propto \exp(-E/kT)$, where $k$ is the Boltzmann constant. The description is nonquantum mechanical in that there is no restriction on the number of particles that can have the same state (the same wavefunction) with an energy $E$. Also, it applies when there are only a few particles compared to the number of possible states, so the likelihood of two particles having the same state becomes negligible. This is generally the case for thermally excited electrons in the conduction band of a semiconductor, where there are many more states than electrons. The kinetic energy distribution of gas molecules in a tank obeys the Boltzmann statistics.

**Cathode** is a negative electrode. It emits electrons or attracts positive charges, that is, cations.

**Debye frequency** is the maximum frequency of lattice vibrations that can exist in a particular crystal. It is the cut-off frequency for lattice vibrations.

**Debye temperature** is a characteristic temperature of a particular crystal above which nearly all the atoms are vibrating in accordance with the kinetic molecular theory, that is, each atom has an average energy (potential + kinetic) of $3kT$ due to atomic vibrations, and the heat capacity is determined by the Dulong–Petit rule.

**Density of states** $g(E)$ is the number of electron states [*e.g.*, wavefunctions, $\psi(n, \ell, m_\ell, m_s)$] per unit energy per unit volume. Thus, $g(E)\,dE$ is the number of states in the energy range $E$ to $(E + dE)$ per unit volume.

**Density of vibrational states** is the number of lattice vibrational modes per unit angular frequency range.

**Dispersion relation** relates the angular frequency $\omega$ and the wavevector $K$ of a wave. In a crystal lattice, the coupling of atomic oscillations leads to a particular relationship between $\omega$ and $K$ which determines the allowed lattice waves and their group velocities. The dispersion relation is specific to the crystal structure, that is, it depends on the lattice, basis, and bonding.

**Effective electron mass** $m_e^*$ represents the inertial resistance of an electron inside a crystal against an acceleration imposed by an external force, such as the applied electric field. If $F_{\text{ext}} = e\mathcal{E}_x$ is the external applied force due to the applied field $\mathcal{E}_x$, then the effective mass $m_e^*$ determines the acceleration $a$ of the electron by $e\mathcal{E}_x = m_e^* a$. This takes into account the effect of the internal fields on the motion of the electron. In vacuum where there are no internal fields, $m_e^*$ is the mass in vacuum $m_e$.

**Fermi–Dirac statistics** determines the probability of an electron occupying a state at an energy level $E$. This takes into account that a collection of electrons must obey the Pauli exclusion principle. The Fermi–Dirac function quantifies this probability via $f(E) = 1/\{1 + \exp[(E - E_F)/kT]\}$, where $E_F$ is the Fermi energy.

**Fermi energy** is the maximum energy of the electrons in a metal at 0 K.

**Field emission** is the tunneling of an electron from the surface of a metal into vacuum, due to the application of a strong electric field (typically $\mathcal{E} > 10^9$ V m$^{-1}$).

**Group velocity** is the velocity at which traveling waves carry energy. If $\omega$ is the angular frequency and $K$ is the wavevector of a wave, then the group velocity $v_g = d\omega/dK$.

**Harmonic oscillator** is an oscillating system, for example, two masses joined by a spring, that can be described by *simple harmonic motion.* In quantum mechanics, the energy of a harmonic oscillator is quantized and can only increase or decrease by a discrete amount $\hbar\omega$. The minimum energy of a harmonic oscillator is not zero but $\frac{1}{2}\hbar\omega$ (see **zero-point energy**).

**Lattice wave** is a wave in a crystal due to coupled oscillations of the atoms. Lattice waves may be traveling or stationary waves.

**Linear combination of atomic orbitals (LCAO)** is a method for obtaining the electron wavefunction in the molecule from a linear combination of individual atomic wavefunctions. For example, when two H atoms $A$ and $B$ come together, the electron wavefunctions, based on LCAO, are

$$\psi_a = \psi_{1s}(A) + \psi_{1s}(B)$$

$$\psi_b = \psi_{1s}(A) - \psi_{1s}(B)$$

where $\psi_{1s}(A)$ and $\psi_{1s}(B)$ are atomic wavefunctions centered around the H atoms $A$ and $B$, respectively. The $\psi_a$ and $\psi_b$ represent molecular orbital wavefunctions for the electron; they reflect the behavior of the electron, or its probability distribution, in the molecule.

**Mode** or **state of lattice vibration** is a distinct, independent way in which a crystal lattice can vibrate with its own particular frequency $\omega$ and wavevector $K$. There are only a finite number of vibrational modes in a crystal.

**Molecular orbital wavefunction,** or simply molecular orbital, is a wavefunction for an electron within a system of two or more nuclei (*e.g.*, molecule). A molecular orbital determines the probability distribution of the electron within the molecule, just as the atomic orbital determines the electron's probability distribution within the atom. A molecular orbital can take two electrons with opposite spins.

**Orbital** is a region of space in an atom or molecule where an electron with a given energy may be found. An orbit, which is a well-defined path for an electron, cannot be used to describe the whereabouts of the electron in an atom or molecule because the electron has a probability distribution. Orbitals are generally represented by a surface within which the total probability is high, for example, 90 percent.

**Orbital wavefunction,** or simply orbital, describes the spatial dependence of the electron. The orbital is $\psi(r, \theta, \phi)$, which depends on $n$, $\ell$, and $m_\ell$, and the spin dependence $m_s$ is excluded.

**Phonon** is a quantum of lattice vibrational energy of magnitude $\hbar\omega$, where $\omega$ is the vibrational angular frequency. A phonon has a momentum $\hbar K$ where $K$ is the wavevector of the lattice wave.

**Seebeck effect** is the development of a built-in potential difference across a material as a result of a temperature gradient. If $dV$ is the built-in potential across a temperature difference $dT$, then the Seebeck coefficient $S$ is defined as $S = dV/dT$. The coefficient gauges the magnitude of the Seebeck effect. Only the net Seebeck voltage difference between different metals can be measured. The principle of the thermocouple is based on the Seebeck effect.

**State** is a possible wavefunction for the electron that defines its spatial (orbital) and spin properties, for example, $\psi(n, \ell, m_\ell, m_s)$ is a state of the electron. From the Schrödinger equation, each state corresponds to a certain electron energy $E$. We thus speak of a state with energy $E$, state of energy $E$, or even an energy state. Generally there may be more than one state $\psi$ with the same energy $E$.

**Thermionic emission** is the emission of electrons from the surface of a heated metal.

**Work function** is the minimum energy needed to free an electron from the metal at a temperature of absolute zero. It is the energy separation of the Fermi level from the vacuum level.

**Zero-point energy** is the minimum energy of a harmonic oscillator $\frac{1}{2}\hbar\omega$. Even at 0 K, an oscillator in quantum mechanics will have a finite amount of energy which is its zero-point energy. Heisenberg's uncertainty principle does not allow a harmonic oscillator to have zero energy because that would mean no uncertainty in the momentum and consequently an infinite uncertainty in space ($\Delta p_x \, \Delta x > \hbar$).

# QUESTIONS AND PROBLEMS

**4.1 Phase of an atomic orbital**

*a.* What is the functional form of a $1s$ wavefunction $\psi(r)$? Sketch schematically the atomic wavefunction $\psi_{1s}(r)$ as a function of distance from the nucleus.

*b.* What is the total wavefunction $\Psi_{1s}(r, t)$?

*c.* What is meant by two wavefunctions $\Psi_{1s}(A)$ and $\Psi_{1s}(B)$ that are out of phase?

*d.* Sketch schematically the two wavefunctions $\Psi_{1s}(A)$ and $\Psi_{1s}(B)$ at one instant.

**4.2 Molecular orbitals and atomic orbitals** Consider a linear chain of four identical atoms representing a hypothetical molecule. Suppose that each atomic wavefunction is a $1s$ wavefunction. This system of identical atoms has a center of symmetry $O$ with respect to the center of the molecule (midway between the second and the third atom), and all molecular wavefunctions must be either symmetric or antisymmetric about $O$.

*a.* Using the LCAO principle, sketch the possible molecular orbitals.

*b.* Sketch the probability distributions $|\psi|^2$.

*c.* If more nodes in the wavefunction lead to greater energies, order the energies of the molecular orbitals.

Note: The electron wavefunctions, and the related probability distributions, in a simple potential energy well that are shown in Figure 3.15 can be used as a rough *guide* toward finding the appropriate molecular wavefunctions in the four-atom symmetric molecule. For example, if we were to smooth the electron potential energy in the four-atom molecule into a constant potential energy, that is, generate a potential energy well, we should be able to modify or distort, without flipping, the molecular orbitals to somewhat resemble $\psi_1$ to $\psi_4$ sketched in Figure 3.15. Consider also that the number of nodes increases from none for $\psi_1$ to three for $\psi_4$ in Figure 3.15.

**4.3 Diamond and tin** Germanium, silicon, and diamond have the same crystal structure, that of diamond. Bonding in each case involves $sp^3$ hybridization. The bonding energy decreases as we go from diamond to Si to Ge, as noted in Table 4.7.

*a.* What would you expect for the bandgap of diamond? How does it compare with the experimental value of 5.5 eV?

*b.* Tin has a tetragonal crystal structure, which makes it different than its group members, diamond, silicon, and germanium.

1. Is it a metal or a semiconductor?
2. What experiments do you think would expose its semiconductor properties?

Table 4.7

| Property | Diamond | Silicon | Germanium | Tin |
|---|---|---|---|---|
| Melting temperature, °C | 3800 | 1417 | 937 | 232 |
| Covalent radius, nm | 0.077 | 0.117 | 0.122 | 0.146 |
| Bond energy, eV | 3.60 | 1.84 | 1.7 | 1.2 |
| First ionization energy, eV | 11.26 | 8.15 | 7.88 | 7.33 |
| Bandgap, eV | ? | 1.12 | 0.67 | ? |

**4.4** **Compound III–V Semiconductors** Indium as an element is a metal. It has a valency of III. Sb as an element is a metal and has a valency of V. InSb is a semiconductor, with each atom bonding to four neighbors, just like in silicon. Explain how this is possible and why InSb is a semiconductor and not a metal alloy. (Consider the electronic structure and $sp^3$ hybridization for each atom.)

**4.5** **Compound II–VI semiconductors** CdTe is a semiconductor, with each atom bonding to four neighbors, just like in silicon. In terms of covalent bonding and the positions of Cd and Te in the Periodic Table, explain how this is possible. Would you expect the bonding in CdTe to have more ionic character than that in III–V semiconductors?

***4.6** **Density of states for a two-dimensional electron gas** Consider a two-dimensional electron gas in which the electrons are restricted to move freely within a square area $a^2$ in the $xy$ plane. Following the procedure in Section 4.5, show that the density of states $g(E)$ is constant (independent of energy).

**4.7** **Fermi energy of Cu** The Fermi energy of electrons in copper at room temperature is 7.0 eV. The electron drift mobility in copper, from Hall effect measurements, is 33 $cm^2\ V^{-1}\ s^{-1}$.

*a.* What is the speed $v_F$ of conduction electrons with energies around $E_F$ in copper? By how many times is this larger than the average thermal speed $v_{thermal}$ of electrons, if they behaved like an ideal gas (Maxwell–Boltzmann statistics)? Why is $v_F$ much larger than $v_{thermal}$?

*b.* What is the De Broglie wavelength of these electrons? Will the electrons get diffracted by the lattice planes in copper, given that interplanar separation in Cu = 2.09 Å? (Solution guide: Diffraction of waves occurs when $2d \sin\theta = \lambda$, which is the Bragg condition. Find the relationship between $\lambda$ and $d$ that results in $\sin\theta > 1$ and hence no diffraction.)

*c.* Calculate the mean free path of electrons at $E_F$ and comment.

**4.8** **Free electron model, Fermi energy, and density of states** Na and Au both are valency I metals; that is, each atom donates one electron to the sea of conduction electrons. Calculate the Fermi energy (in eV) of each at 300 K and 0 K. Calculate the mean speed of all the conduction electrons and also the speed of electrons at $E_F$ for each metal. Calculate the density of states as states per eV $cm^{-3}$ at the Fermi energy and also at the center of the band, to be taken at $(E_F + \Phi)/2$. (See Table 4.1 for $\Phi$.)

**4.9** **Fermi energy and electron concentration** Consider the metals in Table 4.8 from Groups I, II, and III in the Periodic Table. Calculate the Fermi energies at absolute zero, and compare the values with the experimental values. What is your conclusion?

Table 4.8

| Metal | Group | $M_{at}$ | Density (g $cm^{-3}$) | $E_F$ (eV) [Calculated] | $E_F$ (eV) [Experiment] |
|---|---|---|---|---|---|
| Cu | I | 63.55 | 8.96 | — | 6.5 |
| Zn | II | 65.38 | 7.14 | — | 11.0 |
| Al | III | 27 | 2.70 | — | 11.8 |

**4.10 Temperature dependence of the Fermi energy**

*a.* Given that the Fermi energy for Cu is 7.0 eV at absolute zero, calculate the $E_F$ at 300 K. What is the percentage change in $E_F$ and what is your conclusion?

*b.* Given the Fermi energy for Cu at absolute zero, calculate the average energy and mean speed per conduction electron at absolute zero and 300 K, and comment.

**4.11 X-ray emission spectrum from sodium** Structure of the Na atom is [Ne]$3s^1$. Figure 4.59a shows the formation of the $3s$ and $3p$ energy bands in Na as a function of internuclear separation. Figure 4.59b shows the X-ray emission spectrum (called the $L$-band) from crystalline sodium in the soft X-ray range as explained in Example 4.6.

*a.* From Figure 4.59a, estimate the nearest neighbor equilibrium separation between Na atoms in the crystal if some electrons in the $3s$ band spill over into the states in the $3p$ band.

*b.* Explain the origin of the X-ray emission band in Figure 4.59b and the reason for calling it the $L$-band.

*c.* What is the Fermi energy of the electrons in Na from Figure 4.59b?

*d.* Taking the valency of Na to be I, what is the expected Fermi energy and how does it compare with that in part (c)?

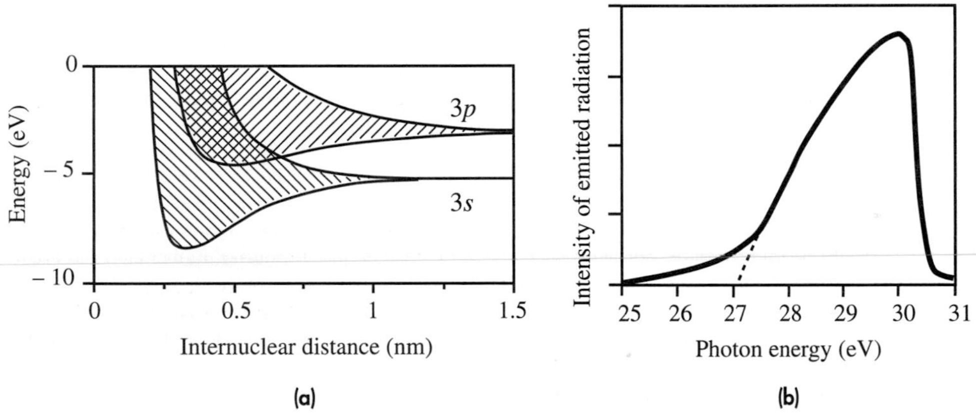


**Figure 4.59**
(a) Energy band formation in sodium.
(b) $L$-emission band of X-rays from sodium.
| SOURCE: (b) Data extracted from W. M. Cadt and D. H. Tomboulian, *Phys. Rev.*, **59,** 1941, p. 381.

**4.12 Conductivity of metals in the free electron model** Consider the general expression for the conductivity of metals in terms of the density of states $g(E_F)$ at $E_F$ given by

$$\sigma = \tfrac{1}{3}e^2 v_F^2 \tau g(E_F)$$

Show that within the free electron theory, this reduces to $\sigma = e^2 n\tau/m_e$, the Drude expression.

**4.13 Mean free path of conduction electrons in a metal** Show that within the free electron theory, the mean free path $\ell$ and conductivity $\sigma$ are related by

*Mean free path and conductivity in the free electron model*

$$\sigma = \frac{e^2}{3^{1/3}\pi^{2/3}\hbar}\ell n^{2/3} = 7.87 \times 10^{-5}\ell n^{2/3}$$

Calculate $\ell$ for Cu and Au, given each metal's resistivity of 17 nΩ m and 22 nΩ m, respectively, and that each has a valency of I.

*4.14 **Low-temperature heat capacity of metals** The heat capacity of conduction electrons in a metal is proportional to the temperature. The overall heat capacity of a metal is determined by the lattice heat capacity, except at the lowest temperatures. If $\delta E_t$ is the increase in the total energy of the conduction electrons (per unit volume) and $\delta T$ is the increase in the temperature of the metal as a result of heat addition, $E_t$ has been calculated as follows:

$$E_t = \int_0^\infty E g(E) f(E)\, dE = E_t(0) + \left(\frac{\pi^2}{4}\right)\frac{n(kT)^2}{E_{FO}}$$

where $E_t(0)$is the total energy per unit volume at 0 K, $n$ is the concentration of conduction electrons, and $E_{FO}$ is the Fermi energy at 0 K. Show that the heat capacity per unit volume due to conduction electrons in the free electron model of metals is

*Heat capacity of conduction electrons*

$$C_e = \frac{\pi^2}{2}\left(\frac{nk^2}{E_{FO}}\right)T = \gamma T \qquad [4.84]$$

where $\gamma = (\pi^2/2)(nk^2/E_{FO})$. Calculate $C_e$ for Cu, and then using the Debye equation for the lattice heat capacity, find $C_v$ for Cu at 10 K. Compare the two values and comment. What is the comparison at room temperature? (Note: $C_{\text{volume}} = C_{\text{molar}}(\rho/M_{\text{at}})$, where $\rho$ is the density in g cm$^{-3}$, $C_{\text{volume}}$ is in J K$^{-1}$ cm$^{-3}$, and $M_{\text{at}}$ is the atomic mass in g mol$^{-1}$.)

*4.15 **Secondary emission and photomultiplier tubes** When an energetic (high velocity) projectile electron collides with a material with a low work function, it can cause electron emission from the surface. This phenomenon is called **secondary emission.** It is fruitfully utilized in photomultiplier tubes as illustrated in Figure 4.60. The tube is evacuated and has a photocathode for receiving photons as a signal. An incoming photon causes photoemission of an electron from the photocathode material. The electron is then accelerated by a positive voltage applied to an electrode called a dynode which has a work function that easily allows secondary emission. When the accelerated electron strikes dynode $D_1$, it can release several electrons. All these electrons, the original and the secondary electrons, are then accelerated by the more positive voltage applied to dynode $D_2$. On impact with $D_2$, further electrons are released by secondary emission. The secondary emission process continues at each dynode stage until the final electrode, called the anode, is reached whereupon all the electrons are collected which results in a signal. Typical applications for photomultiplier tubes are in X-ray and nuclear medical instruments

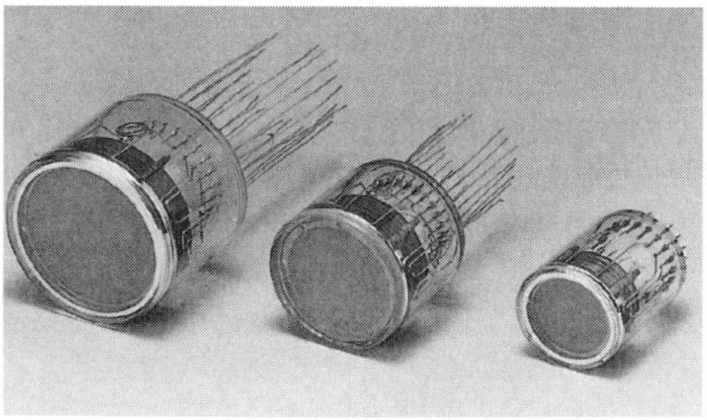

Photomultiplier tubes.
| SOURCE: Courtesy of Hamamatsu.

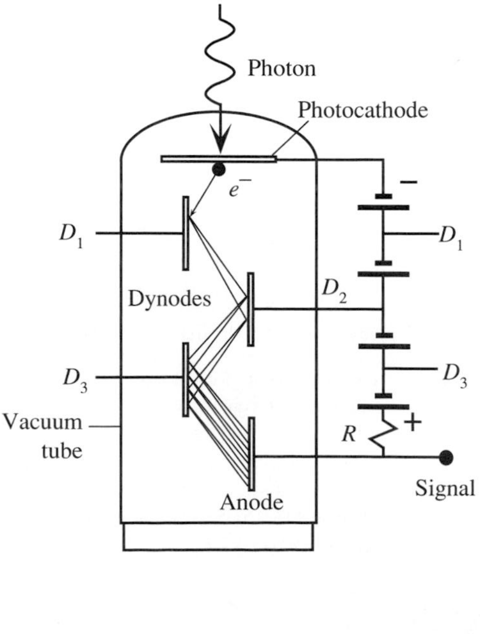


**Figure 4.60** The photomultiplier tube.

(X-ray CT scanner, positron CT scanner, gamma camera, etc.), radiation measuring instruments (*e.g.*, radon counter), X-ray diffractometers, and radiation measurement in high-energy physics research.

A particular photomultiplier tube has the following properties. The photocathode is made of a semiconductor-type material with $E_g \approx 1$ eV, an electron affinity $\chi$ of 0.4 eV, and a quantum efficiency of 20 percent at 400 nm. *Quantum efficiency* is defined as the number of photoemitted electrons per absorbed photon. The diameter of the photocathode is 18 mm. There are 10 dynode electrodes and an applied voltage of 1250 V between the photocathode and anode. Assume that this voltage is equally distributed among all the electrodes.

*a.* What is the longest threshold wavelength for the phototube?

*b.* What is the maximum kinetic energy of the emitted electron if the photocathode is illuminated with a 400 nm radiation?

*c.* What is the emission current from the photocathode at 400 nm illumination per unit intensity of radiation?

*d.* What is the *KE* of the electron as it strikes the first dynode electrode?

*e.* It has been found that the tube has a gain of $10^6$ electrons per incident photon. What is the average number of secondary electrons released at each dynode?

**4.16** **Thermoelectric effects and $E_F$** Consider a thermocouple pair that consists of gold and aluminum. One junction is at 100 °C and the other is at 0 °C. A voltmeter (with a very large input resistance) is inserted into the aluminum wire. Use the properties of Au and Al in Table 4.3 to estimate the emf registered by the voltmeter and identify the positive end.

**4.17** **The thermocouple equation** Although inputting the measured emf for $V$ in the thermocouple equation $V = a\Delta T + b(\Delta T)^2$ leads to a quadratic equation, which in principle can be solved for $\Delta T$, in general $\Delta T$ is related to the measured emf via

$$\Delta T = a_1 V + a_2 V^2 + a_3 V^3 + \cdots$$

with the coefficients $a_1$, $a_2$, etc., determined for each pair of TCs. By carrying out a Taylor's expansion of the TC equation, find the first two coefficients $a_1$ and $a_2$. Using an emf table for the K-type thermocouple or Figure 4.33, evaluate $a_1$ and $a_2$.

**4.18** **Thermionic emission** A vacuum tube is required to have a cathode operating at 800 °C and providing an emission (saturation) current of 10 A. What should be the surface area of the cathode for the two materials in Table 4.9? What should be the operating temperature for the Th on W cathode, if it is to have the same surface area as the oxide-coated cathode?

**Table 4.9**

| | $B_e$ (A m$^{-2}$ K$^{-2}$) | $\Phi$ (eV) |
|---|---|---|
| Th on W | $3 \times 10^4$ | 2.6 |
| Oxide coating | 100 | 1 |

**4.19** **Field-assisted emission in MOS devices** Metal-oxide-semiconductor (MOS) transistors in microelectronics have a metal gate on an $SiO_2$ insulating layer on the surface of a doped Si crystal. Consider this as a parallel plate capacitor. Suppose the gate is an Al electrode of area 50 $\mu$m × 50 $\mu$m and has a voltage of 10 V with respect to the Si crystal. Consider two thicknesses for the $SiO_2$, (a) 100 Å and (b) 40 Å, where (1 Å = $10^{-10}$ m). The work function of Al is 4.2 eV, but this refers to electron emission into vacuum, whereas in this case, the electron is emitted into the oxide. The potential energy barrier $\Phi_B$ between Al and $SiO_2$ is about 3.1 eV, and the field-emission current density is given by Equation 4.46a and b. Calculate the field-emission current for the two cases. For simplicity, take $m_e$ to be the electron mass in free space. What is your conclusion?

**4.20** **CNTs and field emission** The electric field at the tip of a sharp emitter is much greater than the "applied field," $\mathcal{E}_o$. The applied field is simply defined as $V_G/d$ where $d$ is the distance from the cathode tip to the gate or the grid; it represents the average nearly uniform field that would exist if the tip were replaced by a flat surface so that the cathode and the gate would almost constitute a parallel plate capacitor. The tip experiences an effective field $\mathcal{E}$ that is much greater than $\mathcal{E}_o$, which is expressed by a **field enhancement factor** $\beta$ that depends on the geometry of the cathode–gate emitter, and the shape of the emitter; $\mathcal{E} = \beta\mathcal{E}_o$. Further, we can take $\Phi_{\text{eff}}^{1/2}\Phi \approx \Phi^{3/2}$ in Equation 4.46. The final expression for the field-emission current density then becomes

*Fowler–Nordheim field emission current*

$$J = \frac{1.5 \times 10^{-6}}{\Phi}\beta^2\mathcal{E}_o^2 \exp\left(\frac{10.4}{\Phi^{1/2}}\right)\exp\left(-\frac{6.44 \times 10^7 \Phi^{3/2}}{\beta\mathcal{E}_o}\right) \quad [4.85]$$

where $J$ is in A cm$^{-2}$, $\mathcal{E}_o$ is in V cm$^{-1}$ and $\Phi$ is in eV. For a particular CNT emitter, $\Phi = 4.9$ eV. Estimate the applied field required to achieve a field-emission current density of 100 mA cm$^{-2}$ in the absence of field enhancement ($\beta = 1$) and with a field enhancement of $\beta = 800$ (typical value for a CNT emitter).

**4.21** **Nordheim–Fowler field emission in an FED** Table 4.10 shows the results of I–V measurements on a Motorola FED microemitter. By a suitable plot show that the I–V follows the Nordheim–Fowler emission characteristics.

**Table 4.10** Tests on a Motorola FED micro field emitter

| $V_G$ (V) | 40.0 | 42 | 44 | 46 | 48 | 50 | 52 | 53.8 | 56.2 | 58.2 | 60.4 |
|---|---|---|---|---|---|---|---|---|---|---|---|
| $I_{\text{emission}}$ (μA) | 0.40 | 2.14 | 9.40 | 20.4 | 34.1 | 61 | 93.8 | 142.5 | 202 | 279 | 367 |

**4.22** **Lattice waves and heat capacity**

*a.* Consider an aluminum sample. The nearest separation $2R$ (2 × atomic radius) between the Al–Al atoms in the crystal is 0.286 nm. Taking $a$ to be $2R$, and given the sound velocity in Al as 5100 m s$^{-1}$, calculate the force constant $\beta$ in Equation 4.66. Use the group velocity $v_g$ from the actual dispersion relation, Equation 4.55, to calculate the "sound velocity" at wavelengths of $\Lambda$ = 1mm, 1 $\mu$m, and 1 nm. What is your conclusion?

*b.* Aluminum has a Debye temperature of 394 K. Calculate its specific heat at 30 °C (Darwin, Australia) and at −30 °C (January, Resolute Nunavut, Canada).

*c.* Calculate the specific heat capacity of a germanium crystal at 25 °C and compare it with the experimental value in Table 4.5.

**4.23** **Specific heat capacity of GaAs and InSb**

*a.* The Debye temperature $T_D$ of GaAs is 344 K. Calculate its specific heat capacity at 300 K and at 30 °C.

*b.* For InSb, $T_D$ = 203 K. Calculate the room temperature specific heat capacity of InSb and compare it with the value expected from the Dulong–Petit rule ($T > T_D$).

**4.24** **Thermal conductivity**

*a.* Given that silicon has a Young's modulus of about 110 GPa and a density of 2.3 g cm$^{-3}$, calculate the mean free path of phonons in Si at room temperature.

*b.* Diamond has the same crystal structure as Si but has a very large thermal conductivity, about 1000 W m$^{-1}$ K$^{-1}$ at room temperature. Given that diamond has a specific heat capacity $c_s$ of 0.50 J K$^{-1}$ g$^{-1}$, Young's modulus $Y$ of 830 GPa, and density $\rho$ of 0.35 g cm$^{-3}$, calculate the mean free path of phonons in diamond.

*c.* GaAs has a thermal conductivity of 200 W m$^{-1}$ K$^{-1}$ at 100 K and 80 W m$^{-1}$ K$^{-1}$ at 200 K. Calculate its thermal conductivity at 25 °C and compare with the experimental value of 44 W m$^{-1}$ K$^{-1}$. (Hint: Take $\kappa \propto T^{-n}$ in the temperature region of interest; see Figure 4.48.)

***4.25** **Overlapping bands** Consider Cu and Ni with their density of states as schematically sketched in Figure 4.61. Both have overlapping $3d$ and $4s$ bands, but the $3d$ band is very narrow compared to the $4s$ band. In the case of Cu the band is full, whereas in Ni, it is only partially filled.

*a.* In Cu, do the electrons in the $3d$ band contribute to electrical conduction? Explain.

*b.* In Ni, do electrons in both bands contribute to conduction? Explain.

*c.* Do electrons have the same effective mass in the two bands? Explain.

*d.* Can an electron in the $4s$ band with energy around $E_F$ become scattered into the $3d$ band as a result of a scattering process? Consider both metals.

*e.* Scattering of electrons from the $4s$ band to the $3d$ band and vice versa can be viewed as an additional scattering process. How would you expect the resistivity of Ni to compare with that of Cu, even though Ni has two valence electrons and nearly the same density as Cu? In which case would you expect a stronger temperature dependence for the resistivity?

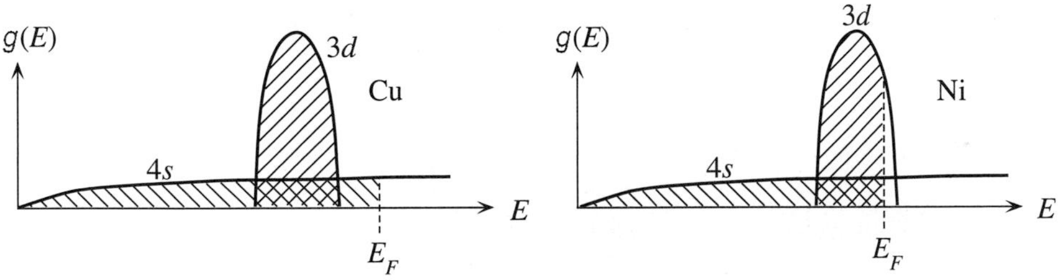


**Figure 4.61** Density of states and electron filling in Cu and Ni.

***4.26** **Overlapping bands at $E_F$ and higher resistivity** Figure 4.61 shows the density of states for Cu (or Ag) and Ni (or Pd). The $d$ band in Cu is filled, and only electrons at $E_F$ in the $s$ band make a contribution to the conductivity. In Ni, on the other hand, there are electrons at $E_F$ both in the $s$ and $d$ bands. The $d$ band is narrow compared with the $s$ band, and the electron's effective mass in this $d$ band is large; for simplicity, we will assume $m_e^*$ is "infinite" in this band. Consequently, the $d$-band electrons cannot be accelerated by the field (infinite $m_e^*$), have a negligible drift mobility, and make no contribution to the conductivity. Electrons in the $s$ band can become scattered by phonons into the $d$ band, and hence become relatively immobile until they are scattered back into the $s$ band when they can drift again. Consider Ni and one particular conduction electron at $E_F$ starting in the $s$ band. Sketch schematically the magnitude of the velocity gained $|v_x - u_x|$ from the field $E_x$ as a function of time for 10 scattering events; $v_x$ and $u_x$ are the instantaneous and initial velocities, and $|v_x - u_x|$ increases linearly with time, as the electron accelerates in the $s$ band and then drops to zero upon scattering. If $\tau_{ss}$ is the mean time for $s$ to $s$-band scattering, $\tau_{sd}$ is for $s$-band to $d$-band scattering, $\tau_{ds}$ is for $d$-band to $s$-band scattering, assume the following sequence of 10 events in your sketch: $\tau_{ss}, \tau_{ss}, \tau_{sd}, \tau_{ds}, \tau_{ss}, \tau_{sd}, \tau_{ds}, \tau_{ss}, \tau_{sd}, \tau_{ds}$. What would a similar sketch look like for Cu? Suppose that we wish to apply Equation 4.27. What does $g(E_F)$ and $\tau$ represent? What is the most important factor that makes Ni more resistive than Cu? Consider Matthiessen's rule. (Note: There are also electron spin related effects on the resistivity of Ni, but for simplicity these have been neglected.)

**4.27** **Grüneisen's law** Al and Cu both have metallic bonding and the same crystal structure. Assuming that the Grüneisen's parameter $\gamma$ for Al is the same as that for Cu, $\gamma = 0.23$, estimate the linear expansion coefficient $\lambda$ of Al, given that its bulk modulus $K = 75$ GPa, $c_s = 900$ J K$^{-1}$ kg$^{-1}$, and $\rho = 2.7$ g cm$^{-3}$. Compare your estimate with the experimental value of $23.5 \times 10^{-6}$ K$^{-1}$.

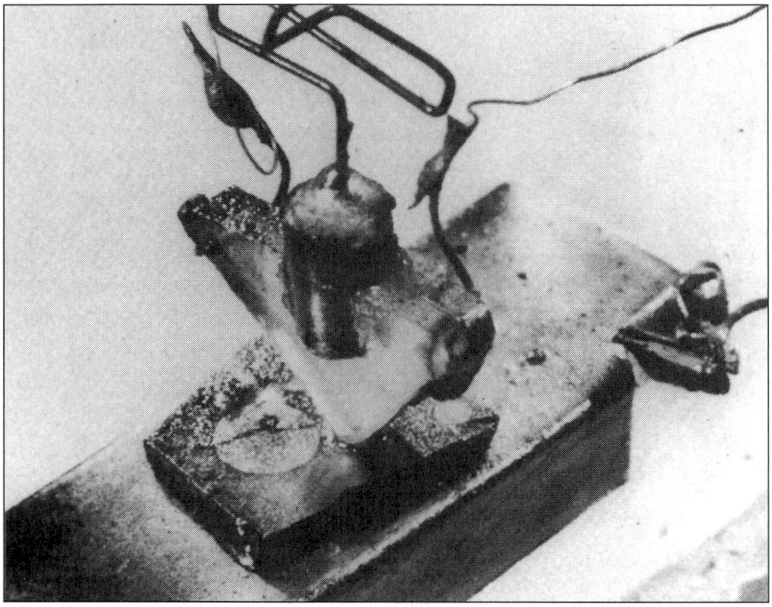

First point-contact transistor invented at Bell Labs.
| SOURCE: Courtesy of Bell Labs.

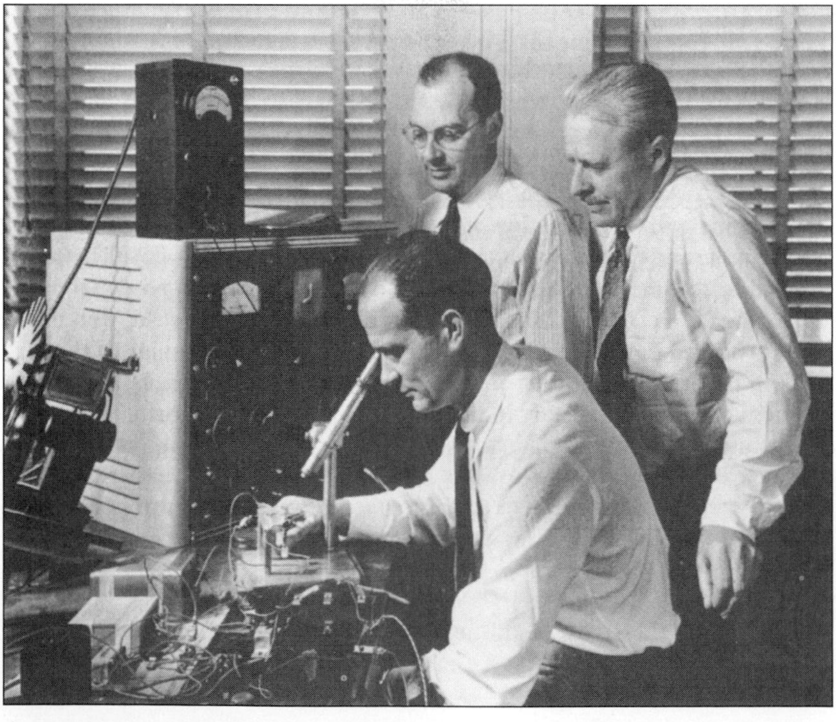

The three inventors of the transistor: William Shockley (seated), John Bardeen (left), and Walter Brattain (right) in 1948; the three inventors shared the Nobel prize in 1956.
| SOURCE: Courtesy of Bell Labs.

CHAPTER

# 5

# Semiconductors

In this chapter we develop a basic understanding of the properties of intrinsic and extrinsic semiconductors. Although most of our discussions and examples will be based on Si, the ideas are applicable to Ge and to the compound semiconductors such as GaAs, InP, and others. By intrinsic Si we mean an ideal perfect crystal of Si that has no impurities or crystal defects such as dislocations and grain boundaries. The crystal thus consists of Si atoms perfectly bonded to each other in the diamond structure. At temperatures above absolute zero, we know that the Si atoms in the crystal lattice will be vibrating with a distribution of energies. Even though the average energy of the vibrations is at most $3kT$ and incapable of breaking the Si–Si bond, a few of the lattice vibrations in certain crystal regions may nonetheless be sufficiently energetic to "rupture" a Si–Si bond. When a Si–Si bond is broken, a "free" electron is created that can wander around the crystal and also contribute to electrical conduction in the presence of an applied field. The broken bond has a missing electron that causes this region to be positively charged. The vacancy left behind by the missing electron in the bonding orbital is called a **hole.** An electron in a neighboring bond can readily tunnel into this broken bond and fill it, thereby effectively causing the hole to be displaced to the original position of the tunneling electron. By electron tunneling from a neighboring bond, holes are therefore also free to wander around the crystal and also contribute to electrical conduction in the presence of an applied field. In an intrinsic semiconductor, the number of thermally generated electrons is equal to the number of holes (broken bonds). In an extrinsic semiconductor, impurities are added to the semiconductor that can contribute either excess electrons or excess holes. For example, when an impurity such as arsenic is added to Si, each As atom acts as a donor and contributes a free electron to the crystal. Since these electrons do not come from broken bonds, the numbers of electrons and holes are not equal in an extrinsic semiconductor, and the As-doped Si in this example will have excess electrons. It will be an $n$-type Si since electrical conduction will be mainly due to the motion of electrons. It is also possible to obtain a $p$-type Si crystal in which hole concentration is in excess of the electron concentration due to, for example, boron doping.

# 5.1 INTRINSIC SEMICONDUCTORS

## 5.1.1 SILICON CRYSTAL AND ENERGY BAND DIAGRAM

The electronic configuration of an isolated Si atom is [Ne]$3s^2p^2$. However, in the vicinity of other atoms, the $3s$ and $3p$ energy levels are so close that the interactions result in the *four* orbitals $\psi(3s)$, $\psi(3p_x)$, $\psi(3p_y)$, and $\psi(3p_z)$ mixing together to form *four* new hybrid orbitals (called $\psi_{\text{hyb}}$) that are symmetrically directed as far away from each other as possible (toward the corners of a tetrahedron). In two dimensions, we can simply view the orbitals pictorially as in Figure 5.1a. The four hybrid orbitals, $\psi_{\text{hyb}}$, each have one electron so that they are half-occupied. Therefore, a $\psi_{\text{hyb}}$ orbital of one Si atom can overlap a $\psi_{\text{hyb}}$ orbital of a neighboring Si atom to form a covalent bond with two spin-paired electrons. In this manner one Si atom bonds with four other Si atoms by overlapping the half-occupied $\psi_{\text{hyb}}$ orbitals, as illustrated in Figure 5.1b.

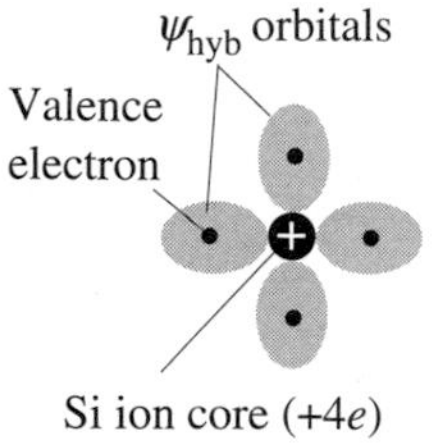


(a)

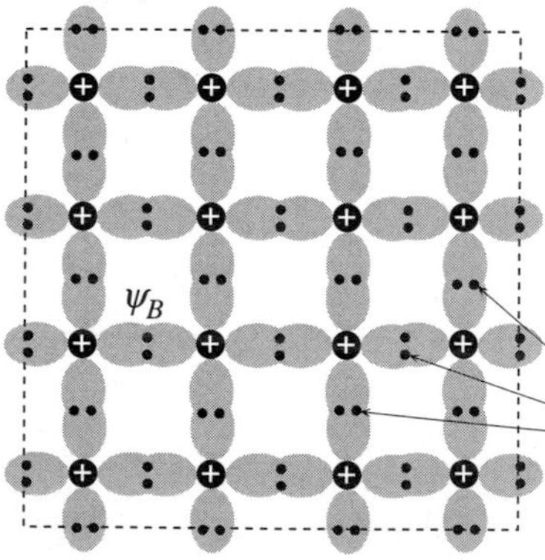


(b)

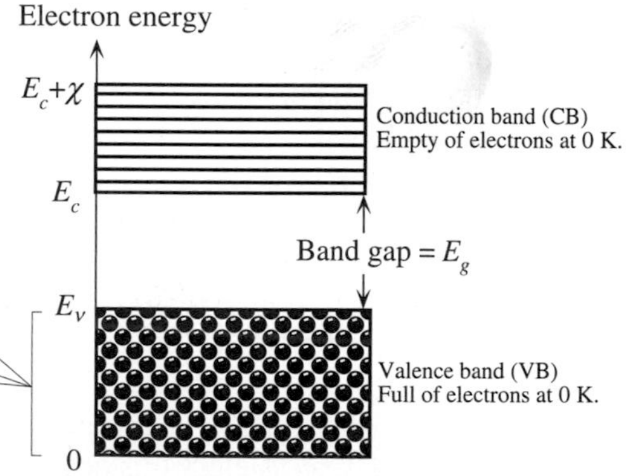


(c)

**Figure 5.1**

(a) A simplified two-dimensional illustration of a Si atom with four hybrid orbitals $\psi_{\text{hyb}}$. Each orbital has one electron.

(b) A simplified two-dimensional view of a region of the Si crystal showing covalent bonds.

(c) The energy band diagram at absolute zero of temperature.

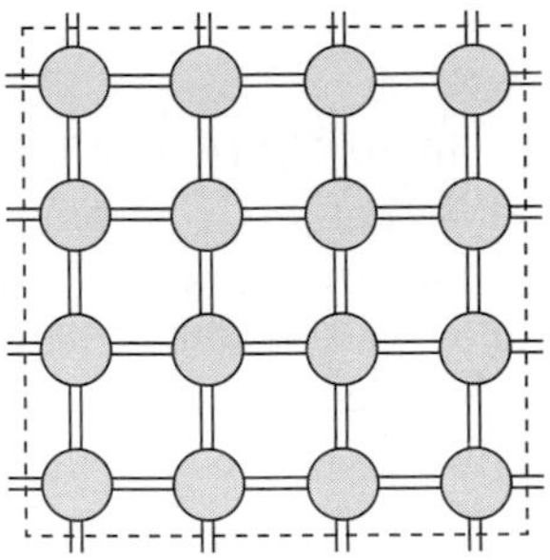

**Figure 5.2** A two-dimensional pictorial view of the Si crystal showing covalent bonds as two lines where each line is a valence electron.

Each Si–Si bond corresponds to a bonding orbital, $\psi_B$, obtained by overlapping two neighboring $\psi_{\text{hyb}}$ orbitals. Each bonding orbital ($\psi_B$) has two spin-paired electrons and is therefore *full*. Neighboring Si atoms can also form covalent bonds with other Si atoms, thus forming a three-dimensional network of Si atoms. The resulting structure is the Si crystal in which each Si atom bonds with four Si atoms in a tetrahedral arrangement. The crystal structure is that of a *diamond,* which was described in Chapter 1. We can imagine the Si crystal in two dimensions as depicted in Figure 5.1b. The electrons in the covalent bonds are the valence electrons.

The energy band diagram of the silicon crystal is shown in Figure 5.1c.[1] The vertical axis is the electron energy in the crystal. The valence band (VB) contains those electronic states that correspond to the overlap of bonding orbitals ($\psi_B$). Since all the bonding orbitals ($\psi_B$) are full with valence electrons in the crystal, the VB is also full with these valence electrons at a temperature of absolute zero. The conduction band (CB) contains electronic states that are at higher energies, those corresponding to the overlap of antibonding orbitals. The CB is separated from the VB by an energy gap $E_g$, called the **bandgap.** The energy level $E_v$ marks the top of the VB and $E_c$ marks the bottom of the CB. The energy distance from $E_c$ to the vacuum level, the width of the CB, is called the **electron affinity** $\chi$. The general energy band diagram in Figure 5.1c applies to all crystalline semiconductors with appropriate changes in the energies.

The electrons shown in the VB in Figure 5.1c are those in the covalent bonds between the Si atoms in Figure 5.1b. An electron in the VB, however, is not localized to an atomic site but extends throughout the whole solid. Although the electrons appear localized in Figure 5.1b, at the bonding orbitals between the Si atoms this is not, in fact, true. In the crystal, the electrons can tunnel from one bond to another and exchange places. If we were to work out the wavefunction of a valence electron in the Si crystal, we would find that it extends throughout the whole solid. This means that the electrons in the covalent bonds are indistinguishable. We cannot label an electron from the start and say that the electron is in the covalent bond between these two atoms.

We can crudely represent the silicon crystal in two dimensions as shown in Figure 5.2. Each covalent bond between Si atoms is represented by two lines corresponding to two spin-paired electrons. Each line represents a valence electron.

---

[1] The formation of energy bands in the silicon crystal was described in detail in Chapter 4.

## 5.1.2 ELECTRONS AND HOLES

The only empty electronic states in the silicon crystal are in the CB (Figure 5.1c). An electron placed in the CB is free to move around the crystal and also respond to an applied electric field because there are plenty of neighboring empty energy levels. An electron in the CB can easily gain energy from the field and move to higher energy levels because these states are empty. Generally we can treat an electron in the CB as if it were free within the crystal with certain modifications to its mass, as explained later in Section 5.1.3.

Since the only empty states are in the CB, the excitation of an electron from the VB requires a minimum energy of $E_g$. Figure 5.3a shows what happens when a photon of energy $h\nu > E_g$ is incident on an electron in the VB. This electron absorbs the incident photon and gains sufficient energy to surmount the energy gap $E_g$ and reach the CB. Consequently, a free electron and a "hole," corresponding to a missing electron in the VB, are created. In some semiconductors such as Si and Ge, the photon absorption process also involves lattice vibrations (vibrations of the Si atoms), which we have not shown in Figure 5.3b.

Although in this specific example a photon of energy $h\nu > E_g$ creates an electron–hole pair, this is not necessary. In fact, in the absence of radiation, there is an electron–hole generation process going on in the sample as a result of **thermal generation.** Due to thermal energy, the atoms in the crystal are constantly vibrating, which corresponds to the bonds between the Si atoms being periodically deformed. In a certain region, the atoms, at some instant, may be moving in such a way that a bond becomes overstretched, as pictorially depicted in Figure 5.4. This will result in the overstretched bond rupturing and hence releasing an electron into the CB (the electron effectively

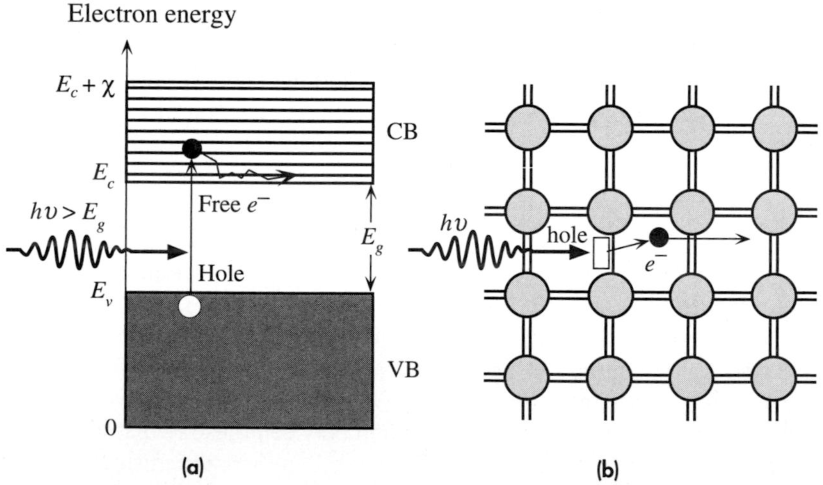


**Figure 5.3**
(a) A photon with an energy greater than $E_g$ can excite an electron from the VB to the CB.
(b) When a photon breaks a Si–Si bond, a free electron and a hole in the Si–Si bond are created.

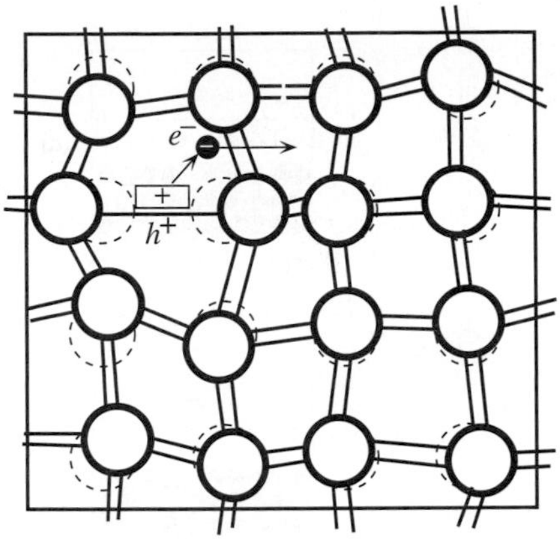


**Figure 5.4** Thermal vibrations of atoms can break bonds and thereby create electron–hole pairs.

becomes "free"). The empty electronic state of the missing electron in the bond is what we call a **hole** in the valence band. The free electron, which is in the CB, can wander around the crystal and contribute to the electrical conduction when an electric field is applied. The region remaining around the hole in the VB is positively charged because a charge of $-e$ has been removed from an otherwise neutral region of the crystal. This hole, denoted as $h^+$, can also wander around the crystal as if it were free. This is because an electron in a neighboring bond can "jump," that is, tunnel, into the hole to fill the vacant electronic state at this site and thereby create a hole at its original position. This is effectively equivalent to the hole being displaced in the opposite direction, as illustrated in Figure 5.5a. This single step can reoccur, causing the hole to be further displaced. As a result, the hole moves around the crystal as if it were a free positively charged entity, as pictured in Figure 5.5a to d. Its motion is quite independent from that of the original electron. When an electric field is applied, the hole will drift in the direction of the field and hence contribute to electrical conduction. It is now apparent that there are essentially two types of charge carriers in semiconductors: *electrons* and *holes.* A hole is effectively an empty electronic state in the VB that behaves as if it were a positively charged "particle" free to respond to an applied electric field.

When a wandering electron in the CB meets a hole in the VB, the electron has found an empty state of lower energy and therefore occupies the hole. The electron falls from the CB to the VB to fill the hole, as depicted in Figure 5.5e and f. This is called **recombination** and results in the annihilation of an electron in the CB and a hole in the VB. The excess energy of the electron falling from CB to VB in certain semiconductors such as GaAs and InP is emitted as a photon. In Si and Ge the excess energy is lost as lattice vibrations (heat).

It must be emphasized that the illustrations in Figure 5.5 are pedagogical pictorial visualizations of hole motion based on classical notions and cannot be taken too seriously, as discussed in more advanced texts (see also Section 5.11). We should remember that the electron has a wavefunction in the crystal that is extended and not localized, as the pictures in Figure 5.5 imply. Further, the hole is a concept that corresponds to an empty valence band wavefunction that normally has an electron. Again, we cannot localize the hole to a particular site, as the pictures in Figure 5.5 imply.

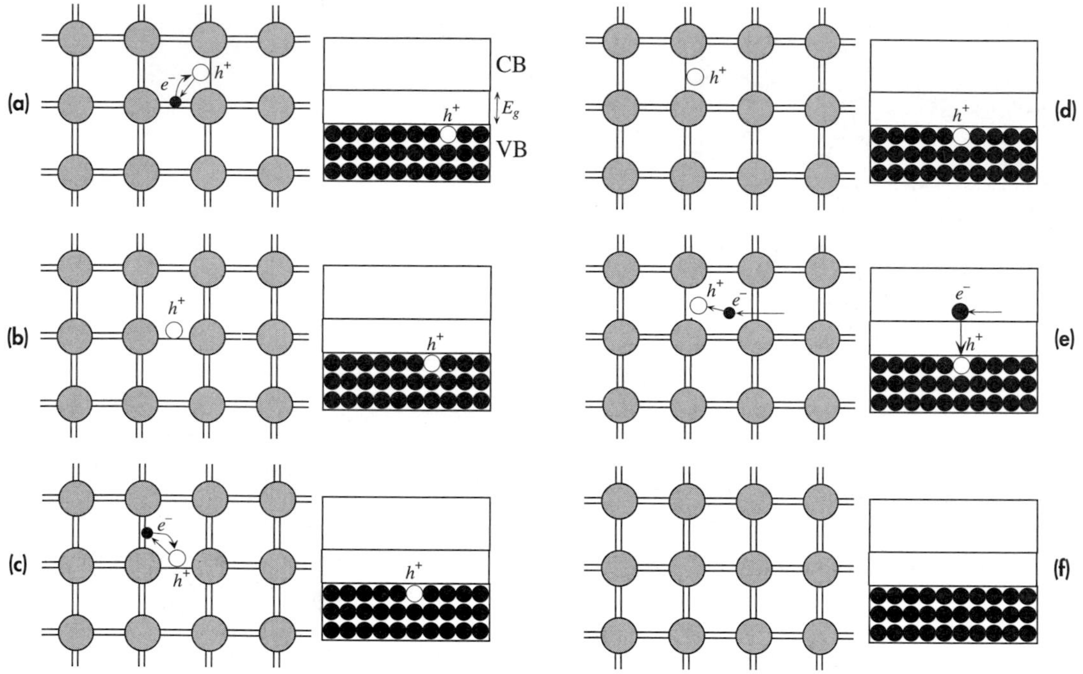


**Figure 5.5** A pictorial illustration of a hole in the valence band wandering around the crystal due to the tunneling of electrons from neighboring bonds.

### 5.1.3 Conduction in Semiconductors

When an electric field is applied across a semiconductor as shown in Figure 5.6, the energy bands bend. The total electron energy $E$ is $KE + PE$, but now there is an additional electrostatic $PE$ contribution that is not constant in an applied electric field. A uniform electric field $\mathcal{E}_x$ implies a linearly decreasing potential $V(x)$, by virtue of $(dV/dx) = -\mathcal{E}_x$, that is, $V = -Ax + B$. This means that the $PE$, $-eV(x)$, of the electron is now $eAx - eB$, which increases linearly across the sample. All the energy levels and hence the energy bands must therefore tilt up in the $x$ direction, as shown in Figure 5.6, in the presence of an applied field.

Under the action of $\mathcal{E}_x$, the electron in the CB moves to the left and immediately starts gaining energy from the field. When the electron collides with a thermal vibration of a Si atom, it loses some of this energy and thus "falls" down in energy in the CB. After the collision, the electron starts to accelerate again, until the next collision, and so on. We recognize this process as the drift of the electron in an applied field, as illustrated in Figure 5.6. The drift velocity $v_{de}$ of the electron is $\mu_e\mathcal{E}_x$ where $\mu_e$ is the drift mobility of the electron. In a similar fashion, the holes in the VB also drift in an applied field, but here the drift is along the field. Notice that when a hole gains energy, it moves "down" in the VB because the potential energy of the hole is of opposite sign to that of the electron.

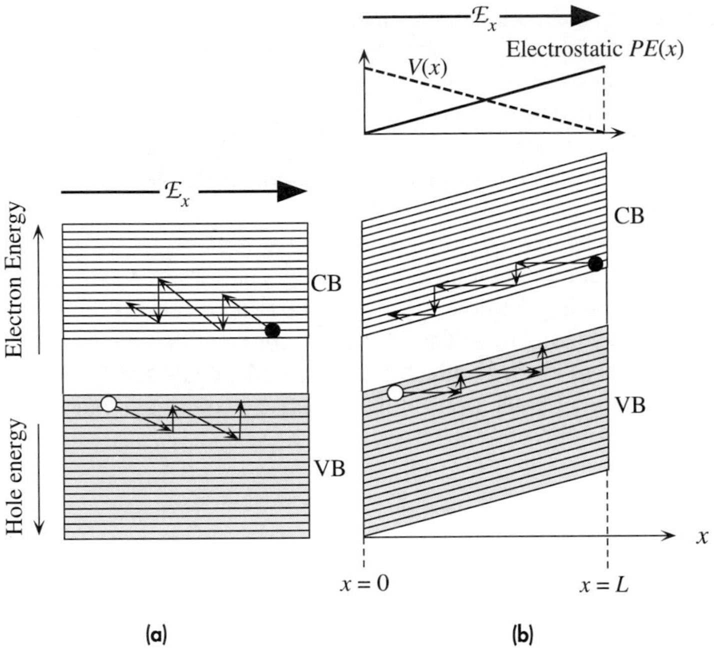


**Figure 5.6** When an electric field is applied, electrons in the CB and holes in the VB can drift and contribute to the conductivity.
(a) A simplified illustration of drift in $\mathcal{E}_x$.
(b) Applied field bends the energy bands since the electrostatic *PE* of the electron is $-eV(x)$ and $V(x)$ decreases in the direction of $\mathcal{E}_x$, whereas *PE* increases.

Since both electrons and holes contribute to electrical conduction, we may write the current density $J$, from its definition, as

$$J = env_{de} + epv_{dh} \qquad [5.1]$$

where $n$ is the electron concentration in the CB, $p$ is the hole concentration in the VB, and $v_{de}$ and $v_{dh}$ are the drift velocities of electrons and holes in response to an applied electric field $\mathcal{E}_x$, Thus,

*Electron and hole drift velocities*

$$v_{de} = \mu_e \mathcal{E}_x \qquad \text{and} \qquad v_{dh} = \mu_h \mathcal{E}_x \qquad [5.2]$$

where $\mu_e$ and $\mu_h$ are the electron and hole drift mobilities. In Chapter 2 we derived the drift mobility $\mu_e$ of the electrons in a conductor as

$$\mu_e = \frac{e\tau_e}{m_e} \qquad [5.3]$$

where $\tau_e$ is the mean free time between scattering events and $m_e$ is the electronic mass. The ideas on electron motion in metals can also be applied to the electron motion in the CB of a semiconductor to rederive Equation 5.3. We must, however, use an effective mass $m_e^*$ for the electron in the crystal rather than the mass $m_e$ in free space. A "free" electron in a crystal is not entirely free because as it moves it interacts with the potential energy (*PE*) of the ions in the solid and therefore experiences various internal forces. The effective mass $m_e^*$ accounts for these internal forces in such a way that we can relate the acceleration $a$ of the electron in the CB to an external force $F_{\text{ext}}$ (*e.g.*, $-e\mathcal{E}_x$) by $F_{\text{ext}} = m_e^* a$ just as we do for the electron in vacuum by $F_{\text{ext}} = m_e a$. In applying the

$F_{ext} = m_e^* a$ type of description to the motion of the electron, we are assuming, of course, that the effective mass of the electron can be calculated or measured experimentally. It is important to remark that the true behavior is governed by the solution of the Schrödinger equation in a periodic lattice (crystal) from which it can be shown that we can indeed describe the inertial resistance of the electron to acceleration in terms of an effective mass $m_e^*$. The effective mass depends on the interaction of the electron with its environment within the crystal.

We can now speculate on whether the hole can also have a mass. As long as we view mass as resistance to acceleration, that is, inertia, there is no reason why the hole should not have a mass. Accelerating the hole means accelerating electrons tunneling from bond to bond in the opposite direction. Therefore it is apparent that the hole will have a nonzero finite inertial mass because otherwise the smallest external force will impart an infinite acceleration to it. If we represent the effective mass of the hole in the VB by $m_h^*$, then the hole drift mobility will be

$$\mu_h = \frac{e\tau_h}{m_h^*} \qquad [5.4]$$

where $\tau_h$ is the mean free time between scattering events for holes.

Taking Equation 5.1 for the current density further, we can write the **conductivity of a semiconductor** as

*Conductivity of a semiconductor*

$$\sigma = en\mu_e + ep\mu_h \qquad [5.5]$$

where $n$ and $p$ are the electron and hole concentrations in the CB and VB, respectively. This is a general equation valid for all semiconductors.

### 5.1.4 Electron and Hole Concentrations

The general equation for the conductivity of a semiconductor, Equation 5.5, depends on $n$, the electron concentration, and $p$, the hole concentration. How do we determine these quantities? We follow the procedure schematically shown in Figure 5.7a to d in which the density of states is multiplied by the probability of a state being occupied and integrated over the entire CB for $n$ and over the entire VB for $p$.

We define $g_{cb}(E)$ as the **density of states** in the CB, that is, the number of states per unit energy per unit volume. The probability of finding an electron in a state with energy $E$ is given by the Fermi–Dirac function $f(E)$, which is discussed in Chapter 4. Then $g_{cb}(E)f(E)$ is the actual number of electrons per unit energy per unit volume $n_E(E)$ in the CB. Thus,

$$n_E\, dE = g_{cb}(E) f(E)\, dE$$

is the number of electrons in the energy range $E$ to $E + dE$. Integrating this from the bottom ($E_c$) to the top ($E_c + \chi$) of the CB gives the electron concentration $n$, number of electrons per unit volume, in the CB. In other words,

$$n = \int_{E_c}^{E_c+\chi} n_E(E)\, dE = \int_{E_c}^{E_c+\chi} g_{cb}(E) f(E)\, dE$$

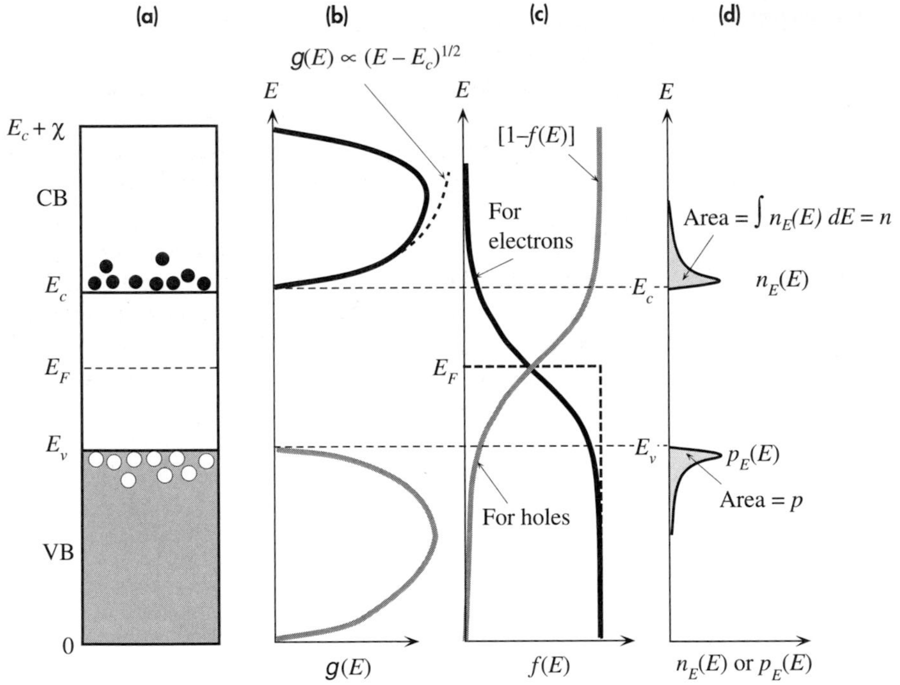


**Figure 5.7**
(a) Energy band diagram.
(b) Density of states (number of states per unit energy per unit volume).
(c) Fermi–Dirac probability function (probability of occupancy of a state).
(d) The product of $g(E)$ and $f(E)$ is the energy density of electrons in the CB (number of electrons per unit energy per unit volume). The area under $n_E(E)$ versus $E$ is the electron concentration.

We will assume that $(E_c - E_F) \gg kT$ (*i.e.*, $E_F$ is at least a few $kT$ below $E_c$) so that

$$f(E) \approx \exp[-(E - E_F)/kT]$$

We are thus replacing Fermi–Dirac statistics by Boltzmann statistics and thereby inherently assuming that the number of electrons in the CB is far less than the number of states in this band.

Further, we will take the upper limit to be $E = \infty$ rather than $E_c + \chi$ since $f(E)$ decays rapidly with energy so that $g_{cb}(E)f(E) \to 0$ near the top of the band. Furthermore, since $g_{cb}(E)f(E)$ is significant only close to $E_c$, we can use

*Density of states in conduction band*

$$g_{cb}(E) = \frac{(\pi 8\sqrt{2})m_e^{*3/2}}{h^3}(E - E_c)^{1/2}$$

for an electron in a three-dimensional *PE* well without having to consider the exact form of $g_{cb}(E)$ across the whole band. Thus

$$n \approx \frac{(\pi 8\sqrt{2})m_e^{*3/2}}{h^3}\int_{E_c}^{\infty}(E - E_c)^{1/2}\exp\left[-\frac{(E - E_F)}{kT}\right]dE$$

which leads to

*Electron concentration in CB*

$$n = N_c \exp\left[-\frac{(E_c - E_F)}{kT}\right] \qquad [5.6]$$

where

*Effective density of states at CB edge*

$$N_c = 2\left(\frac{2\pi m_e^* kT}{h^2}\right)^{3/2} \qquad [5.7]$$

The result of the integration in Equation 5.6 seems to be simple, but it is an approximation as it assumes that $(E_c - E_F) \gg kT$. $N_c$ is a temperature-dependent constant, called the **effective density of states at the CB edge.** Equation 5.6 can be interpreted as follows. If we take all the states in the conduction band and replace them with an effective concentration $N_c$ (number of states per unit volume) at $E_c$ and then multiply this simply by the Boltzmann probability function, $f(E_c) = \exp[-(E_c - E_F)/kT]$, we obtain the concentration of electrons at $E_c$, that is, in the conduction band. $N_c$ is thus an effective density of states at the CB band edge.

We can carry out a similar analysis for the concentration of holes in the VB. Multiplying the density of states $g_{\text{vb}}(E)$ in the VB with the probability of occupancy by a hole $[1 - f(E)]$, that is, the probability that an electron is absent, gives $p_E$, the hole concentration per unit energy. Integrating this over the VB gives the hole concentration

$$p = \int_0^{E_v} p_E \, dE = \int_0^{E_v} g_{\text{vb}}(E)[(1 - f(E)] \, dE$$

With the assumption that $E_F$ is a few $kT$ above $E_v$, the integration simplifies to

*Hole concentration in VB*

$$p = N_v \exp\left[-\frac{(E_F - E_v)}{kT}\right] \qquad [5.8]$$

where $N_v$ is the effective density of states at the VB edge and is given by

*Effective density of states at VB edge*

$$N_v = 2\left(\frac{2\pi m_h^* kT}{h^2}\right)^{3/2} \qquad [5.9]$$

We can now see the virtues of studying the density of states $g(E)$ as a function of energy $E$ and the Fermi–Dirac function $f(E)$. Both were central factors in deriving the expressions for $n$ and $p$. There are no specific assumptions in our derivations, except for $E_F$ being a few $kT$ away from the band edges, which means that Equations 5.6 and 5.8 are generally valid.

The general equations that determine the free electron and hole concentrations are thus given by Equations 5.6 and 5.8. It is interesting to consider the product $np$,

$$np = N_c \exp\left[-\frac{(E_c - E_F)}{kT}\right] N_v \exp\left[-\frac{(E_F - E_v)}{kT}\right] = N_c N_v \exp\left[-\frac{(E_c - E_v)}{kT}\right]$$

or

$$np = N_c N_v \exp\left(-\frac{E_g}{kT}\right) \qquad [5.10]$$

where $E_g = E_c - E_v$ is the bandgap energy. First, we note that this is a general expression in which the right-hand side, $N_c N_v \exp(-E_g/kT)$, is a constant that depends on the temperature and the material properties, for example, $E_g$, and not on the position of the Fermi level. In the special case of an intrinsic semiconductor, $n = p$, which we can denote as $n_i$, the **intrinsic concentration,** so that $N_c N_v \exp(-E_g/kT)$ must be $n_i^2$. From Equation 5.10 we therefore have

$$np = n_i^2 = N_c N_v \exp\left(-\frac{E_g}{kT}\right) \quad [5.11]$$

*Mass action law*

This is a general equation that is valid as long as we have thermal equilibrium. External excitation, such as photogeneration, is excluded. It states that the product $np$ is a temperature-dependent constant. If we somehow increase the electron concentration, then we inevitably reduce the hole concentration. The constant $n_i$ has a special significance because it represents the free electron and hole concentrations in the intrinsic material.

An **intrinsic semiconductor** is a pure semiconductor crystal in which the electron and hole concentrations are equal. By pure we mean virtually no impurities in the crystal. We should also exclude crystal defects that may capture carriers of one sign and thus result in unequal electron and hole concentrations. Clearly in a pure semiconductor, electrons and holes are generated in pairs by thermal excitation across the bandgap. It must be emphasized that Equation 5.11 is generally valid and therefore applies to both intrinsic and nonintrinsic ($n \neq p$) semiconductors.

When an electron and hole meet in the crystal, they "recombine." The electron falls in energy and occupies the empty electronic state that the hole represents. Consequently, the broken bond is "repaired," but we losc two free charge carriers. **Recombination** of an electron and hole results in their annihilation. In a semiconductor we therefore have thermal generation of electron–hole pairs by thermal excitation from the VB to the CB, and we also have recombination of electron–hole pairs that removes them from their conduction and valence bands, respectively. The rate of recombination $R$ will be proportional to the number of electrons and also to the number of holes. Thus

$$R \propto np$$

The rate of generation $G$ will depend on how many electrons are available for excitation at $E_v$, that is, $N_v$; how many empty states are available at $E_c$, that is, $N_c$; and the probability that the electron will make the transition, that is, $\exp(-E_g/kT)$, so that

$$G \propto N_c N_v \exp\left(-\frac{E_g}{kT}\right)$$

Since in thermal equilibrium we have no continuous increase in $n$ or $p$, we must have the rate of generation equal to the rate of recombination, that is, $G = R$. This is equivalent to Equation 5.11.

In sketching the diagrams in Figure 5.7a to d to illustrate the derivation of the expressions for $n$ and $p$ (in Equations 5.6 and 5.8), we assumed that the Fermi level $E_F$ is somewhere around the middle of the energy bandgap. This was not an assumption in the mathematical derivations but only in the sketches. From Equations 5.6 and 5.8 we also note that the position of Fermi level is important in determining the electron and hole concentrations. It serves as a "mathematical crank" to determine $n$ and $p$.

We first consider an intrinsic semiconductor, $n = p = n_i$. Setting $p = n_i$ in Equation 5.8, we can solve for the Fermi energy in the intrinsic semiconductor, $E_{Fi}$, that is,

$$N_v \exp\left[-\frac{(E_{Fi} - E_v)}{kT}\right] = (N_c N_v)^{1/2} \exp\left(-\frac{E_g}{2kT}\right)$$

which leads to

*Fermi energy in intrinsic semiconductor*

$$E_{Fi} = E_v + \frac{1}{2}E_g - \frac{1}{2}kT \ln\left(\frac{N_c}{N_v}\right) \qquad [5.12]$$

Furthermore, substituting the proper expressions for $N_c$ and $N_v$ we get

*Fermi energy in intrinsic semiconductor*

$$E_{Fi} = E_v + \frac{1}{2}E_g - \frac{3}{4}kT \ln\left(\frac{m_e^*}{m_h^*}\right) \qquad [5.13]$$

It is apparent from these equations that if $N_c = N_v$ or $m_e^* = m_h^*$, then

$$E_{Fi} = E_v + \frac{1}{2}E_g$$

that is, $E_{Fi}$ is right in the middle of the energy gap. Normally, however, the effective masses will not be equal and the Fermi level will be slightly shifted down from midgap by an amount $\frac{3}{4}kT \ln(m_e^*/m_h^*)$, which is quite small compared with $\frac{1}{2}E_g$. For Si and Ge, the hole effective mass (for density of states) is slightly smaller than the electron effective mass, so $E_{Fi}$ is slightly below the midgap.

The condition $np = n_i^2$ means that if we can somehow increase the electron concentration in the CB over the intrinsic value—for example, by adding impurities into the Si crystal that donate additional electrons to the CB—we will then have $n > p$. The semiconductor is then called ***n*-type.** The Fermi level must be closer to $E_c$ than $E_v$, so that

$$E_c - E_F < E_F - E_v$$

and Equations 5.6 and 5.8 yield $n > p$. The $np$ product always yields $n_i^2$ in thermal equilibrium in the absence of external excitation, for example, illumination.

It is also possible to have an excess of holes in the VB over electrons in the CB, for example, by adding impurities that remove electrons from the VB and thereby generate holes. In that case $E_F$ is closer to $E_v$ than to $E_c$. A semiconductor in which $p > n$ is called a ***p*-type semiconductor.** The general band diagrams with the appropriate Fermi levels for intrinsic, $n$-type, and $p$-type semiconductors (*e.g.*, *i*-Si, *n*-Si, and *p*-Si, respectively) are illustrated in Figure 5.8a to c.

It is apparent that if we know where $E_F$ is, then we have effectively determined $n$ and $p$ by virtue of Equations 5.6 and 5.8. We can view $E_F$ as a *material property* that is related to the concentration of charge carriers that contribute to electrical conduction. Its significance, however, goes beyond $n$ and $p$. It also determines the energy needed to remove an electron from the semiconductor. The energy difference between the vacuum level (where the electron is free) and $E_F$ is the **work function** $\Phi$ of the semiconductor, the energy required to remove an electron even though there are no electrons at $E_F$ in a semiconductor.

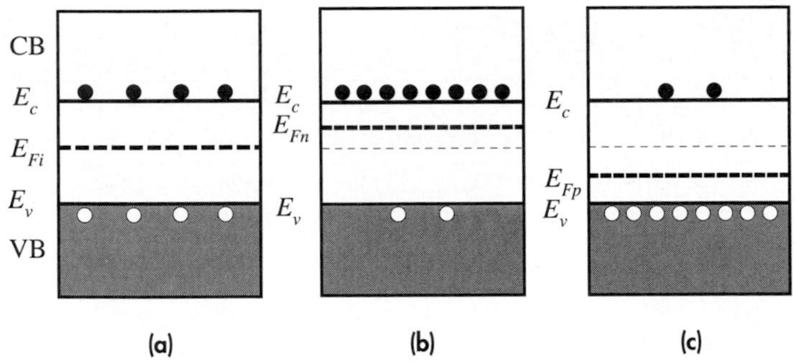


**Figure 5.8** Energy band diagrams for (a) intrinsic, (b) *n*-type, and (c) *p*-type semiconductors.
In all cases, $np = n_i^2$.

The Fermi level can also be interpreted in terms of the potential energy per electron for electrical work similar to the interpretation of electrostatic *PE*. Just as $e\,\Delta V$ is the electrical work involved in taking a charge $e$ across a potential difference $\Delta V$, any difference in $E_F$ in going from one end of a material (or system) to another is available to do an amount $\Delta E_F$ of external work. A corollary to this is that if electrical work is done on the material, for example, by passing a current through it, then the Fermi level is not uniform in the material. $\Delta E_F$ then represents the work done per electron. For a material in thermal equilibrium and not subject to any external excitation such as illumination or connections to a voltage supply, the Fermi level in the material must therefore be uniform, $\Delta E_F = 0$.

What is the average energy of an electron in the conduction band of a semiconductor? Also, what is the mean speed of an electron in the conduction band? We note that the concentration of electrons with energies $E$ to $E + dE$ is $n_E(E)\,dE$ or $g_{cb}(E)f(E)\,dE$. Thus the average energy of electrons in the CB, by definition of the mean, is

$$\overline{E}_{CB} = \frac{1}{n}\int_{CB} E\,g_{cb}(E)f(E)\,dE$$

where the integration must be over the CB. Substituting the proper expressions for $g_{cb}(E)$ and $f(E)$ in the integrand and carrying out the integration from $E_c$ to the top of the band, we find the very simple result that

*Average electron energy in CB*

$$\overline{E}_{CB} = E_c + \frac{3}{2}kT \tag{5.14}$$

Thus, an electron in the conduction band has an average energy of $\frac{3}{2}kT$ *above* $E_c$. Since we know that an electron at $E_c$ is "free" in the crystal, $\frac{3}{2}kT$ must be its average kinetic energy.

This is just like the average kinetic energy of gas atoms (such as He atoms) in a tank assuming that the atoms (or the "particles") do not interact, that is, they are independent. We know from the kinetic theory that the statistics of a collection of independent gas atoms obeys the classical Maxwell–Boltzmann description with an average energy given by $\frac{3}{2}kT$. We should also recall that the description of electron statistics in a metal involves the Fermi–Dirac function, which is based on the Pauli exclusion principle. In a metal the average energy of the conduction electron is $\frac{3}{5}E_F$ and, for all practical purposes, temperature independent. We see that the collective electron behavior is completely different in the two solids. We can explain the difference by noting that the conduction band in a

**Table 5.1** Selected typical properties of Ge, Si, and GaAs at 300 K

| | $E_g$ (eV) | $\chi$ (eV) | $N_c$ ($cm^{-3}$) | $N_v$ ($cm^{-3}$) | $n_i$ ($cm^{-3}$) | $\mu_e$ ($cm^2\ V^{-1}\ s^{-1}$) | $\mu_h$ ($cm^2\ V^{-1}\ s^{-1}$) | $m_e^*/m_e$ | $m_h^*/m_e$ | $\varepsilon_r$ |
|---|---|---|---|---|---|---|---|---|---|---|
| Ge | 0.66 | 4.13 | $1.04 \times 10^{19}$ | $6.0 \times 10^{18}$ | $2.3 \times 10^{13}$ | 3900 | 1900 | 0.12*a*<br>0.56*b* | 0.23*a*<br>0.40*b* | 16 |
| Si | 1.10 | 4.01 | $2.8 \times 10^{19}$ | $1.2 \times 10^{19}$ | $1.0 \times 10^{10}$ | 1350 | 450 | 0.26*a*<br>1.08*b* | 0.38*a*<br>0.60*b* | 11.9 |
| GaAs | 1.42 | 4.07 | $4.7 \times 10^{17}$ | $7 \times 10^{18}$ | $2.1 \times 10^{6}$ | 8500 | 400 | 0.067*a,b* | 0.40*a*<br>0.50*b* | 13.1 |

NOTE: Effective mass related to conductivity (labeled *a*) is different than that for density of states (labeled *b*). In numerous textbooks, $n_i$ is taken as $1.45 \times 10^{10}$ $cm^{-3}$ and is therefore the most widely used value of $n_i$ for Si, though the correct value is actually $1.0 \times 10^{10}$ $cm^{-3}$. (M. A. Green, *J. Appl. Phys.*, **67**, 2944, 1990.)

semiconductor is only scarcely populated by electrons, which means that there are many more electronic states than electrons and thus the likelihood of two electrons trying to occupy the same electronic state is practically nil. We can then neglect the Pauli exclusion principle and use the Boltzmann statistics. This is not the case for metals where the number of conduction electrons and the number of states are comparable in magnitude.

Table 5.1 is a comparative table of some of the properties of the important semiconductors, Ge, Si, and GaAs.

**EXAMPLE 5.1** **INTRINSIC CONCENTRATION AND CONDUCTIVITY OF Si** Given that the density of states related effective masses of electrons and holes in Si are approximately $1.08m_e$ and $0.60m_e$, respectively, and the electron and hole drift mobilities at room temperature are 1350 and 450 $cm^2\ V^{-1}\ s^{-1}$, respectively, calculate the intrinsic concentration and intrinsic resistivity of Si.

**SOLUTION**

We simply calculate the effective density of states $N_c$ and $N_v$ by

$$N_c = 2\left(\frac{2\pi m_e^* kT}{h^2}\right)^{3/2} \quad \text{and} \quad N_v = 2\left(\frac{2\pi m_h^* kT}{h^2}\right)^{3/2}$$

Thus

$$N_c = 2\left[\frac{2\pi(1.08 \times 9.1 \times 10^{-31}\ \text{kg})(1.38 \times 10^{-23}\ \text{J K}^{-1})\ (300\ \text{K})}{(6.63 \times 10^{-34}\ \text{J s})^2}\right]^{3/2}$$

$$= 2.81 \times 10^{25}\ \text{m}^{-3} \quad \text{or} \quad 2.81 \times 10^{19}\ \text{cm}^{-3}$$

and

$$N_v = 2\left[\frac{2\pi(0.60 \times 9.1 \times 10^{-31}\ \text{kg})(1.38 \times 10^{-23}\ \text{J K}^{-1})(300\ \text{K})}{(6.63 \times 10^{-34}\ \text{J s})^2}\right]^{3/2}$$

$$= 1.16 \times 10^{25}\ \text{m}^{-3} \quad \text{or} \quad 1.16 \times 10^{19}\ \text{cm}^{-3}$$

The intrinsic concentration is

$$n_i = (N_c N_v)^{1/2} \exp\left(-\frac{E_g}{2kT}\right)$$

so that

$$n_i = [(2.81 \times 10^{19}\ \text{cm}^{-3})(1.16 \times 10^{19}\ \text{cm}^{-3})]^{1/2} \exp\left[-\frac{(1.10\ \text{eV})}{2(300\ \text{K})(8.62 \times 10^{-5}\ \text{eV K}^{-1})}\right]$$
$$= 1.0 \times 10^{10}\,\text{cm}^{-3}$$

The conductivity is

$$\sigma = en\mu_e + ep\mu_h = en_i(\mu_e + \mu_h)$$

that is,

$$\sigma = (1.6 \times 10^{-19}\ \text{C})(1.0 \times 10^{10}\ \text{cm}^{-3})(1350 + 450\ \text{cm}^2\ \text{V}^{-1}\ \text{s}^{-1})$$
$$= 2.9 \times 10^{-6}\ \Omega^{-1}\ \text{cm}^{-1}$$

The resistivity is

$$\rho = \frac{1}{\sigma} = 3.5 \times 10^5\ \Omega\ \text{cm}$$

Although we calculated $n_i = 1.0 \times 10^{10}$ cm$^{-3}$, the most widely used $n_i$ value in the literature has been $1.45 \times 10^{10}$ cm$^{-3}$. The difference arises from a number of factors but, most importantly, from what exact value of the effective hole mass should be used in calculating $N_v$. Henceforth we will simply use[2] $n_i = 1.0 \times 10^{10}$ cm$^{-3}$, which seems to be the "true" value.

---

**EXAMPLE 5.2**

**MEAN SPEED OF ELECTRONS IN THE CB** Estimate the mean speed of electrons in the conduction band of Si at 300 K. If $a$ is the magnitude of lattice vibrations, then the kinetic theory predicts $a^2 \propto T$; or stated differently, the mean energy associated with lattice vibrations (proportional to $a^2$) increases with $kT$. Given the temperature dependence of the mean speed of electrons in the CB, what should be the temperature dependence of the drift mobility? The effective mass of an electron in the conduction band is $0.26m_e$.

**SOLUTION**

The fact that the average $KE$, $\frac{1}{2}m_e^* v_e^2$, of an electron in the CB of a semiconductor is $\frac{3}{2}kT$ means that the effective mean speed $v_e$ must be

$$v_e = \left(\frac{3kT}{m_e^*}\right)^{1/2} = \left[\frac{(3 \times 1.38 \times 10^{-23} \times 300)}{(0.26 \times 9.1 \times 10^{-31})}\right]^{1/2} = 2.3 \times 10^5\ \text{m s}^{-1}$$

The effective mean speed $v_e$ is called the **thermal velocity** $v_{\text{th}}$ of the electron.

The mean free time $\tau$ of the electron between scattering events due to thermal vibrations of the atoms is inversely proportional to both the mean speed $v_e$ of the electron and the scattering cross section of the thermal vibrations, that is,

$$\tau \propto \frac{1}{v_e(\pi a^2)}$$

where $a$ is the amplitude of the atomic thermal vibrations. But, $v_e \propto T^{1/2}$ and $(\pi a^2) \propto kT$, so that $\tau \propto T^{-3/2}$ and consequently $\mu_e \propto T^{-3/2}$.

Experimentally $\mu_e$ is not exactly proportional to $T^{-3/2}$ but to $T^{-2.4}$, a higher power index. The effective mass used in the density of states calculations is actually different than that used in transport calculations such as the mean speed, drift mobility, and so on.

---

[2] The correct value appears to be $1.0 \times 10^{10}$ cm$^{-3}$ as discussed by M. A. Green (*J. Appl. Phys.*, **67**, 2944, 1990) and A. B. Sproul and M. A. Green (*J. Appl. Phys.*, **70**, 846, 1991).

## 5.2 EXTRINSIC SEMICONDUCTORS

By introducing small amounts of impurities into an otherwise pure Si crystal, it is possible to obtain a semiconductor in which the concentration of carriers of one polarity is much in excess of the other type. Such semiconductors are referred to as **extrinsic semiconductors** vis-à-vis the intrinsic case of a pure and perfect crystal. For example, by adding pentavalent impurities, such as arsenic, which have a valency of more than four, we can obtain a semiconductor in which the electron concentration is much larger than the hole concentration. In this case we will have an *n*-type semiconductor. If we add trivalent impurities, such as boron, which have a valency of less than four, then we find that we have an excess of holes over electrons. We now have a *p*-type semiconductor. How do impurities change the concentrations of holes and electrons in a semiconductor?

### 5.2.1 *n*-TYPE DOPING

Consider what happens when small amounts of a pentavalent (valency of 5) element from Group V in the Periodic Table, such as As, P, Sb, are introduced into a pure Si crystal. We only add small amounts (*e.g.*, one impurity atom for every million host atoms) because we wish to surround each impurity atom by millions of Si atoms, thereby forcing the impurity atoms to bond with Si atoms in the same diamond crystal structure. Arsenic has five valence electrons, whereas Si has four. Thus when an As atom bonds with four Si atoms, it has one electron left unbonded. It cannot find a bond to go into, so it is left orbiting around the As atom, as illustrated in Figure 5.9. The $As^+$ ionic center with an electron $e^-$ orbiting it is just like a hydrogen atom in a silicon environment. We can easily calculate how much energy is required to free this electron away from the As site, thereby ionizing the As impurity. Had this been a hydrogen atom in free space, the energy required to remove the electron from its ground state (at $n = 1$) to far away from the positive center would have been given by $-E_n$ with $n = 1$. The binding energy of the electron in the H atom is thus

$$E_b = -E_1 = \frac{m_e e^4}{8\varepsilon_o^2 h^2} = 13.6 \text{ eV}$$

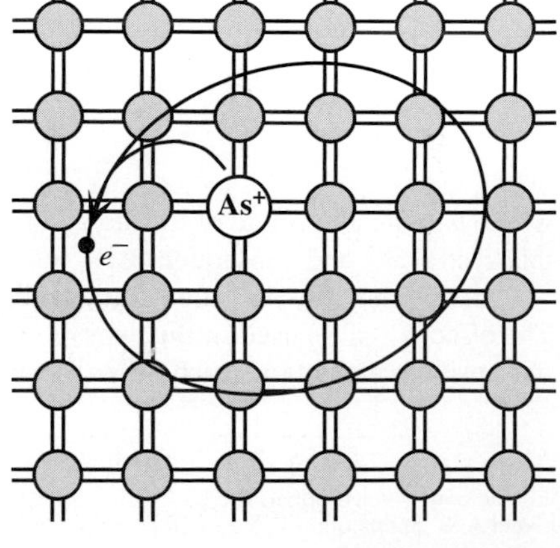


**Figure 5.9** Arsenic-doped Si crystal.
The four valence electrons of As allow it to bond just like Si, but the fifth electron is left orbiting the As site. The energy required to release the free fifth electron into the CB is very small.

If we wish to apply this to the electron around an $As^+$ core in the Si crystal environment, we must use $\varepsilon_r\varepsilon_o$ instead of $\varepsilon_o$, where $\varepsilon_r$ is the relative permittivity of silicon, and also the effective mass of the electron $m_e^*$ in the silicon crystal. Thus, the binding energy of the electron to the $As^+$ site in the Si crystal is

*Electron binding energy at a donor*

$$E_b^{\text{Si}} = \frac{m_e^* e^4}{8\varepsilon_r^2\varepsilon_o^2 h^2} = (13.6\text{ eV})\left(\frac{m_e^*}{m_e}\right)\left(\frac{1}{\varepsilon_r^2}\right) \quad [5.15]$$

With $\varepsilon_r = 11.9$ and $m_e^* \approx \frac{1}{3}m_e$ for silicon, we find $E_b^{\text{Si}} = 0.032$ eV, which is comparable with the average thermal energy of atomic vibrations at room temperature, $\sim 3kT$ ($\sim 0.07$ eV). Thus, the fifth valence electron can be readily freed by thermal vibrations of the Si lattice. The electron will then be "free" in the semiconductor, or, in other words, it will be in the CB. The energy required to excite the electron to the CB is therefore 0.032 eV. The addition of As atoms introduces localized electronic states at the As sites because the fifth electron has a localized wavefunction, of the hydrogenic type, around $As^+$. The energy $E_d$ of these states is 0.032 eV below $E_c$ because this is how much energy is required to take the electron away into the CB. Thermal excitation by the lattice vibrations at room temperature is sufficient to ionize the As atom, that is, excite the electron from $E_d$ into the CB. This process creates free electrons but immobile $As^+$ ions, as shown in the energy band diagram of an $n$-type semiconductor in Figure 5.10. Because the As atom donates an electron into the CB, it is called a **donor atom.** $E_d$ is the electron energy around the donor atom. $E_d$ is close to $E_c$, so the spare fifth electron from the dopant can be readily donated to the CB. If $N_d$ is the donor atom concentration in the crystal, then provided that $N_d \gg n_i$, at room temperature the electron concentration in the CB will be nearly equal to $N_d$, that is $n \approx N_d$. The hole concentration will be $p = n_i^2/N_d$, which is less than the intrinsic concentration because a few of the large number of electrons in the CB recombine with holes in the VB so as to maintain $np = n_i^2$. The conductivity will then be

*n-type conductivity*

$$\sigma = eN_d\mu_e + e\left(\frac{n_i^2}{N_d}\right)\mu_h \approx eN_d\mu_e \quad [5.16]$$

At low temperatures, however, not all the donors will be ionized and we need to know the probability, denoted as $f_d(E_d)$, of finding an electron in a state with energy

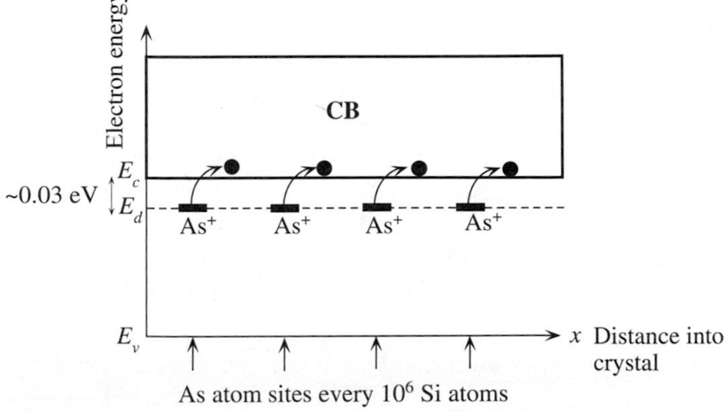


**Figure 5.10** Energy band diagram for an $n$-type Si doped with 1 ppm As.
There are donor energy levels just below $E_c$ around $As^+$ sites.

$E_d$ at a donor. This probability function is similar to the Fermi–Dirac function $f(E_d)$ except that it has a factor of $\frac{1}{2}$ multiplying the exponential term,

*Occupation probability at a donor*

$$f_d(E_d) = \frac{1}{1 + \dfrac{1}{2}\exp\left[\dfrac{(E_d - E_F)}{kT}\right]} \quad [5.17]$$

The factor $\frac{1}{2}$ is due to the fact that the electron state at the donor can take an electron with spin either up or down but not both[3] (once the donor has been occupied, a second electron cannot enter this site). Thus, the number of ionized donors at a temperature $T$ is given by

$$N_d^+ = N_d \times (\text{probability of not finding an electron at } E_d)$$

$$= N_d[1 - f_d(E_d)]$$

$$= \frac{N_d}{1 + 2\exp\left[\dfrac{(E_F - E_d)}{kT}\right]} \quad [5.18]$$

## 5.2.2 *p*-TYPE DOPING

We saw that introducing a pentavalent atom into a Si crystal results in $n$-type doping because the fifth electron cannot go into a bond and escapes from the donor into the CB by thermal excitation. By similar arguments, we should anticipate that doping a Si crystal with a trivalent atom (valency of 3) such as B, Al, Ga, or In will result in a $p$-type Si crystal. We consider doping Si with small amounts of B as shown in Figure 5.11a. Because B has only three valence electrons, when it shares them with four neighboring Si atoms, one of the bonds has a missing electron, which of course is a hole. A nearby electron can tunnel into this hole and displace the hole further away from the boron atom. As the hole moves away, it gets attracted by the negative charge left behind on the boron atom and therefore takes an orbit around the $B^-$ ion, as shown in Figure 5.11b. The binding energy of this hole to the $B^-$ ion can be calculated using the hydrogenic atom analogy as in the $n$-type Si case. This binding energy turns out to be very small, $\sim$0.05 eV, so at room temperature the thermal vibrations of the lattice can free the hole away from the $B^-$ site. A free hole, we recall, exists in the VB. The escape of the hole from the $B^-$ site involves the B atom *accepting* an electron from a neighboring Si–Si bond (from the VB), which effectively results in the hole being displaced away and its eventual escape to freedom in the VB. The B atom introduced into the Si crystal therefore acts as an electron acceptor and, because of this, it is called an **acceptor impurity.** The electron accepted by the B atom comes from a nearby bond. On the energy band diagram, an electron leaves the VB and gets accepted by a B atom, which becomes negatively charged. This process leaves a hole in the VB that is free to wander away, as illustrated in Figure 5.12.

It is apparent that doping a silicon crystal with a trivalent impurity results in a $p$-type material. We have many more holes than electrons for electrical conduction

[3] The proof can be found in advanced solid-state physics texts.

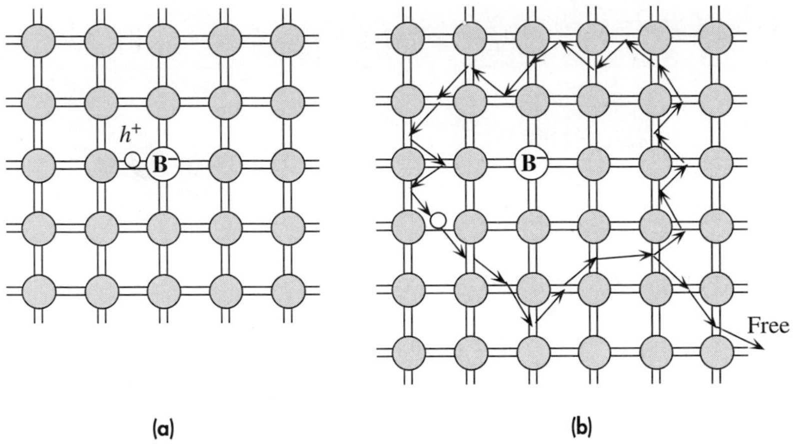


**Figure 5.11** Boron-doped Si crystal.
B has only three valence electrons. When it substitutes for a Si atom, one of its bonds has an electron missing and therefore a hole, as shown in (a). The hole orbits around the $B^-$ site by the tunneling of electrons from neighboring bonds, as shown in (b). Eventually, thermally vibrating Si atoms provide enough energy to free the hole from the $B^-$ site into the VB, as shown.

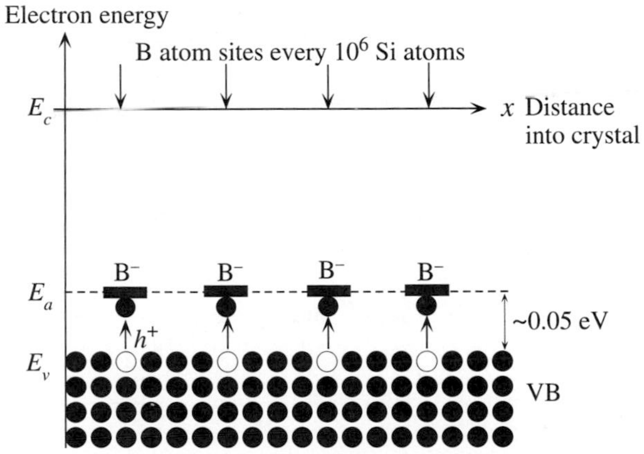


**Figure 5.12** Energy band diagram for a *p*-type Si doped with 1 ppm B.
There are acceptor energy levels $E_a$ just above $E_v$ around $B^-$ sites. These acceptor levels accept electrons from the VB and therefore create holes in the VB.

since the negatively charged B atoms are immobile and hence cannot contribute to the conductivity. If the concentration of acceptor impurities $N_a$ in the crystal is much greater than the intrinsic concentration $n_i$, then at room temperature all the acceptors would have been ionized and thus $p \approx N_a$. The electron concentration is then determined by the mass action law, $n = n_i^2/N_a$, which is much smaller than $p$, and consequently the conductivity is simply given by $\sigma = eN_a\mu_h$.

Typical ionization energies for donor and acceptor atoms in the silicon crystal are summarized in Table 5.2.

**Table 5.2** Examples of donor and acceptor ionization energies (eV) in Si

| Donors | | | Acceptors | | |
|---|---|---|---|---|---|
| P | As | Sb | B | Al | Ga |
| 0.045 | 0.054 | 0.039 | 0.045 | 0.057 | 0.072 |

### 5.2.3 COMPENSATION DOPING

What happens when a semiconductor contains both donors and acceptors? **Compensation doping** is a term used to describe the doping of a semiconductor with both donors and acceptors to control the properties. For example, a $p$-type semiconductor doped with $N_a$ acceptors can be converted to an $n$-type semiconductor by simply adding donors until the concentration $N_d$ exceeds $N_a$. The effect of donors compensates for the effect of acceptors and vice versa. The electron concentration is then given by $N_d - N_a$ provided the latter is larger than $n_i$. When both acceptors and donors are present, what essentially happens is that electrons from donors recombine with the holes from the acceptors so that the mass action law $np = n_i^2$ is obeyed. Remember that we cannot simultaneously increase the electron and hole concentrations because that leads to an increase in the recombination rate that returns the electron and hole concentrations to satisfy $np = n_i^2$. When an acceptor atom accepts a valence band electron, a hole is created in the VB. This hole then recombines with an electron from the CB. Suppose that we have more donors than acceptors. If we take the initial electron concentration as $n = N_d$, then the recombination between the electrons from the donors and $N_a$ holes generated by $N_a$ acceptors results in the electron concentration reduced by $N_a$ to $n = N_d - N_a$. By a similar argument, if we have more acceptors than donors, the hole concentration becomes $p = N_a - N_d$, with electrons from $N_d$ donors recombining with holes from $N_a$ acceptors. Thus there are two compensation effects:

*Compensation doping*

1. More donors: $N_d - N_a \gg n_i \qquad n = (N_d - N_a) \quad \text{and} \quad p = \dfrac{n_i^2}{(N_d - N_a)}$

2. More acceptors: $N_a - N_d \gg n_i \qquad p = (N_a - N_d) \quad \text{and} \quad n = \dfrac{n_i^2}{(N_a - N_d)}$

These arguments assume that the temperature is sufficiently high for donors and acceptors to have been ionized. This will be the case at room temperature. At low temperatures, we have to consider donor and acceptor statistics and the charge neutrality of the whole crystal, as in Example 5.8.

**EXAMPLE 5.3** **RESISTIVITY OF INTRINSIC AND DOPED Si** Find the resistance of a 1 cm$^3$ pure silicon crystal. What is the resistance when the crystal is doped with arsenic if the doping is 1 in $10^9$, that is, 1 part per billion (ppb) (note that this doping corresponds to one foreigner living in China)? Given data: Atomic concentration in silicon is $5 \times 10^{22}$ cm$^{-3}$, $n_i = 1.0 \times 10^{10}$ cm$^{-3}$, $\mu_e = 1350$ cm$^2$ V$^{-1}$ s$^{-1}$, and $\mu_h = 450$ cm$^2$ V$^{-1}$ s$^{-1}$.

**SOLUTION**

For the intrinsic case, we apply

$$\sigma = en\mu_e + ep\mu_h = en(\mu_e + \mu_h)$$

so

$$\sigma = (1.6 \times 10^{-19}\ \text{C})(1.0 \times 10^{10}\ \text{cm}^{-3})(1350 + 450\ \text{cm}^2\ \text{V}^{-1}\ \text{s}^{-1})$$
$$= 2.88 \times 10^{-6}\ \Omega^{-1}\ \text{cm}^{-1}$$

Since $L = 1$ cm and $A = 1\ \text{cm}^2$, the resistance is

$$R = \frac{L}{\sigma A} = \frac{1}{\sigma} = 3.47 \times 10^5\ \Omega \qquad \text{or} \qquad 347\ \text{k}\Omega$$

When the crystal is doped with 1 in $10^9$, then

$$N_d = \frac{N_{\text{Si}}}{10^9} = \frac{5 \times 10^{22}}{10^9} = 5 \times 10^{13}\ \text{cm}^{-3}$$

At room temperature all the donors are ionized, so

$$n = N_d = 5 \times 10^{13}\ \text{cm}^{-3}$$

The hole concentration is

$$p = \frac{n_i^2}{N_d} = \frac{(1.0 \times 10^{10})^2}{(5 \times 10^{13})} = 2.0 \times 10^6\ \text{cm}^{-3} \ll n_i$$

Therefore,

$$\sigma = en\mu_e = (1.6 \times 10^{-19}\ \text{C})(5 \times 10^{13}\ \text{cm}^{-3})(1350\ \text{cm}^2\ \text{V}^{-1}\ \text{s}^{-1})$$
$$= 1.08 \times 10^{-2}\ \Omega^{-1}\ \text{cm}^{-1}$$

Further,

$$R = \frac{L}{\sigma A} = \frac{1}{\sigma} = 92.6\ \Omega$$

Notice the drastic fall in the resistance when the crystal is doped with only 1 in $10^9$ atoms.

Doping the silicon crystal with boron instead of arsenic, but still in amounts of 1 in $10^9$, means that $N_a = 5 \times 10^{13}\ \text{cm}^{-3}$, which results in a conductivity of

$$\sigma = ep\mu_h = (1.6 \times 10^{-19}\ \text{C})(5 \times 10^{13}\ \text{cm}^{-3})(450\ \text{cm}^2\ \text{V}^{-1}\ \text{s}^{-1})$$
$$= 3.6 \times 10^{-3}\ \Omega^{-1}\ \text{cm}^{-1}$$

Therefore,

$$R = \frac{L}{\sigma A} = \frac{1}{\sigma} = 278\ \Omega$$

The reason for a higher resistance with $p$-type doping compared with the same amount of $n$-type doping is that $\mu_h < \mu_e$.

---

**EXAMPLE 5.4**

**COMPENSATION DOPING** An $n$-type Si semiconductor containing $10^{16}$ phosphorus (donor) atoms $\text{cm}^{-3}$ has been doped with $10^{17}$ boron (acceptor) atoms $\text{cm}^{-3}$. Calculate the electron and hole concentrations in this semiconductor.

**SOLUTION**

This semiconductor has been compensation doped with excess acceptors over donors, so

$$N_a - N_d = 10^{17} - 10^{16} = 9 \times 10^{16}\ \text{cm}^{-3}$$

This is much larger than the intrinsic concentration $n_i = 1.0 \times 10^{10}$ cm$^{-3}$ at room temperature, so

$$p = N_a - N_d = 9 \times 10^{16}\ \text{cm}^{-3}$$

The electron concentration

$$n = \frac{n_i^2}{p} = \frac{(1.0 \times 10^{10}\ \text{cm}^{-3})^2}{(9 \times 10^{16}\ \text{cm}^{-3})} = 1.1 \times 10^3\ \text{cm}^{-3}$$

Clearly, the electron concentration and hence its contribution to electrical conduction is completely negligible compared with the hole concentration. Thus, by excessive boron doping, the $n$-type semiconductor has been converted to a $p$-type semiconductor.

**EXAMPLE 5.5**

**THE FERMI LEVEL IN *n*- AND *p*-TYPE Si** An $n$-type Si wafer has been doped uniformly with $10^{16}$ antimony (Sb) atoms cm$^{-3}$. Calculate the position of the Fermi energy with respect to the Fermi energy $E_{Fi}$ in intrinsic Si. The above $n$-type Si sample is further doped with $2 \times 10^{17}$ boron atoms cm$^{-3}$. Calculate the position of the Fermi energy with respect to the Fermi energy $E_{Fi}$ in intrinsic Si. (Assume that $T = 300$ K, and $kT = 0.0259$ eV.)

**SOLUTION**

Sb gives $n$-type doping with $N_d = 10^{16}$ cm$^{-3}$, and since $N_d \gg n_i$ $(= 1.0 \times 10^{10}$ cm$^{-3})$, we have

$$n = N_d = 10^{16}\ \text{cm}^{-3}$$

For intrinsic Si,

$$n_i = N_c \exp\left[-\frac{(E_c - E_{Fi})}{kT}\right]$$

whereas for doped Si,

$$n = N_c \exp\left[-\frac{(E_c - E_{Fn})}{kT}\right] = N_d$$

where $E_{Fi}$ and $E_{Fn}$ are the Fermi energies in the intrinsic and $n$-type Si. Dividing the two expressions,

$$\frac{N_d}{n_i} = \exp\left[\frac{(E_{Fn} - E_{Fi})}{kT}\right]$$

so that

$$E_{Fn} - E_{Fi} = kT \ln\left(\frac{N_d}{n_i}\right) = (0.0259\ \text{eV}) \ln\left(\frac{10^{16}}{1.0 \times 10^{10}}\right) = 0.36\ \text{eV}$$

When the wafer is further doped with boron, the acceptor concentration is

$$N_a = 2 \times 10^{17}\ \text{cm}^{-3} > N_d = 10^{16}\ \text{cm}^{-3}$$

The semiconductor is compensation doped and compensation converts the semiconductor to $p$-type Si. Thus

$$p = N_a - N_d = (2 \times 10^{17} - 10^{16}) = 1.9 \times 10^{17}\ \text{cm}^{-3}$$

For intrinsic Si,

$$n_i = N_v \exp\left[-\frac{(E_{Fi} - E_v)}{kT}\right]$$

whereas for doped Si,

$$p = N_v \exp\left[-\frac{(E_{Fp} - E_v)}{kT}\right] = N_a - N_d$$

where $E_{Fi}$ and $E_{Fp}$ are the Fermi energies in the intrinsic and $p$-type Si, respectively. Dividing the two expressions,

$$\frac{p}{n_i} = \exp\left[-\frac{(E_{Fp} - E_{Fi})}{kT}\right]$$

so that

$$E_{Fp} - E_{Fi} = -kT \ln\left(\frac{p}{n_i}\right) = -(0.0259 \text{ eV}) \ln\left(\frac{1.9 \times 10^{17}}{1.0 \times 10^{10}}\right)$$

$$= -0.43 \text{ eV}$$

**EXAMPLE 5.6**

**ENERGY BAND DIAGRAM OF AN $n$-TYPE SEMICONDUCTOR CONNECTED TO A VOLTAGE SUPPLY** Consider the energy band diagram for an $n$-type semiconductor that is connected to a voltage supply of $V$ and is carrying a current. The applied voltage drops uniformly along the semiconductor, so the electrons in the semiconductor now also have an imposed electrostatic potential energy that decreases toward the positive terminal, as depicted in Figure 5.13. The whole band structure, the CB and the VB, therefore tilts. When an electron drifts from $A$ toward

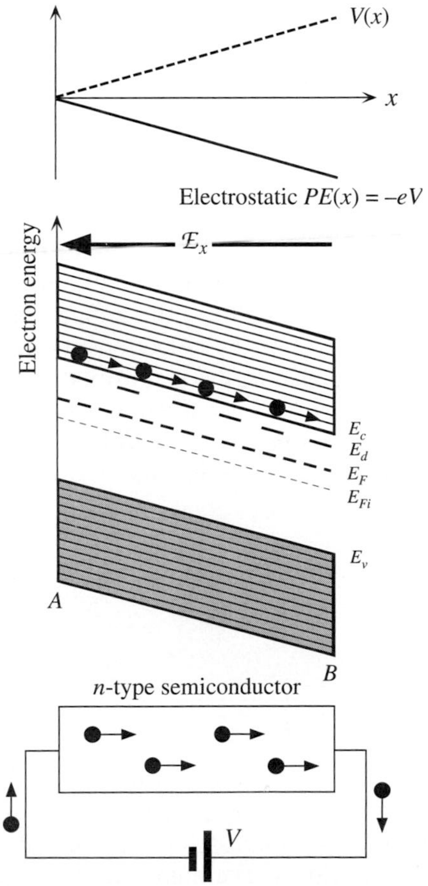


**Figure 5.13** Energy band diagram of an $n$-type semiconductor connected to a voltage supply of $V$ volts.

The whole energy diagram tilts because the electron now also has an electrostatic potential energy.

$B$, its $PE$ decreases because it is approaching the positive terminal. The Fermi level $E_F$ is above that for the intrinsic case, $E_{Fi}$.

We should remember that an important property of the Fermi level is that a change in $E_F$ within a system is available externally to do electrical work. As a corollary we note that when electrical work is done on the system, for example, when a battery is connected to a semiconductor, then $E_F$ is not uniform throughout the whole system. A change in $E_F$ within a system $\Delta E_F$ is equivalent to electrical work per electron or $eV$. $E_F$ therefore follows the electrostatic $PE$ behavior, and the change in $E_F$ from one end to the other, $E_F(A) - E_F(B)$, is just $eV$, the energy expended in taking an electron through the semiconductor, as shown in Figure 5.13. Electron concentration in the semiconductor is uniform, so $E_c - E_F$ must be constant from one end to the other. Thus the CB, VB, and $E_F$ all bend by the same amount.

## 5.3 TEMPERATURE DEPENDENCE OF CONDUCTIVITY

So far we have been calculating conductivities and resistivities of doped semiconductors at room temperature by simply assuming that $n \approx N_d$ for $n$-type and $p \approx N_a$ for $p$-type doping, with the proviso that the concentration of dopants is much greater than the intrinsic concentration $n_i$. To obtain the conductivity at other temperatures we have to consider two factors: the temperature dependence of the carrier concentration and the drift mobility.

### 5.3.1 CARRIER CONCENTRATION TEMPERATURE DEPENDENCE

Consider an $n$-type semiconductor doped with $N_d$ donors per unit volume where $N_d \gg n_i$. We take the semiconductor down to very low temperatures until its conductivity is practically nil. At this temperature, the donors will *not* be ionized because the thermal vibrational energy is insufficiently small. As the temperature is increased, some of the donors become ionized and donate their electrons to the CB, as shown in Figure 5.14a. The Si–Si bond breaking, that is, thermal excitation from $E_v$ to $E_c$, is unlikely because it takes too much energy. Since the donor ionization energy $\Delta E = E_c - E_d$ is very small ($\ll E_g$), thermal generation involves exciting electrons from $E_d$ to $E_c$. The electron concentration at low temperatures is given by the expression

$$n = \left(\frac{1}{2}N_c N_d\right)^{1/2} \exp\left(-\frac{\Delta E}{2kT}\right) \qquad [5.19]$$

similar to the intrinsic case, that is,

$$n = (N_c N_v)^{1/2} \exp\left(-\frac{E_g}{2kT}\right) \qquad [5.20]$$

Equation 5.20 is valid when thermal generation occurs across the bandgap $E_g$ from $E_v$ to $E_c$. Equation 5.19 is the counterpart of Equation 5.20 taking into account that at low temperatures the excitation is from $E_d$ to $E_c$ (across $\Delta E$) and that instead of $N_v$, we have $N_d$ as the number of available electrons. The numerical factor $\frac{1}{2}$ in

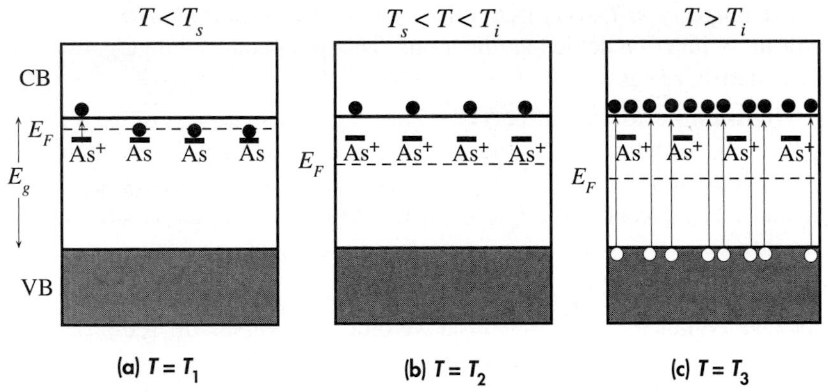


**Figure 5.14**
(a) Below $T_s$, the electron concentration is controlled by the ionization of the donors.
(b) Between $T_s$ and $T_i$, the electron concentration is equal to the concentration of donors since they would all have ionized.
(c) At high temperatures, thermally generated electrons from the VB exceed the number of electrons from ionized donors and the semiconductor behaves as if intrinsic.

Equation 5.19 arises because donor occupation statistics is different by this factor from the usual Fermi–Dirac function, as mentioned earlier.

As the temperature is increased further, eventually all the donors become ionized and the electron concentration is equal to the donor concentration, that is, $n = N_d$, as depicted in Figure 5.14b. This state of affairs remains unchanged until very high temperatures are reached, when thermal generation across the bandgap begins to dominate. At very high temperatures, thermal vibrations of the atoms will be so strong that many Si–Si bonds will be broken and thermal generation across $E_g$ will dominate. The electron concentration in the CB will then be mainly due to thermal excitation from the VB to the CB, as illustrated in Figure 5.14c. But this process also generates an equal concentration of holes in the VB. Accordingly, the semiconductor behaves as if it were intrinsic. The electron concentration at these temperatures will therefore be equal to the intrinsic concentration $n_i$, which is given by Equation 5.20.

The dependence of the electron concentration on temperature thus has three regions:

1. **Low-temperature range ($T < T_s$).** The increase in temperature at these low temperatures ionizes more and more donors. The donor ionization continues until we reach a temperature $T_s$, called the **saturation temperature,** when all donors have been ionized and we have saturation in the concentration of ionized donors. The electron concentration is given by Equation 5.19. This temperature range is often referred to as the **ionization range.**

2. **Medium-temperature range ($T_s < T < T_i$).** Since nearly all the donors have been ionized in this range, $n = N_d$. This condition remains unchanged until $T = T_i$, when $n_i$, which is temperature dependent, becomes equal to $N_d$. It is this

temperature range $T_s < T < T_i$ that utilizes the $n$-type doping properties of the semiconductor in $pn$ junction device applications. This temperature range is often referred to as the **extrinsic range.**

**3. High-temperature range ($T > T_i$).** The concentration of electrons generated by thermal excitation across the bandgap $n_i$ is now much larger than $N_d$, so the electron concentration $n = n_i(T)$. Furthermore, as excitation occurs from the VB to the CB, the hole concentration $p = n$. This temperature range is referred to as the **intrinsic range.**

Figure 5.15 shows the behavior of the electron concentration with temperature in an $n$-type semiconductor. By convention we plot ln($n$) versus the reciprocal temperature $T^{-1}$. At low temperatures, ln($n$) versus $T^{-1}$ is almost a straight line with a slope $-(\Delta E/2k)$, since the temperature dependence of $N_c^{1/2}(\propto T^{3/4})$ is negligible compared with the $\exp(-\Delta E/2kT)$ part in Equation 5.19. In the high-temperature range, however, the slope is quite steep and almost $-E_g/2k$ since Equation 5.20 implies that

$$n \propto T^{3/2} \exp\left(-\frac{E_g}{2kT}\right)$$

and the exponential part again dominates over the $T^{3/2}$ part. In the intermediate range, $n$ is equal to $N_d$ and practically independent of the temperature.

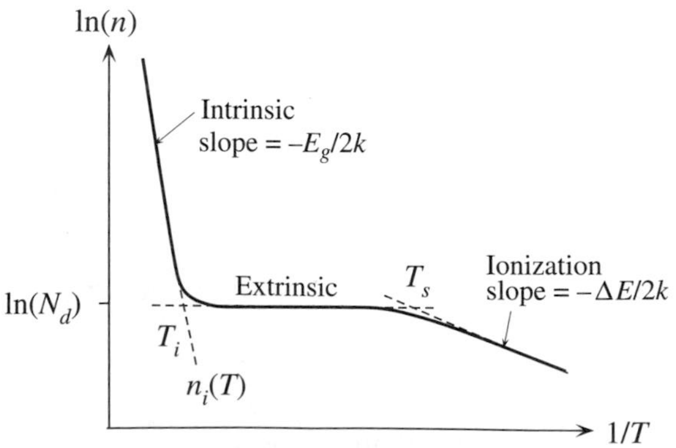


**Figure 5.15** The temperature dependence of the electron concentration in an $n$-type semiconductor.

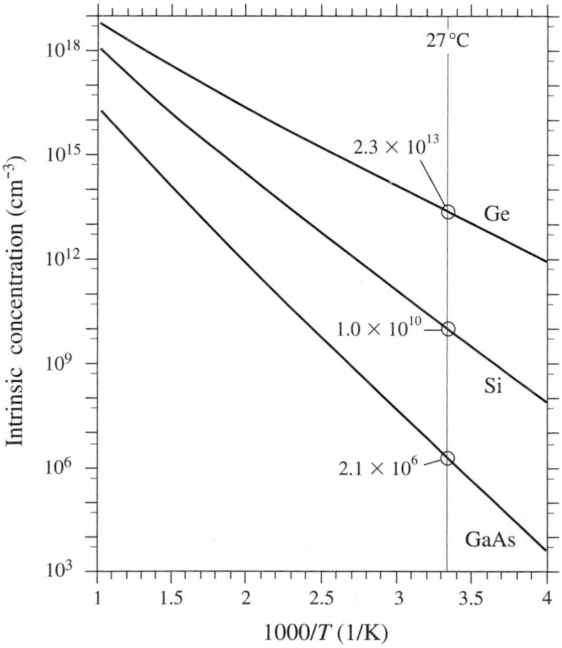


**Figure 5.16** The temperature dependence of the intrinsic concentration.

Figure 5.16 displays the temperature dependence of the intrinsic concentration in Ge, Si, and GaAs as $\log(n_i)$ versus $1/T$ where the slope of the lines is, of course, a measure of the bandgap energy $E_g$. The $\log(n_i)$ versus $1/T$ graphs can be used to find, for example, whether the dopant concentration at a given temperature is more than the intrinsic concentration. As we will find out in Chapter 6, the reverse saturation current in a *pn* junction diode depends on $n_i^2$, so Figure 5.16 also indicates how this saturation current varies with temperature.

**EXAMPLE 5.7**

**SATURATION AND INTRINSIC TEMPERATURES** An *n*-type Si sample has been doped with $10^{15}$ phosphorus atoms $\text{cm}^{-3}$. The donor energy level for P in Si is 0.045 eV below the conduction band edge energy.

*a.* Estimate the temperature above which the sample behaves as if intrinsic.

*b.* Estimate the lowest temperature above which most of the donors are ionized.

**SOLUTION**

Remember that $n_i(T)$ is highly temperature dependent, as shown in Figure 5.16 so that as $T$ increases, eventually at $T \approx T_i$, $n_i$ becomes comparable to $N_d$. Beyond $T_i$, $n_i(T > T_i) \gg N_d$. Thus we need to solve

$$n_i(T_i) = N_d = 10^{15}\ \text{cm}^{-3}$$

From the $\log(n_i)$ versus $10^3/T$ graph for Si in Figure 5.16, when $n_i = 10^{15}\ \text{cm}^{-3}$, $(10^3/T_i) \approx 1.85$, giving $T_i \approx 541$ K or 268 °C.

We will assume that most of the donors are ionized, say at $T \approx T_s$, where the extrinsic and the extrapolated ionization lines intersect in Figure 5.15:

$$n = \left(\frac{1}{2}N_c N_d\right)^{1/2} \exp\left(-\frac{\Delta E}{2kT_s}\right) \approx N_d$$

This is the temperature at which the ionization behavior intersects the extrinsic region. In the above equation, $N_d = 10^{15}\ \text{cm}^{-3}$, $\Delta E = 0.045$ eV, and $N_c \propto T^{3/2}$, that is,

$$N_c(T_s) = N_c(300\ \text{K})\left(\frac{T_s}{300}\right)^{3/2}$$

Clearly, then, the equation can only be solved numerically. Similar equations occur in a wide range of physical problems where one term has the strongest temperature dependence. Here, $\exp(-\Delta E/kT_s)$ has the strongest temperature dependence. First assume $N_c$ is that at 300 K, $N_c = 2.8 \times 10^{19}\ \text{cm}^{-3}$, and evaluate $T_s$,

$$T_s = \frac{\Delta E}{k \ln\left(\dfrac{N_c}{2N_d}\right)} = \frac{0.045\ \text{eV}}{(8.62 \times 10^{-5}\ \text{eV K}^{-1}) \ln\left[\dfrac{2.8 \times 10^{19}\ \text{cm}^{-3}}{2(1.0 \times 10^{15}\ \text{cm}^{-3})}\right]} = 54.7\ \text{K}$$

At $T = 54.7$ K,

$$N_c(54.7\ \text{K}) = N_c(300\ \text{K})\left(\frac{54.7}{300}\right)^{3/2} = 2.18 \times 10^{18}\ \text{cm}^{-3}$$

With this new $N_c$ at a lower temperature, the improved $T_s$ is 74.6 K. Since we only need an estimate of $T_s$, the extrinsic range of this semiconductor is therefore from about 75 to 541 K or −198 to about 268 °C.

**EXAMPLE 5.8**

*Electron concentration in the ionization region*

**TEMPERATURE DEPENDENCE OF THE ELECTRON CONCENTRATION** By considering the mass action law, charge neutrality within the crystal, and occupation statistics of electronic states, we can show that at the lowest temperatures the electron concentration in an $n$-type semiconductor is given by

$$n = \left(\frac{1}{2}N_c N_d\right)^{1/2} \exp\left(-\frac{\Delta E}{2kT}\right)$$

where $\Delta E = E_c - E_d$. Furthermore, at the lowest temperatures, the Fermi energy is midway between $E_d$ and $E_c$.

There are only a few physical principles that must be considered to arrive at the effect of doping on the electron and hole concentrations. For an $n$-type semiconductor, these are

1. **Charge carrier statistics.**

$$n = N_c \exp\left[-\frac{(E_c - E_F)}{kT}\right] \quad (1)$$

2. **Mass action law.**

$$np = n_i^2 \quad (2)$$

3. **Electrical neutrality of the crystal.** We must have the same number of positive and negative charges:

$$p + N_d^+ = n \quad (3)$$

where $N_d^+$ is the concentration of *ionized* donors.

4. **Statistics of ionization of the dopants.**

$$N_d^+ = N_d \times \text{(probability of not finding an electron at } E_d) = N_d[1 - f_d(E_d)]$$

$$= \frac{N_d}{1 + 2\exp\left[\frac{(E_F - E_d)}{kT}\right]} \quad (4)$$

Solving Equations 1 to 4 for $n$ will give the dependence of $n$ on $T$ and $N_d$. For example, from the mass action law, Equation 2, and the charge neutrality condition, Equation 3, we get

$$\frac{n_i^2}{n} + N_d^+ = n$$

This is a quadratic equation in $n$. Solving this equation gives

$$n = \frac{1}{2}(N_d^+) + \left[\frac{1}{4}(N_d^+)^2 + n_i^2\right]^{1/2}$$

Clearly, this equation should give the behavior of $n$ as a function of $T$ and $N_d$ when we also consider the statistics in Equation 4. In the low-temperature region ($T < T_s$), $n_i^2$ is negligible in the expression for $n$ and we have

$$n = N_d^+ = \frac{N_d}{1 + 2\exp\left[\frac{(E_F - E_d)}{kT}\right]} \approx \frac{1}{2}N_d \exp\left[-\frac{(E_F - E_d)}{kT}\right]$$

But the statistical description in Equation 1 is generally valid, so multiplying the low-temperature region equation by Equation 1 and taking the square root eliminates $E_F$ from the expression, giving

$$n = \left(\frac{1}{2}N_c N_d\right)^{1/2} \exp\left[-\frac{(E_c - E_d)}{2kT}\right]$$

To find the location of the Fermi energy, consider the general expression

$$n = N_c \exp\left[-\frac{(E_c - E_F)}{kT}\right]$$

which must now correspond to $n$ at low temperatures. Equating the two and rearranging to obtain $E_F$ we find

$$E_F = \frac{E_c + E_d}{2} + \frac{1}{2}kT \ln\left(\frac{N_d}{2N_c}\right)$$

which puts the Fermi energy near the middle of $\Delta E = E_c - E_d$ at low temperatures.

### 5.3.2 DRIFT MOBILITY: TEMPERATURE AND IMPURITY DEPENDENCE

The temperature dependence of the drift mobility follows two distinctly different temperature variations. In the high-temperature region, it is observed that the drift mobility is limited by scattering from lattice vibrations. As the magnitude of atomic vibrations increases with temperature, the drift mobility decreases in the fashion $\mu \propto T^{-3/2}$. However, at low temperatures the lattice vibrations are not sufficiently strong to be the major limitation to the mobility of the electrons. It is observed that at low temperatures the scattering of electrons by ionized impurities is the major mobility limiting mechanism and $\mu \propto T^{3/2}$, as we will show below.

We recall from Chapter 2 that the electron drift mobility $\mu$ depends on the mean free time $\tau$ between scattering events via

$$\mu = \frac{e\tau}{m_e^*} \qquad [5.21]$$

in which

$$\tau = \frac{1}{Sv_{\text{th}}N_s} \qquad [5.22]$$

where $S$ is the cross-sectional area of the scatterer; $v_{\text{th}}$ is the mean speed of the electrons, called the **thermal velocity;** and $N_s$ is the number of scatterers per unit volume. If $a$ is the amplitude of the atomic vibrations about the equilibrium, then $S = \pi a^2$. As the temperature increases, so does the amplitude $a$ of the lattice vibrations following $a^2 \propto T$ behavior, as shown in Chapter 2. An electron in the CB is free to wander around and therefore has only *KE*. We also know that the mean kinetic energy per electron in the CB is $\frac{3}{2}kT$, just as if the kinetic molecular theory could be applied to all those electrons in the CB. Therefore,

$$\frac{1}{2}m_e^* v_{\text{th}}^2 = \frac{3}{2}kT$$

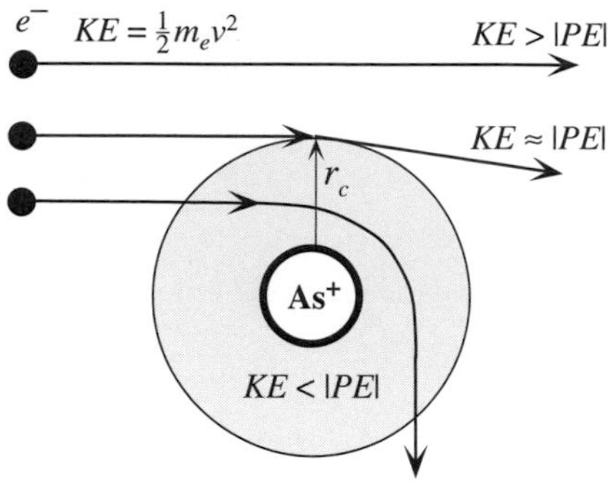


**Figure 5.17** Scattering of electrons by an ionized impurity.

so that $v_{\text{th}} \propto T^{1/2}$. Thus the mean time $\tau_L$ between scattering events from lattice vibrations is

$$\tau_L = \frac{1}{(\pi a^2)v_{\text{th}}N_s} \propto \frac{1}{(T)(T^{1/2})} \propto T^{-3/2}$$

*Lattice-scattering-limited mobility*

which leads to a **lattice vibration scattering limited mobility,** denoted as $\mu_L$, of the form

$$\mu_L \propto T^{-3/2} \tag{5.23}$$

At low temperatures, scattering of electrons by thermal vibrations of the lattice will not be as strong as the electron scattering brought about by ionized donor impurities. As an electron passes by an ionized donor $As^+$, it is attracted and thus deflected from its straight path, as schematically shown in Figure 5.17. This type of scattering of an electron is what limits the drift mobility at low temperatures.

The *PE* of an electron at a distance $r$ from an $As^+$ ion is due to the Coulombic attraction, and its magnitude is given by

$$|PE| = \frac{e^2}{4\pi\varepsilon_o\varepsilon_r r}$$

If the *KE* of the electron approaching an $As^+$ ion is larger than its *PE* at distance $r$ from $As^+$, then the electron will essentially continue without feeling the *PE* and therefore without being deflected, and we can say that it has not been scattered. Effectively, due to its high *KE*, the electron does not feel the Coulombic pull of the donor. On the other hand, if the *KE* of the electron is less than its *PE* at $r$ from $As^+$, then the *PE* of the Coulombic interaction will be so strong that the electron will be strongly deflected. This is illustrated in Figure 5.17. The critical radius $r_c$ corresponds to the case when the electron is just scattered, which is when $KE \approx |PE(r_c)|$. But average $KE = \frac{3}{2}kT$, so at $r = r_c$

$$\frac{3}{2}kT = |PE(r_c)| = \frac{e^2}{4\pi\varepsilon_o\varepsilon_r r_c}$$

from which $r_c = e^2/(6\pi\varepsilon_o\varepsilon_r kT)$. As the temperature increases, the scattering radius decreases. The scattering cross section $S = \pi r_c^2$ is thus given by

$$S = \frac{\pi e^4}{(6\pi\varepsilon_o\varepsilon_r kT)^2} \propto T^{-2}$$

Incorporating $v_{\text{th}} \propto T^{1/2}$ as well, the temperature dependence of the mean scattering time $\tau_I$ between impurities, from Equation 5.22, must be

$$\tau_I = \frac{1}{S v_{\text{th}} N_I} \propto \frac{1}{(T^{-2})(T^{1/2})N_I} \propto \frac{T^{3/2}}{N_I}$$

where $N_I$ is the concentration of ionized impurities (all ionized impurities including donors and acceptors). Consequently, the **ionized impurity scattering limited mobility** from Equation 5.21 is

*Ionized impurity scattering limited mobility*

$$\mu_I \propto \frac{T^{3/2}}{N_I} \quad \textbf{[5.24]}$$

Note also that $\mu_I$ decreases with increasing ionized dopant concentration $N_I$, which itself may be temperature dependent. Indeed, at the lowest temperatures, below the saturation temperature $T_s$, $N_I$ will be strongly temperature dependent because not all the donors would have been fully ionized.

The overall temperature dependence of the drift mobility is then, simply, the reciprocal additions of the $\mu_I$ and $\mu_L$ by virtue of Matthiessen's rule, that is,

*Effective mobility*

$$\frac{1}{\mu_e} = \frac{1}{\mu_I} + \frac{1}{\mu_L} \quad \textbf{[5.25]}$$

so the scattering process having the lowest mobility determines the overall (effective) drift mobility.

The experimental temperature dependence of the electron drift mobility in both Ge and Si is shown in Figure 5.18 as a log–log plot for various donor concentrations. The slope on this plot corresponds to the index $n$ in $\mu_e \propto T^n$. The simple theoretical sketches in the insets show how $\mu_L$ and $\mu_I$ from Equations 5.23 and 5.24 depend on the temperature. For Ge, at low doping concentrations (*e.g.*, $N_d = 10^{13}\ \text{cm}^{-3}$), the experiments indicate a $\mu_e \propto T^{-1.5}$ type of behavior, which is in agreement with $\mu_e$ determined by $\mu_L$ in Equation 5.23. Curves for Si at low-level doping ($\mu_I$ negligible)

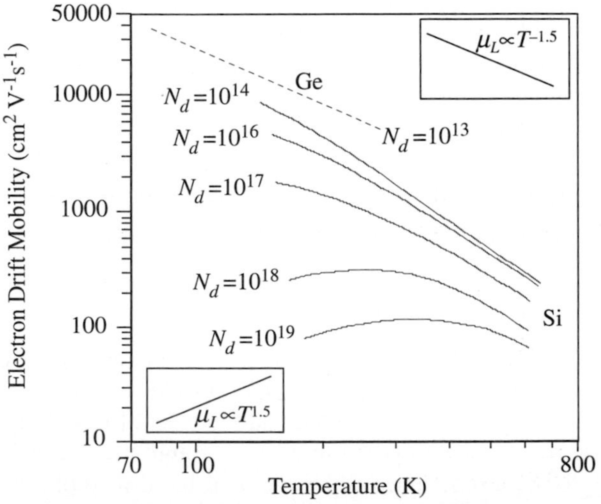


**Figure 5.18** Log–log plot of drift mobility versus temperature for *n*-type Ge and *n*-type Si samples. Various donor concentrations for Si are shown. $N_d$ are in $\text{cm}^{-3}$. The upper right inset is the simple theory for lattice limited mobility, whereas the lower left inset is the simple theory for impurity scattering limited mobility.

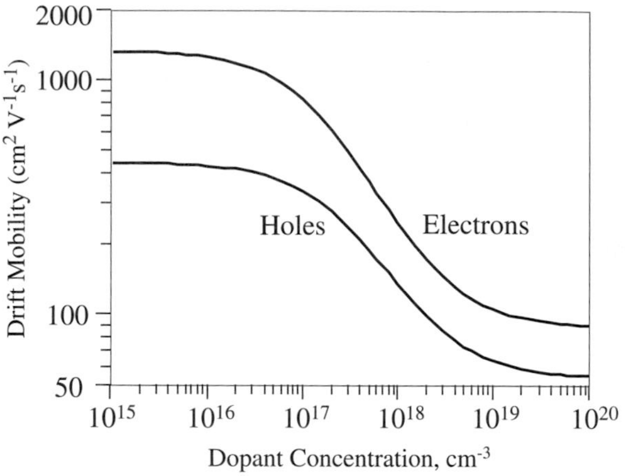


**Figure 5.19** The variation of the drift mobility with dopant concentration in Si for electrons and holes at 300 K.

at high temperatures, however, exhibit a $\mu_e \propto T^{-2.5}$ type of behavior rather than $T^{-1.5}$, which can be accounted for in a more rigorous theory. As the donor concentration increases, the drift mobility decreases by virtue of $\mu_I$ getting smaller. At the highest doping concentrations and at low temperatures, the electron drift mobility in Si exhibits almost a $\mu_e \propto T^{3/2}$ type of behavior. Similar arguments can be extended to the temperature dependence of the hole drift mobility.

The dependences of the room temperature electron and hole drift mobilities on the dopant concentration for Si are shown in Figure 5.19 where, as expected, past a certain amount of impurity addition, the drift mobility is overwhelmingly controlled by $\mu_I$ in Equation 5.25.

### 5.3.3 Conductivity Temperature Dependence

The conductivity of an extrinsic semiconductor doped with donors depends on the electron concentration and the drift mobility, both of which have been determined above. At the lowest temperatures in the ionization range, the electron concentration depends exponentially on the temperature by virtue of

*Electron concentration in ionization region*

$$n = \left(\frac{1}{2}N_cN_d\right)^{1/2} \exp\left[-\frac{(E_c - E_d)}{2kT}\right]$$

which then also dominates the temperature dependence of the conductivity. In the intrinsic range at the highest temperatures, the conductivity is dominated by the temperature dependence of $n_i$ since

$$\sigma = en_i(\mu_e + \mu_h)$$

and $n_i$ is an exponential function of temperature in contrast to $\mu \propto T^{-3/2}$. In the extrinsic temperature range, $n = N_d$ and is constant, so the conductivity follows the temperature dependence of the drift mobility. Figure 5.20 shows schematically the semilogarithmic plot of the conductivity against the reciprocal temperature where through the extrinsic range $\sigma$ exhibits a broad "S" due to the temperature dependence of the drift mobility.

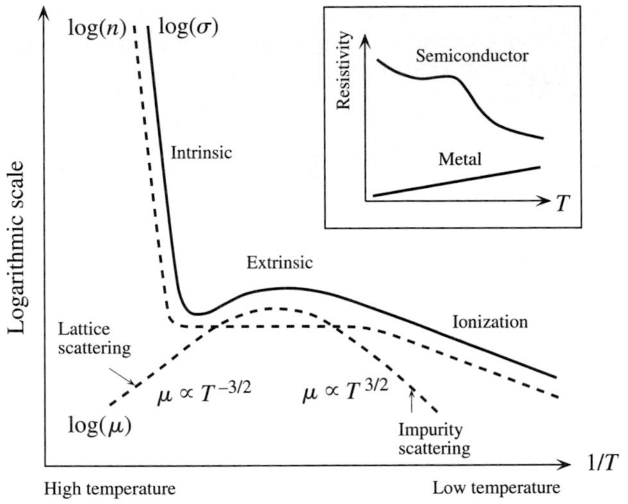


**Figure 5.20** Schematic illustration of the temperature dependence of electrical conductivity for a doped (*n*-type) semiconductor.

## COMPENSATION-DOPED Si

**EXAMPLE 5.9**

*a.* A Si sample has been doped with $10^{17}$ arsenic atoms $cm^{-3}$. Calculate the conductivity of the sample at 27 °C (300 K) and at 127 °C (400 K).

*b.* The above *n*-type Si sample is further doped with $9 \times 10^{16}$ boron atoms $cm^{-3}$. Calculate the conductivity of the sample at 27 °C and 127 °C.

### SOLUTION

*a.* The arsenic dopant concentration, $N_d = 10^{17}$ cm$^{-3}$, is much larger than the intrinsic concentration $n_i$, which means that $n = N_d$ and $p = (n_i^2/N_d) \ll n$ and can be neglected. Thus $n = 10^{17}$ cm$^{-3}$ and the electron drift mobility at $N_d = 10^{17}$ cm$^{-3}$ is 800 cm$^2$ V$^{-1}$ s$^{-1}$ from the drift mobility versus dopant concentration graph in Figure 5.19, so

$$\sigma = en\mu_e + ep\mu_h = eN_d\mu_e$$

$$= (1.6 \times 10^{-19}\text{ C})(10^{17}\text{ cm}^{-3})(800\text{ cm}^2\text{ V}^{-1}\text{ s}^{-1}) = 12.8\ \Omega^{-1}\text{ cm}^{-1}$$

At $T = 127$ °C = 400 K,

$$\mu_e \approx 420\text{ cm}^2\text{ V}^{-1}\text{ s}^{-1}$$

(from the $\mu_e$ versus $T$ graph in Figure 5.18). Thus

$$\sigma = eN_d\mu_e = 6.72\ \Omega^{-1}\text{ cm}^{-1}$$

*b.* With further doping we have $N_a = 9 \times 10^{16}$ cm$^{-3}$, so from the compensation effect

$$N_d - N_a = 1 \times 10^{17} - 9 \times 10^{16} = 10^{16}\text{ cm}^{-3}$$

Since $N_d - N_a \gg n_i$, we have an *n*-type material with $n = N_d - N_a = 10^{16}$ cm$^{-3}$. But the drift mobility now is about ~ 600 cm$^2$ V$^{-1}$ s$^{-1}$ because, even though $N_d - N_a$ is now $10^{16}$ cm$^{-3}$ and not $10^{17}$ cm$^{-3}$, all the donors and acceptors are still ionized and hence still scatter the charge carriers. The recombination of electrons from the donors and holes from the acceptors does not alter the fact that at room temperature all the dopants will be ionized.

Effectively, the compensation effect is as if all electrons from the donors were being accepted by the acceptors. Although with compensation doping the net electron concentration is $n = N_d - N_a$, the drift mobility scattering is determined by $(N_d + N_a)$, which in this case is $10^{17} + 9 \times 10^{16}$ cm$^{-3}$ = $1.9 \times 10^{17}$ cm$^{-3}$, which gives an electron drift mobility of ~600 cm$^2$ V$^{-1}$ s$^{-1}$ at 300 K and ~400 cm$^2$ V$^{-1}$ s$^{-1}$ at 400 K. Then, neglecting the hole concentration $p = n_i^2/(N_d - N_a)$, we have

At 300 K, $$\sigma = e(N_d - N_a)\mu_e \approx (1.6 \times 10^{-19}\ \text{C})(10^{16}\ \text{cm}^{-3})(600\ \text{cm}^2\ \text{V}^{-1}\ \text{s}^{-1}) = 0.96\ \Omega^{-1}\ \text{cm}^{-1}$$

At 400 K, $$\sigma = e(N_d - N_a)\mu_e \approx (1.6 \times 10^{-19}\ \text{C})(10^{16}\ \text{cm}^{-3})(400\ \text{cm}^2\ \text{V}^{-1}\ \text{s}^{-1}) = 0.64\ \Omega^{-1}\ \text{cm}^{-1}$$

### 5.3.4 DEGENERATE AND NONDEGENERATE SEMICONDUCTORS

The general exponential expression for the concentration of electron in the CB,

$$n \approx N_c \exp\left[-\frac{(E_c - E_F)}{kT}\right] \quad \text{[5.26]}$$

is based on replacing Fermi–Dirac statistics with Boltzmann statistics, which is only valid when $E_c$ is several $kT$ above $E_F$. In other words, we assumed that the number of states in the CB far exceeds the number of electrons there, so the likelihood of two electrons trying to occupy the same state is almost nil. This means that the Pauli exclusion principle can be neglected and the electron statistics can be described by the Boltzmann statistics. $N_c$ is a measure of the density of states in the CB. The Boltzmann expression for $n$ is valid only when $n \ll N_c$. Those semiconductors for which $n \ll N_c$ and $p \ll N_v$ are termed **nondegenerate semiconductors.** They essentially follow all the discussions above and exhibit all the normal semiconductor properties outlined above.

When the semiconductor has been excessively doped with donors, then $n$ may be so large, typically $10^{19}$–$10^{20}$ cm$^{-3}$, that it may be comparable to or greater than $N_c$. In that case the Pauli exclusion principle becomes important in the electron statistics and we have to use the Fermi–Dirac statistics. Equation 5.26 for $n$ is then no longer valid. Such a semiconductor exhibits properties that are more metal-like than semiconductor-like; for example, the resistivity follows $\rho \propto T$. Semiconductors that have $n > N_c$ or $p > N_v$ are called **degenerate semiconductors.**

The large carrier concentration in a degenerate semiconductor is due to its heavy doping. For example, as the donor concentration in an $n$-type semiconductor is increased, at sufficiently high doping levels, the donor atoms become so close to each other that their orbitals overlap to form a narrow energy band that overlaps and becomes part of the conduction band. $E_c$ is therefore slightly shifted down and $E_g$ becomes slightly narrower. The valence electrons from the donors fill the band from $E_c$. This situation is reminiscent of the valence electrons filling overlapping energy bands in a metal. In a degenerate $n$-type semiconductor, the Fermi level is therefore within the CB, or above $E_c$ just like $E_F$ is within the band in a metal. The

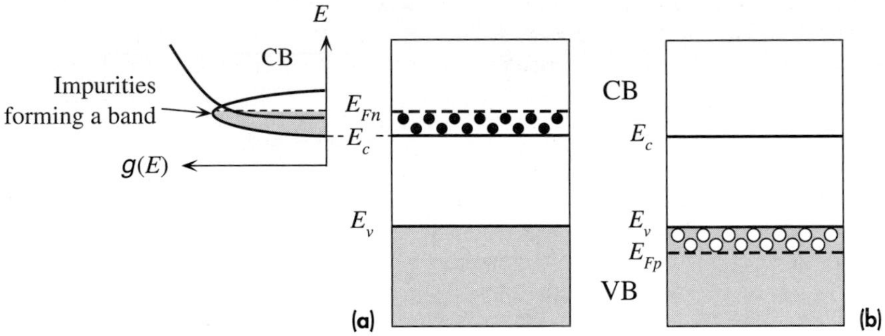


**Figure 5.21**
(a) Degenerate *n*-type semiconductor. Large number of donors form a band that overlaps the CB.
(b) Degenerate *p*-type semiconductor.

majority of the states between $E_c$ and $E_F$ are full of electrons as indicated in Figure 5.21. In the case of a *p*-type degenerate semiconductor, the Fermi level lies in the VB below $E_v$. It should be emphasized that one cannot simply assume that $n = N_d$ or $p = N_a$ in a degenerate semiconductor because the dopant concentration is so large that they interact with each other. Not all dopants are able to become ionized, and the carrier concentration eventually reaches a saturation typically around $\sim 10^{20}$ cm$^{-3}$. Furthermore, the mass action law $np = n_i^2$ is not valid for degenerate semiconductors.

Degenerate semiconductors have many important uses. For example, they are used in laser diodes, zener diodes, and ohmic contacts in ICs, and as metal gates in many microelectronic MOS devices.

## 5.4 RECOMBINATION AND MINORITY CARRIER INJECTION

### 5.4.1 DIRECT AND INDIRECT RECOMBINATION

Above absolute zero of temperature, the thermal excitation of electrons from the VB to the CB continuously generates free electron–hole pairs. It should be apparent that in equilibrium there should be some annihilation mechanism that returns the electron from the CB down to an empty state (a hole) in the VB. When a free electron, wandering around in the CB of a crystal, "meets" a hole, it falls into this low-energy empty electronic state and fills it. This process is called **recombination.** Intuitively, recombination corresponds to the free electron finding an incomplete bond with a missing electron. The electron then enters and completes this bond. The free electron in the CB and the free hole in the VB are consequently annihilated. On the energy band diagram, the recombination process is represented by returning the electron from the CB (where it is free) into a hole in the VB (where it is in a bond). Figure 5.22

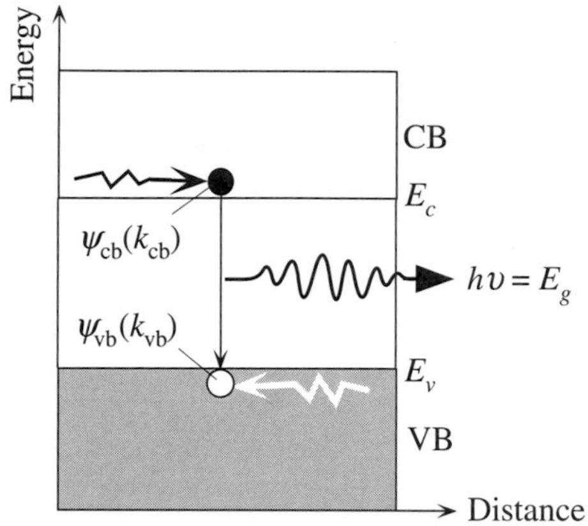


**Figure 5.22** Direct recombination in GaAs. $k_{cb} = k_{vb}$ so that momentum conservation is satisfied.

shows a direct recombination mechanism, for example, as it occurs in GaAs, in which a free electron recombines with a free hole when they meet at one location in the crystal. The excess energy of the electron is lost as a photon of energy $h\nu = E_g$. In fact, it is this type of recombination that results in the emitted light from light emitting diodes (LEDs).

The recombination process between an electron and a hole, like every other process in nature, must obey the momentum conservation law. The wavefunction of an electron in the CB, $\psi_{cb}(k_{cb})$, will have a certain momentum $\hbar k_{cb}$ associated with the wavevector $k_{cb}$ and, similarly, the electron wavefunction $\psi_{vb}(k_{vb})$ in the VB will have a momentum $\hbar k_{vb}$ associated with the wavevector $k_{vb}$. Conservation of linear momentum during recombination requires that when the electron drops from the CB to the VB, its wavevector should remain the same, $k_{vb} = k_{cb}$. For the elemental semiconductors, Si and Ge, the electronic states $\psi_{vb}(k_{vb})$ with $k_{vb} = k_{cb}$ are right in the middle of the VB and are therefore fully occupied. Consequently, there are no empty states in the VB that can satisfy $k_{vb} = k_{cb}$, and so direct recombination in Si and Ge is next to impossible. For some compound semiconductors, such as GaAs and InSb, for example, the states with $k_{vb} = k_{cb}$ are right at the top of the valence band, so they are essentially empty (contain holes). Consequently, an electron in the CB of GaAs can drop down to an empty electronic state at the top of the VB and maintain $k_{vb} = k_{cb}$. Thus **direct recombination** is highly probable in GaAs, and it is this very reason that makes GaAs an LED material.

In elemental semiconductor crystals, for example, in Si and Ge, electrons and holes usually recombine through recombination centers. A recombination center increases the probability of recombination because it can "take up" any momentum difference between a hole and electron. The process essentially involves a third body, which may be an impurity atom or a crystal defect. The electron is captured by the recombination center and thus becomes localized at this site. It is "held" at the center until some hole arrives and recombines with it. In the energy band diagram picture shown in Figure 5.23a, the recombination center provides a localized electronic state below $E_c$ in the bandgap, which is at a certain location in the crystal. When an electron approaches the center, it is captured. The electron is then localized and bound to this

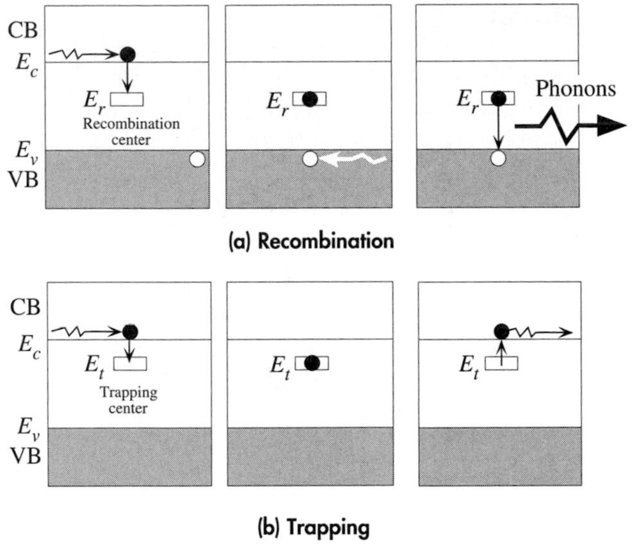


**Figure 5.23** Recombination and trapping.
(a) Recombination in Si via a recombination center that has a localized energy level at $E_r$ in the bandgap, usually near the middle.
(b) Trapping and detrapping of electrons by trapping centers. A trapping center has a localized energy level in the bandgap.

center and "waits" there for a hole with which it can recombine. In this recombination process, the energy of the electron is usually lost to lattice vibrations (as "sound") via the "recoiling" of the third body. Emitted lattice vibrations are called phonons. A **phonon** is a quantum of energy associated with atomic vibrations in the crystal analogous to the photon.

Typical recombination centers, besides the donor and acceptor impurities, might be metallic impurities and crystal defects such as dislocations, vacancies, or interstitials. Each has its own peculiar behavior in aiding recombination, which will not be described here.

It is instructive to mention briefly the phenomenon of charge carrier **trapping** since in many devices this can be the main limiting factor on the performance. An electron in the conduction band can be captured by a localized state, just like a recombination center, located in the bandgap, as shown in Figure 5.23b. The electron falls into the trapping center at $E_t$ and becomes temporarily removed from the CB. At a later time, due to an incident energetic lattice vibration, it becomes excited back into the CB and is available for conduction again. Thus trapping involves the temporary removal of the electron from the CB, whereas in the case of recombination, the electron is permanently removed from the CB since the capture is followed by recombination with a hole. We can view a trap as essentially being a flaw in the crystal that results in the creation of a localized electronic state, around the flaw site, with an energy in the bandgap. A charge carrier passing by the flaw can be captured and lose its freedom. The flaw can be an impurity or a crystal imperfection in the same way as a recombination center. The only difference is that when a charge carrier is captured at a recombination site, it has no possibility of escaping again because the center aids recombination. Although Figure 5.23b illustrates an electron trap, similar arguments also apply to hole traps, which are normally closer to $E_v$. In general, flaws and defects that give localized states near the middle of the bandgap tend to act as recombination centers.

## 5.4.2 MINORITY CARRIER LIFETIME

Consider what happens when an $n$-type semiconductor, doped with $5 \times 10^{16}$ cm$^{-3}$ donors, is uniformly illuminated with appropriate wavelength light to photogenerate electron–hole pairs (EHPs), as shown in Figure 5.24. We will now define thermal equilibrium majority and minority carrier concentrations in an extrinsic semiconductor. In general, the subscript $n$ or $p$ is used to denote the type of semiconductor, and $o$ to refer to thermal equilibrium in the dark.

In an $n$-type semiconductor, electrons are the majority carriers and holes are the minority carriers

$n_{no}$ is defined as the **majority carrier concentration** (electron concentration in an $n$-type semiconductor) in thermal equilibrium in the dark. These electrons, constituting the majority carriers, are thermally ionized from the donors.

$p_{no}$ is termed the **minority carrier concentration** (hole concentration in an $n$-type semiconductor) in thermal equilibrium in the dark. These holes that constitute the minority carriers are thermally generated across the bandgap.

In both cases the subscript *no* refers to an $n$-type semiconductor and thermal equilibrium conditions, respectively. Thermal equilibrium means that the mass action law is obeyed and $n_{no}p_{no} = n_i^2$.

When we illuminate the semiconductor, we create *excess* EHPs by photogeneration. Suppose that the electron and hole concentrations at any instant are denoted by $n_n$ and $p_n$, which are defined as the *instantaneous* majority (electron) and minority (hole) concentrations, respectively. At any instant and at any location in the semiconductor, we define the departure from the equilibrium by **excess concentrations** as follows:

$\Delta n_n$ is the *excess* electron (majority carrier) concentration: $\Delta n_n = n_n - n_{no}$

$\Delta p_n$ is the *excess* hole (minority carrier) concentration: $\Delta p_n = p_n - p_{no}$

Under illumination, at any instant, therefore

$$n_n = n_{no} + \Delta n_n \qquad \text{and} \qquad p_n = p_{no} + \Delta p_n$$

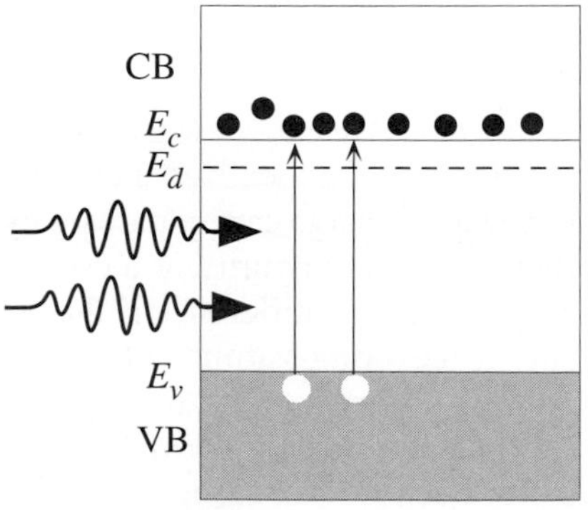


**Figure 5.24** Low-level photoinjection into an $n$-type semiconductor in which $\Delta n_n < n_{no}$.

Photoexcitation creates EHPs or an equal number of electrons and holes, as shown in Figure 5.24, which means that

$$\Delta p_n = \Delta n_n$$

and obviously the mass action law is not obeyed: $n_n p_n \neq n_i^2$. It is worth remembering that

$$\frac{dn_n}{dt} = \frac{d\Delta n_n}{dt} \quad \text{and} \quad \frac{dp_n}{dt} = \frac{d\Delta p_n}{dt}$$

since $n_{no}$ and $p_{no}$ depend only on temperature.

Let us assume that we have "weak" illumination, which causes, say, only a 10 percent change in $n_{no}$, that is,

$$\Delta n_n = 0.1 n_{no} = 0.5 \times 10^{16}\,\text{cm}^{-3}$$

Then

$$\Delta p_n = \Delta n_n = 0.5 \times 10^{16}\,\text{cm}^{-3}$$

Figure 5.25 shows a single-axis plot of the majority ($n_n$) and minority ($p_n$) concentrations in the dark and in light. The scale is logarithmic to allow large orders of magnitude changes to be recorded. Under illumination, the minority carrier concentration is

$$p_n = p_{no} + \Delta p_n = 2.0 \times 10^3 + 0.5 \times 10^{16} \approx 0.5 \times 10^{16} = \Delta p_n$$

That is, $p_n \approx \Delta p_n$, which shows that although $n_n$ changes by only 10 percent, $p_n$ changes *drastically,* that is, by a factor of $\sim 10^{12}$.

Figure 5.26 shows a pictorial view of what is happening inside an $n$-type semiconductor when light is switched on at a certain time and then later switched off again. Obviously when the light is switched off, the condition $p_n = \Delta p_n$ (state $B$ in Figure 5.26) must eventually revert back to the dark case (state $A$) where $p_n = p_{no}$. In other words, the excess minority carriers $\Delta p_n$ and excess majority carriers $\Delta n_n$ must

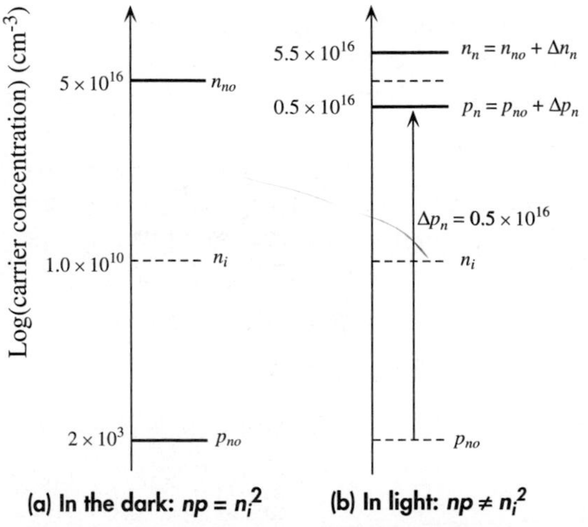


**Figure 5.25** Low-level injection in an $n$-type semiconductor does not significantly affect $n_n$ but drastically affects the minority carrier concentration $p_n$.

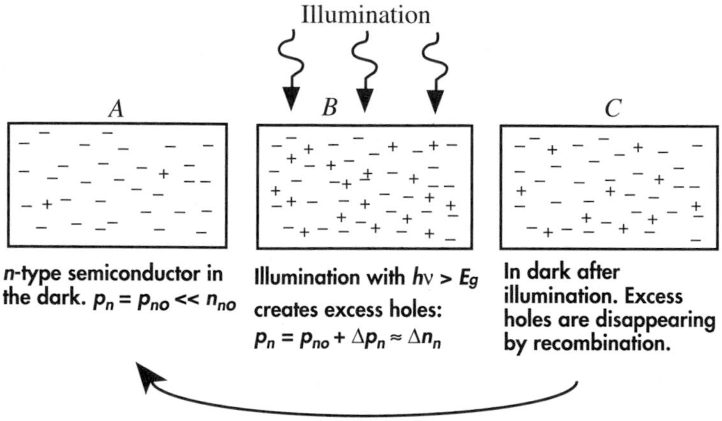


**Figure 5.26** Illumination of an *n*-type semiconductor results in excess electron and hole concentrations.
After the illumination, the recombination process restores equilibrium; the excess electrons and holes simply recombine.

be removed. This removal occurs by recombination. Excess holes recombine with the electrons available and disappear. This, however, takes time because the electrons and holes have to find each other. In order to describe the rate of recombination, we introduce a temporal quantity, denoted by $\tau_h$ and called the **minority carrier lifetime (mean recombination time),** which is defined as follows: $\tau_h$ is the average time a hole exists in the VB from its generation to its recombination, that is, the mean time the hole is free before recombining with an electron. An alternative and equivalent definition is that $1/\tau_h$ is the average probability per unit time that a hole will recombine with an electron. We must remember that the recombination process occurs through recombination centers, so the recombination time $\tau_h$ will depend on the concentration of these centers and their effectiveness in capturing the minority carriers. Once a minority carrier has been captured by a recombination center, there are many majority carriers available to recombine with it, so $\tau_h$ in an indirect process is independent of the majority carrier concentration. This is the reason for defining the recombination time as a minority carrier lifetime.

If the minority carrier recombination time is, say, 10 s, and if there are some 1000 excess holes, then it is clear that these excess holes will be disappearing at a rate of $1000/10 \text{ s} = 100$ per second. The rate of recombination of excess minority carriers is simply $\Delta p_n/\tau_h$. At any instant, therefore,

Rate of increase in excess hole concentration = Rate of photogeneration − Rate of recombination of excess holes

*Excess minority carrier concentration*

If $G_{\text{ph}}$ is the rate of photogeneration, then clearly the net rate of change of $\Delta p_n$ is

$$\frac{d\Delta p_n}{dt} = G_{\text{ph}} - \frac{\Delta p_n}{\tau_h} \qquad [5.27]$$

This is a general expression that describes the time evolution of the excess minority carrier concentration given the photogeneration rate $G_{\text{ph}}$, the minority carrier lifetime $\tau_h$, and the initial condition at $t = 0$. The only assumption is weak injection ($\Delta p_n < n_{no}$).

We should note that the recombination time $\tau_h$ depends on the semiconductor material, impurities, crystal defects, temperature, and so forth, and there is no typical value to quote. It can be anywhere from nanoseconds to seconds. Later it will be shown that certain applications require a short $\tau_h$, as in fast switching of $pn$ junctions, whereas others require a long $\tau_h$, for example, persistent luminescence.

**EXAMPLE 5.10**

**PHOTORESPONSE TIME** Sketch the hole concentration when a step illumination is applied to an $n$-type semiconductor at time $t = 0$ and switched off at time $t = t_{\text{off}}(\gg \tau_h)$.

**SOLUTION**

We use Equation 5.27 with $G_{\text{ph}} = \text{constant}$ in $0 \le t \le t_{\text{off}}$. Since Equation 5.27 is a first-order differential equation, integrating it we simply find

$$\ln\left[G_{\text{ph}} - \left(\frac{\Delta p_n}{\tau_h}\right)\right] = -\frac{t}{\tau_h} + C_1$$

where $C_1$ is the integration constant. At $t = 0$, $\Delta p_n = 0$, so $C_1 = \ln G_{\text{ph}}$. Therefore the solution is

$$\Delta p_n(t) = \tau_h G_{\text{ph}}\left[1 - \exp\left(-\frac{t}{\tau_h}\right)\right] \qquad 0 \le t < t_{\text{off}} \tag{5.28}$$

We see that as soon as the illumination is turned on, the minority carrier concentration rises exponentially toward its steady-state value $\Delta p_n(\infty) = \tau_h G_{\text{ph}}$. This is reached after a time $t > \tau_h$.

At the instant the illumination is switched off, we assume that $t_{\text{off}} \gg \tau_h$ so that from Equation 5.28,

$$\Delta p_n(t_{\text{off}}) = \tau_h G_{\text{ph}}$$

We can define $t'$ to be the time measured from $t = t_{\text{off}}$, that is, $t' = t - t_{\text{off}}$. Then

$$\Delta p_n(t' = 0) = \tau_h G_{\text{ph}}$$

Solving Equation 5.27 with $G_{\text{ph}} = 0$ in $t > t_{\text{off}}$ or $t' > 0$, we get

$$\Delta p_n(t') = \Delta p_n(0)\exp\left(-\frac{t'}{\tau_h}\right)$$

where $\Delta p_n(0)$ is actually an integration constant that is equivalent to the boundary condition on $\Delta p_n$ at $t' = 0$. Putting $t' = 0$ and $\Delta p_n = \tau_h G_{\text{ph}}$ gives

$$\Delta p_n(t') = \tau_h G_{\text{ph}}\exp\left(-\frac{t'}{\tau_h}\right) \tag{5.29}$$

We see that the excess minority carrier concentration decays exponentially from the instant the light is switched off with a time constant equal to the minority carrier recombination time. The time evolution of the minority carrier concentration is sketched in Figure 5.27.

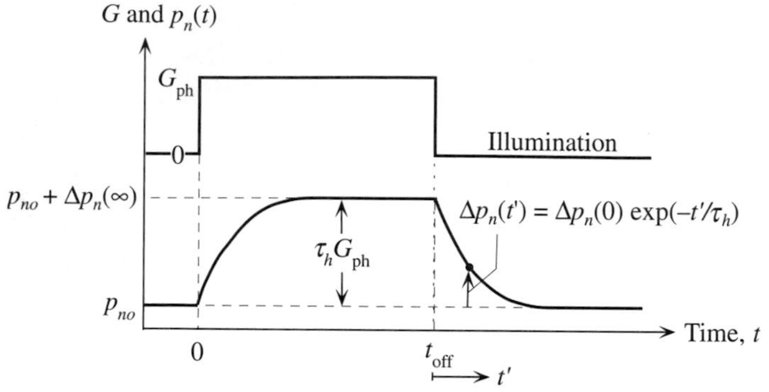


**Figure 5.27** Illumination is switched on at time $t = 0$ and then off at $t = t_{off}$.
The excess minority carrier concentration $\Delta p_n(t)$ rises exponentially to its steady-state value with a time constant $\tau_h$. From $t_{off}$, the excess minority carrier concentration decays exponentially to its equilibrium value.

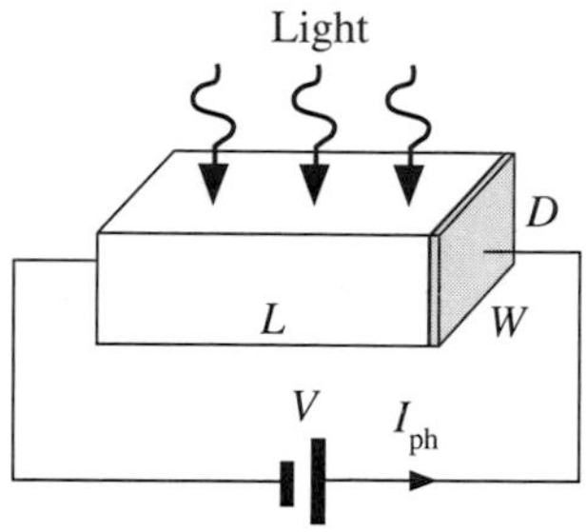


**Figure 5.28** A semiconductor slab of length $L$, width $W$, and depth $D$ is illuminated with light of wavelength $\lambda$. $I_{ph}$ is the steady-state photocurrent.

**EXAMPLE 5.11**

**PHOTOCONDUCTIVITY** Suppose that a direct bandgap semiconductor with no traps is illuminated with light of intensity $I(\lambda)$ and wavelength $\lambda$ that will cause photogeneration as shown in Figure 5.28. The area of illumination is $A = (L \times W)$, and the thickness (depth) of the semiconductor is $D$. If $\eta$ is the quantum efficiency (number of free EHPs generated per absorbed photon) and $\tau$ is the recombination lifetime of the photogenerated carriers, show that the **steady-state photoconductivity,** defined as

$$\Delta\sigma = \sigma(\text{in light}) - \sigma(\text{in dark})$$

is given by

*Steady-state photo-conductivity*

$$\Delta\sigma = \frac{e\eta I\lambda\tau(\mu_e + \mu_h)}{hcD} \qquad [5.30]$$

A photoconductive cell has a CdS crystal 1 mm long, 1 mm wide, and 0.1 mm thick with electrical contacts at the end, so the receiving area of radiation is 1 mm$^2$, whereas the area of each contact is 0.1 mm$^2$. The cell is illuminated with a blue radiation of wavelength 450 nm and intensity 1 mW/cm$^2$. For unity quantum efficiency and an electron recombination time of 1 ms, calculate

*a.* The number of EHPs generated per second

*b.* The photoconductivity of the sample

*c.* The photocurrent produced if 50 V is applied to the sample

Note that a CdS photoconductor is a direct bandgap semiconductor with an energy gap $E_g = 2.6$ eV, electron mobility $\mu_e = 0.034$ m$^2$ V$^{-1}$ s$^{-1}$, and hole mobility $\mu_h = 0.0018$ m$^2$ V$^{-1}$ s$^{-1}$.

**SOLUTION**

If $\Gamma_{ph}$ is the number of photons arriving per unit area per unit second (the photon flux), then $\Gamma_{ph} = I/h\nu$ where $I$ is the light intensity (energy flowing per unit area per second) and $h\nu$ is the energy per photon. The quantum efficiency $\eta$ is defined as the number of free EHPs generated per absorbed photon. Thus, the number of EHPs generated *per unit volume per second,* the photogeneration rate per unit volume $G_{ph}$ is given by

$$G_{ph} = \frac{\eta A \Gamma_{ph}}{AD} = \frac{\eta\left(\frac{I}{h\nu}\right)}{D} = \frac{\eta I \lambda}{hcD}$$

In the steady state,

$$\frac{d\Delta n}{dt} = G_{ph} - \frac{\Delta n}{\tau} = 0$$

so

$$\Delta n = \tau G_{ph} = \frac{\tau \eta I \lambda}{hcD}$$

But, by definition,

$$\Delta\sigma = e\mu_e\,\Delta n + e\mu_h\,\Delta p = e\,\Delta n(\mu_e + \mu_h)$$

since electrons and holes are generated in pairs, $\Delta n = \Delta p$. Thus, substituting for $\Delta n$ in the $\Delta\sigma$ expression, we get Equation 5.30:

$$\Delta\sigma = \frac{e\eta I\lambda\tau(\mu_e + \mu_h)}{hcD}$$

*a.* The photogeneration rate per unit time is not $G_{ph}$, which is per unit time per unit volume. We define $EHP_{ph}$ as the total number of EHPs photogenerated per unit time in the whole volume $(AD)$. Thus

$$\begin{aligned} EHP_{ph} &= \text{Total photogeneration rate} \\ &= (AD)G_{ph} = (AD)\frac{\eta I\lambda}{hcD} = \frac{A\eta I\lambda}{hc} \\ &= [(10^{-3} \times 10^{-3}\ \text{m}^2)(1)(10^{-3} \times 10^4\ \text{J s}^{-1}\ \text{m}^{-2})(450 \times 10^{-9}\ \text{m})] \\ &\quad \div [(6.63 \times 10^{-34}\ \text{J s})(3 \times 10^8\ \text{m s}^{-1})] \\ &= 2.26 \times 10^{13}\ \text{EHP s}^{-1} \end{aligned}$$

*b.* From Equation 5.30,

$$\Delta\sigma = \frac{e\eta I\lambda\tau(\mu_e + \mu_h)}{hcD}$$

That is

$$\Delta\sigma = \frac{(1.6 \times 10^{-19}\ \text{C})(1)(10^{-3} \times 10^4\ \text{J s}^{-1}\ \text{m}^{-2})(450 \times 10^{-9}\ \text{m})(1 \times 10^{-3}\ \text{s})(0.0358\ \text{m}^2\ \text{V}^{-1}\ \text{s}^{-1})}{(6.63 \times 10^{-34}\ \text{J s})(3 \times 10^8\ \text{ms}^{-1})(0.1 \times 10^{-3}\ \text{m})}$$

$$= 1.30\ \Omega^{-1}\text{m}^{-1}$$

c. Photocurrent density will be

$$\Delta J = \mathcal{E}\,\Delta\sigma = (1.30\ \Omega^{-1}\ \text{m}^{-1})(50\ \text{V}/10^{-3}\ \text{m}) = 6.50 \times 10^4\ \text{A m}^{-2}$$

Thus the photocurrent

$$\Delta I = A\,\Delta J = (10^{-3} \times 0.1 \times 10^{-3}\ \text{m}^2)(6.50 \times 10^4\ \text{A m}^{-2})$$
$$= 6.5 \times 10^{-3}\ \text{A} \quad \text{or} \quad 6.5\ \text{mA}$$

We assumed that all the incident radiation is absorbed.

## 5.5 DIFFUSION AND CONDUCTION EQUATIONS, AND RANDOM MOTION

It is well known that, by virtue of their random motion, gas particles diffuse from high-concentration regions to low-concentration regions. When a perfume bottle is opened at one end of a room, the molecules diffuse out from the bottle and, after a while, can be smelled at the other end of the room. Whenever there is a concentration gradient of particles, there is a net diffusional motion of particles in the direction of decreasing concentration. The origin of diffusion lies in the random motion of particles. To quantify particle flow, we define the **particle flux** $\Gamma$ just like current, as the number of particles (not charges) crossing unit area per unit time. Thus if $\Delta N$ particles cross an area $A$ in time $\Delta t$, then, by definition, the particle flux is

*Definition of particle flux*

$$\Gamma = \frac{\Delta N}{A\,\Delta t} \quad \textbf{[5.31]}$$

Clearly if the particles are charged with a charge $Q$ ($-e$ for electrons and $+e$ for holes), then the electric current density $J$, which is basically a charge flux, is related to the particle flux $\Gamma$ by

*Definition of current density*

$$J = Q\Gamma \quad \textbf{[5.32]}$$

Suppose that the electron concentration at some time $t$ in a semiconductor decreases in the $x$ direction and has the profile $n(x, t)$ shown in Figure 5.29a. This may have been achieved, for example, by photogeneration at one end of a semiconductor. We will assume that the electron concentration changes only in the $x$ direction so that the diffusion of electrons can be simplified to a one-dimensional problem as depicted in Figure 5.29a. We know that in the absence of an electric field, the electron motion is random and involves scattering from lattice vibrations and impurities. Suppose that $\ell$ is the mean free path in the $x$ direction and $\tau$ is the mean free time between the scattering events. The electron moves a mean distance $\ell$ in the $+x$ or $-x$ direction and then it is scattered and changes direction. Its mean speed along $x$ is $v_x = \ell/\tau$. Let us evaluate the flow of electrons in the $+x$ and $-x$ directions through the plane at $x_o$ and hence find the net flow in the $+x$ direction.

We can divide the $x$ axis into hypothetical segments of length $\ell$ so that each segment corresponds to a mean free path. Going across a segment, the electron experiences one scattering process. Consider what happens during one mean free time, the time it takes for the electrons to move across a segment toward the left or right. Half of the electrons in $(x_o - \ell)$ would be moving toward $x_o$ and the other half away from $x_o$, and in time $\tau$ half of them will reach $x_o$ and cross as shown in Figure 5.29b. If $n_1$ is the concentration of electrons at $x_o - \frac{1}{2}\ell$, then the number of electrons moving toward the right to

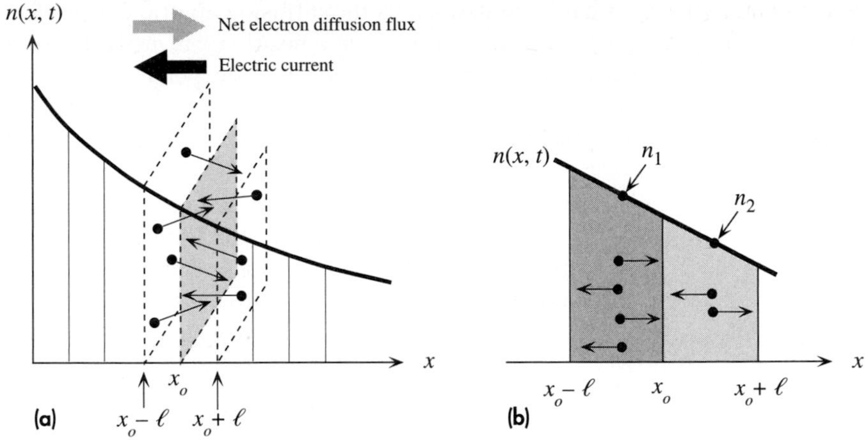


**Figure 5.29**

(a) Arbitrary electron concentration $n(x, t)$ profile in a semiconductor. There is a net diffusion (flux) of electrons from higher to lower concentrations.

(b) Expanded view of two adjacent sections at $x_o$. There are more electrons crossing $x_o$ coming from the left ($x_o - \ell$) than coming from the right ($x_o + \ell$).

cross $x_o$ is $\frac{1}{2}n_1 A\ell$ where $A$ is the cross-sectional area and hence $A\ell$ is the volume of the segment. Similarly half of the electrons in $(x_o + \ell)$ would be moving toward the left and in time $\tau$ would reach $x_o$. Their number is $\frac{1}{2}n_2 A\ell$ where $n_2$ is the concentration at $x_o + \frac{1}{2}\ell$. The net number of electrons crossing $x_o$ per unit time per unit area in the $+x$ direction is the electron flux $\Gamma_e$,

$$\Gamma_e = \frac{\frac{1}{2}n_1 A\ell - \frac{1}{2}n_2 A\ell}{A\tau}$$

that is,

$$\Gamma_e = -\frac{\ell}{2\tau}(n_2 - n_1) \qquad \textbf{[5.33]}$$

As far as calculus of variations is concerned, the mean free path $\ell$ is small, so we can calculate $n_2 - n_1$ from the concentration gradient using

$$n_2 - n_1 \approx \left(\frac{dn}{dx}\right)\Delta x = \left(\frac{dn}{dx}\right)\ell$$

We can now write the flux in Equation 5.33 in terms of the concentration gradient as

$$\Gamma_e = -\frac{\ell^2}{2\tau}\left(\frac{dn}{dx}\right)$$

or

$$\Gamma_e = -D_e\frac{dn}{dx} \qquad \textbf{[5.34]}$$

*Fick's first law*

where the quantity $(\ell^2/2\tau)$ has been defined as the diffusion coefficient of electrons and denoted by $D_e$. Thus, the net electron flux $\Gamma_e$ at a position $x$ is proportional to the concentration gradient and the diffusion coefficient. The steeper this gradient, the larger the flux $\Gamma_e$. In fact, we can view the concentration gradient $dn/dx$ as the driving force for the diffusion flux, just like the electric field $-(dV/dx)$ is the driving force for the electric current: $J = \sigma \mathcal{E} = -\sigma(dV/dx)$.

Equation 5.34 is called **Fick's first law** and represents the relationship between the net particle flux and the driving force, which is the concentration gradient. It is the counterpart of Ohm's law for diffusion. $D_e$ has the dimensions of m$^2$ s$^{-1}$ and is a measure of how readily the particles (in this case, electrons) diffuse in the medium. Note that Equation 5.34 gives the electron flux $\Gamma_e$ at a position $x$ where the electron concentration gradient is $dn/dx$. Since from Figure 5.29, the slope $dn/dx$ is a negative number, $\Gamma_e$ in Equation 5.34 comes out positive, which indicates that the flux is in the positive $x$ direction. The electric current (conventional current) due to the diffusion of electrons to the right will be in the negative direction by virtue of Equation 5.32. Representing this electric current density due to diffusion as $J_{D,e}$ we can write

*Electron diffusion current density*

$$J_{D,e} = -e\Gamma_e = eD_e\frac{dn}{dx} \tag{5.35}$$

In the case of a hole concentration gradient, as shown in Figure 5.30, the hole flux $\Gamma_h(x)$ is given by

$$\Gamma_h = -D_h\frac{dp}{dx}$$

where $D_h$ is the hole diffusion coefficient. Putting in a negative number for the slope $dp/dx$, as shown in Figure 5.30, results in a positive hole flux (in the positive $x$ direction), which in turn implies a diffusion current density toward the right. The current density due to hole diffusion is given by

*Hole diffusion current density*

$$J_{D,h} = e\Gamma_h = -eD_h\frac{dp}{dx} \tag{5.36}$$

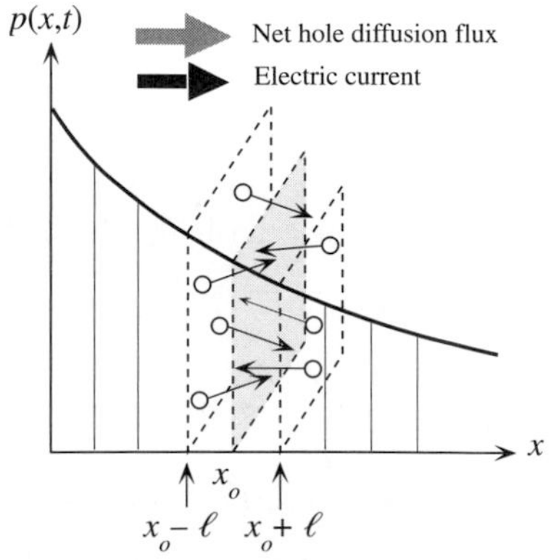


**Figure 5.30** Arbitrary hole concentration $p(x, t)$ profile in a semiconductor.
There is a net diffusion (flux) of holes from higher to lower concentrations. There are more holes crossing $x_o$ coming from the left $(x_o - \ell)$ than coming from the right $(x_o + \ell)$.

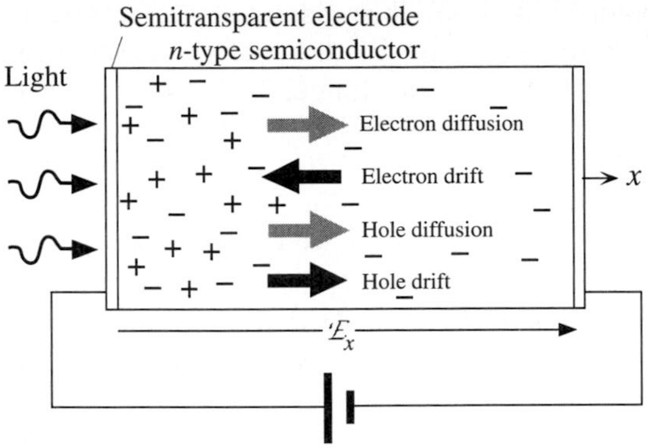


**Figure 5.31** When there is an electric field and also a concentration gradient, charge carriers move both by diffusion and drift.

Suppose that there is also a positive electric field $\mathcal{E}_x$ acting along $+x$ in Figures 5.29 and 5.30. A practical example is shown in Figure 5.31 in which a semiconductor is sandwiched between two electrodes, the left one semitransparent. By connecting a battery to the electrodes, an applied field of $\mathcal{E}_x$ is set up in the semiconductor along $+x$. The left electrode is continuously illuminated, so excess EHPs are generated at this surface that give rise to concentration gradients in $n$ and $p$. The applied field imposes an electrical force on the charges, which then try to drift. Holes drift toward the right and electrons toward the left. Charge motion then involves both drift and diffusion. The total current density due to the electrons drifting, driven by $\mathcal{E}_x$, and also diffusing, driven by $dn/dx$, is then given by adding Equation 5.35 to the usual electron drift current density,

*Total electron current due to drift and diffusion*

$$J_e = en\mu_e\mathcal{E}_x + eD_e\frac{dn}{dx} \qquad [5.37]$$

We note that as $\mathcal{E}_x$ is along $x$, so is the drift current (first term), but the diffusion current (second term) is actually in the opposite direction by virtue of a negative $dn/dx$.

Similarly, the hole current due to holes drifting and diffusing, Equation 5.36, is given by

*Total hole current due to drift and diffusion*

$$J_h = ep\mu_h\mathcal{E}_x - eD_h\frac{dp}{dx} \qquad [5.38]$$

In this case the drift and diffusion currents are in the same direction.

We mentioned that the diffusion coefficient is a measure of the ease with which the diffusing charge carriers move in the medium. But drift mobility is also a measure of the ease with which the charge carriers move in the medium. The two quantities are related through the **Einstein relation,**

*Einstein relation*

$$\frac{D_e}{\mu_e} = \frac{kT}{e} \quad \text{and} \quad \frac{D_h}{\mu_h} = \frac{kT}{e} \qquad [5.39]$$

In other words, the diffusion coefficient is proportional to the temperature and mobility. This is a reasonable expectation since increasing the temperature will increase the mean speed and thus accelerate diffusion. The randomizing effect against diffusion in one particular direction is introduced by the scattering of the carriers from lattice vibrations, impurities, and so forth, so that the longer the mean free path between scattering events, the larger the diffusion coefficient. This is examined in Example 5.12.

We equated the diffusion coefficient $D$ to $\ell^2/2\tau$ in Equation 5.34. Our analysis, as represented in Figure 5.29, is oversimplified because we simply assumed that all electrons move a distance $\ell$ before scattering and all are free for a time $\tau$. We essentially assumed that all those at a distance $\ell$ from $x_o$ and moving toward $x_o$ cross the plane exactly in time $\tau$. This assumption is not entirely true because scattering is a stochastic process and consequently not all electrons moving toward $x_o$ will cross it even in the segment of thickness $\ell$. A rigorous statistical analysis shows that the diffusion coefficient is given by

*Diffusion coefficient*

$$D = \frac{\ell^2}{\tau} \qquad [5.40]$$

**EXAMPLE 5.12** **THE EINSTEIN RELATION** Using the relation between the drift mobility and the mean free time $\tau$ between scattering events and the expression for the diffusion coefficient $D = \ell^2/\tau$, derive the Einstein relation for electrons.

**SOLUTION**

In one dimension, for example, along $x$, the diffusion coefficient for electrons is given by $D_e = \ell^2/\tau$ where $\ell$ is the mean free path along $x$ and $\tau$ is the mean free time between scattering events for electrons. The mean free path $\ell = v_x\tau$, where $v_x$ is the mean (or effective) speed of the electrons along $x$. Thus,

$$D_e = v_x^2\tau$$

In the conduction band and in one dimension, the mean $KE$ of electrons is $\frac{1}{2}kT$, so $\frac{1}{2}kT = \frac{1}{2}m_e^* v_x^2$ where $m_e^*$ is the effective mass of the electron in the CB. This gives

$$v_x^2 = \frac{kT}{m_e^*}$$

Substituting for $v_x$ in the $D_e$ equation, we get,

$$D_e = \frac{kT\tau}{m_e^*} = \frac{kT}{e}\left(\frac{e\tau}{m_e^*}\right)$$

Further, we know from Chapter 2 that the electron drift mobility $\mu_e$ is related to the mean free time $\tau$ via $\mu_e = e\tau/m_e^*$, so we can substitute for $\tau$ to obtain

$$D_e = \frac{kT}{e}\mu_e$$

which is the Einstein relation. We assumed that Boltzmann statistics, that is, $v_x^2 = kT/m_e^*$ is applicable, which, of course, is true for the conduction band electrons in a semiconductor but not for the conduction electrons in a metal. Thus, the Einstein relation is only valid for electrons and holes in a nondegenerate semiconductor and certainly not valid for electrons in a metal.

EXAMPLE 5.13

**DIFFUSION COEFFICIENT OF ELECTRONS IN Si** Calculate the diffusion coefficient of electrons at 27 °C in $n$-type Si doped with $10^{15}$ As atoms $cm^{-3}$.

**SOLUTION**

From the $\mu_e$ versus dopant concentration graph, the electron drift mobility $\mu_e$ with $10^{15}$ $cm^{-3}$ of dopants is about 1300 $cm^2\ V^{-1}\ s^{-1}$, so

$$D_e = \frac{\mu_e kT}{e} = (1300\ \text{cm}^2\ \text{V}^{-1}\ \text{s}^{-1})(0.0259\ \text{V}) = 33.7\ \text{cm}^2\ \text{s}^{-1}$$

EXAMPLE 5.14

**BUILT-IN POTENTIAL DUE TO DOPING VARIATION** Suppose that due to a variation in the amount of donor doping in a semiconductor, the electron concentration is nonuniform across the semiconductor, that is, $n = n(x)$. What will be the potential difference between two points in the semiconductor where the electron concentrations are $n_1$ and $n_2$? If the donor profile in an $n$-type semiconductor is $N(x) = N_o \exp(-x/b)$, where $b$ is a characteristic of the exponential doping profile, evaluate the built-in field $\mathcal{E}_x$. What is your conclusion?

**SOLUTION**

Consider a nonuniformly doped $n$-type semiconductor in which immediately after doping the donor concentration, and hence the electron concentration, decreases toward the right. Initially, the sample is neutral everywhere. The electrons will immediately diffuse from higher- to lower-concentration regions. But this diffusion accumulates *excess* electrons in the right region and exposes the positively charged donors in the left region, as depicted in Figure 5.32. The electric field between the accumulated negative charges and the exposed donors prevents further accumulation. Equilibrium is reached when the diffusion toward the right is just balanced by the drift of electrons toward the left. The total current in the sample must be zero (it is an open circuit),

$$J_e = en\mu_e\mathcal{E}_x + eD_e\frac{dn}{dx} = 0$$

But the field is related to the potential difference by $\mathcal{E}_x = -(dV/dx)$, so

$$-en\mu_e\frac{dV}{dx} + eD_e\frac{dn}{dx} = 0$$

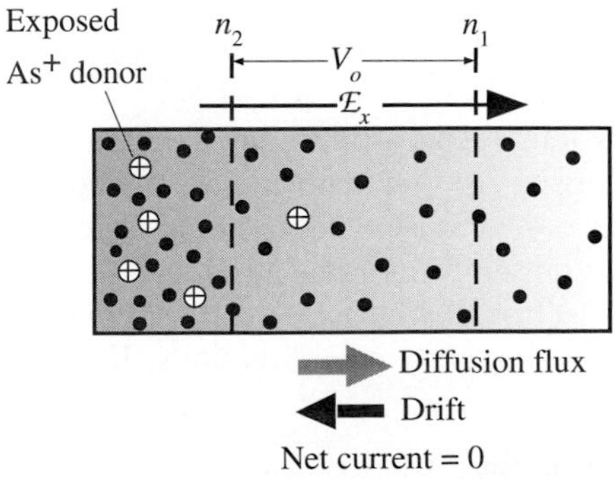


**Figure 5.32** Nonuniform doping profile results in electron diffusion toward the less concentrated regions.
This exposes positively charged donors and sets up a built-in field $\mathcal{E}_x$. In the steady state, the diffusion of electrons toward the right is balanced by their drift toward the left.

We can now use the Einstein relation $D_e/\mu_e = kT/e$ to eliminate $D_e$ and $\mu_e$ and then cancel $dx$ and integrate the equation,

$$\int_{V_1}^{V_2} dV = \frac{kT}{e}\int_{n_1}^{n_2}\frac{dn}{n}$$

Integrating, we obtain the potential difference between points 1 and 2,

*Built-in potential and concentration*

$$V_2 - V_1 = \frac{kT}{e}\ln\left(\frac{n_2}{n_1}\right) \qquad \textbf{[5.41]}$$

To find the built-in field, we will assume that (and this is a reasonable assumption) the diffusion of electrons toward the right has not drastically upset the original $n(x) = N_d(x)$ variation because the field builds up quickly to establish equilibrium. Thus

$$n(x) \approx N_d(x) = N_o \exp\left(-\frac{x}{b}\right)$$

Substituting into the equation for $J_e = 0$, and again using the Einstein relation, we obtain $\mathcal{E}_x$ as

*Built-in field*

$$\mathcal{E}_x = \frac{kT}{be} \qquad \textbf{[5.42]}$$

Note: As a result of the fabrication process, the base region of a bipolar transistor has nonuniform doping, which can be approximated by an exponential $N_d(x)$. The resulting electric field $\mathcal{E}_x$ in Equation 5.42 acts to drift minority carriers faster and therefore speeds up the transistor operation as discussed in Chapter 6.

## 5.6 CONTINUITY EQUATION[4]

### 5.6.1 TIME-DEPENDENT CONTINUITY EQUATION

Many semiconductor devices operate on the principle that excess charge carriers are injected into a semiconductor by external means such as illumination or an applied voltage. The injection of carriers upsets the equilibrium concentration. To determine the carrier concentration at any point at any instant we need to solve the **continuity equation,** which is based on accounting for the total charge at that location in the semiconductor. Consider an $n$-type semiconductor slab as shown in Figure 5.33 in which the hole concentration has been upset along the $x$ axis from its equilibrium value $p_{no}$ by some external means.

Consider an infinitesimally thin elemental volume $A\,\delta x$ as in Figure 5.33 in which the hole concentration is $p_n(x, t)$. The current density at $x$ due to holes flowing into the volume is $J_h$ and that due to holes flowing out at $x + \delta x$ is $J_h + \delta J_h$. There is a change in the hole current density $J_h$; that is, $J_h(x, t)$ is not uniform along $x$. (Recall that the total current will also have a component due to electrons.) We assume that $J_h(x, t)$ and $p_n(x, t)$ do not change across the cross section along the $y$ or $z$ directions. If $\delta J_h$ is

[4] This section may be skipped without loss of continuity. (No pun intended.)

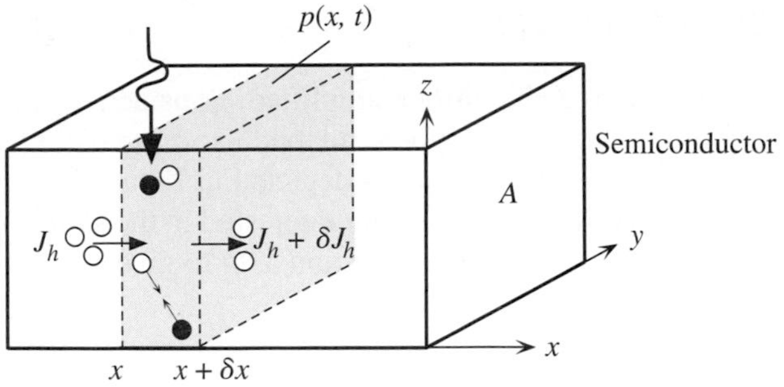


**Figure 5.33** Consider an elemental volume $A\,\delta x$ in which the hole concentration is $p(x, t)$.

*negative,* then the current leaving the volume is less than that entering the volume, which leads to an increase in the hole concentration in $A\,\delta x$. Thus,

$$\frac{1}{A\,\delta x}\left(\frac{-A\,\delta J_h}{e}\right) = \text{Rate of increase in hole concentration due to the change in } J_h \qquad [5.43]$$

The negative sign ensures that negative $\delta J_h$ leads to an increase in $p_n$. Recombination taking place in $A\,\delta x$ removes holes from this volume. In addition, there may also be photogeneration at $x$ at time $t$. Thus,

The *net* rate of increase in the hole concentration $p_n$ in $A\,\delta x$
= Rate of increase due to decrease in $J_h$ − Rate of recombination + Rate of photogeneration

*Continuity equation for holes*

$$\frac{\partial p_n}{\partial t} = -\frac{1}{e}\left(\frac{\partial J_h}{\partial x}\right) - \frac{p_n - p_{no}}{\tau_h} + G_{\text{ph}} \qquad [5.44]$$

where $\tau_h$ is the hole recombination time (lifetime), $G_{\text{ph}}$ is the photogeneration rate at $x$ at time $t$, and we used $\partial J_h/\partial x$ for $\delta J_h/\delta x$ since $J_h$ depends on $x$ and $t$.

Equation 5.44 is called the **continuity equation** for holes. The current density $J_h$ is given by diffusion and drift components in Equations 5.37 and 5.38. There is a similar expression for electrons as well, but the negative sign multiplying $\partial J_e/\partial x$ is changed to positive (the charge $e$ is negative for electrons).

The solutions of the continuity equation depend on the initial and boundary conditions. Many device scientists and engineers have solved Equation 5.44 for various semiconductor problems to characterize the behavior of devices. In most cases numerical solutions are necessary as analytical solutions are not mathematically tractable. As a simple example, consider uniform illumination of the surface of a semiconductor with suitable electrodes at its end as in Figure 5.28. Photogeneration and current density do not vary with distance along the sample length, so $\partial J_h/\partial x = 0$. If $\Delta p_n$ is the excess concentration, $\Delta p_n = p_n - p_{no}$, then the time derivative of $p_n$ in Equation 5.44 is the same as $\Delta p_n$. Thus, the continuity equation becomes

*Continuity equation with uniform photo-generation*

$$\frac{\partial \Delta p_n}{\partial t} = -\frac{\Delta p_n}{\tau_h} + G_{\text{ph}} \qquad [5.45]$$

which is identical to the semiquantitatively derived Equation 5.27 from which photoconductivity was calculated in Example 5.11.

## 5.6.2 STEADY-STATE CONTINUITY EQUATION

For certain problems, the continuity equation can be further simplified. Consider, for example, the continuous illumination of one end of an $n$-type semiconductor slab by light that is absorbed in a very small thickness $x_o$ at the surface as depicted in Figure 5.34a. There is no bulk photogeneration, so $G_{\text{ph}} = 0$. Suppose we are interested in the **steady-state** behavior; then the time derivative would be zero in Equation 5.44 to give,

*Steady-state continuity equation for holes*

$$\frac{1}{e}\left(\frac{\partial J_h}{\partial x}\right) = -\frac{p_n - p_{no}}{\tau_h} \qquad [5.46]$$

The hole current density $J_h$ would have diffusion and drift components. If we assume that the electric field is very small, we can use Equation 5.38 with $\mathcal{E} \approx 0$ in Equation 5.46. Further, since the excess concentration $\Delta p_n(x) = p_n(x) - p_{no}$, we obtain,

*Steady-state continuity equation with $\mathcal{E} = 0$*

$$\frac{d^2 \Delta p_n}{dx^2} = \frac{\Delta p_n}{L_h^2} \qquad [5.47]$$

where, by definition, $L_h = \sqrt{D_h \tau_h}$ and is called the **diffusion length of holes.** Equation 5.47 describes the **steady-state** behavior of minority carrier concentration in a semiconductor under time-invariant excitation. When the appropriate boundary conditions are also included, its solution gives the *spatial* dependence of the excess minority carrier concentration $\Delta p_n(x)$.

In Figure 5.34a, both excess electrons and holes are photogenerated at the surface, but the percentage increase in the concentration of holes is much more dramatic since

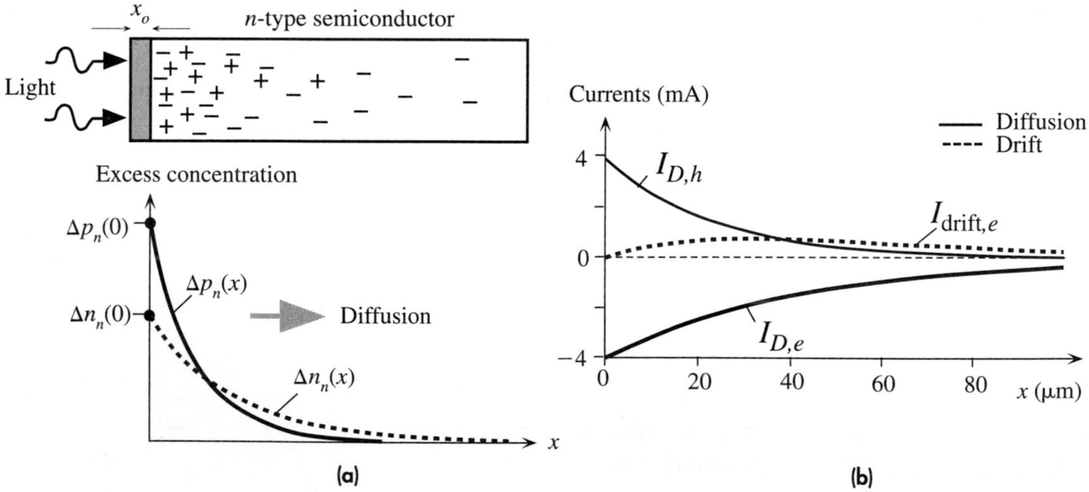


**Figure 5.34**

(a) Steady-state excess carrier concentration profiles in an $n$-type semiconductor that is continuously illuminated at one end.

(b) Majority and minority carrier current components in open circuit. Total current is zero.

$p_{no} \ll n_{no}$. We will assume **weak injection,** that is, $\Delta p_n \ll n_{no}$. Suppose that illumination is such that it causes the excess hole concentration at $x = 0$ to be $\Delta p_n(0)$. As holes diffuse toward the right, they meet electrons and recombine as a result of which the hole concentration $p_n(x)$ decays with distance into the semiconductor. If the bar is very long, then far away from the injection end we would expect $p_n$ to be equal to the thermal equilibrium concentration $p_{no}$. The solution of Equation 5.47 with these boundary conditions shows that $\Delta p_n(x)$ decays *exponentially* as

*Minority carrier concentration, long bar*

$$\Delta p_n(x) = \Delta p_n(0)\exp\left(-\frac{x}{L_h}\right) \quad [5.48]$$

This decay in the hole concentration results in a hole diffusion current $I_{D,h}(x)$ that has the same spatial dependence. Thus, if $A$ is the cross-sectional area, the hole current is

*Hole diffusion current*

$$I_h \approx I_{D,h} = -AeD_h\frac{dp_n(x)}{dx} = \frac{AeD_h}{L_h}\Delta p_n(0)\exp\left(-\frac{x}{L_h}\right) \quad [5.49]$$

We find $\Delta p_n(0)$ as follows. Under steady state, the holes generated per unit time in $x_o$ must be removed by the hole current (at $x = 0$) at the *same* rate. Thus,

$$Ax_oG_{\text{ph}} = \frac{1}{e}I_{D,h}(0) = \frac{AD_h}{L_h}\Delta p_n(0)$$

or

$$\Delta p_n(0) = x_oG_{\text{ph}}\left(\frac{\tau_h}{D_h}\right)^{1/2} \quad [5.50]$$

Similarly, electrons photogenerated in $x_o$ diffuse toward the bulk, but their diffusion coefficient $D_e$ and length $L_e$ are larger than those for holes. The excess electron concentration $\Delta n_n$ decays as

*Majority carrier concentration, long bar*

$$\Delta n_n(x) = \Delta n_n(0)\exp\left(-\frac{x}{L_e}\right) \quad [5.51]$$

where $L_e = \sqrt{D_e\tau_h}$ and $\Delta n_n(x)$ decays more slowly than $\Delta p_n(x)$ as $L_e > L_h$. (Note that $\tau_e = \tau_h$.) The electron diffusion current $I_{D,e}$ is

*Electron diffusion current*

$$I_{D,e} = AeD_e\frac{dn_n(x)}{dx} = -\frac{AeD_e}{L_e}\Delta n_n(0)\exp\left(-\frac{x}{L_e}\right) \quad [5.52]$$

The field at the surface is zero. Under steady state, the electrons generated per unit time in $x_o$ must be removed by the electron current at the *same* rate. Thus, similarly to Equation 5.50,

$$\Delta n_n(0) = x_oG_{\text{ph}}\left(\frac{\tau_h}{D_e}\right)^{1/2} \quad [5.53]$$

so that

$$\frac{\Delta p_n(0)}{\Delta n_n(0)} = \left(\frac{D_e}{D_h}\right)^{1/2} \quad [5.54]$$

which is greater than unity for Si.

**Table 5.3** Currents in an infinite slab illuminated at one end for weak injection near the surface

| Currents at | Minority Diffusion $I_{D,h}$ (mA) | Minority Drift $I_{\text{drift},h}$ (mA) | Majority Diffusion $I_{D,e}$ (mA) | Majority Drift $I_{\text{drift},e}$ (mA) | Field $\mathcal{E}$ (V cm$^{-1}$) |
|---|---|---|---|---|---|
| $x = 0$ | 3.94 | 0 | −3.94 | 0 | 0 |
| $x = L_e$ | 0.70 | 0.0022 | −1.45 | 0.75 | 0.035 |

It is apparent that the hole and electron diffusion currents are in *opposite* directions. At the surface, the electron and hole diffusion currents are equal and opposite, so the total current is zero. As apparent from Equations 5.49 and 5.52, the hole diffusion current decays more rapidly than the electron diffusion current, so there must be some electron drift to keep the total current zero. The electrons are majority carriers which means that even a small field can cause a marked majority carrier drift current. If $I_{\text{drift},e}$ is the electron drift current, then in an open circuit the total current $I = I_{D,h} + I_{D,e} + I_{\text{drift},e} = 0$, so

*Electron drift current*

$$I_{\text{drift},e} = -I_{D,h} - I_{D,e} \quad [5.55]$$

The electron drift current increases with distance, so the total current $I$ at every location is zero. It must be emphasized that there must be some field $\mathcal{E}$ in the sample, however small, to provide the necessary drift to balance the currents to zero. The field can be found from $I_{\text{drift},e} \approx Aen_{no}\mu_e\mathcal{E}$, inasmuch as $n_{no}$ does not change significantly (weak injection),

*Electric field*

$$\mathcal{E} = \frac{I_{\text{drift},e}}{Aen_{no}\mu_e} \quad [5.56]$$

The hole drift current due to this field is

*Hole drift current*

$$I_{\text{drift},h} = Ae\mu_h p_n(x)\mathcal{E} \quad [5.57]$$

and it will be negligibly small as $p_n \ll n_{no}$.

We can use actual values to gauge magnitudes. Suppose that $A = 1$ mm$^2$ and $N_d = 10^{16}$ cm$^{-3}$ so that $n_{no} = N_d = 10^{16}$ cm$^{-3}$ and $p_{no} = n_i^2/N_d = 1 \times 10^4$ cm$^{-3}$. The light intensity is adjusted to yield $\Delta p_n(0) = 0.05n_{no} = 5 \times 10^{14}$ cm$^{-3}$: *weak injection.* Typical values at 300 K for the material properties in this $N_d$-doped $n$-type Si would be $\tau_h = 480$ ns, $\mu_e = 1350$ cm$^2$ V$^{-1}$ s$^{-1}$, $D_e = 34.9$ cm$^2$ s$^{-1}$, $L_e = 0.0041$ cm $=$ 41 μm, $\mu_h = 450$ cm$^2$ V$^{-1}$ s$^{-1}$, $D_h = 11.6$ cm$^2$ s$^{-1}$, $L_h = 0.0024$ cm $=$ 24 μm. We can now calculate each current term using the Equations 5.49, 5.52, 5.55 and 5.57 above as shown in Figure 5.34b. The actual values at two locations, $x = 0$ and $x = L_e = 41$ μm, are shown in Table 5.3.[5]

[5] The reader may have observed that the currents in Table 5.3 do not add exactly to zero. The analysis here is only approximate and, further, it was based on neglecting the hole drift current and taking the field as nearly zero to use Equation 5.47 in deriving the carrier concentration profiles. Note that hole drift current is much smaller than the other current components.

**EXAMPLE 5.15**

**INFINITELY LONG SEMICONDUCTOR ILLUMINATED AT ONE END** Find the minority carrier concentration profile $p_n(x)$ in an infinite $n$-type semiconductor that is illuminated continuously at one end as in Figure 5.34. Assume that photogeneration occurs near the surface. Show that the mean distance diffused by the minority carriers before recombination is $L_h$.

**SOLUTION**

Continuous illumination means that we have steady-state conditions and thus Equation 5.47 can be used. The general solution of this second-order differential equation is

$$\Delta p_n(x) = A \exp\left(-\frac{x}{L_h}\right) + B \exp\left(\frac{x}{L_h}\right) \qquad [5.58]$$

where $A$ and $B$ are constants that have to be found from the boundary conditions. For an infinite bar, at $x = \infty$, $\Delta p_n(\infty) = 0$ gives $B = 0$. At $x = 0$, $\Delta p_n = \Delta p_n(0)$ so $A = \Delta p_n(0)$. Thus, the excess (photoinjected) hole concentration at position $x$ is

$$\Delta p_n(x) = \Delta p_n(0) \exp\left(-\frac{x}{L_h}\right) \qquad [5.59]$$

which is shown in Figure 5.34a. To find the mean position of the photoinjected holes, we use the definition of the "mean," that is,

$$\bar{x} = \frac{\int_0^\infty x\, \Delta p_n(x)\, dx}{\int_0^\infty \Delta p_n(x)\, dx}$$

Substituting for $\Delta p_n(x)$ from Equation 5.59 and carrying out the integration gives $\bar{x} = L_h$. We conclude that the **diffusion length** $L_h$ is the average distance diffused by the minority carriers before recombination. As a corollary, we should infer that $1/L_h$ is the mean probability per unit distance that the hole recombines with an electron.

## 5.7 OPTICAL ABSORPTION

We have already seen that a photon of energy $h\nu$ greater than $E_g$ can be absorbed in a semiconductor, resulting in the excitation of an electron from the valence band to the conduction band, as illustrated in Figure 5.35. The average energy of electrons in the conduction band is $\frac{3}{2}kT$ above $E_c$ (average kinetic energy is $\frac{3}{2}kT$), which means that the electrons are very close to $E_c$. If the photon energy is much larger than the bandgap energy $E_g$, then the excited electron is not near $E_c$ and has to lose the extra energy $h\nu - E_g$ to reach thermal equilibrium. The excess energy $h\nu - E_g$ is lost to lattice vibrations as heat as the electron is scattered from one atomic vibration to another. This process is called **thermalization.** If, on the other hand, the photon energy $h\nu$ is less than the bandgap energy, the photon will not be absorbed and we can say that the semiconductor is transparent to wavelengths longer than $hc/E_g$ provided that there are no energy states in the bandgap. There, of course, will be reflections occurring at the air/semiconductor surface due to the change in the refractive index.

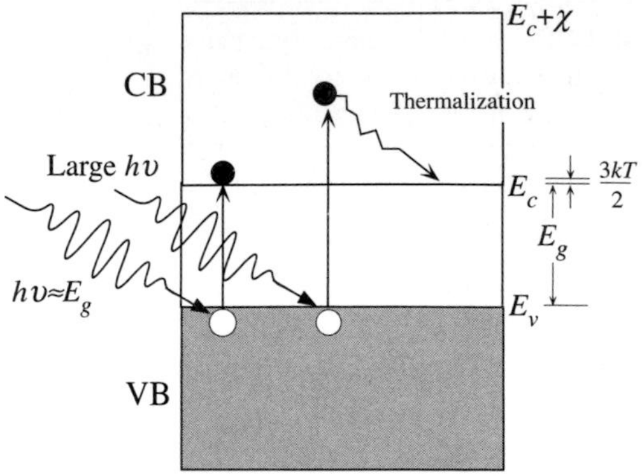


**Figure 5.35** Optical absorption generates electron–hole pairs.
Energetic electrons must lose their excess energy to lattice vibrations until their average energy is $\frac{3}{2}kT$ in the CB.

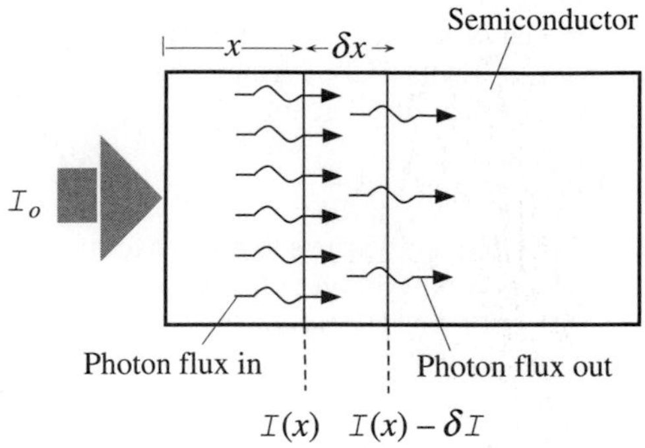


**Figure 5.36** Absorption of photons within a small elemental volume of width $\delta x$.

Suppose that $I_o$ is the intensity of a beam of photons incident on a semiconductor material. Thus, $I_o$ is the energy incident per unit area per unit time. If $\Gamma_{\text{ph}}$ is the photon flux, then

$$I_o = h\nu\Gamma_{\text{ph}}$$

When the photon energy is greater than $E_g$, photons from the incident radiation will be absorbed by the semiconductor. The absorption of photons requires the excitation of valence band electrons, and there are only so many of them with the right energy *per unit volume.* Consequently, absorption depends on the thickness of the semiconductor. Suppose that $I(x)$ is the light intensity at $x$ and $\delta I$ is the change in the light intensity in the small elemental volume of thickness $\delta x$ at $x$ due to photon absorption, as illustrated in Figure 5.36. Then $\delta I$ will depend on the number of photons arriving at this volume $I(x)$ and the thickness $\delta x$. Thus

$$\delta I = -\alpha I \,\delta x$$

where $\alpha$ is a proportionality constant that depends on the photon energy and hence wavelength, that is, $\alpha = \alpha(\lambda)$. The negative sign ensures that $\delta I$ is a reduction. The constant $\alpha$ as defined by this equation is called the **absorption coefficient** of the semiconductor. It is therefore defined by

*Definition of absorption coefficient*

$$\alpha = -\frac{\delta I}{I\,\delta x} \qquad [5.60]$$

which has the dimensions of length$^{-1}$ (m$^{-1}$).

When we integrate Equation 5.60 for illumination with constant wavelength light, we get the **Beer–Lambert law,** the transmitted intensity decreases exponentially with the thickness,

*Beer–Lambert law*

$$I(x) = I_o \exp(-\alpha x) \qquad [5.61]$$

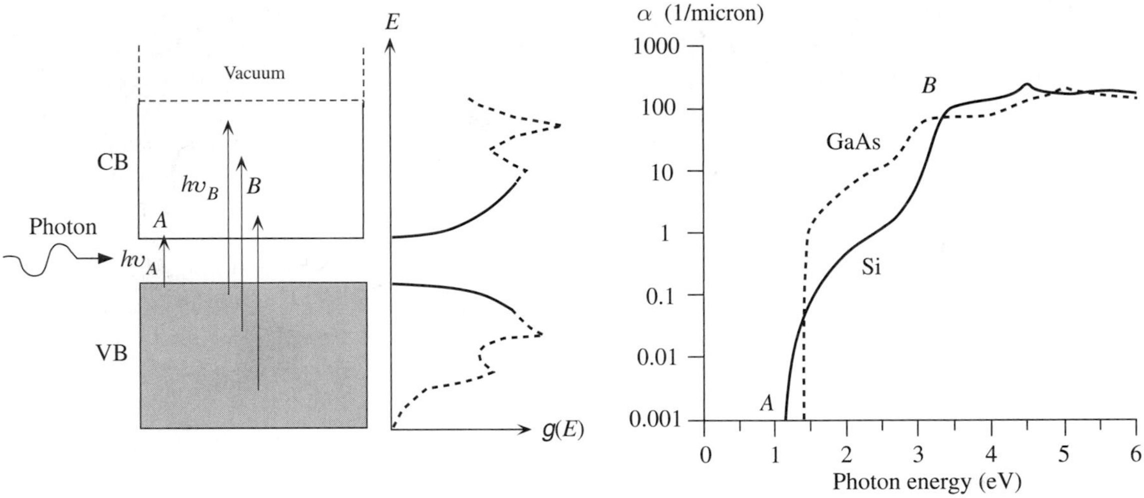


**Figure 5.37** The absorption coefficient $\alpha$ depends on the photon energy $h\nu$ and hence on the wavelength. Density of states increases from band edges and usually exhibits peaks and troughs. Generally $\alpha$ increases with the photon energy greater than $E_g$ because more energetic photons can excite electrons from populated regions of the VB to numerous available states deep in the CB.

As apparent from Equation 5.61, over a distance $x = 1/\alpha$, the light intensity falls to a value $0.37\mathcal{I}_o$; that is, it decreases by 63 percent. This distance over which 67 percent of the photons are absorbed is called the **penetration depth,** denoted by $\delta = 1/\alpha$.

The absorption coefficient depends on the photon absorption processes occurring in the semiconductor. In the case of **band-to-band (interband) absorption,** $\alpha$ increases rapidly with the photon energy $h\nu$ above $E_g$ as shown for Si ($E_g = 1.1$ eV) and GaAs ($E_g = 1.42$ eV) in Figure 5.37. Notice that $\alpha$ is plotted on a logarithmic scale. The general trend of the $\alpha$ versus $h\nu$ behavior can be intuitively understood from the density of states diagram also shown in the same figure.

Density of states $g(E)$ represents the number of states per unit energy per unit volume. We assume that the VB states are filled and the CB states are empty since the number of electrons in the CB is much smaller than the number of states in this band ($n \ll N_c$). The photon absorption process increases when there are more VB states available as more electrons can be excited. We also need available CB states into which the electrons can be excited, otherwise the electrons cannot find empty states to fill. The probability of photon absorption depends on both the density of VB states *and* the density of CB states. For photons of energy $h\nu_A = E_g$, the absorption can only occur from $E_v$ to $E_c$ where the VB and CB densities of states are low and thus the absorption coefficient is small, which is illustrated as $A$ in Figure 5.37. For photon energies $h\nu_B$, which can take electrons from very roughly the middle region of the VB to the middle of the CB, the densities of states are large and $\alpha$ is also large as indicated by $B$ in Figure 5.37. Furthermore, there are more choices of excitation for the $h\nu_B$ photon as illustrated by the three arrows in the figure. At even higher photon energies, photon absorption can of course excite electrons from the VB into vacuum. In reality, the density of states $g(E)$ of a real crystalline semiconductor is much more complicated with various sharp peaks and troughs on the density of states function, shown as dashed curves in $g(E)$ in Figure 5.37, particularly away from the band edges. In addition, the absorption process has to satisfy the conservation of momentum and quantum mechanical transition rules which means that certain transitions from the CB to the VB will be more favorable than others. For example, GaAs is a **direct bandgap** semiconductor, so photon absorption can lead directly to the excitation of an electron from the CB to the VB for photon energies just above $E_g$ just as direct recombination of an electron and hole results in photon emission. Si is an **indirect bandgap** semiconductor. Just as direct electron and hole recombination is not possible in silicon, the electron excitation from states near $E_v$ to states near $E_c$ must be accompanied by the emission or absorption of lattice vibrations, and hence the absorption is less efficient; $\alpha$ versus $h\nu$ for GaAs rises more sharply than that for Si above $E_g$ as apparent in Figure 5.37. At sufficiently high photon energies, it is possible to excite electrons directly from the VB to the CB in Si and this gives the sharp rise in $\alpha$ versus $h\nu$ before $B$ in Figure 5.37. (Band-to-band absorption is further discussed in Chapter 9.)

**EXAMPLE 5.16** **PHOTOCONDUCTIVITY OF A THIN SLAB** Modify the photoconductivity expression

$$\Delta\sigma = \frac{e\eta I_o\lambda\tau(\mu_e + \mu_h)}{hcD}$$

derived for a direct bandgap semiconductor in Figure 5.28 to take into account that some of the light intensity is transmitted through the material.

**SOLUTION**

If we assume that all the photons are absorbed (there is no transmitted light intensity), then the photoconductivity expression is

$$\Delta\sigma = \frac{e\eta I_o\lambda\tau(\mu_e + \mu_h)}{hcD}$$

But, in reality, $I_o \exp(-\alpha D)$ is the transmitted intensity through the specimen with thickness $D$, so absorption is determined by the intensity lost in the material $I_o[1 - \exp(-\alpha D)]$, which means that $\Delta\sigma$ must be accordingly scaled down to

$$\Delta\sigma = \frac{e\eta I_o[1 - \exp(-\alpha D)]\lambda\tau(\mu_e + \mu_h)}{hcD}$$

**EXAMPLE 5.17** **PHOTOGENERATION IN GaAs AND THERMALIZATION** Suppose that a GaAs sample is illuminated with a 50 mW HeNe laser beam (wavelength 632.8 nm) on its surface. Calculate how much power is dissipated as heat in the sample during thermalization. Give your answer as mW. The energy bandgap $E_g$ of GaAs is 1.42 eV.

**SOLUTION**

Suppose $P_L$ is the power in the laser beam; then $P_L = IA$, where $I$ is the intensity of the beam and $A$ is the area of incidence. The photon flux, photons arriving per unit area per unit time, is

$$\Gamma_{\text{ph}} = \frac{I}{h\nu} = \frac{P_L}{Ah\nu}$$

so the number of EHPs generated per unit time is

$$\frac{dN}{dt} = \Gamma_{\text{ph}} A = \frac{P_L}{h\nu}$$

These carriers *thermalize*—lose their excess energy as lattice vibrations (heat) via collisions with the lattice—so eventually their average kinetic energy becomes $\frac{3}{2}kT$ above $E_g$ as depicted in Figure 5.35. Remember that we assume that electrons in the CB are nearly free, so they must obey the kinetic theory and hence have an average kinetic energy of $\frac{3}{2}kT$. The average energy of the electron is then $E_g + \frac{3}{2}kT \approx 1.46$ eV. The excess energy

$$\Delta E = h\nu - \left(E_g + \frac{3}{2}kT\right)$$

is lost to the lattice as heat, that is, lattice vibrations. Since each electron loses an amount of energy $\Delta E$ as heat, the heat power generated is

$$P_H = \left(\frac{dN}{dt}\right)\Delta E = \left(\frac{P_L}{h\nu}\right)(\Delta E)$$

The incoming photon has an energy $h\nu = hc/\lambda = 1.96$ eV, so

$$P_H = \frac{(50\text{ mW})(1.96\text{ eV} - 1.46\text{ eV})}{1.96\text{ eV}} = 12.76\text{ mW}$$

Notice that in this example, and also in Figure 5.35, we have assigned the excess energy $\Delta E = h\nu - E_g - \frac{3}{2}kT$ to the electron rather than share it between the electron and the hole that is photogenerated. This assumption depends on the ratio of the electron and hole effective masses, and hence depends on the semiconductor material. It is approximately true in GaAs because the electron is much lighter than the hole, almost 10 times, and consequently the absorbed photon is able to "impart" a much higher kinetic energy to the electron than to the hole; $h\nu - E_g$ is used in the photogeneration, and the remainder goes to impart kinetic energy to the photogenerated electron hole pair.

## 5.8 PIEZORESISTIVITY

When a mechanical stress is applied to a semiconductor sample, as shown in Figure 5.38a, it is found that the resistivity of the semiconductor changes by an amount that depends on the stress.[6] **Piezoresistivity** is the change in the resistivity of a semiconductor (indeed, any material), due to an applied stress. **Elastoresistivity** refers to the change in the resistivity due to an induced strain in the substance. Since the application of stress invariably leads to strain, piezoresistivity and elastoresistivity refer to

[6] *Mechanical stress* is defined as the applied force per unit area, $\sigma_m = F/A$, and the resulting strain $\varepsilon_m$ is the fractional change in the length of a sample caused by $\sigma_m$; $\varepsilon_m = \delta L/L$, where $L$ is the sample length. The two are related through the elastic modulus $Y$; $\sigma_m = Y\varepsilon_m$. Subscript $m$ is used to distinguish the stress $\sigma_m$ and strain $\varepsilon_m$ from the conductivity $\sigma$ and permittivity $\varepsilon$.

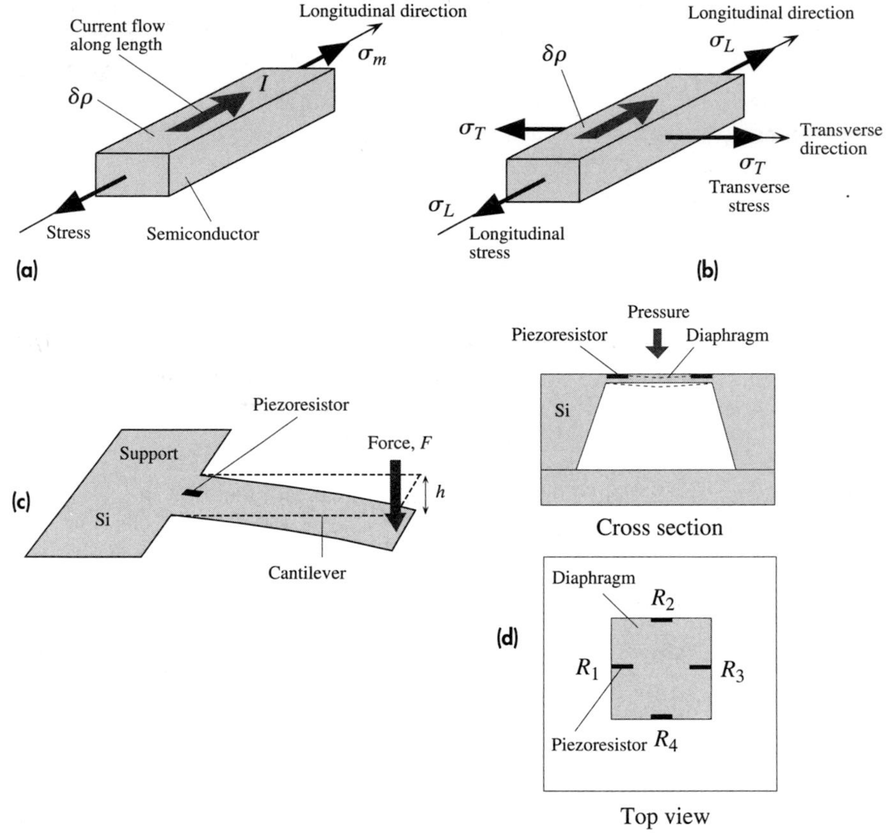


**Figure 5.38** Piezoresistivity and its applications.
(a) Stress $\sigma_m$ along the current (longitudinal) direction changes the resistivity by $\delta\rho$.
(b) Stresses $\sigma_L$ and $\sigma_T$ cause a resistivity change.
(c) A force applied to a cantilever bends it. A piezoresistor at the support end (where the stress is large) measures the stress, which is proportional to the force.
(d) A pressure sensor has four piezoresistors $R_1$, $R_2$, $R_3$, $R_4$ embedded in a diaphragm. The pressure bends the diaphragm, which generates stresses that are sensed by the four piezoresistors.

the same phenomenon. Piezoresistivity is fruitfully utilized in a variety of useful sensor applications such as force, pressure and strain gauges, accelerometers, and microphones.

The change in the resistivity may be due to a change in the concentration of carriers or due to a change in the drift mobility of the carriers, both of which can be modified by a strain in the crystal. Typically, in an extrinsic or doped semiconductor, the concentration of carriers does not change as significantly as the drift mobility; the piezoresistivity is then associated with the change in the mobility. For example, in an $n$-type Si, the change in the electron mobility $\mu_e$ with mechanical strain $\varepsilon_m$, $d\mu_e/d\varepsilon_m$, is of the order of $10^5$ cm$^2$ V$^{-1}$ s$^{-1}$, so that a strain of 0.015 percent will result in a change in the mobility that is about 1 percent, and a similar change in the resistivity, which is readily measurable. In this case, the change in the mobility $\mu_e$ is due to the induced strain changing the effective mass $m_e^*$ which then modifies $\mu_e$. (Recall that $\mu_e = e\tau/m_e^*$, where $\tau$ is the mean scattering time.)

The change in the resistivity $\delta\rho$ has been shown to be proportional to the induced strain in the crystal and hence proportional to the applied stress $\sigma_m$. The fractional change $\delta\rho/\rho$ can be written as

$$\frac{\delta\rho}{\rho} = \pi\sigma_m \qquad [5.62]$$

*Piezoresistivity*

where $\pi$ is a constant called the **piezoresistive coefficient;** $\pi$ has the units of 1/stress, *e.g.*, $m^2/N$ or 1/Pa. The piezoresistive coefficient $\pi$ depends on the type of doping, $p$-or $n$-type; the dopant concentration; the temperature; and the crystallographic direction. A stress along a certain direction in a crystal, for example, along the length of a semiconductor crystal, will change the resistivity not only in the same direction but also in transverse directions. We know from elementary mechanics that a strain in one direction is accompanied by a transverse strain, as implied by the Poisson ratio, so it is not unexpected that a stress in one direction will also modify the resistivity in a transverse direction. Thus, the change in the resistivity of a semiconductor in a "longitudinal" direction, taken as the direction of current flow, is due to stresses in the longitudinal and transverse directions. If $\sigma_L$ is the stress along a longitudinal direction, the direction of current flow, and $\sigma_T$ is the stress along a transverse direction, as in Figure 5.38b, then, generally, the fractional change in the resistivity along the current flow direction (longitudinal direction) is given by

$$\frac{\delta\rho}{\rho} = \pi_L\sigma_L + \pi_T\sigma_T \qquad [5.63]$$

*Piezoresistivity*

where $\pi_L$ is the piezoresistive coefficient along a longitudinal direction (different for $p$- and $n$-type Si), and $\pi_T$ is the piezoresistive coefficient in the transverse direction.

The piezoresistive effect is actually more complicated than what we have implied. In reality, we have to consider six types of stresses, three uniaxial stresses along the $x$, $y$, and $z$ directions (*e.g.*, trying to pull the crystal along in three independent directions) and three shear stresses (*e.g.*, trying to shear the crystal in three independent ways). In very simple terms, a change in the resistivity $(\delta\rho/\rho)_i$ along a particular direction $i$ (an arbitrary direction) can be induced by a stress $\sigma_j$ along another direction $j$ (which may or may not be identical to $i$). The two, $(\delta\rho/\rho)_i$ and $\sigma_j$, are then related through a piezoresistivity coefficient denoted by $\pi_{ij}$. Consequently, the full description of piezoresistivity involves tensors, and the piezoresistivity coefficients $\pi_{ij}$ form the elements of this tensor; a treatment beyond the scope of this book. Nonetheless, it is useful to be able to calculate $\pi_L$ and $\pi_T$ from various tabulated piezoresistivity coefficients $\pi_{ij}$, without having to learn tensors. It turns out that it is sufficient to identify three *principal piezoresistive coefficients* to describe the piezoresistive effect in cubic crystals, which are denoted as $\pi_{11}$, $\pi_{12}$, and $\pi_{44}$. From the latter set we can easily calculate $\pi_L$ and $\pi_T$ for a crystallographic direction of interest; the relevant equations can be found in advanced textbooks.

Advances in silicon fabrication technologies and micromachining (ability to fabricate micromechanical structures) have now enabled various piezoresistive silicon microsensors to be developed that have a wide range of useful applications. Figure 5.38c shows a very simple Si microcantilever in which an applied force $F$ to the free end bends the cantilever; the tip of the cantilever is deflected by a distance $h$. According to elementary mechanics, this deflection induces a maximum stress $\sigma_m$ that is at the surface, at the support end, of the cantilever. A properly placed piezoresistor at this end can be used to measure this stress $\sigma_m$, and hence the deflection or the force. The piezoresistor is implanted by selectively diffusing dopants into the Si cantilever at the support end. Obviously, we need to relate the deflection $h$ of the cantilever tip to the stress $\sigma_m$, which is well described in mechanics. In addition, $h$ is proportional to the applied force $F$ through a factor that depends on the elastic modulus and the geometry of the cantilever. Thus, we can measure both the displacement ($h$) and force ($F$).

Another useful application is in pressure sensors, which are commercially available. Again, the structure is fabricated from Si. A very thin elastic membrane, called a *diaphragm,* has four piezoresistors embedded, by appropriate dopant diffusion, on its surface as shown in Figure 5.38d. Under pressure, the Si diaphragm deforms elastically, and the stresses that are generated by this deformation cause the resistance of the piezoresistors to change. There are four piezoresistors because the four are connected in a Wheatstone bridge arrangement for better signal detection. The diaphragm area is typically 1 mm × 1 mm, and the thickness is 20 μm. There is no doubt that recent advances in micromachining have made piezoresistivity an important topic for a variety of sensor applications.

**EXAMPLE 5.18** **PIEZORESISTIVE STRAIN GAUGE** Suppose that we apply a stress $\sigma_L$ along the length, taken along the [110] direction, of a $p$-type silicon crystal sample. We will measure the resistivity along this direction by passing a current along the length and measuring the voltage drop between two fixed points as in Figure 5.38a. The stress $\sigma_L$ along the length will result in a strain $\varepsilon_L$ along the same length given by $\varepsilon_L = \sigma_L/Y$, where $Y$ is the elastic modulus. From Equation 5.63 the change in the resistivity is

$$\frac{\Delta\rho}{\rho} = \pi_L\sigma_L + \pi_T\sigma_T = \pi_L Y\varepsilon_L$$

where we have ignored the presence of any transverse stresses; $\sigma_T \approx 0$. These transverse stresses depend on how the piezoresistor is used, that is, whether it is allowed to contract laterally. If the resistor cannot contract, it must be experiencing a transverse stress. In any event, for the particular direction of interest, [110], the Poisson ratio is very small (less than 0.1), and we can simply neglect any $\sigma_T$. Clearly, we can find the strain $\varepsilon_L$ from the measurement of $\Delta\rho/\rho$, which is the principle of the strain gauge. The **gauge factor** $G$ of a strain gauge measures the sensitivity of the gauge in terms of the fractional change in the resistance per unit strain,

*Semiconductor strain gauge*

$$G = \frac{\left(\dfrac{\Delta R}{R}\right)}{\left(\dfrac{\Delta L}{L}\right)} \approx \frac{\left(\dfrac{\Delta\rho}{\rho}\right)}{\varepsilon_L} \approx Y\pi_L$$

where we have assumed that $\Delta R$ is dominated by $\Delta\rho$, since the effects from geometric changes in the sample shape can be ignored compared with the piezoresistive effect in semiconductors.

Using typical values for a $p$-type Si piezoresistor which has a length along [110], $Y \approx 170$ GPa, $\pi_L \approx 72 \times 10^{-11}$ Pa$^{-1}$, we find $G \approx 122$. This is much greater than $G \approx 1.7$ for metal resistor–based strain gauges. In most metals, the fractional change in the resistance $\Delta R/R$ is due to the geometric effect, the sample becoming elongated and narrower, whereas in semiconductors it is due to the piezoresistive effect.

## 5.9 SCHOTTKY JUNCTION

### 5.9.1 Schottky Diode

We consider what happens when a metal and an $n$-type semiconductor are brought into contact. In practice, this process is frequently carried out by the evaporation of a metal onto the surface of a semiconductor crystal in vacuum.

The energy band diagrams for the metal and the semiconductor are shown in Figure 5.39. The work function, denoted as $\Phi$, is the energy difference between the vacuum level and the Fermi level. The vacuum level defines the energy where the electron is free from that particular solid and where the electron has zero $KE$.

For the metal, the work function $\Phi_m$ is the minimum energy required to remove an electron from the solid. In the metal there are electrons at the Fermi level $E_{Fm}$, but in the

John Bardeen, Walter Schottky, and Walter Brattain. Walter H. Schottky (1886–1976) obtained his PhD from the University of Berlin in 1912. He made many distinct contributions to physical electronics. He invented the screen grid vacuum tube in 1915, and the tetrode vacuum tube in 1919 while at Siemens. The Schottky junction theory was formulated in 1938. He also made distinct contributions to thermal and shot noise in devices. His book *Thermodynamik* was published in 1929 and included an explanation of the Schottky defect (Chapter 1).

| SOURCE: AIP Emilio Segre Visual Archives, Brattain Collection.

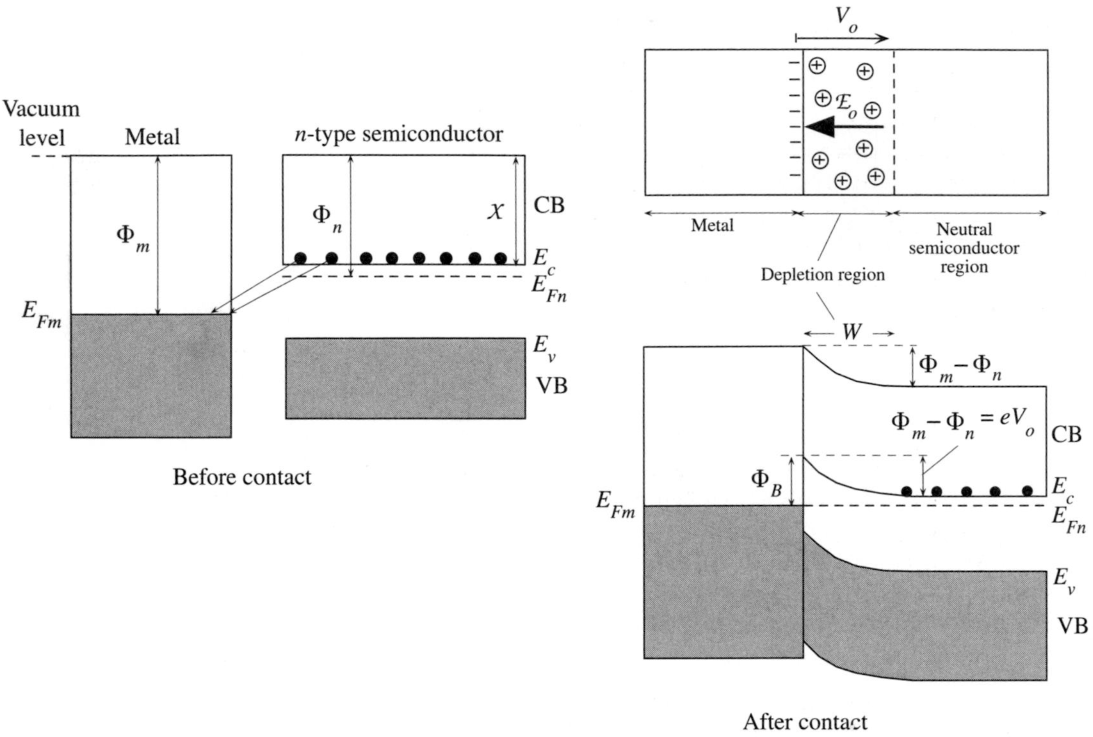


**Figure 5.39** Formation of a Schottky junction between a metal and an $n$-type semiconductor when $\Phi_m > \Phi_n$.

semiconductor there are none at $E_{Fn}$. Nonetheless, the semiconductor work function $\Phi_n$ still represents the energy required to remove an electron from the semiconductor. It may be thought that the minimum energy required to remove an electron from the semiconductor is simply the electron affinity $\chi$, but this is not so. Thermal equilibrium requires that only a certain fraction of all the electrons in the semiconductor should be in the CB at a given temperature. When an electron is removed from the conduction band, then thermal equilibrium can be maintained only if an electron is excited from the VB to CB, which involves absorbing heat (energy) from the environment; thus it takes more energy than simply $\chi$. We will not derive the effective thermal energy required to remove an electron but state that, as for a metal, this is equal to $\Phi_n$, even though there are no electrons at $E_{Fn}$. In fact, the thermionic emission of electrons from a heated semiconductor is also described by the Richardson–Dushman expression in Equation 4.37 but with $\Phi$ representing the work function of the semiconductor, $\Phi_n$ in the present $n$-type case. (In contrast, the minimum *photon energy* required to remove an electron from a semiconductor above absolute zero would be the electron affinity.)

We assume that $\Phi_m > \Phi_n$, the work function of the metal is greater than that of the semiconductor. When the two solids come into contact, the more energetic electrons in the CB of the semiconductor can readily tunnel into the metal in search of lower empty energy levels (just above $E_{Fm}$) and accumulate near the surface of the metal, as illustrated in Figure 5.39. Electrons tunneling from the semiconductor leave behind an electron-depleted region of width $W$ in which there are exposed positively charged donors, in other words, net positive space charge. The contact potential, called the **built-in potential** $V_o$, therefore develops between the metal and the semiconductor. There is obviously also a **built-in electric field** $\mathcal{E}_o$ from the positive charges to the negative charges on the metal surface. Eventually this built-in potential reaches a value that prevents further accumulation of electrons at the metal surface and an equilibrium is reached. The value of the built-in voltage $V_o$ is the same as that in the metal–metal junction case in Chapter 4, namely, $(\Phi_m - \Phi_n)/e$. The **depletion region** has been depleted of free carriers (electrons) and hence contains the exposed positive donors. This region thus constitutes a **space charge layer** (SCL) in which there is a nonuniform internal field directed from the semiconductor to the metal surface. The maximum value of this built-in field is denoted as $\mathcal{E}_o$ and occurs right at the metal–semiconductor junction (this is where there are a maximum number of field lines from positive to negative charges).

The Fermi level throughout the whole solid, the metal and semiconductor in contact, must be uniform in equilibrium. Otherwise, a change in the Fermi level $\Delta E_F$ going from one end to the other end will be available to do external (electrical) work. Thus, $E_{Fm}$ and $E_{Fn}$ line up. The $W$ region, however, has been depleted of electrons, so in this region $E_c - E_{Fn}$ must increase so that $n$ decreases. The bands must bend to increase $E_c - E_{Fn}$ toward the junction, as depicted in Figure 5.39. Far away from the junction, we, of course, still have an $n$-type semiconductor. The bending is just enough for the vacuum level to be continuous and changing by $\Phi_m - \Phi_n$ from the semiconductor to the metal, as this much energy is needed to take an electron across from the semiconductor to the metal. The $PE$ barrier for electrons moving from the metal to the semiconductor is called the **Schottky barrier height** $\Phi_B$, which is given by

*Schottky barrier*

$$\Phi_B = \Phi_m - \chi = eV_o + (E_c - E_{Fn}) \qquad [5.64]$$

which is greater than $eV_o$.

Under open circuit conditions, there is no net current flowing through the metal–semiconductor junction. The number of electrons thermally emitted over the $PE$ barrier $\Phi_B$ from the metal to the semiconductor is equal to the number of electrons thermally emitted over $eV_o$ from the semiconductor to the metal. Emission probability depends on the $PE$ barrier for emission through the Boltzmann factor. There are two current components due to electrons flowing through the junction. The current due to electrons being thermally emitted from the metal to the CB of the semiconductor is

$$J_1 = C_1 \exp\left(-\frac{\Phi_B}{kT}\right) \qquad [5.65]$$

where $C_1$ is some constant, whereas the current due to electrons being thermally emitted from the CB of the semiconductor to the metal is

$$J_2 = C_2 \exp\left(-\frac{eV_o}{kT}\right) \qquad [5.66]$$

where $C_2$ is some constant different than $C_1$.

In equilibrium, that is, open circuit conditions in the dark, the currents are equal but in the reverse directions:

$$J_{\text{open circuit}} = J_2 - J_1 = 0$$

Under forward bias conditions, the semiconductor side is connected to the negative terminal, as depicted schematically in Figure 5.40a. Since the depletion region $W$ has a much larger resistance than the neutral $n$-region (outside $W$) and the metal side, nearly all the voltage drop is across the depletion region. The applied bias is in the opposite direction to the built-in voltage $V_o$. Thus $V_o$ is reduced to $V_o - V$. $\Phi_B$ remains unchanged. The semiconductor band diagram outside the depletion region has been effectively shifted up with respect to the metal side by an amount $eV$ because

$$PE = \text{Charge} \times \text{Voltage}$$

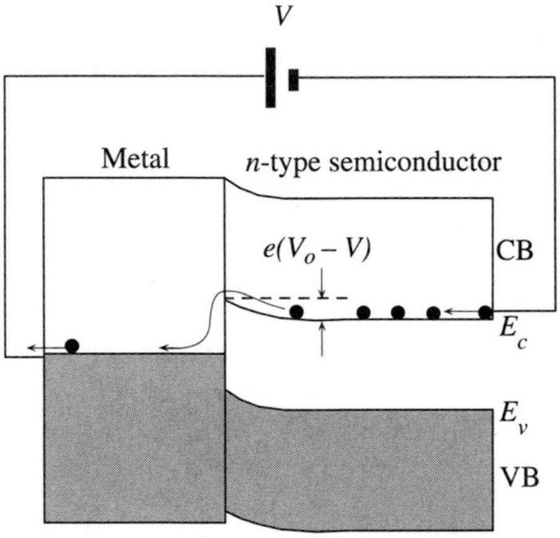


**(a) Forward-biased Schottky junction. Electrons in the CB of the semiconductor can easily overcome the small *PE* barrier to enter the metal.**

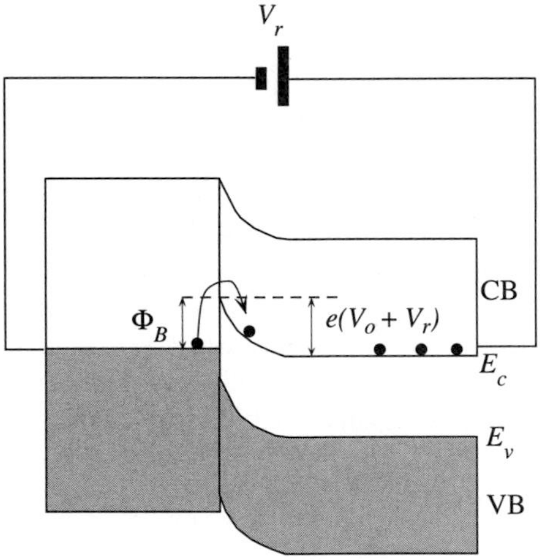


**(b) Reverse-biased Schottky junction. Electrons in the metal cannot easily overcome the *PE* barrier $\Phi_B$ to enter the semiconductor.**

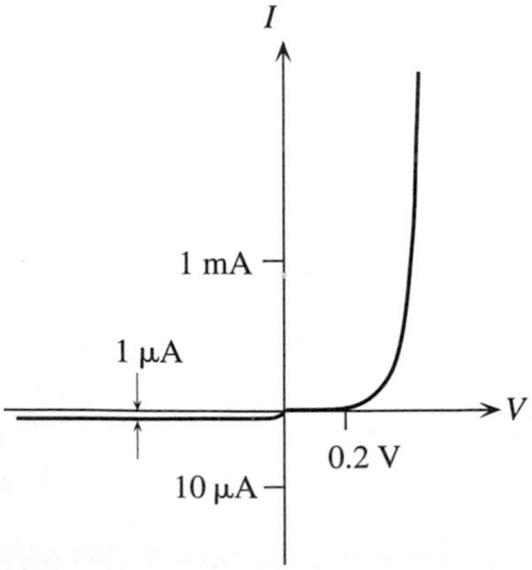


**(c) *I*–*V* characteristics of a Schottky junction exhibits rectifying properties (negative current axis is in microamps).**

**Figure 5.40** The Schottky junction.

The charge is negative but so is the voltage connected to the semiconductor, as shown in Figure 5.40a.

The *PE* barrier for thermal emission of electrons from the semiconductor to the metal is now $e(V_o - V)$. The electrons in the CB can now readily overcome the *PE* barrier to the metal.

The current $J_2^{\text{for}}$, due to the electron emission from the semiconductor to the metal, is now

$$J_2^{\text{for}} = C_2 \exp\left[-\frac{e(V_o - V)}{kT}\right] \tag{5.67}$$

Since $\Phi_B$ is the same, $J_1$ remains unchanged. The net current is then

$$J = J_2^{\text{for}} - J_1 = C_2 \exp\left[-\frac{e(V_o - V)}{kT}\right] - C_2 \exp\left(-\frac{eV_o}{kT}\right)$$

or

$$J = C_2 \exp\left(-\frac{eV_o}{kT}\right)\left[\exp\left(\frac{eV}{kT}\right) - 1\right]$$

giving

*Schottky junction*

$$J = J_o\left[\exp\left(\frac{eV}{kT}\right) - 1\right] \tag{5.68}$$

where $J_o$ is a constant that depends on the material and surface properties of the two solids. In fact, examination of the above steps shows that $J_o$ is also $J_1$ in Equation 5.65.

When the Schottky junction is reverse biased, then the positive terminal is connected to the semiconductor, as illustrated in Figure 5.40b. The applied voltage $V_r$ drops across the depletion region since this region has very few carriers and is highly resistive. The built-in voltage $V_o$ thus increases to $V_o + V_r$. Effectively, the semiconductor band diagram is shifted down with respect to the metal side because the charge is negative but the voltage is positive and $PE = \text{Charge} \times \text{Voltage}$. The *PE* barrier for thermal emission of electrons from the CB to the metal becomes $e(V_o + V_r)$, which means that the corresponding current component becomes

$$J_2^{\text{rev}} = C_2 \exp\left[-\frac{e(V_o + V_r)}{kT}\right] \ll J_1 \tag{5.69}$$

Since generally $V_o$ is typically a fraction of a volt and the reverse bias is more than a few volts, $J_2^{\text{rev}} \ll J_1$ and the reverse bias current is essentially limited by $J_1$ only and is very small. Thus, under reverse bias conditions, the current is primarily due to the thermal emission of electrons over the barrier $\Phi_B$ from the metal to the CB of the semiconductor as determined by Equation 5.65. Figure 5.40c illustrates the *I–V* characteristics of a typical Schottky junction. The *I–V* characteristics exhibit rectifying properties, and the device is called a **Schottky diode.**

Equation 5.68, which is derived for forward bias conditions, is also valid under reverse bias by making $V$ negative, that is, $V = -V_r$. Furthermore, it turns out to be applicable not only to Schottky-type metal–semiconductor junctions but also to junctions between a $p$-type and an $n$-type semiconductor, $pn$ junctions, as we will show in Chapter 6. Under a forward bias $V_f$, which is greater than 25 mV at room temperature, the forward current is simply

*Schottky junction forward bias*

$$J_f = J_o \exp\left(\frac{eV_f}{kT}\right) \qquad V_f > \frac{kT}{e} \tag{5.70}$$

It should be mentioned that it is also possible to obtain a Schottky junction between a metal and a $p$-type semiconductor. This arises when $\Phi_m < \Phi_p$, where $\Phi_p$ is the work function for the $p$-type semiconductor.

### 5.9.2 Schottky Junction Solar Cell

The built-in field in the depletion region of the Schottky junction allows this type of device to function as a photovoltaic device and also as a photodetector. We consider a Schottky device that has a thin metal film (usually Au) deposited onto an $n$-type semiconductor. The energy band diagram is shown in Figure 5.41. The metal is sufficiently thin (~10 nm) to allow light to reach the semiconductor.

For photon energies greater than $E_g$, EHPs are generated in the depletion region in the semiconductor, as indicated in Figure 5.41. The field in this region separates the EHPs and drifts the electrons toward the semiconductor and holes toward the metal. When an electron reaches the neutral $n$-region, there is now one extra electron there and therefore an additional negative charge. This end therefore becomes more negative with respect to the situation in the dark or the equilibrium situation. When a hole reaches the metal, it recombines with an electron and reduces the effective charge there by one electron, thus making it more positive relative to its dark state. Under open circuit conditions, therefore, a voltage develops across the Schottky junction device with the metal end positive and semiconductor end negative.

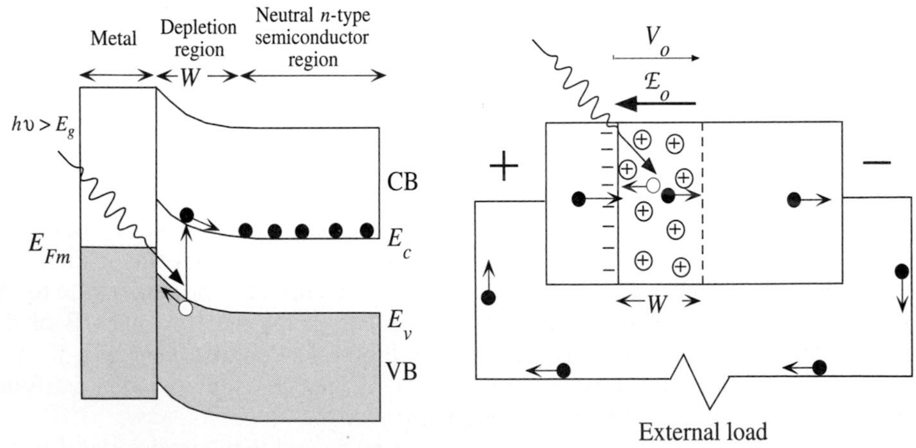


**Figure 5.41** The principle of the Schottky junction solar cell.

The photovoltaic explanation in terms of the energy band diagram is simple. At the point of photogeneration, the electron finds itself at a *PE* slope as $E_c$ is decreasing toward the semiconductor, as shown in Figure 5.41. It has no option but to roll down the slope just as a ball that is let go on a slope would roll down the slope to decrease its gravitational *PE*. Recall that there are many more empty states in the CB than electrons, so there is nothing to prevent the electron from rolling down the CB in search of lower energy. When the electron reaches the neutral region (flat $E_c$ region), it upsets the equilibrium there. There is now an additional electron in the CB and this side acquires a negative charge. If we remember that hole energy increases downward on the energy band diagram, then similar arguments also apply to the photogenerated hole in the VB, which rolls down its own *PE* slope to reach the surface of the metal and recombine with an electron there.

If the device is connected to an external load, then the extra electron in the neutral $n$-region is conducted through the external leads, through the load, toward the metal side, where it replenishes the lost electron in the metal. As long as photons are generating EHPs, the flow of electrons around the external circuit will continue and there will be photon energy to electrical energy conversion. Sometimes it is useful to think of the neutral $n$-type semiconductor region as a "conductor," an extension of the external wire (except that the $n$-type semiconductor has a higher resistivity). As soon as the photogenerated electron crosses the depletion region, it reaches a conductor and is conducted around the external circuit to the metal side to replenish the lost electron there.

For photon energies less than $E_g$, the device can still respond, providing that the $h\nu$ can excite an electron from $E_{Fm}$ in the metal over the *PE* barrier $\Phi_B$ into the CB, from where the electron will roll down toward the neutral $n$-region. In this case, $h\nu$ must only be greater than $\Phi_B$.

If the Schottky junction diode is reverse-biased, as shown in Figure 5.42, then the reverse bias $V_r$ increases the built-in potential $V_o$ to $V_o + V_r$ ($V_r \gg V_o$). The internal field increases to substantially high values. This has the advantage of increasing the drift velocity of the EHPs ($v_d = \mu_d \mathcal{E}$) in the depletion region and therefore

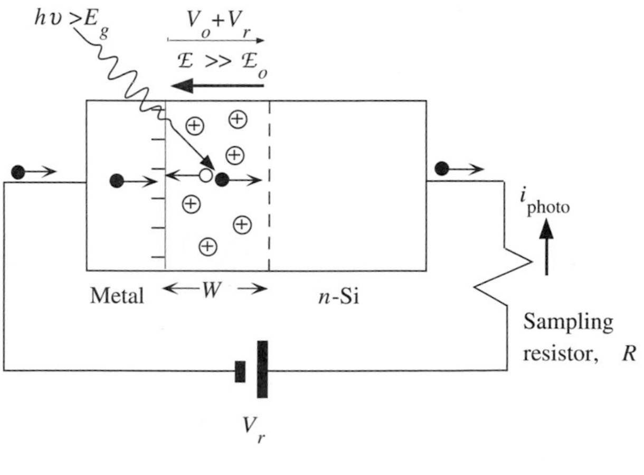


**Figure 5.42** Reverse-biased Schottky photodiodes are frequently used as fast photodetectors.

shortening the transit time required to cross the depletion width. The device responds faster and is useful as a fast photodetector. The photocurrent $i_{photo}$ in the external circuit is due to the drift of photogenerated carriers in the depletion region and can be readily measured.

---

**EXAMPLE 5.19** **THE SCHOTTKY DIODE** The reverse saturation current $J_o$ in the Schottky junction, as expressed in Equation 5.68, is the same current that is given by the Richardson–Dushman equation for thermionic emission over a potential barrier $\Phi(= \Phi_B)$ derived in Chapter 4. $J_o$ is given by

*Reverse saturation current in Schottky junction*

$$J_o = B_e T^2 \exp\left(-\frac{\Phi_B}{kT}\right)$$

where $B_e$ is the effective Richardson constant that depends on the characteristics of the metal–semiconductor junction. $B_e$ for metal–semiconductor junctions, among other factors, depends on the density of states related effective mass of the thermally emitted carriers in the semiconductor. For example, for a metal to $n$-Si junction, $B_e$ is about 110 A cm$^{-2}$ K$^{-2}$, and for a metal to $p$-Si junction, which involves holes, $B_e$ is about 30 A cm$^{-2}$ K$^{-2}$.

*a.* Consider a Schottky junction diode between W (tungsten) and $n$-Si, doped with $10^{16}$ donors cm$^{-3}$. The cross-sectional area is 1 mm$^2$. Given that the electron affinity $\chi$ of Si is 4.01 eV and the work function of W is 4.55 eV, what is the theoretical barrier height $\Phi_B$ from the metal to the semiconductor?

*b.* What is the built-in voltage $V_o$ with no applied bias?

*c.* Given that the experimental barrier height $\Phi_B$ is about 0.66 eV, what is the reverse saturation current and the current when there is a forward bias of 0.2 V across the diode?

**SOLUTION**

*a.* From Figure 5.39, it is clear that the barrier height $\Phi_B$ is

$$\Phi_B = \Phi_m - \chi = 4.55 \text{ eV} - 4.01 \text{ eV} = 0.54 \text{ eV}$$

The experimental value is around 0.66 eV, which is greater than the theoretical value due to various effects at the metal–semiconductor interface arising from dangling bonds, defects, and so forth. For example, dangling bonds give rise to what are called *surface states* within the bandgap of the semiconductor that can capture electrons and modify the Schottky energy band diagram. (The energy band diagram in Figure 5.39 represents an ideal junction with no surface states.) Further, in some cases, such as Pt on $n$-Si, the experimental value can be lower than the theoretical value.

*b.* We can find $E_c - E_{Fn}$ in Figure 5.39 from

$$n = N_d = N_c \exp\left(-\frac{E_c - E_{Fn}}{kT}\right)$$

$$10^{16} \text{ cm}^{-3} = (2.8 \times 10^{19} \text{ cm}^{-3}) \exp\left(-\frac{E_c - E_{Fn}}{0.026 \text{ eV}}\right)$$

which gives $\Delta E = E_c - E_{Fn} = 0.206$ eV. Thus, the built-in potential $V_o$ can be found from

$$V_o = \frac{\Phi_B}{e} - \frac{E_c - E_{Fn}}{e} = 0.54 \text{ V} - 0.206 \text{ V} = 0.33 \text{ V}$$

*c.* If $A$ is the cross-sectional area, 0.01 cm$^2$, taking $B_e$ to be 110 A K$^{-2}$ cm$^{-2}$, and using the experimental value for the barrier height $\Phi_B$, the saturation current is

$$I_o = AB_eT^2 \exp\left(-\frac{\Phi_B}{kT}\right) = (0.01)(110)(300^2)\exp\left(-\frac{0.66 \text{ eV}}{0.026 \text{ eV}}\right)$$

$$= 9.36 \times 10^{-7} \text{ A} \quad \text{or} \quad 0.94\,\mu\text{A}$$

When the applied voltage is $V_f$, the forward current $I_f$ is

$$I_f = I_o\left[\exp\left(\frac{V_f}{kT}\right) - 1\right] = (0.94\,\mu\text{A})\left[\exp\left(\frac{0.2}{0.026}\right) - 1\right] = 2.0 \text{ mA}$$

## 5.10 OHMIC CONTACTS AND THERMOELECTRIC COOLERS

An **ohmic contact** is a junction between a metal and a semiconductor that does not limit the current flow. The current is essentially limited by the resistance of the semiconductor outside the contact region rather than the thermal emission rate of carriers across a potential barrier at the contact. In the Schottky diode, the *I–V* characteristics were determined by the thermal emission rate of carriers across the contact. It should be mentioned that, contrary to intuition, when we talk about an ohmic contact, we do not generally infer a linear *I–V* characteristic for the ohmic contact itself. We only imply that the contact does not limit the current flow.

Figure 5.43 shows the formation of an ohmic contact between a metal and an $n$-type semiconductor. The work function of the metal $\Phi_m$ is smaller than the work function $\Phi_n$ of the semiconductor. There are more energetic electrons in the metal than

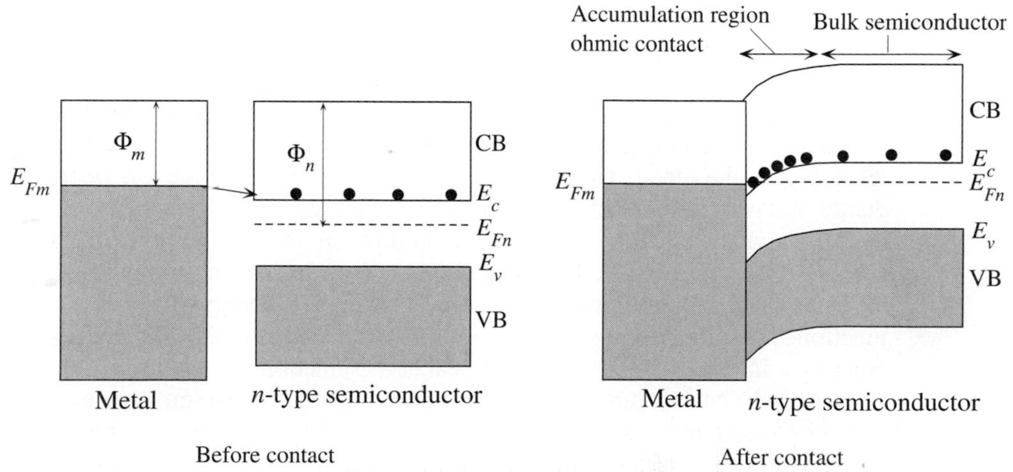


**Figure 5.43** When a metal with a smaller work function than an *n*-type semiconductor is put into contact with the *n*-type semiconductor, the resulting junction is an ohmic contact in the sense that it does not limit the current flow.

in the CB, which means that the electrons (around $E_{Fm}$) tunnel into the semiconductor in search of lower energy levels, which they find around $E_c$, as indicated in Figure 5.43. Consequently, many electrons pile in the CB of the semiconductor near the junction. Equilibrium is reached when the accumulated electrons in the CB of the semiconductor prevent further electrons tunneling from the metal. Put more rigorously, equilibrium is reached when the Fermi level is uniform across the whole system from one end to the other.

The semiconductor region near the junction in which there are excess electrons is called the **accumulation region.** To show the increase in $n$, we draw the semiconductor energy bands bending downward to decrease $E_c - E_{Fn}$, which increases $n$. Going from the far end of the metal to the far end of the semiconductor, there are always conduction electrons. In sharp contrast, the depletion region of the Schottky junction separates the conduction electrons in the metal from those in the semiconductor. It can be seen from the contact in Figure 5.43 that the conduction electrons immediately on either side of the junction (at $E_{Fm}$ and $E_c$) have about the same energy and therefore there is no barrier involved when they cross the junction in either direction under the influence of an applied field.

It is clear that the excess electrons in the accumulation region increase the conductivity of the semiconductor in this region. When a voltage is applied to the structure, the voltage drops across the higher resistance region, which is the bulk semiconductor region. Both the metal and the accumulation region have comparatively high concentrations of electrons compared with the bulk of the semiconductor. The current is therefore determined by the resistance of the bulk region. The current density is then simply $J = \sigma \mathcal{E}$ where $\sigma$ is the conductivity of the semiconductor in the bulk and $\mathcal{E}$ is the applied field in this region.

One of the interesting and important applications of semiconductors is in **thermoelectric,** or **Peltier,** devices, which enable small volumes to be cooled by direct currents. Whenever a dc current flows through a contact between two dissimilar materials, heat is either released or absorbed in the contact region, depending on the direction of the current. Suppose that there is a dc current flowing from an $n$-type semiconductor to a metal through an ohmic contact, as depicted in Figure 5.44a. Then electrons are flowing from the metal to the CB of the semiconductor. We only consider the contact region where the Peltier effect occurs. Current is carried by electrons near the Fermi level $E_{Fm}$ in the metal. These electrons then cross over into the CB of the semiconductor and when they reach the end of the contact region, their energy is $E_c$ plus average $KE$ (which is $\frac{3}{2}kT$). There is therefore an increase in the average energy $(PE + KE)$ per electron in the contact region. The electron must therefore absorb heat from the environment (lattice vibrations) to gain this energy as it drifts through the junction. Thus, the passage of an electron from the metal to the CB of an $n$-type semiconductor involves the absorption of heat at the junction.

When the current direction is from the metal to the $n$-type semiconductor, the electrons flow from the CB of the semiconductor to the Fermi level of the metal as they pass through the contact. Since $E_{Fm}$ is lower than $E_c$, the passing electron has to lose energy, which it does to lattice vibrations as heat. Thus, the passage of a CB electron from the $n$-type semiconductor to the metal involves the release of heat at the junction, as indicated in Figure 5.44b.

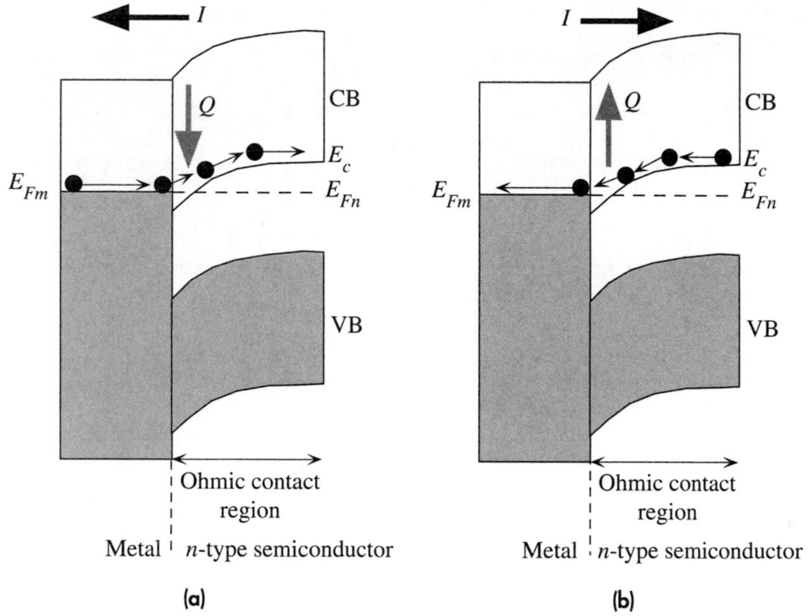


**Figure 5.44**

(a) Current from an *n*-type semiconductor to the metal results in heat absorption at the junction.

(b) Current from the metal to an *n*-type semiconductor results in heat release at the junction.

It is apparent that depending on the direction of the current flow through a junction between a metal and an *n*-type semiconductor, heat is either absorbed or released at the junction. Although we considered current flow between a metal and an *n*-type semiconductor through an ohmic contact, this thermoelectric effect is a general phenomenon that occurs at a junction between any two dissimilar materials. It is called the **Peltier effect** after its discoverer. In the case of metal–*p*-type semiconductor junctions, heat is absorbed for current flowing from the metal to the *p*-type semiconductor and heat is released in the other direction. Thermoelectric effects occurring at metal–semiconductor junctions are summarized in Figure 5.45. It is important not to confuse the Peltier effect with the Joule heating of the semiconductor and the metal. Joule heating, which we simply call $I^2R$ (or $J^2\rho$) heating, arises from the finite resistivity of the material. It is due to the conduction electrons losing their energy gained from the field to lattice vibrations when they become scattered by such vibrations, as discussed in Chapter 2.

It is self-evident that when a current flows through a semiconductor sample with metal contacts at its ends, as depicted in Figure 5.45, one of the contacts will always absorb heat and the other will always release heat. The contact where heat is absorbed will be cooled and is called the cold junction, whereas the other contact, where heat is released, will warm up and is called the hot junction. One can use the cold junction to

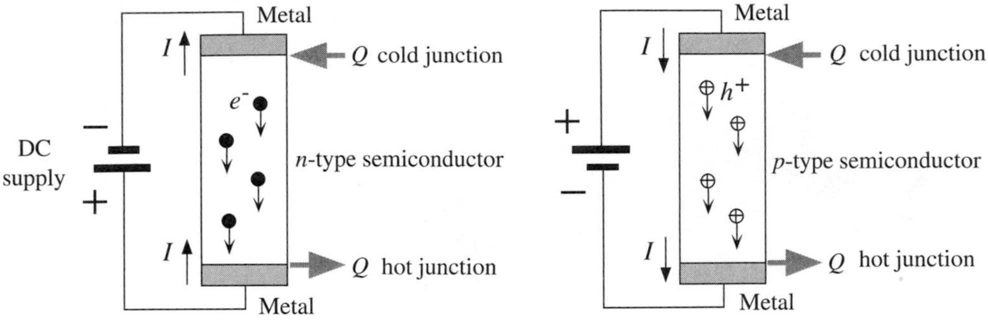


**Figure 5.45** When a dc current is passed through a semiconductor to which metal contacts have been made, one junction absorbs heat and cools (the cold junction) and the other releases heat and warms (the hot junction).

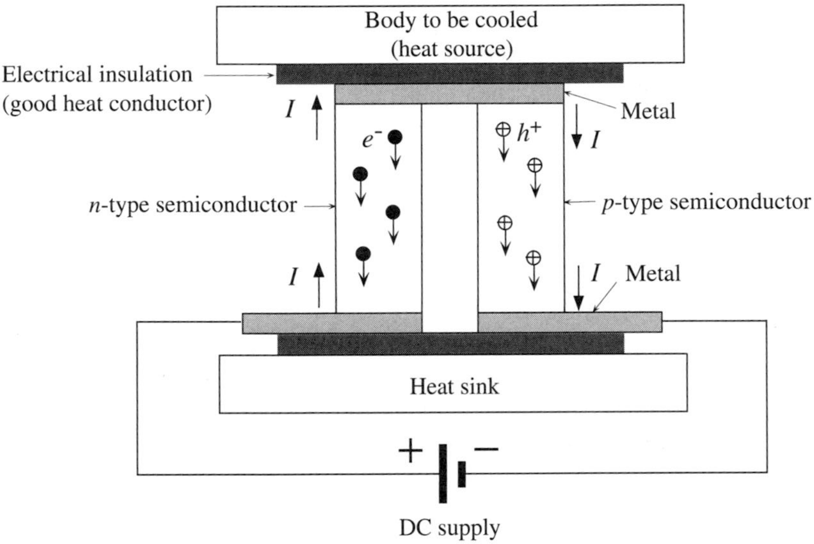


**Figure 5.46** Cross section of a typical thermoelectric cooler.

cool another body, providing that the heat generated at the hot junction can be removed from the semiconductor sufficiently quickly to reduce its conduction through the semiconductor to the cold junction. Furthermore, there will always be the Joule heating ($I^2R$) of the whole semiconductor sample since the bulk will always have a finite resistance.

A simplified schematic diagram of a practical single-element thermoelectric cooling device is shown in Figure 5.46. It uses two semiconductors, one $n$-type and the other $p$-type, each with ohmic contacts. The current direction therefore has opposite thermoelectric effects. On one side, the semiconductors share the same metal

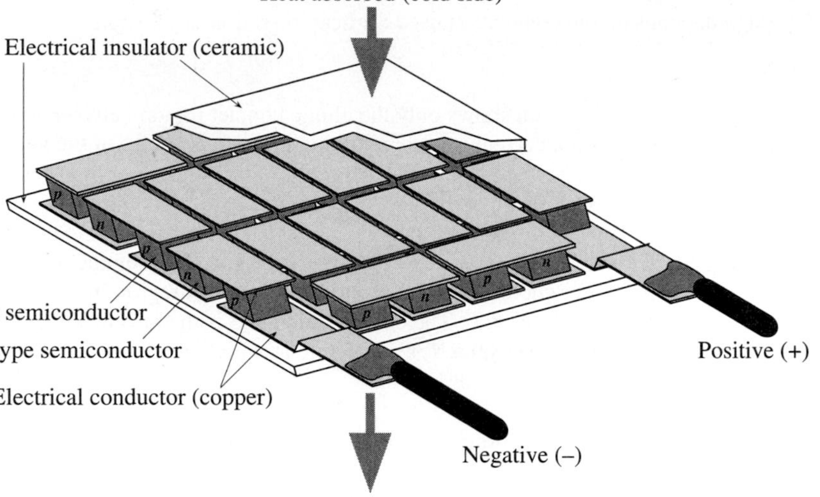


**Figure 5.47** Typical structure of a commercial thermoelectric cooler.

electrode. Effectively, the structure is an *n*-type and a *p*-type semiconductor connected in series through a common metal electrode. Typically, either $Bi_2Te_3$, $Bi_2Se_3$, or $Sb_2Te_3$ is used as the semiconductor material with copper usually as the metal electrode.

The current flowing through the *n*-type semiconductor to the common metal electrode causes heat absorption, which cools this junction and hence the metal. The same current then enters the *p*-type semiconductor and causes heat absorption at this junction, which cools the same metal electrode. Thus the common metal electrode is cooled at both ends. The other ends of the semiconductors are hot junctions. They are connected to a large heat sink to remove the heat and thus prevent heat conduction through the semiconductors toward the cold junctions. The other face of the common metal electrode is in contact, through a thin ceramic plate (electrical insulator but thermal conductor), with the body to be cooled. In commercial Peltier devices, many of these elements are connected in series, as illustrated in Figure 5.47, to increase the cooling efficiency.

---

**EXAMPLE 5.20**

**THE PELTIER COEFFICIENT** Consider the motion of electrons across an ohmic contact between a metal and an *n*-type semiconductor and hence show that the rate of heat generation $Q'$ at the contact is approximately

$$Q' = \pm \Pi I$$

where $\Pi$, called the **Peltier coefficient** between the two materials, is given by

$$\Pi = \frac{1}{e}\left[(E_c - E_{Fn}) + \frac{3}{2}kT\right]$$

where $E_c - E_{Fn}$ is the energy separation of $E_c$ from the Fermi level in the $n$-type semiconductor. The sign depends on the convention used for heat liberation or absorption.

**SOLUTION**

We consider Figure 5.44a, which shows only the ohmic contact region between a metal and an $n$-type semiconductor when a current is passing through it. The majority of the applied voltage drops across the bulk of the semiconductor because the contact region, or the accumulation region, has an accumulation of electrons in the CB. The current is limited by the bulk resistance of the semiconductor. Thus, in the contact region we can take the Fermi level to be almost undisturbed and hence uniform, $E_{Fm} \approx E_{Fn}$. In the bulk of the metal, a conduction electron is at around $E_{Fm}$ (same as $E_{Fn}$), whereas just at the end of the contact region in the semiconductor it is at $E_c$ plus an average $KE$ of $\frac{3}{2}kT$. The energy difference is the heat absorbed per electron going through the contact region. Since $I/e$ is the rate at which electrons are flowing through the contact,

$$\text{Rate of energy absorption} = \left[\left(E_c + \frac{3}{2}kT\right) - E_{Fm}\right]\left(\frac{I}{e}\right)$$

or

$$Q' = \left[\frac{(E_c - E_{Fn}) + \frac{3}{2}kT}{e}\right] I = \Pi I$$

so the Peltier coefficient is approximately given by the term in the square brackets. A more rigorous analysis gives $\Pi$ as

$$\Pi = \frac{1}{e}[(E_c - E_{Fn}) + 2kT]$$

---

# ADDITIONAL TOPICS

## 5.11 DIRECT AND INDIRECT BANDGAP SEMICONDUCTORS

***E–k Diagrams*** We know from quantum mechanics that when the electron is within a potential well of size $L$, its energy is quantized and given by

$$E_n = \frac{(\hbar k_n)^2}{2m_e}$$

where the wavevector $k_n$ is essentially a quantum number determined by

$$k_n = \frac{n\pi}{L}$$

where $n = 1, 2, 3, \ldots$ The energy increases parabolically with the wavevector $k_n$. We also know that the electron momentum is given by $\hbar k_n$. This description can be used to represent the behavior of electrons in a metal within which their average potential energy can be taken to be roughly zero. In other words, we take $V(x) = 0$ within the metal crystal and $V(x)$ to be large [*e.g.*, $V(x) = V_o$] outside so that the electron is contained within the metal. This is the **nearly free electron model** of a metal that has been quite successful in interpreting many of the properties. Indeed, we were able to calculate the density of states $g(E)$ based on the three-dimensional potential well problem. It is quite obvious that this model is too simple since it does not take into account the actual variation of the electron potential energy in the crystal.

The potential energy of the electron depends on its location within the crystal and is periodic due to the regular arrangement of the atoms. How does a periodic potential energy affect the relationship between $E$ and $k$? It will no longer simply be $E_n = (\hbar k_n)^2/2m_e$.

To find the energy of the electron in a crystal, we need to solve the Schrödinger equation for a periodic potential energy function in three dimensions. We first consider the hypothetical one-dimensional crystal shown in Figure 5.48. The electron potential energy functions for each atom add to give an overall potential energy function $V(x)$, which is clearly periodic in $x$ with the periodicity of the crystal $a$. Thus,

*Periodic potential energy*

$$V(x) = V(x + a) = V(x + 2a) = \cdots \qquad [5.71]$$

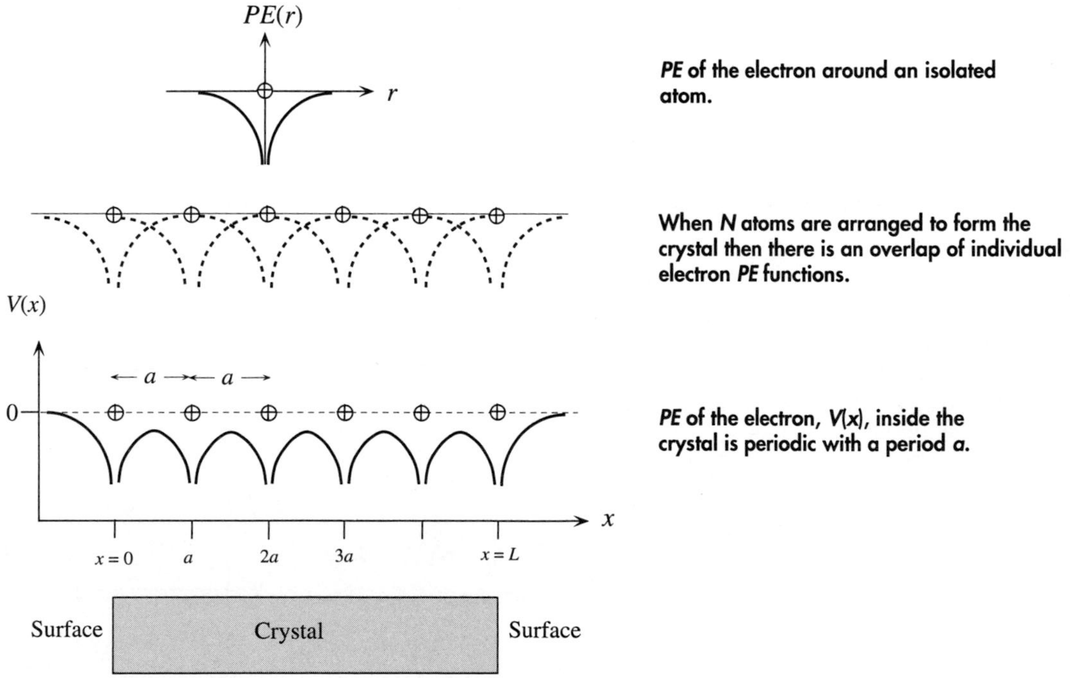


**Figure 5.48** The electron potential energy (*PE*), *V*(*x*), inside the crystal is periodic with the same periodicity *a* as that of the crystal. Far away outside the crystal, by choice, $V = 0$ (the electron is free and $PE = 0$).

and so on. Our task is therefore to solve the Schrödinger equation

*Schrödinger equation*

$$\frac{d^2\psi}{dx^2} + \frac{2m_e}{\hbar^2}[E - V(x)]\psi = 0 \quad [5.72]$$

subject to the condition that the potential energy $V(x)$ is periodic in $a$, that is,

*Periodic potential*

$$V(x) = V(x + ma) \qquad m = 1, 2, 3, \ldots \quad [5.73]$$

The solution of Equation 5.72 will give the electron wavefunction in the crystal and hence the electron energy. Since $V(x)$ is periodic, we should expect, by intuition at least, the solution $\psi(x)$ to be periodic. It turns out that the solutions to Equation 5.72, which are called **Bloch wavefunctions,** are of the form

*Bloch wavefunction*

$$\psi_k(x) = U_k(x)\exp(jkx) \quad [5.74]$$

where $U_k(x)$ is a periodic function that depends on $V(x)$ and has the same periodicity $a$ as $V(x)$. The term $\exp(jkx)$, of course, represents a traveling wave. We should remember that we have to multiply this by $\exp(-jEt/\hbar)$, where $E$ is the energy, to get the overall wavefunction $\Psi(x, t)$. Thus the electron wavefunction in the crystal is a traveling wave that is modulated by $U_k(x)$.

There are many such Bloch wavefunction solutions to the one-dimensional crystal, each identified with a particular $k$ value, say $k_n$, which acts as a kind of quantum number. Each $\psi_k(x)$ solution corresponds to a particular $k_n$ and represents a state with an energy $E_k$. The dependence of the energy $E_k$ on the wavevector $k$ is what we call the $E$–$k$ diagram. Figure 5.49 shows a typical $E$–$k$ diagram for the hypothetical one-dimensional solid for $k$ values in the range $-\pi/a$ to $+\pi/a$. Just as $\hbar k$ is the momentum of a free electron, $\hbar k$ for the Bloch electron is the momentum involved in its interaction with external fields, for example, those involved in the photon absorption process. Indeed, the rate of change of $\hbar k$ is the externally applied force $F_{ext}$ on the electron such as that due to an electric field ($F_{ext} = e\mathcal{E}$). Thus, for the electron within

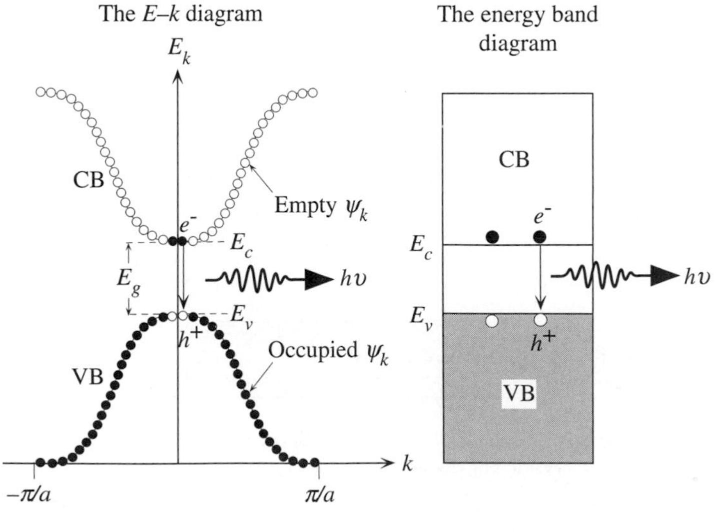


**Figure 5.49** The $E$–$k$ diagram of a direct bandgap semiconductor such as GaAs.
The $E$–$k$ curve consists of many discrete points, each corresponding to a possible state, wavefunction $\psi_k(x)$, that is allowed to exist in the crystal. The points are so close that we normally draw the $E$–$k$ relationship as a continuous curve. In the energy range $E_v$ to $E_c$, there are no points [$\psi_k(x)$ solutions].

the crystal,

$$\frac{d(\hbar k)}{dt} = F_{\text{ext}}$$

and consequently we call $\hbar k$ the **crystal momentum** of the electron.[7]

Inasmuch as the momentum of the electron in the $x$ direction in the crystal is given by $\hbar k$, the $E$–$k$ diagram is an **energy versus crystal momentum plot.** The states $\psi_k(x)$ in the lower $E$–$k$ curve constitute the wavefunctions for the valence electrons and thus correspond to the states in the VB. Those in the upper $E$–$k$ curve, on the other hand, correspond to the states in the conduction band (CB) since they have higher energies. All the valence electrons at absolute zero of temperature therefore fill the states, particular $k_n$ values, in the lower $E$–$k$ diagram.

It should be emphasized that an $E$–$k$ curve consists of many discrete points, each corresponding to a possible state, wavefunction $\psi_k(x)$, that is allowed to exist in the crystal. The points are so close that we draw the $E$–$k$ relationship as a continuous curve. It is clear from the $E$–$k$ diagram that there is a range of energies, from $E_v$ to $E_c$, for which there are no solutions to the Schrödinger equation and hence there are no $\psi_k(x)$ with energies in $E_v$ to $E_c$. Furthermore, we also note that the $E$–$k$ behavior is not a simple parabolic relationship except near the bottom of the CB and the top of the VB.

Above absolute zero of temperature, due to thermal excitation, however, some of the electrons from the top of the valence band will be excited to the bottom of the conduction band. According to the $E$–$k$ diagram in Figure 5.49, when an electron and hole recombine, the electron simply drops from the bottom of the CB to the top of the VB without any change in its $k$ value, so this transition is quite acceptable in terms of momentum conservation. We should recall that the momentum of the emitted photon is negligible compared with the momentum of the electron. The $E$–$k$ diagram in Figure 5.49 is therefore for a **direct bandgap semiconductor.**

The simple $E$–$k$ diagram sketched in Figure 5.49 is for the hypothetical one-dimensional crystal in which each atom simply bonds with two neighbors. In real crystals, we have a three-dimensional arrangement of atoms with $V(x, y, z)$ showing periodicity in more than one direction. The $E$–$k$ curves are then not as simple as that in Figure 5.49 and often show unusual features. The $E$–$k$ diagram for GaAs, which is shown in Figure 5.50a, as it turns out, has main features that are quite similar to that sketched in Figure 5.49. GaAs is therefore a direct bandgap semiconductor in which electron–hole pairs can recombine directly and emit a photon. It is quite apparent that light emitting devices use direct bandgap semiconductors to make use of direct recombination.

---

[7] The actual momentum of the electron, however, is not $\hbar k$ because

$$\frac{d(\hbar k)}{dt} \neq F_{\text{external}} + F_{\text{internal}}$$

where $F_{\text{external}} + F_{\text{internal}}$ are all forces acting on the electron. The true momentum $p_e$ satisfies

$$\frac{dp_e}{dt} = F_{\text{external}} + F_{\text{internal}}$$

However, as we are interested in interactions with external forces such as an applied field, we treat $\hbar k$ as if it were the momentum of the electron in the crystal and use the name **crystal momentum.**

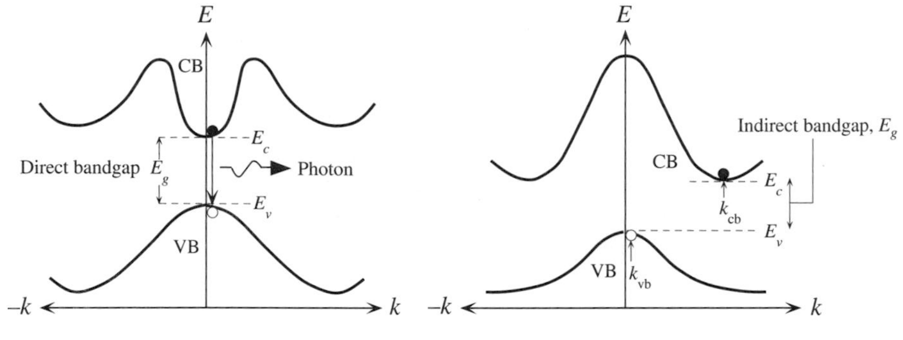


(a) GaAs (b) Si

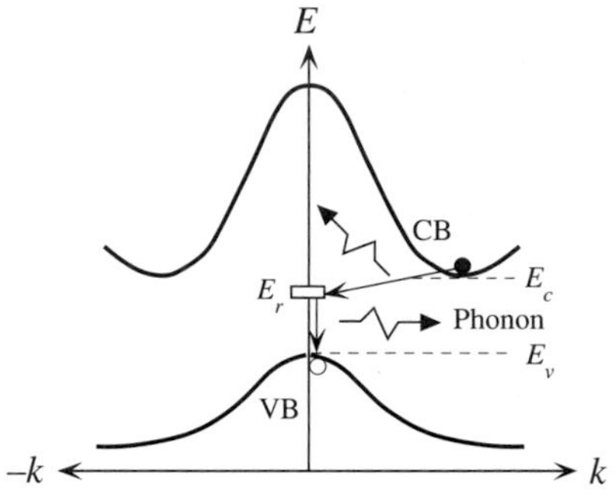


(c) Si with a recombination center

**Figure 5.50**
(a) In GaAs the minimum of the CB is directly above the maximum of the VB. GaAs is therefore a direct bandgap semiconductor.
(b) In Si, the minimum of the CB is displaced from the maximum of the VB and Si is an indirect bandgap semiconductor.
(c) Recombination of an electron and a hole in Si involves a recombination center.

In the case of Si, the diamond crystal structure leads to an $E$–$k$ diagram that has the essential features depicted in Figure 5.50b. We notice that the minimum of the CB is not directly above the maximum of the VB. An electron at the bottom of the CB therefore cannot recombine directly with a hole at the top of the VB because, for the electron to fall down to the top of the VB, its momentum must change from $k_{\text{cb}}$ to $k_{\text{vb}}$, which is not allowed by the law of conservation of momentum. Thus direct electron–hole recombination does not take place in Si and Ge. The recombination process in these elemental semiconductors occurs via a recombination center at an energy level $E_r$. The electron is captured by the defect at $E_r$, from where it can fall down into the top of the VB. The indirect recombination process is illustrated in Figure 5.50c. The energy of the electron is lost by the emission of phonons, that is, lattice vibrations. The $E$–$k$ diagram in Figure 5.50b for Si is an example of an **indirect bandgap semiconductor.**

In some indirect bandgap semiconductors such as GaP, the recombination of the electron with a hole at certain recombination centers results in photon emission. The $E$–$k$ diagram is similar to that shown in Figure 5.50c except that the recombination centers at $E_r$ are generated by the purposeful addition of nitrogen impurities to GaP. The electron transition from $E_r$ to $E_v$ involves photon emission.

***Electron Motion and Drift*** We can understand the response of a conduction band electron to an applied external force, for example, an applied field, by examining the $E$–$k$ diagram. Again, for simplicity, we consider the one-dimensional crystal. The

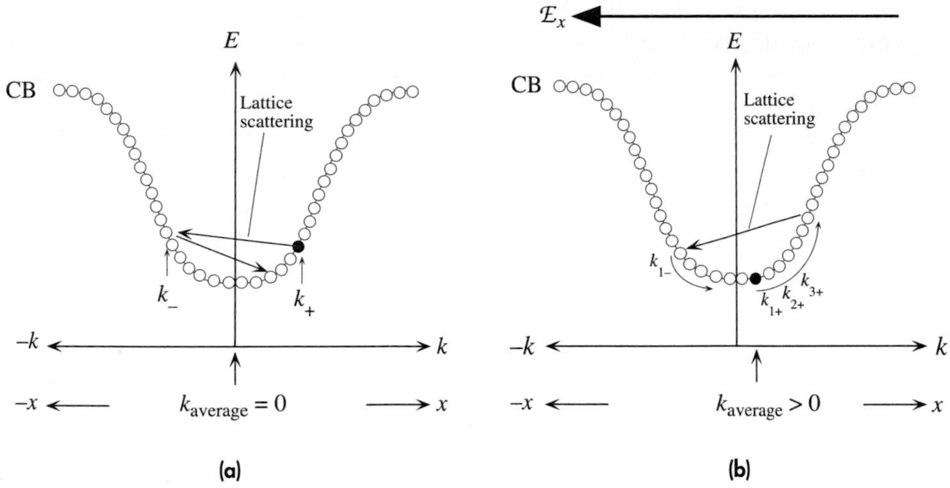


**Figure 5.51**

(a) In the absence of a field, over a long time, the average of all $k$ values is zero; there is no net momentum in any one particular direction.

(b) In the presence of a field in the $-x$ direction, the electron accelerates in the $+x$ direction increasing its $k$ value along $x$ until it is scattered to a random $k$ value. Over a long time, the average of all $k$ values is along the $+x$ direction. Thus the electron drifts along $+x$.

electron is wandering around the crystal quite randomly due to scattering from lattice vibrations. Thus the electron moves with a certain $k$ value in the $+x$ direction, say $k_+$, as illustrated in the $E$–$k$ diagram of Figure 5.51a. When it is scattered by a lattice vibration, its $k$ value changes, perhaps to $k_-$, which is also shown in Figure 5.51a. This process of $k$ changing randomly from one scattering to another scattering process continues all the time, so over a long time the average value of $k$ is zero; that is, average $k_+$ is the same as average $k_-$.

When an electric field is applied, say in the $-x$ direction, then the electron gains momentum in the $+x$ direction from the force of the field $e\mathcal{E}_x$. With time, while the electron is not scattered, it moves up in the $E$–$k$ diagram from $k_{1+}$ to $k_{2+}$ to $k_{3+}$ and so on until a lattice vibration randomly scatters the electron to say $k_{1-}$ (or to some other random $k$ value) as shown in Figure 5.51b. Over a long time, the average of all $k_+$ is no longer equal to the average of all $k_-$ and there is a net momentum in the $+x$ direction, which is tantamount to a drift in the same direction.

***Effective Mass*** The usual definition of inertial mass of a particle in classical physics is based on

$$\text{Force} = \text{Mass} \times \text{Acceleration}$$

$$F = ma$$

When we treat the electron as a wave within the semiconductor crystal, we have to determine whether we can still, in some way, use the convenient classical $F = ma$ relation to describe the motion of an electron under an applied force such as $e\mathcal{E}_x$ and, if so, what the apparent mass of the electron in the crystal should be.

We will evaluate the velocity and acceleration of the electron in the CB in response to an electric field $\mathcal{E}_x$ along $-x$ that imposes an external force $F_{\text{ext}} = e\mathcal{E}_x$ in the $+x$ direction, as shown in Figure 5.51b. Our treatment will make use of the quantum mechanical $E$–$k$ diagram.

Since we are treating the electron as a wave, we have to evaluate the group velocity $v_g$, which, by definition, is $v_g = d\omega/dk$. We know that the time dependence of the wavefunction is $\exp(-jEt/\hbar)$ where the energy $E = \hbar\omega$ ($\omega$ is an "angular frequency" associated with the wave motion of the electron). Both $E$ and $\omega$ depend on $k$. Thus, the group velocity is

*Electron's group velocity*

$$v_g = \frac{1}{\hbar}\frac{dE}{dk} \qquad [5.75]$$

Thus the group velocity is determined by the **gradient** of the $E$–$k$ curve. In the presence of an electric field, the electron experiences a force $F_{\text{ext}} = e\mathcal{E}_x$ from which it gains energy and moves up in the $E$–$k$ diagram until, later on, it collides with a lattice vibration, as shown in Figure 5.51b. During a small time interval $\delta t$ between collisions, the electron moves a distance $v_g\,\delta t$ and hence gains energy $\delta E$, which is

$$\delta E = F_{\text{ext}} v_g\,\delta t \qquad [5.76]$$

To find the acceleration of the electron and the effective mass, we somehow have to put this equation into a form that looks like $F_{\text{ext}} = m_e a$, where $a$ is the acceleration. From Equation 5.76, the relationship between the external force and energy is

$$F_{\text{ext}} = \frac{1}{v_g}\frac{dE}{dt} = \hbar\frac{dk}{dt} \qquad [5.77]$$

where we used Equation 5.75 for $v_g$ in Equation 5.76. Equation 5.77 is the reason for interpreting $\hbar k$ as the **crystal momentum** inasmuch as the rate of change of $\hbar k$ is the externally applied force.

The acceleration $a$ is defined as $dv_g/dt$. We can use Equation 5.75,

$$a = \frac{dv_g}{dt} = \frac{d\left[\dfrac{1}{\hbar}\dfrac{dE}{dk}\right]}{dt} = \frac{1}{\hbar}\frac{d^2E}{dk^2}\frac{dk}{dt} \qquad [5.78]$$

From Equation 5.78, we can substitute for $dk/dt$ in Equation 5.77, which is then a relationship between $F_{\text{ext}}$ and $a$ of the form

*External force and acceleration*

$$F_{\text{ext}} = \frac{\hbar^2}{\left[\dfrac{d^2E}{dk^2}\right]}a \qquad [5.79]$$

We know that the response of a free electron to the external force is $F_{\text{ext}} = m_e a$, where $m_e$ is its mass in vacuum. Therefore it is quite clear from Equation 5.79 that the **effective mass** of the electron in the crystal is

*Effective mass*

$$m_e^* = \hbar^2\left[\frac{d^2E}{dk^2}\right]^{-1} \qquad [5.80]$$

Thus, the electron responds to an external force and moves as if its mass were given by Equation 5.80. The effective mass obviously depends on the $E$–$k$ relationship, which in turn depends on the crystal symmetry and the nature of bonding between the atoms. Its value is different for electrons in the CB and for those in the VB, and moreover, it depends on the energy of the electron since it is related to the curvature of the $E$–$k$ behavior $(d^2E/dk^2)$. Further, it is clear from Equation 5.80 that the effective mass is a quantum mechanical quantity inasmuch as the $E$–$k$ behavior is a direct consequence of the application of quantum mechanics (the Schrödinger equation) to the electron in the crystal.

It is interesting that, according to Equation 5.80, when the $E$–$k$ curve is a downward concave as at the top of a band (*e.g.*, Figure 5.49), the effective mass of an electron at these energies in a band is then negative. What does a negative effective mass mean? When the electron moves up on the $E$–$k$ curve by gaining energy from the field, it actually decelerates, that is, moves more slowly. Its acceleration is therefore in the opposite direction to an electron at the bottom of the band. Electrons in the CB are at the bottom of a band, so their effective masses are positive quantities. At the top of a valence band, however, we have plenty of electrons. These electrons have negative effective masses and under the action of a field, they decelerate. Put differently, they accelerate in the opposite direction to the applied external force $F_{ext}$. It turns out that we can describe the collective motion of these electrons near the top of a band by considering the motion of a few holes with positive masses.

It should be mentioned that Equation 5.80 defines the meaning of the effective mass in quantum mechanical terms. Its usefulness as a concept lies in the fact that we can measure it experimentally, for example, by cyclotron resonance experiments, and have actual values for it. This means we can simply replace $m_e$ by $m_e^*$ in equations that describe the effect of an external force on electron transport in semiconductors.

***Holes*** To understand the concept of a hole, we consider the $E$–$k$ curve corresponding to energies in the VB, as shown in Figure 5.52a. If all the states are filled, then there are no empty states for the electrons to move into and consequently an electron cannot gain energy from the field. For each electron moving in the positive $x$ direction with a momentum $\hbar k_+$, there is a corresponding electron with an equal and opposite momentum $\hbar k_-$, so there is no net motion. For example, the electron at $b$ is moving toward the

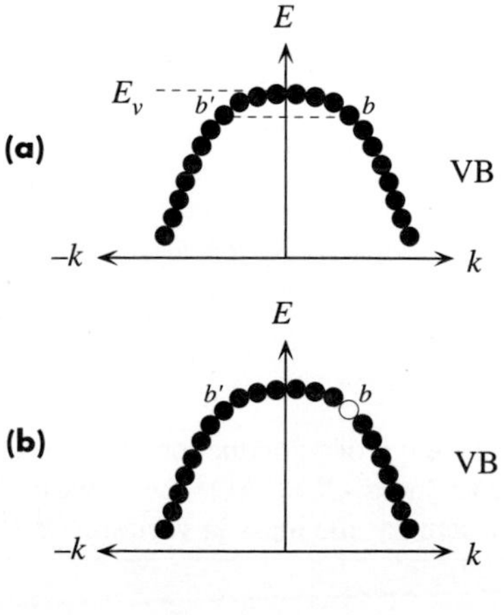


**Figure 5.52**

(a) In a full valence band, there is no net contribution to the current. There are equal numbers of electrons (*e.g.*, at $b$ and $b'$) with opposite momenta.

(b) If there is an empty state (*hole*) at $b$ at the top of the band, then the electron at $b'$ contributes to the current.

right with $k_{+b}$, but its effect is canceled by that at $b'$ moving toward the left with $k_{-b'}$. This cancellation of momenta by electron pairs applies to all the electrons since the VB is assumed to be full. Thus, a full VB cannot contribute to the electric current.

Suppose that one of the states, labeled as $b$ in Figure 5.52b, near the top of the valence band has a missing electron, or a hole, because the electron normally at $b$ has been removed by some means of excitation to the conduction band. It is immediately obvious that the motion of the electron at $b'$ toward the left, that is, $k_{-b'}$, is now *not* canceled, which means that this electron makes a net contribution to the current. We realize that the reason the presence of a hole makes conduction possible is the fact that the momenta of all the VB electrons are canceled except that at $b'$. It is also clear that in reaching this conclusion, we had to consider all the electrons in the valence band.

Let us maintain strict sign rules so that quantities such as the field ($\mathcal{E}_x$), group velocity ($v_g$), and acceleration ($a$) along the $+x$ direction are positive and those along the $-x$ direction are negative. If $\mathcal{E}_x$ is along the $+x$ direction, then the acceleration of a *free* electron from force/mass is $[(-e)(\mathcal{E}_x)]/m_e$, which is negative and along $-x$ as we expect. Similarly, an electron at the bottom of the CB has a positive effective mass and an acceleration that is also negative. Our treatment of conduction in metals by electrons in Chapter 2 inherently assumed that electrons accelerated in the opposite direction to the applied field, that is, positive effective mass.

However, the electrons at the top of the VB have a negative effective mass, which we can write as $-|m_e^*|$. The acceleration $a$ of the electron at $b'$ contributing to the current is

$$a = \frac{-e\mathcal{E}_x}{-|m_e^*|} = \frac{+e\mathcal{E}_x}{+|m_e^*|}$$

which is positive, $a$ along $\mathcal{E}_x$. This means that the acceleration of an electron with a negative effective mass at the top of a VB is equivalent to the acceleration of a positive charge $+e$ with an effective mass $|m_e^*|$. Put differently, we therefore can equivalently describe current conduction by the motion of the hole alone by assigning to it a positive charge and a positive effective mass.

**EXAMPLE 5.21**

**EFFECTIVE MASS** Show that the effective mass of a free electron is the same as its mass in vacuum.

**SOLUTION**

The expression for the energy of a free electron is

$$E = \frac{(\hbar k)^2}{2m_e}$$

The effective mass, by definition, is given by

$$m_e^* = \hbar^2\left[\frac{d^2E}{dk^2}\right]^{-1}$$

Substituting $E = (\hbar k)^2/2m_e$ we get $m_e^* = m_e$. Since the energy of a conduction electron in a metal, within the nearly free electron model, will also have an energy $E = (\hbar k)^2/2m_e$, we can show that the effective mass of the electron in a metal is the same as the mass in vacuum.

**EXAMPLE 5.22**

**CURRENT DUE TO A MISSING ELECTRON IN THE VB** First, let us consider a completely full valence band that contains, say, $N$ electrons. $N/2$ of these are moving with momentum in the $+x$, and $N/2$ in the $-x$ direction. Suppose that the crystal is unit volume. An electron with charge $-e$ moving with a group velocity $\mathbf{v}_{gi}$ contributes to the current by an amount $-e\mathbf{v}_{gi}$. We can determine the current density $\mathbf{J}_N$ due to the motion of all the electrons ($N$ of them) in the band,

$$\mathbf{J}_N = -e\sum_{i=1}^{N}\mathbf{v}_{gi} = 0$$

$\mathbf{J}_N$ is zero because for each value of $\mathbf{v}_{gi}$, there is a corresponding velocity equal in magnitude but opposite in direction ($b$ and $b'$ in Figure 5.52a). Our conclusion from this is that the contribution to the current density from a full valence band is nil, as we expect.

Suppose now that the $j$th electron is missing ($b$ in Figure 5.52b). The net current density is due to $N - 1$ electrons in the band, so

$$\mathbf{J}_{N-1} = -e\sum_{i=1,i\neq j}^{N}\mathbf{v}_{gi} \qquad [5.81]$$

where the summation is for $i = 1$ to $N$ and $i \neq j$ ($j$th electron is missing). We can write the sum as summation to $N$ including the $j$th electron and minus the missing $j$th electron contribution,

$$\mathbf{J}_{N-1} = -e\sum_{i=1}^{N}\mathbf{v}_{gi} - (-e\mathbf{v}_{gj})$$

that is,

$$\mathbf{J}_{N-1} = +e\mathbf{v}_{gj} \qquad [5.82]$$

where we used $\mathbf{J}_N = 0$. We see that when there is a missing electron, there is a net current due to that empty state ($j$th). The current appears as the motion of a charge $+e$ with a velocity $\mathbf{v}_{gj}$, where $\mathbf{v}_{gj}$ is the group velocity of the missing electron. In other words, the current is due to the motion of a positive charge $+e$ at the site of the missing electron at $k_j$, which is what we call a hole. One should note that Equation 5.81 describes the current by considering the motions of *all* the $N - 1$ electrons, whereas Equation 5.82 describes the same current by simply considering the missing electron as if it were a positively charged particle ($+e$) moving with a velocity equal to that of the missing electron. Equation 5.82 is the convenient description universally adopted for a valence band containing missing electrons.

## 5.12 INDIRECT RECOMBINATION

We consider the recombination of minority carriers in an extrinsic indirect bandgap semiconductor such as Si or Ge. As an example, we consider the recombination of electrons in a $p$-type semiconductor. In an indirect bandgap semiconductor, the recombination mechanism involves a recombination center, a third body that may be a crystal defect or an impurity, in the recombination process to satisfy the requirements of conservation of momentum. We can view the recombination process as follows. Recombination occurs when an electron is captured by the recombination center at the energy level $E_r$. As soon as the electron is captured, it will recombine with a hole because holes are abundant in a $p$-type semiconductor. In other words, since there are many majority carriers, the limitation on the rate of recombination is the actual capture of the minority carrier by the center. Thus, if $\tau_e$ is the electron recombination time, since the electrons will have to be captured by the centers, $\tau_e$ is given by

$$\tau_e = \frac{1}{S_r N_r v_{\text{th}}} \quad [5.83]$$

where $S_r$ is the capture (or recombination) cross section of the center, $N_r$ is the concentration of centers, and $v_{\text{th}}$ is the mean speed of the electron that you may take as its effective thermal velocity.

Equation 5.83 is valid under small injection conditions, that is, $p_{po} \gg n_p$. There is a more general treatment of indirect recombination called the Shockley–Read statistics of indirect recombination and generation, which is treated in more advanced semiconductor physics textbooks. That theory eventually arrives at Equation 5.83 for low-level injection conditions. We derived Equation 5.83 from a purely physical reasoning.

Gold is frequently added to silicon to aid recombination. It is found that the minority carrier recombination time is inversely proportional to the gold concentration, following Equation 5.83.

## 5.13 AMORPHOUS SEMICONDUCTORS

Up to now we have been dealing with crystalline semiconductors, those crystals that have perfect periodicity and are practically flawless unless purposefully doped for use in device applications. They are used in numerous solid-state devices including large-area solar cells. Today's microprocessor uses a single crystal of silicon that contains millions of transistors; indeed, we are heading for the 1-billion-transistor chip. There are, however, various applications in electronics that require inexpensive large-area devices to be fabricated and hence require a semiconductor material that can be prepared in a large area. In other applications, the semiconductor material is required to be deposited as a film on a flexible substrate for use as a sensor. Best known examples of large-area devices are flat panel displays based on thin-film transistors (TFTs), inexpensive solar cells, photoconductor drums (for printing and photocopying), image sensors, and newly developed X-ray image detectors. Many of these applications typically use hydrogenated amorphous silicon, a-Si:H.

A distinctive property of an electron in a crystalline solid is that its wavefunction is a traveling wave, a Bloch wave, $\psi_k$, as in Equation 5.74. The Bloch wavefunction is a consequence of the periodicity of an electron's potential energy $PE$, $V(x)$, within the crystal. One can view the electron's motion as tunneling through the periodic potential energy hills. The wavefunctions $\psi_k$ form **extended states** because they *extend* throughout the whole crystal. The electron belongs to the whole crystal, and there is an equal probability of finding an electron in any unit cell. The wavevector $k$ in this traveling wave $\psi_k$ acts as a quantum number. There are many discrete $k_n$ values, which form a nearly continuous set of $k$ values (see Figure 5.49). We can describe the interaction of the electron with an external force, or with photons and phonons, by assigning a momentum $\hbar k$ to the electron, which is called the electron's crystal momentum.

The electron's wavefunction $\psi_k$ is frequently scattered by lattice vibrations (or by defects or impurities) from one $k$-value to another, *e.g.*, from $\psi_k$ to $\psi_{k'}$. The scattering of the wavefunction imposes a mean free path $\ell$ on the electron's motion, that is, a mean distance over which a wave can travel without being scattering. Over the distance $\ell$, the wavefunction is coherent, that is, well defined and predictable as a traveling Bloch wave; $\ell$ is also known as the coherence length of the wavefunction. The mobility is determined by the mean free path $\ell$, which at room temperature is typically of the order of several hundreds of mean interatomic separations. The crystal periodicity and the unit cell atomic structure control the types of Bloch wave solutions one can obtain to the Schrödinger equation. The solutions allow the electron energy $E$ to be examined as a function of $k$ (or momentum $\hbar k$) and these $E - k$ diagrams categorize crystalline semiconductors into two classes: direct bandgap (GaAs type) and indirect bandgap (Si type) semiconductors.

**Hydrogenated amorphous** silicon (a-Si:H) is the noncrystalline form of silicon in which the structure has no long-range order but only short-range order; that is, we can only identify the nearest neighbors of a given atom. Each Si atom has four neighbors as in the crystal, but there is no periodicity or long-range order as illustrated in Figure 1.59. Without the hydrogen, pure a-Si would have dangling bonds. In such a structure sometimes a Si atom would not be able to find a fourth neighboring Si atom to bond with and will be left with a dangling bond as in Figure 1.59b. The hydrogen in the structure ($\sim$10 percent) passivates (*i.e.*, neutralizes) the unsatisfied ("dangling") bonds inherent in a noncrystalline structure and so reduces the density of dangling bonds or defects. a-Si:H belongs to a class of solids called **amorphous semiconductors** that do not follow typical crystalline concepts such as Bloch wavefunctions. First, due to the lack of periodicity, we cannot describe the electron as a Bloch wave. Consequently, we cannot use a wavevector $k$, and hence $\hbar k$, to describe the electron's motion. These semiconductors however do have a short-range order and also possess an energy bandgap that separates a conduction band and a valence band. A window glass has a noncrystalline structure but also has a bandgap, which makes it transparent. Photons with energies less than the bandgap energy can pass through the window glass.

The examination of the structure of a-Si:H in Figure 1.59c should make it apparent that the potential energy $V(x)$ of the electron in this noncrystalline structure fluctuates randomly from site to site. In some cases, the local changes in $V(x)$ can be quite strong, forming effective local *PE* wells (obviously finite wells). Such fluctuations in the *PE* within the solid can capture or trap electrons, that is, localize electrons at certain spatial locations. A localized electron will have a wavefunction that resembles the wavefunction in the hydrogen atom, so the probability of finding the electron is localized to the site. Such locations that can trap electrons, give them localized wavefunctions, are called **localized states.** The amorphous structure also has electrons that possess extended wavefunctions; that is, they belong to the whole solid. These extended wavefunctions are distinctly different than those in the crystal because they have very short coherence lengths due to the random potential fluctuations; the electron is scattered from site to site and hence the mean free path is of the order of a few atomic spacings. The extended wavefunction has random phase fluctuations. Figure 5.53 compares localized and extended wavefunctions in an amorphous semiconductor.

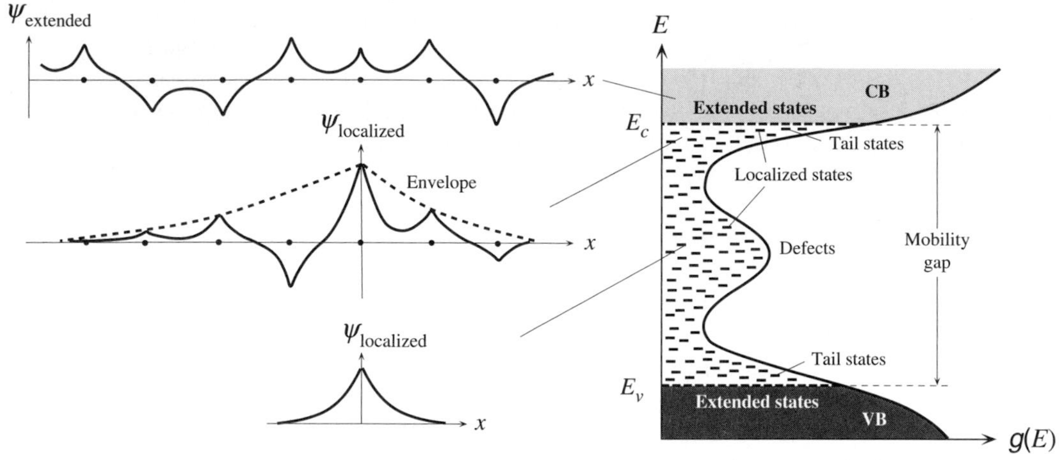


**Figure 5.53** Schematic representation of the density of states $g(E)$ versus energy $E$ for an amorphous semiconductor and the associated electron wavefunctions for an electron in the extended and localized states.

Electronic properties of all amorphous semiconductors can be explained in terms of the energy distribution of their density of states (DOS) function, $g(E)$. The DOS function has well-defined energies $E_v$ and $E_c$ that separate extended states from localized states as in Figure 5.53. There is a distribution of localized states, called **tail states** below $E_c$ and above $E_v$. The usual **bandgap** $E_c - E_v$ is called the **mobility gap.** The reason is that there is a change in the character of charge transport, and hence in the carrier mobility, in going from extended states above $E_c$ to localized states below $E_c$.

Electron transport above $E_c$ in the conduction band is dominated by scattering from random potential fluctuations arising from the disordered nature of the structure. The electrons are scattered so frequently that their effective mobility is much less than what it is in crystalline Si: $\mu_e$ in a-Si:H is typically 5–10 $cm^2\ V^{-1}\ s^{-1}$ whereas it is 1400 $cm^2\ V^{-1}\ s^{-1}$ in a single crystal Si. Electron transport below $E_c$, on the other hand, requires an electron to jump, or hop, from one localized state to another, aided by thermal vibrations of the lattice, in an analogous way to the diffusion of an interstitial impurity in a crystal. We know from Chapter 1 that the jump or diffusion of the impurity is a thermally activated process because it relies on the thermal vibrations of all the crystal atoms to occasionally give the impurity enough energy to make that jump. The electron's mobility associated with this type of hopping motion among localized states is thermally activated, and its value is small. Thus, there is a change in the electron mobility across $E_c$, which is called the conduction band **mobility edge.**

The localized states (frequently simply called *traps*) between $E_v$ and $E_c$ have a profound effect on the overall electronic properties. The tail localized states are a direct result of the structural disorder that is inherent in noncrystalline solids, variations in the bond angles and length. Various prominent peaks and features in the DOS within the mobility gap have been associated with possible structural defects, such as under- and overcoordinated atoms in the structure, dangling bonds, and dopants. Electrons that drift in the conduction band can fall into localized states and become immobilized (trapped) for a while. Thus, electron transport in a-Si:H occurs by multiple trapping in shallow localized states. The effective electron drift mobility in a-Si:H is therefore reduced to ~1 cm$^2$ V$^{-1}$ s$^{-1}$. Low drift mobilities obviously prevent the use of amorphous semiconductor materials in high-speed or high-gain electronic applications. Nonetheless, low-speed electronics is just as important as high-speed electronics in the electronics market in such applications as flat panel displays, solar cells, and image sensors. A low-speed flat panel display made from hydrogenated amorphous silicon (a-Si:H) TFTs costs very roughly the same as a high-speed crystalline Si microchip that runs the CPU.

## CD Selected Topics and Solved Problems

| **Selected Topics** | **Solved Problems** |
|---|---|
| Hall Effect in Semiconductors | Piezoresistance: Pressure Sensors and Strain Gauges |
| Transferred Electron Devices: Gunn Effect | Hall Effect |
| Elements of Photoconductivity | Ionization Region in Doped Semiconductors |
| Thermoelectric Effects in Semiconductors: Voltage Drift in Semiconductor Devices | Compensation Doping of Semiconductors |
| | Electron-Hole Recombination in Semiconductors and Photoconductivity |

## DEFINING TERMS

**Acceptor atoms** are dopants that have one less valency than the host atom. They therefore accept electrons from the VB and thereby create holes in the VB, which leads to a $p > n$ and hence to a *p*-type semiconductor.

**Average energy** of an electron in the CB is $\frac{3}{2}kT$ as if the electrons were obeying Maxwell–Boltzmann statistics. This is only true for a nondegenerate semiconductor.

**Bloch wave** refers to an electron wavefunction of the form $\psi_k = U_k(x)\exp(jkx)$, which is a traveling wave that is modulated by a function $U_k(x)$ that has the periodicity of the crystal. The Bloch wavefunction is a consequence of the periodicity of an electron's potential energy within the crystal.

**Compensated semiconductor** contains both donors and acceptors in the same crystal region that compensate for each other's effects. For example, if there are more donors than acceptors, $N_d > N_a$, then some of the electrons released by donors are captured by acceptors and the net effect is that $N_d - N_a$ number of electrons per unit volume are left in the CB.

**Conduction band** (CB) is a band of energies for the electron in a semiconductor where it can gain energy from an applied field and drift and thereby contribute to electrical conduction. The electron in the CB behaves as if it were a "free" particle with an effective mass $m_e^*$.

**Degenerate semiconductor** has so many dopants that the electron concentration in the CB, or hole concentration in the VB, is comparable with the density of states in the band. Consequently, the Pauli exclusion principle is significant and Fermi–Dirac statistics must be used. The Fermi level is either in the CB for a $n^+$-type degenerate or in the VB for a $p^+$-type degenerate semiconductor. The superscript + indicates a heavily doped semiconductor.

**Diffusion** is a random process by which particles move from high-concentration regions to low-concentration regions.

**Donor atoms** are dopants that have a valency one more than the host atom. They therefore donate electrons to the CB and thereby create electrons in the CB, which leads to $n > p$ and hence to an *n*-type semiconductor.

**Effective density of states ($N_c$) at the CB edge** is a quantity that represents all the states in the CB per unit volume as if they were all at $E_c$. Similarly, $N_v$ at the

VB edge is quantity that represents all the states in the VB per unit volume as if they were all at $E_v$.

**Effective mass** ($m_e^*$) of an electron is a quantum mechanical quantity that behaves like the inertial mass in classical mechanics, $F = ma$, in that it measures the object's inertial resistance to acceleration. It relates the acceleration $a$ of an electron in a crystal to the applied external force $F_{\text{ext}}$ by $F_{\text{ext}} = m_e^* a$. The external force is most commonly the force of an electric field $e\mathcal{E}$ and excludes all internal forces within the crystal.

**Einstein relation** relates the diffusion coefficient $D$ and the drift mobility $\mu$ of a given species of charge carriers through $(D/\mu) = (kT/e)$.

**Electron affinity** ($\chi$) is the energy required to remove an electron from $E_c$ to the vacuum level.

**Energy** of the electron in the crystal, whether in the CB or VB, depends on its momentum $\hbar k$ through the $E$–$k$ behavior determined by the Schrödinger equation. $E$–$k$ behavior is most conveniently represented graphically through $E$–$k$ diagrams. For example, for an electron at the bottom of the CB, $E$ increases as $(\hbar k)^2/m_e^*$ where $\hbar k$ is the momentum and $m_e^*$ is the effective mass of the electron, which is determined from the $E$–$k$ behavior.

**Excess carrier concentration** is the excess concentration *above* the thermal equilibrium value. Excess carriers are generated by an external excitation such as photogeneration.

**Extended state** refers to an electron wavefunction $\psi_k$ whose magnitude does not decay with distance; that is, it is extended in the crystal. An extended wavefunction of an electron in a *crystal* is a **Bloch wave,** that is, $\psi_k = U_k(x)\exp(jkx)$, which is a traveling wave that is modulated by a function $U_k(x)$ that has the periodicity of the crystal. There is an equal probability of finding an electron in any unit cell of the crystal. Scattering of an electron in the crystal by lattice vibrations or impurities, etc., corresponds to the electron being scattered from one $\psi_k$ to another $\psi_{k'}$, *i.e.* a change in the wavevector from $\mathbf{k}$ to $\mathbf{k}'$. Valence and conduction bands in a crystal have extended states.

**Extrinsic semiconductor** is a semiconductor that has been doped so that the concentration of one type of charge carrier far exceeds that of the other. Adding donor impurities releases electrons into the CB and $n$ far exceeds $p$; thus, the semiconductor becomes $n$-type.

**Fermi energy or level** ($E_F$) may be defined in several equivalent ways. The Fermi level is the energy level corresponding to the energy required to remove an electron from the semiconductor; there need not be any actual electrons at this energy level. The energy needed to remove an electron defines the work function $\Phi$. We can define the Fermi level to be $\Phi$ below the vacuum level. $E_F$ can also be defined as that energy value below which all states are full and above which all states are empty at absolute zero of temperature. $E_F$ can also be defined through a difference. A difference in the Fermi energy $\Delta E_F$ in a system is the external electrical work done per electron either on the system or by the system such as electrical work done when a charge $e$ moves through an electrostatic $PE$ difference is $e\Delta V$. It can be viewed as a fundamental material property.

**Intrinsic carrier concentration** ($n_i$) is the electron concentration in the CB of an intrinsic semiconductor. The hole concentration in the VB is equal to the electron concentration.

**Intrinsic semiconductor** has an equal number of electrons and holes due to thermal generation across the bandgap $E_g$. It corresponds to a pure semiconductor crystal in which there are no impurities or crystal defects.

**Ionization energy** is the energy required to ionize an atom, for example, to remove an electron.

**Ionized impurity scattering limited mobility** is the mobility of the electrons when their motion is limited by scattering from the ionized impurities in the semiconductor (*e.g.*, donors and acceptors).

***k*** is the wavevector of the electron's wavefunction. In a crystal the electron wavefunction, $\psi_k(x)$ is a *modulated traveling wave* of the form

$$\psi_k(x) = U_k(x)\exp(jkx)$$

where $k$ is the wavevector and $U_k(x)$ is a periodic function that depends on the $PE$ of interaction between the electron and the lattice atoms. $k$ identifies all possible states $\psi_k(x)$ that are allowed to exist in the crystal. $\hbar k$ is called the *crystal momentum* of the electron as its rate of change is the externally applied force to the electron, $d(\hbar k)/dt = F_{\text{external}}$.

**Lattice-scattering-limited mobility** is the mobility of the electrons when their motion is limited by scattering from thermal vibrations of the lattice atoms.

**Localized state** refers to an electron wavefunction $\psi_{\text{localized}}$ whose magnitude, or the envelope of the wavefunction, decays with distance, which localizes the electron to a spatial region in the semiconductor. For example, a $1s$-type wavefunction of the form $\psi_{\text{localized}} \propto \exp(-\alpha r)$, where $r$ is the distance measured from some center at $r = 0$, and $\alpha$ is a positive constant, would represent a localized state centered at $r = 0$.

**Majority carriers** are electrons in an $n$-type and holes in a $p$-type semiconductor.

**Mass action law** in semiconductor science refers to the law $np = n_i^2$, which is valid under thermal equilibrium conditions and in the absence of external biases and illumination.

**Minority carrier diffusion length** ($L$) is the mean distance a minority carrier diffuses before recombination, $L = \sqrt{D\tau}$, where $D$ is the diffusion coefficient and $\tau$ is the minority carrier lifetime.

**Minority carrier lifetime** ($\tau$) is the mean time for a minority carrier to disappear by recombination. $1/\tau$ is the mean probability per unit time that a minority carrier recombines with a majority carrier.

**Minority carriers** are electrons in a $p$-type and holes in an $n$-type semiconductor.

**Nondegenerate semiconductor** has electrons in the CB and holes in the VB that obey Boltzmann statistics. Put differently, the electron concentration $n$ in the CB is much less than the effective density of states $N_c$ and similarly $p \ll N_v$. It refers to a semiconductor that has not been heavily doped so that these conditions are maintained; typically, doping concentrations are less than $10^{18}$ cm$^{-3}$.

**Ohmic contact** is a contact that can supply charge carriers to a semiconductor at a rate determined by charge transport through the semiconductor and not by the contact properties itself. Thus the current is limited by the conductivity of the semiconductor and not by the contact.

**Peltier effect** is the phenomenon of heat absorption or liberation at the contact between two dissimilar materials as a result of a dc current passing through the junction. The rate of heat generation $Q'$ is proportional to the dc current $I$ passing through the contact so that $Q' = +\Pi I$, where $\Pi$ is called the Peltier coefficient and the sign depends on whether heat is absorbed or released.

**Phonon** is a quantum of energy associated with the vibrations of the atoms in the crystal, analogous to the photon. A phonon has an energy $\hbar\omega$ where $\omega$ is the frequency of the lattice vibration.

**Photoconductivity** is the change in the conductivity from dark to light, $\sigma_{\text{light}} - \sigma_{\text{dark}}$.

**Photogeneration** is the excitation of an electron into the CB by the absorption of a photon. If the photon is absorbed by an electron in the VB, then its excitation to the CB will generate an EHP.

**Photoinjection** is the photogeneration of carriers in the semiconductor by illumination. Photogeneration may be VB to CB excitation, in which case electrons and holes are generated in pairs.

**Piezoresistivity** is the change in the resistivity of a semiconductor due to an applied mechanical stress $\sigma_m$. **Elastoresistivity** refers to the change in the resistivity due to an induced strain in the substance. Application of stress normally leads to strain, so piezoresistivity and elastoresistivity refer to the same phenomenon. In simple terms, the change in the resistivity may be due to a change in the concentration of carriers or due to a change in the drift mobility of the carriers. The fractional change in the resistivity $\delta\rho/\rho$ is proportional to the applied stress $\sigma_m$, and the proportionality constant is called the **piezoresistive coefficient** $\pi$ (1/Pa units), which is a tensor quantity because a stress in one direction in a crystal can alter the resistivity in another direction.

**Recombination of an electron–hole pair** involves an electron in the CB falling down in energy into an empty state (hole) in the VB to occupy it. The result is the annihilation of an EHP. Recombination is direct when the electron falls directly down into an empty state in the VB as in GaAs. Recombination is indirect if the electron is first captured locally by a defect or an impurity, called a recombination center, and from there it falls down into an empty state (hole) in the VB as in Si and Ge.

**Schottky junction** is a contact between a metal and a semiconductor that has rectifying properties. For a metal/$n$-type semiconductor junction, electrons on the metal side have to overcome a potential energy barrier $\Phi_B$ to enter the conduction band of the semiconductor, whereas the conduction electrons in the semiconductor have to overcome a smaller barrier $eV_o$ to enter the metal. Forward bias decreases $eV_o$ and thereby greatly encourages electron emissions over the barrier $e(V_o - V)$. Under reverse bias, electrons have to overcome $\Phi_B$ and the current is very small.

**Thermal equilibrium carrier concentrations** are those electron and hole concentrations that are solely determined by the statistics of the carriers and the density of states in the band. Thermal equilibrium concentrations obey the mass action law, $np = n_i^2$.

**Thermal velocity** ($v_{\text{th}}$) of an electron in the CB is its mean (or effective) speed in the semiconductor as it moves around in the crystal. For a nondegenerate semiconductor, it can be obtained simply from $\frac{1}{2}m_e^* v_{\text{th}}^2 = \frac{3}{2}kT$

**Vacuum level** is the energy level where the *PE* of the electron and the *KE* of the electron are both zero. It defines the energy level where the electron is just free from the solid.

**Valence band (VB)** is a band of energies for the electrons in bonds in a semiconductor. The valence band is made of all those states (wavefunctions) that constitute the bonding between the atoms in the crystal. At absolute zero of temperature, the VB is full of all the bonding electrons of the atoms. When an electron is excited to the CB, this leaves behind an empty state, which is called a hole. It carries a positive charge and behaves as if it were a "free" positively charged entity with an effective mass of $m_h^*$. It moves around the VB by having a neighboring electron tunnel into the unoccupied state.

**Work function** ($\Phi$) is the energy required to remove an electron from the solid to the vacuum level.

# QUESTIONS AND PROBLEMS

**5.1 Bandgap and photodetection**

*a.* Determine the maximum value of the energy gap that a semiconductor, used as a photoconductor, can have if it is to be sensitive to yellow light (600 nm).

*b.* A photodetector whose area is $5 \times 10^{-2}$ cm$^2$ is irradiated with yellow light whose intensity is 2 mW cm$^{-2}$. Assuming that each photon generates one electron–hole pair, calculate the number of pairs generated per second.

*c.* From the known energy gap of the semiconductor GaAs ($E_g = 1.42$ eV), calculate the primary wavelength of photons emitted from this crystal as a result of electron–hole recombination.

*d.* Is the above wavelength visible?

*e.* Will a silicon photodetector be sensitive to the radiation from a GaAs laser? Why?

**5.2 Intrinsic Ge** Using the values of the density of states effective masses $m_e^*$ and $m_h^*$ in Table 5.1, calculate the intrinsic concentration in Ge. What is $n_i$ if you use $N_c$ and $N_v$ from Table 5.1? Calculate the intrinsic resistivity of Ge at 300 K.

**5.3 Fermi level in intrinsic semiconductors** Using the values of the density of states effective masses $m_e^*$ and $m_h^*$ in Table 5.1, find the position of the Fermi energy in intrinsic Si, Ge, and GaAs with respect to the middle of the bandgap ($E_g/2$).

**5.4 Extrinsic Si** A Si crystal has been doped with P. The donor concentration is $10^{15}$ cm$^{-3}$. Find the conductivity and resistivity of the crystal.

**5.5 Extrinsic Si** Find the concentration of acceptors required for a $p$-type Si crystal to have a resistivity of 1 Ω cm.

**5.6 Minimum conductivity**

*a.* Consider the conductivity of a semiconductor, $\sigma = en\mu_e + ep\mu_h$. Will doping always increase the conductivity?

*b.* Show that the minimum conductivity for Si is obtained when it is $p$-type doped such that the hole concentration is

$$p_m = n_i \sqrt{\frac{\mu_e}{\mu_h}}$$

and the corresponding minimum conductivity (maximum resistivity) is

$$\sigma_{\min} = 2en_i\sqrt{\mu_e \mu_h}$$

*c.* Calculate $p_m$ and $\sigma_{\min}$ for Si and compare with intrinsic values.

**5.7** **Extrinsic $p$-Si** A Si crystal is to be doped $p$-type with B acceptors. The hole drift mobility $\mu_h$ depends on the total concentration of ionized dopants $N_{\text{dopant}}$, in this case acceptors only, as

$$\mu_h \approx 54.3 + \frac{407}{1 + 3.745 \times 10^{-18} N_{\text{dopant}}} \quad \text{cm}^2\ \text{V}^{-1}\ \text{s}^{-1}$$

where $N_{\text{dopant}}$ is in cm$^{-3}$. Find the required concentration of B doping for the resistivity to be 0.1 Ω cm.

**5.8** **Thermal velocity and mean free path in GaAs** Given that the electron effective mass $m_e^*$ for the GaAs is $0.067m_e$, calculate the thermal velocity of the conduction band (CB) electrons. The electron drift mobility $\mu_e$ depends on the mean free time $\tau_e$ between electron scattering events (between electrons and lattice vibrations). Given $\mu_e = e\tau_e/m_e^*$, and $\mu_e = 8500$ cm$^2$ V$^{-1}$ s$^{-1}$ for GaAs, calculate $\tau_e$, and hence the mean free path $\ell$ of CB electrons. How many unit cells is $\ell$ if the lattice constant $a$ of GaAs is 0.565 nm? Calculate the drift velocity $v_d = \mu_e \mathcal{E}$ of the CB electrons in an applied field $\mathcal{E}$ of $10^4$ V m$^{-1}$. What is your conclusion?

**5.9** **Compensation doping in Si**

*a.* A Si wafer has been doped $n$-type with $10^{17}$ As atoms cm$^{-3}$.

1. Calculate the conductivity of the sample at 27 °C.
2. Where is the Fermi level in this sample at 27 °C with respect to the Fermi level ($E_{Fi}$) in intrinsic Si?
3. Calculate the conductivity of the sample at 127 °C.

*b.* The above $n$-type Si sample is further doped with $9 \times 10^{16}$ boron atoms ($p$-type dopant) per centimeter cubed.

1. Calculate the conductivity of the sample at 27 °C.
2. Where is the Fermi level in this sample with respect to the Fermi level in the sample in (*a*) at 27 °C? Is this an $n$-type or $p$-type Si?

**5.10** **Temperature dependence of conductivity** An $n$-type Si sample has been doped with $10^{15}$ phosphorus atoms cm$^{-3}$. The donor energy level for P in Si is 0.045 eV below the conduction band edge energy.

*a.* Calculate the room temperature conductivity of the sample.

*b.* Estimate the temperature above which the sample behaves as if intrinsic.

*c.* Estimate to within 20 percent the lowest temperature above which all the donors are ionized.

*d.* Sketch schematically the dependence of the electron concentration in the conduction band on the temperature as log($n$) versus $1/T$, and mark the various important regions and critical temperatures. For each region draw an energy band diagram that clearly shows from where the electrons are excited into the conduction band.

*e.* Sketch schematically the dependence of the conductivity on the temperature as log($\sigma$) versus $1/T$ and mark the various critical temperatures and other relevant information.

***5.11** **Ionization at low temperatures in doped semiconductors** Consider an $n$-type semiconductor. The probability that a donor level $E_d$ is occupied by an electron is

$$f_d = \frac{1}{1 + \dfrac{1}{g}\exp\left(\dfrac{E_d - E_F}{kT}\right)} \qquad [5.84]$$

*Probability of donor occupancy*

where $k$ is the Boltzmann constant, $T$ is the temperature, $E_F$ is the Fermi energy, and $g$ is a constant called the degeneracy factor; in Si, $g = 2$ for donors, and for the occupation statistics of acceptors $g = 4$. Show that

*Electron concentration in extrinsic semiconductors*

$$n^2 + \frac{nN_c}{g\exp\left(\frac{\Delta E}{kT}\right)} - \frac{N_d N_c}{g\exp\left(\frac{\Delta E}{kT}\right)} = 0 \qquad \textbf{[5.85]}$$

where $n$ is the electron concentration in the conduction band, $N_c$ is the effective density of states at the conduction band edge, $N_d$ is the donor concentration, and $\Delta E = E_c - E_d$ is the ionization energy of the donors. Show that Equation 5.85 at low temperatures is equivalent to Equation 5.19. Consider a $p$-type Si sample that has been doped with $10^{15}$ gallium (Ga) atoms cm$^{-3}$. The acceptor energy level for Ga in Si is 0.065 eV above the valence band edge energy, $E_v$. Estimate the lowest temperature (°C) above which 90 percent of the acceptors are ionized by assuming that the acceptor degeneracy factor $g = 4$.

**5.12** **Compensation doping in $n$-type Si** An $n$-type Si sample has been doped with $1 \times 10^{17}$ phosphorus (P) atoms cm$^{-3}$. The drift mobilities of holes and electrons in Si at 300 K depend on the total concentration of dopants $N_{\text{dopant}}$ (cm$^{-3}$) as follows:

*Electron drift mobility*

$$\mu_e \approx 88 + \frac{1252}{1 + 6.984 \times 10^{-18} N_{\text{dopant}}} \quad \text{cm}^2\ \text{V}^{-1}\ \text{s}^{-1}$$

and

*Hole drift mobility*

$$\mu_h \approx 54.3 + \frac{407}{1 + 3.745 \times 10^{-18} N_{\text{dopant}}} \quad \text{cm}^2\ \text{V}^{-1}\ \text{s}^{-1}$$

*a.* Calculate the room temperature conductivity of the sample.

*b.* Calculate the necessary acceptor doping (*i.e.*, $N_a$) that is required to make this sample $p$-type with approximately the same conductivity.

**5.13** **GaAs** Ga has a valency of III and As has V. When Ga and As atoms are brought together to form the GaAs crystal, as depicted in Figure 5.54, the three valence electrons in each Ga and the five valence electrons in each As are all shared to form four covalent bonds per atom. In the GaAs crystal with some $10^{23}$ or so equal numbers of Ga and As atoms, we have an average of four valence electrons per atom, whether Ga or As, so we would expect the bonding to be similar to that in the Si crystal: four bonds per atom. The crystal structure, however, is not that of diamond but rather that of zinc blende (Chapter 1).

*a.* What is the average number of valence electrons per atom for a pair of Ga and As atoms and in the GaAs crystal?

*b.* What will happen if Se or Te, from Group VI, are substituted for an As atom in the GaAs crystal?

*c.* What will happen if Zn or Cd, from Group II, are substituted for a Ga atom in the GaAs crystal?

*d.* What will happen if Si, from Group IV, is substituted for an As atom in the GaAs crystal?

*e.* What will happen if Si, from Group IV, is substituted for a Ga atom in the GaAs crystal? What do you think **amphoteric dopant** means?

*f.* Based on the discussion of GaAs, what do you think the crystal structures of the III–V compound semiconductors AlAs, GaP, InAs, InP, and InSb will be?

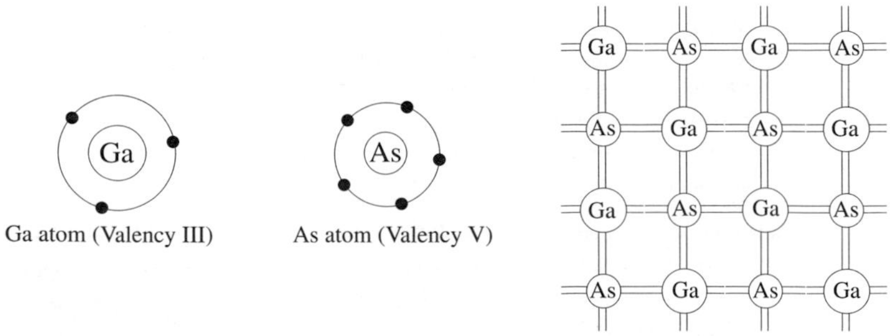


**Figure 5.54** The GaAs crystal structure in two dimensions.

Average number of valence electrons per atom is four. Each Ga atom covalently bonds with four neighboring As atoms and vice versa.

**5.14** **Doped GaAs** Consider the GaAs crystal at 300 K.

*a.* Calculate the intrinsic conductivity and resistivity.

*b.* In a sample containing only $10^{15}$ cm$^{-3}$ ionized donors, where is the Fermi level? What is the conductivity of the sample?

*c.* In a sample containing $10^{15}$ cm$^{-3}$ ionized donors and $9 \times 10^{14}$ cm$^{-3}$ ionized acceptors, what is the free hole concentration?

**5.15** **Varshni equation and the change in the bandgap with temperature** The Varshni equation describes the change in the energy bandgap $E_g$ of a semiconductor with temperature $T$ in terms of

$$E_g = E_{go} - \frac{AT^2}{B+T}$$

*Varshni equation*

where $E_{go}$ is the bandgap at $T = 0$ K, and $A$ and $B$ are material-specific constants. For example, for GaAs, $E_{go} = 1.519$ eV, $A = 5.405 \times 10^{-4}$ eV K$^{-1}$, $B = 204$ K, so that at $T = 300$ K, $E_g = 1.42$ eV. Show that

$$\frac{dE_g}{dT} = -\frac{AT(T+2B)}{(B+T)^2} = -\frac{(E_{go} - E_g)}{T}\left(\frac{T+2B}{T+B}\right)$$

*Bandgap shift with temperature*

What is $dE_g/dT$ for GaAs? The Varshni equation can be used to calculate the shift in the peak emission wavelength of a light emitting diode (LED) with temperature or the cutoff wavelength of a detector. If the emitted photon energy from an electron and hole recombination is $h\nu \approx E_g + kT$, find the shift in the emitted wavelength from 27 °C down to −30 °C from a GaAs LED.

**5.16** **Degenerate semiconductor** Consider the general exponential expression for the concentration of electrons in the CB,

$$n = N_c \exp\left[-\frac{(E_c - E_F)}{kT}\right]$$

and the mass action law, $np = n_i^2$. What happens when the doping level is such that $n$ approaches $N_c$ and exceeds it? Can you still use the above expressions for $n$ and $p$?

Consider an $n$-type Si that has been heavily doped and the electron concentration in the CB is $10^{20}$ cm$^{-3}$. Where is the Fermi level? Can you use $np = n_i^2$ to find the hole concentration? What is its resistivity? How does this compare with a typical metal? What use is such a semiconductor?

**5.17** **Photoconductivity and speed** Consider two $p$-type Si samples both doped with $10^{15}$ B atoms cm$^{-3}$. Both have identical dimensions of length $L$ (1 mm), width $W$ (1 mm), and depth (thickness) $D$ (0.1 mm). One sample, labeled $A$, has an electron lifetime of 1 $\mu$s whereas the other, labeled $B$, has an electron lifetime of 5 $\mu$s.

*a.* At time $t = 0$, a laser light of wavelength 750 nm is switched on to illuminate the surface ($L \times W$) of both the samples. The incident laser light intensity on both samples is 10 mW cm$^{-2}$. At time $t = 50$ $\mu$s, the laser is switched off. Sketch the time evolution of the minority carrier concentration for both samples on the same axes.

*b.* What is the photocurrent (current due to illumination alone) if each sample is connected to a 1 V battery?

***5.18** **Hall effect in semiconductors** The Hall effect in a semiconductor sample involves not only the electron and hole concentrations $n$ and $p$, respectively, but also the electron and hole drift mobilities $\mu_e$ and $\mu_h$. The Hall coefficient of a semiconductor is (see Chapter 2)

$$R_H = \frac{p - nb^2}{e(p+nb)^2} \quad [5.86]$$

*Hall coefficient of a semiconductor*

where $b = \mu_e/\mu_h$.

*a.* Given the mass action law $np = n_i^2$, find $n$ for maximum $|R_H|$ (negative and positive $R_H$). Assume that the drift mobilities remain relatively unaffected as $n$ changes (due to doping). Given the electron and hole drift mobilities $\mu_e = 1350$ cm$^2$ V$^{-1}$ s$^{-1}$ and $\mu_h = 450$ cm$^2$ V$^{-1}$ s$^{-1}$ for silicon, determine $n$ for maximum $|R_H|$ in terms of $n_i$.

*b.* Taking $b = 3$, plot $R_H$ as a function of electron concentration $n/n_i$ from 0.01 to 10.

*c.* Show that, when $n \gg n_i$, $R_H = -1/en$ and when $n \ll n_i$, $R_H = +1/ep$.

**5.19** **Hall effect in semiconductors** Most Hall-effect high-sensitivity sensors typically use III–V semiconductors, such as GaAs, InAs, InSb. Hall-effect integrated circuits with integrated amplifiers, on the other hand, use Si. Consider nearly intrinsic samples in which $n \approx p \approx n_i$, and calculate $R_H$ for each using the data in Table 5.4. What is your conclusion? Which sensor would exhibit the worst temperature drift? (Consider the bandgap, and drift in $n_i$.)

**Table 5.4** Hall effect in selected semiconductors

| | $E_g$(eV) | $n_i$(cm$^{-3}$) | $\mu_e$ (cm$^2$ V$^{-1}$ s$^{-1}$) | $\mu_h$ (cm$^2$ V$^{-1}$ s$^{-1}$) | $b$ | $R_H$ (m$^3$ A$^{-1}$ s$^{-1}$) |
|---|---|---|---|---|---|---|
| Si | 1.10 | $1 \times 10^{10}$ | 1,350 | 450 | 3 | −312 |
| GaAs | 1.42 | $2 \times 10^{6}$ | 8,500 | 400 | ? | ? |
| InAs | 0.36 | $1 \times 10^{15}$ | 33,000 | 460 | ? | ? |
| InSb | 0.17 | $2 \times 10^{16}$ | 78,000 | 850 | ? | ? |

***5.20** **Compound semiconductor devices** Silicon and germanium crystalline semiconductors are what are called elemental Group IV semiconductors. It is possible to have compound semiconductors from atoms in Groups III and V. For example, GaAs is a compound semiconductor that has Ga from Group III and As from Group V, so in the crystalline structure we have an "effective" or "mean" valency of IV per atom and the solid behaves like a semiconductor. Similarly GaSb (gallium antimonide) would be a III–V type semiconductor. Provided we have a stoichiometric compound, the semiconductor will be ideally intrinsic. If, however, there is an excess of Sb atoms in the solid GaSb, then we will have nonstoichiometry and the semiconductor will be extrinsic. In this case, excess Sb atoms will act as donors in the GaSb structure. There are many useful compound semiconductors, the most important of which is GaAs. Some can be doped both $n$- and $p$-type, but many are one type only. For example, ZnO is a II–VI compound semiconductor with a direct bandgap of 3.2 eV, but unfortunately, due to the presence of excess Zn, it is naturally $n$-type and cannot be doped to $p$-type.

*a.* GaSb (gallium antimonide) is an interesting direct bandgap semiconductor with an energy bandgap $E_g = 0.67$ eV, almost equal to that of germanium. It can be used as an light emitting diode (LED) or laser diode material. What would be the wavelength of emission from a GaSb LED? Will this be visible?

*b.* Calculate the intrinsic conductivity of GaSb at 300 K taking $N_c = 2.3 \times 10^{19}$ cm$^{-3}$, $N_v = 6.1 \times 10^{19}$ cm$^{-3}$, $\mu_e = 5000$ cm$^2$ V$^{-1}$ s$^{-1}$, and $\mu_h = 1000$ cm$^2$ V$^{-1}$ s$^{-1}$. Compare with the intrinsic conductivity of Ge.

*c.* Excess Sb atoms will make gallium antimonide nonstoichiometric, that is, GaSb$_{1+\delta}$, which will result in an extrinsic semiconductor. Given that the density of GaSb is 5.4 g cm$^{-3}$, calculate $\delta$ (excess Sb) that will result in GaSb having a conductivity of 100 $\Omega^{-1}$ cm$^{-1}$. Will this be an $n$- or $p$-type semiconductor? You may assume that the drift mobilities are relatively unaffected by the doping.

**5.21** **Excess minority carrier concentration** Consider an $n$-type semiconductor and weak injection conditions. Assume that the minority carrier recombination time $\tau_h$ is constant (independent of injection—hence the weak injection assumption). The rate of change of the instantaneous hole concentration $\partial p_n/\partial t$ due to recombination is given by

*Recombination rate*

$$\frac{\partial p_n}{\partial t} = -\frac{p_n}{\tau_h} \quad [5.87]$$

The net rate of increase (change) in $p_n$ is the sum of the total generation rate $G$ and the rate of change due to recombination, that is,

Excess carries under uniform photogeneration and recombination

$$\frac{dp_n}{dt} = G - \frac{p_n}{\tau_h} \quad [5.88]$$

By separating the generation term $G$ into thermal generation $G_o$ and photogeneration $G_{\text{ph}}$ and considering the dark condition as one possible solution, show that

$$\frac{d\Delta p_n}{dt} = G_{\text{ph}} - \frac{\Delta p_n}{\tau_h} \quad [5.89]$$

How does your derivation compare with Equation 5.27? What are the assumptions inherent in Equation 5.89?

***5.22 Direct recombination and GaAs** Consider recombination in a direct bandgap $p$-type semiconductor, *e.g.*, GaAs doped with an acceptor concentration $N_a$. The recombination involves a direct meeting of an electron–hole pair as depicted in Figure 5.22. Suppose that excess electrons and holes have been injected (*e.g.*, by photoexcitation), and that $\Delta n_p$ is the excess electron concentration and $\Delta p_p$ is the excess hole concentration. Assume $\Delta n_p$ is controlled by recombination and thermal generation only; that is, recombination is the equilibrium storing mechanism. The recombination rate will be proportional to $n_p p_p$, and the thermal generation rate will be proportional to $n_{po} p_{po}$. In the dark, in equilibrium, thermal generation rate is equal to the recombination rate. The latter is proportional to $n_{no} p_{po}$. The rate of change of $\Delta n_p$ is

Recombination rate

$$\frac{\partial \Delta n_p}{\partial t} = -B[n_p p_p - n_{po} p_{po}] \quad [5.90]$$

where $B$ is a proportionality constant, called the **direct recombination capture coefficient.** The **recombination lifetime** $\tau_r$ is *defined* by

Definition of recombination lifetime

$$\frac{\partial \Delta n_p}{\partial t} = -\frac{\Delta n_p}{\tau_r} \quad [5.91]$$

*a.* Show that for *low-level injection*, $n_{po} \ll \Delta n_p \ll p_{po}$, $\tau_r$ is constant and given by

Low injection recombination time

$$\tau_r = \frac{1}{B p_{po}} = \frac{1}{B N_a} \quad [5.92]$$

*b.* Show that under *high-level injection*, $\Delta n_p \gg p_{po}$,

High injection

$$\frac{\partial \Delta n_p}{\partial t} \approx -B \Delta p_p \, \Delta n_p = -B(\Delta n_p)^2 \quad [5.93]$$

so that the recombination lifetime $\tau_r$ is now given by

High-injection recombination time

$$\tau_r = \frac{1}{B \Delta p_p} = \frac{1}{B \Delta n_p} \quad [5.94]$$

that is, the lifetime $\tau_r$ is inversely proportional to the injected carrier concentration.

*c.* Consider what happens in the presence of photogeneration at a rate $G_{\text{ph}}$ (electron–hole pairs per unit volume per unit time). Steady state will be reached when the photogeneration rate and recombination rate become equal. That is,

Steady-state photogeneration rate

$$G_{\text{ph}} = \left(\frac{\partial \Delta n_p}{\partial t}\right)_{\text{recombination}} = B[n_p p_p - n_{po} p_{po}]$$

A photoconductive film of $n$-type GaAs doped with $10^{13}$ cm$^{-3}$ donors is 2 mm long ($L$), 1 mm wide ($W$), and 5 μm thick ($D$). The sample has electrodes attached to its ends (electrode area is therefore 1 mm × 5 μm) which are connected to a 1 V supply through an ammeter. The GaAs photoconductor is uniformly illuminated over the surface area 2 mm × 1 mm with a 1 mW laser radiation of wavelength $\lambda = 840$ nm (infrared). The recombination coefficient $B$ for GaAs is $7.21 \times 10^{-16}$ m$^3$ s$^{-1}$. At $\lambda = 840$ nm, the absorption coefficient is about $5 \times 10^3$ cm$^{-1}$. Calculate the photocurrent $I_{\text{photo}}$ and the electrical power dissipated as Joule heating in the sample. What will be the power dissipated as heat in the sample in an open circuit, where $I = 0$?

**5.23 Piezoresistivity application to deflection and force measurement** Consider the cantilever in Figure 5.38c. Suppose we apply a force $F$ to the free end, which results in a deflection $h$ of the tip of the cantilever from its horizontal equilibrium position. The maximum stress $\sigma_m$ is induced at the support end of the cantilever, at its surface where the piezoresistor is embedded to measure the stress. When the cantilever is bent, there is a tensile or longitudinal stress $\sigma_L$ on the surface because the top surface is extended and the bottom surface is contracted. If $L$, $W$, and $D$ are respectively the length, width, and thickness of the cantilever, then the relationships between the force $F$ and deflection $h$, and the maximum stress $\sigma_L$ are

*Cantilever equations*

$$\sigma_L(\max) = \frac{3YDh}{2L^2} \qquad \text{and} \qquad F = \frac{WD^3Y}{4L^3}h$$

where $Y$ is the elastic (Young's) modulus. A particular Si cantilever has a length ($L$) of 500 μm, width ($W$) of 100 μm, and thickness ($D$) of 10 μm. Given $Y = 170$ GPa, and that the piezoresistor embedded in the cantilever is along the [110] direction with $\pi_L \approx 72 \times 10^{-11}$ Pa$^{-1}$, find the percentage change in the resistance, $\Delta R/R$, of the piezoresistor when the deflection is 0.1 μm. What is the force that would give this deflection? (Neglect the transverse stresses on the piezoresistor.) How does the design choice for the length $L$ of the cantilever depend on whether one is interested in measuring the deflection $h$ or the force $F$? (Note: $\sigma_L$ depends on the distance $x$ from the support end; it decreases with $x$. Assume that the length of the piezoresistor is very short compared with $L$ so that $\sigma_L$ does not change significantly along its length.)

**5.24 Schottky junction**

*a.* Consider a Schottky junction diode between Au and $n$-Si, doped with $10^{16}$ donors cm$^{-3}$. The cross-sectional area is 1 mm$^2$. Given the work function of Au as 5.1 eV, what is the theoretical barrier height $\Phi_B$ from the metal to the semiconductor?

*b.* Given that the experimental barrier height $\Phi_B$ is about 0.8 eV, what is the reverse saturation current and the current when there is a forward bias of 0.3 V across the diode? (Use Equation 4.37.)

**5.25 Schottky junction** Consider a Schottky junction diode between Al and $n$-Si, doped with $5 \times 10^{16}$ donors cm$^{-3}$. The cross-sectional area is 1 mm$^2$. Given that the electron affinity $\chi$ of Si is 4.01 eV and the work function of Al is 4.28 eV, what is the theoretical barrier height $\Phi_B$ from the metal to the semiconductor? What is the built-in voltage? If the experimental barrier height $\Phi_B$ is about 0.6 eV, what is the reverse saturation current and the current when there is a forward bias of 0.2 V across the diode? Take $B_e = 110$ A cm$^{-2}$ K$^{-2}$.

**5.26 Schottky and ohmic contacts** Consider an $n$-type Si sample doped with $10^{16}$ donors cm$^{-3}$. The length $L$ is 100 $\mu$m; the cross-sectional area $A$ is 10 μm × 10 μm. The two ends of the sample are labeled as $B$ and $C$. The electron affinity ($\chi$) of Si is 4.01 eV and the work functions $\Phi$ of four potential metals for contacts at $B$ and $C$ are listed in Table 5.5.

**Table 5.5** Work functions in eV

| Cs | Li | Al | Au |
|---|---|---|---|
| 1.8 | 2.5 | 4.25 | 5.0 |

*a.* Ideally, which metals will result in a Schottky contact?

*b.* Ideally, which metals will result in an ohmic contact?

*c.* Sketch the $I$–$V$ characteristics when both $B$ and $C$ are ohmic contacts. What is the relationship between $I$ and $V$?

*d.* Sketch the $I$–$V$ characteristics when $B$ is ohmic and $C$ is a Schottky junction. What is the relationship between $I$ and $V$?

*e.* Sketch the $I$–$V$ characteristics when both $B$ and $C$ are Schottky contacts. What is the relationship between $I$ and $V$?

**5.27** **Peltier effect and electrical contacts** Consider the Schottky junction and the ohmic contact shown in Figures 5.39 and 5.43 between a metal and $n$-type semiconductor.

*a.* Is the Peltier effect similar in both contacts?

*b.* Is the sign in $Q' = \pm\Pi I$ the same for both contacts?

*c.* Which junction would you choose for a thermoelectric cooler? Give reasons.

***5.28** **Peltier coolers and figure of merit (FOM)** Consider the thermoelectric effect shown in Figure 5.45 in which a semiconductor has two contacts at its ends and is conducting an electric current $I$. We assume that the cold junction is at a temperature $T_c$ and the hot junction is at $T_h$ and that there is a temperature difference of $\Delta T = T_h - T_c$ between the two ends of the semiconductor. The current $I$ flowing through the cold junction absorbs Peltier heat at a rate $Q'_P$, given by

$$Q'_P = \Pi I \quad [5.95]$$

where $\Pi$ is the Peltier coefficient for the junction between the metal and semiconductor. The current $I$ flowing through the semiconductor generates heat due to the Joule heating of the semiconductor. The rate of Joule heat generated through the bulk of the semiconductor is

$$Q'_J = \left(\frac{L}{\sigma A}\right) I^2 \quad [5.96]$$

We assume that half of this heat flows to the cold junction.

In addition there is heat flow from the hot to the cold junction through the semiconductor, given by the thermal conduction equation

$$Q'_{TC} = \left(\frac{A\kappa}{L}\right) \Delta T \quad [5.97]$$

The net rate of heat absorption (cooling rate) at the cold junction is then

$$Q'_{\text{net cool}} = Q'_P - \frac{1}{2}Q'_J - Q'_{TC} \quad [5.98]$$

By substituting from Equations 5.95 to 5.97 into Equation 5.98, obtain the net cooling rate in terms of the current $I$. Then by differentiating $Q'_{\text{net cool}}$ with respect to current, show that maximum cooling is

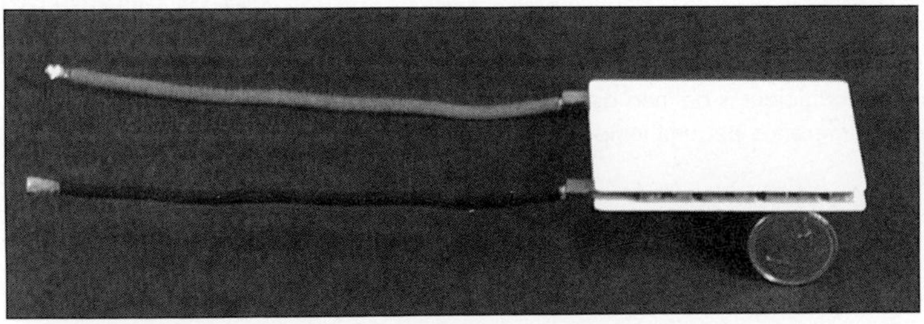

A commercial thermoelectric cooler (by Melcor); an example of the Peltier effect. The device area is 5.5 cm × 5.5 cm (approximately 2.2 inches × 2.2 inches). Its maximum current is 14 A; maximum heat pump ability is 67 W; maximum temperature difference between the hot and cold surfaces is 67 °C.

**Table 5.6**

| Material | $\Pi$ (V) | $\rho$ ($\Omega$ m) | $\kappa$ (W m$^{-1}$K$^{-1}$) | FOM |
|---|---|---|---|---|
| $n$-$Bi_2Te_3$ | $6.0 \times 10^{-2}$ | $10^{-5}$ | 1.70 | |
| $p$-$Bi_2Te_3$ | $7.0 \times 10^{-2}$ | $10^{-5}$ | 1.45 | |
| Cu | $5.5 \times 10^{-4}$ | $1.7 \times 10^{-8}$ | 390 | |
| W | $3.3 \times 10^{-4}$ | $5.5 \times 10^{-8}$ | 167 | |

obtained when the current is

$$I_m = \left(\frac{A}{L}\right)\Pi\sigma \qquad [5.99]$$

and the maximum cooling rate is

*Maximum cooling rate*

$$Q'_{\text{max cool}} = \frac{A}{L}\left[\frac{1}{2}\Pi^2\sigma - \kappa\,\Delta T\right] \qquad [5.100]$$

Under steady-state operating conditions, the temperature difference $\Delta T$ reaches a steady-state value and the net cooling rate at the junction is then zero ($\Delta T$ is constant). From Equation 5.100 show that the maximum temperature difference achievable is

*Maximum temperature difference*

$$\Delta T_{\text{max}} = \frac{1}{2}\frac{\Pi^2\sigma}{\kappa} \qquad [5.101]$$

The quantity $\Pi^2\sigma/\kappa$ is defined as the **figure of merit** (FOM) for the semiconductor as it determines the maximum $\Delta T$ achievable. The same expression also applies to metals, though we will not derive it here.

Use Table 5.6 to determine the FOM for various materials listed therein and discuss the significance of your calculations. Would you recommend a thermoelectric cooler based on a metal-to-metal junction?

***5.29 Seebeck coefficient of semiconductors and thermal drift in semiconductor devices** Consider an $n$-type semiconductor that has a temperature gradient across it. The right end is hot and the left end is cold, as depicted in Figure 5.55. There are more energetic electrons in the hot region than in the cold region. Consequently, electron diffusion occurs from hot to cold regions, which immediately exposes negatively charged donors in the hot region and therefore builds up an internal field and a built-in voltage, as shown in Figure 5.55. Eventually an equilibrium is reached when the diffusion of electrons is balanced by their drift driven by the built-in field. The net current must be zero. The Seebeck coefficient (or thermoelectric power) $S$ measures

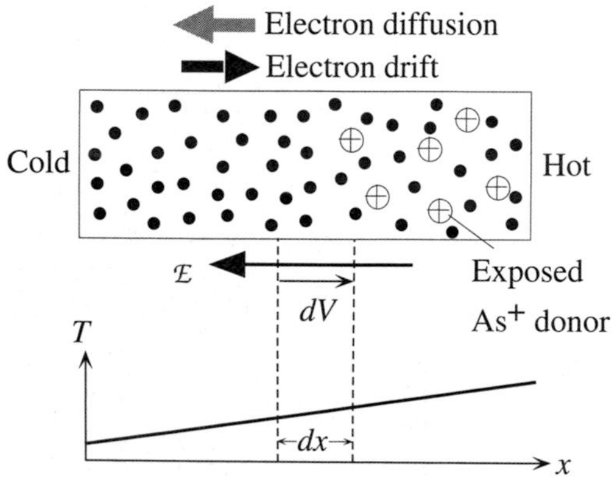


**Figure 5.55** In the presence of a temperature gradient, there is an internal field and a voltage difference.
The Seebeck coefficient is defined as $dV/dT$, the potential difference per unit temperature difference.

this effect in terms of the voltage developed as a result of an applied temperature gradient as

$$S = \frac{dV}{dT} \quad [5.102]$$

*a.* How is the Seebeck effect in a $p$-type semiconductor different than that for an $n$-type semiconductor when both are placed in the same temperature gradient in Figure 5.55? Recall that the sign of the Seebeck coefficient is the polarity of the voltage at the cold end with respect to the hot end (see Section 4.8.2).

*b.* Given that for an $n$-type semiconductor,

*Seebeck coefficient n-type semiconductor*

$$S_n = -\frac{k}{e}\left[2 + \frac{(E_c - E_F)}{kT}\right] \quad [5.103]$$

what are typical magnitudes for $S_n$ in Si doped with $10^{14}$ and $10^{16}$ donors $cm^{-3}$? What is the significance of $S_n$ at the semiconductor device level?

*c.* Consider a $pn$ junction Si device that has the $p$-side doped with $10^{18}$ acceptors $cm^{-3}$ and the $n$-side doped with $10^{14}$ donors $cm^{-3}$. Suppose that this $pn$ junction forms the input stage of an op amp with a large gain, say 100. What will be the output signal if a small thermal fluctuation gives rise to a 1 °C temperature difference across the $pn$ junction?

**5.30 Photogeneration and carrier kinetic energies** Figure 5.35 shows what happens when a photon with energy $h\nu > E_g$ is absorbed in GaAs to photogenerate an electron and a hole. The figure shows that the electron has a higher kinetic energy ($KE$), which is the excess energy above $E_c$, than the hole, since the hole is almost at $E_v$. The reason is that the electron effective mass in GaAs is almost 10 times less than the hole effective mass, so the photogenerated electron has a much higher $KE$. When an electron and hole are photogenerated in a direct bandgap semiconductor, they have the same **k** vector. Energy conservation requires that the photon energy $h\nu$ divides according to

*Photogeneration*

$$h\nu = E_g + \frac{(\hbar k)^2}{2m_e^*} + \frac{(\hbar k)^2}{2m_h^*}$$

where $k$ is the wavevector of the electron and hole and $m_e^*$ and $m_h^*$ are the effective masses of the electron and hole, respectively.

*a.* What is the ratio of the electron to hole $KE$s right after photogeneration?

*b.* If the incoming photon has an energy of 2.0 eV, and $E_g = 1.42$ eV for GaAs, calculate the $KE$s of the electron and the hole in eV, and calculate to which energy levels they have been excited with respect to their band edges.

*c.* Explain why the electron and hole wavevector $k$ should be approximately the same right after photogeneration. Consider $k_{\text{photon}}$ for the photon, and the momentum conservation.

William Shockley and his group celebrate Shockley's Nobel prize in 1956. First left, sitting, is G. E. Moore (chairman emeritus of Intel), standing fourth from right is R. N. Noyce, inventor of the integrated circuit, and standing at the extreme right is J. T. Last.

SOURCE: P. K. Bondyopadhyay, "W = Shockley, the Transistor Pioneer—Portrait of an Inventive Genius," *Proceedings IEEE*, vol. 86, no. 1, January 1998, p. 202, figure 16 (Courtesy of IEEE.)

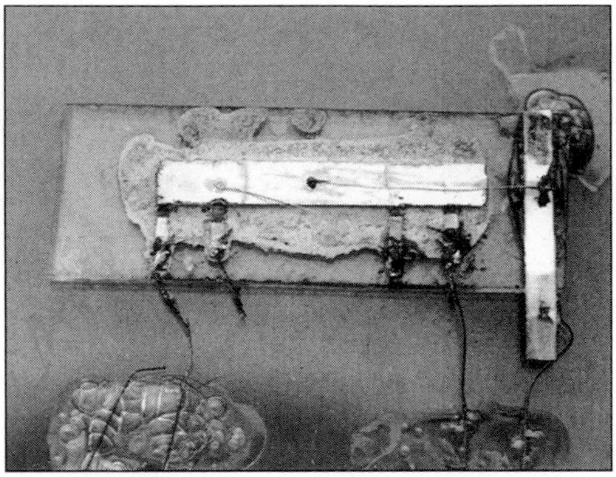

The first monolithic integrated circuit, about the size of a fingertip, was documented and developed at Texas Instruments by Jack Kilby in 1958; he won the 2000 Nobel prize in physics for his contribution to the development of the first integrated circuit. The IC was a chip of a single Ge crystal containing one transistor, one capacitor, and one resistor. Left: Jack Kilby holding his IC (photo, 1998). Right. The photo of the chip.

SOURCE: Courtesy of Texas Instruments.

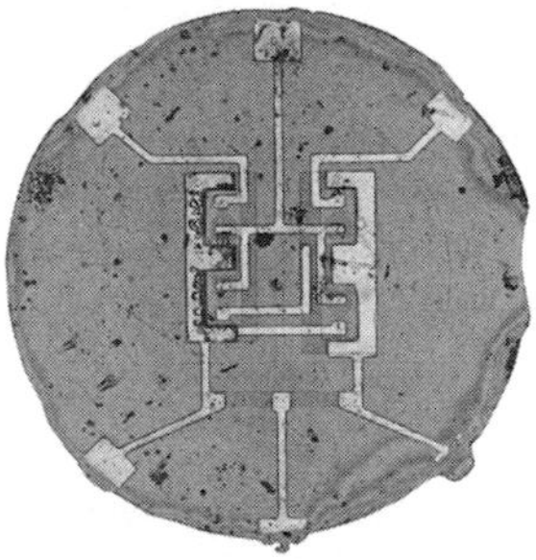

Robert Noyce and Jean Hoerni (a Swiss physicist) were responsible for the invention of the first planar IC at Fairchild (1961). The planar fabrication process was the key to the success of their IC. The photograph is that of the first logic chip at Fairchild.

SOURCE: Courtesy of Fairchild Semiconductor.

Left to right: Andrew Grove, Robert Noyce (1927–1990), and Gordon Moore, who founded Intel in 1968. Andrew Grove's book *Physics and Technology of Semiconductor Devices* (Wiley, 1967) was one of the classic texts on devices in the sixties and seventies. "Moore's law" that started as a rough rule in 1965 states that the number of transistors in a chip will double every 18 months; Moore updated it in 1995 to every couple of years.

SOURCE: Courtesy of Intel.

CHAPTER

# 6

# Semiconductor Devices

Most diodes are essentially *pn* junctions fabricated by forming a contact between a *p*-type and an *n*-type semiconductor. The junction possesses rectifying properties in that a current in one direction can flow quite easily whereas in the other direction it is limited by a leakage current that is generally very small. A transistor is a three-terminal solid-state device in which a current flowing between two electrodes is controlled by the voltage between the third and one of the other terminals. Transistors are capable of providing current and voltage gains thereby enabling weak signals to be amplified. Transistors can also be used as switches just like electromagnetic relays. Indeed, the whole microcomputer industry is based on transistor switches. The majority of the transistors in microelectronics are of essentially two types: **bipolar junction transistors** (BJTs) and **field effect transistors** (FETs). The appreciation of the underlying principles of the *pn* junction is essential to understanding the operation of not only the bipolar transistor but also a variety of related devices. The central fundamental concept is the **minority carrier injection** as purported by William Shockley in his explanations of the transistor operation. Field effect transistors operate on a totally different principle than BJTs. Their characteristics arise from the effect of the applied field on a conducting channel between two terminals. The last two decades have seen enormous advances and developments in optoelectronic and photonic devices which we now take for granted, the best examples being **light emitting diodes** (LEDs), **semiconductor lasers, photodetectors,** and **solar cells.** Nearly all these devices are based on *pn* junction principles. The present chapter takes the semiconductor concepts developed in Chapter 5 to device level applications, from the basic *pn* junction to heterojunction laser diodes.

# 6.1 IDEAL *pn* JUNCTION

## 6.1.1 NO APPLIED BIAS: OPEN CIRCUIT

Consider what happens when one side of a sample of Si is doped $n$-type and the other $p$-type, as shown in Figure 6.1a. We assume that there is an abrupt discontinuity between the $p$- and $n$-regions, which we call the **metallurgical junction** and label as M in Figure 6.1a, where the fixed (immobile) ionized donors and the free electrons (in the conduction band, CB) in the $n$-region and fixed ionized acceptors and holes (in the valence band, VB) in the $p$-region are also shown.

Due to the hole concentration gradient from the $p$-side, where $p = p_{po}$, to the $n$-side, where $p = p_{no}$, holes diffuse toward the right. Similarly the electron concentration

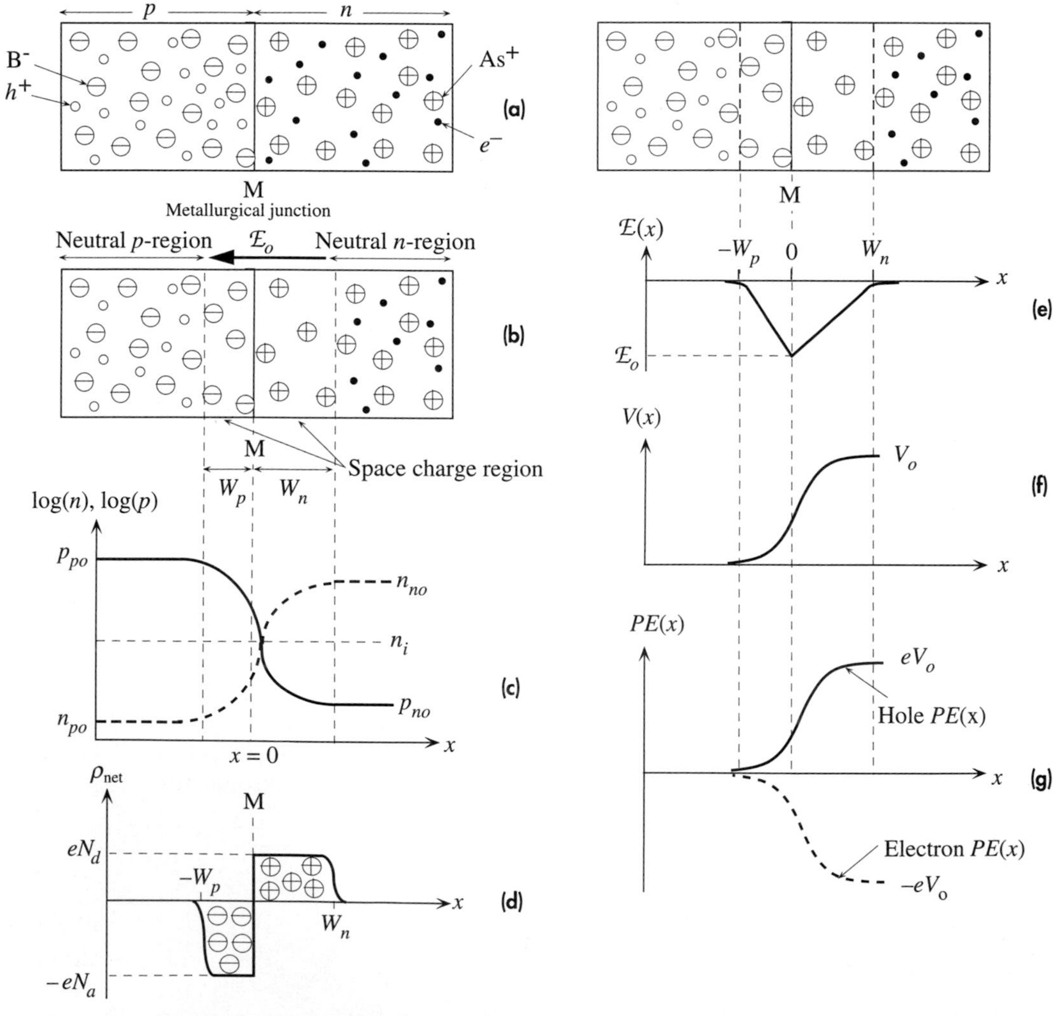


**Figure 6.1** Properties of the *pn* junction.

gradient drives the electrons by diffusion toward the left. Holes diffusing and entering the $n$-side recombine with the electrons in the $n$-side near the junction. Similarly, electrons diffusing and entering the $p$-side recombine with holes in the $p$-side near the junction. The junction region consequently becomes depleted of free carriers in comparison with the bulk $p$- and $n$-regions far away from the junction. Note that we must, under equilibrium conditions (*e.g.*, no applied bias or photoexcitation), have $pn = n_i^2$ everywhere. Electrons leaving the $n$-side near the junction M leave behind exposed positively charged donor ions, say $As^+$, of concentration $N_d$. Similarly, holes leaving the $p$-region near M expose negatively charged acceptor ions, say $B^-$, of concentration $N_a$. There is therefore a **space charge layer** (SCL) around M. Figure 6.1b shows the **depletion region,** or the space charge layer, around M, whereas Figure 6.1c illustrates the hole and electron concentration profiles in which the vertical concentration scale is logarithmic. The depletion region is also called the transition region.

It is clear that there is an internal electric field $\mathcal{E}_o$ from positive ions to negative ions, that is, in the $-x$ direction, that tries to drift the holes back into the $p$-region and electrons back into the $n$-region. This field drives the holes in the opposite direction to their diffusion. As shown in Figure 6.1b, $\mathcal{E}_o$ imposes a drift force on holes in the $-x$ direction, whereas the hole diffusion flux is in the $+x$ direction. A similar situation also applies for electrons with the electric field attempting to drift the electrons against diffusion from the $n$-region to the $p$-region. It is apparent that as more and more holes diffuse toward the right, and electrons toward the left, the internal field around M will increase until eventually an "equilibrium" is reached when the rate of holes diffusing toward the right is just balanced by holes drifting back to the left, driven by the field $\mathcal{E}_o$. The electron diffusion and drift fluxes will also be balanced in equilibrium.

For uniformly doped $p$- and $n$-regions, the net space charge density $\rho_{\text{net}}(x)$ across the semiconductor will be as shown in Figure 6.1d. (Why are the edges rounded?) The net space charge density $\rho_{\text{net}}$ is negative and equal to $-eN_a$ in the SCL from $x = -W_p$ to $x = 0$ (where we take M to be) and then positive and equal to $+eN_d$ from $x = 0$ to $W_n$. The total charge on the left-hand side must be equal to that on the right-hand side for overall charge neutrality, so

*Depletion widths*

$$N_a W_p = N_d W_n \qquad [6.1]$$

In Figure 6.1, we arbitrarily assumed that the donor concentration is less than the acceptor concentration, $N_d < N_a$. From Equation 6.1 this implies that $W_n > W_p$; that is, the depletion region penetrates the $n$-side, the lightly doped side, more than the $p$-side, the heavily doped side. Indeed, if $N_a \gg N_d$, then the depletion region is almost entirely on the $n$-side. We generally indicate heavily doped regions with the plus sign as a superscript, that is, $p^+$.

The electric field $\mathcal{E}(x)$ and the net space charge density $\rho_{\text{net}}(x)$ at a point are related in electrostatics[1] by

*Field and net space charge density*

$$\frac{d\mathcal{E}}{dx} = \frac{\rho_{\text{net}}(x)}{\varepsilon}$$

[1] This is called **Gauss's law in point** form and comes from Gauss's law in electrostatics. Gauss's law is discussed in Section 7.5.

where $\varepsilon = \varepsilon_o\varepsilon_r$ is the permittivity of the medium and $\varepsilon_o$ and $\varepsilon_r$ are the absolute permittivity and relative permittivity of the semiconductor material. We can thus integrate $\rho_{\text{net}}(x)$ across the diode and thus determine the electric field $\mathcal{E}(x)$, that is,

*Field in depletion region*

$$\mathcal{E}(x) = \frac{1}{\varepsilon}\int_{-W_p}^{x} \rho_{\text{net}}(x)\,dx \qquad [6.2]$$

The variation of the electric field across the *pn* junction is shown in Figure 6.1e. The negative field means that it is in the $-x$ direction. Note that $\mathcal{E}(x)$ reaches a maximum value $\mathcal{E}_o$ at the metallurgical junction M.

The potential $V(x)$ at any point $x$ can be found by integrating the electric field since by definition $\mathcal{E} = -dV/dx$. Taking the potential on the $p$-side far away from M as zero (we have no applied voltage), which is an arbitrary reference level, then $V(x)$ increases in the depletion region toward the $n$-side, as indicated in Figure 6.1f. Its functional form can be determined by integrating Equation 6.2, which is, of course, a parabola. Notice that on the $n$-side the potential reaches $V_o$, which is called the **built-in potential.**

The fact that we are considering an abrupt *pn* junction means that $\rho_{\text{net}}(x)$ can simply be described by step functions, as displayed in Figure 6.1d. Using the step form of $\rho_{\text{net}}(x)$ in Figure 6.1d in the integration of Equation 6.2 gives the electric field at M as

*Built-in field*

$$\mathcal{E}_o = -\frac{eN_dW_n}{\varepsilon} = -\frac{eN_aW_p}{\varepsilon} \qquad [6.3]$$

where $\varepsilon = \varepsilon_o\varepsilon_r$. We can integrate the expression for $\mathcal{E}(x)$ in Figure 6.1e to evaluate the potential $V(x)$ and thus find $V_o$ by putting in $x = W_n$. The graphical representation of this integration is the step from Figure 6.1e to f. The result is

*Built-in voltage*

$$V_o = -\frac{1}{2}\mathcal{E}_oW_o = \frac{eN_aN_dW_o^2}{2\varepsilon(N_a + N_d)} \qquad [6.4]$$

where $W_o = W_n + W_p$ is the total width of the depletion region under a zero applied voltage. If we know $W_o$, then $W_n$ or $W_p$ follows readily from Equation 6.1. Equation 6.4 is a relationship between the built-in voltage $V_o$ and the depletion region width $W_o$. If we know $V_o$, we can calculate $W_o$.

The simplest way to relate $V_o$ to the doping parameters is to make use of the fact that in the system consisting of $p$- and $n$-type semiconductors joined together, in equilibrium, Boltzmann statistics[2] demands that the concentrations $n_1$ and $n_2$ of carriers at potential energies $E_1$ and $E_2$ are related by

$$\frac{n_2}{n_1} = \exp\left[-\frac{(E_2 - E_1)}{kT}\right]$$

where $E = qV$, where $q$ is the charge of the carrier. Considering electrons ($q = -e$), we see from Figure 6.1g that $E = 0$ on the $p$-side far away from M where $n = n_{po}$, and

[2] We use Boltzmann statistics, that is, $n(E) \propto \exp(-E/kT)$, because the concentration of electrons in the conduction band, whether on the $n$-side or $p$-side, is never so large that the Pauli exclusion principle becomes important. As long as the carrier concentration in the conduction band is much smaller than $N_c$, we can use Boltzmann statistics.

$E = -eV_o$ on the $n$-side away from M where $n = n_{no}$. Thus

$$\frac{n_{po}}{n_{no}} = \exp\left(-\frac{eV_o}{kT}\right) \quad [6.5a]$$

*Boltzmann statistics for electrons*

This shows that $V_o$ depends on $n_{no}$ and $n_{po}$ and hence on $N_d$ and $N_a$. The corresponding equation for hole concentrations is clearly

$$\frac{p_{no}}{p_{po}} = \exp\left(-\frac{eV_o}{kT}\right) \quad [6.5b]$$

Thus, rearranging Equations 6.5a and b we obtain

$$V_o = \frac{kT}{e}\ln\left(\frac{n_{no}}{n_{po}}\right) \quad \text{and} \quad V_o = \frac{kT}{e}\ln\left(\frac{p_{po}}{p_{no}}\right)$$

We can now write $p_{po}$ and $p_{no}$ in terms of the dopant concentrations inasmuch as $p_{po} = N_a$ and

$$p_{no} = \frac{n_i^2}{n_{no}} = \frac{n_i^2}{N_d}$$

so $V_o$ becomes

$$V_o = \frac{kT}{e}\ln\left(\frac{N_a N_d}{n_i^2}\right) \quad [6.6]$$

*Built-in voltage*

Clearly $V_o$ has been conveniently related to the dopant and material properties via $N_a$, $N_d$, and $n_i^2$. The built-in voltage ($V_o$) is the voltage across a *pn* junction, going from $p$- to $n$-type semiconductor, in an open circuit. It is *not* the voltage across the diode, which is made up of $V_o$ as well as the contact potentials at the metal-to-semiconductor junctions at the electrodes. If we add $V_o$ and the contact potentials at the electroded ends, we will find zero.

Once we know the built-in potential from Equation 6.6, we can then calculate the width of the depletion region from Equation 6.4, namely

$$W_o = \left[\frac{2\varepsilon(N_a + N_d)V_o}{eN_a N_d}\right]^{1/2} \quad [6.7]$$

*Depletion region width*

Notice that the depletion width $W_o \propto V_o^{1/2}$. This results in the capacitance of the depletion region being voltage dependent, as we will see in Section 6.3.

**EXAMPLE 6.1**

**THE BUILT-IN POTENTIALS FOR Ge, Si, AND GaAs *pn* JUNCTIONS** A *pn* junction diode has a concentration of $10^{16}$ acceptor atoms cm$^{-3}$ on the $p$-side and a concentration of $10^{17}$ donor atoms cm$^{-3}$ on the $n$-side. What will be the built-in potential for the semiconductor materials Ge, Si, and GaAs?

**SOLUTION**

The built-in potential is given by Equation 6.6, which requires the knowledge of the intrinsic concentration for each semiconductor. From Chapter 5 we can tabulate the following at 300 K:

| Semiconductor | $E_g$ (eV) | $n_i$ ($cm^{-3}$) | $V_o$ (V) |
|---|---|---|---|
| Ge | 0.7 | $2.40 \times 10^{13}$ | 0.37 |
| Si | 1.1 | $1.0 \times 10^{10}$ | 0.78 |
| GaAs | 1.4 | $2.1 \times 10^{6}$ | 1.21 |

Using

$$V_o = \left(\frac{kT}{e}\right) \ln\left(\frac{N_d N_a}{n_i^2}\right)$$

for Si with $N_d = 10^{17}$ $cm^{-3}$ and $N_a = 10^{16}$ $cm^{-3}$, $kT/e = 0.0259$ V at 300 K, and $n_i = 1.0 \times 10^{10}$ $cm^{-3}$, we obtain

$$V_o = (0.0259 \text{ V}) \ln\left[\frac{(10^{17})(10^{16})}{(1.0 \times 10^{10})^2}\right] = 0.775 \text{ V}$$

The results for all three semiconductors are summarized in the last column of the table in this example.

**EXAMPLE 6.2** **THE $p^+n$ JUNCTION** Consider a $p^+n$ junction, which has a heavily doped $p$-side relative to the $n$-side, that is, $N_a \gg N_d$. Since the amount of charge $Q$ on both sides of the metallurgical junction must be the same (so that the junction is overall neutral)

$$Q = eN_aW_p = eN_dW_n$$

it is clear that the depletion region essentially extends into the $n$-side. According to Equation 6.7, when $N_d \ll N_a$, the width is

$$W_o = \left[\frac{2\varepsilon V_o}{eN_d}\right]^{1/2}$$

What is the depletion width for a $pn$ junction Si diode that has been doped with $10^{18}$ acceptor atoms $cm^{-3}$ on the $p$-side and $10^{16}$ donor atoms $cm^{-3}$ on the $n$-side?

**SOLUTION**

To apply the above equation for $W_o$, we need the built-in potential, which is

$$V_o = \left(\frac{kT}{e}\right) \ln\left(\frac{N_d N_a}{n_i^2}\right) = (0.0259 \text{ V}) \ln\left[\frac{(10^{16})(10^{18})}{(1.0 \times 10^{10})^2}\right] = 0.835 \text{ V}$$

Then with $N_d = 10^{16}$ $cm^{-3}$, that is, $10^{22}$ $m^{-3}$, $V_o = 0.835$ V, and $\varepsilon_r = 11.9$ in the equation for $W_o$

$$W_o = \left[\frac{2\varepsilon V_o}{eN_d}\right]^{1/2} = \left[\frac{2(11.9)(8.85 \times 10^{-12})(0.835)}{(1.6 \times 10^{-19})(10^{22})}\right]^{1/2}$$

$$= 3.32 \times 10^{-7} \text{ m} \quad \text{or} \quad 0.33 \text{ μm}$$

Nearly all of this region (99 percent of it) is on the $n$-side.

EXAMPLE 6.3

**BUILT-IN VOLTAGE** There is a rigorous derivation of the built-in voltage across a *pn* junction. Inasmuch as in equilibrium there is no net current through the *pn* junction, drift of holes due to the built-in field $\mathcal{E}(x)$ must be just balanced by their diffusion due to the concentration gradient $dp/dx$. We can thus set the total electron and hole current densities (drift + diffusion) through the depletion region to zero. Considering holes alone, from Equation 5.38,

$$J_{\text{hole}}(x) = ep(x)\mu_h\mathcal{E}(x) - eD_h\frac{dp}{dx} = 0$$

The electric field is defined by $\mathcal{E} = -dV/dx$, so substituting we find,

$$-ep\mu_h\, dV - eD_h\, dp = 0$$

We can now use the *Einstein relation* $D_h/\mu_h = kT/e$ to get

$$-ep\, dV - kT\, dp = 0$$

We can integrate this equation. According to Figure 6.1, in the *p*-side, $p = p_{po}$, $V = 0$, and in the *n*-side, $p = p_{no}$, $V = V_o$, thus,

$$\int_0^{V_o} dV + \frac{kT}{e}\int_{p_{po}}^{p_{no}} \frac{dp}{p} = 0$$

that is,

$$V_o + \frac{kT}{e}[\ln(p_{no}) - \ln(p_{po})] = 0$$

giving

$$V_o = \frac{kT}{e}\ln\left(\frac{p_{po}}{p_{no}}\right)$$

which is the same as Equation 6.5b and hence leads to Equation 6.6.

## 6.1.2 FORWARD BIAS: DIFFUSION CURRENT

Consider what happens when a battery is connected across a *pn* junction so that the positive terminal of the battery is attached to the *p*-side and the negative terminal to the *n*-side. Suppose that the applied voltage is *V*. It is apparent that the negative polarity of the supply will reduce the potential barrier $V_o$ by *V*, as shown in Figure 6.2a. The reason for this is that the bulk regions outside the depletion width have high conductivities due to plenty of majority carriers in the bulk, in comparison with the depletion region in which there are mainly immobile ions. Thus, the applied voltage drops mostly across the depletion width *W*. Consequently, *V* directly opposes $V_o$ and the potential barrier against diffusion is reduced to $(V_o - V)$, as depicted in Figure 6.2b. This has drastic consequences because the probability that a hole will surmount this potential barrier and diffuse to the right now becomes proportional to $\exp[-e(V_o - V)/kT]$. In other words, the applied voltage effectively reduces the built-in potential and hence the built-in field, which acts against diffusion. Consequently many holes can now diffuse across the depletion region and enter the *n*-side. This results in the **injection of excess minority carriers,** holes, into the *n*-region. Similarly, excess electrons can now diffuse toward the *p*-side and enter this region and thereby become injected minority carriers.

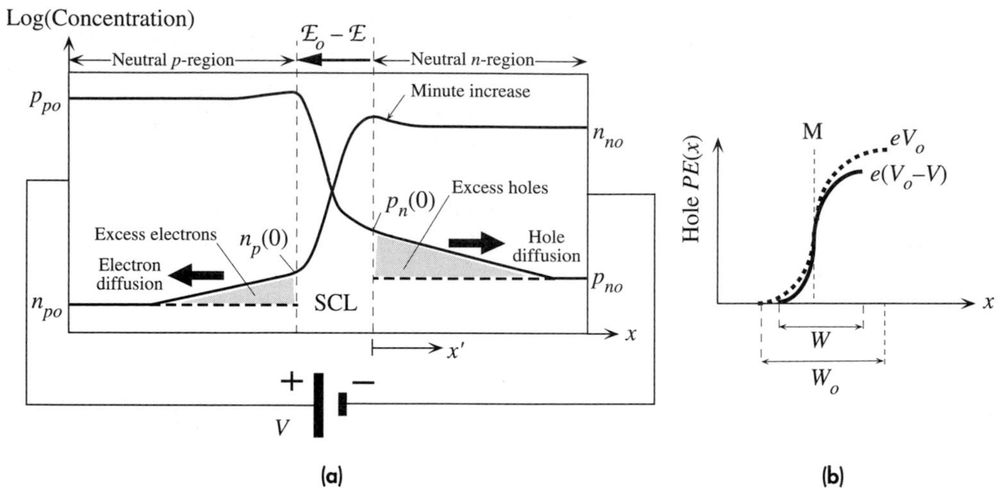


**Figure 6.2** Forward-biased *pn* junction and the injection of minority carriers.
(a) Carrier concentration profiles across the device under forward bias.
(b) The hole potential energy with and without an applied bias. *W* is the width of the SCL with forward bias.

The hole concentration

$$p_n(0) = p_n(x' = 0)$$

just outside the depletion region at $x' = 0$ ($x'$ is measured from $W_n$) is due to the excess of holes diffusing as a result of the reduction in the built-in potential barrier. This concentration $p_n(0)$ is determined by the probability of surmounting the new potential energy barrier $e(V_o - V)$,

$$p_n(0) = p_{po} \exp\left[-\frac{e(V_o - V)}{kT}\right] \tag{6.8}$$

This follows directly from the Boltzmann equation, by virtue of the hole potential energy rising by $e(V_o - V)$ from $x = -W_p$ to $x = W_n$, as indicated in Figure 6.2b, and at the same time the hole concentration falling from $p_{po}$ to $p_n(0)$. By dividing Equation 6.8 by Equation 6.5b, we obtain the effect of the applied voltage directly, which shows how the voltage $V$ determines the amount of excess holes diffusing and arriving at the *n*-region. Equation 6.8 divided by Equation 6.5b is

*Law of the junction*

$$p_n(0) = p_{no} \exp\left(\frac{eV}{kT}\right) \tag{6.9}$$

which is called the **law of the junction.** Equation 6.9 is an important equation that we will use again in dealing with *pn* junction devices. It describes the effect of the applied voltage $V$ on the injected minority carrier concentration just outside the depletion region $p_n(0)$. Obviously, with no applied voltage, $V = 0$ and $p_n(0) = p_{no}$, which is exactly what we expect.

Injected holes diffuse in the $n$-region and eventually recombine with electrons in this region as there are many electrons in the $n$-side. Those electrons lost by recombination are readily replenished by the negative terminal of the battery connected to this side. The current due to holes diffusing in the $n$-region can be sustained because more holes can be supplied by the $p$-region, which itself can be replenished by the positive terminal of the battery.

Electrons are similarly injected from the $n$-side to the $p$-side. The electron concentration $n_p(0)$ just outside the depletion region at $x = -W_p$ is given by the equivalent of Equation 6.9 for electrons, that is,

*Law of the junction*

$$n_p(0) = n_{po} \exp\left(\frac{eV}{kT}\right) \quad [6.10]$$

In the $p$-region, the injected electrons diffuse toward the positive terminal looking to be collected. As they diffuse they recombine with some of the many holes in this region. Those holes lost by recombination can be readily replenished by the positive terminal of the battery connected to this side. The current due to the diffusion of electrons in the $p$-side can be maintained by the supply of electrons from the $n$-side, which itself can be replenished by the negative terminal of the battery. It is apparent that an electric current can be maintained through a *pn* junction under forward bias, and that the current flow, surprisingly, seems to be due to the **diffusion of minority carriers.** There is, however, some drift of majority carriers as well.

If the lengths of the $p$- and $n$-regions are longer than the minority carrier diffusion lengths, then we will be justified to expect the hole concentration $p_n(x')$ on the $n$-side to fall exponentially toward the thermal equilibrium value $p_{no}$, that is,

*Excess minority carrier profile*

$$\Delta p_n(x') = \Delta p_n(0) \exp\left(-\frac{x'}{L_h}\right) \quad [6.11]$$

where

*Excess minority carrier concentration*

$$\Delta p_n(x') = p_n(x') - p_{no}$$

is the excess carrier distribution and $L_h$ is the **hole diffusion length,** defined by $L_h = \sqrt{D_h \tau_h}$ in which $\tau_h$ is the mean hole recombination lifetime (minority carrier lifetime) in the $n$-region. We base Equation 6.11 on our experience with the minority carrier injection in Chapter 5.[3]

The hole **diffusion current density** $J_{D,\text{hole}}$ is therefore

$$J_{D,\text{hole}} = -eD_h \frac{dp_n(x')}{dx'} = -eD_h \frac{d\Delta p_n(x')}{dx'}$$

that is,

$$J_{D,\text{hole}} = \left(\frac{eD_h}{L_h}\right) \Delta p_n(0) \exp\left(-\frac{x'}{L_h}\right)$$

[3] This is simply the solution of the continuity equation in the absence of an electric field, which is discussed in Chapter 5. Equation 6.11 is identical to Equation 5.48.

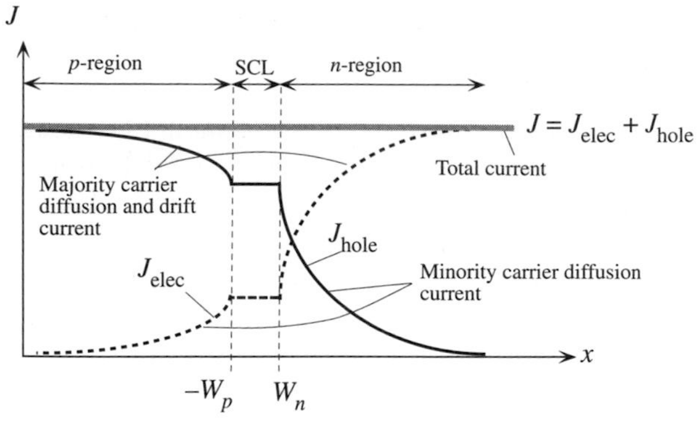


**Figure 6.3** The total current anywhere in the device is constant. Just outside the depletion region, it is due to the diffusion of minority carriers.

Although this equation shows that the hole diffusion current depends on location, the total current at any location is the sum of hole and electron contributions, which is independent of $x$, as indicated in Figure 6.3. The decrease in the minority carrier diffusion current with $x'$ is made up by the increase in the current due to the drift of the majority carriers, as schematically shown in Figure 6.3. The field in the neutral region is not totally zero but a small value, just sufficient to drift the huge number of majority carriers there.

At $x' = 0$, just outside the depletion region, the hole diffusion current is

$$J_{D,\text{hole}} = \left(\frac{eD_h}{L_h}\right)\Delta p_n(0)$$

We can now use the law of the junction to substitute for $\Delta p_n(0)$ in terms of the applied voltage $V$. Writing

$$\Delta p_n(0) = p_n(0) - p_{no} = p_{no}\left[\exp\left(\frac{eV}{kT}\right) - 1\right]$$

and substituting in $J_{D,\text{hole}}$, we get

*Hole diffusion current in n-side*

$$J_{D,\text{hole}} = \left(\frac{eD_h p_{no}}{L_h}\right)\left[\exp\left(\frac{eV}{kT}\right) - 1\right]$$

Thermal equilibrium hole concentration $p_{no}$ is related to the donor concentration by

$$p_{no} = \frac{n_i^2}{n_{no}} = \frac{n_i^2}{N_d}$$

Thus,

*Hole diffusion current in n-side*

$$J_{D,\text{hole}} = \left(\frac{eD_h n_i^2}{L_h N_d}\right)\left[\exp\left(\frac{eV}{kT}\right) - 1\right]$$

There is a similar expression for the electron diffusion current density $J_{D,\text{elec}}$ in the $p$-region. We will assume (quite reasonably) that the electron and hole currents do not change across the depletion region because, in general, the width of this region is narrow (reality is not quite like the schematic sketches in Figures 6.2 and 6.3). The electron current at $x = -W_p$ is the same as that at $x = W_n$. The total current density is then simply given by $J_{D,\text{hole}} + J_{D,\text{elec}}$, that is,

$$J = \left(\frac{eD_h}{L_h N_d} + \frac{eD_e}{L_e N_a}\right) n_i^2 \left[\exp\left(\frac{eV}{kT}\right) - 1\right]$$

or

$$J = J_{so}\left[\exp\left(\frac{eV}{kT}\right) - 1\right] \qquad [6.12]$$

*Ideal diode (Shockley) equation*

This is the familiar diode equation with

$$J_{so} = \left[\left(\frac{eD_h}{L_h N_d}\right) + \left(\frac{eD_e}{L_e N_a}\right)\right] n_i^2$$

*Reverse saturation current*

It is frequently called the **Shockley equation.** The constant $J_{so}$ depends not only on the doping, $N_d$ and $N_a$, but also on the material via $n_i$, $D_h$, $D_e$, $L_h$, and $L_e$. It is known as the **reverse saturation current density,** as explained below. Writing

$$n_i^2 = (N_c N_v)\exp\left(-\frac{eV_g}{kT}\right)$$

*Intrinsic concentration*

where $V_g = E_g/e$ is the bandgap energy expressed in volts, we can write Equation 6.12 as

$$J = \left(\frac{eD_h}{L_h N_d} + \frac{eD_e}{L_e N_a}\right)\left[(N_c N_v)\exp\left(-\frac{eV_g}{kT}\right)\right]\left[\exp\left(\frac{eV}{kT}\right) - 1\right]$$

that is,

$$J = J_1 \exp\left(-\frac{eV_g}{kT}\right)\left[\exp\left(\frac{eV}{kT}\right) - 1\right]$$

or

$$J = J_1 \exp\left[\frac{e(V - V_g)}{kT}\right] \quad \text{for} \quad \frac{eV}{kT} \gg 1 \qquad [6.13]$$

*Diode current and bandgap energy*

where

$$J_1 = \left(\frac{eD_h}{L_h N_d} + \frac{eD_e}{L_e N_a}\right)(N_c N_v)$$

is a new constant.

The significance of Equation 6.13 is that it reflects the dependence of *I–V* characteristics on the bandgap (via $V_g$), as displayed in Figure 6.4 for the three important semiconductors, Ge, Si, and GaAs. Notice that the voltage across the *pn* junction for an appreciable current of say ~0.1 mA is about 0.2 V for Ge, 0.6 V for Si, and 0.9 V for GaAs.

The diode equation, Equation 6.12, was derived by assuming that the lengths of the *p* and *n* regions outside the depletion region are long in comparison with the diffusion lengths $L_h$ and $L_e$. Suppose that $\ell_p$ is the length of the *p*-side outside the depletion region

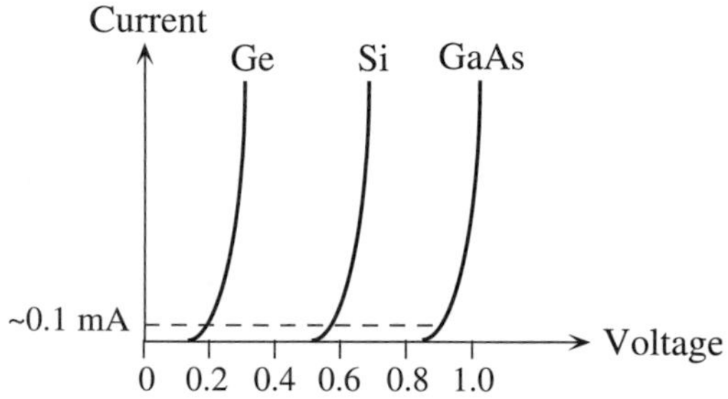


**Figure 6.4** Schematic sketch of the *I–V* characteristics of Ge, Si, and GaAs *pn* junctions.

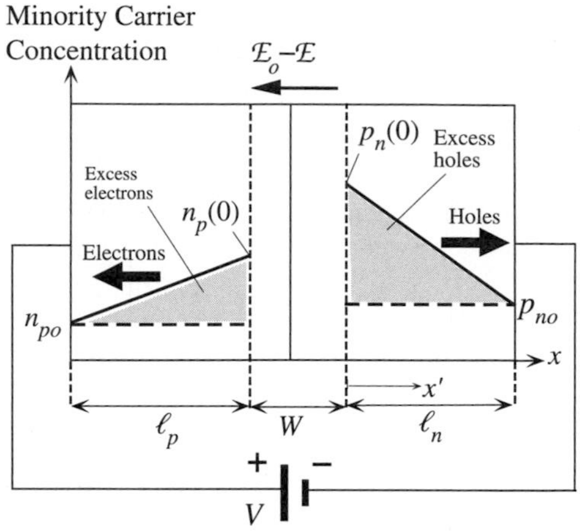


**Figure 6.5** Minority carrier injection and diffusion in a short diode.

and $\ell_n$ is that of the $n$-side outside the depletion region. If $\ell_p$ and $\ell_n$ are shorter than the diffusion lengths $L_e$ and $L_h$, respectively, then we have what is called a **short diode** and consequently the minority carrier distribution profiles fall almost linearly with distance from the depletion region, as depicted in Figure 6.5. This can be readily proved by solving the continuity equation, but an intuitive explanation makes it clear. At $x' = 0$, the minority carrier concentration is determined by the law of the junction, whereas at the battery terminal there can be no excess carriers as the battery will simply collect these. Since the length of the neutral region is shorter than the diffusion length, there are practically no holes lost by recombination, and therefore the hole flow is expected to be uniform across $\ell_n$. This can be so only if the driving force for diffusion, the concentration gradient, is linear.

The excess minority carrier gradient is

$$\frac{d\Delta p_n(x')}{dx'} = -\frac{[p_n(0) - p_{no}]}{\ell_n}$$

The current density $J_{D,\text{hole}}$ due to the injection and diffusion of holes in the $n$-region as a result of forward bias is

$$J_{D,\text{hole}} = -eD_h \frac{d\Delta p_n(x')}{dx'} = eD_h \frac{[p_n(0) - p_{no}]}{\ell_n}$$

We can now use the law of the junction

$$p_n(0) = p_{no} \exp\left(\frac{eV}{kT}\right)$$

for $p_n(0)$ in the above equation and also obtain a similar equation for electrons diffusing in the $p$-region and then sum the two for the total current $J$,

*Short diode*

$$J = \left(\frac{eD_h}{\ell_n N_d} + \frac{eD_e}{\ell_p N_a}\right) n_i^2 \left[\exp\left(\frac{eV}{kT}\right) - 1\right] \quad \textbf{[6.14]}$$

It is clear that this expression is identical to that of a long diode, that is, Equation 6.12, if in the latter we replace the diffusion lengths $L_h$ and $L_e$ by the lengths $\ell_n$ and $\ell_p$ of the *n*- and *p*-regions outside the SCL.

### 6.1.3 FORWARD BIAS: RECOMBINATION AND TOTAL CURRENT

So far we have assumed that, under a forward bias, the minority carriers diffusing and recombining in the neutral regions are supplied by the external current. However, some of the minority carriers will recombine in the depletion region. The external current must therefore also supply the carriers lost in the recombination process in the SCL. Consider for simplicity a symmetrical *pn* junction as in Figure 6.6 under forward bias. At the metallurgical junction at the center *C*, the hole and electron concentrations are $p_M$ and $n_M$ and are equal. We can find the SCL recombination current by considering electrons recombining in the *p*-side in $W_p$ and holes recombining in the *n*-side in $W_n$ as shown by the shaded areas *ABC* and *BCD*, respectively, in Figure 6.6. Suppose that the **mean hole recombination time** in $W_n$ is $\tau_h$ and **mean electron recombination time** in $W_p$ is $\tau_e$. The rate at which the electrons in *ABC* are recombining is the area *ABC* (nearly all injected electrons) divided by $\tau_e$. The electrons are replenished by the diode current. Similarly, the rate at which holes in *BCD* are recombining is the area *BCD* divided by $\tau_h$. Thus, the recombination current density is

$$J_{\text{recom}} = \frac{eABC}{\tau_e} + \frac{eBCD}{\tau_h}$$

We can evaluate the areas *ABC* and *BCD* by taking them as triangles, $ABC \approx \frac{1}{2}W_p n_M$, etc., so that

$$J_{\text{recom}} \approx \frac{e\frac{1}{2}W_p n_M}{\tau_e} + \frac{e\frac{1}{2}W_n p_M}{\tau_h}$$

Under steady-state and equilibrium conditions, assuming a nondegenerate semiconductor, we can use Boltzmann statistics to relate these concentrations to the potential

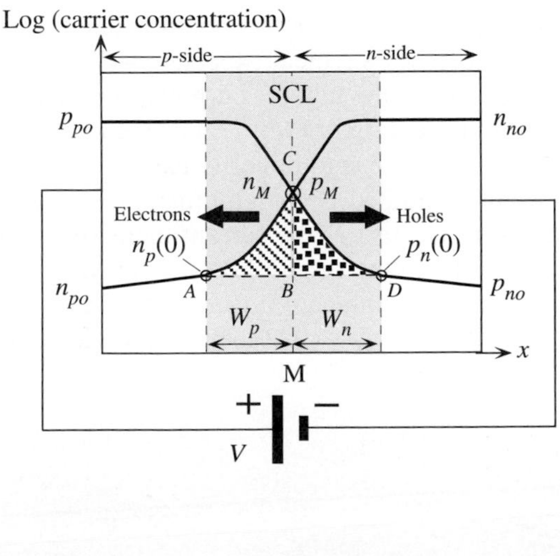


**Figure 6.6** Forward-biased *pn* junction and the injection of carriers and their recombination in SCL.

energy. At $A$, the potential is zero and at $M$ it is $\frac{1}{2}e(V_o - V)$, so

$$\frac{p_M}{p_{po}} = \exp\left[-\frac{e(V_o - V)}{2kT}\right]$$

Since $V_o$ depends on dopant concentrations and $n_i$ as in Equation 6.6 and further $p_{po} = N_a$, we can simplify this equation to

$$p_M = n_i \exp\left(\frac{eV}{2kT}\right)$$

This means that the recombination current for $V > kT/e$ is given by

*Recombination current*

$$J_{\text{recom}} = \frac{en_i}{2}\left(\frac{W_p}{\tau_e} + \frac{W_n}{\tau_h}\right)\exp\left(\frac{eV}{2kT}\right) \quad [6.15]$$

From a better quantitative analysis, the expression for the recombination current can be shown to be[4]

*Recombination current*

$$J_{\text{recom}} = J_{ro}\left[\exp(eV/2kT) - 1\right] \quad [6.16]$$

where $J_{ro}$ is the preexponential constant in Equation 6.15.

Equation 6.15 is the current that supplies the carriers that recombine in the depletion region. The total current into the diode will supply carriers for minority carrier diffusion in the neutral regions and recombination in the space charge layer, so it will be the sum of Equations 6.12 and 6.15. For $V > kT/e$,

*Total diode current = diffusion + recombination*

$$J = J_{so}\exp\left(\frac{eV}{kT}\right) + J_{ro}\exp\left(\frac{eV}{2kT}\right)$$

This expression is often lumped into a single exponential as

*The diode equation for $V > kT/e$*

$$J = J_o\exp\left(\frac{eV}{\eta kT}\right) \quad [6.17]$$

where $J_o$ is a new constant and $\eta$ is an **ideality factor,** which is 1 when the current is due to minority carrier diffusion in the neutral regions and 2 when it is due to recombination in the space charge layer. Figure 6.7 shows typical expected *I–V* characteristics of *pn* junction Ge, Si, and GaAs diodes. At the highest currents, invariably, the bulk resistances of the neutral regions limit the current (why?). For Ge diodes, typically $\eta = 1$ and the overall *I–V* characteristics are due to minority carrier diffusion. In the case of GaAs, $\eta \approx 2$ and the current is limited by recombination in the space charge layer. For Si, typically, $\eta$ changes from 2 to 1 as the current increases, indicating that both processes play an important role. In the case of heavily doped Si diodes, heavy doping leads to short minority carrier recombination times and the current is controlled by recombination in the space charge layer so that the $\eta = 2$ region extends all the way to the onset of bulk resistance limitation.

[4] This is generally proved in advanced texts.

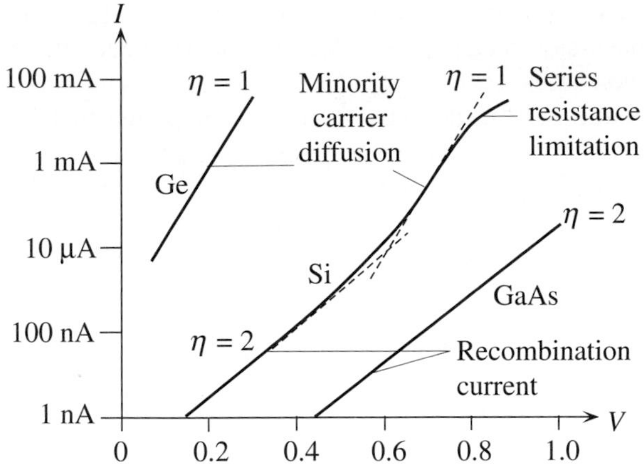


**Figure 6.7** Schematic sketch of typical *I–V* characteristics of Ge, Si, and GaAs *pn* junctions as log(*I*) versus *V*.
The slope indicates $e/(\eta kT)$.

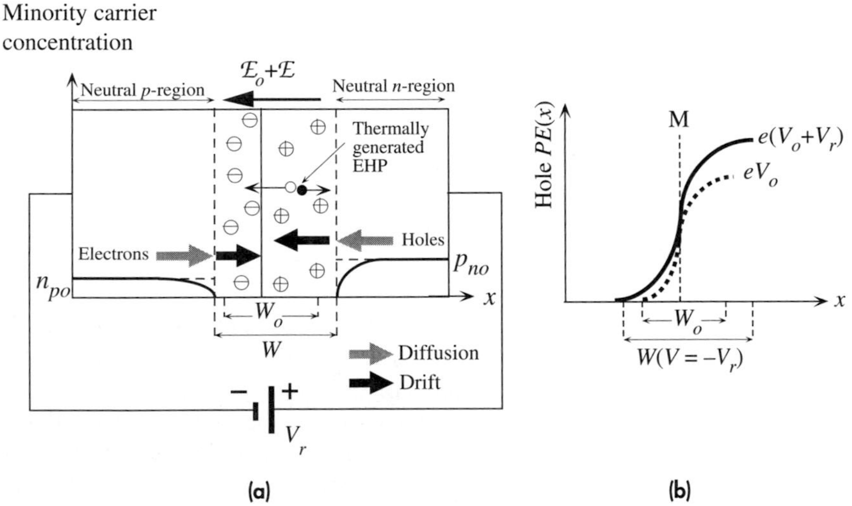


**Figure 6.8** Reverse-biased *pn* junction.
(a) Minority carrier profiles and the origin of the reverse current.
(b) Hole *PE* across the junction under reverse bias.

## 6.1.4 REVERSE BIAS

When a *pn* junction is reverse-biased, as shown in Figure 6.8a, the applied voltage, as before, drops mainly across the depletion region, that is, the space charge layer (SCL), which becomes wider. The negative terminal will attract the holes in the *p*-side to move away from the SCL, which results in more exposed negative acceptor ions and thus a wider SCL. Similarly, the positive terminal will attract electrons away from the SCL, which exposes more positively charged donors. The depletion width on the *n*-side also widens. The movement of electrons in the *n*-region toward the positive battery terminal cannot be sustained because there is no electron supply to this $n$-side. The $p$-side cannot supply electrons to the $n$-side because it has almost none. However, there is a small reverse current due to two causes.

The applied voltage increases the built-in potential barrier, as depicted in Figure 6.8b. The electric field in the SCL is larger than the built-in internal field $\mathcal{E}_o$. The small number of holes on the $n$-side near the SCL become extracted and swept by the field across the SCL over to the $p$-side. This small current can be maintained by the diffusion of holes from the $n$-side bulk to the SCL boundary.

Assume that the reverse bias $V_r > kT/e = 25$ mV. The hole concentration $p_n(0)$ just outside the SCL is nearly zero by the law of the junction, Equation 6.9, whereas the hole concentration in the bulk (or near the negative terminal) is the equilibrium concentration $p_{no}$, which is small. There is therefore a small concentration gradient and hence a small hole diffusion current toward the SCL as shown in Figure 6.8a. Similarly, there is a small electron diffusion current from bulk $p$-side to the SCL. Within the SCL, these carriers are drifted by the field. This minority carrier diffusion current is essentially the Shockley model. The reverse current is given by Equation 6.12 with a negative voltage which leads to a diode current density of $-J_{so}$ called the **reverse saturation current density.** The value of $J_{so}$ depends only on the material via $n_i$, $\mu_h$, $\mu_e$, dopant concentrations, but not on the voltage ($V_r > kT/e$). Furthermore, as $J_{so}$ depends on $n_i^2$, it is strongly temperature dependent. In some books it is stated that the causes of reverse current are the thermal generation of minority carriers in the neutral region within a diffusion length to the SCL, the diffusion of these carriers to the SCL, and their subsequent drift through the SCL. This description, in essence, is identical to the Shockley model we just described.

The thermal generation of electron–hole pairs (EHPs) in the SCL, as shown in Figure 6.8a, can also contribute to the observed reverse current since the internal field in this layer will separate the electron and hole and drift them toward the neutral regions. This drift will result in an external current in addition to the reverse current due to the diffusion of minority carriers. The theoretical evaluation of SCL generation current involves an in-depth knowledge of the charge carrier generation processes via recombination centers, which is discussed in advanced texts. Suppose that $\tau_g$ is the **mean time to generate an electron–hole pair** by virtue of the thermal vibrations of the lattice; $\tau_g$ is also called the **mean thermal generation time.** Given $\tau_g$, the rate of thermal generation per unit volume must be $n_i/\tau_g$ because it takes on average $\tau_g$ seconds to create $n_i$ number of EHPs per unit volume. Furthermore, since $WA$, where $A$ is the cross-sectional area, is the volume of the depletion region, the rate of EHP, or charge carrier, generation is $(AWn_i)/\tau_g$. Both holes and electrons drift in the SCL each contributing equally to the current. The observed current density must be $e(Wn_i)/\tau_g$. Therefore the reverse current density component due to thermal generation of EHPs within the SCL should be given by

*EHP thermal generation in SCL*

$$J_{\text{gen}} = \frac{eWn_i}{\tau_g} \qquad [6.18]$$

The reverse bias widens the width $W$ of the depletion layer and hence increases $J_{\text{gen}}$. The total reverse current density $J_{\text{rev}}$ is the sum of the diffusion and generation

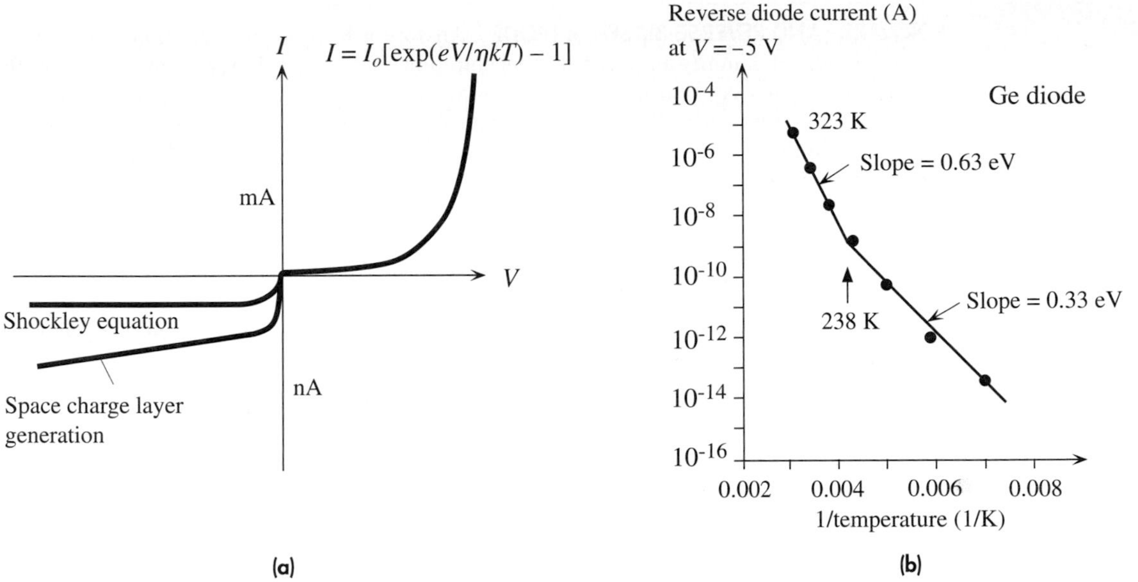


**Figure 6.9**

(a) Forward and reverse *I–V* characteristics of a *pn* junction (the positive and negative current axes have different scales and hence the discontinuity at the origin).

(b) Reverse diode current in a Ge *pn* junction as a function of temperature in a $\ln(I_{rev})$ versus $1/T$ plot. Above 238 K, $I_{rev}$ is controlled by $n_i^2$, and below 238 K, it is controlled by $n_i$. The vertical axis is a logarithmic scale with actual current values.

| SOURCE: (b) From D. Scansen and S. O. Kasap, *Cnd. J. Physics*, **70**, 1070, 1992.

components,

$$J_{\text{rev}} = \left(\frac{eD_h}{L_h N_d} + \frac{eD_e}{L_e N_a}\right)n_i^2 + \frac{eWn_i}{\tau_g} \qquad [6.19]$$

*Total reverse current*

which is shown schematically in Figure 6.9a. The thermal generation component $J_{\text{gen}}$ in Equation 6.18 increases with reverse bias $V_r$ because the SCL width $W$ increases with $V_r$.

The terms in the reverse current in Equation 6.19 are predominantly controlled by $n_i^2$ and $n_i$. Their relative importance depends not only on the semiconductor properties but also on the temperature since $n_i \propto \exp(-E_g/2kT)$. Figure 6.9b shows the reverse current $I_{\text{rev}}$ in dark in a Ge *pn* junction (a photodiode) plotted as $\ln(I_{\text{rev}})$ versus $1/T$ to highlight the two different processes in Equation 6.19. The measurements in Figure 6.9b show that above 238 K, $I_{\text{rev}}$ is controlled by $n_i^2$ because the slope of $\ln(I_{\text{rev}})$ versus $1/T$ yields an $E_g$ of approximately 0.63 eV, close to the expected $E_g$ of about 0.66 eV in Ge. Below 238 K, $I_{\text{rev}}$ is controlled by $n_i$ because the slope of $\ln(I_{\text{rev}})$ versus $1/T$ is equivalent to $E_g/2$ of approximately 0.33 eV. In this range, the reverse current is due to EHP generation in the SCL via defects and impurities (recombination centers).

**EXAMPLE 6.4** **FORWARD- AND REVERSE-BIASED Si DIODE** An abrupt Si $p^+n$ junction diode has a cross-sectional area of 1 mm$^2$, an acceptor concentration of $5 \times 10^{18}$ boron atoms cm$^{-3}$ on the $p$-side, and a donor concentration of $10^{16}$ arsenic atoms cm$^{-3}$ on the $n$-side. The lifetime of holes in the $n$-region is 417 ns, whereas that of electrons in the $p$-region is 5 ns due to a greater concentration of impurities (recombination centers) on that side. Mean thermal generation lifetime ($\tau_g$) is about 1 μs. The lengths of the $p$- and $n$-regions are 5 and 100 microns, respectively.

*a.* Calculate the minority diffusion lengths and determine what type of a diode this is.

*b.* What is the built-in potential across the junction?

*c.* What is the current when there is a forward bias of 0.6 V across the diode at 27 °C? Assume that the current is by minority carrier diffusion.

*d.* Estimate the forward current at 100 °C when the voltage across the diode remains at 0.6 V. Assume that the temperature dependence of $n_i$ dominates over those of $D$, $L$, and $\mu$.

*e.* What is the reverse current when the diode is reverse-biased by a voltage $V_r = 5$ V?

**SOLUTION**

The general expression for the diffusion length is $L = \sqrt{D\tau}$ where $D$ is the diffusion coefficient and $\tau$ is the carrier lifetime. $D$ is related to the carrier mobility $\mu$ via the Einstein relationship $D/\mu = kT/e$. We therefore need to know $\mu$ to calculate $D$ and hence $L$. Electrons diffuse in the $p$-region and holes in the $n$-region, so we need $\mu_e$ in the presence of $N_a$ acceptors and $\mu_h$ in the presence of $N_d$ donors. From the drift mobility, $\mu$ versus dopant concentration in Figure 5.19, we have the following:

| | | |
|---|---|---|
| With | $N_a = 5 \times 10^{18}$ cm$^{-3}$ | $\mu_e \approx 120$ cm$^2$ V$^{-1}$ s$^{-1}$ |
| With | $N_d = 10^{16}$ cm$^{-3}$ | $\mu_h \approx 440$ cm$^2$ V$^{-1}$ s$^{-1}$ |

Thus

$$D_e = \frac{kT\mu_e}{e} \approx (0.0259 \text{ V})(120 \text{ cm}^2 \text{ V}^{-1} \text{ s}^{-1}) = 3.10 \text{ cm}^2 \text{ s}^{-1}$$

$$D_h = \frac{kT\mu_h}{e} \approx (0.0259 \text{ V})(440 \text{ cm}^2 \text{ V}^{-1} \text{ s}^{-1}) = 11.39 \text{ cm}^2 \text{ s}^{-1}$$

Diffusion lengths are

$$L_e = \sqrt{D_e\tau_e} = \sqrt{[(3.10 \text{ cm}^2 \text{ s}^{-1})(5 \times 10^{-9} \text{ s})]}$$

$$= 1.2 \times 10^{-4} \text{ cm} \qquad \text{or} \qquad 1.2\ \mu\text{m} < 5\ \mu\text{m}$$

$$L_h = \sqrt{D_h\tau_h} = \sqrt{[(11.39 \text{ cm}^2 \text{ s}^{-1})(417 \times 10^{-9} \text{ s})]}$$

$$= 21.8 \times 10^{-4} \text{ cm} \qquad \text{or} \qquad 21.8\ \mu\text{m} < 100\ \mu\text{m}$$

We therefore have a long diode. The built-in potential is

$$V_o = \left(\frac{kT}{e}\right)\ln\left(\frac{N_dN_a}{n_i^2}\right) = (0.0259 \text{ V}) \ln\left[\frac{(5 \times 10^{18} \times 10^{16})}{(1.0 \times 10^{10})^2}\right] = 0.877 \text{ V}$$

To calculate the forward current when $V = 0.6$ V, we need to evaluate both the diffusion and recombination components to the current. It is likely that the diffusion component will exceed the recombination component at this forward bias (this can be easily verified). Assuming that the forward current is due to minority carrier diffusion in neutral regions,

$$I = I_{so}\left[\exp\left(\frac{eV}{kT}\right) - 1\right] \approx I_{so}\exp\left(\frac{eV}{kT}\right) \qquad \text{for } V \gg \frac{kT}{e} \qquad (= 0.0259 \text{ V})$$

where

$$I_{so} = AJ_{so} = Aen_i^2\left[\left(\frac{D_h}{L_hN_d}\right) + \left(\frac{D_e}{L_eN_a}\right)\right] \approx \frac{Aen_i^2D_h}{L_hN_d}$$

as $N_a \gg N_d$. In other words, the current is mainly due to the diffusion of holes in the $n$-region. Thus,

$$I_{so} = \frac{(0.01 \text{ cm}^2)(1.6 \times 10^{-19} \text{ C})(1.0 \times 10^{10} \text{ cm}^{-3})^2(11.39 \text{ cm}^2 \text{ s}^{-1})}{(21.8 \times 10^{-4} \text{ cm})(10^{16} \text{ cm}^{-3})}$$

$$= 8.36 \times 10^{-14} \text{ A} \qquad \text{or} \qquad 0.084 \text{ pA}$$

Then the diode current is

$$I \approx I_{so}\exp\left(\frac{eV}{kT}\right) = (8.36 \times 10^{-14} \text{ A})\exp\left[\frac{(0.6 \text{ V})}{(0.0259 \text{ V})}\right]$$

$$= 0.96 \times 10^{-3} \text{ A} \qquad \text{or} \qquad 0.96 \text{ mA}$$

We note that when a forward bias of 0.6 V is applied, the built-in potential is reduced from 0.877 V to 0.256 V, which encourages minority carrier injection, that is, diffusion of holes from $p$- to $n$-side and electrons from $n$- to $p$-side. To find the current at 100 °C, first we assume that $I_{so} \propto n_i^2$. Then at $T = 273 + 100 = 373$ K, $n_i \approx 1.0 \times 10^{12}$ cm$^{-3}$ (approximately from $n_i$ versus $1/T$ graph in Figure 5.16), so

$$I_{so}(373 \text{ K}) \approx I_{so}(300 \text{ K})\left[\frac{n_i(373 \text{ K})}{n_i(300 \text{ K})}\right]^2$$

$$\approx (8.36 \times 10^{-14})\left(\frac{1.0 \times 10^{12}}{1.0 \times 10^{10}}\right)^2 = 8.36 \times 10^{-10}\text{A} \qquad \text{or} \qquad 0.836 \text{ nA}$$

At 100 °C, the forward current with 0.6 V across the diode is

$$I = I_{so}\exp\left(\frac{eV}{kT}\right) = (8.36 \times 10^{-10} \text{ A})\exp\left[\frac{(0.6 \text{ V})(300 \text{ K})}{(0.0259 \text{ V})(373 \text{ K})}\right] = 0.10 \text{ A}$$

When a reverse bias of $V_r$ is applied, the potential difference across the depletion region becomes $V_o + V_r$ and the width $W$ of the depletion region is

$$W = \left[\frac{2\varepsilon(V_o + V_r)}{eN_d}\right]^{1/2} = \left[\frac{2(11.9)(8.85 \times 10^{-12})(0.877 + 5)}{(1.6 \times 10^{-19})(10^{22})}\right]^{1/2}$$

$$= 0.88 \times 10^{-6} \text{ m} \qquad \text{or} \qquad 0.88\ \mu\text{m}$$

The thermal generation current with $V_r = 5$ V is

$$I_{\text{gen}} = \frac{eAWn_i}{\tau_g} = \frac{(1.6 \times 10^{-19} \text{ C})(0.01 \text{ cm}^2)(0.88 \times 10^{-4} \text{ cm})(1.0 \times 10^{10} \text{ cm}^{-3})}{(10^{-6} \text{ s})}$$

$$= 1.41 \times 10^{-9} \text{ A} \qquad \text{or} \qquad 1.4 \text{ nA}$$

This thermal generation current is much greater than the reverse saturation current $I_{so}(= 0.084 \text{ pA})$. The reverse current is therefore dominated by $I_{\text{gen}}$ and it is 1.4 nA.

## 6.2 *pn* JUNCTION BAND DIAGRAM

### 6.2.1 OPEN CIRCUIT

Figure 6.10a shows the energy band diagrams for a *p*-type and an *n*-type semiconductor of the same material (same $E_g$) when the semiconductors are isolated from each other. In the *p*-type material the Fermi level $E_{Fp}$ is $\Phi_p$ below the vacuum level and is close to $E_v$. In the *n*-type material the Fermi level $E_{Fn}$ is $\Phi_n$ below the vacuum level and is close to $E_c$. The separation $E_c - E_{Fn}$ determines the electron concentration $n_{no}$ in the *n*-type and $E_{Fp} - E_v$ determines the hole concentration $p_{po}$, in the *p*-type semiconductor under thermal equilibrium conditions.

An important property of the Fermi energy $E_F$ is that in a system in equilibrium, the Fermi level must be spatially continuous. A difference in Fermi levels $\Delta E_F$ is equivalent to electrical work $eV$, which is either done on the system or extracted from the system. When the two semiconductors are brought together, as in Figure 6.10b, the Fermi level must be uniform through the two materials and the junction at M, which marks the position of the metallurgical junction. Far away from M, in the bulk of the *n*-type semiconductor, we should still have an *n*-type semiconductor and $E_c - E_{Fn}$ should be the same as before. Similarly, $E_{Fp} - E_v$ far away from M inside the *p*-type material should also be the same as before. These features are sketched in Figure 6.10b keeping $E_{Fp}$ and $E_{Fn}$ the same through the whole system and, of course, keeping the bandgap $E_c - E_v$ the same. Clearly, to draw the energy band diagram, we have to bend the bands $E_c$ and $E_v$ around the junction at M because $E_c$ on the *n*-side is close to $E_{Fn}$ whereas on the *p*-side it is far away from $E_{Fp}$. How do bands bend and what does it mean?

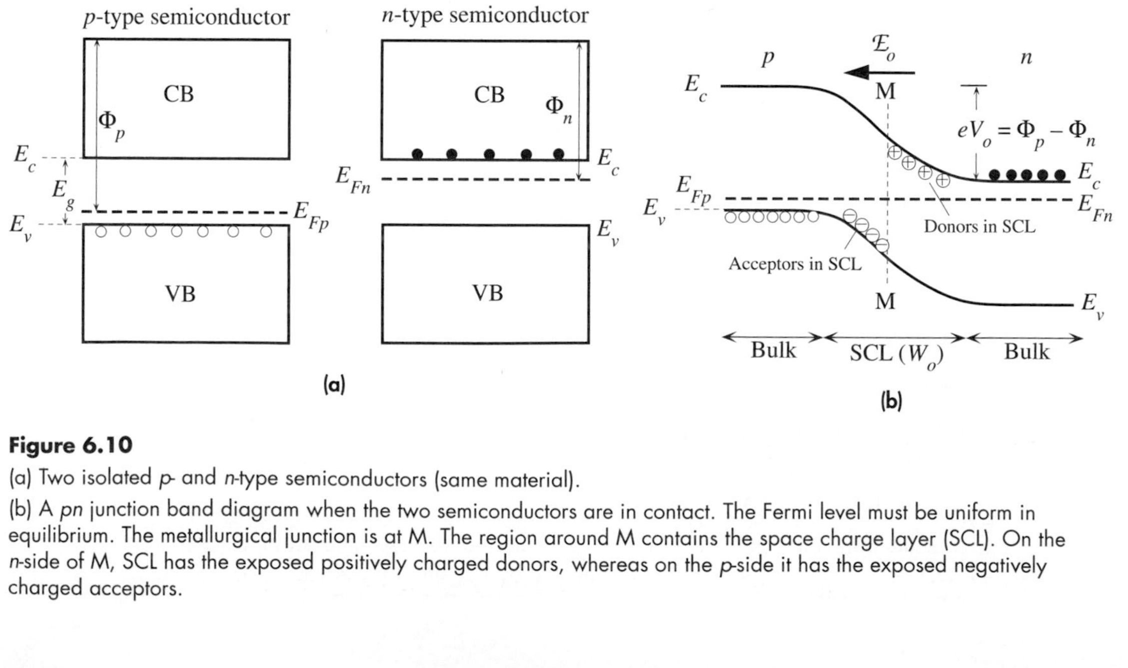


**Figure 6.10**

(a) Two isolated *p*- and *n*-type semiconductors (same material).

(b) A *pn* junction band diagram when the two semiconductors are in contact. The Fermi level must be uniform in equilibrium. The metallurgical junction is at M. The region around M contains the space charge layer (SCL). On the *n*-side of M, SCL has the exposed positively charged donors, whereas on the *p*-side it has the exposed negatively charged acceptors.

As soon as the two semiconductors are brought together to form the junction, electrons diffuse from the *n*-side to the *p*-side and as they do so they deplete the *n*-side near the junction. Thus $E_c$ must move away from $E_{Fn}$ toward M, which is exactly what is sketched in Figure 6.10b. Holes diffuse from the *p*-side to the *n*-side and the loss of holes in the *p*-type material near the junction means that $E_v$ moves away from $E_{Fp}$ toward M, which is also in the figure.

Furthermore, as electrons and holes diffuse toward each other, most of them recombine and disappear around M, which leads to the formation of a depletion region or the space charge layer, as we saw in Figure 6.1. The electrostatic potential energy ($PE$) of the electron decreases from 0 inside the *p*-region to $-eV_o$ inside the *n*-region, as shown in Figure 6.1g. The total energy of the electron must therefore decrease going from the *p*- to the *n*-region by an amount $eV_o$. In other words, the electron in the *n*-side at $E_c$ must overcome a *PE* barrier to go over to $E_c$ in the *p*-side. This *PE* barrier is $eV_o$, where $V_o$ is the built-in potential that we evaluated in Section 6.1. Band bending around M therefore accounts not only for the variation of electron and hole concentrations in this region but also for the effect of the built-in potential (and hence the built-in field as the two are related).

In Figure 6.10b we have also schematically sketched in the positive donor (at $E_d$) and the negative acceptor (at $E_a$) charges in the SCL around M to emphasize that there are exposed charges near M. These charges are, of course, immobile and, generally, they are not shown in band diagrams. It should be noted that in the SCL region, marked as $W_o$, the Fermi level is close to neither $E_c$ nor $E_v$, compared with the bulk semiconductor regions. This means that both *n* and *p* in this zone are much less than their bulk values $n_{no}$ and $p_{po}$. The metallurgical junction zone has been depleted of carriers compared with the bulk. Any applied voltage must therefore drop across the SCL.

### 6.2.2 FORWARD AND REVERSE BIAS

The energy band diagram of the *pn* junction under open circuit conditions is shown in Figure 6.11a. There is no net current, so the diffusion current of electrons from the *n*- to *p*-side is balanced by the electron drift current from the *p*- to *n*-side driven by the built-in field $\mathcal{E}_o$. Similar arguments apply to holes. The probability that an electron diffuses from $E_c$ in the *n*-side to $E_c$ in the *p*-side determines the diffusion current density $J_{\text{diff}}$. The probability of overcoming the *PE* barrier is proportional to $\exp(-eV_o/kT)$. Therefore, under zero bias,

$$J_{\text{diff}}(0) = B\exp\left(-\frac{eV_o}{kT}\right) \tag{6.20}$$

$$J_{\text{net}}(0) = J_{\text{diff}}(0) + J_{\text{drift}}(0) = 0 \tag{6.21}$$

where $B$ is a proportionality constant and $J_{\text{drift}}(0)$ is the current due to the drift of electrons by $\mathcal{E}_o$. Clearly $J_{\text{drift}}(0) = -J_{\text{diff}}(0)$; that is, drift is in the opposite direction to diffusion.

When the *pn* junction is forward-biased, the majority of the applied voltage drops across the depletion region, so the applied voltage is in opposition to the built-in potential $V_o$. Figure 6.11b shows the effect of forward bias, which is to reduce the *PE*

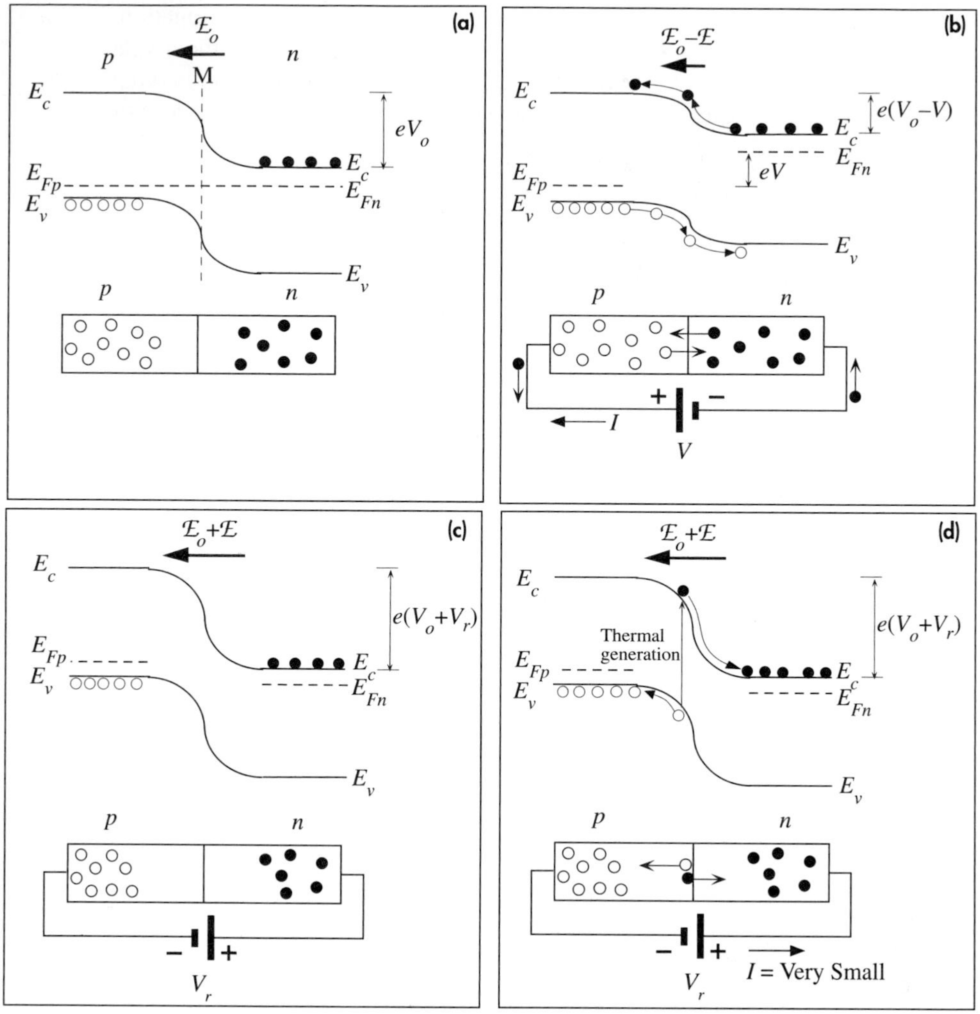


**Figure 6.11** Energy band diagrams for a *pn* junction: (a) open circuit, (b) forward bias, (c) reverse bias conditions, (d) thermal generation of electron–hole pairs in the depletion region results in a small reverse current.

barrier from $eV_o$ to $e(V_o - V)$. The electrons at $E_c$ in the $n$-side can now readily overcome the *PE* barrier and diffuse to the $p$-side. The diffusing electrons from the $n$-side can be replenished easily by the negative terminal of the battery connected to this side. Similarly holes can now diffuse from the $p$- to $n$-side. The positive terminal of the battery can replenish those holes diffusing away from the $p$-side. There is therefore a current flow through the junction and around the circuit.

The probability that an electron at $E_c$ in the $n$-side overcomes the new *PE* barrier and diffuses to $E_c$ in the $p$-side is now proportional to $\exp[-e(V_o - V)/kT]$. The latter increases enormously even for small forward voltages. The new diffusion current due to electrons diffusing from the *n*- to *p*-side is

$$J_{\text{diff}}(V) = B\exp\left[-\frac{e(V_o - V)}{kT}\right]$$

There is still a drift current due to electrons being drifted by the new field $\mathcal{E}_o - \mathcal{E}$ ($\mathcal{E}$ is the applied field) in the SCL. This drift current now has the value $J_{\text{drift}}(V)$. The net current is the diode current under forward bias

$$J = J_{\text{diff}}(V) + J_{\text{drift}}(V)$$

$J_{\text{drift}}(V)$ is difficult to evaluate. As a first approximation we can assume that although $\mathcal{E}_o$ has decreased to $\mathcal{E}_o - \mathcal{E}$, there is, however, an increase in the electron concentration in the SCL due to diffusion so that we can approximately take $J_{\text{drift}}(V)$ to remain the same as $J_{\text{drift}}(0)$. Thus

$$J \approx J_{\text{diff}}(V) + J_{\text{drift}}(0) = B\exp\left[-\frac{e(V_o - V)}{kT}\right] - B\exp\left(-\frac{eV_o}{kT}\right)$$

Factoring leads to

$$J \approx B\exp\left(-\frac{eV_o}{kT}\right)\left[\exp\left(\frac{eV}{kT}\right) - 1\right]$$

We should also add to this the hole contribution, which has a similar form with a different constant *B*. The diode current–voltage relationship then becomes the familiar diode equation,

$$J = J_o\left[\exp\left(\frac{eV}{kT}\right) - 1\right]$$

*pn Junction I–V characteristics*

where $J_o$ is a temperature-dependent constant.[5]

When a reverse bias, $V = -V_r$, is applied to the *pn* junction, the voltage again drops across the SCL. In this case, however, $V_r$ adds to the built-in potential $V_o$, so the *PE* barrier becomes $e(V_o + V_r)$, as shown in Figure 6.11c. The field in the SCL at M increases to $\mathcal{E}_o + \mathcal{E}$, where $\mathcal{E}$ is the applied field.

The diffusion current due to electrons diffusing from $E_c$ in the *n*-side to $E_c$ in the *p*-side is now almost negligible because it is proportional to $\exp[-e(V_o + V_r)/kT]$, which rapidly becomes very small with $V_r$. There is, however, a small reverse current arising from the drift component. When an electron–hole pair (EHP) is thermally generated in the SCL, as shown in Figure 6.11d, the field here separates the pair. The electron falls down the *PE* hill, down to $E_c$, in the *n*-side to be collected by the battery. Similarly the hole falls down its own *PE* hill (energy increases downward for holes) to make it to the *p*-side. The process of falling down a *PE* hill is the same process as being driven by a field, in this case by $\mathcal{E}_o + \mathcal{E}$. Under reverse bias conditions, there is therefore a small reverse current that depends on the rate of thermal generation of EHPs in the SCL. An electron in the *p*-side that is thermally generated within a diffusion length

[5] The derivation is similar to that for the Schottky diode, but there were more assumptions here.

$L_e$ to the SCL can diffuse to the SCL and consequently can become drifted by the field, that is, roll down the *PE* hill in Figure 6.11d. Such minority carrier thermal generation in neutral regions can also give rise to a small reverse current.

**EXAMPLE 6.5** **THE BUILT-IN VOLTAGE $V_o$ FROM THE ENERGY BAND DIAGRAM** The energy band treatment allows a simple way to calculate $V_o$. When the junction is formed in Figure 6.10 from a to b, $E_{Fp}$ and $E_{Fn}$ must shift and line up. Using the energy band diagrams in this figure and semiconductor equations for $n$ and $p$, derive an expression for the built-in voltage $V_o$ in terms of the material and doping properties $N_d$, $N_a$, and $n_i$.

**SOLUTION**

The shift in $E_{Fp}$ and $E_{Fn}$ to line up is clearly $\Phi_p - \Phi_n$, the work function difference. Thus the *PE* barrier $eV_o$ is $\Phi_p - \Phi_n$. From Figure 6.10, we have

$$eV_o = \Phi_p - \Phi_n = (E_c - E_{Fp}) - (E_c - E_{Fn})$$

But on the $p$- and $n$-sides, the electron concentrations in thermal equilibrium are given by

$$n_{po} = N_c \exp\left[-\frac{(E_c - E_{Fp})}{kT}\right]$$

$$n_{no} = N_c \exp\left[-\frac{(E_c - E_{Fn})}{kT}\right]$$

From these equations, we can now substitute for $(E_c - E_{Fp})$ and $(E_c - E_{Fn})$ in the expression for $eV_o$. The $N_c$ cancel and we obtain

$$eV_o = kT \ln\left(\frac{n_{no}}{n_{po}}\right)$$

Since $n_{po} = n_i^2/N_a$ and $n_{no} = N_d$, we readily obtain the built-in potential $V_o$,

*Built-in voltage*

$$V_o = \left(\frac{kT}{e}\right) \ln\left[\frac{(N_a N_d)}{n_i^2}\right]$$

## 6.3 DEPLETION LAYER CAPACITANCE OF THE *pn* JUNCTION

It is apparent that the depletion region of a *pn* junction has positive and negative charges separated over a distance $W$ similar to a parallel plate capacitor. The stored charge in the depletion region, however, unlike the case of a parallel plate capacitor, does not depend linearly on the voltage. It is useful to define an incremental capacitance that relates the incremental charge stored to an incremental voltage change across the *pn* junction.

The width of the depletion region is given by

*Depletion region width*

$$W = \left[\frac{2\varepsilon(N_a + N_d)(V_o - V)}{eN_a N_d}\right]^{1/2} \quad [6.22]$$

where, for forward bias, $V$ is positive, which reduces $V_o$, and, for reverse bias, $V$ is negative, so $V_o$ is increased. We are interested in obtaining the capacitance of the

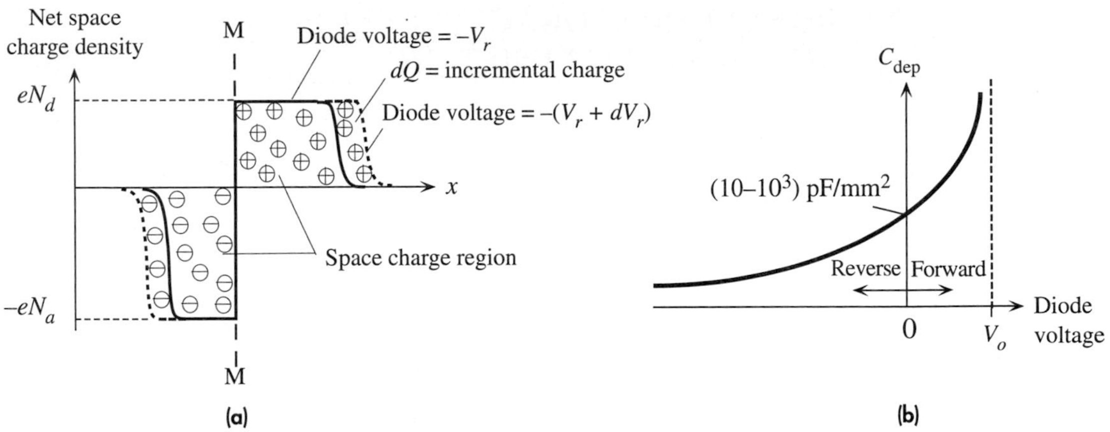


**Figure 6.12** The depletion region behaves like a capacitor.
(a) The charge in the depletion region depends on the applied voltage just as in a capacitor. A reverse bias example is shown.
(b) The incremental capacitance of the depletion region increases with forward bias and decreases with reverse bias. Its value is typically in the range of picofarads per mm[2] of device area.

depletion region under dynamic conditions, that is, when $V$ is a function of time. When the applied voltage $V$ changes by $dV$, to $V + dV$, then $W$ also changes via Equation 6.22, and as a result, the amount of charge in the depletion region becomes $Q + dQ$, as shown in Figure 6.12a for the reverse bias case, that is, $V = -V_r$ and $dV = -dV_r$. The **depletion layer capacitance** $C_{dep}$ is defined by

*Definition of depletion layer capacitance*

$$C_{dep} = \left| \frac{dQ}{dV} \right| \quad [6.23]$$

where the amount of charge (on any one side of the depletion layer) is

$$|Q| = eN_d W_n A = eN_a W_p A$$

and $W = W_n + W_p$. We can therefore substitute for $W$ in Equation 6.22 in terms of $Q$ and then differentiate it to obtain $dQ/dV$. The final result for the depletion capacitance is

*Depletion capacitance*

$$C_{dep} = \frac{\varepsilon A}{W} = \frac{A}{(V_o - V)^{1/2}} \left[ \frac{e\varepsilon (N_a N_d)}{2(N_a + N_d)} \right]^{1/2} \quad [6.24]$$

We should note that $C_{dep}$ is given by the same expression as that for the parallel plate capacitor, $\varepsilon A/W$, but with $W$ being voltage dependent by virtue of Equation 6.22. The $C_{dep} - V$ behavior is sketched in Figure 6.12b. Notice that $C_{dep}$ decreases with increasing reverse bias, which is expected since the separation of the charges increases via $W \propto (V_o + V_r)^{1/2}$. The capacitance $C_{dep}$ is present under both forward and reverse bias conditions.

The voltage dependence of the depletion capacitance is utilized in **varactor diodes** (varicaps), which are employed as voltage-dependent capacitors in tuning circuits. A varactor diode is reverse biased to prevent conduction, and its depletion capacitance is varied by the magnitude of the reverse bias.

## 6.4 DIFFUSION (STORAGE) CAPACITANCE AND DYNAMIC RESISTANCE

The diffusion or storage capacitance arises under forward bias only. As shown in Figure 6.2a, when the $p^+n$ junction is forward biased, we have stored a positive charge on the $n$-side by the continuous injection and diffusion of minority carriers. Similarly, a negative charge has been stored on the $p^+$-side by electron injection, but the magnitude of this negative charge is small for the $p^+n$ junction. When the applied voltage is increased from $V$ to $V + dV$, as shown in Figure 6.13, then $p_n(0)$ changes from $p_n(0)$ to $p_n'(0)$. If $dQ$ is the additional minority carrier charge injected into the $n$-side, as a result of a small increase $dV$ in $V$, then the incremental **storage** or **diffusion capacitance** $C_{\text{diff}}$ is defined as $C_{\text{diff}} = dQ/dV$. At voltage $V$, the injected positive charge $Q$ on the $n$-side is disappearing by recombination at a rate $Q/\tau_h$, where $\tau_h$ is the minority carrier lifetime. The diode current $I$ is therefore $Q/\tau_h$, from which

*Injected minority carrier charge*

$$Q = \tau_h I = \tau_h I_o\left[\exp\left(\frac{eV}{kT}\right) - 1\right] \qquad [6.25]$$

Thus,

*Diffusion capacitance*

$$C_{\text{diff}} = \frac{dQ}{dV} = \frac{\tau_h e I}{kT} = \frac{\tau_h I(\text{mA})}{25} \qquad [6.26]$$

where we used $e/kT \approx 1/0.025$ at room temperature. Generally the value of the diffusion capacitance, typically in the nanofarads range, far exceeds that of the depletion layer capacitance.

Suppose that the voltage $V$ across the diode is increased by an infinitesimally small amount $dV$, as shown in an exaggerated way in Figure 6.14. This gives rise to a small increase $dI$ in the diode current. We define the **dynamic** or **incremental resistance** $r_d$ of the diode as $dV/dI$, so

*Dynamic/incremental resistance*

$$r_d = \frac{dV}{dI} = \frac{kT}{eI} = \frac{25}{I(\text{mA})} \qquad [6.27]$$

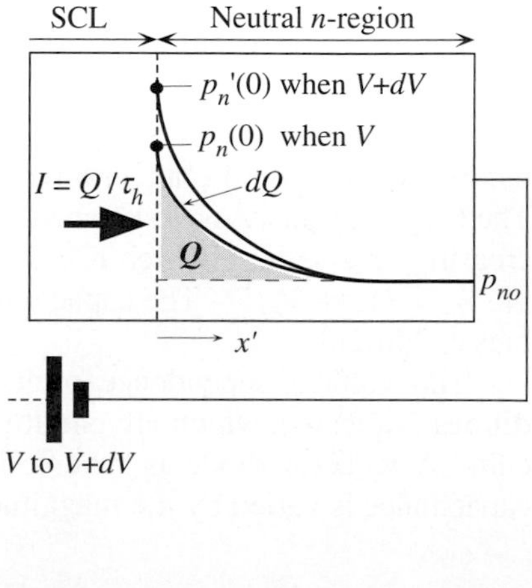


**Figure 6.13** Consider the injection of holes into the $n$-side during forward bias.
Storage or diffusion capacitance arises because when the diode voltage increases from $V$ to $V + dV$, more minority carriers are injected and more minority carrier charge is stored in the $n$-region.

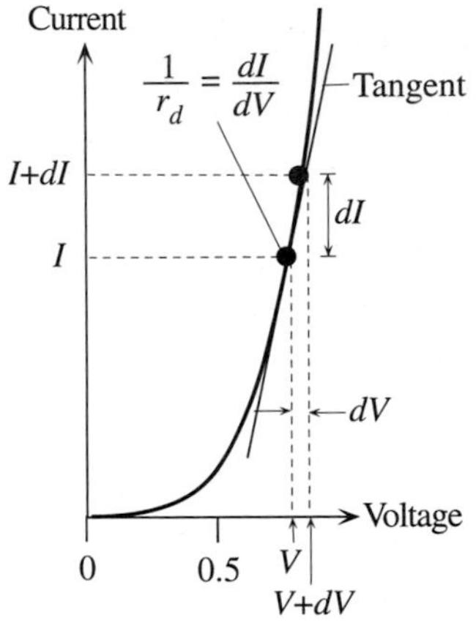


**Figure 6.14** The dynamic resistance of the diode is defined as $dV/dI$, which is the inverse of the tangent at $I$.

The dynamic resistance is therefore the inverse of the slope of the $I$–$V$ characteristics at a point and hence depends on the current $I$. It relates the changes in the diode current and voltage arising from the **diode action** alone, by which we mean the modulation of the rate of minority carrier diffusion by the diode voltage. We could have equivalently defined a dynamic conductance by

*Dynamic conductance*

$$g_d = \frac{dI}{dV} = \frac{1}{r_d}$$

From Equations 6.26 and 6.27 we have

$$r_d C_{\text{diff}} = \tau_h \tag{6.28}$$

The dynamic resistance $r_d$ and diffusion capacitance $C_{\text{diff}}$ of a diode determine its response to small ac signals under forward bias conditions. By *small* we usually mean voltages smaller than the thermal voltage $kT/e$ or 25 mV at room temperature. For small ac signals we can simply represent a forward-biased diode as a resistance $r_d$ in parallel with a capacitance $C_{\text{diff}}$.

**EXAMPLE 6.6**

**INCREMENTAL RESISTANCE AND CAPACITANCE** An abrupt Si $p^+n$ junction diode of cross-sectional area ($A$) 1 mm$^2$ with an acceptor concentration of $5 \times 10^{18}$ boron atoms cm$^{-3}$ on the $p$-side and a donor concentration of $10^{16}$ arsenic atoms cm$^{-3}$ on the $n$-side is forward-biased to carry a current of 5 mA. The lifetime of holes in the $n$-region is 417 ns, whereas that of electrons in the $p$-region is 5 ns. What are the small-signal ac resistance, incremental storage, and depletion capacitances of the diode?

**SOLUTION**

This is the same diode we considered in Example 6.4 for which the built-in potential was 0.877 V and $I_{so} = 0.0836$ pA. The current through the diode is 5 mA. Thus

$$I = I_{so} \exp\left(\frac{eV}{kT}\right) \quad \text{or} \quad V = \left(\frac{kT}{e}\right) \ln\left(\frac{I}{I_{so}}\right) = (0.0259) \ln\left(\frac{5 \times 10^{-3}}{0.0836 \times 10^{-12}}\right) = 0.643 \text{ V}$$

The dynamic diode resistance is given by

$$r_d = \frac{25}{I\,(\text{mA})} = \frac{25}{5} = 5\ \Omega$$

The depletion capacitance per unit area with $N_a \gg N_d$ is

$$C_{\text{dep}} = A\left[\frac{e\varepsilon(N_a N_d)}{2(N_a + N_d)(V_o - V)}\right]^{1/2} \approx A\left[\frac{e\varepsilon N_d}{2(V_o - V)}\right]^{1/2}$$

At $V = 0.643$ V, with $V_o = 0.877$ V, $N_d = 10^{22}$ m$^{-3}$, $\varepsilon_r = 11.9$, and $A = 10^{-6}$ m$^2$, the above equation gives

$$C_{\text{dep}} = 10^{-6}\left[\frac{(1.6 \times 10^{-19})(11.9)(8.85 \times 10^{-12})(10^{22})}{2(0.877 - 0.643)}\right]^{1/2}$$

$$= 6.0 \times 10^{-10}\ \text{F} \qquad \text{or} \qquad 600\ \text{pF}$$

The incremental diffusion capacitance $C_{\text{diff}}$ due to holes injected and stored in the $n$-region is

$$C_{\text{diff}} = \frac{\tau_h I\,(\text{mA})}{25} = \frac{(417 \times 10^{-9})(5)}{25} = 8.3 \times 10^{-8}\ \text{F} \qquad \text{or} \qquad 83\ \text{nF}$$

Clearly the diffusion capacitance (83 nF) that arises during forward bias completely overwhelms the depletion capacitance (600 pF).

We note that there is also a diffusion capacitance due to electrons injected and stored in the $p$-region. However, electron lifetime in the $p$-region is very short (here 5 ns), so the value of this capacitance is much smaller than that due to holes in the $n$-region. In calculating the diffusion capacitance, we normally consider the minority carriers that have the longest recombination lifetime, here $\tau_h$. These are the carriers that take a long time to disappear by recombination when the bias is suddenly switched off.

## 6.5 REVERSE BREAKDOWN: AVALANCHE AND ZENER BREAKDOWN

The reverse voltage across a *pn* junction cannot be increased without limit. Eventually the *pn* junction breaks down either by the Avalanche or Zener breakdown mechanisms, which lead to large reverse currents, as shown in Figure 6.15. In the $V = -V_{\text{br}}$ region, the reverse current increases dramatically with the reverse bias. If unlimited, the large

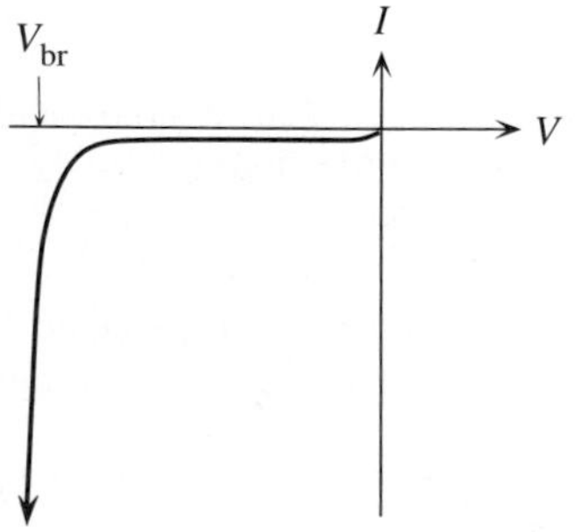


**Figure 6.15** Reverse *I–V*, characteristics of a *pn* junction.

reverse current will increase the power dissipated, which in turn raises the temperature of the device, which leads to a further increase in the reverse current and so on. If the temperature does not burn out the device, for example, by melting the contacts, then the breakdown is recoverable. If the current is limited by an external resistance to a value within the power dissipation specifications, then there is no reason why the device cannot operate under breakdown conditions.

### 6.5.1 Avalanche Breakdown

As the reverse bias increases, the field in the SCL can become so large that an electron drifting in this region can gain sufficient kinetic energy to impact on a Si atom and ionize it, or break a Si–Si bond. The phenomenon by which a drifting electron gains sufficient energy from the field to ionize a host crystal atom by bombardment is termed **impact ionization.** The accelerated electron must gain at least an energy equal to $E_g$ as impact ionization breaks a Si–Si bond, which is tantamount to exciting an electron from the valence band to the conduction band. Thus an additional electron–hole pair is created by this process.

Consider what happens when a thermally generated electron just inside the SCL in the $p$-side is accelerated by the field. The electron accelerates and gains sufficient energy to collide with a host Si atom and release an EHP by impact ionization, as depicted in Figure 6.16. It will lose at least $E_g$ amount of energy, but it can accelerate and head for another ionizing collision further along the depletion region until it reaches the neutral $n$-region. The EHPs generated by impact ionization themselves can now be accelerated by the field and will themselves give rise to further EHPs by ionizing collisions and so on, leading to an **avalanche effect.** One initial carrier can thus create many carriers in the SCL through an avalanche of impact ionizations.

If the reverse current in the SCL in the absence of impact ionization is $I_o$, then due to the avalanche of ionizing collisions in the SCL, the reverse current becomes $MI_o$ where $M$ is the multiplication. It is the net number of carriers generated by the avalanche effect per carrier in the SCL. Impact ionization depends strongly on the electric field. Small increases in the reverse bias can lead to dramatic increases in the

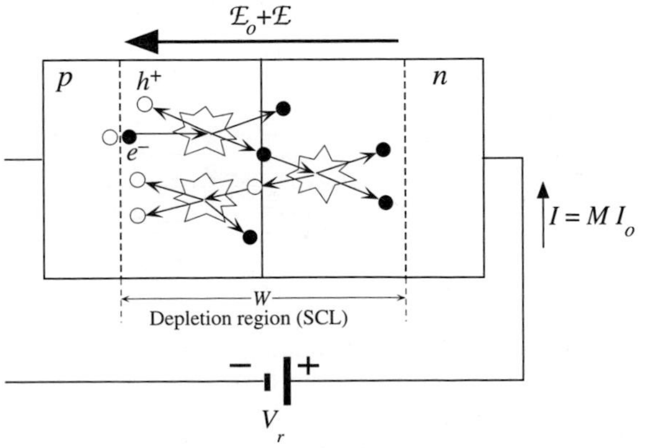


**Figure 6.16** Avalanche breakdown by impact ionization.

multiplication process. Typically

$$M = \frac{1}{1 - \left(\frac{V_r}{V_{br}}\right)^n} \qquad [6.29]$$

where $V_r$ is the reverse bias, $V_{br}$ is the breakdown voltage, and $n$ is an index in the range 3 to 5. It is clear that the reverse current $MI_o$ increases sharply with $V_r$ near $V_{br}$, as depicted in Figure 6.15. Indeed, the voltage across a diode under reverse breakdown remains around $V_{br}$ for very large current variations (several orders of magnitude). If the reverse current under breakdown is limited by an appropriate external resistor $R$, as shown in Figure 6.17, to prevent destructive power dissipation in the diode, then the voltage across the diode remains approximately at $V_{br}$. Thus, as long as $V_r > V_{br}$, the diode clamps the voltage between $A$ and $B$ to approximately $V_{br}$. The reverse current in the circuit is then $(V_r - V_{br})/R$.

Since the electric field in the SCL depends on the width of the depletion region $W$, which in turn depends on the doping parameters, $V_{br}$ also depends on the doping, as discussed in Example 6.7.

### 6.5.2 ZENER BREAKDOWN

Heavily doped *pn* junctions have narrow depletion widths, which lead to large electric fields within this region. When a reverse bias is applied to a *pn* junction, the energy band diagram of the *n*-side can be viewed as being lowered with respect to the *p*-side, as depicted in Figure 6.18. For a sufficient reverse bias (typically less than 10 V), $E_c$

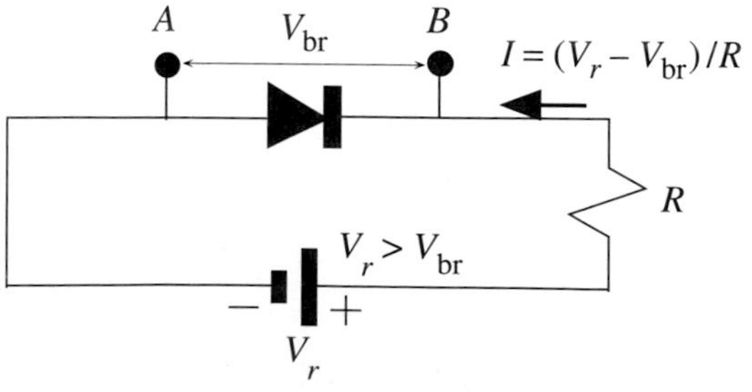


**Figure 6.17** If the reverse breakdown current when $V_r > V_{br}$ is limited by an external resistance $R$ to prevent destructive power dissipation, then the diode can be used to clamp the voltage between $A$ and $B$ to remain approximately $V_{br}$.

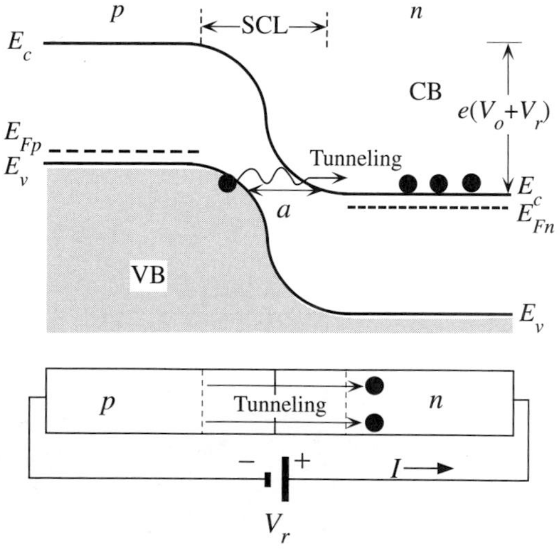


**Figure 6.18** Zener breakdown involves electrons tunneling from the VB of *p*-side to the CB of *n*-side when the reverse bias reduces $E_c$ to line up with $E_v$.

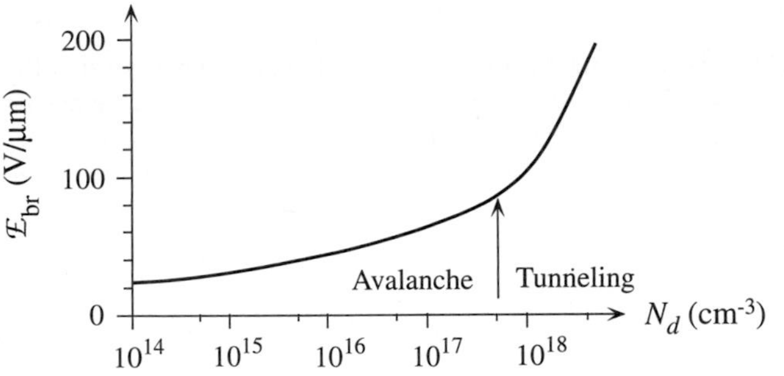


**Figure 6.19** The breakdown field $\mathcal{E}_{br}$ in the depletion layer for the onset of reverse breakdown versus doping concentration $N_d$ in the lightly doped region in a one-sided ($p^+n$ or $pn^+$) abrupt $pn$ junction.
Avalanche and tunneling mechanisms are separated by the arrow.
SOURCE: Data extracted from M. Sze and G. Gibbons, *Solid State Electronics*, 9, no. 831, 1966.

on the $n$-side may be lowered to be below $E_v$ on the $p$-side. This means that electrons at the top of the VB in the $p$-side are now at the same energy level as the empty states in the CB in the $n$-side. As the separation between the VB and CB narrows, shown as $a$ ($< W$), the electrons easily tunnel from the VB in the $p$-side to the CB in the $n$-side, which leads to a current. This process is called the **Zener effect.** As there are many electrons in the VB and many empty states in the CB, the tunneling current can be substantial. The reverse voltage $V_r$, which starts the tunneling current and hence the Zener breakdown, is clearly that which lowers $E_c$ on the $n$-side to be below $E_v$ on the $p$-side and thereby gives a separation that encourages tunneling. In nonquantum mechanical terms, one may intuitively view the Zener effect as the strong electric field in the depletion region ripping out some of those electrons in the Si–Si bonds and thereby releasing them for conduction.

Figure 6.19 shows the dependence of the breakdown field $\mathcal{E}_{br}$ in the depletion region for the onset of avalanche or Zener breakdown in a one-sided ($p^+n$ or $pn^+$) abrupt junction on the dopant concentration $N_d$ in the lightly doped side. At high fields, the tunneling becomes the dominant reverse breakdown mechanism.

---

**EXAMPLE 6.7**

**AVALANCHE BREAKDOWN** Consider a uniformly doped abrupt $p^+n$ junction ($N_a \gg N_d$) reverse biased by $V = -V_r$.

*a.* What is the relationship between the depletion width $W$ and the potential difference ($V_o + V_r$) across $W$?

*b.* If avalanche breakdown occurs when the maximum field in the depletion region $\mathcal{E}_o$ reaches the breakdown field $\mathcal{E}_{br}$, show that the breakdown voltage $V_{br}$ ($\gg V_o$) is then given by

$$V_{br} = \frac{\varepsilon \mathcal{E}_{br}^2}{2eN_d}$$

*c.* An abrupt Si $p^+n$ junction has boron doping of $10^{19}$ cm$^{-3}$ on the $p$-side and phosphorus doping of $10^{16}$ cm$^{-3}$ on the $n$-side. The dependence of the avalanche breakdown field on the impurity concentration is shown in Figure 6.19.
1. What is the reverse breakdown voltage of this Si diode?
2. Calculate the reverse breakdown voltage when the phosphorus doping is increased to $10^{17}$ cm$^{-3}$.

**SOLUTION**

One can assume that all the applied reverse bias drops across the depletion layer so that the new voltage across $W$ is now $V_o + V_r$. We have to integrate $d\mathcal{E}/dx = \rho_{\text{net}}/\varepsilon$ as before across $W$ to find the maximum field. The most important fact to remember here is that the *pn* junction equations relating $W$, $\mathcal{E}_o$, $V_o$, $N_o$, $N_d$, and so on remain the same but with $V_o$ replaced with $V_o + V_r$ since the applied reverse bias of $V_r$ increases $V_o$ to $V_o + V_r$. Then from Equation 6.4,

$$W^2 = \frac{2\varepsilon(V_o + V_r)(N_a^{-1} + N_d^{-1})}{e} \approx \frac{2\varepsilon(V_o + V_r)}{eN_d}$$

*Maximum field and reverse bias*

since $N_a \gg N_d$. The maximum field that corresponds to the breakdown field $\mathcal{E}_{\text{br}}$ is given by

$$\mathcal{E}_o = -\frac{2(V_o + V_r)}{W}$$

*Breakdown voltage and doping*

Thus, from these two equations we can eliminate $W$ and obtain $V_{\text{br}} = V_r$ as

$$V_{\text{br}} = \frac{\varepsilon \mathcal{E}_{\text{br}}^2}{2eN_d}$$

Given $N_a \gg N_d$ we have a $p^+n$ junction with $N_d = 10^{16}$ cm$^{-3}$. The depletion region extends into the *n*-region, so the maximum field actually occurs in the *n*-region. Here the breakdown field $\mathcal{E}_{\text{br}}$ depends on the doping level as given in the graph of the critical field at breakdown $\mathcal{E}_{\text{br}}$ versus doping concentration $N_d$ in Figure 6.19. Taking $\mathcal{E}_{\text{br}} \approx 40$ V/$\mu$m or $4.0 \times 10^5$ V cm$^{-1}$ at $N_d = 10^{16}$ cm$^{-3}$ and using the above equation for $V_{\text{br}}$, we get $V_{\text{br}} = 53$ V.

When $N_d = 10^{17}$ cm$^{-3}$, $\mathcal{E}_{\text{br}}$ from the graph is about $6 \times 10^5$ V cm$^{-1}$, which leads to $V_{\text{br}} = 11.8$ V.

## 6.6 BIPOLAR TRANSISTOR (BJT)

### 6.6.1 COMMON BASE (CB) DC CHARACTERISTICS

As an example, we will consider the *pnp* bipolar junction transistor (BJT) whose basic structure is shown in Figure 6.20a. The *pnp* transistor has three differently doped semiconductor regions. These regions of different doping occur within the same single crystal by the variation of acceptor and donor concentrations resulting from the fabrication process. The most heavily doped *p*-region ($p^+$) is called the **emitter.** In contact with this region is the lightly doped *n*-region, which is called the **base.** The next region is the *p*-type doped **collector.** The base region has the most narrow width for reasons discussed below. Although the three regions in Figure 6.20a have identical cross-sectional areas, in practice, due to the fabrication process, the cross-sectional area increases from the emitter to the collector and the collector region has an extended width. For simplicity, we will assume that the cross-sectional area is uniform, as in Figure 6.20a.

The *pnp* BJT connected as shown in Figure 6.20b is said to be operating under normal and active conditions, which means that the base–emitter (BE) junction is forward biased and the base–collector (BC) junction is reverse biased. The circuit in

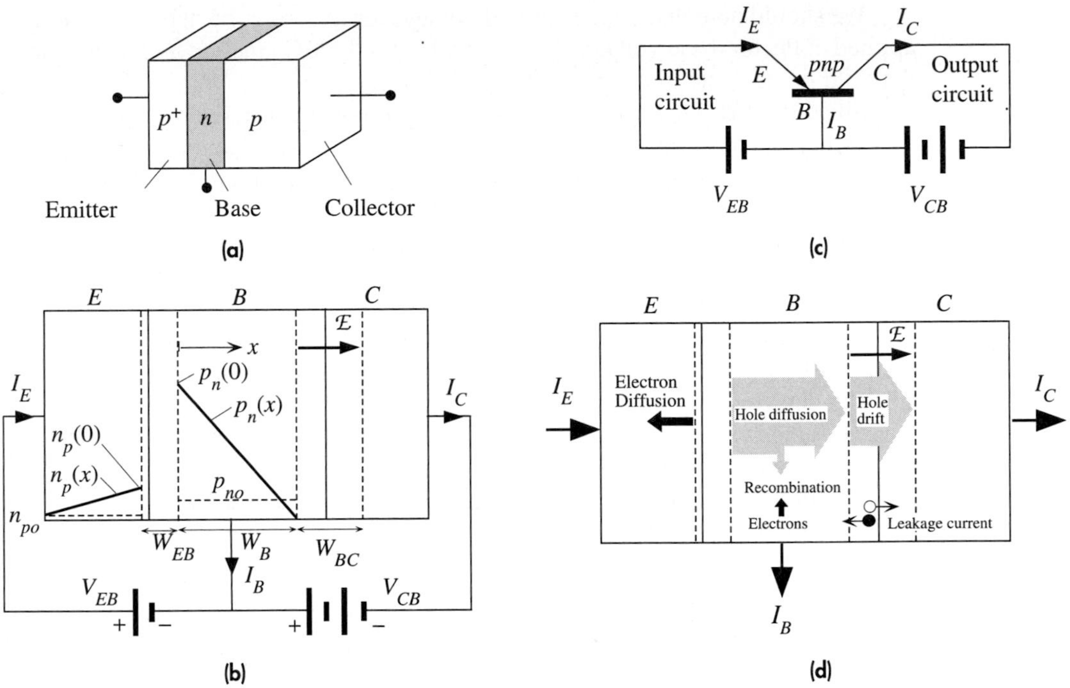


**Figure 6.20**
(a) A schematic illustration of the *pnp* bipolar transistor with three differently doped regions.
(b) The *pnp* bipolar operated under normal and active conditions.
(c) The CB configuration with input and output circuits identified.
(d) The illustration of various current components under normal and active conditions.

Figure 6.20b, in which the base is common to both the collector and emitter bias voltages, is known as the common base (CB) configuration.[6] Figure 6.20c shows the CB transistor circuit with the BJT represented by its circuit symbol. The arrow identifies the emitter junction and points in the direction of current flow when the EB junction is forward biased. Figure 6.20c also identifies the emitter circuit, where $V_{EB}$ is connected, as the input circuit. The collector circuit, where $V_{CB}$ is connected, is the output circuit.

The base–emitter junction is simply called the **emitter junction** and the base-collector junction is called the **collector junction.** As the emitter is heavily doped, the base–emitter depletion region $W_{EB}$ extends almost entirely into the base. Generally, the base and collector regions have comparable doping, so the base–collector depletion region $W_{BC}$ extends to both sides. The width of the neutral base region outside the depletion regions is labeled as $W_B$. All these parameters are shown and defined in Figure 6.20b.

[6] CB should not be confused with the conduction band abbreviation.

We should note that all the applied voltages drop across the depletion widths. The applied collector–base voltage $V_{CB}$ reverse biases the BC junction and hence increases the field in the depletion region at the collector junction.

Since the EB junction is forward-biased, minority carriers are then injected into the emitter and base exactly as they are in the forward-biased diode. Holes are injected into the base and electrons into the emitter, as depicted in Figure 6.20d. Hole injection into the base, however, far exceeds the electron injection into the emitter because the emitter is heavily doped. We can then assume that the emitter current is almost entirely due to holes injected from the emitter into the base. Thus, when forward biased, the emitter "emits," that is, injects holes into the base.

Injected holes into the base must diffuse toward the collector junction because there is a hole concentration gradient in the base. Hole concentration $p_n(W_B)$ just outside the depletion region at the collector junction is negligibly small because the increased field sweeps nearly all the holes here across the junction into the collector (the collector junction is reverse biased).

The hole concentration $p_n(0)$ in the base just outside the emitter junction depletion region is given by the law of the junction. Measuring $x$ from this point (Figure 6.20b),

$$p_n(0) = p_{no} \exp\left(\frac{eV_{EB}}{kT}\right) \qquad [6.30]$$

whereas at the collector end, $x = W_B$, $p_n(W_B) \approx 0$.

If no holes are lost by recombination in the base, then all the injected holes diffuse to the collector junction. There is no field in the base to drift the holes. Their motion is by diffusion. When they reach the collector junction, they are quickly swept across into the collector by the internal field $\mathcal{E}$ in $W_{BC}$. It is apparent that all the injected holes from the emitter become collected by the collector. The collector current is then the same as the emitter current. The only difference is that the emitter current flows across a smaller voltage difference $V_{EB}$, whereas the collector current flows through a larger voltage difference $V_{CB}$. This means a *net gain in power* from the emitter circuit to the collector circuit.

Since the current in the base is by diffusion, to evaluate the emitter and collector currents we must know the hole concentration gradient at $x = 0$ and $x = W_B$ and therefore we must know the hole concentration profile $p_n(x)$ across the base.[7] In the first instance, we can approximate the $p_n(x)$ profile in the base as a straight line from $p_n(0)$ to $p_n(W_B) = 0$, as shown in Figure 6.20b. This is only true in the absence of any recombination in the base as in the short diode case. The emitter current is then

$$I_E = -eAD_h\left(\frac{dp_n}{dx}\right)_{x=0} = eAD_h\frac{p_n(0)}{W_B}$$

[7] The actual concentration profile can be calculated by solving the steady-state continuity equation, which can be found in more advanced texts.

We can substitute for $p_n(0)$ from Equation 6.30 to obtain

$$I_E = \frac{eAD_h p_{no}}{W_B} \exp\left(\frac{eV_{EB}}{kT}\right) \quad [6.31]$$

*Emitter current*

It is apparent that $I_E$ is determined by $V_{EB}$, the forward bias applied across the EB junction, and the base width $W_B$. In the absence of recombination, the collector current is the same as the emitter current, $I_C = I_E$. The control of the collector current $I_C$ in the output (collector) circuit by $V_{EB}$ in the input (emitter) circuit is what constitutes the **transistor action.** The common base circuit has a **power gain** because $I_C$ in the output in Figure 6.20c flows around a larger voltage difference $V_{CB}$ compared with $I_E$ in the input, which flows across $V_{EB}$ (about 0.6 V).

The ratio of the collector current $I_C$ to the emitter current $I_E$ is defined as the **CB current gain** or **current transfer ratio** $\alpha$ of the transistor,

$$\alpha = \frac{I_C}{I_E} \quad [6.32]$$

*Definition of CB current gain*

Typically, $\alpha$ is less than unity, in the range 0.99–0.999, due to two reasons. First is the limitation due to the emitter injection efficiency. When the *BE* junction is forward-biased, holes are injected from the emitter into the base, giving an emitter current $I_{E(\text{hole})}$, and electrons are injected from the base into the emitter, giving an emitter current $I_{E(\text{electron})}$. The total emitter current is, therefore,

$$I_E = I_{E(\text{hole})} + I_{E(\text{electron})}$$

*Total emitter current*

Only the holes injected into the base are useful in giving a collector current because only they can reach the collector. The emitter injection efficiency is defined as

$$\gamma = \frac{I_{E(\text{hole})}}{I_{E(\text{hole})} + I_{E(\text{electron})}} = \frac{1}{1 + \dfrac{I_{E(\text{electron})}}{I_{E(\text{hole})}}} \quad [6.33]$$

*Emitter injection efficiency*

Consequently, the collector current, which depends on $I_{E(\text{hole})}$ only, is less than the emitter current. We would like $\gamma$ to be as close to unity as possible; $I_{E(\text{hole})} \gg I_{E(\text{electron})}$. $\gamma$ can be readily calculated for the forward-biased *pn* junction current equations as shown in Example 6.9.

Secondly, a small number of the diffusing holes in the narrow base inevitably become lost by recombination with the large number of electrons present in this region as depicted in Figure 6.20d. Thus, a fraction of $I_{E(\text{hole})}$ is lost in the base due to recombination, which further reduces the collector current. We define the **base transport factor** $\alpha_T$ as

$$\alpha_T = \frac{I_C}{I_{E(\text{hole})}} = \frac{I_C}{\gamma I_E} \quad [6.34]$$

*Base transport factor*

If the emitter were a perfect injector, $I_E = I_{E(\text{hole})}$, then the current gain $\alpha$ would be $\alpha_T$. If $\tau_h$ is the hole (minority carrier) lifetime in the base, then $1/\tau_h$ is the probability per unit time that a hole will recombine and disappear. We also know that in time $t$, a particle diffuses a distance $x$, given by $x = \sqrt{2Dt}$ where $D$ is the diffusion coefficient. The time $\tau_t$ it takes for a hole to diffuse across $W_B$ is then given by

*Base minority carrier transit time*

$$\tau_t = \frac{W_B^2}{2D_h} \quad [6.35]$$

This diffusion time is called the **transit time** of the minority carriers across the base.

The probability of recombination in time $\tau_t$ is then $\tau_t/\tau_h$. The probability of not recombining and therefore diffusing across is $(1 - \tau_t/\tau_h)$. Since $I_{E(\text{hole})}$ represents the holes entering the base per unit time, $I_{E(\text{hole})}(1 - \tau_t/\tau_h)$ represents the number of holes leaving the base per unit time (without recombining) which is the collector current $I_C$. Substituting for $I_C$ and $I_{E(\text{hole})}$ in Equation 6.34 gives the base transport factor $\alpha_T$,

*Base transport factor*

$$\alpha_T = \frac{I_C}{I_{E(\text{hole})}} = 1 - \frac{\tau_t}{\tau_h} \quad [6.36]$$

Using Equations 6.32, 6.34, and 6.36 we can find the total **CB current gain** $\alpha$:

*CB current gain*

$$\alpha = \alpha_T \gamma = \left(1 - \frac{\tau_t}{\tau_h}\right)\gamma \quad [6.37]$$

The recombination of holes with electrons in the base means that the base must be replenished with electrons, which are supplied by the external battery in the form of a small base current $I_B$, as shown in Figure 6.20d. In addition, the base current also has to supply the electrons injected from the base into the emitter, that is, $I_{E(\text{electron})}$, and shown as electron diffusion in the emitter in Figure 6.20d. The number of holes entering the base per unit time is represented by $I_{E(\text{hole})}$, and the number recombining per unit time is then $I_{E(\text{hole})}(\tau_t/\tau_h)$. Thus, $I_B$ is

*Base current*

$$I_B = \left(\frac{\tau_t}{\tau_h}\right) I_{E(\text{hole})} + I_{E(\text{electron})} = \gamma \frac{\tau_t}{\tau_h} I_E + (1 - \gamma) I_E \quad [6.38]$$

which further simplifies to $I_E - I_C$; the difference between the emitter current and the collector current is the base current. (This is exactly what we expect from Kirchoff's current law.)

The ratio of the collector current to the base current is defined as the **current gain** $\beta$ of the transistor.[8] By using Equations 6.32, 6.37, and 6.38, we can relate $\beta$ to $\alpha$:

*Base-to-collector current gain*

$$\beta = \frac{I_C}{I_B} = \frac{\alpha}{1 - \alpha} \approx \frac{\gamma \tau_h}{\tau_t} \quad [6.39]$$

The base–collector junction in Figure 6.20b is reverse biased, which leads to a leakage current into the collector terminal even in the absence of an emitter current. This leakage current is due to thermally generated electron–hole pairs in the depletion region $W_{BC}$ being drifted by the internal field, as schematically illustrated in Figure 6.20d.

[8] $\beta$ is a useful parameter when the transistor is used in what is called the common emitter (CE) configuration, in which the input current is made to flow into the base of the transistor, and the collector current is made to flow in the output circuit.

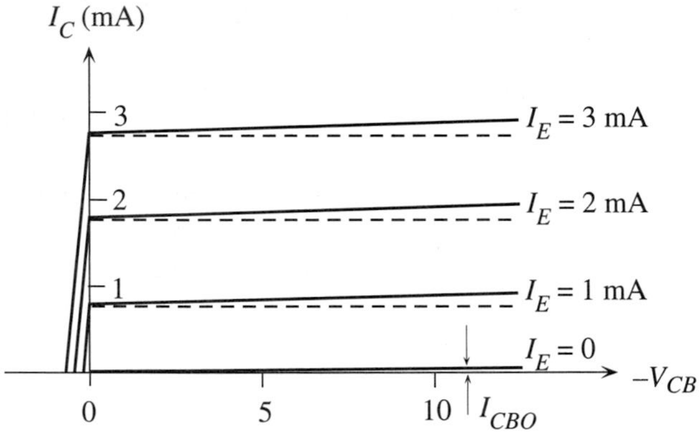


**Figure 6.21** DC *I–V* characteristics of the *pnp* bipolar transistor (exaggerated to highlight various effects).

Suppose that we open circuit the emitter ($I_E = 0$). Then the collector current is simply the leakage current, denoted by $I_{CBO}$. The base current is then $-I_{CBO}$ (flowing out from the base terminal). In the presence of an emitter current $I_E$, we have

*Active region collector current*

$$I_C = \alpha I_E + I_{CBO} \quad [6.40]$$

*Active region base current*

$$I_B = (1 - \alpha)I_E - I_{CBO} \quad [6.41]$$

Equations 6.40 and 6.41 give the collector and base currents in terms of the input current $I_E$, which in turn depends on $V_{EB}$. They only hold when the collector junction is reverse biased and the emitter junction is forward biased, which is defined as the **active region** of the BJT. It should be emphasized that what constitutes the transistor action is the control of $I_E$, and hence $I_C$, by $V_{EB}$.

The dc characteristics of the CB-connected BJT as in Figure 6.20b are normally represented by plotting the collector current $I_C$ as a function of $V_{CB}$ for various fixed values of the emitter current. A typical example of such dc characteristics for a *pnp* transistor is illustrated in Figure 6.21. The following characteristics are apparent. The collector current when $I_E = 0$ is the CB junction leakage current $I_{CBO}$, typically a fraction of a microampere. As long as the collector is negatively biased with respect to the base, the CB junction is reverse biased and the collector current is given by $I_C = \alpha I_E + I_{CBO}$, which is close to the emitter current when $I_E \gg I_{CBO}$. When the polarity of $V_{CB}$ is changed, the CB junction becomes forward biased. The collector junction is then like a forward biased diode and the collector current is the difference between the forward biased CB junction current and the forward biased EB junction current. As they are in opposite directions, they subtract.

We note that $I_C$ increases slightly with the magnitude of $V_{CB}$ even when $I_E$ is constant. In our treatment $I_C$ did not directly depend on $V_{CB}$, which simply reverse biased the collector junction to collect the diffusing holes. In our discussions we assumed that the base width $W_B$ does not depend on the applied voltages. This is only approximately true. Suppose that we increase the reverse bias $V_{CB}$ (for example, from $-5$ to $-10$ V). Then the base–collector depletion width $W_{BC}$ also increases, as schematically depicted in Figure 6.22. Consequently the base width $W_B$ gets slightly narrower, which leads to a slightly shorter base transit time $\tau_t$. The base transport factor $\alpha_T$ in Equation 6.36 and

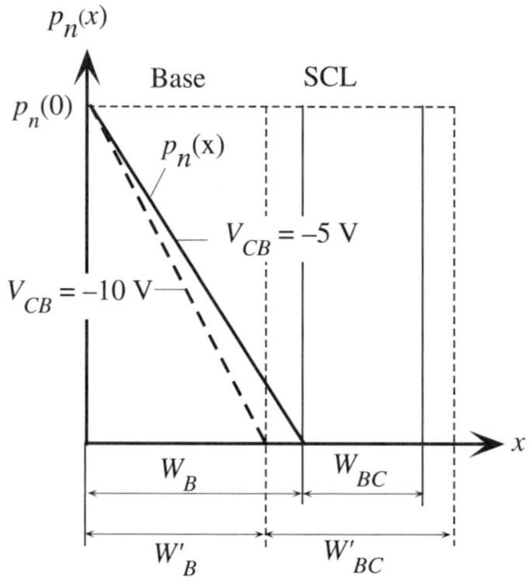


**Figure 6.22** The Early effect.
When the BC reverse bias increases, the depletion width $W_{BC}$ increases to $W'_{BC}$, which reduces the base width $W_B$ to $W'_B$. As $p_n(0)$ is constant (constant $V_{EB}$), the minority carrier concentration gradient becomes steeper and the collector current, $I_C$ increases.

hence $\alpha$ are then slightly larger, which leads to a small increase in $I_C$. The modulation of the base width $W_B$ by $V_{CB}$ is not very strong, which means that the slopes of the $I_C - V_{CB}$ lines at a fixed $I_E$ are very small in Figure 6.21. The base width modulation by $V_{CB}$ is called the **Early effect.**

**EXAMPLE 6.8** **A *pnp* TRANSISTOR** Consider a *pnp* Si BJT that has the following properties. The emitter region mean acceptor doping is $2 \times 10^{18}$ cm$^{-3}$, the base region mean donor doping is $1 \times 10^{16}$ cm$^{-3}$, and the collector region mean acceptor doping is $1 \times 10^{16}$ cm$^{-3}$. The hole drift mobility in the base is 400 cm$^2$ V$^{-1}$ s$^{-1}$, and the electron drift mobility in the emitter is 200 cm$^2$ V$^{-1}$ s$^{-1}$. The transistor emitter and base neutral region widths are about 2 μm each when the transistor is under normal operating conditions, that is, when the EB junction is forward-biased and the BC junction is reverse-biased. The effective cross-sectional area of the device is 0.02 mm$^2$. The hole lifetime in the base is approximately 400 ns. Assume that the emitter has 100 percent injection efficiency, $\gamma = 1$. Calculate the CB current transfer ratio $\alpha$ and the current gain $\beta$. What is the emitter–base voltage if the emitter current is 1 mA?

**SOLUTION**

The hole drift mobility $\mu_h = 400$ cm$^2$ V$^{-1}$ s$^{-1}$ (minority carriers in the base). From the Einstein relationship we can easily find the diffusion coefficient of holes,

$$D_h = \left(\frac{kT}{e}\right)\mu_h = (0.0259 \text{ V})(400 \text{ cm}^2 \text{ V}^{-1} \text{ s}^{-1}) = 10.36 \text{ cm}^2 \text{ s}^{-1}$$

The minority carrier transit time $\tau_t$ across the base is

$$\tau_t = \frac{W_B^2}{2D_h} = \frac{(2 \times 10^{-4} \text{ cm})^2}{2(10.36 \text{ cm}^2 \text{ s}^{-1})} = 1.93 \times 10^{-9} \text{ s} \qquad \text{or} \qquad 1.93 \text{ ns}$$

The base transport factor and hence the CB current gain is

$$\alpha = \gamma\alpha_T = 1 - \frac{\tau_t}{\tau_h} = 1 - \frac{1.93 \times 10^{-9} \text{ s}}{400 \times 10^{-9} \text{ s}} = 0.99517$$

The current gain $\beta$ of the transistor is

$$\beta = \frac{\alpha}{1-\alpha} = \frac{0.99517}{1-0.99517} = 206.2$$

The emitter current is due to holes diffusing in the base ($\gamma = 1$),

$$I_E = I_{EO}\exp\left(\frac{eV_{EB}}{kT}\right)$$

where

$$I_{EO} = \frac{eAD_h p_{no}}{W_B} = \frac{eAD_h n_i^2}{N_d W_B}$$

$$= \frac{(1.6\times10^{-19}\text{ C})(0.02\times10^{-2}\text{ cm}^2)(10.36\text{ cm s}^{-1})(1.0\times10^{10}\text{ cm}^{-3})^2}{(1\times10^{16}\text{ cm}^{-3})(2\times10^{-4}\text{ cm})}$$

$$= 1.66\times10^{-14}\text{ A}$$

Thus,

$$V_{EB} = \frac{kT}{e}\ln\left(\frac{I_E}{I_{EO}}\right) = (0.0259\text{ V})\ln\left(\frac{1\times10^{-3}\text{ A}}{1.66\times10^{-14}\text{ A}}\right) = 0.64\text{ V}$$

The major assumption is $\gamma = 1$, which is generally not true, as shown in Example 6.9. The actual $\alpha$ and hence $\beta$ will be smaller due to less than 100 percent emitter injection. Note also that $W_B$ is the *neutral region width*, that is, the region of base outside the depletion regions. It is not difficult to calculate the depletion layer widths within the base, which are about 0.2 μm on the emitter side and roughly about 0.7 μm on the collector side, so that the total base width junction to junction is $2 + 0.2 + 0.7 = 2.9$ μm.

The transit time of minority carriers across the base is $\tau_t$. If the input signal changes before the minority carriers have diffused across the base, then the collector current cannot respond to the changes in the input. Thus, if the frequency of the input signal is greater than $1/\tau_t$, the minority carriers will not have time to transit the base and the collector current will remain unmodulated by the input signal. One can set the upper frequency limit at $\sim 1/\tau_t$ which is 518 MHz.

## EMITTER INJECTION EFFICIENCY $\gamma$

**EXAMPLE 6.9**

*a.* Consider a *pnp* transistor with the parameters as defined in Figure 6.20. Show that the **injection efficiency of the emitter**, defined as

$$\gamma = \frac{\text{Emitter current due to minority carriers injected into the base}}{\text{Total emitter current}}$$

is given by

$$\gamma = \frac{1}{1 + \dfrac{N_d W_B \mu_{e(\text{emitter})}}{N_a W_E \mu_{h(\text{base})}}}$$

*b.* How would you modify the CB current gain $\alpha$ to include the emitter injection efficiency?

*c.* Calculate the emitter injection efficiency for the *pnp* transistor in Example 6.8, which has an acceptor doping of $2\times10^{18}$ cm$^{-3}$ in the emitter, donor doping of $1\times10^{16}$ cm$^{-3}$ in the base, emitter and base neutral region widths of 2 μm, and a minority carrier lifetime of 400 ns in the base. What are its $\alpha$ and $\beta$ taking into account the emitter injection efficiency?

**SOLUTION**

When the BE junction is forward biased, holes are injected into the base, giving an emitter current $I_{E(\text{hole})}$, and electrons are injected into the emitter, giving an emitter current $I_{E(\text{electron})}$. The total emitter current is therefore

$$I_E = I_{E(\text{hole})} + I_{E(\text{electron})}$$

Only the holes injected into the base are useful in giving a collector current because only they can reach the collector. Injection efficiency is defined as

*Emitter injection efficiency definition*

$$\gamma = \frac{I_{E(\text{hole})}}{I_{E(\text{hole})} + I_{E(\text{electron})}} = \frac{1}{1 + \dfrac{I_{E(\text{electron})}}{I_{E(\text{hole})}}}$$

But, provided that $W_E$ and $W_B$ are shorter than minority carrier diffusion lengths,

$$I_{E(\text{hole})} = \frac{eAD_{h(\text{base})}n_i^2}{N_d W_B}\exp\left(\frac{eV_{EB}}{kT}\right) \quad \text{and} \quad I_{E(\text{electron})} = \frac{eAD_{e(\text{emitter})}n_i^2}{N_a W_E}\exp\left(\frac{eV_{EB}}{kT}\right)$$

When we substitute into the definition of $\gamma$ and use $D = \mu kT/e$, we obtain

*Emitter injection efficiency*

$$\gamma = \frac{1}{1 + \dfrac{N_d W_B \mu_{e(\text{emitter})}}{N_a W_E \mu_{h(\text{base})}}}$$

The hole component of the emitter current is given as $\gamma I_E$. Of this, a fraction $\alpha_T = (1 - \tau_t/\tau_h)$ will give a collector current. Thus, the emitter-to-collector current transfer ratio $\alpha$, taking into account the emitter injection efficiency, is

*Emitter-to-collector current transfer ratio*

$$\alpha = \gamma\left(1 - \frac{\tau_t}{\tau_h}\right)$$

In the emitter, $N_{a(\text{emitter})} = 2 \times 10^{18}$ cm$^{-3}$ and $\mu_{e(\text{emitter})} = 200$ cm$^2$ V$^{-1}$ s$^{-1}$, and in the base, $N_{d(\text{base})} = 1 \times 10^{16}$ cm$^{-3}$ and $\mu_{h(\text{base})} = 400$ cm$^2$ V$^{-1}$ s$^{-1}$. The emitter injection efficiency is

$$\gamma = \frac{1}{1 + \dfrac{(1 \times 10^{16})(2)(200)}{(2 \times 10^{18})(2)(400)}} = 0.99751$$

The transit time $\tau_t = W_B^2/2D_h = 1.93 \times 10^{-9}$ s (as before), so the overall $\alpha$ is

$$\alpha = 0.99751\left(1 - \frac{1.93 \times 10^{-9}}{400 \times 10^{-9}}\right) = 0.99269$$

and the overall $\beta$ is

$$\beta = \frac{\alpha}{(1 - \alpha)} = 135.8$$

The same transistor with 100 percent emitter injection in Example 6.8 had a $\beta$ of 206. It is clear that the emitter injection efficiency $\gamma$ and the base transport factor $\alpha_T$ have comparable impacts in controlling the overall gain in the example. We neglected the recombination of electrons and holes in the EB depletion region. In fact, if we were to also consider this recombination component of the emitter current, $I_{E(\text{hole})}$ would have to be even smaller compared with the total $I_E$, which would make $\gamma$ and hence $\beta$ even lower.

## 6.6.2 COMMON BASE AMPLIFIER

According to Equation 6.31 the emitter current depends exponentially on $V_{EB}$,

$$I_E = I_{EO} \exp\left(\frac{eV_{EB}}{kT}\right) \qquad [6.42]$$

It is therefore apparent that small changes in $V_{EB}$ lead to large changes in $I_E$. Since $I_C \approx I_E$, we see that small variations in $V_{EB}$ cause large changes in $I_C$ in the collector circuit. This can be fruitfully used to obtain voltage amplification as shown in Figure 6.23. The battery $V_{CC}$, through $R_C$, provides a reverse bias for the base–collector junction. The dc voltage $V_{EE}$ forward biases the EB junction, which means that it provides a dc current $I_E$. The input signal is the ac voltage $v_{eb}$ applied in series with the dc bias voltage $V_{EE}$ to the EB junction. The applied signal $v_{eb}$ modulates the total voltage $V_{EB}$ across the EB junction and hence, by virtue of Equation 6.30, modulates the injected hole concentration $p_n(0)$ up and down about the dc value determined by $V_{EE}$ as depicted in Figure 6.23. This variation in $p_n(0)$ alters the concentration gradient and therefore gives rise to a change in $I_E$, and hence a nearly identical change in $I_C$. The change in the collector current can be converted to a voltage change by using a resistor $R_C$ in the collector circuit as shown in Figure 6.23. However, the output is commonly taken between the collector, and the base and this voltage $V_{CB}$ is

$$V_{CB} = -V_{CC} + R_C I_C$$

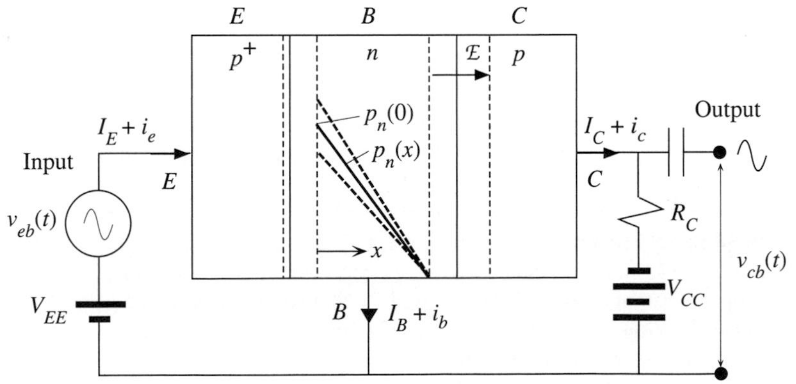


**Figure 6.23** A *pnp* transistor operated in the active region in the common base amplifier configuration.
The applied (input) signal $v_{eb}$ modulates the dc voltage across the EB junction and hence modulates the injected hole concentration up and down about the dc value $p_n(0)$. The solid line shows $p_n(x)$ when only the dc bias $V_{EE}$ is present. The dashed lines show how $p_n(x)$ is modulated up and down by the signal $v_{eb}$ superimposed on $V_{EE}$.

Increasing the emitter–base voltage $V_{EB}$ (by increasing $v_{eb}$) increases $I_C$, which increases $V_{CB}$. Since we are interested in ac signals, that voltage variation across CB is tapped out through a dc blocking capacitor in Figure 6.23.

For simplicity we will assume that changes $\delta V_{EB}$ and $\delta I_E$ in the dc values of $V_{EB}$ and $I_E$ are small, which means that $\delta V_{EB}$ and $\delta I_E$ can be related by differentiating Equation 6.42. We are hence tacitly assuming an operation under small signals. Further, we will take the changes to represent the ac signal magnitudes, $v_{eb} = \delta V_{EB}$, $i_e = \delta I_E$, $i_c = \delta I_C \approx \delta I_E \approx i_e$, $v_{cb} = \delta V_{CB}$.

The output signal voltage $v_{cb}$ corresponds to the change in $V_{CB}$,

$$v_{cb} = \delta V_{CB} = R_C\,\delta I_C = R_C\,\delta I_E$$

The variation in the emitter current $\delta I_E$ depends on the variation $\delta V_{EB}$ in $V_{EB}$, which can be determined by differentiating Equation 6.42,

$$\frac{\delta I_E}{\delta V_{EB}} = \frac{e}{kT} I_E$$

By definition, $\delta V_{EB}$ is the input signal $v_{eb}$. The change $\delta I_E$ in $I_E$ is the input signal current ($i_e$) flowing into the emitter as a result of $\delta V_{EB}$. Therefore the quantity $\delta V_{EB}/\delta I_E$ represents an input resistance $r_e$ seen by the source $v_{eb}$.

*Input resistance*

$$r_e = \frac{\delta V_{EB}}{\delta I_E} = \frac{kT}{eI_E} = \frac{25}{I_E(\text{mA})} \qquad [6.43]$$

The output signal is then

$$v_{cb} = R_C\,\delta I_E = R_C \frac{v_{eb}}{r_e}$$

so the voltage amplification is

*CB voltage gain*

$$A_V = \frac{v_{cb}}{v_{eb}} = \frac{R_C}{r_e} \qquad [6.44]$$

To obtain a voltage gain we obviously need $R_C > r_e$, which is invariably the case by the appropriate choice of $I_E$, hence $r_e$, and $R_C$. For example, when the BJT is biased so that $I_E$ is 10 mA and $r_e$ is 2.5 Ω, and if $R_C$ is chosen to be 50 Ω, then the gain is 20.

**EXAMPLE 6.10** **A COMMON BASE AMPLIFIER** Consider a *pnp* Si BJT that has been connected as in Figure 6.23. The BJT has a $\beta = 135$ and has been biased to operate with a 5 mA collector current. What is the small-signal input resistance? What is the required $R_C$ that will provide a voltage gain of 20? What is the base current? What should be the $V_{CC}$ in Figure 6.23? Suppose $V_{CC} = -6$ V, what is the largest swing in the output voltage $V_{CB}$ in Figure 6.23 as the input signal is increased and decreased about the bias point $V_{EE}$, taken as 0.65 V?

**SOLUTION**

The emitter and collector currents are approximately the same. From Equation 6.43,

$$r_e = \frac{25}{I_E\ (\text{mA})} = \frac{25}{5} = 5\ \Omega$$

The voltage gain $A_V$ from Equation 6.44 is

$$A_V = \frac{R_C}{r_e} \qquad \text{or} \qquad 20 = \frac{R_C}{5\ \Omega}$$

so a gain of 20 requires $R_C = 100\ \Omega$.

$$\text{Base current } I_B = \frac{I_C}{\beta} = \frac{5\ \text{mA}}{135} = 0.037\ \text{mA} \qquad \text{or} \qquad 37\ \mu\text{A}$$

There is a dc voltage across $R_C$ given by $I_C R_C = (0.005\ \text{A})(100\ \Omega) = 0.5\ \text{V}$. $V_{CC}$ has to provide the latter voltage across $R_C$ and also a sufficient voltage to keep the BC junction reverse biased at all times under normal operation. Let us set $V_{CC} = -6$ V. Thus, in the absence of any input signal $v_{eb}$, $V_{CB}$ is set to $-6\ \text{V} + 0.5\ \text{V} = -5.5\ \text{V}$. As we increase the signal $v_{eb}$, $V_{EB}$ and hence $I_C$ increase until the point $C$ becomes nearly zero,[9] that is, $V_{CB} = 0$, which occurs when $I_C$ is maximum at $I_{C\max} = |V_{CC}|/R_C$ or 60 mA. As $v_{eb}$ decreases, so does $V_{EB}$ and hence $I_C$. Eventually $I_C$ will simply become zero, and point $C$ will be at $-6$ V, so $V_{CB} = V_{CC}$. Thus, $V_{CB}$ can only swing from $-5.5$ V to 0 V (for increasing input until $I_C = I_{C\max}$), or from $-5.5$ to $-6$ V (for decreasing input until $I_C = 0$).

### 6.6.3 COMMON EMITTER (CE) DC CHARACTERISTICS

An *npn* bipolar transistor when connected in the common emitter (CE) configuration has the emitter common to both the input and output circuits, as shown in Figure 6.24a. The dc voltage $V_{BE}$ forward biases the BE junction and thereby injects electrons as minority carriers into the base. These electrons diffuse to the collector junction where the field $\mathcal{E}$ sweeps them into the collector to constitute the collector current $I_C$. $V_{BE}$ controls the current $I_E$ and hence $I_B$ and $I_C$. The advantage of the CE configuration is that the **input current** is the current flowing between the ac source and the base, which is the base current $I_B$. This current is much smaller than the emitter current by about a factor of $\beta$. The output current is the current flowing between $V_{CE}$ and the collector, which is $I_C$. In the CE configuration, the dc voltage $V_{CE}$ must be greater than $V_{BE}$ to reverse bias the collector junction and collect the diffusing electrons in the base.

The dc characteristics of the BJT in the CE configuration are normally given as $I_C$ versus $V_{CE}$ for various values of fixed base currents $I_B$, as shown in Figure 6.24b. The characteristics can be readily understood by Equations 6.40 and 6.41. We should note that, in practice, we are essentially adjusting $V_{BE}$ to obtain the desired $I_B$ because, by Equation 6.41,

$$I_B = (1 - \alpha)I_E - I_{CBO}$$

and $I_E$ depends on $V_{BE}$ via Equation 6.42.

Increasing $I_B$ requires increasing $V_{BE}$, which increases $I_C$. Using Equations 6.40 and 6.41, we can obtain $I_C$ in terms of $I_B$ alone,

$$I_C = \beta I_B + \frac{1}{(1 - \alpha)} I_{CBO}$$

[9] Various saturation effects are ignored in this approximate discussion.

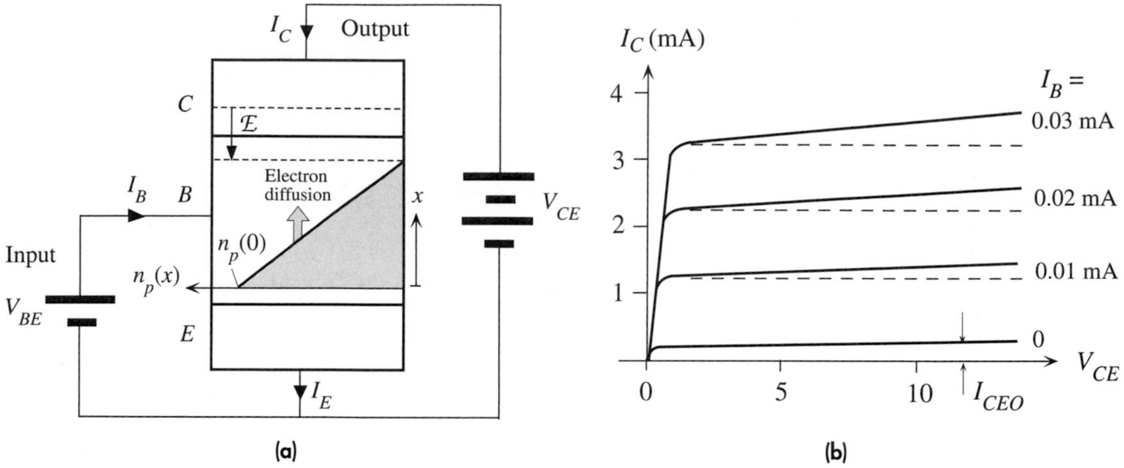


**Figure 6.24**
(a) An *npn* transistor operated in the active region in the common emitter configuration. The input current is the current that flows between $V_{BE}$ and the base which is $I_B$.
(b) DC *I–V* characteristics of the *npn* bipolar transistor in the CE configuration. (Exaggerated to highlight various effects.)

*Active region collector current*

or

$$I_C = \beta I_B + I_{CEO} \quad [6.45]$$

where

$$I_{CEO} = \frac{I_{CBO}}{(1-\alpha)} \approx \beta I_{CBO}$$

is the leakage current into the collector when the base is open circuited. This is much larger in the CE circuit than in the CB configuration.

Even when $I_B$ is kept constant, $I_C$ still exhibits a small increase with $V_{CE}$, which, according to Equation 6.45, indicates an increase in the current gain $\beta$ with $V_{CE}$. This is due to the Early effect or modulation of the base width by $V_{CB}$, shown in Figure 6.22. Increasing $V_{CE}$ increases $V_{CB}$, which increases $W_{BC}$, reduces $W_B$, and hence shortens $\tau_t$. The resulting effect is a larger $\beta$ ($\approx \tau_h/\tau_t$).

When $V_{CE}$ is less than $V_{BE}$, the collector junction becomes forward biased and Equation 6.45 is not valid. The collector current is then the difference between forward currents of emitter and collector junctions. The transistor operating in this region is said to be **saturated.**

## 6.6.4 Low-Frequency Small-Signal Model

The *npn* bipolar transistor in the CE (common emitter) amplifier configuration is shown in Figure 6.25. The input circuit has a dc bias $V_{BB}$ to forward bias the base–emitter (BE) junction and the output circuit has a dc voltage $V_{CC}$ (larger than $V_{BB}$) to reverse bias the base–collector (BC) junction through a collector resistor $R_C$.

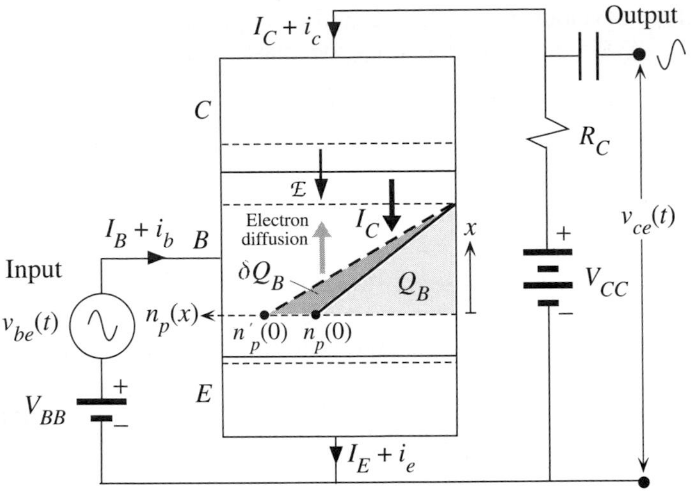


**Figure 6.25** An *npn* transistor operated in the active region in the common emitter amplifier configuration.
The applied signal $v_{be}$ modulates the dc voltage across the BE junction and hence modulates the injected electron concentration up and down about the dc value $n_p(0)$. The solid line shows $n_p(x)$ when only the dc bias $V_{BB}$ is present. The dashed line shows how $n_p(x)$ is modulated up by a positive small signal $v_{be}$ superimposed on $V_{BB}$.

The actual reverse bias voltage across the BC junction is $V_{CE} - V_{BE}$, where $V_{CE}$ is

$$V_{CE} = V_{CC} - I_C R_C$$

An input signal in the form of a small ac signal $v_{be}$ is applied in series with the bias voltage $V_{BB}$ and modulates the voltage $V_{BE}$ across the BE junction about its dc value $V_{BB}$. The varying voltage across the BE modulates $n_p(0)$ up and down about its dc value, which leads to a varying emitter current and hence to an almost identically varying collector current in the output circuit. The variation in the collector current is converted to an output voltage signal by the collector resistance $R_C$. Note that increasing $V_{BE}$ increases $I_C$, which leads to a decrease in $V_{CE}$. Thus, the output voltage is 180° out of phase with the input voltage.

Since the BE junction is forward-biased, the relationship between $I_E$ and $V_{BE}$ is exponential,

*Emitter current and $V_{BE}$*

$$I_E = I_{EO} \exp\left(\frac{eV_{BE}}{kT}\right) \qquad [6.46]$$

where $I_{EO}$ is a constant. We can differentiate this expression to relate small variations in $I_E$ and $V_{BE}$ as in the presence of small signals superimposed on dc values. For small signals, we have $v_{be} = \delta V_{BE}$, $i_b = \delta I_B$, $i_e = \delta I_E$, $i_c = \delta I_C$. Then from Equation 6.45 we see that $\delta I_C = \beta\,\delta I_B$, so $i_c = \beta i_b$. Since $\alpha \approx 1$, $i_e \approx i_c$.

What is the advantage of the CE circuit over the common base (CB) configuration? First, the input current is the base current, which is about a factor of $\beta$ smaller than the emitter current. The ac input resistance of the CE circuit is therefore a factor of $\beta$ higher than that of the CB circuit. This means that the amplifier does not load the ac source; the input resistance of the amplifier is much greater than the internal (or output) resistance of the ac source at the input. The small-signal input resistance $r_{be}$ is

*Input resistance*

$$r_{be} = \frac{v_{be}}{i_b} = \frac{\delta V_{BE}}{\delta I_B} \approx \beta\frac{\delta V_{BE}}{\delta I_E} = \frac{\beta kT}{eI_E} \approx \frac{\beta 25}{I_C(\text{mA})} \qquad [6.47]$$

where we differentiated Equation 6.46.

The output ac signal $v_{ce}$ develops across the CE and is tapped out through a capacitor. Since $V_{CE} = V_{CC} - I_C R_C$, as $I_C$ increases, $V_{CE}$ decreases. Thus,

$$v_{ce} = \delta V_{CE} = -R_C\,\delta I_C = -R_C i_c$$

The voltage amplification is

*Voltage gain*

$$A_V = \frac{v_{ce}}{v_{be}} = \frac{-R_C i_c}{r_{be} i_b} = \frac{-R_C \beta}{r_{be}} \approx -\frac{R_C I_C(\text{mA})}{25} \qquad [6.48]$$

which is the same as that in the CB configuration. However, in the CE configuration the output to input current ratio $i_c/i_b = \beta$, whereas this is almost unity in the CB configuration. Consequently, the CE configuration provides a greater power amplification, which is the second advantage of the CE circuit.

The input signal $v_{be}$ gives rise to an output current $i_c$. This input voltage to output current conversion is defined in a parameter called the **mutual conductance,** or **transconductance, $g_m$.**

*Transconductance*

$$g_m = \frac{i_c}{v_{be}} \approx \frac{\delta I_E}{\delta V_{BE}} = \frac{I_E(\text{mA})}{25} = \frac{1}{r_e} \qquad [6.49]$$

The voltage amplification of the CE amplifier is then

*Voltage gain*

$$A_V = -g_m R_C \qquad [6.50]$$

We generally find it convenient to use a small-signal equivalent circuit for the low-frequency behavior of a BJT in the CE configuration. Between the base and emitter, the applied ac source voltage $v_s$ sees only an input resistance of $r_{be}$, as shown in Figure 6.26. To underline the importance of the transistor input resistance, the output (or the internal) resistance $R_s$ of the ac source is also shown. In the output circuit there is a voltage-controlled current source $i_c$ which generates a current of $g_m v_{be}$. The current $i_c$ passes through the load (or collector) resistance $R_C$ across which the voltage signal develops. As we are only interested in ac signals, the batteries are taken as a short-circuit path for the ac current, which means that the internal resistances of the batteries are taken as zero. This model, of course, is valid only under normal and active operating conditions and small signals about dc values, and at low frequencies.

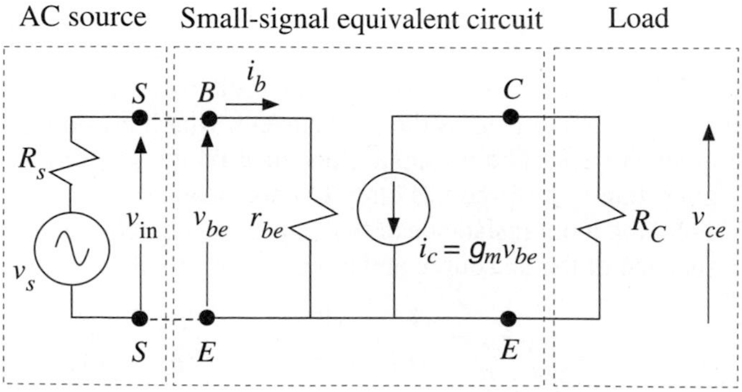


**Figure 6.26** Low-frequency small-signal simplified equivalent circuit of the bipolar transistor in the CE configuration with a load resistor $R_C$ in the collector circuit.

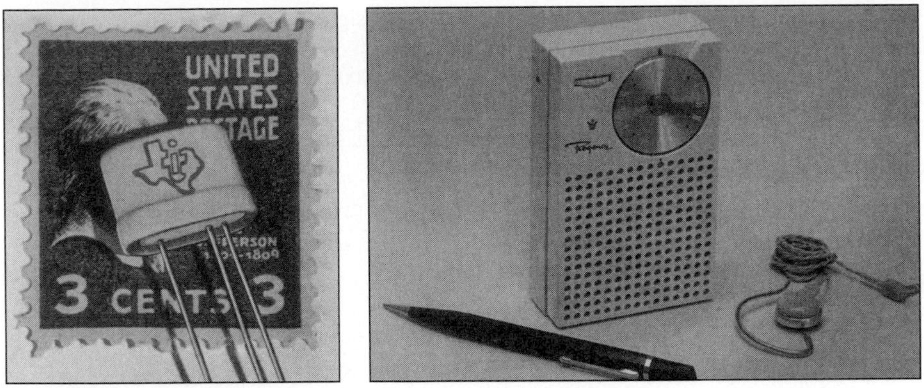


Left: The first commercial Si transistor from Texas Instruments (1954). Right: The first transistor pocket radio (1954). It had four Ge *npn* transistors.

| SOURCE: Courtesy of Texas Instruments.

The bipolar transistor general dc current equation $I_C = \beta I_B$, where $\beta \approx \tau_h/\tau_t$ is a material-dependent constant, implies that the ac small-signal collector current is

$$\delta I_C = \beta\, \delta I_B \qquad \text{or} \qquad i_c = \beta i_b$$

Thus the CE dc and ac small-signal current gains are the same. This is a reasonable approximation in the low-frequency range, typically at frequencies below $1/\tau_h$. It is useful to have a relationship between $\beta$, $g_m$, and $r_{be}$. Using Equations 6.47 and 6.49, we have

$$\beta = g_m r_{be} \tag{6.51}$$

*$\beta$ at low frequencies*

In transistor data books, the dc current gain $I_C/I_B$ is denoted as $h_{FE}$ whereas the small-signal ac current gain $i_c/i_b$ is denoted as $h_{fe}$. Except at high frequencies, $h_{fe} \approx h_{FE}$.

**EXAMPLE 6.11**

**CE LOW-FREQUENCY SMALL-SIGNAL EQUIVALENT CIRCUIT** Consider a BJT with a $\beta$ of 100, used in a CE amplifier in which the collector current is 2.5 mA and $R_C$ is 1 kΩ. If the ac source has an rms voltage of 1 mV and an output resistance $R_s$ of 50 Ω, what is the rms output voltage? What is the input and output power and the overall power amplification?

**SOLUTION**

As the collector current is 2.5 mA, the input resistance and the transconductance are

$$r_{be} = \frac{\beta 25}{I_C(\text{mA})} = \frac{(100)(25)}{2.5} = 1000\ \Omega$$

and

$$g_m = \frac{I_C(\text{mA})}{25} = \frac{2.5}{25} = 0.1\ \text{A/V}$$

The *magnitude* of the voltage gain of the BJT small-signal equivalent circuit is

$$A_V = \frac{v_{\text{ce}}}{v_{be}} = g_m R_C = (0.1)(1000) = 100$$

When the ac source is connected to the *B* and *E* terminals (Figure 6.26), the input resistance $r_{be}$ of the BJT loads the ac source, so $v_{be}$ across BE is

$$v_{be} = v_s \frac{r_{be}}{(r_{be} + R_s)} = (1\text{ mV})\frac{1000\ \Omega}{(1000\ \Omega + 50\ \Omega)} = 0.952\text{ mV}$$

The output voltage (rms) is, therefore,

$$v_{\text{ce}} = A_V v_{be} = 100(0.952\text{ mV}) = 95.2\text{ mV}$$

The loading effect makes the output less than 100 mV. To reduce the loading of the ac source, we need to increase $r_{be}$, *i.e.*, reduce the collector current, but that also reduces the gain. So to keep the gain the same, we need to reduce $I_C$ and increase $R_C$. However, $R_C$ cannot be increased indefinitely because $R_C$ itself is loaded by the input of the next stage and, in addition, there is an incremental resistance between the collector and emitter terminals (typically $\sim$100 k$\Omega$) that shunts $R_C$ (not shown in Figure 6.26).

The power amplification of the CE BJT itself is

$$A_P = \frac{i_c v_{\text{ce}}}{i_b v_{be}} = \beta A_V = (100)(100) = 10{,}000$$

The input power into the BE terminals is

$$P_{\text{in}} = v_{be} i_b = \frac{v_{be}^2}{r_{be}} = \frac{(0.952 \times 10^{-3}\text{ V})^2}{1000\ \Omega} = 9.06 \times 10^{-10}\text{ W} \quad \text{or} \quad 0.906\text{ nW}$$

The output power is

$$P_{\text{out}} = P_{\text{in}} A_P = (9.06 \times 10^{-10})(10{,}000) = 9.06 \times 10^{-6}\text{ W} \quad \text{or} \quad 9.06\ \mu\text{W}$$

---

# 6.7 JUNCTION FIELD EFFECT TRANSISTOR (JFET)

## 6.7.1 GENERAL PRINCIPLES

The basic structure of the junction field effect transistor (JFET) with an *n*-type channel (*n*-channel) is depicted in Figure 6.27a. An *n*-type semiconductor slab is provided with contacts at its ends to pass current through it. These terminals are called **source** (*S*) and **drain** (*D*). Two of the opposite faces of the *n*-type semiconductor are heavily *p*-type doped to some small depth so that an *n*-type channel is formed between the source and drain terminals, as shown in Figure 6.27a. The two $p^+$ regions are normally electrically connected and are called the **gate** (*G*). As the gate is heavily doped, the depletion layers extend almost entirely into the *n*-channel, as shown in Figure 6.27. For simplicity we will assume that the two gate regions are identical (both $p^+$ type) and that the doping in the *n*-type semiconductor is uniform. We will define the *n*-channel to be the region of conducting *n*-type material contained between the two depletion layers.

The basic and idealized symmetric structure in Figure 6.27a is useful in explaining the principle of operation as discussed later but does not truly represent

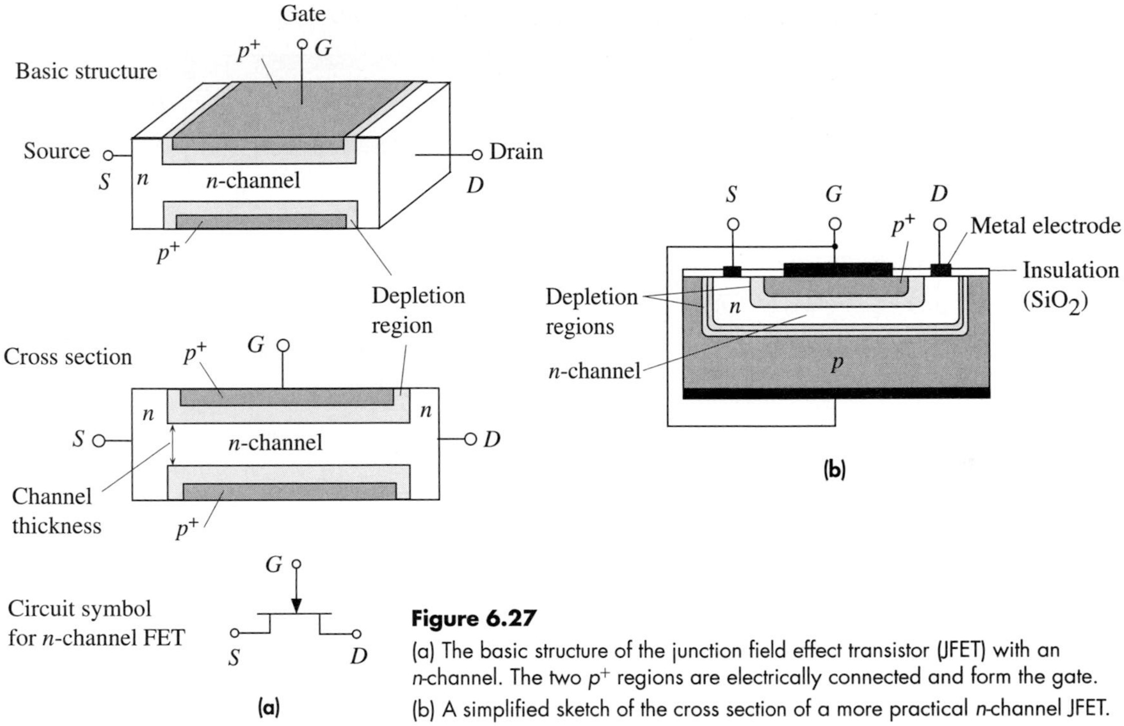


**Figure 6.27**
(a) The basic structure of the junction field effect transistor (JFET) with an *n*-channel. The two $p^+$ regions are electrically connected and form the gate.
(b) A simplified sketch of the cross section of a more practical *n*-channel JFET.

the structure of a typical practical device. A simplified schematic sketch of the cross section of a more practical device (as, for example, fabricated by the planar technology) is shown in Figure 6.27b where it is apparent that the two gate regions do not have identical doping and that, except for one of the gates, all contacts are on one surface.

We first consider the behavior of the JFET with the gate and source shorted ($V_{GS} = 0$), as shown in Figure 6.28a. The resistance between $S$ and $D$ is essentially the resistance of the conducting $n$-channel between $A$ and $B$, $R_{AB}$. When a positive voltage is applied to $D$ with respect to $S$ ($V_{DS} > 0$), then a current flows from $D$ to $S$, which is called the **drain current** $I_D$. There is a voltage drop along the channel, between $A$ and $B$, as indicated in Figure 6.28a. The voltage in the $n$-channel is zero at $A$ and $V_{DS}$ at $B$. As the voltage along the $n$-channel is positive, the $p^+n$ junctions between the gates and the $n$-channel become progressively more reverse-biased from $A$ to $B$. Consequently the depletion layers extend more into the channel and thereby decrease the thickness of the conducting channel from $A$ to $B$.

Increasing $V_{DS}$ increases the widths of the depletion layers, which penetrate more into the channel and hence result in more channel narrowing toward the drain. The resistance of the $n$-channel $R_{AB}$ therefore increases with $V_{DS}$. The drain current therefore does not increase linearly with $V_{DS}$ but falls below it because

$$I_D = \frac{V_{DS}}{R_{AB}}$$

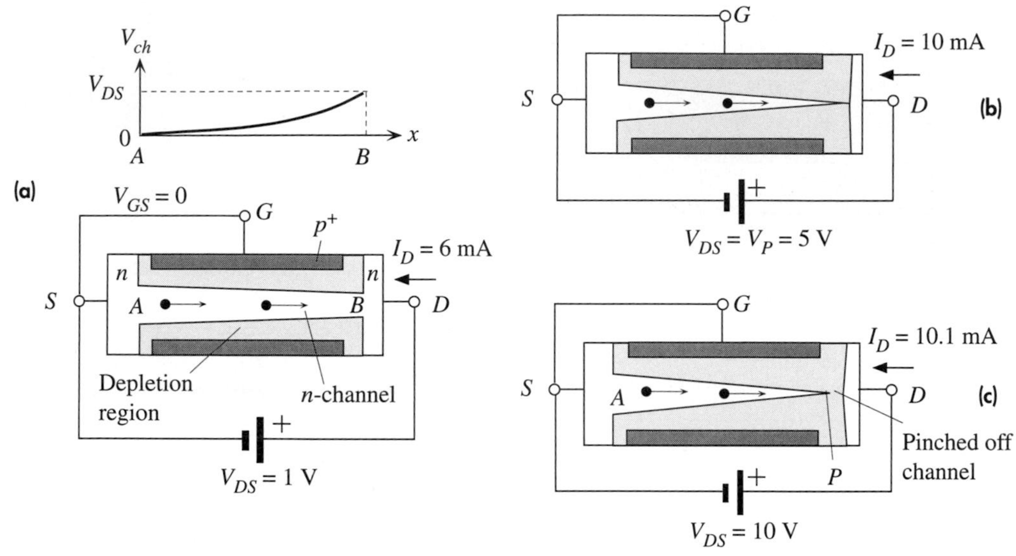


**Figure 6.28**

(a) The gate and source are shorted ($V_{GS} = 0$) and $V_{DS}$ is small.

(b) $V_{DS}$ has increased to a value that allows the two depletion layers to just touch, when $V_{DS} = V_P$ ($= 5$ V) and the $p^+n$ junction voltage at the drain end, $V_{GD} = -V_{DS} = -V_P = -5$ V.

(c) $V_{DS}$ is large ($V_{DS} > V_P$), so a short length of the channel is pinched off.

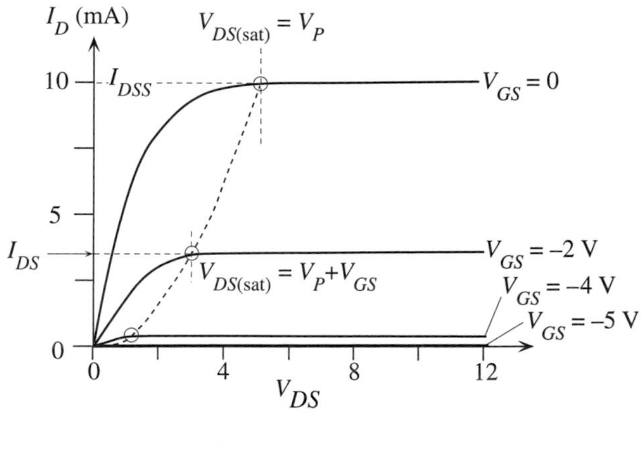


**Figure 6.29** Typical $I_D$ versus $V_{DS}$ characteristics of a JFET for various fixed gate voltages $V_{GS}$.

and $R_{AB}$ increases with $V_{DS}$. Thus $I_D$ versus $V_{DS}$ exhibits a sublinear behavior, as shown in the $V_{DS} < 5$ V region in Figure 6.29.

As $V_{DS}$ increases further, the depletion layers extend more into the channel and eventually, when $V_{DS} = V_P$ ($= 5$ V), the two depletion layers around $B$ meet at point $P$ at the drain end of the channel, as depicted in Figure 6.28b. The channel is then said to be "pinched off" by the two depletion layers. The voltage $V_P$ is called the **pinch-off voltage.** It is equal to the magnitude of reverse bias needed across the $p^+n$ junctions to

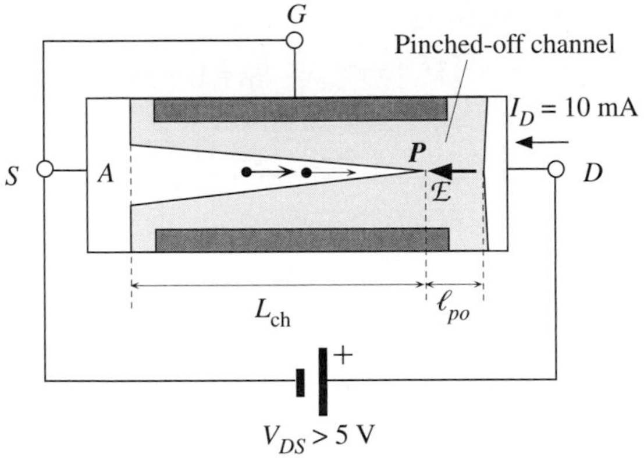


**Figure 6.30** The pinched-off channel and conduction for $V_{DS} > V_P$ (= 5 V).

make them just touch at the drain end. Since the actual bias voltage across the $p^+n$ junctions at the drain end ($B$) is $V_{GD}$, the pinch-off occurs whenever

*Pinch-off condition*

$$V_{GD} = -V_P \quad [6.52]$$

In the present case, gate to source is shorted, $V_{GS} = 0$, so $V_{GD} = -V_{DS}$ and pinch-off occurs when $V_{DS} = V_P$ (5 V). The drain current from pinch-off onwards, as shown in Figure 6.29, does not increase significantly with $V_{DS}$ for reasons given below. Beyond $V_{DS} = V_P$, there is a short pinched-off channel of length $\ell_{po}$.

The pinched-off channel is a reverse-biased depletion region that separates the drain from the $n$-channel, as depicted in Figure 6.30. There is a very strong electric field $\mathcal{E}$ in this pinched-off region in the $D$ to $S$ direction. This field is the vector sum of the fields from positive donors to negative acceptors in the depletion regions of the channel and the gate on the drain side. Electrons in the $n$-channel drift toward $P$, and when they arrive at $P$, they are swept across the pinched-off channel by $\mathcal{E}$. This process is similar to minority carriers in the base of a BJT reaching the collector junction depletion region, where the internal field there sweeps them across the depletion layer into the collector. Consequently the drain current is actually determined by the resistance of the conducting $n$-channel over $L_{ch}$ from $A$ to $P$ in Figure 6.30 and not by the pinched-off channel.

As $V_{DS}$ increases, most of the additional voltage simply drops across $\ell_{po}$ as this region is depleted of carriers and hence highly resistive. Point $P$, where the depletion layers first meet, moves slightly toward $A$, thereby slightly reducing the channel length $L_{ch}$. Point $P$ must still be at a potential $V_P$ because it is this potential that just makes the depletion layers touch. Thus the voltage drop across $L_{ch}$ remains as $V_P$. Beyond pinch-off then

$$I_D = \frac{V_P}{R_{AP}} \qquad (V_{DS} > V_P)$$

Since $R_{AP}$ is determined by $L_{ch}$, which decreases slightly with $V_{DS}$, $I_D$ increases slightly with $V_{DS}$. In many cases, $I_D$ is conveniently taken to be saturated at a value $I_{DSS}$ for $V_{DS} > V_P$. Typical $I_D$ versus $V_{DS}$ behavior is shown in Figure 6.29.

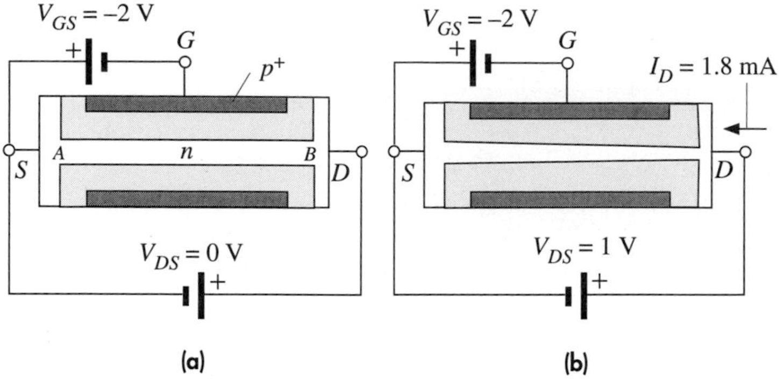


**Figure 6.31**

(a) The JFET with a negative $V_{GS}$ voltage has a narrower *n*-channel at the start.

(b) Compared to the $V_{GS} = 0$ case, the same $V_{DS}$ gives less $I_D$ as the channel is narrower.

(c) The channel is pinched off at $V_{DS} = 3$ V sooner than the $V_{GS} = 0$ case, where it was $V_{DS} = 5$ V.

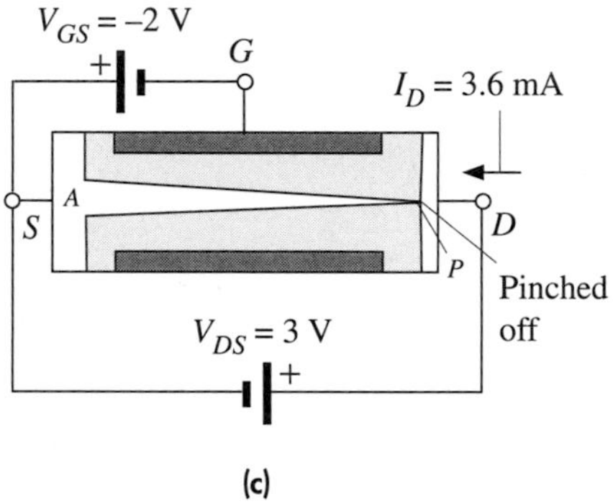


We now consider what happens when a negative voltage, say $V_{GS} = -2$ V, is applied to the gate with respect to the source, as shown in Figure 6.31a with $V_{DS} = 0$. The $p^+n$ junctions are now reverse-biased from the start, the channel is narrower, and the channel resistance is now larger than in the $V_{GS} = 0$ case. The drain current that flows when a small $V_{DS}$ is applied, as in Figure 6.31b, is now smaller than in the $V_{GS} = 0$ case as apparent in Figure 6.29. The $p^+n$ junctions are now progressively more reverse-biased from $V_{GS}$ at the source end to $V_{GD} = V_{GS} - V_{DS}$ at the drain end. We therefore need a smaller $V_{DS}$ (= 3 V) to pinch off the channel, as shown in Figure 6.31c. When $V_{DS} = 3$ V, the $G$ to $D$ voltage $V_{GD}$ across the $p^+n$ junctions at the drain end is $-5$ V, which is $-V_P$, so the channel becomes pinched off. Beyond pinch-off, $I_D$ is nearly saturated just as in the $V_{GS} = 0$ case, but its magnitude is obviously smaller as the thickness of the channel at $A$ is smaller; compare Figures 6.28 and 6.31. In the presence of $V_{GS}$, the pinch-off occurs at $V_{DS} = V_{DS(\text{sat})}$, and from Equation 6.52.

*Pinch-off condition*

$$V_{DS(\text{sat})} = V_P + V_{GS} \quad [6.53]$$

where $V_{GS}$ is a negative voltage (reducing $V_P$). Beyond pinch-off when $V_{DS} > V_{DS(\text{sat})}$, the point $P$ where the channel is *just pinched* still remains at potential $V_{DS(\text{sat})}$, given by Equation 6.53.

For $V_{DS} > V_{DS(\text{sat})}$, $I_D$ becomes nearly saturated at a value denoted as $I_{DS}$, which is indicated in Figure 6.29. When $G$ and $S$ are shorted ($V_{GS} = 0$), $I_{DS}$ is called $I_{DSS}$ (which stands for $I_{DS}$ with shorted gate to source). Beyond pinch-off, with negative $V_{GS}$, $I_{DS}$ is

$$I_D \approx I_{DS} \approx \frac{V_{DS(\text{sat})}}{R_{AP}(V_{GS})} = \frac{V_P + V_{GS}}{R_{AP}(V_{GS})} \qquad V_{DS} > V_{DS(\text{sat})} \tag{6.54}$$

where $R_{AP}$ $(V_{GS})$ is the effective resistance of the conducting $n$-channel from $A$ to $P$ (Figure 6.31b), which depends on the channel thickness and hence on $V_{GS}$. The resistance increases with more negative gate voltage as this increases the reverse bias across the $p^+n$ junctions, which leads to the narrowing of the channel. For example, when $V_{GS} = -4$ V, the channel thickness at $A$ becomes narrower than in the case with $V_{GS} = -2$ V, thereby increasing the resistance, $R_{AP}$, of the conducting channel and therefore decreasing $I_{DS}$. Further, there is also a reduction in the drain current by virtue of $V_{DS(\text{sat})}$ decreasing with negative $V_{GS}$, as apparent in Equation 6.54. Figure 6.29 shows the effect of the gate voltage on the $I_D$ versus $V_{DS}$ behavior. The two effects, that from $V_{DS(\text{sat})}$ and that from $R_{AP}$ $(V_{GS})$ in Equation 6.54, lead to $I_{DS}$ almost decreasing parabolically with $-V_{GS}$.

When the gate voltage is such that $V_{GS} = -V_P\,(= -5\text{ V})$ with the source and drain shorted ($V_{DS} = 0$), then the two depletion layers touch over the entire channel length and the whole channel is closed, as illustrated in Figure 6.32. The channel is said to be off. The only drain current that flows when a $V_{DS}$ is applied is due to the thermally generated carriers in the depletion layers. This current is very small.

Figure 6.29 summarizes the full $I_D$ versus $V_{DS}$ characteristics of the $n$-channel JFET at various gate voltages $V_{GS}$. It is apparent that $I_{DS}$ is relatively independent of $V_{DS}$ and that it is controlled by the gate voltage $V_{GS}$, as expected by Equation 6.54. This is analogous to the BJT in which the collector current $I_C$ is controlled by the base–emitter bias voltage $V_{BE}$. Figure 6.33a shows the dependence of $I_{DS}$ on the gate voltage $V_{GS}$. The transistor action is the control of the drain current $I_{DS}$, in the drain–source (output) circuit by the voltage $V_{GS}$ in the gate–source (input circuit), as shown in Figure 6.33b. This control is only possible if $V_{DS} > V_{DS(\text{sat})}$. When $V_{GS} = -V_p$, the drain current is nearly zero because the channel has been totally pinched off. This gate–source voltage is denoted by $V_{GS(\text{off})}$ as the drain current has been switched off. Furthermore, we should note that as $V_{GS}$ reverse biases the $p^+n$ junction, the current into the gate $I_G$ is the reverse leakage current of these junctions. It is usually very small. In some JFETs, $I_G$ is as low as a fraction of a nanoampere. We should also note that the circuit symbol for the JFET, as shown in Figure 6.27a, has an arrow to identify the gate and the $pn$ junction direction.

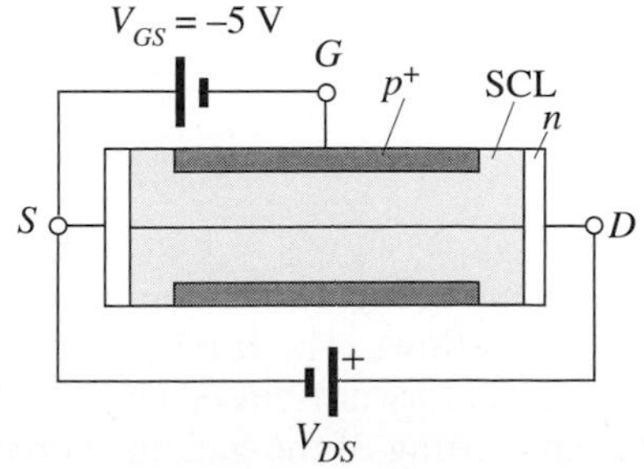


**Figure 6.32** When $V_{GS} = -5$ V, the depletion layers close the whole channel from the start, at $V_{DS} = 0$. As $V_{DS}$ is increased, there is a very small drain current, which is the small reverse leakage current due to thermal generation of carriers in the depletion layers.

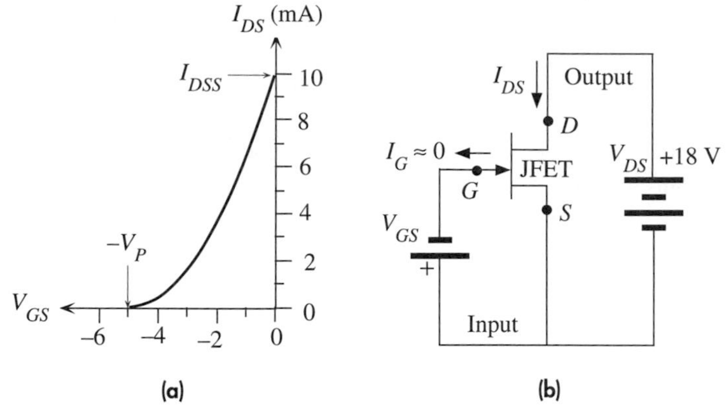


**Figure 6.33**
(a) Typical $I_{DS}$ versus $V_{GS}$ characteristics of a JFET.
(b) The dc circuit where $V_{GS}$ in the gate–source circuit (input) controls the drain current $I_{DS}$ in the drain–source (output) circuit in which $V_{DS}$ is kept constant and large ($V_{DS} > V_P$).

Is there a convenient relationship between $I_{DS}$ and $V_{GS}$? If we calculate the effective resistance $R_{AP}$ of the $n$-channel between $A$ and $P$, we can obtain its dependence on the channel thickness, and thus on the widths of the depletion layers and hence on $V_{GS}$. We can then find $I_{DS}$ from Equation 6.54. It turns out that a simple parabolic dependence seems to represent the data reasonably well,

*Beyond pinch-off*

$$I_{DS} = I_{DSS}\left[1 - \left(\frac{V_{GS}}{V_{GS(\text{off})}}\right)\right]^2 \qquad [6.55]$$

where $I_{DSS}$ is the drain current when $V_{GS} = 0$ (Figure 6.33) and $V_{GS(\text{off})}$ is defined as $-V_P$, that is, that gate–source voltage that just pinches off the channel. The pinch-off voltage $V_P$ here is a positive quantity because it was introduced through $V_{DS(\text{sat})}$. $V_{GS(\text{off})}$ however is negative, $-V_P$. We should note two important facts about the JFET. Its name originates from the effect that modulating the electric field in the reverse-biased depletion layers (by changing $V_{GS}$) varies the depletion layer penetration into the channel and hence the resistance of the channel. The transistor action hence can be thought of as being based on a **field effect.** Since there is a $p^+n$ junction between the gate and the channel, the name has become JFET. This junction in reverse bias provides the isolation between the gate and channel.

Secondly, the region beyond pinch-off, where Equations 6.54 and 6.55 hold, is commonly called the **current saturation region,** as well as **constant current region** and **pentode region.** The term **saturation** should not be confused with similar terms used for saturation effects in bipolar transistors. A saturated BJT cannot be used as an amplifier, but JFETs are invariably used as amplifiers in the saturated current region.

### 6.7.2 JFET Amplifier

The transistor action in the JFET is the control of $I_{DS}$ by $V_{GS}$, as shown in Figure 6.33. The input circuit is therefore the gate–source circuit containing $V_{GS}$ and the output circuit is the drain–source circuit in which the drain current $I_{DS}$ flows. The JFET is almost never used with the $pn$ junction between the gate and channel forward-biased ($V_{GS} > 0$) as this would lead to a very large gate current and near shorting of the gate to source

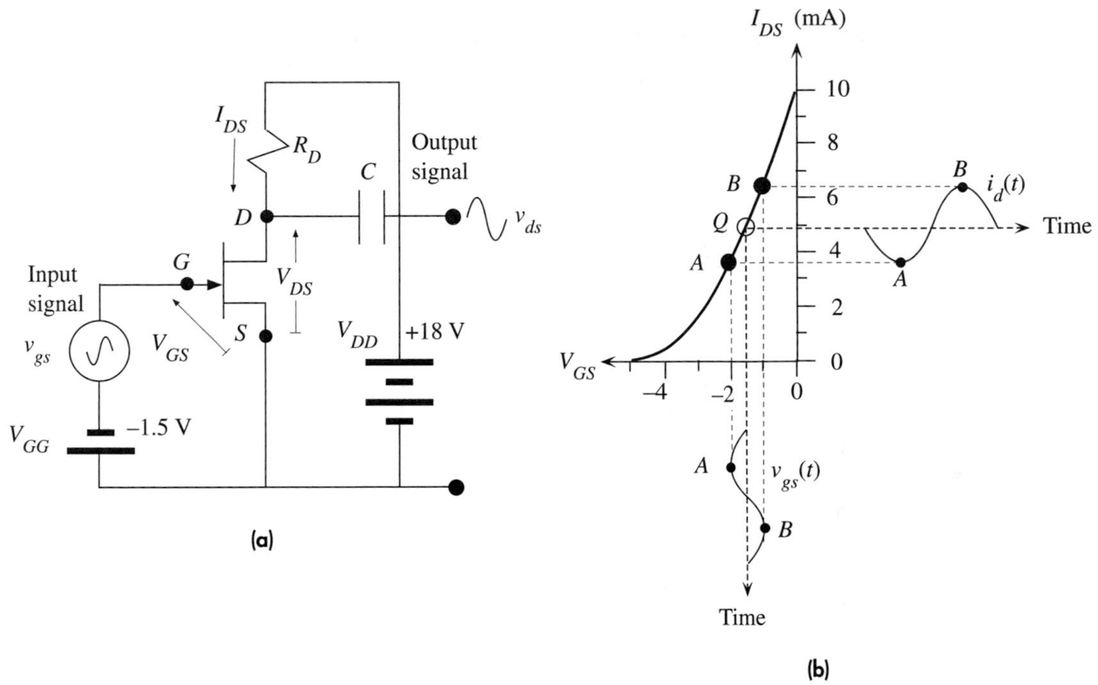


**Figure 6.34**
(a) Common source (CS) ac amplifier using a JFET.
(b) Explanation of how $I_D$ is modulated by the signal $v_{gs}$ in series with the dc bias voltage $V_{GG}$.

voltage. With $V_{GS}$ limited to negative voltages, the maximum current in the output circuit can only be $I_{DSS}$, as shown in Figure 6.33a. The maximum input voltage $V_{GS}$ should therefore give an $I_{DS}$ less than $I_{DSS}$.

Figure 6.34a shows a simplified illustration of a typical JFET voltage amplifier. As the source is common to both the input and output circuits, this is called a **common source** (CS) **amplifier.** The input signal is the ac source $v_{gs}$ connected in series with a negative dc bias voltage $V_{GG}$ of $-1.5$ V in the *GS* circuit. First we will find out what happens when there is no ac signal in the circuit ($v_{gs} = 0$). The dc supply ($-1.5$ V) in the input provides a negative dc voltage to the gate and therefore gives a dc current $I_{DS}$ in the output circuit (less than $I_{DSS}$). Figure 6.34b shows that when $V_{GS} = -1.5$ V, point $Q$ on the $I_{DS}$ versus $V_{GS}$ characteristics gives $I_{DS} = 4.9$ mA. Point $Q$, which determines the dc operation, is called the **quiescent point.**

The ac source $v_{gs}$ is connected in series with the negative dc bias voltage $V_{GS}$. It therefore modulates $V_{GS}$ up and down about $-1.5$ V with time, as shown in Figure 6.34b. Suppose that $v_{gs}$ varies sinusoidally between $-0.5$ V and $+0.5$ V. Then, as shown in Figure 6.34b when $v_{gs}$ is $-0.5$ V (point $A$), $V_{GS} = -2.0$ V and the drain current is given by point $A$ on the $I_{DS}$–$V_{GS}$ curve and is about 3.6 mA. When $v_{gs}$ is $+0.5$ V (point $B$), then $V_{GS} = -1.0$ V and the drain current is given by point $B$ on the $I_{DS}$–$V_{GS}$ curve and is about 6.4 mA. The input variation from $-0.5$ V to $+0.5$ V has thus been

**Table 6.1** Voltage and current in the common source amplifier of Figure 6.34a

| $v_{gs}$ (V) | $V_{GS}$ (V) | $I_{DS}$ (mA) | $i_d$ (mA) | $V_{DS} = V_{DD} - I_{DS}R_D$ | $v_{ds}$ (V) | Voltage Gain | Comment |
|---|---|---|---|---|---|---|---|
| 0 | −1.5 | 4.9 | 0 | 8.2 | 0 | | dc conditions, point $Q$ |
| −0.5 | −2.0 | 3.6 | −1.3 | 10.8 | +2.6 | −5.2 | Point $A$ |
| +0.5 | −1.0 | 6.4 | +1.5 | 5.2 | −3.0 | −6 | Point $B$ |

NOTE: $V_{DD} = 18$ V and $R_D = 2000\ \Omega$.

converted to a drain current variation from 3.6 mA to 6.4 mA as indicated in Figure 6.34b. We could have just as easily calculated the drain current from Equation 6.55. Table 6.1 summarizes what happens to the drain current as the ac input voltage is varied about zero.

The change in the drain current with respect to its dc value is the output signal current denoted as $i_d$. Thus at $A$,

$$i_d = 3.6 - 4.9 = -1.3 \text{ mA}$$

and at $B$,

$$i_d = 6.4 - 4.9 = 1.5 \text{ mA}$$

The variation in the output current is not quite symmetric as that in the input signal $v_{gs}$ because the $I_{DS}$–$V_{GS}$ relationship, Equation 6.55, is not linear.

The drain current variations in the $DS$ circuit are converted to voltage variations by the resistance $R_D$. The voltage across $DS$ is

$$V_{DS} = V_{DD} - I_{DS} R_D \qquad [6.56]$$

where $V_{DD}$ is the bias battery voltage in the $DS$ circuit. Thus, variations in $I_{DS}$ result in variations in $V_{DS}$ that are in the opposite direction or 180° out of phase. The ac output voltage between $D$ and $S$ is tapped out through a capacitor $C$, as shown in Figure 6.34a. The capacitor $C$ simply blocks the dc. Suppose that $R_D = 2000\ \Omega$ and $V_{DD} = 18$ V, then using Equation 6.56 we can calculate the dc value of $V_{DS}$ and also the minimum and maximum values of $V_{DS}$, as shown in Table 6.1.

It is apparent that as $v_{gs}$ varies from −0.5 V, at $A$, to +0.5 V, at $B$, $V_{DS}$ varies from 10.8 V to 5.2 V, respectively. The change in $V_{DS}$ with respect to dc is what constitutes the output signal $v_{ds}$, as only the ac is tapped out. From Equation 6.56, the change in $V_{DS}$ is related to the change in $I_{DS}$ by

$$v_{ds} = -R_D i_d \qquad [6.57]$$

Thus the output, $v_{ds}$, changes from −3.0 V to 2.6 V. The peak-to-peak voltage amplification is

$$A_{V(\text{pk-pk})} = \frac{\Delta V_{DS}}{\Delta V_{GS}} = \frac{v_{ds(\text{pk-pk})}}{v_{gs(\text{pk-pk})}} = \frac{-3\text{ V} - (2.6\text{ V})}{0.5\text{ V} - (-0.5\text{ V})} = -5.6$$

The negative sign represents the fact that the output and input voltages are out of phase by 180°. This can also be seen from Table 6.1 where a negative $v_{gs}$ results in a positive $v_{ds}$. Even though the ac input signal $v_{gs}$ is symmetric about zero, ±0.5 V, the ac output signal $v_{ds}$ is not symmetric, which is due to the $I_{DS}$ versus $V_{GS}$ curve being nonlinear, and thus varies between −3.0 V and 2.6 V. If we were to calculate the voltage amplification for the most negative input signal, we would find −5.2, whereas for the most positive input signal, it would be −6. The peak-to-peak voltage amplification, which was −5.6, represents a mean gain taking both negative and positive input signals into account.

The amplification can of course be increased by increasing $R_D$, but we must maintain $V_{DS}$ at all times above $V_{DS(\text{sat})}$ (beyond pinch-off) to ensure that the drain current $I_{DS}$ in the output circuit is only controlled by $V_{GS}$ in the input circuit.

When the signals are small about dc values, we can use differentials to represent small signals. For example, $v_{gs} = \delta V_{GS}$, $i_d = \delta I_{DS}$, $v_{ds} = \delta V_{DS}$, and so on. The variation $\delta I_{DS}$ due to $\delta V_{GS}$ about the dc value may be used to define a **mutual transconductance** $g_m$ (sometimes denoted as $g_{fs}$) for the JFET,

*Definition of JFET transconductance*

$$g_m = \frac{dI_{DS}}{dV_{GS}} \approx \frac{\delta I_{DS}}{\delta V_{GS}} = \frac{i_d}{v_{gs}}$$

This transconductance can be found by differentiating Equation 6.55,

*JFET transconductance*

$$g_m = \frac{dI_{DS}}{dV_{GS}} = -\frac{2I_{DSS}}{V_{GS(\text{off})}}\left[1 - \left(\frac{V_{GS}}{V_{GS(\text{off})}}\right)\right] = -\frac{2[I_{DSS}I_{DS}]^{1/2}}{V_{GS(\text{off})}} \qquad [6.58]$$

The output signal current is

$$i_d = g_m v_{gs}$$

so using Equation 6.57, the small-signal voltage amplification is

*Small-signal voltage gain*

$$A_V = \frac{v_{ds}}{v_{gs}} = \frac{-R_D(g_m v_{gs})}{v_{gs}} = -g_m R_D \qquad [6.59]$$

Equation 6.59 is only valid under small-signal conditions in which the variations about the dc values are small compared with the dc values themselves. The negative sign indicates that $v_{ds}$ and $v_{gs}$ are 180° out of phase.

**EXAMPLE 6.12**

**THE JFET AMPLIFIER** Consider the *n*-channel JFET common source amplifier shown in Figure 6.34a. The JFET has an $I_{DSS}$ of 10 mA and a pinch-off voltage $V_P$ of 5 V as in Figure 6.34b. Suppose that the gate dc bias voltage supply $V_{GG} = -1.5$ V, the drain circuit supply $V_{DD} = 18$ V, and $R_D = 2000\ \Omega$. What is the voltage amplification for small signals? How does this compare with the peak-to-peak amplification of −5.6 found for an input signal that had a peak-to-peak value of 1 V ?

**SOLUTION**

We first calculate the operating conditions at the bias point with no ac signals. This corresponds to point $Q$ in Figure 6.34b. The dc bias voltage $V_{GS}$ across the gate to source is −1.5 V. The resulting dc drain current $I_{DS}$ can be calculated from Equation 6.55 with $V_{GS(\text{off})} = -V_P = -5$ V:

$$I_{DS} = I_{DSS}\left[1 - \left(\frac{V_{GS}}{V_{GS(\text{off})}}\right)\right]^2 = (10\text{ mA})\left[1 - \left(\frac{-1.5}{-5}\right)\right]^2 = 4.9\text{ mA}$$

The transconductance at this dc current (at $Q$) is given by Equation 6.58,

$$g_m = -\frac{2(I_{DSS}I_{DS})^{1/2}}{V_{GS(\text{off})}} = -\frac{2[(10\times10^{-3})(4.9\times10^{-3})]^{1/2}}{-5} = 2.8\times10^{-3}\text{ A/V}$$

The voltage amplification of small signals about point $Q$ is

$$A_V = -g_m R_D = -(2.8\times10^{-3})(2000) = -5.6$$

This turns out to be the same as the peak-to-peak voltage amplification we calculated in Table 6.1. When the input ac signal $v_{gs}$ varies between $-0.5$ and $+0.5$ V, as in Table 6.1, the output signal is not symmetric. It varies between $-3$ V and 2.8 V, so the voltage gain depends on the input signal. The amplifier is then said to exhibit **nonlinearity.**

## 6.8 METAL-OXIDE-SEMICONDUCTOR FIELD EFFECT TRANSISTOR (MOSFET)

### 6.8.1 FIELD EFFECT AND INVERSION

The metal-oxide-semiconductor field effect transistor is based on the effect of a field penetrating into a semiconductor. Its operation can be understood by first considering a parallel plate capacitor with metal electrodes and a vacuum as insulation in between, as shown in Figure 6.35a. When a voltage $V$ is applied between the plates, charges $+Q$ and $-Q$ (where $Q = CV$) appear on the plates and there is an electric field given by $\mathcal{E} = V/L$. The origins of these charges are the conduction electrons for $-Q$ and exposed positively charged metal ions for $+Q$. Metallic bonding is based on all the valence electrons forming a sea of conduction electrons and permeating the space between metal ions that are fixed at crystal lattice sites. Since the electrons are mobile, they are readily displaced by the field. Thus in the lower plate $\mathcal{E}$ displaces some of the conduction electrons to the surface to form $-Q$. In the top plate $\mathcal{E}$ displaces some electrons from the surface into the bulk to expose positively charged metal ions to form $+Q$.

Suppose that the plate area is 1 cm$^2$ and spacing is 0.1 $\mu$m and that we apply 2 V across it. The capacitance $C$ is 8.85 nF and the magnitude of charge $Q$ on each plate is $1.77\times10^{-8}$ C, which corresponds to $1.1\times10^{11}$ electrons. A typical metal such as copper has something like $1.9\times10^{15}$ atoms per cm$^2$ on the surface. Thus, there will be that number of positive metal ions and electrons on the surface (assuming one conduction electron per atom). The charges $+Q$ and $-Q$ can therefore be generated by the electrons and metal ions at the surface alone. For example, if one in every $1.7\times10^4$ electrons on the surface moves one atomic spacing ($\sim$0.3 nm) into the bulk, then the surface will have a charge of $+Q$ due to exposed positive metal ions. It is clear that, for all practical purposes, the electric field does not penetrate into the metal and terminates at the metal surface.

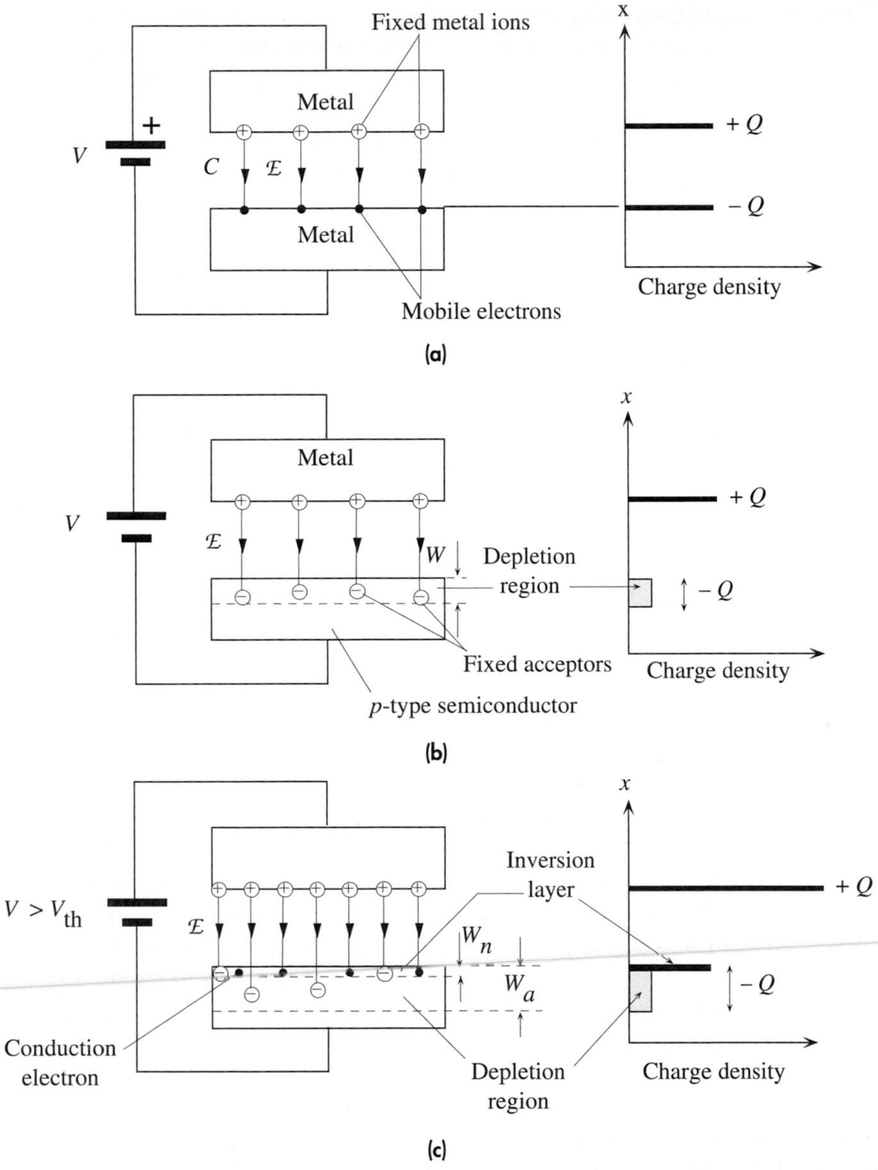


**Figure 6.35** The field effect.
(a) In a metal-air-metal capacitor, all the charges reside on the surface.
(b) Illustration of field penetration into a *p*-type semiconductor.
(c) As the field increases, eventually when $V > V_{th}$, an inversion layer is created near the surface in which there are conduction electrons.

The same is not true when one of the electrodes is a semiconductor, as shown in Figure 6.35b where the structure now is of the metal-insulator-semiconductor type. Suppose that we replace the lower metal in Figure 6.35a with a $p$-type semiconductor with an acceptor concentration of $10^{15}$ cm$^{-3}$. The number of acceptor atoms on the surface[10] is $1 \times 10^{10}$ cm$^{-2}$. We may assume that at room temperature all the acceptors are ionized and thus negatively charged. It is immediately apparent that we do not have a sufficient number of negative acceptors at the surface to generate the charge $-Q$. We must therefore also expose negative acceptors in the bulk, which means that the field must penetrate into the semiconductor. Holes in the surface region of the semiconductor become repelled toward the bulk and thereby expose more negative acceptors. We can estimate the width $W$ into which the field penetrates since the total negative charge exposed $eAWN_a$ must be $Q$. We find that $W$ is of the order of 1 $\mu$m, which is something like 4000 atomic layers. Our conclusion is that the field penetrates into a semiconductor by an amount that depends on the doping concentration.

The penetrating field into the semiconductor drifts away most of the holes in this region and thereby exposes negatively charged acceptors to make up the charge $-Q$. The region into which the field penetrates has lost holes and is therefore depleted of its equilibrium concentration of holes. We refer to this region as a **depletion layer.** As long as $p > n$ even though $p \ll N_a$, this still has $p$-type characteristics as holes are in the majority.

If the voltage increases further, $-Q$ also increases, as the field becomes stronger and penetrates more into the semiconductor but eventually it becomes more difficult to make up the charge $-Q$ by simply extending the depletion layer width $W$ into the bulk. It becomes possible (and more favorable) to attract conduction electrons into the depletion layer and form a thin electron layer of width $W_n$ near the surface. The charge $-Q$ is now made up of the fixed negative charge of acceptors in $W_a$ and of conduction electrons in $W_n$, as shown in Figure 6.35c. Further increases in the voltage do not change the width $W_a$ of the depletion layer but simply increase the electron concentration in $W_n$. Where do these electrons come from as the semiconductor is doped $p$-type? Some are attracted into the depletion layer from the bulk, where they were minority carriers. But most are thermally generated by the breaking of Si–Si bonds (*i.e.*, across the bandgap) in the depleted layer. Thermal generation in the depletion layer generates electron–hole pairs that become separated by the field. The holes are then drifted by the field into the bulk and the electrons toward the surface. Recombination of the thermally generated electrons and holes with other carriers is greatly reduced because the depletion layer has so few carriers. Since the electron concentration in the electron layer exceeds the hole concentration and this layer is within a normally $p$-type semiconductor, we call this an **inversion layer.**

It is now apparent that increasing the field in the metal-insulator-semiconductor device first creates a depletion layer and then an inversion layer at the surface when the voltage exceeds some threshold value $V_{th}$ . This is the basic principle of the field effect device. As long as $V > V_{th}$, any increase in the field and hence $-Q$ leads to more electrons in the inversion layer, whereas the width of the depletion layer $W_a$ and hence the quantity

[10] Surface concentration of atoms (atoms per unit area) can be found from $n_{surf} \approx (n_{bulk})^{2/3}$.

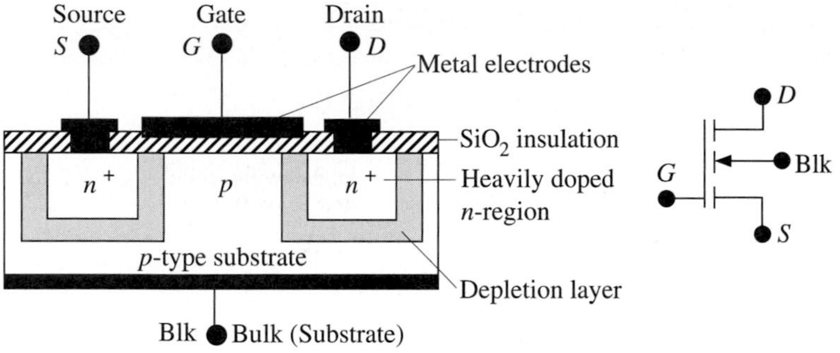


**Figure 6.36** The basic structure of the enhancement MOSFET and its circuit symbol.

of fixed negative charge remain constant. The insulator between the metal and the semiconductor, that is, a vacuum in Figure 6.35, is typically $SiO_2$ in many devices.

## 6.8.2 ENHANCEMENT MOSFET

Figure 6.36 shows the basic structure of an enhancement $n$-channel MOSFET device (NMOSFET). A metal-insulator-semiconductor structure is formed between a $p$-type Si substrate and an aluminum electrode, which is called the gate ($G$). The insulator is the $SiO_2$ oxide grown during fabrication. There are two $n^+$ doped regions at the ends of the MOS device that form the source ($S$) and drain ($D$). A metal contact is also made to the $p$-type Si substrate (or the bulk), which in many devices is connected to the source terminal as shown in Figure 6.36. Further, many MOSFETs have a degenerately doped polycrystalline Si material as the gate that serves the same function as the metal electrode.

With no voltage applied to the gate, $S$ to $D$ is an $n^+pn^+$ structure that is always reverse-biased whatever the polarity of the source to drain voltage. However, if the substrate (bulk) is connected to the source, a negative $V_{DS}$ will forward bias the $n^+p$ junction between the drain and the substrate. As the $n$-channel MOSFET device is not normally used with a negative $V_{DS}$, we will not consider this polarity.

When a positive voltage less than $V_{\text{th}}$ is applied to the gate, $V_{GS} < V_{\text{th}}$ , as shown in Figure 6.37a, the $p$-type semiconductor under the gate develops a depletion layer as a result of the expulsion of holes into the bulk, just as in Figure 6.35b. Since $S$ and $D$ are isolated by a low-conductivity $p$-doped region that has a depletion layer from $S$ to $D$, no current can flow for any positive $V_{DS}$.

With $V_{DS} = 0$, as soon as $V_{GS}$ is increased beyond the threshold voltage $V_{\text{th}}$, an $n$-channel inversion layer is formed within the depletion layer under the gate and immediately below the surface, as shown in Figure 6.37b. This $n$-channel links the two $n^+$ regions of source and drain. We then have a continuous $n$-type material with electrons as mobile carriers between the source and drain. When a small $V_{DS}$ is applied, a drain current $I_D$ flows that is limited by the resistance of the $n$-channel $R_{n\text{-ch}}$:

$$I_D = \frac{V_{DS}}{R_{n\text{-ch}}} \tag{6.60}$$

Thus, $I_D$ initially increases with $V_{DS}$ almost linearly, as shown in Figure 6.37b.

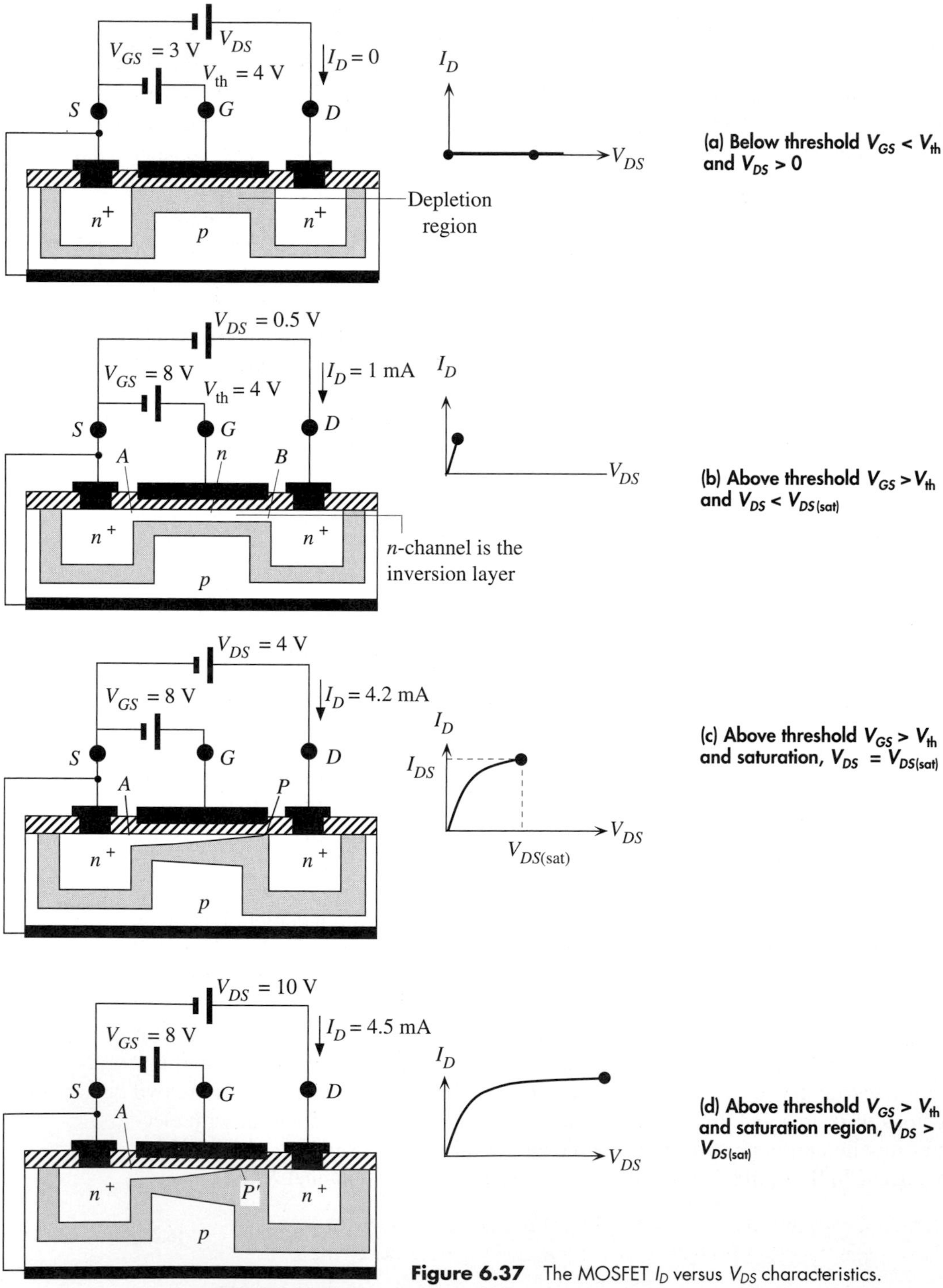


**Figure 6.37** The MOSFET $I_D$ versus $V_{DS}$ characteristics.

The voltage variation along the channel is from zero at $A$ (source end) to $V_{DS}$ at $B$ (drain end). The gate to the $n$-channel voltage is then $V_{GS}$ at $A$ and $V_{GD} = V_{GS} - V_{DS}$ at $B$. Thus point $A$ depends only on $V_{GS}$ and remains undisturbed by $V_{DS}$. As $V_{DS}$ increases, the voltage at $B$ ($V_{GD}$) decreases and thereby causes less inversion. This means that the channel gets narrower from $A$ to $B$ and its resistance $R_{n\text{-ch}}$, increases with $V_{DS}$. $I_D$ versus $V_{DS}$ then falls increasingly below the $I_D \propto V_{DS}$ line. Eventually when the gate to $n$-channel voltage at $B$ decreases to just below $V_{\text{th}}$, the inversion layer at $B$ disappears and a depletion layer is exposed, as illustrated in Figure 6.37c. The $n$-channel becomes pinched off at this point $P$. This occurs when $V_{DS} = V_{DS(\text{sat})}$, satisfying

$$V_{GD} = V_{GS} - V_{DS(\text{sat})} = V_{\text{th}} \qquad [6.61]$$

It is apparent that the whole process of the narrowing of the $n$-channel and its eventual pinch-off is similar to the operation of the $n$-channel JFET. When the drifting electrons in the $n$-channel reach $P$, the large electric field within the very narrow depletion layer at $P$ sweeps the electrons across into the $n^+$ drain. The current is limited by the supply of electrons from the $n$-channel to the depletion layer at $P$, which means that it is limited by the effective resistance of the $n$-channel between $A$ and $P$.

When $V_{DS}$ exceeds $V_{DS(\text{sat})}$, the additional $V_{DS}$ drops mainly across the highly resistive depletion layer at $P$, which extends slightly to $P'$ toward $A$, as shown in Figure 6.37d. At $P'$, the gate to channel voltage must still be just $V_{\text{th}}$ as this is the voltage required to just pinch off the channel and just eliminate inversion. The widening of the depletion layer (from $B$ to $P'$) at the drain end with $V_{DS}$, however, is small compared with the channel length $AB$. The resistance of the channel from $A$ to $P'$ does not change significantly with increasing $V_{DS}$, which means that the drain current is then nearly saturated at $I_{DS}$,

$$I_D \approx I_{DS} \approx \frac{V_{DS(\text{sat})}}{R_{AP'n\text{-ch}}} \qquad V_{DS} > V_{DS(\text{sat})} \qquad [6.62]$$

As $V_{DS(\text{sat})}$ depends on $V_{GS}$, so does $I_{DS}$. The overall $I_{DS}$ versus $V_{DS}$ characteristics for various fixed gate voltages $V_{GS}$ of a typical enhancement MOSFET is shown in Figure 6.38a. It can be seen that there is only a slight increase in $I_{DS}$ with $V_{DS}$ beyond $V_{DS(\text{sat})}$. The $I_{DS}$ versus $V_{GS}$ when $V_{DS} > V_{DS(\text{sat})}$ characteristics are shown in Figure 6.38b. It is apparent that as long as $V_{DS} > V_{DS(\text{sat})}$, the saturated drain current $I_{DS}$ in the source–drain (or output) circuit is almost totally controlled by the gate voltage $V_{GS}$ in the source–gate (or input) circuit. This is what constitutes the MOSFET action. Variations in $V_{GS}$ then lead to variations in the drain current $I_{DS}$ (just as in the JFET), which forms the basis of the MOSFET amplifier. The term *enhancement* refers to the fact that a gate voltage exceeding $V_{\text{th}}$ is required to enhance a conducting channel between the source and drain. This contrasts with the JFET where the gate voltage depletes the channel and decreases the drain current.

The experimental relationship between $I_{DS}$ and $V_{GS}$ (when $V_{DS} > V_{DS(\text{sat})}$) has been found to be best described by a parabolic equation similar to that for the JFET, except that now $V_{GS}$ enhances the channel when $V_{GS} > V_{\text{th}}$ so $I_{DS}$ exists only when $V_{GS} > V_{\text{th}}$,

*Enhancement NMOSFET*

$$I_{DS} = K(V_{GS} - V_{\text{th}})^2 \qquad [6.63]$$

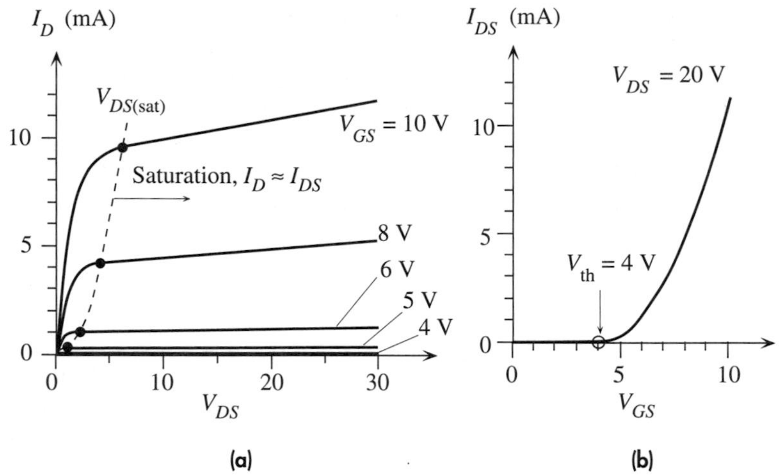


**Figure 6.38**
(a) Typical $I_D$ versus $V_{DS}$ characteristics of an enhancement MOSFET ($V_{th} = 4$ V) for various fixed gate voltages $V_{GS}$.
(b) Dependence of $I_{DS}$ on $V_{GS}$ at a given $V_{DS}$ ($> V_{DS\,(sat)}$).

where $K$ is a constant. For an ideal MOSFET, it can be expressed as

*Enhancement NMOSFET constant*

$$K = \frac{Z\mu_e\varepsilon}{2Lt_{ox}}$$

where $\mu_e$ is the electron drift mobility in the channel, $L$ and $Z$ are the length and width of the gate controlling the channel, and $\varepsilon$ and $t_{ox}$ are the permittivity ($\varepsilon_r\varepsilon_o$) and thickness of the oxide insulation under the gate. According to Equation 6.63, $I_{DS}$ is independent of $V_{DS}$. The shallow slopes of the $I_D$ versus $V_{DS}$ lines beyond $V_{DS(sat)}$ in Figure 6.38a can be accounted for by writing Equation 6.63 as

*Enhancement NMOSFET*

$$I_{DS} = K(V_{GS} - V_{th})^2(1 + \lambda V_{DS}) \qquad [6.64]$$

where $\lambda$ is a constant that is typically 0.01 $V^{-1}$. If we extend the $I_{DS}$ versus $V_{DS}$ lines, they intersect the $-V_{DS}$ axis at $1/\lambda$, which is called the **Early voltage.** It should be apparent that $I_{DSS}$, which is $I_{DS}$ with the gate and source shorted ($V_{GS} = 0$), is zero and is not a useful quantity in describing the behavior of the enhancement MOSFET.

**EXAMPLE 6.13** **THE ENHANCEMENT NMOSFET** A particular enhancement NMOS transistor has a gate with a width ($Z$) of 50 $\mu$m, length ($L$) of 10 $\mu$m, and $SiO_2$ thickness of 450 Å. The relative permittivity of $SiO_2$ is 3.9. The $p$-type bulk is doped with $10^{16}$ acceptors cm$^{-3}$. Its threshold voltage is 4 V. Estimate the drain current when $V_{GS} = 8$ V and $V_{DS} = 20$ V, given $\lambda = 0.01$. Due to the strong scattering of electrons near the crystal surface assume that the electron drift mobility $\mu_e$ in the channel is half the drift mobility in the bulk.

**SOLUTION**

Since $V_{DS} > V_{th}$, we can assume that the drain current is saturated and we can use the $I_{DS}$ versus $V_{GS}$ relationship in Equation 6.64,

$$I_{DS} = K(V_{GS} - V_{th})^2(1 + \lambda V_{DS})$$

where

$$K = \frac{Z\mu_e\varepsilon}{2Lt_{ox}}$$

The electron mobility in the bulk when $N_a = 10^{16}$ cm$^{-3}$ is 1300 cm$^2$ V$^{-1}$ s$^{-1}$ (Chapter 5). Thus

$$K = \frac{Z\mu_e \varepsilon_r \varepsilon_0}{2Lt_{ox}} = \frac{(50 \times 10^{-6})\left(\frac{1}{2} \times 1300 \times 10^{-4}\right)(3.9 \times 8.85 \times 10^{-12})}{2(10 \times 10^{-6})(450 \times 10^{-10})} = 0.000125$$

When $V_{GS} = 8$ V and $V_{DS} = 20$ V, with $\lambda = 0.01$, we have

$$I_{DS} = 0.000125(8-4)^2[1 + (0.01)(20)] = 0.0024 \text{ A} \quad \text{or} \quad 2.4 \text{ mA}$$

## 6.8.3 THRESHOLD VOLTAGE

The threshold voltage is an important parameter in MOSFET devices. Its control in device fabrication is therefore essential. Figure 6.39a shows an idealized MOS structure where all the electric field lines from the metal pass through the oxide and penetrate the $p$-type semiconductor. The charge $-Q$ is made up of fixed negative acceptors in a surface region of $W_a$ and of conduction electrons in the inversion layer at the surface, as shown in Figure 6.39a. The voltage drop across the MOS structure, however,

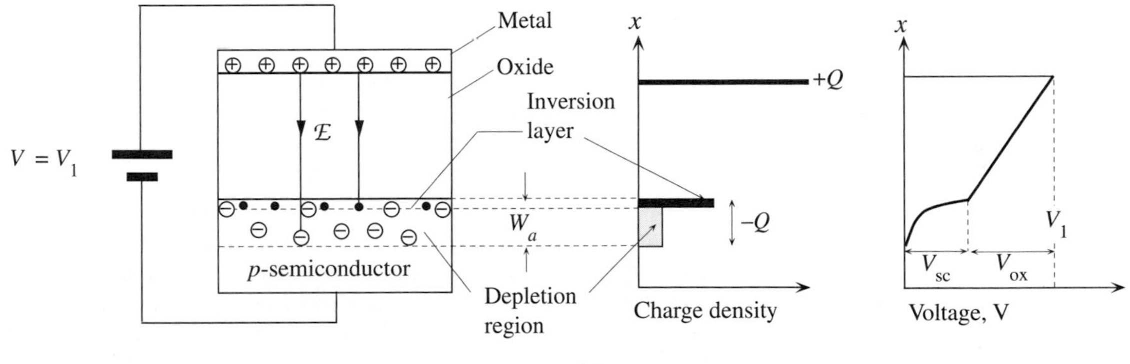


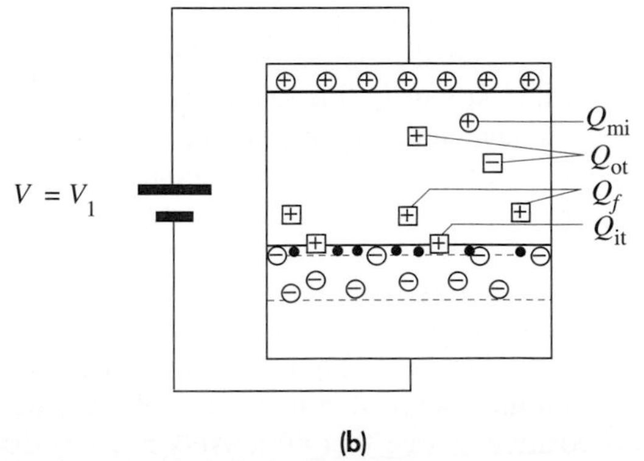


**Figure 6.39**
(a) The threshold voltage and the ideal MOS structure.
(b) In practice, there are several charges in the oxide and at the oxide–semiconductor interface that affect the threshold voltage: $Q_{mi}$ = mobile ionic charge (e.g., $Na^+$), $Q_{ot}$ = trapped oxide charge, $Q_f$ = fixed oxide charge, and $Q_{it}$ = charge trapped at the interface.

is not uniform. As the field penetrates the semiconductor, there is a voltage drop $V_{sc}$ across the field penetration region of the semiconductor by virtue of $\mathcal{E} = -dV/dx$, as shown in Figure 6.39a. The field terminates on both electrons in the inversion layer and acceptors in $W_a$, so within the semiconductor $\mathcal{E}$ is not uniform and therefore the voltage drop is not constant. But the field in the oxide is uniform, as we assumed there were no charges inside the oxide. The voltage drop across the oxide is constant and is $V_{ox}$, as shown in Figure 6.39a. As the applied voltage is $V_1$, we must have $V_{sc} + V_{ox} = V_1$. The actual voltage drop $V_{sc}$ across the semiconductor determines the condition for inversion. We can show this as follows. If the acceptor doping concentration is $10^{16}$ cm$^{-3}$, then the Fermi level $E_F$ in the bulk of the $p$-type semiconductor must be 0.347 eV below $E_{Fi}$ in intrinsic Si. To make the surface $n$-type we need to shift $E_F$ at the surface to go just above $E_{Fi}$. Thus we need to shift $E_F$ from bulk to surface by at least 0.347 eV. We have to bend the energy band by 0.347 eV at the surface. Since the voltage drop across the semiconductor is $V_{sc}$ and the corresponding electrostatic $PE$ change is $eV_{sc}$, this must be 0.347 eV or $V_{sc} = 0.347$ V. The gate voltage for the start of inversion will then be $V_{ox} + 0.347$ V. By inversion, however, we generally infer that the electron concentration at the surface is comparable to the hole concentration in the bulk. This means that we actually have to shift $E_F$ above $E_{Fi}$ by another 0.347 eV, so the gate threshold voltage $V_{th}$ must be $V_{ox} + 0.694$ V.

In practice there are a number of other important effects that must be considered in evaluating the threshold voltage. Invariably there are charges both within the oxide and at the oxide–semiconductor interface that alter the field penetration into the semiconductor and hence the threshold voltage needed at the gate to cause inversion. Some of these are depicted in Figure 6.39b and can be qualitatively summarized as follows.

There may be some mobile ions within the $SiO_2$, such as alkaline ions ($Na^+$, $K^+$), which are denoted as $Q_{mi}$ in Figure 6.39b. These may be introduced unintentionally, for example, during cleaning and etching processes in the fabrication. In addition there may be various trapped (immobile) charges within the oxide $Q_{ot}$ due to structural defects, for example, an interstitial $Si^+$. Frequently these oxide trapped charges are created as a result of radiation damage (irradiation by X-rays or other high-energy beams). They can be reduced by annealing the device.

A significant number of fixed positive charges ($Q_f$) exist in the oxide region close to the interface. They are believed to originate from the nonstoichiometry of the oxide near the oxide–semiconductor interface. They are generally attributed to positively charged $Si^+$ ions. During the oxidation process, a Si atom is removed from the Si surface to react with the oxygen diffusing in through the oxide. When the oxidation process is stopped suddenly, there are unfulfilled Si ions in this region. $Q_f$ depends on the crystal orientation and on the oxidation and annealing processes. The semiconductor to oxide interface itself is a sudden change in the structure from crystalline Si to amorphous oxide. The semiconductor surface itself will have various defects, as discussed in Chapter 1. There is some inevitable mismatch between the two structures at the interface, and consequently there are broken bonds, dangling bonds, point defects such as vacancies and $Si^+$, and other defects at this interface that trap charges (*e.g.*, holes). All these interface charges are represented as $Q_{it}$ in Figure 6.39b. $Q_{it}$ depends not only on the crystal orientation but also on the chemical composition of the interface. Both $Q_f$ and $Q_{it}$ overall represent a positive charge that effectively reduces the gate voltage needed for inversion. They are smaller for the (100) surface than the (111) surface, so (100) is the preferred surface for the Si MOS device.

In addition to various charges in the oxide and at the interface shown in Figure 6.39b, there will also be a voltage difference, denoted as $V_{FB}$, between the semiconductor surface and the metal surface, even in the absence of an applied voltage. $V_{FB}$ arises from the work function difference between the metal and the $p$-type semiconductor, as discussed in Chapter 4. The metal work function is generally smaller than the semiconductor work function, which means that the semiconductor surface will have an accumulation of electrons and the metal surface will have positive charges (exposed metal ions). The gate voltage needed for inversion will therefore also depend on $V_{FB}$. Since $V_{FB}$ is normally positive and $Q_{\text{f}}$ and $Q_{\text{it}}$ are also positive, there may already be an inversion layer formed at the semiconductor surface even without a positive gate voltage. The fabrication of an enhancement MOSFET then requires special fabrication procedures, such as ion implantation, to obtain a positive and predictable $V_{\text{th}}$.

The simplest way to control the threshold gate voltage is to provide a separate electrode to the bulk of an enhancement MOSFET, as shown in Figure 6.36, and to apply a bias voltage to the bulk with respect to the source to obtain the desired $V_{\text{th}}$ between the gate and source. This technique has the disadvantage of requiring an additional bias supply for the bulk and also adjusting the bulk to source voltage almost individually for each MOSFET.

### 6.8.4 ION IMPLANTED MOS TRANSISTORS AND POLY-SI GATES

The most accurate method of controlling the threshold voltage is by ion implantation, as the number of ions that are implanted into a device and their location can be closely controlled. Furthermore, ion implantation can also provide a self-alignment of the edges of the gate electrode with the source and drain regions. In the case of an $n$-channel enhancement MOSFET, it is generally desirable to keep the $p$-type doping in the bulk low to avoid small $V_{DS}$ for reverse breakdown between the drain and the bulk (see Figure 6.36). Consequently, the surface, in practice, already has an inversion layer (without any gate voltage) due to various fixed positive charges residing in the oxide and at the interface, as shown in Figure 6.39b (positive $Q_{\text{f}}$ and $Q_{\text{it}}$ and $V_{FB}$). It then becomes necessary to implant the surface region under the gate with boron acceptors to remove the electrons and restore this region to a $p$-type behavior.

The ion implantation process is carried out in a vacuum where the required impurity ions are generated and then accelerated toward the device. The energy of the arriving ions and hence their penetration into the device can be readily controlled. Typically, the device is implanted with $B$ acceptors under the gate oxide, as shown in Figure 6.40. The distribution of implanted acceptors as a function of distance into the device from the surface of the oxide is also shown in the figure. The position of the peak depends on the energy of the ions and hence on the accelerating voltage. The peak of the concentration of implanted acceptors is made to occur just below the surface of the semiconductor. Since ion implantation involves the impact of energetic ions with the crystal structure, it results in the inevitable generation of various defects within the implanted region. The defects are almost totally eliminated by annealing the device at an

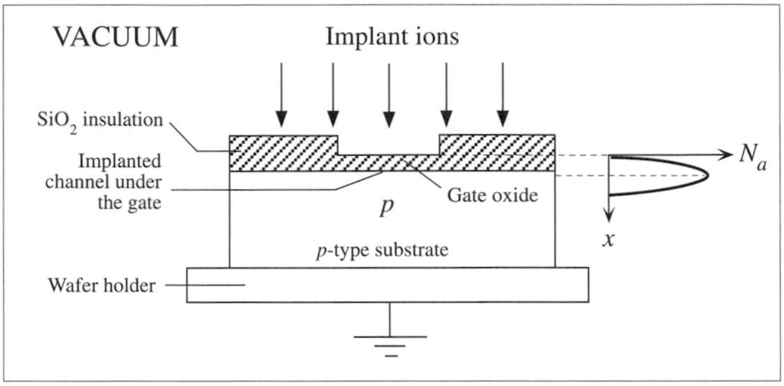


**Figure 6.40** Schematic illustration of ion implantation for the control of $V_{th}$.

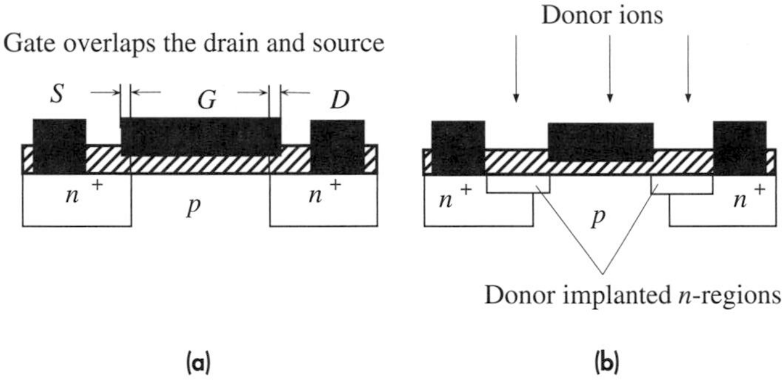


**Figure 6.41**
(a) There is an overlap of the gate electrode with the source and drain regions and hence additional capacitance between the gate and drain.
(b) $n^+$-type ion implantation extends the drain and source to line up with the gate.

elevated temperature. Annealing also broadens the acceptor implanted region as a result of increased diffusion of implanted acceptors.

Ion implantation also has the advantage of providing self-alignment of the drain and source with the edges of the gate electrode. In a MOS transistor, it is important that the gate electrode extends all the way from the source to the drain regions so that the channel formed under the gate can link the two regions; otherwise, an incomplete channel will be formed. To avoid the possibility of forming an incomplete channel, it is necessary to allow for some overlap, as shown in Figure 6.41a, between the gate and source and drain regions because of various tolerances and variations involved in the fabrication of a MOSFET by conventional masking and diffusional techniques. The overlap, however, results in additional capacitances between the gate and source and the gate and drain and adversely affects the high-frequency (or transient) response of the device. It is therefore desirable to align the edges of the gate electrode with the source and drain regions. Suppose that the gate electrode is made narrower so that it does not extend all the way between the source and drain regions, as shown in Figure 6.41b. If the device is now ion implanted with donors, then donor ions passing through the thin oxide will extend the $n^+$ regions up to the edges of the gate and thereby align the drain and source with the edges of the gate. The thick metal gate is practically impervious to the arriving donor ions.

Another method of controlling $V_{th}$ is to use silicon instead of Al for the gate electrode. This technique is called **silicon gate technology.** Typically, the silicon for the

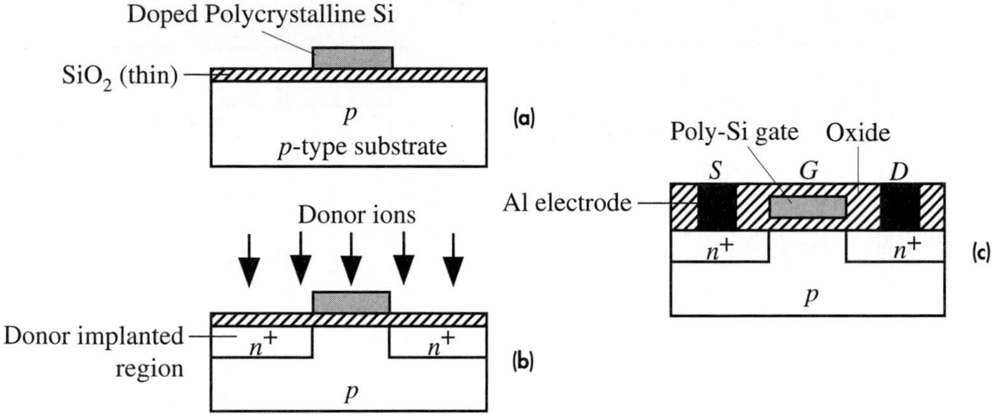


**Figure 6.42** The poly-Si gate technology.
(a) Poly-Si is deposited onto the oxide, and the areas outside the gate dimensions are etched away.
(b) The poly-Si gate acts as a mask during ion implantation of donors to form the $n^+$ source and drain regions.
(c) A simplified schematic sketch of the final poly-Si MOS transistor.

gate is vacuum deposited (*e.g.*, by chemical vapor deposition using silane gas) onto the oxide, as shown in Figure 6.42. As the oxide is noncrystalline, the Si gate is polycrystalline (rather than a single crystal) and is therefore called a **poly-Si gate.** Normally it is heavily doped to ensure that it has sufficiently low resistivity to avoid *RC* time constant limitations in charging and discharging the gate capacitance during transient or ac operations. The advantage of the poly-Si gate is that its work function depends on the doping (type and concentration) and can be controlled so that $V_{FB}$ and hence $V_{\text{th}}$ can also be controlled. There are also additional advantages in using the poly-Si gate. For example, it can be raised to high temperatures (Al melts at 660 °C). It can be used as a mask over the gate region of the semiconductor during the formation of the source and drain regions. If ion implantation is used to deposit donors into the semiconductor, then the $n^+$ source and drain regions are self-aligned with the poly-Si gate, as shown in Figure 6.42.

# 6.9 LIGHT EMITTING DIODES (LED)

## 6.9.1 LED PRINCIPLES

**A light emitting diode** (LED) is essentially a *pn* junction diode typically made from a direct bandgap semiconductor, for example, GaAs, in which the electron–hole pair (EHP) recombination results in the emission of a photon. The emitted photon energy $h\nu$ is approximately equal to the bandgap energy $E_g$. Figure 6.43a shows the energy band diagram of an unbiased $pn^+$ junction device in which the $n$-side is more heavily doped than the $p$-side. The Fermi level $E_F$ is uniform through the device, which is a requirement of equilibrium with no applied bias. The depletion region extends mainly into the $p$-side. There is a *PE* barrier $eV_o$ from $E_c$ on the $n$-side to $E_c$ on the $p$-side

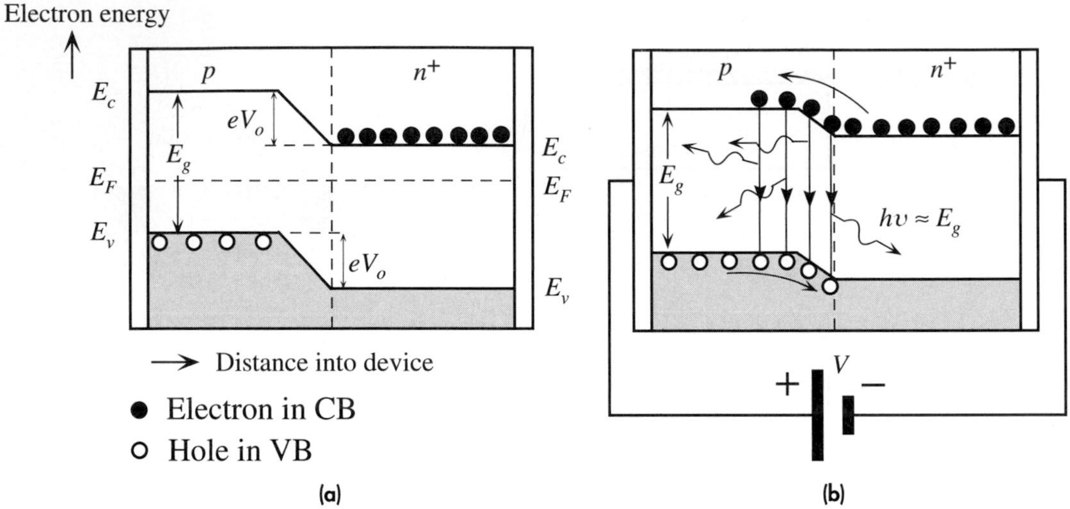


**Figure 6.43** Energy band diagram of a *pn* (heavily *n*-type doped) junction.
(a) No bias voltage.
(b) With forward bias *V*. Recombination around the junction and within the diffusion length of the electrons in the *p*-side leads to photon emission.

where $V_o$ is the *built-in voltage*. The *PE* barrier $eV_o$ prevents the diffusion of electrons from the $n$-side to the $p$-side.

When a forward bias $V$ is applied, the built-in potential $V_o$ is reduced to $V_o - V$, which then allows the electrons from the $n^+$-side to diffuse, that is, become injected, into the $p$-side as depicted in Figure 6.43b. The hole injection component from $p$ into the $n^+$-side is much smaller than the electron injection component from the $n^+$-side to the $p$-side. The recombination of injected electrons in the depletion region and within a volume extending over the electron diffusion length $L_e$ in the $p$-side leads to photon emission. The phenomenon of light emission from the EHP recombination as a result of minority carrier injection is called **injection electroluminescence.** Due to the statistical nature of the recombination process between electrons and holes, the emitted photons are in random directions; they result from spontaneous emission processes. The LED structure has to be such that the emitted photons can escape the device without being reabsorbed by the semiconductor material. This means the $p$-side has to be sufficiently narrow or we have to use *heterostructure* devices as discussed below.

One very simple LED structure is shown in Figure 6.44. First a doped semiconductor layer is grown on a suitable substrate (GaAs or GaP). The growth is done **epitaxially;** that is, the crystal of the new layer is grown to follow the structure of the substrate crystal. The **substrate** is essentially a sufficiently thick crystal that serves as a mechanical support for the *pn* junction device (the doped layers) and can be of different crystal. The $pn^+$ junction is formed by growing another epitaxial layer but doped $p$-type. Those photons that are emitted toward the $n$-side become either absorbed or reflected back at the substrate interface depending on the substrate thickness and the exact structure of the LED. If the epitaxial layer and the substrate crystals have different

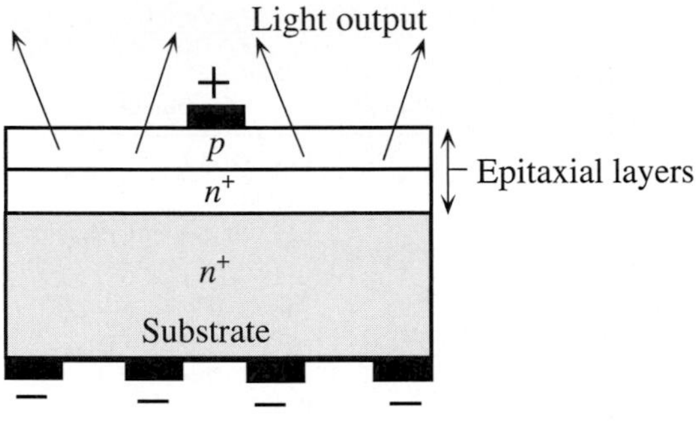


**Figure 6.44** A schematic illustration of one possible LED device structure. First an $n^+$ layer is epitaxially grown on a substrate. A thin $p$ layer is then epitaxially grown on the first layer.

crystal lattice parameters, then there is a lattice mismatch between the two crystal structures. This causes lattice strain in the LED layer and hence leads to crystal defects. Such crystal defects encourage radiationless EHP recombinations. That is, a defect acts as a recombination center. Such defects are reduced by lattice matching the LED epitaxial layer to the substrate crystal. It is therefore important to lattice match the LED layer to the substrate crystal. For example, one of the AlGaAs alloys is a direct bandgap semiconductor that has a bandgap in the red-emission region. It can be grown on GaAs substrates with excellent lattice match which results in high-efficiency LED devices.

There are various direct bandgap semiconductor materials that can be readily doped to make commercial *pn* junction LEDs which emit radiation in the red and infrared range of wavelengths. An important class of commercial semiconductor materials that covers the visible spectrum is the **III–V ternary alloys** based on alloying GaAs and GaP and denoted as $GaAs_{1-y}P_y$. In this compound, As and P atoms from Group V are distributed randomly at normal As sites in the GaAs crystal structure. When $y < 0.45$, the alloy $GaAs_{1-y}P_y$ is a direct bandgap semiconductor and hence the EHP recombination process is direct as depicted in Figure 6.45a. The rate of recombination is directly proportional to the product of electron and hole concentrations. The emitted wavelengths range from about 630 nm, red, for $y = 0.45$ ($GaAs_{0.55}P_{0.45}$) to 870 nm for $y = 0$ (GaAs).

$GaAs_{1-y}P_y$ alloys (which include GaP) with $y > 0.45$ are indirect bandgap semiconductors. The EHP recombination processes occur through recombination centers and involve lattice vibrations rather than photon emission. However, if we add

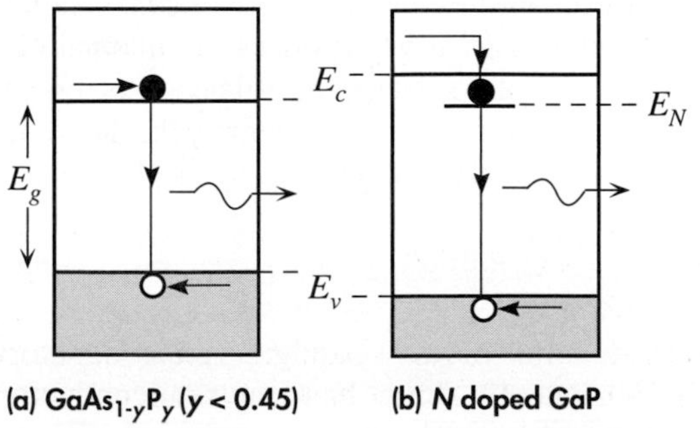


**Figure 6.45**
(a) Photon emission in a direct bandgap semiconductor.
(b) GaP is an indirect bandgap semiconductor. When it is doped with nitrogen, there is an electron recombination center at $E_N$. Direct recombination between a captured electron at $E_N$ and a hole emits a photon.

**Table 6.2** Selected LED semiconductor materials

| Semiconductor Active Layer | Structure | D or I | $\lambda$ (nm) | $\eta_{\text{external}}$ (%) | Comments |
|---|---|---|---|---|---|
| GaAs | DH | D | 870–900 | 10 | Infrared (IR) |
| $Al_xGa_{1-x}As$ $(0 < x < 0.4)$ | DH | D | 640–870 | 3–20 | Red to IR |
| $In_{1-x}Ga_xAs_yP_{1-y}$ $(y \approx 2.20x, 0 < x < 0.47)$ | DH | D | 1–1.6 μm | >10 | LEDs in communications |
| $In_{0.49}Al_xGa_{0.51-x}P$ | DH | D | 590–630 | >10 | Amber, green, red; high luminous intensity |
| InGaN/GaN quantum well | QW | D | 450–530 | 5–20 | Blue to green |
| $GaAs_{1-y}P_y$ $(y < 0.45)$ | HJ | D | 630–870 | < 1 | Red to IR |
| $GaAs_{1-y}P_y$ $(y > 0.45)$ (N or Zn, O doping) | HJ | I | 560–700 | < 1 | Red, orange, yellow |
| SiC | HJ | I | 460–470 | 0.02 | Blue, low efficiency |
| GaP (Zn) | HJ | I | 700 | 2–3 | Red |
| GaP (N) | HJ | I | 565 | < 1 | Green |

NOTE: Optical communication channels are at 850 nm (local network) and at 1.3 and 1.55 μm (long distance). D = direct bandgap, I = indirect bandgap. $\eta_{\text{external}}$ is typical and may vary substantially depending on the device structure. DH = double heterostructure, HJ = homojunction, QW = quantum well.

**isoelectronic impurities** such as nitrogen (in the same Group V as P) into the semiconductor crystal, then some of these N atoms substitute for P atoms. Since N and P have the same valency, N atoms substituting for P atoms form the same number of bonds and do not act as donors or acceptors. The electronic cores of N and P, however, are different. The positive nucleus of N is less shielded by electrons compared with that of the P atom. This means that a conduction electron in the neighborhood of a N atom will be attracted and may become captured at this site. N atoms therefore introduce localized energy levels, or electron traps, $E_N$ near the conduction band (CB) edge as depicted in Figure 6.45b. When a conduction electron is captured at $E_N$, it can attract a hole (in the valence band) in its vicinity by Coulombic attraction and eventually recombine with it directly and emit a photon. The emitted photon energy is only slightly less than $E_g$ as $E_N$ is typically close to $E_c$. As the recombination process depends on N doping, it is not as efficient as direct recombination. Thus, the efficiency of LEDs from N doped indirect bandgap $GaAs_{1-y}P_y$ semiconductors is less than those from direct bandgap semiconductors. Nitrogen doped indirect bandgap $GaAs_{1-y}P_y$ alloys are widely used in inexpensive green, yellow, and orange LEDs.

The **external efficiency** $\eta_{\text{external}}$ of an LED quantifies the efficiency of conversion of electric energy into an emitted external optical energy. It incorporates the internal efficiency of the radiative recombination process and the subsequent efficiency of photon extraction from the device. The input of electric power into an LED is simply the diode current and diode voltage product ($IV$). If $P_{\text{out}}$ is the optical power emitted by the device, then

*External efficiency*

$$\eta_{\text{external}} = \frac{P_{\text{out}}(\text{optical})}{IV} \times 100\% \qquad [6.65]$$

and some typical values are listed in Table 6.2. For indirect bandgap semiconductors, $\eta_{\text{external}}$ are generally less than 1 percent, whereas for direct bandgap semiconductors with the right device structure, $\eta_{\text{external}}$ can be substantial.

## 6.9.2 HETEROJUNCTION HIGH-INTENSITY LEDS

A *pn* junction between two differently doped semiconductors that are of the same material, that is, the same bandgap $E_g$, is called a **homojunction.** A junction between two different bandgap semiconductors is called a **heterojunction.** A semiconductor device structure that has junctions between different bandgap materials is called a **heterostructure device.**

LED constructions for increasing the intensity of the output light make use of the double heterostructure. Figure 6.46a shows a **double-heterostructure** (DH) device based on two junctions between different semiconductor materials with different bandgaps. In this case the semiconductors are AlGaAs with $E_g \approx 2$ eV and GaAs with $E_g \approx 1.4$ eV. The double heterostructure in Figure 6.46a has an $n^+p$ heterojunction between $n^+$-AlGaAs and $p$-GaAs. There is another heterojunction between $p$-GaAs and $p$-AlGaAs. The $p$-GaAs region is a thin layer, typically a fraction of a micron, and it is lightly doped.

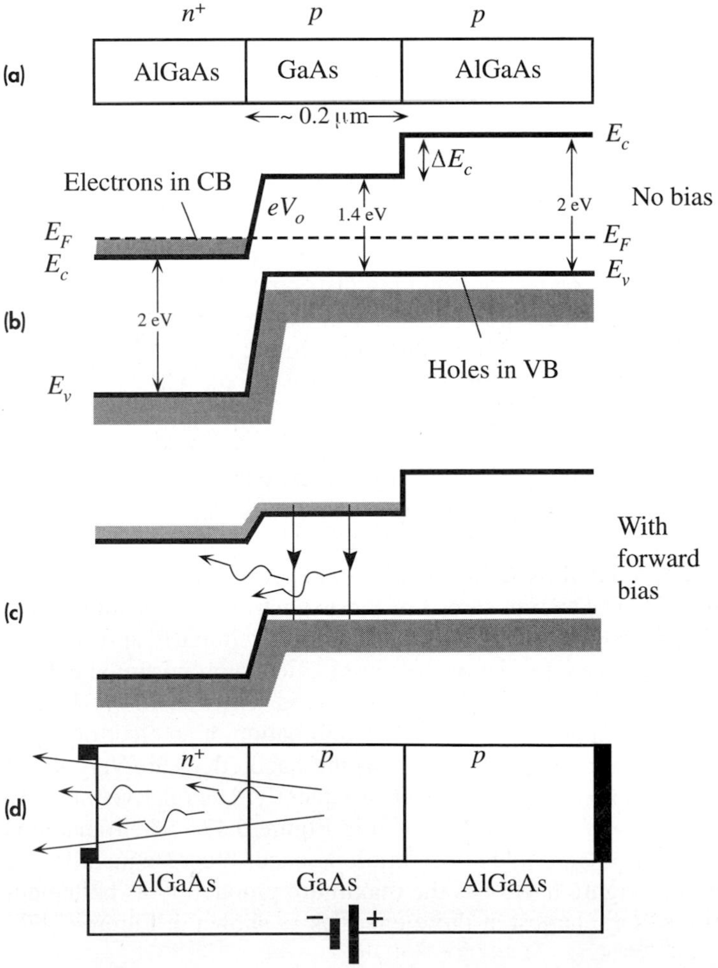


**Figure 6.46**

(a) A double heterostructure diode has two junctions which are between two different bandgap semiconductors (GaAs and AlGaAs).

(b) A simplified energy band diagram with exaggerated features. $E_F$ must be uniform.

(c) Forward-biased simplified energy band diagram.

(d) Forward-biased LED. Schematic illustration of photons escaping reabsorption in the AlGaAs layer and being emitted from the device.

The simplified energy band diagram for the whole device in the absence of an applied voltage is shown in Figure 6.46b. The Fermi level $E_F$ is continuous throughout the whole structure. There is a potential energy barrier $eV_o$ for electrons in the CB of $n^+$-AlGaAs against diffusion into $p$-GaAs. There is a bandgap change at the junction between $p$-GaAs and $p$-AlGaAs which results in a step change $\Delta E_c$ in $E_c$ between the two conduction bands of $p$-GaAs and $p$-AlGaAs. This $\Delta E_c$ is effectively a *potential energy barrier* that prevents any electrons in the CB in $p$-GaAs passing to the CB of $p$-AlGaAs. (There is also a step change $\Delta E_v$ in $E_v$, but this is small and is not shown.)

When a forward bias is applied, most of this voltage drops between the $n^+$-AlGaAs and $p$-GaAs and reduces the potential energy barrier $eV_o$, just as in the normal $pn$ junction. This allows electrons in the CB of $n^+$-AlGaAs to be injected into $p$-GaAs as shown in Figure 6.46c. These electrons, however, are *confined* to the CB of $p$-GaAs since there is a barrier $\Delta E_c$ between $p$-GaAs and $p$-AlGaAs. The wide bandgap AlGaAs layers therefore act as **confining layers** that restrict injected electrons to the $p$-GaAs layer. The recombination of injected electrons and the holes already present in this $p$-GaAs layer results in spontaneous photon emission. Since the bandgap $E_g$ of AlGaAs is greater than GaAs, the emitted photons do not get reabsorbed as they escape the active region and can reach the surface of the device as depicted in Figure 6.46d. Since light is also not absorbed in $p$-AlGaAs, it can be reflected to increase the light output.

### 6.9.3 LED CHARACTERISTICS

The energy of an emitted photon from an LED is not simply equal to the bandgap energy $E_g$ because electrons in the conduction band are distributed in energy and so are the holes in the valence band (VB). Figure 6.47a and b illustrate the energy band diagram and the energy distributions of electrons and holes in the CB and VB, respectively. The electron concentration as a function of energy in the CB is given by $g(E)f(E)$ where $g(E)$ is the density of states and $f(E)$ is the Fermi–Dirac function (probability of finding an electron in a state with energy $E$). The product $g(E)f(E)$ represents the electron concentration per unit energy or the concentration in energy and is plotted along the horizontal axis in Figure 6.47b. There is a similar energy distribution for holes in the VB.

The electron concentration in the CB as a function of energy is asymmetrical and has a peak at $\frac{1}{2}kT$ above $E_c$. The energy spread of these electrons is typically $\sim 2kT$ from $E_c$ as shown in Figure 6.47b. The hole concentration is similarly spread from $E_v$ in the valence band. Recall the rate of direct recombination is proportional to both the electron and hole concentrations at the energies involved. The transition which is identified as 1 in Figure 6.47a involves the direct recombination of an electron at $E_c$ and a hole at $E_v$. But the carrier concentrations near the band edges are very small and hence this type of recombination does not occur frequently. The relative intensity of light at this photon energy $h\nu_1$ is small as shown in Figure 6.47c. The transitions that involve the largest electron and hole concentrations occur most frequently. For example, the transition 2 in Figure 6.47a has the maximum probability as both electron and hole concentrations are largest at these energies as shown in Figure 6.47b.

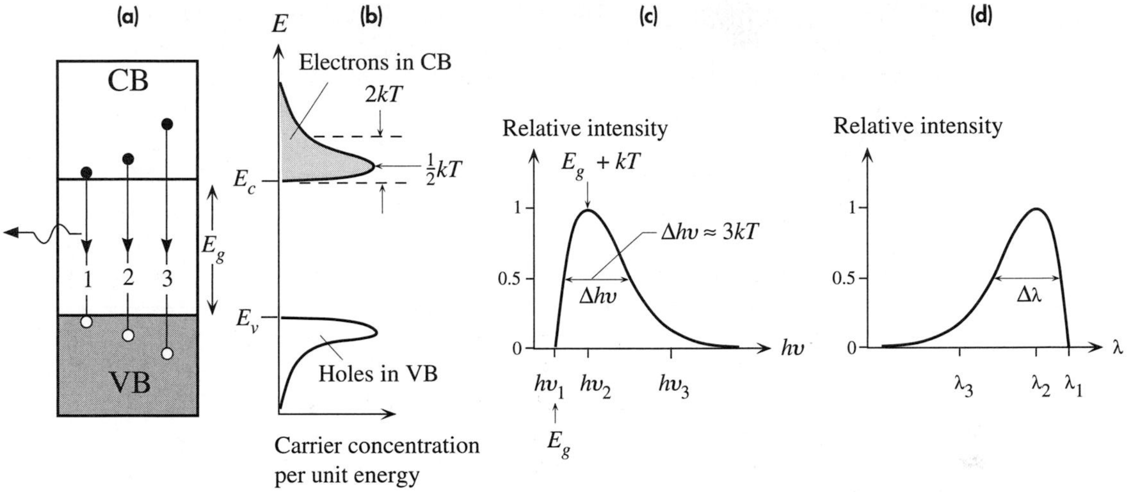


**Figure 6.47**
(a) Energy band diagram with possible recombination paths.
(b) Energy distribution of electrons in the CB and holes in the VB. The highest electron concentration is $\frac{1}{2}kT$ above $E_c$.
(c) The relative light intensity as a function of photon energy based on (b).
(d) Relative intensity as a function of wavelength in the output spectrum based on (b) and (c).

The relative intensity of light corresponding to this transition energy $h\nu_2$ is then maximum, or close to maximum, as indicated in Figure 6.47c.[11] The transitions marked as 3 in Figure 6.47a that emit relatively high energy photons $h\nu_3$ involve energetic electrons and holes whose concentrations are small as apparent in Figure 6.47b. Thus, the light intensity at these relatively high photon energies is small. The fall in light intensity with photon energy is shown in Figure 6.47c. The relative light intensity versus photon energy characteristic of the output spectrum is shown in Figure 6.47c and represents an important LED characteristic. Given the spectrum in Figure 6.47c we can also obtain the relative light intensity versus wavelength characteristic as shown in Figure 6.47d since $\lambda = c/\nu$. The **linewidth** of the output spectrum, $\Delta\nu$ or $\Delta\lambda$, is defined as the width between half-intensity points as shown in Figure 6.47c and d.

The wavelength for the peak intensity and the linewidth $\Delta\lambda$ of the emitted spectrum are obviously related to the energy distributions of the electrons and holes in the conduction and valence bands and therefore to the density of states in these bands. The photon energy for the peak emission is roughly $E_g + kT$ inasmuch as it corresponds to peak-to-peak transitions in the energy distributions of the electrons and holes in Figure 6.47b. The linewidth $\Delta(h\nu)$ of the output radiation between the half intensity points is approximately $3kT$ as shown in Figure 6.47c. It is relatively straightforward to calculate the corresponding spectral linewidth $\Delta\lambda$ in terms of wavelength as explained in Example 6.14.

[11] The intensity is not necessarily maximum when both the electron and hole concentrations are maximum, but it will be close.

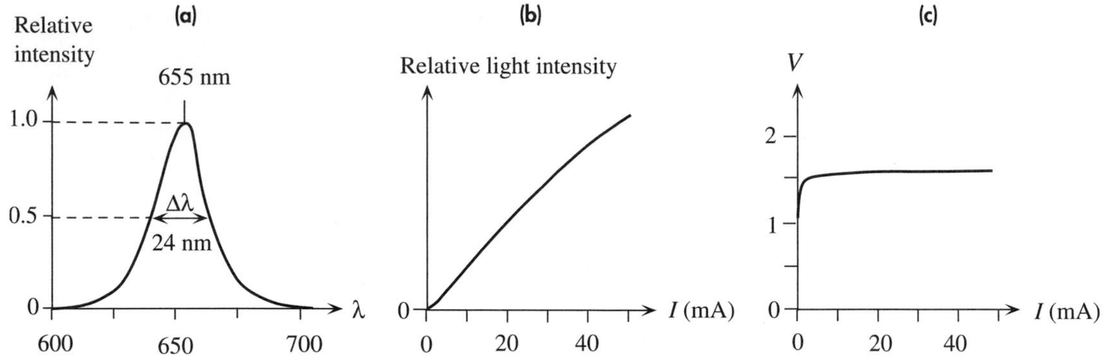


**Figure 6.48**
(a) A typical output spectrum from a red GaAsP LED.
(b) Typical output light power versus forward current.
(c) Typical $I$–$V$ characteristics of a red LED. The turn-on voltage is around 1.5 V.

The output spectrum, or the relative intensity versus wavelength characteristics, from an LED depends not only on the semiconductor material but also on the structure of the *pn* junction diode, including the dopant concentration levels. The spectrum in Figure 6.47d represents an idealized spectrum without including the effects of heavy doping on the energy bands and the reabsorption of some of the photons.

Typical characteristics of a red LED (655 nm), as an example, are shown in Figure 6.48a to c. The output spectrum in Figure 6.48a exhibits less asymmetry than the idealized spectrum in Figure 6.47d. The width of the spectrum is about 24 nm, which corresponds to a width of about $2.7kT$ in the energy distribution of the emitted photons. As the LED current increases so does the injected minority carrier concentration, and thus the rate of recombination and hence the output light intensity. The increase in the output light power is not however linear with the LED current as apparent in Figure 6.48b. At high current levels, a strong injection of minority carriers leads to the recombination time depending on the injected carrier concentration and hence on the current itself; this leads to a nonlinear recombination rate with current. Typical current–voltage characteristics are shown in Figure 6.48c where it can be seen that the **turn-on,** or **cut-in, voltage** is about 1.5 V from which point the current increases very steeply with voltage. The turn-on voltage depends on the semiconductor and generally increases with the energy bandgap $E_g$. For example, typically, for a blue LED it is about 3.5–4.5 V, for a yellow LED it is about 2 V, and for a GaAs infrared LED it is around 1 V.

**EXAMPLE 6.14**

**SPECTRAL LINEWIDTH OF LEDS** We know that a spread in the output wavelengths is related to a spread in the emitted photon energies as depicted in Figure 6.47. The emitted photon energy $E_{ph} = hc/\lambda$ and the spread in the photon energies $\Delta E_{ph} = \Delta(h\nu) \approx 3kT$ between the half-intensity points as shown in Figure 6.47c. Show that the corresponding **linewidth** $\Delta\lambda$ between the *half-intensity points* in the output spectrum is

*LED spectral linewidth*

$$\Delta\lambda = \lambda^2 \frac{3kT}{hc} \qquad [6.66]$$

What is the spectral linewidth of an optical communications LED operating at 1550 nm and at 300 K?

**SOLUTION**

First consider the relationship between the photon frequency $\nu$ and $\lambda$,

$$\lambda = \frac{c}{\nu} = \frac{hc}{h\nu}$$

in which $h\nu$ is the photon energy. We can differentiate this,

$$\frac{d\lambda}{d(h\nu)} = -\frac{hc}{(h\nu)^2} = -\frac{\lambda^2}{hc}$$

The negative sign implies that increasing the photon energy decreases the wavelength. We are only interested in changes or spreads; thus $\Delta\lambda/\Delta(h\nu) \approx |d\lambda/d(h\nu)|$,

$$\Delta\lambda = \frac{\lambda^2}{hc}\Delta(h\nu) = \frac{\lambda^2}{hc}3kT$$

where we used $\Delta(h\nu) = 3kT$, and obtained Equation 6.66. We can substitute $\lambda = 1550\,\text{nm}$ and $T = 300$ K to calculate the linewidth of the 1550 nm LED:

$$\Delta\lambda = \lambda^2\frac{3kT}{hc} = (1550 \times 10^{-9})^2\frac{3(1.38 \times 10^{-23})(300)}{(6.626 \times 10^{-34})(3 \times 10^8)}$$

$$= 1.50 \times 10^{-7}\ \text{m} \quad \text{or} \quad 150\ \text{nm}$$

The spectral linewidth of an LED output is due to the spread in the photon energies, which is fundamentally about $3kT$. The only option for decreasing $\Delta\lambda$ at a given wavelength is to reduce the temperature. The output spectrum of a laser, on the other hand, has a much narrower linewidth. A single-mode laser can have an output linewidth less than 1 nm.

# 6.10 SOLAR CELLS

## 6.10.1 PHOTOVOLTAIC DEVICE PRINCIPLES

A simplified schematic diagram of a typical solar cell is shown in Figure 6.49. Consider a *pn* junction with a very narrow and more heavily doped $n$-region. The illumination is through the thin $n$-side. The depletion region ($W$) or the space charge layer (SCL) extends primarily into the $p$-side. There is a built-in field $\mathcal{E}_o$ in this depletion layer. The electrodes attached to the $n$-side must allow illumination to enter the device and at the same time result in a small series resistance. They are deposited on the $n$-side to form an array of **finger electrodes** on the surface as depicted in Figure 6.50. A thin **antireflection coating** on the surface (not shown in the figure) reduces reflections and allows more light to enter the device.

As the $n$-side is very narrow, most of the photons are absorbed within the depletion region ($W$) and within the neutral $p$-side ($\ell_p$) and photogenerate EHPs in these regions. EHPs photogenerated in the depletion region are immediately separated by the built-in field $\mathcal{E}_o$ which drifts them apart. The electron drifts and reaches the neutral $n^+$-side whereupon it makes this region negative by an amount of charge $-e$. Similarly,

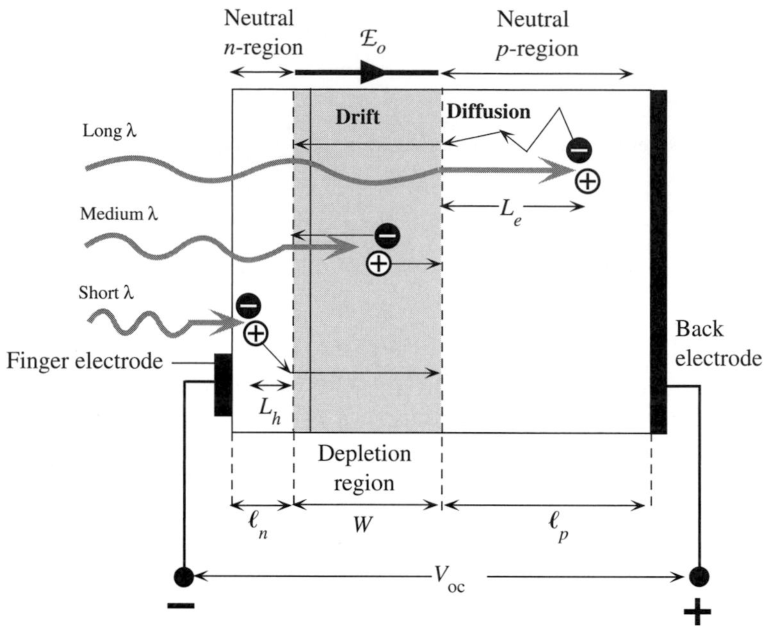


**Figure 6.49** The principle of operation of the solar cell (exaggerated features to highlight principles).

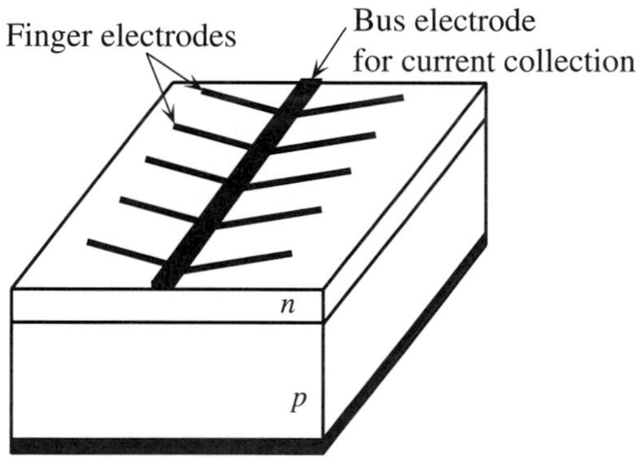


**Figure 6.50** Finger electrodes on the surface of a solar cell reduce the series resistance.

the hole drifts and reaches the neutral $p$-side and thereby makes this side positive. Consequently an **open circuit voltage** develops between the terminals of the device with the $p$-side positive with respect to the $n$-side. If an external load is connected, then the excess electron in the $n$-side can travel around the external circuit, do work, and reach the $p$-side to recombine with the excess hole there. It is important to realize that without the internal field $\mathcal{E}_o$ it is not possible to drift apart the photogenerated EHPs and accumulate excess electrons on the $n$-side and excess holes on the $p$-side.

The EHPs photogenerated by long-wavelength photons that are absorbed in the neutral $p$-side diffuse around in this region as there is no electric field. If the recombination lifetime of the electron is $\tau_e$, it diffuses a mean distance $L_e = \sqrt{2D_e\tau_e}$ where $D_e$ is its diffusion coefficient in the $p$-side. Those electrons within a distance $L_e$ to the depletion region can readily diffuse and reach this region whereupon they become drifted

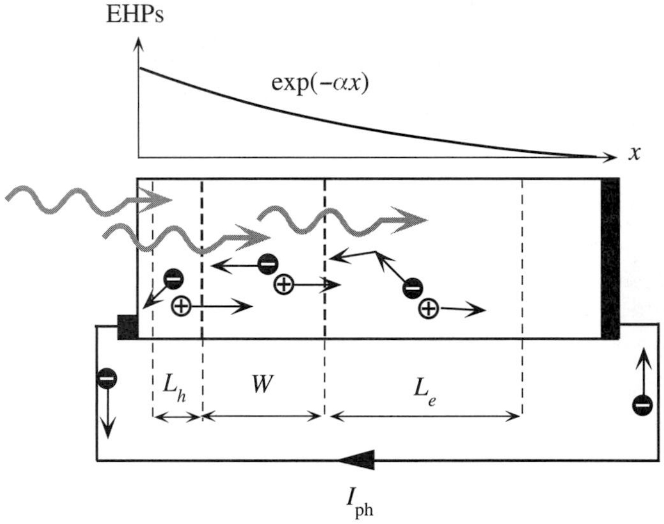


**Figure 6.51** Photogenerated carriers within the volume $L_h + W + L_e$ give rise to a photocurrent $I_{ph}$.
The variation in the photogenerated EHP concentration with distance is also shown where $\alpha$ is the absorption coefficient at the wavelength of interest.

by $\mathcal{E}_o$ to the $n$-side as shown in Figure 6.49. Consequently only those EHPs photogenerated within the minority carrier diffusion length $L_e$ to the depletion layer can contribute to the photovoltaic effect. Again the importance of the built-in field $\mathcal{E}_o$ is apparent. Once an electron diffuses to the depletion region, it is swept over to the $n$-side by $\mathcal{E}_o$ to give an additional negative charge there. Holes left behind in the $p$-side contribute a net positive charge to this region. Those photogenerated EHPs further away from the depletion region than $L_e$ are lost by recombination. It is therefore important to have the minority carrier diffusion length $L_e$ be as long as possible. This is the reason for choosing this side of a Si $pn$ junction to be $p$-type which makes electrons the minority carriers; the electron diffusion length in Si is longer than the hole diffusion length. The same ideas also apply to EHPs photogenerated by short-wavelength photons absorbed in the $n$-side. Those holes photogenerated within a diffusion length $L_h$ can reach the depletion layer and become swept across to the $p$-side. The photogeneration of EHPs that contributes to the photovoltaic effect therefore occurs in a volume covering $L_h + W + L_e$. If the terminals of the device are shorted, as in Figure 6.51, then the excess electron in the $n$-side can flow through the external circuit to neutralize the excess hole in the $p$-side. This current due to the flow of the photogenerated carriers is called the **photocurrent.**

Under a steady-state operation, there can be no net current through an *open circuit* solar cell. This means the photocurrent inside the device due to the flow of photogenerated carriers must be exactly balanced by a flow of carriers in the opposite direction. The latter carriers are minority carriers that become injected by the appearance of the photovoltaic voltage across the $pn$ junction as in a normal diode. This is not shown in Figure 6.49.

EHPs photogenerated by energetic photons absorbed in the $n$-side near the surface region or outside the diffusion length $L_h$ to the depletion layer are lost by recombination as the lifetime in the $n$-side is generally very short (due to heavy doping). The $n$-side is therefore made very thin, typically less than 0.2 μm. Indeed, the length $\ell_n$ of

Solar cell inventors at Bell Labs (left to right): Gerald Pearson, Daryl Chapin, and Calvin Fuller. They are checking a Si solar cell sample for the amount of voltage produced (1954).

SOURCE: Courtesy of Bell Labs, Lucent Technologies.

Helios is a solar cell–powered airplane that is remotely piloted. It has been able to fly as high as about 30 km during the day. Its wingspan is 9 m. It has fuel cells to fly at night.

SOURCE: Courtesy of NASA, Dryden Flight Center.

*pn* Junction Si solar cells at work. Honda's two-seated Dream car is powered by photovoltaics. The Honda Dream was first to finish 3,010 km in four days in the 1996 World Solar Challenge.

SOURCE: Courtesy of Centre for Photovoltaic Engineering, University of New South Wales, Sydney, Australia.

the $n$-side may be shorter than the hole diffusion length $L_h$. The EHPs photogenerated very near the surface of the $n$-side, however, disappear by recombination due to various surface defects acting as recombination centers as discussed below.

At long wavelengths, around 1–1.2 μm, the absorption coefficient $\alpha$ of Si is small and the *absorption depth* $(1/\alpha)$ is typically greater than 100 μm. To capture these long-wavelength photons, we therefore need a thick $p$-side and at the same time a long minority carrier diffusion length $L_e$. Typically the $p$-side is 200–500 μm and $L_e$ tends to be shorter than this.

Crystalline silicon has a bandgap of 1.1 eV which corresponds to a threshold wavelength of 1.1 μm. The incident energy in the wavelength region greater than 1.1 μm is then wasted; this is not a negligible amount (~25 percent). The worst part of the efficiency limitation however comes from the high-energy photons becoming absorbed near the crystal surface and being lost by recombination in the surface region. Crystal surfaces and interfaces contain a high concentration of recombination centers which facilitate the recombination of photogenerated EHPs near the surface. Losses due to EHP recombinations near or at the surface can be as high as 40 percent. These combined effects bring the efficiency down to about 45 percent. In addition, the antireflection coating is not perfect, which reduces the total collected photons by a factor of about 0.8–0.9. When we also include the limitations of the photovoltaic action itself (discussed below), the upper limit to a photovoltaic device that uses a single crystal of Si is about 24–26 percent at room temperature.

Consider an ideal $pn$ junction photovoltaic device connected to a resistive load $R$ as shown in Figure 6.52a. Note that $I$ and $V$ in the figure define the convention for the direction of positive current and positive voltage. If the load is a short circuit, then the only current in the circuit is that generated by the incident light. This is the photocurrent $I_{ph}$ shown in Figure 6.52b which depends on the number of EHPs photogenerated within the volume enclosing the depletion region ($W$) and the diffusion lengths to the depletion region (Figure 6.51). The greater is the light intensity, the

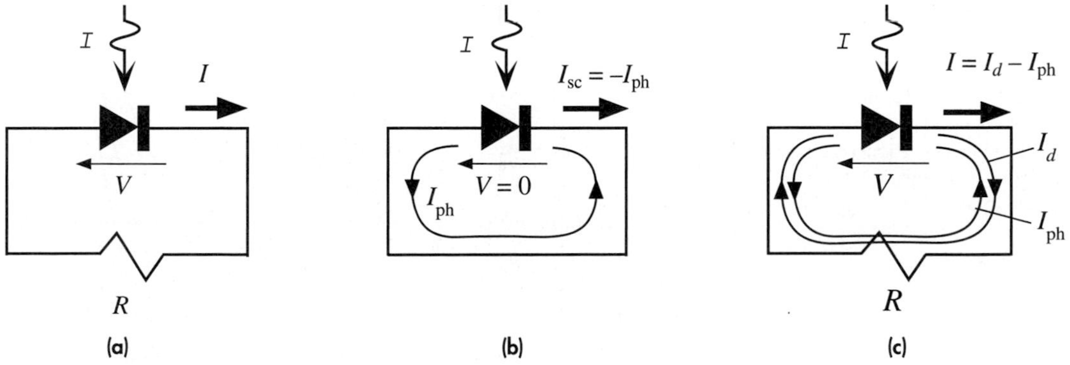


**Figure 6.52**

(a) The solar cell connected to an external load $R$ and the convention for the definitions of positive voltage and positive current.

(b) The solar cell in short circuit. The current is the photocurrent $I_{ph}$.

(c) The solar cell driving an external load $R$. There is a voltage $V$ and current $I$ in the circuit.

higher is the photogeneration rate and the larger is $I_{ph}$. If $\mathcal{I}$ is the light intensity, then the **short circuit current** is

*Short circuit solar cell current in light*

$$I_{sc} = -I_{ph} = -K\mathcal{I} \quad [6.67]$$

where $K$ is a constant that depends on the particular device. The photocurrent does not depend on the voltage across the $pn$ junction because there is always some internal field to drift the photogenerated EHP. We exclude the secondary effect of the voltage modulating the width of the depletion region. The photocurrent $I_{ph}$ therefore flows even when there is not a voltage across the device.

If $R$ is not a short circuit, then a positive voltage $V$ appears across the $pn$ junction as a result of the current passing through it as shown in Figure 6.52c. This voltage reduces the built-in potential of the $pn$ junction and hence leads to minority carrier injection and diffusion just as it would in a normal diode. Thus, in addition to $I_{ph}$ there is also a forward diode current $I_d$ in the circuit as shown in Figure 6.52c which arises from the voltage developed across $R$. Since $I_d$ is due to the normal $pn$ junction behavior, it is given by the diode characteristics,

$$I_d = I_o\left[\exp\left(\frac{eV}{\eta kT}\right) - 1\right]$$

where $I_o$ is the "reverse saturation current" and $\eta$ is the ideality factor ($\eta = 1 - 2$). In an open circuit, the net current is zero. This means that the photocurrent $I_{ph}$ develops just enough photovoltaic voltage $V_{oc}$ to generate a diode current $I_d = I_{ph}$.

Thus the **total current** through the solar cell, as shown in Figure 6.52c, is

*Solar cell I–V*

$$I = -I_{ph} + I_o\left[\exp\left(\frac{eV}{\eta kT}\right) - 1\right] \quad [6.68]$$

The overall $I$–$V$ characteristics of a typical Si solar cell are shown in Figure 6.53. It can be seen that it corresponds to the normal dark characteristics being shifted down

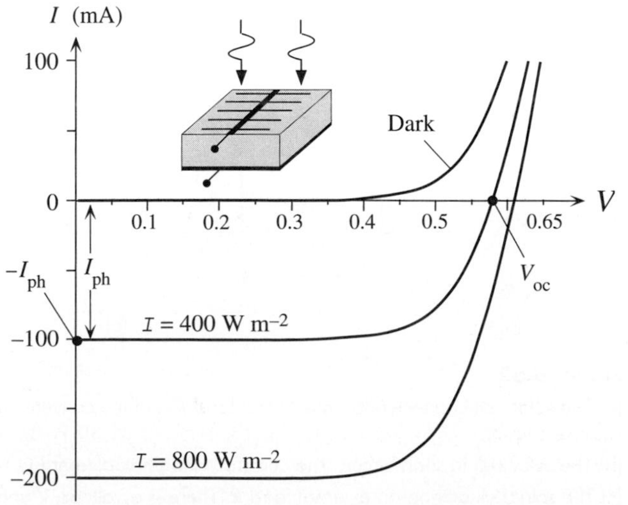


**Figure 6.53** Typical $I$–$V$ characteristics of a Si solar cell.
The short circuit current is $I_{ph}$ and the open circuit voltage is $V_{oc}$. The $I$–$V$ curves for positive current require an external bias voltage. Photovoltaic operation is always in the negative current region.

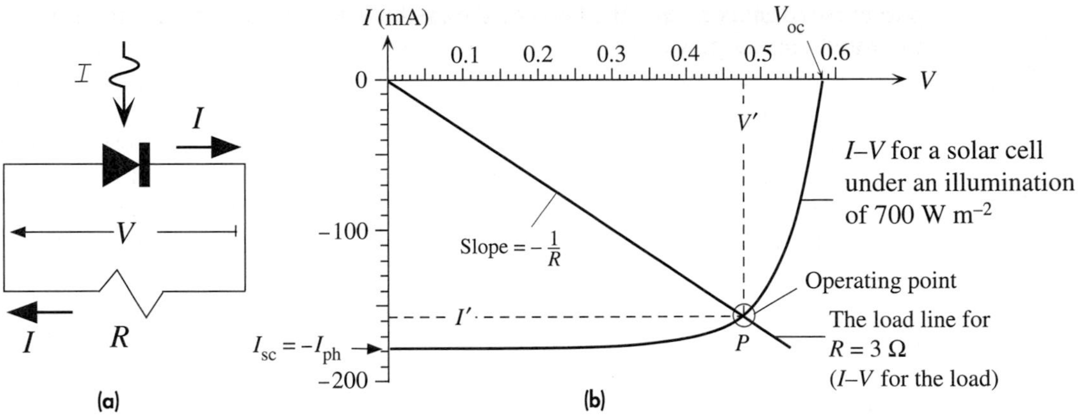


**Figure 6.54**
(a) When a solar cell drives a load $R$, $R$ has the same voltage as the solar cell but the current through it is in the opposite direction to the convention that current flows from high to low potential.
(b) The current $I'$ and voltage $V'$ in the circuit of (a) can be found from a load line construction. Point $P$ is the operating point ($I'$, $V'$). The load line is for $R = 3\ \Omega$.

by the photocurrent $I_{ph}$, which depends on the light intensity $\mathcal{I}$. The open circuit output voltage $V_{oc}$, of the solar cell is given by the point where the $I$–$V$ curve cuts the $V$ axis ($I = 0$). It is apparent that although it depends on the light intensity, its value typically lies in the range 0.5–0.7 V.

Equation 6.68 gives the $I$–$V$ characteristics of the solar cell. When the solar cell is connected to a load as in Figure 6.54a, the load has the same voltage as the solar cell and carries the same current. But the current $I$ through $R$ is now in the opposite direction to the convention that current flows from high to low potential. Thus, as shown in Figure 6.54a,

$$I = -\frac{V}{R} \qquad [6.69]$$

*The load line*

The actual current $I'$ and voltage $V'$ in the circuit must satisfy both the $I$–$V$ characteristics of the solar cell, Equation 6.68, and that of the load, Equation 6.69. We can find $I'$ and $V'$ by solving these two equations simultaneously or using a graphical solution. $I'$ and $V'$ in the solar cell circuit are most easily found by using a **load line construction.** The $I$–$V$ characteristics of the load in Equation 6.69 is a straight line with a negative slope $-1/R$. This is called the **load line** and is shown in Figure 6.54b along with the $I$–$V$ characteristics of the solar cell under a given intensity of illumination. The load line cuts the solar cell characteristic at $P$ where the load and the solar cell have the same current and voltage $I'$ and $V'$. Point $P$ therefore satisfies both Equations 6.68 and 6.69 and thus represents the **operating point of the circuit.**

The **power delivered** to the load is $P_{out} = I'V'$, which is the area of the rectangle bound by the $I$ and $V$ axes and the dashed lines shown in Figure 6.54b. Maximum power is delivered to the load when this rectangular area is maximized (by changing $R$ or the intensity of illumination), when $I' = I_m$ and $V' = V_m$. Since the maximum possible current is $I_{sc}$ and the maximum possible voltage is $V_{oc}$, $I_{sc}V_{oc}$ represents the desirable goal in power delivery for a given solar cell. Therefore it makes sense to compare the maximum power output $I_m V_m$ with $I_{sc}V_{oc}$. The **fill factor** FF, which is a figure of merit for the solar cell, is defined as

*Definition of fill factor*

$$\mathrm{FF} = \frac{I_m V_m}{I_{sc} V_{oc}} \tag{6.70}$$

The FF is a measure of the closeness of the solar cell *I–V* curve to the rectangular shape (the ideal shape). It is clearly advantageous to have the FF as close to unity as possible, but the exponential *pn* junction properties prevent this. Typically FF values are in the range 70–85 percent and depend on the device material and structure.

**EXAMPLE 6.15**

**A SOLAR CELL DRIVING A RESISTIVE LOAD** Consider the solar cell in Figure 6.54 that is driving a load of 3 Ω. This cell has an area of 3 cm × 3 cm and is illuminated with light of intensity 700 W m$^{-2}$. Find the current and voltage in the circuit. Find the power delivered to the load, the efficiency of the solar cell in this circuit, and the fill factor of the solar cell.

**SOLUTION**

The $I$–$V$ characteristic of the load in Figure 6.54a, is the load line in Equation 6.69; that is, $I = -V/(3\ \Omega)$. The line is drawn in Figure 6.54b with a slope $1/(3\ \Omega)$. It cuts the $I$–$V$ characteristics of the solar cell at $I' = 157$ mA and $V' = 0.475$ V as apparent in Figure 6.54b, which are the current and voltage, respectively, in the photovoltaic circuit of Figure 6.54a. The power delivered to the load is

$$P_{out} = I'V' = (157 \times 10^{-3})(0.475\ \mathrm{V}) = 0.0746\ \mathrm{W} \qquad \text{or} \qquad 74.6\ \mathrm{mW}$$

The input of sunlight power is

$$P_{in} = (\text{Light intensity})(\text{Surface area}) = (700\ \mathrm{W\ m^{-2}})(0.03\ \mathrm{m})^2 = 0.63\ \mathrm{W}$$

The efficiency is

$$\eta_{photovoltaic} = (100\%)\frac{P_{out}}{P_{in}} = (100\%)\frac{(0.0746\ \mathrm{W})}{(0.63\ \mathrm{W})} = 11.8\%$$

This will increase if the load is adjusted to extract the maximum power from the solar cell, but the increase will be small as the rectangular area $I'V'$ in Figure 6.54b is already quite close to the maximum.

The fill factor can also be calculated since point $P$ in Figure 6.54b is close to the optimum operation, maximum output power, in which the rectangular area $I'V'$ is maximum:

$$\mathrm{FF} = \frac{I_m V_m}{I_{sc}V_{oc}} \approx \frac{I'V'}{I_{sc}V_{oc}} = \frac{(157\ \mathrm{mA})(0.475\ \mathrm{V})}{(178\ \mathrm{mA})(0.58\ \mathrm{V})} = 0.722 \qquad \text{or} \qquad 72\%$$

**EXAMPLE 6.16**

**OPEN CIRCUIT VOLTAGE AND ILLUMINATION** A solar cell under an illumination of 500 W m$^{-2}$ has a short circuit current $I_{sc}$ of 150 mA and an open circuit output voltage $V_{oc}$ of 0.530 V. What are the short circuit current and open circuit voltage when the light intensity is doubled? Assume $\eta = 1.5$, a typical value for various Si *pn* junctions.

**SOLUTION**

The general $I$–$V$ characteristic under illumination is given by Equation 6.68. Setting $I = 0$ for open circuit,

*Open circuit condition*

$$I = -I_{\text{ph}} + I_o\left[\exp\left(\frac{eV_{\text{oc}}}{\eta kT}\right) - 1\right] = 0$$

Assuming that $V_{\text{oc}} \gg \eta kT/e$, rearranging the above equation we can find $V_{\text{oc}}$,

*Open circuit output voltage*

$$V_{\text{oc}} = \frac{\eta kT}{e}\ln\left(\frac{I_{\text{ph}}}{I_o}\right)$$

The photocurrent $I_{\text{ph}}$ depends on the light intensity $\mathcal{I}$ via $I_{\text{ph}} = K\mathcal{I}$, where $K$ is a constant. Thus, at a given temperature, the change in $V_{\text{oc}}$ is

*Open circuit voltage and light intensity*

$$V_{\text{oc}2} - V_{\text{oc}1} = \frac{\eta kT}{e}\ln\left(\frac{I_{\text{ph}2}}{I_{\text{ph}1}}\right) = \frac{\eta kT}{e}\ln\left(\frac{\mathcal{I}_2}{\mathcal{I}_1}\right)$$

The short circuit current is the photocurrent, so at double the intensity this is

$$I_{\text{sc}2} = I_{\text{sc}1}\left(\frac{\mathcal{I}_2}{\mathcal{I}_1}\right) = (150 \text{ mA})(2) = 300 \text{ mA}$$

Assuming $\eta = 1.5$, the new open circuit voltage is

$$V_{\text{oc}2} = V_{\text{oc}1} + \frac{\eta kT}{e}\ln\left(\frac{\mathcal{I}_2}{\mathcal{I}_1}\right) = 0.530 \text{ V} + (1.5)(0.026)\ln(2) = 0.557 \text{ V}$$

This is a 5 percent increase compared with the 100 percent increase in illumination and the short circuit current.

## 6.10.2 SERIES AND SHUNT RESISTANCE

Practical solar cells can deviate substantially from the ideal *pn* junction solar cell behavior depicted in Figure 6.53 due to a number of reasons. Consider an illuminated *pn* junction driving a load resistance $R_L$ and assume that photogeneration takes place in the depletion region. As shown in Figure 6.55, the photogenerated electrons have to traverse a surface semiconductor region to reach the nearest finger electrode. All these electron paths in the $n$-layer surface region to finger electrodes introduce an **effective series resistance** $R_s$ into the photovoltaic circuit. If the finger electrodes are thin, then the resistance of the electrodes themselves will further increase $R_s$. There is also a series resistance due to the neutral $p$-region, but this is generally small compared with the resistance of the electron paths to the finger electrodes.

Figure 6.56a shows the equivalent circuit of an ideal *pn* junction solar cell. The photogeneration process is represented by a *constant current generator* $I_{\text{ph}}$, which generates a current that is proportional to the light intensity. The flow of photogenerated carriers across the junction gives rise to a photovoltaic voltage difference $V$ across the junction, and this voltage leads to the normal diode current $I_d = I_o[\exp(eV/\eta kT) - 1]$. This diode current $I_d$ is represented by an ideal *pn* junction diode in the circuit as shown in Figure 6.56a. As apparent, $I_{\text{ph}}$ and $I_d$ are in opposite directions ($I_{\text{ph}}$ is "up"

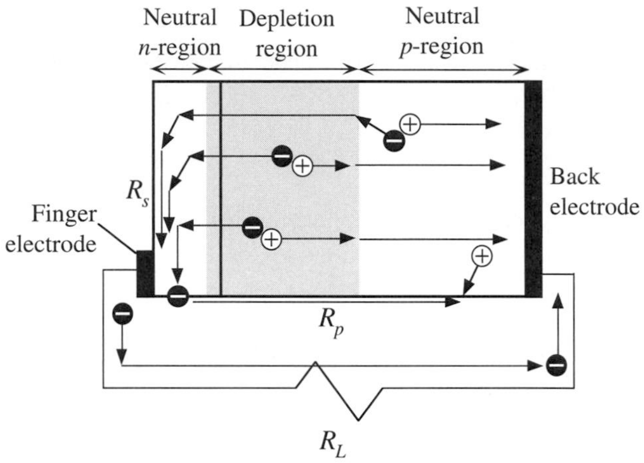


**Figure 6.55** Series and shunt resistances and various fates of photogenerated EHPs.

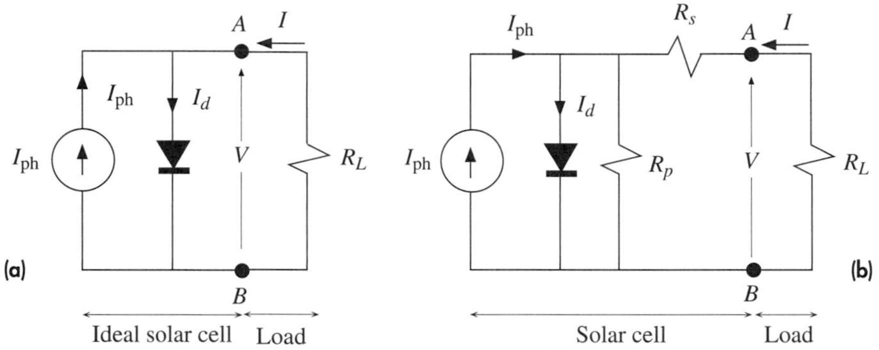


**Figure 6.56** The equivalent circuit of a solar cell.
(a) Ideal *pn* junction solar cell.
(b) Parallel and series resistances $R_s$ and $R_p$.

and $I_d$ is "down"), so in an open circuit the photovoltaic voltage is such that $I_{ph}$ and $I_d$ have the same magnitude and cancel each other. By convention, positive current $I$ at the output terminal is normally taken to flow into the terminal and is given by Equation 6.68. (In reality, of course, the solar cell current is negative, as in Figure 6.53, which represents a current that is flowing out into the load.)

Figure 6.56b shows the equivalent circuit of a more practical solar cell. The **series resistance** $R_s$ in Figure 6.56b gives rise to a voltage drop and therefore prevents the ideal photovoltaic voltage from developing at the output between $A$ and $B$ when a current is drawn. A fraction (usually small) of the photogenerated carriers can also flow through the crystal surfaces (edges of the device) or through *grain boundaries in polycrystalline devices* instead of flowing though the external load $R_L$. These effects that prevent photogenerated carriers from flowing in the external circuit can be represented by an effective internal **shunt** or **parallel resistance** $R_p$ that diverts the photocurrent away from the load $R_L$. Typically $R_p$ is less important than $R_s$ in overall

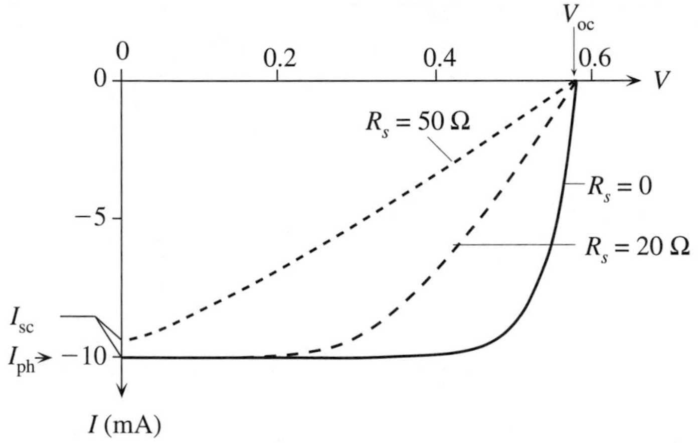


**Figure 6.57** The series resistance broadens the *I–V* curve and reduces the maximum available power and hence the overall efficiency of the solar cell.

The example is a Si solar cell with $\eta \approx 1.5$ and $I_o \approx 3 \times 10^{-6}$ mA. Illumination is such that the photocurrent $I_{ph} = 10$ mA.

device behavior, unless the device is highly polycrystalline and the current component flowing through grain boundaries is not negligible.

The series resistance $R_s$ can significantly deteriorate the solar cell performance as illustrated in Figure 6.57 where $R_s = 0$ is the best solar cell case. It is apparent that the available maximum output power decreases with the series resistance which therefore reduces the cell efficiency. Notice also that when $R_s$ is sufficiently large, it limits the short circuit current. Similarly, low shunt resistance values, due to extensive defects in the material, also reduce the efficiency. The difference is that although $R_s$ does not affect the open circuit voltage $V_{oc}$, low $R_p$ leads to a reduced $V_{oc}$.

### 6.10.3 Solar Cell Materials, Devices, and Efficiencies

Most solar cells use crystalline silicon because silicon-based semiconductor fabrication is now a mature technology that enables cost-effective devices to be manufactured. Typical Si-based solar cell efficiencies range from about 18 percent for polycrystalline to 22–24 percent in high-efficiency single-crystal devices that have special structures to absorb as many of the incident photons as possible. Solar cells fabricated by making a *pn* junction in the same crystal are called *homojunctions*. The best Si homojunction solar cell efficiencies are about 24 percent for expensive single-crystal passivated emitter rear locally diffused (PERL) cells.[12] The PERL and similar cells have a textured surface that is an array of "inverted pyramids" etched into the surface to capture as much of the incoming light as possible as depicted in Figure 6.58. Normal reflections from a flat crystal surface lead to a loss of light, whereas reflections inside the pyramid allow a second or even a third chance for absorption. Further, after refraction, photons would be entering the semiconductor at oblique angles which means that they will be absorbed in the useful photogeneration volume, that is, within the electron diffusion length of the depletion layer as shown in Figure 6.58.

[12] Much of the pioneering work for high-efficiency PERL solar cells was done by Martin Green and coworkers at the University of New South Wales.

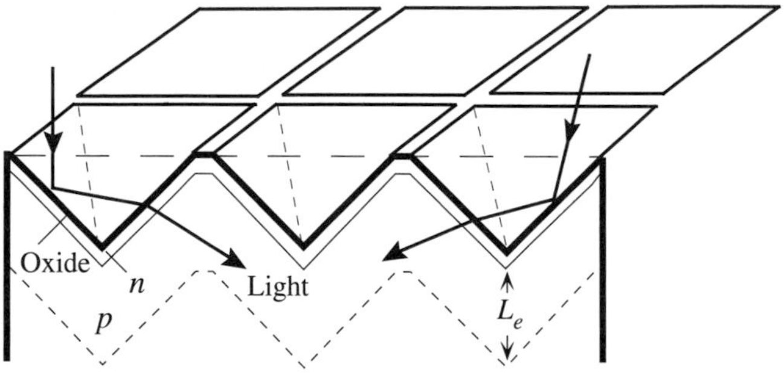


**Figure 6.58** An inverted pyramid textured surface substantially reduces reflection losses and increases absorption probability in the device.

Table 6.3 summarizes some typical characteristics of various solar cells. GaAs and Si solar cells have comparable efficiencies though theoretically GaAs with a higher bandgap is supposed to have a better efficiency. The largest factors reducing the efficiency of a Si solar cell are the unabsorbed photons with $h\nu < E_g$ and short wavelength photons absorbed near the surface. Both these factors are improved if tandem cell structures or heterojunctions are used.

There are a number of III–V semiconductor alloys that can be prepared with different bandgaps but with the same lattice constant. Heterojunctions (junctions between different materials) from these semiconductors have negligible interface defects. AlGaAs has a wider bandgap than GaAs and would allow most solar photons to pass through. If we use a thin AlGaAs layer on a GaAs *pn* junction, as shown in Figure 6.59, then this layer passivates the surface defects normally present in a homojunction GaAs cell. The AlGaAs window layer therefore overcomes the surface recombination limitation and improves the cell efficiency (such cells have efficiencies of about 24 percent).

**Table 6.3** Typical characteristics of various solar cells at room temperature under AM1.5 illumination of 1000 W m$^{-2}$

| Semiconductor | $E_g$ (eV) | $V_{oc}$ (V) | $J_{sc}$ (mA cm$^{-2}$) | FF | $\eta$ (%) | Comments |
|---|---|---|---|---|---|---|
| Si, single crystal | 1.1 | 0.5–0.7 | 42 | 0.7–0.8 | 16–24 | Single crystal, PERL |
| Si, polycrystalline | 1.1 | 0.5–0.65 | 38 | 0.7–0.8 | 12–19 | |
| Amorphous Si:Ge:H film | | | | | 8–13 | Amorphous film with tandem structure, convenient large-area fabrication |
| GaAs, single crystal | 1.42 | 1.02 | 28 | 0.85 | 24–25 | |
| GaAlAs/GaAs, tandem | | 1.03 | 27.9 | 0.864 | 24.8 | Different bandgap materials in tandem increases absorption efficiency |
| GaInP/GaAs, tandem | | 2.5 | 14 | 0.86 | 25–30 | Different bandgap materials in tandem increases absorption efficiency |
| CdTe, thin film | 1.5 | 0.84 | 26 | 0.75 | 15–16 | |
| InP, single crystal | 1.34 | 0.87 | 29 | 0.85 | 21–22 | |
| $CuInSe_2$ | 1.0 | | | | 12–13 | |

NOTE: AM1.5 refers to a solar illumination of "Air Mass 1.5," which represents solar radiation falling on the Earth's surface with a total intensity (or irradiance) of 1000 W m$^{-2}$. AM1.5 is widely used for comparing solar cells.

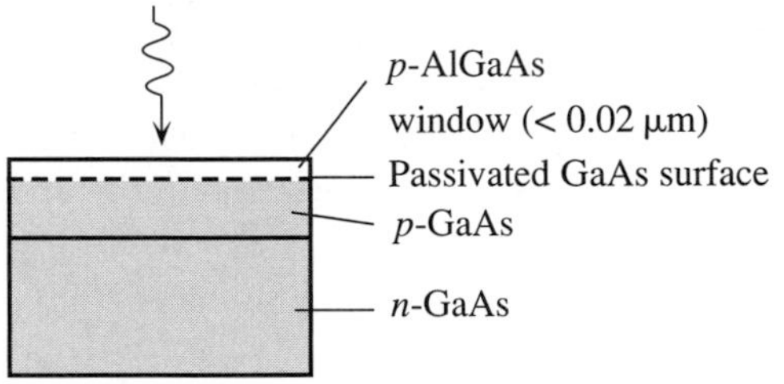


**Figure 6.59** AlGaAs window layer on GaAs passivates the surface states and thereby increases the photogeneration efficiency.

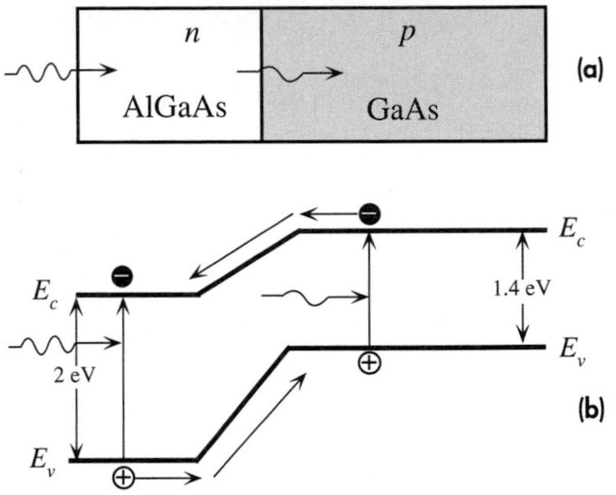


**Figure 6.60** A heterojunction solar cell between two different bandgap semiconductors (GaAs and AlGaAs).

Heterojunctions between different bandgap III–V semiconductors that are lattice matched offer the potential of developing high-efficiency solar cells. The simplest single heterojunction example, shown in Figure 6.60, consists of a *pn* junction using a wider bandgap $n$-AlGaAs with $p$-GaAs. Energetic photons ($h\nu > 2$ eV) are absorbed in AlGaAs, whereas those with energies less than 2 eV but greater than 1.4 eV are absorbed in the GaAs layer. In more sophisticated cells, the bandgap of AlGaAs is graded slowly from the surface by varying the composition of the AlGaAs layer.

Tandem or cascaded cells use two or more cells in tandem or in cascade to increase the absorbed photons from the incident light as illustrated in Figure 6.61. The first cell is made from a wider bandgap ($E_{g1}$) material and only absorbs photons with $h\nu > E_{g1}$. The second cell with bandgap $E_{g2}$ absorbs photons that pass the first cell and have $h\nu > E_{g2}$. The whole structure can be grown within a single crystal by using lattice-matched crystalline layers leading to a monolithic tandem cell. If, in addition, light concentrators are also used, the efficiency can be further increased. For example, a GaAs–GaSb tandem cell operating under a 100-sun condition, that is, 100 times that of ordinary sunlight, have exhibited an efficiency of about 34 percent. Tandem cells have been used in thin-film a-Si:H (hydrogenated amorphous Si) *pin* ($p$-type, intrinsic, and $n$-type structure) solar cells to obtain efficiencies up to about 12 percent. These tandem cells have a-Si:H and a-Si:Ge:H cells and are easily fabricated in large areas.

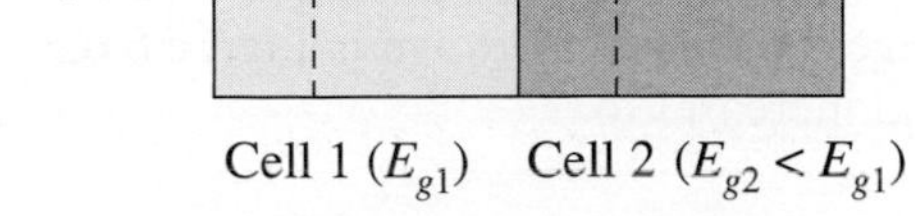


**Figure 6.61** A tandem cell. Cell 1 has a wider bandgap and absorbs energetic photons with $h\nu > E_{g1}$. Cell 2 absorbs photons that pass through cell 1 and have $h\nu > E_{g2}$.

# ADDITIONAL TOPICS

## 6.11 *pin* DIODES, PHOTODIODES, AND SOLAR CELLS

The *pin* Si diode is a device that has a structure with three distinct layers: a heavily doped thin $p^+$-type layer, a relatively thick intrinsic ($i$-Si) layer, and a heavily doped thin $n^+$-type layer, as shown in Figure 6.62a. For simplicity we will assume that the $i$-layer is truly intrinsic, or at least doped so lightly compared with $p^+$ and $n^+$ layers that it behaves almost as if intrinsic. The intrinsic layer is much wider than the $p^+$ and $n^+$ regions, typically 5–50 $\mu$m depending on the particular application. When the structure is first formed, holes diffuse from the $p^+$-side and electrons from the $n^+$-side into the $i$-Si layer where they recombine and disappear. This leaves behind a thin layer of exposed negatively charged acceptor ions in the $p^+$-side and a thin layer of exposed positively charged donor ions in the $n^+$-side as shown in Figure 6.22b. The two charges are separated by the $i$-Si layer of thickness $W$. There is a uniform built-in field $\mathcal{E}_o$ in the $i$-Si layer from the exposed positive ions to the exposed negative ions as illustrated in Figure 6.22c. (Since there is no net space charge in the $i$-layer, from $d\mathcal{E}/dx = \rho/\varepsilon_o\varepsilon_r = 0$, the field must be uniform.) In contrast, the built-in field in the depletion layer of a *pn* junction is not uniform. With no applied bias, the equilibrium is maintained by the built-in field $\mathcal{E}_o$ which prevents further diffusion of majority carriers from the $p^+$ and $n^+$ layers into the $i$-Si layer. A hole that manages to diffuse from the $p^+$-side into the $i$-layer is drifted back by $\mathcal{E}_o$, so the net current is zero. As in the *pn* junction, there is also a built-in potential $V_o$ from the edge of the $p^+$-side depletion region to the edge of the $n^+$-side depletion region. $V_o$ (like $\mathcal{E}_o$) provides a potential barrier against further net diffusion of holes and electrons into the $i$-layer and maintains the equilibrium in the open circuit (net current being zero) as in the *pn* junction. It is apparent from Figure 6.62c that, in the absence of an applied voltage, $\mathcal{E}_o = V_o/W$.

One of the distinct advantages of *pin* diodes is that the depletion layer capacitance is very small and independent of the voltage. The separation of two very thin layers of negative and positive charges by a fixed distance, width $W$ of the $i$-Si layer, is the same as that in a parallel plate capacitor. The **junction or depletion layer capacitance** of the *pin* diode is simply given by

*Junction capacitance of pin*

$$C_{\text{dep}} = \frac{\varepsilon_o\varepsilon_r A}{W} \qquad [6.70]$$

where $A$ is the cross-sectional area and $\varepsilon_o\varepsilon_r$ is the permittivity of the semiconductor (Si), respectively. Further, since the width $W$ of the $i$-Si layer is fixed by the structure, the junction capacitance does not depend on the applied voltage in contrast to that of the *pn* junction. $C_{\text{dep}}$ is typically of the order of a picofarad in fast *pin* photodiodes, so with a 50 Ω resistor, the $RC_{\text{dep}}$ time constant is about 50 ps.

When a reverse bias voltage $V_r$ is applied across the *pin* device, it drops almost entirely across the width of the $i$-Si layer. The depletion layer widths of the thin sheets of acceptor and donor charges in the $p^+$ and $n^+$ sides are negligible compared with $W$. The reverse bias $V_r$ increases the built-in voltage to $V_o + V_r$ as shown in Figure 6.62d. The field $\mathcal{E}$ in the $i$-Si layer is still uniform and increases to

*Reverse-biased pin*

$$\mathcal{E} = \mathcal{E}_o + \frac{V_r}{W} \approx \frac{V_r}{W} \qquad (V_r \gg V_o) \qquad [6.71]$$

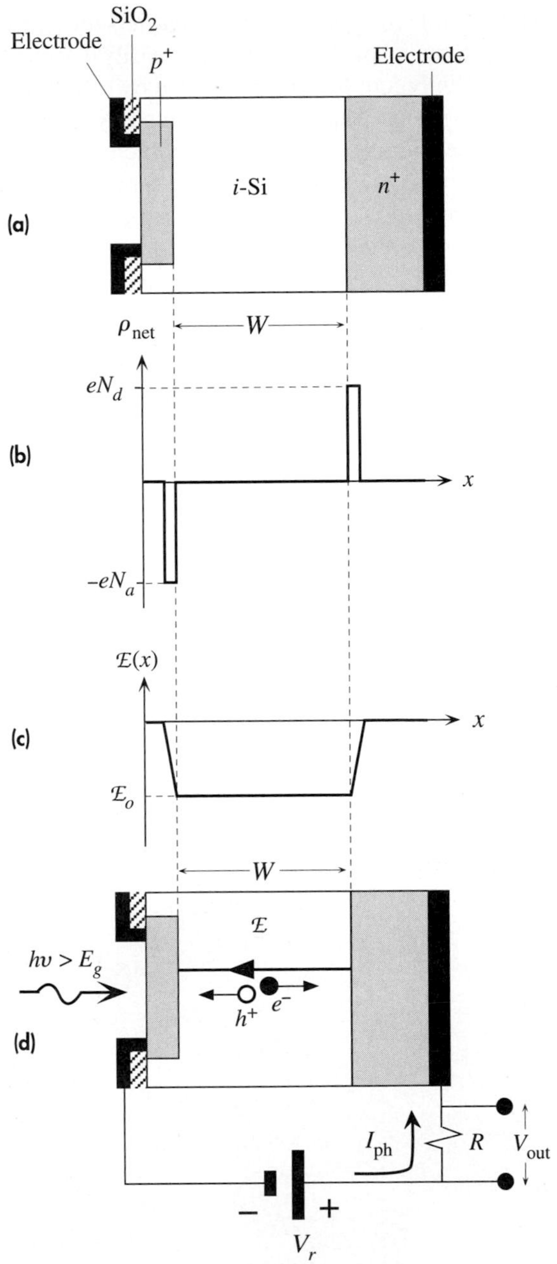


**Figure 6.62**
(a) The schematic structure of an idealized *pin* photodiode.
(b) The net space charge density across the photodiode.
(c) The built-in field across the diode.
(d) The *pin* photodiode in photodetection is reverse-biased.

Since the width of the $i$-layer in a *pin* device is typically much larger than the depletion layer width in an ordinary *pn* junction, the *pin* devices usually have higher breakdown voltages, which makes them useful where high breakdown voltages are required.

In *pin* photodetectors, the *pin* structure is designed so that photon absorption occurs primarily over the $i$-Si layer. The photogenerated electron–hole pairs (EHPs) in the $i$-Si layer are then separated by the field $\mathcal{E}$ and drifted toward the $n^+$ and $p^+$ sides, respectively, as illustrated in Figure 6.62d. While the photogenerated carriers are drifting through the *i*-Si layer, they give rise to an external photocurrent which is easily detected as a voltage across a small sampling resistor *R* in Figure 6.62d (or detected by a current-to-voltage converter). The response time of the *pin* photodiode is determined by the transit times of the photogenerated carriers across the width *W* of the *i*-Si layer. Increasing *W* allows more photons to be absorbed, which increases the output signal per input light intensity, but it slows down the speed of response because carrier transit times become longer.

The simple *pn* junction photodiode has two major drawbacks. Its junction or depletion layer capacitance is not sufficiently small to allow photodetection at high modulation frequencies. This is an RC time constant limitation. Secondly, its depletion layer is at most a few microns. This means that at long wavelengths where the penetration depth is greater than the depletion layer width, the majority of photons are absorbed outside the depletion layer where there is no field to separate the EHPs and drift them. The photodetector efficiency is correspondingly low at these long wavelengths. These problems are substantially reduced in the *pin* photodiode.[13] The *pin* photovoltaic devices, such as a-Si:H solar cells, are designed to have the photogeneration occur in the *i*-layer as in the case of photodetectors. Obviously, there is no external applied bias, and the built-in field $\mathcal{E}_o$ separates the EHPs and drives the photocurrent.

## 6.12 SEMICONDUCTOR OPTICAL AMPLIFIERS AND LASERS

All practical semiconductor laser diodes are double heterostructures (DH) whose energy band diagrams are similar to the LED diagram in Figure 6.46. The energy band diagram of a forward biased DH laser diode is shown in Figure 6.63a and b.

Izuo Hayashi and Morton Panish at Bell Labs (1971) were able to design the first semiconductor laser that operated continuously at room temperature. (Notice the similarity of the energy band diagram on the chalkboard with that in Figure 6.63.)

SOURCE: Courtesy of Bell Labs, Lucent Technologies.

[13] The *pin* photodiode was invented by J. Nishizawa and his research group in Japan in 1950.

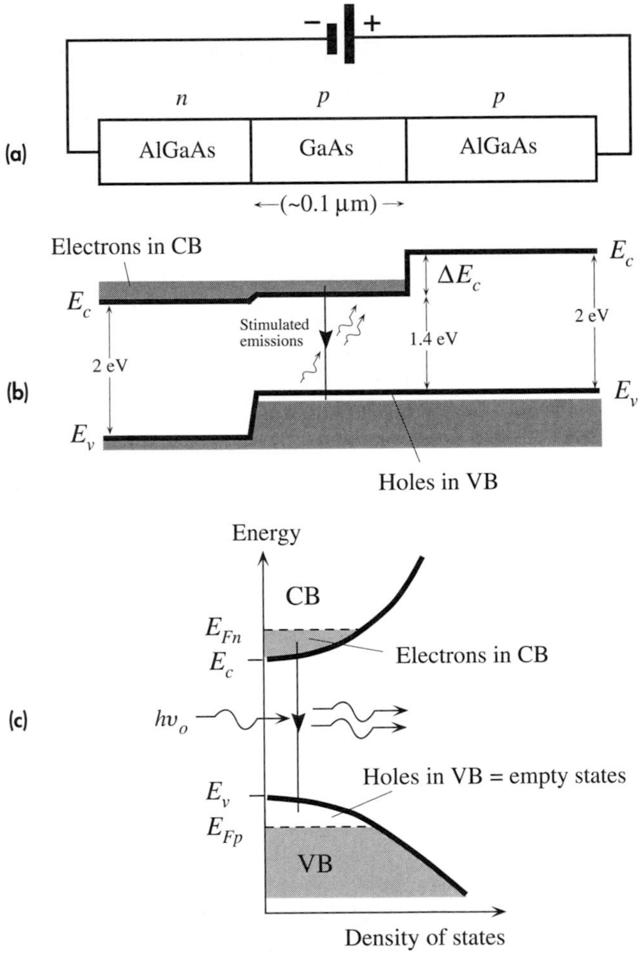


**Figure 6.63**
(a) A double heterostructure diode has two junctions which are between two different bandgap semiconductors (GaAs and AlGaAs).
(b) Simplified energy band diagram under a large forward bias. Lasing recombination takes place in the *p*-GaAs layer, the *active layer*.
(c) The density of states and energy distribution of electrons and holes in the conduction and valence bands in the active layer.

In this case the semiconductors are AlGaAs with $E_g \approx 2$ eV and GaAs with $E_g \approx 1.4$ eV. The $p$-GaAs region is a thin layer, typically 0.1–0.2 $\mu$m, and constitutes the **active layer** in which stimulated emissions take place. Both $p$-GaAs and $p$-AlGaAs are heavily $p$-type doped and are degenerate with the Fermi level $E_{Fp}$ in the valence band. When a sufficiently large forward bias is applied, $E_c$ of $n$-AlGaAs moves very close to the $E_c$ of $p$-GaAs which leads to a large injection of electrons in the CB of $n$-AlGaAs into $p$-GaAs as shown in Figure 6.63b. In fact, with a sufficient large forward bias, $E_c$ of AlGaAs can be moved above the $E_c$ of GaAs, which causes an enormous electron injection from $n$-AlGaAs into the CB of $p$-GaAs. These injected electrons, however, are *confined* to the CB of $p$-GaAs since there is a barrier $\Delta E_c$ between $p$-GaAs and $p$-AlGaAs due to the change in the bandgap.

The $p$-GaAs layer is degenerately doped. Thus, the top of its valence band (VB) is full of holes, or it has all the electronic states *empty* above the Fermi level $E_{Fp}$

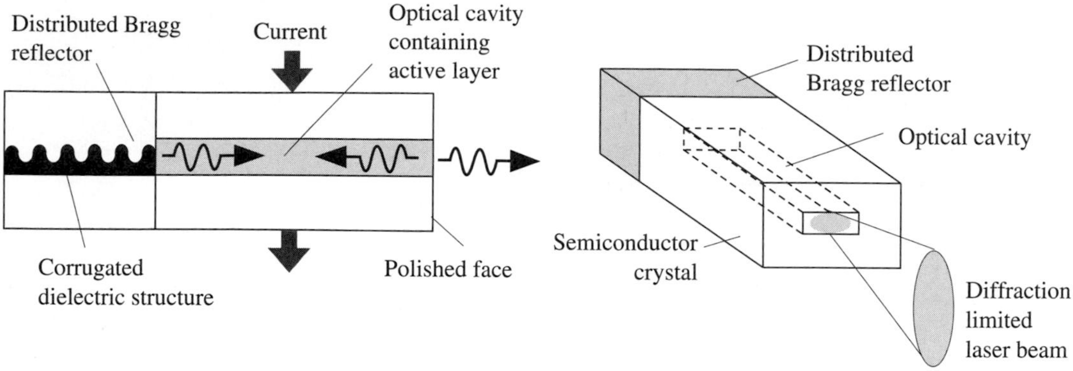


**Figure 6.64** Semiconductor lasers have an optical cavity to build up the required electromagnetic oscillations. In this example, one end of the cavity has a Bragg distributed reflector, a reflection grating, that reflects only certain wavelengths back into the cavity.

in this layer. The large forward bias injects a very large concentration of electrons from $n$-AlGaAs into the conduction band of $p$-GaAs. Consequently, as shown in Figure 6.63c, there is a large concentration of electrons in the CB and totally empty states at the top of the VB, which means that there is a *population inversion.* An incoming photon with an energy $h\nu_o$ just above $E_g$ can stimulate a conduction electron in the $p$-GaAs layer to fall down from the CB to the VB and emit a photon by *stimulated emission* as depicted in Figure 6.63c. Such a transition is a photon-stimulated electron–hole recombination, or a lasing recombination. Thus, an avalanche of stimulated emissions in the active layer provides an **optical amplification** of photons with $h\nu_o$ in this layer. The amplification depends on the extent of population inversion and hence on the diode forward current. The device operates as a **semiconductor optical amplifier** which amplifies an optical signal that is passed through the active layer. There is a threshold current below which there is no stimulated emission and no optical amplification.

To construct a **semiconductor laser** with a self-sustained lasing emission we have to incorporate the active layer into an *optical cavity* just as in the case of the HeNe laser in Chapter 3. The optical cavity with reflecting ends, reflects the coherent photons back and forward and encourages their constructive interference within the cavity as depicted in Figure 6.64. This leads to a buildup of high-energy electromagnetic oscillations in the cavity. Some of this electromagnetic energy in the cavity is tapped out as output radiation by having one end of the cavity as partially reflecting. For example, one type of optical cavity, as shown in Figure 6.64, has a special reflector, called a **Bragg distributed reflector** (BDR), at one end to reflect only certain wavelengths back into the cavity.[14] A BDR is a periodic corrugated

[14] Partial reflections of waves from the corrugations in the DBR can interfere constructively and constitute a reflected wave only for certain wavelengths, called *Bragg wavelengths*, that are related to the periodicity of the corrugations. A DBR acts like a reflection grating in optics.

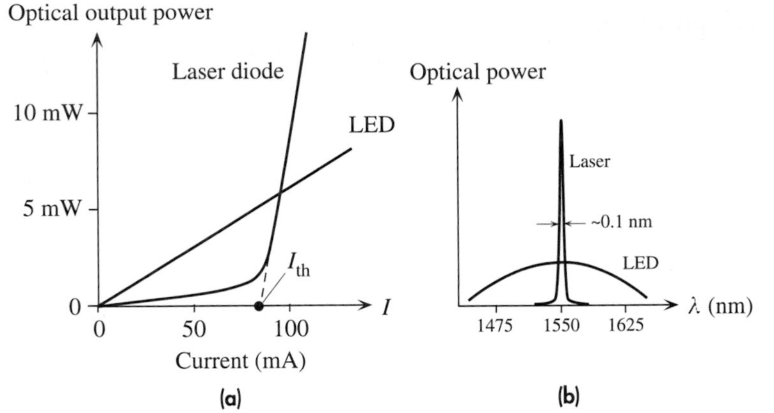


**Figure 6.65**
(a) Typical optical power output versus forward current for a laser diode and an LED.
(b) Comparison of spectral output characteristics.

structure, like a reflection grating, etched in a semiconductor that reflects only certain wavelengths that are related to the corrugation periodicity. This Bragg reflector has a corrugation periodicity such that it reflects only one desirable wavelength that falls within the optical gain of the active region. This wavelength selective reflection leads to only one possible electromagnetic radiation mode existing in the cavity, which leads to a very narrow output spectrum: a *single-mode output,* that is, only one peak in the output spectrum shown in Figure 3.43. Semiconductor lasers that operate with only one mode in the radiation output are called **single-mode** or **single-frequency lasers;** the spectral linewidth of a single-mode laser output is typically ~0.1 nm, which should be compared with an LED spectral width of 150 nm operating at a 1550 nm emission.

The double heterostructure has further advantages. Wider bandgap semiconductors generally have lower refractive indices, which means AlGaAs has a lower refractive index than that of GaAs. The change in the refractive index defines an optical dielectric waveguide that confines the photons to the active region of the optical cavity and thereby reduces photon losses and increases the photon concentration. This increase in the photon concentration increases the rate of stimulated emissions and the efficiency of the laser.

To achieve the necessary stimulated emissions from a laser diode and build up the necessary optical oscillations in the cavity (to overcome all the optical losses) the current must exceed a certain **threshold current** $I_{th}$ as shown in Figure 6.65a. The optical power output at a current $I$ is then very roughly proportional to $I - I_{th}$. There is still some weak optical power output below $I_{th}$, but this is simply due to spontaneous recombinations of injected electrons and holes in the active layer; the laser diode behaves like a "poor" LED below $I_{th}$. The output light from an LED however increases almost in proportion to the diode current. Figure 6.65b compares the output spectrum from the two devices. Remember that the output light from the laser diode is *coherent radiation,* whereas that from an LED is a stream of incoherent photons.

## CD Selected Topics and Solved Problems

**Selected Topics**

The *pn* Junction: Diffusion or Drift? Fick or Ohm?
Shot Noise Generated by the *pn* Junction
Voltage Drift in Semiconductor Devices due to Thermoelectric Effects
Transistor Switches: Why the Saturated Collector–Emitter Voltage is 0.2 V
Semiconductor Device Fabrication (Overview)
Photolithography and Minimum Line Width in Semiconductor Fabrication
Depletion MOSFET Fundamentals
High-Frequency Small-Signal BJT Model

**Solved Problems**

*pn* Junction: The Shockley Model
Recombination Current and I–V Characteristics of a *pn* Junction Diode
Design of a *pn* Junction Diode
Bipolar Junction Transistors at Low Frequencies: Principles and Solved Problems
BJT and Nonuniform Base Doping Effect
Junction Field Effect Transistor (JFET)
Enhancement MOSFET and CS Amplifier
LED Emission Wavelength and Temperature

## DEFINING TERMS

**Accumulation** occurs when an applied voltage to the gate (or metal electrode) of a MOS device causes the semiconductor under the oxide to have a greater number of majority carriers than the equilibrium value. Majority carriers have been accumulated at the surface of the semiconductor under the oxide.

**Active device** is a device that exhibits gain (current or voltage, or both) and has a directional electronic function. Transistors are active devices, whereas resistors, capacitors, and inductors are passive devices.

**Antireflection coating** reduces light reflection from a surface.

**Avalanche breakdown** is the enormous increase in the reverse current in a *pn* junction when the applied reverse field is sufficiently high to cause the generation of electron–hole pairs by impact ionization in the space charge layer.

**Base width modulation (the Early effect)** is the modulation of the base width by the voltage appearing across the base–collector junction. An increase in the base to collector voltage increases the collector junction depletion layer width, which results in the narrowing of the base width.

**Bipolar junction transistor (BJT)** is a transistor whose normal operation is based on the injection of carriers from the emitter into the base region, where they become minority carriers, and their subsequent diffusion to the collector, where they give rise to a collector current. The voltage between the base and the emitter controls the collector current.

**Built-in field** is the internal electric field in the depletion region of a *pn* junction that is maximum at the metallurgical junction. It is due to exposed negative acceptors on the *p*-side and positive donors on the *n*-side of the junction.

**Built-in voltage** ($V_o$) is the voltage across a *pn* junction, going from a *p*- to *n*-type semiconductor, in an open circuit.

**Channel** is the conducting strip between the source and drain regions of a MOSFET.

**Chip** is a piece (or a volume) of a semiconductor crystal that contains many integrated active and passive components to implement a circuit.

**Collector junction** is the metallurgical junction between the base and the collector of a bipolar transistor.

**Critical electric field** is the field in the space charge (or depletion) region at reverse breakdown (avalanche or Zener).

**Depletion layer** (or **space charge layer, SCL**) is a region around the metallurgical junction where recombination of electrons and holes has depleted this region of its large number of equilibrium majority carriers.

**Depletion (space charge) layer capacitance** is the incremental capacitance ($dQ/dV$) due to the change in the exposed dopant charges in the depletion layer as a result of the change in the voltage across the *pn* junction.

**Diffusion** is the flow of particles of a given species from high- to low-concentration regions by virtue of their random thermal motions.

**Diffusion (storage) capacitance** is the *pn* junction capacitance due to the diffusion and storage of minority carriers in the neutral regions when a forward bias is applied.

**Dynamic (incremental) resistance** $r_d$ of a diode is the change in the voltage across the diode per unit change in the current through the diode $r_d = dV/dI$. It is the low-frequency ac resistance of the diode. *Dynamic conductance* $g_d$ is the reciprocal dynamic resistance: $g_d = 1/r_d$.

**Emitter junction** is the metallurgical junction between the emitter and the base.

**Enhancement MOSFET** is a MOSFET device that needs a gate to source voltage above the threshold voltage to form a conducting channel between the source and the drain. In the absence of a gate voltage, there is no conduction between the source and drain. In its usual mode of operation, the gate voltage enhances the conductance of the source to drain inversion layer and increases the drain current.

**Epitaxial layer** is a thin layer of crystal that has been grown on the surface of another crystal which is usually a substrate, a mechanical support for the new crystal layer. The atoms of the new layer bond to follow the crystal pattern of the substrate, so the crystal structure of the epitaxial layer is matched with the crystal structure of the substrate.

**External quantum efficiency** is the optical power emitted from a light emitting device per unit electric input power.

**Field effect transistor (FET)** is a transistor whose normal operation is based on controlling the conductance of a channel between two electrodes by the application of an external field. The effect of the applied field is to control the current flow. The current is due to majority carrier drift from the source to the drain and is controlled by the voltage applied to the gate.

**Fill factor** (FF) is a figure of merit for a solar cell that represents, as a percentage, the maximum power $I_m V_m$ available to an external load as a fraction of the *ideal* theoretical power determined by the product of the short circuit current $I_{sc}$ and the open circuit voltage $V_{oc}$: $\text{FF} = (I_m V_m)/(I_{sc} V_{oc})$.

**Forward bias** is the application of an external voltage to a *pn* junction such that the positive terminal is connected to the *p*-side and the negative to the *n*-side. The applied voltage reduces the built-in potential.

**Heterojunction** is a junction between different semiconductor materials, for example, between GaAs and AlGaAs ternary alloy. There may or may not be a change in the doping.

**Homojunction** is a junction between differently doped regions of the same semiconducting material, for example, a *pn* junction in the same silicon crystal; there is no change in the bandgap energy $E_g$.

**Impact ionization** is the process by which a high electric field accelerates a free charge carrier (electron in the CB), which then impacts with a Si–Si bond to generate a free electron–hole pair. The impact excites an electron from $E_v$ to $E_c$.

**Integrated circuit (IC)** is a chip of a semiconductor crystal in which many active and passive components have been miniaturized and integrated together to form a sophisticated circuit.

**Inversion** occurs when an applied voltage to the gate (or metal electrode) of a MOS device causes the semiconductor under the oxide to develop a conducting layer (or a channel) at the surface of the semiconductor. The conducting layer has opposite polarity carriers to the bulk semiconductor and hence is termed an inversion layer.

**Ion implantation** is a process that is used to bombard a sample in a vacuum with ions of a given species of atom. First the dopant atoms are ionized in a vacuum and then accelerated by applying voltage differences to impinge on a sample to be doped. The sample is grounded to neutralize the implanted ions.

**Isoelectronic impurity** atom has the same valency as the host atom.

**Law of the junction** relates the injected minority carrier concentration just outside the depletion layer to the applied voltage. For holes in the $n$-side, it is

$$p_n(0) = p_{no}\exp\left(\frac{eV}{kT}\right)$$

where $p_n(0)$ is the hole concentration just outside the depletion layer.

**Linewidth** is the width of the intensity versus wavelength spectrum, usually between the half-intensity points, emitted from a light emitting device.

**Long diode** is a $pn$ junction with neutral regions longer than the minority carrier diffusion lengths.

**Metallurgical junction** is where there is an effective junction between the $p$-type and $n$-type doped regions in the crystal. It is where the donor and acceptor concentrations are equal or where there is a transition from $n$- to $p$-type doping.

**Metal-oxide-semiconductor transistor (MOST)** is a field effect transistor in which the conductance between the source and drain is controlled by the voltage supplied to the gate electrode, which is insulated from the channel by an oxide layer.

**Minority carrier injection** is the flow of electrons into the $p$-side and holes into the $n$-side of a $pn$ junction when a voltage is applied to reduce the built-in voltage across the junction.

**MOS** is short for a metal-insulator-semiconductor structure in which the insulator is typically silicon oxide. It can also be a different type of dielectric; for example, it can be the nitride $Si_3N_4$.

**NMOS** is an enhancement type $n$-channel MOSFET.

**Passive device** or component is a device that exhibits no gain and no directional function. Resistors, capacitors, and inductors are passive components.

**Photocurrent** is the current generated by a light-receiving device when it is illuminated.

**Pinch-off voltage** is the gate to source voltage needed to just pinch off the conducting channel between the source and drain with no source to drain voltage applied. It is also the source to drain voltage that just pinches off the channel when the gate and source are shorted. Beyond pinch-off, the drain current is almost constant and controlled by $V_{GS}$.

**PMOS** is an enhancement type $p$-channel MOSFET.

**Poly-Si gate** is short for a polycrystalline and highly doped Si gate.

**Recombination current** flows under forward bias to replenish the carriers recombining in the space charge (depletion) layer. Typically, it is described by $I = I_{ro}[\exp(eV/2kT) - 1]$.

**Reverse bias** is the application of an external voltage to a $pn$ junction such that the positive terminal is connected to the $n$-side and the negative to the $p$-side. The applied voltage increases the built-in potential.

**Reverse saturation current** is the reverse current that would flow in a reverse-biased ideal $pn$ junction obeying the Shockley equation.

**Shockley diode equation** relates the diode current to the diode voltage through $I = I_o[\exp(eV/kT) - 1]$. It is based on the injection and diffusion of injected minority carriers by the application of a forward bias.

**Short diode** is a $pn$ junction in which the neutral regions are shorter than the minority carrier diffusion lengths.

**Small-signal equivalent circuit** of a transistor replaces the transistor with an equivalent circuit that consists of resistances, capacitances, and dependent sources (current or voltage). The equivalent circuit represents the transistor behavior under small-signal ac conditions. The batteries are replaced with short circuits (or their internal resistances). Small signals imply small variations about dc values.

**Substrate** is a single mechanical support that carries active and passive devices. For example, in integrated circuit technology, typically, many integrated circuits are fabricated on a single silicon crystal wafer that serves as the substrate.

**Thermal generation current** is the current that flows in a reverse-biased $pn$ junction as a result of the thermal generation of electron–hole pairs in the depletion layer that become separated and swept across by the built-in field.

**Threshold voltage** is the gate voltage needed to establish a conducting channel between the source and drain of an enhancement MOST (metal-oxide-semiconductor transistor).

**Transistor** is a three-terminal solid-state device in which a current flowing between two electrodes is controlled by the voltage between the third and one of the other terminals or by a current flowing into the third terminal.

**Turn-on,** or **cut-in,** voltage of a diode is the voltage beyond which there is a substantial increase in the current. The turn-on voltage of a Si diode is about 0.6 V whereas it is about 1 V for a GaAs LED. The turn-on voltage of a *pn* junction diode depends on the bandgap of the semiconductor and the device structure.

**Zener breakdown** is the enormous increase in the reverse current in a *pn* junction when the applied voltage is sufficient to cause the tunneling of electrons from the valence band in the *p*-side to the conduction band in the *n*-side. Zener breakdown occurs in *pn* junctions that are heavily doped on both sides so that the depletion layer width is narrow.

# QUESTIONS AND PROBLEMS

**6.1** **The *pn* junction** Consider an abrupt Si $pn^+$ junction that has $10^{15}$ acceptors cm$^{-3}$ on the *p*-side and $10^{19}$ donors on the *n*-side. The minority carrier recombination times are $\tau_e = 490$ ns for electrons in the *p*-side and $\tau_h = 2.5$ ns for holes in the *n*-side. The cross-sectional area is 1 mm$^2$. Assuming a long diode, calculate the current $I$ through the diode at room temperature when the voltage $V$ across it is 0.6 V. What are $V/I$ and the incremental resistance ($r_d$) of the diode and why are they different?

***6.2** **The Si *pn* junction** Consider a long *pn* junction diode with an acceptor doping $N_a$ of $10^{18}$ cm$^{-3}$ on the *p*-side and donor concentration of $N_d$ on the *n*-side. The diode is forward-biased and has a voltage of 0.6 V across it. The diode cross-sectional area is 1 mm$^2$. The minority carrier recombination time $\tau$ depends on the dopant concentration $N_{\text{dopant}}$(cm$^{-3}$) through the following approximate relation

$$\tau = \frac{5 \times 10^{-7}}{(1 + 2 \times 10^{-17} N_{\text{dopant}})}$$

*a.* Suppose that $N_d = 10^{15}$ cm$^{-3}$. Then the depletion layer extends essentially into the *n*-side and we have to consider minority carrier recombination time $\tau_h$ in this region. Calculate the diffusion and recombination contributions to the total diode current. What is your conclusion?

*b.* Suppose that $N_d = N_a = 10^{18}$ cm$^{-3}$. Then $W$ extends equally to both sides and, further, $\tau_e = \tau_h$. Calculate the diffusion and recombination contributions to the diode current. What is your conclusion?

**6.3** **Junction capacitance of a *pn* junction** The capacitance ($C$) of a reverse-biased abrupt Si $p^+n$ junction has been measured as a function of the reverse bias voltage $V_r$ as listed in Table 6.4. The *pn* junction cross-sectional area is 500 μm × 500 μm. By plotting $1/C^2$ versus $V_r$, obtain the built-in potential $V_o$ and the donor concentration $N_d$ in the *n*-region. What is $N_a$?

**Table 6.4** Capacitance at various values of reverse bias ($V_r$)

| $V_r$ (V) | 1 | 2 | 3 | 5 | 10 | 15 | 20 |
|---|---|---|---|---|---|---|---|
| $C$ (pF) | 38.3 | 30.7 | 26.4 | 21.3 | 15.6 | 12.9 | 11.3 |

**6.4** **Temperature dependence of diode properties**

*a.* Consider the reverse current in a *pn* junction. Show that

$$\frac{\delta I_{\text{rev}}}{I_{\text{rev}}} \approx \left(\frac{E_g}{\eta kT}\right)\frac{\delta T}{T}$$

where $\eta = 2$ for Si and GaAs, in which thermal generation in the depletion layer dominates the reverse current, and $\eta = 1$ for Ge, in which the reverse current is due to minority carrier diffusion to the depletion layer. It is assumed that $E_g \gg kT$ at room temperature. Order the semiconductors Ge, Si, and GaAs according to the sensitivity of the reverse current to temperature.

*b.* Consider a forward-biased *pn* junction carrying a *constant* current *I*. Show that the change in the voltage across the *pn* junction per unit change in the temperature is given by

$$\frac{dV}{dT} = -\left(\frac{V_g - V}{T}\right)$$

where $V_g = E_g/e$ is the energy gap expressed in volts. Calculate typical values for $dV/dT$ for Ge, Si, and GaAs assuming that, typically, $V = 0.2$ V for Ge, 0.6 V for Si, and 0.9 V for GaAs. What is your conclusion? Can one assume that, typically, $dV/dT \approx -2$ mV °C$^{-1}$ for these diodes?

**6.5** **Avalanche breakdown** Consider a Si $p^+n$ junction diode that is required to have an avalanche breakdown voltage of 25 V. Given the breakdown field $\mathcal{E}_{\text{br}}$ in Figure 6.19, what should be the donor doping concentration?

**6.6** **Design of a *pn* junction diode** Design an abrupt Si $pn^+$ junction that has a reverse breakdown voltage of 100 V and provides a current of 10 mA when the voltage across it is 0.6 V. Assume that, if $N_{\text{dopant}}$ is in cm$^{-3}$, the minority carrier recombination time is given by

$$\tau = \frac{5 \times 10^{-7}}{(1 + 2 \times 10^{-17} N_{\text{dopant}})}$$

Mention any assumptions made.

**6.7** **Minority carrier profiles (the hyperbolic functions)** Consider a *pnp* BJT under normal operating conditions in which the EB junction is forward-biased and the BC junction is reverse-biased. The field in the neutral base region outside the depletion layers can be assumed to be negligibly small. The continuity equation for holes $p_n(x)$ in the *n*-type base region is

$$D_h \frac{d^2 p_n}{dx^2} - \frac{p_n - p_{no}}{\tau_h} = 0 \qquad \textbf{[6.71]}$$

where $p_n(x)$ is the hole concentration at $x$ from just outside the depletion region and $p_{no}$ and $\tau_h$ are the equilibrium hole concentration and hole recombination lifetime in the base.

*a.* What are the boundary conditions at $x = 0$ and $x = W_B$, just outside the collector region depletion layer? (Consider the law of the junction.)

*b.* Show that the following expression for $p_n(x)$ is a solution of the continuity equation

$$p_n(x) = p_{no}\left[\exp\left(\frac{eV}{kT}\right) - 1\right]\left[\frac{\sinh\left(\dfrac{W_B - x}{L_h}\right)}{\sinh\left(\dfrac{W_B}{L_h}\right)}\right] + p_{no}\left[1 - \frac{\sinh\left(\dfrac{x}{L_h}\right)}{\sinh\left(\dfrac{W_B}{L_h}\right)}\right] \qquad \textbf{[6.72]}$$

where $V = V_{EB}$ and $L_h = \sqrt{D_h \tau_h}$.

*c.* Show that Equation 6.72 satisfies the boundary conditions.

***6.8** **The *pnp* bipolar transistor** Consider a *pnp* transistor in a common base configuration and under normal operating conditions. The emitter–base junction is forward-biased and the base–collector junction is reverse-biased. The emitter, base, and collector dopant concentrations are $N_{a(E)}$, $N_{d(B)}$, and $N_{a(C)}$, respectively, where $N_{a(E)} \gg N_{d(B)} \geq N_{a(C)}$. For simplicity, assume uniform doping in all the regions. The base and emitter widths are $W_B$ and $W_E$, respectively, both much shorter than the minority carrier diffusion lengths, $L_h$ and $L_e$. The minority carrier lifetime in the base is the hole recombination time $\tau_h$. The minority carrier mobility in the base and emitter are denoted by $\mu_h$ and $\mu_e$, respectively.

The minority carrier concentration profile in the base can be represented by Equation 6.72.

*a.* Assuming that the emitter injection efficiency is unity show that

1. $$I_E \approx \frac{eAD_h n_i^2 \coth\left(\frac{W_B}{L_h}\right)}{L_h N_{d(B)}} \exp\left(\frac{eV_{EB}}{kT}\right)$$

2. $$I_C \approx \frac{eAD_h n_i^2 \operatorname{cosech}\left(\frac{W_B}{L_h}\right)}{L_h N_{d(B)}} \exp\left(\frac{eV_{EB}}{kT}\right)$$

3. $$\alpha \approx \operatorname{sech}\left(\frac{W_B}{L_h}\right)$$

4. $$\beta \approx \frac{\tau_h}{\tau_t} \quad \text{where} \quad \tau_t = \frac{W_B^2}{2D_h} \quad \text{is the base transit time.}$$

*b.* Consider the total emitter current $I_E$ through the EB junction, which has diffusion and recombination components as follows:

$$I_E = I_{E(so)} \exp\left(\frac{eV_{EB}}{kT}\right) + I_{E(ro)} \exp\left(\frac{eV_{EB}}{2kT}\right)$$

Only the hole component of the diffusion current (first term) can contribute to the collector current. Show that when $N_{a(E)} \gg N_{d(B)}$, the emitter injection efficiency $\gamma$ is given by

$$\gamma \approx \left[1 + \frac{I_{E(ro)}}{I_{E(so)}} \exp\left(-\frac{eV_{EB}}{2kT}\right)\right]^{-1}$$

How does $\gamma < 1$ modify the expressions derived in part (*a*)? What is your conclusion (consider small and large emitter currents, or $V_{EB} = 0.4$ and 0.7 V)?

**6.9** **Characteristics of an *npn* Si BJT** Consider an idealized silicon *npn* bipolar transistor with the properties in Table 6.5. Assume uniform doping in each region. The emitter and base widths are between metallurgical junctions (not neutral regions). The cross-sectional area is 100 μm × 100 μm. The transistor is biased to operate in the normal active mode. The base–emitter forward bias voltage is 0.6 V and the reverse bias base–collector voltage is 18 V.

**Table 6.5** Properties of an *npn* BJT

| Emitter Width | Emitter Doping | Hole Lifetime in Emitter | Base Width | Base Doping | Electron Lifetime in Base | Collector Doping |
|---|---|---|---|---|---|---|
| 10 μm | $1 \times 10^{18}$ cm$^{-3}$ | 10 ns | 5 μm | $1 \times 10^{16}$ cm$^{-3}$ | 200 ns | $1 \times 10^{16}$ cm$^{-3}$ |

*a.* Calculate the depletion layer width extending from the collector into the base and also from the emitter into the base. What is the width of the neutral base region?

*b.* Calculate $\alpha$ and hence $\beta$ for this transistor, assuming unity emitter injection efficiency. How do $\alpha$ and $\beta$ change with $V_{CB}$?

*c.* What is the emitter injection efficiency and what are $\alpha$ and $\beta$, taking into account that the emitter injection efficiency is not unity?

*d.* What are the emitter, collector, and base currents?

*e.* What is the collector current when $V_{CB} = 19$ V but $V_{EB} = 0.6$ V? What is the incremental collector output resistance defined as $\Delta V_{CB}/\Delta I_C$?

***6.10 Bandgap narrowing and emitter injection efficiency** Heavy doping in semiconductors leads to what is called *bandgap narrowing* which is an effective narrowing of the bandgap $E_g$. If $\Delta E_g$ is the reduction in the bandgap, then for an $n$-type semiconductor, according to Lanyon and Tuft (1979),

*Bandgap narrowing*

$$\Delta E_g(\text{meV}) = 22.5\left(\frac{n}{10^{18}}\right)^{1/2}$$

where $n$ (in cm$^{-3}$) is the concentration of majority carriers which is equal to the dopant concentration if they are all ionized (for example, at room temperature). The new effective intrinsic concentration $n_{i\text{eff}}$ due to the reduced bandgap is given by

*Bandgap narrowing*

$$n_{i\text{eff}}^2 = N_c N_v \exp\left[-\frac{(E_g - \Delta E_g)}{kT}\right] = n_i^2 \exp\left(\frac{\Delta E_g}{kT}\right)$$

where $n_i$ is the intrinsic concentration in the absence of emitter bandgap narrowing.

The equilibrium electron and hole concentrations $n_{no}$ and $p_{no}$, respectively, obey

*Mass action law with bandgap narrowing*

$$n_{no} p_{no} = n_{i\text{eff}}^2$$

where $n_{no} = N_d$ since nearly all donors would be ionized at room temperature.

Consider a Si *npn* bipolar transistor operating under normal active conditions with the base–emitter forward biased, and the base–collector reverse biased. The transistor has narrow emitter and base regions. The emitter neutral region width $W_E$ is 1 μm, and the donor doping is $10^{19}$ cm$^{-3}$. The width $W_B$ of the neutral base region is 1 μm, and the acceptor doping is $10^{17}$ cm$^{-3}$. Assume that $W_E$ and $W_B$ are less than the minority carrier diffusion lengths in the emitter and the base.

*a.* Obtain an expression for the emitter injection efficiency taking into account the emitter bandgap narrowing effect above.

*b.* Calculate the emitter injection efficiency with and without the emitter bandgap narrowing.

*c.* Calculate the common emitter current gain $\beta$ with and without the emitter bandgap narrowing effect given a perfect base transport factor ($\alpha_T = 1$).

**6.11 The JFET pinch-off voltage** Consider the symmetric $n$-channel JFET shown in Figure 6.66. The width of each depletion region extending into the $n$-channel is $W$. The thickness, or depth, of the channel, defined between the two metallurgical junctions, is $2a$. Assuming an abrupt *pn* junction and $V_{DS} = 0$, show that when the gate to source voltage is $-V_p$ the channel is pinched off where

$$V_p = \frac{a^2 e N_d}{2\varepsilon} - V_o$$

**Figure 6.66** A symmetric JFET.

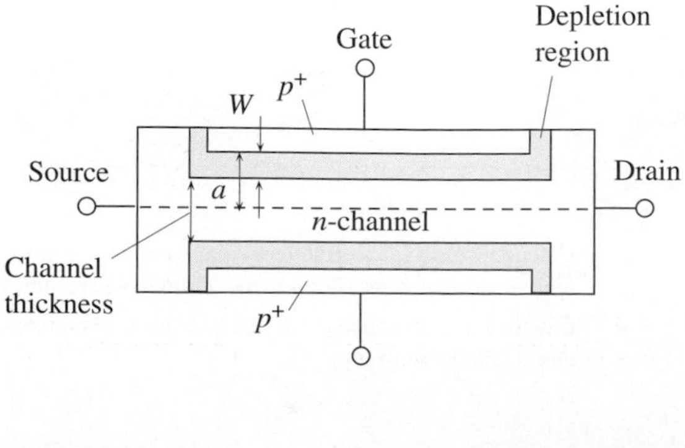

where $V_o$ is the built-in potential between $p^+n$ junction and $N_d$ is the donor concentration of the channel.

Calculate the pinch-off voltage of a JFET that has an acceptor concentration of $10^{19}$ cm$^{-3}$ in the $p^+$ gate, a channel donor doping of $10^{16}$ cm$^{-3}$, and a channel thickness (depth) $2a$ of 2 μm.

**6.12** **The JFET** Consider an $n$-channel JFET that has a symmetric $p^+n$ gate–channel structure as shown in Figures 6.27a and 6.66. Let $L$ be the gate length, $Z$ the gate width, and $2a$ the channel thickness. The pinch-off voltage is given by Question 6.11. The drain saturation current $I_{DSS}$ is the drain current when $V_{GS} = 0$. This occurs when $V_{DS} = V_{DS(\text{sat})} = V_P$ (Figure 6.29), so $I_{DSS} = V_P G_{\text{ch}}$, where $G_{\text{ch}}$ is the conductance of the channel between the source and the pinched-off point (Figure 6.30). Taking into account the shape of the channel at pinch-off, if $G_{\text{ch}}$ is about one-third of the conductance of the free or unmodulated (rectangular) channel, show that

$$I_{DSS} = V_P\left[\frac{1}{3}\frac{(e\mu_e N_d)(2a)Z}{L}\right]$$

A particular $n$-channel JFET with a symmetric $p^+n$ gate–channel structure has a pinch-off voltage of 3.9 V and an $I_{DSS}$ of 5.5 mA. If the gate and channel dopant concentrations are $N_a = 10^{19}$ cm$^{-3}$ and $N_d = 10^{15}$ cm$^{-3}$, respectively, find the channel thickness $2a$ and $Z/L$. If $L = 10$ μm, what is $Z$? What is the gate–source capacitance when the JFET has no voltage supplies connected to it?

**6.13** **The JFET amplifier** Consider an $n$-channel JFET that has a pinch-off voltage ($V_P$) of 5 V and $I_{DSS} = 10$ mA. It is used in a common source configuration as in Figure 6.34a in which the gate to source bias voltage ($V_{GS}$) is $-1.5$ V. Suppose that $V_{DD} = 25$ V.

*a.* If a small-signal voltage gain of 10 is needed, what should be the drain resistance ($R_D$)? What is $V_{DS}$?

*b.* If an ac signal of 3 V peak-to-peak is applied to the gate in series with the dc bias voltage, what will be the ac output voltage peak-to-peak? What is the voltage gain for positive and negative input signals? What is your conclusion?

**6.14** **The enhancement NMOSFET amplifier** Consider an $n$-channel Si enhancement NMOS transistor that has a gate width ($Z$) of 150 μm, channel length ($L$) of 10 μm, and oxide thickness ($t_{\text{ox}}$) of 500 Å. The channel has $\mu_e = 700$ cm$^2$ V$^{-1}$ s$^{-1}$ and the threshold voltage ($V_{\text{th}}$) is 2 V ($\varepsilon_r = 3.9$ for $SiO_2$).

*a.* Calculate the drain current when $V_{GS} = 5$ V and $V_{DS} = 5$ V and assuming $\lambda = 0.01$.

*b.* What is the small-signal voltage gain if the NMOSFET is connected as a common source amplifier, as shown in Figure 6.67, with a drain resistance $R_D$ of 2.2 kΩ, the gate biased at 5 V with respect to

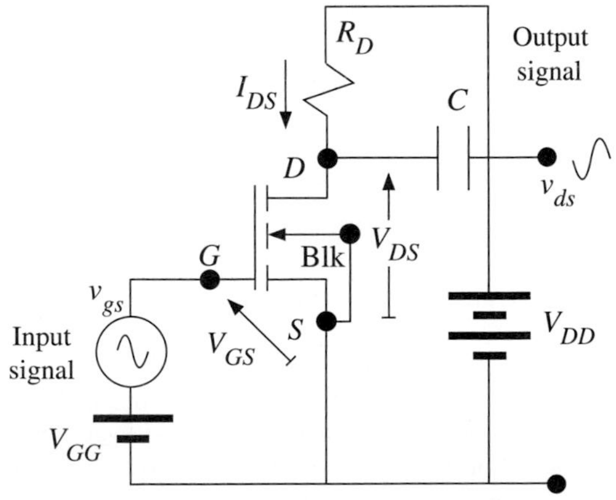


**Figure 6.67** NMOSFET amplifier.

source ($V_{GG} = 5$ V) and $V_{DD}$ is such that $V_{DS} = 5$ V? What is $V_{DD}$? What will happen if the drain supply is smaller?

*c.* Estimate the most positive and negative input signal voltages that can be amplified if $V_{DD}$ is fixed at the above value in part (*b*).

*d.* What factors will lead to a higher voltage amplification?

***6.15 Ultimate limits to device performance**

*a.* Consider the speed of operation of an $n$-channel FET-type device. The time required for an electron to transit from the source to the drain is $\tau_t = L/v_d$, where $L$ is the channel length and $v_d$ is the drift velocity. This transit time can be shortened by shortening $L$ and increasing $v_d$. As the field increase, the drift velocity eventually saturates at about $v_{d\text{sat}} = 10^5$ m s$^{-1}$ when the field in the channel is equal to $\mathcal{E}_c \approx 10^6$ V m$^{-1}$. A short $\tau_t$ requires a field that is at least $\mathcal{E}_c$.

1. What is the change in the *PE* of an electron when it traverses the channel length $L$ from source to drain if the voltage difference is $V_{DS}$?
2. This energy must be greater than the energy due to thermal fluctuations, which is of the order of $kT$. Otherwise, electrons would be brought in and out of the drain due to thermal fluctuations. Given the minimum field and $V_{DS}$, what is the minimum channel length and hence the minimum transit time?

*b.* Heisenberg's uncertainty principle relates the energy and the time duration in which that energy is possessed through a relationship of the form (Chapter 3) $\Delta E \, \Delta t > \hbar$. Given that during the transit of the electron from the source to the drain its energy changes by $eV_{DS}$, what is the shortest transit time $\tau$ satisfying Heisenberg's uncertainty principle? How does it compare with your calculation in part (*a*)?

*c.* How does electron tunneling limit the thickness of the gate oxide and the channel length in a MOSFET? What would be typical distances for tunneling to be effective? (Consider Example 3.10.)

**6.16 Energy distribution of electrons in the conduction band of a semiconductor and LED emission spectrum**

*a.* Consider the energy distribution of electrons $n_E(E)$ in the conduction band (CB). Assuming that the density of state $g_{\text{cb}}(E) \propto (E - E_c)^{1/2}$ and using Boltzmann statistics $f(E) \approx \exp[-(E - E_F)/kT]$, show that the energy distribution of the electrons in the CB can be written as

$$n_x(x) = Cx^{1/2}\exp(-x)$$

where $x = (E - E_c)/kT$ is electron energy in terms of $kT$ measured from $E_c$, and $C$ is a temperature-dependent constant (independent of $E$).

*b.* Setting arbitrarily $C = 1$, plot $n_x$ versus $x$. Where is the maximum, and what is the full width at half maximum (FWHM), *i.e.*, between half maximum points?

*c.* Show that the average electron energy in the CB is $\frac{3}{2}kT$, by using the definition of the average,

$$x_{\text{average}} = \frac{\int_0^\infty x n_x \, dx}{\int_0^\infty n_x \, dx}$$

where the integration is from $x = 0$ ($E_c$) to say $x = 10$ (far away from $E_c$ where $n_x \to 0$). You need to use a numerical integration.

*d.* Show that the maximum in the energy distribution is at $x = \frac{1}{2}$ or at $E_{\text{max}} = \frac{1}{2}kT$ above $E_c$.

*e.* Consider the recombination of electrons and holes in GaAs. The recombination involves the emission of a photon. Given that both electron and hole concentrations have energy distributions in the conduction and valence bands, respectively, sketch schematically the expected light intensity emitted from electron and hole recombinations against the photon energy. What is your conclusion?

**6.17** **LED output spectrum** Given that the width of the relative light intensity between half-intensity points versus photon energy spectrum of an LED is typically $\sim 3kT$, what is the linewidth $\Delta\lambda$ in the output spectrum in terms of the peak emission wavelength? Calculate the spectral linewidth $\Delta\lambda$ of the output radiation from a green LED emitting at 570 nm at 300 K.

**6.18** **LED output wavelength variations** Show that the change in the emitted wavelength $\lambda$ with temperature $T$ from an LED is approximately given by

$$\frac{d\lambda}{dT} \approx -\frac{hc}{E_g^2}\left(\frac{dE_g}{dT}\right)$$

where $E_g$ is the bandgap. Consider a GaAs LED. The bandgap of GaAs at 300 K is 1.42 eV which changes (decreases) with temperature as $dE_g/dT = -4.5 \times 10^{-4}$ eV K$^{-1}$. What is the change in the emitted wavelength if the temperature change is 10 °C?

**6.19** **Linewidth of direct recombination LEDs** Experiments carried out on various direct bandgap semiconductor LEDs give the output spectral linewidth (between half-intensity points) listed in Table 6.6. Since wavelength $\lambda = hc/E_{ph}$, where $E_{ph} = h\nu$ is the photon energy, we know that the spread in the wavelength is related to a spread in the photon energy,

$$\Delta\lambda \approx \frac{hc}{E_{ph}^2}\Delta E_{ph}$$

Suppose that we write $E_{ph} = hc/\lambda$ and $\Delta E_{ph} = \Delta(h\nu) \approx nkT$ where $n$ is a numerical constant. Show that,

$$\Delta\lambda = \frac{nkT}{hc}\lambda^2$$

*LED output spectrum linewidth*

and by appropriately plotting the data in Table 6.6 find $n$.

**Table 6.6** Linewidth $\Delta\lambda_{1/2}$ between half-points in the output spectrum (intensity versus wavelength) of GaAs and AlGaAs LEDs

| | Peak wavelength of emission $\lambda$ (nm) | | | | | | | |
|---|---|---|---|---|---|---|---|---|
| | **650** | **810** | **820** | **890** | **950** | **1150** | **1270** | **1500** |
| $\Delta\lambda_{1/2}$ (nm) | 22 | 36 | 40 | 50 | 55 | 90 | 110 | 150 |
| Material (direct $E_g$) | AlGaAs | AlGaAs | AlGaAs | GaAs | GaAs | InGaAsP | InGaAsP | InGaAsP |

**6.20** **AlGaAs LED emitter** An AlGaAs LED emitter for use in a local optical fiber network has the output spectrum shown in Figure 6.68. It is designed for peak emission at 820 nm at 25 °C.

*a.* What is the linewidth $\Delta\lambda$ between half power points at temperatures −40 °C, 25 °C, and 85 °C? Given these three temperatures, plot $\Delta\lambda$ and $T$ (in K) and find the empirical relationship between $\Delta\lambda$ and $T$. How does this compare with $\Delta(h\nu) \approx 2.5kT$ to $3kT$?

*b.* Why does the peak emission wavelength increase with temperature?

*c.* What is the bandgap of AlGaAs in this LED?

*d.* The bandgap $E_g$ of the ternary alloys $Al_xGa_{1-x}As$ follows the empirical expression

$$E_g(\text{eV}) = 1.424 + 1.266x + 0.266x^2$$

What is the composition of the AlGaAs in this LED?

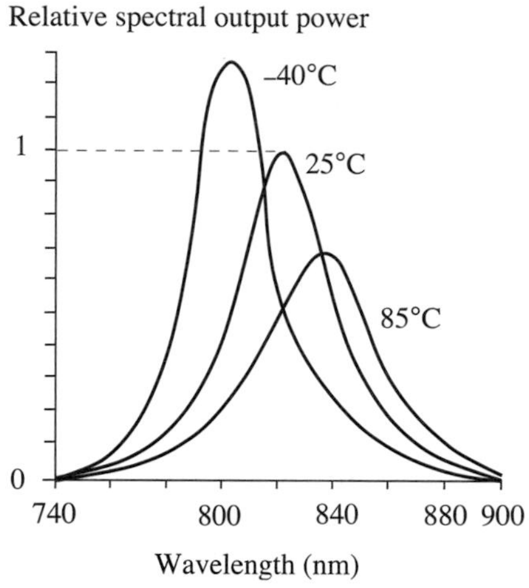


**Figure 6.68** The output spectrum from an AlGaAs LED.
Values are normalized to peak emission at 25 °C.

**6.21 Solar cell driving a load**

*a.* A Si solar cell of area 2.5 cm × 2.5 cm is connected to drive a load $R$ as in Figure 6.54a. It has the $I$–$V$ characteristics in Figure 6.53. Suppose that the load is 2 Ω and it is used under a light intensity of 800 W m$^{-2}$. What are the current and voltage in the curcuit? What is the power delivered to the load? What is the efficiency of the solar cell in this circuit?

*b.* What should the load be to obtain maximum power transfer from the solar cell to the load at 800 W m$^{-2}$ illumination? What is this load at 400 W m$^{-2}$?

*c.* Consider using a number of such solar cells to drive a calculator that needs a minimum of 3 V and draws 50 mA at 3–4 V. It is to be used at a light intensity of about 400 W m$^{-2}$. How many solar cells would you need and how would you connect them?

**6.22 Open circuit voltage** A solar cell under an illumination of 1000 W m$^{-2}$ has a short circuit current $I_{sc}$ of 50 mA and an open circuit output voltage $V_{oc}$ of 0.65 V. What are the short circuit current and open circuit voltages when the light intensity is halved?

**6.23 Maximum power from a solar cell** Suppose that the power delivered by a solar cell, $P = IV$, is maximum when $I = I_m$ and $V = V_m$. Suppose that we define normalized voltage and current for maximum power as

*Normalized solar cell voltage and current*

$$v = \frac{V_m}{\eta V_T} \qquad \text{and} \qquad i = \frac{I_m}{I_{sc}}$$

where $\eta$ is the ideality factor, $V_T = kT/e$ is called the thermal voltage (0.026 V at 300 K), and $I_{sc} = -I_{ph}$. Suppose that $v_{oc} = V_{oc}/(\eta V_T)$ is the normalized open circuit voltage. Under illumination with the solar cell delivering power with $V > \eta V_T$,

*Power delivered by solar cell*

$$P = IV = \left[-I_{ph} + I_o \exp\left(\frac{V}{\eta V_T}\right)\right] V$$

One can differentiate $P = IV$ with respect to $V$, set it to zero for maximum power, and find expressions for $I_m$ and $V_m$ for maximum power. One can then use the open circuit condition ($I = 0$) to relate $V_{oc}$ to $I_o$. Show that maximum power occurs when

*Maximum power delivery*

$$v = v_{oc} - \ln(v + 1) \qquad \text{and} \qquad i = 1 - \exp[-(v_{oc} - v)]$$

Consider a solar cell with $\eta = 1.5$, $V_{oc} = 0.60$ V, and $I_{ph} = 35$ mA, with an area of 1 cm$^2$. Find $i$ and $v$, and hence the current $I_m$ and voltage $V_m$ for maximum power. (Note: Solve the first equation numerically or graphically to find $v \approx 12.76$.) What is the fill factor?

**6.24** **Series resistance** The series resistance causes a voltage drop when a current is drawn from a solar cell. By convention, the positive current is taken to flow into the device. (If calculations yield a negative value, it means that, physically, the current is flowing out, which is the actual case under illumination.) If $V$ is the actual voltage across the solar cell output (accessed by the user), then the voltage across the diode is $V - IR_s$. The solar cell equation becomes

$$I = -I_{ph} + I_d = -I_{ph} + I_o \exp\left(\frac{e(V - IR_s)}{\eta kT}\right)$$

*Solar cell with series resistance*

Plot $I$ versus $V$ for a Si solar cell that has $\eta = 1.5$ and $I_o = 3 \times 10^{-6}$ mA, for an illumination such that $I_{ph} = 10$ mA for $R_s = 0$, 20 and 50 Ω. What is your conclusion?

**6.25** **Shunt resistance** Consider the shunt resistance $R_p$ of a solar cell. Whenever there is a voltage $V$ at the terminals of the solar cell, the shunt resistance draws a current $V/R_p$. Thus, the total current as seen at the terminals (and flowing in by convention) is

$$I = -I_{ph} + I_d + \frac{V}{R_p} = -I_{ph} + I_o \exp\left(\frac{eV}{\eta kT}\right) + \frac{V}{R_p} = 0$$

*Solar cell with shunt resistance*

Plot $I$ versus $V$ for a polycrystalline Si solar cell that has $\eta = 1.5$ and $I_o = 3 \times 10^{-6}$ mA, for an illumination such that $I_{ph} = 10$ mA. Use $R_p = \infty$, 1000, 100 Ω. What is your conclusion?

***6.26** **Series connected solar cells** Consider two identical solar cells connected in series. There are two $R_s$ in series and two *pn* junctions in series. If $I$ is the total current through the devices, then the voltage across one *pn* junction is $V_d = \frac{1}{2}[V - I(2R_s)]$ so that the current $I$ flowing into the combined solar cells is

$$I \approx -I_{ph} + I_o \exp\left[\frac{V - I(2R_s)}{2\eta V_T}\right] \qquad V_d > \eta\left(\frac{kT}{e}\right)$$

*Two solar cells in series*

where $V_T = kT/e$ is the thermal voltage. Rearranging, for two cells in series,

$$V = 2\eta V_T \ln\left(\frac{I + I_{ph}}{I_o}\right) + 2R_s I$$

*Two solar cells in series*

whereas for one cell,

$$V = \eta V_T \ln\left(\frac{I + I_{ph}}{I_o}\right) + R_s I$$

*One solar cell*

Suppose that the cells have the properties $I_o = 25 \times 10^{-6}$ mA, $\eta = 1.5$, $R_s = 20$ Ω, and both are subjected to the same illumination so that $I_{ph} = 10$ mA. Plot the individual $I$–$V$ characteristics and the $I$–$V$ characteristics of the two cells in series. Find the maximum power that can be delivered by one cell and two cells in series. Find the corresponding voltage and current at the maximum power point.

**6.27** **A solar cell used in Eskimo Point** The intensity of light arriving at a point on Earth, where the solar latitude is $\alpha$ can be approximated by the Meinel and Meinel equation:

$$I = 1.353(0.7)^{(\operatorname{cosec}\alpha)^{0.678}} \quad \text{kW m}^{-2}$$

where $\operatorname{cosec}\alpha = 1/(\sin\alpha)$. The solar latitude $\alpha$ is the angle between the sun's rays and the horizon. Around September 23 and March 22, the sun's rays arrive parallel to the plane of the equator. What is the maximum power available for a photovoltaic device panel of area 1 m$^2$ if its efficiency of conversion is 10 percent?

A manufacturer's characterization tests on a particular Si *pn* junction solar cell at 27 °C specifies an open circuit output voltage of 0.45 V and a short circuit current of 400 mA when illuminated directly with a light of intensity 1 kW m$^{-2}$. The fill factor for the solar cell is 0.73. This solar cell is to be used in a portable equipment application near Eskimo Point (Canada) at a geographical latitude ($\phi$) of 63°. Calculate the open circuit output voltage and the maximum available power when the solar cell is used at noon on September 23 when the temperature is around −10 °C. What is the maximum current this solar cell can supply to an electronic equipment? What is your conclusion? (Note: $\alpha + \phi = \pi/2$, and assume $\eta = 1$, and that $I_o \propto n_i^2$)

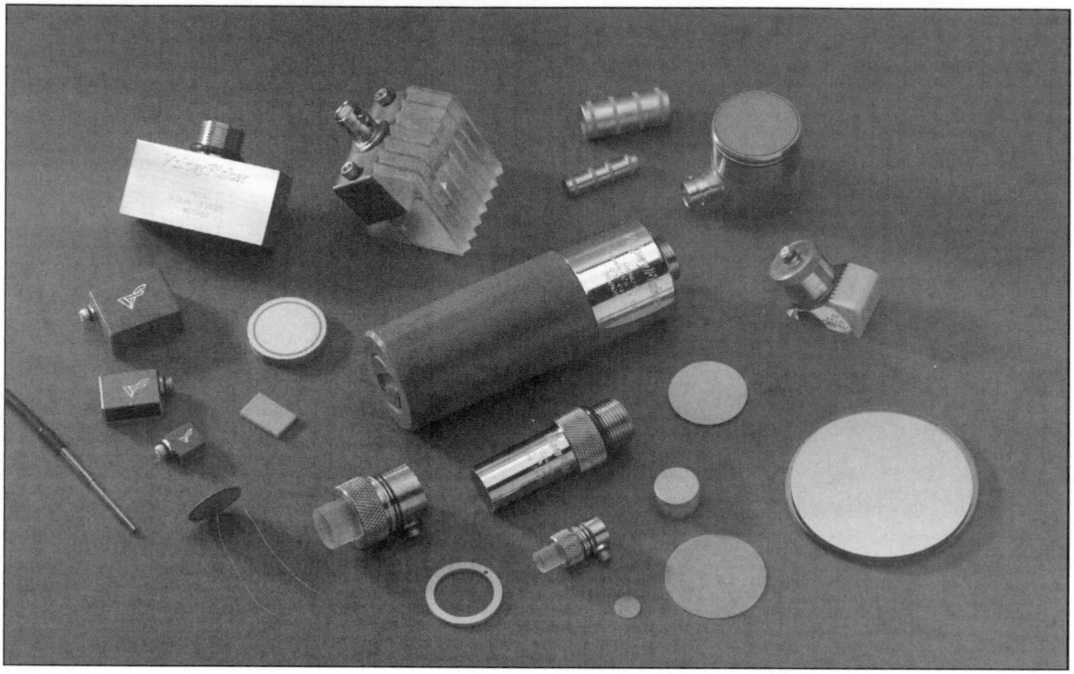

A selection of ultrasonic transducers (piezoelectric effect devices).

SOURCE: Courtesy of Valpey Fisher.

An HV capacitor bushing being subjected to mains-frequency overvoltage. The photo is one of prolonged exposure, recording multiple surface flashovers.

SOURCE: Courtesy of Dr. Simon Rowland, UMIST, England.

CHAPTER

# 7

# Dielectric Materials and Insulation

The familiar parallel plate capacitor equation with free space as an insulator is given by

$$C = \frac{\varepsilon_o A}{d}$$

where $\varepsilon_o$ is the absolute permittivity, $A$ is the plate area, and $d$ is the separation between the plates. If there is a material medium between the plates, then the capacitance, the charge storage ability per unit voltage, increases by a factor of $\varepsilon_r$, where $\varepsilon_r$ is called the **dielectric constant** of the medium or its **relative permittivity.** The increase in the capacitance is due to the **polarization** of the medium in which positive and negative charges are displaced with respect to their equilibrium positions. The opposite surfaces of the dielectric medium acquire opposite surface charge densities that are related to the amount of polarization in the material. An important concept in dielectric theory is that of an **electric dipole moment** $p$, which is a measure of the electrostatic effects of a pair of opposite charges $+Q$ and $-Q$ separated by a finite distance $a$, and so is defined by

$$p = Qa$$

Although the net charge is zero, this entity still gives rise to an electric field in space and also interacts with an electric field from other sources. The relative permittivity is a material property that is frequency dependent. Some capacitors are designed to work at low frequencies, whereas others have a wide frequency range. Furthermore, even though they are regarded as energy storage devices, all practical capacitors exhibit some losses when used in an electric circuit. These losses are no different than $I^2R$ losses in a resistor carrying a current. The power dissipation in a practical capacitor depends on the frequency, and for some applications it can be an important factor. A defining property of a dielectric medium is not only its ability to increase capacitance but also, and equally important, its insulating behavior or low conductivity so that the charges are not conducted from one plate of the capacitor to the other through the dielectric. Dielectric materials often serve to insulate current-carrying conductors or conductors at different voltages. Why can we not simply use air as insulation between high-voltage conductors? When the electric field inside an insulator exceeds a critical field called the **dielectric strength,** the medium suffers dielectric breakdown and a large discharge current flows through the dielectric. Some 40 percent of utility generator failures are linked to insulation failures in the generator. Dielectric breakdown is probably one of the oldest electrical engineering problems and that which has been most widely studied and never fully explained.

# 7.1 MATTER POLARIZATION AND RELATIVE PERMITTIVITY

## 7.1.1 RELATIVE PERMITTIVITY: DEFINITION

We first consider a parallel plate capacitor with vacuum as the dielectric medium between the plates, as shown in Figure 7.1a. The plates are connected to a constant voltage supply $V$. Let $Q_o$ be the charge on the plates. This charge can be easily measured. The capacitance $C_o$ of the parallel plate capacitor in free space, as in Figure 7.1a, is defined by

*Definition of capacitance*

$$C_o = \frac{Q_o}{V} \qquad [7.1]$$

The electric field, directed from high to low potential, is defined by the gradient of the potential $\mathcal{E} = -dV/dx$. Thus, the electric field $\mathcal{E}$ between the plates is just $V/d$ where $d$ is the separation of the plates.

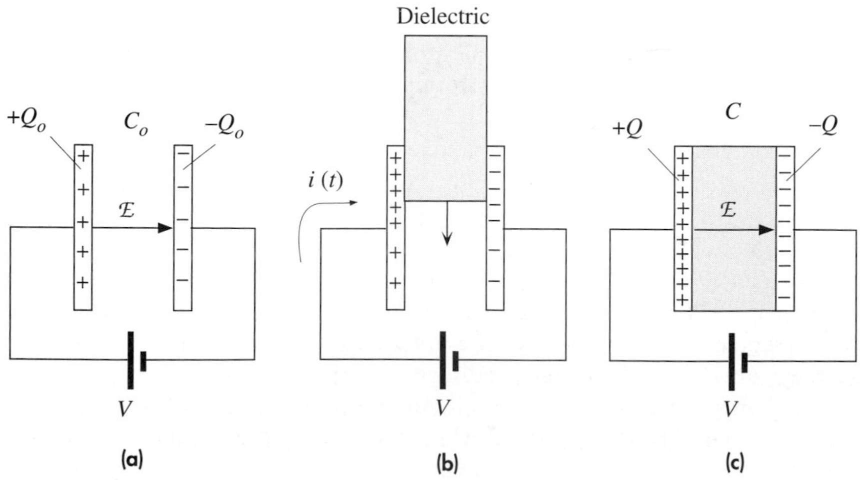


**Figure 7.1**
(a) Parallel plate capacitor with free space between the plates.
(b) As a slab of insulating material is inserted between the plates, there is an external current flow indicating that more charge is stored on the plates.
(c) The capacitance has been increased due to the insertion of a medium between the plates.

Consider now what happens when a dielectric slab (a slab of any nonconducting material) is inserted into this parallel plate capacitor, as shown in Figure 7.1b and c with $V$ kept the same. During the insertion of the dielectric slab, there is an external current flow that indicates that there is additional charge being stored on the plates. The charge on the electrodes increases from $Q_o$ to $Q$. We can easily measure the extra charge $Q - Q_o$ flowing from the battery to the plates by integrating the observed current in the circuit during the process of insertion, as shown in Figure 7.1b. Because there is now a greater amount of charge stored on the plates, the capacitance of the system in Figure 7.1c is larger than that in Figure 7.1a by the ratio $Q$ to $Q_o$. The **relative permittivity** (or the **dielectric constant**) $\varepsilon_r$ is defined to reflect this increase in the capacitance or the charge storage ability by virtue of having a dielectric medium. If $C$ is the capacitance with the dielectric medium as in Figure 7.1c, then by definition

$$\varepsilon_r = \frac{Q}{Q_o} = \frac{C}{C_o} \tag{7.2}$$

*Definition of relative permittivity*

The increase in the stored charge is due to the polarization of the dielectric by the applied field, as explained below. It is important to remember that when the dielectric medium is inserted, the electric field remains unchanged, provided that the insulator fills the whole space between the plates as shown in Figure 7.1c. The voltage $V$ remains the same and therefore so does the gradient $V/d$, which means that $\mathcal{E}$ remains constant.

### 7.1.2 DIPOLE MOMENT AND ELECTRONIC POLARIZATION

An electrical dipole moment is simply a separation between a negative and positive charge of equal magnitude $Q$ as shown in Figure 7.2. If **a** is the vector from the negative to the positive charge, the **electric dipole moment** is defined as a vector by

$$\mathbf{p} = Q\mathbf{a} \tag{7.3}$$

*Definition of dipole moment*

The region that contains the $+Q$ and $-Q$ charges has zero net charge. Unless the two charge centers coincide, this region will nonetheless, by virtue of the definition in Equation 7.3, contain a dipole moment.

The net charge within a neutral atom is zero. Furthermore, on average, the center of negative charge of the electrons coincides with the positive nuclear charge, which means that the atom has no net dipole moment, as indicated in Figure 7.3a. However, when this atom is placed in an external electric field, it will develop an induced dipole moment. The electrons, being much lighter than the positive nucleus, become easily displaced by the field, which results in the separation of the negative charge center

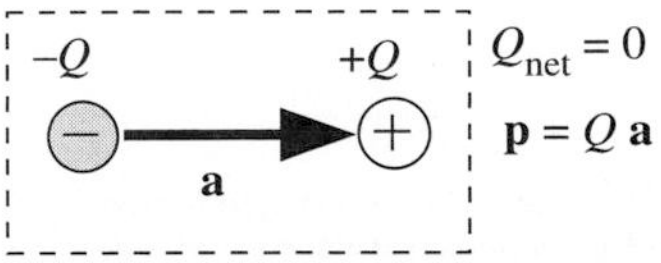


**Figure 7.2** The definition of electric dipole moment.

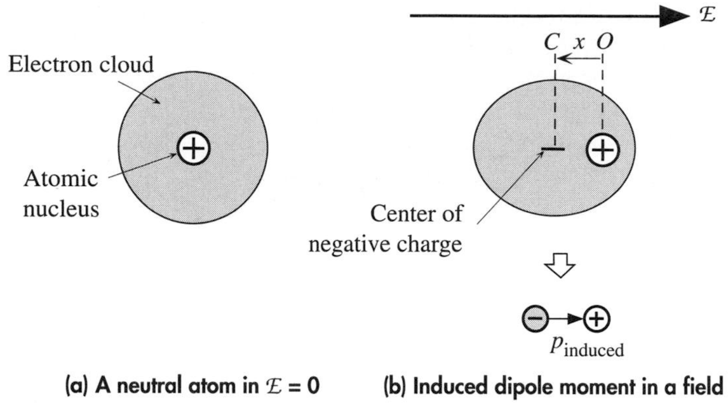


**Figure 7.3** The origin of electronic polarization.

from the positive charge center, as shown in Figure 7.3b. This separation of negative and positive charges and the resulting induced dipole moment are termed **polarization.** An atom is said to be **polarized** if it possesses an effective dipole moment, that is, if there is a separation between the centers of negative and positive charge distributions.

The induced dipole moment depends on the electric field causing it. We define a quantity called the **polarizability** $\alpha$ to relate the induced dipole moment $p_{\text{induced}}$ to the field $\mathcal{E}$ causing it,

*Definition of polarizability*

$$p_{\text{induced}} = \alpha\mathcal{E} \tag{7.4}$$

where $\alpha$ is a coefficient called the polarizability of the atom. It depends on the polarization mechanism. Since the polarization of a neutral atom involves the displacement of electrons, $\alpha$ is called **electronic polarization** and denoted as $\alpha_e$. Inasmuch as the electrons in an atom are not rigidly fixed, all atoms possess a certain amount of electronic polarizability.

In the absence of an electric field, the center of mass of the orbital motions of the electrons coincides with the positively charged nucleus and the electronic dipole moment is zero. Suppose that the atom has $Z$ number of electrons orbiting the nucleus and all the electrons are contained within a certain spherical region. When an electric field $\mathcal{E}$ is applied, the light electrons become displaced in the opposite direction to $\mathcal{E}$, so their center of mass $C$ is shifted by some distance $x$ with respect to the nucleus $O$, which we take to be the origin, as shown in Figure 7.3b. As the electrons are "pushed" away by the applied field, the Coulombic attraction between the electrons and nuclear charge "pulls in" the electrons. The force on the electrons, due to $\mathcal{E}$, trying to separate them away from the nuclear charge is $Ze\mathcal{E}$. The restoring force $F_r$, which is the Coulombic attractive force between the electrons and the nucleus, can be taken to be proportional to the displacement $x$, provided that the latter is small.[1] The restoring force $F_r$ is obviously zero when $C$ coincides with $O\,(x = 0)$. We can write

*Restoring force*

$$F_r = -\beta x$$

[1] It may be noticed that even if $F_r$ is a complicated function of $x$, it can still be expanded in a series in terms of powers of $x$, that is, $x$, $x^2$, $x^3$, and so on, and for small $x$ only the $x$ term is significant, $F_r = -\beta x$.

where $\beta$ is a constant and the negative sign indicates that $F_r$ is always directed toward the nucleus $O$ (Figure 7.3b). In equilibrium, the net force on the negative charge is zero or

$$Ze\mathcal{E} = \beta x$$

from which $x$ is known. Therefore the **magnitude** of the induced electronic dipole moment $p_e$ is given by

$$p_e = (Ze)x = \left(\frac{Z^2e^2}{\beta}\right)\mathcal{E} \quad [7.5]$$

*Electronic polarization*

As expected, $p_e$ is proportional to the applied field. The electronic dipole moment in Equation 7.5 is valid under static conditions, that is, when the electric field is a dc field. Suppose that we suddenly remove the applied electric field polarizing the atom. There is then only the restoring force $-\beta x$, which always acts to pull the electrons toward the nucleus $O$. The equation of motion of the negative charge center is then (from force = mass × acceleration)

*Equation for simple harmonic motion*

$$-\beta x = Zm_e\frac{d^2x}{dt^2}$$

Thus the displacement at any time is

$$x(t) = x_o \cos(\omega_o t)$$

where

$$\omega_o = \left(\frac{\beta}{Zm_e}\right)^{1/2} \quad [7.6]$$

*Electronic polarization resonance frequency*

is the oscillation frequency of the center of mass of the electron cloud about the nucleus and $x_o$ is the displacement before the removal of the field. After the removal of the field, the electronic charge cloud executes simple harmonic motion about the nucleus with a natural frequency determined by Equation 7.6; $\omega_o$ is called the **electronic polarization resonance frequency.**[2] It is analogous to a mass on a spring being pulled and let go. The system then executes simple harmonic motion. The oscillations of course die out with time. In the atomic case, a sinusoidal displacement implics that the electronic charge cloud has an acceleration

$$\frac{d^2x}{dt^2} = -x_o\omega_o^2\cos(\omega_o t)$$

It is well known from classical electromagnetism that an accelerating charge radiates electromagnetic energy just like a radio antenna. Consequently the oscillating charge

[2] The term *natural frequency* refers to a system's characteristic frequency of oscillation when it is excited. A mass attached to a spring and then let go will execute simple harmonic motion with a certain natural frequency $\omega_o$. If we then decide to oscillate this mass with an applied force, the maximum energy transfer will occur when the applied force has the same frequency as $\omega_o$; the system will be put in resonance. $\omega_o$ is also a *resonant frequency*. Strictly, $\omega = 2\pi f$ is the angular frequency and $f$ is the frequency. It is quite common to simply refer to $\omega$ as a frequency because the literature is dominated by $\omega$; the meaning should be obvious within context.

cloud loses energy, and thus its amplitude of oscillation decreases. (Recall that the average energy is proportional to the square of the amplitude of the displacement.)

From the expression derived for $p_e$ in Equation 7.5, we can find the electronic polarizability $\alpha_e$ from Equation 7.4,

*Static electronic polarizability*

$$\alpha_e = \frac{Ze^2}{m_e\omega_o^2} \quad [7.7]$$

**EXAMPLE 7.1**

**ELECTRONIC POLARIZABILITY** Consider the electronic polarizability of inert gas atoms. These atoms have closed shells. Their electronic polarizabilities are listed in Table 7.1. For each type of atom calculate the electronic polarization resonant frequency $f_o = \omega_o/2\pi$, and plot $\alpha_e$ and $f_o$ against the number of electrons $Z$ in the atom. What is your conclusion?

**SOLUTION**

We can use Equation 7.7 to calculate the resonant frequency $f_o = \omega_o/2\pi$. Taking Ar,

$$\omega_o = \left(\frac{Ze^2}{\alpha_e m_e}\right)^{1/2} = \left[\frac{(18)(1.6\times10^{-19})^2}{(1.7\times10^{-40})(9.1\times10^{-31})}\right]^{1/2} = 5.46\times10^{16}\ \text{rad s}^{-1}$$

**Table 7.1** Electronic polarizability $\alpha_e$ dependence on $Z$ for the inert element atoms

| | Atom | | | | | |
|---|---|---|---|---|---|---|
| | **He** | **Ne** | **Ar** | **Kr** | **Xe** | **Rn*** |
| $Z$ | 2 | 10 | 18 | 36 | 56 | |
| $\alpha_e \times 10^{-40}$ (F m$^2$) | 0.18 | 0.45 | 1.7 | 2.7 | 4.4 | 5.9 |
| $f_o \times 10^{15}$ (Hz) | 8.90 | 12.6 | 8.69 | 9.76 | 9.36 | 10.2 |

*Rn (radon) gas is radioactive.

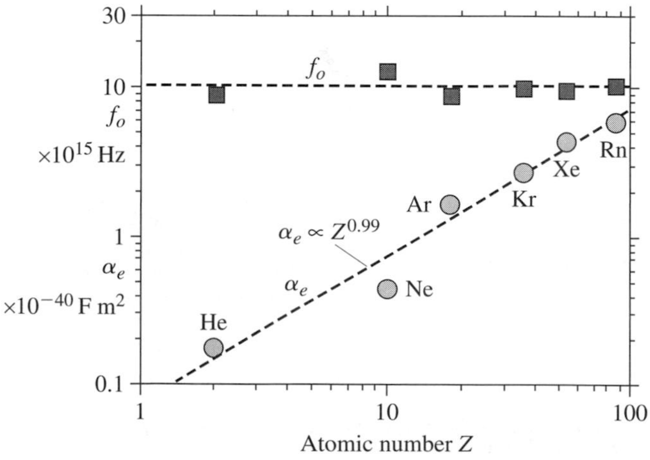


**Figure 7.4** Electronic polarizability and its resonance frequency versus the number of electrons in the atom ($Z$).
The dashed line is the best-fit line.

so that

$$f_o = \frac{\omega_o}{2\pi} = 8.69 \times 10^{15} \text{ Hz}$$

which is listed in Table 7.1, among other $f_o$ calculations for the other atoms. Such frequencies correspond to the field oscillations in UV light, that is, at optical frequencies. For all practical purposes, electronic polarization occurs very rapidly, that is, on a time scale $1/f_o$ or $10^{-15}$ s, and we can take the static polarizability $\alpha_e$ to remain the same up to optical frequencies.[3]

Figure 7.4 shows the dependence of $\alpha_e$ and $f_o$ on the number of electrons $Z$. It is apparent that $\alpha_e$ is nearly linearly proportional to $Z$, whereas $f_o$ is very roughly constant. It is left as an exercise to show that $\beta$ increases with $Z$, which is reasonable since the restoring force was defined as the total force between *all* the electrons and the nucleus when the electrons are displaced.

## 7.1.3 POLARIZATION VECTOR P

When a material is placed in an electric field, the atoms and the molecules of the material become polarized, so we have a distribution of dipole moments in the material. We can visualize this effect with the insertion of the dielectric slab into the parallel plate capacitor, as depicted in Figure 7.5a. The placement of the dielectric slab into an electric field polarizes the molecules in the material. The induced dipole moments all point in the direction of the field. Consider the polarized medium alone, as shown in Figure 7.5b. In the bulk of the material, the dipoles are aligned head to tail. Every positive charge has a negative charge next to it and vice versa. There is therefore no net charge within the bulk. But the positive charges of the dipoles appearing at the right-hand face are not canceled by negative charges of any dipoles at this face. There is therefore a surface charge $+Q_P$ on the right-hand face that results from the polarization of the medium. Similarly, there is a negative charge $-Q_P$ with the same magnitude appearing on the left-hand face due to the negative charges of the dipoles at this face. We see that charges $+Q_P$ and $-Q_P$ appear on the opposite surfaces of a material when it becomes polarized in an electric field, as shown in Figure 7.5c. These charges are **bound** and are a direct result of the polarization of the molecules. They are termed **surface polarization charges.** Figure 7.5c emphasizes this aspect of dielectric behavior in an electric field by showing the dielectric and its polarization charges only.

We represent the polarization of a medium by a quantity called **polarization P,** which is defined as the total dipole moment per unit volume,

*Definition of polarization vector*

$$\mathbf{P} = \frac{1}{\text{Volume}}[\mathbf{p}_1 + \mathbf{p}_2 + \cdots + \mathbf{p}_N] \qquad \textbf{[7.8a]}$$

where $\mathbf{p}_1, \mathbf{p}_2, \ldots, \mathbf{p}_N$ are the dipole moments induced at $N$ molecules in the volume. If $\mathbf{p}_{\text{av}}$ is the average dipole moment per molecule, then an equivalent definition of **P** is

*Definition of polarization vector*

$$\mathbf{P} = N\mathbf{p}_{\text{av}} \qquad \textbf{[7.8b]}$$

[3] Electronic polarization at optical frequencies controls the optical properties such as the refractive index, as will be covered in Chapter 9.

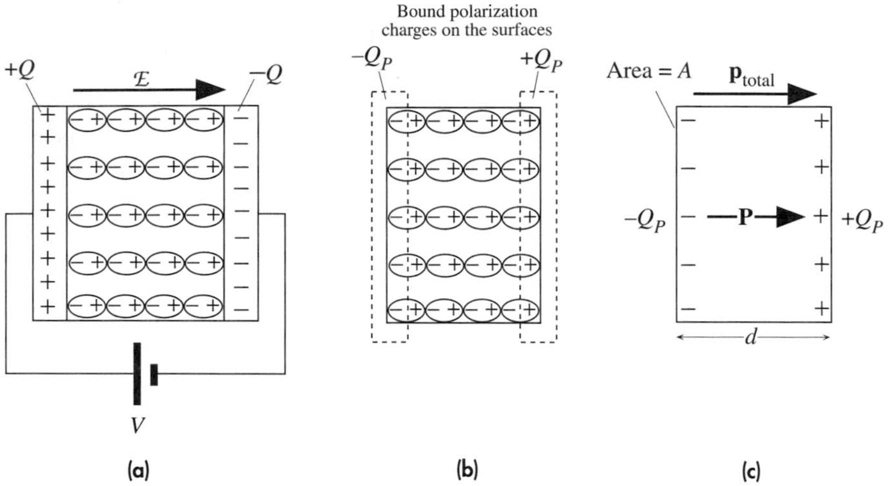


**Figure 7.5**

(a) When a dielectric is placed in an electric field, bound polarization charges appear on the opposite surfaces.

(b) The origin of these polarization charges is the polarization of the molecules of the medium.

(c) We can represent the whole dielectric in terms of its surface polarization charges $+Q_P$ and $-Q_P$.

where $N$ is the number of molecules per unit volume. There is an important relationship, given below, between $\mathbf{P}$ and the polarization charges $Q_P$ on the surfaces of the dielectric. It should be emphasized for future discussions that if polarization arises from the effect of the applied field, as shown in Figure 7.5a, which is usually the case, $\mathbf{p}_{av}$ must be the *average dipole moment per atom in the direction of the applied field.* In that case we often also denote $\mathbf{p}_{av}$ as the induced average dipole moment per molecule $\mathbf{p}_{induced}$.

To calculate the polarization $\mathbf{P}$ for the polarized dielectric in Figure 7.5b, we need to sum all the dipoles in the medium and divide by the volume $Ad$, as in Equation 7.8a. However, the polarized medium can be simply represented as in Figure 7.5c in terms of surface charge $+Q_P$ and $-Q_P$, which are separated by the thickness distance $d$. We can view this arrangement as one big dipole moment $p_{total}$ from $-Q_P$ to $+Q_P$. Thus

$$p_{total} = Q_P d$$

Since the polarization is defined as the total dipole moment per unit volume, the magnitude of $\mathbf{P}$ is

*Polarization and bound surface charge density*

$$P = \frac{p_{total}}{\text{Volume}} = \frac{Q_P d}{Ad} = \frac{Q_P}{A}$$

But $Q_P/A$ is the **surface polarization charge density** $\sigma_P$, so

$$P = \sigma_p \tag{7.9a}$$

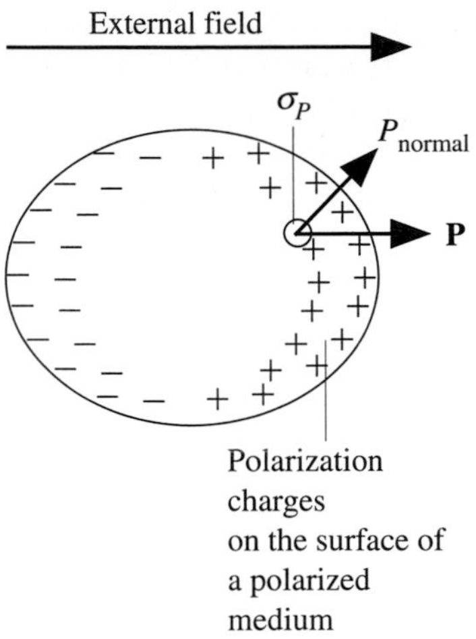


**Figure 7.6** Polarization charge density on the surface of a polarized medium is related to the normal component of the polarization vector.

Polarization is a vector and Equation 7.9a only gives its magnitude. For the rectangular slab in Figure 7.5c, the direction of $P$ is normal to the surface. For $+\sigma_p$ (right face), it comes out from the surface and for $-\sigma_P$ (left face), it is directed into the surface. Although Equation 7.9a is derived for one specific geometry, the rectangular slab, it can be generalized as follows. *The charge per unit area appearing on the surface of a polarized medium is equal to the component of the polarization vector normal to this surface.* If $P_{\text{normal}}$ is the component of **P** normal to the surface where the polarization charge density is $\sigma_P$, as shown in Figure 7.6, then,

*Polarization and bound surface charge density*

$$P_{\text{normal}} = \sigma_p \qquad [7.9b]$$

The polarization $P$ induced in a dielectric medium when it is placed in an electric field depends on the field itself. The induced dipole moment per molecule within the medium depends on the electric field by virtue of Equation 7.4. To express the dependence of $P$ on the field $\mathcal{E}$, we define a quantity called the **electric susceptibility** $\chi_e$ by

*Definition of electric susceptibility*

$$P = \chi_e \varepsilon_o \mathcal{E} \qquad [7.10]$$

Equation 7.10 shows an *effect* $P$ due to a *cause* $\mathcal{E}$ and the quantity $\chi_e$ relates the effect to its cause. Put differently, $\chi_e$ acts as a proportionality constant. It may depend on the field itself, in which case the effect is nonlinearly related to the cause. Further, electronic polarizability is defined by

$$p_{\text{induced}} = \alpha_e \mathcal{E}$$

so

$$P = Np_{\text{induced}} = N\alpha_e \mathcal{E}$$

where $N$ is the number of molecules per unit volume. Then from Equation 7.10, $\chi_e$ and $\alpha_e$ are related by

*Electric susceptibility and polarization*

$$\chi_e = \frac{1}{\varepsilon_o} N\alpha_e \qquad [7.11]$$

It is important to recognize the difference between *free* and *polarization* (or *bound*) charges. The charges stored on the metal plates in Figure 7.5a are free because they result from the motion of free electrons in the metal. For example both $Q_o$ and $Q$, before and after the dielectric insertion in Figure 7.1, are free charges that arrive on the plates from the battery. The polarization charges $+Q_P$ and $-Q_P$, on the other hand, are bound to the molecules. They cannot move within the dielectric or on its surface.

The field $\mathcal{E}$ *before* the dielectric was inserted (Figure 7.1a) is given by

$$\mathcal{E} = \frac{V}{d} = \frac{Q_o}{C_o d} = \frac{Q_o}{\varepsilon_o A} = \frac{\sigma_o}{\varepsilon_o} \qquad [7.12]$$

where $\sigma_o = Q_o/A$ is the **free surface charge density** without any dielectric medium between the plates, as in Figure 7.1a.

*After* the insertion of the dielectric, this field remains the same $V/d$, but the free charges on the plates are different. The free surface charge on the plates is now $Q$. In addition there are bound polarization charges on the dielectric surfaces next to the plates, as shown in Figure 7.5a. It is apparent that the flow of current during the insertion of the dielectric, Figure 7.1b, is due to the additional free charges $Q - Q_o$ needed on the capacitor plates to neutralize the opposite polarity polarization charges $Q_P$ appearing on the dielectric surfaces. The total charge (see Figure 7.5a) due to that on the plate plus that appearing on the dielectric surface, $Q - Q_P$, must be the same as before, $Q_o$, so that the field, as given by Equation 7.12, does not change inside the dielectric, that is,

$$Q - Q_P = Q_o$$

or

$$Q = Q_o + Q_P$$

Dividing by $A$, defining $\sigma = Q/A$ as the free surface charge density on the plates with the dielectric inserted, and using Equation 7.12, we obtain

$$\sigma = \varepsilon_o \mathcal{E} + \sigma_P$$

Since $\sigma_P = P$ and $P = \chi_e \varepsilon_o \mathcal{E}$, Equations 7.9 and 7.10, we can eliminate $\sigma_P$ to obtain

$$\sigma = \varepsilon_o (1 + \chi_e)\mathcal{E}$$

From the definition of the relative permittivity in Equation 7.2 we have

$$\varepsilon_r = \frac{Q}{Q_o} = \frac{\sigma}{\sigma_o}$$

*Relative permittivity and electric susceptibility*

so substituting for $\sigma$ and using Equation 7.12 we obtain

$$\varepsilon_r = 1 + \chi_e \qquad [7.13]$$

In terms of electronic polarization, from Equation 7.11, this is

*Relative permittivity and polarizability*

$$\varepsilon_r = 1 + \frac{N\alpha_e}{\varepsilon_o} \qquad [7.14]$$

The significance of Equation 7.14 is that it relates the microscopic polarization mechanism that determines $\alpha_e$ to the macroscopic property $\varepsilon_r$.

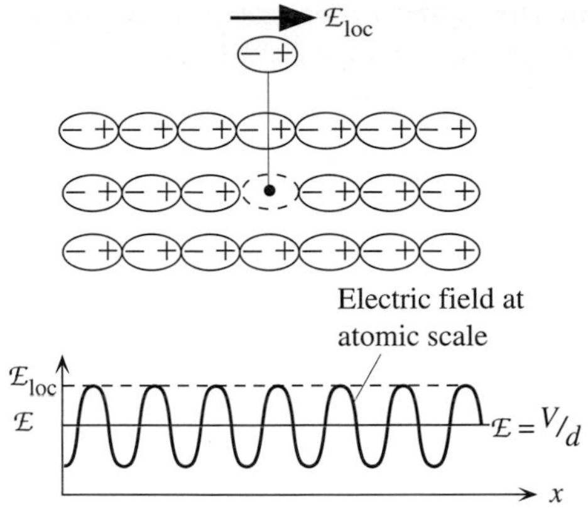


**Figure 7.7** The electric field inside a polarized dielectric at the atomic scale is not uniform. The local field is the actual field that acts on a molecule. It can be calculated by removing that molecule and evaluating the field at that point from the charges on the plates and the dipoles surrounding the point.

## 7.1.4 LOCAL FIELD $\mathcal{E}_{loc}$ AND CLAUSIUS–MOSSOTTI EQUATION

Equation 7.14, which relates $\varepsilon_r$ to electronic polarizability $\alpha_e$ is only approximate because it assumes that the field acting on an individual atom or molecule is the field $\mathcal{E}$, which is assumed to be uniform within the dielectric. In other words, the induced polarization, $p_{induced} \propto \mathcal{E}$. However, the induced polarization depends on the actual field experienced by the molecule. It is apparent from Figure 7.5a that there are polarized molecules within the dielectric with their negative and positive charges separated so that the field is not constant *on the atomic scale* as we move through the dielectric. This is depicted in Figure 7.7. The field experienced by an individual molecule is actually different than $\mathcal{E}$, which represents the average field in the dielectric. As soon as the dielectric becomes polarized, the field at some arbitrary point depends not only on the charges on the plates ($Q$) but also on the orientations of all the other dipoles around this point in the dielectric. When averaged over some distance, say a few thousand molecules, this field becomes $\mathcal{E}$, as shown in Figure 7.7.

The actual field experienced by a molecule in a dielectric is defined as the **local field** and denoted by $\mathcal{E}_{loc}$. It depends not only on the free charges on the plates but also on the arrangement of all the polarized molecules around this point. In evaluating $\mathcal{E}_{loc}$ we simply remove the molecule from this point and calculate the field at this point coming from all sources, including neighboring polarized molecules, as visualized in Figure 7.7. $\mathcal{E}_{loc}$ will depend on the amount of polarization the material has experienced. The greater the polarization, the greater is the local field because there are bigger dipoles around this point. $\mathcal{E}_{loc}$ depends on the arrangement of polarized molecules around the point of interest and hence depends on the crystal structure. In the simplest case of a material with a cubic crystal structure, or a liquid (no crystal structure), the local field $\mathcal{E}_{loc}$ acting on a molecule increases with polarization as[4]

$$\mathcal{E}_{loc} = \mathcal{E} + \frac{1}{3\varepsilon_o} P \qquad [7.15]$$

*Lorentz local field in dielectrics*

[4] This field is called the **Lorentz field** and the proof, though not difficult, is not necessary for the present introductory treatment of dielectrics. This local field expression does not apply to dipolar dielectrics discussed in Section 7.3.2.

Equation 7.15 is called the **Lorentz field.** The induced polarization in the molecule now depends on this local field $\mathcal{E}_{\text{loc}}$ rather than the average field $\mathcal{E}$. Thus

$$p_{\text{induced}} = \alpha_e \mathcal{E}_{\text{loc}}$$

The fundamental definition of electric susceptibility by the equation

$$P = \chi_e \varepsilon_o \mathcal{E}$$

is unchanged, which means that $\varepsilon_r = 1 + \chi_e$, Equation 7.13, remains intact. The polarization is defined by $P = Np_{\text{induced}}$, and $p_{\text{induced}}$ can be related to $\mathcal{E}_{\text{loc}}$ and hence to $\mathcal{E}$ and $P$. Then

$$P = (\varepsilon_r - 1)\varepsilon_o \mathcal{E}$$

can be used to eliminate $\mathcal{E}$ and $P$ and obtain a relationship between $\varepsilon_r$ and $\alpha_e$. This is the **Clausius–Mossotti equation,**

*Clausius–Mossotti equation*

$$\frac{\varepsilon_r - 1}{\varepsilon_r + 2} = \frac{N\alpha_e}{3\varepsilon_o} \qquad [7.16]$$

This equation allows the calculation of the macroscopic property $\varepsilon_r$ from microscopic polarization phenomena, namely, $\alpha_e$.

**EXAMPLE 7.2**

**ELECTRONIC POLARIZABILITY OF A VAN DER WAALS SOLID** The electronic polarizability of the Ar atom is $1.7 \times 10^{-40}$ F m$^2$. What is the static dielectric constant of solid Ar (below 84 K) if its density is 1.8 g cm$^{-3}$?

**SOLUTION**

To calculate $\varepsilon_r$ we need the number of Ar atoms per unit volume $N$ from the density $d$. If $M_{\text{at}} = 39.95$ is the relative atomic mass of Ar and $N_A$ is Avogadro's number, then

$$N = \frac{N_A d}{M_{\text{at}}} = \frac{(6.02 \times 10^{23}\ \text{mol}^{-1})(1.8\ \text{g cm}^{-3})}{(39.95\ \text{g mol}^{-1})} = 2.71 \times 10^{22}\ \text{cm}^{-3}$$

with $N = 2.71 \times 10^{28}$ m$^{-3}$ and $\alpha_e = 1.7 \times 10^{-40}$ F m$^2$, we have

$$\varepsilon_r = 1 + \frac{N\alpha_e}{\varepsilon_o} = 1 + \frac{(2.71 \times 10^{28})(1.7 \times 10^{-40})}{(8.85 \times 10^{-12})} = 1.52$$

If we use the Clausius–Mossotti equation, we get

$$\varepsilon_r = \frac{1 + \dfrac{2N\alpha_e}{3\varepsilon_o}}{1 - \dfrac{N\alpha_e}{3\varepsilon_o}} = 1.63$$

The two values are different by about 7 percent. The simple relationship in Equation 7.14 underestimates the relative permittivity.

## 7.2 ELECTRONIC POLARIZATION: COVALENT SOLIDS

When a field is applied to a solid substance, the constituent atoms or molecules become polarized, as we visualized in Figure 7.5a. The electron clouds within each atom become shifted by the field, and this gives rise to **electronic polarization.** This type of electronic polarization within an atom, however, is quite small compared with the polarization due to the valence electrons in the covalent bonds within the solid. For example, in crystalline silicon, there are electrons shared with neighboring Si atoms in covalent bonds, as shown in Figure 7.8a. These valence electrons form bonds (*i.e.*, become shared) between the Si atoms because they are already loosely bound to their parent atoms. If this were not the case, the solid would be a van der Waals solid with atoms held together by secondary bonds (*e.g.*, solid Ar below 83.8 K). In the covalent solid, the valence electrons therefore are not rigidly tied to the ionic cores left in the Si atoms. Although intuitively we often view these valence electrons as living in covalent bonds between the ionic Si cores, they nonetheless belong to the whole crystal because they can tunnel from bond to bond and exchange places with each other. We refer to their wavefunctions as delocalized, that is, not localized to any particular Si atom. When an electric field is applied, the negative charge distribution associated with these valence electrons becomes readily shifted with respect to the positive charges of the ionic Si cores, as depicted in Figure 7.8b and the crystal exhibits polarization, or develops a polarization vector. One can appreciate the greater flexibility of electrons in covalent bonds compared with those in individual ionic cores by comparing the energy involved in freeing each. It takes perhaps 1–2 eV to break a covalent bond to free the valence electron, but it takes more than 10 eV to free an electron from an individual ionic Si core. Thus, the valence electrons in the bonds readily respond to an applied field and become displaced. This type of electronic polarization, due to the displacement of electrons in covalent bonds, is responsible for the large dielectric constants of covalent crystals. For example $\varepsilon_r = 11.9$ for the Si crystal and $\varepsilon_r = 16$ for the Ge crystal.

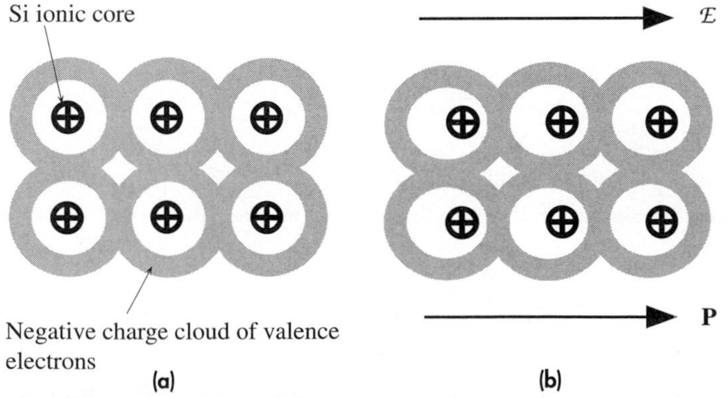


**Figure 7.8**
(a) Valence electrons in covalent bonds in the absence of an applied field.
(b) When an electric field is applied to a covalent solid, the valence electrons in the covalent bonds are shifted very easily with respect to the positive ionic cores. The whole solid becomes polarized due to the collective shift in the negative charge distribution of the valence electrons.

**EXAMPLE 7.3** **ELECTRONIC POLARIZABILITY OF COVALENT SOLIDS** Consider a pure Si crystal that has $\varepsilon_r = 11.9$.

*a.* What is the electronic polarizability due to valence electrons per Si atom (if one could portion the observed crystal polarization to individual atoms)?

*b.* Suppose that a Si crystal sample is electroded on opposite faces and has a voltage applied across it. By how much is the local field greater than the applied field?

*c.* What is the resonant frequency $f_o$ corresponding to $\omega_o$?

From the density of the Si crystal, the number of Si atoms per unit volume, $N$, is given as $5 \times 10^{28}\ \text{m}^{-3}$.

**SOLUTION**

*a.* Given the number of Si atoms, we can apply the Clausius–Mossotti equation to find $\alpha_e$

$$\alpha_e = \frac{3\varepsilon_o}{N}\frac{\varepsilon_r - 1}{\varepsilon_r + 2} = \frac{3(8.85 \times 10^{-12})}{(5 \times 10^{28})}\frac{11.9 - 1}{11.9 + 2} = 4.17 \times 10^{-40}\ \text{F m}^2$$

This is larger, for example, than the electronic polarizability of an isolated Ar atom, which has more electrons. If we were to take the inner electrons in each Si atom as very roughly representing Ne, we would expect their contribution to the overall electronic polarizability to be roughly the same as the Ne atom, which is $0.45 \times 10^{-40}\ \text{F m}^2$.

*b.* The local field is

$$\mathcal{E}_{\text{loc}} = \mathcal{E} + \frac{1}{3\varepsilon_o}P$$

But, by definition,

$$P = \chi_e\varepsilon_o\mathcal{E} = (\varepsilon_r - 1)\varepsilon_o\mathcal{E}$$

Substituting for $P$,

$$\mathcal{E}_{\text{loc}} = \mathcal{E} + \frac{1}{3}(\varepsilon_r - 1)\mathcal{E}$$

so the local field with respect to the applied field is

$$\frac{\mathcal{E}_{\text{loc}}}{\mathcal{E}} = \frac{1}{3}(\varepsilon_r + 2) = 4.63$$

The local field is a factor of 4.63 greater than the applied field.

*c.* Since polarization is due to valence electrons and there are four per Si atom, we can use Equation 7.7,

$$\omega_o = \left(\frac{Ze^2}{m_e\alpha_e}\right)^{1/2} = \left[\frac{4(1.6 \times 10^{-19})^2}{(9.1 \times 10^{-31})(4.17 \times 10^{-40})}\right]^{1/2} = 1.65 \times 10^{16}\ \text{rad s}^{-1}$$

The corresponding resonant frequency is $\omega_o/2\pi$ or $2.6 \times 10^{15}$ Hz, which is typically associated with electromagnetic waves of wavelength in the ultraviolet region.

## 7.3 POLARIZATION MECHANISMS

In addition to electronic polarization, we can identify a number of other polarization mechanisms that may also contribute to the relative permittivity.

### 7.3.1 IONIC POLARIZATION

This type of polarization occurs in ionic crystals such as NaCl, KCl, and LiBr. The ionic crystal has distinctly identifiable ions, for example, $Na^+$ and $Cl^-$, located at well-defined lattice sites, so each pair of oppositely charged neighboring ions has a dipole moment. As an example, we consider the one-dimensional NaCl crystal depicted as a chain of alternating $Na^+$ and $Cl^-$ ions in Figure 7.9a. In the absence of an applied field, the solid has no net polarization because the dipole moments of equal magnitude are lined up head to head and tail to tail so that the net dipole moment is zero. The dipole moment $p_+$ in the positive $x$ direction has the same magnitude as $p_-$ in the negative $x$ direction, so the net dipole moment

$$p_{\text{net}} = p_+ - p_- = 0$$

In the presence of a field $\mathcal{E}$ along the $x$ direction, however, the $Cl^-$ ions are pushed in the $-x$ direction and the $Na^+$ ions in the $+x$ direction about their equilibrium positions. Consequently, the dipole moment $p_+$ in the $+x$ direction *increases* to $p'_+$ and the dipole moment $p_-$ *decreases* to $p'_-$, as shown in Figure 7.9b. The net dipole moment is now no longer zero. The net dipole moment, or the average dipole moment, per ion pair is now $(p'_+ - p'_-)$, which depends on the electric field $\mathcal{E}$. Thus the induced average dipole moment per ion pair $p_{\text{av}}$ depends on the field $\mathcal{E}$. The ionic polarizability $\alpha_i$ is defined in terms of the local field experienced by the ions,

$$p_{\text{av}} = \alpha_i \mathcal{E}_{\text{loc}} \qquad [7.17]$$

*Ionic polarizability*

The larger the $\alpha_i$, the greater the induced dipole moment. Generally, $\alpha_i$ is larger than the electronic polarizability $\alpha_e$ by a factor of 10 or more, which leads to ionic solids having large dielectric constants. The polarization $P$ exhibited by the ionic solid

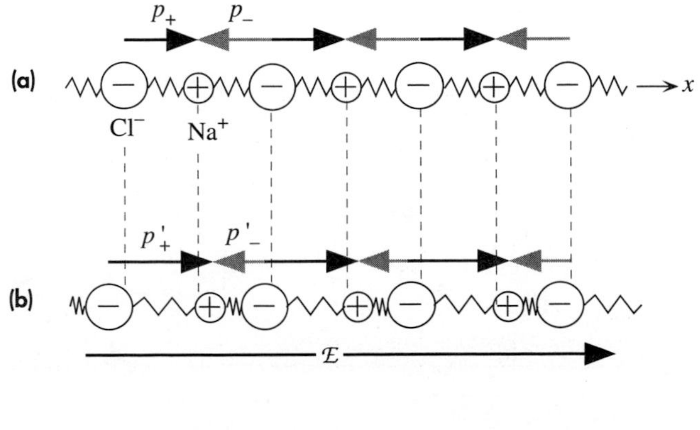


**Figure 7.9**

(a) A NaCl chain in the NaCl crystal without an applied field. Average or net dipole moment per ion is zero.

(b) In the presence of an applied field, the ions become slightly displaced, which leads to a net average dipole moment per ion.

is therefore given by

$$P = N_i p_{av} = N_i \alpha_i \mathcal{E}_{loc}$$

where $N_i$ is the number of ion pairs per unit volume. By relating the local field to $\mathcal{E}$ and using

$$P = (\varepsilon_r - 1)\varepsilon_o \mathcal{E}$$

*Clausius–Mossotti equation for ionic polarization*

we can again obtain the Clausius–Mossotti equation, but now due to ionic polarization,

$$\frac{\varepsilon_r - 1}{\varepsilon_r + 2} = \frac{1}{3\varepsilon_o} N_i \alpha_i \qquad [7.18]$$

Each ion also has a core of electrons that become displaced in the presence of an applied field with respect to their positive nuclei and therefore also contribute to the polarization of the solid. This electronic polarization simply adds to the ionic polarization. Its magnitude is invariably much smaller than the ionic contribution in these solids.

### 7.3.2 Orientational (Dipolar) Polarization

Certain molecules possess permanent dipole moments. For example, the HCl molecule shown in Figure 7.10a has a permanent dipole moment $p_o$ from the $Cl^-$ ion to the $H^+$ ion. In the liquid or gas phases, these molecules, in the absence of an electric field, are randomly oriented as a result of thermal agitation, as shown in Figure 7.10b. When an electric field $\mathcal{E}$ is applied, $\mathcal{E}$ tries to align the dipoles parallel to itself, as depicted in Figure 7.10c. The $Cl^-$ and $H^+$ charges experience forces in opposite directions. But the nearly rigid bond between $Cl^-$ and $H^+$ holds them together, which means that the

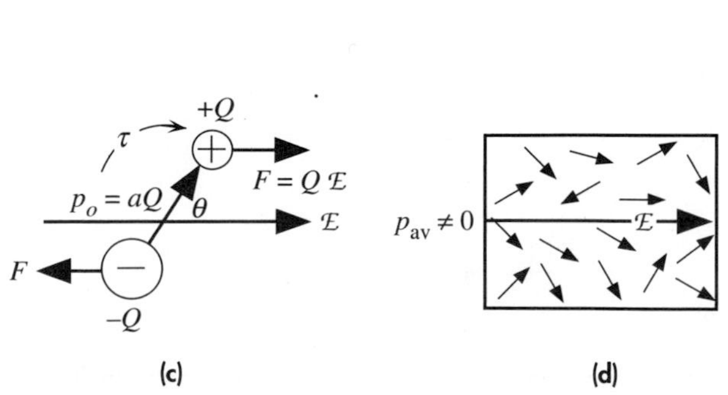


**Figure 7.10**
(a) A HCl molecule possesses a permanent dipole moment $p_o$.
(b) In the absence of a field, thermal agitation of the molecules results in zero net average dipole moment per molecule.
(c) A dipole such as HCl placed in a field experiences a torque that tries to rotate it to align $p_o$ with the field $\mathcal{E}$.
(d) In the presence of an applied field, the dipoles try to rotate to align with the field against thermal agitation. There is now a net average dipole moment per molecule along the field.

molecule experiences a torque $\tau$ about its center of mass.[5] This torque acts to rotate the molecule to align $p_o$ with $\mathcal{E}$. If all the molecules were to simply rotate and align with the field, the polarization of the solid would be

$$P = Np_o$$

where $N$ is the number of molecules per unit volume. However, due to their thermal energy, the molecules move around randomly and collide with each other and with the walls of the container. These collisions destroy the dipole alignments. Thus the thermal energy tries to randomize the orientations of the dipole moments. A snapshot of the dipoles in the material in the presence of a field can be pictured as in Figure 7.10d in which the dipoles have different orientations. There is, nonetheless, a net average dipole moment per molecule $p_{av}$ that is finite and directed along the field. Thus the material exhibits net polarization, which leads to a dielectric constant that is determined by this **orientational polarization.**

To find the induced average dipole moment $p_{av}$ along $\mathcal{E}$, we need to know the average potential energy $E_{dip}$ of a dipole placed in a field $\mathcal{E}$ and how this compares with the average thermal energy $\frac{5}{2}kT$ per molecule as in the present case of five degrees of freedom. $E_{dip}$ represents the average external work done by the field in aligning the dipoles with the field. If $\frac{5}{2}kT$ is much greater than $E_{dip}$, then the average thermal energy of collisions will prevent any dipole alignment with the field. If, however, $E_{dip}$ is much greater than $\frac{5}{2}kT$, then the thermal energy is insufficient to destroy the dipole alignments.

A dipole at an angle $\theta$ to the field experiences a torque $\tau$ that tries to rotate it, as shown in Figure 7.10c. Work done $dW$ by the field in rotating the dipole by $d\theta$ is $\tau\, d\theta$ (as in $F\,dx$). This work $dW$ represents a small change $dE$ in the potential energy of the dipole. No work is done if the dipole is already aligned with $\mathcal{E}$, when $\theta = 0$, which corresponds to the minimum in *PE*. On the other hand, maximum work is done when the torque has to rotate the dipole from $\theta = 180°$ to $\theta = 0°$ (either clockwise or counterclockwise, it doesn't matter). The torque experienced by the dipole, according to Figure 7.10c, is given by

*Torque on a dipole*

$$\tau = (F \sin\theta)a \qquad \text{or} \qquad \mathcal{E}p_o \sin\theta$$

where

$$p_o = aQ$$

If we take $PE = 0$ when $\theta = 0$, then the maximum *PE* is when $\theta = 180°$, or

$$E_{max} = \int_0^{\pi} p_o \mathcal{E} \sin\theta\, d\theta = 2p_o\mathcal{E}$$

The average dipole potential energy is then $\frac{1}{2}E_{max}$ or $p_o\mathcal{E}$. For orientational polarization to be effective, this energy must be greater than the average thermal energy. The average dipole moment $p_{av}$ along $\mathcal{E}$ is directly proportional to the magnitude of $p_o$ itself and also proportional to the average dipole energy to average thermal energy

[5] The oppositely directed forces also slightly stretch the $Cl^-$–$H^+$ bond, but we neglect this effect.

ratio, that is,

$$p_{av} \propto p_o \frac{p_o \mathcal{E}}{\frac{5}{2}kT}$$

If we were to do the calculation properly using Boltzmann statistics for the distribution of dipole energies among the molecules, that is, the probability that the dipole has an energy $E$ is proportional to $\exp(-E/kT)$, then we would find that when $p_o E < kT$ (generally the case),

*Average dipole moment in orientational polarization*

$$p_{av} = \frac{1}{3}\frac{p_o^2 \mathcal{E}}{kT} \qquad [7.19]$$

It turns out that the intuitively derived expression for $p_{av}$ is roughly the same as Equation 7.19. Strictly, of course, we should use the local field acting on each molecule, in which case $\mathcal{E}$ is simply replaced by $\mathcal{E}_{loc}$. From Equation 7.19 we can define a **dipolar orientational polarizability** $\alpha_d$ per molecule by

*Dipolar orientational polarizability*

$$\alpha_d = \frac{1}{3}\frac{p_o^2}{kT} \qquad [7.20]$$

It is apparent that, in contrast to the electronic and ionic polarization, dipolar orientational polarization is strongly temperature dependent. $\alpha_d$ decreases with temperature, which means that the relative permittivity $\varepsilon_r$ also decreases with temperature. Dipolar orientational polarization is normally exhibited by polar liquids (*e.g.*, water, alcohol, acetone, and various electrolytes) and polar gases (*e.g.*, gaseous HCl and steam). It can also occur in solids if there are permanent dipoles within the solid structure, even if dipolar rotation involves a discrete jump of an ion from one site to another, such as in various glasses.

### 7.3.3 INTERFACIAL POLARIZATION

**Interfacial polarization** occurs whenever there is an accumulation of charge at an interface between two materials or between two regions within a material. The simplest example is interfacial polarization due to the accumulation of charges in the dielectric near one of the electrodes, as depicted in Figure 7.11a and b. Invariably materials, however perfect, contain crystal defects, impurities, and various mobile charge carriers such as electrons (*e.g.*, from donor-type impurities), holes, or ionized host or impurity ions. In the particular example in Figure 7.11a, the material has an equal number of positive ions and negative ions, but the positive ions are assumed to be far more mobile. For example, if present, the $H^+$ ion (which is a proton) and the $Li^+$ ion in ceramics and glasses are more mobile than negative ions in the structure because they are relatively small. Under the presence of an applied field, these positive ions migrate to the negative electrode. The positive ions, however, cannot leave the dielectric and enter the crystal structure of the metal electrode. They therefore simply pile up at the interface and give rise to a positive space charge near the electrode. These positive charges at the interface attract more electrons to the negative electrode. This additional charge on the electrode,

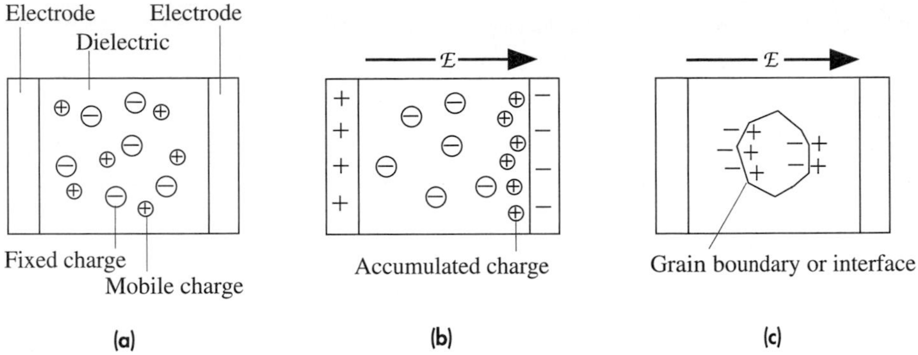


**Figure 7.11**

(a) A crystal with equal number of mobile positive ions and fixed negative ions. In the absence of a field, there is no net separation between all the positive charges and all the negative charges.

(b) In the presence of an applied field, the mobile positive ions migrate toward the negative electrode and accumulate there. There is now an overall separation between the negative charges and positive charges in the dielectric. The dielectric therefore exhibits interfacial polarization.

(c) Grain boundaries and interfaces between different materials frequently give rise to interfacial polarization.

of course, appears as an increase in the dielectric constant. The term **interfacial polarization** arises because the positive charges accumulating at the interface and the remainder of negative charges in the bulk together constitute dipole moments that appear in the polarization vector **P** (**P** sums all the dipoles within the material per unit volume).

Another typical interfacial polarization mechanism is the trapping of electrons or holes at defects at the crystal surface, at the interface between the crystal and the electrode. In this case we can view the positive charges in Figure 7.11a as holes and negative charges as immobile ionized acceptors. We assume that the contacts are blocking and do not allow electrons or holes to be injected, that is, exchanged between the electrodes and the dielectric. In the presence of a field, the holes drift to the negative electrode and become trapped in defects at the interface, as in Figure 7.11b.

Grain boundaries frequently lead to interfacial polarization as they can trap charges migrating under the influence of an applied field, as indicated in Figure 7.11c. Dipoles between the trapped charges increase the polarization vector. Interfaces also arise in heterogeneous dielectric materials, for example, when there is a dispersed phase within a continuous phase. The principle is then the same as schematically illustrated in Figure 7.11c.

## 7.3.4 Total Polarization

In the presence of electronic, ionic, and dipolar polarization mechanisms, the average induced dipole moment per molecule will be the sum of all the contributions in terms of the local field,

*Total induced dipole moment*

$$p_{av} = \alpha_e \mathcal{E}_{loc} + \alpha_i \mathcal{E}_{loc} + \alpha_d \mathcal{E}_{loc}$$

**Table 7.2** Typical examples of polarization mechanisms

| Example | Polarization | Static $\varepsilon_r$ | Comment |
|---|---|---|---|
| Ar gas | Electronic | 1.0005 | Small $N$ in gases: $\varepsilon_r \approx 1$ |
| Ar liquid ($T < 87.3$ K) | Electronic | 1.53 | van der Waals bonding |
| Si crystal | Electronic polarization due to valence electrons | 11.9 | Covalent solid; bond polarization |
| NaCl crystal | Ionic | 5.90 | Ionic crystalline solid |
| CsCl crystal | Ionic | 7.20 | Ionic crystalline solid |
| Water | Orientational | 80 | Dipolar liquid |
| Nitromethane (27 °C) | Orientational | 34 | Dipolar liquid |
| PVC (polyvinyl chloride) | Orientational | 7 | Dipole orientations partly hindered in the solid |

Each effect adds linearly to the net dipole moment per molecule, a fact verified by experiments. Interfacial polarization cannot be simply added to the above equation as $\alpha_{if}\mathcal{E}_{\text{loc}}$ because it occurs at interfaces and cannot be put into an average polarization per molecule in the bulk. Further, the fields are not well defined at the interfaces. In addition, we *cannot* use the simple Lorentz local field approximation for dipolar materials. That is, the Clausius–Mossotti equation does not work with dipolar dielectrics and the calculation of the local field is quite complicated. The dielectric constant $\varepsilon_r$ under **electronic** and **ionic polarizations,** however, can be obtained from

*Clausius–Mossotti equation*

$$\frac{\varepsilon_r - 1}{\varepsilon_r + 2} = \frac{1}{3\varepsilon_o}(N_e\alpha_e + N_i\alpha_i) \tag{7.21}$$

Table 7.2 summarizes the various polarization mechanisms and the corresponding static (or very low frequency) dielectric constant. Typical examples where one mechanism dominates over others are also listed.

**EXAMPLE 7.4** **IONIC AND ELECTRONIC POLARIZABILITY** Consider the CsCl crystal which has one $Cs^+$–$Cl^-$ pair per unit cell and a lattice parameter $a$ of 0.412 nm. The electronic polarizability of $Cs^+$ and $Cl^-$ ions is $3.35 \times 10^{-40}$ F m$^2$ and $3.40 \times 10^{-40}$ F m$^2$, respectively, and the mean ionic polarizability per ion pair is $6 \times 10^{-40}$ F m$^2$. What is the dielectric constant at low frequencies and that at optical frequencies?

**SOLUTION**

The CsCl structure has one cation ($Cs^+$) and one anion ($Cl^-$) in the unit cell. Given the lattice parameter $a = 0.412 \times 10^{-9}$ m, the number of ion pairs $N_i$ per unit volume is $1/a^3 = 1/(0.412 \times 10^{-9}\text{ m})^3 = 1.43 \times 10^{28}$ m$^{-3}$. $N_i$ is also the concentration of cations and anions individually. From the Clausius–Mossotti equation,

$$\frac{\varepsilon_r - 1}{\varepsilon_r + 2} = \frac{1}{3\varepsilon_o}[N_i\alpha_e(\text{Cs}^+) + N_i\alpha_e(\text{Cl}^-) + N_i\alpha_i]$$

That is,

$$\frac{\varepsilon_r - 1}{\varepsilon_r + 2} = \frac{(1.43 \times 10^{28}\ \text{m}^{-3})(3.35 \times 10^{-40} + 3.40 \times 10^{-40} + 6 \times 10^{-40}\ \text{F m}^2)}{3(8.85 \times 10^{-12}\ \text{F m}^{-1})}$$

Solving for $\varepsilon_r$, we find $\varepsilon_r = 7.56$.

At high frequencies—that is, near-optical frequencies—the ionic polarization is too sluggish to allow ionic polarization to contribute to $\varepsilon_r$. Thus, $\varepsilon_{rop}$, relative permittivity at optical frequencies, is given by

$$\frac{\varepsilon_{rop} - 1}{\varepsilon_{rop} + 2} = \frac{1}{3\varepsilon_o}[N_i\alpha_e(\text{Cs}^+) + N_i\alpha_e(\text{Cl}^-)]$$

That is,

$$\frac{\varepsilon_{rop} - 1}{\varepsilon_{rop} + 2} = \frac{(1.43 \times 10^{28}\ \text{m}^{-3})(3.35 \times 10^{-40} + 3.40 \times 10^{-40}\ \text{F m}^2)}{3(8.85 \times 10^{-12}\ \text{F m}^{-1})}$$

Solving for $\varepsilon_{rop}$, we find $\varepsilon_{rop} = 2.71$. Note that experimental values are $\varepsilon_r = 7.20$ at low frequencies and $\varepsilon_{rop} = 2.62$ at high frequencies, very close to calculated values.

## 7.4 FREQUENCY DEPENDENCE: DIELECTRIC CONSTANT AND DIELECTRIC LOSS

### 7.4.1 DIELECTRIC LOSS

The static dielectric constant is an effect of polarization under dc conditions. When the applied field, or the voltage across a parallel plate capacitor, is a sinusoidal signal, then the polarization of the medium under these ac conditions leads to an ac dielectric constant that is generally different than the static case. As an example we will consider orientational polarization involving dipolar molecules. The sinusoidally varying field changes magnitude and direction continuously, and it tries to line up the dipoles one way and then the other way and so on. If the instantaneous induced dipole moment $p$ per molecule can instantaneously follow the field variations, then at any instant

$$p = \alpha_d \mathcal{E} \qquad [7.22]$$

and the polarizability $\alpha_d$ has its expected maximum value from dc conditions, that is,

$$\alpha_d = \frac{p_o^2}{3kT} \qquad [7.23]$$

There are two factors opposing the immediate alignment of the dipoles with the field. First is that thermal agitation tries to randomize the dipole orientations. Collisions in the gas phase, random jolting from lattice vibrations in the liquid and solid phases, for example, aid the randomization of the dipole orientations. Second, the molecules rotate in a viscous medium by virtue of their interactions with neighbors, which is particularly strong in the liquid and solid states and means that the dipoles cannot respond instantaneously to the changes in the applied field. If the field changes too

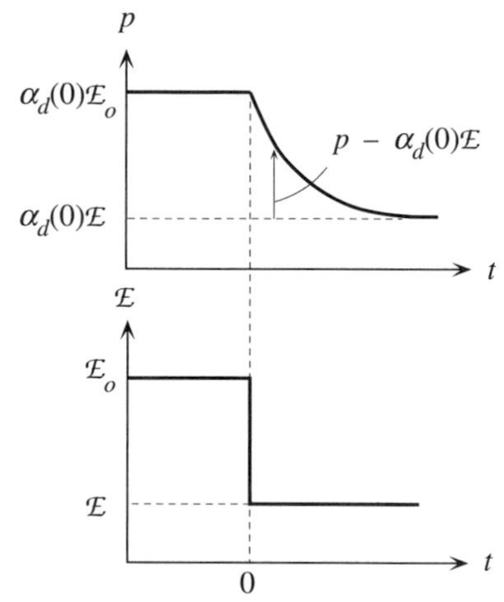


**Figure 7.12** The applied dc field is suddenly changed from $\mathcal{E}_o$ to $\mathcal{E}$ at time $t = 0$.
The induced dipole moment $p$ has to decrease from $\alpha_d(0)\mathcal{E}_o$ to a final value of $\alpha_d(0)\mathcal{E}$. The decrease is achieved by random collisions of molecules in the gas.

rapidly, then the dipoles cannot follow the field and, as a consequence, remain randomly oriented. At high frequencies, therefore, $\alpha_d$ will be zero as the field cannot induce a dipole moment. At low frequencies, of course, the dipoles can respond rapidly to follow the field and $\alpha_d$ has its maximum value. It is clear that $\alpha_d$ changes from its maximum value in Equation 7.23 to zero as the frequency of the field is increased. We need to find the behavior of $\alpha_d$ as a function of frequency $\omega$ so that we can determine the dielectric constant $\varepsilon_r$ by the Clausius–Mossotti equation.

Suppose that after a prolonged application, corresponding to dc conditions, the applied field across the dipolar gaseous medium is suddenly decreased from $\mathcal{E}_o$ to $\mathcal{E}$ at a time we define as zero, as shown in Figure 7.12. The field $\mathcal{E}$ is smaller than $\mathcal{E}_o$, so the induced dc dipole moment per molecule should be smaller and given by $\alpha_d(0)\mathcal{E}$ where $\alpha_d(0)$ is $\alpha_d$ at $\omega = 0$, dc conditions. Therefore, the induced dipole moment per molecule has to decrease, or *relax,* from $\alpha_d(0)\mathcal{E}_o$ to $\alpha_d(0)\mathcal{E}$. In a gas medium the molecules would be moving around randomly and their collisions with each other and the walls of the container randomize the induced dipole per molecule. Thus the decrease, or the **relaxation process,** in the induced dipole moment is achieved by random collisions. Assuming that $\tau$ is the average time, called the **relaxation time,** between molecular collisions, then this is the mean time it takes per molecule to randomize the induced dipole moment. If $p$ is the instantaneous induced dipole moment, then $p - \alpha_d(0)\mathcal{E}$ is the *excess* dipole moment, which must eventually disappear to zero through random collisions as $t \to \infty$. It would take an average $\tau$ seconds to eliminate the excess dipole moment $p - \alpha_d(0)\mathcal{E}$. The rate at which the induced dipole moment is changing is then $-[p - \alpha_d(0)\mathcal{E}]/\tau$, where the negative sign represents a decrease. Thus,

*Dipolar relaxation equation*

$$\frac{dp}{dt} = -\frac{p - \alpha_d(0)\mathcal{E}}{\tau} \qquad [7.24]$$

Although we did not derive Equation 7.24 rigorously, it is nonetheless a good first-order description of the behavior of the induced dipole moment per molecule in a dipolar medium. Equation 7.24 can be used to obtain the dipolar polarizability under ac conditions. For an ac field, we would write

$$\mathcal{E} = \mathcal{E}_o \sin(\omega t)$$

and solve Equation 7.24, but in engineering we prefer to use an exponential representation for the field

*Applied field*

$$\mathcal{E} = \mathcal{E}_o \exp(j\omega t)$$

as in ac voltages. In this case the impedance of a capacitor $C$ and an inductor $L$ become $1/j\omega C$ and $j\omega L$, where $j$ represents a phase shift of 90°. With $\mathcal{E} = \mathcal{E}_o \exp(j\omega t)$ in Equation 7.24, we have

*Dipole relaxation equation*

$$\frac{dp}{dt} = -\frac{p}{\tau} + \frac{\alpha_d(0)}{\tau}\mathcal{E}_o \exp(j\omega t) \quad [7.25]$$

Solving this we find the induced dipole moment as

$$p = \alpha_d(\omega)\mathcal{E}_o \exp(j\omega t)$$

where $\alpha_d(\omega)$ is given by

*Orientational polarizability and frequency*

$$\alpha_d(\omega) = \frac{\alpha_d(0)}{1 + j\omega\tau} \quad [7.26]$$

and represents the orientational polarizability under ac field conditions. Polarizability $\alpha_d(\omega)$ is a complex number that indicates that $p$ and $\mathcal{E}$ are out of phase.[6] Put differently, if $N$ is the number of molecules per unit volume, $P = Np$ and $\mathcal{E}$ are out of phase, as indicated in Figure 7.13a. At low frequencies, $\omega\tau \ll 1$, $\alpha_d(\omega)$ is nearly $\alpha_d(0)$, and $p$ is in phase with $\mathcal{E}$. The rate of relaxation $1/\tau$ is much faster than the frequency of the field or the rate at which the polarization is being changed; $p$ then closely follows $\mathcal{E}$. At very high frequencies, $\omega\tau \gg 1$, the rate of relaxation $1/\tau$ is much slower than the frequency of the field and $p$ can no longer follow the variations in the field.

We can easily obtain the dielectric constant $\varepsilon_r$ from $\alpha_d(\omega)$ by using Equation 7.14, which then leads to a complex number for $\varepsilon_r$ since $\alpha_d$ itself is a complex number. By convention, we generally write the **complex dielectric constant** as

*Complex relative permittivity*

$$\varepsilon_r = \varepsilon_r' - j\varepsilon_r'' \quad [7.27]$$

where $\varepsilon_r'$ is the real part and $\varepsilon_r''$ is the imaginary part, both being frequency dependent, as shown in Figure 7.13b. The real part $\varepsilon_r'$ decreases from its maximum value $\varepsilon_r'(0)$, corresponding to $\alpha_d(0)$, to 1 at high frequencies when $\alpha_d = 0$ as $\omega \to \infty$ in Equation 7.26. The imaginary part $\varepsilon_r''(\omega)$ is zero at low and high frequencies but peaks when $\omega\tau = 1$ or when $\omega = 1/\tau$. The real part $\varepsilon_r'$ represents the relative permittivity that we would use in calculating the capacitance, as for example in $C = \varepsilon_r\varepsilon_o A/d$. The imaginary part $\varepsilon_r''(\omega)$ represents the energy lost in the dielectric medium as the dipoles are oriented against random collisions one way and then the other way and so on by the field. Consider

[6] The polarization $P$ lags behind $\mathcal{E}$ by some angle $\phi$, that is determined by Equation 7.26 as shown in Figure 7.13.

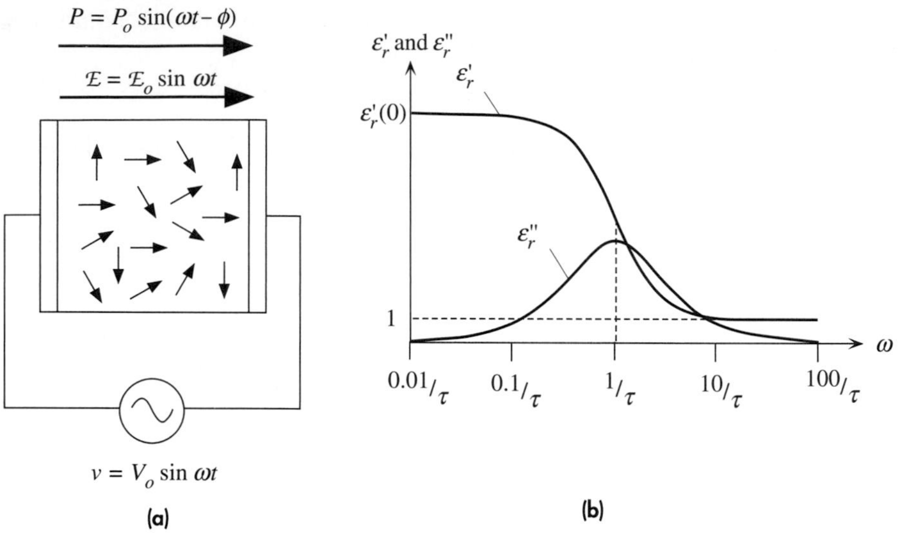


**Figure 7.13**
(a) An ac field is applied to a dipolar medium. The polarization $P$ ($P = Np$) is out of phase with the ac field.
(b) The relative permittivity is a complex number with real ($\varepsilon_r'$) and imaginary ($\varepsilon_r''$) parts that exhibit relaxation at $\omega \approx 1/\tau$.

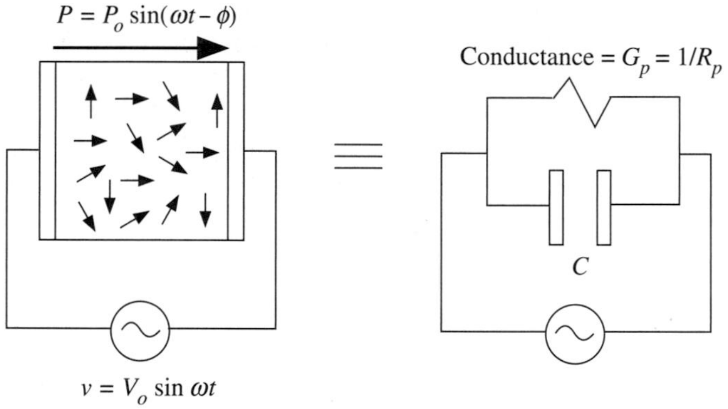


**Figure 7.14** The dielectric medium behaves like an ideal (lossless) capacitor of capacitance $C$, which is in parallel with a conductance $G_P$.

the capacitor in Figure 7.14, which has this dielectric medium between the plates. Then the admittance $Y$, *i.e.*, the reciprocal of impedance of this capacitor, with $\varepsilon_r$ given in Equation 7.27 is

*Admittance of a parallel plate capacitor*

$$Y = \frac{j\omega A\varepsilon_o\varepsilon_r(\omega)}{d} = \frac{j\omega A\varepsilon_o\varepsilon_r'(\omega)}{d} + \frac{\omega A\varepsilon_o\varepsilon_r''(\omega)}{d}$$

which can be written as

$$Y = j\omega C + G_P \quad [7.28]$$

where

$$C = \frac{A\varepsilon_o\varepsilon'_r}{d} \quad [7.29]$$

*Equivalent ideal capacitance*

and

$$G_P = \frac{\omega A\varepsilon_o\varepsilon''_r}{d} \quad [7.30]$$

*Equivalent parallel conductance*

is a real number just as if we had a conductive medium with some conductance $G_P$ or resistance $1/G_P$. The admittance of the dielectric medium according to Equation 7.28 is a parallel combination of an ideal, or lossless, capacitor $C$, with a relative permittivity $\varepsilon'_r$, and a resistance of $R_P = 1/G_P$ as indicated in Figure 7.14. Thus the dielectric medium behaves as if $C_o$ and $R_P$ were in parallel. There is no real electric power dissipated in $C$, but there is indeed real power dissipated in $R_P$ because

$$\text{Input power} = IV = YV^2 = j\omega CV^2 + \frac{V^2}{R_P}$$

and the second term is real. Thus the power dissipated in the dielectric medium is related to $\varepsilon''_r$ and peaks when $\omega = 1/\tau$. The rate of energy storage by the field is determined by $\omega$ whereas the rate of energy transfer to molecular collisions is determined by $1/\tau$. When $\omega = 1/\tau$, the two processes, energy storage by the field and energy transfer to random collisions, are then occurring at the same rate, and hence energy is being transferred to heat most efficiently. The peak in $\varepsilon''_r$ versus $\omega$ is called a **relaxation peak,** which is at a frequency when the dipole relaxations are at the right rate for maximum power dissipation. This process is known as **dielectric resonance.**

According to Equation 7.28, the magnitude of $G_P$ and hence the energy loss is determined by $\varepsilon''_r$. In engineering applications of dielectrics in capacitors, we would like to minimize $\varepsilon''_r$ for a given $\varepsilon'_r$. We define the relative magnitude of $\varepsilon''_r$ with respect to $\varepsilon'_r$ through a quantity, $\tan\delta$, called the **loss tangent** (or **loss factor**), as

$$\tan\delta = \frac{\varepsilon''_r}{\varepsilon'_r} \quad [7.31]$$

*Loss tangent*

which is frequency dependent and peaks just beyond $\omega = 1/\tau$. The actual value of $1/\tau$ depends on the material, but typically for liquid and solid media it is in the gigahertz range, that is, microwave frequencies. We can easily find the energy per unit time—power—dissipated as dielectric loss in the medium. The resistance $R_P$ represents the dielectric loss, so

$$W_{\text{vol}} = \frac{\text{Power loss}}{\text{Volume}} = \frac{V^2}{R_P} \times \frac{1}{dA} = \frac{V^2}{\dfrac{d}{\omega A\varepsilon_o\varepsilon''_r}} \times \frac{1}{dA} = \frac{V^2}{d^2}\omega\varepsilon_o\varepsilon''_r$$

Using Equation 7.31 and $\mathcal{E} = V/d$, we obtain

$$W_{\text{vol}} = \omega\mathcal{E}^2\varepsilon_o\varepsilon'_r \tan\delta \quad [7.32]$$

*Dielectric loss per unit volume*

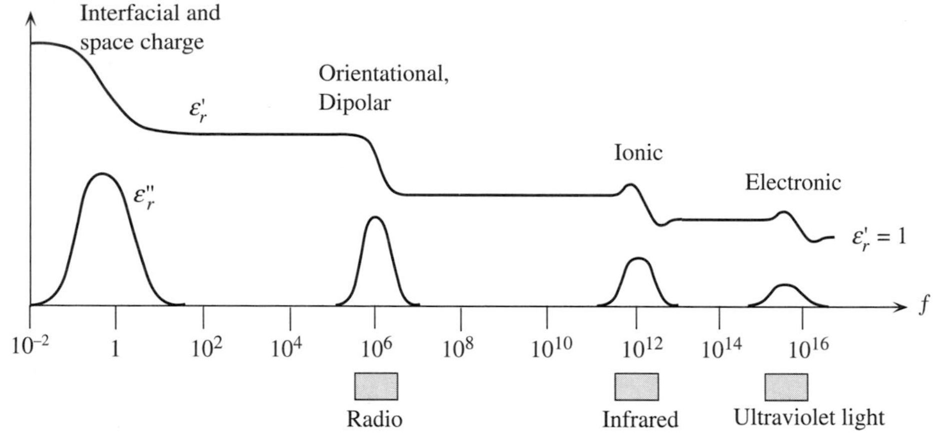


**Figure 7.15** The frequency dependence of the real and imaginary parts of the dielectric constant in the presence of interfacial, orientational, ionic, and electronic polarization mechanisms.

Equation 7.32 represents the power dissipated per unit volume in the polarization mechanism: energy lost per unit time to random molecular collisions as heat. It is clear that dielectric loss is influenced by three factors: $\omega$, $\mathcal{E}$, and tan $\delta$.

Although we considered only orientational polarization, in general a dielectric medium will also exhibit other polarization mechanisms and certainly electronic polarization since there will always be electron clouds around individual atoms, or electrons in covalent bonds. If we were to consider the ionic polarizability in ionic solids, we would also find $\alpha_I$ to be frequency dependent and a complex number. In this case, lattice vibrations in the crystal, typically at frequencies $\omega_I$ in the infrared region of the electromagnetic spectrum, will dissipate the energy stored in the induced dipole moments just as energy was dissipated by molecular collisions in the gaseous dipolar medium. Thus, the energy loss will be greatest when the frequency of the polarizing field is the same as the lattice vibration frequency, $\omega = \omega_I$, which tries to randomize the polarization.

We can represent the general features of the frequency dependence of the real and imaginary parts of the dielectric constant as in Figure 7.15. Although the figure shows distinctive peaks in $\varepsilon_r''$ and transition features in $\varepsilon_r'$, in reality these peaks and various features are broader. First, there is no single well-defined lattice vibration frequency but instead an allowed range of frequencies just as in solids where there is an allowed range of energies for the electron. Moreover, the polarization effects depend on the crystal orientation. In the case of polycrystalline materials, various peaks in different directions overlap to exhibit a broadened overall peak. At low frequencies the interfacial or space charge polarization features are even broader because there can be a number of conduction mechanisms (different species of charge carriers and different carrier mobilities) for the charges to accumulate at interfaces, each having its own speed. Orientational polarization, especially in many liquid dielectrics at room temperature, typically takes place at radio to microwave frequencies. In some polymeric materials, this type of polarization involves a limited rotation of dipolar side groups

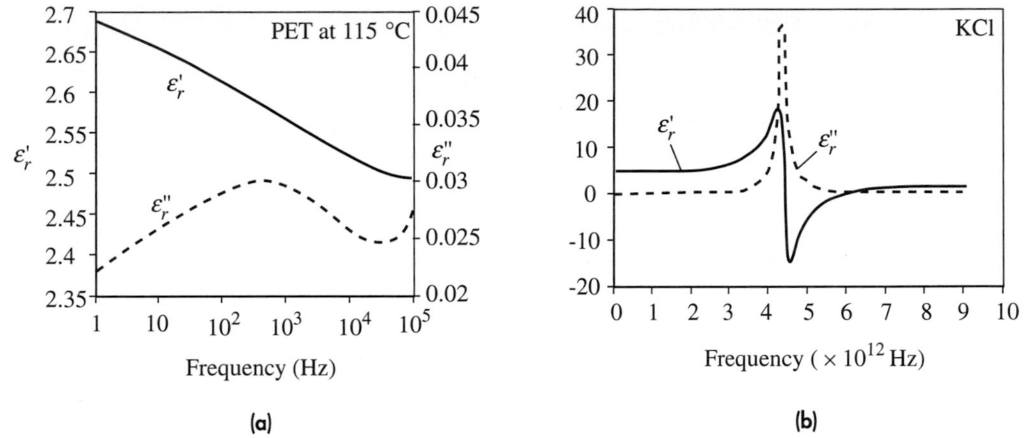


**Figure 7.16** Real and imaginary parts of the dielectric constant, $\varepsilon_r'$ and $\varepsilon_r''$, versus frequency for (a) a polymer, PET, at 115 °C and (b) an ionic crystal, KCl, at room temperature.
Both exhibit relaxation peaks but for different reasons.

SOURCE: Data for (a) from author's own experiments using a dielectric analyzer (DEA), (b) from C. Smart, G. R. Wilkinson, A. M. Karo, and J. R. Hardy, International Conference on Lattice Dynamics, Copenhagen, 1963, as quoted by D. H. Martin, "The Study of the Vibration of Crystal Lattices by Far Infra-Red Spectroscopy," *Advances in Physics,* 14, no. 53–56, 1965, pp. 39–100.

attached to the polymeric chain and can occur at much lower frequencies depending on the temperature. Figure 7.16 shows two typical examples of dielectric behavior, $\varepsilon_r'$ and $\varepsilon_r''$ as a function of frequency, for a polymer (PET) and an ionic crystal (KCl). Both exhibit loss peaks, peaks in $\varepsilon_r''$ versus frequency, but for different reasons. The particular polymer, PET (a polyester), exhibits orientational polarization due to dipolar side groups, whereas KCl exhibits ionic polarization due to the displacement of $K^+$ and $Cl^-$ ions. The frequency of the loss peak in the case of orientational polarization is highly temperature dependent. For the PET example in Figure 7.16 at 115 °C, the peak occurs at around 400 Hz, even below typical radio frequencies.

**EXAMPLE 7.5**

**DIELECTRIC LOSS PER UNIT CAPACITANCE AND THE LOSS ANGLE $\delta$** Obtain the dielectric loss per unit capacitance in a capacitor in terms of the loss tangent. Obtain the phase difference between the current through the capacitor and that through $R_P$. What is the significance of $\delta$?

**SOLUTION**

We consider the equivalent circuit in Figure 7.14. The power loss in the capacitor is due to $R_P$. If $V$ is the rms value of the voltage across the capacitor, then the power dissipated per unit capacitance $W_{cap}$ is

$$W_{cap} = \frac{V^2}{R_P} \times \frac{1}{C} = V^2 \frac{\omega\varepsilon_o\varepsilon_r'' A}{d} \times \frac{d}{\varepsilon_o\varepsilon_r' A} = V^2 \frac{\omega\varepsilon_r''}{\varepsilon_r'}$$

or

$$W_{cap} = V^2\omega \tan\delta$$

**Table 7.3** Dielectric properties of three insulators

| | $f = 60$ Hz | | | $f = 1$ MHz | | |
|---|---|---|---|---|---|---|
| **Material** | $\varepsilon_r'$ | **tan** $\delta$ | $\omega$ **tan** $\delta$ | $\varepsilon_r'$ | **tan** $\delta$ | $\omega$ **tan** $\delta$ |
| Polycarbonate | 3.17 | $9 \times 10^{-4}$ | 0.34 | 2.96 | $1 \times 10^{-2}$ | $6.2 \times 10^4$ |
| Silicone rubber | 3.7 | $2.25 \times 10^{-2}$ | 8.48 | 3.4 | $4 \times 10^{-3}$ | $2.5 \times 10^4$ |
| Epoxy with mineral filler | 5 | $4.7 \times 10^{-2}$ | 17.7 | 3.4 | $3 \times 10^{-2}$ | $18 \times 10^4$ |

As tan $\delta$ is frequency dependent and peaks at some frequency, so does the power dissipated per unit capacitance. A clear design objective would be to keep $W_{\text{cap}}$ as small as possible. Further, for a given voltage, $W_{\text{cap}}$ does not depend on the dielectric geometry. For a given voltage and capacitance, we therefore cannot reduce the power dissipation by simply changing the dimensions of the dielectric.

Consider the rms currents through $R_P$ and $C$, $I_{\text{loss}}$ and $I_{\text{cap}}$ respectively, and their ratio,[7]

$$\frac{I_{\text{loss}}}{I_{\text{cap}}} = \frac{V}{R_P} \times \frac{\dfrac{1}{j\omega C}}{V} = \frac{\omega\varepsilon_o\varepsilon_r'' A}{d} \times \frac{d}{j\omega\varepsilon_o\varepsilon_r' A} = -j \tan\delta$$

As expected, the two are 90° out of phase ($-j$) and the loss current (through $R_P$) is a factor, tan $\delta$, of the capacitive current (through $C$). The ratio of $I_{\text{cap}}$ and the total current, $I_{\text{total}} = I_{\text{cap}} + I_{\text{loss}}$, is

$$\frac{I_{\text{cap}}}{I_{\text{total}}} = \frac{I_{\text{cap}}}{I_{\text{cap}} + I_{\text{loss}}} = \frac{1}{1 + \dfrac{I_{\text{loss}}}{I_{\text{cap}}}} = \frac{1}{1 - j\tan\delta}$$

The phase angle between $I_{\text{cap}}$ and $I_{\text{total}}$ is determined by the negative of the phase of the denominator term $(1 - j\tan\delta)$. Thus the phase angle between $I_{\text{cap}}$ and $I_{\text{total}}$ is $\delta$, where $I_{\text{cap}}$ leads $I_{\text{total}}$ by $\delta$. $\delta$ is also called the **loss angle.** When the loss angle is zero, $I_{\text{cap}}$ and $I_{\text{total}}$ are equal and there is no loss in the dielectric.

**EXAMPLE 7.6** **DIELECTRIC LOSS PER UNIT CAPACITANCE** Consider the three dielectric materials listed in Table 7.3 with their dielectric constant $\varepsilon_r'$ (usually simply stated as $\varepsilon_r$) and loss factors tan $\delta$. At a given voltage, which dielectric will have the lowest power dissipation per unit capacitance at 60 Hz? Is this also true at 1 MHz?

**SOLUTION**

The power dissipated at a given voltage per unit capacitance depends only on $\omega$ tan $\delta$, so we do not need to use $\varepsilon_r'$. Calculating $\omega$ tan $\delta$ or $(2\pi f)$ tan $\delta$, we find the values listed in the table at 60 Hz and 1 MHz. At 60 Hz, polycarbonate has the lowest power dissipation per unit capacitance, but at 1 MHz it is silicone rubber.

[7] These currents are phasors, each with a rms magnitude and phase angle.

**Table 7.4** Dielectric loss per unit volume for two insulators ($\kappa$ is the thermal conductivity)

| | $f = 60$ Hz | | | $f = 1$ MHz | | | |
|---|---|---|---|---|---|---|---|
| **Material** | $\varepsilon_r'$ | $\tan\delta$ | Loss ($\text{mW cm}^{-3}$) | $\varepsilon_r'$ | $\tan\delta$ | Loss ($\text{W cm}^{-3}$) | $\kappa$ ($\text{W cm}^{-1}\text{ K}^{-1}$) |
| XLPE | 2.3 | $3 \times 10^{-4}$ | 0.230 | 2.3 | $4 \times 10^{-4}$ | 5.12 | 0.005 |
| Alumina | 8.5 | $1 \times 10^{-3}$ | 2.84 | 8.5 | $1 \times 10^{-3}$ | 47.3 | 0.33 |

**EXAMPLE 7.7**

**DIELECTRIC LOSS AND FREQUENCY** Calculate the heat generated per second due to dielectric loss per cm$^3$ of cross-linked polyethylene, XLPE (typical power cable insulator), and alumina, $Al_2O_3$ (typical substrate in thin- and thick-film electronics), at 60 Hz and 1 MHz at a field of 100 kV cm$^{-1}$. Their properties are given in Table 7.4. What is your conclusion?

**SOLUTION**

The power dissipated per unit volume is

$$W_{\text{vol}} = (2\pi f)\mathcal{E}^2\varepsilon_o\varepsilon_r' \tan\delta$$

We can calculate $W_{\text{vol}}$ by substituting the properties of individual dielectrics at the given frequency $f$. For example, for XLPE at 60 Hz,

$$W_{\text{vol}} = (2\pi 60 \text{ Hz})(100 \times 10^3 \times 10^2 \text{ V m}^{-1})^2(8.85 \times 10^{-12} \text{ F m}^{-1})(2.3)(3 \times 10^{-4})$$
$$= 230 \text{ W m}^{-3}$$

We can convert this into per cm$^3$ by

$$W'_{\text{vol}} = \frac{W_{\text{vol}}}{10^6} = 0.230 \text{ mW cm}^{-3}$$

which is shown in Table 7.4.

From similar calculations we can obtain the heat generated per second per cm$^3$ as shown in Table 7.4. The heats at 60 Hz are small. The thermal conductivity of the insulation and its connecting electrodes can remove the heat without substantially increasing the temperature of the insulation. At 1 MHz, the heats generated are not trivial. One has to remove 5.12 W of heat from 1 cm$^3$ of XLPE and 47.3 W from 1 cm$^3$ of alumina. The thermal conductivity $\kappa$ of XLPE is about 0.005 W cm$^{-1}$ K$^{-1}$, whereas that of alumina is almost 100 times larger, 0.33 W cm$^{-1}$ K$^{-1}$. The poor thermal conductivity of polyethylene means that 5.12 W of heat cannot be conducted away easily and it will raise the temperature of the insulation until dielectric breakdown ensues. In the case of alumina, 47.3 W of heat will substantially increase the temperature. *Dielectric loss is the mechanism by which microwave ovens heat food.* Dielectric heating at high frequencies is used in industrial applications such as heating plastics and drying wood.

## 7.4.2 DEBYE EQUATIONS, COLE–COLE PLOTS, AND EQUIVALENT SERIES CIRCUIT

Consider a dipolar dielectric in which there are both orientational and electronic polarizations, $\alpha_d$ and $\alpha_e$, respectively, contributing to the overall polarizability. Electronic polarization $\alpha_e$ will be independent of frequency over the typical frequency range of operation of a dipolar dielectric, well below optical frequencies. At high frequencies, orientational polarization will be too sluggish too respond, $\alpha_d = 0$, and the $\varepsilon_r$ will be $\varepsilon_{r\infty}$. (The subscript "infinity" simply means high frequencies where orientational polarization is negligible.) The dielectric constant and polarizabilities are generally related through[8]

*Dielectric constant of a dipolar material*

$$\varepsilon_r = 1 + \frac{N}{\varepsilon_o}\alpha_e + \frac{N}{\varepsilon_o}\alpha_d(\omega) = \varepsilon_{r\infty} + \frac{N}{\varepsilon_o}\alpha_d(\omega)$$

where we have combined 1 and $\alpha_e$ terms to represent the high frequency $\varepsilon_r$ as $\varepsilon_{r\infty}$. Further $N\alpha_d(0)/\varepsilon_o$ determines the contribution of orientational polarization to the static dielectric constant $\varepsilon_{r\text{dc}}$, so that $N\alpha_d(0)/\varepsilon_o$ is simply $(\varepsilon_{r\text{dc}} - \varepsilon_{r\infty})$. Substituting for the frequency dependence of $\alpha_d(\omega)$ from Equation 7.26, and writing $\varepsilon_r$ in terms of real and imaginary parts,

*Dipolar dielectric constant*

$$\varepsilon_r' - j\varepsilon_r'' = \varepsilon_{r\infty} + \frac{N}{\varepsilon_o}\frac{\alpha_d(0)}{1 + j\omega\tau} = \varepsilon_{r\infty} + \frac{(\varepsilon_{r\text{dc}} - \varepsilon_{r\infty})}{1 + j\omega\tau} \qquad [7.33]$$

We can eliminate the complex denominator by multiplying both the denominator and numerator of the right-hand side by $1 - j\omega\tau$ and equate real and imaginary parts to obtain what are known as **Debye equations:**

*Debye equations for real and imaginary parts*

$$\varepsilon_r' = \varepsilon_{r\infty} + \frac{\varepsilon_{r\text{dc}} - \varepsilon_{r\infty}}{1 + (\omega\tau)^2} \qquad [7.34a]$$

*Debye equations for real and imaginary parts*

and

$$\varepsilon_r'' = \frac{(\varepsilon_{r\text{dc}} - \varepsilon_{r\infty})(\omega\tau)}{1 + (\omega\tau)^2} \qquad [7.34b]$$

Equations 7.34a and b reflect the behavior of $\varepsilon_r'$ and $\varepsilon_r''$ as a function of frequency shown in Figure 7.13b. The imaginary part $\varepsilon_r''$ that represents the dielectric loss exhibits a peak at $\omega = 1/\tau$ which is called a **Debye loss peak.** Many dipolar gases and some liquids with dipolar molecules exhibit this type of behavior. In the case of solids the peak is typically much broader because we cannot represent the losses in terms of just one single well-defined relaxation time $\tau$; the relaxation in the solid is usually represented by a distribution of relaxation times. Further, the simple relaxation process that is described in Equation 7.25 assumes that the dipoles do not influence each other either through their electric fields or through their interactions with the lattice; that is, they are not coupled. In solids, the dipoles can also couple, which complicates the relaxation process. Nonetheless, there are also many solids whose dielectric relaxation can be approximated by a nearly Debye relaxation or by slightly modifying Equation 7.33.

In dielectric studies of materials it is quite common to find a plot of the imaginary part ($\varepsilon_r''$) versus the real part ($\varepsilon_r'$) as a function of frequency $\omega$. Such plots are called **Cole–Cole plots** after their originators. The Debye equations 7.34a and b obviously

[8] This simple relationship is used because the Lorentz local field equation does not apply in dipolar dielectrics and the local field problem is particularly complicated in these dielectrics.

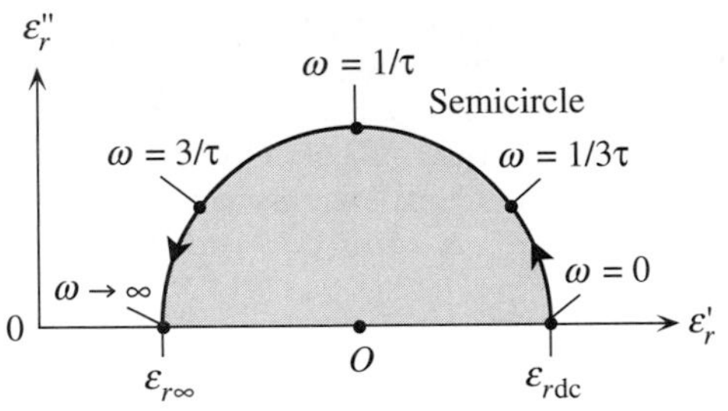


**Figure 7.17** Cole–Cole plot is a plot of $\varepsilon_r''$ versus $\varepsilon_r'$ as a function of frequency $\omega$.
As the frequency is changed from low to high, the plot traces out a semicircle.

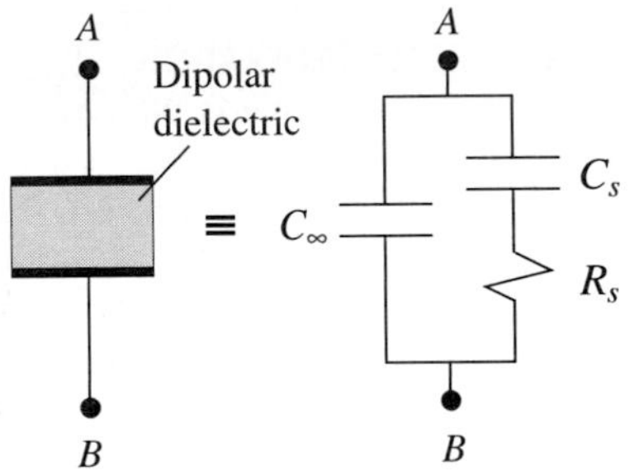


**Figure 7.18** A capacitor with a dipolar dielectric and its equivalent circuit in terms of an ideal Debye relaxation.

provide the necessary values for $\varepsilon_r'$ and $\varepsilon_r''$ to be plotted for the present simple dipolar relaxation mechanism that has only a single relaxation time $\tau$. In fact, by simply putting in $\tau = 1$ second, we can calculate and plot $\varepsilon_r''$ versus $\varepsilon_r'$ for $\omega = 0$ (dc) to $\omega \to \infty$ as shown in Figure 7.17. The result is a *semicircle.* While for certain substances, such as gases and some liquids, the Cole–Cole plots do indeed generate a semicircle, for many dielectrics, the curve is typically flattened and asymmetric, and not a semicircle.[9]

The Debye equations lead to a particular *RC* circuit representation of a dielectric material that is quite useful. Suppose that we have a resistance $R_s$ in series with a capacitor $C_s$, both of which are in parallel with the capacitor $C_\infty$ as in Figure 7.18. If we were to write down the equivalent admittance of this circuit, we would find that it corresponds to Equation 7.33, that is, the Debye equation. (The circuit mathematics is straightforward and is not reproduced here.) The reader may wonder why this circuit is different than the general model shown in Figure 7.14. Any series $R_s$ and $C_s$ circuit can be transformed to be equivalent to a parallel $R_p$ and $C_p$ (or $G_p$ and $C$ in Figure 7.14) circuit as is well known in circuit theory; the relationships between the elements depend on the frequency. Many electrolytic capacitors are frequently represented by an equivalent series $R_s$ and $C_s$ circuit as in Figure 7.18. If $A$ is the area and $d$ is the thickness of a parallel plate capacitor with a dipolar dielectric, then

*Equivalent circuit of a Debye dielectric*

$$C_\infty = \frac{\varepsilon_o \varepsilon_{r\infty} A}{d} \qquad C_s = \frac{\varepsilon_o(\varepsilon_{rdc} - \varepsilon_{r\infty})A}{d} \qquad \text{and} \qquad R_s = \frac{\tau}{C_s} \tag{7.35}$$

Notice that in this circuit model, $R_s$, $C_s$, and $C_\infty$ do not depend on the frequency, which is only true for an ideal Debye dielectric, that with a single relaxation time $\tau$.

[9] The departure is simply due to the fact that a simple relaxation process with a single relaxation time cannot describe the dielectric behavior accurately. (A good overview of non-Debye relaxations is given by Andrew Jonscher in *J. Phys D*, **32,** R57, 1999.)

**EXAMPLE 7.8**

**NEARLY DEBYE RELAXATION** There are some dielectric solids that exhibit nearly Debye relaxation. One example is the $La_{0.7}Sr_{0.3}MnO_3$ ceramic whose relaxation peak and Cole–Cole plots are similar to those shown in Figures 7.13b and 7.17,[10] especially in the high-frequency range past the resonance peak. $La_{0.7}Sr_{0.3}MnO_3$'s low frequency ($\varepsilon_{r\text{dc}}$) and high frequency ($\varepsilon_{r\infty}$) dielectric constants are 3.6 and 2.58, respectively, where *low* and *high* refer, respectively, to frequencies far below and above the Debye relaxation peak, *i.e.*, $\varepsilon_{r\text{dc}}$ and $\varepsilon_{r\infty}$. The Debye loss peak occurs at 6 kHz. Calculate $\varepsilon_r'$ and the dielectric loss factor tan $\delta$ at 29 kHz.

**SOLUTION**

The loss peak occurs when $\omega_o = 1/\tau$, so that $\tau = 1/\omega_o = 1/(2\pi 6000) = 26.5$ μs. We can now calculate the real and imaginary parts of $\varepsilon_r$ at 29 kHz,

$$\varepsilon_r' = \varepsilon_{r\infty} + \frac{\varepsilon_{r\text{dc}} - \varepsilon_{r\infty}}{1 + (\omega\tau)^2} = 2.58 + \frac{3.6 - 2.58}{1 + [(2\pi)(29 \times 10^3)(26.5 \times 10^{-6})]^2} = 2.62$$

$$\varepsilon_r'' = \frac{(\varepsilon_{r\text{dc}} - \varepsilon_{r\infty})(\omega\tau)}{1 + (\omega\tau)^2} = \frac{(3.6 - 2.58)[(2\pi)(29 \times 10^3)(26.5 \times 10^{-6})]}{1 + [(2\pi)(29 \times 10^3)(26.5 \times 10^{-6})]^2} = 0.202$$

and hence

$$\tan\delta = \frac{\varepsilon_r''}{\varepsilon_r'} = \frac{0.202}{2.62} = 0.077$$

which is close to the experimental value of 0.084.

This example was a special case of nearly Debye relaxation. Debye equations have been modified over the years to account for the broad relaxation peaks that have been observed, particularly in polymers, by writing the complex $\varepsilon_r$ as

*Non-Debye relaxation*

$$\varepsilon_r = \varepsilon_{r\infty} + \frac{\varepsilon_{r\text{dc}} - \varepsilon_{r\infty}}{[1 + (j\omega\tau)^\alpha]^\beta} \qquad \textbf{[7.36]}$$

where $\alpha$ and $\beta$ are constants, typically less than unity (setting $\alpha = \beta = 1$ generates the Debye equations). Such equations are useful in engineering for predicting $\varepsilon_r$ at any frequency from a few known values at various frequencies, as highlighted in this simple nearly Debye example. Further, if $\tau$ dependence on the temperature $T$ is known (often $\tau$ is thermally activated), then we can predict $\varepsilon_r$ at any $\omega$ and $T$.

## 7.5 GAUSS'S LAW AND BOUNDARY CONDITIONS

An important fundamental theorem in electrostatics is Gauss's law, which relates the integration of the electric field over a surface to the total charge enclosed. It can be derived from Coulomb's law, or the latter can be derived from Gauss's law. Suppose $\mathcal{E}_n$ is the electric field normal to a small surface area $dA$ on a closed surface, as shown in Figure 7.19; then summing $\mathcal{E}_n\, dA$ products over the whole surface gives total net charge $Q_{\text{total}}$ inside it,

*Gauss's law*

$$\oint_{\text{Surface}} \mathcal{E}_n\, dA = \frac{Q_{\text{total}}}{\varepsilon_o} \qquad \textbf{[7.37]}$$

[10] Z. C. Xia *et al., J. Phys. Cond. Matter,* **13,** 4359, 2001. The origin of the dipolar activity in this ceramic is quite complex and involves an electron hopping (jumping) from a $Mn^{3+}$ to $Mn^{4+}$ ion; we do not need the physical details in the example.

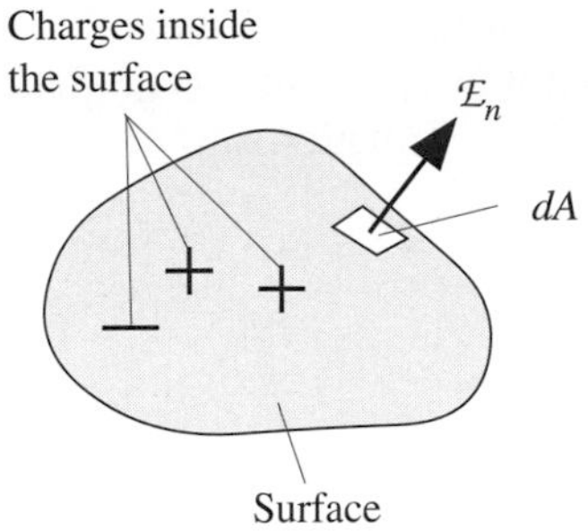


**Figure 7.19** Gauss's law.
The surface integral of the electric field normal to the surface is the total charge enclosed. The field is positive if it is coming out, negative if it is going into the surface.

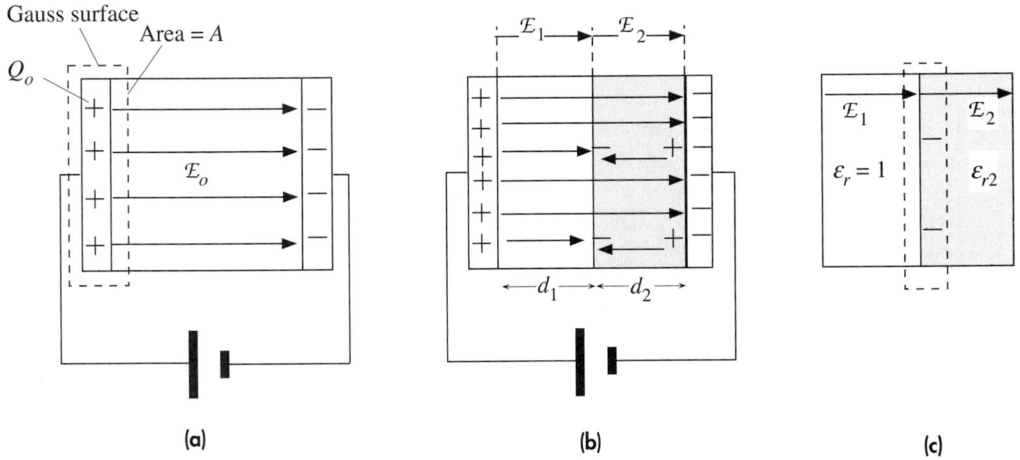


**Figure 7.20**
(a) The Gauss surface is a very thin rectangular surface just surrounding the positive electrode and enclosing the positive charges $Q_o$. The field cuts only the face just inside the capacitor.
(b) A solid dielectric occupies part of the distance between the plates. The vacuum (air)–dielectric boundary is parallel to the plates and normal to the fields $\mathcal{E}_1$ and $\mathcal{E}_2$.
(c) A thin rectangular Gauss surface at the boundary encloses the negative polarization charges.

where the circle on the integral sign represents integrating over the whole surface (any shape) enclosing the charges constituting $Q_{\text{total}}$ as shown in Figure 7.19. The total charge $Q_{\text{total}}$ includes *all charges,* both free charges and bound polarization charges. Gauss's law is one of the most useful laws for calculating electric fields in electrostatics, more so than the Coulomb law with which the reader is probably more familiar. The surface can be of any shape as long as it contains the charges. We generally choose convenient surfaces to simplify the integral in Equation 7.37, and these convenient surfaces are called Gauss surfaces. It should be noted from Figure 7.19 that the field $\mathcal{E}_n$ is coming *out* from the surface.

As an example, we can consider the field in the parallel plate capacitor in Figure 7.20a with no dielectric medium. We draw a thin rectangular Gauss surface (a hypothetical surface) just enclosing the positive electrode that contains the free charges $+Q_o$ on the plate. The field $\mathcal{E}_o$ is normal to the inner face (area $A$) of the Gauss surface.

Further, we can assume that $\mathcal{E}_o$ is uniform across the plate surface, which means that the integral of $\mathcal{E}_n\, dA$ in Equation 7.37 over the surface is simply $\mathcal{E}_o A$. There is no field on the other faces of this rectangular Gauss surface. Then from Equation 7.37,

$$\mathcal{E}_o A = \frac{Q_o}{\varepsilon_o}$$

which gives

$$\mathcal{E}_o = \frac{\sigma_o}{\varepsilon_o} \quad [7.38]$$

where

$$\sigma_o = \frac{Q_o}{A}$$

is the free surface charge density. This is the same as the field we calculated using $\mathcal{E}_o = V/d$ and $Q_o = CV$.

An important application of Gauss's law is determining what happens at boundaries between dielectric materials. The simplest example is the insertion of a dielectric slab to only partially fill the distance between the plates, as shown in Figure 7.20b. The applied voltage remains the same, but the field is no longer uniform between the plates. There is an air–dielectric boundary. The field is different in the air and dielectric regions. Suppose that the field is $\mathcal{E}_1$ in the air region and $\mathcal{E}_2$ in the dielectric region. Both these fields are normal to the boundary by the choice of the dielectric shape (faces parallel to the plates). As a result of polarization, bound surface charges $+A\sigma_P$ and $-A\sigma_P$ appear on the surfaces of the dielectric slab, as shown in Figure 7.20b, where $\sigma_P = P$, the polarization in the dielectric. We draw a very narrow rectangular Gauss surface that encompasses the air–dielectric interface and hence the surface polarization charges $-A\sigma_P$ as shown in Figure 7.20c. The field coming *in* at the left face in air is $\mathcal{E}_1$ (taken as negative) and the field coming *out* at the right face in the dielectric is $\mathcal{E}_2$. The surface integral $\mathcal{E}_n\, dA$ and Gauss's law become

$$\mathcal{E}_2 A - \mathcal{E}_1 A = \frac{-(A\sigma_P)}{\varepsilon_o}$$

or

$$\mathcal{E}_1 = \mathcal{E}_2 + \frac{P}{\varepsilon_o}$$

The polarization $P$ and the field $\mathcal{E}_2$ in the dielectric are related by

$$P = \varepsilon_o \chi_{e2} \mathcal{E}_2$$

or

$$P = \varepsilon_o(\varepsilon_{r2} - 1)\mathcal{E}_2$$

where $\chi_{e2}$ is the electrical susceptibility and $\varepsilon_{r2}$ is the relative permittivity of the inserted dielectric. Then, substituting for $P$, we can relate $\mathcal{E}_1$ and $\mathcal{E}_2$,

$$\mathcal{E}_1 = \mathcal{E}_2 + (\varepsilon_{r2} - 1)\mathcal{E}_2$$

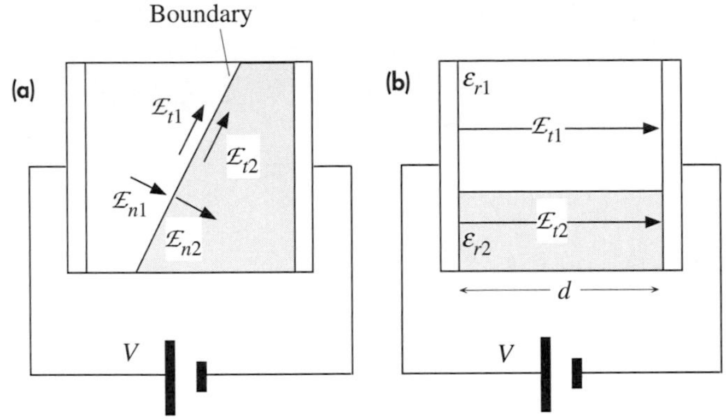


**Figure 7.21**
(a) Boundary conditions between dielectrics.
(b) The case for $\mathcal{E}_{t1} = \mathcal{E}_{t2}$.

or

$$\mathcal{E}_1 = \varepsilon_{r2}\mathcal{E}_2$$

The field in the air part is $\mathcal{E}_1$ and the relative permittivity is 1. The example in Figure 7.20b involved a boundary between air (vacuum) and a dielectric solid, and the boundary was parallel to the plates and hence normal to the fields $\mathcal{E}_1$ and $\mathcal{E}_2$. A more general expression can be shown to relate the normal components of the electric field, shown as $\mathcal{E}_{n1}$ and $\mathcal{E}_{n2}$ in Figure 7.21a, on either side of a boundary by

*General boundary condition*

$$\varepsilon_{r1}\mathcal{E}_{n1} = \varepsilon_{r2}\mathcal{E}_{n2} \qquad [7.39]$$

There is a second boundary condition that relates the tangential components of the electric field, shown as $\mathcal{E}_{t1}$ and $\mathcal{E}_{t2}$ in Figure 7.21a, on either side of a boundary. These tangential fields must be equal.

*General boundary condition*

$$\mathcal{E}_{t1} = \mathcal{E}_{t2} \qquad [7.40]$$

We can readily appreciate this boundary condition by examining the fields in a parallel plate capacitor, which has two dielectrics longitudinally filling the space between the plates but with a boundary parallel to the field, as shown in Figure 7.21b. The field in each, $\mathcal{E}_{t1}$ and $\mathcal{E}_{t2}$, is parallel to the boundary. The voltage across each longitudinal dielectric slab is the same, and since $\mathcal{E} = dV/dx$, the field in each is the same, $\mathcal{E}_{t1} = \mathcal{E}_{t2} = V/d$.

The above boundary conditions are widely used in explaining dielectric behavior when boundaries are involved. For example, consider a small disk-shaped cavity within a solid dielectric between two electrodes, as depicted in Figure 7.22. The disk-shaped cavity has its face perpendicular to the electric field. Suppose that the dielectric length $d$ is 1 cm and the cavity size is on the scale of micrometers. The average field within the dielectric will still be close to $V/d$ because in integrating the field $\mathcal{E}(x)$ to find the voltage across the dielectric, the contribution from a tiny distance of a few microns will be negligible compared with contributions coming over the rest of the 1 cm. But the field within the cavity will not be the same as the average field $\mathcal{E}_1$ in the dielectric. If $\varepsilon_{r1} = 5$ for the dielectric medium and the cavity has air, then at the cavity face we have

$$\varepsilon_{r2}\mathcal{E}_2 = \varepsilon_{r1}\mathcal{E}_1$$

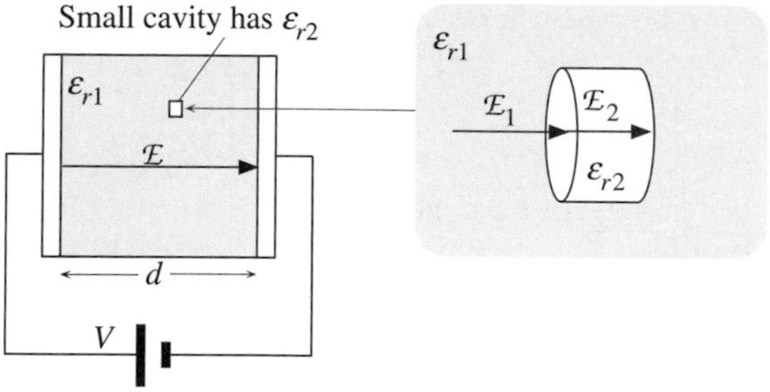


**Figure 7.22** Field in the cavity is higher than the field in the solid.

which gives

$$\mathcal{E}_2 = 5\left(\frac{V}{d}\right)$$

Air insulation in a 100 micron (0.1 mm) thick cavity breaks down when $\mathcal{E}_2$ is typically 100 kV cm$^{-1}$. From $\mathcal{E}_2 = 5(V/d)$, a voltage of 20 kV will result in the breakdown of air in the cavity and hence a discharge current. This is called a **partial discharge** as only a partial breakdown of the insulation, that in the cavity, has occurred between the electrodes. Under an ac voltage, the discharge in the cavity can often be sustained by the capacitive current through the surrounding dielectric. Without this cavity, the dielectric would accept a greater voltage across it, which in this case is typically greater than 100 kV.

**EXAMPLE 7.9** **FIELD INSIDE A THIN DIELECTRIC WITHIN A SECOND DIELECTRIC** When the dielectric fills the whole space between the plates of a capacitor, the net field within the dielectric is the same as before, $\mathcal{E} = V/d$. Explain what happens when a dielectric slab of thickness $t \ll d$ is inserted in the middle of the space between the plates, as shown in Figure 7.23. What is the field inside the dielectric?

**SOLUTION**

The problem is illustrated in Figure 7.23 and has symmetry in that the field in air on either side of the dielectric is the same and $\mathcal{E}_1$. The boundary conditions give

$$\varepsilon_{r1}\mathcal{E}_1 = \varepsilon_{r2}\mathcal{E}_2$$

**Figure 7.23** A thin slab of dielectric is placed in the middle of a parallel plate capacitor.
The field inside the thin slab is $\mathcal{E}_2$.

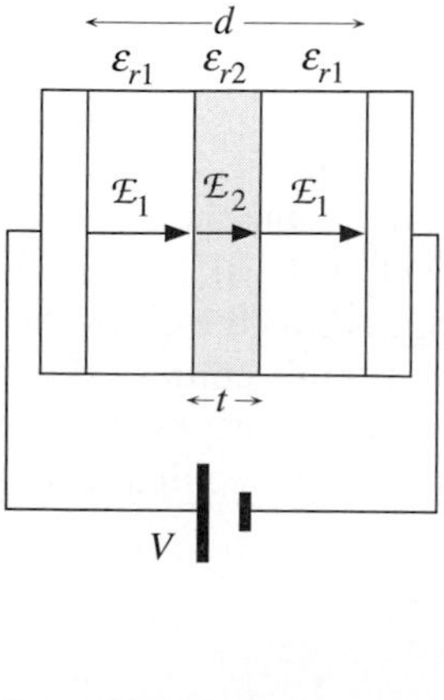

Further, the integral of the field from one plate to the other must be $V$ because $dV/dx = \mathcal{E}$. Examining Figure 7.23, we see that the integration is

$$\mathcal{E}_1(d - t) + \mathcal{E}_2 t = V$$

We now have to eliminate $\mathcal{E}_1$ between the previous two equations and obtain $\mathcal{E}_2$, which can be done by algebraic manipulation,

$$\mathcal{E}_2 = \frac{\varepsilon_{r1}}{\varepsilon_{r2} - \dfrac{t}{d}(\varepsilon_{r2} - \varepsilon_{r1})}\left(\frac{V}{d}\right) \qquad [7.41]$$

If $t \ll d$, then this approximates to

$$\mathcal{E}_2 = \frac{\varepsilon_{r1}}{\varepsilon_{r2}}\left(\frac{V}{d}\right) \quad \text{and} \quad \mathcal{E}_1 = \left(\frac{V}{d}\right) \quad (t \ll d) \qquad [7.42]$$

Clearly $\mathcal{E}_1$ in the air space remains the same as the applied field $V/d$. Since $\varepsilon_{r1} = 1$ (air) and $\varepsilon_{r2} > 1$, $\mathcal{E}_2$ in the thin dielectric slab is smaller than the applied field $V/d$. On the other hand, if we have air space between two dielectric slabs, then the field in this air space will be greater than the field inside the two dielectric slabs. Indeed, if the applied voltage is sufficiently large, the field in the air gap can cause dielectric breakdown of this region.

**EXAMPLE 7.10**

**GAUSS'S LAW WITHIN A DIELECTRIC AND FREE CHARGES** Gauss's law in Equation 7.37 contains the total charge $Q_{\text{total}}$, enclosed within the surface. Generally, these enclosed charges are free charges $Q_{\text{free}}$, due to the free carriers on the electrode, and bound charges $Q_P$, due to polarization charges on the dielectric surface. Apply Gauss's law using a Gaussian rectangular surface enclosing the left electrode and the dielectric surface in Figure 7.24. Show that the electric field $\mathcal{E}$ in the dielectric can be expressed in terms of free charges only, $Q_{\text{free}}$, through

*Free charges and field in a dielectric*

$$\oint_{\text{Surface}} \mathcal{E}_n\, dA = \frac{Q_{\text{free}}}{\varepsilon_o \varepsilon_r} \qquad [7.43]$$

where $\varepsilon_r$ is the relative permittivity of the dielectric medium.

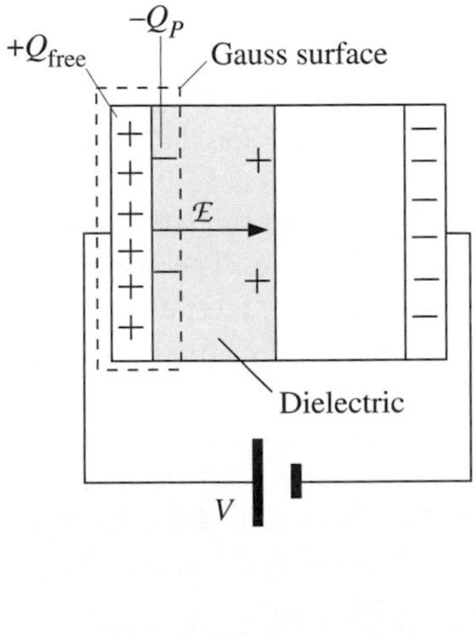


**Figure 7.24** A convenient Gauss surface for calculating the field inside the dielectric is a very thin rectangular surface enclosing the surface of the dielectric.
The total charges enclosed are the free charges on the electrodes and the polarization charges on the surface of the dielectric.

**SOLUTION**

We apply Gauss's law to a hypothetical rectangular surface enclosing the left electrode and the dielectric surface. The field $\mathcal{E}$ in the dielectric is normal and outwards at the Gauss surface in Figure 7.24. Thus $\mathcal{E}_n = \mathcal{E}$ in the left-hand side of Equation 7.37.

$$\varepsilon_o A\mathcal{E} = Q_{\text{total}} = Q_{\text{free}} - Q_P = Q_{\text{free}} - AP = Q_{\text{free}} - A\varepsilon_o(\varepsilon_r - 1)\mathcal{E}$$

where we have used $P = \varepsilon_o(\varepsilon_r - 1)\mathcal{E}$. Rearranging,

$$\varepsilon_o\varepsilon_r A\mathcal{E} = Q_{\text{free}}$$

Since $A\mathcal{E}$ is effectively the surface integral of $\mathcal{E}_n$, the above corresponds to writing Gauss's law in a dielectric in terms of free charges as

$$\oint_{\text{Surface}} \mathcal{E}_n\, dA = \frac{Q_{\text{free}}}{\varepsilon_o\varepsilon_r}$$

The above equation assumes that polarization $P$ and $\mathcal{E}$ are linearly related,

$$P = \varepsilon_o(\varepsilon_r - 1)\mathcal{E}$$

We note that if we only use free charges in Gauss's law, then we simply multiply $\varepsilon_o$ by the dielectric constant of the medium. The above proof is by no means a rigorous derivation.

## 7.6 DIELECTRIC STRENGTH AND INSULATION BREAKDOWN

### 7.6.1 DIELECTRIC STRENGTH: DEFINITION

A defining property of a dielectric medium is not only its ability to increase capacitance but also, and equally important, its insulating behavior or low conductivity so that the charges are not simply conducted from one plate of the capacitor to the other through the dielectric. Dielectric materials are widely used as insulating media between conductors at different voltages to prevent the ionization of air and hence current flashovers between conductors. The voltage across a dielectric material and hence the field within it cannot, however, be increased without limit. Eventually a voltage is reached that causes a substantial current to flow between the electrodes, which appears as a short between the electrodes and leads to what is called **dielectric breakdown.** In gaseous and many liquid dielectrics, the breakdown does not generally permanently damage the material. This means that if the voltage causing breakdown is removed, then the dielectric can again sustain voltages until the voltage is sufficiently high to cause breakdown again. In solid dielectrics the breakdown process invariably leads to the formation of a permanent conducting channel and hence to permanent damage. The **dielectric strength** $\mathcal{E}_{\text{br}}$ is the maximum field that can be applied to an insulating medium without causing dielectric breakdown. Beyond $\mathcal{E}_{\text{br}}$, dielectric breakdown takes place. The dielectric strength of solids depends on a number of factors besides simply the molecular structure, such as the impurities in the material, microstructural defects (*e.g.*, microvoids), sample geometry, nature of the electrodes, temperature, and ambient conditions (*e.g.*, humidity), as well as the duration and frequency of the applied field. Dielectric strength

**Table 7.5** Dielectric strength; typical values at room temperature and 1 atm

| Dielectric Medium | Dielectric Strength | Comments |
|---|---|---|
| Atmosphere at 1 atm pressure | 31.7 kV cm$^{-1}$ at 60 Hz | 1 cm gap. Breakdown by electron avalanche by impact ionization. |
| $SF_6$ gas | 79.3 kV cm$^{-1}$ at 60 Hz | Used in high-voltage circuit breakers to avoid discharges. |
| Polybutene | >138 kV cm$^{-1}$ at 60 Hz | Liquid dielectric used as oil filler and HV pipe cables. |
| Transformer oil | 128 kV cm$^{-1}$ at 60 Hz | |
| Amorphous silicon dioxide ($SiO_2$) in MOS technology | 10 MV cm$^{-1}$ dc | Very thin oxide films without defects. Intrinsic breakdown limit. |
| Borosilicate glass | 10 MV cm$^{-1}$ duration of 10 μs<br>6 MV cm$^{-1}$ duration of 30 s | Intrinsic breakdown.<br>Thermal breakdown. |
| Polypropylene | 295–314 kV cm$^{-1}$ | Likely to be thermal breakdown or electrical treeing. |

is different under dc and ac conditions. There are also **aging effects** that slowly degrade the properties of the insulator and reduce the dielectric strength. For engineers involved in insulation, the dielectric strength of solids is therefore one of the most difficult parameters to interpret and use. For example, the breakdown field also depends on the thickness of the insulation because thicker insulators have more volume and hence a greater probability of containing a microstructural defect (*e.g.*, a microcavity) that can initiate a dielectric breakdown. Table 7.5 shows some typical dielectric strengths for various dielectrics used in electrical insulation. Unpressurized gases have lower breakdown strengths than liquids and solids.

## 7.6.2 DIELECTRIC BREAKDOWN AND PARTIAL DISCHARGES: GASES

Due to cosmic radiation, there are always a few free electrons in a gas. If the field is sufficiently large, then one of these electrons can be accelerated to sufficiently large kinetic energies to impact ionize a neutral gas molecule and produce an additional free electron and a positively charged gas ion. Both the first and liberated electrons are now available to accelerate in the field again and further impact ionize more neutral gas molecules, and so on. Thus, an avalanche of impact ionization processes creates many free electrons and positive gas ions in the gas, which give rise to a discharge current between the electrodes. The process is similar to avalanche breakdown in a reverse-biased *pn* junction. The breakdown in gases depends on the pressure. The concentration of gas molecules is greater at higher pressures. This means that the mean separation between molecules, and, hence, the mean free path of a free electron, is shorter. Shorter mean free paths inhibit the free electrons from accelerating to reach impact ionization energies unless the field is increased. Thus, generally, $\mathcal{E}_{br}$ increases with the gas pressure. The 60 Hz breakdown field for an air gap of 1 cm at room temperature and at atmospheric pressure is about 31.7 kV cm$^{-1}$. On the other hand, the gas sulfurhexafluoride, $SF_6$, has

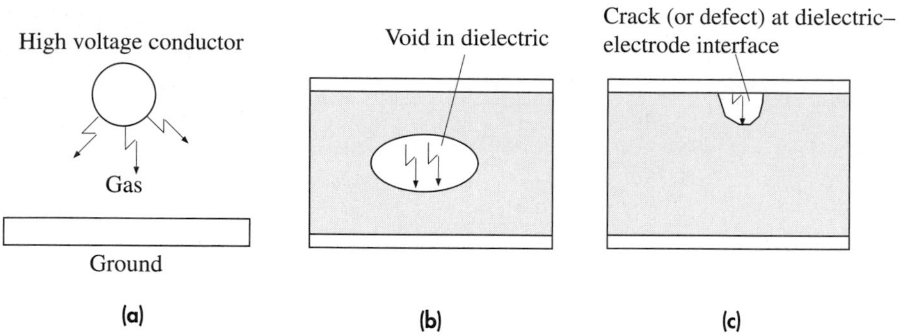


**Figure 7.25**

(a) The field is greatest on the surface of the cylindrical conductor facing the ground. If the voltage is sufficiently large, this field gives rise to a corona discharge.

(b) The field in a void within a solid can easily cause partial discharge.

(c) The field in the crack at the solid–metal interface can also lead to a partial discharge.

a dielectric strength of 79.3 kV $cm^{-1}$ and an even higher strength when pressurized. $SF_6$ is therefore used instead of air in high-voltage circuit breakers.

A **partial discharge** occurs when only a local region of the dielectric is exhibiting discharge, so the discharge does not directly connect the two electrodes. For example, for the cylindrical conductor carrying a high voltage above a grounded plate, as in Figure 7.25a, the electric field is greatest on the surface of the conductor facing the ground. This field initiates discharge locally in this region because the field is sufficiently high to give rise to an electron avalanche effect. Away from the conductor, however, the field is not sufficiently strong to continue the electron avalanche discharge. This type of local discharge in high field regions is termed **corona discharge.** Voids and cracks occurring within solid dielectrics and discontinuities at the dielectric–electrode interface can also lead to partial discharges as the field in these voids is higher than the average field in the dielectric, and, further, the dielectric strength in the gas (*e.g.*, atmosphere) in the void is less than that of the continuous solid insulation. Figure 7.25b and c depict two examples of partial discharges occurring in voids, one inside the solid (perhaps an air or gas bubble introduced during the processing of the dielectric) and the other (perhaps in the form of a crack) at the solid–electrode interface. In practice, a variety of factors can lead to microvoids and microcavities inside solids as well as at interfaces. Partial discharges in these voids physically and chemically erode the surrounding dielectric region and lead to an overall deterioration of the dielectric strength. If uncontrolled, they can eventually give rise to a major breakdown.

### 7.6.3 DIELECTRIC BREAKDOWN: LIQUIDS

The processes that lead to the breakdown of insulation in liquids are not as clear as the electron avalanche effect in gases. In impure liquids with small conductive particles in suspension, it is believed that these impurities coalesce end to end to form a conducting bridge between the electrodes and thereby give rise to discharge. In some liquids, the discharge initiates as partial discharges in gas bubbles entrapped in the liquid. These partial discharges can locally raise the temperature and vaporize more of the liquid and hence increase the size of the bubble. The eventual discharge can be a series of partial discharges in entrapped gas bubbles. Moisture absorption and absorption of gases from the ambient generally deteriorate the dielectric strength. Oxidation of certain liquids, such as oils, with time produces more acidic and hence higher conductivity inclusions or regions that eventually give discharge. In some liquids, the discharge involves the emission of a large number of electrons from the electrode into the liquid due to field emission at high fields. This is a discharge process by electrode injection.

### 7.6.4 DIELECTRIC BREAKDOWN: SOLIDS

There are various major mechanisms that can lead to dielectric breakdown in solids. The most likely mechanism depends on the dielectric material's condition and sometimes on extrinsic factors such as the ambient conditions, moisture absorption being a typical example.

***Intrinsic Breakdown or Electronic Breakdown*** The most common type of electronic breakdown is an **electron avalanche breakdown.** A free electron in the conduction band (CB) of a dielectric in the presence of a large field can be accelerated to sufficiently large energies to collide with and ionize a host atom of the solid. The electron gains an energy $e\mathcal{E}_{br}\ell$ when it moves a distance $\ell$ under an applied field $\mathcal{E}_{br}$. If this energy is greater than the bandgap energy $E_g$, then the electron, as a result of a collision with the lattice vibrations, can excite an electron from the valence band to the conduction band, that is, break a bond. Both the primary and the released electron can further impact ionize other host atoms and thereby generate an electron avalanche effect that leads to a substantial current. The initial conduction electrons for the avalanche are either present in the CB or are injected from the metal into the CB as a result of field-assisted thermal emission from the Fermi energy in the metal to the CB in the dielectric. Taking typical values, $E_g \approx 5$ eV and $\ell$ to be of the order of the mean free path for lattice scattering, say $\sim$50 nm, one finds $\mathcal{E}_{br} \approx 1$ MV cm$^{-1}$. Obviously, $\mathcal{E}_{br}$ depends on the choice of $\ell$, but its order of magnitude indicates voltages that are quite large. This type of breakdown represents an upper theoretical limit that is probably approached by only certain dielectrics—those that have practically no defects. Usually, microstructural defects lead to a lower dielectric strength than the limit indicated by intrinsic breakdown. Silicon dioxide ($SiO_2$) films with practically no structural defects in present MOS (metal-oxide-semiconductor) capacitors (as in the gates of MOSFETs) probably exhibit an intrinsic breakdown.

If dielectric breakdown does not occur by an electron avalanche effect (perhaps due to short mean free paths in the insulator), then another insulation breakdown mechanism is the enormous increase in the injection of electrons from the metal electrode into the insulator at very high fields as a result of field-assisted emission.[11] It has

[11] The emission of electrons by tunneling from an electrode in the presence of a large field was treated in Chapter 4 as Fowler–Nordheim field emission.

been proposed that insulation breakdown under short durations in some thin polymer films is due to tunneling injection.

***Thermal Breakdown*** Finite conductivity of the insulation means that there is Joule heat $\sigma\mathcal{E}^2$ being released within the solid. Further, at high frequencies, the dielectric loss, $V^2\omega \tan\delta$, becomes especially significant. For example, the work done by the external field in rotating the dipoles is transferred more frequently to random molecular collisions as heat as the frequency of the field increases. Both conduction and dielectric losses therefore generate heat within the dielectric. If this heat cannot be removed from the solid sufficiently quickly by thermal conduction (or by other means), then the temperature of the dielectric will increase. The increase in the temperature invariably increases the conductivity of an insulator. The increase in the conductivity then leads to more Joule heating and hence further rises in the temperature and so on. If the heat cannot be conducted away to limit the temperature, then the result is a thermal runaway condition in which the temperature and the current increase until a discharge occurs through various sections of the solid. As a consequence of sample inhomogeneities, frequently thermal runaway is severe in certain parts of the solid that become hot spots and suffer local melting and physical and chemical erosion. Hot spots are those local regions or inhomogeneities where $\sigma$ or $\varepsilon_r''$ is larger or where the thermal conductivity is too poor to remove the heat generated. Local breakdown at various hot spots eventually leads to a conducting channel connecting the opposite electrodes and hence to a dielectric breakdown. Since it takes time to raise the temperature of the dielectric, due to the heat capacity, this breakdown process has a marked thermal lag. The time to achieve thermal breakdown depends on the heat generated, and hence on $\mathcal{E}^2$. Conversely, this means that the dielectric strength $\mathcal{E}_{br}$ depends on the duration of application of the field. For example, at 70 °C, pyrex has an $\mathcal{E}_{br}$ of typically 9 MV cm$^{-1}$ if the applied field duration is kept short, not more than 1 ms or so. If the field is kept for 30 s, then the breakdown field is only 2.5 MV cm$^{-1}$. Dielectric breakdown in various ceramics and glasses at high frequencies has been attributed directly to thermal breakdown. A characteristic feature of thermal breakdown is not only the thermal lag, the time dependence, but also the temperature dependence. Thermal breakdown is facilitated by increasing the temperature of the dielectric, which means that $\mathcal{E}_{br}$ decreases with temperature.

***Electromechanical Breakdown and Electrofracture*** A dielectric medium between oppositely charged electrodes experiences compressional forces because the opposite charges $+Q$ and $-Q$ on the plates attract each other, as depicted in Figure 7.26. As the voltage increases, so does the compressive load, and the dielectric becomes squeezed, or the thickness $d$ gets smaller. At each stage, the increase in the compressive load is normally balanced by the elastic deformation of the insulation to a new smaller thickness. However, if the elastic modulus is sufficiently small, then compressive loads cannot be simply balanced by the elasticity of the solid, and there is a mechanical runaway for the following reasons. The decrease in $d$, due to the compressive load, leads to a higher field ($\mathcal{E} = V/d$) and also to more charges on the electrodes ($Q = CV$, $C = \varepsilon_o\varepsilon_r A/d$). This in turn leads to a greater compressive load, which further decreases $d$, and so on, until the shear stresses within the insulation cause the insulation to flow plastically (for example, by viscous deformation). Eventually, the insulation breaks down. In addition, the increase in $\mathcal{E}$ as $d$ gets

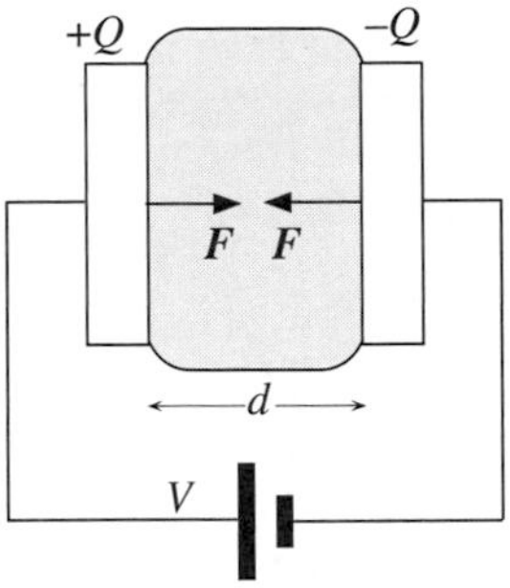


**Figure 7.26** An exaggerated schematic illustration of a soft dielectric medium experiencing strong compressive forces due to the applied voltage.

smaller results in more Joule ($\sigma \mathcal{E}^2$) and dielectric-loss heating ($\omega \mathcal{E}^2 \tan \delta$) in the dielectric, which increases the temperature and hence lowers the elastic modulus and viscosity, thereby further deteriorating the mechanical stability. It is also possible for the field during the mechanical deformation of the dielectric to reach the thermal breakdown field, in which case the dielectric failure is not truly a mechanical breakdown mechanism though initiated by mechanical deformations. Another possibility is the initiation and growth of internal cracks (perhaps filamentary cracks) by internal stresses around inhomogeneous regions inside the dielectric. For example, an imperfection or a tiny cavity experiences shear stresses and also large local electric fields. Combined effects of both large shear stresses and large electric fields eventually lead to crack propagation and mechanical and, hence, dielectric failure. This type of process is sometimes called **electrofracture.** It is generally believed that certain thermoplastic polymers suffer from electromechanical dielectric breakdown, especially close to their softening temperatures. Polyethylene and polyisobutylene have been cited as examples.

***Internal Discharges*** These are partial discharges that take place in microstructural voids, cracks, or pores within the dielectric where the gas atmosphere (usually air) has lower dielectric strength. A porous ceramic, for example, would experience partial discharges if the applied field is sufficiently large. The discharge current in a void,

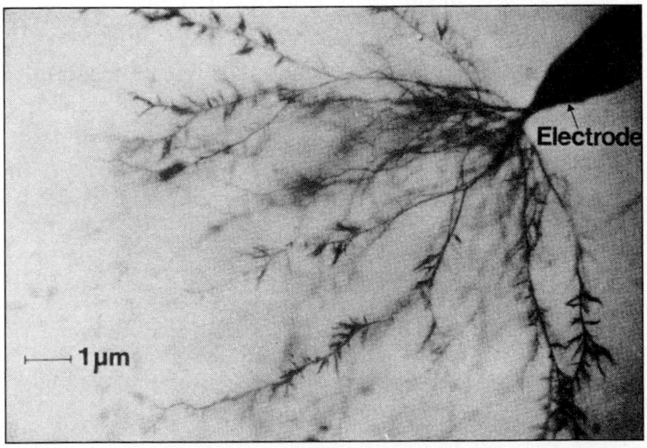


Electrical breakdown by *treeing* (formation of discharge channels) in a low-density polyethylene insulation when a 50 Hz, 20 kV (rms) voltage is applied for 200 minutes to an electrode embedded in the insulation.

SOURCE: J. W. Billing and D. J. Groves, *Proceedings of the Institution of Electrical Engineers*, **212,** 1451, 1974.

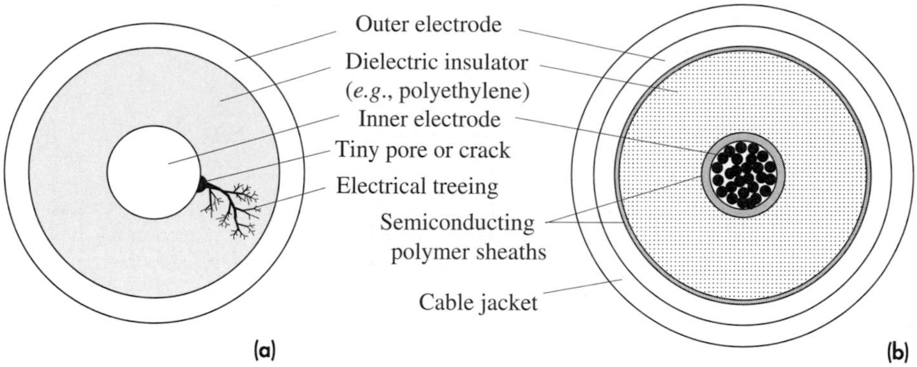


**Figure 7.27**

(a) A schematic illustration of electrical treeing breakdown in a high-voltage coaxial cable that was initiated by a partial discharge in the void at the inner conductor–dielectric interface.

(b) A schematic diagram of a typical high-voltage coaxial cable with semiconducting polymer layers around the inner conductor and around the outer surface of the dielectric.

such as those in Figure 7.25b and c, can be easily sustained under ac conditions, which accounts for the severity of this type of breakdown mechanism under ac conditions. Initially, the pore size (or the number of pores) may be small and the partial discharge insignificant, but with time the partial discharge erodes the internal surfaces of the void. Partial discharges can locally melt the insulator and can easily cause chemical transformations. Eventually, and usually, an **electrical tree** type of discharge develops from a partial discharge that has been eroding the dielectric, as depicted in Figure 7.27a for a high-voltage cable in which there is a tiny void at the interface between the dielectric and the inner conductor (generated perhaps by the differential thermal expansion of the electrode and polymeric insulation). The erosion of the dielectric by the partial discharge propagates like a branching tree. The "tree branches" are erosion channels—hollow filaments of various sizes—in which gaseous discharge takes place and forms a conducting channel during operation.

In the case of a coaxial high-voltage cable in Figure 7.27a, the dielectric is usually a polymer, polyethylene (PE) being one of the most popular. The electric field is maximum at the surface of the inner conductor, which is the reason for the initiation of most electrical trees near this surface. Electrical treeing is substantially controlled by having semiconductive polymer layers or sheaths surrounding the inner conductor and the outer surface of the insulator, as shown in Figure 7.27b. For flexibility, the inner conductor is frequently multicored, or stranded, rather than solid. Due to the extrusion process used to draw the insulation, the semiconductive polymer sheaths are bonded to the insulation. There are therefore practically no microvoids at the interfaces between the insulator and the semiconducting sheath. Further, these semiconducting polymer sheaths are sufficiently conductive to become "part of the electrodes." Both the conductor and the adjacent semiconductor are roughly at the same voltage, which means that there is no breakdown in the semiconductor–conductor interfaces. There is normally an outer jacket (*e.g.*, PVC) to protect the cable.

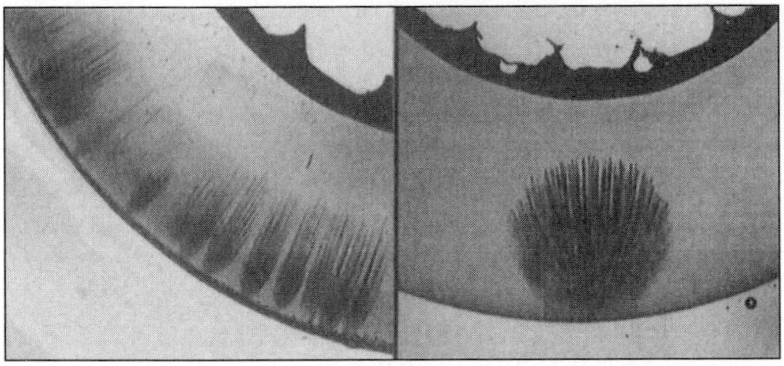

Some typical water trees found in field aged cables. Left: Trees in a cable with tape and graphite insulation. Right: Trees in a cable with strippable insulation.

SOURCE: P. Werellius, P. Tharning, R. Eriksson, B. Holmgren. J. Gafvert, "Dielectric Spectroscopy for Diagnosis of Water Tree Deterioration in XLPE Cables," *IEEE Transactions on Dielectrics and Electrical Insulation,* vol. **8,** February 2001, p. 34, figure 10 (© IEEE, 2001).

***Insulation Aging*** It is well recognized that during service, the properties of an insulating material become degraded and eventually dielectric breakdown occurs at a field below that predicted by experiments on fresh forms of the insulation. **Aging** is a term used to describe, in a general sense, the deterioration in the properties of the insulation. Aging therefore determines the useful life of the insulation. There are many factors that either directly or indirectly affect the properties and performance of an insulator in service. Even in the absence of an electric field, the insulation will experience physical and chemical aging whereby its physical and chemical properties change considerably. An insulation that is subjected to temperature and mechanical stress variations can develop structural defects, such as microcracks, which are quite damaging to the dielectric strength, as mentioned above. Irradiation by ionizing radiation such as X-rays, exposure to severe ambient conditions such as excessive humidity, ozone, and many other external conditions, through various chemical processes, deteriorate the chemical structure and properties of an insulator. This is generally much more severe for polymers than ceramics, but it is not practical to use a solid ceramic insulation in a coaxial power cable. Oxidation of a polymeric insulation with time is another form of chemical aging and is well-known to degrade the insulation performance. This is the reason for adding various antioxidants into semicrystalline polymers for use in insulation. The chemical aging processes are generally accelerated with temperature. In service, the insulation also experiences electrical aging as a result of the effects of the field on the properties of the insulation. For example, dc fields can disassociate and transport various ions in the structure and thereby slowly change the structure and properties of the insulation. Electrical trees develop as a result of electrical aging because, in service, the ac field gives rise to continual partial discharges in an internal or surface microcavity, which then erodes the region around it and slowly grows like a branching tree. In well-manufactured insulation systems, electrical treeing has been substantially reduced or eliminated from microvoids. A form of electrical aging that is currently in vogue is **water treeing,** which eventually leads to electrical treeing. The definition of a water tree, as viewed under an optical microscope, is a diffused bushy (or broccoli) type growth that consists of millions of microscopic voids (per $mm^3$) containing water or aqueous electrolyte. They invariably occur in moist environments and are relatively nonconducting, which means that they do not themselves lead to a direct discharge.

***External Discharges*** There are many examples where the surface of the insulation becomes contaminated by ambient conditions such as excessive moisture, deposition

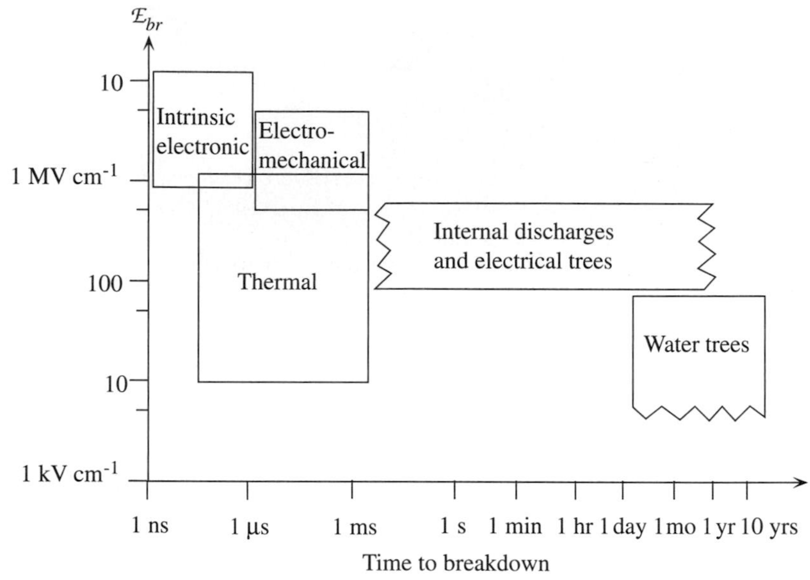


**Figure 7.28** Time to breakdown and the field at breakdown $\mathcal{E}_{br}$ are interrelated and depend on the mechanism that causes the insulation breakdown.

External discharges have been excluded.

SOURCE: Based on L. A. Dissado and J. C. Fothergill, *Electrical Degradation and Breakdown in Polymers*, United Kingdom: Peter Peregrinus Ltd. for IEE, 1992, p. 63.

of pollutants, dirt, dust, and salt spraying. Eventually the contaminated surface develops sufficient conductance to allow discharge between the electrodes at a field below the normal breakdown strength of the insulator. This type of dielectric breakdown over the surface of the insulation is termed **surface tracking.**

It is apparent that there are a number of dielectric breakdown mechanisms and the one that causes eventual breakdown depends not only on the properties and quality of the material but also on the operating conditions, environmental factors being no less important. Figure 7.28 provides an illustrative diagram showing the relationship between the breakdown field and the time to breakdown. An insulation that can withstand large fields for a very short duration will break down at a lower field if the duration of the field increases. The breakdown mechanism is also likely to change from being intrinsic to being, perhaps, thermal. When insulation breakdown occurs in times beyond a few days, it is generally attributed to the degradation of the insulation, which eventually leads to a breakdown through, most probably, electrical treeing. It is also apparent that it is not possible to clearly identify a specific dielectric breakdown mechanism for a given material.

**EXAMPLE 7.11** **DIELECTRIC BREAKDOWN IN A COAXIAL CABLE** Consider the coaxial cable in Figure 7.29 with $a$ and $b$ defining the radii of the inner and outer conductors.

*a.* Using Gauss's law, find the capacitance of the coaxial cable.

*b.* What is the electric field at $r$ from the center of the cable $(r > a)$? Where is the field maximum?

*c.* Consider two candidate materials for the dielectric insulation: cross-linked polyethylene, (XLPE) and silicone rubber. Suppose that the inner conductor diameter is 5 mm and the insulation thickness is also 5 mm. What is the voltage that will cause dielectric breakdown in each insulator?

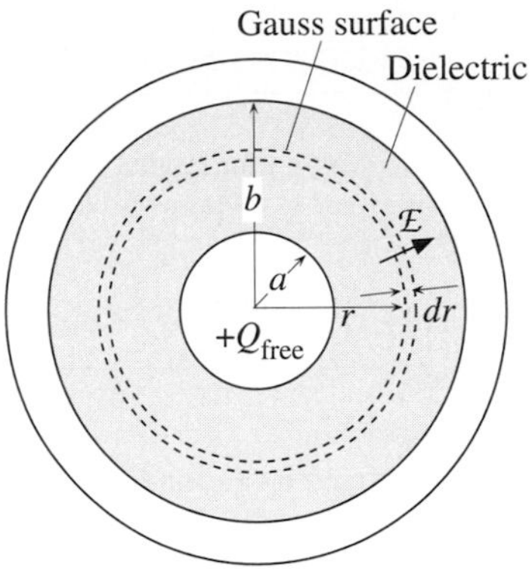


**Figure 7.29** A schematic diagram for the calculation of the capacitance of a coaxial cable and the field at point *r* from the axis.

Consider an infinitesimally thin cylindrical shell of radius *r* and thickness *dr* in the dielectric and concentrically around the inner conductor. This surface is chosen as the Gauss surface. The voltage across the dielectric thickness *dr* is *dV*. The field $\mathcal{E} = -dV/dr$.

*d.* What typical voltage will initiate a partial discharge in a small air pore (perhaps formed during mechanical and thermal stressing) at the inner conductor–insulator interface? Assume that the breakdown field for air at 1 atm and gap spacing around 0.1 mm is about 100 kV cm$^{-1}$.

**SOLUTION**

Consider a cylindrical shell of thickness $dr$ of the dielectric as shown in Figure 7.29. Suppose that the voltage across the shell thickness is $dV$. Then the field $\mathcal{E}$ at $r$ is $-dV/dr$ (this is the definition of $\mathcal{E}$). Suppose that $Q_{free}$ is the free charge on the inner conductor. We take a Gauss surface that is a cylinder of radius $r$ and concentric with the inner conductor as depicted in Figure 7.29. The surface area $A$ of this cylinder is $2\pi rL$ where $L$ is the length of the cable. The field at the surface, at distance $r$, is $\mathcal{E}$, which is normal to $A$ and coming out of $A$. Then from Equation 7.43

$$\mathcal{E}(2\pi rL) = \frac{Q_{free}}{\varepsilon_o \varepsilon_r} \qquad [7.44]$$

Thus

$$-\frac{dV}{dr} = \frac{Q_{free}}{\varepsilon_o \varepsilon_r 2\pi rL}$$

This can be integrated from $r = a$, where the voltage is $V$, to $b$, where $V = 0$. Then

$$V = \frac{Q_{free}}{\varepsilon_o \varepsilon_r 2\pi L} \ln\left(\frac{b}{a}\right) \qquad [7.45]$$

We can obtain the capacitance of the coaxial cable from $C_{coax} = Q_{free}/V$, which is

*Capacitance of a coaxial cable*

$$C_{coax} = \frac{\varepsilon_o \varepsilon_r 2\pi L}{\ln\left(\frac{b}{a}\right)} \qquad [7.46]$$

The capacitance per unit length can be calculated using $a = 2.5$ mm and

$$b = a + \text{Thickness} = 7.5 \text{ mm}$$

and the appropriate dielectric constants, $\varepsilon_r = 2.3$ for XLPE and 3.7 for silicone rubber. The values are around 100–200 pF per meter, as listed in the fourth column in Table 7.6.

**Table 7.6** Dielectric insulation candidates for a coaxial cable

| Dielectric | $\varepsilon_r$ (60 Hz) | Strength (60 Hz) (kV cm$^{-1}$) | $C$ (60 Hz) (pF m$^{-1}$) | Breakdown Voltage (kV) | Voltage for Partial Discharge in a Microvoid (kV) |
|---|---|---|---|---|---|
| XLPE | 2.3 | 217 | 116 | 59.6 | 11.9 |
| Silicone rubber | 3.7 | 158 | 187 | 43.4 | 7.4 |

The electric field $\mathcal{E}$ follows directly when we substitute for $Q_{\text{free}}$ from Equation 7.45 into Equation 7.44,

*Field in a coaxial cable*

$$\mathcal{E} = \frac{V}{r \ln\left(\dfrac{b}{a}\right)} \quad [7.47]$$

Equation 7.47 is valid for $r$ from $a$ to $b$ (there is no field within the conductors). The field is maximum where $r = a$,

*Maximum field in a coaxial cable*

$$\mathcal{E}_{\max} = \frac{V}{a \ln\left(\dfrac{b}{a}\right)} \quad [7.48]$$

The breakdown voltage $V_{\text{br}}$ is reached when this maximum field $\mathcal{E}_{\max}$ reaches the dielectric strength or the breakdown field $\mathcal{E}_{\text{br}}$

*Breakdown voltage*

$$V_{\text{br}} = \mathcal{E}_{\text{br}} a \ln\left(\frac{b}{a}\right) \quad [7.49]$$

The breakdown voltages calculated from Equation 7.49 are listed in the fifth column in Table 7.6. Although the values are high, it must be remembered that, due to a number of other factors such as insulation aging, one cannot expect the cable to withstand these voltages forever.

If there is an air cavity or bubble at the inner conductor to dielectric surface, then the field in this gaseous space will be $\mathcal{E}_{\text{air}} \approx \varepsilon_r \mathcal{E}_{\max}$, where $\mathcal{E}_{\max}$ is the field at $r = a$. Air breakdown occurs when

$$\mathcal{E}_{\text{air}} = \mathcal{E}_{\text{air-br}} = 100 \text{ kV cm}^{-1}$$

at 1 atm and 25 °C for a 0.1 mm gap. Then $\mathcal{E}_{\max} \approx \mathcal{E}_{\text{air-br}}/\varepsilon_r$. The corresponding voltage from Equation 7.48 is

$$V_{\text{air-br}} \approx \frac{\mathcal{E}_{\text{air-br}}}{\varepsilon_r} a \ln\left(\frac{b}{a}\right)$$

The voltages for partial discharges for the two coaxial cables are shown in the sixth column of Table 7.6. It should be noted that these voltages will only give partial discharges contained within microvoids and will not normally lead to the immediate breakdown of the insulation. The partial discharges erode the cavities and also release vapor from the polymer that accumulates in the cavities. Thus, gaseous content and pressure in a cavity will change as the partial discharge continues. For example, the pressure buildup will increase the breakdown field and elevate the voltage for partial breakdown. Eventual degradation is likely to lead to electrical treeing.

We should also note that the actual field in the air cavity depends on the shape of the cavity, and the above treatment is only valid for a thin disk-like cavity lying perpendicular to the field (see Section 7.9, Additional Topics).

# 7.7 CAPACITOR DIELECTRIC MATERIALS

## 7.7.1 Typical Capacitor Constructions

The selection criteria of dielectric materials for capacitors depend on the capacitance value, frequency of application, maximum tolerable loss, and maximum working voltage, with size and cost being additional external constraints. Requirements for high-voltage power capacitors are distinctly different than those used in small integrated circuits. Large capacitance values are more easily obtained at low frequencies because low-frequency polarization mechanisms such as interfacial and dipolar polarization make a substantial contribution to the dielectric constant. At high frequencies, it becomes more difficult to achieve large capacitances and at the same time maintain acceptable low dielectric loss, inasmuch as the dielectric loss per unit volume is $\varepsilon_o \varepsilon_r' \omega \mathcal{E}^2 \tan \delta$.

The bar-chart diagrams in Figures 7.30 and 7.31 provide some typical examples of dielectrics for a range of capacitance values and for a range of usable frequencies. For example, electrolytic dielectrics characteristically provide capacitances between one to thousands of microfarads, but their frequency response is typically limited to below

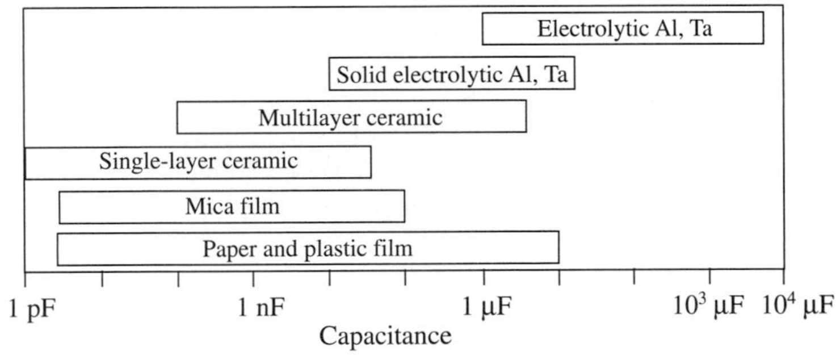


**Figure 7.30** Examples of dielectrics that can be used for various capacitance values.

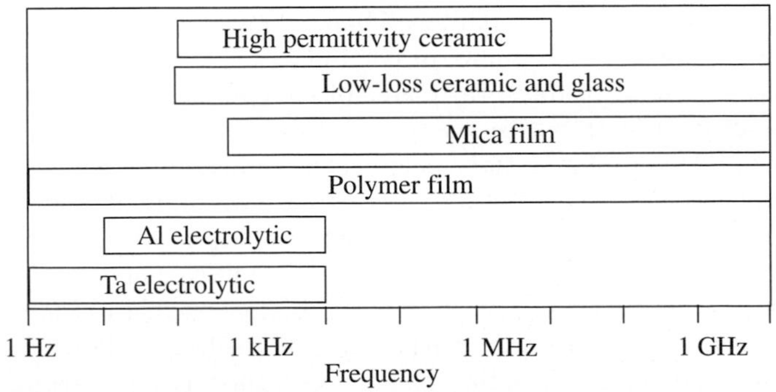


**Figure 7.31** Examples of dielectrics that can be used in various frequency ranges.

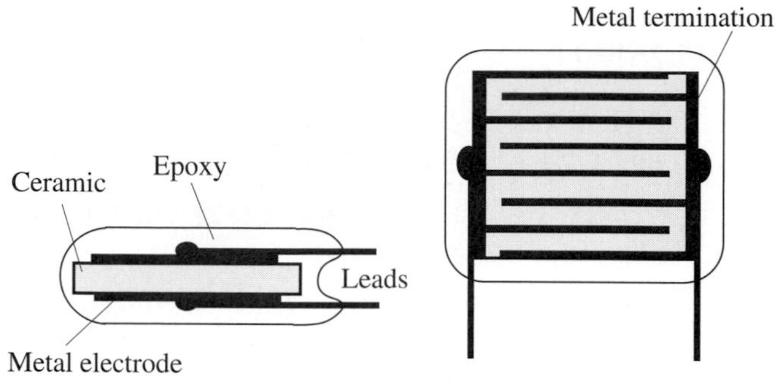


(a) Single-layer ceramic capacitor (e.g., disk capacitors)

(b) Multilayer ceramic capacitor (stacked ceramic layers)

**Figure 7.32** Single- and multilayer dielectric capacitors.

10 kHz. On the other hand, polymeric film capacitors typically have values less than 10 $\mu$F but a frequency response that is flat well into the gigahertz range.

We can understand the principles utilized in capacitor design from the capacitance of a parallel plate capacitor,

$$C = \frac{\varepsilon_o \varepsilon_r A}{d} \quad [7.50]$$

where $\varepsilon_r$ infers $\varepsilon_r'$. Large capacitances can be achieved by using high $\varepsilon_r$ dielectrics, thin dielectrics, and large areas. There are various commercial ceramics, usually a mixture of various oxides or ferroelectric ceramics, that have high dielectric constants, ranging up to several thousands. These are typically called high-$K$ (or high-$\kappa$), where $K$ (or $\kappa$) stands for the relative permittivity. A ceramic dielectric with $\varepsilon_r = 10$, $d$ of perhaps 10 $\mu$m, and an area of 1 cm$^2$ has a capacitance of 885 pF. Figure 7.32a shows a typical single-layer ceramic capacitor. The thin ceramic disk or plate has suitable metal electrodes, and the whole structure has been encapsulated in an epoxy by dipping it in a thermosetting resin. The epoxy coating prevents moisture from degrading the dielectric properties of the ceramic (increasing $\varepsilon_r''$ and the loss, tan $\delta$). One way to increase the capacitance is to connect $N$ number of these in parallel, and this is done in a space-efficient way by using the multilayer ceramic structure shown in Figure 7.32b. In this case there are $N$ electroded dielectric layers. Each ceramic has offset metal electrodes that align with the opposite sides of the plate and make contact with the metal terminations on these sides. The result is $N$ number of parallel plate capacitors. There is therefore an effective use of volume as the surface area of the component stays the same but the height increases to at least $Nd$. By using multilayer ceramic structures, capacitances up to a few hundred microfarads have been recently obtained.

Many wide-frequency-range capacitors utilize **polymeric thin films** for two reasons. Although $\varepsilon_r$ is typically 2 to 3 (less than those for many ceramics), it is constant over a wide frequency range. The dielectric loss $\varepsilon_o \varepsilon_r \omega \mathcal{E}^2$ tan $\delta$ becomes significant at high frequencies and polymers have low tan $\delta$ values. Low $\varepsilon_r$ values mean that one has to find a space-efficient way of constructing polymer film capacitors. One method is shown in Figure 7.33a and b for constructing a metallized film polymer capacitor. Two polymeric

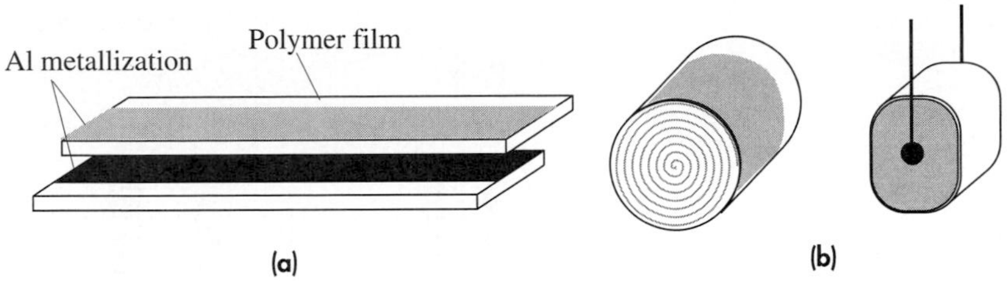


**Figure 7.33** Two polymer tapes in (a), each with a metallized film electrode on the surface (offset from each other), can be rolled together (like a Swiss roll) to obtain a polymer film capacitor as in (b).
As the two separate metal films are lined at opposite edges, electroding is done over the whole side surface.

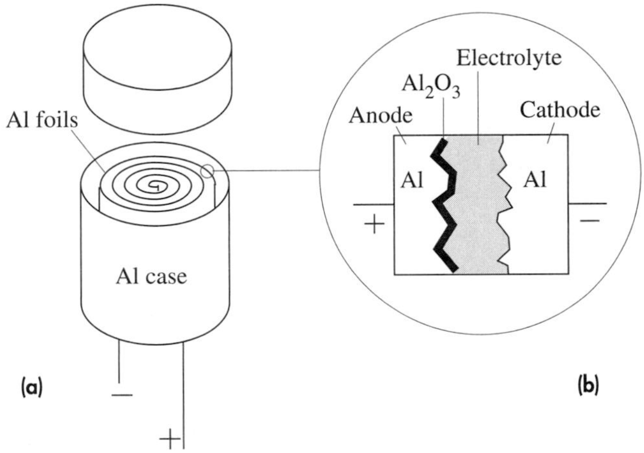


**Figure 7.34** Aluminum electrolytic capacitor.

tapes have metallized electrodes (typically vacuum deposited Al) on one surface, leaving a margin on one side. These metal film electrodes have been offset in opposite directions so that they line up with the opposite sides of the tapes. The two tapes together are rolled up (like a Swiss-roll cake) and the opposite sides are electroded using suitable conducting glues or other means. The concept is therefore similar to the multilayer ceramic capacitor except that the layers are rolled up to form a circular cross section. It is also possible to cut and stack the layers as in the multilayer ceramic construction.

**Electrolytic capacitors** provide large values of capacitance while maintaining a tolerable size. There are various types of electrolytic capacitors. In aluminum electrolytic capacitors, the metal electrodes are two Al foils, typically 50–100 $\mu$m thick, that are separated by a porous paper medium soaked with a liquid electrolyte. The two foils together are wound into a cylindrical form and held within a cylindrical case, as shown in Figure 7.34a. Contrary to intuition, the paper-soaked electrolyte is not the dielectric. The dielectric medium is the thin alumina $Al_2O_3$ layer grown on the roughened surface of one of the foils, as shown in Figure 7.34b. This foil is then called the anode (+ terminal). Both Al foils are etched to obtain rough surfaces, which increases the surface area

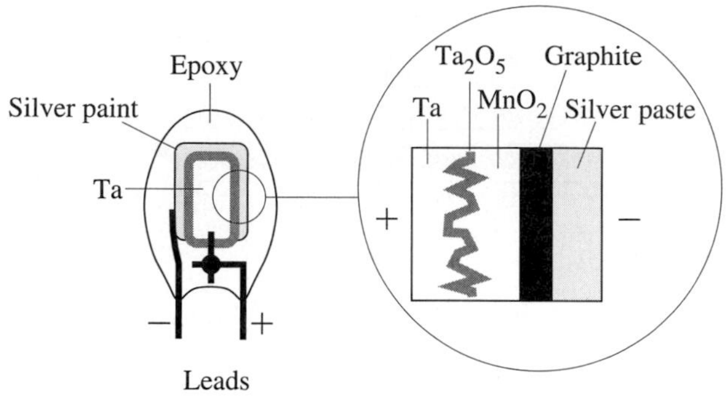


**Figure 7.35** Solid electrolyte tantalum capacitor.
(a) A cross section without fine detail.
(b) An enlarged section through the Ta capacitor.

compared with smooth surfaces. The capacitor is called electrolytic because the $Al_2O_3$ layer is grown electrolytically on one of the foils and is typically 0.1 μm in thickness. This small thickness and the large surface area are responsible for the large capacitance. The electrolyte is conducting and serves to heal local minor breakdowns in the $Al_2O_2$ by an electrolytic reaction, provided that the anode has been positively biased. The capacitive behavior is due to the Al/($Al_2O_3$)/electrolyte structure. Furthermore, Al/$Al_2O_3$ contact is like a metal to *p*-type semiconductor contact and has rectifying properties. It must be reverse-biased to prevent charge injection into the $Al_2O_3$ and hence conduction through the capacitor. Thus the Al must be connected to the positive terminal, which makes it the anode. When the electrolytic Al capacitor in Figure 7.34b is oppositely biased, it becomes conducting.

Electrolytic capacitors using liquid electrolytes tend to dry up over a long period, which is a disadvantage. **Solid electrolyte tantalum capacitors** overcome the drying-up problem by using a solid electrolyte. The structure of a typical solid Ta capacitor is shown in Figure 7.35a and b. The anode (+ electrode) is a porous (sintered) Ta pellet that has the surface anodized to obtain a thin surface layer of tantalum pentoxide, $Ta_2O_5$, which is the dielectric medium (with $\varepsilon_r' = 28$). The Ta pellet with $Ta_2O_5$ is then coated with a thick solid electrolyte, in this case $MnO_2$. Subsequently, graphite and silver paste layers are applied. Leads are then attached and the whole construction is molded into a resin chip. Solid tantalum capacitors are widely used in numerous electronics applications due to their small size, temperature and time stability, and high reliability.

## 7.7.2 DIELECTRICS: COMPARISON

The **capacitance per unit volume** $C_{vol}$, which characterizes the **volume efficiency** of a dielectric, can be obtained by dividing $C$ by $Ad$,

*Capacitance per unit volume*

$$C_{vol} = \frac{\varepsilon_o \varepsilon_r}{d^2} \qquad [7.51]$$

It is clear that large capacitances require high dielectric constants and thin dielectrics. We should note that $d$ appears as $d^2$, so the importance of $d$ cannot be understated.

**Table 7.7** Comparison of dielectrics for capacitor applications

| | Capacitor Name | | | | | |
|---|---|---|---|---|---|---|
| | **Polypropylene** | **Polyester** | **Mica** | **Aluminum, Electrolytic** | **Tantalum, Electrolytic, Solid** | **High-*K* Ceramic** |
| Dielectric | Polymer film | Polymer film | Mica | Anodized $Al_2O_3$ film | Anodized $Ta_2O_5$ film | X7R $BaTiO_3$ base |
| $\varepsilon_r'$ | 2.2–2.3 | 3.2–3.3 | 6.9 | 8.5 | 27 | 2000 |
| $\tan\delta$ | $4 \times 10^{-4}$ | $4 \times 10^{-3}$ | $2 \times 10^{-4}$ | 0.05–0.1 | 0.01 | 0.01 |
| $\mathcal{E}_{br}$ (kV mm$^{-1}$) dc | 100–350 | 100–300 | 50–300 | 400–1000 | 300–600 | 10 |
| $d$ (typical minimum) (μm) | 3–4 | 1 | 2–3 | 0.1 | 0.1 | 10 |
| $C_{vol}$ (μF cm$^{-3}$) | 2 | 30 | 15 | 7500* | 24,000* | 180 |
| $R_p = 1/G_p$(kΩ) for $C = 1$ μF, $f = 1$ kHz | 400 | 40 | 800 | 1.5–3 | 16 | 16 |
| $E_{vol}$ (mJ cm$^{-3}$)† | 10 | 15 | 8 | 1000 | 1200 | 100 |
| Polarization | Electronic | Electronic and dipolar | Ionic | Ionic | Ionic | Large ionic displacement |

* Proper volumetric calculations must also consider the volumes of electrodes and the electrolyte necessary for these dielectrics to work; hence the number would have to be decreased.

† $E_{vol}$ depends very sensitively on $\mathcal{E}_{br}$ and the choice of $\eta$; hence it can vary substantially.

NOTES: Values are typical. Assume $\eta = 3$. The table is for comparison purposes only. Breakdown fields are typical dc values and can vary substantially, by at least an order of magnitude; $\mathcal{E}_{br}$ depends on the thickness, material quality, and the duration of the applied voltage. Polyester is PET, or polyethylene terephthalate. Mica is potassium aluminosilicate, a muscovite crystal. X7R is the name of a particular $BaTiO_3$-based ceramic solid solution.

Although mica has a higher $\varepsilon_r$ than polymer films, the latter can be made quite thin, a few microns, which leads to a greater capacitance per unit volume. The reason that electrolytic aluminum capacitors can achieve large capacitance per unit volume is that $d$ can be made very thin over a large surface area by using the liquid electrolyte to heal minor local dielectric breakdowns. Table 7.7 shows a selection of dielectric materials for capacitor applications and compares the "volume efficiency" $C_{vol}$ based on a typical minimum thickness that a convenient process can handle. It is apparent that, compared with polymeric films, ceramics have substantial volume efficiency as a result of large dielectric constants (high-$K$ ceramics) in some cases and as a consequence of a thin dielectric thickness in other cases ($Al_2O_3$).

Another engineering consideration in selecting a dielectric is the working voltage. Although $d$ can be decreased to obtain large capacitances per unit volume, this also decreases the working voltage. The maximum voltage that can be applied to a capacitor depends on the breakdown field of the dielectric medium $\mathcal{E}_{br}$, which itself is a highly variable quantity. A safe working voltage must be some safety factor $\eta$ less than the breakdown voltage $\mathcal{E}_{br}d$. Thus, if $V_m$ is the maximum safe working voltage, then the maximum energy that can be stored per unit volume is given by

*Maximum energy per unit volume*

$$E_{vol} = \frac{1}{2}CV_m^2 \times \frac{1}{Ad} = \frac{\varepsilon_o\varepsilon_r'}{2\eta^2}\mathcal{E}_{br}^2 \qquad [7.52]$$

It is clear that both $\varepsilon_r'$ and $\mathcal{E}_{br}$ of the dielectric are significant in determining the energy storage ability of the capacitor. Moreover, at the maximum working voltage, the rate of dielectric loss per unit volume in the capacitor becomes

*Dielectric loss per unit volume*

$$W_{vol} = \frac{\mathcal{E}_{br}^2}{\eta^2}\omega\varepsilon_o\varepsilon_r' \tan\delta \qquad [7.53]$$

Those materials that have relatively higher tan $\delta$ exhibit greater dielectric losses. Although dielectric losses may be small at low frequencies, at high frequencies they become quite significant. Table 7.7 compares the energy storage efficiency $E_{vol}$ and tan $\delta$ for various dielectrics. It seems that ceramics have a better energy storage efficiency than polymers. High-$K$ ceramics tend to have large tan $\delta$ values and suffer from greater dielectric loss. Polystyrene and polypropylene have particularly low tan $\delta$ as the polarization mechanism is due to electronic polarization and the dielectric losses are the least. Indeed, polystyrene and polypropylene capacitors have found applications in high-quality audio electronics. Equations 7.52 and 7.53 should be used with care, because the breakdown field $\mathcal{E}_{br}$ can depend on the thickness $d$, among many other factors, including the quality of the dielectric material. For example, for polypropylene insulation, $\mathcal{E}_{br}$ is typically quoted as roughly 50 kV mm$^{-1}$ (500 kV cm$^{-1}$), whereas for thin films (*e.g.*, 25 μm), over short durations, $\mathcal{E}_{br}$ can be as high as 200 kV mm$^{-1}$. Further, in some cases, $\mathcal{E}_{br}$ is more suitably defined in terms of the maximum allowable leakage current, that is, a field at which the dielectric is sufficiently conducting.

The temperature stability of a capacitor is determined by the temperature dependences of $\varepsilon_r'$ and tan $\delta$, which are controlled by the dominant polarization mechanism. For example, polar polymers have permanent dipole groups attached to the polymer chains as in polyethyleneterephthalate (PET). In the absence of an applied field, these dipoles are randomly oriented and also restricted in their rotations by neighboring chains, as depicted in Figure 7.36a. In the presence of an applied dc field, as in Figure 7.36b, some very limited rotation enables partial dipolar (orientational) polarization to take place. Typically, at room temperature, dipolar contribution to $\varepsilon_r$ under ac conditions, however, is small because restricted and hindered rotation prevents the dipoles to closely follow the ac field. Close to the softening temperature of the polymer, the molecular motions become easier and, further, there is more volume between chains for the dipoles to rotate. The dipolar side groups and polarized chains become capable of responding to the field. They can align with the field and also follow the field variations, as shown in Figure 7.36c. Dipolar contribution to $\varepsilon_r$ is substantial even at high frequencies. Both $\varepsilon_r'$ and tan $\delta$ therefore increase with temperature. Thus, polar polymers exhibit temperature dependent $\varepsilon_r$ and tan $\delta$, which reflect in the properties of the capacitor.

On the other hand, in nonpolar polymers such as polystyrene and polypropylene, the polarization is due to electronic polarization and $\varepsilon_r$ and tan $\delta$ remain relatively constant. Thus polystyrene and polypropylene capacitors are more stable compared with PET (polyester) capacitors. The change in the capacitance with temperature is measured by the **temperature coefficient of capacitance** (TCC), which is defined as the fractional (or percentage) change in the capacitance per unit temperature change. The temperature controls not only $\varepsilon_r$ but also the linear expansion of the dielectric,

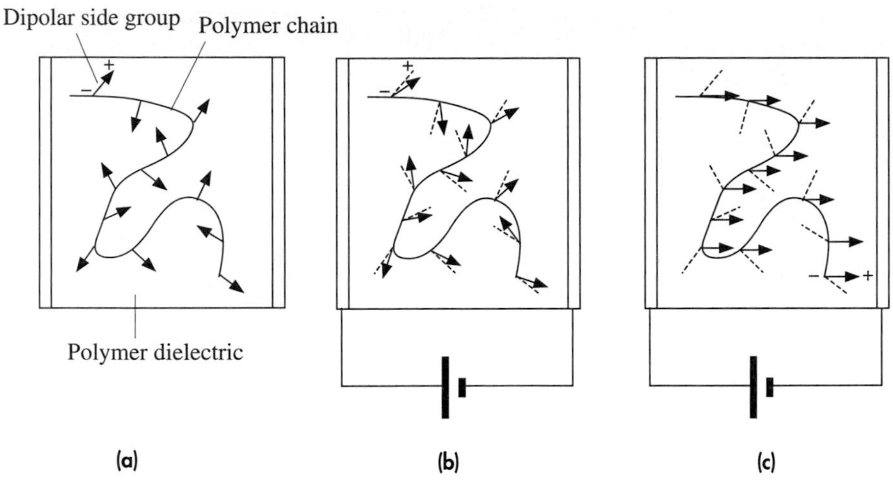


**Figure 7.36**

(a) A polymer dielectric that has dipolar side groups attached to the polymer chains. With no applied field, the dipoles are randomly oriented.

(b) In the presence of an applied field, some very limited rotation enables dipolar polarization to take place.

(c) Near the softening temperature of the polymer, the molecular motions are rapid and there is also sufficient volume between chains for the dipoles to align with the field. The dipolar contribution to $\varepsilon_r$ is substantial, even at high frequencies.

which changes the dimensions $A$ and $d$. For example, polystyrene, polycarbonate, and mica capacitors are particularly stable with small TCC values. Plastic capacitors are typically limited to operations well below their melting temperatures, which is one of their main drawbacks. The specified operating temperature, for example, from −55 °C to 125 °C, for many of the ceramic capacitors is often a limitation of the epoxy coating of the capacitor rather than the actual limitation of the ceramic material. In many capacitors, the working voltage has to be derated for operation at high temperatures and high frequencies because $\mathcal{E}_{br}$ decreases with ambient temperature and the frequency of the applied field. For example, a 1000 V dc polypropylene capacitor will have a substantially lower ac working voltage, *e.g.*, 100 V at 10 kHz.

**EXAMPLE 7.12**

**DIELECTRIC LOSS AND EQUIVALENT CIRCUIT OF A POLYESTER CAPACITOR AT 1 kHz** Figure 7.37 shows the temperature dependence of $\varepsilon_r'$ and $\tan\delta$ for a polyester film. Calculate the equivalent circuit at 25 °C at 1 kHz for a 560 pF PET capacitor that uses a 0.5 micron thick polyester film. What happens to these values at 100 °C?

**SOLUTION**

From Figure 7.37 at 25 °C, $\varepsilon_r' = 2.60$ and $\tan\delta \approx 0.002$. The capacitance $C$ at 25 °C is given as 560 pF. The equivalent parallel conductance $G_P$, representing the dielectric loss, is given by

$$G_P = \frac{\omega A \varepsilon_o \varepsilon_r' \tan\delta}{d} = \omega C \tan\delta$$

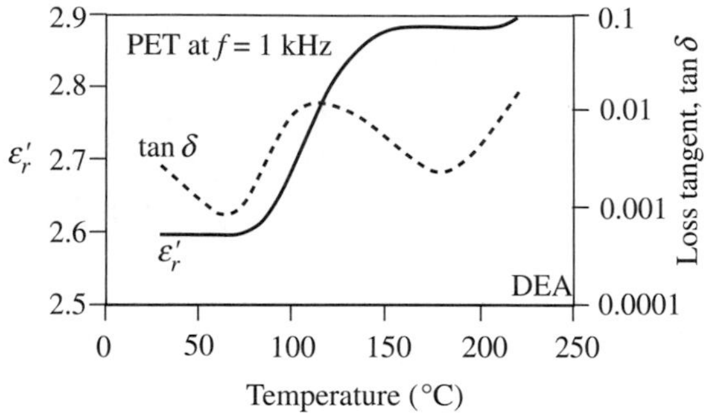


**Figure 7.37** Real part of the dielectric constant $\varepsilon_r'$ and loss tangent, tan $\delta$, at 1 kHz versus temperature for PET.

SOURCE: Data obtained by Kasap and Maeda (1995) using a dielectric analyzer (DEA).

Substituting

$$\omega = 2\pi f = 2000\pi$$

and $\tan\delta = 0.002$, we get

$$G_P = (2000\pi)(560 \times 10^{-12})(0.002) = 7.04 \times 10^{-9}\frac{1}{\Omega}$$

This is equivalent to a resistance of 142 MΩ. The equivalent circuit is an ideal (lossless) capacitor of 560 pF in parallel with a 142 MΩ resistance (this resistance value decreases with the frequency).

At 100 °C, $\varepsilon_r' = 2.69$ and $\tan\delta \approx 0.01$, so the new capacitance is

$$C_{100\,°\text{C}} = C_{25\,°\text{C}}\frac{\varepsilon_r(100\,°\text{C})}{\varepsilon_r(25\,°\text{C})} = (560\text{ pF})\frac{2.69}{2.60} = 579\text{ pF}$$

The equivalent parallel conductance at 100 °C is

$$G_P = (2000\pi)(579 \times 10^{-12})(0.01) = 3.64 \times 10^{-8}\frac{1}{\Omega}$$

This is equivalent to a resistance of 27.5 MΩ. The equivalent circuit is an ideal (lossless) capacitor of 579 pF in parallel with a 27.5 MΩ resistance.

## 7.8 PIEZOELECTRICITY, FERROELECTRICITY, AND PYROELECTRICITY

### 7.8.1 PIEZOELECTRICITY

Certain crystals, for example, quartz (crystalline $SiO_2$) and $BaTiO_3$, become polarized when they are mechanically stressed. Charges appear on the surfaces of the crystal, as depicted in Figure 7.38a and b. Appearance of surface charges leads to a voltage difference between the two surfaces of the crystal. The same crystals also exhibit mechanical strain or distortion when they experience an electric field, as shown in Figure 7.38c and d. The direction of mechanical deformation (*e.g.*, extension or compression) depends on

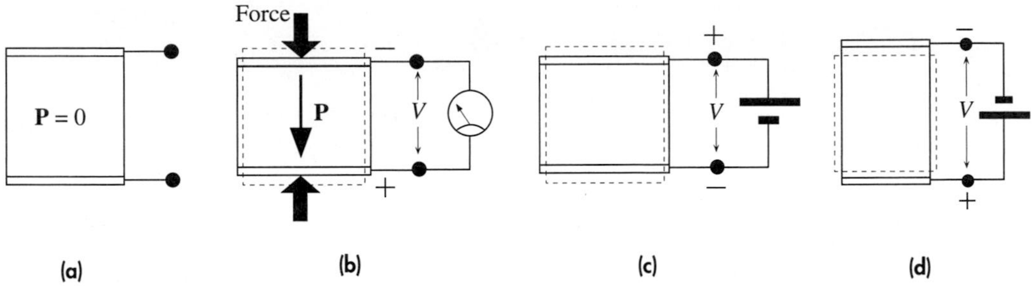


**Figure 7.38** The piezoelectric effect.
(a) A piezoelectric crystal with no applied stress or field.
(b) The crystal is strained by an applied force that induces polarization in the crystal and generates surface charges.
(c) An applied field causes the crystal to become strained. In this case the field compresses the crystal.
(d) The strain changes direction with the applied field and now the crystal is extended.

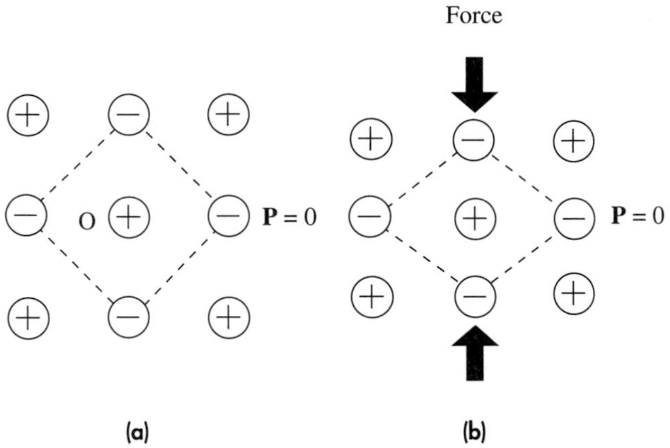


**Figure 7.39** A NaCl-type cubic unit cell has a center of symmetry.
(a) In the absence of an applied force, the centers of mass for positive and negative ions coincide.
(b) This situation does not change when the crystal is strained by an applied force.

the direction of the applied field, or the polarity of the applied voltage. The two effects are complementary and define **piezoelectricity.**

Only certain crystals can exhibit piezoelectricity because the phenomenon requires a special crystal structure—that which has no center of symmetry. Consider a NaCl-type cubic unit cell in Figure 7.39a. We can describe the whole crystal behavior by examining the properties of the unit cell. This unit cell has a **center of symmetry** at O because if we draw a vector from O to any charge and then draw the reverse vector, we will find the same type of charge. Indeed, any point on any charge is a center of symmetry. Many similar cubic crystals (not all) possess a center of symmetry. When unstressed, the center of mass of the negative charges at the corners of the unit cell coincides with the positive charge at the center, as shown in Figure 7.39a. There is therefore no net polarization in the unit cell and $\mathbf{P} = 0$. Under stress, the unit cell becomes strained, as shown in Figure 7.39b, but the center of mass of the negative charges still coincides with the positive charge and the net polarization is still zero. Thus, the strained crystal

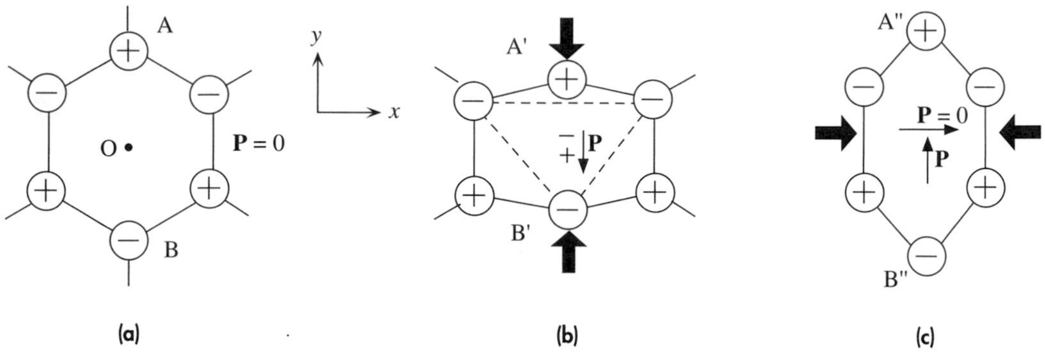


**Figure 7.40** A hexagonal unit cell has no center of symmetry.
(a) In the absence of an applied force, the centers of mass for positive and negative ions coincide.
(b) Under an applied force in the $y$ direction, the centers of mass for positive and negative ions are shifted, which results in a net dipole moment, **P**, along $y$.
(c) When the force is along a different direction, along $x$, there may not be a resulting net dipole moment in that direction though there may be a net **P** along a different direction ($y$).

still has $\mathbf{P} = 0$. This result is generally true for all crystals that have a center of symmetry. The centers of mass of negative and positive charges in the unit cell remain coincident when the crystal is strained.

Piezoelectric crystals have no center of symmetry. For example, the hexagonal unit cell shown in Figure 7.40a exhibits no center of symmetry. If we draw a vector from point O to any charge and then reverse the vector, we will find an opposite charge. The unit cell is said to be **noncentrosymmetric.** When unstressed, as shown in Figure 7.40a, the center of mass of the negative charges coincides with the center of mass of the positive charges, both at O. However, when the unit cell is stressed, as shown in Figure 7.40b, the positive charge at A and the negative charge at B both become displaced inwards to A′ and B′, respectively. The two centers of mass therefore become shifted and there is now a net polarization **P**. Thus, an applied stress produces a net polarization **P** in the unit cell, and in this case **P** appears to be in the same direction as the applied stress, along $y$.

The direction of the induced polarization depends on the direction of the applied stress. When the same unit cell in Figure 7.40a is stressed along $x$, as illustrated in Figure 7.40c, there is no induced dipole moment along this direction because there is no net displacement of the centers of mass in the $x$ direction. However, the stress causes the atoms A and B to be displaced outwards to A″ and B″, respectively, and this results in the shift of the centers of mass away from each other along $y$. In this case, an applied stress along $x$ results in an induced polarization along $y$. Generally, an applied stress in one direction can give rise to induced polarization in other crystal directions. Suppose that $T_j$ is the applied mechanical stress along some $j$ direction and $P_i$ is the induced polarization along some $i$ direction; then the two are linearly related by

*Piezoelectric effect*

$$P_i = d_{ij} T_j \qquad [7.54]$$

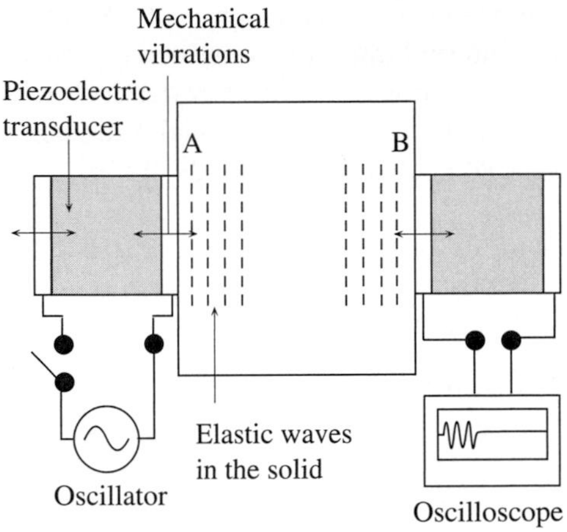


**Figure 7.41** Piezoelectric transducers are widely used to generate ultrasonic waves in solids and also to detect such mechanical waves. The transducer on the left is excited from an ac source and vibrates mechanically. These vibrations are coupled to the solid and generate elastic waves. When the waves reach the other end, they mechanically vibrate the transducer on the right, which converts the vibrations to an electrical signal.

where $d_{ij}$ are called the **piezoelectric coefficients.** Reversing the stress reverses the polarization. Although we did not specifically consider shear stresses in Figure 7.40, they, as well as tensile stresses, can also induce a net polarization, which means that $T$ in Equation 7.54 can also represent shear stresses. The converse piezoelectric effect is that between an induced strain $S_j$ along $j$ and an applied electric field $\mathcal{E}_i$ along $i$,

*Converse piezoelectric effect*

$$S_j = d_{ij}\mathcal{E}_i \quad [7.55]$$

The coefficients $d_{ij}$ in Equations 7.54 and 7.55 are the same.[12]

As apparent from the foregoing discussions and Figure 7.38, piezoelectric crystals are essentially electromechanical transducers because they convert an electrical signal, an electric field, to a mechanical signal, strain, and vice versa. They are used in many engineering applications that involve electromechanical conversions, as in ultrasonic transducers, microphones, accelerometers, and so forth. Piezoelectric transducers are widely used to generate ultrasonic waves in solids and also to detect such mechanical waves, as illustrated in Figure 7.41. The transducer is simply a piezoelectric crystal, for example, quartz, that is appropriately cut and electroded to generate the desired types of mechanical vibrations (*e.g.*, longitudinal or transverse vibrations). The transducer on the left is attached to the surface A of the solid under examination, as shown in Figure 7.41. It is excited from an ac source, which means that it mechanically vibrates. These vibrations are coupled to the solid by a proper coupling medium (typically grease) and generate mechanical waves or elastic waves that propagate away from A. They are called **ultrasonic waves** as their frequencies are typically above the audible range. When the waves reach the other end, B, they mechanically vibrate the transducer attached to B, which converts the vibrations to an electrical signal that can readily be displayed on an oscilloscope. In this trivial example, one can easily measure the time it takes for elastic waves to travel in the solid from A to B and hence determine the ultrasonic velocity of the waves since the distance AB is

[12] The equivalence of the coefficients in Equations 7.54 and 7.55 can be shown by using thermodynamics and is not considered in this textbook. For rigorous piezoelectric definitions see IEEE Standard 176-1987 (*IEEE Trans. on Ultrasonics, Ferroelectrics and Frequency Control*, September 1996).

known. From the ultrasonic velocity one can determine the elastic constants (Young's modulus) of the solid. Furthermore, if there are internal imperfections such as cracks in the solid, then they reflect or scatter the ultrasonic waves. These reflections can lead to echoes that can be detected by suitably located transducers. Such ultrasonic testing methods are widely used for nondestructive evaluations of solids in mechanical engineering.

It is clear that an important engineering factor in the use of piezoelectric transducers is the electromechanical coupling between electrical and mechanical energies. The **electromechanical coupling factor** $k$ is defined in terms of $k^2$ by

*Electromechanical coupling factor*

$$k^2 = \frac{\text{Electrical energy converted to mechanical energy}}{\text{Input of electrical energy}} \quad [7.56a]$$

or equivalently by

*Electromechanical coupling factor*

$$k^2 = \frac{\text{Mechanical energy converted to electrical energy}}{\text{Input of mechanical energy}} \quad [7.56b]$$

Table 7.8 summarizes some typical piezoelectric materials with some applications. The so-called PZT ceramics are widely used in many piezoelectric applications. PZT stands for lead zirconate titanate and the ceramic is a solid solution of lead zirconate, $PbZrO_3$, and lead titanate, $PbTiO_3$, so its composition is $PbTi_{1-x}Zr_xO_3$ where $x$ is determined by the extent of the solid solution but typically is around 0.5. PZT piezoelectric components are manufactured by sintering, which is a characteristic ceramic manufacturing process in which PZT powders are placed in a mold and subjected to a pressure at high temperatures. During sintering the ceramic powders are fused through interdiffusion. The final properties depend not only on the composition of the solid solution but also on the manufacturing process, which controls the average grain size or polycrystallinity. Electrodes are deposited onto the final ceramic component, which is then poled by the application of a temporary electric field to induce it to become

**Table 7.8** Piezoelectric materials and some typical values for $d$ and $k$

| Crystal | $d$ (m V$^{-1}$) | $k$ | Comment |
|---|---|---|---|
| Quartz (crystal $SiO_2$) | $2.3 \times 10^{-12}$ | 0.1 | Crystal oscillators, ultrasonic transducers, delay lines, filters |
| Rochelle salt ($NaKC_4H_4O_6 \cdot 4H_2O$) | $350 \times 10^{-12}$ | 0.78 | |
| Barium titanate ($BaTiO_3$) | $190 \times 10^{-12}$ | 0.49 | Accelerometers |
| PZT, lead zirconate titanate ($PbTi_{1-x}Zr_xO_3$) | $480 \times 10^{-12}$ | 0.72 | Wide range of applications including earphones, microphones, spark generators (gas lighters, car ignition), displacement transducers, accelerometers |
| Polyvinylidene fluoride (PVDF) | $18 \times 10^{-12}$ | — | Must be poled; heated, put in an electric field and then cooled. Large area and inexpensive |

piezoelectric. **Poling** refers to the application of a temporary electric field, generally at an elevated temperature, to align the polarizations of various grains and thereby develop piezoelectric behavior.

**EXAMPLE 7.13**

**PIEZOELECTRIC SPARK GENERATOR** The piezoelectric spark generator, as used in various applications such as lighters and car ignitions, operates by stressing a piezoelectric crystal to generate a high voltage which is discharged through a spark gap in air as schematically shown in Figure 7.42a. Consider a piezoelectric sample in the form of a cylinder as in Figure 7.42a. Suppose that the piezoelectric coefficient $d = 250 \times 10^{-12}$ m V$^{-1}$ and $\varepsilon_r = 1000$. The piezoelectric cylinder has a length of 10 mm and a diameter of 3 mm. The spark gap is in air and has a breakdown voltage of about 3.5 kV. What is the force required to spark the gap? Is this a realistic force?

**SOLUTION**

We need to express the induced voltage in terms of the applied force. If the applied stress is $T$, then the induced polarization $P$ is

$$P = dT = d\frac{F}{A}$$

Induced polarization $P$ leads to induced surface polarization charges given by $Q = AP$. If $C$ is the capacitance, then the induced voltage is

$$V = \frac{Q}{C} = \frac{AP}{\left(\dfrac{\varepsilon_o\varepsilon_r A}{L}\right)} = \frac{LP}{\varepsilon_o\varepsilon_r} = \frac{L\left(d\dfrac{F}{A}\right)}{\varepsilon_o\varepsilon_r} = \frac{dLF}{\varepsilon_o\varepsilon_r A}$$

Therefore, the required force is

$$F = \frac{\varepsilon_o\varepsilon_r AV}{dL} = \frac{(8.85 \times 10^{-12} \times 1000)\pi(1.5 \times 10^{-3})^2(3500)}{(250 \times 10^{-12})(10 \times 10^{-3})} = 87.6 \text{ N}$$

This force can be applied by squeezing by hand an appropriate lever arrangement; it is the weight of 9 kg. The force must be applied quickly because the piezoelectric charge generated

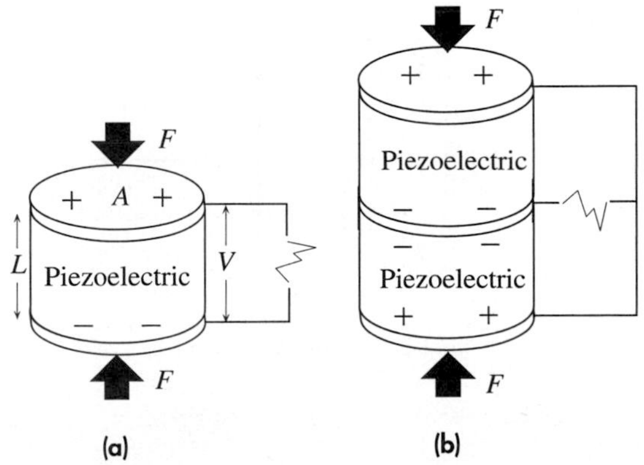


**Figure 7.42** The piezoelectric spark generator.

will leak away (or become neutralized) if the charge is generated too slowly; many spark igniters use mechanical impact. The *energy* in the spark depends on the amount of charge generated. This can increase by using two piezoelectric crystals back to back as in Figure 7.42b, which is a more practical arrangement for a spark generator. The induced voltage per unit force $V/F$ is proportional to $d/(\varepsilon_o\varepsilon_r)$ which is called the **piezoelectric voltage coefficient.** In general, if an applied stress $T = F/A$ induces a field $\mathcal{E} = V/L$ in a piezoelectric crystal, then the effect is related to the cause by the piezoelectric voltage coefficient $g$,

*Piezoelectric voltage coefficient*

$$\mathcal{E} = gT \tag{7.57}$$

It is left as an exercise to show that $g = d/(\varepsilon_o\varepsilon_r)$.

## 7.8.2 PIEZOELECTRICITY: QUARTZ OSCILLATORS AND FILTERS

One of the most important applications of the piezoelectric quartz crystal in electronics is in the frequency control of oscillators and filters. Consider a suitably cut thin plate of a quartz crystal that has thin gold electrodes on the opposite faces. Suppose that we set up mechanical vibrations in the crystal by connecting the electrodes to an ac source, as in Figure 7.43a. It is possible to set up a mechanical resonance, or mechanical standing waves, in the crystal if the wavelength $\lambda$ of the waves and the length $\ell$ along which the waves are traveling satisfy the condition for standing waves:

*Mechanical standing waves*

$$n\left(\frac{1}{2}\lambda\right) = \ell \tag{7.58}$$

where $n$ is an integer.

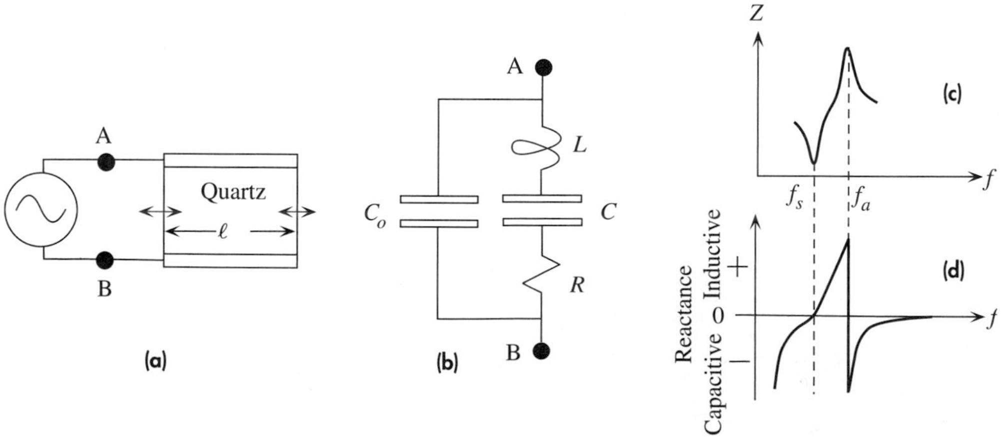


**Figure 7.43** When a suitably cut quartz crystal with electrodes is excited by an ac voltage as in (a), it behaves as if it has the equivalent circuit in (b).
(c) and (d) The magnitude of the impedance $Z$ and reactance (both between A and B) versus frequency, neglecting losses.

The frequency of these mechanical vibrations $f_s$ is given by $f_s = v/\lambda$, where $v$ is the velocity of the waves in the medium and $\lambda$ is the wavelength. These mechanical vibrations in quartz experience very small losses and therefore have a high-quality factor $Q$, which means that resonance can only be set up if the frequency of the excitation, the electrical frequency, is close to $f_s$. Because of the coupling of energy between the electrical excitation and mechanical vibrations through the piezoelectric effect, mechanical vibrations appear like a series $LCR$ circuit to the ac source, as shown in Figure 7.43b. This $LCR$ series circuit has an impedance that is minimum at the **mechanical resonant frequency** $f_s$, given by

*Mechanical resonant frequency*

$$f_s = \frac{1}{2\pi\sqrt{LC}} \quad [7.59]$$

In this series $LCR$ circuit, $L$ represents the mass of the transducer, $C$ the stiffness, and $R$ the losses or mechanical damping. Since the quartz crystal has electrodes at opposite faces, there is, in addition, the parallel plate capacitance $C_o$ between the electrodes. Thus, the whole equivalent circuit is $C_o$ in *parallel* with $LCR$, as in Figure 7.43b. As far as $L$ is concerned, $C_o$ and $C$ are in series. There is a second higher resonant frequency $f_a$, called the **antiresonant frequency,** that is due to $L$ resonating with $C$ and $C_o$ in series,

*Antiresonant frequency*

$$f_a = \frac{1}{2\pi\sqrt{LC'}} \quad [7.60]$$

where

$$\frac{1}{C'} = \frac{1}{C_o} + \frac{1}{C}$$

The impedance between the terminals of the quartz crystal has the frequency dependence shown in Figure 7.43c. The two frequencies $f_s$ and $f_a$ are called the series and

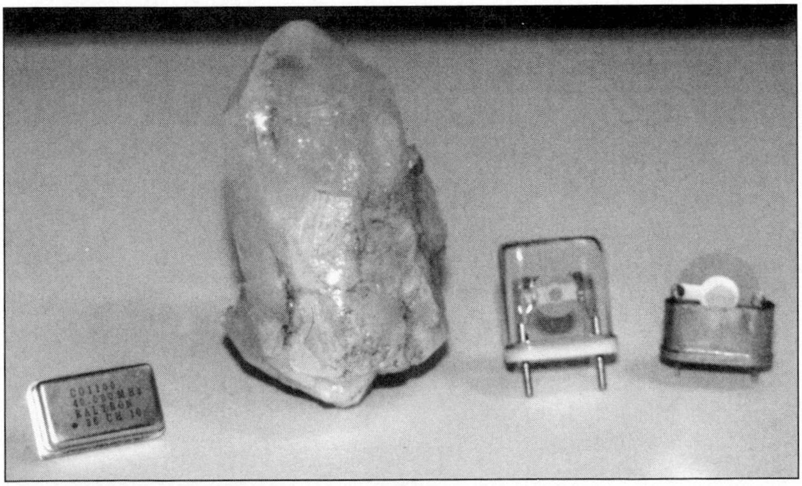

Various quartz crystal "oscillators." Left to right: Raltron 40 MHz; a natural quartz crystal (South Dakota); Phillips 27 MHz; a cutaway view of a typical crystal oscillator.

parallel resonant frequencies, respectively. It is apparent that around $f_a$, the crystal behaves like a filter with a high $Q$ value. If we were to examine the reactance of the crystal, whether it is behaving capacitively or inductively, we would find the behavior in Figure 7.43d, where positive reactance refers to an inductive and negative reactance to a capacitive behavior. Between $f_s$ and $f_a$ the crystal behaves inductively, and capacitively outside this range. Indeed, between $f_s$ and $f_a$ the response of the transducer is controlled by the mass of the crystal. This property has been utilized by electrical engineers in designing quartz oscillators.

In quartz oscillators, the crystal is invariably used in one of two modes. First, it can be used at $f_s$ where it behaves as a resistance of $R$ without any reactance. The circuit is designed so that oscillations can take place only when the crystal in the circuit exhibits no reactance or phase change—in other words, at $f_s$. Outside this frequency, the crystal introduces reactance or phase changes that do not lead to sustained oscillations. In a different mode of operation, the oscillator circuit is designed to make use of the **inductance** of the crystal just above $f_s$. Oscillations are maintained close to $f_s$ because even very large changes in the inductance result in small changes in the frequency between $f_s$ and $f_a$.

**EXAMPLE 7.14** **THE QUARTZ CRYSTAL AND ITS EQUIVALENT CIRCUIT** From the following equivalent definition of the coupling coefficient,

$$k^2 = \frac{\text{Mechanical energy stored}}{\text{Total energy stored}}$$

show that

$$k^2 = 1 - \frac{f_s^2}{f_a^2}$$

Given that typically for an X-cut quartz crystal, $k = 0.1$, what is $f_a$ for $f_s = 1$ MHz? What is your conclusion?

**SOLUTION**

$C$ represents the mechanical mass where the mechanical energy is stored, whereas $C_o$ is where the electrical energy is stored. If $V$ is the applied voltage, then

$$k^2 = \frac{\text{Mechanical energy stored}}{\text{Total energy stored}} = \frac{\frac{1}{2}CV^2}{\frac{1}{2}CV^2 + \frac{1}{2}C_oV^2} = \frac{C}{C + C_o} = 1 - \frac{f_s^2}{f_a^2}$$

Rearranging this equation, we find

$$f_a = \frac{f_s}{\sqrt{1-k^2}} = \frac{1\text{ MHz}}{\sqrt{1-(0.1)^2}} = 1.005\text{ MHz}$$

Thus, $f_a - f_s$ is only 5 kHz. The two frequencies $f_s$ and $f_a$ in Figure 7.43d are very close. An oscillator designed to oscillate at $f_s$, that is, at 1 MHz, therefore, cannot drift far (for example, a few kHz) because that would change the reactance enormously, which would upset the oscillation conditions.

**EXAMPLE 7.15**

**QUARTZ CRYSTAL AND ITS INDUCTANCE** A typical 1 MHz quartz crystal has the following properties:

$$f_s = 1 \text{ MHz} \qquad f_a = 1.0025 \text{ MHz} \qquad C_o = 5 \text{ pF} \qquad R = 20 \ \Omega$$

What are $C$ and $L$ in the equivalent circuit of the crystal? What is the quality factor $Q$ of the crystal, given that

$$Q = \frac{1}{2\pi f_s RC}$$

**SOLUTION**

The expression for $f_s$ is

$$f_s = \frac{1}{2\pi \sqrt{LC}}$$

From the expression for $f_a$, we have

$$f_a = \frac{1}{2\pi \sqrt{LC'}} = \frac{1}{2\pi \sqrt{L \dfrac{CC_o}{C + C_o}}}$$

Dividing $f_a$ by $f_s$ eliminates $L$, and we get

$$\frac{f_a}{f_s} = \sqrt{\frac{C + C_o}{C_o}}$$

so that $C$ is

$$C = C_o \left[ \left( \frac{f_a}{f_s} \right)^2 - 1 \right] = (5 \text{ pF})(1.0025^2 - 1) = 0.025 \text{ pF}$$

Thus

$$L = \frac{1}{C(2\pi f_s)^2} = \frac{1}{0.025 \times 10^{-12}(2\pi 10^6)^2} = 1.01 \text{ H}$$

This is a substantial inductance, and the enormous increase in the inductive reactance above $f_s$ is intuitively apparent. The quality factor

$$Q = \frac{1}{2\pi f_s RC} = 3.18 \times 10^5$$

is very large.

### 7.8.3 FERROELECTRIC AND PYROELECTRIC CRYSTALS

Certain crystals are permanently polarized even in the absence of an applied field. The crystal already possesses a finite polarization vector due to the separation of positive and negative charges in the crystal. These crystals are called **ferroelectric.**[13] Barium

[13] In analogy with the ferromagnetic crystals that already possess magnetization.

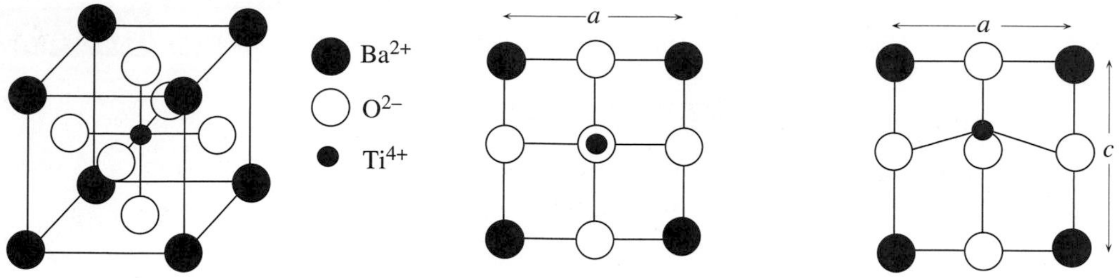


(a) $BaTiO_3$ cubic crystal structure above 130 °C

(b) $BaTiO_3$ cubic structure above 130 °C

(c) $BaTiO_3$ tetragonal structure below 130 °C

**Figure 7.44** $BaTiO_3$ has different crystal structures above and below 130 °C that lead to different dielectric properties.

titanate ($BaTiO_3$) is probably the best cited example. Above approximately 130 °C, the crystal structure of $BaTiO_3$ has a cubic unit cell, as shown in Figure 7.44a. The centers of mass of the negative charges ($O^{2-}$) and the positive charges, $Ba^{2+}$ and $Ti^{4+}$, coincide at the $Ti^{4+}$ ion, as shown in Figure 7.44b. There is therefore no net polarization and $\mathbf{P} = 0$. Above 130 °C, therefore, the barium titanate crystal exhibits no permanent polarization and is not ferroelectric. However, below 130 °C, the structure of barium titanate is tetragonal, as shown in Figure 7.44c, in which the $Ti^{4+}$ atom is not located at the center of mass of the negative charges. The crystal is therefore polarized by the separation of the centers of mass of the negative and positive charges. The crystal possesses a finite polarization vector **P** and is ferroelectric. The critical temperature above which ferroelectric property is lost, in this case 130 °C, is called the **Curie temperature** ($T_C$). Below the Curie temperature, the whole crystal becomes spontaneously polarized. The onset of spontaneous polarization is accompanied by the distortion of the crystal structure, as shown by the change from Figure 7.44b to Figure 7.44c. The spontaneous displacement of the $Ti^{4+}$ ion below the Curie temperature elongates the cubic structure, which becomes tetragonal. It is important to emphasize that we have only described an observation and not the reasons for the spontaneous polarization of the whole crystal. The development of the permanent dipole moment below the Curie temperature involves long-range interactions between the ions outside the simple unit cell pictured in Figure 7.44. The energy of the crystal is lower when the $Ti^{4+}$ ion in each unit cell is slightly displaced along the $c$ direction, as in Figure 7.44c, which generates a dipole moment in each unit cell. The interaction energy of these dipoles when all are aligned in the same direction lowers the energy of the whole crystal. It should be mentioned that the distortion of the crystal that takes place when spontaneous polarization occurs just below $T_C$ is very small relative to the dimensions of the unit cell. For $BaTiO_3$, for example, $c/a$ is 1.01 and the displacement of the $Ti^{4+}$ ion from the center is only 0.012 nm, compared with $a = 0.4$ nm.

An important and technologically useful characteristic of a ferroelectric crystal is its ability to be poled. Above 130 °C there is no permanent polarization in the crystal. If we apply a temporary field $\mathcal{E}$ and let the crystal cool to below 130 °C, we can induce the spontaneous polarization **P** to develop along the field direction. In other words, we would define the $c$ axis by imposing a temporary external field. This process is called poling. The $c$ axis is the polar axis along which **P** develops. It is also called the

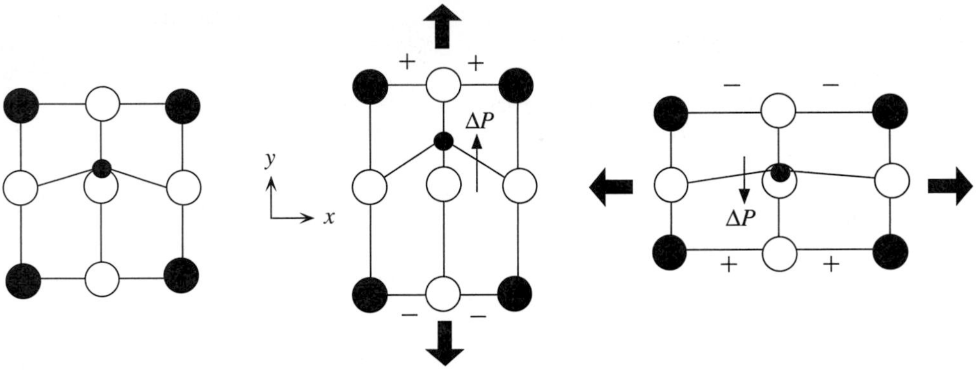


(a) $BaTiO_3$ tetragonal structure below 130 °C

(b) $BaTiO_3$ crystal under stress along *y*

(c) $BaTiO_3$ crystal under stress along *x*

**Figure 7.45** Piezoelectric properties of $BaTiO_3$ below its Curie temperature.

**ferroelectric axis.** Since below the Curie temperature the ferroelectric crystal already has a permanent polarization, it is not possible to use the expression

$$P = \varepsilon_o(\varepsilon_r - 1)\mathcal{E}$$

to define a relative permittivity. Suppose that we use a ferroelectric crystal as a dielectric medium between two parallel plates. Since any change $\Delta P$ normal to the plates changes the stored charge, what is of significance to the observer is the change in the polarization. We can appreciate this by noting that $C = Q/V$ is not a good definition of capacitance if there are already charges on the plates, even in the absence of voltage.[14] We then prefer a definition of $C$ based on $\Delta Q/\Delta V$ where $\Delta Q$ is the change in stored charge due to a change $\Delta V$ in the voltage. Similarly, we define the relative permittivity $\varepsilon_r$ in this case in terms of the change $\Delta P$ in $P$ induced by $\Delta\mathcal{E}$ in the field $\mathcal{E}$,

$$\Delta P = \varepsilon_o(\varepsilon_r - 1)\,\Delta\mathcal{E}$$

An applied field along the *a* axis can displace the $Ti^{4+}$ ion more easily than that along the *c* axis, and experiments show that $\varepsilon_r \approx 4100$ along *a* is much greater than $\varepsilon_r \approx 160$ along *c*. Because of their large dielectric constants, ferroelectric ceramics are used as high-*K* dielectrics in capacitors.

All ferroelectric crystals are also piezoelectric, but the reverse is not true: not all piezoelectric crystals are ferroelectric. When a stress along *y* is applied to the $BaTiO_3$ crystal in Figure 7.45a, the crystal is stretched along *y*, as a result of which the $Ti^{4+}$ atom becomes displaced, as shown in Figure 7.45b. There is, however, no shift in the center of mass of the negative charges, which means that there is a change $\Delta P$ in the polarization vector along *y*. Thus, the applied stress induces a change in the polarization, which is a piezoelectric effect. If the stress is along *x*, as illustrated in Figure 7.45c, then the change in the polarization is along *y*. In both cases, $\Delta P$ is proportional to the stress, which is a characteristic of the piezoelectric effect.

[14] A finite $Q$ on the plates of a capacitor when $V = 0$ implies an infinite capacitance, $C = \infty$. However, $C = dQ/dV$ definition avoids this infinity.

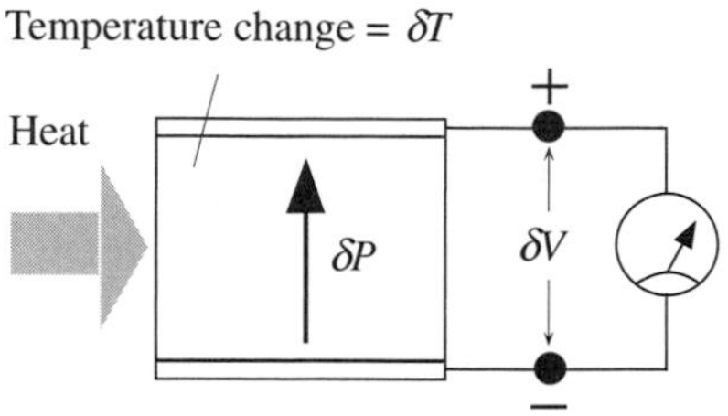


**Figure 7.46** The heat absorbed by the crystal increases the temperature by $\delta T$, which induces a change $\delta P$ in the polarization.
This is the pyroelectric effect. The change $\delta P$ gives rise to a change $\delta V$ in the voltage that can be measured.

The barium titanate crystal in Figure 7.44 is also said to be pyroelectric because when the temperature increases, the crystal expands and the relative distances of ions change. The $Ti^{4+}$ ion becomes shifted, which results in a change in the polarization. Thus, a temperature change $\delta T$ induces a change $\delta P$ in the polarization of the crystal. This is called **pyroelectricity,** which is illustrated in Figure 7.46. The magnitude of this effect is quantized by the **pyroelectric coefficient** $p$, which is defined by

*Pyroelectric coefficient*

$$p = \frac{dP}{dT} \qquad [7.61]$$

A few typical pyroelectric crystals and their pyroelectric coefficients are listed in Table 7.9. Very small temperature changes, even in thousandths of degrees, in the material can develop voltages that can be readily measured. For example, for a PZT-type pyroelectric ceramic in Table 7.9, taking $\delta T = 10^{-3}$ K and $p \approx 380 \times 10^{-6}$, we find $\delta P = 3.8 \times 10^{-7}$ C m$^{-2}$. From

$$\delta P = \varepsilon_o(\varepsilon_r - 1)\,\delta \mathcal{E}$$

with $\varepsilon_r = 290$, we find

$$\delta \mathcal{E} = 148 \text{ V m}^{-1}$$

If the distance between the faces of the ceramic where the charges are developed is 0.1 mm, then

$$\delta V = 0.0148 \text{ V} \qquad \text{or} \qquad 15 \text{ mV}$$

**Table 7.9** Some pyroelectric (and also ferroelectric) crystals and typical properties

| Material | $\varepsilon_r'$ | tan $\delta$ | Pyroelectric Coefficient ($\times 10^{-6}$ C m$^{-2}$ K$^{-1}$) | Curie Temperature (°C) |
|---|---|---|---|---|
| $BaTiO_3$ | 4100 ⊥ polar axis; 160 //polar axis | $7 \times 10^{-3}$ | 20 | 130 |
| $LiTaO_3$ | 47 | $5 \times 10^{-3}$ | 230 | 610 |
| PZT modified for pyroelectric | 290 | $2.7 \times 10^{-3}$ | 380 | 230 |
| PVDF, polymer | 12 | 0.01 | 27 | 80 |

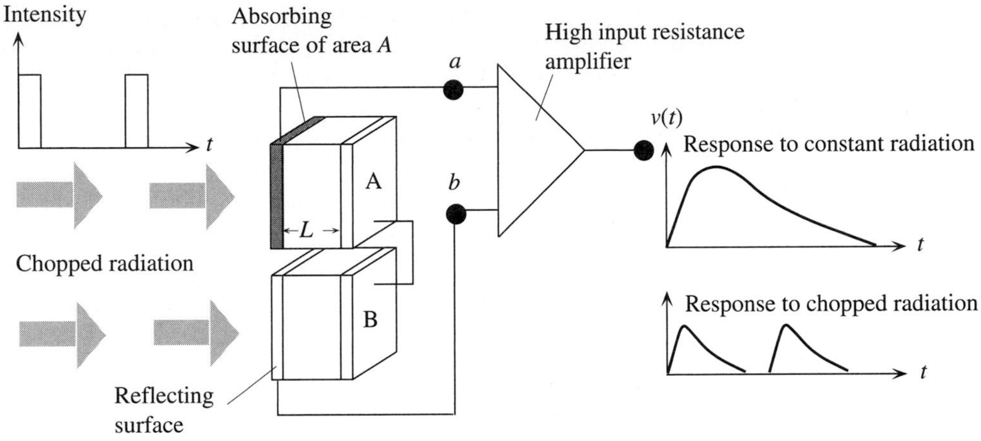


**Figure 7.47** The pyroelectric detector.
Radiation is absorbed in the detecting element, A, which generates a pyroelectric voltage that is measured by the amplifier. The second element, B, has a reflecting electrode and does not absorb the radiation. It is a dummy element that compensates for the piezoelectric effects. Piezoelectric effects generate equal voltages in both A and B, which cancel each other across *a* and *b*, the input of the amplifier.

which can be readily measured. Pyroelectric crystals are widely used as infrared detectors. Any infrared radiation that can raise the temperature of the crystal even by a thousandth of a degree can be detected. For example, many intruder alarms use pyroelectric detectors because as the human or animal intruder passes by the view of detector, the infrared radiation from the warm body raises the temperature of the pyroelectric detector, which generates a voltage that actuates an alarm.

Figure 7.47 shows a simplified schematic circuit for a pyroelectric radiation detector. The detecting element, labeled A, is actually a thin crystal or ceramic (or even a polymer) of a pyroelectric material that has electrodes on opposite faces. Pyroelectric materials are also piezoelectric and therefore also sensitive to stresses. Thus, pressure fluctuations, for example, vibrations from the detector mount or sound waves, interfere with the response of the detector to radiation alone. These can be compensated for by having a second dummy detector B that has a reflecting coating and is subjected to the same vibrations (air and mount), as depicted in Figure 7.47. Thus, there are two elements in the detector, one with an absorbing surface, detecting element A, and the other with a reflecting surface, compensating element B. Stress fluctuations give rise to the same piezoelectric voltage in both, which then cancel each other between *a* and *b* at the input of the amplifier. When radiation is incident, then only the detecting element absorbs the radiation, becomes warmer, and hence generates a pyroelectric voltage. This voltage appears directly across *a* and *b*. As the incident radiation warms the detecting element and increases its temperature, the pyroelectric voltage increases with time. Eventually the temperature reaches a steady-state value determined by heat losses from the element. We therefore expect the pyroelectric voltage to reach a constant value as well. However, the problem is that a constant pyroelectric voltage cannot be sustained because the surface charges slowly become neutralized or leak away.

The constant radiation is therefore normally chopped to subject the detector to periodic bursts of radiation, as shown in Figure 7.47. The pyroelectric voltage is then a changing function of time, which is readily measured and related to the power in the incident radiation.

Many pyroelectric applications refer to a pyroelectric current that is generated by the temperature rise. There is another way to look at the pyroelectric phenomenon instead of considering the induced pyroelectric voltage that is created across the crystal (Figure 7.46). The induced polarization $\delta P$ in a small time interval $\delta t$, due to the change $\delta T$ in the temperature, generates an induced polarization charge density $\delta P$ on the crystal's surfaces. This charge density $\delta P$ flows in a time interval $\delta t$, and hence generates an *induced polarization current density* $J_p$ to flow, *i.e.*,

*Pyroelectric current density*

$$J_p = \frac{dP}{dt} = p\frac{dT}{dt} \tag{7.62}$$

$J_p$ in Equation 7.62 is called the **pyroelectric current density** and depends on the rate of change of the temperature $dT/dt$ brought about by the absorption of radiation.

Most pyroelectric detectors are characterized by their **current responsivity** $\mathcal{R}_I$ defined as the pyroelectric current generated per unit input radiation power,

*Pyroelectric current responsivity*

$$\mathcal{R}_I = \frac{\text{Pyroelectric current generated}}{\text{Input radiation power}} = \frac{J_p}{\mathcal{I}} \tag{7.63}$$

where $\mathcal{I}$ is the radiation intensity (W m$^{-2}$); $\mathcal{R}_I$ is quoted in A W$^{-1}$. If the pyroelectric current generated by the crystal flows into the self-capacitance of the crystal itself (no external resistors or capacitors connected, and the voltmeter is an ideal meter), it charges the self-capacitance to generate the observed voltage $\delta V$ in Figure 7.46. The **pyroelectric voltage responsivity** $\mathcal{R}_V$ is defined similarly to Equation 7.63 but considers the voltage that is developed upon receiving the input radiation:

*Pyroelectric voltage responsivity*

$$\mathcal{R}_V = \frac{\text{Pyroelectric output voltage generated}}{\text{Input radiation power}} \tag{7.64}$$

The output voltage that is generated depends not only on the pyroelectric crystal's dielectric properties, but also on the input impedance of the amplifier, and can be quite complicated. A typical commercial $LiTaO_3$ pyroelectric detector has a current responsivity of 0.1–1 μA/W.

**EXAMPLE 7.16** **A PYROELECTRIC RADIATION DETECTOR** Consider the radiation detector in Figure 7.47 but with a single element A. Suppose that the radiation is chopped so that the radiation is passed to the detector for a time $\Delta t$ seconds every $\tau$ seconds, where $\Delta t \ll \tau$. If $\Delta t$ is sufficiently small, then the temperature rise $\Delta T$ is small and hence the heat losses are negligible during $\Delta t$. Using the heat capacity to find the temperature change during $\Delta t$, relate the magnitude of the voltage $\Delta V$ to the incident radiation intensity $\mathcal{I}$. What is your conclusion?

Consider a PZT-type pyroelectric material with a density of about 7 g cm$^{-3}$ and a specific heat capacity of about 380 J K$^{-1}$ kg$^{-1}$. If $\Delta t = 0.2$ s and the minimum voltage that can be detected above the background noise is 1 mV, what is the minimum radiation intensity that can be measured?

**SOLUTION**

Suppose that the radiation of intensity $I$ is received during a time interval $\Delta t$ and delivers an amount of energy $\Delta H$ to the pyroelectric detector. This energy $\Delta H$, in the absence of any heat losses, increases the temperature by $\Delta T$. If $c$ is the specific heat capacity (heat capacity per unit mass) and $\rho$ is the density,

$$\Delta H = (AL\rho)c\,\Delta T$$

where $A$ is the surface area and $L$ the thickness of the detector. The change in the polarization $\Delta P$ is

$$\Delta P = p\,\Delta T = \frac{p\,\Delta H}{AL\rho c}$$

The change in the surface charge $\Delta Q$ is

$$\Delta Q = A\,\Delta P = \frac{p\,\Delta H}{L\rho c}$$

This change in the surface charge gives a voltage change $\Delta V$ across the electrodes of the detector. If $C = \varepsilon_o\varepsilon_r A/L$ is the capacitance of the pyroelectric crystal,

$$\Delta V = \frac{\Delta Q}{C} = \frac{p\,\Delta H}{L\rho c} \times \frac{L}{\varepsilon_o\varepsilon_r A} = \frac{p\,\Delta H}{A\rho c\varepsilon_o\varepsilon_r}$$

The absorbed energy (heat) $\Delta H$ during $\Delta t$ depends on the intensity of incident radiation. Incident intensity $I$ is the energy arriving per unit area per unit time. In time $\Delta t$, $I$ delivers an energy $\Delta H = IA\,\Delta t$. Substituting for $\Delta H$ in the expression for $\Delta V$, we find

*Pyroelectric detector output voltage*

$$\Delta V = \frac{pI\,\Delta t}{\rho c\varepsilon_r\varepsilon_o} = \left(\frac{p}{\rho c\varepsilon_r\varepsilon_o}\right) I\,\Delta t \qquad [7.65]$$

The parameters in the parentheses are material properties and reflect the "goodness" of the pyroelectric material for the application. We should emphasize that in deriving Equation 7.65 we did not consider any heat losses that will prevent the rise of the temperature indefinitely. If $\Delta t$ is short, then the temperature change will be small and heat losses negligible.

For a PZT-type pyroelectric, we can take $p = 380 \times 10^{-6}$ C m$^{-2}$ K$^{-1}$, $\varepsilon_r = 290$, $c =$ 380 J K$^{-1}$ kg$^{-1}$, and $\rho = 7 \times 10^3$ kg m$^{-3}$, and then from Equation 7.65 with $\Delta V = 0.001$ V and $\Delta t = 0.2$ s, we have

$$I = \left(\frac{p}{\rho c\varepsilon_o\varepsilon_r}\right)^{-1} \frac{\Delta V}{\Delta t} = \left(\frac{380 \times 10^{-6}}{(7000)(380)(290)(8.85 \times 10^{-12})}\right)^{-1} \frac{0.001}{0.2}$$

$$= 0.090 \text{ W m}^{-2} \quad \text{or} \quad 9\,\mu\text{W cm}^{-2}$$

We have assumed that all the incident radiation $I$ is absorbed by the pyroelectric crystal. In practice, only a fraction $\eta$ (called the *emissivity* of the surface), that is, $\eta I$, will be absorbed instead of $I$. We also assumed that the output voltage $\Delta V$ is developed totally across the pyroelectric element capacitance; that is, the amplifier's input impedance (parallel combination of its input capacitance and resistance) is negligible compared with that of the pyroelectric crystal. As stated, we also neglected all heat losses from the pyroelectric crystal so that the absorbed radiation simply increases the crystal's temperature. These simplifying assumptions lead to the maximum signal $\Delta V$ that can be generated from a given input radiation signal $I$ as stated in Equation 7.65. It is left as an exercise to show that Equation 7.65 can also be easily derived by starting from Equation 7.62 for the pyroelectric current density $J_p$, and have $J_p$ charge up the capacitance $C = \varepsilon_o\varepsilon_r A/L$ of the crystal.

# ADDITIONAL TOPICS

## 7.9 ELECTRIC DISPLACEMENT AND DEPOLARIZATION FIELD

***Electric Displacement (D) and Free Charges*** Consider a parallel plate capacitor with free space between the plates, as shown in Figure 7.48a, which has been charged to a voltage $V_o$ by connecting it to a battery of voltage $V_o$. The battery has been suddenly removed, which has left the free positive and negative charges $Q_{\text{free}}$ on the plates. These charges are free in the sense that they can be conducted away. An ideal electrometer (with no leakage current) measures the total charge on the positive plate (or voltage of the positive plate with respect to the negative plate). The voltage across the plates is $V_o$ and the capacitance is $C_o$. The field in the free space between the plates is

*Electric field without dielectric*

$$\mathcal{E}_o = \frac{Q_{\text{free}}}{\varepsilon_o A} = \frac{V_o}{d} \quad [7.66]$$

where $d$ is the separation of the plates.

When we insert a dielectric to fit between the plates, the field polarizes the dielectric and polarization charges $-Q_P$ and $+Q_P$ appear on the left and right surfaces of the dielectric, as shown in Figure 7.48b. As there is no battery to supply more free charges, the net charge on the left plate (positive plate) becomes $Q_{\text{free}} - Q_P$. Similarly the net negative charge on the right plate becomes $-Q_{\text{free}} + Q_P$. The field inside the dielectric is no longer $\mathcal{E}_o$ but less because induced polarization charges have the opposite polarity to the original free charges and the net charge on each plate has been reduced. The new field can be found by applying Gauss's law. Consider a Gauss surface just enclosing the left plate and the surface region of the dielectric with its negative polarization charges, as shown in Figure 7.49. Then Gauss's law gives

*Gauss's law with dielectric*

$$\oint_{\text{Surface}} \varepsilon_o \mathcal{E}\, dA = Q_{\text{total}} = Q_{\text{free}} - Q_P \quad [7.67]$$

where $A$ is the plate area (same as dielectric surface area) and we take the field $\mathcal{E}$ to be normal to the surface area $dA$, as indicated in Figure 7.49. If the polarization charge is

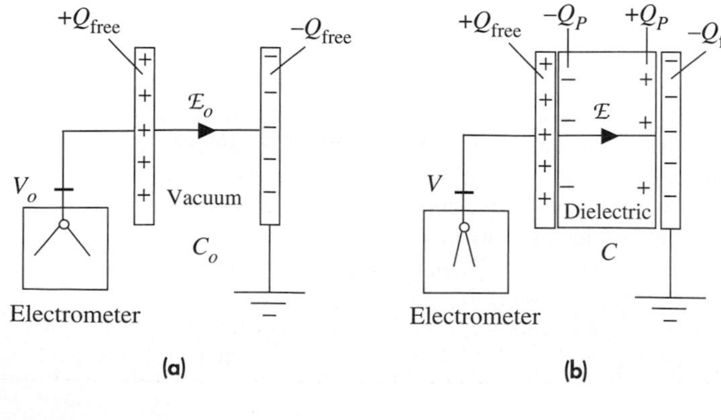


**Figure 7.48**
(a) Parallel plate capacitor with free space between plates that has been charged to a voltage $V_o$. There is no battery to maintain the voltage constant across the capacitor. The electrometer measures the voltage difference across the plates and, in principle, does not affect the measurement.
(b) After the insertion of the dielectric, the voltage difference is $V$, less than $V_o$, and the field in the dielectric is $\mathcal{E}$ less than $\mathcal{E}_o$.

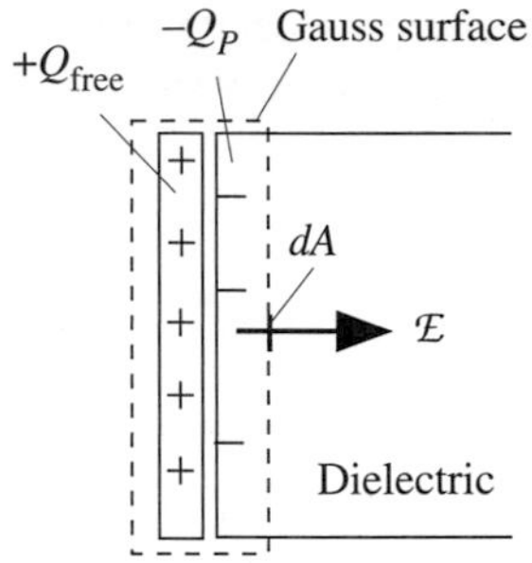


**Figure 7.49** A Gauss surface just around the left plate and within the dielectric, encompassing both $+Q_{free}$ and $-Q_P$.

$dQ_P$ over a small surface area $dA$ of the dielectric, then the polarization charge density $\sigma_P$ at this point is defined as

$$\sigma_P = \frac{dQ_P}{dA}$$

For uniform polarization, the charge distribution is $Q_P/A$, as we have used previously. Since $\sigma_P = P$, where $P$ is the polarization vector, we can write

$$P = \frac{dQ_P}{dA}$$

and therefore express $Q_P$ as

$$Q_P = \oint_{\text{Surface}} P\, dA \qquad \textbf{[7.68]}$$

We can now substitute for $Q_P$ in Equation 7.67 and take this term to the left-hand side to add the two surface integrals. The right-hand side is left with only $Q_{free}$. Thus,

$$\oint_{\text{Surface}} (\varepsilon_o \mathcal{E} + P)\, dA = Q_{free} \qquad \textbf{[7.69]}$$

What is important here is that the surface integration of the quantity $\varepsilon_o \mathcal{E} + P$ is always equal to the total free charges on the surface. Whatever the dielectric material, this integral is always $Q_{free}$. It becomes convenient to define $\varepsilon_o \mathcal{E} + P$ as a usable quantity, called the **electric displacement** and denoted as $D$, that is,

*Definition of electric displacement*

$$D = \varepsilon_o \mathcal{E} + P \qquad \textbf{[7.70]}$$

Then, Gauss's law in terms of free charges alone in Equation 7.69 becomes

*Gauss's law for free charges*

$$\oint_{\text{Surface}} D\, dA = Q_{free} \qquad \textbf{[7.71]}$$

In Equation 7.71 we take $D$ to be normal to the surface area $dA$ as in the case of $\mathcal{E}$ in Gauss's law. Equation 7.71 provides a convenient way to calculate the electric displacement $D$, from which one should be able to determine the field. We should note that, in general, $\mathcal{E}$ is a vector and so is $P$, so the definition in Equation 7.70 is strictly in terms of vectors. Inasmuch as the electric displacement depends only on free charges, as a vector it starts at negative free charges and finishes on positive free charges.

Equation 7.71 for $D$ defines it in terms of $\mathcal{E}$ and $P$, but we can express $D$ in terms of the field $\mathcal{E}$ in the dielectric alone. The polarization $P$ and $\mathcal{E}$ are related by the definition of the relative permittivity $\varepsilon_r$,

$$P = \varepsilon_o(\varepsilon_r - 1)\mathcal{E}$$

Substituting for $P$ in Equation 7.70 and rearranging, we find that $D$ is simply given by

*Electric displacement and the field*

$$D = \varepsilon_o\varepsilon_r\mathcal{E} \qquad [7.72]$$

We should note that this simple equation applies in an isotropic medium where the field along one direction, for example, $x$, does not generate polarization along a different direction, for example, $y$. In those cases, Equation 7.72 takes a tensor form whose mathematics is beyond the scope of this book.

We can now apply Equation 7.71 for a Gauss surface surrounding the left plate,

$$D = \frac{Q_{\text{free}}}{A} = \varepsilon_o\mathcal{E}_o \qquad [7.73]$$

where we used Equation 7.66 to replace $Q_{\text{free}}$. Thus $D$ does not change when we insert the dielectric because the same free charges are still on the plates (they cannot be conducted away anywhere). The new field $\mathcal{E}$ between the plates after the insertion of the dielectric is

$$\mathcal{E} = \frac{1}{\varepsilon_o\varepsilon_r}D = \frac{1}{\varepsilon_r}\mathcal{E}_o \qquad [7.74]$$

The original field is reduced by the polarization of the dielectric. We should recall that the field does *not* change in the case where the parallel plate capacitor is connected to a battery that keeps the voltage constant across the plates and supplies additional free charges ($\Delta Q_{\text{free}}$) to make up for the induced opposite-polarity polarization charges.

Gauss's law in Equation 7.71 in terms of $D$ and the enclosed free charges $Q_{\text{free}}$ can also be written in terms of the field $\mathcal{E}$, but including the relative permittivity, because $D$ and $\mathcal{E}$ are related by Equation 7.72. Using Equation 7.72, Equation 7.71 becomes

*Gauss's law for free charges*

$$\oint_{\text{Surface}} \varepsilon_o\varepsilon_r\mathcal{E}\, dA = Q_{\text{free}}$$

For an isotropic medium where $\varepsilon_r$ is the same everywhere,

*Gauss's law in an isotropic dielectric*

$$\oint_{\text{Surface}} \mathcal{E}\, dA = \frac{Q_{\text{free}}}{\varepsilon_o\varepsilon_r} \qquad [7.75]$$

As before, $\mathcal{E}$ in the surface integral is taken as normal to $dA$ everywhere. Equation 7.75 is a convenient way of evaluating the field from the free charges alone, given the dielectric constant of the medium.

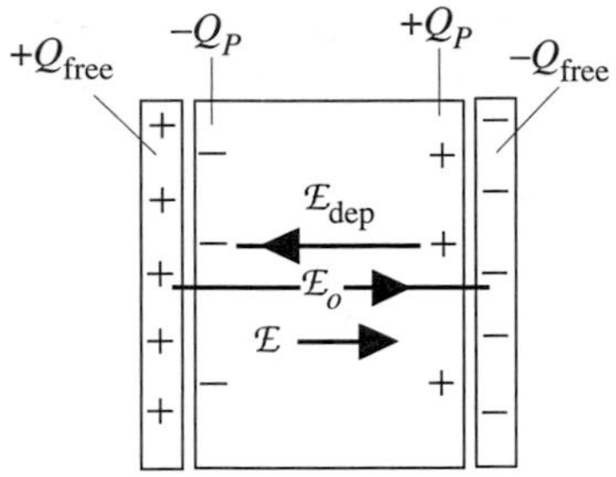


**Figure 7.50** The field inside the dielectric can be considered to be the sum of the field due to the free charges ($Q_{\text{free}}$) and a field due to the polarization of the dielectric, called the depolarization field.

***The Depolarizing Field*** We can view the field $\mathcal{E}$ as arising from two electric fields: that due to the free charges $\mathcal{E}_o$ and that due to the polarization charges, denoted as $\mathcal{E}_{\text{dep}}$. These two fields are indicated in Figure 7.50. $\mathcal{E}_o$ is called the **applied field** as it is due to the free charges that have been put on the plates. It starts and ends at free charges on the plates. The field due to polarization charges starts and ends at these bound charges and is in the *opposite* direction to the $\mathcal{E}_o$. Although $\mathcal{E}_o$ polarizes the molecules of the medium, $\mathcal{E}_{\text{dep}}$, being in the opposite direction, tries to depolarize the medium. It is called the **depolarizing field** (and hence the subscript). Thus the field inside the medium is

$$\mathcal{E} = \mathcal{E}_o - \mathcal{E}_{\text{dep}} \qquad [7.76]$$

The depolarizing field depends on the amount of polarization since it is determined by $+Q_P$ and $-Q_P$. For the dielectric plate in Figure 7.50, we know the field $\mathcal{E}$ is $\mathcal{E}_o/\varepsilon_r$, so we can eliminate $\mathcal{E}_o$ in Equation 7.76 and relate $\mathcal{E}_{\text{dep}}$ directly to $\mathcal{E}$,

$$\mathcal{E}_{\text{dep}} = \mathcal{E}(\varepsilon_r - 1)$$

However, the polarization $P$ is related to the field $\mathcal{E}$ by

$$P = \varepsilon_o(\varepsilon_r - 1)\mathcal{E}$$

which means that the depolarization field is

*Depolarizing field in a dielectric plate*

$$\mathcal{E}_{\text{dep}} = \frac{1}{\varepsilon_o}P \qquad [7.77]$$

As we expected, the depolarizing field is proportional to the polarization $P$. We should emphasize that $\mathcal{E}_{\text{dep}}$ is in the *opposite direction* to $\mathcal{E}$ and $P$ and Equation 7.77 is for magnitudes only. If we write it as a vector equation, then we must introduce a negative sign to give $\mathcal{E}_{\text{dep}}$ a direction opposite to that of $P$. Moreover, the relationship in Equation 7.77 is special to the dielectric plate geometry in Figure 7.50. In general, the depolarizing field is still proportional to the polarization, as in Equation 7.77, but it is given by

*Depolarizing field in a dielectric*

$$\mathcal{E}_{\text{dep}} = \frac{N_{\text{dep}}}{\varepsilon_o}P \qquad [7.78]$$

where $N_{\text{dep}}$ is a numerical factor called the **depolarization factor.** It takes into account the shape of the dielectric and the variation in the polarization within the medium. For

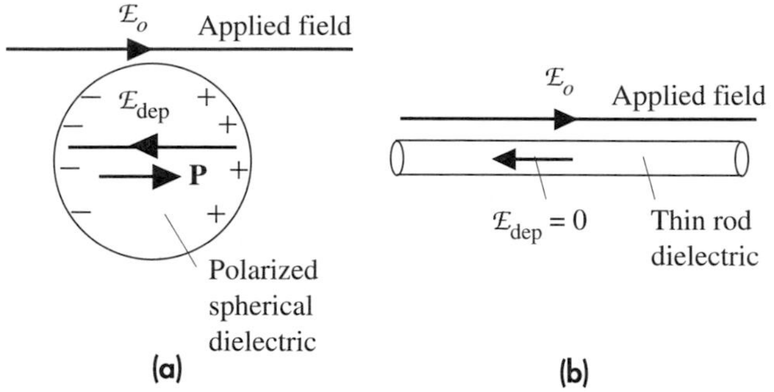


**Figure 7.51**

(a) Polarization and the depolarizing field in a spherical-shaped dielectric placed in an applied field.

(b) Depolarization field in a thin rod placed in an applied field is nearly zero.

a dielectric plate placed perpendicularly to an external field, $N_{dep} = 1$, as we found in Equation 7.77. For the spherical dielectric medium as in Figure 7.51a, $N_{dep} = \frac{1}{3}$. For a long thin dielectric rod placed with its axis along the applied field, as in Figure 7.51b, $N_{dep} \approx 0$ and becomes exactly zero as the diameter shrinks to zero. $N_{dep}$ is always between 0 and 1. If we know $N_{dep}$, we can determine the field inside the dielectric, for example, in a small spherical cavity within an insulation given the external field.

## 7.10 LOCAL FIELD AND THE LORENTZ EQUATION

When a dielectric medium is placed in an electric field, it becomes polarized and there is a macroscopic, or an average, field $\mathcal{E}$ in the medium. The actual field at an atom, called the **local field** $\mathcal{E}_{loc}$, however, is not the same as the average field as illustrated in Figure 7.7.

Consider a dielectric plate polarized by placing it between the plates of a capacitor as shown in Figure 7.52a. The macroscopic field $\mathcal{E}$ in the dielectric is given by the applied field $\mathcal{E}_o$ due to the free charges $Q_{free}$ on the plates, and the depolarization field due to $P$, or polarization charges on the dielectric plate surfaces $A$. Since we have a plate

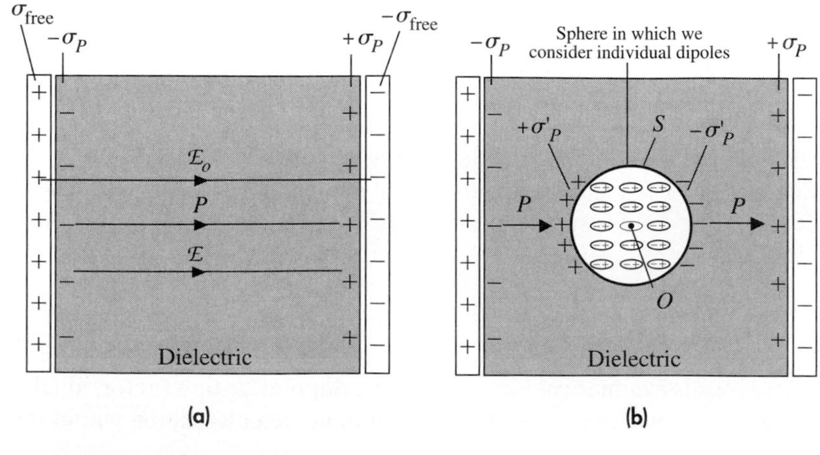


**Figure 7.52**

(a) The macroscopic field $\mathcal{E}$ is determined by the applied field $\mathcal{E}_o$ and the depolarization field due to $P$.

(b) Calculation of the local field involves making a hypothetical spherical cavity $S$ inside the dielectric. This produces polarization surface charges on the inside surface $S$ of the cavity. The effects of the dipoles inside the cavity are treated individually.

dielectric, the depolarization field is $P/\varepsilon_o$, so

$$\mathcal{E} = \mathcal{E}_o - \mathcal{E}_{\text{dep}} = \mathcal{E}_o - \frac{1}{\varepsilon_o}P$$

Consider the field at some atomic site, point $O$, but with the atom itself removed. We evaluate the field at $O$ coming from all the charges except the atom at $O$ itself since we are looking at the field experienced by this atom (the atom cannot become polarized by its own field). We then cut a (hypothetical) spherical cavity $S$ centered at $O$ and consider the atomic polarizations individually within the spherical cavity. In other words, the effects of the dipoles in the cavity are treated separately from the remaining dielectric medium which is now left with a spherical cavity. This remaining dielectric is considered as a continuous medium but with a spherical cavity. Its dielectric property is represented by its polarization vector $P$. Because of the cavity, we must now put polarization charges on the inner surface $S$ of this cavity as illustrated in Figure 7.52b. This may seem surprising, but we should remember that we are treating the effects of the atomic dipoles within the cavity individually and separately by cutting out a spherical cavity from the medium and thereby introducing a surface $S$.

The field at $O$ comes from four sources:

1. Free charges $Q_{\text{free}}$ on the electrodes, represented by $\mathcal{E}_o$.
2. Polarization charges on the plate surfaces $A$, represented by $\mathcal{E}_{\text{dep}}$.
3. Polarization charges on the inner surface of the spherical cavity $S$, represented by $\mathcal{E}_S$.
4. Individual dipoles within the cavity, represented by $\mathcal{E}_{\text{dipoles}}$.

Thus,

$$\mathcal{E}_{\text{loc}} = \mathcal{E}_o + \mathcal{E}_{\text{dep}} + \mathcal{E}_S + \mathcal{E}_{\text{dipoles}}$$

Since the first two terms make up the macroscopic field, we can write this as

*Local field in a crystal*

$$\mathcal{E}_{\text{loc}} = \mathcal{E} + \mathcal{E}_S + \mathcal{E}_{\text{dipoles}}$$

The field from the individual dipoles surrounding $O$ depends on the positions of these atomic dipoles which depend on the crystal structure. For cubic crystals, amorphous solids (*e.g.*, glasses), or liquids, effects of these dipoles around $O$ cancel each other and $\mathcal{E}_{\text{dipoles}} = 0$. Thus,

*Local field in a cubic crystal or a non-crystalline material*

$$\mathcal{E}_{\text{loc}} = \mathcal{E} + \mathcal{E}_S \qquad [7.79]$$

We are then left with evaluating the field due to polarization charges on the inner surface $S$ of the cavity. This field comes from polarization charges on the surface $S$. Consider a thin spherical shell on surface $S$ as shown in Figure 7.53 which makes an angle $\theta$ with $O$. The radius of this shell is $a \sin\theta$, whereas its width (or thickness) is $a\,d\theta$. The surface area $dS$ is then $(2\pi a \sin\theta)(a\,d\theta)$. The polarization charge $dQ_P$ on this spherical shell surface is $P_n\,dS$ where $P_n$ is the polarization vector normal to the surface $dS$. Thus,

$$dQ_P = P_n\,dS = (P\cos\theta)(2\pi a \sin\theta)(a\,d\theta)$$

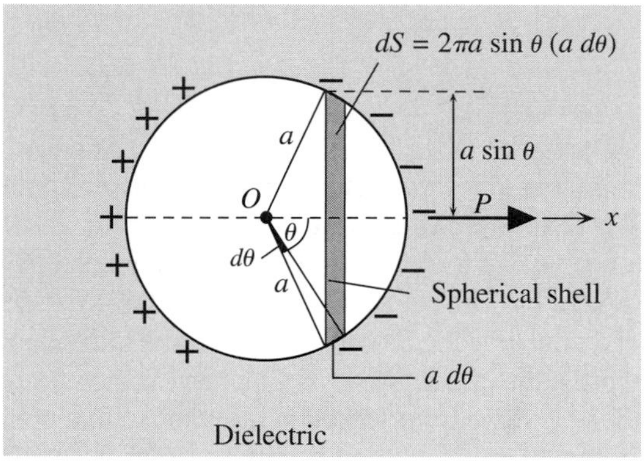


**Figure 7.53** Calculation of the field due to polarization charges on the inner surface $S$ of the spherical cavity.
Consider a spherical shell of radius $a$. The surface area is $dS = 2\pi a \sin\theta\,(a\,d\theta)$.

But the field at $O$ from $dQ_P$ is given from electrostatics as

$$d\mathcal{E}_S = \frac{dQ_P}{4\pi\varepsilon_o a^2} = \frac{(P\cos\theta)(2\pi a\,\sin\theta)(a\,d\theta)}{4\pi\varepsilon_o a^2}$$

To find the total field coming from the whole surface $S$ we have to integrate $d\mathcal{E}_S$ from $\theta = 0$ to $\theta = \pi$,

$$\mathcal{E}_S = \int_0^{\pi} \frac{(P\cos\theta)(\sin\theta)}{2\varepsilon_o}\,d\theta$$

which integrates to

*Field in a spherical cavity*

$$\mathcal{E}_S = \frac{1}{3\varepsilon_o}P \tag{7.80}$$

*Local field in a cubic crystal or noncrystalline material*

The local field by Equation 7.79 is

$$\mathcal{E}_{loc} = \mathcal{E} + \frac{1}{3\varepsilon_o}P \tag{7.81}$$

Equation 7.81 is the **Lorentz relation** for the local field in terms of the polarization $P$ of the medium and is valid for cubic crystals and noncrystalline materials, such as glasses. It does *not* apply to dipolar dielectrics in which the local field can be quite complicated.

## 7.11 DIPOLAR POLARIZATION

Consider a medium with molecules that have permanent dipole moments. Each permanent dipole moment is $p_o$. In the presence of an electric field the dipoles try to align perfectly with the field, but random thermal collisions, *i.e.*, thermal agitation, act against this perfect alignment. A molecule that manages to rotate and align with the field finds itself later colliding with another molecule and losing its alignment. We are interested in the mean dipole moment in the presence of an applied field taking into

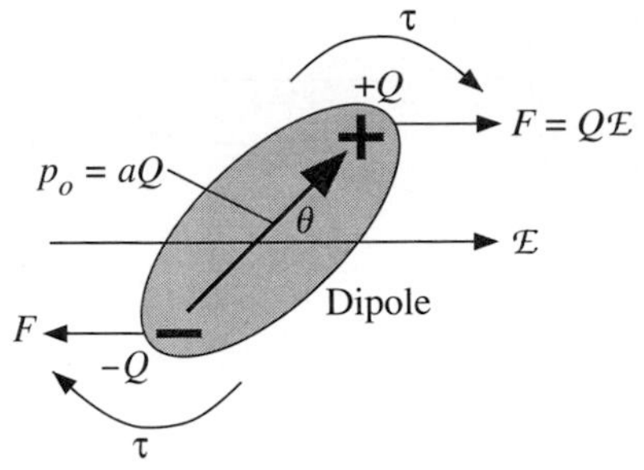


**Figure 7.54** In the presence of an applied field a dipole tries to rotate to align with the field against thermal agitation.

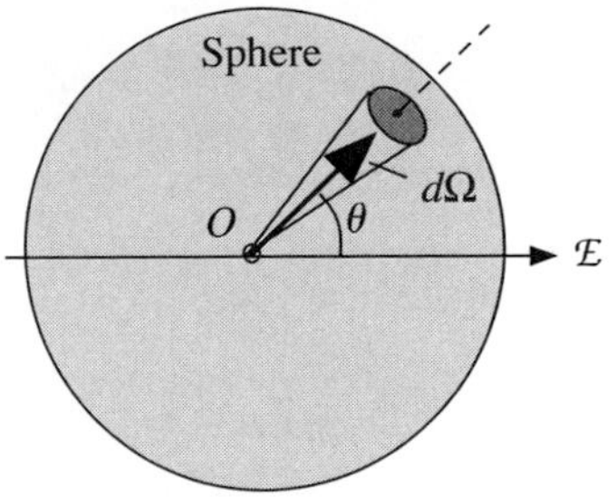


**Figure 7.55** The dipole is pointing within a solid angle $d\Omega$.

account the thermal energies of the molecules and their random collisions. We will assume that the probability that a molecule has an energy $E$ is given by the Boltzmann factor, $\exp(-E/kT)$.

Consider an arbitrary dipolar molecule in an electric field as in Figure 7.54 with its dipole moment $p_o$ at an angle $\theta$ with the field $\mathcal{E}$. The torque experienced by the dipole is given by $\tau = (F \sin\theta)a$ or $\mathcal{E}p_o \sin\theta$ where $p_o = aQ$. The potential energy $E$ at an angle $\theta$ is given by integrating $\tau\, d\theta$,

*Potential energy of a dipole at an angle θ*

$$E = \int_0^\theta p_o\mathcal{E} \sin\theta\, d\theta = -p_o\mathcal{E}\cos\theta + p_o\mathcal{E}$$

Inasmuch as the $PE$ depends on the orientation, there is a certain probability of finding a dipole oriented at this angle as determined by the Boltzmann distribution. The fraction $f$ of molecules oriented at $\theta$ is proportional to $\exp(-E/kT)$,

*Boltzmann distribution*

$$f \propto \exp\left(\frac{p_o\mathcal{E}\cos\theta}{kT}\right) \qquad [7.82]$$

The initial orientation of the dipole should be considered in three dimensions and not as in the two-dimensional illustration in Figure 7.54. In three dimensions we use solid angles, and the fraction $f$ then represents the fraction of molecules pointing in a direction defined by a small solid angle $d\Omega$ as shown in Figure 7.55. The whole sphere around the dipole corresponds to a solid angle of $4\pi$. Furthermore, we need to find the average dipole moment along $\mathcal{E}$ as this will be the induced net dipole moment by the field. The dipole moment along $\mathcal{E}$ is $p_o\cos\theta$. Then from the definition of the average

$$p_{\text{av}} = \frac{\int_0^{4\pi}(p_o\cos\theta)f\,d\Omega}{\int_0^{4\pi} f\,d\Omega} \qquad [7.83]$$

where $f$ is the Boltzmann factor given in Equation 7.82 and depends on $\mathcal{E}$ and $\theta$. The final result of this integration is a special function called the **Langevin function** which

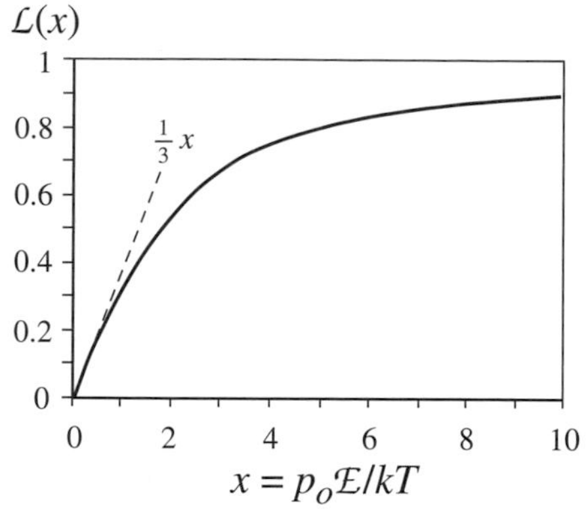


**Figure 7.56** The Langevin function.

*Average dipole moment and the Langevin function*

is denoted as $\mathcal{L}(x)$ where $x$ is the argument of the function (not the $x$ coordinate). The integration of Equation 7.83 then gives

$$p_{\text{av}} = p_o\mathcal{L}(x) \qquad \text{and} \qquad x = \frac{\mathcal{E}}{kT} \tag{7.84}$$

The behavior of the Langevin function is shown in Figure 7.56. At the highest fields $\mathcal{L}(x)$ tends toward saturation at unity. Then, $p_{\text{av}} = p_o$, which corresponds to nearly all the dipoles aligning with the field, so increasing the field cannot increase $p_{\text{av}}$ anymore. In the low field region, $p_{\text{av}}$ increases linearly with the field. In practice, the applied fields are such that all dipolar polarizations fall into this linear behavior region where the Langevin function $\mathcal{L}(x) \approx \frac{1}{3}x$. Then Equation 7.84 becomes

*Average induced dipole in orientational polarization*

$$p_{\text{av}} = \frac{1}{3}\frac{p_o^2\mathcal{E}}{kT} \tag{7.85}$$

The **dipolar** or **orientational polarizability** is then simply

*Dipolar or orientational polarizability*

$$\alpha_d = \frac{1}{3}\frac{p_o^2}{kT} \tag{7.86}$$

## 7.12 IONIC POLARIZATION AND DIELECTRIC RESONANCE

In ionic polarization, as shown in Figure 7.9, the applied field displaces the positive and negative ions in opposite directions, which results in a net dipole moment per ion, called the *induced dipole moment* $p_i$ per ion. We can calculate the ionic polarizability $\alpha_i$ and the ionic contribution to the relative permittivity as a function of frequency by applying an ac field of the form $\mathcal{E} = \mathcal{E}_o \exp(j\omega t)$.

Consider two oppositely charged neighboring ions, *e.g.*, $Na^+$ and $Cl^-$, which experience forces $Q\mathcal{E}$ in opposite directions where $Q$ is the magnitude of the ionic charge of each ion as shown in Figure 7.57. The bond between the ions becomes

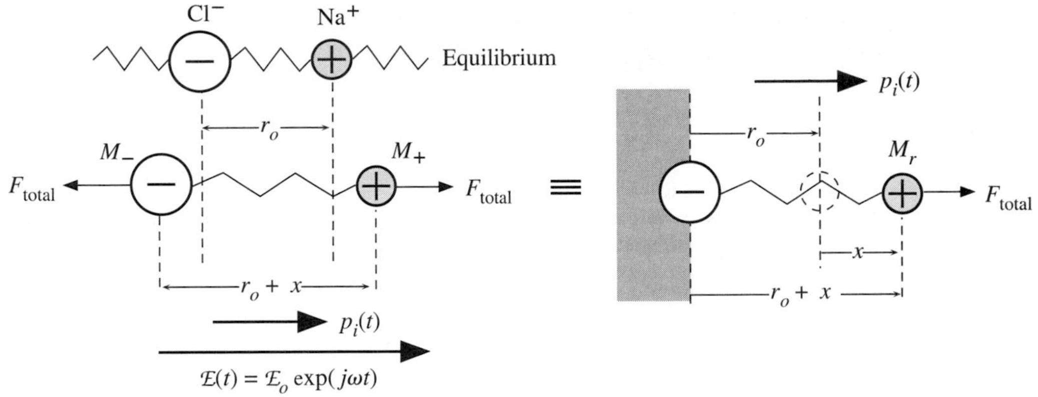


**Figure 7.57** Consider a pair of oppositely charged ions. In the presence of an applied field $\mathcal{E}$ along x, the $Na^+$ and $Cl^-$ ions are displaced from each other by a distance $x$. The net average (or induced) dipole moment is $p_i$.

stretched, and the two ions become displaced from the equilibrium separation $r_o$ to a new separation $r_o + x$ as depicted in Figure 7.57. The force $F = Q\mathcal{E}$ of the applied field is the polarizing force, which causes the relative displacement. We take $F$ to be along the $x$ direction. The applied force is resisted by a **restoring force** $F_r$ that is due to the stretching of the bond (Hooke's law) and is proportional to the amount of bond stretching, *i.e.*, $F_r = -\beta x$ where $\beta$ is the **spring constant** associated with the ionic bond (easily calculated from the potential energy curve of the bond), and the negative sign ensures that $F_r$ is directed in the opposite direction to the applied force. Thus, the net force acting on the ions is $Q\mathcal{E} - \beta x$. As the ions are oscillated by the applied force, they couple some of the energy in the applied field to lattice vibrations and this energy is then lost as heat (lattice vibrations) in the crystal. As in classical mechanics, this type of energy loss through a coupling mechanism can be represented as a **frictional force** (force associated with losses) $F_{\text{loss}}$ that acts against the effect of the applied force. This frictional force is proportional to the velocity of the ions or $dx/dt$, so it is written as $F_{\text{loss}} = -\gamma(dx/dt)$ where $\gamma$ is a proportionality constant that depends on the exact mechanism for the energy loss from the field, and the negative sign ensures that it is opposing the applied field. The total (net) force on the ions is

$$F_{\text{total}} = F + F_r + F_{\text{loss}} = Q\mathcal{E} - \beta x - \gamma\frac{dx}{dt}$$

*Total force*

Normally we would examine the equations of motion (Newton's second law) under forced oscillation for each ion separately, and then we would use the results to find the overall extension $x$. An equivalent procedure (as well known in mechanics) is to keep one ion stationary and allow the other one to oscillate with a reduced mass $M_r$, which is $M_r = (M_+M_-)/(M_+ + M_-)$ where $M_+$ and $M_-$ are the masses of $Na^+$ and $Cl^-$ ions, respectively. For example, we can simply examine the oscillations of the $Na^+$ -ion within the reference frame of the $Cl^-$ -ion (kept "stationary") and attach a reduced mass $M_r$ to $Na^+$ as depicted in Figure 7.57. Then Newton's second law gives

*Forced oscillations of $Na^+$–$Cl^-$ ion pair*

$$M_r \frac{d^2x}{dt^2} = Q\mathcal{E} - \beta x - \gamma \frac{dx}{dt} \tag{7.87}$$

It is convenient to put $M_r$ and $\beta$ together into a new constant $\omega_I$ which represents the **resonant** or **natural angular frequency** of the ionic bond, or the natural oscillations when the applied force is removed. Defining $\omega_I = (\beta/M_r)^{1/2}$ and $\gamma_I$ as $\gamma$ per unit reduced mass, *i.e.*, $\gamma_I = \gamma/M_r$,we have

*Forced dipole oscillator, ionic polarization*

$$\frac{d^2x}{dt^2} + \gamma_I \frac{dx}{dt} + \omega_I^2 x = \frac{Q}{M_r}\mathcal{E}_o \exp(j\omega t) \tag{7.88}$$

Equation 7.88 is a second-order differential equation for the induced displacement $x$ of a pair of neighboring ions about the equilibrium separation as a result of an applied force $Q\mathcal{E}$. It is called the *forced oscillator* equation and is well known in mechanics. (The same equation would describe the damped motion of a ball attached to a spring in a viscous medium and oscillated by an applied force.) The solution to Equation 7.88 will give the displacement $x = x_o \exp(j\omega t)$, which will have the same time dependence as $\mathcal{E}$ but *phase shifted;* that is, $x_o$ will be a complex number. The *relative* displacement of the ions from the equilibrium gives rise to a *net* or **induced polarization** $p_i = Qx$. Thus Equation 7.88 can be multiplied by $Q$ to represent the forced oscillations of the induced dipole. Equation 7.88 is also called the **Lorentz dipole oscillator model.**

The induced dipole $p_i$ will also be phase shifted with respect to the applied force $Q\mathcal{E}$. When we divide $p_i$ by the applied field $\mathcal{E}$, we get the **ionic polarizability** $\alpha_i$, given by

*Ionic polarizability*

$$\alpha_i = \frac{p_i}{\mathcal{E}} = \frac{Qx}{\mathcal{E}} = \frac{Q^2}{M_r\left(\omega_I^2 - \omega^2 + j\gamma_I\omega\right)} \tag{7.89}$$

It can be seen that the polarizability is also a complex number as we expect; there is a phase shift between $\mathcal{E}$ and induced $p_i$. It therefore has real $\alpha_i'$ and imaginary $\alpha_i''$ parts and can be written as $\alpha_i = \alpha_i' - j\alpha_i''$. We note that, by convention, the imaginary part is written with a minus sign to keep $\alpha_i''$ as a positive quantity. Further, when $\omega = 0$, under dc conditions, the ionic polarizability $\alpha_i(0)$ from Equation 7.89 is

*DC ionic polarizability*

$$\alpha_i(0) = \frac{Q^2}{M_r\omega_I^2} \tag{7.90}$$

The dc polarizability is a real quantity as there can be no phase shift under dc conditions. We can then write the ionic polarizability in Equation 7.89 in terms of the normalized frequency $(\omega/\omega_I)$ as

*AC ionic polarizability*

$$\alpha_i(\omega) = \frac{\alpha_i(0)}{\left[1 - \left(\dfrac{\omega}{\omega_I}\right)^2 + j\left(\dfrac{\gamma_I}{\omega_I}\right)\left(\dfrac{\omega}{\omega_I}\right)\right]} \tag{7.91}$$

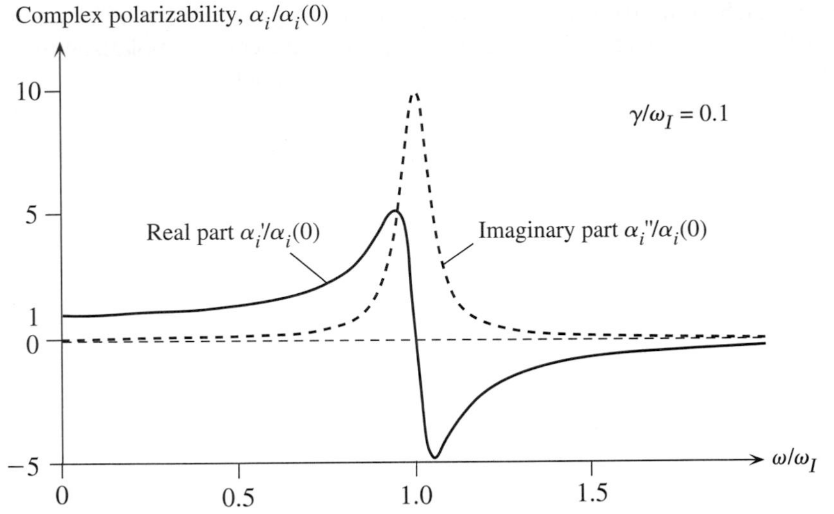


**Figure 7.58** A schematic representation of the frequency dependence of the real and imaginary parts of normalized polarizability $\alpha_i/\alpha_i(0)$ versus $\omega/\omega_I$.

The dependences of the real and imaginary parts of $\alpha_i$ on the frequency of the field are shown in Figure 7.58 in terms of the normalized frequency $(\omega/\omega_I)$ for one particular value of the loss factor, $\gamma_I = 0.1\omega_I$. Note that $\alpha_i''$ peaks at a frequency very close to the ionic bond resonant frequency $\omega_I$ (it is exactly $\omega_I$ when $\gamma_I = 0$). The sharpness and magnitude of the $\alpha_i''$ peak depends on the loss factor $\gamma_I$. The peak is sharper and higher for smaller $\gamma_I$. Notice that $\alpha_i'$ is nearly constant at frequencies lower than $\omega_I$. Indeed, in a dc field, $\alpha_i' = \alpha_i(0)$. But, through $\omega_I$, $\alpha_i'$ shows a rapid change from positive to negative values and then it tends toward zero for frequencies greater than $\omega_I$.

Zero or negative $\alpha_i'$ should not be disconcerting since the actual magnitude of the polarizability is $|\alpha_i| = (\alpha_i'^2 + \alpha_i''^2)^{1/2}$, which is always positive through $\omega_I$ and maximum at $\omega_I$. The phase of $\alpha_i$ however changes through $\omega_I$. The phase of $\alpha_i$, and hence the phase of the polarization with respect to the field, are zero at low frequencies $(\omega \ll \omega_I)$. As the frequency increases, the polarization lags behind the field and the phase of $\alpha_i$ becomes more negative. At $\omega = \omega_I$, the polarization lags behind the field by 90°. However, the rate of change of polarization is in phase with the field oscillations, which leads to a maximum energy transfer. At high frequencies, well above $\omega_I$, the ions cannot respond to the rapidly changing field and the coupling between the field and the ions is negligible. The peak in the $\alpha_i''$ versus $\omega$ behavior around $\omega = \omega_I$ is what is called the **dielectric resonance peak,** and in this particular case it is called the **ionic polarization relaxation peak** and is due to the strong coupling of the applied field with the natural vibrations of the ionic bond at $\omega = \omega_I$.

The resulting relative permittivity $\varepsilon_r$ can be found from the Clausius–Mossotti equation. But we also have to consider the electronic polarizability $\alpha_e$ of the two types of ions since this type of polarization operates up to optical frequencies $(\omega \gg \omega_I)$, which means that

*Dielectric constant of an ionic solid*

$$\frac{\varepsilon_r(\omega) - 1}{\varepsilon_r(\omega) + 2} = \frac{N_i}{3\varepsilon_o}[\alpha_i + \alpha_{e+} + \alpha_{e-}] \quad [7.92]$$

where $N_i$ is the concentrations of negative and positive ion pairs (assuming an equal number of positive and negative ions), and $\alpha_{e+}$ and $\alpha_{e-}$ are the electronic polarizabilities of the negative and positive ion species, respectively. Inasmuch as $\alpha_i$ is a complex quantity, so is the relative permittivity $\varepsilon_r(\omega)$. We can express Equation 7.92 differently by noting that at very high frequencies, $\omega \gg \omega_I$, $\alpha_i = 0$, and the relative permittivity is then denoted as $\varepsilon_{rop}$. Equation 7.92 then becomes

*Dispersion relation for ionic polarization*

$$\frac{\varepsilon_r(\omega) - 1}{\varepsilon_r(\omega) + 2} - \frac{\varepsilon_{rop} - 1}{\varepsilon_{rop} + 2} = \frac{N_i \alpha_i}{3\varepsilon_o} = \frac{N_i Q^2}{3\varepsilon_o M_r \left(\omega_I^2 - \omega^2 + j\gamma_I \omega\right)} \qquad [7.93]$$

This is called the **dielectric dispersion relation** between the relative permittivity, due to ionic polarization, and the frequency of the electric field. Figure 7.16b shows the behavior of $\varepsilon_r(\omega)$ with frequency for KCl where $\varepsilon_r''$ peaks at $\omega = \omega_I = 2\pi(4.5 \times 10^{12})$ rad s$^{-1}$ and $\varepsilon_r'$ exhibits sharp changes around this frequency. It is clear that as $\omega$ gets close to $\omega_I$, there are rapid changes in $\varepsilon_r(\omega)$. The resonant frequencies ($\omega_I$) for ionic polarization relaxations are typically in the infrared frequency range, and the "applied" field in the crystal is then due to a propagating electromagnetic (EM) wave rather than an ac applied field between two external electrodes placed on the crystal.[15]

It should be mentioned that electronic polarization can also be described by the Lorentz oscillator model, and can also be represented by Equation 7.91 if we appropriately replace $\alpha_i$ by $\alpha_e$ and interpret $\omega_I$ and $\gamma_I$ as the resonant frequency and loss factor involved in electronic polarization.

**EXAMPLE 7.17** **IONIC POLARIZATION RESONANCE IN KCl** Consider a KCl crystal which has the FCC crystal structure and the following properties. The optical dielectric constant is 2.19, the dc dielectric constant is 4.84, and the lattice parameter $a$ is 0.629 nm. Calculate the dc ionic polarizability $\alpha_i(0)$. Estimate the ionic resonance absorption frequency and compare the value with the experimentally observed resonance at $4.5 \times 10^{12}$ Hz in Figure 7.16b. The atomic masses of K and Cl are 39.09 and 35.45 g mol$^{-1}$, respectively.

**SOLUTION**

At optical frequencies the dielectric constant $\varepsilon_{rop}$ is determined by electronic polarization. At low frequencies and under dc conditions, the dielectric constant $\varepsilon_{rdc}$ is determined by both electronic and ionic polarization. If $N_i$ is the concentration of negative and positive ion pairs, then equation 7.93 becomes

$$\frac{\varepsilon_{rdc} - 1}{\varepsilon_{rdc} + 2} = \frac{\varepsilon_{rop} - 1}{\varepsilon_{rop} + 2} + \frac{1}{3\varepsilon_o} N_i \alpha_i(0)$$

There are four negative and positive ion pairs per unit cell, and the cell dimension is $a$. The concentration of negative and positive ion pairs $N_i$ is

$$N_i = \frac{4}{a^3} = \frac{4}{(0.629 \times 10^{-9}\ \text{m})^3} = 1.61 \times 10^{28}\ \text{m}^{-3}$$

[15] More rigorous theories of ionic polarization would consider the interactions of a propagating electromagnetic wave with various phonon modes within the crystal, which is beyond the scope of this book.

Substituting $\varepsilon_{r\text{dc}} = 4.84$ and $\varepsilon_{rop} = 2.19$ and $N_i$ in Equation 7.93

$$\alpha_i(0) = \frac{3\varepsilon_o}{N_i}\left[\frac{\varepsilon_{r\text{dc}} - 1}{\varepsilon_{r\text{dc}} + 2} - \frac{\varepsilon_{r\text{op}} - 1}{\varepsilon_{r\text{op}} + 2}\right] = \frac{3(8.85 \times 10^{-12})}{1.61 \times 10^{28}}\left[\frac{4.84 - 1}{4.84 + 2} - \frac{2.19 - 1}{2.19 + 2}\right]$$

we find

$$\alpha_i(0) = 4.58 \times 10^{-40} \text{ F m}^2$$

The relationship between $\alpha_i(0)$ and the resonance absorption frequency involves the reduced mass $M_r$ of the $K^+$–$Cl^-$ ion pair,

$$M_r = \frac{M_+ M_-}{M_+ + M_-} = \frac{(39.09)(35.45)(10^{-3})}{(39.09 + 35.45)(6.022 \times 10^{23})} = 3.09 \times 10^{-26} \text{ kg}$$

At $\omega = 0$, the polarizability is given by Equation 7.90, so the resonance absorption frequency $\omega_I$ is

$$\omega_I = \left[\frac{Q^2}{M_r \alpha_i(0)}\right]^{1/2} = \left[\frac{(1.6 \times 10^{-19})^2}{(3.09 \times 10^{-26})(4.58 \times 10^{-40})}\right]^{1/2} = 4.26 \times 10^{13} \text{ rad s}^{-1}$$

or

$$f_I = \frac{\omega_I}{2\pi} = 6.8 \times 10^{12} \text{ Hz}$$

This is about a factor of 1.5 greater than the observed resonance absorption frequency of $4.5 \times 10^{12}$ Hz. Typically one accounts for the difference by noting that the actual ionic charges may not be exactly $+e$ on $K^+$ and $-e$ on $Cl^-$, but $Q$ is effectively $0.76e$. Taking $Q = 0.76e$ makes $f_I = 5.15 \times 10^{12}$ Hz, only 14 percent greater than the observed value. A closer agreement can be obtained by refining the simple theory and considering how many effective dipoles there are in the unit cell along the direction of the applied field.

## 7.13 DIELECTRIC MIXTURES AND HETEROGENEOUS MEDIA

Many dielectrics are composite materials; that is, they are mixtures of two or more different types of dielectric materials with different relative permittivities and loss factors. The simplest example is a porous dielectric which has small air pores randomly dispersed within the bulk of the material as shown in Figure 7.59a (analogous to a random raisin pudding). Another example would be a dielectric material composed of two distinctly different phases that are randomly mixed, as shown in Figure 7.59b, somewhat like a Swiss cheese that has air bubbles. We often need to find the overall or the **effective dielectric constant** $\varepsilon_{r\text{eff}}$ of the mixture, which is not a trivial problem.[16] This overall $\varepsilon_{r\text{eff}}$ can then be used to treat the mixture as if it were one dielectric substance with this particular dielectric constant; for example, the capacitance can be calculated

[16] The theories that try to represent a heterogeneous medium in terms of effective quantities are called *effective medium theories* (or approximations). The theory of finding an effective dielectric constant of a mixture has intrigued many famous scientists in the past. Over the years, many quite complicated mixture rules have been developed, and there is no shortage of formulas in this field. Many engineers however still tend to use simple empirical rules to model a composite dielectric. The primary reason is that many theoretical mixture rules depend on the exact knowledge of the geometrical shapes, sizes, and distributions of the mixed phases.

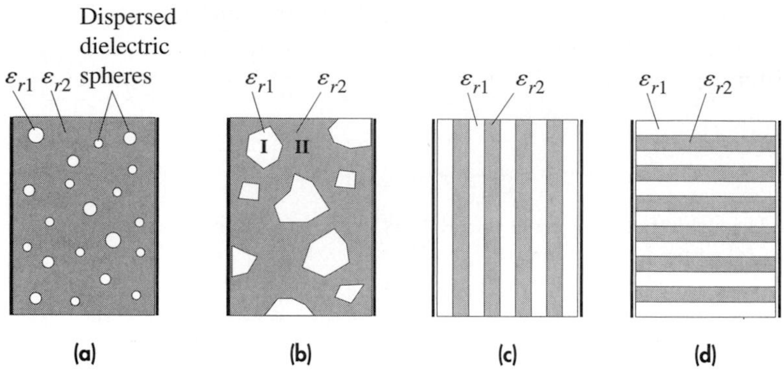


**Figure 7.59** Heterogeneous dielectric media examples.
(a) Dispersed dielectric spheres in a dielectric matrix.
(b) A heterogeneous medium with two distinct phases I and II.
(c) Series mixture rule.
(d) Parallel mixture rule.

from $C = \varepsilon_o \varepsilon_{reff} A/d$ by simply using $\varepsilon_{reff}$. It should be emphasized that if mixing occurs at the atomic level so that the material is essentially a *solid solution,* then, in principle, the Clausius–Mossotti equation can be used in which we simply add the polarizabilities of each species of atoms or ions weighted by their concentration. (We did this for CsCl in Example 7.4.) The present problem examines **heterogeneous materials,** and hence excludes such solid solutions.

The theoretical treatment of mixtures can be quite complicated since one has to consider not only individual dielectric properties but also the geometrical shapes, sizes, and distributions of the two (or more) phases present in the composite material. In many cases, empirical rules that have been shown to work have been used to predict $\varepsilon_{reff}$. Consider a heterogeneous dielectric that has two mixed phases I and II with dielectric constants $\varepsilon_{r1}$ and $\varepsilon_{r2}$, and volume fractions $v_1$ and $v_2$, respectively, $(v_1 + v_2 = 1)$ as in Figure 7.59b. One simple and useful mixture rule is

*Generalized mixture rule*

$$\varepsilon_{reff}^n = v_1 \varepsilon_{r1}^n + v_2 \varepsilon_{r2}^n \quad \textbf{[7.94]}$$

where $n$ is an index (a constant), usually determined empirically, that depends on the type of mixture. If we have a parallel stack of plates of I and II in alternating (or in random) sequence between the two electrodes, this would be like many series-connected dielectrics and $n$ would be $-1$. If the phases are in parallel as plates of I and II stacked on top of each other, as shown in Figure 7.59d, then $n$ is 1. As $n$ approaches 0, Equation 7.94 can be shown to be equivalent to a **logarithmic mixture** rule:

*Lichtenecker formula*

$$\ln \varepsilon_{reff} = v_2 \ln \varepsilon_{r1} + v_2 \ln \varepsilon_{r2} \quad \textbf{[7.95]}$$

which is known as the **Lichtenecker formula** (1926). Although its scientific basis is not strong, it has shown remarkable applicability to various heterogeneous media; perhaps due to the fact that it is a kind of compromise between the two extreme limits of series and parallel mixtures.

There is one particular mixture rule for dispersed dielectric spheres (with $\varepsilon_{r1}$), such as air pores, in a continuous dielectric matrix (with $\varepsilon_{r2}$), that works quite well for volume fractions up to about 20 percent, called the **Maxwell–Garnett formula**

$$\frac{\varepsilon_{\text{reff}} - \varepsilon_{r2}}{\varepsilon_{\text{reff}} + 2\varepsilon_{r2}} = v_1 \frac{\varepsilon_{r1} - \varepsilon_{r2}}{\varepsilon_{r1} + 2\varepsilon_{r2}} \qquad [7.96]$$

*Maxwell–Garnett formula*

The Maxwell–Garnett equation can predict the effective dielectric constant of many different types of dielectrics that have dispersed pores. There are other mixture rules, but the above are some of the common types.

**EXAMPLE 7.18**

**LOW-$\kappa$ POROUS DIELECTRICS FOR MICROELECTRONICS** It was mentioned in Chapter 2 that today's high transistor density ICs have multilayers of metal interconnect lines that are separated by an **interlayer dielectric** (ILD). The speed of the chip (as limited by the RC time constant) depends on the overall interconnect capacitance, which depends on the relative permittivity $\varepsilon_{r\text{ILD}}$ of the ILD. The traditional ILD material has been $SiO_2$ with $\varepsilon_r = 3.9$. There is much research interest in finding suitable low-$\kappa$ (also called low-$k$) materials for such ILD applications, especially in ultralarge-scale integration (ULSI). What is the required porosity in $SiO_2$ if its effective relative permittivity is to be 2.5?

**SOLUTION**

The Maxwell–Garnett equation is particularly useful for such porous media calculations. Substituting $\varepsilon_{r2} = 3.9$, $\varepsilon_{r1} = 1$ (air pores), and setting $\varepsilon_{\text{reff}} = 2.5$ in Equation 7.96 we have

$$\frac{2.5 - 3.9}{2.5 + 2(3.9)} = v_1 \frac{1 - 3.9}{1 + 2(3.9)}$$

and solving gives

$$v_1 = 0.412, \qquad \text{or} \qquad 41\% \text{ porosity}$$

Such porosity is achievable but it may have side effects such as poorer mechanical properties and lower breakdown voltage. Note that the Lichtenecker formula gives 32.6 percent porosity. As apparent from this example, there is a distinct advantage in starting with a dielectric that has a low initial $\varepsilon_r$, and then using porosity to lower $\varepsilon_r$ further. For example, if we start with $\varepsilon_r = 3$, then the same 41 percent porosity will yield $\varepsilon_{\text{reff}} = 2.05$. Many polymeric materials have $\varepsilon_r$ values $\sim 2.5$ and have been candidate materials for low-$\kappa$ ILD applications in microelectronics.

## CD Selected Topics and Solved Problems

| Selected Topics | Solved Problems |
|---|---|
| Static Dielectric Constant of Materials | Static Electronic Polarizability |
| Piezoelectric Materials and Devices: Elementary Concepts | Relative Permittivity of an Ionic Crystal at Low and Optical Frequencies |
| Real and Imaginary Dielectric Constant | Piezoelectric Coefficients |
| Conduction in Solid Insulating Materials | Piezoelectric Spark Generator |
| | Electric Field in Coaxial Cables: Double Layer Insulation for Controlling the Maximum Field |

## DEFINING TERMS

**Boundary conditions** relate the normal and tangential components of the electric field next to the boundary. The tangential component must be continuous through the boundary. Suppose that $\mathcal{E}_{n1}$ is the normal component of the field in medium 1 at the boundary and $\varepsilon_{r1}$ is the relative permittivity in medium 1. Using a similar notation for medium 2, then the boundary condition is $\varepsilon_{r1}\mathcal{E}_{n1} = \varepsilon_{r2}\mathcal{E}_{n2}$.

**Clausius–Mossotti equation** relates the dielectric constant ($\varepsilon_r$), a macroscopic property, to the polarizability ($\alpha$), a microscopic property.

**Complex relative permittivity** ($\varepsilon_r' + j\varepsilon_r''$) has a real part ($\varepsilon_r'$) that determines the charge storage ability and an imaginary part ($\varepsilon_r''$) that determines the energy losses in the material as a result of the polarization mechanism. The real part determines the capacitance through $C = \varepsilon_o\varepsilon_r' A/d$ and the imaginary part determines the electric power dissipation per unit volume as heat by $\mathcal{E}^2\omega\varepsilon_o\varepsilon_r''$.

**Corona discharge** is a local discharge in a gaseous atmosphere where the field is sufficiently high to cause dielectric breakdown, for example, by avalanche ionization.

**Curie temperature** $T_C$ is the temperature above which ferroelectricity disappears, that is, the spontaneous polarization of the crystal is lost.

**Debye equations** attempt to describe the frequency response of the complex relative permittivity $\varepsilon_r' + j\varepsilon_r''$ of a dipolar medium through the use of a single relaxation time $\tau$ to describe the sluggishness of the dipoles driven by the external ac field.

**Dielectric** is a material in which energy can be stored by the polarization of the molecules. It is a material that increases the capacitance or charge storage ability of a capacitor. Ideally, it is a nonconductor of electrical charge so that an applied field does not cause a flow of charge but instead relative displacement of opposite charges and hence polarization of the medium.

**Dielectric loss** is the electrical energy lost as heat in the polarization process in the presence of an applied ac field. The energy is absorbed from the ac voltage and converted to heat during the polarization of the molecules. It should not be confused with conduction loss $\sigma\mathcal{E}^2$ or $V^2/R$.

**Dielectric strength** is the maximum field ($\mathcal{E}_{\text{br}}$) that can be sustained in a dielectric beyond which dielectric breakdown ensues; that is, there is a large conduction current through the dielectric shorting the plates.

**Dipolar (orientational) polarization** arises when randomly oriented polar molecules in a dielectric are rotated and aligned by the application of a field so as to give rise to a net average dipole moment per molecule. In the absence of the field, the dipoles (polar molecules) are randomly oriented and there is no average dipole moment per molecule. In the presence of the field, the dipoles are rotated, some partially and some fully, to align with the field and hence give rise to a net dipole moment per molecule.

**Dipolar relaxation equation** describes the time response of the induced dipole moment per molecule in a dipolar material in the presence of a time-dependent applied field. The response of the dipoles depends on their relaxation time, which is the mean time required to dissipate the stored electrostatic energy in the dipole alignment to heat through lattice vibrations or molecular collisions.

**Dipole relaxation (dielectric resonance)** occurs when the frequency of the applied ac field is such that there is maximum energy transfer from the ac voltage source to heat in the dielectric through the alternating polarization and depolarization of the molecules by the ac field. The stored electrostatic energy is dissipated through molecular collisions and lattice vibrations (in solids). The peak occurs when the angular frequency of the ac field is the reciprocal of the relaxation time.

**Electric dipole moment** exists when a positive charge $+Q$ is separated from a negative charge $-Q$. Even though the net charge is zero, there is nonetheless an electric dipole moment $\mathbf{p}$ given by $\mathbf{p} = Q\mathbf{x}$ where $\mathbf{x}$ is the distance vector from $-Q$ to $+Q$. Just as two charges exert a Coulombic force on each other, two dipoles also exert a force on each other that depends on the magnitudes of the dipoles, their separation, and orientation.

**Electric susceptibility** ($\chi_e$) is a material quantity that measures the extent of polarization in the material per unit field. It relates the amount of polarization $P$ at a point in the dielectric to the field $\mathcal{E}$ at that point via $P = \chi_e \varepsilon_o \mathcal{E}$. If $\varepsilon_r$ is the relative permittivity, then $\chi_e = \varepsilon_r - 1$. Vacuum has no electric susceptibility.

**Electromechanical breakdown and electrofracture** are breakdown processes that directly or indirectly involve electric field–induced mechanical weakening, for example, crack propagation, or mechanical deformation that eventually lead to dielectric breakdown.

**Electronic bond polarization** is the displacement of valence electrons in the bonds in covalent solids (*e.g.*, Ge, Si). It is a collective displacement of the electrons in the bonds with respect to the positive nuclei.

**Electronic polarization** is the displacement of the electron cloud of an atom with respect to the positive nucleus. Its contribution to the relative permittivity of a solid is usually small.

**External discharges** are discharges or shorting currents over the surface of the insulator when the conductance of the surface increases as a result of surface contamination, for example, excessive moisture, deposition of pollutants, dirt, dust, and salt spraying. Eventually the contaminated surface develops sufficient conductance to allow discharge between the electrodes at a field below the normal breakdown strength of the insulator. Dielectric breakdown over the surface of an insulation is termed **surface tracking.**

**Ferroelectricity** is the occurrence of spontaneous polarization in certain crystals such as barium titanate ($BaTiO_3$). Ferroelectric crystals have a permanent polarization **P** as a result of spontaneous polarization. The direction of **P** can be defined by the application of an external field.

**Gauss's law** is a fundamental law of physics that relates the surface integral of the electric field over a closed (hypothetical) surface to the sum of all the charges enclosed within the surface. If $\mathcal{E}_n$ is the field normal to a small surface area $dA$ and $Q_{\text{total}}$ is the enclosed total charge, then over the whole closed surface $\varepsilon_o \oint \mathcal{E}_n \, dA = Q_{\text{total}}$.

**Induced polarization** is the polarization of a molecule as a result of its placement in an electric field. The induced polarization is along the direction of the field. If the molecule is already polar, then induced polarization is the additional polarization that arises due to the applied field alone and it is directed along the field.

**Insulation aging** is a term used to describe the physical and chemical deterioration in the properties of the insulation so that its dielectric breakdown characteristics worsen with time. Aging therefore determines the useful life of the insulation.

**Interfacial polarization** occurs whenever there is an accumulation of charge at an interface between two materials or between two regions within a material. Grain boundaries and electrodes are regions where charges generally accumulate and give rise to this type of polarization.

**Internal discharges** are partial discharges that take place in microstructural voids, cracks, or pores within the dielectric where the gas atmosphere (usually air) has lower dielectric strength. A porous ceramic, for example, would experience partial discharges if the field is sufficiently large. Initially, the pore size (or the number of pores) may be small and the partial discharge insignificant, but with time the partial discharge erodes the internal surfaces of the void. Eventually (and usually) an *electrical tree* type of discharge develops from a partial discharge that has been eroding the dielectric. The erosion of the dielectric by the partial discharge propagates like a branching tree. The "tree branches" are erosion channels, filaments of various sizes, in which gaseous discharge takes place and forms a conducting channel during operation.

**Intrinsic breakdown** or **electronic breakdown** commonly involves the avalanche multiplication of electrons (and holes in solids) by impact ionization in the presence of high electric fields. The large number of free carriers generated by the avalanche of impact ionizations leads to a runaway current between the electrodes and hence to insulation breakdown.

**Ionic polarization** is the relative displacement of oppositely charged ions in an ionic crystal that results in the polarization of the whole material. Typically, ionic polarization is important in ionic crystals below the infrared wavelengths.

**Local field** ($\mathcal{E}_{\text{loc}}$) is the true field experienced by a molecule in a dielectric that arises from the free charges on the plates and all the induced dipoles surrounding the molecule. The true field at a molecule is not simply the applied field $(V/d)$ because of the field of the neighboring induced dipoles.

**Loss tangent** or **tan** $\delta$ is the ratio of the dielectric constant's imaginary part to the real part, $\varepsilon_r''/\varepsilon_r'$. The angle $\delta$ is the phase angle between the capacitive current and the total current. If there is no dielectric loss, then the two currents are the same and $\delta = 0$.

**Partial discharge** occurs when only a local region of the dielectric is exhibiting discharge, so the discharge does not directly connect the two electrodes.

**Piezoelectric material** has a noncentrosymmetric crystal structure that leads to the generation of a polarization vector $P$, or charges on the crystal surfaces, upon the application of a mechanical stress. When strained, a piezoelectric crystal develops an internal field and therefore exhibits a voltage difference between two of its faces.

**PLZT,** lead lanthanum zirconate titanate, is a PZT-type material with lanthanum occupying the Pb site.

**Polarizability** $(\alpha)$ is the ability of an atom or molecule to become polarized in the presence of an electric field. It is induced polarization in the molecule per unit field along the field direction.

**Polarization** is the separation of positive and negative charges in a system so that there is a net electric dipole moment per unit volume.

**Polarization vector (P)** measures the extent of polarization in a unit volume of dielectric matter. It is the vector sum of dielectric dipoles per unit volume. If **p** is the average dipole moment per molecule and $n$ is the number of molecules per unit volume, then $\mathbf{P} = n\mathbf{p}$. In a polarized dielectric matter (*e.g.*, in an electric field), the bound surface charge density $\sigma_p$ due to polarization is equal to the normal component of **P** at that point, $\sigma_p = P_{\text{normal}}$.

**Poling** is the application of a temporary electric field to a piezoelectric (or ferroelectric) material, generally at an elevated temperature, to align the polarizations of various grains and thereby develop piezoelectric behavior.

**Pyroelectric** material is a polar dielectric (such as barium titanate) in which a temperature change $\Delta T$ induces a proportional change $\Delta P$ in the polarization, that is, $\Delta P = p\,\Delta T$, where $p$ is the pyroelectric coefficient of the crystal.

**PZT** is a general acronym for the lead zirconate titanate ($PbZrO_3$-$PbTiO_3$ or $PbTi_{0.48}Zr_{0.52}O_3$) family of crystals.

***Q*-factor** or **quality factor** for an impedance is the ratio of its reactance to its resistance. The $Q$-factor of a capacitor is $X_c/R_p$ where $X_c = 1/\omega C$ and $R_p$ is the equivalent parallel resistance that represents the dielectric and conduction losses. The $Q$-factor of a resonant circuit measures the circuit's peak response at the resonant frequency and also its bandwidth. The greater the $Q$, the higher the peak response and the narrower the bandwidth. For a series $RLC$ resonant circuit,

$$Q = \frac{\omega_o L}{R} = \frac{1}{\omega_o CR}$$

where $\omega_o$ is the resonant angular frequency, $\omega_o = 1/\sqrt{LC}$. The width of the resonant response curve between half-power points is $\Delta\omega = \omega_o/Q$.

**Relative permittivity** $(\varepsilon_r)$ or **dielectric constant** of a dielectric is the fractional increase in the stored charge per unit voltage on the capacitor plates due to the presence of the dielectric between the plates (the whole space between the plates is assumed to be filled). Alternatively, we can define it as the fractional increase in the capacitance of a capacitor when the insulation between the plates is changed from a vacuum to a dielectric material, keeping the geometry the same.

**Relaxation time** $(\tau)$ is a characteristic time that determines the sluggishness of the dipole response to an applied field. It is the mean time for the dipole to lose its alignment with the field due to its random interactions with the other molecules through molecular collisions, lattice vibrations, and so forth.

**Surface tracking** is an *external dielectric breakdown* that occurs over the surface of the insulation.

**Temperature coefficient of capacitance** (TCC) is the fractional change in the capacitance per unit temperature change.

**Thermal breakdown** is a breakdown process that involves thermal runaway, which leads to a runaway current or discharge between the electrodes. If the heat generated by dielectric loss, due to $\varepsilon_r''$, or Joule heating, due to finite $\sigma$, cannot be removed sufficiently rapidly, then the temperature of the dielectric rises, which increases the conductivity and the dielectric loss. The increases in $\varepsilon_r''$ and $\sigma$ lead to more heat generation and a further rise in the temperature, so thermal runaway ensues, followed by either a large shorting current or local thermal decomposition of the insulation accompanied by a partial discharge in this region.

**Transducer** is a device that converts electrical energy into another form of usable energy or vice versa. For example, piezoelectric transducers convert electrical energy to mechanical energy and vice versa.

# QUESTIONS AND PROBLEMS

**7.1** **Relative permittivity and polarizability**

*a.* Show that the local field is given by

$$\mathcal{E}_{\text{loc}} = \mathcal{E}\left(\frac{\varepsilon_r + 2}{3}\right)$$

*Local field*

*b.* Amorphous selenium (a-Se) is a high-resistivity semiconductor that has a density of approximately 4.3 g cm$^{-3}$ and an atomic number and mass of 34 and 78.96, respectively. Its relative permittivity at 1 kHz has been measured to be 6.7. Calculate the relative magnitude of the local field in a-Se. Calculate the polarizability per Se atom in the structure. What type of polarization is this? How will $\varepsilon_r$ depend on the frequency?

*c.* If the electronic polarizability of an isolated atom is given by

$$\alpha_e \approx 4\pi\varepsilon_o r_o^3$$

where $r_o$ is the radius of the atom, then calculate the electronic polarizability of an isolated Se atom, which has $r_o = 0.12$ nm, and compare your result with that for an atom in a-Se. Why is there a difference?

**7.2** **Electronic polarization and $SF_6$** Because of its high dielectric strength, $SF_6$ (sulfur hexafluoride) gas is widely used as an insulator and a dielectric in HV applications such as HV transformers, switches, circuit breakers, transmission lines, and even HV capacitors. The $SF_6$ gas at 1 atm and at room temperature has a dielectric constant of 1.0015. The number of $SF_6$ molecules per unit volume $N$ can be found by the gas law, $P = (N/N_A)RT$. Calculate the electronic polarizability $\alpha_e$ of the $SF_6$ molecule. How does $\alpha_e$ compare with the $\alpha_e$ versus $Z$ line in Figure 7.4? (Note: The $SF_6$ molecule has no net dipole. Assume that the overall polarizability of $SF_6$ is due to electronic polarization.)

**7.3** **Electronic polarization in liquid xenon** Liquid xenon has been used in radiation detectors. The density of the liquid is 3.0 g cm$^{-3}$. What is the relative permittivity of liquid xenon given its electronic polarizability in Table 7.1? (The experimental $\varepsilon_r$ is 1.96.)

**7.4** **Relative permittivity, bond strength, bandgap, and refractive index** Diamond, silicon, and germanium are covalent solids with the same crystal structure. Their relative permittivities are shown in Table 7.10.

*a.* Explain why $\varepsilon_r$ increases from diamond to germanium.

*b.* Calculate the polarizability per atom in each crystal and then plot polarizability against the elastic modulus $Y$ (Young's modulus). Should there be a correlation?

*c.* Plot the polarizability from part (*b*) against the bandgap energy $E_g$. Is there a relationship?

*d.* Show that the refractive index $n$ is $\sqrt{\varepsilon_r}$. When does this relationship hold and when does it fail?

*e.* Would your conclusions apply to ionic crystals such as NaCl?

**Table 7.10** Properties of diamond, Si, and Ge

| | $\varepsilon_r$ | $M_{at}$ | Density (g cm$^{-3}$) | $\alpha_e$ | $Y$ (GPa) | $E_g$ (eV) | $n$ |
|---|---|---|---|---|---|---|---|
| Diamond | 5.8 | 12 | 3.52 | | 827 | 5.5 | 2.42 |
| Si | 11.9 | 28.09 | 2.33 | | 190 | 1.12 | 3.45 |
| Ge | 16 | 72.61 | 5.32 | | 75.8 | 0.67 | 4.09 |

**7.5** **Dipolar liquids** Given the static dielectric constant of water as 80, its high-frequency dielectric constant (due to electronic polarization) as 4, and its density as 1 g cm$^{-3}$, calculate the permanent dipole moment $p_o$ per water molecule assuming that it is the orientational and electronic polarization of individual molecules that gives rise to the dielectric constant. Use both the simple relationship in Equation 7.14 where the local field is the same as the macroscopic field and also the Clausius–Mossotti equation and compare your results with the permanent dipole moment of the water molecule which is $6.1 \times 10^{-30}$ C m. What is your conclusion? What is $\varepsilon_r$ calculated from the Clausius–Mossotti equation taking the true $p_o$ ($6.1 \times 10^{-30}$ C m) of a water molecule? (Note: Static dielectric constant is due to both orientational and electronic polarization. The Clausius–Mossotti equation does not apply to dipolar materials because the local field is not described by the Lorentz field.)

**7.6** **Dielectric constant of water vapor or steam** The isolated water molecule has a permanent dipole $p_o$ of $6.1 \times 10^{-30}$ C m. The electronic polarizability $\alpha_e$ of the water molecule under dc conditions is about $4 \times 10^{-40}$ C m. What is the dielectric constant of steam at a pressure of 10 atm ($10 \times 10^5$ Pa) and at a temperature of 400 °C? [Note: The number of water molecules per unit volume $N$ can be found from the simple gas law, $P = (N/N_A)RT$. The Clausius–Mossotti equation does not apply to orientational polarization. Since $N$ is small, use Equation 7.14.]

**7.7** **Dipole moment in a nonuniform electric field** Figure 7.60 shows an electric dipole moment $p$ in a nonuniform electric field. Suppose the gradient of the field is $d\mathcal{E}/dx$ at the dipole $p$, and the dipole is oriented to be along the direction of increasing $\mathcal{E}$ as in Figure 7.60. Show that the *net force* acting on this dipole is given by

*Net force on a dipole*

$$F = p\frac{d\mathcal{E}}{dx}$$

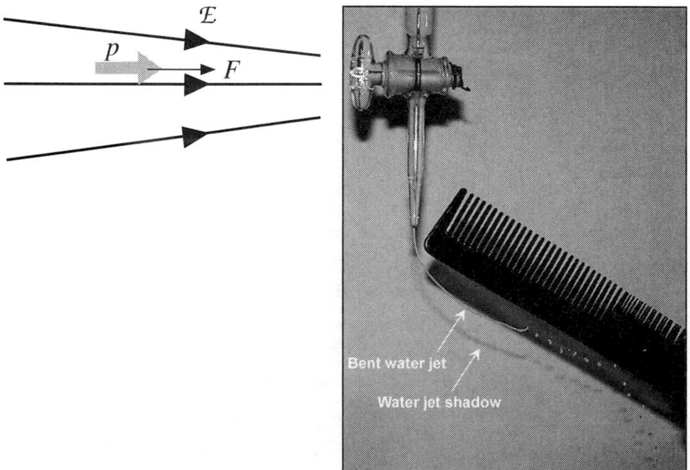


**Figure 7.60**
Left: A dipole moment in a nonuniform field experiences a net force $F$ that depends on the dipole moment $p$ and the field gradient $d\mathcal{E}/dx$.
Right: When a charged comb (by combing hair) is brought close to a water jet, the field from the comb polarizes the liquid by orientational polarization. The induced polarization vector $P$ and hence the liquid is attracted to the comb where the field is higher.

Which direction is the force? What happens to this net force when the dipole moment is facing the direction of decreasing field? Given that a dipole normally also experiences a torque as described in Section 7.3.2, explain qualitatively what happens to a randomly placed dipole in a nonuniform electric field. Explain the experimental observation of bending a flow of water by a nonuniform field from a charged comb as shown in the photograph in Figure 7.60? (Remember that a dielectric medium placed in a field develops polarization $P$ directed along the field.)

**7.8** **Ionic and electronic polarization** Consider a CsBr crystal that has the CsCl unit cell crystal structure (one $Cs^+$–$Br^-$ pair per unit cell) with a lattice parameter ($a$) of 0.430 nm. The electronic polarizability of $Cs^+$ and $Br^-$ ions are $3.35 \times 10^{-40}$ F m$^2$ and $4.5 \times 10^{-40}$ F m$^2$, respectively, and the mean ionic polarizability per ion pair is $5.8 \times 10^{-40}$ F m$^2$. What is the low-frequency dielectric constant and that at optical frequencies?

**7.9** **Electronic and ionic polarization in KCl** KCl has the same crystal structure as NaCl. KCl's lattice parameter is 0.629 nm. The ionic polarizability per ion pair (per $K^+$–$Cl^-$ ion) is $4.58 \times 10^{-40}$ F m$^2$. The electronic polarizability of $K^+$ is $1.26 \times 10^{-40}$ F m$^2$ and that of $Cl^-$ is $3.41 \times 10^{-40}$ F m$^2$. Calculate the dielectric constant under dc operation and at optical frequencies. Experimental values are 4.84 and 2.19.

**7.10** **Debye relaxation** We will test the Debye equations for approximately calculating the real and imaginary parts of the dielectric constant of water just above the freezing point at 0.2 °C. Assume the following values in the Debye equations for water: $\varepsilon_{rdc} = 87.46$ (dc), $\varepsilon_{r\infty} = 4.87$ (at $f = 300$ GHz well beyond the relaxation peak), and $\tau = 1/\omega_o = (2\pi 9.18 \text{ GHz})^{-1} = 0.017$ ns. Calculate the real and imaginary, $\varepsilon_r'$ and $\varepsilon_r''$, parts of $\varepsilon_r$ for water at frequencies in Table 7.11, and plot both the experimental values and your calculations on a linear–log plot (frequency on the log axis). What is your conclusion? (Note: It is possible to obtain a better agreement by using two relaxation times or using more sophisticated models.)

**Table 7.11** Dielectric properties of water at 0.2 °C

| | *f* (GHz) | | | | | | | | | | | | |
|---|---|---|---|---|---|---|---|---|---|---|---|---|---|
| | **0.3** | **0.5** | **1** | **1.5** | **3** | **5** | **9.18** | **10** | **20** | **40** | **70** | **100** | **300** |
| $\varepsilon_r'$ | 87.46 | 87.25 | 86.61 | 85.34 | 76.20 | 68.19 | 46.13 | 42.35 | 19.69 | 10.16 | 7.20 | 6.14 | 4.87 |
| $\varepsilon_r''$ | 2.60 | 4.50 | 8.85 | 13.18 | 24.28 | 34.53 | 40.55 | 40.24 | 30.23 | 17.68 | 11.15 | 8.31 | 3.68 |

I SOURCE: Data extracted from R. Buchner et al., *Chem. Phys. Letts*, **306,** 57, 1999.

***7.11** **Debye and non-Debye relaxation and Cole–Cole plots** Consider the Debye equation

$$\varepsilon_r = \varepsilon_{r\infty} + \frac{\varepsilon_{rdc} - \varepsilon_{r\infty}}{1 + j\omega\tau}$$

*Debye relaxation*

and also the **generalized dielectric relaxation** equation, which "stretches" (broadens) the Debye function,

$$\varepsilon_r = \varepsilon_{r\infty} + \frac{\varepsilon_{rdc} - \varepsilon_{r\infty}}{[1 + (j\omega\tau)^\alpha]^\beta}$$

*Generalized dielectric relaxation*

Take $\tau = 1$, $\varepsilon_{rdc} = 5$, $\varepsilon_{r\infty} = 2$, and $\alpha = 0.8$, and $\beta = 1$. Plot the real and imaginary parts of $\varepsilon_r$ versus frequency (on a log scale) for both functions from $\omega = 0, 0.1/\tau, 1/3\tau, 1/\tau, 3/\tau$, and $10\tau$. For the same $\omega$ values, plot $\varepsilon_r''$ versus $\varepsilon_r'$ (Cole–Cole plot) for both functions using a graph in which the $x$ and $y$ axes have the same divisions. What is your conclusion?

**7.12** **Equivalent circuit of a polyester capacitor** Consider a 1 nF polyester capacitor that has a polymer (PET) film thickness of 1 μm. Calculate the equivalent circuit of this capacitor at 50 °C and at 120 °C for operation at 1 kHz. What is your conclusion?

**7.13** **Student microwaves mashed potatoes** A microwave oven uses electromagnetic waves at 2.48 GHz to heat food by dielectric loss, that is, making use of $\varepsilon_r''$ of the food material, which normally has substantial water content. An undergraduate student microwaves 10 $\text{cm}^3$ of mashed potatoes in 60 seconds. The microwave generates an rms field of $\mathcal{E}_{\text{rms}}$ of 200 V $\text{cm}^{-1}$ in mashed potatoes. At 2.48 GHz, mashed potatoes have $\varepsilon_r'' = 21$. Calculate the average power dissipated per $\text{cm}^3$, and also the total energy dissipated heating the food. (Note: You can use $\mathcal{E}_{\text{rms}}$ instead of $E$ in Equation 7.32.)

**7.14** **Dielectric loss per unit capacitance** Consider the three dielectric materials listed in Table 7.12 with the real and imaginary dielectric constants $\varepsilon_r'$ and $\varepsilon_r''$. At a given voltage, which dielectric will have the lowest power dissipation per unit capacitance at 1 kHz and at an operating temperature of 50 °C? Is this also true at 120 °C?

**Table 7.12** Dielectric properties of three insulators at 1 kHz

| | $T$ = 50 °C | | $T$ = 120 °C | |
|---|---|---|---|---|
| **Material** | $\varepsilon_r'$ | $\varepsilon_r''$ | $\varepsilon_r'$ | $\varepsilon_r''$ |
| Polycarbonate | 2.47 | 0.003 | 2.535 | 0.003 |
| PET | 2.58 | 0.003 | 2.75 | 0.027 |
| PEEK | 2.24 | 0.003 | 2.25 | 0.003 |

| SOURCE: Data taken using a DEA by Kasap and Nomura (1995).

**7.15** **Parallel and series equivalent circuits** Figure 7.61 shows simplified parallel and series equivalent circuits for a capacitor. The elements $R_p$ and $C_p$ in the parallel circuit and the elements $R_s$ and $C_s$ in the series circuit are related. We can write down the impedance $Z_{AB}$ between the terminals $A$ and $B$ for both the circuits, and then equate $Z_{AB}$(parallel) = $Z_{AB}$(series). Show that

*Equivalent series resistance and capacitance*

$$R_s = \frac{R_p}{1 + (\omega R_p C_p)^2} \quad \text{and} \quad C_s = C_p\left[1 + \frac{1}{(\omega R_p C_p)^2}\right]$$

and similarly by considering the admittance (1/impedance),

*Equivalent series resistance and capacitance*

$$R_p = R_s\left[1 + \frac{1}{(\omega R_s C_s)^2}\right] \quad \text{and} \quad C_p = \frac{C_s}{1 + (\omega R_s C_s)^2}$$

A 10 nF capacitor operating at 1 MHz has a parallel equivalent resistance of 100 kΩ. What are $C_s$ and $R_s$?

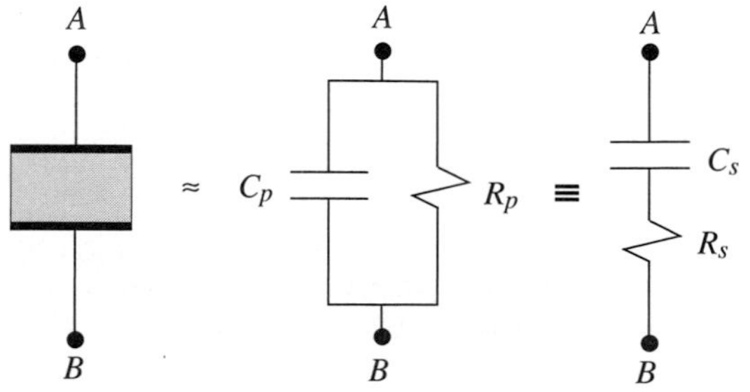


**Figure 7.61** An equivalent parallel $R_p$ and $C_p$ circuit is equivalent to a series $R_s$ and $C_s$ circuit. The elements $R_p$ and $C_p$ in the parallel circuit are related to the elements $R_s$ and $C_s$ in the series circuit.

**7.16** **Tantalum capacitors** Electrolytic capacitors tend to be modeled by a series $R_s + j\omega C_s$ equivalent circuit. A nominal 22 μF Ta capacitor (22 μF at low frequencies) has the following properties at 10 kHz:

$\varepsilon_r' \approx 20$ (at this frequency), $\tan\delta \approx 0.05$, dielectric thickness $d = 0.16$ μm, effective area $A = 150$ cm$^2$. Calculate $C_p$, $R_p$, $C_s$, and $R_s$.

**7.17** **Tantalum versus niobium oxide capacitors** Niobium oxide ($Nb_2O_5$) is a competing dielectric to $Ta_2O_5$ (the dielectric in the tantalum capacitor). The dielectric constants are 41 for $Nb_2O_5$ and 27 for $Ta_2O_5$. For operation at the same voltage, the $Ta_2O_5$ thickness is 0.17 μm, and that of $Nb_2O_5$ is 0.25 μm. Explain why the niobium oxide capacitor is superior (or inferior) to the Ta capacitor. (Use a quantitative argument, such as the capacitance per unit volume.) What other factors would you consider if you were choosing between the two?

***7.18** **TCC of a polyester capacitor** Consider the parallel plate capacitor equation

$$C = \frac{\varepsilon_o \varepsilon_r xy}{z}$$

where $\varepsilon_r$ is the relative permittivity (or $\varepsilon_r'$), $x$ and $y$ are the side lengths of the dielectric so that $xy$ is the area $A$, and $z$ is the thickness of the dielectric. The quantities $\varepsilon_r$, $x$, $y$, and $z$ change with temperature. By differentiating this equation with respect to temperature, show that the **temperature coefficient of capacitance** (TCC) is

*Temperature coefficient of capacitance*

$$\text{TCC} = \frac{1}{C}\frac{dC}{dT} = \frac{1}{\varepsilon_r}\frac{d\varepsilon_r}{dT} + \lambda$$

where $\lambda$ is the linear expansion coefficient defined by

$$\lambda = \frac{1}{L}\frac{dL}{dT}$$

where $L$ stands for any length of the material ($x$, $y$, or $z$). Assume that the dielectric is isotropic and $\lambda$ is the same in all directions. Using $\varepsilon_r'$ versus $T$ behavior in Figure 7.62 and taking $\lambda = 50 \times 10^{-6}$ K$^{-1}$ as a typical value for polymers, predict the TCC at room temperature and at 10 kHz.

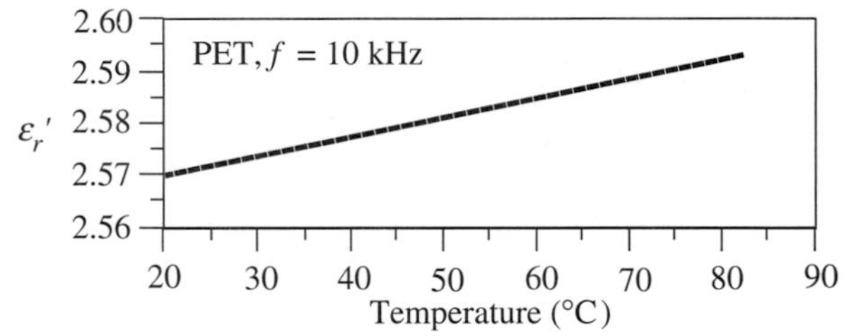


**Figure 7.62** Temperature dependence of $\varepsilon_r'$ at 10 kHz.
SOURCE: Data taken by Kasap and Maeda (1995).

**7.19** **Dielectric breakdown of gases and Paschen curves** Dielectric breakdown in gases typically involves the avalanche ionization of the gas molecules by energetic electrons accelerated by the applied field. The mean free path between collisions must be sufficiently long to allow the electrons to gain sufficient energy from the field to impact ionize the gas molecules. The breakdown voltage $V_{br}$ between two electrodes depends on the distance $d$ between the electrodes as well as the gas pressure $P$, as shown in Figure 7.63. $V_{br}$ versus $Pd$ plots are called **Paschen curves.** We consider gaseous insulation, air and $SF_6$, in an HV switch.

*a.* What is the breakdown voltage between two electrodes of a switch separated by a 5 mm gap with air at 1 atm when the gaseous insulation is air and when it is $SF_6$?

*b.* What are the breakdown voltages in the two cases when the pressure is 10 times greater? What is your conclusion?

*c.* At what pressure is the breakdown voltage a minimum?

*d.* What air gap spacing $d$ at 1 atm gives the minimum breakdown voltage?

*e.* What would be the reasons for preferring gaseous insulation over liquid or solid insulation?

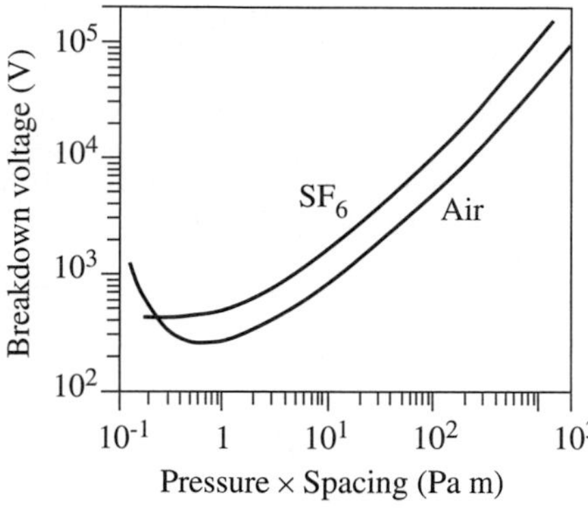


**Figure 7.63** Breakdown voltage versus (pressure × electrode spacing) (Paschen curves).

***7.20** **Capacitor design** Consider a nonpolarized 100 nF capacitor design at 60 Hz operation. Note that there are three candidate dielectrics, as listed in Table 7.13.

*a.* Calculate the volume of the 100 nF capacitor for each dielectric, given that they are to be used under low voltages and each dielectric has its minimum fabrication thickness. Which one has the smallest volume?

*b.* How is the volume affected if the capacitor is to be used at a 500 V application and the maximum field in the dielectric must be a factor of 2 less than the dielectric strength? Which one has the smallest volume?

*c.* At a 500 V application, what is the power dissipated in each capacitor at 60 Hz operation? Which one has the lowest dissipation?

**Table 7.13** Comparison of dielectric properties at 60 Hz (typical values)

| | Polymer Film PET | Ceramic $TiO_2$ | High-*K* Ceramic ($BaTiO_3$ based) |
|---|---|---|---|
| Name | Polyester | Polycrystalline titania | X7R |
| $\varepsilon_r'$ | 3.2 | 90 | 1800 |
| tan $\delta$ | $5 \times 10^{-3}$ | $4 \times 10^{-4}$ | $5 \times 10^{-2}$ |
| $\mathcal{E}_{br}$(kV cm$^{-1}$) | 150 | 50 | 100 |
| Typical minimum thickness | 1–2 μm | 10 μm | 10 μm |

***7.21** **Dielectric breakdown in a coaxial cable** Consider a coaxial underwater high-voltage cable as in Figure 7.64a. The current flowing through the inner conductor generates heat, which has to flow through the dielectric insulation to the outer conductor where it will be carried away by conduction and convection. We will assume that steady state has been reached and the inner conductor is carrying a dc current $I$. Heat generated per unit second $Q' = dQ/dt$ by Joule heating of the inner conductor is

*Rate of heat generation*

$$Q' = RI^2 = \frac{\rho L I^2}{\pi a^2} \qquad [7.97]$$

where $\rho$ is the resistivity, $a$ the radius of the conductor, and $L$ the cable length.

This heat flows radially out from the inner conductor through the dielectric insulator to the outer conductor, then to the ambient. This heat flow is by thermal conduction through the dielectric. The rate of heat flow $Q'$ depends on the temperature difference $T_i - T_o$ between the inner and outer conductors;

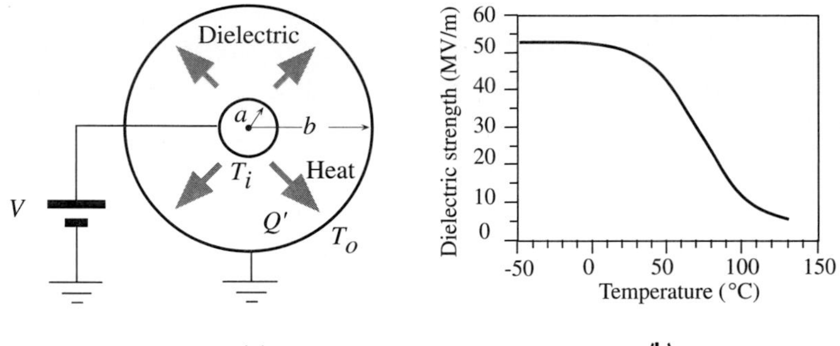


(a) (b)

**Figure 7.64**
(a) The Joule heat generated in the core conductor flows outward radially through the dielectric material.
(b) Typical temperature dependence of the dielectric strength of a polyethylene-based polymeric insulation.

on the sample geometry ($a$, $b$, and $L$); and on the thermal conductivity $\kappa$ of the dielectric. From elementary thermal conduction theory, this is given by

*Rate of heat conduction*

$$Q' = (T_i - T_o)\frac{2\pi\kappa L}{\ln\left(\dfrac{b}{a}\right)} \qquad [7.98]$$

The inner core temperature $T_i$ rises until, in the steady state, the rate of Joule heat generation by the electric current in Equation 7.97 is just removed by the rate of thermal conduction through the dielectric insulation, given by Equation 7.98.

*a.* Show that the inner conductor temperature is

*Steady-state inner conductor temperature*

$$T_i = T_o + \frac{\rho I^2}{2\pi^2 a^2 \kappa}\ln\left(\frac{b}{a}\right) \qquad [7.99]$$

*b.* The breakdown occurs at the maximum field point, which is at $r = a$, just outside the inner conductor and is given by (see Example 7.11).

*Maximum field in a coaxial cable*

$$\mathcal{E}_{\max} = \frac{V}{a\ln\left(\dfrac{b}{a}\right)} \qquad [7.100]$$

The dielectric breakdown occurs when $\mathcal{E}_{\max}$ reaches the dielectric strength $\mathcal{E}_{br}$. However the dielectric strength $\mathcal{E}_{br}$ for many polymeric insulation materials depends on the temperature, and generally it decreases with temperature, as shown for a typical example in Figure 7.64b. If the load current $I$ increases, then more heat $Q'$ is generated per second and this leads to a higher inner core temperature $T_i$ by virtue of Equation 7.99. The increase in $T_i$ with $I$ eventually lowers $\mathcal{E}_{br}$ so much that it becomes equal to $\mathcal{E}_{\max}$ and the insulation breaks down (thermal breakdown). Suppose that a certain coaxial cable has an aluminum inner conductor of diameter 10 mm and resistivity 27 nΩ m. The insulation is 3 mm thick and is a polyethylene-based polymer whose long-term dc dielectric strength is shown in Figure 7.64b. Suppose that the cable is carrying a voltage of 40 kV and the outer shield temperature is the ambient temperature, 25 °C. Given that the thermal conductivity of the polymer is about 0.3 W K$^{-1}$ m$^{-1}$, at what dc current will the cable fail?

*c.* Rederive $T_i$ in Equation 7.99 by considering that $\rho$ depends on the temperature as $\rho = \rho_o[1 + \alpha_o(T - T_o)]$ (Chapter 2). Recalculate the maximum current in $b$ given that $\alpha_o = 3.9 \times 10^{-3}$ °C$^{-1}$ at 25 °C.

**7.22** **Piezoelectricity** Consider a quartz crystal and a PZT ceramic filter both designed for operation at $f_s = 1$ MHz. What is the bandwidth of each? Given Young's modulus ($Y$), density ($\rho$) for each, and that the filter is a disk with electrodes and is oscillating radially, what is the diameter of the disk for each material? For quartz, $Y = 80$ GPa and $\rho = 2.65$ g cm$^{-3}$. For PZT, $Y = 70$ GPa and $\rho = 7.7$ g m$^{-3}$.

Assume that the velocity of mechanical oscillations in the crystal is $v = \sqrt{Y/\rho}$ and the wavelength $\lambda = v/f_s$. Consider only the fundamental mode ($n = 1$).

**7.23** **Piezoelectric voltage coefficient** The application of a stress $T$ to a piezoelectric crystal leads to a polarization $P$ and hence to an electric field $\mathcal{E}$ in the crystal such that

*Piezoelectric voltage coefficient*

$$\mathcal{E} = gT$$

where $g$ is the *piezoelectric voltage coefficient.* If $\varepsilon_o\varepsilon_r$ is the permittivity of the crystal, show that

$$g = \frac{d}{\varepsilon_o \varepsilon_r}$$

A $BaTiO_3$ sample, along a certain direction (called 3), has $d = 190$ pC $N^{-1}$, and its $\varepsilon_r \approx 1900$ along this direction. What do you expect for its $g$ coefficient for this direction and how does this compare with the measured value of approximately 0.013 $m^2\ C^{-1}$?

**7.24** **Piezoelectricity and the piezoelectric bender**

*a.* Consider using a piezoelectric material in an application as a mechanical positioner where the displacements are expected to be small (as in a scanning tunneling microscope). For the piezoelectric plate shown in Figure 7.65a, we will take $L = 20$ mm, $W = 10$ mm, and $D$ (thickness) = 0.25 mm. Under an applied voltage of $V$, the plate changes length, width, and thickness according to the piezoelectric coefficients $d_{ij}$, relating the applied field along $i$ to the resulting strain along $j$.

Suppose we define direction 3 along the thickness $D$ and direction 1 along the length $L$, as shown in Figure 7.65a. Show that the changes in the thickness and length are

*Piezoelectric effects*

$$\delta D = d_{33}\, V$$

$$\delta L = \left(\frac{L}{D}\right) d_{31}\, V$$

Given $d_{33} \approx 500 \times 10^{-12}$ m $V^{-1}$ and $d_{31} \approx -250 \times 10^{-12}$ m $V^{-1}$, calculate the changes in the length and thickness for an applied voltage of 100 V. What is your conclusion?

*b.* Consider two oppositely poled and joined ceramic plates, A and B, forming a bimorph, as shown in Figure 7.65b. This piezoelectric bimorph is mounted as a cantilever; one end is fixed and the other end is free to move. Oppositely poled means that the electric field elongates A and contracts B, and the two relative motions *bend* the plate. The displacement $h$ of the tip of the cantilever is given by

*Piezoelectric bending*

$$h = \frac{3}{2} d_{31} \left(\frac{L}{D}\right)^2 V$$

What is the deflection of the cantilever for an applied voltage of 100 V? What is your conclusion?

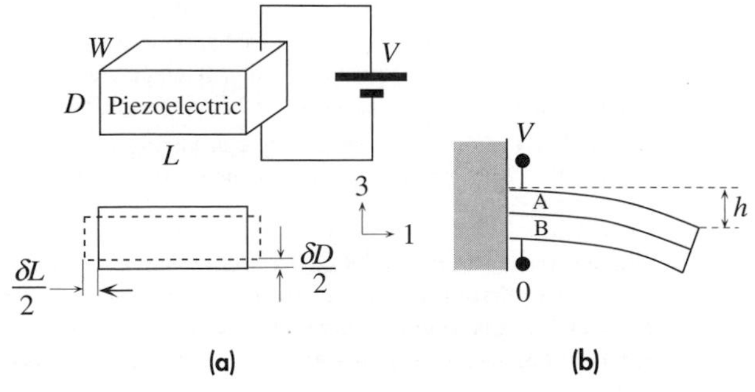


**Figure 7.65**
(a) A mechanical positioner using a piezoelectric plate under an applied voltage of $V$.
(b) A cantilever-type piezoelectric bender. An applied voltage bends the cantilever.

**7.25** **Piezoelectricity** The wavelength $\lambda$ of mechanical oscillations in a piezoelectric slab satisfies

$$n\left(\frac{1}{2}\lambda\right) = L$$

where $n$ is an integer, $L$ is the length of the slab along which mechanical oscillations are set up, and the wavelength $\lambda$ is determined by the frequency $f$ and velocity $v$ of the waves. The ultrasonic wave velocity $v$ depends on Young's modulus $Y$ as

$$v = \left(\frac{Y}{\rho}\right)^{1/2}$$

where $\rho$ is the density. For quartz, $Y = 80$ GPa and $\rho = 2.65$ g cm$^{-3}$. Considering the fundamental mode ($n = 1$), what are practical dimensions for crystal oscillators operating at 1 kHz and 1 MHz?

**7.26** **Pyroelectric detectors** Consider two different radiation detectors using PZT and PVDF as pyroelectric materials whose properties are summarized in Table 7.14. The receiving area is 4 mm$^2$. The thicknesses of the PZT ceramic and the PVDF polymer film are 0.1 mm and 0.005 mm, respectively. In both cases the incident radiation is chopped periodically to allow the radiation to pass for a duration of 0.05 s.

*a.* Calculate the magnitude of the output voltage for each detector if both receive a radiation of intensity 10 μW cm$^{-2}$. What is the corresponding current in the circuit? In practice, what would limit the magnitude of the output voltage?

*b.* What is the minimum detectable radiation intensity if the minimum detectable signal voltage is 10 nV?

**Table 7.14** Properties of PZT and PVDF

| | $\varepsilon_r'$ | Pyroelectric Coefficient ($\times 10^{-6}$ C m$^{-2}$ K$^{-1}$) | Density (g cm$^{-3}$) | Heat Capacity (J K$^{-1}$ g$^{-1}$) |
|---|---|---|---|---|
| PZT | 290 | 380 | 7.7 | 0.3 |
| PVDF | 12 | 27 | 1.76 | 1.3 |

**7.27** **$LiTaO_3$ pyroelectric detector** $LiTaO_3$ (lithium tantalate) detectors are available commercially. $LiTaO_3$ has the following properties: pyroelectric coefficient $p \approx 200 \times 10^{-6}$ C m$^{-2}$ K$^{-1}$, density $\rho = 7.5$ g cm$^{-3}$, specific heat capacity $c_s = 0.43$ J K$^{-1}$ g$^{-1}$. A particular detector has a cylindrical crystal with a diameter of 10 mm and thickness of 0.2 mm. Suppose we chop the input radiation and allow the radiation to fall on the detector for short periods of time. Each input radiation pulse has a duration of $\Delta t = 10$ ms. (The time between the radiation pulses is long, so consider only the response of the detector to a single pulse of radiation.) Suppose that all the incident radiation is absorbed. If the input radiation has an intensity of 10 μW cm$^{-2}$, calculate the pyroelectric current, and the maximum possible output voltage that can be generated assuming that the input impedance of the amplifier is sufficiently large to be negligible. What is the current responsivity of this detector? What are the major assumptions in your calculation of the voltage signal?

***7.28** **Pyroelectric detectors** Consider a typical pyroelectric radiation detector circuit as shown in Figure 7.66. The FET circuit acts as a voltage follower (source follower). The resistance $R_1$ represents the input resistance of the FET in parallel with a bias resistance that is usually inserted between the gate and source. $C_1$ is the overall input capacitance of the FET including any stray capacitance but excluding the capacitance of the pyroelectric detector. Suppose that the incident radiation intensity is constant and equal to $I$. Emissivity $\eta$ of a surface characterizes what fraction of the incident radiation that is absorbed? $\eta I$ is the energy absorbed per unit area per unit time. Some of the absorbed energy will increase

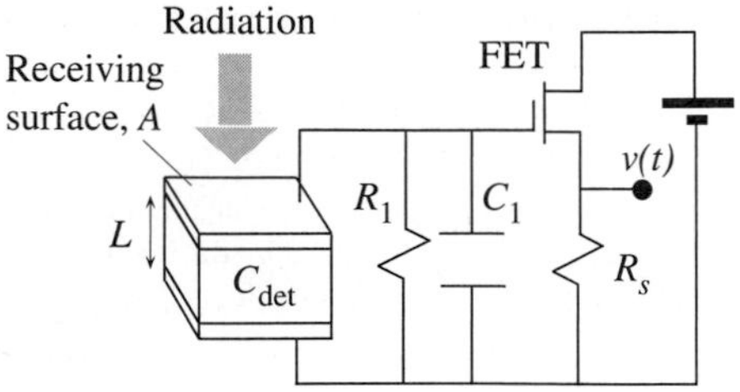


**Figure 7.66** A pyroelectric detector with an FET voltage follower circuit.

the temperature of the detector and some of it will be lost to surroundings by thermal conduction and convection. Let the detector receiving area be $A$, thickness be $L$, density be $\rho$, and specific heat capacity (heat capacity per unit mass) be $c$. The heat losses will be proportional to the temperature difference between the detector temperature $T$ and the ambient temperature $T_o$, as well as the surface area $A$ (much greater than $L$). Energy balance requires that

Rate of increase in the internal energy (heat content) of the detector
= Rate of energy absorption − Rate of heat losses

that is,

$$(AL\rho)c\frac{dT}{dt} = A\eta I - KA(T - T_o)$$

where $K$ is a constant of proportionality that represents the heat losses and hence depends on the thermal conductivity $\kappa$. If the heat loss involves pure thermal conduction from the detector surface to the detector base (detector mount), then $K = \kappa/L$. In practice, this is generally not the case and $K = \kappa/L$ is an oversimplification.

*a.* Show that the temperature of the detector rises exponentially as

*Detector temperature*

$$T = T_o + \frac{\eta I}{K}\left[1 - \exp\left(-\frac{t}{\tau_{th}}\right)\right]$$

where $\tau_{th}$ is a **thermal time constant** defined by $\tau_{th} = L\rho c/K$. Further show that for very small $K$, this equation simplifies to

$$T = T_o + \frac{\eta I}{L\rho c}t$$

*b.* Show that temperature change $dT$ in time $dt$ leads to a pyroelectric current $i_p$ given by

*Pyroelectric current*

$$i_p = Ap\frac{dT}{dt} = \frac{Ap\eta I}{L\rho c}\exp\left(-\frac{t}{\tau_{th}}\right)$$

where $p$ is the pyroelectric coefficient. What is the initial current?

*c.* The voltage across the FET and hence the output voltage $v(t)$ is given by

*Pyroelectric detector output voltage*

$$v(t) = V_o\left[\exp\left(-\frac{t}{\tau_{th}}\right) - \exp\left(-\frac{t}{\tau_{el}}\right)\right]$$

where $V_o$ is a constant and $\tau_{el}$ is the **electrical time constant** given by $R_1C_t$, where $C_t$, total capacitance, is $(C_1 + C_{det})$, where $C_{det}$ is the capacitance of the detector. Consider a particular PZT pyroelectric detector with an area of 1 mm$^2$ and a thickness of 0.05 mm. Suppose that this PZT has $\varepsilon_r = 250$, $\rho = 7.7$ g cm$^{-3}$, $c = 0.3$ J K$^{-1}$ g$^{-1}$, and $\kappa = 1.5$ W K$^{-1}$ m$^{-1}$. The detector is connected to an FET circuit that has $R_1 = 10$ MΩ and $C_1 = 3$ pF. Taking the thermal conduction loss constant $K$ as $\kappa/L$, and $\eta = 1$, calculate $\tau_{th}$ and $\tau_{el}$. Sketch schematically the output voltage. What is your conclusion?

**7.29** **Spark generator design** Design a PLZT piezoelectric spark generator using two back-to-back PLZT crystals that provide a 60 μJ spark in an air gap of 0.5 mm from a force of 50 N. At 1 atm in an air gap of 0.5 mm, the breakdown voltage is about 3000 V. The design will need to specify the dimensions of the crystal and the dielectric constant. Assume that the piezoelectric voltage coefficient is 0.023 V m $N^{-1}$.

**7.30** **Ionic polarization resonance in CsCl** Consider a CsCl crystal which has the following properties. The optical dielectric constant is 2.62, the dc dielectric constant is 7.20, and the lattice parameter $a$ is 0.412 nm. There is only one ion pair ($Cs^+$–$Cl^-$) in the cubic-type unit cell. Calculate (estimate) the ionic resonance absorption frequency and compare the value with the experimentally observed resonance at $3.1 \times 10^{12}$ Hz. What effective value of $Q$ would bring the calculated value to within 10 percent of the experimental value?

**7.31** **Low-$\kappa$ porous dielectrics for microelectronics** Interconnect technologies need lower $\varepsilon_r$ interlayer dielectrics (ILDs) to minimize the interconnect capacitances. These materials are called **low-$\kappa$ dielectrics.**

*a.* Consider fluorinated silicon dioxide, also known as fluorosilicate glass (FSG). Its $\varepsilon_r$ is 3.2. What would be the effective dielectric constant if the ILD is 40 percent porous?

*b.* What should be the starting $\varepsilon_r$ if we need an effective $\varepsilon_r$ less than 2 and the porosity cannot exceed 40 percent?

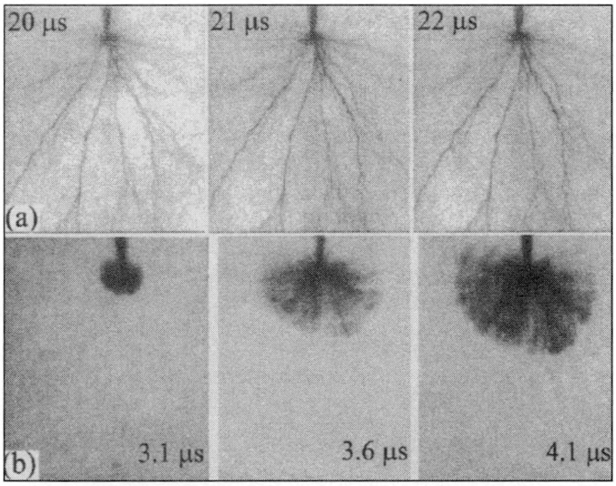


Tree and bush type electrical discharge structures. (a) Voltage $V =$ 160 kV, gap spacing $d = 0.06$ m at various times. (b) Dense bush discharge structure, $V = 300$ kV, $d = 0.06$ m at various times.

SOURCE: V. Lopatin, M. D. Noskov, R. Badent, K. Kist, A. J. Swab, "Positive Discharge Development in Insulating Oil: Optical Observation and Simulation," *IEEE Trans. on Dielec. and Elec. Insulation,* vol. 5, no. 2, 1998, p. 251, figure 2. (© IEEE, 1998)

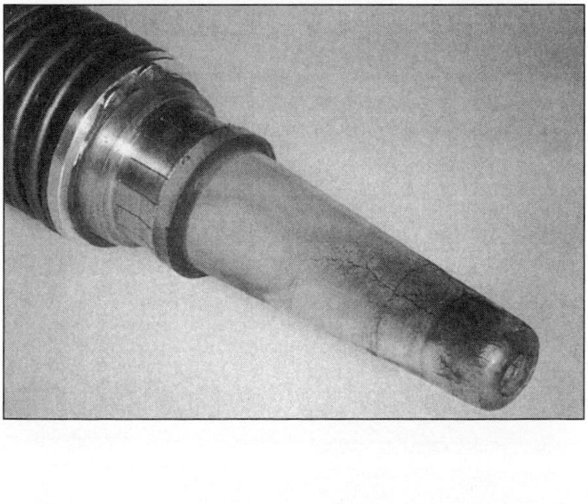

Coaxial cable connector with traces of corona discharge; electrical treeing.

SOURCE: M. Mayer and G. H. Schröder, "Coaxial 30 kV Connectors for the RG220/U Cable: 20 Years of Operational Experience," *IEEE Electrical Insulation Magazine,* vol. **16,** March/April 2000, p. 11, figure 6. (© IEEE, 2000)

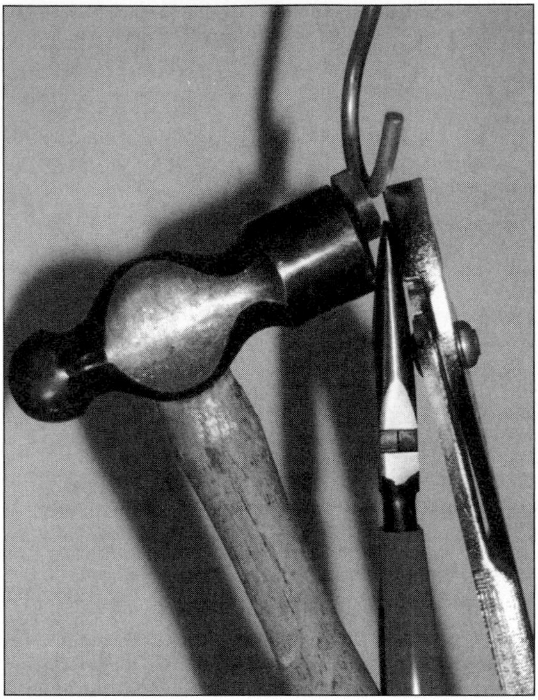

This small neodymium-iron-boron permanent magnet (diameter about the same as one-cent coin) is capable of lifting up to 10 pounds. Nd-Fe-B magnets typically have large $(BH)_{max}$ values (200–275 kJ $m^{-3}$).

In 1986 J. George Bednorz (right) and K. Alex Müller, at IBM Research Laboratories in Zurich, discovered that a copper oxide based ceramic-type compound (La-Ba-Cu-O) which normally has high resistivity becomes superconducting when cooled below 35 K. This Nobel prize winning discovery opened a new era of high-temperature-superconductivity research; now there are various ceramic compounds that are superconducting above the liquid nitrogen (an inexpensive cryogen) temperature (77 K).

| SOURCE: IBM Zürich Research Laboratories.

CHAPTER

# 8

# Magnetic Properties and Superconductivity

Many electrical engineering devices such as inductors, transformers, rotating machines, and ferrite antennas are based on utilizing the magnetic properties of materials. There are many instances where permanent magnets are also used either on their own or as part of a device such as a rotating machine or a loud speaker. The majority of engineering devices make use of the ferromagnetic and ferrimagnetic properties, which are therefore treated in much more detail than other magnetic properties such as diamagnetism and paramagnetism. Although superconductivity involves the vanishing of the resistivity of a conductor at low temperatures and is normally explained within quantum mechanics, we treat the subject in this chapter because all superconductors are perfect diamagnets and, further, they have present or potential uses that involve magnetic fields. The advent of high-$T_c$ superconductivity, discovered in 1986 by George Bednorz and Alex Müller at IBM Research Laboratories in Zürich, is undoubtedly one of the most significant discoveries over the last 50 years, as popularized in various magazines. High-$T_c$ superconductors are already finding applications in such devices as superconducting solenoids, sensitive magnetometers, and high-Q microwave filters.

## 8.1 MAGNETIZATION OF MATTER

### 8.1.1 Magnetic Dipole Moment

Magnetic properties of materials involve concepts based on the magnetic dipole moment. Consider a current loop, as shown in Figure 8.1, where the circulating current is $I$. This may, for example, be a coil carrying a current. For simplicity we will assume that the current loop lies within a single plane. The area enclosed by the current is $A$. Suppose that $\mathbf{u}_n$ is a unit vector coming out from the area $A$. The direction of $\mathbf{u}_n$ is such that

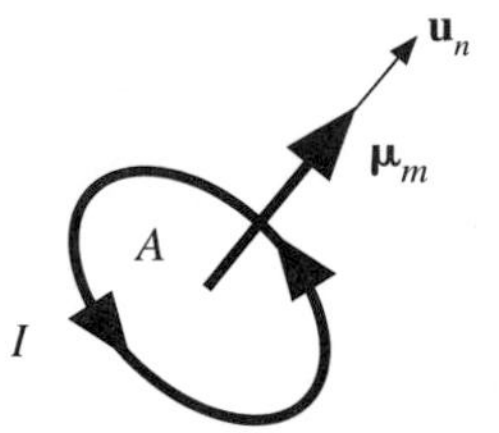


**Figure 8.1** Definition of a magnetic dipole moment.

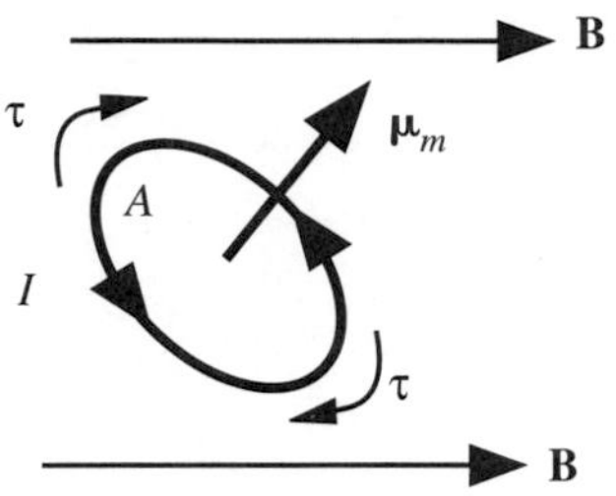


**Figure 8.2** A magnetic dipole moment in an external field experiences a torque.

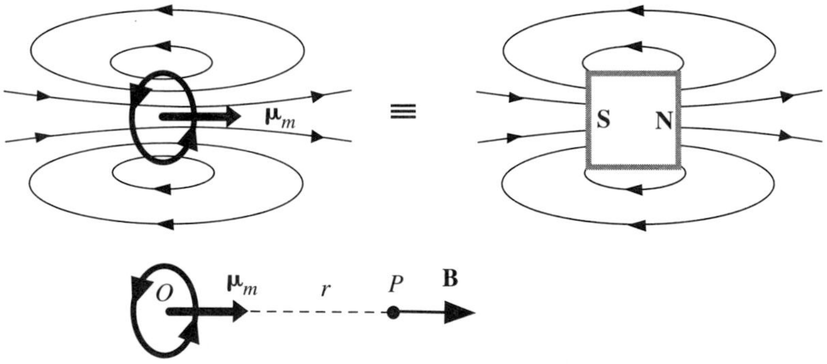


**Figure 8.3** A magnetic dipole moment creates a magnetic field just like a bar magnet.
The field **B** depends on $\boldsymbol{\mu}_m$.

looking along it, the current circulates clockwise. Then the **magnetic dipole moment,** or simply the **magnetic moment** $\boldsymbol{\mu}_m$, is defined by[1]

*Definition of magnetic moment*

$$\boldsymbol{\mu}_m = IA\mathbf{u}_n \tag{8.1}$$

When a magnetic moment is placed in a magnetic field, it experiences a torque that tries to rotate the magnetic moment to align its axis with the magnetic field, as depicted in Figure 8.2. Moreover, since a magnetic moment is a current loop, it gives rise to a magnetic field **B** around it, as shown in Figure 8.3, which is similar to the magnetic field around a bar magnet. We can find the field **B** from the current $I$ and its geometry, which are treated in various physics textbooks. For example, the field **B** at a point $P$ at a distance $r$ along the axis of the coil from the center, as shown in Figure 8.3, is directly proportional to the magnitude of the magnetic moment but inversely proportional to $r^3$, that is, $\mathbf{B} \propto \boldsymbol{\mu}_m / r^3$.

[1] The symbol $\boldsymbol{\mu}$ for the magnetic dipole moment should not be confused with the permeability. Absolute and relative permeabilities will be denoted by $\mu_o$ and $\mu_r$.

### 8.1.2 ATOMIC MAGNETIC MOMENTS

An orbiting electron in an atom behaves much like a current loop and has a magnetic dipole moment associated with it, called the **orbital magnetic moment** ($\mu_{\text{orb}}$), as illustrated in Figure 8.4. If $\omega$ is the angular frequency of the electron, then the current $I$ due to the orbiting electron is

$$I = \text{Charge flowing per unit time} = -\frac{e}{\text{Period}} = -\frac{e\omega}{2\pi}$$

If $r$ is the radius of the orbit, then the magnetic dipole moment is

$$\mu_{\text{orb}} = I(\pi r^2) = -\frac{e\omega r^2}{2}$$

But the velocity $v$ of the electron is $\omega r$ and its orbital angular momentum is

$$L = (m_e v)r = m_e \omega r^2$$

Using this in $\mu_{\text{orb}}$, we get

*Orbital magnetic moment of the electron*

$$\mu_{\text{orb}} = -\frac{e}{2m_e}L \qquad [8.2]$$

We see that the magnetic moment is proportional to the orbital angular momentum through a factor that has the charge to mass ratio of the electron. The numerical factor, in this case $e/2m_e$, relating the angular momentum to the magnetic moment, is called the **gyromagnetic ratio.** The negative sign in Equation 8.2 indicates that $\mu_{\text{orb}}$ is in the opposite direction to $L$ and is due to the negative charge of the electron.

The electron also has an intrinsic angular momentum $S$, that is, spin. The spin of the electron has a **spin magnetic moment,** denoted by $\mu_{\text{spin}}$, but the relationship between $\mu_{\text{spin}}$ and $S$ is not the same as that in Equation 8.2. The gyromagnetic ratio is a factor of 2 greater,

*Spin magnetic moment of the electron*

$$\mu_{\text{spin}} = -\frac{e}{m_e}S \qquad [8.3]$$

The overall magnetic moment of the electron consists of $\boldsymbol{\mu}_{\text{orb}}$ and $\boldsymbol{\mu}_{\text{spin}}$ appropriately added. We cannot simply add them numerically as they are vector quantities. Furthermore, the overall magnetic moment $\boldsymbol{\mu}_{\text{atom}}$ of the atom itself depends on the

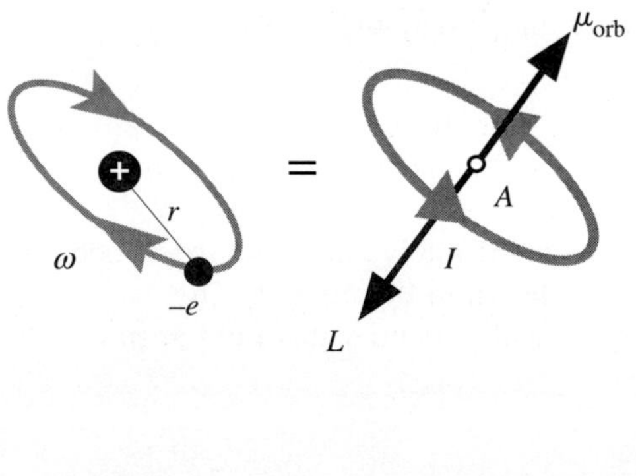


**Figure 8.4** An orbiting electron is equivalent to a magnetic dipole moment $\mu_{\text{orb}}$.

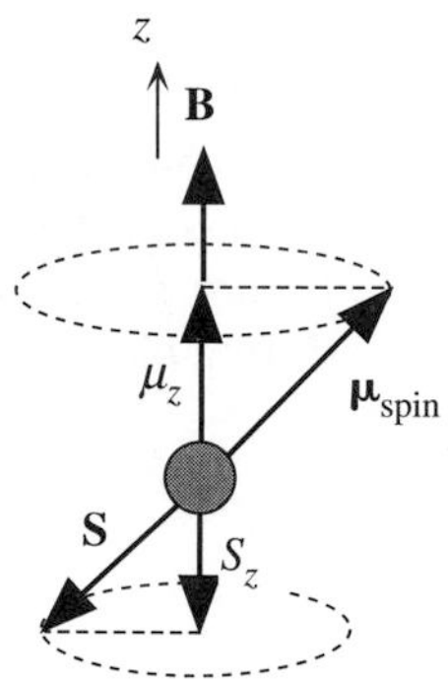


**Figure 8.5** The spin magnetic moment precesses about an external magnetic field along $z$ and has a value $\mu_z$ along $z$.

orbital motions and spins of *all* the electrons. Electrons in closed subshells, however, do not contribute to the overall magnetic moment because for every electron with a given **L** (or **S**), there is another one with an opposite **L** (or **S**). The reason is that the direction of **L** is space quantized by $m_\ell$ and all negative and positive values of $m_\ell$ are occupied in a closed shell. Similarly, there are as many electrons spinning up as there are spinning down, so there is no net electron spin in a closed shell and no net $\boldsymbol{\mu}_{spin}$. Thus, only **unfilled subshells** contribute to the overall magnetic moment of an atom.

Consider an atom that has closed inner shells and a single electron in an $s$ orbital ($\ell = 0$). This means that the orbital magnetic moment is zero and the atom has a magnetic moment due to the spin of the electron alone, $\boldsymbol{\mu}_{atom} = \boldsymbol{\mu}_{spin}$. In the presence of an external magnetic field along the $z$ direction, the magnetic moment cannot simply rotate and align with the field because quantum mechanics requires the spin angular momentum to be space quantized, that is, $S_z$ (the component of **S** along $z$) must be $m_s\hbar$ where $m_s = \pm\frac{1}{2}$ is the spin magnetic quantum number. The torque experienced by the spinning electron causes the spin magnetic moment to precess about the external magnetic field, as shown in Figure 8.5. This precession is such that $S_z = -\frac{1}{2}\hbar$ and leads to an average magnetic moment $\mu_z$ along the field given by Equation 8.3 with $S_z$, that is,

*Magnetic moment along the field*

$$\mu_z = -\frac{e}{m_e}S_z = -\frac{e}{m_e}(m_s\hbar) = \frac{e\hbar}{2m_e} = \beta \qquad [8.4]$$

The quantity $\beta = e\hbar/2m_e$ is called the **Bohr magneton** and has the value $9.27 \times 10^{-24}$ A m$^2$ or J T$^{-1}$.

Thus, the spin of a single electron has a magnetic moment of one Bohr magneton along the field.

### 8.1.3 MAGNETIZATION VECTOR M

Consider a tightly wound long solenoid, ideally infinitely long, with free space (or vacuum) as the medium inside the solenoid, as shown in Figure 8.6a. The magnetic field inside the solenoid is denoted by $\mathbf{B}_o$ to specifically identify this field as in free space.

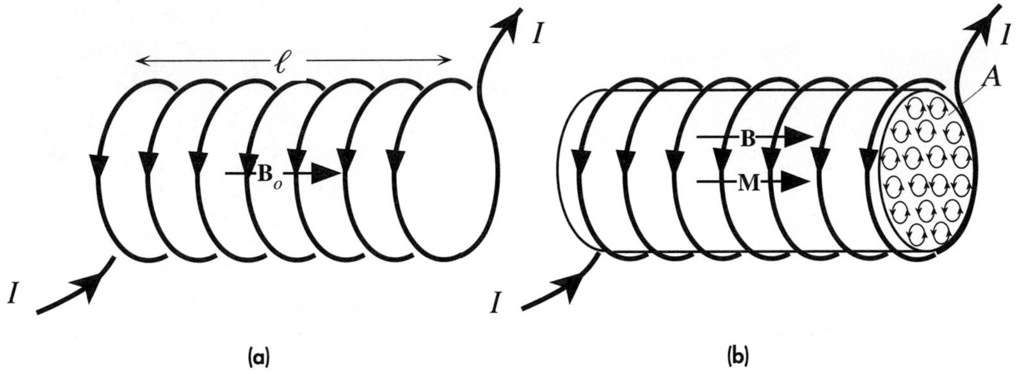


**Figure 8.6**
(a) Consider a long solenoid. With free space as the medium inside, the magnetic field is $\mathbf{B}_o$.
(b) A material medium inserted into the solenoid develops a magnetization **M**.

This field depends on the current $I$ through the solenoid wire and the number of turns per unit length $n$ and is given by[2]

*Free space field inside solenoid*

$$B_o = \mu_o nI = \mu_o I' \quad [8.5]$$

where $I'$ is the current per unit length of the solenoid, that is, $I' = nI$, and $\mu_o$ is the absolute permeability of free space in henries per meter, H $\text{m}^{-1}$.

If we now place a cylindrical material medium to fill the inside of this solenoid, as in Figure 8.6b, we find that the magnetic field has changed. The new magnetic field in the presence of a medium is denoted as **B**. We will take $\mathbf{B}_o$ to be the applied magnetic field into which the material medium is placed.

Each atom of the material responds to the applied field $\mathbf{B}_o$ and develops, or acquires, a net magnetic moment $\boldsymbol{\mu}_m$ along the applied field. We can view each magnetic moment $\boldsymbol{\mu}_m$ as the result of the precession of each atomic magnetic moment about $\mathbf{B}_o$. The medium therefore develops a net magnetic moment along the field and becomes **magnetized.** The magnetic vector **M** describes the extent of magnetization of the medium. **M** is defined as the **magnetic dipole moment per unit volume.** Suppose that there are $N$ atoms in a small volume $\Delta V$ and each atom $i$ has a magnetic moment $\boldsymbol{\mu}_{mi}$ (where $i = 1$ to $N$). Then **M** is defined by

*Magnetization vector*

$$\mathbf{M} = \frac{1}{\Delta V}\sum_{i=1}^{N} \boldsymbol{\mu}_{mi} = n_{\text{at}}\boldsymbol{\mu}_{\text{av}} \quad [8.6]$$

where $n_{\text{at}}$ is the number of atoms per unit volume and $\boldsymbol{\mu}_{\text{av}}$ is the average magnetic moment per atom. We can assume that each atom acquires a magnetic moment $\boldsymbol{\mu}_{\text{av}}$ along $\mathbf{B}_o$. Each of these magnetic moments along $\mathbf{B}_o$ can be viewed as an elementary current loop at the atomic scale, as schematically depicted in Figure 8.6b. These elementary

[2] The proof of this comes out from Ampere's law and can be found in any textbook of electromagnetism.

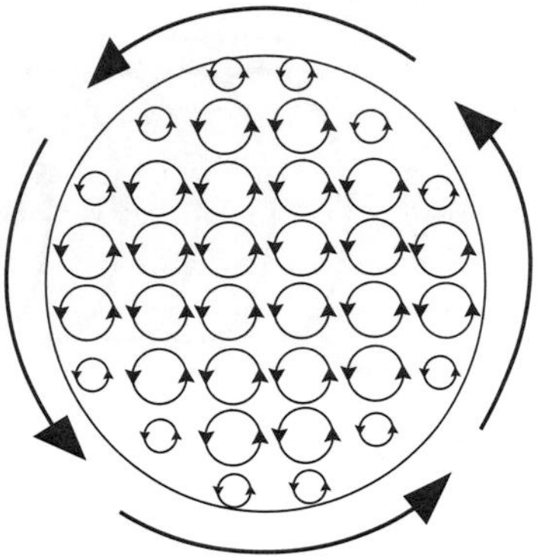


**Figure 8.7** Elementary current loops result in surface currents.
There is no internal current, as adjacent currents on neighboring loops are in opposite directions.

current loops are due to electronic currents within the atom and arise from both orbital and spin motions of the electrons. Each current loop has its current plane normal to $\mathbf{B}_o$.

Consider a cross section of the magnetized medium, as in Figure 8.7. All the elementary current loops in this plane have the current circulation in the same direction inasmuch as each atom acquires the same magnetic moment $\boldsymbol{\mu}_{av}$. All neighboring loops in the bulk have adjacent currents in opposite directions that cancel each other, as apparent in Figure 8.7. Thus, there are no net bulk currents, or internal currents, within the bulk of the material. However, the currents at the surface in the surface loops cannot be canceled and this leads to a net **surface current,** as depicted in Figure 8.7. The surface currents are induced by the magnetization of the medium by the applied magnetic field and therefore depend on the magnetization $M$ of the specimen.

From the definition of $M$, the total magnetic moment of the cylindrical specimen is

$$\text{Total magnetic moment} = M\,(\text{Volume}) = MA\ell$$

Suppose that the magnetization current on the surface per unit length of the specimen is $I_m$. Then the total circulating surface current is $I_m\ell$ and the total magnetic moment of the specimen, by definition, is

$$\text{Total magnetic moment} = (\text{Total current}) \times (\text{Cross-sectional area}) = I_m\ell A$$

*Magnetization and surface currents*

Equating the two total magnetic moments, we find

$$M = I_m \qquad \textbf{[8.7]}$$

We derived this for a particular sample geometry, a cylindrical specimen, in which **M** is along the axis of the cylindrical specimen and $I_m$ flows in a plane perpendicular to **M**. The relationship, however, is more general, as derived in more advanced texts. It should be emphasized that the magnetization current $I_m$ is not due to the flow of free charge carriers, as in a current-carrying copper wire, but due to localized electronic currents within the atoms of the solid at the surface. Equation 8.7 states that we can represent the magnetization of a medium by a surface current per unit length $I_m$ that is equal to $M$.

### 8.1.4 MAGNETIZING FIELD OR MAGNETIC FIELD INTENSITY H

The magnetized specimen in Figure 8.6b placed inside the solenoid develops magnetization currents on the surface. It therefore behaves like a solenoid. We can now regard the solenoid with medium inside, as depicted in Figure 8.8. The magnetic field within the medium now arises from not only the conduction current per unit length $I'$ in the solenoid wires but also from the magnetization current $I_m$ on the surface. The magnetic field $B$ inside the solenoid is now given by the usual solenoid expression but with a current that includes both $I'$ and $I_m$, as shown in Figure 8.8:

$$B = \mu_o(I' + I_m) = B_o + \mu_o M$$

This relationship is generally valid and can be written in vector form as

*Magnetic field in a magnetized medium*

$$\mathbf{B} = \mathbf{B}_o + \mu_o\mathbf{M} \quad [8.8]$$

The field at a point inside a magnetized material is the sum of the applied field $\mathbf{B}_o$ and a contribution from the magnetization $\mathbf{M}$ of the material. The magnetization arises from the application of $\mathbf{B}_o$ due to the current of free carriers in the solenoid wires, called the **conduction current,** which we can externally adjust. It becomes useful to introduce a vector field that represents the effect of the external or conduction current alone. In general, $\mathbf{B} - \mu_o\mathbf{M}$ at a point is the contribution of the external currents alone to the magnetic field at that point inside the material that we called $\mathbf{B}_o$. $\mathbf{B} - \mu_o\mathbf{M}$ represents a magnetizing field because it is the field of the external currents that magnetize the material. The **magnetizing field H** is defined as

*Definition of the magnetizing field*

$$\mathbf{H} = \frac{1}{\mu_o}\mathbf{B} - \mathbf{M} \quad [8.9]$$

or

*Definition of the magnetizing field*

$$\mathbf{H} = \frac{1}{\mu_o}\mathbf{B}_o$$

The magnetizing field is also known as the **magnetic field intensity** and is measured in A m$^{-1}$. The reason for the division by $\mu_o$ is that the resulting vector field **H** becomes simply related to the external conduction currents (through Ampere's law). Since in the solenoid $\mathbf{B}_o$ is $\mu_o nI$, we see that the magnetizing field in a solenoid is

$$H = nI = \text{Total conduction current per unit length} \quad [8.10]$$

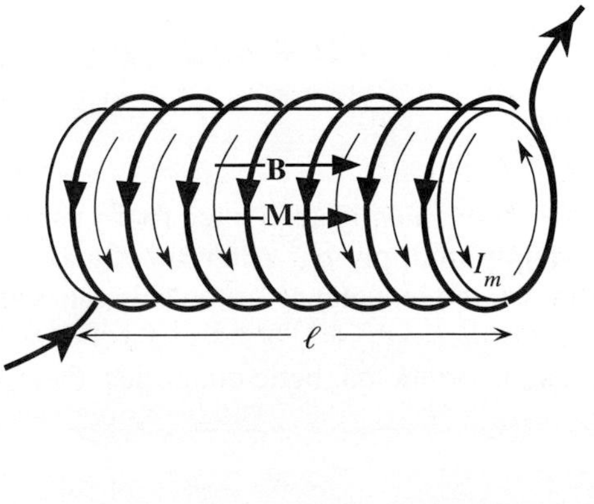


**Figure 8.8** The field **B** in the material inside the solenoid is due to the conduction current $I$ through the wires and the magnetization current $I_m$ on the surface of the magnetized medium, or $\mathbf{B} = \mathbf{B}_o + \mu_o\mathbf{M}$.

It is generally helpful to imagine **H** as the *cause* and **B** as the *effect*. The cause **H** depends only on the external conduction currents, whereas the effect **B** depends on the magnetization **M** of matter.

### 8.1.5 MAGNETIC PERMEABILITY AND MAGNETIC SUSCEPTIBILITY

Suppose that at a point $P$ in a material, the magnetic field is **B** and the magnetizing field is **H**. We let $\mathbf{B}_o$ be the magnetic field at $P$ in the absence of any material (*i.e.*, in free space). The magnetic permeability of the medium at $P$ is defined as the magnetic field per unit magnetizing field,

*Definition of magnetic permeability*

$$\mu = \frac{B}{H} \qquad [8.11]$$

It relates the effect $B$ to the cause $H$ at the same point $P$ inside a material. In simple qualitative terms, $\mu$ represents to what extent a medium is permeable by magnetic fields. Relative permeability $\mu_r$ of a medium is the fractional increase in the magnetic field with respect to the field in free space when a material medium is introduced. For example, suppose that the field in a solenoid with free space in it is $B_o$ but with material inserted it is $B$. Then $\mu_r$ is defined by

*Definition of relative permeability*

$$\mu_r = \frac{B}{B_o} = \frac{B}{\mu_o H} \qquad [8.12]$$

From Equations 8.11 and 8.12, clearly,

$$\mu = \mu_o \mu_r$$

The magnetization **M** produced in a material depends on the net magnetic field **B**. It would be natural to proceed as in dielectrics by relating **M** to **B** analogously to relating $P$ (polarization) to $\mathcal{E}$ (electric field). However, for historic reasons, **M** is related to **H**, the magnetizing field. Suppose that the medium is isotropic (same properties in all directions), then magnetic susceptibility $\chi_m$ of the medium is defined simply by

*Definition of magnetic susceptibility*

$$\mathbf{M} = \chi_m \mathbf{H} \qquad [8.13]$$

This relationship is not obeyed by all magnetic materials. For example, as we will see later, ferromagnetic materials do not obey Equation 8.12. Since the magnetic field

$$\mathbf{B} = \mu_o(\mathbf{H} + \mathbf{M})$$

we have

$$B = \mu_o H + \mu_o M = \mu_o H + \mu_o \chi_m H = \mu_o(1 + \chi_m)H$$

and

*Relative permeability and susceptibility*

$$\mu_r = 1 + \chi_m \qquad [8.14]$$

The presence of a magnetizable material is conveniently accounted for by using the relative permeability $\mu_r$, or $(1 + \chi_m)$, to simply multiply $\mu_o$. Alternatively, one can simply replace $\mu_o$ with $\mu = \mu_o \mu_r$. For example, the inductance of the solenoid with a magnetic medium inside increases by a factor of $\mu_r$.

Table 8.1 provides a summary of various important magnetic quantities, their definitions, and units.

**Table 8.1** Magnetic quantities and their units

| Magnetic Quantity | Symbol | Definition | Units | Comment |
|---|---|---|---|---|
| Magnetic field; magnetic induction | $\mathbf{B}$ | $\mathbf{F} = q\mathbf{v} \times \mathbf{B}$ | T = tesla = webers $m^{-2}$ | Produced by moving charges or currents, acts on moving charges or currents. |
| Magnetic flux | $\Phi$ | $\Delta\Phi = B_{normal}\,\Delta A$ | Wb = weber | $\Delta\Phi$ is flux through $\Delta A$ and $B_{normal}$ is normal to $\Delta A$. Total flux through any closed surface is zero. |
| Magnetic dipole moment | $\boldsymbol{\mu}_m$ | $\mu_m = IA$ | A $m^2$ | Experiences a torque in **B** and a net force in a nonuniform **B**. |
| Bohr magneton | $\beta$ | $\beta = e\hbar/2m_e$ | A $m^2$ or J $T^{-1}$ | Magnetic moment due to the spin of the electron. $\beta = 9.27 \times 10^{-24}$ A $m^2$ |
| Magnetization vector | $\mathbf{M}$ | Magnetic moment per unit volume | A $m^{-1}$ | Net magnetic moment in a material per unit volume. |
| Magnetizing field; magnetic field intensity | $\mathbf{H}$ | $\mathbf{H} = \mathbf{B}/\mu_o - \mathbf{M}$ | A $m^{-1}$ | **H** is due to external conduction currents only and is the cause of **B** in a material. |
| Magnetic susceptibility | $\chi_m$ | $\mathbf{M} = \chi_m\mathbf{H}$ | None | Relates the magnetization of a material to the magnetizing field **H**. |
| Absolute permeability | $\mu_o$ | $c = [\varepsilon_o\mu_o]^{-1/2}$ | H $m^{-1}$ = Wb $m^{-1}$ $A^{-1}$ | A fundamental constant in magnetism. In free space, $\mu_o = B/H$. |
| Relative permeability | $\mu_r$ | $\mu_r = B/\mu_o H$ | None | |
| Magnetic permeability | $\mu$ | $\mu = \mu_o\mu_r$ | H $m^{-1}$ | Not to be confused with magnetic moment. |
| Inductance | $L$ | $L = \Phi_{total}/I$ | H (henries) | Total flux threaded per unit current. |
| Magnetostatic energy density | $E_{vol}$ | $dE_{vol} = H\,dB$ | J $m^{-3}$ | $dE_{vol}$ is the energy required per unit volume in changing $B$ by $dB$. |

**EXAMPLE 8.1**

**AMPERE'S LAW AND THE INDUCTANCE OF A TOROIDAL COIL** Ampere's law provides a relationship between the conduction current $I$ and the magnetic field intensity $H$ threading this current. The conduction current $I$ is the current due to the flow of free charge carriers through a conductor and not due to the magnetization of any medium. Consider an arbitrary closed path $C$ around a conductor carrying a current $I$, as shown in Figure 8.9. The tangential component of $\mathbf{H}$ to the curve $C$ at point $P$ is $H_t$. If $dl$ is an infinitesimally small path length of $C$ at $P$, as shown in Figure 8.9, then the summation of $H_t\,dl$ around the path $C$ gives the conduction current enclosed within $C$. This is **Ampere's law,**

$$\oint_C H_t\,dl = I \qquad [8.15]$$

*Ampere's law*

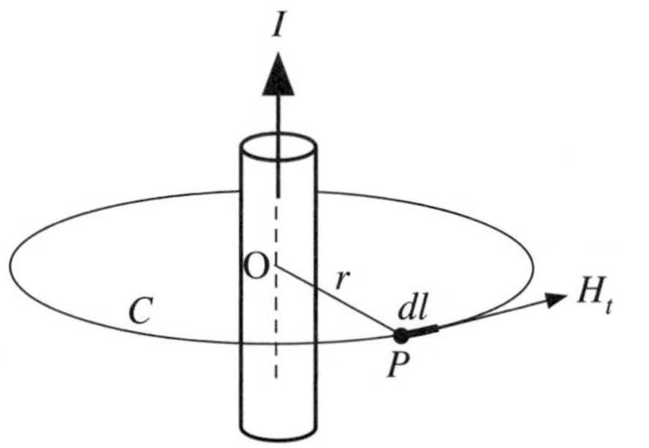


**Figure 8.9** Ampere's circuital law.

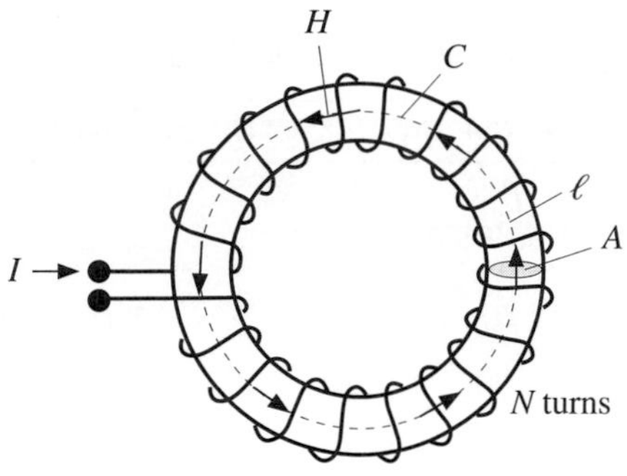


**Figure 8.10** A toroidal coil with $N$ turns.

Consider the toroidal coil with $N$ turns shown in Figure 8.10. First assume that the toroid core is air ($\mu_r \approx 1$). Suppose that the current through the coils is $I$. By symmetry, the magnetic field intensity $H$ inside the toroidal core is the same everywhere and is directed along the circumference. Suppose that $l$ is the length of the mean circumference $C$. The current is linked $N$ times by the circumference $C$, so Equation 8.15 is

$$\oint_C H_t\, dl = H\ell = NI$$

or

$$H = \frac{NI}{\ell}$$

The magnetic field $B_o$ with air as core material is then simply

$$B_o = \mu_o H = \frac{\mu_o NI}{\ell}$$

When the toroidal coil has a magnetic medium with a relative permeability $\mu_r$, the magnetic field intensity is still $H$ because the conduction current $I$ has not changed. But the magnetic field $B$ is now different than $B_o$ and is given by

*Magnetic field inside toroidal coil*

$$B = \mu_o \mu_r H = \frac{\mu_o \mu_r NI}{\ell}$$

If $A$ is the cross-sectional area of the toroid, then the total flux $\Phi$ through the core is $BA$ or $\mu_o \mu_r NAI/\ell$. The current $I$ in Figure 8.10 threads the flux $N$ times. The inductance $L$ of the toroidal coil, by definition, is then

*Inductance of toroidal coil*

$$L = \frac{\text{Total flux threaded}}{\text{Current}} = \frac{N\Phi}{I} = \frac{\mu_o \mu_r N^2 A}{\ell}$$

Having a magnetic material as the toroid core increases the inductance by a factor of $\mu_r$ in the same way a dielectric material increases the capacitance by a factor of $\varepsilon_r$.

**EXAMPLE 8.2** **MAGNETOSTATIC ENERGY PER UNIT VOLUME** Consider a toroidal coil with $N$ turns that is energized from a voltage supply through a rheostat, as shown in Figure 8.11. The core of the toroid may be any material. Suppose that by adjusting the rheostat we increase the current $i$

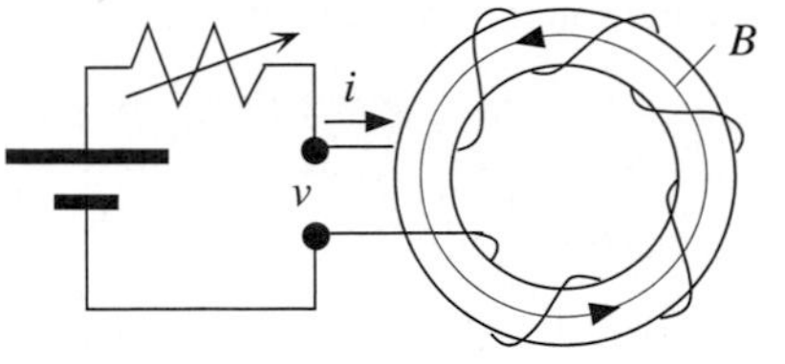


**Figure 8.11** Energy required to magnetize a toroidal coil.

supplied to the coil. The current $i$ produces magnetic flux $\Phi$ in the core, which is $BA$, where $B$ is the magnetic field and $A$ is the cross-sectional area. We can now use Ampere's law for $H$ to relate the current $i$ to $H$, as in Example 8.1. If $\ell$ is the mean circumference, then

$$H\ell = Ni \tag{8.16}$$

The changing current means that the flux is also changing (both increasing). We know from Faraday's law that a changing flux that threads a circuit generates a voltage $v$ in that circuit given by the rate of change of total threaded flux, or $N\Phi$. Lenz's law makes the polarity of the induced voltage oppose the applied voltage. Suppose that in a time interval $\delta t$ seconds, the magnetic field within the core changes by $\delta B$; then $\delta\Phi = A\delta B$ and

$$v = \frac{\delta(\text{Total flux threaded})}{\delta t} = \frac{N\delta\Phi}{\delta t} = NA\frac{\delta B}{\delta t} \tag{8.17}$$

The battery has to supply the current $i$ against this induced voltage $v$, which means that it has to do electrical work $iv$ every second. In other words, the battery has to do work $iv\,\delta t$ in a time interval $\delta t$ to supply the necessary current to increase the magnetic field by $\delta B$. The electric energy $\delta E$ that is input into the coil in time $\delta t$ is then, using Equations 8.16 and 8.17,

$$\delta E = iv\,\delta t = \left(\frac{H\ell}{N}\right)\left(NA\frac{\delta B}{\delta t}\right)\delta t = (A\ell)H\,\delta B$$

This energy $\delta E$ is the work done in increasing the field in the core by $\delta B$. The volume of the toroid is $A\ell$. Therefore, the total energy or work required per unit volume to increase the magnetic field from an initial value $B_1$ to a final value $B_2$ in the toroid is

*Work done per unit volume during magnetization*

$$E_{\text{vol}} = \int_{B_1}^{B_2} H\,dB \tag{8.18}$$

where the integration limits are determined by the initial and final magnetic field. This is the expression for calculating the **energy density** (energy per unit volume) required to change the field from $B_1$ to $B_2$. It should be emphasized that Equation 8.18 is valid for *any medium.* We conclude that an incremental energy density of $dE_{\text{vol}} = H\,dB$ is required to increase the magnetic field by $dB$ at a point in any medium including free space.

We can now consider a core material that we can represent by a *constant* relative permeability $\mu_r$. This means we can exclude those materials that do not have a linear relationship between $B$ and $H$, such as ferromagnetic and ferrimagnetic materials, which we will discuss later. If the core is free space or air, then $\mu_r = 1$.

Suppose that we increase the current in Figure 8.11 from zero to some final value $I$ so that the magnetic field changes from zero to some final value $B$. Since the medium has a constant relative permeability $\mu_r$, we can write

$$B = \mu_r\mu_o H$$

and use this in Equation 8.18 to integrate and find the energy per unit volume needed to establish the field $B$ or field intensity $H$

*Energy density of a magnetic field*

$$E_{\text{vol}} = \frac{1}{2}\mu_r \mu_o H^2 = \frac{B^2}{2\mu_r \mu_o} \qquad [8.19]$$

This is the energy absorbed from the battery per unit volume of core medium to establish the magnetic field. This energy is stored in the magnetic field and is called **magnetostatic energy density.** It is a form of magnetic potential energy. If we were to suddenly remove the battery and short those terminals, the current will continue to flow for a short while (determined by $L/R$) and do external work in heating the resistor. This external work comes from the stored energy in the magnetic field. If the medium is free space, or air, then the energy density is

*Magnetostatic energy density in free space*

$$E_{\text{vol}}(\text{air}) = \frac{1}{2}\mu_o H^2 = \frac{B^2}{2\mu_o}$$

A magnetic field of 2 T corresponds to a magnetostatic energy density of 1.6 MJ $m^{-3}$ or 1.6 J $cm^{-3}$. The energy in a magnetic field of 2 T in a 1 $cm^3$ volume (size of a thimble) has the work ability (potential energy) to raise an average-sized apple by 5 feet. We should note that as long as the core material is linear, that is, $\mu_r$ is independent of the magnetic field itself, magnetostatic energy density can also be written as

*Magnetostatic energy in a linear magnetic medium*

$$E_{\text{vol}} = \frac{1}{2}HB \qquad [8.20]$$

## 8.2 MAGNETIC MATERIAL CLASSIFICATIONS

In general, magnetic materials are classified into five distinct groups: diamagnetic, paramagnetic, ferromagnetic, antiferromagnetic, and ferrimagnetic. Table 8.2 provides a summary of the magnetic properties of these classes of materials.

### 8.2.1 DIAMAGNETISM

Typical diamagnetic materials have a magnetic susceptibility that is negative and small. For example, the silicon crystal is diamagnetic with $\chi_m = -5.2 \times 10^{-6}$. The relative permeability of diamagnetic materials is slightly less than unity. When a diamagnetic substance such as a silicon crystal is placed in a magnetic field, the magnetization vector **M** in the material is in the *opposite* direction to the applied field $\mu_o\mathbf{H}$ and the resulting field **B** within the material is less than $\mu_o\mathbf{H}$. The negative susceptibility can be interpreted as the diamagnetic substance trying to expel the applied field from the material. When a diamagnetic specimen is placed in a nonuniform magnetic field, the magnetization **M** of the material is in the opposite direction to **B** and the specimen experiences a net force toward smaller fields, as depicted in Figure 8.12. A substance exhibits diamagnetism whenever the constituent atoms in the material have closed subshells and shells. This means that each constituent atom has no permanent magnetic moment in the absence of an applied field. Covalent

**Table 8.2** Classification of magnetic materials

| Type | $\chi_m$ (typical values) | $\chi_m$ versus $T$ | Comments and Examples |
|---|---|---|---|
| Diamagnetic | Negative and small ($-10^{-6}$) | $T$ independent | Atoms of the material have closed shells. Organic materials, *e.g.*, many polymers; covalent solids, *e.g.*, Si, Ge, diamond; some ionic solids, *e.g.*, alkalihalides; some metals, *e.g.*, Cu, Ag, Au. |
| | Negative and large ($-1$) | Below a critical temperature | Superconductors |
| Paramagnetic | Positive and small ($10^{-5}$–$10^{-4}$) | Independent of $T$ | Due to the alignment of spins of conduction electrons. Alkali and transition metals. |
| | Positive and small ($10^{-5}$) | Curie or Curie–Weiss law, $\chi_m = C/(T - T_C)$ | Materials in which the constituent atoms have a permanent magnetic moment, *e.g.*, gaseous and liquid oxygen; ferromagnets (Fe), antiferromagnets (Cr), and ferrimagnets ($Fe_3O_4$) at high temperatures. |
| Ferromagnetic | Positive and very large | Ferromagnetic below and paramagnetic above the Curie temperature | May possess a large permanent magnetization even in the absence of an applied field. Some transition and rare earth metals, Fe, Co, Ni, Gd, Dy. |
| Antiferromagnetic | Positive and small | Antiferromagnetic below and paramagnetic above the Néel temperature | Mainly salts and oxides of transition metals, *e.g.*, MnO, NiO, $MnF_2$, and some transition metals, $\alpha$–Cr, Mn. |
| Ferrimagnetic | Positive and very large | Ferrimagnetic below and paramagnetic above the Curie temperature | May possess a large permanent magnetization even in the absence of an applied field. Ferrites. |

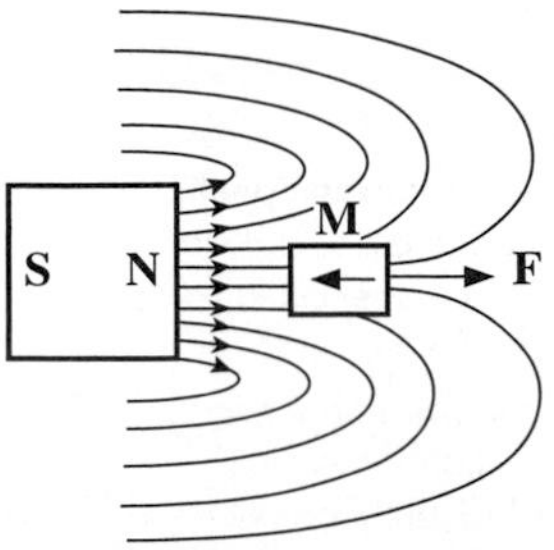


**Figure 8.12** A diamagnetic material placed in a nonuniform magnetic field experiences a force toward smaller fields. This repels the diamagnetic material away from a permanent magnet.

crystals and many ionic crystals are typical diamagnetic materials because the constituent atoms have no unfilled subshells. Superconductors, as we will discuss later, are perfect diamagnets with $\chi_m = -1$ and totally expel the applied field from the material.

### 8.2.2 PARAMAGNETISM

Paramagnetic materials have a small positive magnetic susceptibility. For example, oxygen gas is paramagnetic with $\chi_m = 2.1 \times 10^{-6}$ at atmospheric pressure and room temperature. Each oxygen molecule has a net magnetic dipole moment $\boldsymbol{\mu}_{\text{mol}}$. In the absence of an applied field, these molecular moments are randomly oriented due to the random collisions of the molecules, as depicted in Figure 8.13a. The magnetization of the gas is zero. In the presence of an applied field, the molecular magnetic moments take various alignments with the field, as illustrated in Figure 8.13b. The degree of alignment of $\boldsymbol{\mu}_{\text{mol}}$ with the applied field and hence magnetization **M** increases with the strength of the applied field $\mu_o\mathbf{H}$. Magnetization $M$ typically decreases with increasing temperature because at higher temperatures there are more molecular collisions, which destroy the alignments of molecular magnetic moments with the applied field. When a paramagnetic substance is placed in a nonuniform magnetic field, the induced magnetization **M** is along **B** and there is a net force toward greater fields. For example, when liquid oxygen is poured close to a strong magnet, as depicted in Figure 8.14, the liquid becomes attracted to the magnet.

Many metals are also paramagnetic, such as magnesium with $\chi_m = 1.2 \times 10^{-5}$. The origin of paramagnetism (called **Pauli spin paramagnetism**) in these metals is due to the alignment of the majority of spins of conduction electrons with the field.

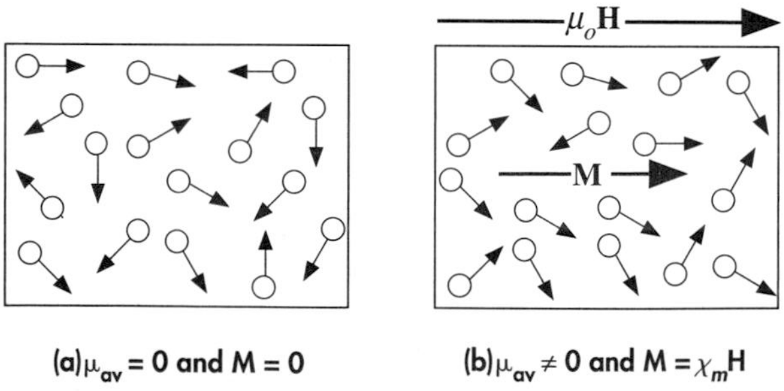


**Figure 8.13**
(a) In a paramagnetic material, each individual atom possesses a permanent magnetic moment, but due to thermal agitation there is no average moment per atom and **M** = 0.
(b) In the presence of an applied field, individual magnetic moments take alignments along the applied field and **M** is finite and along **B**.

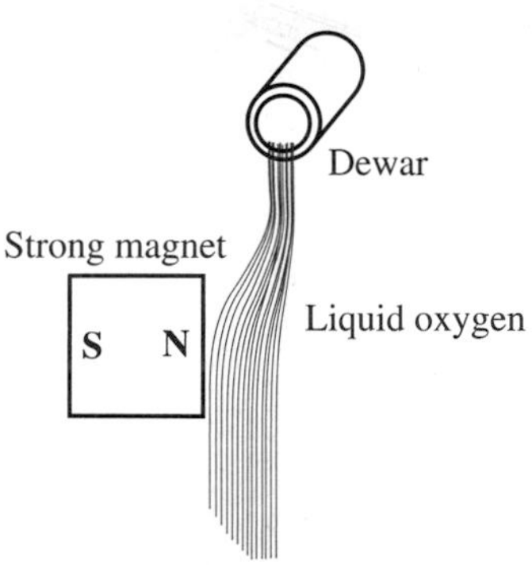


**Figure 8.14** A paramagnetic material placed in a nonuniform magnetic field experiences a force toward greater fields.
This attracts the paramagnetic material (*e.g.*, liquid oxygen) toward a permanent magnet.

### 8.2.3 FERROMAGNETISM

Ferromagnetic materials such as iron can possess large permanent magnetizations even in the absence of an applied magnetic field. The magnetic susceptibility $\chi_m$ is typically positive and very large (even infinite) and, further, depends on the applied field intensity. The relationship between the magnetization **M** and the applied magnetic field $\mu_o\mathbf{H}$ is highly nonlinear. At sufficiently high fields, the magnetization **M** of the ferromagnet saturates. The origin of ferromagnetism is the quantum mechanical exchange interaction (discussed later) between the constituent atoms that results in regions of the material possessing permanent magnetization. Figure 8.15 depicts a region of the Fe crystal, called a **magnetic domain,** that has a net magnetization vector **M** due to the alignment of the magnetic moments of all Fe atoms in this region. This crystal domain has **magnetic ordering** as all the atomic magnetic moments have been aligned parallel to each other. Ferromagnetism occurs below a critical temperature called the Curie temperature $T_C$. At temperatures above $T_C$, ferromagnetism is lost and the material becomes paramagnetic.

### 8.2.4 ANTIFERROMAGNETISM

Antiferromagnetic materials such as chromium have a small but positive susceptibility. They cannot possess any magnetization in the absence of an applied field, in contrast to ferromagnets. Antiferromagnetic materials possess a magnetic ordering in which the magnetic moments of alternating atoms in the crystals align in opposite directions, as schematically depicted in Figure 8.16. The opposite alignments of atomic magnetic moments are due to quantum mechanical exchange forces (described later in Section 8.3). The net result is that in the absence of an applied field, there is no net magnetization. Antiferromagnetism occurs below a critical temperature called the **Néel temperature** $T_N$. Above $T_N$, antiferromagnetic material becomes paramagnetic.

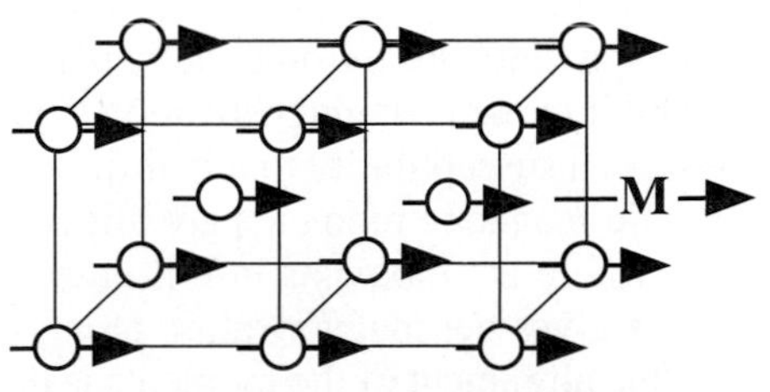


**Figure 8.15** In a magnetized region of a ferromagnetic material such as iron, all the magnetic moments are spontaneously aligned in the same direction.

There is a strong magnetization vector **M** even in the absence of an applied field.

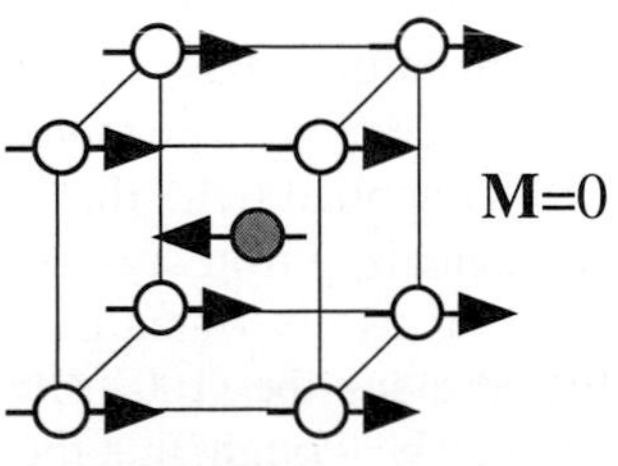


**Figure 8.16** In this antiferromagnetic BCC crystal (Cr), the magnetic moment of the center atom is canceled by the magnetic moments of the corner atoms (one-eighth of the corner atom belongs to the unit cell).

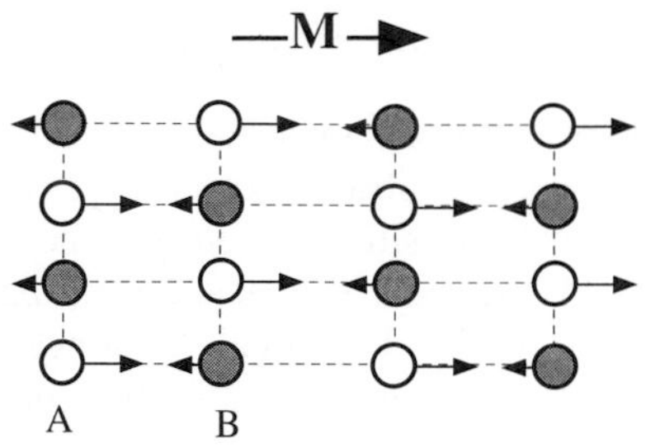


**Figure 8.17** Illustration of magnetic ordering in the ferrimagnetic crystal.
All A atoms have their spins aligned in one direction and all B atoms have their spins aligned in the opposite direction. As the magnetic moment of an A atom is greater than that of a B atom, there is net magnetization **M** in the crystal.

### 8.2.5 FERRIMAGNETISM

Ferrimagnetic materials such as ferrites (*e.g.*, $Fe_3O_4$) exhibit magnetic behavior similar to ferromagnetism below a critical temperature called the Curie temperature $T_C$. Above $T_C$ they become paramagnetic. The origin of ferrimagnetism is based on magnetic ordering, as schematically illustrated in Figure 8.17. All A atoms have their spins aligned in one direction and all B atoms have their spins aligned in the opposite direction. As the magnetic moment of an A atom is greater than that of a B atom, there is net magnetization **M** in the crystal. Unlike the antiferromagnetic case, the oppositely directed magnetic moments have different magnitudes and do not cancel. The net effect is that the crystal can possess magnetization even in the absence of an applied field. Since ferrimagnetic materials are typically nonconducting and therefore do not suffer from eddy current losses, they are widely used in high-frequency electronics applications.

All useful magnetic materials in electrical engineering are invariably ferromagnetic or ferrimagnetic.

## 8.3 FERROMAGNETISM ORIGIN AND THE EXCHANGE INTERACTION

The transition metals iron, cobalt, and nickel are all ferromagnetic at room temperature. The rare earth metals gadolinium and dysprosium are ferromagnetic below room temperature. Ferromagnetic materials can exhibit permanent magnetization even in the absence of an applied field; that is, they possess a susceptibility that is infinite.

In a magnetized iron crystal, all the atomic magnetic moments are aligned in the same direction, as illustrated in Figure 8.15, where the moments in this case have all been aligned along the [100] direction, which gives net magnetization along this direction. It may be thought that the reason for the alignment of the moments is the magnetic forces between the moments, just as bar magnets will tend to align head to tail in an SNSN . . . fashion. This is not, however, the cause, as the magnetic potential energy of interaction is small, indeed smaller than the thermal energy.

The iron atom has the electron structure [Ar]$3d^6 4s^2$. An isolated iron atom has only the $3d$ subshell with four of the five orbitals unfilled. By virtue of Hund's rule, the electrons try to align their spins so that the five $3d$ orbitals contain two paired electrons

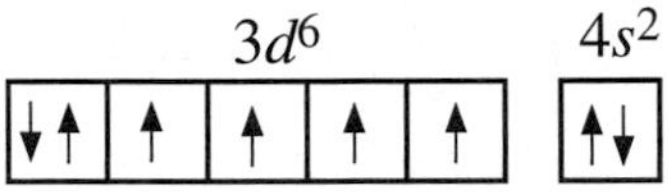


**Figure 8.18** The isolated Fe atom has four unpaired spins and a spin magnetic moment of $4\beta$.

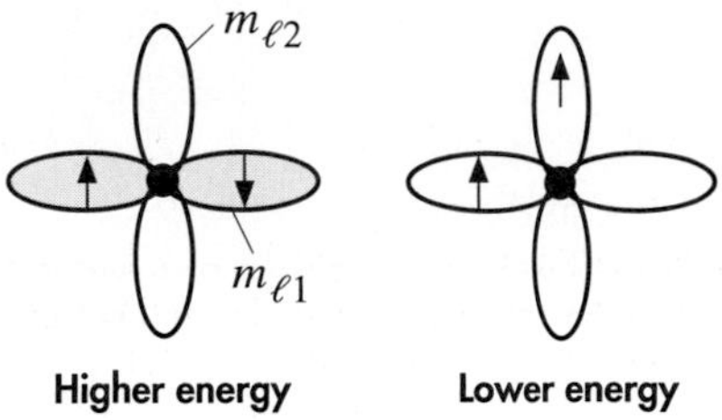


**Figure 8.19** Hund's rule for an atom with many electrons is based on the exchange interaction.

and four unpaired electrons, as in Figure 8.18. The isolated atom has four parallel electron spins and hence a spin magnetic moment of $4\beta$.

The origin of Hund's rule, visualized in Figure 8.19, lies in the fact that when the spins are parallel (same $m_s$), as a requirement of the Pauli exclusion principle, the electrons must occupy orbitals with different $m_\ell$ and hence possess different spatial distributions (recall that $m_\ell$ determines the orientation of an orbit). Different $m_\ell$ values result in a smaller Coulombic repulsion energy between the electrons compared with the case where the electrons have opposite spins (different $m_s$), where they would be in the same orbital (same $m_\ell$), that is, in the same spatial region. It is apparent that even though the interaction energy between the electrons has nothing to do with magnetic forces, it does depend nonetheless on the orientations of their spins ($m_s$), or on their spin magnetic moments, and it is less when the spins are parallel. Two electrons parallel their spins not because of the direct magnetic interaction between the spin magnetic moments but because of the **Pauli exclusion principle** and the **electrostatic interaction energy.** Together they constitute what is known as an **exchange interaction,** which forces two electrons to take $m_s$ and $m_\ell$ values that result in the minimum of electrostatic energy. In an atom, the exchange interaction therefore forces two electrons to take the same $m_s$ but different $m_\ell$ if this can be done within the Pauli exclusion principle. This is the reason an isolated Fe atom has four unpaired spins in the $3d$ subshell.

In the crystal, of course, the outer electrons are no longer strictly confined to their parent Fe atoms, particularly the $4s$ electrons. The electrons now have wavefunctions that belong to the whole solid. Something like Hund's rule also operates at the crystal level for Fe, Co, and Ni. If two $3d$ electrons parallel their spins and occupy different wavefunctions (and hence different negative charge distributions), the resulting mutual Coulombic repulsion between them and also with all the other electrons and the attraction to the positive Fe ions result in an overall reduction of potential energy. This reduction in energy is again due to the exchange interaction and is a direct consequence of the Pauli exclusion principle and the Coulombic forces. Thus, the majority of $3d$ electrons spontaneously parallel their spins without the need for the application of an external magnetic field. The number of electrons that actually parallel their spins depends on the strength of the exchange interaction, and for the iron crystal this turns out to be about 2.2 electrons per atom. Since typically the wavefunctions

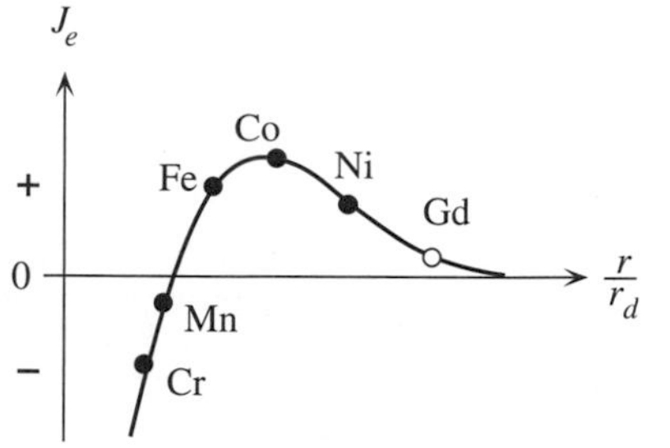


**Figure 8.20** The exchange integral as a function of $r/r_d$, where $r$ is the interatomic distance and $r_d$ the radius of the $d$ orbit (or the average $d$ subshell radius).
Cr to Ni are transition metals. For Gd, the $x$ axis is $r/r_f$, where $r_f$ is the radius of the $f$ orbit.

of the $3d$ electrons in the whole iron crystal show localization around the iron ions, some people prefer to view the $3d$ electrons as spending the majority of their time around Fe atoms, which explains the reason for drawing the magnetized iron crystal as in Figure 8.15.

It may be thought that all solids should follow the example of Fe and become spontaneously ferromagnetic since paralleling spins would result in different spatial distributions of negative charge and probably a reduction in the electrostatic energy, but this is not generally the case at all. We know that, in the case of covalent bonding, the electrons have the lowest energy when the two electrons spin in opposite directions. In covalent bonding in molecules, the exchange interaction does not reduce the energy. Making the electron spins parallel leads to spatial negative charge distributions that result in a net mutual electrostatic repulsion between the positive nuclei.

In the simplest case, for two atoms only, the exchange energy depends on the interatomic separation between two interacting atoms and the relative spins of the two outer electrons (labeled as 1 and 2). From quantum mechanics, the exchange interaction can be represented in terms of an exchange energy $E_{ex}$ as

$$E_{ex} = -2J_e \mathbf{S}_1 \cdot \mathbf{S}_2 \qquad [8.21]$$

where $\mathbf{S}_1$ and $\mathbf{S}_2$ are the spin angular momenta of the two electrons and $J_e$ is a numerical quantity called the **exchange integral** that involves integrating the wavefunctions with the various potential energy interaction terms. It therefore depends on the electrostatic interactions and hence on the interatomic distance. For the majority of solids, $J_e$ is negative, so the exchange energy is negative if $\mathbf{S}_1$ and $\mathbf{S}_2$ are in the opposite directions, that is, the spins are antiparallel (as we found in covalent bonding). This is the antiferromagnetic state. For Fe, Co, and Ni, however, $J_e$ is positive. $E_{ex}$ is then negative if $\mathbf{S}_1$ and $\mathbf{S}_2$ are parallel. Spins of the $3d$ electrons on the Fe atoms therefore spontaneously align in the same direction to reduce the exchange energy. This spontaneous magnetization is the phenomenon of ferromagnetism. Figure 8.20 illustrates how $J_e$ changes with the ratio of interatomic separation to the radius of the $3d$ subshell ($r/r_d$). For the transition metals Fe, Co, and Ni, the $r/r_d$ is such that $J_e$ is positive.[3] In all other cases, it is negative and does not produce ferromagnetic behavior. It should be

[3] According to H. P. Myers, *Introductory Solid State Physics* 2nd ed., London: Taylor and Francis Ltd., 1997, p. 362, there have been no theoretical calculations of the exchange integral $J_e$ for any real magnetic substance.

mentioned that Mn, which is not ferromagnetic, can be alloyed with other elements to increase $r/r_d$ and hence endow ferromagnetism in the alloy.

**EXAMPLE 8.3**

**SATURATION MAGNETIZATION IN IRON** The maximum magnetization, called **saturation magnetization** $M_{sat}$, in iron is about $1.75 \times 10^6$ A m$^{-1}$. This corresponds to all possible net spins aligning parallel to each other. Calculate the effective number of Bohr magnetons per atom that would give $M_{sat}$, given that the density and relative atomic mass of iron are 7.86 g cm$^{-3}$ and 55.85, respectively.

**SOLUTION**

The number of Fe atoms per unit volume is

$$n_{at} = \frac{\rho N_A}{M_{at}} = \frac{(7.86 \times 10^3 \text{ kg m}^{-3})(6.022 \times 10^{23} \text{ mol}^{-1})}{55.85 \times 10^{-3} \text{ kg mol}^{-1}}$$

$$= 8.48 \times 10^{28} \text{ atoms m}^{-3}$$

If each Fe atom contributes $x$ number of net spins, then since each net spin has a magnetic moment of $\beta$, we have,

$$M_{sat} = n_{at}(x\beta)$$

so

$$x = \frac{M_{sat}}{n_{at}\beta} = \frac{1.75 \times 10^6}{(8.48 \times 10^{28})(9.27 \times 10^{-24})} \approx 2.2$$

In the solid, each Fe atom contributes only 2.2 Bohr magnetons to the magnetization even though the isolated Fe atom has 4 Bohr magnetons. There is no orbital contribution to the magnetic moment per atom in the solid because all the outer electrons, $3d$ and $4s$ electrons, can be viewed as belonging to the whole crystal, or being in an energy band, rather than orbiting individual atoms. A $3d$ electron is attracted by various Fe ions in the crystal and therefore does not experience a central force, in contrast to the $3d$ electron in the isolated Fe atom that orbits the nucleus. The orbital momentum in the crystal is said to be quenched.

We should note that when the magnetization is saturated, all atomic magnetic moments are aligned. The resulting magnetic field within the iron specimen in the absence of an applied magnetizing field ($H = 0$) is

$$B_{sat} = \mu_o M_{sat} = 2.2 \text{ T}$$

## 8.4 SATURATION MAGNETIZATION AND CURIE TEMPERATURE

The maximum magnetization in a ferromagnet when all the atomic magnetic moments have been aligned as much as possible is called the saturation magnetization $M_{sat}$. In the iron crystal, for example, this corresponds to each Fe atom with an effective spin magnetic moment of 2.2 Bohr magnetons aligning in the same direction to give a magnetic field $\mu_o M_{sat}$ or 2.2 T. As we increase the temperature, lattice vibrations become more energetic, which leads to a frequent disruption of the alignments of the spins. The spins cannot align perfectly with each other as the temperature increases due to lattice vibrations

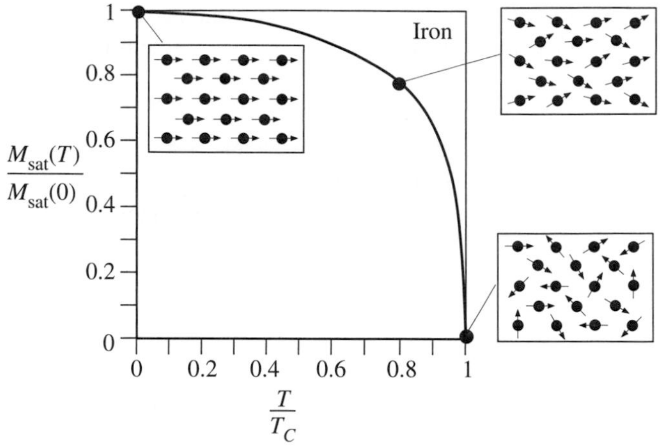


**Figure 8.21** Normalized saturated magnetization versus reduced temperature $T/T_C$ where $T_C$ is the Curie temperature (1043 K).

randomly agitating the individual spins. When an energetic lattice vibration passes through a spin site, the energy in the vibration may be sufficient to disorientate the spin of the atom. The ferromagnetic behavior disappears at a critical temperature called the **Curie temperature,** denoted by $T_C$, when the thermal energy of lattice vibrations in the crystal can overcome the potential energy of the exchange interaction and hence destroy the spin alignments. Above the Curie temperature, the crystal behaves as if it were paramagnetic. The saturation magnetization $M_{sat}$, therefore, decreases from its maximum value $M_{sat}(0)$ at absolute zero of temperature to zero at the Curie temperature. Figure 8.21 shows the dependence of $M_{sat}$ on the temperature when $M_{sat}$ has been normalized to $M_{sat}(0)$ and the temperature is the reduced temperature, that is, $T/T_C$. At $T/T_C = 1$, $M_{sat} = 0$. When plotted in this way, the ferromagnets cobalt and nickel follow closely the observed behavior for iron. We should note that since for iron $T_C = 1043$ K, at room temperature, $T/T_C = 0.29$ and $M_{sat}$ is very close to its value at $M_{sat}(0)$.

Since at the Curie temperature, the thermal energy, of the order of $kT_C$, is sufficient to overcome the energy of the exchange interaction $E_{ex}$ that aligns the spins, we can take $kT_C$ as an order of magnitude estimate of $E_{ex}$. For iron, $E_{ex}$ is ~0.09 eV and for cobalt this is ~0.1 eV.

Table 8.3 summarizes some of the important properties of the ferromagnets Fe, Co, Ni, and Gd (rare earth metal).

**Table 8.3** Properties of the ferromagnets Fe, Co, Ni, and Gd

| | Fe | Co | Ni | Gd |
|---|---|---|---|---|
| Crystal structure | BCC | HCP | FCC | HCP |
| Bohr magnetons per atom | 2.22 | 1.72 | 0.60 | 7.1 |
| $M_{sat}(0)$ (MA m$^{-1}$) | 1.75 | 1.45 | 0.50 | 2.0 |
| $B_{sat} = \mu_o M_{sat}$(T) | 2.2 | 1.82 | 0.64 | 2.5 |
| $T_C$ | 770 °C | 1127 °C | 358 °C | 16 °C |
| | 1043 K | 1400 K | 631 K | 289 K |

# 8.5 MAGNETIC DOMAINS: FERROMAGNETIC MATERIALS

## 8.5.1 MAGNETIC DOMAINS

A single crystal of iron does not necessarily possess a net permanent magnetization in the absence of an applied field. If a magnetized piece of iron is heated to a temperature above its Curie temperature and then allowed to cool in the absence of a magnetic field, it will possess no net magnetization. The reason for the absence of net magnetization is due to the formation of magnetic domains that effectively cancel each other, as discussed below. A **magnetic domain** is a region of the crystal in which all the spin magnetic moments are aligned to produce a magnetic moment in one direction only.

Figure 8.22a shows a single crystal of iron that has a permanent magnetization as a result of ferromagnetism (aligning of all atomic spins). The crystal is like a bar magnet with magnetic field lines around it. As we know, there is potential energy (*PE*), called **magnetostatic energy,** stored in a magnetic field, and we can reduce this energy in the external field by dividing the crystal into two domains where the magnetizations are in the opposite directions, as shown in Figure 8.22b. The external magnetic field lines are reduced and there is now less potential energy stored in the magnetic field. There are only field lines at the ends. This arrangement is energetically favorable because the magnetostatic energy has been reduced by decreasing the external field

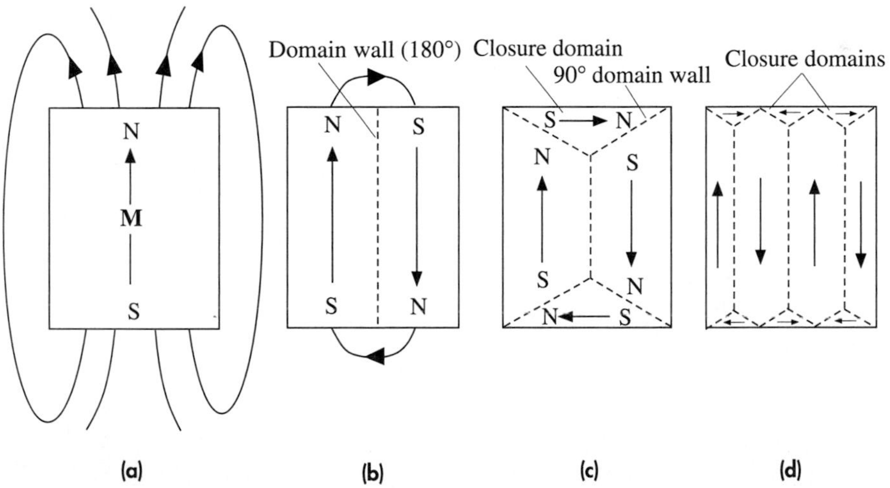


**Figure 8.22**

(a) Magnetized bar of ferromagnet in which there is only one domain and hence an external magnetic field.

(b) Formation of two domains with opposite magnetizations reduces the external field. There are, however, field lines at the ends.

(c) Domains of closure fitting at the ends eliminate the external fields at the ends.

(d) A specimen with several domains and closure domains. There is no external magnetic field and the specimen appears unmagnetized.

lines. However, there is now a boundary, called a **domain wall** (or **Bloch wall**), between the two domains where the magnetization changes from one direction to the opposite direction and hence the atomic spins do, also. It requires energy to rotate the atomic spin through 180° with respect to its neighbor because the exchange energy favors aligning neighboring atomic spins (0°). The wall in Figure 8.22b is a 180° wall inasmuch as the magnetization through the wall is rotated by 180°. It is apparent that the wall region where the neighboring atomic spins change their relative direction (or orientation) from one domain to the neighboring one has higher *PE* than the bulk of the domain, where all the atomic spins are aligned. As we will show below, the domain wall is not simply one atomic spacing but has a finite thickness, which for iron is typically of the order of 0.1 μm, or several hundred atomic spacings. The excess energy in the wall increases with the area of the wall.

The magnetostatic energy associated with the field lines at the ends in Figure 8.22b can be further reduced by eliminating these external field lines by closing the ends with sideway domains with magnetizations at 90°, as shown in Figure 8.22c. These end domains are **closure domains** and have walls that are 90° walls. The magnetization is rotated through 90° through the wall. Although we have reduced the magnetostatic energy, we have increased the potential energy in the walls by adding additional walls. The creation of magnetic domains continues (spontaneously) until the potential energy reduction in creating an additional domain is the same as the increase in creating an additional wall. The specimen then possesses minimum potential energy and is in equilibrium with no net magnetization. Figure 8.22d shows a specimen with several domains and no net magnetization. The sizes, shapes, and distributions of domains depend on a number of factors, including the size and shape of the whole specimen. For iron particles of dimensions less than of the order of 0.01 μm, the increase in the potential energy in creating a domain wall is too costly and these particles are single domains and hence always magnetized.

The magnetization of each domain is normally along one of the preferred directions in which the atomic spin alignments are easiest (the exchange interaction is the strongest). For iron, the magnetization is easiest along any one of six ⟨100⟩ directions (along cube edges), which are called **easy directions.** The domains have magnetizations along these easy directions. The magnetization of the crystal along an applied field occurs, in principle, by the growth of domains with magnetizations (or components of **M**) along the applied field (**H**), as illustrated in Figure 8.23a and b. For simplicity, the magnetizing field is taken along an easy direction. The Bloch wall between the domains A and B migrates toward the right, which enlarges the domain A and shrinks domain B, with the net result that the crystal has an effective magnetization **M** along **H**. The migration of the Bloch wall is caused by the spins in the wall, and also spins in section B adjacent to the wall, being gradually rotated by the applied field (they experience a torque). The magnetization process therefore involves the motions of Bloch walls in the crystal.

## 8.5.2 MAGNETOCRYSTALLINE ANISOTROPY

Ferromagnetic crystals characteristically exhibit magnetic anisotropy, which means that the magnetic properties are different along different crystal directions. In the case of iron (BCC), the spins in a domain are most easily aligned in any of the six [100] type

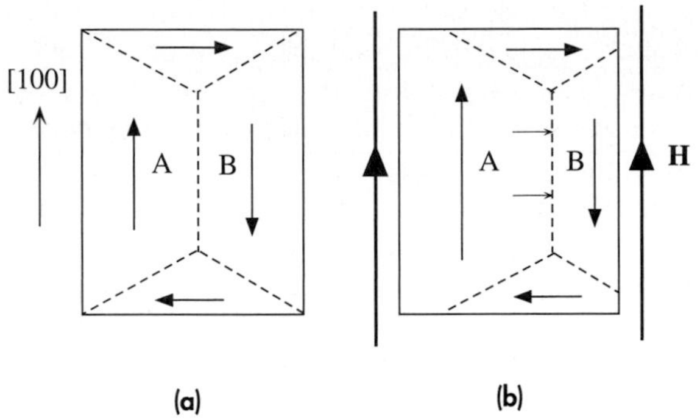


**Figure 8.23**
(a) An unmagnetized crystal of iron in the absence of an applied magnetic field. Domains A and B are the same size and have opposite magnetizations.
(b) When an external magnetic field is applied, the domain wall migrates into domain B, which enlarges A and shrinks B. The result is that the specimen now acquires net magnetization.

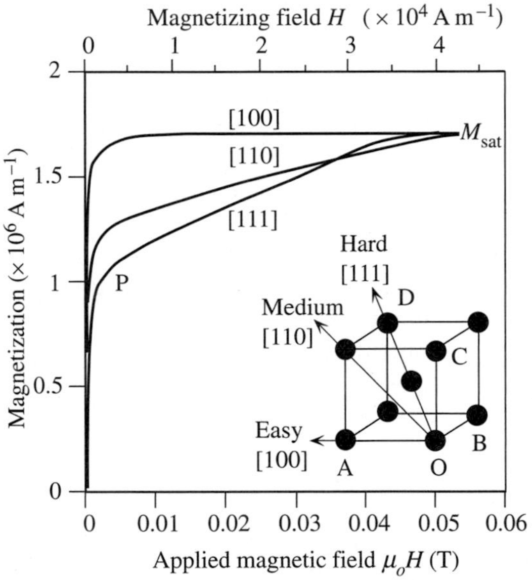


**Figure 8.24** Magnetocrystalline anisotropy in a single iron crystal.
$M$ versus $H$ depends on the crystal direction and is easiest along [100] and hardest along [111].

directions, collectively labeled as ⟨100⟩, and correspond to the six edges of the cubic unit cell. The exchange interactions are such that spin magnetic moments are most easily aligned with each other if they all point in one of the six ⟨100⟩ directions. Thus ⟨100⟩ directions in the iron crystal constitute the easy directions for magnetization. When a magnetizing field **H** along a [100] direction is applied, as illustrated in Figure 8.23a and b, domain walls migrate to allow those domains (*e.g.*, A) with magnetizations along **H** to grow at the expense of those domains (*e.g.*, B) with magnetizations opposing **H**. The observed $M$ versus $H$ behavior is shown in Figure 8.24. Magnetization rapidly increases and saturates with an applied field of less than 0.01 T.

On the other hand, if we want to magnetize the crystal along the [111] direction by applying a field along this direction, then we have to apply a stronger field than that along [100]. This is clearly shown in Figure 8.24, where the resulting magnetization along [111] is smaller than that along [100] for the same magnitude of applied field. Indeed, saturation is reached at an applied field that is about a factor of 4 greater than

**Table 8.4** Exchange interaction, magnetocrystalline anisotropy energy $K$, and saturation magnetostriction coefficient $\lambda_{sat}$

| Material | Crystal | $E_{ex} \approx kT_C$ (meV) | Easy | Hard | $K$ (mJ cm$^{-3}$) | $\lambda_{sat}$ ($\times 10^{-6}$) |
|---|---|---|---|---|---|---|
| Fe | BCC | 90 | <100>; cube edge | <111>; cube diagonal | 48 | 20 [100]<br>−20 [111] |
| Co | HCP | 120 | // to $c$ axis | ⊥ to $c$ axis | 450 | |
| Ni | FCC | 50 | <111>; cube diagonal | <100>; cube edge | 5 | −46 [100]<br>−24 [111] |

NOTE: $K$ is the magnitude of what is called the first anisotropy constant ($K_1$) and is approximately the magnitude of the anisotropy energy. $E_{ex}$ is an estimate from $kT_C$, where $T_C$ is the Curie temperature. All approximate values are from various sources. (Further data can be found in D. Jiles, *Introduction to Magnetism and Magnetic Materials*, London, England: Chapman and Hall, 1991.)

that along [100]. The [111] direction in the iron crystal is consequently known as the **hard direction.** The $M$ versus $H$ behavior along [100], [110], and [111] directions in an iron crystal and the associated anisotropy are shown in Figure 8.24.

When an external field is applied along the diagonal direction OD in Figure 8.24, initially all those domains with **M** along OA, OB, and OC, that is, those with magnetization components along OD, grow by consuming those with **M** in the wrong direction and eventually take over the whole specimen. This is an easy process (similar to the process along [100]) and requires small fields and represents the processes from 0 to P on the magnetization curve for [111] in Figure 8.24. However, from P onwards, the magnetizations in the domains have to be rotated away from their easy directions, that is, from OA, OB, and OC toward OD. This process consumes substantial energy and hence needs much stronger applied fields.

It is apparent that the magnetization of the crystal along [100] needs the least energy, whereas that along [111] consumes the greatest energy. The excess energy required to magnetize a unit volume of a crystal in a particular direction with respect to that in the easy direction is called the **magnetocrystalline anisotropy energy** and is denoted by $K$. For iron, the anisotropy energy is zero for [100] and largest for the [111] direction, about 48 kJ m$^{-3}$ or $3.5 \times 10^{-6}$ eV per atom. For cobalt, which has the HCP crystal structure, the anisotropy energy is at least an order of magnitude greater. Table 8.4 summarizes the easy and hard directions, and the anisotropy energy $K$ for the hard direction.

## 8.5.3 Domain Walls

We recall that the spin magnetic moments rotate across a domain wall. We mentioned that the wall is not simply one atomic spacing wide, as this would mean two neighboring spins being at 180° to each other and hence possessing excessive exchange interaction. A schematic illustration of the structure of a typical 180° Bloch wall, between two domains A and B, is depicted in Figure 8.25. It can be seen that the neighboring spin magnetic moments are rotated gradually, and over several hundred atomic spacings the magnetic moment reaches a rotation of 180°. Exchange

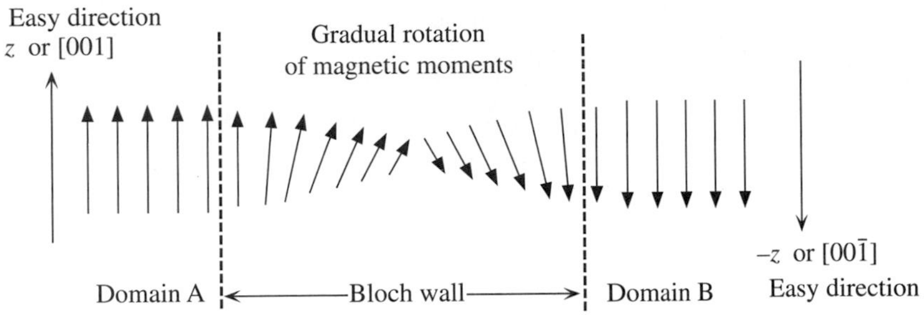


**Figure 8.25** In a Bloch wall, the neighboring spin magnetic moments rotate gradually, and it takes several hundred atomic spacings to rotate the magnetic moment by 180°.

forces between neighboring atomic spins favor very little relative rotation. Had it been left to exchange forces alone, relative rotation of neighboring spins would be so minute that the wall would have to be very thick (infinitely thick) to achieve a 180° rotation.

However, magnetic moments that are oriented away from the easy direction possess excess energy, called the anisotropy energy ($K$). If the wall is thick, then it will contain many magnetic moments rotated away from the easy direction and there would be a substantial anisotropy energy in the wall. Minimum anisotropy energy in the wall is obtained when the magnetic moment changes direction by 180° from the easy direction along $+z$ to that along $-z$ in Figure 8.25 without any intermediate rotations away from $z$. This requires a wall of one atomic spacing. In reality, the wall thickness is a compromise between the exchange energy, demanding a thick wall, and anisotropy energy, demanding a thin wall. The equilibrium wall thickness is that which minimizes the total potential energy, which is the sum of the exchange energy *and* the anisotropy energy within the wall. This thickness turns out to be $\sim$0.1 μm for iron and less for cobalt, in which the anisotropy energy is greater.

**EXAMPLE 8.4**

**MAGNETIC DOMAIN WALL ENERGY AND THICKNESS** The Bloch wall energy and thickness depend on two main factors: the exchange energy $E_{ex}$ (J atom$^{-1}$) and magnetocrystalline energy $K$ (J m$^{-3}$). Suppose that we consider a Bloch wall of unit area, and thickness $\delta$, and calculate the potential energy $U_{wall}$ in this wall due to the exchange energy and the magnetocrystalline anisotropy energy. The spins change by 180° across the thickness $\delta$ of the Bloch wall as in Figure 8.25. The contribution $U_{exchange}$ from the exchange energy arises because it takes energy to rotate one spin with respect to another. If the thickness $\delta$ is large, then the angular change from one spin to the next will be small, and the exchange energy contribution $U_{exchange}$ will also be small. Thus, $U_{exchange}$ is inversely proportional to $\delta$. $U_{exchange}$ is also directly proportional to $E_{ex}$ which gauges the magnitude of this exchange energy; it costs $E_{ex}$ to rotate the two spins 180° to each other. Thus, $U_{exchange} \propto E_{ex}/\delta$.

The anisotropy energy contribution $U_{anisotropy}$ arises from having spins point away from the easy direction. If the thickness $\delta$ is large, there are more and more spin moments that are aligned away from the easy direction, and the anisotropy energy contribution $U_{anisotropy}$ is also large. Thus, $U_{anisotropy}$ is proportional to $\delta$, and also, obviously, to the anisotropy energy $K$ that gauges the magnitude of this energy. Thus, $U_{anisotropy} \propto K\delta$.

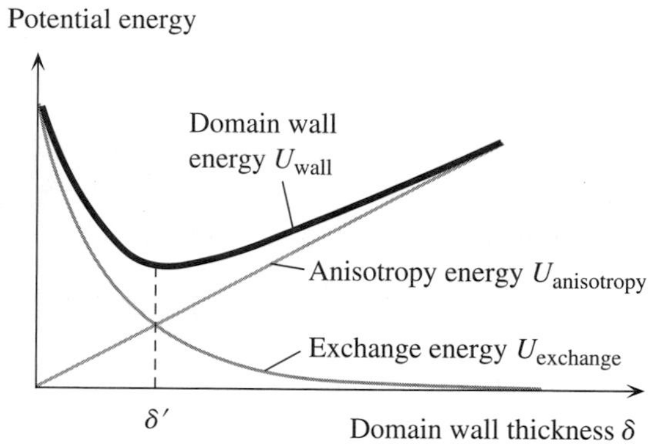


**Figure 8.26** The potential energy of a domain wall depends on the exchange and anisotropy energies.

Figure 8.26 shows the contributions of the exchange and anisotropy energies, $U_{exchange}$ and $U_{anisotropy}$, to the total Bloch wall energy as a function of wall thickness $\delta$. It is clear that exchange and anisotropy energies have opposite (or conflicting) requirements on the wall thickness. There is, however, an optimum thickness $\delta'$ that *minimizes* the Bloch wall energy, that is, a thickness that balances the requirements of exchange and anisotropy forces.

If the interatomic spacing is $a$, then there would be $N = \delta/a$ atomic layers in the wall. Since the spin moment angle changes by 180° across $\delta$, we can calculate the relative spin orientations $(180°/N)$ of adjacent atomic layers, and hence we can find the exact contributions of exchange and anisotropy energies. We do not need the exact mathematics, but the final result is that the potential energy $U_{wall}$ per unit area of the wall is approximately

*Potential energy of a Bloch wall*

$$U_{wall} \approx \frac{\pi^2 E_{ex}}{2a\delta} + K\delta$$

The first term on the right is the exchange energy contribution (proportional to $E_{ex}/\delta$), and the second is the anisotropy energy contribution (proportional to $K\delta$); both have the features we discussed.

Show that the minimum energy occurs when the wall has the thickness

*Bloch wall thickness*

$$\delta' = \left(\frac{\pi^2 E_{ex}}{2aK}\right)^{1/2}$$

Taking $E_{ex} \approx kT_C$, where $T_C$ is the Curie temperature, and for iron, $K \approx 50$ kJ m$^{-3}$, and $a \approx 0.3$ nm, estimate the thickness of a Bloch wall and its energy per unit area.

**SOLUTION**

We can differentiate $U_{wall}$ with respect to $\delta$,

$$\frac{dU_{wall}}{d\delta} = -\frac{\pi^2 E_{ex}}{2a\delta^2} + K$$

and then set it to zero for $\delta = \delta'$ to find,

$$\delta' = \left(\frac{\pi^2 E_{ex}}{2aK}\right)^{1/2}$$

Since $T_C = 1043$ K, $E_{ex} = kT_C = (1.38 \times 10^{-23}\ \text{J K}^{-1})(1043\ \text{K}) = 1.4 \times 10^{-20}$ J, so that

$$\delta' = \left(\frac{\pi^2 E_{ex}}{2aK}\right)^{1/2} = \left[\frac{\pi^2(1.4 \times 10^{-20})}{2(0.3 \times 10^{-9})(50,000)}\right]^{1/2} = 6.8 \times 10^{-8}\text{m} \quad \text{or} \quad 68\ \text{nm}$$

and $$U_{wall} = \frac{\pi^2 E_{ex}}{2a\delta'} + K\delta' = \frac{\pi^2(1.4 \times 10^{-20})}{2(0.3 \times 10^{-9})(6.8 \times 10^{-8})} + (50 \times 10^3)(6.8 \times 10^{-8})$$

$$= 0.007 \text{ J m}^{-2} \quad \text{or} \quad 7 \text{ mJ m}^{-2}$$

A better calculation gives $\delta'$ and $U_{wall}$ as 40 nm and 3 mJ m$^{-2}$, respectively, about the same order of magnitude.[4] The Bloch wall thickness is roughly 70 nm or $\delta/a = 230$ atomic layers. It is left as an exercise to show that when $\delta = \delta'$, the exchange and anisotropy energy contributions are equal.

## 8.5.4 MAGNETOSTRICTION

If we were to strain a ferromagnetic crystal (by applying a suitable stress) along a certain direction, we would change the interatomic spacing not only along this direction but also in other directions and hence change the exchange interactions between the atomic spins. This would lead to a change in the magnetization properties of the crystal. In the converse effect, the magnetization of the crystal generates strains or changes in the physical dimensions of the crystal. For example, in very qualitative terms, when an iron crystal is magnetized along the [111] direction by a strong field, the atomic spins within domains are rotated from their easy directions toward the hard [111] direction. These electron spin rotations involve changes in the electron charge distributions around the atoms and therefore affect the interatomic bonding and hence the interatomic spacing. When an iron crystal is placed in a magnetic field along an easy direction [100], it gets longer along this direction but contracts in the transverse directions [010] and [001], as depicted in Figure 8.27. The reverse is true for nickel. The longitudinal strain $\Delta\ell/\ell$ along the direction of magnetization is called the **magnetostrictive constant,** denoted by $\lambda$. The magnetostrictive constant depends on the crystal direction and may be positive (extension) or negative (contraction). Further, $\lambda$ depends on the applied field and can even change sign as the field is increased; for example, $\lambda$ for iron along the [110] direction is initially positive and then, at higher fields, becomes negative. When the crystal reaches saturation magnetization, $\lambda$ also reaches saturation, called **saturation magnetostriction strain** $\lambda_{sat}$, which is typically $10^{-6}$–$10^{-5}$. Table 8.4 summarizes the $\lambda_{sat}$ values for Fe and Ni along the easy and hard directions. The crystal lattice strain energy associated with magnetostriction is called the **magnetostrictive energy,** which is typically less than the anisotropy energy.

Magnetostriction is responsible for the transformer hum noise one hears near power transformers. As the core of a transformer is magnetized one way and then in the opposite direction under an alternating voltage, the alternating changes in the longitudinal strain vibrate the surrounding environment, air, oil, and so forth, and generate an acoustic noise at twice the main frequency, or 120 Hz, and its harmonics. (Why?)

The magnetostrictive constant can be controlled by alloying metals. For example, $\lambda_{sat}$ along the easy direction for nickel is negative and for iron it is positive, but for the alloy 85% Ni–15% Fe, it is zero. In certain magnetic materials, $\lambda$ can be quite large,

[4] See, for example, D. Jiles, *Introduction to Magnetism and Magnetic Materials*, London, England: Chapman and Hall, 1991.

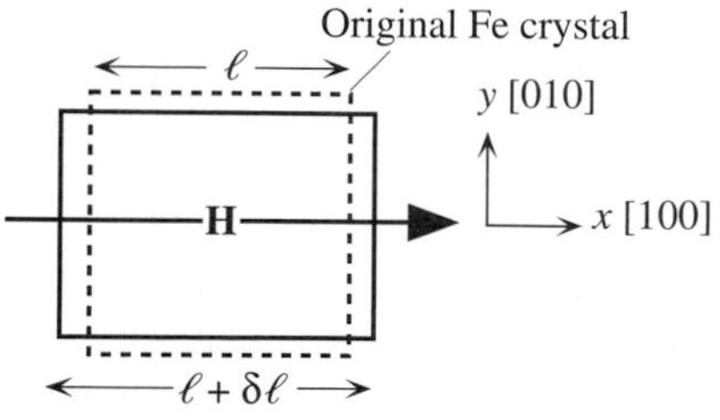


**Figure 8.27** Magnetostriction means that the iron crystal in a magnetic field along x, an easy direction, elongates along x but contracts in the transverse directions (in low fields).

greater than $10^{-4}$, which has opened up new areas of sensor applications based on the magnetostriction effect. For example, it may be possible to develop torque sensors for automotive steering applications by using Co-ferrite type magnetic materials[5] (*e.g.*, CoO–$Fe_2O_3$ or similar compounds) that have $\lambda_{sat}$ of the order of $10^{-4}$.

## 8.5.5 DOMAIN WALL MOTION

The magnetization of a single ferromagnetic crystal involves the motions of domain boundaries to allow the favorably oriented domains to grow at the expense of domains with magnetizations directed away from the field (Figure 8.23). The motion of a domain wall in a crystal is affected by crystal imperfections and impurities and is not smooth. For example, in a 90° Bloch wall, the magnetization changes direction by 90° across the boundary. Due to magnetostriction (Figure 8.27), there is a change in the distortion of the lattice across the 90° boundary, which leads to a complicated strain and hence stress distribution around this boundary. We also know that crystal imperfections such as dislocations and point defects also have strain and stress distributions around them. Domain walls and crystal imperfections therefore interact with each other. Dislocations are line defects that have a substantial volume of strained lattice around them. Figure 8.28 visualizes a dislocation with tensile and compressive strains around it and a domain wall that has a tensile strain on the side of the dislocation. If the wall gets close to the dislocation, the tensile and compressive strains cancel, which results in an unstrained lattice and hence a lower strain energy. This energetically favorable arrangement keeps the domain boundary close to the dislocation. It now takes greater magnetic field to snap away the boundary from the dislocation. Domain walls also interact with nonmagnetic impurities and inclusions. For example, an inclusion that finds itself in a domain becomes magnetized and develops south and north poles, as shown in Figure 8.29a. If the domain wall were to intersect the inclusion and if there were to be two neighboring domains around the inclusion, as in Figure 8.29b, then the magnetostatic energy would be lowered—energetically a favorable event. This reduction in magnetostatic potential energy means that it now takes greater force to move the domain wall past the impurity, as if the wall were "pinned" by the impurity.

The motion of a domain wall in a crystal is therefore not smooth but rather jerky. The wall becomes pinned somewhere by a defect or an impurity and then needs a

[5] See, for example, D. Jiles and C. C. H. Lo, *Sensors and Actuators*, **A106**, 3, 2003.

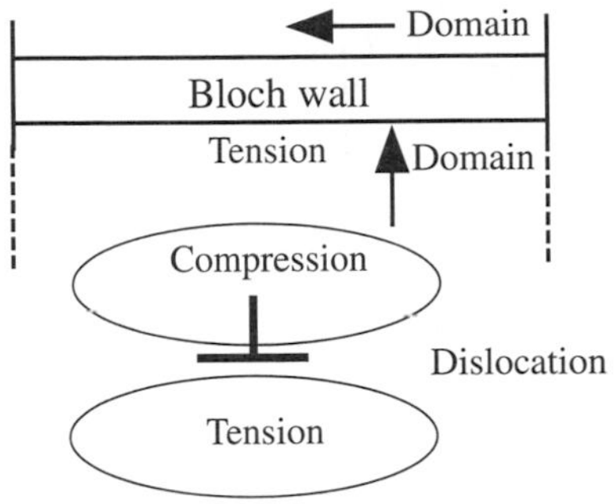


**Figure 8.28** Stress and strain distributions around a dislocation and near a domain wall.

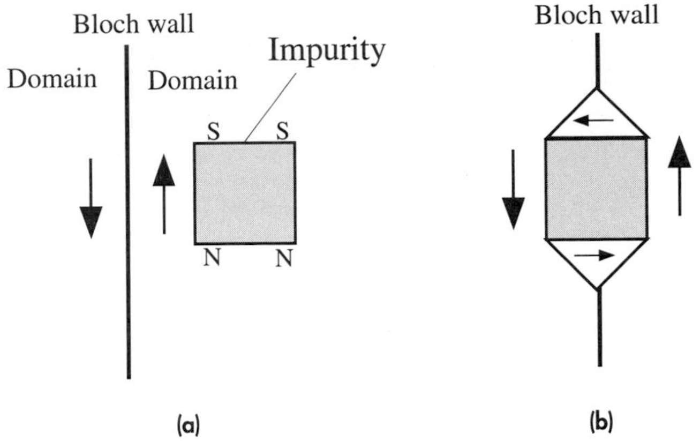


**Figure 8.29** Interaction of a Bloch wall with a nonmagnetic (no permanent magnetization) inclusion.
(a) The inclusion becomes magnetized and there is magnetostatic energy.
(b) This arrangement has lower potential energy and is thus favorable.

greater applied field to break free. Once it snaps off, the domain wall is moved until it is attracted by another type of imperfection, where it is held until the field increases further to snap it away again. Each time the domain wall is snapped loose, lattice vibrations are generated, which means loss of energy as heat. The whole domain wall motion is nonreversible and involves energy losses as heat to the crystal.

## 8.5.6 POLYCRYSTALLINE MATERIALS AND THE $M$ VERSUS $H$ BEHAVIOR

The majority of the magnetic materials used in engineering are polycrystalline and therefore have a microstructure that consists of many grains of various sizes and orientations depending on the preparation and thermal history of the component. In an unmagnetized polycrystalline sample, each crystal grain will possess domains, as depicted in Figure 8.30. The domain structure in each grain will depend on the size and shape of the grain and, to some extent, on the magnetizations in neighboring grains. Although very small grains, perhaps smaller than 0.1 μm, may be single domains, in most cases the majority of the grains will have many domains. Overall, the structure will possess no net magnetization, provided that it was not previously subjected to an applied magnetic field. We can assume that the component was heated to a temperature above the Curie point and then allowed to cool to room temperature without an applied field.

Suppose that we start applying a very small external magnetic field ($\mu_o H$) along some direction, which we can arbitrarily label as $+x$. The domain walls within various grains begin to move small distances, and favorably oriented domains (those with a component of $M$ along $+x$) grow a little larger at the expense of those pointing away from the field, as indicated by point $a$ in Figure 8.31. The domain walls that are pinned by imperfections tend to bow out. There is a very small but net magnetization

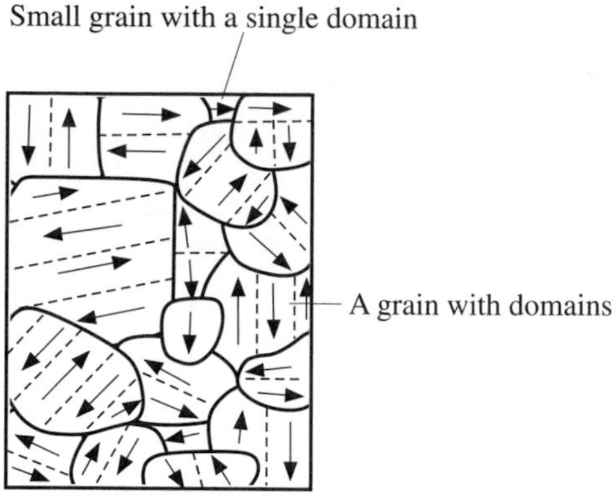


**Figure 8.30** Schematic illustration of magnetic domains in the grains of an unmagnetized polycrystalline iron sample.
Very small grains have single domains.

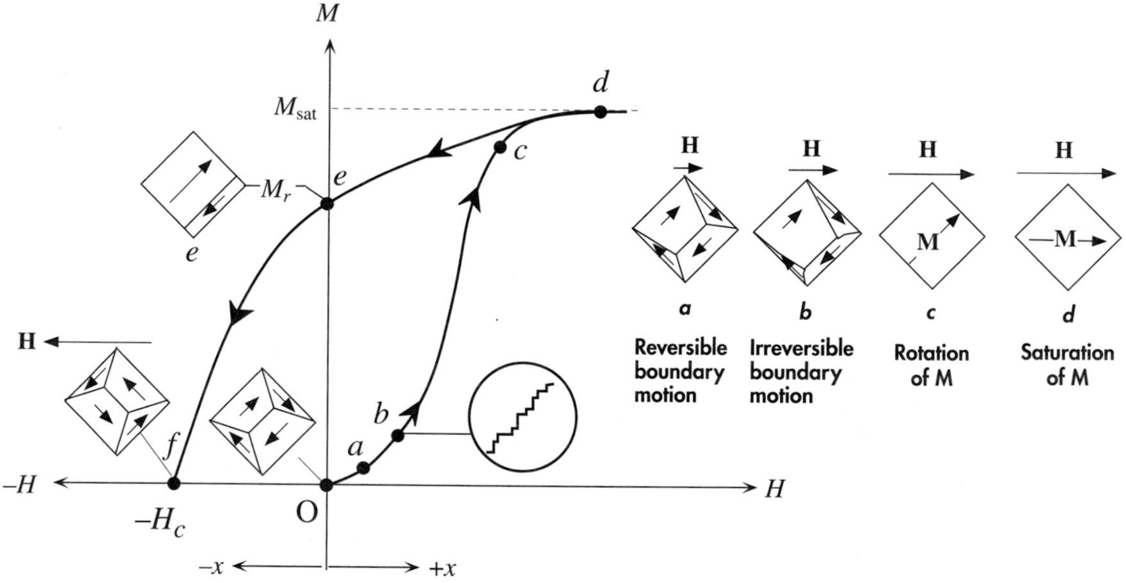


**Figure 8.31** *M* versus *H* behavior of a previously unmagnetized polycrystalline iron specimen.
An example grain in the unmagnetized specimen is that at O.
(a) Under very small fields, the domain boundary motion is reversible.
(b) The boundary motions are irreversible and occur in sudden jerks.
(c) Nearly all the grains are single domains with saturation magnetizations in the easy directions.
(d) Magnetizations in individual grains have to be rotated to align with the field *H*.
(e) When the field is removed, the specimen returns along *d* to *e*.
(f) To demagnetize the specimen, we have to apply a magnetizing field of $H_c$ in the reverse direction.

along the field, as indicated by the O*a* region in the magnetization versus magnetizing field (*M* versus *H*) behavior in Figure 8.31. As we increase the magnetizing field, the domain motions extend larger distances, as shown for point *b* in Figure 8.31, and walls encounter various obstacles such as crystal imperfections, impurities, second phases, and so on, which tend to attract the walls and thereby hinder their motions. A domain wall that is stuck (or pinned) at an imperfection at a given field cannot move until the field increases sufficiently to provide the necessary force to snap the wall free, which then suddenly surges forward to the next obstacle. As a wall suddenly snaps free and shoots forward to the next obstacle, essentially two causes lead to heat generation. Sudden changes in the lattice distortion, due to magnetostriction, create lattice waves that carry off some of the energy. Sudden changes in the magnetization induce eddy currents that dissipate energy via Joule heating (domains have a finite electrical resistance). These processes involve energy conversion to heat and are irreversible. Sudden jerks in the wall motions lead to small jumps in the magnetization of the specimen as the magnetizing field is increased; the phenomenon is known as the **Barkhausen effect.** If we could examine the magnetization precisely with a highly sensitive instrument, we would see jumps in the $M$ versus $H$ behavior, as shown in the inset in Figure 8.31.

As we increase the field, magnetization continues to increase by jerky domain wall motions that enlarge domains with favorably oriented magnetizations and shrink away those with magnetizations pointing away from the applied field. Eventually domain wall motions leave each crystal grain with a single domain and magnetization in one of the easy directions, as indicated by point $c$ in Figure 8.31. Although some grains would be oriented to have the easy direction and hence $M$ along the applied field, the magnetization in many grains will be pointing at some angle to $H$ as shown for point $c$ in Figure 8.31. From then until point $d$, the increase in the applied field forces the magnetization in a grain, such as that at point $c$ to rotate toward the direction of $H$. Eventually the applied field is sufficiently strong to align $M$ along $H$, and the specimen reaches saturation magnetization $M_{sat}$, directed along $H$ or $+x$, as at point $d$ in Figure 8.31.

If we were to decrease and remove the magnetizing field, the magnetization in each grain would rotate to align parallel with the nearest easy direction in that grain. Further, in some grains, additional small domains may develop that reduce the magnetization within that grain, as indicated at point $e$ in Figure 8.31. This process, from point $d$ to point $e$, leaves the specimen with a permanent magnetization, called the **remanent** or **residual magnetization** and denoted by $M_r$.

If we were now to apply a magnetizing field in the reverse direction $-x$, the magnetization of the specimen, still along $+x$, would decrease and eventually, at a sufficiently large applied field $M$ would be zero and the sample would have been totally demagnetized. This is shown as point $f$ in Figure 8.31. The magnetizing field $H_c$ required to totally demagnetize the sample is called the **coercivity** or the **coercive field.** It represents the resistance of the sample to demagnetization. We should note that at point $f$ in Figure 8.31, the sample again has grains with many domains, which means that during the demagnetization process, from point $e$ to point $f$, new domains had to be generated. The demagnetization process invariably involves the nucleation of various domains at various crystal imperfections to cancel the overall magnetization. The treatment of the nucleation of domains is beyond the scope of this book; we will nonetheless, accept it as required process for the demagnetization of the crystal grains.

If we continue to increase the applied magnetic field along $-x$, as illustrated in Figure 8.32a, the process from point $f$ onward becomes similar to that described for magnetization from point $a$ to point $d$ in Figure 8.31 along $+x$ except that it is now directed along $-x$. At point $g$, the sample reaches saturation magnetization along the $-x$ direction. The full $M$ versus $H$ behavior as the magnetizing field is cycled between

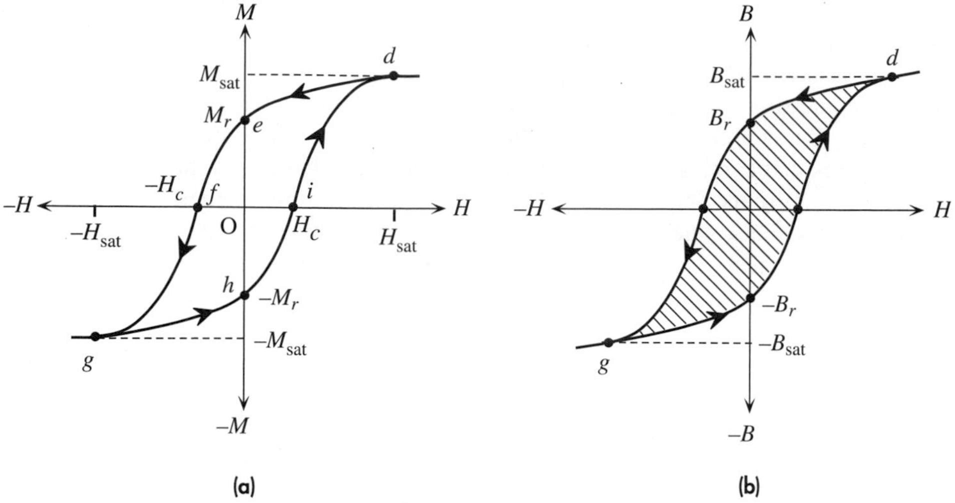


**Figure 8.32**
(a) A typical $M$ versus $H$ hysteresis curve.
(b) The corresponding $B$ versus $H$ hysteresis curve. The shaded area inside the hysteresis loop is the energy loss per unit volume per cycle.

$+x$ to $-x$ has a closed loop shape, shown in Figure 8.32a, called the **hysteresis loop.** We observe that in both $+x$ and $-x$ directions, the magnetization reaches saturation $M_{sat}$ when $H$ reaches $H_{sat}$, and on removing the applied field, the specimen retains an amount of permanent magnetization, represented by points $e$ and $h$ and denoted by $M_r$. The necessary applied field of magnitude $H_c$ that is needed to demagnetize the specimen is the coercivity (or coercive field), which is represented by points $f$ and $i$. The initial magnetization curve, O$abcd$ in Figure 8.31, which starts from an unmagnetized state, is called the **initial magnetization curve.**

We can, of course, monitor the magnetic field $B$ instead of $M$, as in Figure 8.32b, where

$$B = \mu_o M + \mu_o H$$

which leads to a hysteresis loop in the $B$ versus $H$ behavior. The very slight increase in $B$ with $H$ when $M$ is in saturation is due to the permeability of free space $(\mu_o H)$. The area enclosed within the $B$ versus $H$ hysteresis loop, shown as the hatched region in Figure 8.32b, represents the energy dissipated per unit volume per cycle of applied field variation.

Suppose we do not take a magnetic material to saturation but still subject the specimen to a cyclic applied field alternating between the $+x$ and $-x$ directions. Then the hysteresis loop would be different than that when the sample is taken all the way to saturation, as shown in Figure 8.33. The magnetic field in the material does not reach $B_{sat}$ (corresponding to $M_{sat}$) but instead reaches some maximum value $B_m$ when the magnetizing field is $H_m$. There is still a hysteresis effect because the magnetization and demagnetization processes are nonreversible and do not retrace each other. The shape of the hysteresis

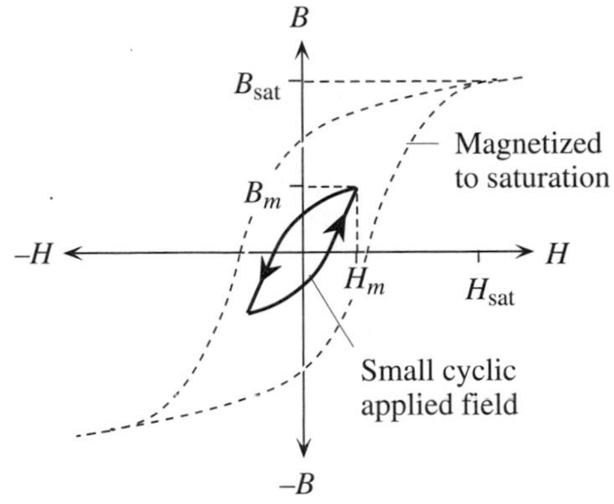


**Figure 8.33** The $B$ versus $H$ hysteresis loop depends on the magnitude of the applied field in addition to the material and sample shape and size.

loop depends on the magnitude of the applied field in addition to the material and sample shape and size. The area enclosed within the loop is still the energy dissipated per unit volume per cycle of applied field oscillation. The hysteresis loop taken to saturation, as in Figure 8.32a and b, is called the **saturation (major) hysteresis loop.** It is apparent from Figure 8.33 that the remanence and coercivity exhibited by the sample depend on the $B$–$H$ loop. The quoted values normally correspond to the saturation hysteresis loop.

Ferrimagnetic materials exhibit properties that closely resemble those of ferromagnetic materials. One can again identify distinct magnetic domains and domain wall motions during magnetization and demagnetization that also lead to $B$–$H$ hysteresis curves with the same characteristic parameters, namely, saturation magnetization ($M_{sat}$ and $B_{sat}$ at $H_{sat}$), remanence ($M_r$ and $B_r$), coercivity ($H_c$), hysteresis loss, and so on.

### 8.5.7 DEMAGNETIZATION

The $B$–$H$ hysteresis curves, as in Figure 8.32b, that are commonly given for magnetic materials represent $B$ versus $H$ behavior observed under repeated cycling. The applied field intensity $H$ is cycled back and forward between the $-x$ and $+x$ directions. If we were to try and demagnetize a specimen with a remanent magnetization at point $e$ in Figure 8.34 by applying a reverse field intensity, then the magnetization would move along from point $e$ to point $f$. If at point $f$ we were to suddenly switch off the applied field, we would find that $B$ does not actually remain zero but recovers along $f$ to point $e'$ and attains some value $B_r'$. The main reason is that small domain wall motions are reversible and as soon as the field is removed, there is some reversible domain wall motion "bouncing back" the magnetization along $f$–$e'$. We can anticipate this recovery and remove the field intensity at some point $f'$ so that the sample recovers along $f'$O and the magnetization ends up being zero. However, to remove the field intensity at point $f'$, we need to know not only the exact $B$–$H$ behavior but also the exact location of point $f'$ (or the recovery behavior). The simplest method to demagnetize the sample is first to cycle $H$ with ample magnitude to reach saturation and then to continue cycling $H$ but with a gradually decreasing magnitude, as depicted in Figure 8.35. As $H$ is cycled with a decreasing magnitude, the sample traces out smaller and smaller $B$–$H$ loops until the $B$–$H$ loops are so small that they end up at the origin when $H$ reaches

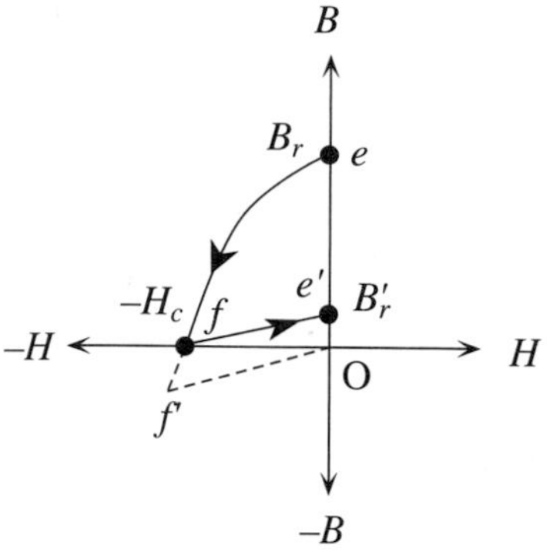


**Figure 8.34** Removal of the demagnetizing field at point *f* does not necessarily result in zero magnetization as the sample recovers along *f–e'*.

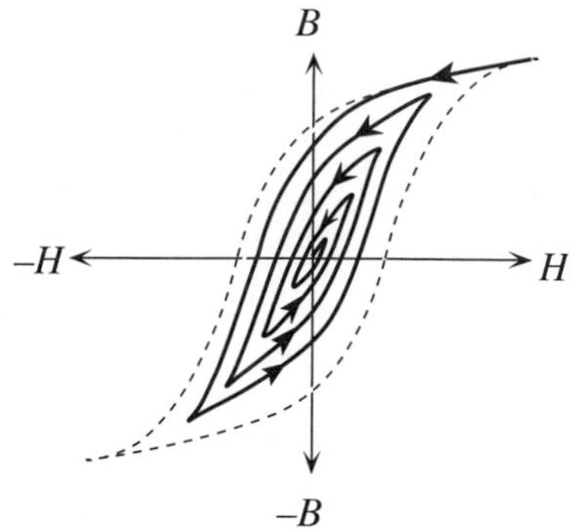


**Figure 8.35** A magnetized specimen can be demagnetized by cycling the field intensity with a decreasing magnitude, that is, tracing out smaller and smaller *B–H* loops until the origin is reached, $H = 0$.

zero. The demagnetization process in Figure 8.35 is commonly known as **deperming.** Undesirable magnetization of various magnetic devices such as recording heads is typically removed by this deperming process (for example, a demagnetizing gun brought close to a magnetized recording head implements deperming by applying a cycled $H$ with decreasing magnitude).

**EXAMPLE 8.5**

**ENERGY DISSIPATED PER UNIT VOLUME AND THE HYSTERESIS LOOP** Consider a toroidal coil with an iron core that is energized from a voltage supply through a rheostat, as shown in Figure 8.11. Suppose that by adjusting the rheostat we can adjust the current $i$ supplied to the coil and hence the magnetizing field $H$ in the core material. $H$ and $i$ are simply related by Ampere's law. However, the magnetic field $B$ in the core is determined by the $B$–$H$ characteristics of the core material. From electromagnetism (see Example 8.2), we know that the battery has to do work $dE_{\text{vol}}$ per unit volume of core material to increase the magnetic field by $dB$, where

$$dE_{\text{vol}} = H\,dB$$

so that the total energy or work involved per unit volume in changing the magnetic field from an initial value $B_1$ to a final value $B_2$ in the core is

*Work done per unit volume during magnetization*

$$E_{\text{vol}} = \int_{B_1}^{B_2} H\,dB \qquad \textbf{[8.22]}$$

where the integration limits are determined by the initial and final magnetic fields.

Equation 8.22 corresponds to the area between the $B$–$H$ curve and the $B$ axis between $B_1$ and $B_2$. Suppose that we take the iron core in the toroid from point $P$ on the hysteresis curve to $Q$, as shown in Figure 8.36. This is a magnetization process for which energy is put into the sample. The work done per unit volume from $P$ to $Q$ is the area $PQRS$, shown as hatched. On returning the sample to the same initial magnetization (same magnetic field $B$ as we had at $P$), taking it from $Q$ to $S$, energy is returned from the core into the electric circuit. This energy per unit volume is the area $QRS$, shown as gray, and is less than $PQRS$ during magnetization. The difference is the energy dissipated in the sample as heat (moving domain walls and so on) and

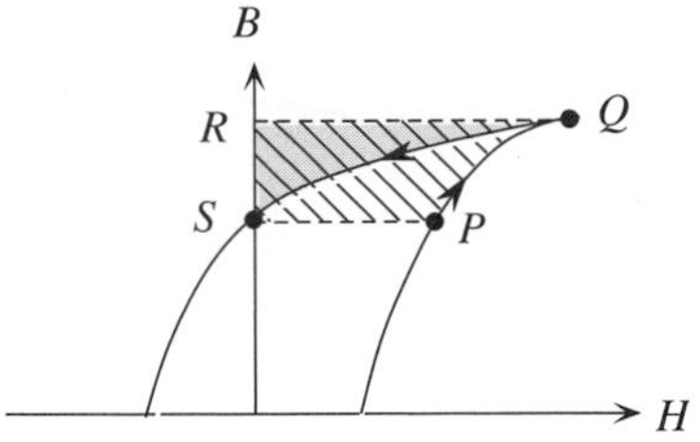


**Figure 8.36** The area between the $B$–$H$ curve and the $B$ axis is the energy absorbed per unit volume in magnetization or released during demagnetization.

corresponds to the hysteresis loop area $PQS$. Over one full cycle, the energy dissipated per unit volume is the total hysteresis loop area.

The hysteresis loop and hence the energy dissipated per unit volume per cycle depend not only on the core material but also on the magnitude of the magnetic field ($B_m$), as apparent in Figure 8.33. For example, for magnetic steels used in transformer cores, the hysteresis **power loss** $P_h$ per unit volume of core is empirically expressed in terms of the maximum magnetic field $B_m$ and the ac frequency $f$ as[6]

*Hysteresis power loss per $m^3$*

$$P_h = KfB_m^n \tag{8.23}$$

where $K$ is a constant that depends on the core material (typically, $K = 150.7$), $f$ is the ac frequency (Hz), $B_m$ is the maximum magnetic field (T) in the core (assumed to be in the range 0.1–1.5 T), and $n = 1.6$. According to Equation 8.23, the hysteresis loss can be decreased by operating the transformer with a reduced magnetic field.

# 8.6 SOFT AND HARD MAGNETIC MATERIALS

## 8.6.1 DEFINITIONS

Based on their $B$–$H$ behavior, engineering materials are typically classified into soft and hard magnetic materials. Their typical $B$–$H$ hysteresis curves are shown in Figure 8.37. Soft magnetic materials are easy to magnetize and demagnetize and hence require relatively low magnetic field intensities. Put differently, their $B$–$H$ loops are narrow, as shown in Figure 8.37. The hysteresis loop has a small area, so the hysteresis power loss per cycle is small. Soft magnetic materials are typically suitable for applications where repeated cycles of magnetization and demagnetization are involved, as in electric motors, transformers, and inductors, where the magnetic field varies cyclically. These applications also require low hysteresis losses, or small hysteresis loop area. Electromagnetic relays that have to be turned on and off require the relay iron to be magnetized and demagnetized and therefore need soft magnetic materials.

Hard magnetic materials, on the other hand, are difficult to magnetize and demagnetize and hence require relatively large magnetic field intensities, as apparent in Figure 8.37. Their $B$–$H$ curves are broad and almost rectangular. They possess relatively large coercivities, which means that they need large applied fields to be demagnetized. The coercive field for hard materials can be millions of times greater than those for soft

[6] This is the power engineers Steinmetz equation for commercial magnetic steels. It has been applied not only to silicon irons (Fe + few percent Si) but also to a wide range of magnetic materials.

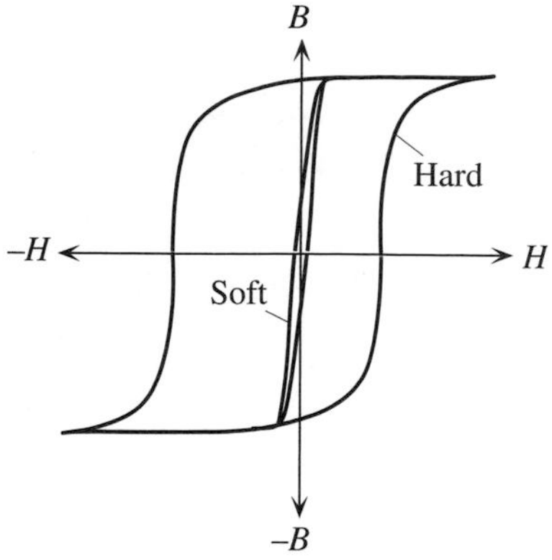


**Figure 8.37** Soft and hard magnetic materials.

magnetic materials. Their characteristics make hard magnetic materials useful as permanent magnets in a variety of applications. It is also clear that the magnetization can be switched from one very persistent direction to another very persistent direction, from $+B_r$ to $-B_r$, by a suitably large magnetizing field intensity. As the coercivity is strong, both the states $+B_r$ and $-B_r$ persist until a suitable (large) magnetic field intensity switches the field from one direction to the other. It is apparent that hard magnetic materials can also be used in magnetic storage of digital data, where the states $+B_r$ and $-B_r$ can be made to represent 1 and 0 (or vice versa).

### 8.6.2 Initial and Maximum Permeability

It is useful to characterize the magnetization of a material by a relative permeability $\mu_r$, since this simplifies magnetic calculations. For example, inductance calculations become straightforward if one could represent the magnetic material by $\mu_r$ alone. But it is clear from Figure 8.38a that

$$\mu_r = \frac{B}{\mu_o H}$$

is not even approximately constant because it depends on the applied field and the magnetic history of the sample. Nonetheless, we still find it useful to specify a relative permeability to compare various materials and even use it in various calculations. The definition $\mu_r = B/(\mu_o H)$ represents the slope of the straight line from the origin O to the point $P$, as shown in Figure 8.38a. This is a maximum when the line becomes a tangent to the $B$–$H$ curve at $P$, as in the figure. Any other line from O to the $B$–$H$ curve that is not a tangent does not yield a maximum relative permeability (the mathematical proof is left to the reader, though the argument is intuitively acceptable from the figure). The **maximum relative permeability,** as defined in Figure 8.38a, is denoted by $\mu_{r,\max}$ and serves as a useful magnetic parameter. The point $P$ in Figure 8.38a that defines the maximum permeability corresponds to what is called the "knee" of the $B$–$H$ curve. Many transformers are designed to operate with the maximum magnetic field in the core reaching this knee point. For pure iron, $\mu_{r,\max}$ is less than $10^4$, but for certain

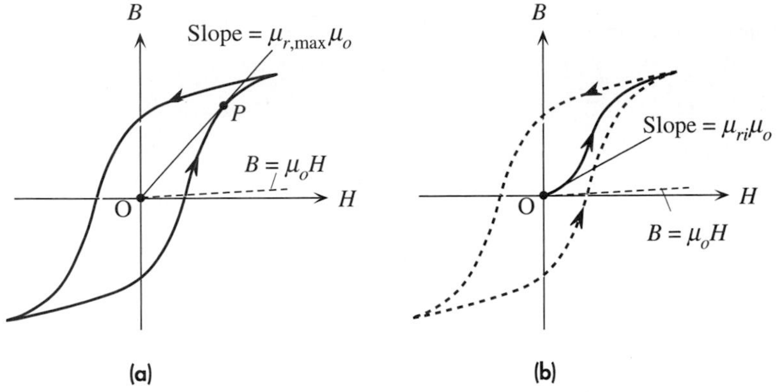


**Figure 8.38** Definitions of (a) maximum permeability and (b) initial permeability.

soft magnetic materials such as supermalloys (a nickel–iron alloy), the values of $\mu_{r,\max}$ can be as high as $10^6$.

**Initial relative permeability,** denoted as $\mu_{ri}$, represents the initial slope of the initial *B* versus *H* curve as the material is first magnetized from an unmagnetized state, as illustrated in Figure 8.38b. This definition is useful for soft magnetic materials that are employed at very low magnetic fields (*e.g.*, small signals in electronics and communications engineering). In practice, weak applied magnetic fields where $\mu_{ri}$ is useful are typically less than $10^{-4}$ T. In contrast, $\mu_{r,\max}$ is useful when the magnetic field in the material is not far removed from saturation. Initial relative permeability of a magnetically soft material can vary by orders of magnitude. For example, $\mu_{ri}$ for iron is 150, whereas for supernumetal-200, a commercial alloy of nickel and iron, it is about $2 \times 10^5$.

## 8.7 SOFT MAGNETIC MATERIALS: EXAMPLES AND USES

Table 8.5 identifies what properties are desirable in soft magnetic materials and also lists some typical examples with various applications. An *ideal* soft magnetic material would have zero coercivity ($H_c$), a very large saturation magnetization ($B_{sat}$), zero remanent magnetization ($B_r$), zero hysteresis loss, and very large $\mu_{r,\max}$ and $\mu_{ri}$. A number of example materials, from pure iron to ferrites, which are ferrimagnetic, are listed in Table 8.5. Pure iron, although soft, is normally not used in electric machines (except in a few specific relay-type applications) because its good conductivity allows large eddy currents to be induced under varying fields. Induced eddy currents in the iron lead to Joule losses ($RI^2$), which are undesirable. The addition of a few percentages of silicon to iron (silicon–iron), known typically as silicon–steels, increases the resistivity and hence reduces the eddy current losses. Silicon–iron is widely used in power transformers and electric machinery.

The nickle–iron alloys with compositions around 77% Ni–23% Fe constitute an important class of soft magnetic materials with low coercivity, low hysteresis losses, and high permeabilities ($\mu_{ri}$ and $\mu_{r,\max}$). High $\mu_{ri}$ makes these alloys particularly useful in low magnetic field applications that are typically found in high-frequency work in

**Table 8.5** Selected soft magnetic materials and some typical values and applications

| Magnetic Material | $\mu_o H_c$ (T) | $B_{sat}$ (T) | $B_r$ (T) | $\mu_{ri}$ | $\mu_{r,max}$ | $W_h$ | Typical Applications |
|---|---|---|---|---|---|---|---|
| Ideal soft | 0 | Large | 0 | Large | Large | 0 | Transformer cores, inductors, electric machines, electromagnet cores, relays, magnetic recording heads. |
| Iron (commercial) grade, 0.2% impurities) | $<10^{-4}$ | 2.2 | <0.1 | 150 | $10^4$ | 250 | Large eddy current losses. Generally not preferred in electric machinery except in some specific applications (*e.g.*, some electromagnets and relays). |
| Silicon iron (Fe: 2–4% Si) | $<10^{-4}$ | 2.0 | 0.5–1 | $10^3$ | $10^4$– $4 \times 10^5$ | 30–100 | Higher resistivity and hence lower eddy current losses. Wide range of electric machinery (*e.g.*, transformers). |
| Supermalloy (79% Ni–15.5% Fe–5% Mo–0.5% Mn) | $2 \times 10^{-7}$ | 0.7–0.8 | <0.1 | $10^5$ | $10^6$ | <0.5 | High permeability, low-loss electric devices, *e.g.*, specialty transformers, magnetic amplifiers. |
| 78 Permalloy (78.5% Ni–21.5% Fe) | $5 \times 10^{-6}$ | 0.86 | <0.1 | $8 \times 10^3$ | $10^5$ | <0.1 | Low-loss electric devices, audio transformers, HF transformers, recording heads, filters. |
| Glassy metals, Fe–Si–B | $2 \times 10^{-6}$ | 1.6 | $<10^{-6}$ | — | $10^5$ | 20 | Low-loss transformer cores. |
| Ferrites, Mn–Zn ferrite | $10^{-5}$ | 0.4 | <0.01 | $2 \times 10^3$ | $5 \times 10^3$ | <0.01 | HF low-loss applications. Low conductivity ensures negligible eddy current losses. HF transformers, inductors (*e.g.*, pot cores, E and U cores), recording heads. |

NOTE: $W_h$ is the hysteresis loss, energy dissipated per unit volume per cycle in hysteresis losses, J m$^{-3}$ cycle$^{-1}$, typically at $B_m = 1$ T.

electronics (*e.g.*, audio and wide-band transformers). They have found many engineering uses in sensitive relays, pulse and wide-band transformers, current transformers, magnetic recording heads, magnetic shielding, and so forth. Alloying iron with nickel increases the resistivity and hence reduces eddy current losses. The magnetocrystalline anisotropy energy is least at these nickel compositions, which leads to easier domain wall motions and hence smaller hysteresis losses. There are a number of commercial nickel–iron alloys whose application depends on the exact composition (which may also have a few percentages of Mo, Cu, or Cr) and the method of preparation (*e.g.*, mechanical rolling). For example, supermalloy (79% Ni–16% Fe–5% Co) has $\mu_{ri} \approx 10^5$, compared with commercial grade iron, which has $\mu_{ri}$ less than $10^3$.

**Amorphous magnetic metals,** as the name implies, have no crystal structure (they only have short-range order) and consequently possess no crystalline imperfections such as grain boundaries and dislocations. They are prepared by rapid solidification of the melt by using special techniques such as melt spinning (as described in Chapter 1). Typically they are thin ribbons by virtue of their preparation method. Since they have no crystal structure, they also have no magnetocrystalline anisotropy energy, which means that all directions are easy. The absence of magnetocrystalline anisotropy and usual crystalline defects which normally impede domain wall motions, leads to low coercivities and hence to soft magnetic properties. The coercivity, however, is not zero inasmuch as there is still some magnetic anisotropy due to the directional nature of the strains frozen in the metal during rapid solidification. By virtue of their disordered structure, these metallic glasses also have higher resistivities and hence they have smaller eddy current losses. Although they are ideally suited for various transformer and electric machinery applications, their limited size and shape, at present, prevent their use in power applications.

**Ferrites** are ferrimagnetic materials that are typically oxides of mixed transition metals, one of which is iron. For example, Mn ferrite is $MnFe_2O_4$ and MgZn ferrite is $Mn_{1-x}Zn_xFe_2O_4$. They are normally insulators and therefore do not suffer from eddy current losses. They are ideal as magnetic materials for high-frequency work where eddy current losses would prevent the use of any material with a reasonable conductivity. Although they can have high initial permeabilities and low losses, they do not possess as large saturation magnetizations as ferromagnets, and further, their useful temperature range (determined by the Curie temperature) is lower. There are many types of commercial ferrites available depending on the application, tolerable losses, and the required upper frequency of operation. MnZn ferrites, for example, have high initial permeabilities (*e.g.*, $2 \times 10^3$) but are only useful up to about 1 MHz, whereas NiZn ferrites have lower initial permeability (*e.g.*, $10^2$) but can be used up to 200 MHz. Generally, the initial permeability in the high-frequency region decreases with frequency.

Garnets are ferrimagnetic materials that are typically used at the highest frequencies that cover the microwave range (1–300 GHz). The yttrium iron garnet, YIG, which is $Y_3Fe_5O_{12}$, is one of the simplest garnets with a very low hysteresis loss at microwave frequencies. Garnets have excellent dielectric properties with high resistivities and hence low losses. The main disadvantages are the low saturation magnetization, which is 0.18 T for YIG, and low Curie temperature, 280 °C for YIG. The compositions of garnets depend on the properties required for the particular microwave application. For example, $Y_{2.1}Gd_{0.98}Fe_5O_{12}$ is a garnet that is used in X-band (8–12 GHz) three-port circulators handling high microwave powers (*e.g.*, peak power 200 kW and average power 200 W).

**AN INDUCTOR WITH A FERRITE CORE** Consider a toroidal coil with a ferrite core. Suppose that the coil has 200 turns and is used in HF work with small signals. The mean diameter of the toroid is 2.5 cm and the core diameter is 0.5 cm. If the core is a MnZn ferrite, what is the approximate inductance of the coil? **EXAMPLE 8.6**

**SOLUTION**

The inductance $L$ of a toroidal coil is given by

$$L = \frac{\mu_{ri}\mu_o N^2 A}{\ell}$$

so

$$L = \frac{(2 \times 10^3)(4\pi \times 10^{-7}\ \text{H m}^{-1})(200)^2\pi\left(\dfrac{0.005}{2}\ \text{m}\right)^2}{(\pi 0.025\ \text{m})} = 0.025\ \text{H} \quad \text{or} \quad 25\ \text{mH}$$

Had the core been air, the inductance would have been $1.26 \times 10^{-5}$ H or 12.6 μH. The main assumption is that $B$ is uniform in the core, and this will be only so if the diameter of the toroid (2.5 cm) is much greater than the core diameter (0.5 cm). Here this ratio is 5 and the calculation is only approximate.

## 8.8 HARD MAGNETIC MATERIALS: EXAMPLES AND USES

An ideal hard magnetic material, as summarized in Table 8.6, has very large coercivity and remanent magnetic field. Further, since they are used as permanent magnets, the energy stored per unit volume in the external magnetic field should be as large as possible since this is the energy available to do work. This energy density (J m$^{-3}$) in the external field depends on the maximum value of the product $BH$ in the second quadrant of the $B$–$H$ characteristics and is denoted as $(BH)_{max}$. It corresponds to the largest rectangular area that fits the $B$–$H$ curve in the second quadrant, as shown in Figure 8.39.

When the size of a ferromagnetic sample falls below a certain critical dimension, of the order of 0.1 μm for cobalt, the whole sample becomes a single domain, as depicted in Figure 8.40, because the cost of energy in generating a domain wall is too high compared with the reduction in external magnetostatic energy. These small particle-like pieces of magnets are called **single domain fine particles.** Their magnetic

**Table 8.6** Hard magnetic materials and typical values

| Magnetic Material | $\mu_o H_c$ (T) | $B_r$ (T) | $(BH)_{max}$ (kJ m$^{-3}$) | Examples and Uses |
|---|---|---|---|---|
| Ideal hard | Large | Large | Large | Permanent magnets in various applications. |
| Alnico (Fe–Al–Ni–Co–Cu) | 0.19 | 0.9 | 50 | Wide range of permanent magnet applications. |
| Alnico (Columnar) | 0.075 | 1.35 | 60 | |
| Strontium ferrite (anisotropic) | 0.3–0.4 | 0.36–0.43 | 24–34 | Starter motors, dc motors, loudspeakers, telephone receivers, various toys. |
| Rare earth cobalt, *e.g.*, $Sm_2Co_{17}$ (sintered) | 0.62–1.1 | 1.1 | 150–240 | Servo motors, stepper motors, couplings, clutches, quality audio headphones. |
| NdFeB magnets | 0.9–1.0 | 1.0–1.2 | 200–275 | Wide range of applications, small motors (*e.g.*, in hand tools), walkman equipment, CD motors, MRI body scanners, computer applications. |
| Hard particles, $\gamma$-$Fe_2O_3$ | 0.03 | 0.2 | | Audio and video tapes, floppy disks. |

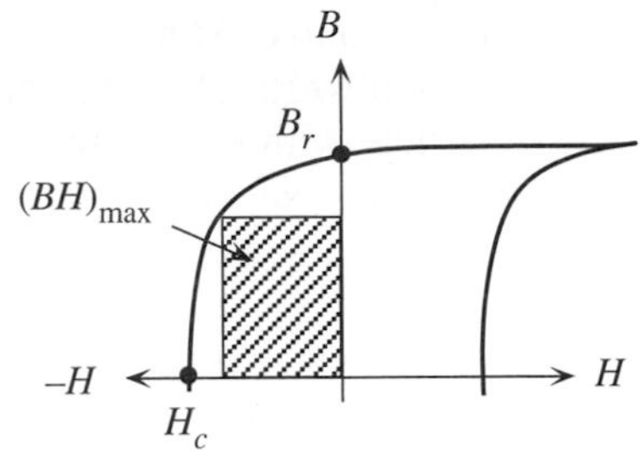


**Figure 8.39** Hard magnetic materials and $(BH)_{max}$.

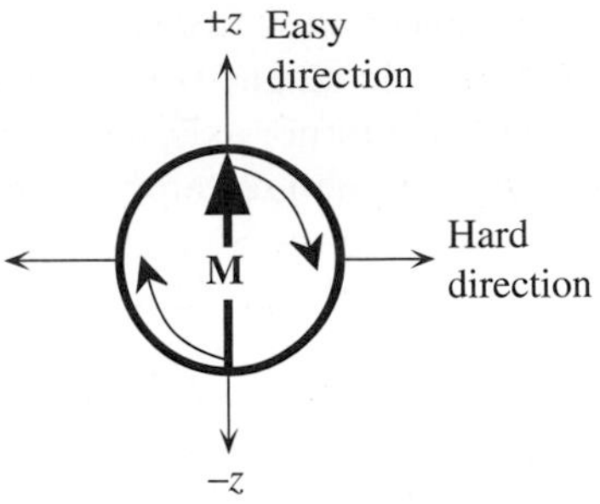


**Figure 8.40** A single domain fine particle.

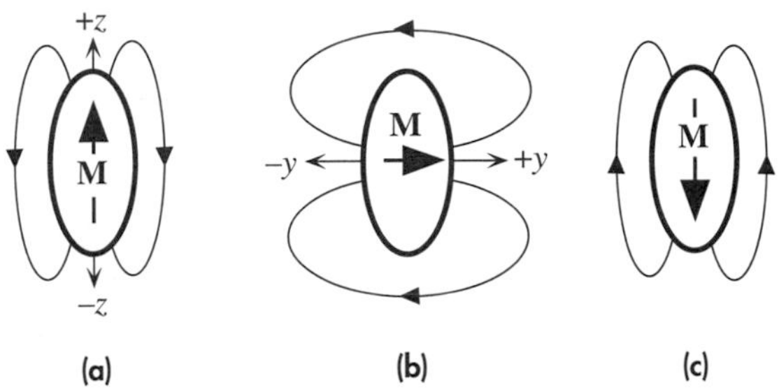


**Figure 8.41** A single domain elongated particle. Due to shape anisotropy, magnetization prefers to be along the long axis as in (a). Work has to be done to change **M** from (a) to (b) to (c).

properties depend not only on the crystal structure of the particle but also on the shape of the particle because different shapes give rise to different external magnetic fields. For a spherical iron particle, the magnetization **M** will be in an easy direction, for example, along [100] taken along $+z$. To reverse the magnetization from $+z$ to $-z$ by an applied field, we have to rotate the spins around past the hard direction, as shown in Figure 8.40, since we cannot generate reverse domains (or move domain walls). The rotation of magnetization involves substantial work due to the magnetocrystalline anisotropic energy, and the result is high coercivity. The higher the magnetocrystalline anisotropy energy, the greater the coercivity. The energy involved in creating a domain wall increases with the magnetocrystalline anisotropy energy. The critical size below which a particle becomes a single domain therefore increases with the crystalline anisotropy. Barium ferrite crystals have the hexagonal structure and hence have a high degree of magnetocrystalline anisotropy. Critical size for single domain barium ferrite particles is about 1–1.5 μm, and the coercivity $\mu_o H_c$ of small particles can be as high as 0.3 T, compared with values 0.02–0.1 T in multidomain barium ferrite pieces.

Particles that are not spherical may even have higher coercivity as a result of shape anisotropy. Consider an ellipsoid (elongated) fine particle, shown in Figure 8.41a. If the magnetization **M** is along the long axis (along $z$), then the potential energy in the external magnetic field is less than if **M** were along the minor axis (along $y$), as compared in Figure 8.41a and b. Thus, we have to do work to rotate **M** from the long to the short axis, or from Figure 8.41a to b. An elongated fine particle therefore has its magnetization along its length, and the effect is called **shape anisotropy.** If we have to reverse the magnetization from $+z$ to $-z$ by applying a reverse field, then we can only do so by rotating the magnetization, as shown in Figure 8.41a to c. **M** has to be rotated around through the minor axis, and this involves substantial work. Thus the coercivity is high. In general, the greater the elongation of the particle with respect to its width, the higher the coercivity. Small spherical Fe–Cr–Co particles have a coercivity $\mu_o H_c$ at most 0.02 T, but elongated and aligned particles can have a coercivity as high as 0.075 T due to shape anisotropy.

High coercivity magnets can be fabricated by having elongated fine particles dispersed by precipitation in a structure. Fine particles will be single domains. Alnico is a popular permanent magnet material that is an alloy of the metals Al, Ni, Co, and Fe (hence the name). Its microstructure consists of fine elongated Fe–Co rich particles, called the $\alpha'$-phase, dispersed in a matrix that is Ni–Al rich and called the $\alpha$-phase. The structure is obtained by an appropriate heat treatment that allows fine $\alpha'$ particles to precipitate out from a solid solution of the alloy. The $\alpha'$ particles are strongly magnetic, whereas the $\alpha$-phase matrix is weakly magnetic. When the heat treatment is carried out in the presence of a strong applied magnetic field, the $\alpha'$ particles that are formed have their elongations (or lengths) and hence their magnetizations along the applied field. The demagnetization process requires the rotations of the magnetizations in single domain elongated $\alpha'$ particles, which is a difficult process (shape anisotropy), and hence the coercivity is high. The main drawback of the Alnico magnet is that the alloy is mechanically hard and brittle and cannot be shaped except by casting or sintering before heat treatment. There are, however, other alloy permanent magnets that can be machined.

A variety of permanent magnets are made by compacting high-coercivity particles by using powder metallurgy (*e.g.*, powder pressing or sintering). The particles are magnetically hard because they are sufficiently small for each to be of single domain or they possess substantial shape anisotropy (elongated particles may be ferromagnetic alloys, *e.g.*, Fe–Co, or various hard ferrites). These are generically called powdered solid permanent magnets. An important class is the **ceramic magnets** that are made by compacting barium ferrite, $BaFe_{12}O_9$, or strontium ferrite, $SrFe_{12}O_9$, particles. The barium ferrite has the hexagonal crystal structure with a large magnetocrystalline anisotropy, which means that barium ferrite particles have high coercivity. The ceramic magnet is typically formed by wet pressing ferrite powder in the presence of a magnetizing field, which allows the easy directions of the particles to be aligned, and then drying and carefully sintering the ceramic. They are used in many low-cost applications.

**Rare earth cobalt** permanent magnets based on samarium–cobalt (Sm–Co) alloys have very high $(BH)_{max}$ values and are widely used in many applications such as dc motors, stepper and servo motors, traveling wave tubes, klystrons, and gyroscopes. The intermetallic compound $SmCo_5$ has a hexagonal crystal structure with high magnetocrystalline anisotropy and hence high coercivity. The $SmCo_5$ powder is pressed in the presence of an applied magnetic field to align the magnetizations of the particles. This is followed by careful sintering to produce a solid powder magnet. The $Sm_2Co_{15}$ magnets are more recent and have particularly high values of $(BH)_{max}$ up to about 240 kJ $m^{-3}$. $Sm_2Co_{15}$ is actually a generic name and the alloy may contain other transition metals substituting for some of the Co atoms.

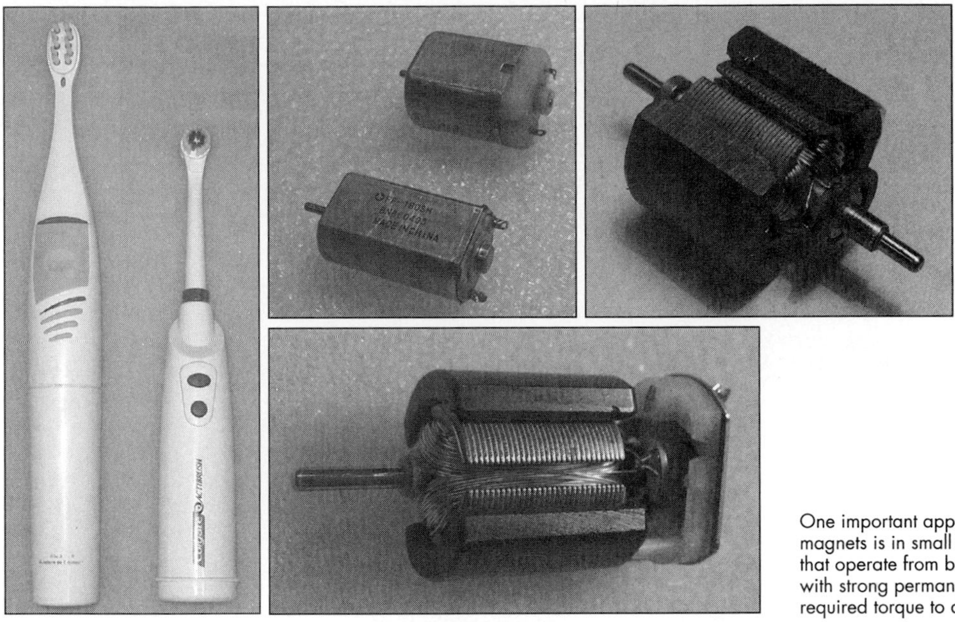

One important application of permanent magnets is in small dc motors. Toothbrushes that operate from batteries use dc motors with strong permanent magnets to get the required torque to drive the brushes.

The more recent **neodymium–iron–boron,** NdFeB, powdered solid magnets can have very large $(HB)_{\max}$ values up to about 275 kJ $m^{-3}$. The tetragonal crystal structure has the easy direction along the long axis and possesses high magnetocrystalline anisotropic energy. This means that we need a substantial amount of work to rotate the magnetization around through the hard direction, and hence the coercivity is also high. The main drawback is the lower Curie temperature, typically around 300 °C, whereas for Alnico and rare earth cobalt magnets, the Curie temperatures are above 700 °C. Another method of preparing NdFeB magnets is by the recrystallization of amorphous NdFeB at an elevated temperature in an applied field. The grains in the recrystallized structure are sufficiently small to be single domain grains and therefore possess high coercivity.

**EXAMPLE 8.7**

**$(BH)_{\max}$ FOR A PERMANENT MAGNET** Consider the permanent magnet in Figure 8.42. There is a small air gap of length $\ell_g$ where there is an external magnetic field that is available to do work. For example, if we were to insert an appropriate coil in the gap and pass a current through the coil, it would rotate as in a moving coil panel meter. Show that the magnetic energy per unit volume stored in the gap is proportional to the maximum value of $BH$. How does $(BH)_{\max}$ vary with the magnetizing field?

**SOLUTION**

Let $\ell_m$ be the mean length of the magnet from one end to the other, as shown in Figure 8.42. We assume that the cross-sectional area $A$ is constant throughout. There are no windings around the magnet and no current, $I = 0$. Ampere's law for $H$ involves integrating $H$ along a closed path or around the mean path length $\ell_m + \ell_g$. Suppose that $H_m$ and $H_g$ are the magnetic

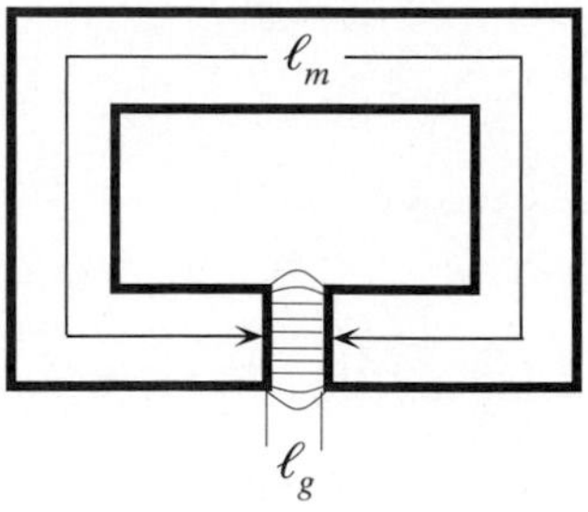


**Figure 8.42** A permanent magnet with a small air gap.

field intensities in the permanent magnet and in the gap, respectively. Then $H\,d\ell$ integrated around $\ell_m + \ell_g$ is

$$\oint H\,d\ell = H_m \ell_m + H_g \ell_g = 0$$

so that

$$H_g = -H_m \frac{\ell_m}{\ell_g}$$

and hence

*B–H for air gap*

$$B_g = -\mu_o \frac{\ell_m}{\ell_g} H_m \qquad [8.24]$$

Equation 8.24 is a relationship between $B_g$ in the gap and $H_m$ in the magnet. In addition, we have the *B–H* relationship for the magnetic material itself between the magnetic field $B_m$ and intensity $H_m$ in the magnet, that is,

*B–H for magnet material*

$$B_m = f(H_m) \qquad [8.25]$$

The magnetic flux in the magnet and in the air gap must be continuous. Since we assumed a uniform cross-sectional area, the continuity of flux across the air gap implies that $B_m = B_g$. Thus we need to equate Equation 8.24 to Equation 8.25. Equation 8.24 is a straight line with a negative slope in a $B_g$ versus $H_m$ plot, as shown in Figure 8.43a. Equation 8.25 is, of course, the *B–H* characteristics of the material. The two intersect at point *P*, as shown in Figure 8.43a, where $B_g = B_m = B'_m$ and $H_m = H'_m$.

We know that there is magnetic energy in the air gap given by

$$E_{\text{mag}} = (\text{Gap volume})(\text{Magnetic energy density in the gap})$$

$$= (A\ell_g)\left(\frac{1}{2}B_g H_g\right) = \frac{1}{2}(A\ell_g)B'_m H'_m \left(\frac{\ell_m}{\ell_g}\right)$$

$$= \frac{1}{2}(A\ell_m)B'_m H'_m$$

*Energy in air gap of a magnet*

$$= \frac{1}{2}(\text{Magnet volume})B'_m H'_m \qquad [8.26]$$

Thus, the external magnetic energy depends on the magnet volume and the *product* of $B'_m$ and $H'_m$ of the magnet characteristics at the operating point *P*. For a given magnet size, the magnetic energy in the gap is proportional to the rectangular area $B'_m H'_m$, $OB'_m PH'_m$ in Figure 8.43a,

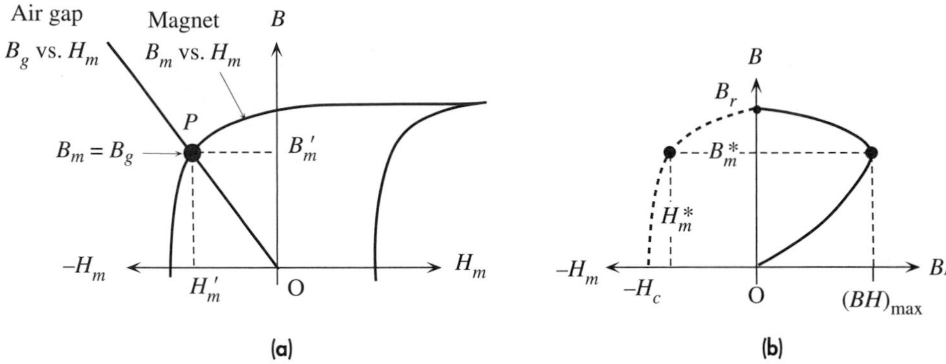


**Figure 8.43**

(a) Point $P$ represents the operating point of the magnet and determines the magnetic field inside and outside the magnet.

(b) Energy density in the gap is proportional to $BH$, and for a given geometry and size of gap, this is a maximum at a particular magnetic field $B^*_m$ or $B^*_g$.

and we have to maximize this area for the best energy extraction. Figure 8.43b shows how the product $BH$ varies with $B$ in a typical magnetic material. $BH$ is maximum at $(BH)_{max}$, when the magnetic field is $B^*_m$ and the field intensity is $H^*_m$. We can appropriately choose the air-gap size to operate at these values, in which case we will be only limited by the $(BH)_{max}$ available for that magnetic material. It is clear that $(BH)_{max}$ is a good figure of merit for comparing hard magnetic materials. According to Table 8.6, we can extract four to five times more work from a rare earth cobalt magnet than from an Alnico magnet of the same size if we were not limited by economics and weight. It should be mentioned that Equation 8.26 is only approximate as it neglects all fringe fields.

# 8.9 SUPERCONDUCTIVITY

## 8.9.1 ZERO RESISTANCE AND THE MEISSNER EFFECT

In 1911 Kamerlingh Onnes at the University of Leiden in Holland observed that when a sample of mercury is cooled to below 4.2 K, its resistivity totally vanishes and the material behaves as a **superconductor,** exhibiting no resistance to current flow. Other experiments since then have shown that there are many such substances, not simply metals, that exhibit superconductivity when cooled below a **critical temperature** $T_c$ that depends on the material. On the other hand, there are also many conductors, including some with the highest conductivities such as silver, gold, and copper, that do not exhibit superconductivity. The resistivity of these **normal conductors** at low temperatures is limited by scattering from impurities and crystal defects and saturates at a finite value determined by the residual resistivity. The two distinctly different types of behavior are depicted in Figure 8.44. Between 1911 and 1986, many different metals and metal alloys had been studied, and the highest

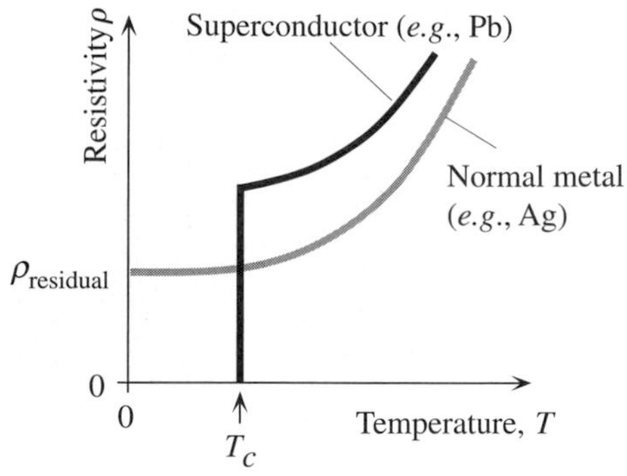


**Figure 8.44** A superconductor such as lead evinces a transition to zero resistivity at a critical temperature $T_c$ (7.2 K for Pb).
A normal conductor such as silver exhibits residual resistivity down to lowest temperatures.

recorded critical temperature was about 23 K in a niobium–germanium compound ($Nb_3Ge$) whose superconductivity was discovered in the early 1970s. In 1986 Bednorz and Müller, at IBM Research Laboratories in Zürich, discovered that a copper oxide–based ceramic-type compound La–Ba–Cu–O, which normally has high resistivity, becomes superconducting when cooled below 35 K. Following this Nobel prize–winning discovery, a variety of copper oxide–based compounds (called cuprate ceramics) have been synthesized and studied. In 1987 it was found that yttrium barium copper oxide (Y–Ba–Cu–O) becomes superconducting at a critical temperature of 95 K, which is above the boiling point of nitrogen (77 K). This discovery was particularly significant because liquid nitrogen is an inexpensive cryogent that is readily liquified and easy to use compared with cryogent liquids that had to be used in the

Superconductivity, zero resistance below a certain critical temperature, was discovered by a Dutch physicist, Heike Kamerlingh Onnes, in 1911. Kamerlingh Onnes and one of his graduate students found that the resistance of frozen mercury simply vanished at 4.15 K; Kamerlingh Onnes won the Nobel prize in 1913.

SOURCE: © Rijksmuseum voor de Geschiedenis der Natuurwetenschappen, courtesy AIP Emilio Segrè Visual Archives.

John Bardeen, Leon N. Cooper, and John Robert Schrieffer, in Nobel prize ceremony (1972). They received the Nobel prize for the explanation of superconductivity in terms of Cooper pairs.
| SOURCE: AIP Emilio Segrè Visual Archives.
"My belief is that the pairing condensation is what Mother Nature had in mind when she created these fascinating high-$T_c$ systems." Robert Schrieffer (1991)

past (liquid helium). At present the highest critical temperature for a superconductor is around 130 K (−143 °C) for Hg–Ba–Ca–Cu–O. These superconductors with $T_c$ above ~30 K are now typically referred as **high-$T_c$ superconductors.** The quest for a near-room-temperature superconductor goes on, with many scientists around the world trying different materials, or synthesizing them, to raise $T_c$ even higher. There are already commercial devices utilizing high-$T_c$ superconductors, for example, thin-film SQUIDs[7] that can accurately measure very small magnetic fluxes, high-Q filters, and resonant cavities in microwave communications.

The vanishing of resistivity is not the only characteristic of a superconductor. A superconductor cannot be viewed simply as a substance that has infinite conductivity below its critical temperature. A superconductor below its critical temperature expels all the magnetic field from the bulk of the sample as if it were a perfectly diamagnetic substance. This phenomenon is known as the **Meissner effect.** Suppose that we place a superconducting material in a magnetic field above $T_c$. The magnetic field lines will penetrate the sample, as we expect for any low $\mu_r$ medium. However, when the superconductor is cooled below $T_c$, it rejects all the magnetic flux in the sample, as depicted in Figure 8.45. The superconductor develops a magnetization $M$ by developing surface currents, such that $M$ and the applied field cancel everywhere inside

| [7] SQUID is a superconducting quantum interference device that can detect very small magnetic fluxes.

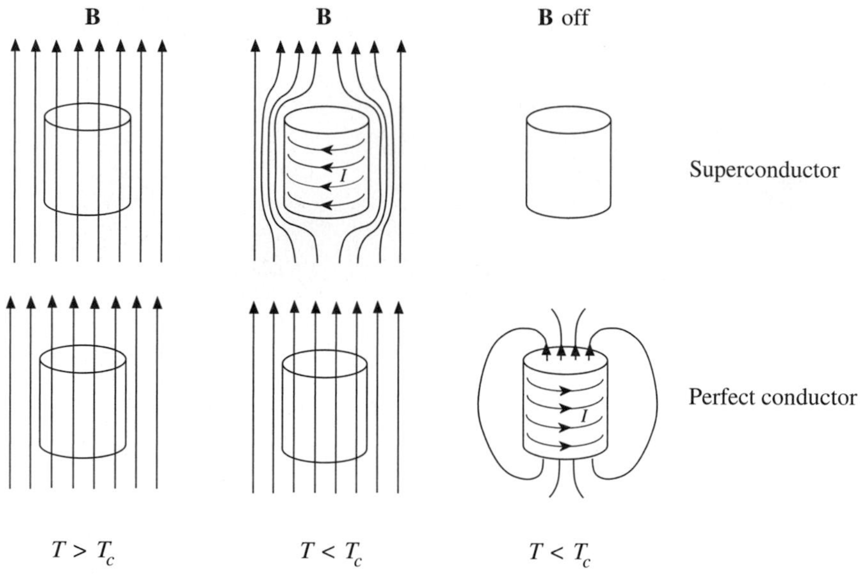


**Figure 8.45** The Meissner effect.
A superconductor cooled below its critical temperature expels all magnetic field lines from the bulk by setting up a surface current. A perfect conductor ($\sigma = \infty$) shows no Meissner effect.

the sample. Put differently $\mu_o M$ is in the *opposite* direction to the applied field and equal to it in magnitude. Thus, below $T_c$ a superconductor is a perfectly diamagnetic substance ($\chi_m = -1$). This should be contrasted with the behavior of a perfect conductor, which only exhibits infinite conductivity, or $\rho = 0$, below $T_c$. If we place a perfect conductor in a magnetic field and then cool it below $T_c$, the magnetic field is not rejected. These two types of behavior are identified in Figure 8.45. If we switch off the field, the field around the superconductor simply disappears. But switching off the field means there is a decreasing applied field. This change in the field induces currents in the perfect conductor by virtue of Faraday's law of induction. These currents generate a magnetic field that opposes the change (Lenz's law); in other words, they generate a field along the same direction as the applied field to reenforce the decreasing field. As the current can be sustained ($\rho = 0$) without Joule dissipation, it keeps on flowing and maintaining the magnetic field. The two final situations are shown in Figure 8.45 and distinguish the Meissner effect, a distinct characteristic of a superconductor, from the behavior of a perfect conductor ($\rho = 0$ only). The photograph showing the levitation of a magnet above the surface of a superconductor (Figure 8.46) is the direct result of the Meissner effect: the exclusion of the magnet's magnetic fields from the interior of the superconductor.

The transition from the normal state to the superconducting state as the temperature falls below the critical temperature has similarities with phase transitions such as solid to liquid or liquid to vapor changes. At the critical temperature, there is a sharp change in the heat capacity as one would observe for any phase change. In the superconducting

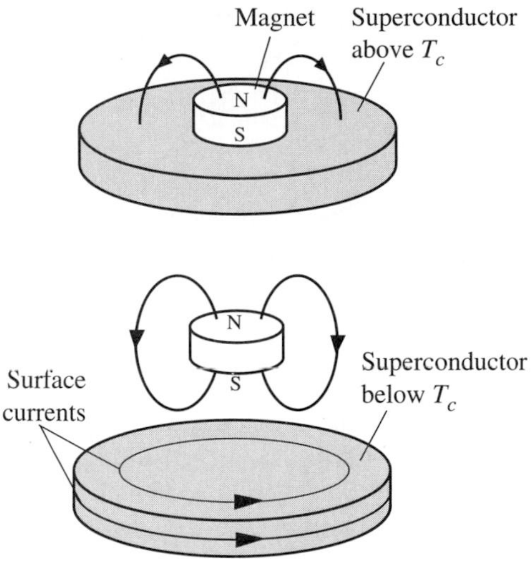

**Figure 8.46**

Left: A magnet over a superconductor becomes levitated. The superconductor is a perfect diamagnet which means that there can be no magnetic field inside the superconductor.

Right: Photograph of a magnet levitating above a superconductor immersed in liquid nitrogen (77 K). This is the Meissner effect.

| SOURCE: Photo courtesy of Professor Paul C. W. Chu.

state, we cannot treat a conduction electron in isolation. The electrons behave collectively and thereby impart the superconducting characteristics to the substance, as discussed later.

## 8.9.2 Type I and Type II Superconductors

The superconductivity below the critical temperature has been observed to disappear in the presence of an applied magnetic field exceeding a critical value denoted by $B_c$. This critical field depends on the temperature and is a characteristic of the material. Figure 8.47 shows the dependence of the critical field on the temperature. The critical field is maximum, $B_c(0)$, when $T = 0$ K (obtained by extrapolation[8]). As long as the applied field is below $B_c$ at that temperature, the material is in the superconducting state, but when the field exceeds $B_c$, the material reverts to the normal state. We know that in the superconducting state, the applied magnetic field lines are expelled from the sample and the phenomenon is called the Meissner effect. The external field, in fact, does penetrate the sample from the surface into the bulk, but the magnitude of this penetrating field decreases exponentially from the surface. If the field at the surface of the sample is $B_o$, then at a distance $x$ from the surface, the field is

[8] There is a third law to thermodynamics that is not as emphasized as the first two laws, which dominate all branches of engineering. That is, one can never reach the absolute zero of temperature.

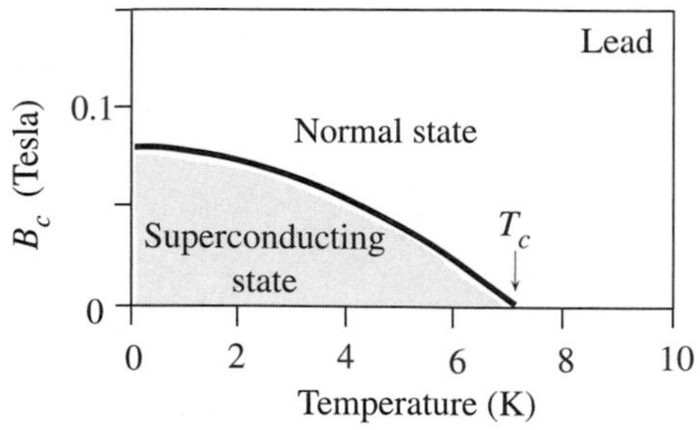


**Figure 8.47** The critical field versus temperature in Type I superconductors.

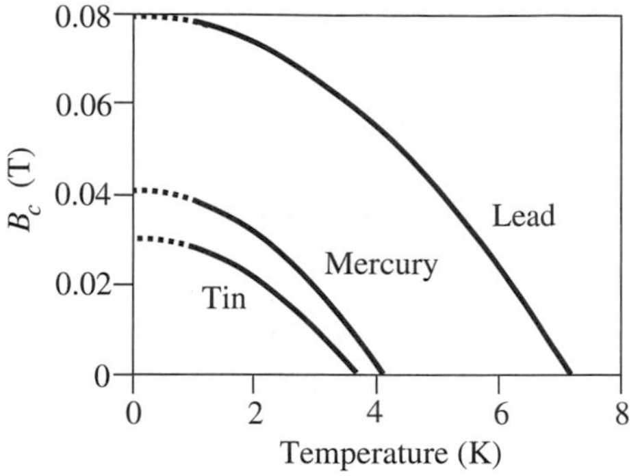


**Figure 8.48** The critical field versus temperature in three examples of Type I superconductors.

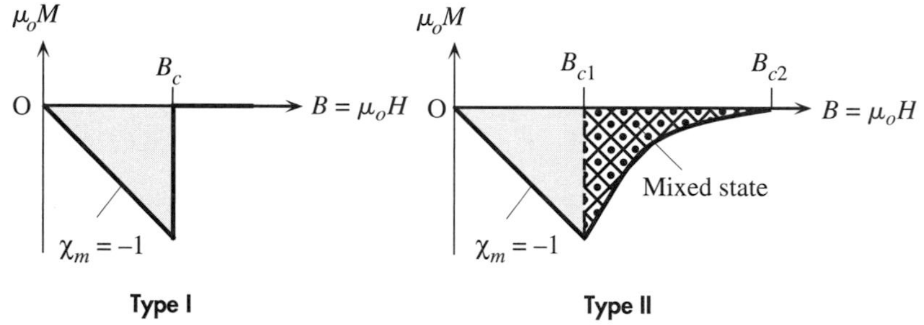


**Figure 8.49** Characteristics of Type I and Type II superconductors. $B = \mu_o H$ is the applied field and $M$ is the overall magnetization of the sample. Field inside the sample, $B_{\text{inside}} = \mu_o H + \mu_o M$, which is zero only for $B < B_c$ (Type I) and $B < B_{c1}$ (Type II).

given by an exponential decay,

$$B(x) = B_o \exp\left(-\frac{x}{\lambda}\right)$$

where $\lambda$ is a "characteristic length" of penetration, called the **penetration depth,** and depends on the temperature and $T_c$ (or the material). At the critical temperature, the penetration length is infinite and any magnetic field can penetrate the sample and destroy the superconducting state. Near absolute zero of temperature, however, typical penetration depths are 10–100 nm. Figure 8.48 shows the $B_c$ versus $T$ behavior for three example superconductors, tin, mercury, and lead.

Superconductors are classified into two types, called Type I and Type II, based on their diamagnetic properties. In Type I superconductors, as the applied magnetic field $B$ increases, so does the opposing magnetization $M$ until the field reaches the critical field $B_c$, whereupon the superconductivity disappears. At that point, the perfect diamagnetic behavior, the Meissner effect, is lost, as illustrated in Figure 8.49. A Type I superconductor below $B_c$ is in the **Meissner state,** where it excludes all the magnetic

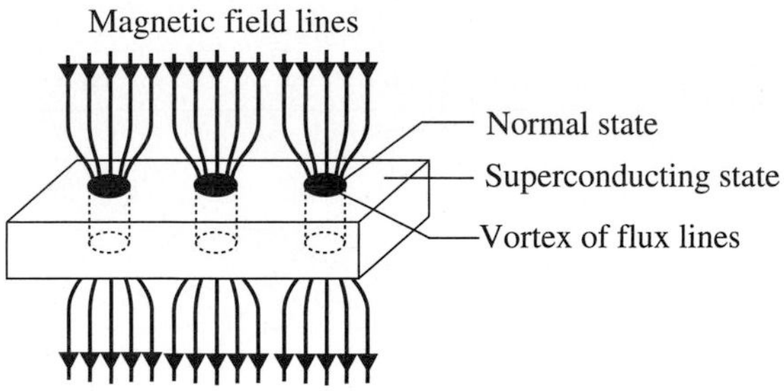


**Figure 8.50** The mixed or vortex state in a Type II superconductor.

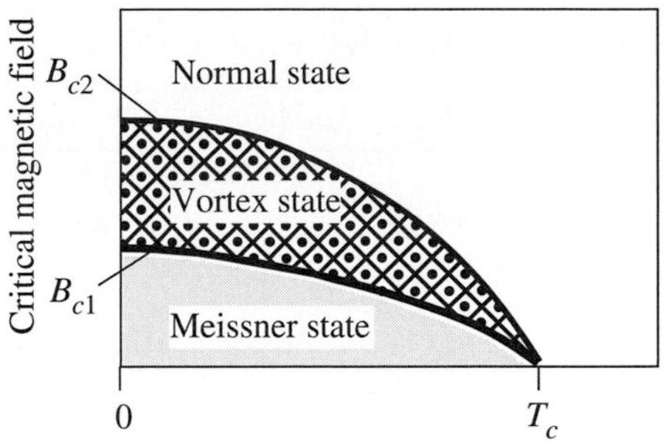


**Figure 8.51** Temperature dependence of $B_{c1}$ and $B_{c2}$.

flux from the interior of the sample. Above $B_c$ it is in the normal state, where the magnetic flux penetrates the sample as it would normally and the conductivity is finite.

In the case of Type II superconductors, the transition does not occur sharply from the Meissner state to the normal state but goes through an intermediate phase in which the applied field is able to pierce through certain local regions of the sample. As the magnetic field increases, initially the sample behaves as a perfect diamagnet exhibiting the Meissner effect and rejecting all the magnetic flux. When the applied field increases beyond a critical field denoted as $B_{c1}$, the **lower critical field,** the magnetic flux lines are no longer totally expelled from the sample. The overall magnetization $M$ in the sample opposes the field, but its magnitude does not cancel the field everywhere. As the field increases, $M$ gets smaller and more flux lines pierce through the sample until at $B_{c2}$, the **upper critical field,** all field lines penetrate the sample and superconductivity disappears. This behavior is shown in Figure 8.49. Type II superconductors therefore have two critical fields $B_{c1}$ and $B_{c2}$.

When the applied field is between $B_{c1}$ and $B_{c2}$, the field lines pierce through the sample through tubular local regions, as pictured in Figure 8.50. The sample develops local small cylindrical (filamentary) regions of normal state in a matrix of superconducting state and the magnetic flux lines go though these filaments of local normal state, as shown in Figure 8.50. The state between $B_{c1}$ and $B_{c2}$ is called the **mixed state** (or **vortex state**) because there are two states—normal and superconducting—mixed in the same sample. The filaments of normal state have finite conductivity and a quantized amount of flux through them. Each filament is a **vortex** of flux lines (hence the name *vortex state*). It should be apparent that there should be currents circulating around the walls of vortices. These circulating currents ensure that the magnetic flux through the superconducting matrix is zero. The sample overall has infinite conductivity due to the superconducting regions. Figure 8.51 shows the dependence of $B_{c1}$ and $B_{c2}$ on the temperature and identifies the regions of Meissner, mixed, and normal states. All engineering applications of superconductors invariably use Type II materials because $B_{c2}$ is typically much greater than $B_c$ found in Type I materials and, furthermore, the critical temperatures of Type II materials are higher than those of Type I. Many superconductors, including the recent high-$T_c$ superconductors, are of Type II. Table 8.7 summarizes the characteristics of selected Type I and Type II superconductors.

**Table 8.7** Examples of Type I and Type II superconductors

| Type I | Sn | Hg | Ta | V | Pb | Nb |
|---|---|---|---|---|---|---|
| $T_c$ (K) | 3.72 | 4.15 | 4.47 | 5.40 | 7.19 | 9.2 |
| $B_c$ (T) | 0.030 | 0.041 | 0.083 | 0.14 | 0.08 | 0.198 |

| Type II | $Nb_3Sn$ | $Nb_3Ge$ | $Ba_{2-x}Br_xCuO_4$ | Y–Ba–Cu–O ($YBa_2Cu_3O_7$) | Bi–Sr–Ca–Cu–O ($Bi_2Sr_2Ca_2Cu_3O_{10}$) | Hg–Ba–Ca–Cu–O |
|---|---|---|---|---|---|---|
| $T_c$ (K) | 18.05 | 23.2 | 30–35 | 93–95 | 122 | 130–135 |
| $B_{c2}$ (Tesla) at 0 K | 24.5 | 38 | ~150 | ~300 | | |
| $J_c$ (A cm$^{-2}$) at 0 K | ~$10^7$ | | | $10^4$–$10^7$ | | |

NOTE: Critical fields are close to absolute zero, obtained by extrapolation. Type I for pure, clean elements.

### 8.9.3 CRITICAL CURRENT DENSITY

Another important characteristic feature of the superconducting state is that when the current density through the sample exceeds a critical value $J_c$, it is found that superconductivity disappears. This is not surprising since the current through the superconductor will itself generate a magnetic field and at sufficiently high current densities, the magnetic field at the surface of the sample will exceed the critical field and extinguish superconductivity. This plausible direct relation between $B_c$ and $J_c$ is only true for Type I superconductors, whereas in Type II superconductors, $J_c$ depends in a complicated way on the interaction between the current and the flux vortices. New high-$T_c$ superconductors have exceedingly high critical fields, as apparent in Table 8.7, that do not seem to necessarily translate to high critical current densities. The critical current density in Type II superconductors depends not only on the temperature and the applied magnetic field but also on the preparation and hence the microstructure (*e.g.*, polycrystallinity) of the superconductor material. Critical current densities in new high-$T_c$ superconductors vary widely with preparation conditions. For example, in Y–Ba–Cu–O, $J_c$ may be greater than $10^7$ A cm$^{-2}$ in some carefully prepared thin films and single crystals but around $10^3$–$10^6$ A cm$^{-2}$ in some of the polycrystalline bulk material (*e.g.*, sintered bulk samples). In $Nb_3Sn$, used in superconducting solenoid magnets, on the other hand, $J_c$ is close to $10^7$ A cm$^{-2}$ at near 0 K.

The critical current density is important in engineering because it limits the total current that can be passed through a superconducting wire or a device. The limits of superconductivity are therefore defined by the critical temperature $T_c$, critical magnetic field $B_c$ (or $B_{c2}$), and critical current density $J_c$. These constitute a surface in a three-dimensional plot, as shown in Figure 8.52, which separates the superconducting state from the normal state. Any operating point ($T_1$, $B_1$, $J_1$) inside this surface is in the superconducting state. When the cuprate ceramic superconductors were first discovered, their $J_c$ values were too low to allow immediate significant applications in engineering.

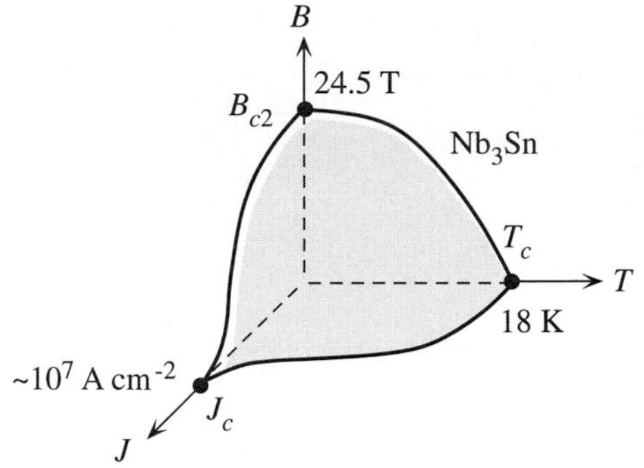


**Figure 8.52** The critical surface for a niobium–tin alloy, which is a Type II superconductor.

Their synthesis over the last 10 years has advanced to a level that we can now benefit from large critical currents and fields. Over the same temperature range, ceramic cuprate superconductors now easily outperform the traditional superconductors. There are already a number of applications of these high-$T_c$ superconductors in the commercial market.

**EXAMPLE 8.8**

**SUPERCONDUCTING SOLENOIDS**[9] Superconducting solenoid magnets can produce very large magnetic fields up to $\sim$15 T or so, whereas the magnetic fields available from a ferromagnetic core solenoid is limited to $\sim$2 T. High field magnets used in magnetic resonance imaging are based on superconducting solenoids wound using a superconducting wire. They are operated around 4 K with expensive liquid helium as the cryogen. These superconducting wires are typically $Nb_3Sn$ or NbTi alloy filaments embedded in a copper matrix. A very large current, several hundred amperes, is passed through the solenoid winding to obtain the necessary high magnetic fields. There is, of course, no Joule heating once the current is flowing in the superconducting state. The main problem is the large forces and hence stresses in the coil due to large currents. Two wires carrying currents in the opposite direction repel each other, and the force is proportional to $I^2$. Thus the magnetic forces between the wires of the coil give rise to outward radial forces trying to "blow open" the solenoid, as depicted in Figure 8.53. The forces between neighboring wires are attractive and hence give rise to compressional forces squeezing the solenoid axially. The solenoid has to have a proper mechanical support structure around it to prevent mechanical fracture and failure due to large forces between the windings. The copper matrix serves as mechanical support to cushion against the stresses as well as a good thermal conductor in the event that superconductivity is inadvertently lost during operation.

Suppose that we have a superconducting solenoid that is 10 cm in diameter and 1 m in length and has 500 turns of $Nb_3Sn$ wire, whose critical field $B_c$ at 4.2 K (liquid He temperature) is about 20 T and critical current density $J_c$ is $3 \times 10^6$ A cm$^{-2}$. What is the current necessary to set up a field of 5 T at the center of a solenoid? What is the approximate energy stored in the

[9] Designing a superconducting solenoid is by no means trivial, and the enthusiastic student is referred to a very readable description given by James D. Doss, *Engineer's Guide to High Temperature Superconductivity,* New York: John Wiley & Sons, 1989, ch. 4. Photographs and descriptions of catastrophic failure in high field solenoids can be found in an article by G. Broebinger, A. Passner, and J. Bevk, "Building World-Record Magnets" in *Scientific American,* June 1995, pp. 59–66.

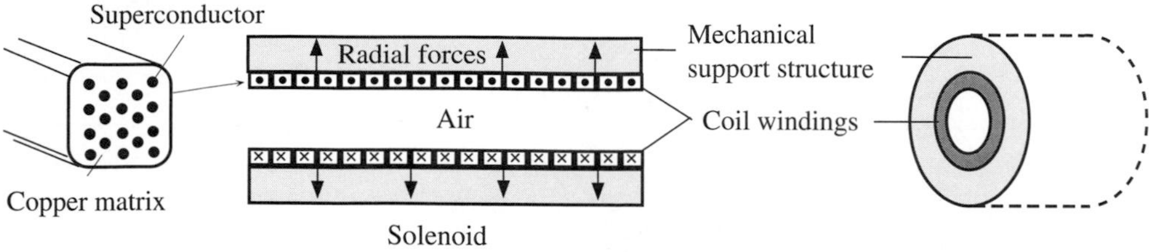


**Figure 8.53** A solenoid carrying a current experiences radial forces pushing the coil apart and axial forces compressing the coil.

Superconducting electromagnets used on MRI. Operates with liquid He, providing a magnetic field 0.5–1.5 T.

SOURCE: Courtesy of IGC Magnet Business group.

solenoid? Assume that the critical current density decreases linearly with the applied field. Further, assume also that the field across the diameter of the solenoid is approximately uniform (field at the windings is the same as that at the center).

**SOLUTION**

We can assume that we have a long solenoid, that is, length (100 cm) $\gg$ diameter (10 cm). The field at the center of a long solenoid is given by

$$B = \frac{\mu_o N I}{\ell}$$

so the current necessary for $B = 5$ T is

$$I = \frac{B\ell}{\mu_o N} = \frac{(5)(1)}{(4\pi \times 10^{-7})(500)} = 7958 \text{ A} \quad \text{or} \quad 7.96 \text{ kA}$$

As the coil is 1 m and there are 500 turns, the coil wire radius must be 1 mm. If all the cross section of the wire were of superconducting medium, then the corresponding current density would be

$$J_{\text{wire}} = \frac{I}{\pi r^2} = \frac{7958}{\pi (0.001)^2} = 2.5 \times 10^9 \text{ A m}^{-2} \quad \text{or} \quad 2.5 \times 10^5 \text{ A cm}^{-2}$$

The actual current density through the superconductors will be greater than this as the wires are embedded in a metal matrix. Suppose that 20 percent by cross-sectional area (and hence as volume percentage) is the superconductor; then the actual current density through the superconductor is

$$J_{\text{super}} = \frac{J_{\text{wire}}}{0.2} = 1.25 \times 10^6 \text{ A cm}^{-2}$$

We now need the critical current density $J_c'$ at a field of 5 T. Assuming $J_c$ decreases linearly with the applied field and vanishes when $B = B_c$, we can find $J_c'$, from linear interpolation

$$J_c' = J_c \frac{B_c - B}{B_c} = (3 \times 10^6 \text{ A cm}^{-2}) \frac{20 \text{ T} - 5 \text{ T}}{20 \text{ T}} = 2.25 \times 10^6 \text{ A cm}^{-2}$$

The actual current density $J_{\text{super}}$ through the superconductors is less than this critical value $J_c'$. We can assume that the superconducting solenoid will operate "safely" (with all other designs correctly implemented). It should be emphasized that accurate and reliable calculations will involve the actual $J_c$–$B_c$–$T_c$ surface, as in Figure 8.52 for the given material.

Since the field in the solenoid is $B = 5$ T, assuming that this is uniform along the axis and the core is air, the energy density or energy per unit volume is

$$E_{\text{vol}} = \frac{B^2}{2\mu_o} = \frac{5^2}{2(4\pi \times 10^{-7})} = 9.95 \times 10^6 \text{ J m}^{-3}$$

so the total energy

$$E = E_{\text{vol}} [\text{volume}] = (9.95 \times 10^6 \text{ J m}^{-3})[(1 \text{ m})(\pi 0.05^2 \text{ m}^2)]$$

$$= 7.81 \times 10^4 \text{ J} \quad \text{or} \quad 78.1 \text{ kJ}$$

If all this energy can be converted to electrical work, it would light a 100 W lamp for 13 min (and if converted to mechanical work, it could lift an 8 ton truck by 1 m).

## 8.10 SUPERCONDUCTIVITY ORIGIN

Although superconductivity was discovered in 1911, the understanding of its origin did not emerge until 1957 when Bardeen, Cooper, and Schrieffer formulated the theory (called the **BCS theory**) in terms of quantum mechanics. The quantum mechanical treatment is certainly beyond the scope of this book, but one can nonetheless grasp an intuitive understanding, as follows. The cardinal idea is that, at sufficiently low temperatures, two oppositely spinning and oppositely traveling electrons can attract each other indirectly through the deformation of the crystal lattice of positive metal ions. The idea is illustrated pictorially in Figure 8.54. The electron 1 distorts the lattice

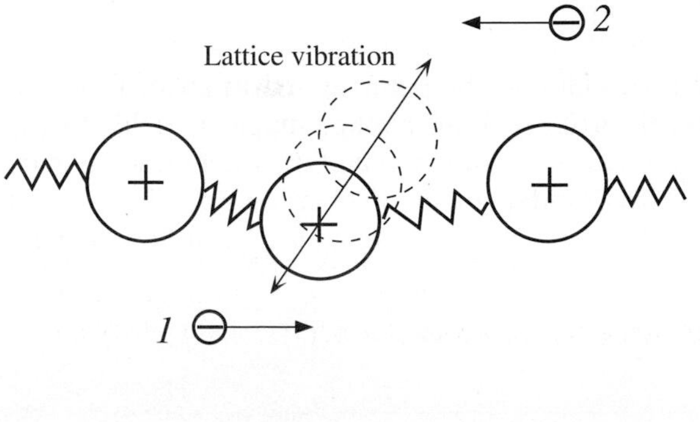


**Figure 8.54** A pictorial and intuitive view of an indirect attraction between two oppositely traveling electrons via lattice distortion and vibration.

around it and changes its vibrations as it passes through this region. Random thermal vibrations of the lattice at low temperatures are not strong enough to randomize this induced lattice distortion and vibration. The vibrations of this distorted region now look differently to another electron, 2, passing by. This second electron feels a "net" attractive force due to the slight displacements of positive metal ions from their equilibrium positions. The two electrons interact indirectly through the deformations and vibrations of the lattice of positive ions. This indirect interaction at sufficiently low temperatures is able to overcome the mutual Coulombic repulsion between the electrons and hence bind the two electrons to each other. The two electrons are called a **Cooper pair.** The intuitive diagram in Figure 8.54, of course, does not even convey the intuition why the spins of the electrons should be opposite. The requirement of opposite spins comes from the formal quantum mechanical theory. The net spin of the Cooper pair is zero and their net linear momentum is also zero. There is a further significance to the pairing of electron spins in the Cooper pair. As a quasi-particle, or an entity, the Cooper pair has no net spin and hence the Cooper pairs do not obey the Fermi–Dirac statistics.[10] They can therefore all "condense" to the *lowest energy* state and possess one single wavefunction that can describe the whole collection of Cooper pairs. All the paired electrons are described collectively by a single coherent wavefunction $\Psi$, which extends over the whole sample. A crystal imperfection cannot simply scatter a single Cooper pair because all the pairs behave as a single entity—like a "huge molecule." Scattering one pair involves scattering all, which is simply not possible. An analogy may help. One can scatter an individual football player running on his own. But if all the team members got together and moved forward arm in arm as a rigid line, then the scattering of any one now is impossible, as the rest will hold him in the line and continue to move forward (don't forget, it's only an analogy!). Superconductivity is said to be a macroscopic manifestation of quantum mechanics. The BCS theory has had good success with traditional superconductors, but there seems to be some doubt about its applicability to the new high-$T_c$ superconductors. There are a number of high-$T_c$ superconductivity theories at present, and the interested student can easily find additional reading on the subject.

# ADDITIONAL TOPICS

## 8.11 ENERGY BAND DIAGRAMS AND MAGNETISM

### 8.11.1 PAULI SPIN PARAMAGNETISM

Consider a paramagnetic metal such as sodium. The paramagnetism arises from the alignment of the spins of conduction electrons with the applied magnetic field. A conduction electron in a metal has an extended wave function and does not orbit any particular metal ion. The conduction electron's magnetic moment arises from the electron spin alone, and $\boldsymbol{\mu}_{\text{spin}}$ is in the opposite direction to the spin; $\boldsymbol{\mu}_{\text{spin}}$ can be either up

[10] In fact, the Cooper pair without a net spin behaves as if is were a **boson** particle.

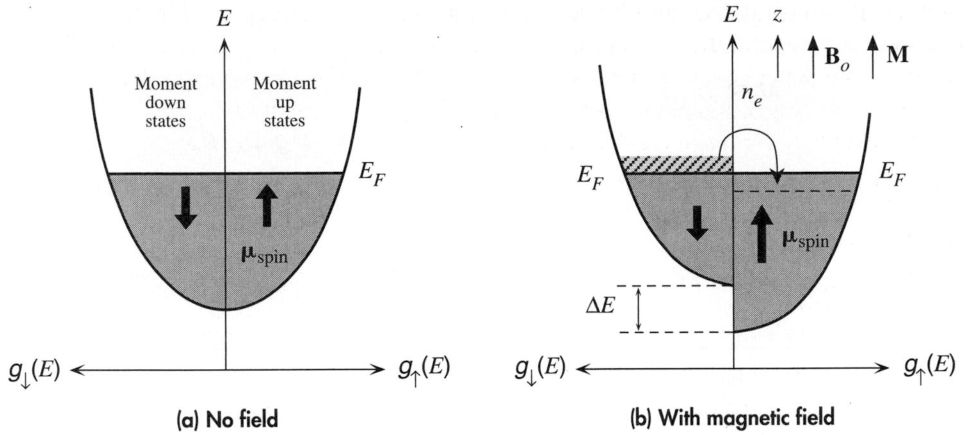


**Figure 8.55** Pauli spin paramagnetism in metals due to conduction electrons.

($m_s = -\frac{1}{2}$) or down ($m_s = +\frac{1}{2}$). In the absence of a magnetic field, the energies of magnetic moment up and down states (or wavefunctions) are the same and there are as many electrons with magnetic moment up as there are with magnetic moment down. Figure 8.55a shows the density of states (number of states per unit energy per unit volume) for states with magnetic moment up ($\uparrow$), denoted as $g_\uparrow(E)$, and for states with magnetic moment down ($\downarrow$), denoted as $g_\downarrow(E)$. Both states have the same energy and both are equally occupied. All energy levels up to the Fermi energy $E_F$ are occupied as shown in Figure 8.55a. Effectively we are viewing the energy band of the metal as two subbands corresponding to magnetic moment up and down bands. The bands overlap in the absence of a field and are indistinguishable.

Consider what happens in the presence of an applied field $B_o$ along the $z$ direction. If a conduction electron's magnetic moment $\mu_z$ is *along* the field (aligned with the field), then it has a lower potential energy. Thus, those electron wavefunctions with a magnetic moment up have lower energy, whereas those wavefunctions with a magnetic moment down have higher energy. In the presence of a field $B_o$, therefore, all states with magnetic moment up, and hence $g_\uparrow(E)$, are lowered in energy by $\beta B_o$ where $\beta$ is the Bohr magneton. All states with magnetic moment down, and hence $g_\downarrow(E)$, are raised by $\beta B_o$. Both shifts are shown in Figure 8.55b. Those electrons with magnetic moment down near $E_F$ in the $g_\downarrow(E)$ band can now find lower energy states in the $g_\uparrow(E)$ band and hence flip their spins and transfer to the $g_\uparrow(E)$ band. There are now more electrons in states with magnetic moment up in the $g_\uparrow(E)$ band than in the $g_\downarrow(E)$ band. When averaged over all conduction electrons there is now a net magnetic moment per conduction electron along the $z$ direction or the applied field.

To find the net magnetic moment per conduction electron we have to find how many electrons transfer from the $g_\downarrow(E)$ band to the $g_\uparrow(E)$ band. The energy separation $\Delta E$ between the magnetic moment down and up states is $2\beta B_o$. All electrons, $n_e$ per unit volume, in the $g_\downarrow(E)$ band around $E_F$ within an energy range $\frac{1}{2}\Delta E$ transfer to the $g_\uparrow(E)$ band. $\Delta E$ is small, so $n_e$ is approximately $g_\downarrow(E_F)(\frac{1}{2}\Delta E)$ or $\frac{1}{2}\, g(E_F)(\frac{1}{2}\Delta E)$ because $g(E_F)$ includes states with spin up and down, that is, $\frac{1}{2}\, g(E_F) = g_\downarrow(E_F)$. The magnetic moment down band decreases by $n_e$ and the magnetic moment up band increases by $n_e$ and the net magnetic moment per unit volume is

$$M \approx 2n_e\mu_z = 2\left[\tfrac{1}{2}\,g(E_F)\left(\tfrac{1}{2}\,\Delta E\right)\right]\beta$$
$$= 2\left[\tfrac{1}{2}\,g(E_F)\left(\tfrac{1}{2}\,2\beta B_o\right)\right]\beta = \beta^2 g(E_F) B_o$$

*Pauli spin para-magnetism*

Using $B_o = \mu_o H$ and the definition $\chi_m = M/H$, the paramagnetic susceptibility is

$$\chi_{\text{para}} \approx \mu_o\beta^2 g(E_F)$$

We see that the density of states at the Fermi level determines the susceptibility.

**EXAMPLE 8.9** **PAULI SPIN PARAMAGNETISM OF SODIUM** The Fermi energy of sodium, $E_F$, is 3.15 eV. Using the density of states $g(E)$ expression for the free conduction electrons in a metal, evaluate the paramagnetic susceptibility of sodium and compare with the experimental value of $9.1 \times 10^{-6}$.

**SOLUTION**

The density of states $g(E)$ in the free electron model is

$$g(E) = (8\pi 2^{1/2})\left(\frac{m_e}{h^2}\right)^{3/2} E^{1/2}$$

We have to evaluate $g(E)$ at the Fermi energy $E = E_F = 3.15$ eV,

$$g(E_F) = (8\pi 2^{1/2})\left(\frac{9.1 \times 10^{-31}}{(6.626 \times 10^{-34})^2}\right)^{3/2} (3.15 \times 1.6 \times 10^{-19})^{1/2} = 7.54 \times 10^{46}\ \text{J}^{-1}\ \text{m}^{-3}$$

Paramagnetic susceptibility is

$$\chi_{\text{para}} = \mu_o\beta^2 g(E_F) = (4\pi \times 10^{-7})(9.27 \times 10^{-24})^2(7.54 \times 10^{46}) = 8.16 \times 10^{-6}$$

We need to subtract the diamagnetic from the calculated paramagnetic susceptibility to obtain the net susceptibility, which would decrease the calculated value slightly. Nonetheless, given the approximate nature of the theory, the calculated value is not far out from the measured value.

## 8.11.2 ENERGY BAND MODEL OF FERROMAGNETISM

The energy band model of paramagnetism can be extended to explain ferromagnetism. Once we start using the energy band model, we are essentially assigning all the valence (outer shell) electrons of the atoms to a collective sharing among *all* the atoms; they no longer belong to their individual parents. These valence electrons now belong to the whole crystal. (The model is also known as the *itinerant electron model*.)

Recall that in a ferromagnetic crystal there is an internal magnetization, even in the absence of an applied field, due to a net number of unpaired spins; that is, overall, the crystal has more electrons with spins up than with spins down. The reason is the exchange energy, which causes the spin magnetic moments of two electrons to line up parallel to each other so that their energy is lowered in much the same way as Hund's

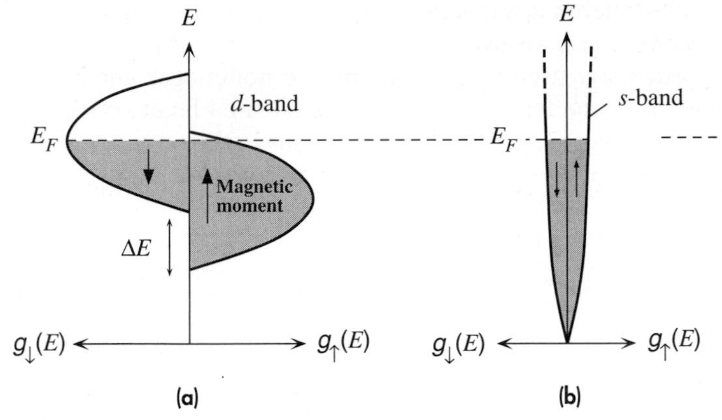


**Figure 8.56** Energy band model of ferromagnetism.
(a) The split *d*-band.
(b) The *s*-band is not affected. The arrows in the bands are spin magnetic moments.

rule works within an atom. In magnetic metals such as Fe, Ni, and Co, there are two bands of interest, the *s*-band and the *d*-band. The two bands overlap but the *s*-band is much wider. We can represent the density of states for magnetic moment up and magnetic moment down states separately. Consider the *d*-band. The density of states $g_\uparrow(E)$ for magnetic moment up states is lowered by $\Delta E$ with respect to the density of states $g_\downarrow(E)$ for magnetic moment down states due to the exchange energy as shown in Figure 8.56a. The energy lowering $\Delta E$ for the *s*-band can be neglected as in Figure 8.56b. All the states up to the Fermi energy are occupied. For Fe, the *d*-band magnetic moment up states are filled almost to the top of the band (this band is 96 percent full), and magnetic moment down states are filled roughly halfway. Thus, there are many more electrons with moments up than moments down; put differently there are many electrons that have aligned their spins. The spin magnetic moment alignment of electrons is exactly what is needed to generate a net magnetization. (In some books, the spin magnetic moment down band is sketched lower than the spin magnetic moment up band in contrast to Figure 8.56a. Both sketches are correct since both would also result in a net number of electrons having their spins in parallel, and hence a net magnetization within the crystal. Another way to look at it is to realize that there are two bands: one band for the "majority of spins," and another band for the "minority spins.")

The *s*-band is filled up to $E_F$, and there are almost equal numbers of electrons with up and down moments in this band. The ferromagnetic effect arises from the behavior of electrons mainly in the *d*-band. Electrical conduction, on the other hand, is determined by electrons in the *s*-band. The reason is that the *s*-band is very wide compared with the *d*-band, and the electron effective mass in the *s*-band is very small. Thus, electrons have a much higher mobility in the *s*-band than in the *d*-band. When an *s*-electron is scattered (by phonons, impurities, defects, etc.) into the *d*-band, it does not make any significant contribution to conduction because the drift mobility is very small in this band. The spin of the electron cannot be flipped easily in a scattering process. An *s*-electron with its moment down can be easily scattered into the empty states in the corresponding moment-down *d*-band (there are many empty states at $E_F$), but the moment-up electron has no states in the moment-up

$d$-band into which it can be scattered. Conduction occurs by moment-up electrons; these are the *favored* electrons for conduction.

The band model is particularly useful in explaining the noninteger number of Bohr magnetons that give rise to the ferromagnetism. The isolated Fe atom has six $3d$ and two $4s$ electrons or 8 valence electrons. These electrons in the crystal become shared by all the atoms. If $N$ is the number of atoms per unit volume, then one unit volume of crystal has $8N$ valence electrons. $8N$ electrons enter the $s$ and $d$ bands, filling states starting from the lowest energy.[11] The exact distribution of electrons depends on how many states are available at each energy as electrons fill the bands. We simply summarize the results of the filling process that is shown in Figure 8.56 for Fe:

$0.3N$ electrons in the moment-up $s$-band ($N$ states available)

$0.3N$ electrons in the moment-down $s$-band ($N$ states available)

$4.8N$ electrons in the moment-up $d$-band ($5N$ states available)

$2.6N$ electrons in the moment-down $d$-band ($5N$ states available)

To find how many electrons have parallel spin magnetic moments, we simply sum the above, which is $2.2N$ moment-up electrons per unit volume or $2.2N$ Bohr magnetons per unit volume, or 2.2 Bohr magnetons per atom. The saturation magnetization $M_{\text{sat}}$ is then $(2.2N)\beta$ or 2.2 T. There is therefore a natural explanation for a noninteger number of spins per atom in the band model of ferromagnetism.

## 8.12 ANISOTROPIC AND GIANT MAGNETORESISTANCE

In general, **magnetoresistance** refers to the change in the resistance of a material (any material) when it is placed in a magnetic field. When a nonmagnetic metal such as copper is placed in a magnetic field, the change in its resistivity, and hence the sample resistance, is so small that it has no real practical use. When a magnetic metal, such as iron, is placed in a magnetic field, the change in the resistivity depends on the direction of the current flow with respect to the magnetic field. The resistivity $\rho_{//}$ for current flow parallel to the magnetic field decreases, and the resistivity $\rho_{\perp}$, perpendicular to the field, increases by roughly the same amount. The change in the resistivity due to the applied magnetic field is *anisotropic* (depends on the direction) and is called **anisotropic magnetoresistance (AMR).** The change in resistivity is limited to a few percent, but, nonetheless, is still useful. The physical origin of this phenomenon is based on the applied field being able to tilt the orbital angular momenta of the $3d$ electrons as shown in Figure 8.57a. The field rotates the $3d$ orbitals, which changes the scattering of the conduction electrons according to their direction of travel; hence $\rho_{//}$ and $\rho_{\perp}$ are different, as shown in Figure 8.57b.

---

[11] $8N$ is used to emphasize that all these valence electrons belong to the crystal, *i.e.*, $8N \approx 7 \times 10^{24}\ \text{cm}^{-3}$.

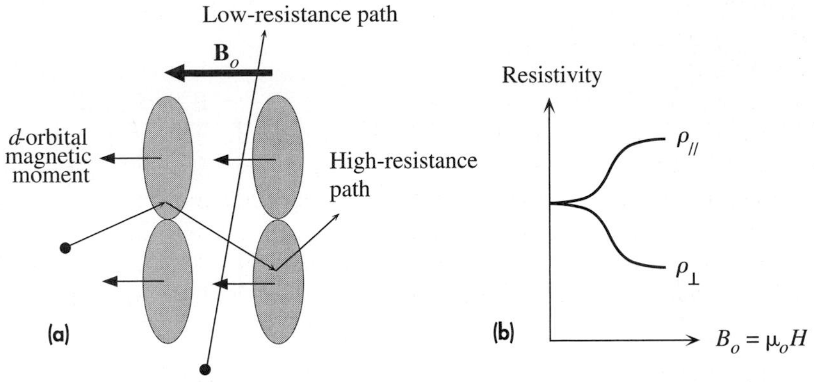


**Figure 8.57**
(a) The origin of anisotropic magnetoresistance (AMR). The electrons traveling along the field experience more scattering than those traveling perpendicular to the field.
(b) Resistivity depends on the current flow direction with respect to the applied magnetic field.

On the other hand, a very large magnetoresistance, called **giant magnetoresistance (GMR),** has been observed in certain special multilayer structures, which exhibit substantial changes in the resistance (*e.g.*, more than 10 percent) when a magnetic field is applied.[12] Even though GMR is a relatively new discovery (1988), it is already widely used in the read heads of hard disk drives. There are also various magnetic field sensors based on the GMR.

The special multilayer structure in its simplest form has two **ferromagnetic layers** (such as Fe or Co or their alloys, etc.) separated by a nonmagnetic transition metal layer (such as Cu), called the **spacer,** as shown in Figure 8.58a. The magnetic layers are thin (less than 10 nm), and the nonmagnetic layer is even thinner. The magnetizations of the two ferromagnetic layers are not random; they depend on the thickness of the spacer because the two layers are "coupled" indirectly through this thin spacer.[13] In the absence of an external field, two magnetic layers are coupled in such a way that their magnetizations are *antiparallel* or in opposite directions; this arrangement is also called an *antiferromagnetically* coupled configuration. We will use the notation FNA to represent the antiparallel configuration, where N stands for the nonmagnetic metal.

We can apply an external magnetic field to one of the layers and rotate its magnetization so that the two magnetizations are now in parallel as in Figure 8.58c. This parallel configuration is frequently called *ferromagnetically* coupled layers and is denoted as FNF. The two structures have a *giant* difference in their resistances, hence the term giant magnetoresistance. The resistance of the antiparallel FNA in Figure 8.58b structure is much higher than that of the parallel structure FNF in Figure 8.58c.

---

[12] GMR was discovered in the late 1980s by Peter Grünberg (Julich, Germany), and Albert Fert (University of Paris-Sud) and their coworkers. Magnetoresistance itself, however, has been well known, and dates back to Lord Kelvin's experiments in 1857.

[13] The physics of the coupling process between the two magnetic layers is an indirect exchange interaction, the details of which are not needed to understand the basics of the GMR phenomenon.

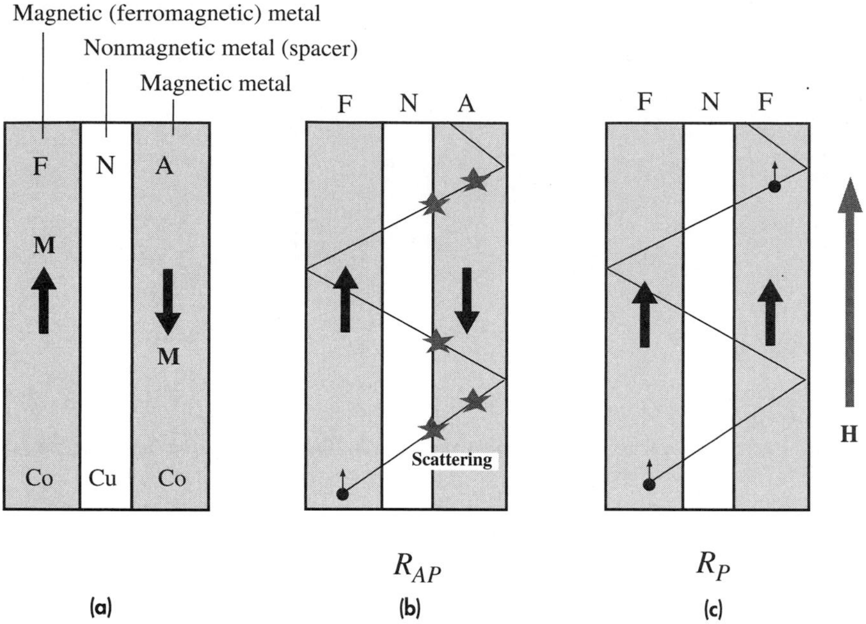


**Figure 8.58** A highly simplified view of the principle of the giant magnetoresistance effect.
(a) The basic trilayer structure.
(b) Antiparallel magnetic layers with high resistance $R_{AP}$.
(c) An external field aligns layers; parallel alignment has a lower resistance $R_P$.

The current flow through this multilayer structure (whether along or perpendicular to the layers) will involve electrons crossing from one layer to another, passing through the interfaces. Recall that it is the electrons around the Fermi energy that are involved in the conduction and that their mean speed is orders of magnitude larger than the drift velocity. The electron trajectories are therefore not parallel to the current flow (and should not be confused with current flow lines).

Consider the antiparallel FNA structure. The magnetic moment up electron in the first magnetic layer is the *favored* conduction electron; that is, it suffers very little scattering. However, when this moment-up electron arrives at the A layer in which the magnetization is reversed, it finds itself with the wrong spin or wrong moment. It is now an *unfavored* electron and is subject to scattering. Thus, the moment-up electron suffers scattering not only in the bulk of A but, more significantly, as it crosses the N-layer into the A-layer, that is, at the interface as in Figure 8.58b. The antiparallel FNA structure therefore has a high resistance, denoted as $R_{AP}$. In contrast, when the magnetizations are parallel, the moment-up electron is the favored electron in both the layers and experiences very little scattering. The resistance $R_P$ of this parallel (FNF) structure is smaller than $R_{AP}$ ($R_P < R_{AP}$). The difference in the resistances $R_P$ and $R_{AP}$ in this simple trilayer is roughly 10 percent or less. But, in multilayered structures, which have a series of alternating magnetic and nonmagnetic layers (*e.g.*, 50 or more magnetic and nonmagnetic alternating layers as in FNANFANFA . . .), the change in the

**Table 8.8** GMR effect in trilayers and multilayers

| Sample | Structure and layer thicknesses | $\Delta R/R_P$ (%) | Temperature (K) |
|---|---|---|---|
| CoFe/CAgCu/CoFe | Trilayer | 4–7 | 300 |
| NiFe/Cu/Co | Trilayer, 10/2.5/2.2 nm (spin valve) | 4.6 | 300 |
| $Co_{90}Fe_{10}/Cu/Co_{90}Fe_{10}$ | Trilayer, 4/2.5/0.8 nm (spin valve) | 7 | 300 |
| $[Co/Cu]_{100}$ | 100 layers of Co/Cu, 1 nm / 1 nm | 80 | 300 |
| $[Co/Co]_{60}$ | 60 layers Co/Cu, 0.8 nm / 0.83 nm | 115 | 4.2 |

SOURCE: Data from P. Grünberg, *Sensors and Actuators*, **A91,** 153, 2001.

resistance can be impressively large, exceeding 100 percent at low temperature and 60–80 percent at room temperature.

The GMR effect is often measured by quoting the change in the resistance with respect to $R_P$,

*Giant magnetoresistance effect*

$$\left(\frac{\Delta R}{R_p}\right)_{\text{GMR}} = \frac{R_{\text{AP}} - R_P}{R_p}$$

Further, the magnetoresistance effect can be measured either by passing a current that flows in the plane of the layers or perpendicular to the plane. Most experiments use the first one, in what is known as **current in plane (CIP)** measurements; but the biggest change, however, is observed for currents perpendicular to the plane of the layers. Table 8.8 summarizes typically reported $\Delta R/R_P$ values for the GMR effect in simple trilayers and multilayers.

The structures with antiparallel and parallel magnetic alignments are obviously two extreme cases. If the angle between the magnetization vectors $\mathbf{M}_1$ and $\mathbf{M}_2$ of the two magnetic layers is $\theta$, then the resistance of the structure depends on $\theta$, with the minimum for $\theta = 0$ (FNF) and the maximum for $\theta = 180°$ (FNA) as shown in Figure 8.59. The fractional change in the resistance depends on $\theta$ as

*GMR and relative magnetizations of magnetic layers*

$$\frac{\Delta R}{R_P} = \left(\frac{\Delta R}{R_P}\right)_{\max} \frac{1 - \cos\theta}{2}$$

As expected, the change is maximum when $\theta = 180°$.

One of the best applications of GMR is in a **spin valve,** in which the current flow is controlled by an external applied magnetic field. Stated differently, the resistance of the valve is controlled by an applied field. Figure 8.60a shows one possible simple spin valve structure. The magnetization of the Co magnetic layer is fixed, that is, *pinned,* by having this layer next to an antiferromagnetic layer, called the *pinning layer*. The exchange interaction between the ferromagnetic Co layer and the antiferromagnetic CoMn layer effectively pins the direction of the Co layer; it takes an enormous field to change the magnetization of the Co layer. A Cu spacer layer separates the Co and the next magnetic FeNi layer. The FeNi layer is called the *free* layer because its magnetization can be changed by an external magnetic field. Normally, in the absence of a

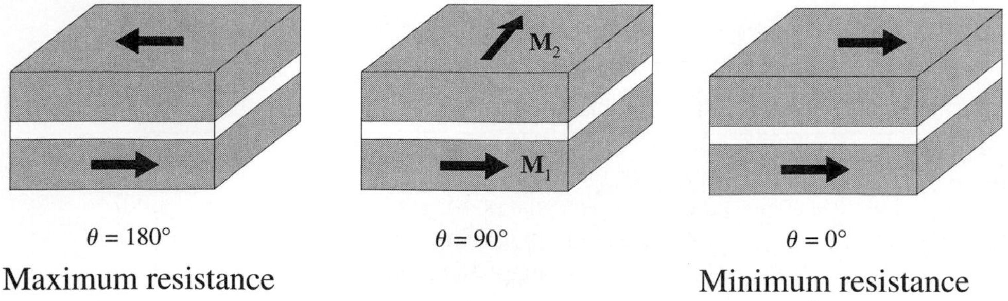


**Figure 8.59** Resistance of the multilayer structure depends on the relative orientations of magnetization in the two magnetic layers.

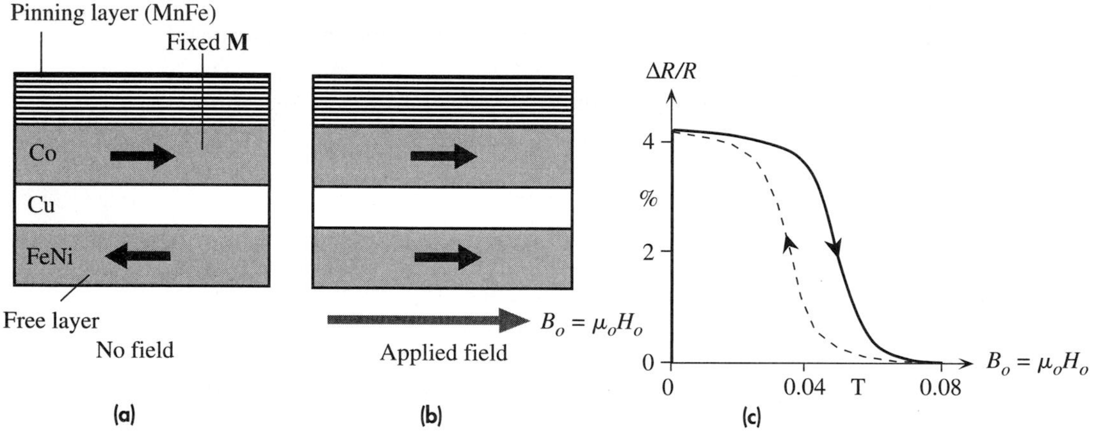


**Figure 8.60** Principle of the spin valve.
(a) No applied field.
(b) Applied field has fully oriented the free-layer magnetization.
(c) Resistance change versus applied magnetic field (schematic) for a FeNi/Cu/FeNi spin valve.

field, the magnetization of the FeNi layer is antiparallel to the Co layer, and the structure has a high resistance $R_{AP}$. An applied external field $B_o = \mu_o H$ can rotate the FeNi layer's magnetization and can easily align FeNi's magnetization fully in parallel with that of Co so that the resistance becomes minimum, *i.e.*, $R_P$ as in Figure 8.60b. It is clear that the external field can be used to control the flow of current through this structure. (The name spin valve reflects the fact that the valve operation relies on the spin of the electrons.) The free layer should be relatively soft to be able to respond to the applied field, whereas the pinned layer should have sufficient coercivity not to lose its magnetization. Figure 8.60c shows a typical magnetoresistance versus applied field characteristics for one particular type of spin valve. The spin valve exhibits hysteresis; that is, the signal $\Delta R$ versus $H$ depends on the direction of magnetization as shown in the figure.

# 8.13 MAGNETIC RECORDING MATERIALS

***General Principles of Magnetic Recording*** Outside electric machinery (mainly rotating machines and transformers), magnetic materials are most widely used in magnetic recording media to store information in either analog or digital form. The deep disappointment of accidently losing valuable stored information on the hard drive of one's computer is well known to most computer users. Magnetic materials in magnetic recording fall into two categories: those used in magnetic heads to record (write), play (read), and erase information, and those used in magnetic media in which the information is stored either permanently or until the next write requirement. The magnetic storage media can be flexible, as in audio and video cassettes and floppy disks, or it can be rigid, as in the hard disk of a computer hard drive. Even though magnetic recording appears in seemingly diverse applications (*e.g.*, audio tape recorders vis-à-vis computer hard drives), the basic principles are nonetheless quite similar.

As a very simple example, we will consider magnetic recording of a signal on an audio tape, as shown schematically in Figure 8.61. The tape is simply a polymer backing tape that has a thin coating of magnetic material on it, as described later. The information is converted into a current signal $i(t)$ that modulates the current around a toroid-type electromagnet with a very small air gap (around 1 μm). This gapped core electromagnet is the **inductive recording head.** The current modulates the magnetic field intensity in the core of the head and hence the field in the gap and around it. The recording of information is achieved by the **fringing magnetic field** around the gap region magnetizing the audio type passing under the head at a constant speed (the tape is usually in contact with the head). As the fringing field changes according to the current signal, so does the magnetization of the audio tape. This means that the electrical signal is stored as a spatial magnetic pattern in the tape. The fringing fields of the recording head modulate the magnetization in the tape in the direction of motion, put differently, along the length of the tape. This type of magnetic information storage is called **longitudinal recording.**

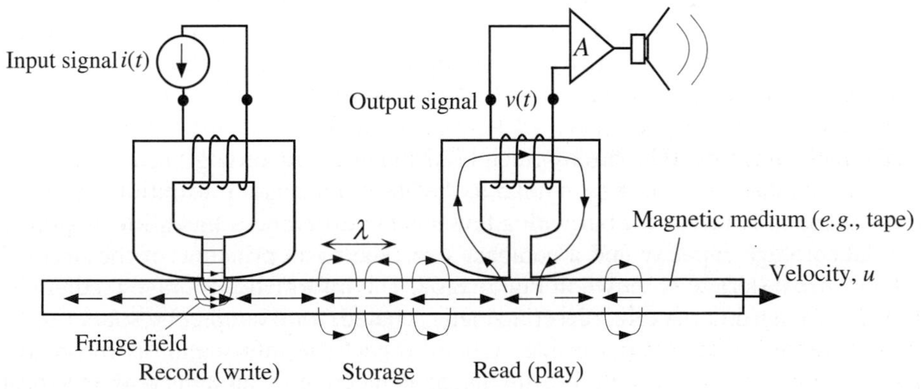


**Figure 8.61** The principle of longitudinal magnetic recording on a flexible medium, for example, magnetic tape in an audio cassette.

The audio tape moves forward and passes under a second head, called the play (or read) head, that converts the spatial variations in the magnetization in the tape into a voltage signal that is amplified and appropriately conditioned for playback, as depicted in Figure 8.61. Of course, the same recording head can also serve as the play head, as is customarily the practice in various general audio recording equipment. The reading process is based on Faraday's law of induction. As the magnetized region in the tape passes under the play head, a portion of the magnetic field from this tape region penetrates into the core and flows around the whole core and hence links the coil. We should recall that magnetic fields prefer to flow in high permeability regions to which they are strongly attracted. The field thus loops around through the core of the head. It does so because the magnetic permeability of the core is very high. As the tape moves past the play head, the field linking the coil changes as different magnetized regions in the tape pass through. The changes in the magnetic flux linking the coil generate a voltage $v(t)$ that is proportional to the strength of the field and hence the magnetization in the tape under the head; the speed of the tape remains the same. Thus the spatial magnetic pattern (information) in the tape is converted into an output voltage signal as the tape is run through under the play head at a constant rate. It should be apparent that the spatial magnetic pattern in the tape is proportional to the current signal $i(t)$, whereas the output signal at the play head is the induced voltage $v(t)$.

Suppose that the input signal has a frequency $f$, or period $1/f$, and the speed of the tape is $u$. Then the magnetic pattern repeats at every $1/f$ seconds. During this time the tape advances by a distance $\Delta x = u/f$. This $\Delta x$ represents a spatial wavelength $\lambda$ that characterizes the repetition of the spatial magnetic pattern that represents the information. The smaller the $\lambda$, the greater the $f$ and hence the greater the information that can be stored. Typical video tapes have $\lambda$ in the submicron range (*e.g.*, 0.75 μm) to be able to store the high density of information in a video signal into a spatial magnetic pattern. The actual recording process in a video cassette recorder is more complicated and involves moving the heads helically across the film, which increases the relative tape speed and hence the induced voltage.

The recording of digital information is straightforward because the information in the form of ones and zeros involves only changes, or no changes, in the direction of magnetization along the tape. In the recording of analog signals, the audio signal is combined with an ac bias signal. However, the analog signal can also be stored as a digital signal by converting it, by an appropriate encoding procedure, to a digital signal.

***Hard Disk Storage*** The basic principle of magnetic recording used in hard disk drives of computers is somewhat similar to the basic schematic illustration for recording on a tape in Figure 8.61, but with a few notable differences that allow high magnetic data storage capacity and a compact size. The basic principle of the magnetic hard disk drive storage is shown in Figure 8.62. The information storage medium is a thin film of magnetic material (described later) coated, for example, by sputtering, on a disk substrate, which rotates inside the hard drive. The information is recorded as magnetization patterns on this thin-film magnetic medium by an inductive write head, similar in principle to the recording head in Figure 8.61. Both the write and the read heads are in a single compact assembly that moves radially across the rotating disk to

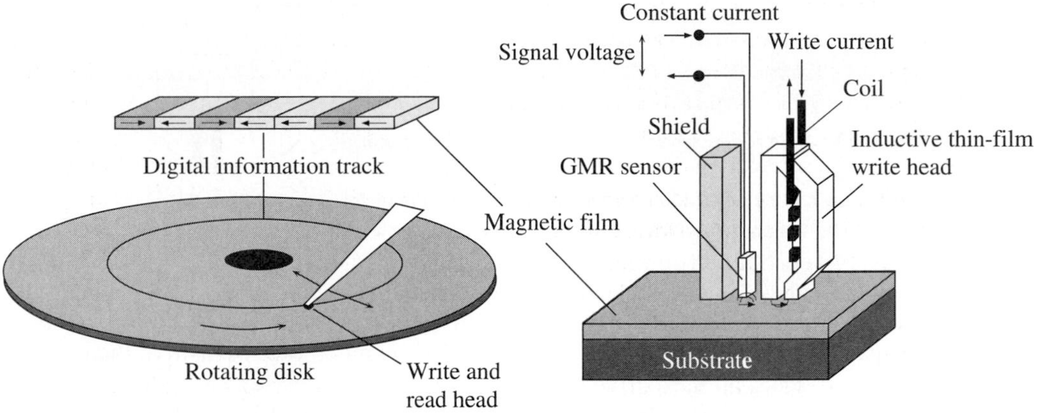


**Figure 8.62** The principle of the hard disk drive magnetic recording.
The write inductive head and the GMR read sensor have been integrated into a single tiny read/write head.

Above: Giant magnetoresistance (GMR) hard disk heads on a U.S. quarter. Left: A small hard disk drive next to a quarter coin—a microdrive.
| SOURCE: Courtesy of IBM.

write or read the information into tracks, called **magnetic bit tracks,** on the magnetic medium. The total area storage density depends on the information density in the track and the track density on the disk. The read head is not an inductive head (as in Figure 8.61) but a tiny giant **magnetoresistance (GMR) sensor** whose resistance depends on an external magnetic field, as explained in Section 8.12. In this case, the field that influences the GMR sensor comes from that of the magnetized region of the disk that is under the GMR sensor. The principle of the GMR is shown in Figure 8.60. The GMR sensor is a multilayered thin-film device whose resistance changes by roughly 10 percent or so in response to an applied field. This change in the resistance generates the read signal. Normally a constant current is passed through the GMR sensor, and the read signal is the voltage variation across the sensor; this voltage is due to the resistance variation induced by the field from the magnetization pattern under the sensor.

There are two important reasons for using a GMR sensor instead of a conventional inductive read head. First is that the GMR sensor is so much smaller than the inductive head that it can probe a much smaller region of the magnetic medium; we can therefore squeeze more information into a given area on the magnetic storage medium. A

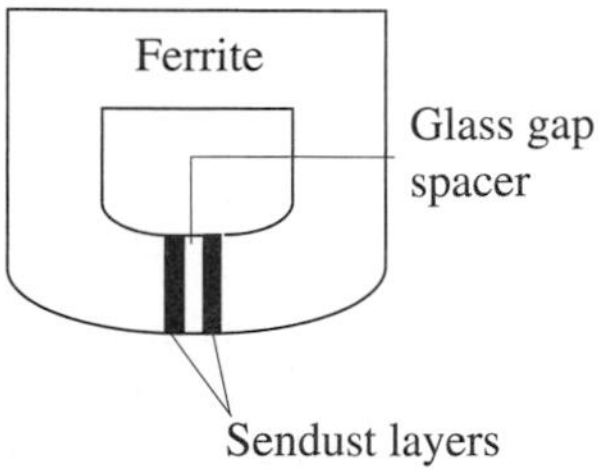


**Figure 8.63** A simplified schematic illustration of a MIG (metal-in-gap) head.
The ferrite core has the poles coated with a ferromagnetic soft metal to enhance the head performance.

typical GMR sensor has a width that is something like 50 nm (~1000 times thinner than the human hair). Second, for the same size, GMR is much more sensitive than the inductive head. Thus, all hard drive read heads are tiny GMR sensors as indicated in Figure 8.62. The inductive write head is normally a **thin-film head,** which has a very small width. Consequently, the information can be written into a very small area on the magnetic storage medium. Usually the thin-film write head and the GMR sensor are integrated into the same structure for convenient write and read operations. The aforementioned basic principles still govern the operation of current magnetic hard drive storage devices.[14]

***Recording Head Materials*** The material for the recording head must be magnetically soft so that its magnetization easily follows the input signal (current $i$ or magnetic field intensity $H$). At the same time, it must provide a strong fringing magnetic field at the gap to magnetize the audio tape, that is, overcome the coercivity of the tape. This requires high saturation magnetization. Thus, the recording head needs small coercivity and large saturation magnetization, which requires soft magnetic materials with as large relative permeabilities as possible.

Typical materials that are used in recording heads are permalloys (Ni–Fe alloys), Sendust (Fe–Al–Si alloy), some sintered soft ferrites (*e.g.*, MnZn and NiZn ferrites), and, more recently, various magnetic amorphous metals such as CoZrNb alloys. Typically, metal-based heads (from permalloy, Sendust, or related materials) are made of laminated metal sheets (with thin insulation between them) to suppress eddy current losses at high frequencies. For high-frequency recording, generally ferrite heads are preferred since ferrites are insulators and suffer no eddy current losses. Ferrites however have low saturation magnetizations and require magnetic storage media of low coercivity. The main problem in ferrite recording heads is that the corners of the poles at the air gap become saturated first. Once saturated, the field around the gap is not proportional to the input current signal, and this degrades the quality of recording. This is overcome by coating the pole faces with a high magnetization metal alloy such as Sendust, or, more recently, a magnetic amorphous metal (*e.g.*, CoZrNb), as depicted in Figure 8.63. Since the magnetic metal alloy is only at the tips of the head, the eddy current losses are still small. This type of head where the poles of the ferrite core have a metal coating is called a metal-in-gap (MIG) head and is widely used in various

[14] One highly recommended book on magnetic recording is R. L. Comstock, *Introduction to Magnetism and Magnetic Recording*, New York: Wiley, 1999. See also R. L. Comstock, "Modern Magnetic Materials in Data Storage," *J. Mater. Sci: Mater. Elecron.* **12,** 509, 2002.

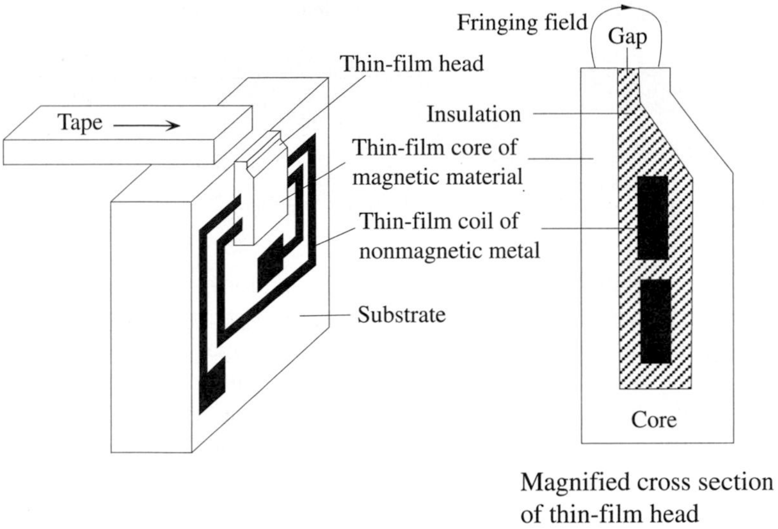


**Figure 8.64** A highly simplified schematic illustration of the principle of a thin-film head.

recording applications. The gap distance itself also influences the extent of the fringing field around it and hence the field penetrating into the magnetic tape. The smaller the gap, the greater the fringing. The necessary fringing fields for proper recording on a tape require gap sizes around 1 μm or less.

More recently, recording head devices have been fabricated using thin films of various ferromagnetic metals or ferrite alloys that have sufficiently small eddy current losses to be useable at high frequencies. A highly simplified illustration of the principle of a thin-film head is shown in Figure 8.64. The head is manufactured by using typical thin-film deposition techniques, such as sputtering of the metal film in a vacuum chamber, photolithography, or some other method. The magnetic core is in the form of a thin film whose thickness is a few microns and whose width is about the same as the tape. The gap at the end of the core has the same width as the core, but its spacing is very small (*e.g.*, 0.25 μm) and generates the necessary fringing field. A spiral-type coil made by depositing a nonmagnetic metal thin film threads the core. The magnetic core is like a U-shaped core that is threaded by the metal strips of the coil. If the core is a metallic material, the coil metal is appropriately insulated from it by thin films of insulation.

***Magnetic Storage Media Materials*** The properties of magnetic storage media such as magnetic tapes, floppy disks, and hard disks used in various magnetic recording applications (audio, video, digital, etc.) must be such that they are able to retain the spatial magnetization pattern written on them after they have passed the recording head. This requires high remanent magnetization $M_r$. High remanent magnetization is also important in the reading process because the magnetic flux that induces voltages in the read head depends on this remanent magnetization, given a particular speed of motion under the read head. Thus the read operation requires media with high $M_r$.

Further, it should be difficult to undesirably erase the magnetic information on the tape by demagnetizing it under stray fields, and this requires high coercivity $H_c$. A

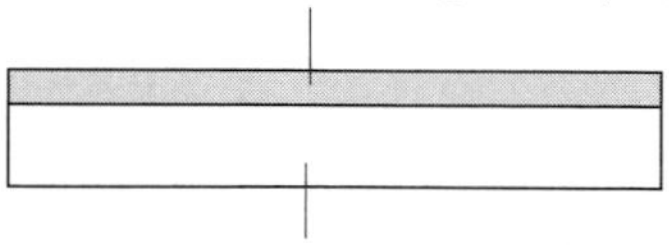


**Figure 8.65** A magnetic tape is typically a magnetic coating on a flexible polymer (*e.g.*, PET) sheet in the form of a tape.

strong magnet passed over a floppy disk can destroy the information stored in it. The coercivity therefore determines the stability of the recording. The coercivity cannot be too high, however, as this would prevent the writing operation, that is, magnetization, under the recording head. One therefore has to find a compromise that allows the information to be written and at the same time retained without ease of demagnetization.

These two requirements, high $M_r$ and medium-to-high $H_c$, lead to a choice of medium to hard magnetic materials as magnetic storage media. Typical flexible storage media (*e.g.*, audio or video tapes) use particulate coatings on flexible polymeric sheets or tapes, as pictured schematically in Figure 8.65. Elongated particles of various magnetic materials are magnetically hard due to a combination of two factors. First, these particles tend to be single domains and are hard due to the magnetocrystalline anisotropy energy. Second, they are also elongated, have a greater length to width ratio (aspect ratio), which means they are also hard due to shape anisotropy; they prefer to be magnetized along the length.

Typical particulate matter used in coatings are $\gamma$-$Fe_2O_3$, Co-modified $\gamma$-$Fe_2O_3$ or Co($\gamma$-$Fe_2O_3$), $CrO_2$, and metallic particles (Fe), as summarized in Table 8.9. The overall magnetic properties of the particulate coating depend not only on the properties of the individual particles (which are hard) but also on the concentration of particles as well as their distribution in the coating. For example, as the packing density of particles increases, the saturation magnetization $M_{sat}$ (total magnetic moment per unit volume) also increases, which is desirable, but the coercivity worsens. The concentrations of particles in the coating are typically between $5 \times 10^{14}$ cm$^{-3}$ (*e.g.*, floppy disk) and

**Table 8.9** Selected examples of flexible magnetic storage media based on coatings of particulate matter: typical values

| Particulate Matter | Typical Application | $\mu_o M_r$ (T) | $\mu_o H_c$ (T) | Comments |
|---|---|---|---|---|
| $\gamma$-$Fe_2O_3$ | Audio tape (Type I) | 0.16 | 0.036 | Widely used particles. |
| $\gamma$-$Fe_2O_3$ | Floppy disk | 0.07 | 0.03 | |
| Co($\gamma$-$Fe_2O_3$) | Video tape | 0.13 | 0.07 | Cobalt-impregnated $\gamma$-$Fe_2O_3$ particles. |
| $CrO_2$ | Audio tape (Type II) | 0.16 | 0.05 | More expensive than $\gamma$-$Fe_2O_3$. |
| $CrO_2$ | Video tape | 0.14 | 0.06 | |
| Fe | Audio tape (Type IV) | 0.30 | 0.11 | High coercivity and magnetization. To avoid corrosion, the particles have to be treated (expensive). |

$5 \times 10^{15}$ cm$^{-3}$ (*e.g.*, video tape), which are sufficient to provide the necessary remanent field and maintain adequate coercivity.

The brown gamma iron oxide, $\gamma$-$Fe_2O_3$, is a metastable form of iron oxide that is ferrimagnetic and is prepared synthetically. Cobalt-treated $\gamma$-$Fe_2O_3$ particles have a small percentage of Co impregnated into the surface of the particles, which improves the magnetic hardness. Cobalt-impregnated $\gamma$-$Fe_2O_3$ particles are used in various video tapes. All these particles in Table 8.9 are needle shaped (elongated rod-like shapes) with length-to-diameter ratios greater than 5, which makes them substantially hard as a result of shape anisotropy. The needle-like particles are typically 0.3–0.6 μm in length and 0.05–0.1 μm in diameter. The particles are initially mixed into a lacquer-like resin binder that is then coated onto a thin polyester backing tape. When the resin coating solidifies, it forms a magnetic coating stuck on the backing tape. Typically between 20–40 percent of this magnetic coating is actually due to the magnetic particles.

Another form of magnetic storage medium is in the form of magnetic thin films deposited onto various hard substrates or even on a flexible plastic tape as in some video tapes. The hard disk in the hard drive of a computer, for example, is typically an aluminum disk that has a thin magnetic film (*e.g.*, CoPtCr) coated onto it. The deposition of the magnetic thin film may involve vacuum deposition techniques (*e.g.*, electron beam evaporation or sputtering) or electroplating. Typical film thicknesses are less than 50 nm. The advantage of using a thin-film coating is that they are solid films of a magnetic material, that is, almost 100 percent dense, whereas in a particulate medium, the packing density of magnetic particles is 20–40 percent. Consequently, thin magnetic films have higher saturation and remanent magnetizations, which enable a smaller area of the thin film to be used for storing the same information as that in a flexible medium. Thus there is an increase in the stored information density—a distinct advantage. Table 8.10 lists the characteristics of a few selected thin magnetic films used as magnetic storage media. Most thin films are alloys of Co because Co has a high degree of magnetocrystalline anisotropy and hence good coercivity $H_c$. Alloying Co with Cr provides good corrosion resistance and increases $H_c$. Alloying with Pt or Ta also increases $H_c$. The desired film properties can usually be obtained by alloying Co with other elements and optimizing the deposition conditions; this is an ongoing research area. The current commercial interest is to increase the storage density even

**Table 8.10** Selected examples of thin films in magnetic storage media: typical values

| Thin Film | Typical Deposition | $\mu_o M_s$ (T) | $\mu_o H_c$ (T) | Comment and Typical or Potential Application |
|---|---|---|---|---|
| Co and rare earth | Sputtering in vacuum | 0.7–0.8 | 0.05–0.07 | Longitudinal magnetic recording media |
| Co($\gamma$-$Fe_2O_3$) | Sputtering in vacuum | 0.3 | 0.07–0.08 | Longitudinal magnetic recording media |
| CoNiP | Electroplating | 1 | 0.1 | Longitudinal magnetic recording media, hard disks |
| CoCr alloys | Sputtering in vacuum | 0.3–0.7 | 0.05–0.3 | Longitudinal and perpendicular magnetic recording media, hard disks |
| CoPtCrB | Sputtering in vacuum | 0.3–0.5 | 0.25–0.6 | Longitudinal and perpendicular magnetic recording media, hard disks |

further by using perpendicular magnetic recording in contrast to longitudinal recording. In perpendicular recording, the local magnetizations in the thin film are perpendicular to the surface of the film.

The magnetic coating on some video tapes may be in the form of a thin film deposited by vacuum evaporation of the magnetic material using an electron beam to heat it. Some recent video tapes have CoNi thin-film coatings that are evaporated by an electron beam onto a polyester (PET) tape.

## 8.14 JOSEPHSON EFFECT

The Josephson junction is a junction between two superconductors that are separated by a thin insulator (a few nanometers thick) as depicted in Figure 8.66. If the insulating barrier is sufficiently thin, then there is a probability that the Cooper pairs can tunnel across the junction. The wavefunction $\Psi$ of the Cooper pair, however, changes phase by $\theta$ when it tunnels through the junction, not unexpected as the pair goes through a potential barrier. The maximum superconducting current $I_c$ that can flow through this weak link depends on not only the thickness and area (size) of the insulator but also on the superconductor materials and the temperature. The current $I$, or the *supercurrent,* through the junction due to Cooper pair tunneling is determined by the phase angle $\theta$,

*Josephson junction supercurrent*

$$I = I_c \sin\theta \qquad [8.27]$$

where $I_c$ is the maximum current or the critical current. If the current through the junction is controlled by an external circuit, then the tunneling Cooper pairs on either side of the junction (in the superconductors) adjust their respective phases to maintain the phase change to satisfy Equation 8.27. If we plot the $I$–$V$ characteristics of this junction as in Figure 8.67, we would find that for $I < I_c$, the behavior follows the vertical *OC* line with no voltage across the junction.

If the current through the junction exceeds $I_c$, then the Cooper pairs cannot tunnel through the insulator because Equation 8.27 cannot be satisfied. There is still a current through the junction, but it is due to the tunneling of normal, that is, single electrons as represented by the curve *OABD* in Figure 8.67. Thus, the current switches from point *C* to point *B* and then follows the normal tunneling curve *B* to *D*. At point *B*, a *voltage*

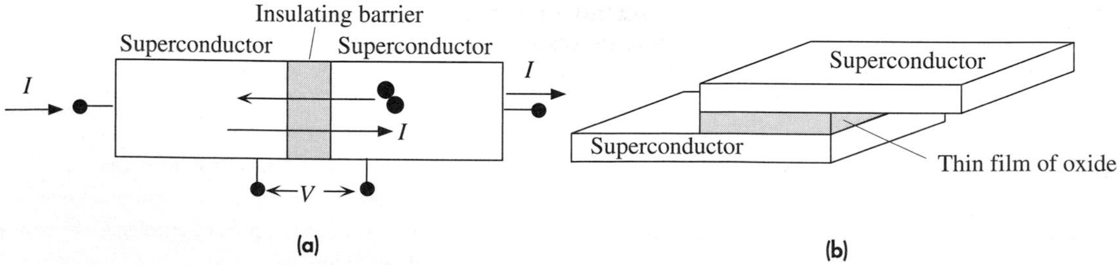


**Figure 8.66**
(a) A Josephson junction is a junction between two superconductors separated by a thin insulator.
(b) In practice, thin-film technology is used to fabricate a Josephson junction.

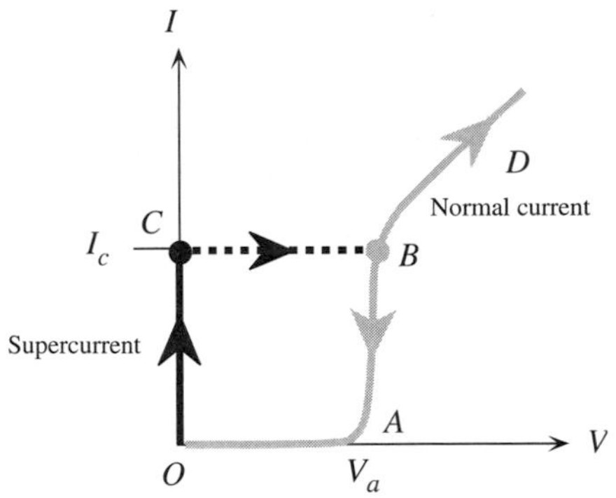


**Figure 8.67** *I–V* characteristics of a Josephson junction for positive currents when the current is controlled by an external circuit.

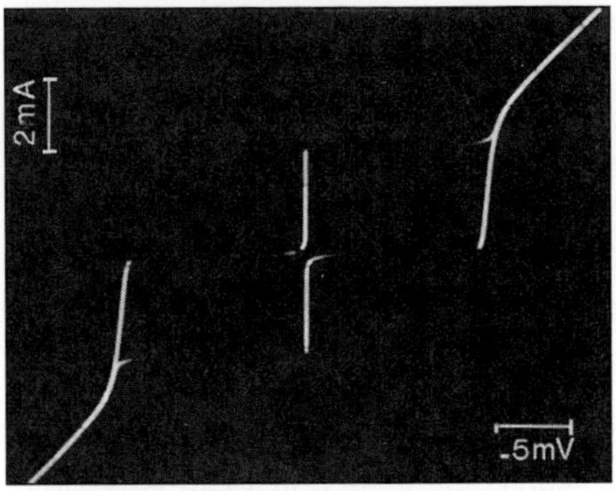


*I–V* characteristics of a Sn-SnO-Sn Josephson junction at $T = 1.52$ K.

SOURCE: E. P. Balsano, G. Paterno, A. Barone, P. Rissman, and M. Russo, "Temperature dependence of the maximum (dc) Josephson current" Phys. Rev. B, vol. 10, p. 1882, Figure 2.


develops across the junction and increases with the current. The normal tunneling current in the range *OA* is negligible and rises suddenly when the voltage exceeds $V_a$. The reason is that a certain amount of voltage (corresponding to a potential energy $eV_a$) is needed to provide the necessary energy to disassociate the tunneling single electron from its Cooper pair. It is apparent that the thin insulation acts as a weak superconductor or as a **weak link** in the superconductor; weak with regard to the currents that can flow in the superconductor itself. The *I–V* characteristic in Figure 8.67 is symmetric about *O* (as in the photograph for an actual device), and is called the **dc characteristic of the Josephson junction.** In addition, the *I–V* behavior exhibits hysteresis; that is, if we were to decrease the current, the behavior does not follow *DBC* down to *O*, but follows the *DBA* curve. When the current is decreased nearly to zero, the normal tunneling current switches to the supercurrent. The Josephson junction is bistable; that is, it has two states corresponding to the superconducting state *OC* and normal state *ABD*. Thus, the device behaves as an electronic switch whose switching time, in theory, is determined by tunneling times, in the picoseconds range. In practice the switching time (~10 ps) is limited by the junction capacitance.

If, on the other hand, a dc voltage is applied across the Josephson junction, then the phase change $\theta$ is modulated by the applied voltage. The most interesting and surprising aspect is that the voltage modulates the rate of change of the phase through the barrier, that is,

*Applied voltage modulates phase*

$$\frac{d\theta}{dt} = \frac{2eV}{\hbar}$$

When we integrate this, we find that $\theta$ is time and voltage dependent, so, according to Equation 8.27, the current is a sinusoidal function of time and voltage, that is,

$$I = I_c \sin\left(\theta_o - \frac{2\pi(2eV)t}{h}\right)$$

or

$$I = I_o \sin(2\pi f t)$$

where $I_o$ is a new constant incorporating $\theta_o$ and the frequency of the oscillations of the current is given by

*ac Josephson effect*

$$f = \frac{2eV}{h} \tag{8.28}$$

The Josephson junction therefore generates an oscillating current of frequency $f$ when there is a dc voltage $V$ across it. This is called the **ac Josephson effect,** a remarkable phenomenon originally predicted by Josephson as a graduate student at Cambridge (1962). According to the ac Josephson effect, the junction generates an ac current at a frequency of $2e/h$ Hz per volt or 483.6 MHz per microvolt. Furthermore, the frequency of the current has nothing to do with the material properties of the junction but is only determined by the applied voltage through $e$ and $h$. The ac Josephson effect has been adopted to define the voltage standard: One volt is the voltage that, when applied to a Josephson junction, will generate an ac current and hence an electromagnetic radiation of frequency 483,597.9 GHz.

## 8.15 FLUX QUANTIZATION

Consider a ring of a superconducting material above its $T_c$. Suppose that the ring is immersed in magnetic flux lines from a magnet placed above it as shown in Figure 8.68a. When we cool the ring to below $T_c$, the magnetic flux lines are excluded from the ring itself, due to the Meissner effect, but they go through the hole, as shown in Figure 8.68b. If we now remove the magnet, we may think that the magnetic flux lines simply disappear, but this is not the case. A persistent current is set up on the inside surface of the superconducting ring that flows to maintain the flux constant in the hollow. This supercurrent generates flux lines in the hollow as if to replace those taken away by the removal of the magnet, as depicted in Figure 8.68c. Since the current can flow indefinitely in the ring, the overall effect is that the magnetic flux is *trapped* within the ring. Indeed, if we were to bring back the magnet, the current in the ring would disappear to ensure that the magnetic flux in the hollow remains unchanged. The origin of

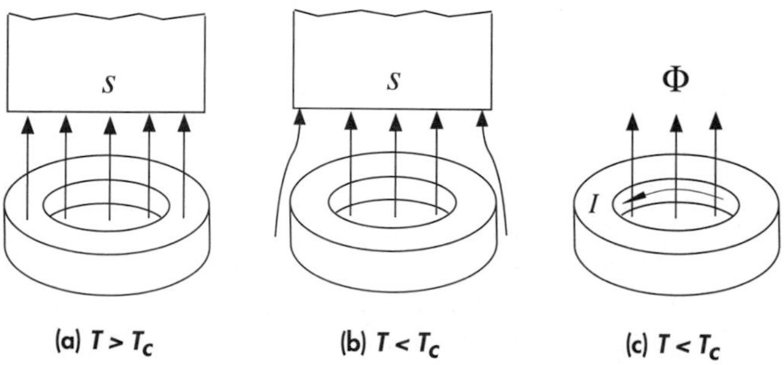


**Figure 8.68**
(a) Above $T_c$, the flux lines enter the ring.
(b) The ring and magnet are cooled through $T_c$. The flux lines do not enter the superconducting ring but stay in the hole.
(c) Removing the magnet does not change the flux in the hole.

flux trapping can be appreciated by considering what would happen if the flux were allowed to change, that is, $d\Phi/dt \neq 0$. A changing flux would induce a voltage $V = -d\Phi/dt$ around the ring that would drive an infinite current $I = V/R$ where $R = 0$. This is not possible, and hence the flux cannot change, which means we must have $d\Phi/dt = 0$. One should also note that there can be no electric field inside a superconductor because

$$\mathcal{E} = \frac{J}{\sigma} = 0$$

since the conductivity $\sigma$ is infinite.

What would happen if we have a superconducting ring (below $T_c$) that initially had no flux in the hole? If we were to bring a magnet to it, then the flux lines would now be excluded from both the ring itself and also the hole since the trapped flux within the ring is zero.

It turns out that the trapped flux $\Phi$ inside the ring is quantized by virtue of superconductivity being a quantum phenomenon. The smallest quantized amount of flux is called the **magnetic flux quantum** and is given by $h/2e$ or $2.0679 \times 10^{-15}$ Wb. The flux $\Phi$ in the ring is an integer multiple $n$ of this quantum,

$$\Phi = n\frac{h}{2e} \qquad [8.29]$$

*Trapped flux is quantized*

## CD Selected Topics and Solved Problems

| Selected Topics | Solved Problems |
|---|---|
| Atomic Diamagnetism | Diamagnetism: Examples |
| Atomic Paramagnetism | |
| Ferrimagnetism and Ferrites | |

## DEFINING TERMS

**Antiferromagnetic materials** have crystals in which alternating permanent atomic spin magnetic moments are equal in magnitude but point in opposite directions (antiparallel), which leads to no net magnetization of the crystal.

**Bloch wall** is a magnetic domain wall.

**Bohr magneton** ($\beta$) is a useful elementary unit of magnetic moment on the atomic scale. It is equal to the magnetic moment of one electron spin along an applied magnetic field $\beta = e\hbar/2m_e$.

**Coercivity** or **coercive field** ($H_c$) measures the ability of a magnetized material to resist demagnetization. It is the required reverse applied field that would remove any remanent magnetization, that is, demagnetize the material.

**Cooper pair** is a quasi-particle formed by the mutual attraction of two electrons with opposite spins and opposite linear momenta below a critical temperature. It has a charge of $-2e$ and a mass of $2m_e$ but no net spin. It does *not* obey Fermi–Dirac statistics. The electrons are held together by the induced distortions and vibrations of the lattice of positive metal ions with which the electrons interact.

**Critical magnetic field** ($B_c$) is the maximum field that can be applied to a superconductor without destroying the superconducting behavior. $B_c$ decreases from its maximum value at absolute zero to zero at $T_c$.

**Critical temperature** ($T_c$) is a temperature that separates the superconducting state from the normal state. Above $T_c$, the substance is in the normal state with a finite resistivity, but below $T_c$, it is in the superconducting state with zero resistivity.

**Curie temperature** ($T_C$) is the critical temperature at which the ferromagnetic and ferrimagnetic properties are lost. Above the Curie temperature, the material behaves as if it were paramagnetic.

**Diamagnetic material** has a negative magnetic susceptibility and reduces or repels applied magnetic fields. Superconductors are perfect diamagnets that repel the applied field. Many substances possess weak diamagnetism, so the applied field is slightly decreased within the material.

**Domain wall** is a region between two neighboring magnetic domains of differing orientations of magnetization.

**Domain wall energy** is the excess energy in the domain wall as a result of the gradual orientations of the neighboring spin magnetic moments of atoms through the wall region. It is the excess energy due to the excess exchange interaction energy, magnetocrystalline anisotropy energy, and magnetostrictive energy in the wall region.

**Easy direction** is the crystal direction along which the atomic magnetic moments (due to spin) are spontaneously and most easily aligned. Exchange interaction energy is lowest (hence favorable) when the alignment of atomic spin magnetic moments is in this direction in the crystal. For the iron crystal, it is one of the six [100] directions (cube edge).

**Eddy current loss** is the Joule energy loss ($I^2R$) in a ferromagnetic material subjected to changing magnetic fields (in ac fields). The varying magnetic field induces voltages in the ferromagnetic material that drive currents (called eddy currents) that generate Joule heating due to $I^2R$.

**Eddy currents** are the induced conduction currents flowing in a ferromagnetic material as a result of varying (ac) magnetic fields.

**Exchange interaction energy** ($E_{ex}$) is a kind of Coulombic interaction energy between two neighboring electrons and positive metal ions that depends on the relative spin orientations of the electrons as a consequence of the Pauli exclusion principle. Its exact origin is quantum mechanical. Qualitatively, different spins lead to different electron wavefunctions, different negative charge distributions, and hence different Coulombic interactions. In ferromagnetic crystals, $E_{ex}$ is negative when the neighboring electron spins are parallel.

**Ferrimagnetic materials** possess crystals that contain two sets of atomic magnetic moments that oppose each other, but one set has greater strength and therefore there is a net magnetization of the crystal. An unmagnetized ferrimagnetic substance normally has many magnetic domains whose magnetization vectors add to give no overall magnetization.

**Ferrites** are ferrimagnetic materials that are ceramics with insulating properties. They are therefore used in HF applications where eddy current losses are significant. Their general composition is (MO)($Fe_2O_3$), where M is typically a divalent metal. For magnetically soft ferrites, M is typically Fe, Mn, Zn, or Ni, whereas for magnetically hard ferrites, M is typically Sr or Ba. Hard ferrites such as $BaOFe_2O_3$ have the hexagonal crystal structure with a high degree of magnetocrystalline anisotropy and therefore possess high coercivity (difficult to demagnetize).

**Ferromagnetic materials** have the ability to possess large permanent magnetizations even in the absence of an applied field. An unmagnetized ferromagnetic material normally has many magnetic domains whose magnetization vectors add to give no overall magnetization. However, in a sufficiently strong magnetizing field, the whole ferromagnetic substance becomes one magnetic domain in which all the atomic spin magnetic moments are aligned to give a large magnetization along the field. Some of this magnetization is retained even after the removal of the field.

**Giant magnetoresistance** (GMR) is the large change in the resistance of a special multilayer structure when a magnetic field is applied; the simplest structure usually consists of two thin ferromagnetic layers (*e.g.*, Fe) sandwiching an even thinner nonmagnetic metal ( *e.g.*, Cu).

**Hard direction** is the crystal direction along which it is hardest to align the atomic spin magnetic moments relative to the easy direction. Exchange interaction energy $E_{ex}$ favors the easy direction most ($E_{ex}$ is most negative) and favors the hard direction least ($E_{ex}$ is least negative).

**Hard magnetic materials** characteristically have high remanent magnetizations ($B_r$) and high coercivities ($H_c$), so once magnetized, they are difficult to demagnetize. They are suitable for permanent magnet applications. They have broad $B$–$H$ hysteresis loops.

**Hard magnetic particles** are small particles of various shapes that have high coercivity due to having a single magnetic domain with high magnetocrystalline anisotropy energy, or possessing substantial shape anisotropy (aspect ratio—length-to-width ratio).

**Hysteresis loop** is the magnetization ($M$) versus applied magnetic field intensity ($H$) or $B$ versus $H$ behavior of a ferromagnetic (or ferrimagnetic) substance through one cycle as it is repeatedly magnetized and demagnetized.

**Hysteresis loss** is the energy loss involved in magnetizing and demagnetizing a ferromagnetic (or ferrimagnetic) substance. It arises from various energy losses involved in the irreversible motions of the domain walls. Hysteresis loss per unit volume of specimen is the area of the $B$–$H$ hysteresis loop.

**Initial permeability** ($\mu_{ri}\mu_o$) is the initial slope of the $B$ versus $H$ characteristic of an unmagnetized ferromagnetic (or ferrimagnetic) material and typically represents the magnetic permeability under very small applied magnetic fields. Initial relative permeability ($\mu_{ri}$) is the relative permeability of an unmagnetized ferromagnetic (or ferrimagnetic) material under very small applied fields.

**Magnetic dipole moment** ($\boldsymbol{\mu}_m$) is defined as $I A\mathbf{u}_n$, where $I$ is the current flowing in a circuit loop of area $A$ and $\mathbf{u}_n$ is the unit vector in the direction of an advance of a screw when it is turned in the direction of the circulating current. Qualitatively, it measures the strength of the magnetic field created by a current loop and also the extent of interaction of the current loop with an externally applied magnetic field. $\boldsymbol{\mu}_m$ is normal to the surface of the loop. Magnetic moment in a magnetic field experiences a torque that tries to rotate $\boldsymbol{\mu}_m$ to align it with the field. In a nonuniform field, the magnetic moment experiences a force that attracts it to a greater field.

**Magnetic domain** is a region of a ferromagnetic (or ferrimagnetic) crystal that has spontaneous magnetization, that is, magnetization in the absence of an applied field, due to the alignment of all magnetic moments in that region.

**Magnetic field, magnetic induction,** or **magnetic flux density (B)** is a field that is generated by a current-carrying conductor that produces a force on a current-carrying conductor elsewhere. Equivalently, we can define it as the field generated by a moving charge that acts to produce a force on a moving charge elsewhere. The force is called the Lorentz force and is given by $\mathbf{F} = q\mathbf{v} \times \mathbf{B}$ where $\mathbf{v}$ is the velocity of the particle with charge $q$. The magnetic field $\mathbf{B}$ in a material is the sum of the applied field $\mu_o\mathbf{H}$, and that due to the magnetization of the material $\mu_o\mathbf{M}$, that is, $\mathbf{B} = \mu_o(\mathbf{H} + \mathbf{M})$.

**Magnetic field intensity** or **magnetizing field (H)** gauges the magnetic strength of external conduction currents (*e.g.*, currents flowing in the windings) in the absence of a material medium. It excludes the magnetization currents that become induced on the surfaces of any material placed in a magnetic field. $\mu_o H$ is the magnetic field in free space and is considered to be the *applied magnetic field*. The terms *intensity* or *strength* distinguish **H** from **B**, which is simply called the magnetic field.

**Magnetic flux** ($\Phi$) represents to what extent magnetic field lines are flowing through a given area perpendicular to the field lines. If $\delta A$ is a small area perpendicular to the magnetic field $B$ and $B$ is constant over $\delta A$, then the flux $\delta\Phi$ through $\delta A$ is defined by $\delta\Phi = B\,\delta A$. Total flux through any closed surface is zero.

**Magnetic permeability** ($\mu$) is the magnetic field generated per unit magnetizing field, that is, $\mu = B/H$. Permeability gauges the effectiveness of a medium in generating as much magnetic field as possible per unit magnetizing field. Permeability of free space is the absolute permeability $\mu_o$, which is the magnetic field generated in a vacuum per unit magnetizing field.

**Magnetic susceptibility** ($\chi_m$) indicates the ease with which the material becomes magnetized under an applied magnetic field. It is the magnetization induced in the material per unit magnetizing field, $\chi_m = M/H$.

**Magnetization** or **magnetization vector (M)** represents the net magnetic moment per unit volume of material. In the presence of a magnetic field, individual atomic moments tend to align with the field, which results in a net magnetization. Magnetization of a specimen can be represented by the flow of currents on the surface over a unit length of the specimen; $M = I_m$, where $I_m$ is the surface magnetization current per unit length.

**Magnetization current** ($I_m$) is a bound current per unit length that exists on the surface of a substance due to its magnetization. It is not, however, due to the flow of free charges but arises in the presence of an applied magnetic field as a result of the orientations of the electronic motions in the constituent atoms. In the bulk, these electronic motions cancel each other and there is no net bulk current, but on the surface, they add to give a bound surface current $I_m$ per unit length, which is equal to the magnetization $M$ of the substance.

**Magnetocrystalline anisotropy** is the anisotropy associated with magnetic properties such as the magnetization in different directions in a ferromagnetic (or ferrimagnetic) crystal. Atomic spins prefer to align along certain directions in the crystal, called easy directions. The direction along which it is most difficult to align the spins is called the hard direction. For example, in the iron crystal, all atomic spins prefer to align along one of the [100] directions (easy directions) and it is most difficult to align the spins along one of the [111] directions (hard directions).

**Magnetocrystalline anisotropy energy** ($K$) is the energy needed to rotate the magnetization of a ferromagnetic (or ferrimagnetic) crystal from its natural easy direction to a hard direction. For example, it takes an energy of about 48 mJ cm$^{-3}$ to rotate the magnetization of an iron crystal from the easy direction [100] to the hard direction [111].

**Magnetoresistance** generally refers to the change in the resistance of a magnetic material when it is placed in a magnetic field. The change in the resistance of a nonmagnetic metal, such as copper, is usually very small. In a magnetic metal, the change in the resistivity due to the applied magnetic field is *anisotropic;* that is, it depends on the direction of current flow with respect to the applied field and is called **anisotropic magnetoresistance (AMR).**

**Magnetostatic energy** is the potential energy stored in an external magnetic field. It takes external work to establish a magnetic field, and this energy is said to be stored in the magnetic field. Magnetic energy per unit volume at a point in free space is given by

$$E_{\text{vol}}(\text{air}) = \frac{1}{2}\mu_o H^2 = \frac{B^2}{2\mu_o}$$

**Magnetostriction** is the change in the length of a ferromagnetic (or ferrimagnetic) crystal as a result of its magnetization. An iron crystal placed in a magnetic field along an easy direction becomes longer along this direction but contracts in the transverse direction.

**Magnetostrictive energy** is the strain energy in the crystal due to magnetostriction, that is, the work done in straining the crystal when it becomes magnetized.

**Maximum relative permeability** ($\mu_{r,\max}$) is the maximum relative permeability of a ferromagnetic (or ferrimagnetic) material.

**Meissner effect** is the repulsion of all magnetic flux from the interior of a superconductor. The superconductor behaves as if it were a perfect diamagnet with $\chi_m = -1$.

**Paramagnetic materials** have a small and positive magnetic susceptibility. In an applied field, they develop a small amount of magnetization in the direction of the applied field, so the magnetic field in the material is slightly greater. They are attracted to a higher magnetic field.

**Relative permeability** ($\mu_r$) measures the magnetic field in a medium with respect to that in a vacuum, $\mu_r = B/\mu_o H$. Since $B$ depends on the magnetization of the medium, $\mu_r$ measures the ease with which the material becomes magnetized.

**Remanence** or **remanent magnetization** ($\mathbf{M}_r$) is the magnetization that remains in a magnetic material after it has been fully magnetized and the magnetizing field has been removed. It measures the ability of a magnetic material to retain a portion of its magnetization after the removal of the applied field. The corresponding magnetic field ($\mu_o \mathbf{M}_r$) is the remanent magnetic field $\mathbf{B}_r$.

**Saturation magnetization** is the maximum magnetization that can be obtained in a ferromagnetic crystal at a given temperature when all the magnetic moments have been aligned in the direction of the applied field, when there is a single magnetic domain with its magnetization **M** along the applied field.

**Shape anisotropy** is the anisotropy in magnetic properties associated with the shape of the ferromagnetic (or ferrimagnetic) substance. A crystal rod that is thin and long prefers to have its magnetization **M** along the length (long axis) of the rod because this direction of magnetization creates less external magnetic fields and leads to less external magnetostatic energy compared with the case when **M** is along the width (short axis) of the rod. Reversing the magnetization involves rotating **M** through the width of the rod, where the external magnetic field and hence magnetostatic energy are large, and requires large substantial work. It is therefore difficult to rotate magnetization around from the long axis to the short axis.

**Soft magnetic materials** characteristically have high saturation magnetizations ($B_{sat}$) but low saturation magnetizing fields ($H_{sat}$) and low coercivities ($H_c$), so they can be magnetized and demagnetized easily. They have tall and narrow *B–H* hysteresis loops.

**Superconductivity** is a phenomenon in which a substance loses all resistance to current flow (acquires zero resistivity) and also exhibits the Meissner effect (becomes a perfect diamagnet).

**Type I superconductors** have a single critical field ($B_c$) above which the superconducting behavior is totally lost.

**Type II superconductors** have a lower ($B_{c1}$) and an upper ($B_{c2}$) critical field. Below $B_{c1}$, the substance is in the superconducting phase with Meissner effect; all magnetic flux is excluded from the interior. Between $B_{c1}$ and $B_{c2}$, magnetic flux lines pierce through local filamentary regions of the superconductor, which behave normally. Above $B_{c2}$, the superconductor reverts to normal behavior.

# QUESTIONS AND PROBLEMS

**8.1** **Inductance of a long solenoid** Consider the very long (ideally infinitely long) solenoid shown in Figure 8.69. If $r$ is the radius of the core and $\ell$ is the length of the solenoid, then $\ell \gg r$. The total number of turns is $N$ and the number of turns per unit length is $n = N/\ell$. The current through the coil wires is $I$. Apply Ampere's law around $C$, which is the rectangular circuit $PQRS$, and show that

$$B \approx \mu_o \mu_r n I$$

Further, show that the inductance is

$$L \approx \mu_o \mu_r n^2 V_{\text{core}}$$

*Inductance of a long solenoid*

where $V_{\text{core}}$ is the volume of the core. How would you increase the inductance of a long solenoid?

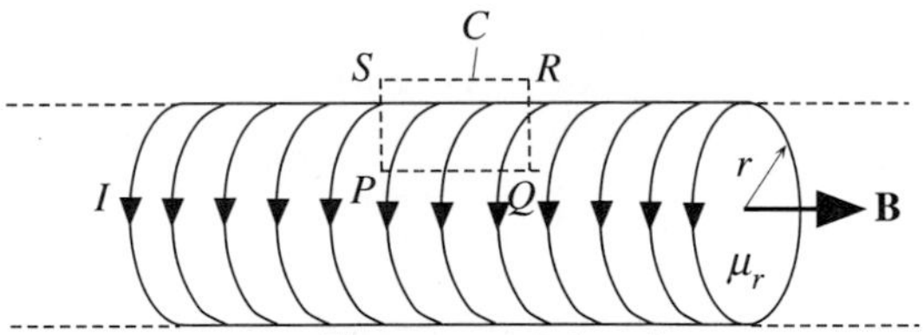


**Figure 8.69**

What is the approximate inductance of an air-cored solenoid with a diameter of 1 cm, length of 20 cm, and 500 turns? What is the magnetic field inside the solenoid and the energy stored in the whole solenoid when the current is 1 A? What happens to these values if the core medium has a relative permeability $\mu_r$ of 600?

**8.2** **Magnetization** Consider a long solenoid with a core that is an iron alloy (see Problem 8.1 for the relevant formulas). Suppose that the diameter of the solenoid is 2 cm and the length of the solenoid is 20 cm. The number of turns on the solenoid is 200. The current is increased until the core is magnetized to saturation at about $I = 2$ A and the saturated magnetic field is 1.5 T.

*a.* What is the magnetic field intensity at the center of the solenoid and the applied magnetic field, $\mu_o H$, for saturation?

*b.* What is the saturation magnetization $M_{sat}$ of this iron alloy?

*c.* What is the total magnetization current on the surface of the magnetized iron alloy specimen?

*d.* If we were to remove the iron-alloy core and attempt to obtain the same magnetic field of 1.5 T inside the solenoid, how much current would we need? Is there a practical way of doing this?

**8.3** **Paramagnetic and diamagnetic materials** Consider bismuth with $\chi_m = -16.6 \times 10^{-5}$ and aluminum with $\chi_m = 2.3 \times 10^{-5}$. Suppose that we subject each sample to an applied magnetic field $B_o$ of 1 T applied in the $+x$ direction. What is the magnetization **M** and the equivalent magnetic field $\mu_o M$ in each sample? Which is paramagnetic and which is diamagnetic?

**8.4** **Mass and molar susceptibilities** Sometimes magnetic susceptibilities are reported as molar or mass susceptibilities. **Mass susceptibility** (in m$^3$ kg$^{-1}$) is $\chi_m/\rho$ where $\rho$ is the density. **Molar susceptibility** (in m$^3$ mol$^{-1}$) is $\chi_m(M_{at}/\rho)$ where $M_{at}$ is the atomic mass. Terbium (Tb) has a magnetic molar susceptibility of 2.0 cm$^3$ mol$^{-1}$. Tb has a density of 8.2 g cm$^{-3}$ and an atomic mass of 158.93 g mol$^{-1}$. What is its susceptibility, mass susceptibility and relative permeability? What is the magnetization in the sample in an applied magnetic field of 2 T?

**8.5** **Pauli spin paramagnetism** Paramagnetism in metals depends on the number of conduction electrons that can flip their spins and align with the applied magnetic field. These electrons are near the Fermi level $E_F$, and their number is determined by the density of states $g(E_F)$ at $E_F$. Since each electron has a spin magnetic moment of $\beta$, paramagnetic susceptibility can be shown to be given by

*Pauli spin paramagnetism*

$$\chi_{para} \approx \mu_o \beta^2 g(E_F)$$

where the density of states is given by Equation 4.10. The Fermi energy of calcium $E_F$ is 4.68 eV. Evaluate the paramagnetic susceptibility of calcium and compare with the experimental value of $1.9 \times 10^{-5}$.

**8.6** **Ferromagnetism and the exchange interaction** Consider dysprosium (Dy), which is a rare earth metal with a density of 8.54 g cm$^{-3}$ and atomic mass of 162.50 g mol$^{-1}$. The isolated atom has the electron structure [Xe]$4f^{10}6s^2$. What is the spin magnetic moment in the isolated atom in terms of number of Bohr magnetons? If the saturation magnetization of Dy near absolute zero of temperature is 2.4 MA m$^{-1}$, what is the effective number of spins per atom in the ferromagnetic state? How does this compare with the number of spins in the isolated atom? What is the order of magnitude for the exchange interaction in eV per atom in Dy if the Curie temperature is 85 K?

**8.7** **Magnetic domain wall energy and thickness** The energy of a Bloch wall depends on two main factors: the exchange energy $E_{ex}$ (J/atom) and magnetocrystalline energy $K$ (J m$^{-3}$). If $a$ is the interatomic distance and $\delta$ is the wall thickness, then it can be shown that the potential energy per unit area of the wall is

*Potential energy of a Bloch wall*

$$U_{wall} = \frac{\pi^2 E_{ex}}{2a\delta} + K\delta$$

Show that the minimum energy occurs when the wall has the thickness

*Bloch wall thickness*

$$\delta' = \left(\frac{\pi^2 E_{ex}}{2aK}\right)^{1/2}$$

and show that when $\delta = \delta'$, the exchange and anisotropy energy contributions are *equal*. Using reasonable values for various parameters, estimate the Bloch energy and wall thickness for Ni. (See Example 8.4.)

***8.8 Toroidal inductor and radio engineers toroidal inductance equation**

*a.* Consider a toroidal coil (Figure 8.10) whose mean circumference is $\ell$ and that has $N$ tightly wound turns around it. Suppose that the diameter of the core is 2a and $\ell \gg a$. By applying Ampere's law, show that if the current through the coil is $I$, then the magnetic field in the core is

$$B = \frac{\mu_o \mu_r N I}{\ell} \qquad [8.30]$$

where $\mu_r$ is the relative permeability of the medium. Why do you need $\ell \gg a$ for this to be valid? Does this equation remain valid if the core cross section is not circular but rectangular, $a \times b$, and $\ell \gg a$ and $b$?

*b.* Show that the inductance of the toroidal coil is

*Toroidal coil inductance*

$$L = \frac{\mu_o \mu_r N^2 A}{\ell} \qquad [8.31]$$

where $A$ is the cross-sectional area of the core.

*c.* Consider a toroidal inductor used in electronics that has a ferrite core size FT-37, that is, round but with a rectangular cross section. The outer diameter is 0.375 in (9.52 mm), the inner diameter is 0.187 in (4.75 mm), and the height of the core is 0.125 in (3.175 mm). The initial relative permeability of the ferrite core is 2000, which corresponds to a ferrite called the 77 Mix. If the inductor has 50 turns, then using Equation 8.31, calculate the approximate inductance of the coil.

*d.* Radio engineers use the following equation to calculate the inductances of toroidal coils,

*Radio engineers inductance equation*

$$L(\text{mH}) = \frac{A_L N^2}{10^6} \qquad [8.32]$$

where $L$ is the inductance in millihenries (mH) and $A_L$ is an inductance parameter, called an **inductance index,** that characterizes the core of the inductor. $A_L$ is supplied by the manufacturers of ferrite cores and is typically quoted as millihenries (mH) per 1000 turns. In using Equation 8.32, one simply substitutes the numerical value of $A_L$ to find $L$ in millihenries. For the FT-37 ferrite toroid with the 77 Mix as the ferrite core, $A_L$ is specified as 884 mH/1000 turns. What is the inductance of the toroidal inductor in part (*c*) from the radio engineers equation in Equation 8.32? What is the percentage difference in values calculated by Equations 8.32 and 8.31? What is your conclusion? (*Comment:* The agreement is not always this close.)

***8.9 A toroidal inductor**

*a.* Equations 8.31 and 8.32 allow the inductance of a toroidal coil in electronics to be calculated. Equation 8.32 is the equation that is used in practice. Consider a toroidal inductor used in electronics that has a ferrite core of size FT-23 that is round but with a rectangular cross section. The outer diameter is 0.230 in (5.842 mm), the inner diameter is 0.120 in (3.05 mm), and the height of the core is 0.06 in (1.5 mm). The ferrite core is a 43-Mix that has an initial relative permeability of 850 and a maximum relative permeability of 3000. The inductance index for this 43-Mix ferrite core of size FT-23 is $A_L = 188$ (mH/1000 turns). If the inductor has 25 turns, then using Equations 8.31 and 8.32, calculate the inductance of the coil under small-signal conditions and comment on the two values.

*b.* The saturation field $B_{sat}$ of the 43-Mix ferrite is 0.2750 T. What will be typical dc currents that will saturate the ferrite core (an estimate calculation is required)? It is not unusual to find such an inductor in an electronic circuit also carrying a dc current. Will your calculation of the inductance remain valid in these circumstances?

*c.* Suppose that the toroidal inductor discussed in parts (a) and (b) is in the vicinity of a very strong magnet that saturates the magnetic field inside the ferrite core. What will be the inductance of the coil?

***8.10 The transformer**

*a.* Consider the transformer shown in Figure 8.70a whose primary is excited by an ac (sinusoidal) voltage of frequency $f$. The current flowing into the primary coil sets up a magnetic flux in the transformer core. By virtue of Faraday's law of induction and Lenz's law, the flux generated in the core is the flux necessary to induce a voltage nearly equal and opposite to the applied voltage. Thus,

$$v = \frac{d(\text{Total flux linked})}{dt} = \frac{NA\,dB}{dt}$$

where $A$ is the cross-sectional area, assumed constant, and $N$ is the number of turns in the primary. Show that if $V_{rms}$ is the rms voltage at the primary ($V_{max} = V_{rms}\sqrt{2}$) and $B_m$ is the maximum magnetic field in the core, then

*Transformer equation*

$$V_{rms} = 4.44\, NAfB_m \qquad [8.33]$$

Transformers are typically operated with $B_m$ at the "knee" of the $B$–$H$ curve, which corresponds roughly to maximum permeability. For transformer irons, $B_m \approx 1.2$ T. Taking $V_{rms} = 120$ V and a transformer core with $A = 10$ cm $\times$ 10 cm, what should $N$ be for the primary winding? If the secondary winding is to generate 240 V, what should be the number of turns for the secondary coil?

*b.* The transformer core will exhibit hysteresis and eddy current losses. The **hysteresis loss** per unit second, as power loss in watts, is given by

*Hysteresis loss*

$$P_h = KfB_m^n V_{core} \qquad [8.34]$$

where $K = 150.7$, $f$ is the ac frequency (Hz), $B_m$ is the maximum magnetic field (T) in the core (assumed to be in the range 0.2–1.5 T), $n = 1.6$, and $V_{core}$ is the volume of the core. The eddy current losses are reduced by laminating the transformer core, as shown in Figure 8.70b. The **eddy current loss** is given by

*Eddy current loss*

$$P_e = 1.65\, f^2 B_m^2 \left(\frac{d^2}{\rho}\right) V_{core} \qquad [8.35]$$

where $d$ is the thickness of the laminated iron sheet in meters (Figure 8.70b) and $\rho$ is its resistivity ($\Omega$ m).

Suppose that the transformer core has a volume of 0.0108 m$^3$ (corresponds to a mean circumference of 1.08 m). If the core is laminated into sheets of thickness 1 mm and the resistivity of the transformer iron is $6 \times 10^{-7}$ $\Omega$ m, calculate both the hysteresis and eddy current losses at $f = 60$ Hz, and comment on their relative magnitudes. How would you reduce each loss?

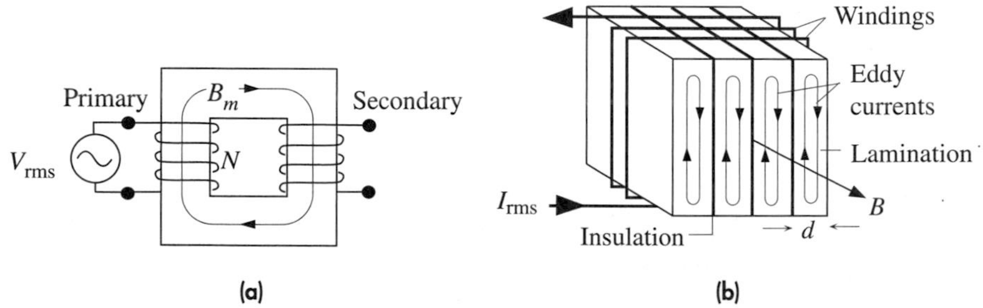


**Figure 8.70**
(a) A transformer with $N$ turns in the primary.
(b) Laminated core reduces eddy current losses.

**8.11** **Losses in a magnetic recording head** Consider eddy current losses in a permalloy magnetic head for audio recording up to 10 kHz. We will use Equation 8.35 for the eddy current losses. Consider a magnetic head weighing 30 g and made from a permalloy with density 8.8 g cm$^{-3}$ and resistivity 6 × 10$^{-7}$ Ω m. The head is to operate at $B_m$ of 0.5 T. If the eddy current losses are not to exceed 1 mW, estimate the thickness of laminations needed. How would you achieve this?

***8.12** **Design of a ferrite antenna for an AM receiver** We consider an AM radio receiver that is to operate over the frequency range 530–1600 kHz. Suppose that the receiving antenna is to be a coil with a ferrite rod as core, as depicted in Figure 8.71. The coil has $N$ turns, its length is $\ell$, and the cross-sectional area is $A$. The inductance $L$ of this coil is tuned with a variable capacitor $C$. The maximum value of $C$ is 265 pF, which with $L$ should correspond to tuning in the lowest frequency at 530 kHz. The coil with the ferrite core receives the EM waves, and the magnetic field of the EM wave permeates the ferrite core and induces a voltage across the coil. This voltage is detected by a sensitive amplifier, and in subsequent electronics it is suitably demodulated. The coil with the ferrite core therefore acts as the antenna of the receiver (ferrite antenna). We will try to find a suitable design for the ferrite coil by carrying out approximate calculations—in practice some trial and error experimentation would also be necessary. We will assume that the inductance of a finite solenoid is

$$L = \frac{\gamma \mu_{ri} \mu_o A N^2}{\ell} \quad [8.36]$$

*Inductance of a solenoid*

where $A$ is the cross-sectional area of the core, $\ell$ is the coil length, $N$ is the number of turns, and $\gamma$ is a geometric factor that accounts for the solenoid coil being of finite length. Assume $\gamma \approx 0.75$. The resonant frequency $f$ of an $LC$ circuit is given by

$$f = \frac{1}{2\pi (LC)^{1/2}} \quad [8.37]$$

*LC circuit resonant frequency*

*a.* If $d$ is the diameter of the enameled wire to be used as the coil winding, then the length $\ell \approx Nd$. If we use an enameled wire of diameter 1 mm, what is the number of coil turns $N$ we need for a ferrite rod given that its diameter is 1 cm and its initial relative permeability is 100?

*b.* Suppose that the magnetic field intensity $H$ of the signal in free space is varying sinusoidally, that is,

$$H = H_m \sin(2\pi f t) \quad [8.38]$$

where $H_m$ is the maximum magnetic field intensity. $H$ is related to the electric field $\mathcal{E}$ at a point by $H = \mathcal{E}/Z_{\text{space}}$, where $Z_{\text{space}}$ is the impedance of free space given by 377 Ω. Show that the induced voltage at the antenna coil is

$$V_m = \frac{\mathcal{E}_m d}{2\pi 377 C f \gamma} \quad [8.39]$$

*Induced voltage across a ferrite antenna*

where $f$ is the frequency of the AM wave and $\mathcal{E}_m$ is the electric field intensity of the AM station at the receiver point. Suppose that the electric field of a local AM station at the receiver is 10 mV m$^{-1}$. What is the voltage induced across the ferrite antenna and can this voltage be detected by an amplifier? Would you use a ferrite rod antenna at short-wave frequencies, given the same $C$ but less $N$?

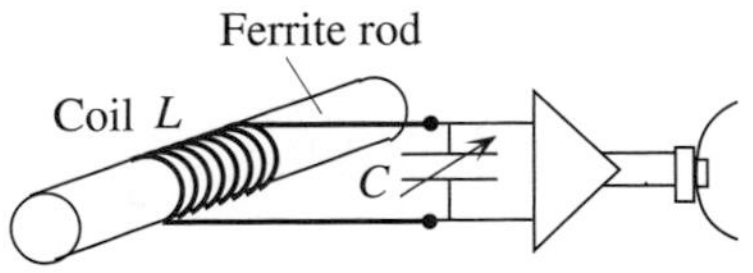


**Figure 8.71** A ferrite antenna of an AM receiver.

***8.13** **A permanent magnet with an air gap** The magnetic field energy in the gap of a permanent magnet is available to do work. Suppose that $B_m$ and $B_g$ are the magnetic field in the magnet and the gap, $H_m$ and $H_g$ are the field intensities in the magnet and the gap, and $V_m$ and $V_g$ are the volumes of the magnet and gap; show that, in terms of magnitudes,

*Magnet and gap relationship*

$$B_g H_g V_g \approx B_m H_m V_m$$

What is the significance of this result?

**8.14 A permanent magnet with an air gap**

*a.* Show that the maximum energy stored in the air gap of a permanent magnet can be written very roughly as

*Energy in gap of a magnet*

$$E_{\text{gap}} \approx \frac{1}{8} B_r H_c V_m$$

where $V_m$ is the volume of the magnet, which is much greater than that of the gap; $B_r$ is the remanent magnetic field; and $H_c$ is the coercivity of the magnet.

*b.* Using Table 8.6, compare the $(BH)_{\max}$ with the product $(\frac{1}{2}H_c)$ $(\frac{1}{2}B_r)$ and comment on the closeness of agreement.

*c.* Calculate the energy in the gap of a rare earth cobalt magnet that has a volume of 0.1 $m^3$. Give an example of typical work (*e.g.*, raising so many apples, each 100 g, by so many meters) that could be done if all this energy could be converted to mechanical work.

**8.15 Weight, cost, and energy of a permanent magnet with an air gap** For a certain application, an energy of 1 kJ is required in the gap of a permanent magnet. There are three candidates, as shown in Table 8.11. Which material will give the lightest magnet? Which will give the cheapest magnet?

**Table 8.11** Three permanent magnet candidates

| Magnet | $(BH)_{\max}$ (kJ $m^{-3}$) | Density (g $cm^{-3}$) | Yesterday's Relative Price (per unit mass) |
|---|---|---|---|
| Alnico | 50 | 7.3 | 1 |
| Rare earth | 200 | 8.2 | 2 |
| Ferrite | 30 | 4.8 | 0.5 |

***8.16 Permanent magnet with yoke and air gap** Consider a permanent magnet bar that has L-shaped ferromagnetic (high permeability) pieces attached to its ends to direct the magnetic field to an air gap as depicted in Figure 8.72. The L-shaped high $\mu_r$ pieces for directing the magnetic field are called **yokes.** Suppose that $A_m$, $A_y$, and $A_g$ are the cross-sectional areas of the magnet, yoke, and gap as indicated in the figure. The lengths of the magnet, yoke, and air gap are $\ell_m$, $y$, and $g$, respectively. The magnet, the two yokes, and the gap can be considered to be all connected end-to-end or in series. Applying Ampere's circuital law for $H$ we can write,

$$H_m \ell_m + 2H_y \ell_y + H_g \ell_g = 0$$

Since all four components, magnet, yokes, and gap, are in series, we can assume that the magnetic flux $\Phi$ through each of them is the same,

$$\Phi = B_m A_m = B_y A_y = B_g A_g$$

*a.* Show that

$$H_m = -\frac{A_m}{\ell_m}\left[\frac{\ell_g}{\mu_o A_g} + \frac{2\ell_y}{\mu_o \mu_{ry} A_y}\right] B_m$$

*b.* What does the equation in part (*a*) represent? Given that $B_m$ and $H_m$ in the magnet must obey the equation in part (*a*), and also the $B$–$H$ characteristic of the magnet material itself, what is your conclusion?

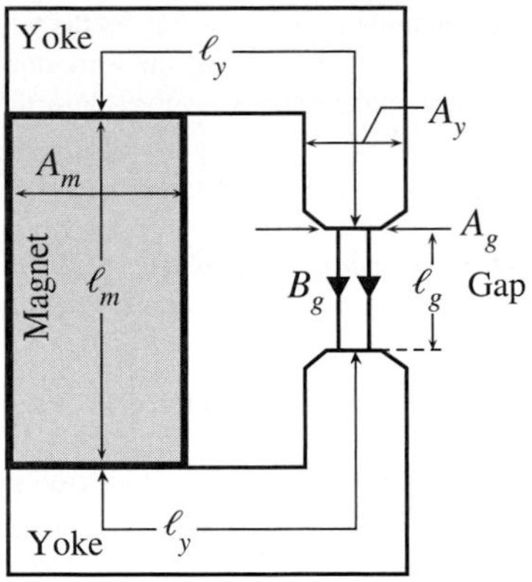


**Figure 8.72** A permanent magnet with two pieces of yoke and an air gap.

*c.* Should the yokes be magnetically hard or soft? Justify your decision.

*d.* Show that if $\mu_{ry}$ is very large ($\mu_{ry} \approx \infty$),

$$H_m = -\frac{1}{\mu_o}\left[\frac{A_m \ell_g}{A_g \ell_m}\right] B_m$$

*e.* If $V_m = A_m \ell_m$ and $V_g = A_g \ell_g$ are the volumes of the magnet and gap, respectively, show that

$$B_g H_g V_g = B_m H_m V_m$$

What is your conclusion (consider the magnetic energy stored in the gap)?

*f.* Consider a rare earth permanent magnet, with a density of 8.2 g cm$^{-3}$, that has a $(BH)_{\max}$ of about 200 kJ m$^{-3}$. Suppose that $(BH)_{\max}$ occurs very roughly at $B_m \approx \frac{1}{2} B_r$ where for this rare earth magnet $B_r \approx 1$ T. Suppose that $A_m \approx A_g$. What is the volume, effective length ($\ell_m$), and mass of the magnet that is needed to store the maximum energy in the gap if $\ell_g = 1$ cm and $A_g = 1000$ cm$^2$? What is the maximum energy in the gap?

**8.17** **Superconductivity and critical current density** Consider two superconducting wires, tin (Sn; Type I) and $Nb_3Sn$ (Type II), each 1 mm in thickness. The magnetic field on the surface of a current-carrying conductor is given by

$$B = \frac{\mu_o I}{2\pi r}$$

*a.* Assuming that Sn wire loses its superconductivity when the field at the surface reaches the critical field (0.2 T), calculate the maximum current and hence the critical current density that can be passed through the Sn wire near absolute zero of temperature.

*b.* Calculate the maximum current and critical current density for the $Nb_3Sn$ wire using the same assumption as in part (*a*) but taking the critical field to be the upper critical field $B_{c2}$, which is 24.5 T at 0 K. How does your calculation of $J_c$ compare with the critical density of about $10^{11}$ A m$^{-2}$ for $Nb_3Sn$ at 0 K?

**8.18** **Magnetic pressure in a solenoid** Consider a long solenoid with an air core. Diametrically opposite windings have oppositely directed currents and, due to the magnetic force, they repel each other. This means that the solenoid experiences a *radial force* $F_r$ that is trying to open up the solenoid, *i.e.*, stretch out the windings as depicted in Figure 8.53. Suppose that $A$ is the surface area of the core (on to which wires are wound). If we decrease the core diameter by $dx$, the volume changes by $dV$. We have to do work $dW$ against the radial magnetic forces $F_r$,

$$dW = F_r\, dx = \left(\frac{F_r}{A}\right) A\, dx = P_r\, dV$$

where $P_r = F_r/A$ is the *radial pressure*, called the **magnetic pressure,** acting on the windings of the solenoid. (This pressure acts to tear apart the solenoid.) Using the fact that the work done against the magnetic forces in changing the volume changes the magnetic energy in the core, show that

*Radial magnetic pressure in a solenoid*

$$P_r = \frac{B^2}{2\mu_o}$$

What is the radial pressure on a solenoid that has a field of 35 T in the core? How many atmospheres is this? What is the equivalent ocean depth that gives the same pressure? What happens to this pressure at 100 T?

***8.19 Enterprising engineers in the high arctic building a superconducting inductor** A current-carrying inductor has energy stored in its magnetic field that can be converted to electrical work. A group of enterprising engineers and scientists living in Resolute in Nunavut (Canada) have decided to build a toroidal inductor to store energy so that this energy can be used to supply a small community of 10 houses each consuming on average 3 kW of energy during the night (6 months). They have discovered a superconductor (Type II) that has a $B_{c2} = 100$ T and a critical current density of $J_c = 5 \times 10^{10}$ A m$^{-2}$ at night temperatures (it is obviously a novel high-$T_c$ superconductor of some sort). Their superconducting wire has a diameter of 5 mm and is available in any desirable length. All the wiring in the community is done by superconductors except where energy needs to be converted to other forms (mechanical, heat, etc.). They have decided on the following design specification for their toroid:

The mean diameter $D_{\text{toroid}}$ of the toroid, $(\frac{1}{2})$ (Outside diameter + Inside diameter), is 10 times longer than the core diameter $D_{\text{core}}$. The field inside the toroid is therefore reasonably uniform to within 10 percent.

The maximum operating magnetic field in the core is 35 T. Fields larger than this can result in mechanical fracture and failure.

Assume that $J_c$ decreases linearly with the magnetic field and that the mechanical engineers in the group can take care of the forces trying to blow open the toroid by building a proper support structure.

Find the size of the toroid (mean diameter and circumference), the number of turns and the length of the superconducting wire they need, the current in the coil, and whether this current is sufficiently below the critical current at that field. Is it feasible?

**8.20 Magnetic storage media**

*a.* Consider the storage of video information (FM signal) on a video tape. Suppose that the maximum signal frequency to be recorded as a spatial magnetic pattern is 10 MHz. The heads helically scan the tape, and the relative velocity of the tape to head is about 10 m s$^{-1}$. What is the minimum spatial wavelength of the stored magnetic pattern (information) on the tape?

*b.* Suppose that the speed of an audio cassette tape in a cassette player is 5 cm s$^{-1}$. If the maximum frequency that needs to be recorded is 20 kHz, what is the minimum spatial wavelength on the tape?

Note: An excellent quantitive description of magnetic recording may be found in R. L. Comstock, *Introduction to Magnetism and Magnetic Recording,* New York: John Wiley & Sons, 1999.

***8.21 Magnetic recording principles** In this "back of an envelope" calculation we consider the principle of operation of a recording head for writing on a magnetic tape (perhaps an audio or a video tape). The recording head has a small gap, of size $g$ (about 1 μm or less), which is much smaller than the mean circumference of the head $\ell$ (perhaps a few millimeters) as shown in Figure 8.73. The coil of this head has $N$ turns and is energized by the signal current $i$. The fringe field intensity $H_f$ at the gap magnetizes the magnetic tape passing under the head. $H_f$ must be greater than the coercivity $H_c$ of the storage medium (tape) to be able to magnetize that region of the tape under the head. Suppose that $H_m$ = magnetic field intensity in the core of the head; $H_g$ = magnetic field intensity in the gap; $H_f$ = fringing field intensity below the gap; $B_m = \mu_r\mu_o H_m$ = magnetic field in the core of the head; $B_g = \mu_o H_g$ = magnetic field in the gap.

The magnetic flux must be continuous through the small gap. Thus, if $A$ is the cross-sectional area,

$$\text{Flux in the core} = AB_m = \text{Flux in the gap} = AB_g \qquad \text{or} \qquad B_g = B_m$$

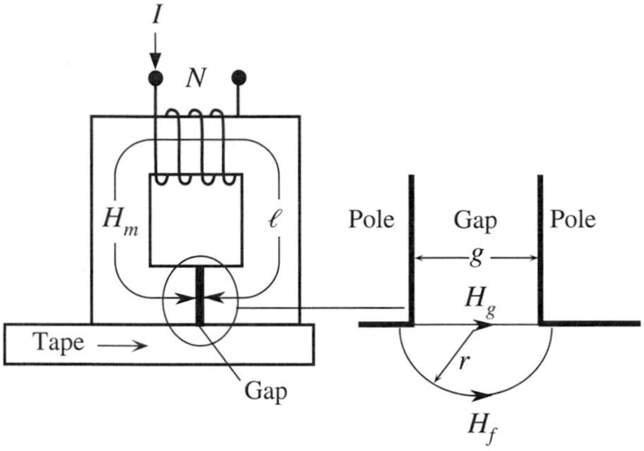


**Figure 8.73** The gap of a recording head and the fringing field for magnetizing the tape.

*a.* Applying Ampere's law for $H$ around the mean circumference, $\ell + g$, show that

$$H_g = \frac{1}{g + \ell/\mu_r} NI$$

*Field in the gap*

*b.* If we apply Ampere's law for $H$ around the semicircle of radius $r$ coming out from the gap into the tape as shown in Figure 8.73 we get

$$H_g g - H_f(\pi r) \approx 0$$

Show that,

$$H_f \approx \frac{\mu_r g}{\pi r(\mu_r g + \ell)} NI$$

*Fringing field for recording on storage media*

*c.* The fringing field must overcome the coercivity of the storage medium. Suppose that the storage medium has $H_c = 50$ kA m$^{-1}$ and we have to determine $NI$ given the head material. Suppose that $\mu_r \approx 10^4$, $g = 1$ μm $= 10^{-6}$ m, $\ell \approx 5$ mm $= 5 \times 10^{-3}$ m, and $r = 1$ μm $= 10^{-6}$ m to record into a depth of 1 μm. What is the minimum $NI$? If the minimum signal current (after amplification) is 5 mA, how many turns do you need for the coil?

*d.* What is the magnetic field $B_m$ in the core? Can you use a ferrite head?

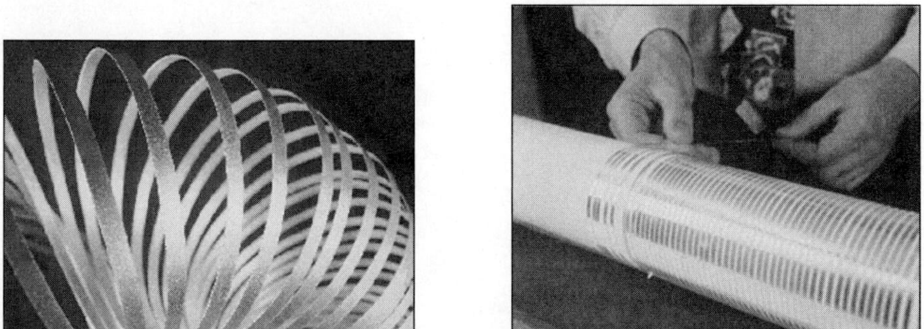

Left: These high-temperature superconductor (HTS) flat tapes are based on $(Bi_{2-x}Pb_x)Sr_2Ca_2Cu_3O_{10-\delta}$(Bi-2223). The tape has an outer surrounding protective metallic sheath. Right: HTS tapes have a major advantage over equivalent-sized metal conductors, in being able to transmit considerably higher power loads. Coils made from HTS tape can be used to create more compact and efficient motors, generators, magnets, transformers, and energy storage devices.

| SOURCE: Courtesy of Australian Superconductors.

Augustin Jean Fresnel (1788–1827) was a French physicist, and a civil engineer for the French government, who was one of the principal proponents of the wave theory of light. He made a number of distinct contributions to optics including the well-known Fresnel lens that was used in lighthouses in the nineteenth century. He fell out with Napoleon in 1815 and was subsequently put into house arrest until the end of Napoleon's reign. During his enforced leisure time he formulated his wave ideas of light into a mathematical theory.

SOURCE: Smithsonian Institution, courtesy of AIP Emilio Segrè Visual Archives.

Christiaan Huygens (1629–1695), a Dutch physicist, explained double refraction of light in calcite in terms of ordinary and extraordinary waves. Christiaan Huygens made many contributions to optics and wrote prolifically on the subject.

SOURCE: Courtesy of Emilo Segrè Visual Archives, American Institute of Physics.

CHAPTER

# 9

# Optical Properties of Materials

The way electromagnetic (EM) radiation interacts with matter depends very much on the wavelength of the EM wave. Many familiar types of EM radiation have wavelengths that range over many orders of magnitude. Although radio waves and X-rays are both EM waves, the two interact in a distinctly different way with matter. We tend to think of "light" as the electromagnetic radiation that we can see, that is, wavelengths in the visible range, typically 400 to 700 nm. However, in many applications, light is also used to describe EM waves that can have somewhat shorter or longer wavelengths such as ultraviolet (UV) and infrared (IR) light. For many practical purposes, it is useful to (arbitrarily) define light as EM waves that have wavelengths shorter than very roughly 100 μm but longer than long-wavelength X-rays, roughly 10 nm. Today's *light wave communications* use EM waves with wavelengths of 1300 and 1550 nm; in the infrared. *Optical properties* of materials are those characteristic properties that determine the interaction of light with matter; the best example being the refractive index $n$ that determines the speed of light in a medium through $v = c/n$, where $v$ is the speed of light in the medium and $c$ is the speed of light in free space. The present chapter examines the key optical properties of matter and how these depend on the material and on the characteristics of the EM wave. The refractive index $n$, for example, depends on the dielectric polarization mechanisms as well as the wavelength $\lambda$. The material's $n$–$\lambda$ behavior is called the **dispersion relation** and is one of the most important characteristics in many optical device applications.

We know from Chapter 3 that, depending on the experiment, we can treat light either as an EM wave, exhibiting typical wave-like properties, or as photons, exhibiting particle-like behavior. In this chapter we will primarily use the wave nature of light, though for absorption of light, the photon interpretation is more appropriate as the photons interact with electrons in the material.

## 9.1 LIGHT WAVES IN A HOMOGENEOUS MEDIUM

We know from well-established experiments that light exhibits typical wave-like properties such as interference and diffraction. We can treat light as an EM wave with time-varying electric and magnetic fields $E_x$ and $B_y$, respectively, which propagate through space in such a way that they are always perpendicular to each other and the direction of propagation $z$ is as depicted in Figure 9.1. The simplest traveling wave is a sinusoidal wave, which, for propagation along $z$, has the general mathematical form,[1]

*Traveling wave along z*

$$E_x = E_o \cos(\omega t - kz + \phi_o) \qquad [9.1]$$

where $E_x$ is the electric field at position $z$ at time $t$; $k$ is the **propagation constant,** or **wavenumber,** given by $2\pi/\lambda$, where $\lambda$ is the wavelength; $\omega$ is the angular frequency; $E_o$ is the amplitude of the wave; and $\phi_o$ is a phase constant which accounts for the fact that at $t = 0$ and $z = 0$, $E_x$ may or may not necessarily be zero depending on the choice of origin. The argument $(\omega t - kz + \phi_o)$ is called the **phase** of the wave and denoted by $\phi$. Equation 9.1 describes a **monochromatic plane wave** of infinite extent traveling in the positive $z$ direction as depicted in Figure 9.2. In any plane perpendicular to the direction of propagation (along $z$), the phase of the wave, according to Equation 9.1, is constant which means that the field in this plane is also constant. A surface over which the phase of a wave is constant is referred to as a **wavefront.** A wavefront of a plane wave is obviously a plane perpendicular to the direction of propagation as shown in Figure 9.2.

We know from electromagnetism that time-varying magnetic fields result in time-varying electric fields (Faraday's law) and vice versa. A time-varying electric field would set up a time-varying magnetic field with the same frequency. According to electromagnetic principles,[2] a traveling electric field $E_x$ as represented by Equation 9.1 would always be accompanied by a traveling magnetic field $B_y$ with the same wave frequency and propagation constant ($\omega$ and $k$) but the directions of the two fields would be orthogonal as in Figure 9.1. Thus, there is a similar traveling wave equation for the magnetic field component $B_y$. We generally describe the interaction of a light wave with a nonconducting matter (conductivity, $\sigma = 0$) through the electric field component $E_x$ rather than $B_y$ because it is the electric field that displaces the electrons in molecules or ions in the crystal and thereby gives rise to the polarization of matter. However, the two fields are linked, as in Figure 9.1, and there is an intimate relationship between the two fields. The **optical field** refers to the electric field $E_x$.

We can also represent a traveling wave using the exponential notation since $\cos\phi = \text{Re}[\exp(j\phi)]$ where Re refers to the real part. We then need to take the real

---

[1] This chapter uses $E$ for the electric field which was reserved for energy in previous chapters. There should be no confusion with $E_g$ that represents the energy bandgap. In addition, $n$ is used to represent the refractive index rather than the electron concentration.

[2] Maxwell's equations formulate electromagnetic phenomena and provide relationships between the electric and magnetic fields and their space and time derivatives. We only need to use a few selected results from Maxwell's equations without delving into their derivations. The *magnetic field B* is also called the magnetic induction or magnetic flux density.

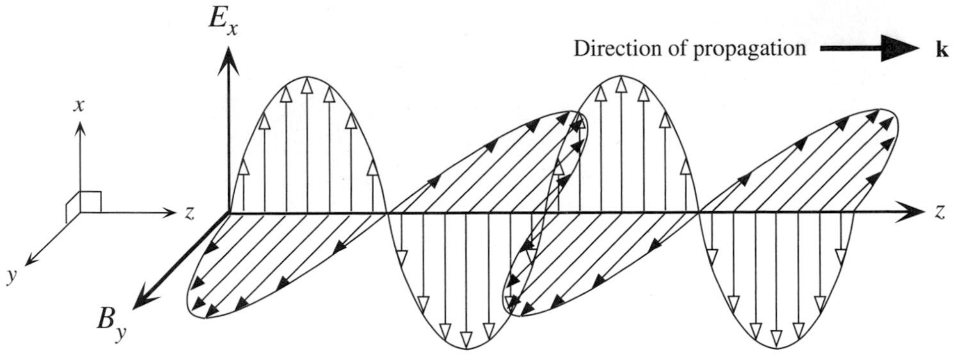


**Figure 9.1** An electromagnetic wave is a traveling wave that has time-varying electric and magnetic fields that are perpendicular to each other and the direction of propagation $z$.

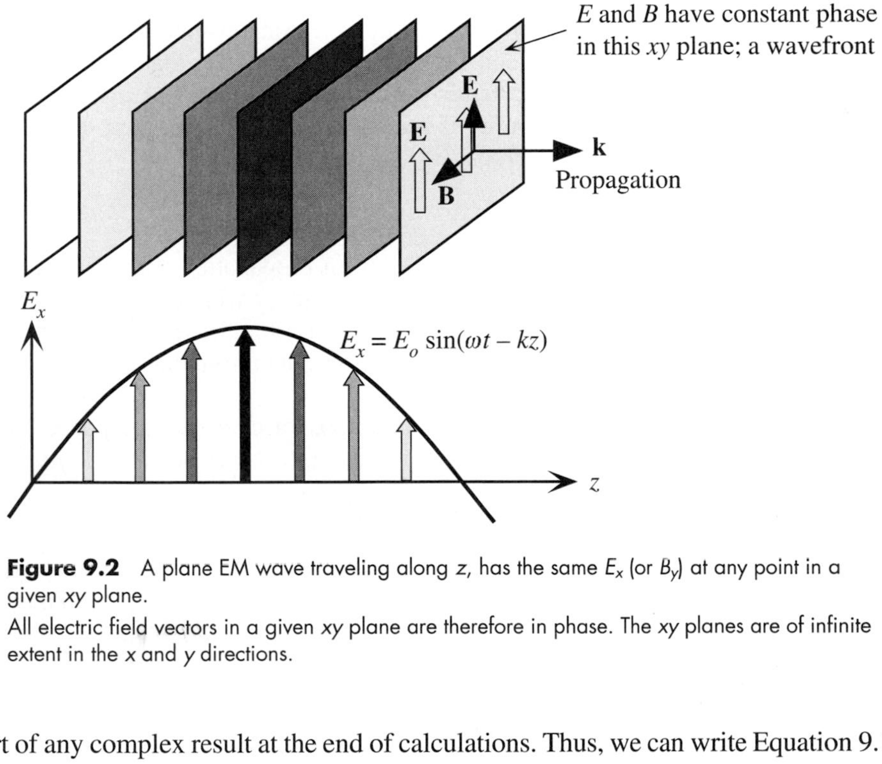


**Figure 9.2** A plane EM wave traveling along $z$, has the same $E_x$ (or $B_y$) at any point in a given $xy$ plane.
All electric field vectors in a given $xy$ plane are therefore in phase. The $xy$ planes are of infinite extent in the $x$ and $y$ directions.

part of any complex result at the end of calculations. Thus, we can write Equation 9.1 as

$$E_x(z, t) = \text{Re}[E_o \exp(j\phi_o) \exp j(\omega t - kz)]$$

or

$$E_x(z, t) = \text{Re}[E_c \exp j(\omega t - kz)] \qquad [9.2]$$

*Traveling wave along z*

where $E_c = E_o \exp(j\phi_o)$ is a complex number that represents the amplitude of the wave and includes the constant phase information $\phi_o$.

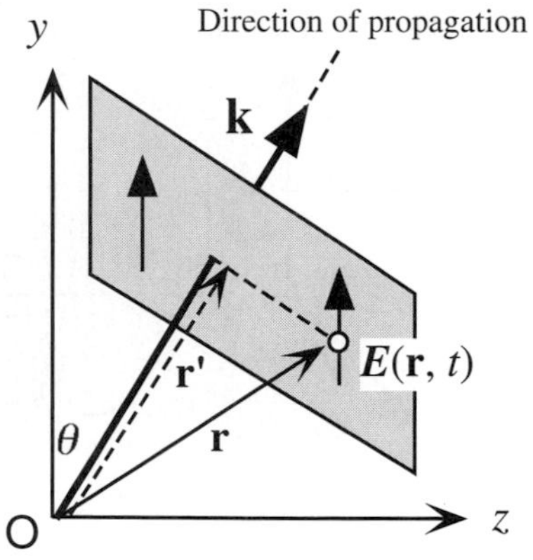


**Figure 9.3** A traveling plane EM wave along a direction **k**.

We indicate the direction of propagation with a vector **k**, called the **wavevector,** whose magnitude is the propagation constant $k = 2\pi/\lambda$. It is clear that **k** is perpendicular to constant phase planes as indicated in Figure 9.2. When the EM wave is propagating along some arbitrary direction **k**, as indicated in Figure 9.3, then the electric field $E(\mathbf{r}, t)$ at a point **r** on a plane perpendicular to **k** is

*Light wave in three dimensions*

$$E(\mathbf{r}, t) = E_o \cos(\omega t - \mathbf{k} \cdot \mathbf{r} + \phi_o) \qquad [9.3]$$

because the dot product $\mathbf{k} \cdot \mathbf{r}$ is along the direction of propagation similar to $kz$. The dot product is the product of **k** and the projection of **r** onto **k** which is $\mathbf{r}'$ in Figure 9.3, so $\mathbf{k} \cdot \mathbf{r} = kr'$. Indeed, if propagation is along $z$, $\mathbf{k} \cdot \mathbf{r}$ becomes $kz$. In general, if **k** has components $k_x$, $k_y$, and $k_z$ along the $x$, $y$, and $z$ directions, then from the definition of the dot product, $\mathbf{k} \cdot \mathbf{r} = k_x x + k_y y + k_z z$.

The time and space evolution of a given phase $\phi$, for example, the phase corresponding to a maximum field, according to Equation 9.1 is described by

$$\phi = \omega t - kz + \phi_o = \text{constant}$$

During a time interval $\delta t$, this constant phase (and hence the maximum field) moves a distance $\delta z$. The phase velocity of this wave is therefore $\delta z/\delta t$. Thus the **phase velocity** $v$ is

*Phase velocity*

$$v = \frac{dz}{dt} = \frac{\omega}{k} = \nu\lambda \qquad [9.4]$$

where $\nu$ is the frequency ($\omega = 2\pi\nu$).

We are frequently interested in the phase difference $\Delta\phi$ at a given time between two points on a wave (Figure 9.1) that are separated by a certain distance. If the wave is traveling along $z$ with a wavevector $k$, as in Equation 9.1, then the phase difference between two points separated by $\Delta z$ is simply $k\,\Delta z$ since $\omega t$ is the same for each point. If this phase difference is 0 or multiples of $2\pi$, then the two points are in phase. Thus, the phase difference $\Delta\phi$ can be expressed as $k\,\Delta z$ or $2\pi\,\Delta z/\lambda$.

## 9.2 REFRACTIVE INDEX

When an EM wave is traveling in a dielectric medium, the oscillating electric field polarizes the molecules of the medium at the frequency of the wave. Intuitively, the EM wave propagation can be considered to be the propagation of this polarization in the medium. The field and the induced molecular dipoles become coupled. The net effect is that the polarization mechanism delays the propagation of the EM wave. The stronger the interaction between the field and the dipoles, the slower is the propagation of the wave. The relative permittivity $\varepsilon_r$ measures the ease with which the medium becomes polarized, and hence it indicates the extent of interaction between the field and the induced dipoles. For an EM wave traveling in a nonmagnetic dielectric medium of relative permittivity $\varepsilon_r$, the phase velocity $\mathsf{v}$ is given by

*Phase velocity in a medium with $\varepsilon_r$*

$$\mathsf{v} = \frac{1}{\sqrt{\varepsilon_r \varepsilon_o \mu_o}} \qquad [9.5]$$

If the frequency $\nu$ is in the optical frequency range, then $\varepsilon_r$ will be due to electronic polarization as ionic polarization will be too sluggish to respond to the field. However, at the infrared frequencies or below, the relative permittivity also includes a significant contribution from ionic polarization and the phase velocity is slower. For an EM wave traveling in free space, $\varepsilon_r = 1$ and $\mathsf{v}_{\text{vacuum}} = 1/\sqrt{\varepsilon_o \mu_o} = c = 3 \times 10^8$ m s$^{-1}$, the velocity of light in a vacuum. The ratio of the speed of light in free space to its speed in a medium is called the **refractive index** $n$ of the medium,

*Definition of refractive index*

$$n = \frac{c}{\mathsf{v}} = \sqrt{\varepsilon_r} \qquad [9.6]$$

Suppose that in free space $k_o$ is the wavevector ($k_o = 2\pi/\lambda_o$) and $\lambda_o$ is the wavelength, then the wavevector $k$ in the medium will be $nk_o$ and the wavelength $\lambda$ will be $\lambda_o/n$. Indeed, we can also define the refractive index in terms of the wavevector $k$ in the medium with respect to that in a vacuum $k_o$,

*Definition of refractive index*

$$n = \frac{k}{k_o} \qquad [9.7]$$

Equation 9.6 is in agreement with our intuition that light propagates more slowly in a denser medium which has a higher refractive index. We should note that the frequency $\nu$ remains the same. The refractive index of a medium is not necessarily the same in all directions. In noncrystalline materials such as glasses and liquids, the material structure is the same in all directions and $n$ does not depend on the direction. The refractive index is then **isotropic.** In crystals, however, the atomic arrangements and interatomic bonding are different along different directions. Crystals, in general, have nonisotropic, or *anisotropic,* properties. Depending on the crystal structure, the relative permittivity $\varepsilon_r$ is different along different crystal directions. This means that, in general, the refractive index $n$ seen by a propagating EM wave in a crystal will depend on the value of $\varepsilon_r$ along the direction of the oscillating electric field (that is, along the direction of polarization). For example, suppose that the wave in Figure 9.1 is traveling along the $z$ direction in a particular crystal with its electric field oscillating along the $x$ direction. If the relative permittivity along this $x$ direction is $\varepsilon_{rx}$, then $n_x = \sqrt{\varepsilon_{rx}}$. The wave therefore propagates with a phase velocity that is $c/n_x$. The variation of $n$ with direction of propagation and the direction of the electric field depends on the particular crystal structure. With the exception of cubic crystals (such as diamond) all crystals exhibit a degree of optical anisotropy which leads to a number of important applications. Typically noncrystalline solids, such as glasses and liquids, and cubic crystals are **optically isotropic;** they possess only one refractive index for all directions.

**EXAMPLE 9.1**

**RELATIVE PERMITTIVITY AND REFRACTIVE INDEX** Relative permittivity $\varepsilon_r$, or the dielectric constant, of materials is frequency dependent and further it depends on crystallographic direction since it is easier to polarize the medium along certain directions in the crystal. Glass has no crystal structure; it is amorphous. The relative permittivity is therefore isotropic but nonetheless frequency dependent.

The relationship $n = \sqrt{\varepsilon_r}$ between the refractive index $n$ and $\varepsilon_r$ must be applied at the same frequency for both $n$ and $\varepsilon_r$. The relative permittivity for many materials can be vastly different at high and low frequencies because different polarization mechanisms operate at these frequencies. At low frequencies all polarization mechanisms present can contribute to $\varepsilon_r$, whereas at optical frequencies only the electronic polarization can respond to the oscillating field. Table 9.1 lists the relative permittivity $\varepsilon_r$(LF) at low frequencies (*e.g.*, 60 Hz or 1 kHz as would be measured for example using a capacitance bridge in the laboratory) for various materials. It then compares $\sqrt{\varepsilon_r(\text{LF})}$ with $n$.

For diamond and silicon there is an excellent agreement between $\sqrt{\varepsilon_r(\text{LF})}$ and $n$. Both are covalent solids in which electronic polarization (electronic bond polarization) is the only polarization mechanism at low and high frequencies. Electronic polarization involves the displacement of light electrons with respect to positive ions of the crystal. This process can readily respond to the field oscillations up to optical or even ultraviolet frequencies.

For AgCl and $SiO_2$, $\sqrt{\varepsilon_r(\text{LF})}$ is larger than $n$ because at low frequencies both of these solids possess a degree of ionic polarization. The bonding has a substantial degree of ionic character which contributes to polarization at frequencies below far-infrared wavelengths. (The AgCl crystal has almost all ionic bonding.) In the case of water, the $\varepsilon_r$(LF) is dominated by orientational or

**Table 9.1** Low-frequency (LF) relative permittivity $\varepsilon_r$(LF) and refractive index $n$

| Material | $\varepsilon_r$ (LF) | $\sqrt{\varepsilon_r(\text{LF})}$ | $n$ (optical) | Comments |
|---|---|---|---|---|
| Diamond | 5.7 | 2.39 | 2.41 (at 590 nm) | Electronic bond polarization up to UV light |
| Si | 11.9 | 3.44 | 3.45 (at 2.15 μm) | Electronic bond polarization up to optical frequencies |
| AgCl | 11.14 | 3.33 | 2.00 (at 1–2 μm) | Ionic polarization contributes to $\varepsilon_r$(LF) |
| $SiO_2$ | 3.84 | 2.00 | 1.46 (at 600 nm) | Ionic polarization contributes to $\varepsilon_r$(LF) |
| Water | 80 | 8.9 | 1.33 (at 600 nm) | Dipolar polarization contributes to $\varepsilon_r$(LF), which is large |

dipolar polarization which is far too sluggish to respond to high-frequency oscillations of the field at optical frequencies.

It is instructive to consider what factors affect $n$. The simplest (and approximate) expression for the relative permittivity is

*Relative permittivity and polarizability*

$$\varepsilon_r \approx 1 + \frac{N\alpha}{\varepsilon_o} \qquad [9.8]$$

where $N$ is the number of molecules per unit volume and $\alpha$ is the polarizability per molecule. Both atomic concentration, or density, and polarizability therefore increase $n$. For example, glasses of given type but with greater density tend to have higher $n$.

## 9.3 DISPERSION: REFRACTIVE INDEX–WAVELENGTH BEHAVIOR

The refractive index of materials in general depends on the frequency, or the wavelength. This wavelength dependence follows directly from the frequency dependence of the relative permittivity $\varepsilon_r$. Figure 9.4 shows what happens to an atom in the presence of an oscillating electric field $E$ which is due to a light wave passing through this location; it may also be due to an applied external field.

In the absence of an electric field and in equilibrium, the center of mass $C$ of the orbital motions of the electrons coincides with the positively charged nucleus at $O$ and the net electric dipole moment is zero as indicated in Figure 9.4a. Suppose that the atom has $Z$ number of electrons orbiting the nucleus and all the electrons are contained

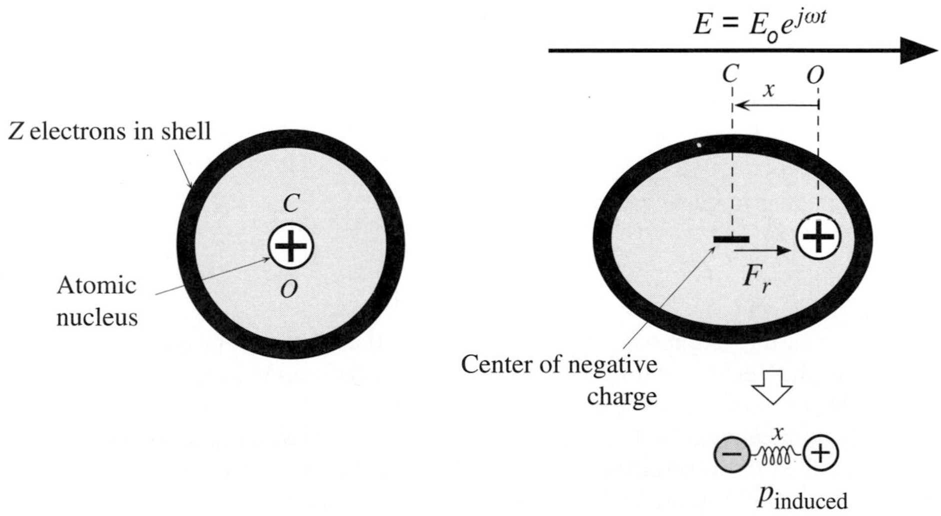


**(a) A neutral atom in $E = 0$.** **(b) Induced dipole moment in a field.**

**Figure 9.4** Electronic polarization of an atom. In the presence of a field in the $+x$ direction, the electrons are displaced in the $-x$ direction (from $O$), and the restoring force is in the $+x$ direction.

within a given shell. In the presence of the electric field $E$, however, the light electrons become displaced in the opposite direction to the field, so their center of mass $C$ is shifted by some distance $x$ with respect to the nucleus $O$ which we take to be the origin as shown in Figure 9.4b. As the electrons are "pushed" away by the applied field, the Coulombic attraction between the electrons and nuclear charge "pulls in" the electrons. The force on the electrons, due to $E$, trying to separate them away from the nuclear charge is $ZeE$. The restoring force $F_r$, which is the Coulombic attractive force between the electrons and the nucleus, can be taken to be proportional to the displacement $x$ provided that the latter is small. The reason is that $F_r = F_r(x)$ can be expanded in powers of $x$, and for small $x$ only the linear term matters. The restoring force $F_r$ is obviously zero when $C$ coincides with $O$ ($x = 0$). We can write $F_r = -\beta x$ where $\beta$ is a constant and the negative sign indicates that $F_r$ is always directed toward the nucleus $O$.

First consider applying a dc field. In equilibrium, the *net* force on the negative charge is zero or $ZeE = \beta x$ from which $x$ is known. Therefore the *magnitude* of the induced electronic dipole moment is given by

*Induced electronic dc dipole moment*

$$p_{\text{induced}} = (Ze)x = \frac{Z^2e^2}{\beta}E \qquad [9.9]$$

As expected $p_{\text{induced}}$ is proportional to the applied field. The electronic dipole moment in Equation 9.9 is valid under static conditions, *i.e.*, when the electric field is a dc field. Suppose that we suddenly remove the applied electric field polarizing the atom. There is then only the restoring force $-\beta x$, which always acts to pull the electrons toward the nucleus $O$. The equation of motion of the negative charge center is then (force = mass × acceleration)

*Simple harmonic motion*

$$-\beta x = Zm_e\frac{d^2x}{dt^2}$$

By solving this differential equation we can show that the displacement at any time is a simple harmonic motion, that is,

$$x(t) = x_o \cos(\omega_o t)$$

where the angular frequency of oscillation $\omega_o$ is

*Natural frequency of the atom*

$$\omega_o = \left(\frac{\beta}{Zm_e}\right)^{1/2} \qquad [9.10]$$

In essence, this is the oscillation frequency of the center of mass of the electron cloud about the nucleus and $x_o$ is the displacement before the removal of the field. After the removal of the field, the electronic charge cloud executes simple harmonic motion about the nucleus with a **natural frequency** $\omega_o$ determined by Equation 9.10; $\omega_o$ is also called the **resonance frequency.** The oscillations, of course, die out with time because there is an inevitable loss of energy from an oscillating charge cloud. An oscillating electron is like an oscillating current and loses energy by radiating EM waves; all accelerating charges emit radiation.

Consider now the presence of an oscillating electric field due to an EM wave passing through the location of this atom as in Figure 9.4b. The applied field oscillates harmonically in the $+x$ and $-x$ directions, that is, $E = E_o \exp(j\omega t)$. This field will drive and oscillate the electrons about the nucleus. There is again a restoring force $F_r$ acting on the displaced electrons trying to bring back the electron shell to its equilibrium placement around the nucleus. For simplicity we will again neglect energy losses. Newton's second law for $Ze$ electrons with mass $Zm_e$ driven by $E$ is given by

*Lorentz oscillator model*

$$Zm_e \frac{d^2x}{dt^2} = -ZeE_o \exp(j\omega t) - \beta x \tag{9.11}$$

The solution of this equation gives the instantaneous displacement $x(t)$ of the center of mass of electrons from the nucleus ($C$ from $O$),

$$x = x(t) = -\frac{eE_o \exp(j\omega t)}{m_e(\omega_o^2 - \omega^2)}$$

The induced electronic dipole moment is then simply given by $p_{\text{induced}} = -(Ze)x$. The negative sign is needed because normally $x$ is measured from negative to positive charge whereas in Figure 9.4b it is measured from the nucleus. By definition, the electronic polarizability $\alpha_e$ is the induced dipole moment per unit electric field,

*Electronic polarizability*

$$\alpha_e = \frac{p_{\text{induced}}}{E} = \frac{Ze^2}{m_e(\omega_o^2 - \omega^2)} \tag{9.12}$$

Thus, the displacement $x$ and hence electronic polarizability $\alpha_e$ increase as $\omega$ increases. Both become very large when $\omega$ approaches the natural frequency $\omega_o$. In practice, charge separation $x$ and hence polarizability $\alpha_e$ do not become infinite at $\omega = \omega_o$ because two factors impose a limit. First, at large $x$, the system is no longer linear and this analysis is not valid. Secondly, there is always some energy loss.

Given that the polarizability is frequency dependent as in Equation 9.12, the effect on the refractive index $n$ is easy to predict. The simplest (and a very rough) relationship between the relative permittivity $\varepsilon_r$ and polarizability $\alpha_e$ is

*Relative permittivity and polarizability*

$$\varepsilon_r = 1 + \frac{N}{\varepsilon_o}\alpha_e$$

where $N$ is the number of atoms per unit volume. Given that the refractive index $n$ is related to $\varepsilon_r$ by $n^2 = \varepsilon_r$, it is clear that $n$ must be frequency dependent, *i.e.*,

*Dispersion relation*

$$n^2 = 1 + \left(\frac{NZe^2}{\varepsilon_o m_e}\right)\frac{1}{\omega_o^2 - \omega^2} \tag{9.13}$$

We can also express this in terms of the wavelength $\lambda$. If $\lambda_o = 2\pi c/\omega_o$ is the resonance wavelength, then Equation 9.13 is equivalent to

*Dispersion relation*

$$n^2 = 1 + \left(\frac{NZe^2}{\varepsilon_o m_e}\right)\left(\frac{\lambda_o}{2\pi c}\right)^2 \frac{\lambda^2}{\lambda^2 - \lambda_o^2} \tag{9.14}$$

This type of relationship between $n$ and the frequency $\omega$, or wavelength $\lambda$, is called the **dispersion relation.** Although the above treatment is grossly simplified, it does nonetheless emphasize that $n$ will always be wavelength dependent and will exhibit a

**Table 9.2** Sellmeier and Cauchy coefficients

| | Sellmeier | | | | | |
|---|---|---|---|---|---|---|
| | $A_1$ | $A_2$ | $A_3$ | $\lambda_1$ (μm) | $\lambda_2$ (μm) | $\lambda_3$ (μm) |
| $SiO_2$ (fused silica) | 0.696749 | 0.408218 | 0.890815 | 0.0690660 | 0.115662 | 9.900559 |
| 86.5% $SiO_2$–13.5% $GeO_2$ | 0.711040 | 0.451885 | 0.704048 | 0.0642700 | 0.129408 | 9.425478 |
| $GeO_2$ | 0.80686642 | 0.71815848 | 0.85416831 | 0.068972606 | 0.15396605 | 11.841931 |
| Sapphire | 1.023798 | 1.058264 | 5.280792 | 0.0614482 | 0.110700 | 17.92656 |
| Diamond | 0.3306 | 4.3356 | — | 0.1750 | 0.1060 | — |

| | Cauchy | | | | |
|---|---|---|---|---|---|
| | Range of $h\nu$ (eV) | $n_{-2}$ (eV$^2$) | $n_0$ | $n_2$ (eV$^{-2}$) | $n_{-4}$ (eV$^{-4}$) |
| Diamond | 0.05–5.47 | $-1.07 \times 10^{-5}$ | 2.378 | $8.01 \times 10^{-3}$ | $1.04 \times 10^{-4}$ |
| Silicon | 0.002–1.08 | $-2.04 \times 10^{-8}$ | 3.4189 | $8.15 \times 10^{-2}$ | $1.25 \times 10^{-2}$ |
| Germanium | 0.002–0.75 | $-1.0 \times 10^{-8}$ | 4.003 | $2.2 \times 10^{-1}$ | $1.4 \times 10^{-1}$ |

SOURCE: Sellmeier coefficients combined from various sources. Cauchy coefficients from D. Y. Smith *et al., J. Phys. CM* **13,** 3883, 2001.

substantial increase as the frequency increases toward a natural frequency of the polarization mechanism. In the above example, we considered the electronic polarization of an isolated atom with a well-defined natural frequency $\omega_o$. In the crystal, however, the atoms interact, and further we also have to consider the valence electrons in the bonds. The overall result is that $n$ is a complicated function of the frequency or the wavelength. One possibility is to assume a number of resonant frequencies, that is, not just $\lambda_o$ but a series of resonant frequencies, $\lambda_1, \lambda_2, \ldots$, and then sum the contributions arising from each with some weighing factor $A_1$, $A_2$, etc.,

*Sellmeier equation*

$$n^2 = 1 + \frac{A_1\lambda^2}{\lambda^2 - \lambda_1^2} + \frac{A_2\lambda^2}{\lambda^2 - \lambda_2^2} + \frac{A_3\lambda^2}{\lambda^2 - \lambda_3^2} + \cdots \quad \textbf{[9.15]}$$

where $A_1$, $A_2$, $A_3$ and $\lambda_1$, $\lambda_2$, and $\lambda_3$ are constants, called **Sellmeier coefficients.**[3] Equation 9.15 turns out to be quite a useful semiempirical expression for calculating $n$ at various wavelengths if the Sellmeier coefficients are known. Higher terms involving $A_4$ and higher $A$ coefficients can generally be neglected in representing $n$ versus $\lambda$ behavior over typical wavelengths of interest. For example, for diamond, we only need the $A_1$ and $A_2$ terms. The Sellmeier coefficients are listed in various optical data handbooks.

There is another well-known useful $n$–$\lambda$ dispersion relation due originally to Cauchy (1836), which has the short form given by

*Cauchy short-form dispersion equation*

$$n = A + \frac{B}{\lambda^2} + \frac{C}{\lambda^4} \quad \textbf{[9.16]}$$

[3] This is also known as the Sellmeier–Herzberger formula.

where $A$, $B$, and $C$ are material specific constants. Typically, the Cauchy equation is used in the visible spectrum for various optical glasses. A more general Cauchy dispersion relation is of the form[4]

*Cauchy dispersion equation in photon energy*

$$n = n_{-2}(h\nu)^{-2} + n_0 + n_2(h\nu)^2 + n_4(h\nu)^4 \quad [9.17]$$

where $h\nu$ is the photon energy, and $n_0, n_{-2}, n_2$, and $n_4$ are constants; values for diamond, Si, and Ge are listed in Table 9.2. The general Cauchy equation is usually applicable over a wide photon energy range.

---

**EXAMPLE 9.2**

**GaAs DISPERSION RELATION** For GaAs, from $\lambda = 0.89$ to $4.1\ \mu\text{m}$, the refractive index is given by the following dispersion relation,

*GaAs dispersion relation*

$$n^2 = 7.10 + \frac{3.78\lambda^2}{\lambda^2 - 0.2767} \quad [9.18]$$

where $\lambda$ is in microns ($\mu$m). What is the refractive index of GaAs for light with a photon energy of 1 eV?

**SOLUTION**

At $h\nu = 1$ eV,

$$\lambda = \frac{hc}{h\nu} = \frac{(6.62 \times 10^{-34}\ \text{J s})(3 \times 10^8\ \text{m s}^{-1})}{(1\ \text{eV} \times 1.6 \times 10^{-19}\ \text{J eV}^{-1})} = 1.24\ \mu\text{m}$$

Thus,

$$n^2 = 7.10 + \frac{3.78\lambda^2}{\lambda^2 - 0.2767} = 7.10 + \frac{3.78(1.24)^2}{(1.24)^2 - 0.2767} = 11.71$$

so that $\quad n = 3.42$

Note that the $n$ versus $\lambda$ expression for GaAs is actually a Sellmeier-type formula because when $\lambda^2 \gg \lambda_1^2$, then $A_1$ can be simply lumped with 1 to give $1 + A_1 = 7.10$.

---

**EXAMPLE 9.3**

**SELLMEIER EQUATION AND DIAMOND** The relevant Sellmeier coefficients for diamond are given in Table 9.2. Calculate its refractive index at 550 nm (green light) to three decimal places.

**SOLUTION**

The Sellmeier dispersion relation for diamond is

$$n^2 = 1 + \frac{0.3306\lambda^2}{\lambda^2 - (175\ \text{nm})^2} + \frac{4.3356\lambda^2}{\lambda^2 - (106\ \text{nm})^2}$$

$$= 1 + \frac{0.3306(550\ \text{nm})^2}{(550\ \text{nm})^2 - (175\ \text{nm})^2} + \frac{4.3356(550\ \text{nm})^2}{(550\ \text{nm})^2 - (106\ \text{nm})^2} = 5.8707$$

So that $\quad n = 2.423$

which is about 0.1 percent different than the experimental value of 2.426.

---

[4] D. Y. Smith *et al.*, *J. Phys. CM* **13,** 3883, 2001.

**EXAMPLE 9.4** **CAUCHY EQUATION AND DIAMOND** Using the Cauchy coefficients for diamond in Table 9.2, calculate the refractive index at 550 nm.

**SOLUTION**

At $\lambda = 550$ nm, the photon energy is

$$h\nu = \frac{hc}{\lambda} = \frac{(6.62 \times 10^{-34}\ \text{J s})(3 \times 10^{8}\ \text{m s}^{-1})}{550 \times 10^{-9}\ \text{m}} \times \frac{1}{1.6 \times 10^{-19}\ \text{J eV}^{-1}} = 2.254\ \text{eV}$$

Using the Cauchy dispersion relation for diamond with coefficients from Table 9.2,

$$\begin{aligned} n &= n_{-2}(h\nu)^{-2} + n_0 + n_2(h\nu)^2 + n_4(h\nu)^4 \\ &= (-1.07 \times 10^{-5})(2.254)^{-2} + 2.378 + (8.01 \times 10^{-3})(2.254)^2 + (1.04 \times 10^{-4})(2.254)^4 \\ &= 2.421 \end{aligned}$$

The difference in $n$ from the value in Example 9.3 is 0.08 percent, and is due to the Cauchy coefficients quoted in Table 9.2 being applicable over a wider wavelength range at the expense of some accuracy.

## 9.4 GROUP VELOCITY AND GROUP INDEX

Since there are no perfect monochromatic waves in practice, we have to consider the way in which a group of waves differing slightly in wavelength will travel along the $z$ direction as depicted in Figure 9.5. When two perfectly harmonic waves of frequencies $\omega - \delta\omega$ and $\omega + \delta\omega$ and wavevectors $k - \delta k$ and $k + \delta k$ interfere, as shown in Figure 9.5, they generate a **wavepacket** which contains an oscillating field at the mean frequency $\omega$ that is amplitude modulated by a slowly varying field of frequency $\delta\omega$. The maximum amplitude moves with a wavevector $\delta k$ and thus with a **group velocity** that is given by $\delta\omega/\delta k$, that is,

*Group velocity*

$$v_g = \frac{d\omega}{dk} \qquad [9.19]$$

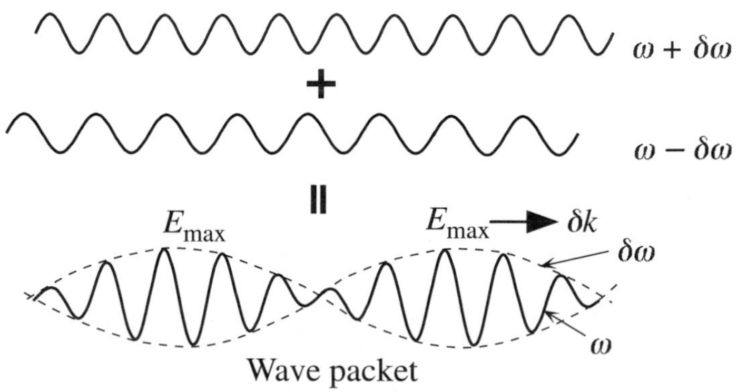


**Figure 9.5** Two slightly different wavelength waves traveling in the same direction result in a wave packet that has an amplitude variation that travels at the group velocity.

The group velocity therefore defines the speed with which energy or information is propagated since it defines the speed of the envelope of the amplitude variation. The maximum electric field in Figure 9.5 advances with a velocity $v_g$, whereas the phase variations in the electric field are propagating at the phase velocity $v$.

Inasmuch as $\omega = vk$ and the phase velocity $v = c/n$, the group velocity in a medium can be readily evaluated from Equation 9.19. In a vacuum, obviously $v$ is simply $c$ and independent of the wavelength or $k$. Thus for waves traveling in a vacuum, $\omega = ck$ and the group velocity is

$$v_g(\text{vacuum}) = \frac{d\omega}{dk} = c = \text{Phase velocity} \tag{9.20}$$

*Group velocity in a vacuum*

On the other hand, suppose that $v$ depends on the wavelength or $k$ by virtue of $n$ being a function of the wavelength as in the case for glasses. Then,

$$\omega = vk = \left[\frac{c}{n(\lambda)}\right]\left(\frac{2\pi}{\lambda}\right) \tag{9.21}$$

where $n = n(\lambda)$ is a function of the wavelength. The group velocity $v_g$ in a medium, from differentiating Equation 9.21 in Equation 9.19, is approximately given by

$$v_g(\text{medium}) = \frac{d\omega}{dk} = \frac{c}{n - \lambda\dfrac{dn}{d\lambda}}$$

This can be written as

$$v_g(\text{medium}) = \frac{c}{N_g} \tag{9.22}$$

*Group velocity in a medium*

where

$$N_g = n - \lambda\frac{dn}{d\lambda} \tag{9.23}$$

*Group index*

is defined as the **group index** of the medium. Equation 9.23 defines the group refractive index $N_g$ of a medium and determines the effect of the medium on the group velocity via Equation 9.22.

In general, for many materials the refractive index $n$ and hence the group index $N_g$ depend on the wavelength of light by virtue of the relative permittivity $\varepsilon_r$ being frequency dependent. Then both the phase velocity $v$ and the group velocity $v_g$ depend on the wavelength and the medium is called a **dispersive medium.** The refractive index $n$ and the group index $N_g$ of pure $SiO_2$ (silica) glass are important parameters in optical fiber design in optical communications. Both of these parameters depend on the wavelength of light as shown in Figure 9.6. Around 1300 nm, $N_g$ is at a minimum which means that for wavelengths close to 1300 nm, $N_g$ is wavelength independent. Thus, light waves with wavelengths around 1300 nm travel with the same group velocity and do not experience dispersion. This phenomenon is significant in the propagation of light in glass fibers used in optical communications.

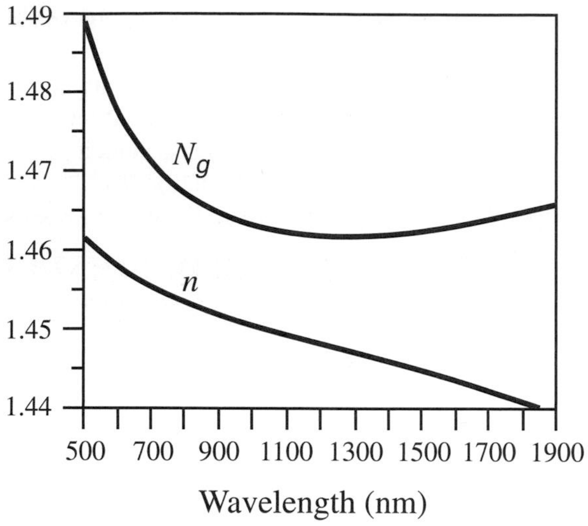


**Figure 9.6** Refractive index $n$ and the group index $N_g$ of pure $SiO_2$ (silica) glass as a function of wavelength.

**EXAMPLE 9.5**

**GROUP VELOCITY** Consider two sinusoidal waves which are close in frequency, that is, waves of frequencies $\omega - \delta\omega$ and $\omega + \delta\omega$ as in Figure 9.5. Their wavevectors will be $k - \delta k$ and $k + \delta k$. The resultant wave will be

$$E_x(z, t) = E_o \cos[(\omega - \delta\omega)t - (k - \delta k)z] + E_o \cos[(\omega + \delta\omega)t - (k + \delta k)z]$$

By using the trigonometric identity $\cos A + \cos B = 2\cos\left[\frac{1}{2}(A - B)\right]\cos\left[\frac{1}{2}(A + B)\right]$ we arrive at

$$E_x(z, t) = 2E_o \cos[(\delta\omega)t - (\delta k)z]\cos[\omega t - kz]$$

As depicted in Figure 9.5, this represents a sinusoidal wave of frequency $\omega$ which is amplitude modulated by a very slowly varying sinusoidal of frequency $\delta\omega$. The system of waves, that is, the modulation, travels along $z$ at a speed determined by the modulating term $\cos[(\delta\omega)t - (\delta k)z]$. The maximum in the field occurs when $[(\delta\omega)t - (\delta k)z] = 2m\pi =$ constant ($m$ is an integer), which travels with a velocity

*Group velocity*

$$\frac{dz}{dt} = \frac{\delta\omega}{\delta k} \quad \text{or} \quad v_g = \frac{d\omega}{dk}$$

This is the group velocity of the waves, as stated in Equation 9.19, since it determines the speed of propagation of the maximum electric field along $z$.

**EXAMPLE 9.6**

**GROUP AND PHASE VELOCITIES** Consider a light wave traveling in a pure $SiO_2$ (silica) glass medium. If the wavelength of light is 1300 nm and the refractive index at this wavelength is 1.447, what is the phase velocity, group index ($N_g$), and group velocity ($v_g$)?

**SOLUTION**

The phase velocity is given by

$$v = \frac{c}{n} = \frac{3 \times 10^8 \text{ m s}^{-1}}{1.447} = 2.073 \times 10^8 \text{ m s}^{-1}$$

From Figure 9.6, at $\lambda = 1300$ nm, $N_g = 1.462$, so

$$v_g = \frac{c}{N_g} = \frac{3 \times 10^8 \text{ m s}^{-1}}{1.462} = 2.052 \times 10^8 \text{ m s}^{-1}$$

The group velocity is ~0.7 percent smaller than the phase velocity.

## 9.5 MAGNETIC FIELD: IRRADIANCE AND POYNTING VECTOR

Although we have considered the electric field component $E_x$ of the EM wave, we should recall that the magnetic field (magnetic induction) component $B_y$ always accompanies $E_x$ in an EM wave propagation. In fact, if $v$ is the phase velocity of an EM wave in an isotropic dielectric medium and $n$ is the refractive index, then according to electromagnetism, at all times and anywhere in an EM wave,[5]

*Fields in an EM wave*

$$E_x = vB_y = \frac{c}{n}B_y \qquad [9.24]$$

where $v = (\varepsilon_o\varepsilon_r\mu_o)^{-1/2}$ and $n = \sqrt{\varepsilon_r}$. Thus, the two fields are simply and intimately related for an EM wave propagating in an isotropic medium. Any process that alters $E_x$ also intimately changes $B_y$ in accordance with Equation 9.24.

As the EM wave propagates in the direction of the wavevector **k** as shown in Figure 9.7, there is an energy flow in this direction. The wave brings with it electromagnetic energy. A small region of space where the electric field is $E_x$ has an energy density, that is, energy per unit volume, given by $\frac{1}{2}\varepsilon_o\varepsilon_r E_x^2$. Similarly, a region of space where the magnetic field is $B_y$ has an energy density $\frac{1}{2}B_y^2/\mu_o$. Since the two fields are related by Equation 9.24, the energy densities in the $E_x$ and $B_y$ fields are the same,

*Energy densities in an EM wave*

$$\frac{1}{2}\varepsilon_o\varepsilon_r E_x^2 = \frac{1}{2\mu_o}B_y^2 \qquad [9.25]$$

The total energy density in the wave is therefore $\varepsilon_o\varepsilon_r E_x^2$. Suppose that an ideal "energy meter" is placed in the path of the EM wave so that the receiving area $A$ of this meter is perpendicular to the direction of propagation. In a time interval $\Delta t$, a portion of the wave of spatial length $v\,\Delta t$ crosses $A$ as shown in Figure 9.7. Thus, a volume $Av\,\Delta t$ of the EM wave crosses $A$ in time $\Delta t$. The energy in this volume consequently becomes received. If $S$ is the EM power flow per unit area,

$$S = \text{Energy flow per unit time per unit area}$$

giving,

$$S = \frac{(Av\,\Delta t)\left(\varepsilon_o\varepsilon_r E_x^2\right)}{A\,\Delta t} = v\varepsilon_o\varepsilon_r E_x^2 = v^2\varepsilon_o\varepsilon_r E_x B_y \qquad [9.26]$$

[5] This is actually a statement of Faraday's law for EM waves. In vector notation it is often expressed as $\omega\mathbf{B} = \mathbf{k} \times \mathbf{E}$.

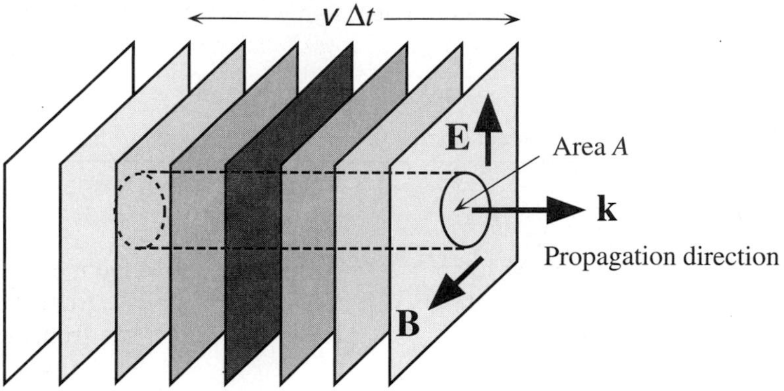


**Figure 9.7** A plane EM wave traveling along **k** crosses an area $A$ at right angles to the direction of propagation. In time $\Delta t$, the energy in the cylindrical volume $Av\,\Delta t$ (shown dashed) flows through $A$.

In an isotropic medium, the energy flow is in the direction of wave propagation. If we use the vectors **E** and **B** to represent the electric and magnetic fields in the EM wave, then the wave propagates in a direction $\mathbf{E} \times \mathbf{B}$, because this direction is perpendicular to both **E** and **B**. The EM power flow per unit area in Equation 9.26 can be written as

*Poynting vector*

$$\mathbf{S} = v^2\varepsilon_o\varepsilon_r\mathbf{E} \times \mathbf{B} \qquad [9.27]$$

where **S**, called the **Poynting vector,** represents the energy flow per unit time per unit area in a direction determined by $\mathbf{E} \times \mathbf{B}$ (direction of propagation). Its magnitude, power flow per unit area, is called the **irradiance.**[6]

The field $E_x$ at the receiver location (say, $z = z_1$) varies sinusoidally which means that the energy flow also varies sinusoidally. The irradiance in Equation 9.26 is the **instantaneous irradiance.** If we write the field as $E_x = E_o \sin(\omega t)$ and then calculate the average irradiance by averaging $S$ over one period, we would find the **average irradiance,**

*Average irradiance (intensity)*

$$I = S_{\text{average}} = \frac{1}{2}v\varepsilon_o\varepsilon_r E_o^2 \qquad [9.28]$$

Since $v = c/n$ and $\varepsilon_r = n^2$ we can write Equation 9.28 as

*Average irradiance (intensity)*

$$I = S_{\text{average}} = \frac{1}{2}c\varepsilon_o n E_o^2$$

$$= (1.33 \times 10^{-3})nE_o^2 \qquad [9.29]$$

The instantaneous irradiance can only be measured if the power meter can respond more quickly than the oscillations of the electric field, and since this is in the

[6] The term *intensity* is widely used and interpreted by many engineers as power flow per unit area even though the strictly correct term is *irradiance*. Many optoelectronic data books simply use intensity to mean irradiance.

optical frequencies range, all practical measurements invariably yield the average irradiance because all detectors have a response rate much slower than the frequency of the wave.

## 9.6 SNELL'S LAW AND TOTAL INTERNAL REFLECTION (TIR)

We consider a traveling plane EM wave in a medium (1) of refractive index $n_1$ propagating toward a medium (2) with a refractive index $n_2$. Constant phase fronts are joined with broken lines, and the wavevector $\mathbf{k}_i$ is perpendicular to the wave fronts as shown in Figure 9.8. When the wave reaches the plane boundary between the two media, a transmitted wave in medium 2 and a reflected wave in medium 1 appear. The transmitted wave is called the **refracted light.** The angles, $\theta_i$, $\theta_t$, $\theta_r$ define the directions of the incident, transmitted, and reflected waves, respectively, with respect to the normal to the boundary plane as shown in Figure 9.8. The wavevectors of the reflected and transmitted waves are denoted as $\mathbf{k}_r$ and $\mathbf{k}_t$, respectively. Since both the incident and reflected waves are in the same medium, the magnitudes of $\mathbf{k}_r$ and $\mathbf{k}_i$ are the same, $k_r = k_i$.

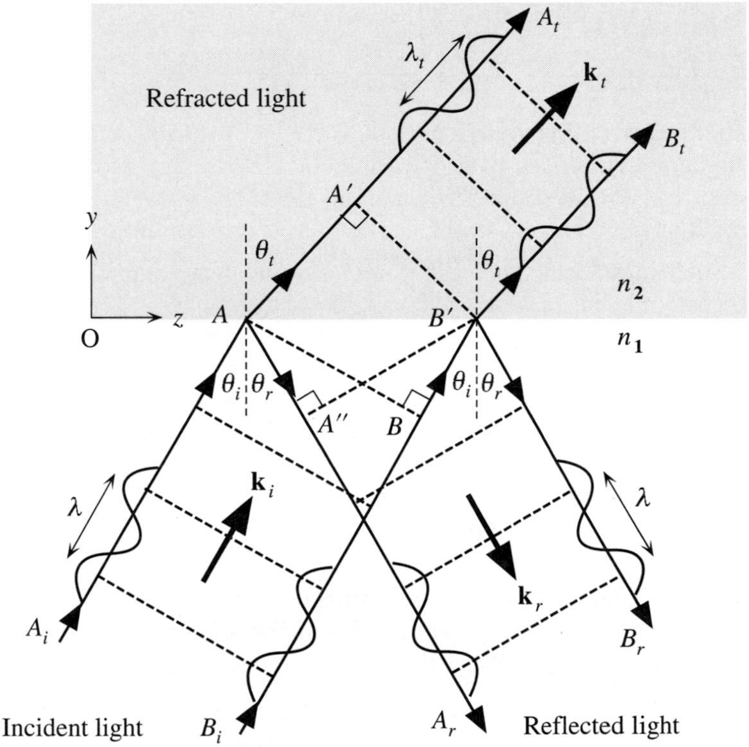


**Figure 9.8** A light wave traveling in a medium with a greater refractive index ($n_1 > n_2$) suffers reflection and refraction at the boundary.

Simple arguments based on constructive interference can be used to show that there can only be one reflected wave that occurs at an angle equal to the incidence angle. The two waves along $A_i$ and $B_i$ are in phase. When these waves are reflected to become waves $A_r$ and $B_r$, then they must still be in phase, otherwise they will interfere destructively and destroy each other. The only way the two waves can stay in phase is if $\theta_r = \theta_i$. All other angles lead to the waves $A_r$ and $B_r$ being out of phase and interfering destructively.

The refracted waves $A_t$ and $B_t$ are propagating in a medium of refracted index $n_2 (< n_1)$ that is different than $n_1$. Hence the waves $A_t$ and $B_t$ have different velocities than $A_i$ and $B_i$. We consider what happens to a wavefront such as $AB$, corresponding perhaps to the maximum field, as it propagates from medium 1 to 2. We recall that the points $A$ and $B$ on this front are always in phase. During the time it takes for the phase $B$ on wave $B_i$ to reach $B'$, phase $A$ on wave $A_t$ has progressed to $A'$. The wavefront $AB$ thus becomes the front $A'B'$ in medium 2. Unless the two waves at $A'$ and $B'$ still have the same phase, there will be no transmitted wave. $A'$ and $B'$ points on the front are only in phase for one particular transmitted angle $\theta_t$.

If it takes time $t$ for the phase at $B$ on wave $B_i$ to reach $B'$, then $BB' = v_1 t = ct/n_1$. During this time $t$, the phase $A$ has progressed to $A'$ where $AA' = v_2 t = ct/n_2$. $A'$ and $B'$ belong to the same front just like $A$ and $B$, so $AB$ is perpendicular to $\mathbf{k}_i$ in medium 1 and $A'B'$ is perpendicular to $\mathbf{k}_t$ in medium 2. From geometrical considerations, $AB' = BB'/\sin\theta_i$ and $AB' = AA'/\sin\theta_t$, so

$$AB' = \frac{v_1 t}{\sin\theta_i} = \frac{v_2 t}{\sin\theta_t}$$

or

*Snell's law*

$$\frac{\sin\theta_i}{\sin\theta_t} = \frac{v_1}{v_2} = \frac{n_2}{n_1} \qquad [9.30]$$

This is **Snell's law**[7] which relates the angles of incidence and refraction to the refractive indices of the media.

If we consider the reflected wave, the wave front $AB$ becomes $A''B'$ in the reflected wave. In time $t$, phase $B$ moves to $B'$ and $A$ moves to $A''$. Since they must still be in phase to constitute the reflected wave, $BB'$ must be equal to $AA''$. Suppose it takes time $t$ for the wavefront $B$ to move to $B'$ (or $A$ to $A''$). Then, since $BB' = AA'' = v_1 t$, from geometrical considerations,

$$AB' = \frac{v_1 t}{\sin\theta_i} = \frac{v_1 t}{\sin\theta_r}$$

so that $\theta_i = \theta_r$. The angles of incidence and reflection are the same.

When $n_1 > n_2$, then obviously the transmitted angle is greater than the incidence angle as apparent in Figure 9.8. When the refraction angle $\theta_t$ reaches 90°, the incidence

[7] Willebrord van Roijen Snell (1581–1626), a Dutch physicist and mathematician, was born in Leiden and eventually became a professor at Leiden University. He obtained his refraction law in 1621 which was published by Réne Descartes in France in 1637; it is not known whether Descartes knew of Snell's law or formulated it independently.

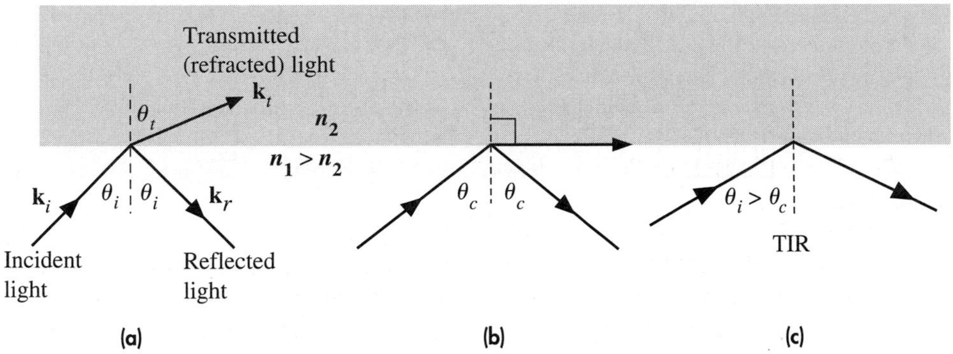


**Figure 9.9** Light wave traveling in a more dense medium strikes a less dense medium. Depending on the incidence angle with respect to $\theta_c$, determined by the ratio of the refractive indices, the wave may be transmitted (refracted) or reflected.
(a) $\theta_i < \theta_c$.
(b) $\theta_i = \theta_c$.
(c) $\theta_i > \theta_c$ and total internal reflection (TIR).

angle is called the **critical angle** $\theta_c$ which is given by

*Critical angle for total internal reflection (TIR)*

$$\sin \theta_c = \frac{n_2}{n_1} \tag{9.31}$$

When the incidence angle $\theta_i$ exceeds $\theta_c$, then there is no transmitted wave but only a reflected wave. The latter phenomenon is called **total internal reflection** (TIR). The effect of increasing the incidence angle is shown in Figure 9.9. It is the TIR phenomenon that leads to the propagation of waves in a dielectric medium surrounded by a medium of smaller refractive index as in optical waveguides (*e.g.*, optical fibers).

**EXAMPLE 9.7**

**OPTICAL FIBERS IN COMMUNICATIONS** Figure 9.10 shows a simplified view of a modern optical communications system. Information is converted into a digital signal (*e.g.*, current pulses) which drives a light emitter such as a semiconductor laser. The light pulses from the emitter are coupled into an **optical fiber,** which acts as a light guide. The optical fiber is a very thin glass fiber [made of silica ($SiO_2$)], almost as thin as your hair, that is able to optically guide the light pulses to their destination. The photodetector at the destination converts the light pulses into an electric signal, which is then decoded into the original information.

The **core** of the optical fiber has a higher refractive index than the surrounding region, which is called the **cladding** as shown in Figure 9.10. Optical fibers for short-distance applications (*e.g.*, communications in local area networks within a large building) usually have a core region that has a diameter of about 100 μm, and the whole fiber would be about 150–200 μm in diameter. The core and cladding refractive indices, $n_1$ and $n_2$, respectively, are normally only 1–3 percent different. The light propagates along the fiber core because light rays experience total internal reflections at the core-cladding interface as shown in Figure 9.10. Only those light rays that can exercise TIR travel along the fiber length and can reach the destination. Consider a fiber with $n_1$(core) = 1.455, and $n_2$(cladding) = 1.440. The critical angle for a ray traveling

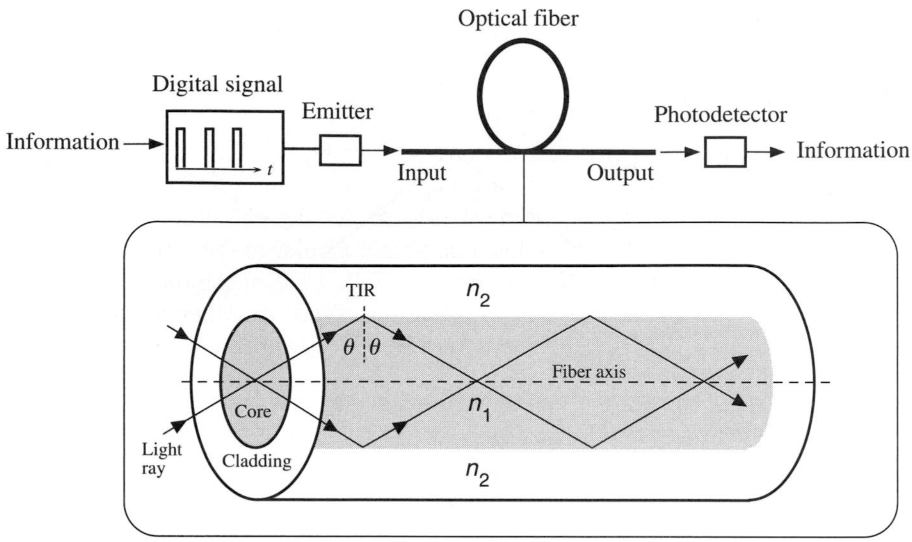


**Figure 9.10** An optical fiber link for transmitting digital information in communications. The fiber core has a higher refractive index, so the light travels along the fiber inside the fiber core by total internal reflection at the core–cladding interface.

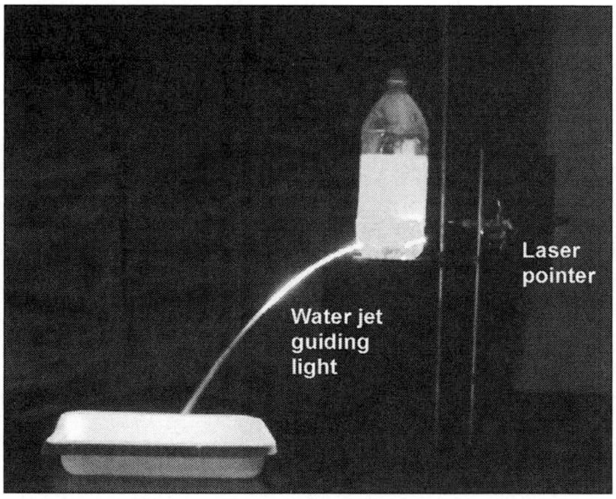


A small hole is made in a plastic bottle full of water to generate a water jet. When the hole is illuminated with a laser beam (from a green laser pointer), the light is guided by total internal reflections along the jet to the tray. Light guiding by a water jet was demonstrated by John Tyndall in 1854 to the Royal Institution. (Water with air bubbles was used to increase the visibility of light, since air bubbles scatter light.)

in the core is

$$\theta_c = \arcsin\left(\frac{n_2}{n_1}\right) = \arcsin\left(\frac{1.440}{1.455}\right) = 81.8^\circ$$

Those light rays that have angles $\theta > \theta_c$ satisfy TIR and can propagate along the fiber.[8] Notice that the ray angles with respect to the fiber axis are less than 8.2°.

[8] The light propagation in an optical fiber is much more complicated than the simple zigzagging of light rays with TIRs at the core–cladding interface. The waves in the core have to satisfy not only TIR but also have to avoid destructive interference so that they are not destroyed as they travel along the guide; see for example, S. O. Kasap, *Optoelectronics and Photonics: Principles and Practices*, Upper Saddle River: Prentice Hall, 2001, chap. 2.

# 9.7 FRESNEL'S EQUATIONS

## 9.7.1 AMPLITUDE REFLECTION AND TRANSMISSION COEFFICIENTS

Although the ray picture with constant phase wave fronts is useful in understanding refraction and reflection, to obtain the magnitude of the reflected and refracted waves and their relative phases, we need to consider the electric field in the light wave. The electric field in the wave must be perpendicular to the direction of propagation as shown in Figure 9.11. We can resolve the field $E_i$ of the incident wave into two components, one in the plane of incidence $E_{i,\parallel}$ and the other perpendicular to the plane of incidence $E_{i,\perp}$. The **plane of incidence** is defined as the plane containing the incident and the reflected rays which in Figure 9.11 corresponds to the plane of the paper.[9] Similarly for both the reflected and transmitted waves, we will have field components parallel and perpendicular to the plane of incidence, *i.e.*, $E_{r,\parallel}$, $E_{r,\perp}$ and $E_{t,\parallel}$, $E_{t,\perp}$.

As apparent from Figure 9.11, the incident, transmitted, and reflected waves all have a wavevector component along the $z$ direction; that is, they have an effective velocity along $z$. The fields $E_{i,\perp}$, $E_{r,\perp}$, and $E_{t,\perp}$ are all perpendicular to the $z$ direction. These waves are called **transverse electric field** (TE) waves. On the other hand, waves with $E_{i,\parallel}$, $E_{r,\parallel}$, and $E_{t,\parallel}$ only have their magnetic field components perpendicular to the $z$ direction and these are called **transverse magnetic field** (TM) waves.

We will describe the incident, reflected, and refracted waves by the exponential representation of a traveling wave, *i.e.*,

*Incident wave*

$$E_i = E_{io} \exp j(\omega t - \mathbf{k}_i \cdot \mathbf{r}) \qquad [9.32]$$

*Reflected wave*

$$E_r = E_{ro} \exp j(\omega t - \mathbf{k}_r \cdot \mathbf{r}) \qquad [9.33]$$

*Transmitted wave*

$$E_t = E_{to} \exp j(\omega t - \mathbf{k}_t \cdot \mathbf{r}) \qquad [9.34]$$

where $\mathbf{r}$ is the position vector; the wavevectors $\mathbf{k}_i$, $\mathbf{k}_r$, and $\mathbf{k}_t$ describe, respectively, the directions of the incident, reflected, and transmitted waves; and $E_{io}$, $E_{ro}$, and $E_{to}$ are the respective amplitudes. Any phase changes such as $\phi_r$ and $\phi_t$ in the reflected and transmitted waves with respect to the phase of the incident wave are incorporated into the complex amplitudes $E_{ro}$ and $E_{to}$. Our objective is to find $E_{ro}$ and $E_{to}$ with respect to $E_{io}$.

We should note that similar equations can be stated for the magnetic field components in the incident, reflected, and transmitted waves, but these will be perpendicular to the corresponding electric fields. The electric and magnetic fields anywhere on the wave must be perpendicular to each other as a requirement of electromagnetic wave theory. This means that with $E_\parallel$ in the EM wave we have a magnetic field $B_\perp$ associated

[9] The definitions of the field components follow those of S. G. Lipson et al., *Optical Physics*, 3rd ed., Cambridge, MA, Cambridge University Press, 1995, and Grant Fowles, *Introduction to Modern Optics*, 2nd ed., New York, Dover Publications, Inc., 1975, whose clear treatments of this subject are highly recommended. The majority of the authors use a different convention which leads to different signs later in the equations; Fresnel's equations are related to the specific electric field directions from which they are derived.

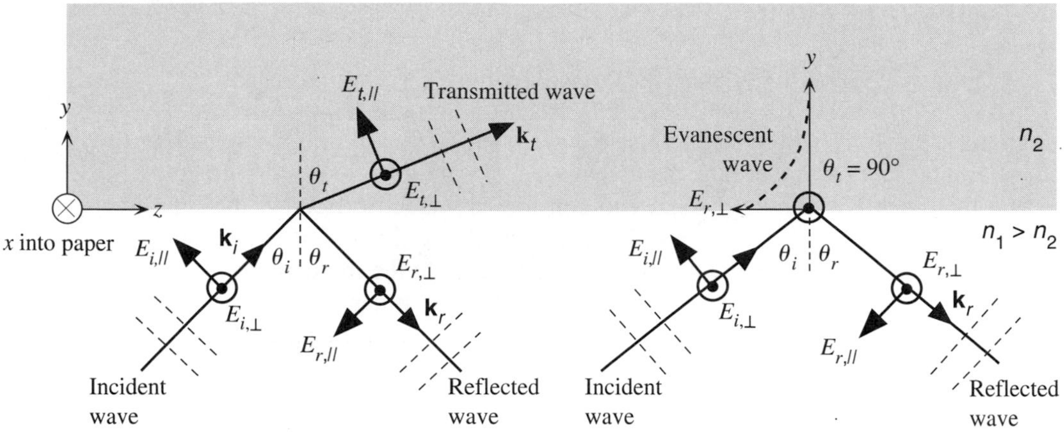


**(a) If $\theta_i < \theta_c$, then some of the wave is transmitted into the less dense medium. Some of the wave is reflected.**

**(b) If $\theta_i > \theta_c$, then the incident wave suffers total internal reflection. There is a decaying evanescent wave into medium 2.**

**Figure 9.11** Light wave traveling in a more dense medium strikes a less dense medium.
The plane of incidence is the plane of the paper and is perpendicular to the flat interface between the two media. The electric field is normal to the direction of propagation. It can be resolved into perpendicular ($\perp$) and parallel ($\|$) components.

with it such that $B_\perp = (n/c)E_\|$. Similarly $E_\perp$ will have a magnetic field $B_\|$ associated with it such that $B_\| = (n/c)E_\perp$.

There are two useful fundamental rules in electromagnetism that govern the behavior of the electric and magnetic fields at a boundary between two dielectric media which we can arbitrarily label as 1 and 2. These rules are called boundary conditions. The first states that the electric field that is tangential to the boundary surface $E_{\text{tangential}}$ must be continuous across the boundary from medium 1 to 2, *i.e.*, at the boundary $y = 0$ in Figure 9.11,

*Boundary condition*

$$E_{\text{tangential}}(1) = E_{\text{tangential}}(2) \qquad [9.35]$$

The second rule is that the tangential component of the magnetic field $B_{\text{tangential}}$ to the boundary must be likewise continuous from medium 1 to 2 provided that the two media are nonmagnetic (relative permeability $\mu_r = 1$),

*Boundary condition*

$$B_{\text{tangential}}(1) = B_{\text{tangential}}(2) \qquad [9.36]$$

Using these boundary conditions for the fields at $y = 0$, and the relationship between the electric and magnetic fields, we can find the reflected and transmitted waves in terms of the incident wave. The boundary conditions can only be satisfied if the reflection and incidence angles are equal, $\theta_r = \theta_i$, and the angles for the transmitted and incident waves obey Snell's law, $n_1 \sin\theta_i = n_2 \sin\theta_r$.

Applying the boundary conditions to the EM wave going from medium 1 to 2, the amplitudes of the reflected and transmitted waves can be readily obtained in terms of

$n_1$, $n_2$, and the incidence angle $\theta_i$ alone.[10] These relationships are called **Fresnel's equations.** If we define $n = n_2/n_1$, as the relative refractive index of medium 2 to that of 1, then the **reflection** and **transmission coefficients** for $E_\perp$ are

$$r_\perp = \frac{E_{r0,\perp}}{E_{i0,\perp}} = \frac{\cos\theta_i - (n^2 - \sin^2\theta_i)^{1/2}}{\cos\theta_i + (n^2 - \sin^2\theta_i)^{1/2}} \quad [9.37]$$

*Reflection coefficient*

and

$$t_\perp = \frac{E_{t0,\perp}}{E_{i0,\perp}} = \frac{2\cos\theta_i}{\cos\theta_i + (n^2 - \sin^2\theta_i)^{1/2}} \quad [9.38]$$

*Transmission coefficient*

There are corresponding coefficients for the $E_\parallel$ fields with corresponding **reflection** and **transmission coefficients** $r_\parallel$ and $t_\parallel$:

$$r_\parallel = \frac{E_{r0,\parallel}}{E_{i0,\parallel}} = \frac{(n^2 - \sin^2\theta_i)^{1/2} - n^2\cos\theta_i}{(n^2 - \sin^2\theta_i)^{1/2} + n^2\cos\theta_i} \quad [9.39]$$

*Reflection coefficient*

$$t_\parallel = \frac{E_{t0,\parallel}}{E_{i0,\parallel}} = \frac{2n\cos\theta_i}{n^2\cos\theta_i + (n^2 - \sin^2\theta_i)^{1/2}} \quad [9.40]$$

*Transmission coefficient*

Further, the reflection and transmission coefficients are related by

$$r_\parallel + nt_\parallel = 1 \qquad \text{and} \qquad r_\perp + 1 = t_\perp \quad [9.41]$$

*Transmission and reflection*

The significance of these equations is that they allow the amplitudes and phases of the reflected and transmitted waves to be determined from the coefficients $r_\perp$, $r_\parallel$, $t_\parallel$, and $t_\perp$. For convenience we take $E_{io}$ to be a real number so that the phase angles of $r_\perp$ and $t_\perp$ correspond to the **phase changes** measured with respect to the incident wave. For example, if $r_\perp$ is a complex quantity, then we can write this as $r_\perp = |r_\perp|\exp(-j\phi_\perp)$ where $|r_\perp|$ and $\phi_\perp$ represent the relative amplitude and phase of the reflected wave with respect to the incident wave for the field perpendicular to the plane of incidence. Of course, when $r_\perp$ is a real quantity, then a positive number represents no phase shift and a negative number is a phase shift of 180° (or $\pi$). As with all waves, a negative sign corresponds to a 180° phase shift. Complex coefficients can only be obtained from Fresnel's equations if the terms under the square roots become negative, and this can only happen when $n < 1$ (or $n_1 > n_2$), and also when $\theta_i > \theta_c$, the critical angle. Thus, phase changes other than 0 or 180° occur only when there is total internal reflection.

Figure 9.12a shows how the magnitudes of the reflection coefficients $|r_\perp|$ and $|r_\parallel|$ vary with the incidence angle $\theta_i$ for a light wave traveling from a more dense medium, $n_1 = 1.44$, to a less dense medium, $n_2 = 1.00$, as predicted by Fresnel's equations. Figure 9.12b shows the changes in the phase of the reflected wave, $\phi_\perp$ and $\phi_\parallel$, with $\theta_i$. The critical angle $\theta_c$ as determined from $\sin\theta_c = n_2/n_1$ in this case is 44°. It is clear that for incidence close to normal (small $\theta_i$), there is no phase change in the reflected wave. For

[10] These equations are readily available in any electromagnetism textbook. Their derivation from the two boundary conditions involves extensive algebraic manipulation which we will not carry out here. The electric and magnetic field components on both sides of the boundary are resolved tangentially to the boundary surface and the boundary conditions are then applied. We then use such relations as $\cos\theta_t = (1 - \sin\theta_t)^{1/2}$ and $\sin\theta_t$ as determined by Snell's law, etc.

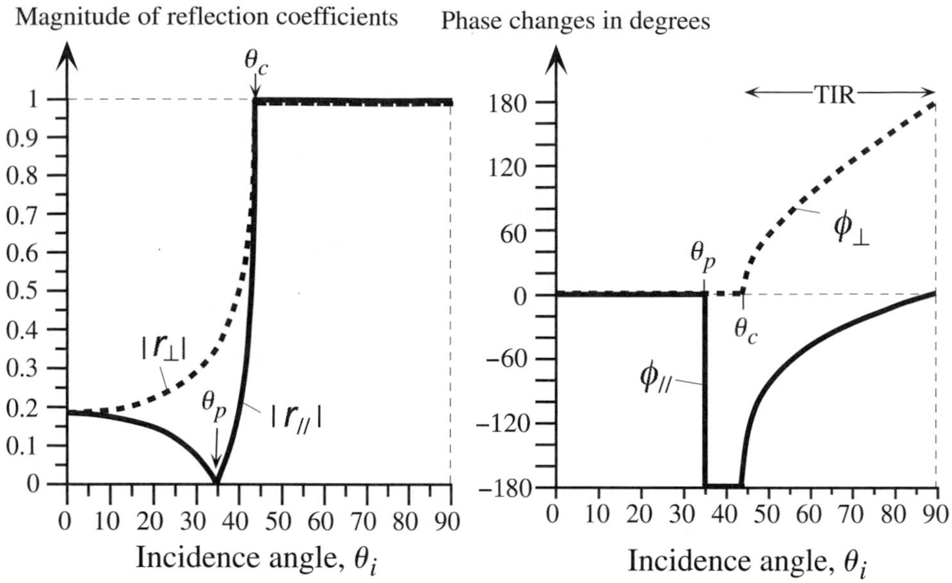


**Figure 9.12** Internal reflection.
(a) Magnitude of the reflection coefficients $r_{\|}$ and $r_{\perp}$ versus the angle of incidence $\theta_i$ for $n_1 = 1.44$ and $n_2 = 1.00$. The critical angle is 44°.
(b) The corresponding phase changes $\phi_{\|}$ and $\phi_{\perp}$ versus incidence angle.

example, putting normal incidence ($\theta_i = 0$) into Fresnel's equations, we find

*Normal incidence*

$$r_{\|} = r_{\perp} = \frac{n_1 - n_2}{n_1 + n_2} \tag{9.42}$$

This is a positive quantity for $n_1 > n_2$ which means that the reflected wave suffers no phase change. This is confirmed by $\phi_{\perp}$ and $\phi_{\|}$ in Figure 9.12b. As the incidence angle increases, eventually $r_{\|}$ becomes zero at an angle of about 35°. We can find this special incidence angle, labeled as $\theta_p$, by solving the Fresnel equation, Equation 9.39, for $r_{\|} = 0$. The field in the reflected wave is then always perpendicular to the plane of incidence and hence well-defined. This special angle is called the **polarization angle** or **Brewster's angle** and from Equation 9.39 is given by

*Brewster's polarization angle*

$$\tan \theta_p = \frac{n_2}{n_1} \tag{9.43}$$

The reflected wave is then said to be **linearly polarized** because it contains *electric field oscillations that are contained within a well-defined plane* which is perpendicular to the plane of incidence and also to the direction of propagation. Electric field oscillations in **unpolarized light,** on the other hand, can be in any one of an infinite number of directions that are perpendicular to the direction of propagation. In linearly polarized light, however, the field oscillations are contained within a well-defined plane. Light emitted from many light sources such as a tungsten light bulb or an LED diode is unpolarized and the field is randomly oriented in a direction that is perpendicular to the direction of propagation.

For incidence angles greater than $\theta_p$ but smaller than $\theta_c$, Fresnel's equation, Equation 9.39, gives a negative number for $r_{\parallel}$ which indicates a phase shift of 180° as shown in $\phi_{\parallel}$ in Figure 9.12b. The magnitudes of both $r_{\parallel}$ and $r_{\perp}$ increase with $\theta_i$ as apparent in Figure 9.12a. At the critical angle and beyond (past 44° in Figure 9.12), *i.e.*, when $\theta_i \geq \theta_c$, the magnitudes of both $r_{\parallel}$ and $r_{\perp}$ go to unity, so the reflected wave has the same amplitude as the incident wave. The incident wave has suffered **total internal reflection** (TIR). When $\theta_i > \theta_c$, in the presence of TIR, the Equations 9.37 to 9.40 are complex quantities because then $\sin\theta_i > n$ and the terms under the square roots become negative. The reflection coefficients become complex quantities of the type $r_{\perp} = 1 \cdot \exp(-j\phi_{\perp})$ and $r_{\parallel} = 1 \cdot \exp(-j\phi_{\parallel})$ with the phase angles $\phi_{\perp}$ and $\phi_{\parallel}$ being other than 0 or 180°. The reflected wave therefore suffers phase changes $\phi_{\perp}$ and $\phi_{\parallel}$ in the components $E_{\perp}$ and $E_{\parallel}$. These phase changes depend on the incidence angle, as apparent in Figure 9.12b, and on $n_1$ and $n_2$.

Examination of Equation 9.37 for $r_{\perp}$ shows that for $\theta_i > \theta_c$, we have $|r_{\perp}| = 1$, but the phase change $\phi_{\perp}$ is given by

$$\tan\left(\frac{1}{2}\phi_{\perp}\right) = \frac{(\sin^2\theta_i - n^2)^{1/2}}{\cos\theta_i} \qquad [9.44]$$

*Phase change in TIR*

For the $E_{\parallel}$ component, the phase change $\phi_{\parallel}$ is given by

$$\tan\left(\frac{1}{2}\phi_{\parallel} + \frac{1}{2}\pi\right) = \frac{(\sin^2\theta_i - n^2)^{1/2}}{n^2\cos\theta_i} \qquad [9.45]$$

*Phase change in TIR*

We can summarize that, in internal reflection ($n_1 > n_2$), the amplitude of the reflected wave from TIR is equal to the amplitude of the incident wave but its phase has shifted by an amount determined by Equations 9.44 and 9.45.[11] The fact that $\phi_{\parallel}$ has an additional $\pi$ shift which makes $\phi_{\parallel}$ negative for $\theta_i > \theta_c$ is due to the choice for the direction of the reflected optical field $E_{r,\parallel}$ in Figure 9.11. (This $\pi$ shift can be ignored if we simply invert $E_{r,\parallel}$.)

The reflection coefficients in Figure 9.12 considered the case in which $n_1 > n_2$. When light approaches the boundary from the higher index side, that is, $n_1 > n_2$, the reflection is said to be **internal reflection** and *at normal incidence there is no phase change*. On the other hand, if light approaches the boundary from the lower index side, that is, $n_1 < n_2$, then it is called **external reflection.** Thus in external reflection light becomes reflected by the surface of an optically denser (higher refractive index) medium. There is an important difference between the two. Figure 9.13 shows how the reflection coefficients $r_{\perp}$ and $r_{\parallel}$ depend on the incidence angle $\theta_i$ for external reflection ($n_1 = 1$ and $n_2 = 1.44$). At normal incidence, both coefficients are negative, which means that *in external reflection at normal incidence there is a phase shift of 180°*. Further, $r_{\parallel}$ goes through zero at the **Brewster angle** $\theta_p$ given by Equation 9.43. At this angle of incidence, the reflected wave is polarized in the $E_{\perp}$ component only. Transmitted light in both internal reflection (when $\theta_i < \theta_c$) and external reflection does not experience a phase shift.

[11] It should be apparent that the concepts and the resulting equations apply to a well-defined linearly polarized light wave.

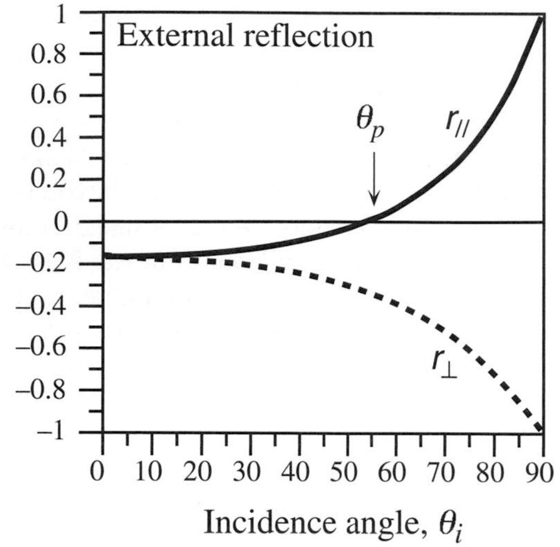


**Figure 9.13** The reflection coefficients $r_{\parallel}$ and $r_{\perp}$ versus angle of incidence $\theta_i$ for $n_1 = 1.00$ and $n_2 = 1.44$.

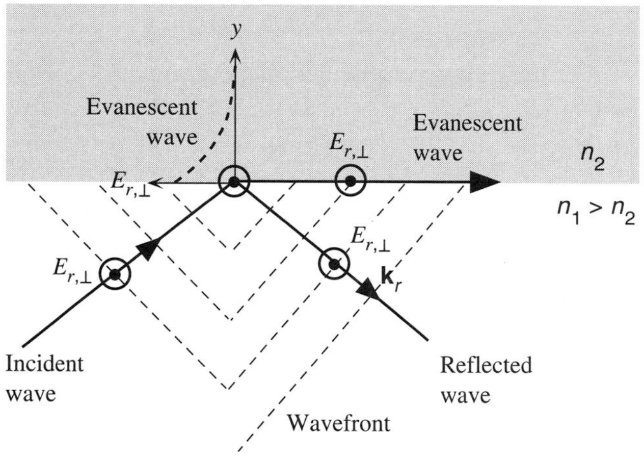


**Figure 9.14** When $\theta_i > \theta_c$, for a plane wave that is reflected, there is an evanescent wave at the boundary propagating along $z$.

What happens to the transmitted wave when $\theta_i > \theta_c$? According to the boundary conditions, there must still be an electric field in medium 2; otherwise, the boundary conditions cannot be satisfied. When $\theta_i > \theta_c$, the field in medium 2 is a wave that travels near the surface of the boundary along the $z$ direction as depicted in Figure 9.14. The wave is called an **evanescent wave** and advances along $z$ with its field decreasing as we move into medium 2, *i.e.*,

*Evanescent wave*

$$E_{t,\perp}(y, z, t) \propto e^{-\alpha_2 y} \exp j(\omega t - k_{iz} z) \qquad [9.46]$$

where $k_{iz} = k_i \sin\theta_i$ is the wavevector of the incident wave along the $z$ axis, and $\alpha_2$ is an **attenuation coefficient** for the electric field penetrating into medium 2,

*Attenuation of evanescent wave*

$$\alpha_2 = \frac{2\pi n_2}{\lambda}\left[\left(\frac{n_1}{n_2}\right)^2 \sin^2\theta_i - 1\right]^{1/2} \qquad [9.47]$$

where $\lambda$ is the free-space wavelength. According to Equation 9.46, the evanescent wave travels along $z$ and has an amplitude that decays exponentially as we move from the boundary into medium 2 (along $y$) as shown in Figure 9.11b. The field of the evanescent wave is $e^{-1}$ in medium 2 when $y = 1/\alpha_2 = \delta$ which is called the **penetration depth.** It is not difficult to show that the evanescent wave is correctly predicted by Snell's law when $\theta_i > \theta_c$. The evanescent wave propagates along the boundary (along $z$) with the same speed as the $z$ component velocity of the incident and reflected waves. In Equations 9.32 to 9.34 we had assumed that the incident and reflected waves were *plane waves,* that is, of infinite extent. If we were to extend the plane wavefronts on the reflected wave, these would cut the boundary as shown in Figure 9.14. The evanescent wave traveling along $z$ can be thought of as arising from these plane wavefronts at the boundary as in Figure 9.14. (The evanescent wave is important in light propagation in optical waveguides such as in optical fibers.) If the incident wave is a narrow beam of light (*e.g.*, from a laser pointer), then the reflected beam would have the same cross section. There would still be an evanescent wave at the boundary, but it would exist only within the cross-sectional area of the reflected beam at the boundary.

## 9.7.2 INTENSITY, REFLECTANCE, AND TRANSMITTANCE

It is frequently necessary to calculate the intensity or irradiance of the reflected and transmitted waves when light traveling in a medium of index $n_1$ is incident at a boundary where the refractive index changes to $n_2$. In some cases we are simply interested in normal incidence where $\theta_i = 0°$. For example, in laser diodes light is reflected from the ends of an optical cavity where there is a change in the refractive index.

**Reflectance** $R$ measures the intensity of the reflected light with respect to that of the incident light and can be defined separately for electric field components parallel and perpendicular to the plane of incidence. The reflectances $R_\perp$ and $R_\parallel$ are defined by

$$R_\perp = \frac{|E_{ro,\perp}|^2}{|E_{io,\perp}|^2} = |r_\perp|^2 \quad \text{and} \quad R_\parallel = \frac{|E_{ro,\parallel}|^2}{|E_{io,\parallel}|^2} = |r_\parallel|^2 \tag{9.48}$$

From Equations 9.37 to 9.40 with normal incidence, these are simply given by

*Reflectance at normal incidence*

$$R = R_\perp = R_\parallel = \left(\frac{n_1 - n_2}{n_1 + n_2}\right)^2 \tag{9.49}$$

Since a glass medium has a refractive index of around 1.5, this means that typically 4 percent of the incident radiation on an air–glass surface will be reflected back.

**Transmittance** $T$ relates the intensity of the transmitted wave to that of the incident wave in a similar fashion to the reflectance. We must, however, consider that the transmitted wave is in a different medium and further its direction with respect to the boundary is also different by virtue of refraction. For normal incidence, the incident and transmitted beams are normal and the transmittances are defined and given by

$$T_\perp = \frac{n_2|E_{to,\perp}|^2}{n_1|E_{io,\perp}|^2} = \left(\frac{n_2}{n_1}\right)|t_\perp|^2 \quad \text{and} \quad T_\parallel = \frac{n_2|E_{to,\parallel}|^2}{n_1|E_{io,\parallel}|^2} = \left(\frac{n_2}{n_1}\right)|t_\parallel|^2 \tag{9.50}$$

*Transmittance at normal incidence*

or

$$T = T_{\perp} = T_{\parallel} = \frac{4n_1n_2}{(n_1+n_2)^2} \qquad [9.51]$$

Further, the fraction of light reflected and fraction transmitted must add to unity. Thus $R + T = 1$.

**EXAMPLE 9.8** **REFLECTION OF LIGHT FROM A LESS DENSE MEDIUM (INTERNAL REFLECTION)** A ray of light which is traveling in a glass medium of refractive index $n_1 = 1.460$ becomes incident on a less dense glass medium of refractive index $n_2 = 1.440$. Suppose that the free-space wavelength ($\lambda$) of the light ray is 1300 nm.

*a.* What should be the minimum incidence angle for TIR?

*b.* What is the phase change in the reflected wave when $\theta_i = 87°$ and when $\theta_i = 90°$?

*c.* What is the penetration depth of the evanescent wave into medium 2 when $\theta_i = 80°$ and when $\theta_i = 90°$?

**SOLUTION**

*a.* The critical angle $\theta_c$ for TIR is given by $\sin\theta_c = n_2/n_1 = 1.440/1.460$, so $\theta_c = 80.51°$.

*b.* Since the incidence angle $\theta_i > \theta_c$, there is a phase shift in the reflected wave. The phase change in $E_{r,\perp}$ is given by $\phi_\perp$. With $n_1 = 1.460$, $n_2 = 1.440$, and $\theta_i = 87°$,

$$\tan\left(\frac{1}{2}\phi_\perp\right) = \frac{(\sin^2\theta_i - n^2)^{1/2}}{\cos\theta_i} = \frac{\left[\sin^2(87°) - \left(\dfrac{1.440}{1.460}\right)^2\right]^{1/2}}{\cos(87°)}$$

$$= 2.989 = \tan\left[\tfrac{1}{2}(143.0°)\right]$$

so the phase change is 143°. For the $E_{r,\parallel}$ component, the phase change is

$$\tan\left(\frac{1}{2}\phi_\parallel + \frac{1}{2}\pi\right) = \frac{(\sin^2\theta_i - n^2)^{1/2}}{n^2\cos\theta_i} = \frac{1}{n^2}\tan\left(\frac{1}{2}\phi_\perp\right)$$

so

$$\tan\left(\frac{1}{2}\phi_\parallel + \frac{1}{2}\pi\right) = \left(\frac{n_1}{n_2}\right)^2\tan\left(\frac{\phi_\perp}{2}\right) = \left(\frac{1.460}{1.440}\right)^2\tan\left[\frac{1}{2}(143°)\right]$$

which gives

$$\phi_\parallel = 143.95° - 180° = -36.05°$$

We can repeat the calculation with $\theta_i = 90°$ to find $\phi_\perp = 180°$ and $\phi_\parallel = 0°$.

Note that as long as $\theta_i > \theta_c$, the magnitude of the reflection coefficients are unity. Only the phase changes.

*c.* The amplitude of the evanescent wave as it penetrates into medium 2 is

$$E_{t,\perp}(y, t) \approx E_{to,\perp}\exp(-\alpha_2 y)$$

We ignore the $z$ dependence, $\exp j(\omega t - k_z z)$, as this only gives a propagating property along $z$. The field strength drops to $e^{-1}$ when $y = 1/\alpha_2 = \delta$, which is called the **penetration**

**depth.** The attenuation constant $\alpha_2$ is

$$\alpha_2 = \frac{2\pi n_2}{\lambda}\left[\left(\frac{n_1}{n_2}\right)^2 \sin^2\theta_i - 1\right]^{1/2}$$

*i.e.*,

$$\alpha_2 = \frac{2\pi(1.440)}{(1300 \times 10^{-9}\ \text{m})}\left[\left(\frac{1.460}{1.440}\right)^2 \sin^2(87°) - 1\right]^{1/2} = 1.104 \times 10^6\ \text{m}^{-1}$$

so the penetration depth is $\delta = 1/\alpha_2 = 1/(1.104 \times 10^6\ \text{m}) = 9.06 \times 10^{-7}$ m, or 0.906 μm. For 90°, repeating the calculation we find $\alpha_2 = 1.164 \times 10^6\ \text{m}^{-1}$, so $\delta = 1/\alpha_2 = 0.859$ μm. We see that the penetration is greater for smaller incidence angles. The values for the refractive indices and wavelength are typical of those values found in optical fiber communications.

---

**REFLECTION AT NORMAL INCIDENCE. INTERNAL AND EXTERNAL REFLECTION** Consider the reflection of light at normal incidence on a boundary between a glass medium of refractive index 1.5 and air of refractive index 1. **EXAMPLE 9.9**

*a.* If light is traveling from air to glass, what is the reflection coefficient and the intensity of the reflected light with respect to that of the incident light?

*b.* If light is traveling from glass to air, what is the reflection coefficient and the intensity of the reflected light with respect to that of the incident light?

*c.* What is the polarization angle in the external reflection in part (*a*)? How would you make a polaroid device that polarizes light based on the polarization angle?

**SOLUTION**

*a.* The light travels in air and becomes partially reflected at the surface of the glass which corresponds to external reflection. Thus $n_1 = 1$ and $n_2 = 1.5$. Then,

$$r_\parallel = r_\perp = \frac{n_1 - n_2}{n_1 + n_2} = \frac{1 - 1.5}{1 + 1.5} = -0.2$$

This is negative which means that there is a 180° phase shift. The reflectance ($R$), which gives the fractional reflected power, is

$$R = r_\parallel^2 = 0.04 \quad \text{or} \quad 4\%$$

*b.* The light travels in glass and becomes partially reflected at the glass–air interface which corresponds to internal reflection. Thus $n_1 = 1.5$ and $n_2 = 1$. Then,

$$r_\parallel = r_\perp = \frac{n_1 - n_2}{n_1 + n_2} = \frac{1.5 - 1}{1.5 + 1} = 0.2$$

There is no phase shift. The reflectance is again 0.04 or 4 percent. In both cases (a) and (b), the amount of reflected light is the same.

*c.* Light is traveling in air and is incident on the glass surface at the polarization angle. Here $n_1 = 1$, $n_2 = 1.5$, and $\tan\theta_p = (n_2/n_1) = 1.5$, so $\theta_p = 56.3°$.

If we were to reflect light from a glass plate keeping the angle of incidence at 56.3°, then the reflected light will be polarized with an electric field component perpendicular to the plane of incidence. The transmitted light will have the field greater in the plane of incidence; that is, it will be partially polarized. By using a stack of glass plates one can increase the polarization of the transmitted light. (This type of *pile-of-plates polarizer* was invented by Dominique F. J. Arago in 1812.)

**EXAMPLE 9.10** **ANTIREFLECTION COATINGS ON SOLAR CELLS** When light is incident on the surface of a semiconductor, it becomes partially reflected. Partial reflection is an important consideration in solar cells where transmitted light energy into the semiconductor device is converted to electric energy. The refractive index of Si is about 3.5 at wavelengths around 700–800 nm. Thus the reflectance with $n_1$(air) = 1 and $n_2$(Si) ≈ 3.5 is

$$R = \left(\frac{n_1 - n_2}{n_1 + n_2}\right)^2 = \left(\frac{1 - 3.5}{1 + 3.5}\right)^2 = 0.309$$

This means that 30 percent of the light is reflected and is not available for conversion to electric energy, a considerable reduction in the efficiency of the solar cell.

However, we can coat the surface of the semiconductor device with a thin layer of a dielectric material such as $Si_3N_4$ (silicon nitride) that has an intermediate refractive index. Figure 9.15 illustrates how the thin dielectric coating reduces the reflected light intensity. In this case $n_1$(air) = 1, $n_2$(coating) ≈ 1.9, and $n_3$(Si) = 3.5. Light is first incident on the air–coating surface, and some of it becomes reflected; this reflected wave is shown as $A$ in Figure 9.15. Wave $A$ has experienced a 180° phase change on reflection as this is an external reflection. The wave that enters and travels in the coating then becomes reflected at the coating–semiconductor surface. This wave, which is shown as $B$, also suffers a 180° phase change since $n_3 > n_2$. When wave $B$ reaches $A$, it has suffered a total delay of traversing the thickness $d$ of the coating twice. The phase difference is equivalent to $k_c(2d)$ where $k_c = 2\pi/\lambda_c$ is the wavevector in the coating and is given by $2\pi/\lambda_c$ where $\lambda_c$ is the wavelength in the coating. Since $\lambda_c = \lambda/n_2$, where $\lambda$ is the free-space wavelength, the phase difference $\Delta\phi$ between $A$ and $B$ is $(2\pi n_2/\lambda)(2d)$. To reduce the reflected light, $A$ and $B$ must interfere destructively, and this requires the phase difference to be $\pi$ or odd multiples of $\pi$, $m\pi$ where $m = 1,3,5,\ldots$ is an odd integer. Thus

$$\left(\frac{2\pi n_2}{\lambda}\right)2d = m\pi \qquad \text{or} \qquad d = m\left(\frac{\lambda}{4n_2}\right)$$

Thus, the thickness of the coating must be multiples of the quarter wavelength in the coating and depends on the wavelength.

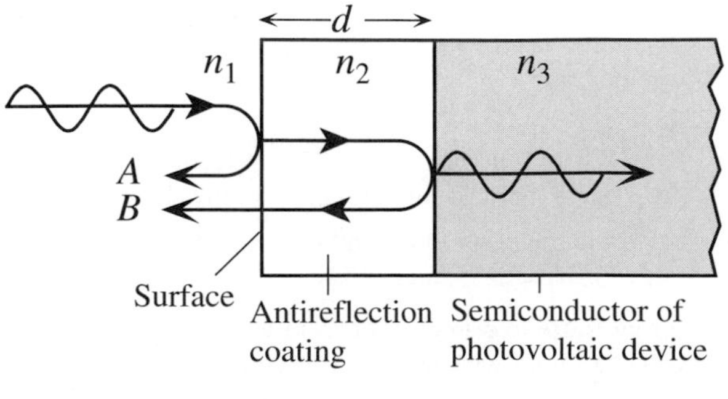


**Figure 9.15** Illustration of how an antireflection coating reduces the reflected light intensity.

To obtain a good degree of destructive interference between waves $A$ and $B$, the two amplitudes must be comparable. It turns out that we need $n_2 = \sqrt{n_1 n_3}$. When $n_2 = \sqrt{n_1 n_3}$, then the reflection coefficient between the air and coating is equal to that between the coating and the semiconductor. In this case we would need $\sqrt{3.5}$ or 1.87. Thus, $Si_3N_4$ is a good choice as an antireflection coating material on Si solar cells.

Taking the wavelength to be 700 nm, $d = (700\text{ nm})/[4(1.9)] = 92.1$ nm or odd multiples of $d$.

**EXAMPLE 9.11**

**DIELECTRIC MIRRORS** A **dielectric mirror** consists of a stack of dielectric layers of alternating refractive indices as schematically illustrated in Figure 9.16 where $n_1$ is smaller than $n_2$. The thickness of each layer is a quarter wavelength or $\lambda_{\text{layer}}/4$, where $\lambda_{\text{layer}}$ is the wavelength of light in that layer, or $\lambda_o/n$, where $\lambda_o$ is the free-space wavelength at which the mirror is required to reflect the incident light and $n$ is the refractive index of the layer. Reflected waves from the interfaces interfere constructively and give rise to a substantial reflected light. If there are a sufficient number of layers, the reflectance can approach unity at the wavelength $\lambda_o$. Figure 9.16 also shows schematically a typical reflectance versus wavelength behavior of a dielectric mirror with many layers.

The reflection coefficient $r_{12}$ for light in layer 1 being reflected at the 1–2 boundary is $r_{12} = (n_1 - n_2)/(n_1 + n_2)$ and is a negative number indicating a $\pi$ phase change. The reflection coefficient for light in layer 2 being reflected at the 2–1 boundary is $r_{21} = (n_2 - n_1)/(n_1 + n_2)$ which is $-r_{12}$ (positive) indicating no phase change. Thus the reflection coefficient alternates in sign through the mirror. Consider two arbitrary waves $A$ and $B$ which are reflected at two consecutive interfaces. The two waves are therefore already out of phase by $\pi$ due to reflections at the different boundaries. Further, wave $B$ travels an additional distance which is twice $(\lambda_2/4)$ before reaching wave $A$ and therefore experiences a phase change equivalent to $2(\lambda_2/4)$ or $\lambda_2/2$, that is, $\pi$. The phase difference between $A$ and $B$ is then $\pi + \pi$ or $2\pi$. Thus waves $A$ and $B$ are in phase and *interfere constructively*. We can similarly show that waves $B$ and $C$ also interfere constructively and so on, so all reflected waves from the consecutive boundaries interfere constructively. After several layers (depending on $n_1$ and $n_2$), the transmitted intensity will be very small and the reflected light intensity will be close to unity. Dielectric mirrors are widely used in modern vertical cavity surface emitting semiconductor lasers.

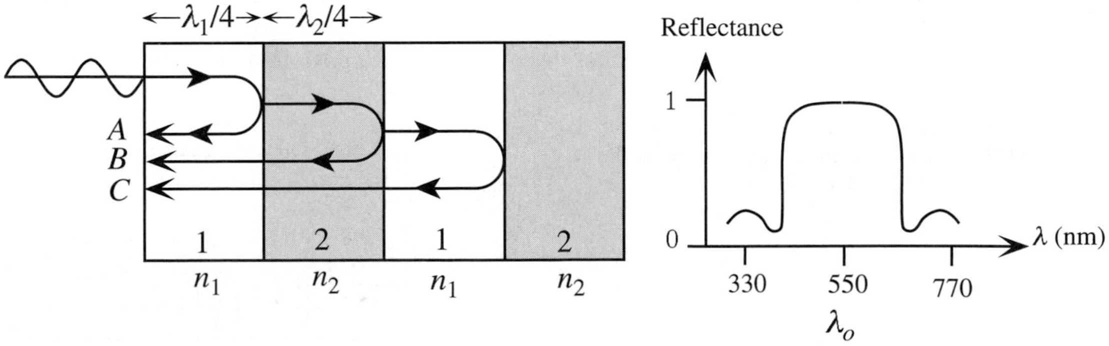


**Figure 9.16** Schematic illustration of the principle of the dielectric mirror with many low and high refractive index layers and its reflectance.

## 9.8 COMPLEX REFRACTIVE INDEX AND LIGHT ABSORPTION

Generally when light propagates through a material, it becomes *attenuated* in the direction of propagation as illustrated in Figure 9.17. We distinguish between *absorption* and *scattering* both of which give rise to a loss of intensity in the regular direction of propagation. In **absorption,** the loss in the power in the propagating EM wave is due to the conversion of light energy to other forms of energy, *e.g.*, lattice vibrations (heat) during the polarization of the molecules of the medium, local vibrations of impurity ions, and excitation of electrons from the valence band to the conduction band. On the other hand, **scattering** is a process by which the energy from a propagating EM wave is redirected as secondary EM waves in various directions away from the original direction of propagation; this is discussed in Section 9.11.

It is instructive to consider what happens when a monochromatic light wave such as

*Lossless propagation*

$$E = E_o \exp j(\omega t - kz) \quad [9.52]$$

is propagating in a dielectric medium. The electric field $E$ in Equation 9.52 is either parallel to $x$ or $y$ since propagation is along $z$. As the wave travels through the medium, the molecules become polarized. This polarization effect is represented by the *relative permittivity* $\varepsilon_r$ of the medium. If there were no losses in the polarization process, then the relative permittivity $\varepsilon_r$ would be a real number and the corresponding refractive index $n = \sqrt{\varepsilon_r}$ would also be a real number. However, we know that there are always some losses in all polarization processes. For example, when the ions of an ionic crystal are displaced from their equilibrium positions by an alternating electric field and made to oscillate, some of the energy from the electric field is coupled and converted to lattice vibrations (intuitively, "sound" and heat). These losses are generally accounted for by describing the whole medium in terms of a **complex relative permittivity** (or **dielectric constant**) $\varepsilon_r$, that is,

*Complex dielectric constant*

$$\varepsilon_r = \varepsilon_r' - j\varepsilon_r'' \quad [9.53]$$

where the real part $\varepsilon_r'$ determines the polarization of the medium with losses ignored and the imaginary part $\varepsilon_r''$ describes the losses in the medium. For a lossless medium, obviously $\varepsilon_r = \varepsilon_r'$. The loss $\varepsilon_r''$ depends on the frequency of the wave and usually peaks at certain natural (resonant) frequencies. If the medium has a finite conductivity

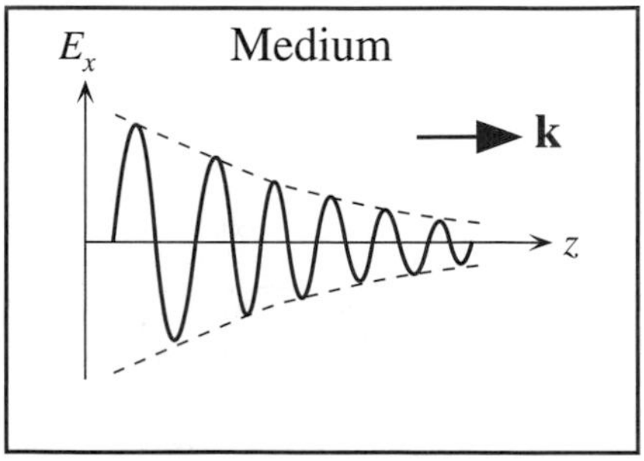


**Figure 9.17** Attenuation of light in the direction of propagation.

(*e.g.*, due to a small number of conduction electrons), then there will be a Joule loss due to the electric field in the wave driving these conduction electrons. This type of light attenuation is called **free carrier absorption.** In such cases, $\varepsilon_r''$ and $\sigma$ are related by

$$\varepsilon_r'' = \frac{\sigma}{\varepsilon_o \omega} \quad [9.54]$$

*Conduction loss*

where $\varepsilon_o$ is the absolute permittivity and $\sigma$ is the conductivity at the frequency of the EM wave. Since $\varepsilon_r$ is a complex quantity, we should also expect to have a *complex refractive index.*

An EM wave that is traveling in a medium and experiencing attenuation due to absorption can be generally described by a **complex propagation constant** $k$, that is,

$$k = k' - jk'' \quad [9.55]$$

*Complex propagation constant*

where $k'$ and $k''$ are the real and imaginary parts. If we put Equation 9.55 into Equation 9.52, we will find the following,

$$E = E_o \exp(-k''z) \exp j(\omega t - k'z) \quad [9.56]$$

*Attenuated propagation*

The amplitude decays exponentially while the wave propagates along $z$. The **real** $k'$ part of the complex propagation constant (wavevector) describes the propagation characteristics, *e.g.*, phase velocity $v = \omega/k'$. The **imaginary** $k''$ part describes the rate of attenuation along $z$. The intensity $I$ at any point along $z$ is

$$I \propto |E|^2 \propto \exp(-2k''z)$$

so the rate of change in the intensity with distance is

$$\frac{dI}{dz} = -2k''I \quad [9.57]$$

*Imaginary part k''*

where the negative sign represents attenuation.

Suppose that $k_o$ is the propagation constant in a vacuum. This is a real quantity as a plane wave suffers no loss in free space. The **complex refractive index** N with real part $n$ and imaginary part $K$ is defined as the ratio of the complex propagation constant in a medium to propagation constant in free space,

$$\mathrm{N} = n - jK = \frac{k}{k_o} = \left(\frac{1}{k_o}\right)[k' - jk''] \quad [9.58a]$$

*Complex refractive index*

*i.e.*,

$$n = \frac{k'}{k_o} \quad \text{and} \quad K = \frac{k''}{k_o} \quad [9.58b]$$

*Refractive index and extinction coefficient*

The real part $n$ is simply and generally called the **refractive index** and $K$ is called the **extinction coefficient.** In the absence of attenuation,

$$k'' = 0 \qquad k = k' \qquad \text{and} \qquad \mathrm{N} = n = \frac{k}{k_o} = \frac{k'}{k_o}$$

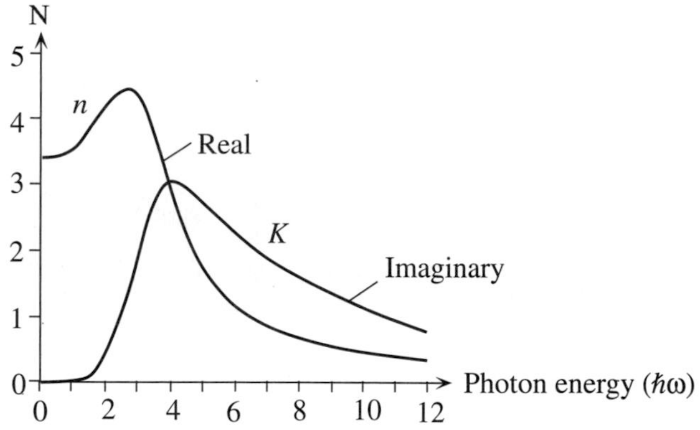


**Figure 9.18** Optical properties of an amorphous silicon film in terms of real ($n$) and imaginary ($K$) parts of the complex refractive index.

We know that in the absence of loss, the relationship between the refractive index $n$ and the relative permittivity $\varepsilon_r$ is $n = \sqrt{\varepsilon_r}$. This relationship is also valid in the presence of loss except that we must use complex refractive index and complex relative permittivity, that is,

*Complex refractive index*

$$\mathrm{N} = n - jK = \sqrt{\varepsilon_r} = \sqrt{\varepsilon_r' - j\varepsilon_r''} \qquad [9.59]$$

By squaring both sides we can relate $n$ and $K$ directly to $\varepsilon_r'$ and $\varepsilon_r''$. The final result is

*Complex refractive index*

$$n^2 - K^2 = \varepsilon_r' \qquad \text{and} \qquad 2nK = \varepsilon_r'' \qquad [9.60]$$

Optical properties of materials are typically reported either by showing the frequency dependences of $n$ and $K$ or $\varepsilon_r'$ and $\varepsilon_r''$. Clearly we can use Equation 9.60 to obtain one set of properties from the other. Figure 9.18 shows the real ($n$) and imaginary ($K$) parts of the complex refractive index of amorphous silicon (noncrystalline form of Si) as a function of photon energy ($h\nu$). For photon energies below the bandgap energy, $K$ is negligible and $n$ is close to 3.5. Both $n$ and $K$ change strongly as the photon energy increases far beyond the bandgap energy.

If we know the frequency dependence of the real part $\varepsilon_r'$ of the relative permittivity of a material, we can also determine the frequency dependence of the imaginary part $\varepsilon_r''$, and vice versa. This may seem remarkable, but it is true provided that we know the frequency dependence of either the real or imaginary part over as wide a range of frequencies as possible (ideally from dc to infinity) and the material is *linear, i.e.*, it has a relative permittivity that is independent of the applied field; the polarization response must be linearly proportional to the applied field.[12] The relationships that relate the real and imaginary parts of the relative permittivity are called **Kramers–Kronig relations.** If $\varepsilon_r'(\omega)$ and $\varepsilon_r''(\omega)$ represent the frequency dependences of the real and imaginary parts, respectively, then one can be determined from the other as depicted schematically in Figure 9.19.

The optical properties $n$ and $K$ can be determined by measuring the reflectance from the surface of a material as a function of polarization and the angle of incidence (based on Fresnel's equations).

[12] In addition the material system should be passive—contain no sources of energy.

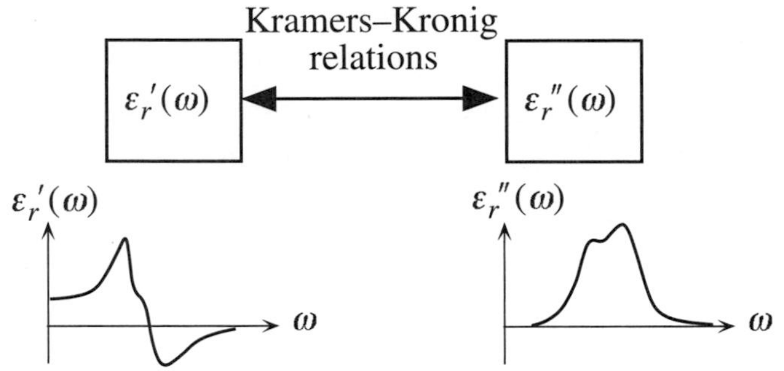


**Figure 9.19** Kramers–Kronig relations allow frequency dependences of the real and imaginary parts of the relative permittivity to be related to each other. The material must be a linear system.

It is instructive to mention that the reflection and transmission coefficients that we derived in Section 9.7 were based on using a real refractive index, that is, neglecting losses. We can still use the reflection and transmission coefficients if we simply use the complex refractive index N instead of $n$. For example, consider a light wave traveling in free space incident on a material at normal incidence ($\theta_i = 90°$). The reflection coefficient is now

$$r = -\frac{\mathrm{N}-1}{\mathrm{N}+1} = -\frac{n - jK - 1}{n - jK + 1} \qquad [9.61]$$

*Reflection coefficient*

The reflectance is then

$$R = \left|\frac{n - jK - 1}{n - jK + 1}\right|^2 = \frac{(n-1)^2 + K^2}{(n+1)^2 + K^2} \qquad [9.62]$$

*Reflectance*

which reduce to the usual forms when the extinction coefficient $K = 0$.

**EXAMPLE 9.12**

**COMPLEX REFRACTIVE INDEX** Spectroscopic ellipsometry measurements on a silicon crystal at a wavelength of 826.6 nm show that the real and imaginary parts of the complex relative permittivity are 13.488 and 0.038, respectively. Find the complex refractive index, the reflectance and the absorption coefficient $\alpha$ at this wavelength, and the phase velocity.

**SOLUTION**

We know that $\varepsilon_r' = 13.488$ and $\varepsilon_r'' = 0.038$. Thus, from Equation 9.60, we have

$$n^2 - K^2 = 13.488 \qquad \text{and} \qquad 2nK = 0.038$$

We can take $K$ from the second equation and substitute for it in the first equation,

$$n^2 - \left(\frac{0.038}{2n}\right)^2 = 13.488$$

This is a quadratic equation in $n^2$ that can be easily solved on a calculator to find $n = 3.67$. Once we know $n$, we can find $K = 0.038/2n = 0.00517$. If we simply take the square root of the real part of $\varepsilon_r$, we would still find $n = 3.67$, because the extinction coefficient $K$ is small. The reflectance of the Si crystal is

$$R = \frac{(n-1)^2 + K^2}{(n+1)^2 + K^2} = \frac{(3.67-1)^2 + 0.00517^2}{(3.67+1)^2 + 0.00517^2} = 0.327$$

which is the same as simply using $(n-1)^2/(n+1)^2 = 0.327$, because $K$ is small.

The absorption coefficient $\alpha$ describes the loss in the light intensity $I$ via $I = I_o \exp(-\alpha z)$. By virtue of Equation 9.57,

$$\alpha = 2k'' = 2k_o K = 2\left(\frac{2\pi}{826.6 \times 10^{-9}}\right)(0.00517) = 7.9 \times 10^4 \text{ m}^{-1}$$

Almost all of this absorption is due to band-to-band absorption (photogeneration of electron–hole pairs).

The phase velocity is given by

$$v = \frac{c}{n} = \frac{3 \times 10^8 \text{ m s}^{-1}}{3.67} = 8.17 \times 10^7 \text{ m s}^{-1}$$

**EXAMPLE 9.13** **COMPLEX REFRACTIVE INDEX OF InP** An InP crystal has a refractive index (real part) $n$ of 3.549 at a wavelength of 620 nm (photon energy of 2 eV). The reflectance of the air–InP crystal surface at this wavelength is 0.317. Calculate the extinction coefficient $K$ and the absorption coefficient $\alpha$ of InP at this wavelength.

**SOLUTION**

The reflectance $R$ is given by

$$R = \frac{(n-1)^2 + K^2}{(n+1)^2 + K^2} \quad \text{or} \quad 0.317 = \frac{(3.549-1)^2 + K^2}{(3.549+1)^2 + K^2}$$

which on solving gives $K = 0.302$.

The absorption coefficient is

$$\alpha = 2k_o K = 2\left(\frac{2\pi}{620 \times 10^{-9}}\right)(0.302) = 6.1 \times 10^6 \text{ m}^{-1}$$

**EXAMPLE 9.14** **FREE CARRIER ABSORPTION COEFFICIENT AND CONDUCTIVITY** Consider a semiconductor sample with a conductivity $\sigma$, and a refractive index $n$. Show that the absorption coefficient due to free carrier absorption (due to conductivity) is given by

$$\alpha = \left(\frac{1}{c\varepsilon_o}\right)\frac{\sigma}{n}$$

An $n$-type Ge has a resistivity of about $5 \times 10^{-3}$ Ω m. Calculate the imaginary part $\varepsilon_r''$ of the relative permittivity at a wavelength of 10 μm where the refractive index is 4. Find the attenuation coefficient $\alpha$ due to free carrier absorption.

**SOLUTION**

*Imaginary relative permittivity and conductivity*

The relationship between the conductivity and the absorption coefficient is given by

$$\varepsilon_r'' = \frac{\sigma}{\varepsilon_o \omega} \tag{9.63}$$

The relationship between the imaginary part $\varepsilon_r''$ of the relative permittivity and the extinction coefficient $K$ is

$$2nK = \varepsilon_r''$$

where $n$ is the refractive index (the real part of N). Since the absorption coefficient from Example 9.13 is

$$\alpha = 2k'' = 2k_o K = 2\left(\frac{2\pi}{\lambda}\right)\left(\frac{\varepsilon_r''}{2n}\right)$$

then

$$\alpha = \left(\frac{\omega}{c}\right)\frac{\varepsilon_r''}{n} \quad [9.64]$$

*Absorption and imaginary relative permittivity*

where $\omega$ is the angular frequency of the EM radiation, $\omega = 2\pi c/\lambda$. Substituting for $\sigma$ in terms of $\varepsilon_r''$ gives

$$\alpha = \left(\frac{1}{c\varepsilon_o}\right)\frac{\sigma}{n} \quad [9.65]$$

*Absorption and conductivity*

The frequency $\omega$ is

$$\omega = \frac{2\pi c}{\lambda} = \left[\frac{2\pi(3\times10^8\ \text{m s}^{-1})}{10\times10^{-6}\ \text{m}}\right] = 1.88\times10^{14}\ \text{rad s}^{-1}$$

The relationship between the conductivity and $\varepsilon_r''$ is given by

$$\varepsilon_r'' = \frac{\sigma}{\varepsilon_o\omega} = \left[\frac{(5\times10^{-3}\ \Omega\ \text{m})^{-1}}{(8.85\times10^{-12}\ \text{F m}^{-1})(1.88\times10^{14}\ \text{rad s}^{-1})}\right]$$

*i.e.,* $\varepsilon_r'' = 0.120$

The absorption coefficient due to free carriers is given by

$$\alpha = \left(\frac{1}{c\varepsilon_o}\right)\frac{\sigma}{n} = \left[\frac{1}{(3\times10^8\ \text{m s}^{-1})(8.85\times10^{-12}\ \text{F m}^{-1})}\right]\frac{(5\times10^{-3}\ \Omega\ \text{m})^{-1}}{4} = 1.9\times10^4\ \text{m}^{-1}$$

**EXAMPLE 9.15**

**COMPLEX REFRACTIVE INDEX AND RESONANCE ABSORPTION** Equation 9.12 is a simple expression for the electronic polarizability $\alpha_e$ due to an oscillating field. It is based on the *Lorentz model* in which there is a restoring force acting against polarization of the atom or the molecule. $\omega_o$ is a *resonant frequency,* or a natural frequency, associated with this type of electronic polarization. The same type of expression will also apply to *ionic polarization,* except that the resonant frequency $\omega_o$ will be lower, and the mass $m_e$ has to be changed to an effective mass of the ions.[13] In practice there will be some loss mechanism that absorbs energy from the oscillating field and dissipates it. For example, in ionic polarization, this would involve energy transfer from light to lattice vibrations. In mechanics it is well known that the loss forces (frictional forces) are always proportional to the velocity $dx/dt$. If we include the energy loss in ac polarization, Equation 9.11 would have an additional term $-\gamma\ dx/dt$ on the right-hand side. If we then follow the same steps to obtain $\alpha_e$, we would find

$$\alpha_e = \frac{Ze^2}{m_e(\omega_o^2 - \omega^2 + j\gamma\omega)} \quad [9.66]$$

*Electronic polarizability with loss*

which is a complex number with real and imaginary parts ($\alpha_e = \alpha_e' - j\alpha_e''$).

[13] Both electronic and ionic polarizabilities have similar expressions. The ionic polarizability in an oscillating field was derived in Chapter 7, and looks almost exactly like Equation 9.66.

Since $\alpha_e$ is a complex quantity, so is $\varepsilon_r$, and hence the refractive index. Consider the simplest relationship between the relative permittivity $\varepsilon_r$ and polarizability $\alpha_e$,

*Relative permittivity*

$$\varepsilon_r = 1 + \frac{N}{\varepsilon_o}\alpha_e \qquad [9.67]$$

where $N$ is the number of atoms per unit volume (or ion pairs per unit volume for ionic polarization). Thus, the relative permittivity is a *complex quantity,* that is $\varepsilon_r = \varepsilon_r' - j\varepsilon_r''$. We can substitute from Equation 9.66 into 9.67, and also use the fact that when $\omega = 0$, $\varepsilon_r = \varepsilon_{rdc}$, to obtain a simple expression for $\varepsilon_r$,

*Complex relative permittivity*

$$\varepsilon_r = 1 + \frac{\varepsilon_{rdc} - 1}{1 - \left(\dfrac{\omega}{\omega_o}\right)^2 + j\dfrac{\gamma\omega}{\omega_o^2}} \qquad [9.68]$$

The relationship between the complex refractive index N and the complex relative permittivity $\varepsilon_r$ is

*Complex refractive index*

$$\mathrm{N} = n - jK = \varepsilon_r^{1/2} = (\varepsilon_r' - j\varepsilon_r'')^{1/2} \qquad [9.69]$$

Suppose for simplicity we consider ionic polarization, and we set $\varepsilon_{rdc} = 9$ and $\gamma = 0.1\omega_o$ (reasonable values for ionic polarization). We can calculate $\varepsilon_r$ from Equation 9.68 for any choice of $\omega/\omega_o$ (or for $\omega$ by taking $\omega_o = 1$), and then calculate N, that is $n$ and $K$. (Our calculator or the math program must be able to handle complex numbers.) Figure 9.20a shows the dependence of $n$ and $K$ on the frequency $\omega/\omega_o$ for the simple Lorentz oscillator model in Equation 9.68. Notice how $n$ and the extinction coefficient $K$ peak close to $\omega = \omega_o$.

The reflectance from Equation 9.62 is plotted in Figure 9.20b as $R$ versus $\omega/\omega_o$. It is apparent that $R$ reaches its maximum value at a frequency slightly above $\omega = \omega_o$, and then remains high until $\omega$ reaches nearly $3\omega_o$; the *reflectance is substantial while absorption is strong*. It may

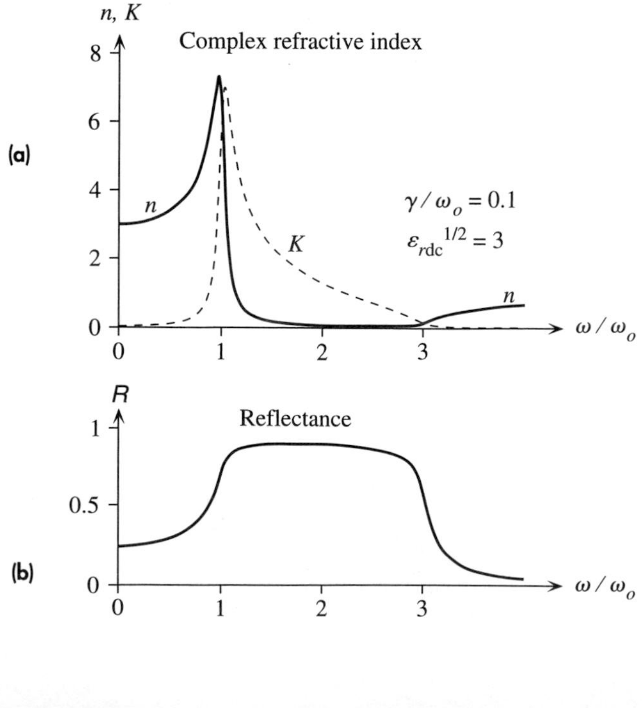


**Figure 9.20**
(a) Refractive index and extinction coefficient versus normalized frequency, $\omega/\omega_o$.
(b) Reflectance versus normalized frequency.

seem strange that the crystal is both highly reflecting and highly absorbing. The light that is incident is strongly reflected, and the light that is inside the crystal becomes strongly absorbed. This phenomenon is known as **infrared reflectance,** and occurs over a band of frequencies, called the **Reststrahlen band;** in the present case from $\omega_o$ to roughly $3\omega_o$.

## 9.9 LATTICE ABSORPTION

In optical absorption, some of the energy from the propagating EM wave is converted to other forms of energy, for example, to heat by the generation of lattice vibrations. There are a number of absorption processes that dissipate the energy from the wave. One important mechanism is called **lattice absorption (Reststrahlen absorption)** and involves the vibrations of the lattice atoms as illustrated in Figure 9.21. The crystal in this example consists of ions, and as an EM wave propagates it displaces the oppositely charged ions in opposite directions and forces them to vibrate at the frequency of the wave. In other words, the medium experiences *ionic polarization.* It is the displacements of these ions that give rise to ionic polarization and its contribution to the relative permittivity $\varepsilon_r$. As the ions and hence the lattice is made to vibrate by the passing EM wave, as shown in Figure 9.21, some energy is coupled into the natural lattice vibrations of the solid. This energy peaks when the frequency of the wave is close to the natural lattice vibration frequencies. Typically these frequencies are in the *infrared region.* Most of the energy is then absorbed from the EM wave and converted to lattice vibrational energy (heat). We associate this absorption with the resonance peak or relaxation peak of ionic polarization loss (imaginary part of the relative permittivity $\varepsilon_r''$).

Figure 9.22 shows the infrared resonance absorption peaks in the extinction coefficient $K$ versus wavelength characteristics of GaAs and CdTe; both crystals have substantial ionic bonding. These absorption peaks in Figure 9.22 are usually called **Reststrahlen bands** because absorption occurs over a band of frequencies (even though the band may be narrow), and in some cases may even have identifiable features. Indeed, if we were to plot the reflectance ($R$) versus wavelength, it would be similar to that shown in Figure 9.20b, and the band would be identified with the high reflectance region.

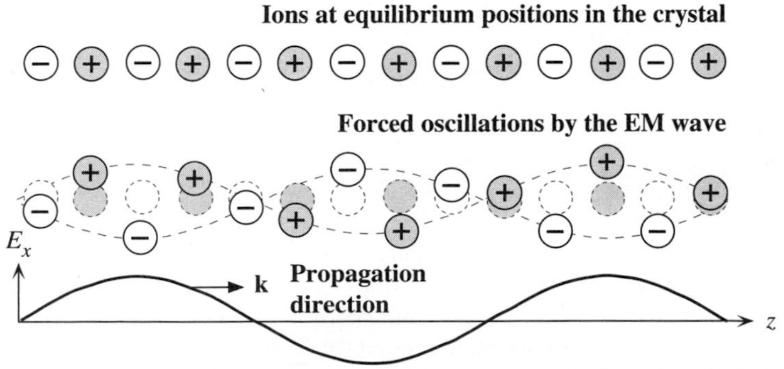


**Figure 9.21** Lattice absorption through a crystal. The field in the EM wave oscillates the ions which consequently generate "mechanical" waves in the crystal; energy is thereby transferred from the wave to lattice vibrations.

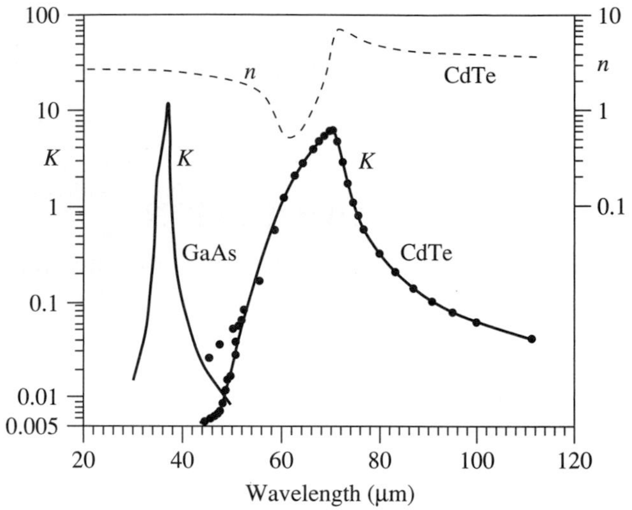


**Figure 9.22** Lattice or Reststrahlen absorption in CdTe and GaAs in terms of the extinction coefficient versus wavelength. For reference, $n$ versus $\lambda$ for CdTe is also shown.

Although Figure 9.21 depicts an ionic solid to visualize absorption due to lattice waves, energy from a passing EM wave can also be absorbed by various ionic impurities in a medium as these charges can couple to the electric field and oscillate. Bonding between an oscillating ion and the neighboring atoms causes the mechanical oscillations of the ion to be coupled to neighboring atoms. This leads to a generation of lattice waves which takes away energy from the EM wave.

**EXAMPLE 9.16** **RESTSTRAHLEN ABSORPTION** Figure 9.22 shows the infrared extinction coefficient $K$ of GaAs and CdTe. Consider CdTe. Calculate the absorption coefficient $\alpha$ and the reflectance $R$ of CdTe at the Reststrahlen peak, and also at 50 μm and at 100 μm. What is your conclusion?

**SOLUTION**

At the resonant peak, $\lambda \approx 72$ μm, $K \approx 6$, and $n \approx 5$, so the corresponding free-space wavevector is

$$k_o = \frac{2\pi}{\lambda} = \frac{2\pi}{72 \times 10^{-6}\text{ m}} = 8.7 \times 10^4 \text{ m}^{-1}$$

The absorption coefficient $\alpha$, by definition, is $2k''$ in Equation 9.57, so

$$\alpha = 2k'' = 2k_o K = 2(8.7 \times 10^4 \text{ m}^{-1})(6) = 1.0 \times 10^6 \text{ m}^{-1}$$

which corresponds to an *absorption depth* $1/\alpha$ of about 1 μm. The reflectance is

$$R = \frac{(n-1)^2 + K^2}{(n+1)^2 + K^2} = \frac{(5-1)^2 + 6^2}{(5+1)^2 + 6^2} = 0.72 \quad \text{or} \quad 72\%$$

Repeating the above calculations at $\lambda = 50$ μm, we get $\alpha = 8.3 \times 10^2 \text{ m}^{-1}$, and $R = 0.11$ or 11 percent. There is a sharp increase in the reflectance from 11 to 72 percent as we approach the resonant peak. At $\lambda = 100$ μm, $\alpha = 6.3 \times 10^3 \text{ m}^{-1}$ and $R = 0.31$ or 31 percent, which is again smaller than the peak reflectance. $R$ is maximum around the Reststrahlen peak.

## 9.10 BAND-TO-BAND ABSORPTION

The photon absorption process for photogeneration, that is, the creation of electron–hole pairs (EHPs), requires the photon energy to be at least equal to the bandgap energy $E_g$ of the semiconductor material to excite an electron from the valence band (VB) to the conduction band (CB). The **upper cut-off wavelength** (or the threshold wavelength) $\lambda_g$ for photogenerative absorption is therefore determined by the bandgap energy $E_g$ of the semiconductor, so $h(c/\lambda_g) = E_g$ or

*Cut-off wavelength and bandgap*

$$\lambda_g(\mu\text{m}) = \frac{1.24}{E_g(\text{eV})} \qquad [9.70]$$

For example, for Si, $E_g = 1.12$ eV and $\lambda_g$ is 1.11 μm whereas for Ge, $E_g = 0.66$ eV and the corresponding $\lambda_g = 1.87$ μm. It is clear that Si photodiodes cannot be used for optical communications at 1.3 and 1.55 μm, whereas Ge photodiodes are commercially available for use at these wavelengths. Table 9.3 lists some typical bandgap energies and the corresponding cut-off wavelengths of various photodiode semiconductor materials.

Incident photons with wavelengths shorter than $\lambda_g$ become absorbed as they travel in the semiconductor, and the light intensity, which is proportional to the number of photons, decays exponentially with distance into the semiconductor. The light intensity $I$ at a distance $x$ from the semiconductor surface is given by

*Absorption coefficient*

$$I(x) = I_o \exp(-\alpha x) \qquad [9.71]$$

where $I_o$ is the intensity of the incident radiation and $\alpha$ is the **absorption coefficient** that depends on the photon energy or wavelength $\lambda$. The absorption coefficient $\alpha$ is a material property. Most of the photon absorption (63%) occurs over a distance $1/\alpha$, and $1/\alpha$ is called the **penetration depth** $\delta$. Figure 9.23 shows the $\alpha$ versus $\lambda$ characteristics of various semiconductors where it is apparent that the behavior of $\alpha$ with the wavelength $\lambda$ depends on the semiconductor material.

Absorption in semiconductors can be understood in terms of the behavior of the electron energy ($E$) with the electron momentum ($\hbar k$) in the crystal, called the **crystal**

**Table 9.3** Bandgap energy $E_g$ at 300 K, cut-off wavelength $\lambda_g$, and type of bandgap (D = direct and I = indirect) for some photodetector materials

| Semiconductor | $E_g$ (eV) | $\lambda_g$ (μm) | Type |
|---|---|---|---|
| InP | 1.35 | 0.91 | D |
| $GaAs_{0.88}Sb_{0.12}$ | 1.15 | 1.08 | D |
| Si | 1.12 | 1.11 | I |
| $In_{0.7}Ga_{0.3}As_{0.64}P_{0.36}$ | 0.89 | 1.4 | D |
| $In_{0.53}Ga_{0.47}As$ | 0.75 | 1.65 | D |
| Ge | 0.66 | 1.87 | I |
| InAs | 0.35 | 3.5 | D |
| InSb | 0.18 | 7 | D |

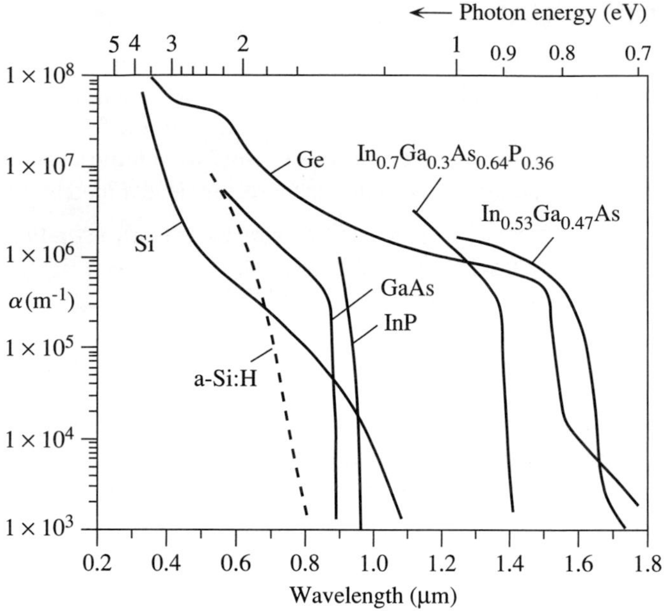


**Figure 9.23** Absorption coefficient $\alpha$ versus wavelength $\lambda$ for various semiconductors.

SOURCE: Data selectively collected and combined from various sources.

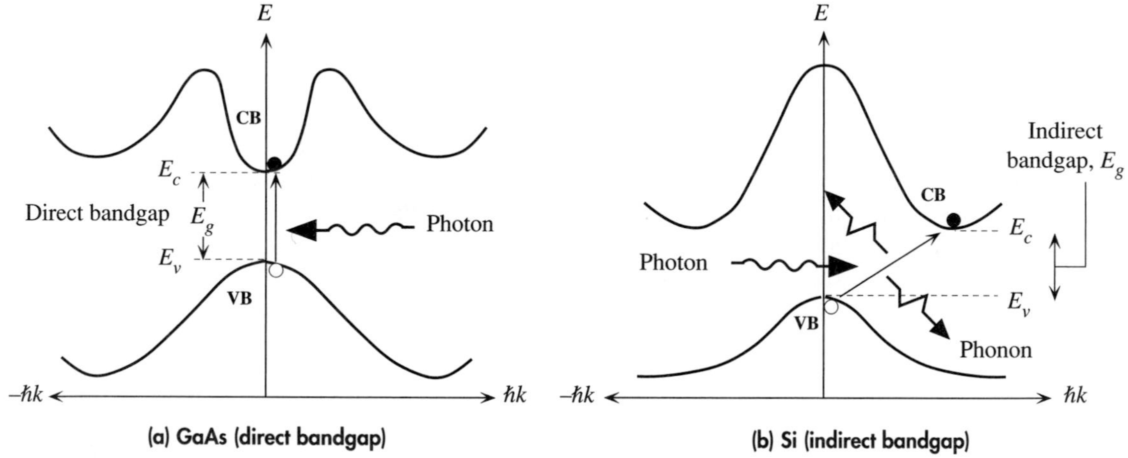


**Figure 9.24** Electron energy $E$ versus crystal momentum $\hbar k$ and photon absorption.
(a) Photon absorption in a direct bandgap semiconductor.
(b) Photon absorption in an indirect bandgap semiconductor (VB = valence band; CB = conduction band).

**momentum.** If $k$ is the wavevector of the electron's wavefunction in the crystal, then the momentum of the electron within the crystal is $\hbar k$. $E$ versus $\hbar k$ behaviors for electrons in the conduction and valence bands of direct and indirect bandgap semiconductors are shown in Figure 9.24a and b, respectively. In **direct bandgap** semiconductors such as III–V semiconductors (*e.g.*, GaAs, InAs, InP, GaP) and in many of their alloys (*e.g.*, InGaAs, GaAsSb) the photon absorption process is a direct process which requires no assistance from lattice vibrations. The photon is absorbed and the electron is excited directly from the valence band to the conduction band without a change in its $k$-vector, or its crystal momentum $\hbar k$, inasmuch as the photon momentum is very small. The change in the electron momentum from the valence to the conduction band is

$$\hbar k_{CB} - \hbar k_{VB} = \text{Photon momentum} \approx 0$$

This process corresponds to a vertical transition on the electron energy ($E$) versus electron momentum ($\hbar k$) diagram as shown in Figure 9.24a. The absorption coefficient of these semiconductors rises sharply with decreasing wavelength from $\lambda_g$ as apparent for GaAs and InP in Figure 9.23.

In **indirect bandgap** semiconductors such as Si and Ge, the photon absorption for photon energies near $E_g$ requires the absorption and emission of lattice vibrations, that is, **phonons,**[14] during the absorption process as shown in Figure 9.24. If $K$ is the wavevector of a lattice wave (lattice vibrations travel in the crystal), then $\hbar K$ represents the momentum associated with such a lattice vibration; that is, $\hbar K$ is a **phonon momentum.** When an electron in the valence band is excited to the conduction band, there is a change in its momentum in the crystal, and this change in the momentum cannot be supplied by the momentum of the incident photon which is very small. Thus, the momentum difference must be balanced by a phonon momentum,

$$\hbar k_{CB} - \hbar k_{VB} = \text{Phonon momentum} = \hbar K$$

The absorption process is said to be **indirect** as it depends on lattice vibrations which in turn depend on the temperature. Since the interaction of a photon with a valence electron needs a third body, a lattice vibration, the probability of photon absorption is not as high as in a direct transition. Furthermore, the cut-off wavelength is not as sharp as for direct bandgap semiconductors. During the absorption process, a phonon may be absorbed or emitted. If $\vartheta$ is the frequency of the lattice vibrations, then the phonon energy is $h\vartheta$. The photon energy is $h\nu$ where $\nu$ is the photon frequency. Conservation of energy requires that

$$h\nu = E_g \pm h\vartheta$$

Thus, the onset of absorption does not exactly coincide with $E_g$, but typically it is very close to $E_g$ inasmuch as $h\vartheta$ is small ($< 0.1$ eV). The absorption coefficient initially rises slowly with decreasing wavelength from about $\lambda_g$ as apparent in Figure 9.23 for Si and Ge.

**FUNDAMENTAL ABSORPTION** A GaAs infrared LED emits at about 860 nm. A Si photodetector is to be used to detect this radiation. What should be the thickness of the Si crystal that absorbs most of this radiation? **EXAMPLE 9.17**

[14] As much as an electromagnetic radiation is quantized in terms of photons, lattice vibrations in the crystal are quantized in terms of phonons. A phonon is a quantum of lattice vibration. If $K$ is the wavevector of a vibrational wave in a crystal lattice and $\omega$ is its angular frequency, then the momentum of the wave is $\hbar K$ and its energy is $\hbar\omega$.

**SOLUTION**

According to Figure 9.23, at $\lambda \approx 0.8\ \mu\text{m}$, Si has $\alpha \approx 6 \times 10^4\ \text{m}^{-1}$, so the absorption depth

$$\delta = \frac{1}{\alpha} = \frac{1}{6 \times 10^4\ \text{m}^{-1}} = 1.7 \times 10^{-5}\ \text{m} \quad \text{or} \quad 17\ \mu\text{m}$$

If the crystal thickness is $\delta$, then 63 percent of the radiation will be absorbed. If the thickness is $2\delta$, then the fraction of absorbed radiation, from Equation 9.71, will be

$$\text{Fraction of absorbed radiation} = 1 - \exp[-\alpha(2\delta)] = 0.86 \quad \text{or} \quad 86\%$$

## 9.11 LIGHT SCATTERING IN MATERIALS

Scattering of an EM wave implies that a portion of the energy in a light beam is directed away from the original direction of propagation as illustrated for a small dielectric particle scattering a light beam in Figure 9.25. There are various types of scattering processes.

Consider what happens when a propagating wave encounters a molecule, or a small dielectric particle (or region), which is smaller than the wavelength. The electric field in the wave polarizes the particle by displacing the lighter electrons with respect to the heavier positive nuclei. The electrons in the molecule couple and oscillate with the electric field in the wave (ac electronic polarization). The oscillation of charge "up" and "down," or the oscillation of the induced dipole, radiates EM waves all around the molecule as depicted in Figure 9.25. We should remember that an oscillating charge is like an alternating current which always radiates EM waves (like an antenna). The net effect is that the incident wave becomes partially reradiated in different directions and hence loses intensity in its original direction of propagation. We may think of the process as the particle absorbing some of the energy via electronic polarization and reradiating it in different directions. It may be thought that the scattered waves constitute a spherical wave emanating from the scattering molecule, but this is not generally the case as the reemitted radiation depends on the shape and polarizability of the molecule in different directions. We assumed a small particle so that at any time the field has no spatial variation through the particle, whose polarization then oscillates with the electric field oscillation. Whenever the size of the scattering region, whether an inhomogeneity or a small

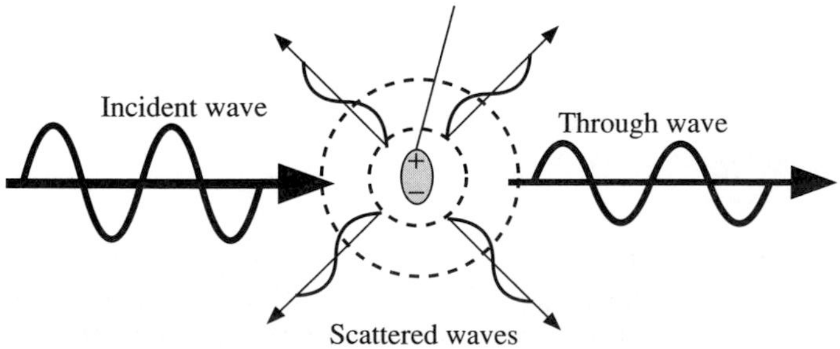


**Figure 9.25** Rayleigh scattering involves the polarization of a small dielectric particle or a region that is much smaller than the light wavelength.
The field forces dipole oscillations in the particle (by polarizing it), which leads to the emission of EM waves in "many" directions so that a portion of the light energy is directed away from the incident beam.

particle or a molecule, is much smaller than the wavelength $\lambda$ of the incident wave, the scattering process is generally termed **Rayleigh scattering.** In this type of scattering, typically the particle size is smaller than one-tenth of the wavelength.

Rayleigh scattering of waves in a medium arises whenever there are small inhomogeneous regions in which the refractive index is different than the medium (which has some average refractive index). This means a local change in the relative permittivity and polarizability. The result is that the small inhomogeneous region acts like a small dielectric particle and scatters the propagating wave in different directions. In the case of optical fibers, dielectric inhomogeneities arise from fluctuations in the relative permittivity that is part of the intrinsic glass structure. As the fiber is drawn by freezing a liquid-like flow, random thermodynamic fluctuations in the composition and structure that occur in the liquid state become frozen into the solid structure. Consequently, the glass fiber has small fluctuations in the relative permittivity which leads to Rayleigh scattering. Nothing can be done to eliminate Rayleigh scattering in glasses as it is part of their intrinsic structure.

It is apparent that the scattering process involves electronic polarization of the molecule or the dielectric particle. We know that this process couples most of the energy at ultraviolet frequencies where the dielectric loss due to electronic polarization is maximum and the loss is due to EM wave radiation. Therefore, as the frequency of light increases, the scattering becomes more severe. In other words, *scattering decreases with increasing wavelength.* For example, blue light which has a shorter wavelength than red light is scattered more strongly by air molecules. When we look at the sun directly, it appears yellow because the blue light has been scattered in the direct light more than the red light. When we look at the sky in any direction but the sun, our eyes receive scattered light which appears blue; hence the sky is blue. At sunrise and sunset, the rays from the sun have to traverse the longest distance through the atmosphere and have the most blue light scattered which gives the sun its red color at these times.

## 9.12 ATTENUATION IN OPTICAL FIBERS

As light propagates through an optical fiber, it becomes attenuated by a number of processes that depend on the wavelength of light. Figure 9.26 shows the attenuation coefficient, as dB per km, of a typical silica-glass-based optical fiber as a function of wavelength. The sharp increase in the attenuation at wavelengths beyond 1.6 μm in the *infrared* region is due to energy absorption by "lattice vibrations" of the constituent ions of the glass material. Fundamentally, energy absorption in this region corresponds to the stretching of the Si–O bonds in ionic polarization induced by the EM wave. Absorption increases with wavelength as we approach the resonance wavelength of the Si–O bond which is around 9 μm. In the case of Ge–O glasses, this is further away, around 11 μm. There is another intrinsic material absorption in the region below 500 nm, not shown in Figure 9.26, which is due to photons exciting electrons from the valence band to the conduction band of the glass.

There is a marked attenuation peak centered at 1.4 μm, and a barely discernible minor peak at about 1.24 μm. These attenuation regions arise from the presence of hydroxyl ions as impurities in the glass structure inasmuch as it is difficult to remove all

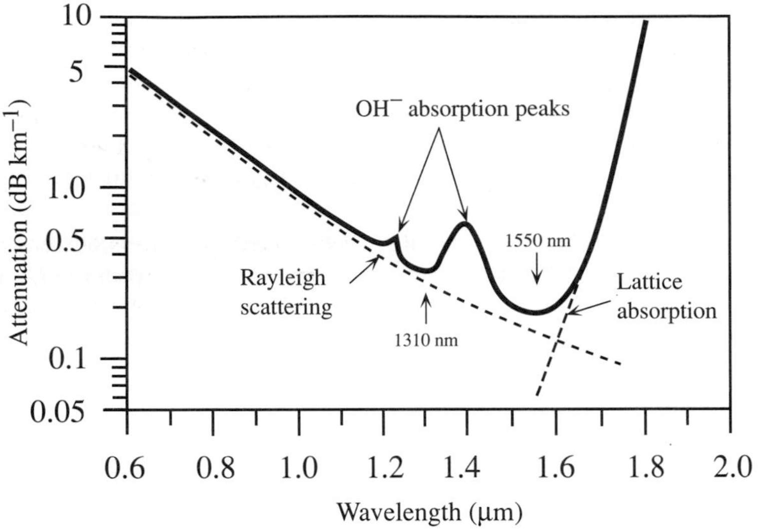


**Figure 9.26** Illustration of typical attenuation versus wavelength characteristics of a silica-based optical fiber.

There are two communications channels at 1310 and 1550 nm.

traces of hydroxyl (water) products during fiber production. Further, hydrogen atoms can easily diffuse into the glass structure at high temperatures during production which leads to the formation of hydrogen bonds in the silica structure and OH ions. Energy is absorbed mainly by the stretching vibrations of the OH bonds within the silica structure which has a fundamental resonance in the infrared region (beyond 2.7 μm) but overtones or harmonics at lower wavelengths (or higher frequencies). The first overtone at around 1.4 μm is the most significant as can be seen in Figure 9.26. The second overtone is around 1 μm, and in high-quality fibers this is negligible. A combination of the first overtone of the OH vibration and the fundamental vibrational frequency of $SiO_2$ gives rise to a minor loss peak at around 1.24 μm. There are two important windows in the attenuation versus wavelength behavior where the attenuation exhibits minima. The window at around 1.3 μm is the region between two neighboring $OH^-$ absorption peaks. This window is widely used in optical communications at 1310 nm. The window at around 1.55 μm is between the first harmonic absorption of $OH^-$ and the infrared lattice absorption tail and represents the lowest attenuation. Current technological drive is to use this window for long-haul communications. It can be seen that it is important to keep the hydroxyl content in the fiber within tolerable levels.

There is a background attenuation process that decreases with wavelength and is due to the Rayleigh scattering of light by the local variations in the refractive index. Glass has a noncrystalline or an amorphous structure which means that there is no long-range order to the arrangement of the atoms but only a short-range order, typically a few bond lengths. The glass structure is as if the structure of the melt has been suddenly frozen. We can only define the number of bonds a given atom in the structure will have. Random variations in the bond angle from atom to atom lead to a disordered structure. There is therefore a random local variation in the density over a few bond lengths which leads to fluctuations in the refractive index over few atomic lengths. These random fluctuations in the refractive index give rise to light scattering and hence light attenuation along the fiber. It should be apparent that since a degree of structural randomness is an intrinsic property of the glass structure, this scattering process is unavoidable and represents the lowest attenuation possible through a glass medium. As one may surmise, attenuation by scattering in a medium is minimum for light propagating through a "perfect" crystal. In this case the only scattering mechanisms will be due to thermodynamic defects (vacancies) and the random thermal vibrations of the lattice atoms.

As mentioned above, the Rayleigh scattering process decreases with wavelength and, according to Lord Rayleigh, it is inversely proportional to $\lambda^4$. The expression for the attenuation $\alpha_R$ in a single component glass due to Rayleigh scattering is approximately given by

*Rayleigh scattering in silica*

$$\alpha_R \approx \frac{8\pi^3}{3\lambda^4}(n^2-1)^2\beta_T k T_f \qquad [9.72]$$

where $\lambda$ is the free-space wavelength, $n$ is the refractive index at the wavelength of interest, $\beta_T$ is the isothermal compressibility (at $T_f$) of the glass, $k$ is the Boltzmann constant, and $T_f$ is a quantity called the *fictive temperature* (roughly the *softening temperature of glass*) where the liquid structure during the cooling of the fiber is frozen to become the glass structure. Fiber is drawn at high temperatures, and as the fiber cools eventually the temperature drops sufficiently for the atomic motions to be so sluggish that the structure becomes essentially "frozen-in" and remains like this even at room temperature. Thus $T_f$ marks the temperature below which the liquid structure is frozen, and hence the density fluctuations are also frozen into the glass structure. It is apparent that Rayleigh scattering represents the lowest attenuation one can achieve using a glass structure. By proper design, the attenuation window at 1.5 μm may be lowered to approach the Rayleigh scattering limit.

---

**EXAMPLE 9.18**

**RAYLEIGH SCATTERING LIMIT** What is the attenuation due to Rayleigh scattering at around the $\lambda = 1.55$ μm window given that pure silica ($SiO_2$) has the following properties: $T_f = 1730$ °C (softening temperature), $\beta_T = 7 \times 10^{-11}$ m$^2$ N$^{-1}$ (at high temperatures), $n = 1.4446$ at 1.5 μm?

**SOLUTION**

We simply calculate the Rayleigh scattering attenuation using

$$\alpha_R \approx \frac{8\pi^3}{3\lambda^4}(n^2-1)^2\beta_T k T_f$$

so

$$\alpha_R \approx \frac{8\pi^3}{3(1.55\times10^{-6})^4}(1.4446^2-1)^2(7\times10^{-11})(1.38\times10^{-23})(1730+273)$$

$$= 3.27\times10^{-5}\ \text{m}^{-1} \quad \text{or} \quad 3.27\times10^{-2}\ \text{km}^{-1}$$

Attenuation in dB per km is then

$$\alpha_{dB} = 4.34\alpha_R = (4.34)(3.27\times10^{-2}\ \text{km}^{-1}) = 0.142\ \text{dB km}^{-1}$$

This represents the lowest possible attenuation for a silica glass fiber at 1.55 μm.

---

## 9.13 LUMINESCENCE, PHOSPHORS, AND WHITE LEDS

We know from our general experience that certain substances, known as *phosphors,* can absorb light and then reemit light even after the excitation light source has been turned off; this is an example of luminescence. In general, **luminescence** is the emission of light by a material, called a **phosphor,** due to the absorption and conversion of energy into electromagnetic radiation as illustrated in Figure 9.27a and b. The luminescent radiation emitted by the phosphor material is considered to be quite separate from the thermal radiation emitted by virtue of its temperature. Luminescence is light emitted by a nonthermal source when it is excited, in contrast to the emission of radiation from a heated object such as the tungsten filament of a light bulb; the latter is called **incandescence.** Typically the emission of light occurs from certain dopants, impurities, or even defects, called **luminescent** or **luminescence centers,** purposefully introduced into a **host matrix,** which may be a crystal or glass as shown in Figure 9.27c. The luminescent center is also called an **activator.** There are many examples of phosphors. For example, in ruby, the $Cr^{3+}$ ions are the luminescent centers in the sapphire ($Al_2O_3$) crystal host. $Cr^{3+}$ ions can absorb UV or violet light and then emit red light. This phosphor system is written as $Al_2O_3$:$Cr^{3+}$. The excitation and emission involves only the $Cr^{3+}$ ion. In other cases, the activator excitation may also involve the host as discussed later.

Luminescence is normally categorized according to the source of excitation energy. **Photoluminescence** involves excitation by photons (light) as in Figure 9.27a. **X-ray luminescence** involves incident X-rays exciting a phosphor to emit light. **Cathodoluminescence,** as shown in Figure 9.27b, is light emission when the excitation is the bombardment of the phosphor with energetic electrons as in TV cathode ray tubes. **Electroluminescence** is light emission due to the passage of an electric current. Electroluminescence in semiconductive materials appears as a result of an excited electron transiting down to the ground energy level, which would correspond to the recombination of an electron and a hole; the excited electron is the conduction band (CB), and its ground state corresponds to a hole in the valence band (VB). The direct electron–hole recombination mechanism generally occurs very quickly. For example, typical minority carrier lifetimes are in the range of nanoseconds, so light emission from a semiconductor stops within nanoseconds after the removal of the excitation. Such quick luminescence processes occurring over a nanosecond time scale or shorter are normally identified as **fluorescence.** The emission of light from a fluorescent tube

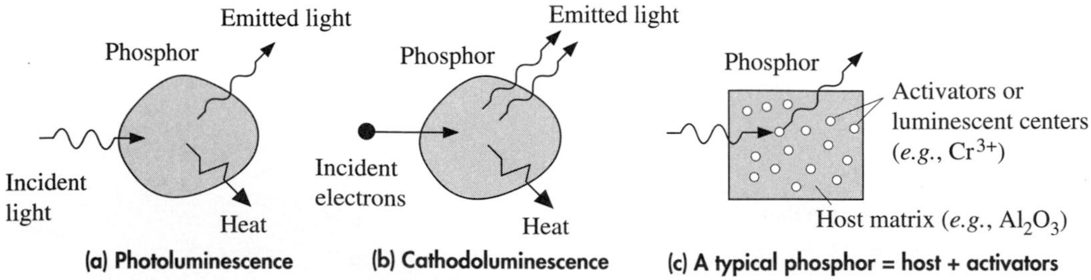


**Figure 9.27** Photoluminescence, cathodoluminescence, and a typical phosphor.

This flashlight uses a white LED instead of an incandescent light bulb. The flashlight can operate continuously for 200 hours and can project an intense spot over 30 ft. White LEDs use a phosphor to generate yellow light from the blue light emitted from the LED's semiconductor chip. The mixture of blue and yellow light appears as white.

is actually a fluorescence process. The tube contains a gas mixture of argon and mercury. The Ar and Hg gas atoms become excited by the electrical discharge process and emit light mainly in the ultraviolet region. This UV light is absorbed by the fluorescent coating on the inside of the tube. The excited activators in the phosphor coating then emit radiation in the visible region. A number of phosphors are used to obtain "white" light from the tube.

There are also phosphors from which light emission may continue for milliseconds to hours after the cessation of excitation. These slow luminescence processes are normally referred to as **phosphorescence** (also known as *afterglow*).

Many phosphors are based on activators doped into a host matrix; for example, $Eu^{3+}$ (europium ion) in a $Y_2O_3$ (yttrium oxide) matrix is a widely used modern phosphor. When excited by UV radiation, it provides an efficient luminescence emission in the red (around 613 nm). It is used as the red-emitting phosphor in color TV tubes and in modern tricolor fluorescent lamps. In very general terms, we can represent the energy of an activator in a host matrix by the highly simplified energy diagram in Figure 9.28.

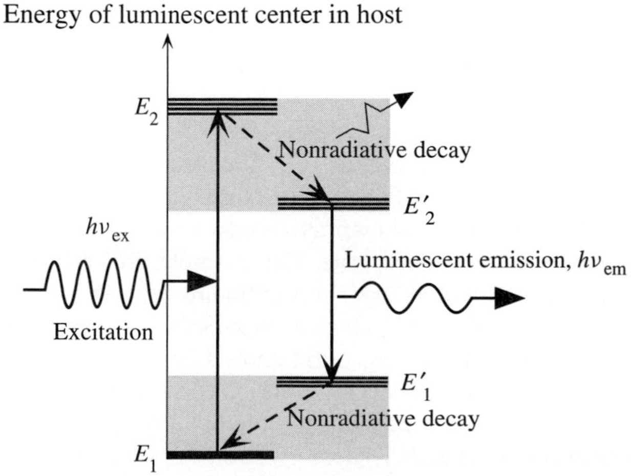


**Figure 9.28** Photoluminescence: light absorption, excitation, nonradiative decay and light emission, and return to the ground state $E_1$.
The energy levels have been displaced horizontally for clarity.

The ground state of the activator is $E_1$. Upon excitation by an incident radiation of suitable energy $h\nu_{ex}$ the activator becomes excited to $E_2$. From this energy level, it decays, or *relaxes,* down relatively quickly (on a time scale of the order of picoseconds) to an energy level $E_2'$ by emitting phonons or lattice vibrations. This type of decay is called *radiationless* or *nonradiative decay.* From $E_2'$, the activator decays down to $E_1'$ by emitting a photon (spontaneous emission), which is the emitted luminescent radiation. The emitted photon energy is $h\nu_{em}$, which is less than the excitation photon energy $h\nu_{ex}$. The return from $E_1'$ to the ground state $E_1$ involves phonon emissions. Further, for some activators, $E_1'$ is either very close to $E_1$, or it is $E_1$.The energy levels such as $E_2$, $E_2'$, $E_1'$, etc., are not well-defined single levels but involve finely spaced multilevels. The higher levels may form multilevel narrow energy "bands." In this example, the activator absorbed the incident radiation and was directly excited, which is known as **activator excitation.** The $Cr^{3+}$ ions in $Al_2O_3$:$Cr^{3+}$ can be excited directly by blue light and would then emit in the red. There are many phosphors in which the excitation involves the host. In **host excitation,** the host matrix absorbs the incident radiation and transfers the energy to the activator, which then becomes excited to $E_2$ in Figure 9.28, and so on. In X-ray phosphors, for example, the X-rays are absorbed by the host, which subsequently transfers the energy to the activators. It is apparent from Figure 9.28 that the emitted radiation ($h\nu_{em}$) has a *longer* wavelength than the exciting radiation ($h\nu_{ex}$), that is, $h\nu_{em} < h\nu_{ex}$. The downshift in the light frequency from absorbed to emitted radiation is called the **Stoke's shift.** It should be emphasized that the energy levels of the activator (as shown in Figure 9.28) also depend on the host, because the internal electric fields within the host crystal act on the activator and shift these levels up and down. The emission characteristics depend firstly on the activator, and secondly on the host.

There are a number of host excitation mechanisms. In one possible process, which involves a semiconductor host, as depicted in Figure 9.29, an incident photon initially excites a valence band (VB) electron to the conduction band (CB). The electron then thermalizes, *i.e.*, loses the excess energy as it collides with lattice vibrations, and falls close to $E_c$, and wanders around in the crystal. In one process, *a* in Figure 9.29, the electron can be captured into an excited state D of a luminescent center or an activator. The electron then falls down in energy to the ground state A of the activator releasing a photon, which is the luminescent emission. The electron at the ground state then recombines with a hole in the VB. Thus the activator acts as a **radiative recombination center.** In some cases D and A may be separate centers representing *donor* and *acceptor*-like centers, hence the labels D and A. In other cases, the radiative recombination center may simply be a single energy level in the bandgap, which is shown as R in Figure 9.29. The electron can emit a photon as it is captured into R, shown as process *b* in Figure 9.29, or emit the photon after it is captured by R, as it recombines with a hole, shown as process *c* in Figure 9.29. Processes *a* and *b* occur in various ZnS-based phosphors. For example, in ZnS:$Cu^+$ phosphors, the activator is $Cu^+$, which has an energy level at A in Figure 9.29. The luminescent emission is enhanced by using a coactivator, such as Al in ZnS:$Cu^+$. Al acts as a shallow donor D, and the luminescence is due to process *a* in Figure 9.29.

There may also be traps in the semiconductor because of various crystal defects, or there may be added impurities. The electron can become captured by a trap at a localized energy level $E_t$ in the bandgap, but close to $E_c$. These electron traps temporarily

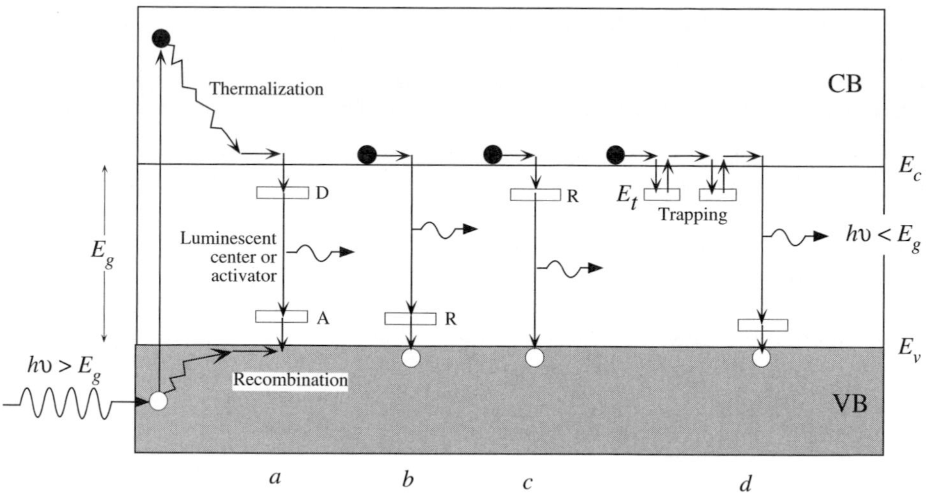


**Figure 9.29** Optical absorption generates an EHP.
Both carriers thermalize. There are a number of recombination processes via a dopant that can result in a luminescent emission.

capture an electron from the conduction band and thereby immobilize it. The time the electron spends trapped at $E_t$ depends on the energy depth of the trap from the conduction band, $E_c - E_t$. After a while a strong lattice vibration returns the electron back into the conduction band (by thermal excitation). The time interval between photogeneration and recombination can be relatively long if the electron remains captured at $E_t$ for a considerable length of time. In fact, the electron may become trapped and detrapped many times before it finally recombines, so the emission of light can persist for a relatively long time after the cessation of excitation (*e.g.*, milliseconds or longer) as indicated by process *d* in Figure 9.29.

It is also possible to excite electrons into the CB by bombarding the material with a high-energy electron beam, which leads to cathodoluminescence. Color CRT displays are typically coated uniformly with three sets of phosphor dots which exhibit cathodoluminescence in the blue, red, and green wavelengths. In electroluminescence, an electric current, either ac or dc, is used to inject electrons into the CB which then recombine with holes and emit light. For example, passing a current through certain semiconducting phosphors such as ZnS doped with Mn causes light emission by electroluminescence. The emission of light from a light emitting diode (LED) is an example of **injection electroluminescence** in which the applied voltage causes charge carrier injection and recombination in a device (diode) that has a junction between a *p*-type and an *n*-type semiconductor.

Zinc sulfide with various activators has been one of the traditional phosphors. The $ZnS:Ag^+$ in which $Ag^+$ is the activator, is still used as a blue emitting phosphor, though in some cases Cd is substituted for some of the Zn. $ZnS:Cu^+$ emits in the green, which is also a useful phosphor. Most modern phosphors, on the other hand, have been based on using rare earth activators in various hosts. For example, $Y_2O_3:Eu^{3+}$ absorbs UV

**Table 9.4** Selected phosphor examples

| Phosphor | Activator | Useful Emission | Example Excitation | Comment or Application |
|---|---|---|---|---|
| $Y_2O_3:Eu^{3+}$ | $Eu^{3+}$ | Red | UV | Fluorescent lamp, color TV |
| $BaMgAl_{10}O_{17}:Eu^{2+}$ | $Eu^{2+}$ | Blue | UV | Fluorescent lamp |
| $CeMgAl_{11}O_{19}:Tb^{3+}$ | $Tb^{3+}$ | Green | UV | Fluorescent lamp |
| $Y_3Al_5O_{12}:Ce^{3+}$ | $Ce^{3+}$ | Yellow | Blue, violet | White LED |
| $Sr_2SiO_4:Eu^{3+}$ | $Eu^{3+}$ | Yellow | Violet | White LED (experimental) |
| $ZnS:Ag^+$ | $Ag^+$ | Blue | Electron beam | Color TV blue phosphor |
| $Zn_{0.68}Cd_{0.32}S:Ag^+$ | $Ag^+$ | Green | Electron beam | Color TV green phosphor |
| $ZnS:Cu^+$ | $Cu^+$ | Green | Electron beam | Color TV green phosphor |

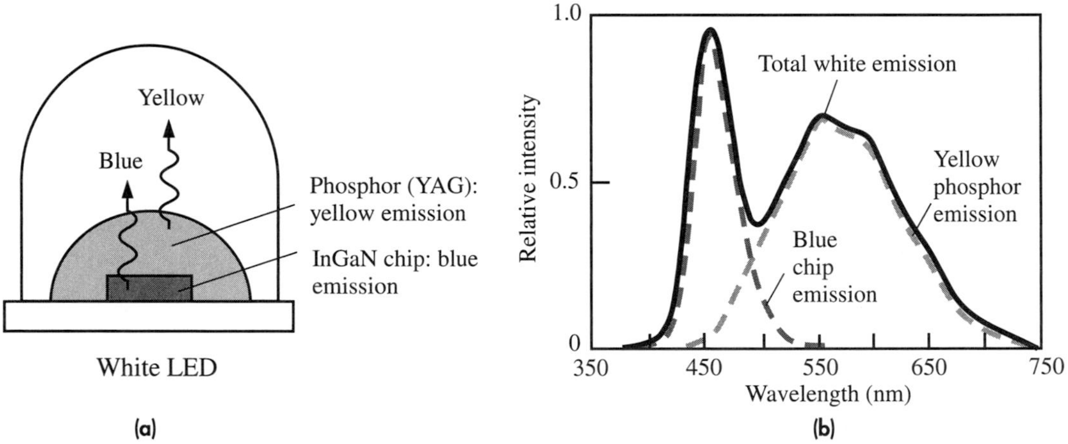


**Figure 9.30**
(a) A typical "white" LED structure.
(b) The spectral distribution of light emitted by a white LED. Blue luminescence is emitted by the GaInN chip and "yellow" phosphorescence or luminescence is produced by a phosphor. The combined spectrum looks "white."

radiation and emits in the red. $Y_3Al_5O_{12}:Ce^{3+}$ absorbs blue light and emits yellow light. Some of the most popular activators are $Eu^{3+}$ for red, $Eu^{2+}$ for blue, and $Tb^{3+}$ for green. Table 9.4 summarizes a number of phosphors commonly used in various applications.

Recent inexpensive white LEDs that have appeared on the market seem to emit white light by emitting a mixture of blue and yellow light which are registered visually by the eye as appearing white. (Yellow consists of red and green mixed together, so mixing blue and yellow generates "white.") The production of white LEDs became possible due to development of bright blue-emitting LEDs based on gallium-indium-nitride (GaInN). The white LED uses a semiconductor chip emitting at a short wavelength (blue, violet, or ultraviolet) and a *phosphor* to convert some of the blue light to yellow light as depicted in Figure 9.30a. The phosphor absorbs light from the diode and undergoes luminescent emission at a longer wavelength. Obviously, the quality and spectral characteristics of the combined emission vary with different designs; Figure 9.30b shows example spectra involved in the blue and yellow emissions and the overall "white" emission from a white LED. Typical phosphors have been based on yttrium-aluminum- ($Y_3Al_5O_{12}$) garnets (YAGs) as the host material. This host is doped with one of the rare earth elements for the activator. Cerium is a common dopant element in YAG phosphors; that is, the phosphor is $Y_3Al_5O_{12}:Ce^{3+}$, which is able to efficiently absorb the blue and emit the yellow. White LEDs are soon expected to challenge the existing incandescent sources for general lighting.

## 9.14 POLARIZATION

A propagating EM wave has its electric and magnetic fields at right angles to the direction of propagation. If we place a $z$ axis along the direction of propagation, then the electric field can be in any direction in the plane perpendicular to the $z$ axis. The term **polarization** of an EM wave describes the behavior of the electric field vector in the EM wave as it propagates through a medium. If the oscillations of the electric field at all times are contained within a well-defined line, then the EM wave is said to be **linearly polarized** as shown in Figure 9.31a. The field vibrations and the direction of propagation ($z$) define a plane of polarization (plane of vibration), so linear polarization implies a wave that is **plane-polarized.** By contrast, if a beam of light has waves with the $E$ field in each in a random direction but perpendicular to $z$, then this light beam is *unpolarized.* A light beam can be linearly polarized by passing the beam

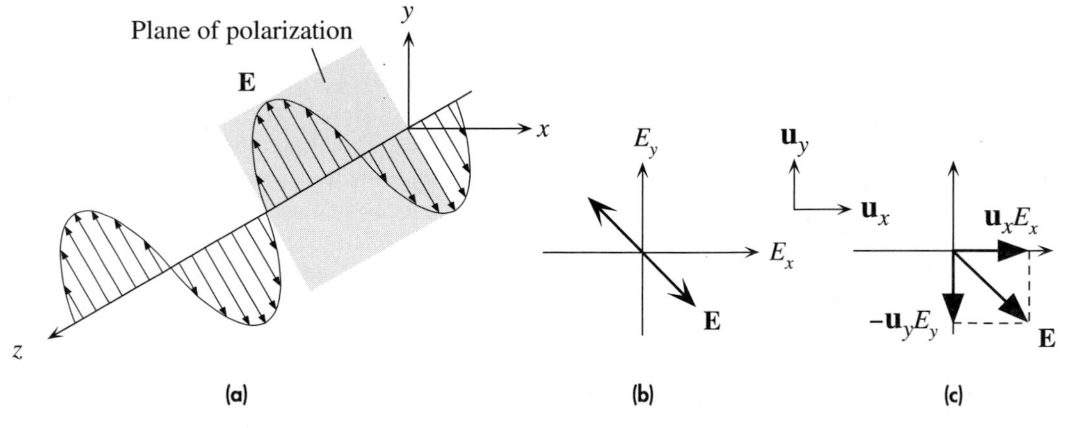


**Figure 9.31**

(a) A linearly polarized wave has its electric field oscillations defined along a line perpendicular to the direction of propagation $z$. The field vector **E** and $z$ define a *plane of polarization.*

(b) The $E$-field oscillations are contained in the plane of polarization.

(c) A linearly polarized light at any instant can be represented by the superposition of two fields $E_x$ and $E_y$ with the right magnitude and phase.

through a *polarizer,* such as a polaroid sheet, a device that only passes electric field oscillations lying on a well-defined plane parallel to its transmission axis.

Suppose that we arbitrarily place the $x$ and $y$ axes and describe the electric field in terms of its components $E_x$ and $E_y$ along $x$ and $y$ (we are justified to do this because $E_x$ and $E_y$ are perpendicular to $z$). To find the electric field in the wave at any space and time location, we add $E_x$ and $E_y$ *vectorially*. Both $E_x$ and $E_y$ can individually be described by a wave equation which must have the same angular frequency $\omega$ and wavenumber $k$. However, we must include a phase difference $\phi$ between the two:

$$E_x = E_{xo} \cos(\omega t - kz) \quad [9.73]$$

and

$$E_y = E_{yo} \cos(\omega t - kz + \phi) \quad [9.74]$$

where $\phi$ is the phase difference between $E_y$ and $E_x$; $\phi$ can arise if one of the components is delayed (retarded).

The linearly polarized wave in Figure 9.31a has the **E** oscillations at $-45°$ to the $x$ axis as shown in Figure 9.31b. We can generate this field by choosing $E_{xo} = E_{yo}$ and $\phi = \pm 180°$ $(\pm\pi)$ in Equations 9.73 and 9.74. Put differently, $E_x$ and $E_y$ have the same magnitude, but they are out of phase by 180°. If $\mathbf{u}_x$ and $\mathbf{u}_y$ are the unit vectors along $x$ and $y$, using $\phi = \pi$ in Equation 9.74, the field in the wave is

$$\mathbf{E} = \mathbf{u}_x E_x + \mathbf{u}_y E_y = \mathbf{u}_x E_{xo} \cos(\omega t - kz) - \mathbf{u}_y E_{yo} \cos(\omega t - kz)$$

or

$$\mathbf{E} = \mathbf{E}_o \cos(\omega t - kz) \quad [9.75]$$

where

$$\mathbf{E}_o = \mathbf{u}_x E_{xo} - \mathbf{u}_y E_{yo} \quad [9.76]$$

Equations 9.75 and 9.76 state that the vector $\mathbf{E}_o$ is at $-45°$ to the $x$ axis and propagates along the $z$ direction.

There are many choices for the behavior of the electric field besides the simple linear polarization in Figure 9.31. For example, if the magnitude of the field vector **E** remains constant but its tip at a given location on $z$ traces out a circle by rotating in a clockwise sense with time, as observed by the receiver of the wave, then the wave is said to be **right circularly polarized**[15] as in Figure 9.32. If the rotation of the tip of **E** is counterclockwise, the wave is said to be **left circularly polarized.** From Equations 9.73 and 9.74, it should be apparent that a right circularly polarized wave has $E_{xo} = E_{yo} = A$ (an amplitude) and $\phi = \pi/2$. This means that,

$$E_x = A \cos(\omega t - kz) \quad [9.77]$$

and

$$E_y = -A \sin(\omega t - kz) \quad [9.78]$$

[15] There is a difference in this definition in optics and engineering. The definition here follows that in optics which is more prevalent in optoelectronics.

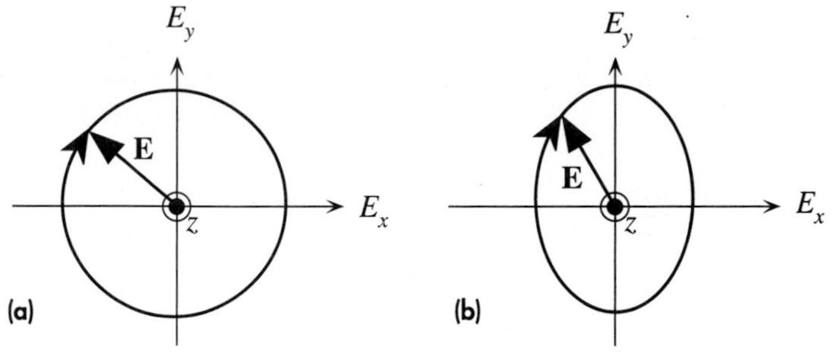


**Figure 9.32**
(a) A *right circularly polarized light* that is traveling along *z* (out of paper). The field vector **E** is always at right angles to *z*, rotates clockwise around *z* with time, and traces out a full circle over one wavelength of distance propagated.
(b) An elliptically polarized light.

It is relatively straightforward to show that Equations 9.77 and 9.78 represent a circle that is

$$E_x^2 + E_y^2 = A^2 \tag{9.79}$$

as shown in Figure 9.32.

When the phase difference $\phi$ is other than 0, $\pm\pi$, or $\pm\pi/2$, the resultant wave is **elliptically polarized** and the tip of the vector in Figure 9.32 traces out an *ellipse*.

## 9.15 OPTICAL ANISOTROPY

An important characteristic of crystals is that many of their properties depend on the crystal direction; that is, crystals are generally anisotropic. The dielectric constant $\varepsilon_r$ depends on electronic polarization which involves the displacement of electrons with respect to positive atomic nuclei. Electronic polarization depends on the crystal direction inasmuch as it is easier to displace electrons along certain crystal directions. *This means that the refractive index n of a crystal depends on the direction of the electric field in the propagating light beam.* Consequently, the velocity of light in a crystal depends on the direction of propagation and on the state of its polarization, *i.e.*, the direction of the electric field. Most noncrystalline materials, such as glasses and liquids, and all cubic crystals are **optically isotropic,** that is, the refractive index is the same in all directions. For all classes of crystals excluding cubic structures, the refractive index depends on the propagation direction and the state of polarization. The result of optical anisotropy is that, except along certain special directions, any unpolarized light ray entering such a crystal breaks into two different rays with different polarizations and phase velocities. When we view an image through a calcite crystal, an optically anisotropic crystal, we see two images, each constituted by light of different polarization passing through the crystal, whereas there is only one image through an optically isotropic crystal as depicted in Figure 9.33. Optically anisotropic crystals are called **birefringent** because an incident light beam may be doubly refracted.

Experiments and theories on "most anisotropic crystals," *i.e.*, those with the highest degree of anisotropy, show that we can describe light propagation in terms of *three* refractive indices, called **principal refractive indices** $n_1$, $n_2$, and $n_3$, along three mutually orthogonal directions in the crystal, say $x$, $y$, and $z$, called **principal axes.** These

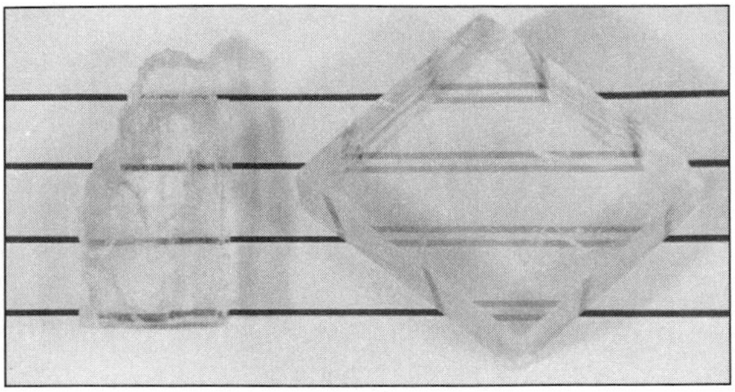

**Figure 9.33** A line viewed through a cubic sodium chloride (halite) crystal (optically isotropic) and a calcite crystal (optically anisotropic).

indices correspond to the polarization state of the EM wave along these axes. In addition, anisotropic crystals may possess one or two optic axes. An **optic axis** is a special direction in the crystal along which the velocity of propagation does *not* depend on the state of polarization. The propagation velocity along the optic axis is the same whatever the polarization of the EM wave.

Crystals that have three distinct principal indices also have *two* optic axes and are called **biaxial crystals.** On the other hand, **uniaxial crystals** have two of their principal indices the same ($n_1 = n_2$) and have only *one* optic axis. Table 9.5 summarizes crystal classifications according to optical anisotropy. Uniaxial crystals, such as quartz, that have $n_3 > n_1$, are called **positive,** and those such as calcite that have $n_3 < n_1$ are called **negative** uniaxial crystals.

**Table 9.5** Principal refractive indices of some optically isotropic and anisotropic crystals (near 589 nm, yellow Na-D line)

| ***Optically Isotropic*** | $n = n_o$ | | |
|---|---|---|---|
| Glass (crown) | 1.510 | | |
| Diamond | 2.417 | | |
| Fluorite ($CaF_2$) | 1.434 | | |
| ***Uniaxial—Positive*** | $n_o$ | $n_e$ | |
| Ice | 1.309 | 1.3105 | |
| Quartz | 1.5442 | 1.5533 | |
| Rutile ($TiO_2$) | 2.616 | 2.903 | |
| ***Uniaxial—Negative*** | $n_o$ | $n_e$ | |
| Calcite ($CaCO_3$) | 1.658 | 1.486 | |
| Tourmaline | 1.669 | 1.638 | |
| Lithium niobate ($LiNbO_3$) | 2.29 | 2.20 | |
| ***Biaxial*** | $n_1$ | $n_2$ | $n_3$ |
| Mica (muscovite) | 1.5601 | 1.5936 | 1.5977 |

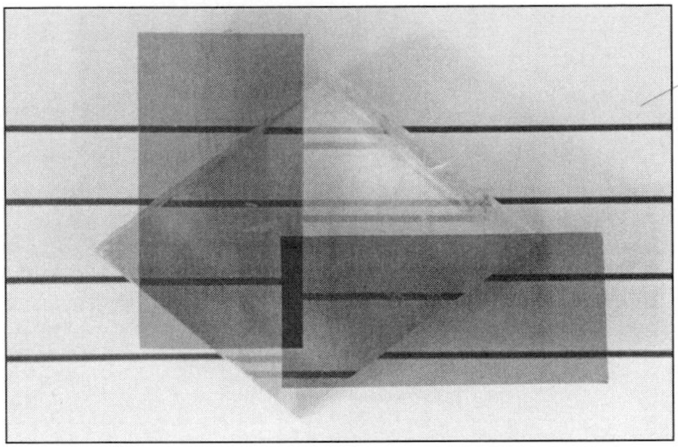

**Figure 9.34** Two polaroid analyzers are placed with their transmission axes, along the long edges, at right angles to each other.

The ordinary ray, undeflected, goes through the left polarizer, whereas the extraordinary wave, deflected, goes through the right polarizer. The two waves therefore have orthogonal polarizations.

## 9.15.1 UNIAXIAL CRYSTALS AND FRESNEL'S OPTICAL INDICATRIX

For our discussions of optical anisotropy, we will consider uniaxial crystals such as calcite and quartz. All experiments and theories lead to the following basic principles.[16]

Any EM wave entering an anisotropic crystal splits into two orthogonal linearly polarized waves that travel with different phase velocities; that is, they experience different refractive indices. These two orthogonally polarized waves in uniaxial crystals are called **ordinary** (*o*) and **extraordinary** (*e*) waves. The *o*-wave has the same phase velocity in all directions and behaves like an ordinary wave in which the field is perpendicular to the phase propagation direction. The *e*-wave has a phase velocity that depends on its direction of propagation and its state of polarization, and further the electric field in the *e*-wave is not necessarily perpendicular to the phase propagation direction. These two waves propagate with the same velocity only along a special direction called the **optic axis.** The *o*-wave is always perpendicularly polarized to the optic axis and obeys the usual Snell's law.

The two images observed through the calcite crystal in Figure 9.33 are due to *o*-waves and *e*-waves being refracted differently, so when they emerge from the crystal they have been separated. Each ray constitutes an image, but the field directions are **orthogonal.** The fact that this is so is easily demonstrated by using two polaroid analyzers with their transmission axes at right angles as in Figure 9.34. If we were to view an object along the optic axis of the crystal, we would not see two images because the two rays would experience the same refractive index.

As mentioned, we can represent the optical properties of a crystal in terms of three refractive indices along three orthogonal axes, the principal axes of the crystal, shown as $x$, $y$, and $z$ in Figure 9.35a. These are special axes along which the polarization vector and the electric field are parallel. (Put differently, the electric displacement[17] **D** and the electric field **E** vectors are parallel.) The refractive indices along these $x$, $y$, and $z$ axes are the principal indices $n_1$, $n_2$, and $n_3$, respectively, for electric

[16] These statements can be proved by solving Maxwell's equations in an anisotropic medium.

[17] Electric displacement **D** at any point is defined by $\mathbf{D} = \varepsilon_o\mathbf{E} + \mathbf{P}$ where **E** is the electric field and **P** is the polarization at that point.

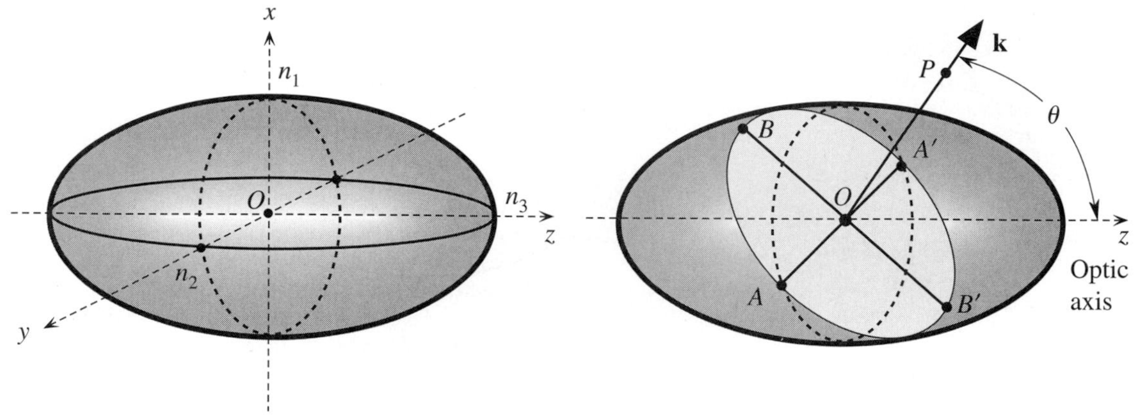


(a) Fresnel's ellipsoid (for $n_1 = n_2 < n_3$; quartz)

(b) An EM wave propagating along *OP* at an angle $\theta$ to optic axis.

**Figure 9.35**

field oscillations along these directions (not to be confused with the wave propagation direction). For example, for a wave with a polarization parallel to the $x$ axes, the refractive index is $n_1$.

The refractive index associated with a particular EM wave in a crystal can be determined by using Fresnel's *refractive index ellipsoid,* called the **optical indicatrix,**[18] which is a refractive index surface placed in the center of the principal axes, as shown in Figure 9.35a, where the $x$, $y$, and $z$ axis intercepts are $n_1$, $n_2$, and $n_3$. If all three indices were the same, $n_1 = n_2 = n_3 = n_o$, we would have a spherical surface and all electric field polarization directions would experience the same refractive index $n_o$. Such a spherical surface would represent an optically isotropic crystal. For positive uniaxial crystals such as quartz, $n_1 = n_2 < n_3$, which is the ellipsoid example shown in Figure 9.35a.

Suppose that we wish to find the refractive indices experienced by a wave traveling with an arbitrary wavevector **k**, which represents the direction of phase propagation. This phase propagation direction is shown as *OP* in Figure 9.35b and is at an angle $\theta$ to the $z$ axis. We place a plane perpendicular to *OP* and passing through the center *O* of the indicatrix. This plane intersects the ellipsoid surface in a curve $ABA'B'$ which is an *ellipse*. The major ($BOB'$) and minor ($AOA'$) axes of this ellipse determine the field oscillation directions and the refractive indices associated with this wave. Put differently, the original wave is now represented by two orthogonally polarized EM waves.

The line $AOA'$, the *minor axis,* corresponds to the polarization of the ordinary wave, and its semiaxis $AA'$ is the refractive index $n_o = n_2$ of this $o$-wave. The electric displacement and the electric field are in the same direction and parallel to $AOA'$. If

[18] There are various names in the literature with various subtle nuances: the Fresnel ellipsoid, optical indicatrix, index ellipsoid, reciprocal ellipsoid, Poinsot ellipsoid, ellipsoid of wave normals.

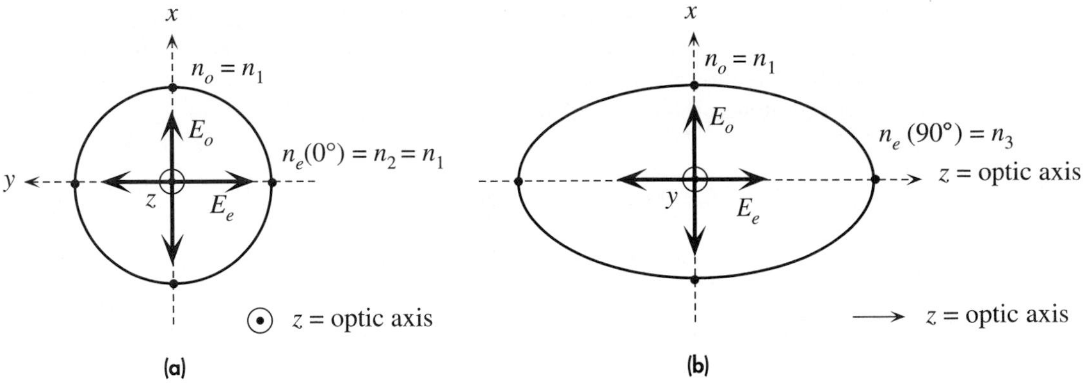


**Figure 9.36** $E_o = E_{o\text{-wave}}$ and $E_e = E_{e\text{-wave}}$.
(a) Wave propagation along the optic axis.
(b) Wave propagation normal to the optic axis.

we were to change the direction of *OP*, we would always find the same minor axis, *i.e.*, $n_o$ is either $n_1$ or $n_2$ whatever the orientation of *OP* (try orientating *OP* to be along $y$ and along $x$). This means that the $o$-wave always experiences the same refractive index in all directions. (The $o$-wave behaves just like an ordinary wave, hence the name.)

The line *BOB*′ in Figure 9.35b, the *major axis,* corresponds to the electric displacement field (**D**) oscillations in the extraordinary wave, and its semiaxis *OB* is the refractive index $n_e(\theta)$ of this $e$-wave. This refractive index is smaller than $n_3$ but greater than $n_2$ $(= n_o)$. The $e$-wave therefore travels more slowly than the $o$-wave in this particular direction and in this crystal. If we change the direction of *OP*, we find that the length of the major axis changes with the *OP* direction. Thus, $n_e(\theta)$ depends on the wave direction $\theta$. As apparent, $n_e = n_o$ when *OP* is along the $z$ axis, that is, when the wave is traveling along $z$ as in Figure 9.36a. This direction is the *optic axis,* and all waves traveling along the optic axis have the same phase velocity whatever their polarization. When the $e$-wave is traveling along the $y$ axis, or along the $x$ axis, $n_e(\theta) = n_3 = n_e$ and the $e$-wave has its slowest phase velocity as shown in Figure 9.36b. Along any *OP* direction that is at an angle $\theta$ to the optic axis, the $e$-wave has a refractive index $n_e(\theta)$ given by

$$\frac{1}{n_e(\theta)^2} = \frac{\cos^2\theta}{n_o^2} + \frac{\sin^2\theta}{n_e^2} \qquad [9.80]$$

*Refractive index of the e-wave*

Clearly, for $\theta = 0°$, $n_e(0°) = n_o$ and for $\theta = 90°$, $n_e(90°) = n_e$.

The major axis *BOB*′ in Figure 9.35b determines the $e$-wave polarization by defining the direction of the displacement vector **D** and not **E**. Although **D** is perpendicular to **k**, this is not true for **E**. The electric field $\mathbf{E}_{e\text{-wave}}$ of the $e$-wave is orthogonal to that of the $o$-wave, and it is in the plane determined by **k** and the optic axis. $\mathbf{E}_{e\text{-wave}}$ is orthogonal to **k** only when the $e$-wave propagates along one of the principal axes. In birefringent crystals it is usual to take the *ray direction* as the direction of energy flow, that is the direction of the Poynting vector (**S**). The $\mathbf{E}_{e\text{-wave}}$ is then orthogonal to the ray direction. For the *o*-wave, the wavefront propagation direction **k** is the same as the energy flow direction **S**. For the *e*-wave, however, the wavefront propagation direction **k** is not the same as the energy flow direction **S**.

### 9.15.2 BIREFRINGENCE OF CALCITE

Consider a calcite crystal ($CaCO_3$) which is a negative uniaxial crystal and also well known for its double refraction. When the surfaces of a calcite crystal have been cleaved, that is, cut along certain crystal planes, the crystal attains a shape that is called a *cleaved form* and the crystal faces are rhombohedrons (parallelogram with 78.08° and 101.92°). A cleaved form of the crystal is called a *calcite rhomb*. A plane of the calcite rhomb that contains the optical axis and is normal to a pair of opposite crystal surfaces is called a *principal section*.

Consider what happens when an unpolarized or natural light enters a calcite crystal at *normal* incidence and thus also normal to a principal section to this surface, but at an angle to the optic axis as shown in Figure 9.37. The ray breaks into ordinary (*o*) and extraordinary (*e*) waves with mutually orthogonal polarizations. The waves propagate in the plane of the principal section as this plane also contains the incident light. The *o*-wave has its field oscillations perpendicular to the optic axis. It obeys Snell's law which means that it enters the crystal undeflected. Thus the direction of *E*-field oscillations must come out of the paper so that it is normal to the optic axis and also to the direction of propagation. The field $E_\perp$ in the *o*-ray is shown as dots, oscillating into and out of the paper.

The *e*-wave has a polarization orthogonal to the *o*-wave and in the principal section. The *e*-wave polarization is in the plane of the paper, indicated as $E_{||}$, in Figure 9.37. It travels with a different velocity and diverges from the *o*-wave. Clearly, the

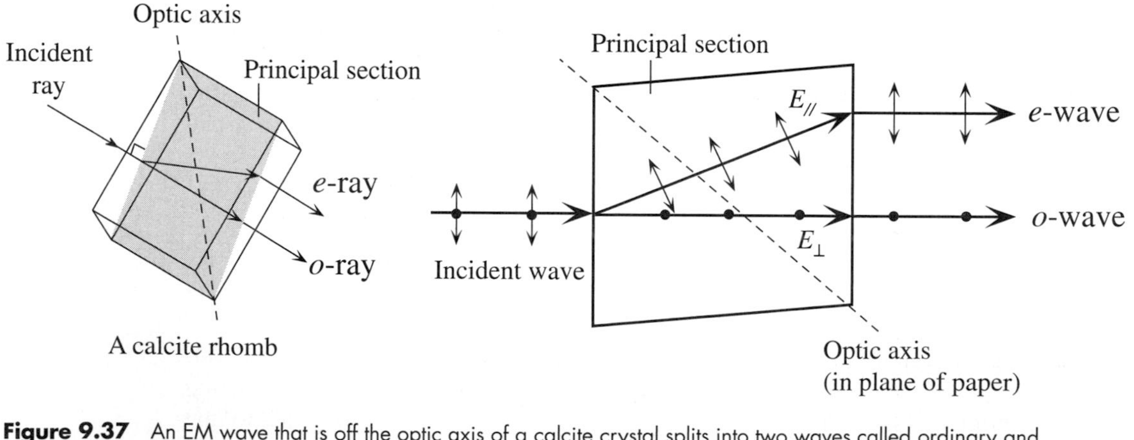


**Figure 9.37** An EM wave that is off the optic axis of a calcite crystal splits into two waves called ordinary and extraordinary waves.
These waves have orthogonal polarizations and travel with different velocities. The o-wave has a polarization that is always perpendicular to the optical axis.

*e*-wave does not obey the usual Snell's law inasmuch as the angle of refraction is not zero. We can determine the *e*-ray direction by noting that the *e*-wave propagates sideways as in Figure 9.37b at right angles to $E_{||}$.

### 9.15.3 DICHROISM

In addition to the variation in the refractive index, some anisotropic crystals also exhibit **dichroism,** a phenomenon in which the optical absorption in a substance depends on the direction of propagation and the state of polarization of the light beam. A dichroic crystal is an optically anisotropic crystal in which either the *e*-wave or the *o*-wave is heavily attenuated (absorbed). This means that a light wave of arbitrary polarization entering a dichroic crystal emerges with a well-defined polarization because the other orthogonal polarization would have been attenuated. Generally dichroism depends on the wavelength of light. For example, in a tourmaline (aluminum borosilicate) crystal, the *o*-wave is much more heavily absorbed with respect to the *e*-wave.

## 9.16 BIREFRINGENT RETARDING PLATES

Consider a positive uniaxial crystal such as a quartz ($n_e > n_o$) plate that has the optic axis (taken along $z$) parallel to the plate faces as in Figure 9.38. Suppose that a *linearly polarized* wave is normally incident on a plate face. If the field **E** is parallel to the optic axis (shown as $E_{||}$), then this wave will travel through the crystal as an *e*-wave with a velocity $c/n_e$ slower than the *o*-wave since $n_e > n_o$. Thus, the optic axis is the "slow axis" for waves polarized parallel to it. If **E** is at right angles to the optic axis (shown as $E_{\perp}$), then this wave will travel with a velocity $c/n_o$, which will be the fastest velocity in the crystal. Thus the axis perpendicular to the optic axis (say $x$) will be the "fast axis" for polarization along this direction. When a light ray enters a crystal at normal incidence to the optic axis and plate surface, then the *o*- and *e*-waves travel along the same direction as shown in Figure 9.38. We can of course resolve a linear polarization at an angle $\alpha$ to $z$ into $E_{\perp}$ and $E_{||}$. The *o*-wave corresponds to the propagation of $E_{\perp}$ and the *e*-wave to the propagation of $E_{||}$ in the crystal. When the light comes out at the

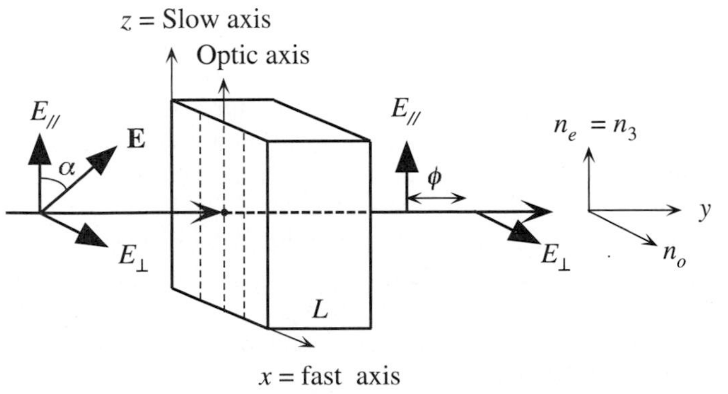


**Figure 9.38** A retarder plate.
The optic axis is parallel to the plate face. The *o*- and *e*-waves travel in the same direction but at different speeds.

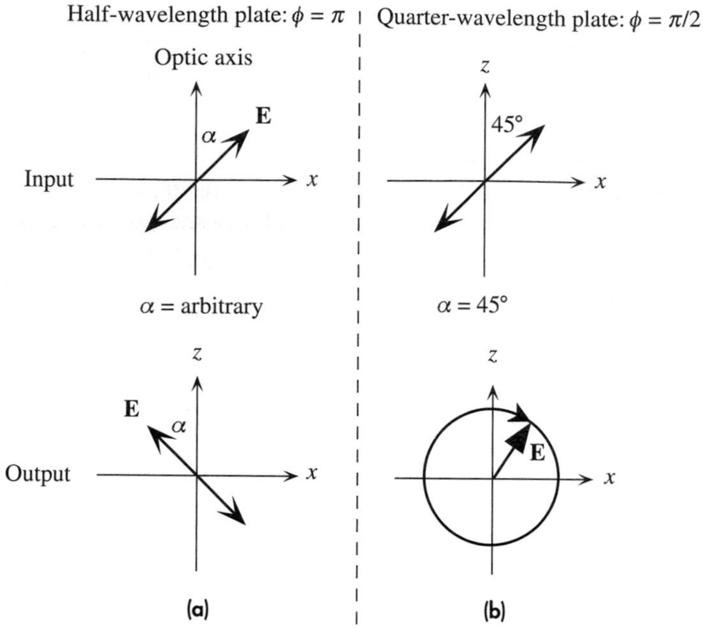


**Figure 9.39** Input and output polarizations of light through (a) a half-wavelength plate and (b) through a quarter-wavelength plate.

opposite face, these two components $E_\perp$ and $E_{||}$ would have been phase shifted by $\phi$. Depending on the initial angle $\alpha$ of **E** and the length of the crystal, which determines the total phase shift $\phi$ through the plate, the emerging beam can have its initial linear polarization rotated, or changed into an elliptically or circularly polarized light as summarized in Figure 9.39.

If $L$ is the thickness of the plate, then the $o$-wave experiences a phase change given by $k_{o\text{-wave}}L$ through the plate where $k_{o\text{-wave}}$ is the wavevector of the $o$-wave; $k_{o\text{-wave}} = (2\pi/\lambda)n_o$, where $\lambda$ is the free-space wavelength. Similarly, the $e$-wave experiences a phase change $(2\pi/\lambda)n_eL$ through the plate. Thus, the phase difference $\phi$ between the orthogonal components $E_\perp$ and $E_{||}$ of the emerging beam is

*Relative phase through retarder plate*

$$\phi = \frac{2\pi}{\lambda}(n_e - n_o)L \qquad [9.81]$$

The phase difference $\phi$ expressed in terms of full wavelengths is called the **retardation** of the plate. For example, a phase difference $\phi$ of 180° is a half-wavelength retardation.

The polarization of the exiting-beam depends on the crystal-type, $(n_e - n_o)$, and the plate thickness $L$. We know that depending on the phase difference $\phi$ between the orthogonal components of the field, the EM wave can be linearly, circularly, or elliptically polarized.

A **half-wave plate retarder** has a thickness $L$ such that the phase difference $\phi$ is $\pi$ or 180°, corresponding to a half wavelength $(\lambda/2)$ of retardation. The result is that $E_{||}$ is delayed by 180° with respect to $E_\perp$. If we add the emerging $E_\perp$ and $E_{||}$ with this phase shift $\phi$, **E** would be at an angle $-\alpha$ to the optic axis and still linearly polarized. **E** has been rotated counterclockwise through $2\alpha$.

A **quarter-wave plate retarder** has a thickness $L$ such that the phase difference $\phi$ is $\pi/2$ or $90°$, corresponding to a quarter wavelength $\frac{1}{4}\lambda$. If we add the emerging $E_\perp$ and $E_{||}$ with this phase shift $\phi$, the emerging light will be elliptically polarized if $0 < \alpha < 45°$ and circularly polarized if $\alpha = 45°$.

**EXAMPLE 9.19**

**QUARTZ HALF-WAVE PLATE** What should be the thickness of a half-wave quartz plate for a wavelength $\lambda \approx 707$ nm given the extraordinary and ordinary refractive indices are $n_o = 1.541$ and $n_e = 1.549$?

**SOLUTION**

Half-wavelength retardation is a phase difference of $\pi$, so from Equation 9.81

$$\phi = \frac{2\pi}{\lambda}(n_e - n_o)L = \pi$$

giving

$$L = \frac{\frac{1}{2}\lambda}{(n_e - n_o)} = \frac{\frac{1}{2}(707 \times 10^{-9}\text{ m})}{(1.549 - 1.541)} = 44.2\ \mu\text{m}$$

This is roughly the thickness of a sheet of paper.

## 9.17 OPTICAL ACTIVITY AND CIRCULAR BIREFRINGENCE

When a linearly polarized light wave is passed through a quartz crystal along its optic axis, it is observed that the emerging wave has its **E**-vector (plane of polarization) rotated, which is illustrated in Figure 9.40. This rotation increases continuously with the distance traveled through the crystal (about 21.7° per mm of quartz). The rotation of the plane of polarization by a substance is called **optical activity.** In very simple intuitive terms, optical activity occurs in materials in which the electron motions induced by the external electromagnetic field follows spiraling or helical paths (orbits).[19] Electrons flowing in helical paths resemble a current flowing in a coil and thus possess a magnetic moment. The optical field in light therefore induces oscillating magnetic moments which can be either parallel or antiparallel to the induced oscillating electric dipoles. Wavelets emitted from these oscillating induced magnetic and electric dipoles interfere to constitute a forward wave that has its optical field rotated either clockwise or counterclockwise.

If $\theta$ is the angle of rotation, then $\theta$ is proportional to the distance $L$ propagated in the optically active medium as depicted in Figure 9.40. For an observer receiving the wave through quartz, the rotation of the plane of polarization may be *clockwise* (to the right) or *counterclockwise* (to the left) which are called *dextrorotatory* and *levorotatory* forms of optical activity. The structure of quartz is such that atomic arrangements spiral around the optic axis either in clockwise or counterclockwise sense. Quartz thus occurs in two distinct crystalline forms, right-handed and left-handed, which exhibit

[19] The explanation of optical activity involves examining both induced magnetic and electric dipole moments which will not be described here in detail.

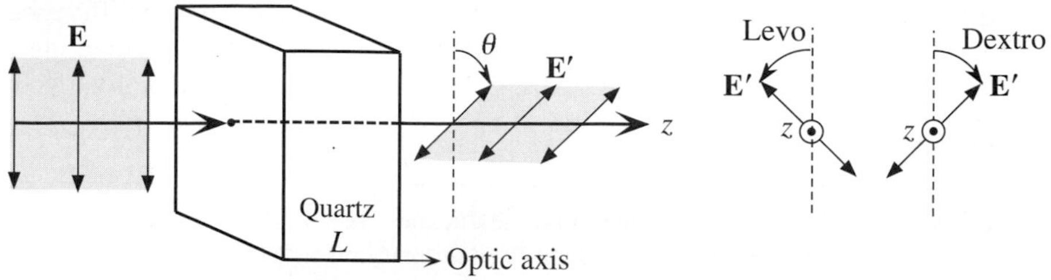


**Figure 9.40** An optically active material such as quartz rotates the plane of polarization of the incident wave: The optical field **E** rotated to **E**′.
If we reflect the wave back into the material, **E**′ rotates back to **E**.

dextrorotatory and levorotatory types of optical activity, respectively. Although we used quartz as an example, there are many substances that are optically active, including various biological substances and even some liquid solutions (*e.g.*, corn syrup) that contain various organic molecules with a rotatory power.

The **specific rotatory power** ($\theta/L$) is defined as the extent of rotation per unit distance traveled in the optically active substance. Specific rotatory power depends on the wavelength. For example, for quartz this is 49° per mm at 400 nm but 17° per mm at 650 nm.

Optical activity can be understood in terms of left and right circularly polarized waves traveling at different velocities in the crystal, *i.e.*, experiencing different refractive indices. Due to the helical twisting of the molecular or atomic arrangements in the crystal, the velocity of a circularly polarized wave depends on whether the optical field rotates clockwise or counterclockwise. A vertically polarized light with a field **E** at the input can be thought of as two right- and left-handed circularly polarized waves $\mathbf{E}_R$ and $\mathbf{E}_L$ that are symmetrical with respect to the $y$ axis, *i.e.*, at any instant $\alpha = \beta$, as shown in Figure 9.41. If they travel at the same velocity through the crystal, then they remain symmetrical with respect to the vertical ($\alpha = \beta$ remains the same) and the resultant is still a vertically polarized light. If, however, these travel at different velocities through a medium, then at the output $\mathbf{E}'_L$ and $\mathbf{E}'_R$ are no longer symmetrical with respect to the vertical, $\alpha' \neq \beta'$, and their resultant is a vector $\mathbf{E}'$ at an angle $\theta$ to the $y$ axis.

Suppose that $n_R$ and $n_L$ are the refractive indices experienced by the right- and left-handed circularly polarized light, respectively. After traversing the crystal length $L$, the phase difference between the two optical fields $\mathbf{E}'_R$ and $\mathbf{E}'_L$ at the output leads to a new optical field $\mathbf{E}'$ that is $\mathbf{E}$ rotated by $\theta$, given by

*Optical activity*

$$\theta = \frac{\pi}{\lambda}(n_L - n_R)L \qquad [9.82]$$

where $\lambda$ is the free-space wavelength. For a left-handed quartz crystal, and for 589 nm light propagation along the optic axis, $n_R = 1.54427$ and $n_L = 1.54420$, which means $\theta$ is about 21.4° per mm of crystal.

In a **circularly birefringent** medium, the right- and left-handed circularly polarized waves propagate with different velocities and experience different refractive indices $n_R$ and $n_L$. Since optically active materials naturally rotate the optical field, it is

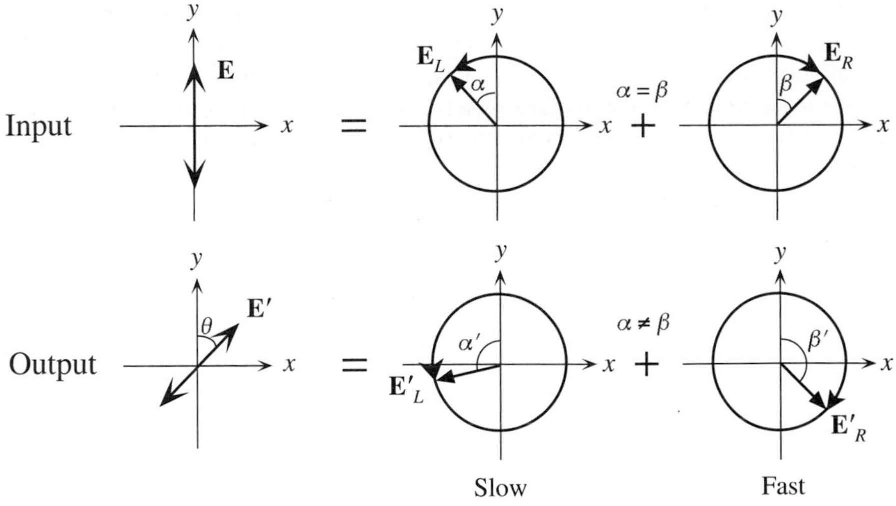


**Figure 9.41** Vertically polarized wave at the input can be thought of as two right- and left-handed circularly polarized waves that are symmetrical; *i.e.*, at any instant $\alpha = \beta$.
If these travel at different velocities through a medium, then at the output they are no longer symmetric with respect to $y$, $\alpha \neq \beta$, and the result is a vector $\mathbf{E}'$ at an angle $\theta$ to $y$.

not unreasonable to expect that a circularly polarized light with its optical field rotating in the same sense as the optical activity will find it easier to travel through the medium. Thus, an optically active medium possesses different refractive indices for right- and left-handed circularly polarized light and exhibits circular birefringence. It should be mentioned that if the direction of the light wave is reversed in Figure 9.40, the ray simply retraces itself and $\mathbf{E}'$ becomes $\mathbf{E}$.

# ADDITIONAL TOPICS

## 9.18 ELECTRO-OPTIC EFFECTS[20]

Electro-optic effects refer to changes in the refractive index of a material induced by the application of an external electric field, which therefore "modulates" the optical properties. We can apply such an external field by placing electrodes on opposite faces of a crystal and connecting these electrodes to a battery. The presence of such a field distorts the electron motions in the atoms or molecules of the substance or distorts the crystal structure resulting in changes in the optical properties. For example, an applied external field can cause an optically isotropic crystal such as GaAs to become birefringent. In this case, the field induces principal axes and an optic axis. Typically changes in the refractive index are small. The frequency of the applied field has to be such that

[20] An extensive discussion and applications of the electro-optic effects may be found in S. O. Kasap, *Optoelectronics and Photonics: Principles and Practices,* Prentice Hall, 2001, Upper Saddle River, NJ, ch. 7.

the field appears static over the time scale it takes for the medium to change its properties, that is, respond, as well as for any light to cross the substance. The electro-optic effects are classified according to first- and second-order effects.

If we were to take the refractive index $n$ to be a function of the applied electric field $E$, that is, $n = n(E)$, we can of course expand this as a Taylor series in $E$. The new refractive index $n'$ is

*Field induced refractive index*

$$n' = n + a_1 E + a_2 E^2 + \cdots \qquad [9.83]$$

where the coefficients $a_1$ and $a_2$ are called the *linear* electro-optic effect and *second-order* electro-optic effect coefficients. Although we would expect even higher terms in the expansion in Equation 9.83, these are generally very small and their effects negligible within the highest practical fields. The change in $n$ due to the first $E$ term is called the **Pockels effect.** The change in $n$ due to the second $E^2$ term is called the **Kerr effect,**[21] and the coefficient $a_2$ is generally written as $\lambda K$ where $K$ is called the Kerr coefficient. Thus, the two effects are

*Pockels effect*

$$\Delta n = a_1 E \qquad [9.84]$$

and

*Kerr effect*

$$\Delta n = a_2 E^2 = (\lambda K) E^2 \qquad [9.85]$$

*All* materials exhibit the Kerr effect. It may be thought that we will always find some (nonzero) value for $a_1$ for all materials, but this is not true and only certain crystalline materials exhibit the Pockels effect. If we apply a field **E** in one direction and then reverse the field and apply −**E**, then according to Equation 9.84, $\Delta n$ should change sign. If the refractive index increases for **E**, it must decrease for −**E**. Reversing the field should *not* lead to an identical effect (the same $\Delta n$). The structure has to respond differently to **E** and −**E**. There must therefore be some *asymmetry* in the structure to distinguish between **E** and −**E**. In a noncrystalline material, $\Delta n$ for **E** would be the same as $\Delta n$ for −**E** as all directions are equivalent in terms of dielectric properties. Thus $a_1 = 0$ for all noncrystalline materials (such as glasses and liquids). Similarly, if the crystal structure has a center of symmetry, then reversing the field direction has an identical effect and $a_1$ is again zero. Only crystals that are **noncentrosymmetric**[22] exhibit the Pockels effect. For example, a NaCl crystal (centrosymmetric) exhibits no Pockels effect, but a GaAs crystal (noncentrosymmetric) does.

The Pockels effect expressed in Equation 9.84 is an oversimplification because in reality we have to consider the effect of an applied field along a particular crystal direction on the refractive index for light with a given propagation direction and polarization. For example, suppose that $x$, $y$, and $z$ are the principal axes of a crystal with refractive indices $n_1$, $n_2$, and $n_3$ along these directions. For an optically isotropic crystal, these would be the same whereas for a uniaxial crystal such as $LiNbO_3$ $n_1 = n_2 \neq n_3$ as depicted in the $xy$ cross section in Figure 9.42a. Suppose that we suitably apply a voltage across a crystal and thereby apply an external dc field $E_a$. In the Pockels effect, the field will modify the

[21] John Kerr (1824–1907) was a Scottish physicist who was a faculty member at Free Church Training College for Teachers, Glasgow (1857–1901) where he set up an optics laboratory and demonstrated the Kerr effect (1875).

[22] A crystal is a center of symmetry about a point $O$, if any atom (or point) with a position vector **r** from $O$ also appears when we invert **r**, that is, take −**r**.

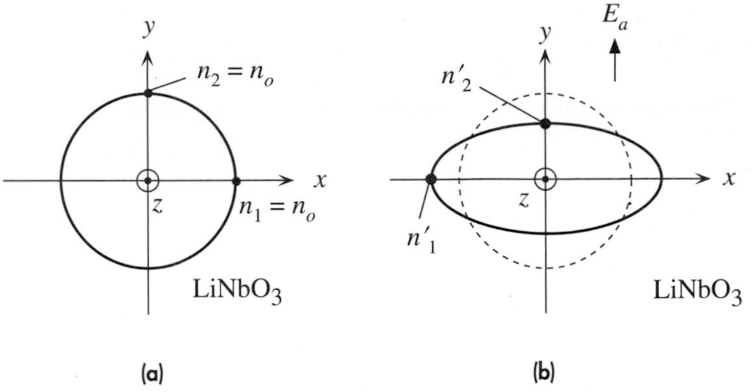


**Figure 9.42**
(a) Cross section of the optical indicatrix with no applied field, $n_1 = n_2 = n_o$.
(b) Applied field along $y$ in $\mathrm{LiNbO_2}$ modifies the indicatrix and changes $n_1$ and $n_2$ to $n'_1$ and $n'_2$.

optical indicatrix. The exact effect depends on the crystal structure. For example, a crystal like GaAs, optically isotropic with a spherical indicatrix, becomes *birefringent* with two different refractive indices. In the case of $\mathrm{LiNbO_3}$ (lithium niobate), which is an optoelectronically important uniaxial crystal, a field $E_a$ along the $y$ direction changes the principal refractive indices $n_1$ and $n_2$ (both equal to $n_o$) to $n'_1$ and $n'_2$ as illustrated in Figure 9.42b. Moreover, in some crystals such as KDP ($\mathrm{KH_2PO_4}$, potassium dihydrogen phosphate), the field $E_a$ along $z$ rotates the principal axes by 45° about $z$ and changes the principal indices. Rotation of principal axes in $\mathrm{LiNbO_3}$ is small and can be neglected.

As an example consider a wave propagating along the $z$ direction (optic axis) in a $\mathrm{LiNbO_3}$ crystal. This wave will experience the same refractive index ($n_1 = n_2 = n_o$) whatever the polarization as in Figure 9.42a. However, in the presence of an applied field $E_a$ parallel to the principal $y$ axis as in Figure 9.42b, the light propagates as two orthogonally polarized waves (parallel to $x$ and $y$) experiencing different refractive indices $n'_1$ and $n'_2$. The applied field thus *induces a birefringence* for light traveling along the $z$ axis. (The field induced rotation of the principal axes in this case, though present, is small and can be neglected.) Before the field $E_a$ is applied, the refractive indices $n_1$ and $n_2$ are both equal to $n_o$. The Pockels effect then gives the new refractive indices $n'_1$ and $n'_2$ in the presence of $E_a$ as

$$n'_1 \approx n_1 + \frac{1}{2}n_1^3 r_{22} E_a \qquad \text{and} \qquad n'_2 \approx n_2 - \frac{1}{2}n_2^3 r_{22} E_a \qquad [9.86]$$

*Pockels effect*

where $r_{22}$ is a constant, called a **Pockels coefficient,** that depends on the crystal structure and the material. The reason for the seemingly unusual subscript notation is that there are more than one constant and these are elements of a tensor that represents the optical response of the crystal to an applied field along a particular direction with respect to the principal axes (the exact theory is more mathematical than intuitive). We therefore have to use the correct Pockels coefficients for the refractive index changes for a given crystal and a given field direction.[23] If the field were along $z$, the Pockels coefficient in Equation 9.86 would be $r_{13}$. Table 9.6 shows some typical values for Pockels coefficients of various crystals.

[23] The reader should not be too concerned with the subscripts but simply interpret them as identifying the right Pockels coefficient value for the particular electro-optic problem at hand.

**Table 9.6** Pockels ($r$) and Kerr ($K$) coefficients in various materials

| Material | Crystal | Indices | Pockels Coefficients $\times 10^{-12}$ m/V | Comment |
|---|---|---|---|---|
| $LiNbO_3$ | Uniaxial | $n_o = 2.272$<br>$n_e = 2.187$ | $r_{13} = 8.6$; $r_{33} = 30.8$<br>$r_{22} = 3.4$; $r_{51} = 28$ | $\lambda \approx 500$ nm |
| KDP | Uniaxial | $n_o = 1.512$<br>$n_e = 1.470$ | $r_{41} = 8.8$; $r_{63} = 10.5$ | $\lambda \approx 546$ nm |
| GaAs | Isotropic | $n_o = 3.6$ | $r_{41} = 1.5$ | $\lambda \approx 546$ nm |

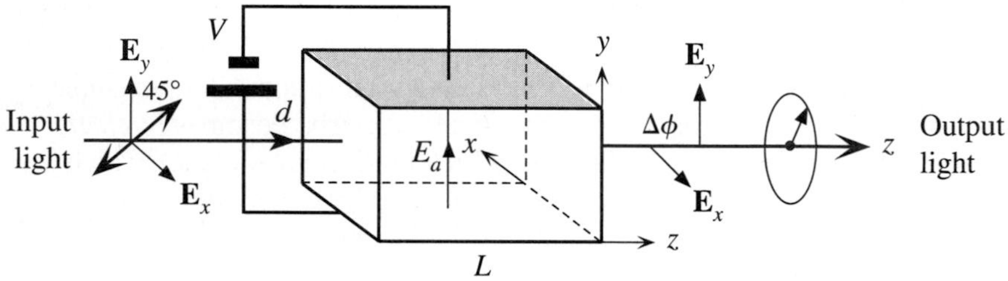


**Figure 9.43** Transverse Pockels cell phase modulator. A linearly polarized input light into an electro-optic crystal emerges as a circularly polarized light.

It is clear that the control of the refractive index by an external applied field (and hence a voltage) is a distinct advantage that enables the phase change through a Pockels crystal to be controlled or modulated; such a **phase modulator** is called a *Pockels cell*. In the *longitudinal Pockels cell phase modulator* the applied field is in the direction of light propagation, whereas in the *transverse phase modulator* the applied field is transverse to the direction of light propagation.

Consider the transverse phase modulator in Figure 9.43. In this example, the applied electric field $E_a = V/d$ is applied parallel to the $y$ direction, normal to the direction of light propagation along $z$. Suppose that the incident beam is linearly polarized (shown as **E**) say at 45° to the $y$ axes. We can represent the incident light in terms of polarizations ($\mathbf{E}_x$ and $\mathbf{E}_y$) along the $x$ and $y$ axes. These components $\mathbf{E}_x$ and $\mathbf{E}_y$ experience refractive indices $n'_1$ and $n'_2$, respectively. Thus, when $\mathbf{E}_x$ traverses the distance $L$, its phase changes by $\phi_1$,

$$\phi_1 = \frac{2\pi n'_1}{\lambda} L = \frac{2\pi L}{\lambda}\left(n_o + \frac{1}{2} n_o^3 r_{22} \frac{V}{d}\right)$$

When the component $\mathbf{E}_y$ traverses the distance $L$, its phase changes by $\phi_2$, given by a similar expression except that $r_{22}$ changes sign. Thus the phase change $\Delta\phi$ between the two field components is

*Transverse Pockels effect*

$$\Delta\phi = \phi_1 - \phi_2 = \frac{2\pi}{\lambda} n_o^3 r_{22} \frac{L}{d} V \qquad [9.87]$$

The applied voltage thus inserts an adjustable phase difference $\Delta\phi$ between the two field components. The polarization state of the output wave can therefore be controlled by the applied voltage and the Pockels cell is a **polarization modulator.** We can change the medium from a quarter-wave to a half-wave plate by simply adjusting $V$. The voltage $V = V_{\lambda/2}$, the **half-wave voltage,** corresponds to $\Delta\phi = \pi$ and generates a half-wave plate.

## CD Selected Topics and Solved Problems

| Selected Topics | Solved Problems |
|---|---|
| Real and Imaginary Dielectric Constant | Fresnel's Equations |
| Optical Dispersion and Absorption | Complex Refractive Index and Light Absorption |
| | Dispersion: Refractive Index versus Wavelength Behavior |

## DEFINING TERMS

**Absorption** is the loss in the power of electromagnetic radiation that is traveling in a medium. The loss is due to the conversion of light energy to other forms of energy, such as lattice vibrations (heat) during the polarization of the molecules of the medium, local vibrations of impurity ions, excitation of electrons from the valence band to the conduction band, and so on.

**Activator** is a luminescent center in a host crystal or glass in which it is excited, by some external excitation such as UV light; following excitation, the activator emits radiation to return to its ground state, or become de-excited.

**Anisotropy (optical)** refers to the fact that the refractive index $n$ of a crystal depends on the direction of propagation of light and on the state of its polarization, that is, the direction of the electric field.

**Antireflection coating** is a thin dielectric layer coated on an optical device or component to reduce the reflection of light and increase the transmitted light intensity.

**Attenuation** is the decrease in the optical power (or irradiance) of a traveling wave in the direction of propagation due to absorption and scattering.

**Attenuation coefficient** $\alpha$ represents the spatial rate of attenuation of an EM wave along the direction of propagation. If $P_o$ is the optical power at some location $O$, and if it is $P$ at a distance $L$ from $O$ along the direction of propagation, then $P = P_o \exp(-\alpha L)$.

**Birefringent crystals** such as calcite are optically anisotropic which leads to an incident light beam becoming separated into ordinary and extraordinary waves with orthogonal polarizations; incident light becomes doubly refracted because these two waves experience different refractive indices $n_o$ and $n_e$.

**Brewster's angle** or **polarization angle** ($\theta_p$) is the angle of incidence that results in the reflected wave having no electric field in the plane of incidence (plane defined by the incident ray and the normal to the surface). The electric field oscillations in the reflected wave are in the plane perpendicular to the plane of incidence.

**Circularly birefringent** medium is a medium in which right and left circularly polarized waves propagate with different velocities and experience different refractive indices $n_R$ and $n_L$.

**Circularly polarized light** is light where the magnitude of the field vector **E** remains constant but its tip at a given location on the direction of propagation traces out a circle by rotating either in a clockwise sense, *right circularly polarized,* with time, as observed by the receiver of the wave, or in a counterclockwise sense, *left circularly polarized.*

**Complex propagation constant** $(k' - jk'')$ describes the propagation characteristics of an electromagnetic wave that is experiencing attenuation as it travels in a lossy medium. If $k = k' - jk''$ is the complex propagation constant, then the electric field component of a plane wave traveling in a lossy medium can be described by

$$E = E_o \exp(-k''z) \exp j(\omega t - k'z)$$

The amplitude decays exponentially while the wave propagates along $z$. The *real* $k'$ part of the complex propagation constant (wavevector) describes the propagation characteristics, that is, the phase velocity $v = \omega/k'$. The *imaginary* $k''$ part describes the rate of attenuation along $z$.

**Complex refractive index** N with real part $n$ and imaginary part $K$ is defined as the ratio of the complex propagation constant $k$ in a medium to propagation constant $k_o$ in free space,

$$\mathrm{N} = n - jK = \frac{k}{k_o} = \left(\frac{1}{k_o}\right)(k' - jk'')$$

The real part $n$ is simply called the refractive index, and $K$ is called the *extinction coefficient.*

**Critical angle** $(\theta_c)$ is the angle of incidence that results in a refracted wave at 90° when the incident wave is traveling in a medium of lower refractive index and is incident at a boundary with a material with a higher refractive index.

**Dielectric mirror** is made from alternating high and low refractive index quarter-wave-thick multilayers such that constructive interference of partially reflected waves gives rise to a high degree of wavelength-selective reflectance.

**Dispersion relation** is a *relationship* between the refractive index $n$ and the wavelength $\lambda$ of the EM wave, $n = n(\lambda)$; the wavelength usually refers to the free-space wavelength. The relationship between the angular frequency $\omega$ and the propagation constant $k$, the $\omega$–$k$ curve, is also called the dispersion relation.

**Dispersive medium** has a refractive index $n$ that depends on the wavelength; that is, $n$ is not a constant.

**Electro-optic effects** refer to changes in the refractive index of a material induced by the application of an external electric field, which therefore "modulates" the optical properties; the applied field is not the electric field of any light wave, but a separate external field.

**Extinction coefficient** is the imaginary part of the complex refractive index $N$.

**Fluorescence** is luminescence that occurs over very short time scales, usually less than $10^{-8}$ seconds (or 10 ns). In fluorescence, the onset and decay of luminescent emission, due to the onset and cessation of excitation of the phosphor, is very short, appearing to be almost instantaneous.

**Fresnel's equations** describe the amplitude and phase relationships between the incident, reflected, and transmitted waves at a dielectric–dielectric interface in terms of the refractive indices of the two media and the angle of incidence.

**Group index** $(N_g)$ represents the factor by which the group velocity of a group of waves in a dielectric medium is reduced with respect to propagation in free space, $N_g = c/v_g$ where $v_g$ is the group velocity.

**Group velocity** $(v_g)$ is the velocity at which energy, or information, is transported by a group of waves; $v_g$ is determined by $d\omega/dk$ whereas phase velocity is determined by $\omega/k$.

**Instantaneous irradiance** is the instantaneous flow of energy per unit time per unit area and is given by the instantaneous value of the Poynting vector **S**.

**Irradiance (average)** is the average flow of energy per unit time per unit area where averaging is typically carried out by the light detector (over many oscillation periods). Average irradiance can also be defined mathematically by the average value of the Poynting vector **S**. The *instantaneous irradiance* can only be measured if the power meter can respond more quickly than the oscillations of the electric field, and since this is in the optical frequencies range, all practical measurements invariably yield the average irradiance.

**Kerr effect** is a second-order effect in which the change in the refractive index $n$ depends on the square of the electric field, that is, $\Delta n = a_2 E^2$, where $a_2$ is a material dependent constant.

**Kramers–Kronig relations** relate the real and imaginary parts of the relative permittivity. If we know the complete frequency dependence of the real part $\varepsilon_r'(\omega)$, using the Kramer–Kronig relation, we can find the frequency dependence of the imaginary part $\varepsilon_r''(\omega)$.

**Luminescence** is the emission of light by a material, called a **phosphor,** due to the absorption and conversion of energy into electromagnetic radiation. Typically the emission of light occurs from certain dopant impurities or even defects, called **luminescent** or **luminescence centers** or **activators** purposefully introduced into a **host matrix,** which may be a crystal or glass, which can accept the activators. **Photoluminescence** involves excitation by photons (light). **Cathodoluminescence** is light emission when the excitation is the bombardment of the phosphor with energetic electrons as in TV cathode ray tubes. **Electroluminescence** is light emission due to the passage of an electric current as in the LED.

**Optic axis** is an axis in the crystal structure along which there is no double refraction for light propagation along this axis.

**Optical activity** is the rotation of the plane of polarization of plane polarized light by a substance such as quartz.

**Optical indicatrix** (Fresnel's ellipsoid) is a refractive index surface placed in the center of the principal axes $x$, $y$, and $z$ of a crystal; the axis intercepts are $n_1$, $n_2$, and $n_3$. We can represent the optical properties of a crystal in terms of three refractive indices along three orthogonal axes, the *principal axes* of the crystal, $x$, $y$, and $z$.

**Phase** of a traveling wave is the quantity $(kx - \omega t)$ which determines the amplitude of the wave at position $x$ and at time $t$ given the propagation constant $k(= 2\pi/\lambda)$ and angular frequency $\omega$. In three dimensions it is the quantity $(\mathbf{k} \cdot \mathbf{r} - \omega t)$ where $\mathbf{k}$ is the wavevector and $\mathbf{r}$ is the position vector.

**Phase velocity** is the rate at which a given phase on a traveling wave advances. It represents the velocity of a given phase rather than the velocity at which information is carried by the wave. Two consecutive peaks of a wave are separated by a wavelength $\lambda$, and it takes a time period $1/\nu$ for one peak to reach the next (or the time separation of two consecutive peaks at one location); then the phase velocity is defined as $v = \lambda\nu$.

**Phosphor** is a substance made of an activator and a host matrix (crystal or glass) that exhibits luminescence upon suitable excitation.

**Phosphorescence** is a slow luminescence process in which luminescent emission occurs well after the cessation of excitation, even after minutes or hours.

**Pockels effect** is a linear change in the refractive index $n$ of a crystal due to an application of an external electric field $E$, other than the field of the light wave, that is, $\Delta n = a_1 E$, where $a_1$ is a constant that depends on the crystal structure.

**Polarization** of an EM wave describes the behavior of the electric field vector in the EM wave as it propagates through a medium. If the oscillations of the electric field at all times are contained within a well-defined line, then the EM wave is said to be *linearly polarized*. The field vibrations and the direction of propagation, *e.g.*, $z$ direction, define a *plane of polarization* (plane of vibration), so linear polarization implies a wave that is plane-polarized.

**Poynting vector (S)** represents the energy flow per unit time per unit area in a direction determined by $\mathbf{E} \times \mathbf{B}$ (direction of propagation), $\mathbf{S} = v^2\varepsilon_o\varepsilon_r\mathbf{E} \times \mathbf{B}$. Its magnitude, power flow per unit area, is called the irradiance.

**Principal axes** of the crystal, normally labeled, $x$, $y$, and $z$, are special axes along which the polarization vector and the electric field are parallel. Put differently, the electric displacement $D$ and the electric field $E$ vectors are parallel. The refractive indices along these $x$, $y$, and $z$ axes are the principal indices $n_1$, $n_2$, and $n_3$, respectively, for electric field oscillations along these directions (not to be confused with the wave propagation direction).

**Reflectance** is the fraction of power in the reflected electromagnetic wave with respect to the incident power.

**Reflection coefficient** is the ratio of the amplitude of the reflected EM wave to that of the incident wave. It can be positive, negative, or a complex number which then represents a phase change.

**Refraction** is a change in the direction of a wave when it enters a medium with a different refractive index. A wave that is incident at a boundary between two media with different refractive indices experiences refraction and changes direction in passing from one to the other medium.

**Refractive index** $n$ of an optical medium is the ratio of the velocity of light in a vacuum to its velocity in the medium $n = c/v$.

**Retarding plates** are optical devices that change the state of polarization of an incident light beam. For example, when a linearly polarized light enters a *quarter-wave plate,* it emerges from the device either as circularly or elliptically polarized light, depending on the angle of the incident electric field with respect to the optic axis of the retarder plate.

**Scattering** is a process by which the energy from a propagating EM wave is redirected as secondary EM waves in various directions away from the original direction of propagation. There are a number of scattering processes. In Rayleigh scattering, fluctuations in the refractive index, inhomogeneities, etc., lead to the scattering of light that decreases with the wavelength as $\lambda^4$.

**Snell's law** is a law that relates the angles of incidence and refraction when an EM wave traveling in one medium becomes refracted as it enters an adjacent medium. If light is traveling in a medium with index $n_1$ is incident on a medium of index $n_2$, and if the angles of incidence and refraction (transmission) are $\theta_i$ and $\theta_t$, then according to Snell's law,

$$\frac{\sin\theta_i}{\sin\theta_t} = \frac{n_2}{n_1}$$

**Specific rotatory power** is defined as the amount of rotation of the optical field in a linearly polarized light per unit distance traveled in the optically active substance.

**Stoke's shift** in luminescence is the shift down in the frequency of the emitted radiation with respect to that of the exciting radiation.

**Total internal reflection** (TIR) is the total reflection of a wave traveling in a medium when it is incident at a boundary with another medium of lower refractive index. The angle of incidence must be greater than the critical angle $\theta_c$ which depends on the refractive indices $\sin\theta_c > n_2/n_1$.

**Transmission coefficient** is the ratio of the amplitude of the transmitted wave to that of the incident wave when the incident wave traveling in a medium meets a boundary with a different medium (different refractive index).

**Transmittance** is the fraction of transmitted intensity when a wave traveling in a medium is incident at a boundary with a different medium (different refractive index).

**Wavefront** is a surface where all the points have the same phase. A wavefront on a plane wave is an infinite plane perpendicular to the direction of propagation.

**Wavenumber** or **propagation constant** is defined as $2\pi/\lambda$ where $\lambda$ is the wavelength. It is the phase shift in the wave over a distance of unit length.

**Wavepacket** is a group of waves with slightly different frequencies traveling together and forming a "group." This wavepacket travels with a group velocity $v_g$ that depends on the slope of $\omega$ versus $k$ characteristics of the wavepacket, *i.e.*, $v_g = d\omega/dk$.

**Wavevector** is a vector denoted as **k** that describes the direction of propagation of a wave and has the magnitude of the wavenumber, $k = 2\pi/\lambda$.

# QUESTIONS AND PROBLEMS

**9.1** **Refractive index and relative permittivity** Using $n = \sqrt{\varepsilon_r}$, calculate the refractive index $n$ of the materials in the table given their low-frequency relative permittivities $\varepsilon_r$ (LF). What is your conclusion?

| | Material | | | |
|---|---|---|---|---|
| | a-Se | Ge | NaCl | MgO |
| $\varepsilon_r$ (LF) | 6.4 | 16.2 | 5.90 | 9.83 |
| $n$ (~ 1–5 μm) | 2.45 | 4.0 | 1.54 | 1.71 |

**9.2** **Refractive index and bandgap** Diamond, silicon, and germanium all have the same diamond unit cell. All three are covalently bonded solids. Their refractive indices ($n$) and energy bandgaps ($E_g$) are shown in the table. ($a$) Plot $n$ versus $E_g$ and ($b$) plot $n^4$ versus $1/E_g$. What is your conclusion? According to **Moss's rule,** very roughly,

*Moss's rule*

$$n^4 E_g \approx K = \text{Constant}$$

What is the value of $K$?

| | Material | | |
|---|---|---|---|
| | **Diamond** | **Silicon** | **Germanium** |
| Bandgap, $E_g$ (eV) | 5 | 1.1 | 0.66 |
| $n$ | 2.4 | 3.46 | 4.0 |

***9.3** **Temperature coefficient of refractive index** Suppose that we could write the relationship between the refractive index $n$ (at frequencies much less than ultraviolet light) and the bandgap $E_g$ of a semiconductor as suggested by Hervé and Vandamme,

$$n^2 = 1 + \left(\frac{A}{E_g + B}\right)^2$$

where $E_g$ is in eV, $A = 13.6$ eV, and $B = 3.4$ eV. ($B$ depends on the incident photon energy.) Temperature dependence in $n$ results from $dE_g/dT$ and $dB/dT$. Show that the temperature coefficient of refractive index (TCRI) is given by,[24]

*Hervé–Vandamme relationship*

$$\text{TCRI} = \frac{1}{n}\cdot\frac{dn}{dT} = -\frac{(n^2-1)^{3/2}}{13.6n^2}\left[\frac{dE_g}{dT} + B'\right]$$

where $B'$ is $dB/dT$. Given that $B' = 2.5 \times 10^{-5}$ eV K$^{-1}$, calculate TCRI for two semiconductors: Si with $n \approx 3.5$ and $dE_g/dT \approx -3 \times 10^{-4}$ eV K$^{-1}$, and AlAs with $n \approx 3.2$ and $dE_g/dT \approx -4 \times 10^{-4}$ eV K$^{-1}$.

**9.4** **Sellmeier dispersion equation** Using the Sellmeier equation and the coefficients in Table 9.2, calculate the refractive index of fused silica ($SiO_2$) and germania ($GeO_2$) at 1550 nm. Which is larger, and why?

**9.5** **Dispersion ($n$ versus $\lambda$) in GaAs** By using the dispersion relation for GaAs, calculate the refractive index $n$ and the group index $N_g$ of GaAs at a wavelength of 1300 nm.

**9.6** **Cauchy dispersion equation** Using the Cauchy coefficients and the general Cauchy equation, calculate the refractive index of a silicon crystal at 200 μm and at 2 μm, over two orders of magnitude wavelength change. What is your conclusion? Would you expect a significant chnge in $n$ for $\hbar\omega > E_g$?

**9.7** **Cauchy dispersion relation for zinc selenide** ZnSe is a II–VI semiconductor and a very useful optical material used in various applications such as optical windows (especially high-power laser windows), lenses, prisms, etc. It transmits over 0.50 to 19 μm. $n$ in the 1–11 μm range described by a Cauchy expression of the form

*ZnSe dispersion relation*

$$n = 2.4365 + \frac{0.0485}{\lambda^2} + \frac{0.0061}{\lambda^4} - 0.0003\lambda^2$$

in which $\lambda$ is in μm. What is ZnSe's refractive index $n$ and group index $N_g$ at 5 μm?

***9.8** **Dispersion ($n$ versus $\lambda$)** Consider an atom in the presence of an oscillating electric field as in Figure 9.4. The applied field oscillates harmonically in the $+x$ and $-x$ directions and is given by $E = E_o \exp(j\omega t)$. The energy losses can be represented by a frictional force whose magnitude is proportional to the velocity

[24] P. J. L. Hervé and L. K. J. Vandamme, *J. Appl. Phys.*, *77*, 5476, 1995 and references therein.

$dx/dt$. If $\gamma$ is the proportionality constant per electron and per unit electron mass, then Newton's second law for $Z$ electrons in the polarized atom is

$$Zm_e \frac{d^2x}{dt^2} = -ZeE_o \exp(j\omega t) - Zm_e\omega_o^2 x - Zm_e\gamma \frac{dx}{dt}$$

where $\omega_o = (\beta/Zm_e)^{1/2}$ is the **natural frequency** of the system composed of $Z$ electrons and a $+Ze$ nucleus and $\beta$ is a force constant for the restoring Coulombic force between the electrons and the nucleus. Show that the electronic polarizability $\alpha_e$ is

*Electronic polarizability*

$$\alpha_e = \frac{p_{\text{induced}}}{E} = \frac{Ze^2}{m_e(\omega_o^2 - \omega^2 + j\gamma\omega)}$$

What does a complex polarizability represent? Since $\alpha_e$ is a complex quantity, so is $\varepsilon_r$ and hence the refractive index. By writing the complex refractive index $\mathrm{N} = \sqrt{\varepsilon_r}$ where $\varepsilon_r$ is related to $\alpha_e$ by the Clausius–Mossotti equation, show that

*Complex refractive index*

$$\frac{\mathrm{N}^2 - 1}{\mathrm{N}^2 + 2} = \frac{NZe^2}{3\varepsilon_o m_e(\omega_o^2 - \omega^2 + j\gamma\omega)}$$

where $N$ is the number of atoms per unit volume. What are your conclusions?

**9.9 Dispersion and diamond** Consider applying the simple electronic polarizability and Clausius–Mossotti equations to diamond. Neglecting losses,

$$\alpha_e = \frac{Ze^2}{m_e(\omega_o^2 - \omega^2)}$$

and

*Dispersion in diamond*

$$\frac{\varepsilon_r - 1}{\varepsilon_r + 2} = \frac{NZe^2}{3\varepsilon_o m_e(\omega_o^2 - \omega^2)}$$

For diamond we can take $Z = 4$ (valence electrons only as these are the most responsive), $N = 1.8 \times 10^{29}$ atoms m$^{-3}$, $\varepsilon_r(\text{DC}) = 5.7$. Find $\omega_o$ and then find the refractive index at $\lambda = 0.5$ μm and 5 μm.

**9.10 Electric and magnetic fields in light** The intensity (irradiance) of the red laser beam from a He–Ne laser in air has been measured to be about 1 mW cm$^{-2}$. What are the magnitudes of the electric and magnetic fields? What are the magnitudes if this 1 mW cm$^{-2}$ beam were in a glass medium with a refractive index $n = 1.45$ and still had the same intensity?

**9.11 Reflection of light from a less dense medium (internal reflection)** A ray of light which is traveling in a glass medium of refractive index $n_1 = 1.450$ becomes incident on a less dense glass medium of refractive index $n_2 = 1.430$. Suppose that the free-space wavelength ($\lambda$) of the light ray is 1 μm.

*a.* What should be the minimum incidence angle for TIR?

*b.* What is the phase change in the reflected wave when $\theta_i = 85°$ and when $\theta_i = 90°$?

*c.* What is the penetration depth of the evanescent wave into medium 2 when $\theta_i = 85°$ and when $\theta_i = 90°$?

**9.12 Internal and external reflection at normal incidence** Consider the reflection of light at normal incidence on a boundary between a GaAs crystal medium of refractive index 3.6 and air of refractive index 1.

*a.* If light is traveling from air to GaAs, what is the reflection coefficient and the intensity of the reflected light in terms of the incident light?

*b.* If light is traveling from GaAs to air, what is the reflection coefficient and the intensity of the reflected light in terms of the incident light?

**9.13 Antireflection coating**

*a.* Consider three dielectric media with flat and parallel boundaries with refractive indices $n_1$, $n_2$, and $n_3$. Show that for normal incidence the reflection coefficient between layers 1 and 2 is the same as that between layers 2 and 3 if $n_2 = \sqrt{n_1 n_3}$. What is the significance of this?

*b.* Consider a Si photodiode that is designed for operation at 900 nm. Given a choice of two possible antireflection coatings, $SiO_2$ with a refractive index of 1.5 and $TiO_2$ with a refractive index of 2.3, which would you use and what would be the thickness of the antireflection coating you chose? The refractive index of Si is 3.5.

**9.14** **Optical fibers in communications** Optical fibers for long-haul applications usually have a core region that has a diameter of about 10 μm, and the whole fiber would be about 125 μm in diameter. The core and cladding refractive indices, $n_1$ and $n_2$, respectively, are normally only 0.3–0.5 percent different. Consider a fiber with $n_1(\text{core}) = 1.4510$, and $n_2(\text{cladding}) = 1.4477$, both at 1550 nm. What is the maximum angle that a light ray can make with the fiber axis if it is still to propagate along the fiber?

**9.15** **Optical fibers in communications** Consider a short-haul optical fiber that has $n_1(\text{core}) = 1.455$ and $n_2(\text{cladding}) = 1.440$ at 870 nm. Assume the core-cladding interface behaves like the flat interface between two infinite media as in Figure 9.11. Consider a ray that is propagating that has an angle of incidence 85° at the core–cladding interface. Can this ray exercise total internal reflection? What would be its penetration depth into the cladding?

**9.16** **Complex refractive index** Spectroscopic ellipsometry measurements on a silicon crystal at a wavelength of 620 nm show that the real and imaginary parts of the complex relative permittivity are 15.2254 and 0.172, respectively. Find the complex refractive index. What is the reflectance and absorption coefficient at this wavelength? What is the phase velocity?

*Free carrier absorption*

**9.17** **Complex refractive index** Spectroscopic ellipsometry measurements on a germanium crystal at a photon energy of 1.5 eV show that the real and imaginary parts of the complex relative permittivity are 21.56 and 2.772, respectively. Find the complex refractive index. What is the reflectance and absorption coefficient at this wavelength? How do your calculations match with the experimental values of $n = 4.653$ and $K = 0.298$, $R = 0.419$ and $\alpha = 4.53 \times 10^6$ m$^{-1}$?

**9.18** **Free carrier absorption** An *n*-type germanium sample has a conductivity of about 300 $\Omega^{-1}$m$^{-1}$. Calculate the imaginary part $\varepsilon_r''$ of the relative permittivity at a wavelength of 20 μm. Find the attenuation coefficient $\alpha$ due to free carrier absorption. The refractive index of germanium at the specified wavelength is $n = 4$.

**9.19** **Reststrahlen absorption in CdTe** Figure 9.22 shows the infrared extinction coefficient $K$ of CdTe. Calculate the absorption coefficient $\alpha$ and the reflectance $R$ of CdTe at 60 μm and 80 μm.

**9.20** **Reststrahlen absorption in GaAs** Figure 9.22 shows the infrared extinction coefficient $K$ of GaAs as a function of wavelength. Optical measurements show that $K$ peaks at $\lambda = 37.1$ μm where $K \approx 11.6$ and $n \approx 6.6$. Calculate the absorption coefficient $\alpha$ and the reflectance $R$ at this wavelength.

**9.21** **Fundamental absorption** Consider the semiconductors in Figure 9.23, and those semiconductors listed in Table 9.3.

*a.* Which semiconductors can be candidates for a photodetector that can detect light in optical communications at 1550 nm?

*b.* For amorphous Si (a-Si), one definition of an *optical gap* is the photon energy that results in an optical absorption coefficient $\alpha$ of $10^4$ cm$^{-1}$. What is the optical gap of a-Si in Figure 9.23?

*c.* Consider a solar cell from crystalline Si. What is the absorption depth of light at 1000 nm, and at 500 nm?

**9.22** **Quartz half-wave plate** What are the possible thicknesses of a half-wave quartz plate for a wavelength $\lambda \approx 1.01$ μm given the extraordinary and ordinary refractive indices are $n_o = 1.534$ and $n_e = 1.543$, respectively?

**9.23** **Pockels cell modulator** What should be the aspect ratio $d/L$ for the transverse $LiNiO_3$ phase modulator in Figure 9.43 that will operate at a free-space wavelength of 1.3 μm and will provide a phase shift $\Delta\phi$ of $\pi$ (half wavelength) between the two field components propagating through the crystal for an applied voltage of 20 V? The Pockels coefficient $r_{22}$ is $3.2 \times 10^{-12}$ m/V and $n_o = 2.2$.

appendix

# A

# Bragg's Diffraction Law and X-ray Diffraction

## Bragg's Diffraction Condition

**X-rays** are electromagnetic (EM) waves with wavelengths typically in the range from 0.01 nm to a few nanometers. This wavelength region is comparable with typical interplanar spacings in crystals. When an X-ray beam impinges on a crystal, the waves in the beam interact with the planes of atoms in the crystal and, as a result, the waves become scattered and the X-ray beam becomes diffracted. An analogy with radio waves may help. Radio waves with wavelengths in the range 1–10 m (short waves and VHF waves) easily interact with objects of comparable size. It is well known that these radio waves become scattered by objects of comparable size such as trees, houses, and buildings. However, long-wave radio waves with wavelengths in kilometers do not become scattered by these objects because the object sizes now are much smaller than the wavelength.

When X-rays strike a crystal, the EM waves penetrate the crystal structure. Each plane of atoms in the crystal reflects a portion of the waves. The reflected waves from different planes then interfere with each other and give rise to a **diffracted beam** which is at a well-defined angle $2\theta$ to the incident beam as depicted in Figure A.1. Some of the incident beam goes through the crystal undiffracted and some of the beam becomes diffracted. Further, the diffracted rays exist only in certain directions. These diffraction directions correspond to well-defined diffraction angles $2\theta$, as defined in Figure A.1. The diffraction angle $2\theta$, the wavelength of the X-rays $\lambda$, and the interplanar separation $d$ of the diffraction planes within the crystal are related through the **Bragg diffraction condition,** that is,

*Bragg's law*

$$2d \sin\theta = n\lambda \qquad n = 1, 2, 3, \ldots \qquad \textbf{[A.1]}$$

**Figure A.1** A schematic illustration of X-ray diffraction by a crystal.
X-rays penetrate the crystal and then become diffracted by a series of atomic planes.

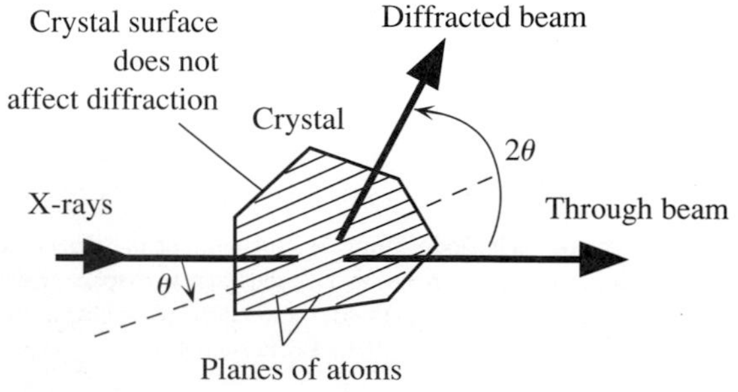

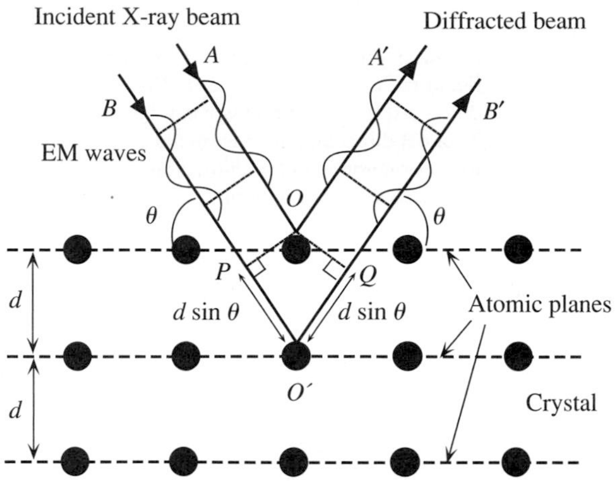


**Figure A.2** Diffraction involves X-ray waves being reflected by various atomic planes in the crystal. These waves interfere constructively to form a diffracted beam only for certain diffraction angles that satisfy the Bragg condition.

Consider X-rays penetrating a crystal structure and becoming reflected by a given set of atomic planes as shown in Figure A.2. We can consider an X-ray beam to be many parallel waves that are in phase. These waves penetrate the crystal structure and become reflected at successive atomic planes. The interplanar separation of these planes is $d$. Waves reflected from adjacent atomic planes interfere constructively to constitute a diffracted beam only when the path difference between the rays is an integer multiple of the wavelength—a requirement of *constructive interference*. This will only be the case for certain directions of reflection. For simplicity, we will consider two waves $A$ and $B$ in an X-ray beam being reflected from two consecutive atomic planes in the crystal. The angle between the X-rays and the atomic planes is $\theta$ as defined in Figure A.2. Initially the waves $A$ and $B$ are in phase. Wave $A$ is reflected from the first plane, whereas wave $B$ is reflected from the second plane. When wave $A$ is reflected at $O$, wave $B$ is at $P$. Wave $B$ becomes reflected from $O'$ on the second plane and then moves along reflected $B'$. Wave $B$ has to travel a further distance, $PO'Q$, equivalent to $2d \sin\theta$ before reaching wave $A$. The path difference between the two reflected waves $A'$ and $B'$ is $PO'Q$ or $2d \sin\theta$. For constructive interference this must be $n\lambda$ where $n$ is an integer. Otherwise the reflected waves will interfere destructively and cancel each other out. Thus the condition for the existence of a diffracted beam is that the path difference between $A'$ and $B'$ should be a multiple of the wavelength $\lambda$; which is Equation A.1. The diffraction condition in Equation A.1 is referred to as **Bragg's law.** The angle $\theta$ is called the **Bragg angle,** whereas $2\theta$ is called the **diffraction angle.** The index $n$ is called the order of diffraction. The incidence angle $\theta$ is the angle between the incident X-ray and the atomic planes within the crystal and not the angle at the actual crystal surface. The crystal surface, whatever shape, does not affect the diffraction process because X-rays penetrate the crystal and then become diffracted by a series of parallel atomic planes. The Bragg diffraction condition has much wider applications than just crystallography; for example, it is of central importance to the operation of modern semiconductor lasers.

## X-ray Diffraction and Study of Crystal Structures

When an X-ray beam is incident on a single crystal, the scattered beam from a given set of planes in the crystal is at an angle $2\theta$ that satisfies the Bragg law. In three dimensions, all directions from the crystal that are at an angle $2\theta$ to the incident beam define a cone as shown in Figure A.3a

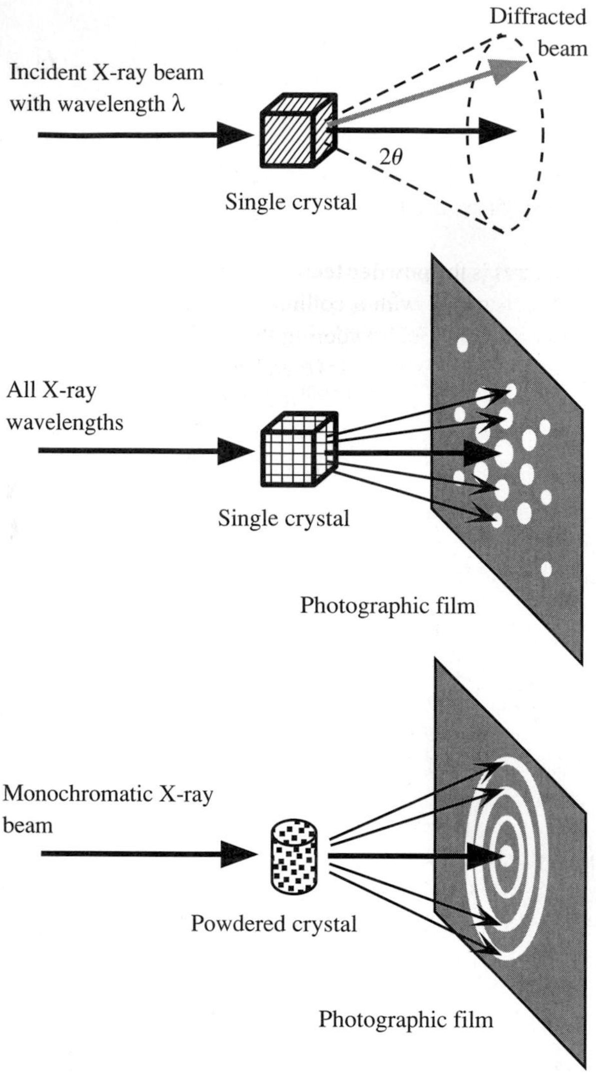


**(a) All 2θ directions around the incident beam define a diffraction cone. The diffracted beam lies on the cone, but its exact direction depends on the exact orientation of the diffraction planes to the incident beam.**

**(b) *Laue technique.* A single crystal is irradiated with a beam of white X-rays. Diffracted X-rays give a spot diffraction pattern on a photographic plate.**

**(c) *Powdered crystal technique.* A sample of powdered crystal is irradiated with a monochromatic (single wavelength) X-ray beam. Diffracted X-rays give diffraction rings on a photographic plate.**

**Figure A.3**

with its apex at the crystal. This is called a *diffraction cone*. There are many such diffraction cones, each corresponding to a different set of diffraction planes with a distinct set of Miller indices ($hkl$). Although all lines lying on a diffraction cone satisfy the Bragg condition, the exact direction of the diffracted beam depends on the exact orientation (or tilt) of the diffracting planes to the incident ray. When a monochromatic X-ray beam is incident on a single crystal, as illustrated in Figure A.3a, the diffracted beam is along one particular direction on the diffraction cone for that set of diffraction planes ($hkl$) with a particular orientation to the incident beam.

The **Laue technique** of studying crystal structures involves irradiating a single crystal with a white X-ray beam that has a wide range of wavelengths. A photographic plate is used to capture the diffraction pattern as shown in Figure A.3b. Effectively we are scanning the wavelength $\lambda$ and picking up diffractions from various (*hkl*) planes each time the Bragg condition is satisfied. Thus, whenever $\lambda$ and $d$ for a particular set of (*hkl*) planes satisfy the Bragg condition, there is a diffraction. The diffraction pattern is a spot pattern where each spot is the result of diffraction from a given set of (*hkl*) planes oriented in a particular way to the incident beam. By using a range of wavelengths we ensure that the required wavelength is available for obtaining diffraction for a given set of planes. The relative positions of the spots are used to determine the crystal structure.

One of the simplest methods for studying crystal structures is the **powder technique** which involves irradiating a powdered crystal, or a polycrystalline sample, with a collimated X-ray beam of known wavelength (monochromatic) as shown in Figure A.3c. Powdering the crystal enables a given set of $(hkl)$ planes to receive the X-rays at many different angles $\theta$ and at many different orientations, or tilts. Put differently, it allows the angle $\theta$ to be scanned for differently oriented crystals. Since all possible crystal orientations are present by virtue powdering, the diffracted rays form diffraction cones and the diffraction pattern developed on a photographic plate has *diffraction rings* as shown in Figure A.3c.

Each diffraction ring in the powder technique in Figure A.3c represents diffraction from a given set of (*hkl*) planes. Whenever the angle $\theta$ satisfies the Bragg law for a given set of atomic planes, with Miller indices (*hkl*) and with an interplanar separation $d_{hkl}$, there is a diffracted beam. An X-ray detector placed at an angle $2\theta$ with respect to the through-beam will register a peak in the detected X-ray intensity, as shown in Figure A.4a. The instrument that allows this type of X-ray diffraction study is called a **diffractometer.** The variation of the detected intensity with the diffraction angle $2\theta$ represents the diffraction pattern of the crystal. The particular diffraction pattern depicted in Figure A.4b is for aluminum, an FCC crystal. Different crystals exhibit different diffraction patterns.

In the case of cubic crystals, the interplanar spacing $d$ is related to the Miller indices of a plane (*hkl*). The separation $d_{hkl}$ between adjacent (*hkl*) planes is given by

*Interplanar separation in cubic crystals*

$$d_{hkl} = \frac{a}{\sqrt{h^2 + k^2 + l^2}} \qquad \textbf{[A.2]}$$

where $a$ is the lattice parameter (side of the cubic unit cell). When we substitute for $d = d_{hkl}$ in the Bragg condition in Equation A.1, square both sides, and rearrange the equation, we find

*Bragg condition for cubic crystals*

$$(\sin\theta)^2 = \frac{n^2\lambda^2}{4a^2}(h^2 + k^2 + l^2) \qquad \textbf{[A.3]}$$

This is essentially **Bragg's law for cubic crystals.** The diffraction angle increases with $(h^2 + k^2 + l^2)$. Higher-order Miller indices, those with greater values of $(h^2 + k^2 + l^2)$, give rise to wider diffraction angles. For example, the diffraction angle for (111) is smaller than that for (200) because $(h^2 + k^2 + l^2)$ is 3 for (111) and 4 for (200). Furthermore, with $\lambda$ and $a$ values that are typically involved in X-ray diffraction, second- and higher-order diffraction peaks, $n = 2, 3, \ldots$, can be ruled out.

In the case of the simple cubic crystal all possible (*hkl*) planes give rise to diffraction peaks with diffraction angles satisfying the Bragg law or Equation A.3. The latter equation therefore defines a diffraction pattern for the simple cubic crystal structure because it generates all the possible values of $2\theta$ for all the planes in the cubic crystal. In the case of FCC and BCC crystals, however, not all (*hkl*) planes give rise to diffraction peaks predicted by Equation A.3. Examination

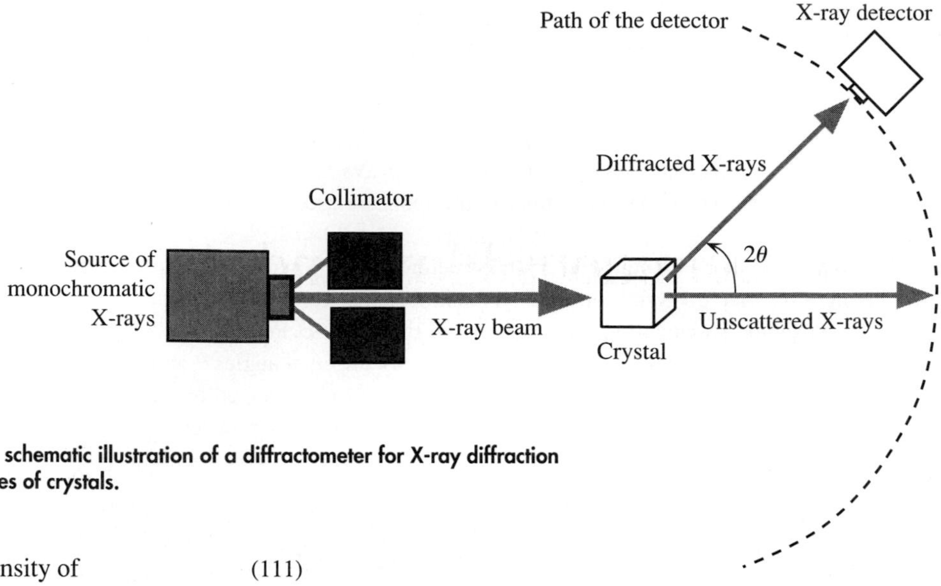


**(a) A schematic illustration of a diffractometer for X-ray diffraction studies of crystals.**

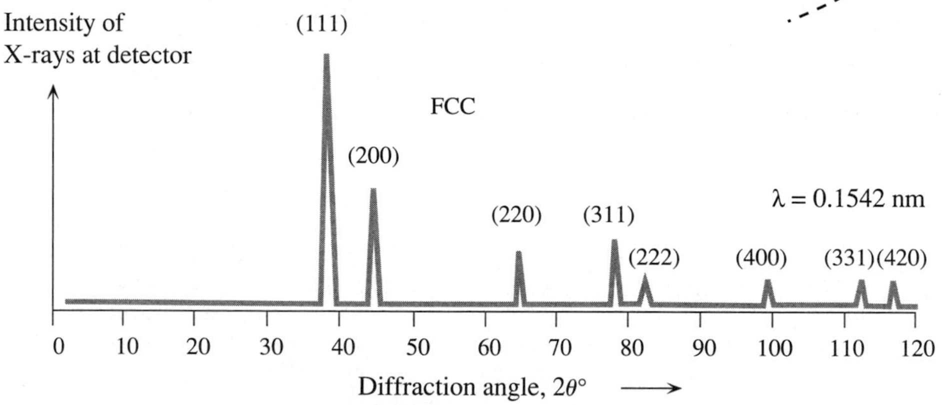


**(b) A schematic diagram illustrating the intensity of X-rays as detected in (a) versus the diffraction angle, 2*θ*, for an FCC crystal (*e.g.*, Al).**

**Figure A.4** A schematic diagram of a diffractometer and the diffraction pattern obtained from an FCC crystal.

of the diffraction pattern in Figure A.4b for an FCC crystal shows that only those planes with Miller indices that are either all odd or all even integers give rise to diffraction peaks. There are no diffractions from those planes with mixed odd and even integers.

The Bragg law for the cubic crystals in Equation A.3 is a necessary diffraction condition but not sufficient because diffraction involves the interaction of EM waves with the electrons in the crystal. To determine whether there will be a diffraction peak from a set of planes in a crystal we also have to consider the distributions of the atoms and their electrons in the crystal. In FCC and BCC structures diffractions from certain planes are missing because the atoms on these planes give rise to out-of-phase reflections.

appendix

# B

# Flux, Luminous Flux, and the Brightness of Radiation

Many optoelectronic light emitting devices are compared by their luminous efficiencies, which requires a knowledge of photometry. **Radiometry** is the science of radiation measurement, for example, the measurement of emitted, absorbed, reflected, transmitted radiation energy; radiation is understood to mean electromagnetic energy in the optical frequency range (ultraviolet, visible, and infrared). **Photometry,** on the other hand, is a subset of radiometry in which radiation is measured with respect to the spectral responsivity of the eye, that is, over the visible spectrum and by taking into account the spectral visual sensitivity of the eye under normal light adapted conditions, *i.e.*, *photopic* conditions.

**Flux** ($\Phi$) in radiometry has three related definitions, **radiant, luminous** and **photon flux,** which correspond to the rate of flow of radiation energy, perceptible visual energy, and photons, respectively. (Notice that, in radiometry, these fluxes are not defined in terms of flow *per unit area.*) For example, **radiant flux** is the energy flow per unit time in the units of Watts. Radiometric quantities, such as *radiant flux* $\Phi_e$, radiant energy flow per unit time, usually have a subscript $e$ and invariably involve energy or power. Radiometric *spectral* quantities, such as *spectral radiant flux* $\Phi_\lambda$, refer to the radiometric quantity per unit wavelength; *i.e.*, $\Phi_\lambda = d\Phi_e/d\lambda$ is the radiant flux per unit wavelength.

**Luminous flux** or **photometric flux** $\Phi_v$, is the visual "brightness" of a source as observed by an average daylight adapted eye and is proportional to the radiant flux (radiation power emitted) of the source and the efficiency of the eye to detect the spectrum of the emitted radiation. While the eye can see a red color source, it cannot see an infrared source, and the luminous flux of the infrared source would be zero. Similarly, the eye is less efficient in the violet than in the green region, and less radiant flux is needed to have a green source at the same luminous flux as the blue source. Luminous flux $\Phi_v$ is measured in **lumens** (lm), and at a particular wavelength it is given by

*Luminous flux in lumens*

$$\Phi_v = \Phi_e \times K \times \eta_{\text{eye}}$$

where $\Phi_e$ is the radiant flux (in Watts), $K$ is a conversion constant (standardized to be 633 lm/W), $\eta_{\text{eye}}$ (also denoted as $V$) is the *luminous efficiency (luminous efficacy)* of the daylight adapted eye, which is unity at 555 nm; $\eta_{\text{eye}}$ depends on the wavelength. By definition, a 1 W light source emitting at 555 nm (green, where $\eta_{\text{eye}} = 1$) emits a luminous flux of 633 lm. The same 1 W light source at 650 nm (red), where $\eta_{\text{eye}} = 0.11$, emits only 70 lm. When we buy a light bulb, we are essentially paying for lumens because it is luminous flux that the eye perceives. A typical 60 W incandescent lamp provides roughly 900 lm. Fluorescent tubes provide more luminous flux output than incandescent lamps for the same electric power input as they have more spectral emission in the visible region and make better use of the eye's spectral sensitivity. Some examples are 100 W incandescent lamps, 1300–1800 lm, depending on the filament operating temperature (hence bulb design), and 25 W compact fluorescent lamps, 1500–1750 lm.

**Luminous efficacy of a light source** (such as a lamp) in the lighting industry is the efficiency with which an electric light source converts the input electric power (W) into an emitted luminous flux (lm). A 100 W light bulb producing 1700 lm has an efficacy of 17 lm/W. While at present the LED efficacies are below those of fluorescent tubes, rapid advances in LED technologies are bringing the expected efficacies to around 50 lm/W or higher. LEDs as solid-state lamps have much longer lifetimes and much higher reliability, and hence are expected to be more economical than incandescent and fluorescent lamps.

From left to right: Michael Faraday, Thomas Henry Huxley, Charles Wheatstone, David Brewster and John Tyndall. Professor Tyndall has been attributed with the first demonstration (1854) of light being guided along a water jet, which is based in total internal reflection.

| SOURCE: Courtesy of AIP Emilio Segrè Visual Archives, Zeleny Collection.

appendix

# C

# Major Symbols and Abbreviations

| | |
|---|---|
| $A$ | area; cross-sectional area; amplification |
| $a$ | lattice parameter; acceleration; amplitude of vibrations; half-channel thickness in a JFET (Ch. 6) |
| $a$ (subscript) | acceptor, *e.g.*, $N_a$ = acceptor concentration ($m^{-3}$) |
| ac | alternating current |
| $a_o$ | Bohr radius (0.0529 nm) |
| $A_V$, $A_P$ | voltage amplification, power amplification |
| APF | atomic packing factor |
| **B**, $B$ | magnetic field vector (T), magnetic field |
| $B$ | frequency bandwidth |
| $B_c$ | critical magnetic field |
| $B_m$ | maximum magnetic field |
| $B_o$, $B_e$ | Richardson–Dushman constant, effective Richardson–Dushman constant |
| BC | base collector |
| BCC | body-centered cubic |
| BE | base emitter |
| BJT | bipolar junction transistor |
| $C$ | capacitance; composition; the Nordheim coefficient (Ω m) |
| $c$ | speed of light ($3 \times 10^8$ m $s^{-1}$); specific heat capacity (J $K^{-1}$ $kg^{-1}$) |
| $C_{dep}$ | depletion layer capacitance |
| $C_m$ | molar heat capacity (J $K^{-1}$ $mol^{-1}$) |
| $C_{diff}$ | diffusion (storage) capacitance of a forward-biased *pn* junction |
| $c_s$ | specific heat capacity (J $K^{-1}$ $kg^{-1}$) |
| $C_v$ | heat capacity per unit volume (J $K^{-1}$ $m^{-3}$) |
| CB | conduction band; common base |
| CE | common emitter |
| CMOS | complementary MOS |
| CN | coordination number |
| CVD | chemical vapor deposition |
| $D$ | diffusion coefficient ($m^2$ $s^{-1}$); thickness; electric displacement (C $m^{-2}$) |
| $d$ | density (kg $m^{-3}$); distance; separation of the atomic planes in a crystal, separation of capacitor plates; piezoelectric coefficient; mean grain size (Ch. 2) |

| | |
|---|---|
| $d$ (subscript) | donor, *e.g.*, $N_d$ = donor concentration ($m^{-3}$) |
| dc | direct current |
| $d_{ij}$ | piezoelectric coefficients |
| $E$ | energy; electric field ($V\ m^{-1}$) (Ch. 9) |
| $E_a$, $E_d$ | acceptor and donor energy levels |
| $E_c$, $E_v$ | conduction band edge, valence band edge |
| $E_{ex}$ | exchange interaction energy |
| $E_F$, $E_{FO}$ | Fermi energy, Fermi energy at 0 K |
| $E_g$ | bandgap energy |
| $E_{mag}$ | magnetic energy |
| $\mathcal{E}$ | electric field ($V\ m^{-1}$) |
| $\mathcal{E}_{br}$ | dielectric strength or breakdown field ($V\ m^{-1}$) |
| $\mathcal{E}_{loc}$ | local electric field |
| $e$ | electronic charge ($1.602 \times 10^{-19}$ C) |
| $e$ (subscript) | electron, *e.g.*, $\mu_e$ = electron drift mobility; electronic |
| eff (subscript) | effective, *e.g.*, $\mu_{eff}$ = effective drift mobility |
| EHP | electron–hole pair |
| EM | electromagnetic |
| EMF (emf) | electromagnetic force (V) |
| $F$ | force (N); function |
| $f$ | frequency; function |
| $f(E)$ | Fermi–Dirac function |
| FCC | face-centered cubic |
| FET | field effect transistor |
| $G$ | rate of generation |
| $G_{ph}$ | rate of photogeneration |
| $G_p$ | parallel conductance ($\Omega^{-1}$) |
| $g(E)$ | density of states |
| $g$ | conductance; transconductance (A/V); piezoelectric voltage coefficient (Ch. 7) |
| $g_d$ | incremental or dynamic conductance (A/V) |
| $g_m$ | mutual transconductance (A/V) |
| $\mathbf{H}$, $H$ | magnetic field intensity (strength), magnetizing field ($A\ m^{-1}$) |
| $h$ | Planck's constant ($6.6261 \times 10^{-34}$ J s) |
| $\hbar$ | Planck's constant divided by $2\pi$ ($\hbar = 1.0546 \times 10^{-34}$ J s) |
| $h$ (subscript) | hole, *e.g.*, $\mu_h$ = hole drift mobility |
| $h_{FE}$, $h_{fe}$ | dc current gain, small-signal (ac) current gain in the common emitter configuration |
| HCP | hexagonal close-packed |
| HF | high frequency |
| $I$ | electric current (A); moment of inertia ($kg\ m^2$) (Ch. 1) |
| $\mathcal{I}$ | light intensity ($W\ m^{-2}$) |
| $I$, $i$ (subscript) | quantity related to ionic polarization |
| $I_{br}$ | breakdown current |
| $I_B$, $I_C$, $I_E$ | base, collector, and emitter currents in a BJT |

| | |
|---|---|
| $i$ | instantaneous current (A); small-signal (ac) current, $i = \delta I$ |
| $i$ (subscript) | intrinsic, *e.g.*, $n_i$ = intrinsic concentration |
| $i_b, i_c, i_e$ | small signal base, collector, and emitter currents ($\delta I_B, \delta I_C, \delta I_E$) in a BJT |
| IC | integrated circuit |
| $J$ | current density (A m$^{-2}$) |
| $\mathbf{J}$ | total angular momentum vector |
| $j$ | imaginary constant: $\sqrt{-1}$ |
| $J_c$ | critical current density (A m$^{-2}$) |
| $J_p$ | pyroelectric current density |
| JFET | junction FET |
| $K$ | spring constant (Ch. 1); phonon wavevector (m$^{-1}$); bulk modulus (Pa); dielectric constant (Ch. 7) |
| $k$ | Boltzmann constant ($k = R/N_A = 1.3807 \times 10^{-23}$ J K$^{-1}$); wavenumber ($k = 2\pi/\lambda$), wavevector (m$^{-1}$); electromechanical coupling factor (Ch. 7) |
| $KE$ | kinetic energy |
| $\mathbf{L}$ | total orbital angular momentum |
| $L$ | length; inductance |
| $\ell$ | length; mean free path; orbital angular momentum quantum number |
| $L_{ch}$ | channel length in a FET |
| $L_e, L_h$ | electron and hole diffusion lengths |
| $\ell_n, \ell_p$ | lengths of the $n$- and $p$-regions outside depletion region in a $pn$ junction |
| ln ($x$) | natural logarithm of $x$ |
| LCAO | linear combination of atomic orbitals |
| $\mathbf{M}, M$ | magnetization vector (A m$^{-1}$), magnetization (A m$^{-1}$) |
| $M$ | multiplication in avalanche effect |
| $M_{at}$ | relative atomic mass; atomic mass; "atomic weight" (g mol$^{-1}$) |
| $M_r$ | remanent or residual magnetization (A m$^{-1}$); reduced mass of two bodies $A$ and $B$, $M_r = M_A M_B/(M_A + M_B)$ |
| $M_{sat}$ | saturation magnetization (A m$^{-1}$) |
| $m$ | mass (kg) |
| $m_e$ | mass of the electron in free space ($9.10939 \times 10^{-31}$ kg) |
| $m_e^*$ | effective mass of the electron in a crystal |
| $m_h^*$ | effective mass of a hole in a crystal |
| $m_\ell$ | magnetic quantum number |
| $m_s$ | spin magnetic quantum number |
| MOS (MOST) | metal-oxide-semiconductor (transistor) |
| MOSFET | metal-oxide-semiconductor FET |
| $N$ | number of atoms or molecules; number of atoms per unit volume (m$^{-3}$) (Chs. 7 and 9); number of turns on a coil (Ch. 8) |
| $N_A$ | Avogadro's number ($6.022 \times 10^{23}$ mol$^{-1}$) |
| $n$ | electron concentration (number per unit volume); atomic concentration; principal quantum number; integer number; refractive index (Ch. 9) |
| $n^+$ | heavily doped $n$-region |
| $n_{at}$ | number of atoms per unit volume |

| | |
|---|---|
| $N_c$, $N_v$ | effective density of states at the conduction and valence band edges ($m^{-3}$) |
| $N_d$, $N_d^+$ | donor and ionized donor concentrations ($m^{-3}$) |
| $n_e$, $n_o$ | refractive index for extraordinary and ordinary waves in a birefringent crystal |
| $n_i$ | intrinsic concentration ($m^{-3}$) |
| $n_{no}$, $p_{po}$ | equilibrium majority carrier concentrations ($m^{-3}$) |
| $n_{po}$, $p_{no}$ | equilibrium minority carrier concentrations ($m^{-3}$) |
| $N_s$ | concentration of electron scattering centers |
| $n_v$ | velocity density function; vacancy concentration ($m^{-3}$) |
| $P$ | probability; pressure (Pa); power (W) or power loss (W) |
| $\mathbf{p}$, $p$ | electric dipole moment (C m) |
| $p$ | hole concentration ($m^{-3}$); momentum (kg m $s^{-1}$); pyroelectric coefficient (C $m^{-2}$ $K^{-1}$) (Ch. 7) |
| $p^+$ | heavily doped $p$-region |
| $p_{\text{av}}$ | average dipole moment per molecule |
| $p_e$ | electron momentum (kg m $s^{-1}$) |
| $PE$ | potential energy |
| $p_{\text{induced}}$ | induced dipole moment (C m) |
| $p_o$ | permanent dipole moment (C m) |
| PET | polyester, polyethylene terephthalate |
| PZT | lead zirconate titanate |
| $Q$ | charge (C); heat (J); quality factor |
| $Q'$ | rate of heat flow (W) |
| $q$ | charge (C); an integer number used in lattice vibrations (Ch. 4) |
| $R$ | gas constant ($N_A k = 8.3145$ J $mol^{-1}$ $K^{-1}$); resistance; radius; reflection coefficient (Ch. 3); rate of recombination (Ch. 5) |
| $\mathsf{R}$ | reflectance (Ch. 9) |
| $\mathcal{R}_I$, $\mathcal{R}_V$ | pyroelectric current and voltage responsivities |
| $\mathbf{r}$ | position vector |
| $r$ | radial distance; radius; interatomic separation; resistance per unit length |
| $\mathsf{r}$ | reflection coefficient (Ch. 9) |
| $R_H$ | Hall coefficient ($m^3$ $C^{-1}$) |
| $r_o$ | bond length, equilibrium separation |
| rms | root mean square |
| $\mathbf{S}$ | total spin momentum, intrinsic angular momentum; Poynting vector (Ch. 9) |
| $S$ | cross-sectional area of a scattering center; Seebeck coefficient, thermoelectric power (V $m^{-1}$); strain (Ch. 7) |
| $S_{\text{band}}$ | number of states per unit volume in the band |
| $S_j$ | strain along direction $j$ |
| SCL | space charge layer |
| $T$ | temperature in Kelvin; transmission coefficient |
| $\mathsf{T}$ | transmittance |
| $t$ | time (s); thickness (m) |
| $\mathsf{t}$ | transmission coefficient |
| $\tan\delta$ | loss tangent |

| | |
|---|---|
| $T_C$ | Curie temperature |
| $T_c$ | critical temperature (K) |
| $T_j$ | mechanical stress along direction $j$ (Pa) |
| TC | thermocouple |
| TCC | temperature coefficient of capacitance ($K^{-1}$) |
| TCR | temperature coefficient of resistivity ($K^{-1}$) |
| $U$ | total internal energy |
| $u$ | mean speed (of electron) (m $s^{-1}$) |
| $V$ | voltage; volume; $PE$ function of the electron, $PE(x)$ |
| $V_{br}$ | breakdown voltage |
| $V_o$ | built-in voltage |
| $V_P$ | pinch-off voltage |
| $V_r$ | reverse bias voltage |
| $v$, $\mathsf{v}$ | velocity (m $s^{-1}$); instantaneous voltage (V) |
| $\overline{v^2}$ | mean square velocity; mean square voltage |
| $v_{dx}$ | drift velocity in the $x$ direction |
| $v_e$, $v_{rms}$ | effective velocity or rms velocity of the electron |
| $v_F$ | Fermi speed |
| $v_g$, $\mathsf{v}_g$ | group velocity |
| $v_{th}$ | thermal velocity |
| VB | valence band |
| $W$ | width; width of depletion layer with applied voltage; dielectric loss |
| $W_o$ | width of depletion region with no applied voltage |
| $W_n$, $W_p$ | width of depletion region on the $n$-side and on the $p$-side with no applied voltage |
| $X$ | atomic fraction |
| $Y$ | admittance ($\Omega^{-1}$); Young's modulus (Pa) |
| $Z$ | impedance ($\Omega$); atomic number, number of electrons in the atom |
| $\alpha$ | polarizability; temperature coefficient of resistivity ($K^{-1}$); absorption coefficient ($m^{-1}$); gain or current transfer ratio from emitter to collector of a BJT |
| $\beta$ | current gain $I_C/I_B$ of a BJT; Bohr magneton ($9.2740 \times 10^{-24}$ J $T^{-1}$); spring constant (Ch. 4) |
| $\beta_S$ | Schottky coefficient |
| $\gamma$ | emitter injection efficiency (Ch. 6); gyromagnetic ratio (Ch. 8); Grüneisen parameter (Ch. 4); loss coefficient in the Lorentz oscillator model |
| $\Gamma$, $\Gamma_{ph}$ | flux ($m^{-2}$ $s^{-1}$), photon flux (photons $m^{-2}$ $s^{-1}$) |
| $\delta$ | small change; skin depth (Ch. 2); loss angle (Ch. 7); domain wall thickness (Ch. 8); penetration depth (Ch. 9) |
| $\Delta$ | change, excess (*e.g.*, $\Delta n$ = excess electron concentration) |
| $\nabla^2$ | $\partial^2/\partial x^2 + \partial^2/\partial y^2 + \partial^2/\partial z^2$ |
| $\varepsilon$ | $\varepsilon_o\varepsilon_r$, permittivity of a medium (C $V^{-1}$ $m^{-1}$ or F $m^{-1}$); elastic strain |
| $\varepsilon_o$ | permittivity of free space or absolute permittivity ($8.8542 \times 10^{-12}$ C $V^{-1}$ $m^{-1}$ or F $m^{-1}$) |

| | |
|---|---|
| $\varepsilon_r$ | relative permittivity or dielectric constant |
| $\eta$ | efficiency; quantum efficiency; ideality factor |
| $\theta$ | angle; an angular spherical coordinate; thermal resistance; angle between a light ray and normal to a surface (Ch. 9) |
| $\kappa$ | thermal conductivity (W $m^{-1}$ $K^{-1}$); dielectric constant |
| $\lambda$ | wavelength (m); thermal coefficient of linear expansion ($K^{-1}$); electron mean free path in the bulk crystal (Ch. 2); characteristic length (Ch. 8) |
| $\boldsymbol{\mu}$, $\mu$ | magnetic dipole moment (A $m^2$) (Ch. 3) |
| $\mu$ | $\mu_o\mu_r$, magnetic permeability (H $m^{-1}$); chemical potential (Ch. 5) |
| $\mu_o$ | absolute permeability ($4\pi \times 10^{-7}$ H $m^{-1}$) |
| $\mu_r$ | relative permeability |
| $\boldsymbol{\mu}_m$, $\mu_m$ | magnetic dipole moment (A $m^2$) (Ch. 8) |
| $\mu_d$ | drift mobility ($m^2$ $V^{-1}$ $s^{-1}$) |
| $\mu_h$, $\mu_e$ | hole drift mobility, electron drift mobility ($m^2$ $V^{-1}$ $s^{-1}$) |
| $\nu$ | frequency (Hz); Poisson's ratio; volume fraction (Ch. 7) |
| $\pi$ | pi, 3.14159. . . ; piezoresistive coefficient ($Pa^{-1}$) |
| $\pi_L$, $\pi_T$ | longitudinal and transverse piezoresistive coefficients ($Pa^{-1}$) |
| $\Pi$ | Peltier coefficient (V) |
| $\rho$ | resistivity ($\Omega$ m); density (kg $m^{-3}$); charge density (C $m^{-3}$) |
| $\rho_E$ | energy density (J $m^{-3}$) |
| $\rho_{net}$ | net space charge density (C $m^{-3}$) |
| $\rho J^2$ | Joule heating per unit volume (W $m^{-3}$) |
| $\sigma$ | electrical conductivity ($\Omega^{-1}$ $m^{-1}$); surface concentration of charge (C $m^{-2}$) (Ch. 7) |
| $\sigma_P$ | polarization charge density (C $m^{-2}$) |
| $\sigma_o$ | free surface charge density (C $m^{-2}$) |
| $\sigma_S$ | Stefan's constant ($5.670 \times 10^{-8}$ W $m^{-2}$ $K^{-4}$) |
| $\tau$ | time constant; mean electron scattering time; relaxation time; torque (N m) |
| $\tau_g$ | mean time to generate an electron–hole pair |
| $\phi$ | angle; an angular spherical coordinate |
| $\Phi$ | work function (J or eV), magnetic flux (Wb) |
| $\Phi_e$ | radiant flux (W) |
| $\Phi_m$ | metal work function (J or eV) |
| $\Phi_n$ | energy required to remove an electron from an *n*-type semiconductor (J or eV) |
| $\Phi_v$ | luminous flux (lumens) |
| $\chi$ | volume fraction; electron affinity; susceptibility ($\chi_e$ is electrical; $\chi_m$ is magnetic) |
| $\Psi(x, t)$ | total wavefunction |
| $\psi(x)$ | spatial dependence of the electron wavefunction under steady-state conditions |
| $\psi_k(x)$ | Bloch wavefunction, electron wavefunction in a crystal |
| $\psi_{hyb}$ | hybrid orbital |
| $\omega$ | angular frequency ($2\pi v$); oscillation frequency (rad $s^{-1}$) |
| $\omega_I$ | ionic polarization resonance frequency (angular) |
| $\omega_o$ | resonance or natural frequency (angular) of an oscillating system. |

appendix

# D

# Elements to Uranium

| Element | Symbol | Z | Atomic Mass (g mol$^{-1}$) | Electronic Structure | Density (g cm$^{-3}$) (*at 0 °C, 1 atm) | Crystal in Solid State |
|---|---|---|---|---|---|---|
| Hydrogen | H | 1 | 1.008 | $1s^1$ | 0.00009* | HCP |
| Helium | He | 2 | 4.003 | $1s^2$ | 0.00018* | FCP |
| Lithium | Li | 3 | 6.941 | $[He]2s^1$ | 0.54 | BCC |
| Beryllium | Be | 4 | 9.012 | $[He]2s^2$ | 1.85 | HCP |
| Boron | B | 5 | 10.81 | $[He]2s^2p^1$ | 2.5 | Rhombohedral |
| Carbon | C | 6 | 12.01 | $[He]2s^2p^2$ | 2.3 | Hexagonal |
| Nitrogen | N | 7 | 14.007 | $[He]2s^2p^3$ | 0.00125* | HCP |
| Oxygen | O | 8 | 16.00 | $[He]2s^2p^4$ | 0.00143* | Monoclinic |
| Fluorine | F | 9 | 18.99 | $[He]2s^2p^5$ | 0.00170* | Monoclinic |
| Neon | Ne | 10 | 20.18 | $[He]2s^2p^6$ | 0.00090* | FCC |
| Sodium | Na | 11 | 22.99 | $[Ne]3s^1$ | 0.97 | BCC |
| Magnesium | Mg | 12 | 24.31 | $[Ne]3s^2$ | 1.74 | HCP |
| Aluminum | Al | 13 | 26.98 | $[Ne]3s^2p^1$ | 2.70 | FCC |
| Silicon | Si | 14 | 28.09 | $[Ne]3s^2p^2$ | 2.33 | Diamond |
| Phosphorus | P | 15 | 30.97 | $[Ne]3s^2p^3$ | 1.82 | Triclinic |
| Sulfur | S | 16 | 32.06 | $[Ne]3s^2p^4$ | 2.0 | Orthorhombic |
| Chlorine | Cl | 17 | 35.45 | $[Ne]3s^2p^5$ | 0.0032* | Orthorhombic |
| Argon | Ar | 18 | 39.95 | $[Ne]3s^2p^6$ | 0.0018* | FCC |
| Potassium | K | 19 | 39.09 | $[Ar]4s^1$ | 0.86 | BCC |
| Calcium | Ca | 20 | 40.08 | $[Ar]4s^2$ | 1.55 | FCC |
| Scandium | Sc | 21 | 44.96 | $[Ar]3d^14s^2$ | 3.0 | HCP |
| Titanium | Ti | 22 | 47.87 | $[Ar]3d^24s^2$ | 4.5 | HCP |
| Vanadium | V | 23 | 50.94 | $[Ar]3d^34s^2$ | 5.8 | BCC |
| Chromium | Cr | 24 | 52.00 | $[Ar]3d^54s^1$ | 7.19 | BCC |
| Manganese | Mn | 25 | 54.95 | $[Ar]3d^54s^2$ | 7.43 | BCC |
| Iron | Fe | 26 | 55.85 | $[Ar]3d^64s^2$ | 7.86 | BCC |
| Cobalt | Co | 27 | 58.93 | $[Ar]3d^74s^2$ | 8.90 | HCP |
| Nickel | Ni | 28 | 58.69 | $[Ar]3d^84s^2$ | 8.90 | FCC |
| Copper | Cu | 29 | 63.55 | $[Ar]3d^{10}4s^1$ | 8.96 | FCC |
| Zinc | Zn | 30 | 65.39 | $[Ar]3d^{10}4s^2$ | 7.14 | HCP |
| Gallium | Ga | 31 | 69.72 | $[Ar]3d^{10}4s^2p^1$ | 5.91 | Orthorhombic |
| Germanium | Ge | 32 | 72.61 | $[Ar]3d^{10}4s^2p^2$ | 5.32 | Diamond |

| Element | Symbol | $Z$ | Atomic Mass (g mol$^{-1}$) | Electronic Structure | Density (g cm$^{-3}$) (*at 0 °C, 1 atm) | Crystal in Solid State |
|---|---|---|---|---|---|---|
| Arsenic | As | 33 | 74.92 | $[Ar]3d^{10}4s^2p^3$ | 5.72 | Rhombohedral |
| Selenium | Se | 34 | 78.96 | $[Ar]3d^{10}4s^2p^4$ | 4.80 | Hexagonal |
| Bromine | Br | 35 | 79.90 | $[Ar]3d^{10}4s^2p^5$ | 3.12 | Orthorhombic |
| Krypton | Kr | 36 | 83.80 | $[Ar]3d^{10}4s^2p^6$ | 3.74 | FCC |
| Rubidium | Rb | 37 | 85.47 | $[Kr]5s^1$ | 1.53 | BCC |
| Strontium | Sr | 38 | 87.62 | $[Kr]5s^2$ | 2.6 | FCC |
| Yttrium | Y | 39 | 88.90 | $[Kr]4d^15s^2$ | 4.5 | HCP |
| Zirconium | Zr | 40 | 91.22 | $[Kr]4d^25s^2$ | 6.50 | HCP |
| Niobium | Nb | 41 | 92.91 | $[Kr]4d^45s^1$ | 8.55 | BCC |
| Molybdenum | Mo | 42 | 95.94 | $[Kr]4d^55s^1$ | 10.2 | BCC |
| Technetium | Tc | 43 | (97.91) | $[Kr]4d^55s^2$ | 11.5 | HCP |
| Ruthenium | Ru | 44 | 101.07 | $[Kr]4d^75s^1$ | 12.2 | HCP |
| Rhodium | Rh | 45 | 102.91 | $[Kr]4d^85s^1$ | 12.4 | FCC |
| Palladium | Pd | 46 | 106.42 | $[Kr]4d^{10}$ | 12.0 | FCC |
| Silver | Ag | 47 | 107.87 | $[Kr]4d^{10}5s^1$ | 10.5 | FCC |
| Cadmium | Cd | 48 | 112.41 | $[Kr]4d^{10}5s^2$ | 8.65 | HCP |
| Indium | In | 49 | 114.82 | $[Kr]4d^{10}5s^2p^1$ | 7.31 | FCT |
| Tin | Sn | 50 | 118.71 | $[Kr]4d^{10}5s^2p^2$ | 7.30 | BCT |
| Antimony | Sb | 51 | 121.75 | $[Kr]4d^{10}5s^2p^3$ | 6.68 | Rhombohedral |
| Tellurium | Te | 52 | 127.60 | $[Kr]4d^{10}5s^2p^4$ | 6.24 | Hexagonal |
| Iodine | I | 53 | 126.91 | $[Kr]4d^{10}5s^2p^5$ | 4.92 | Orthorhombic |
| Xenon | Xe | 54 | 131.29 | $[Kr]4d^{10}5s^2p^6$ | 0.0059* | FCC |
| Cesium | Cs | 55 | 132.90 | $[Xe]6s^1$ | 1.87 | BCC |
| Barium | Ba | 56 | 137.33 | $[Xe]6s^2$ | 3.62 | BCC |
| Lanthanum | La | 57 | 138.91 | $[Xe]5d^16s^2$ | 6.15 | HCP |
| Cerium | Ce | 58 | 140.12 | $[Xe]4f^15d^16s^2$ | 6.77 | FCC |
| Praseodymium | Pr | 59 | 140.91 | $[Xe]4f^36s^2$ | 6.77 | HCP |
| Neodymium | Nd | 60 | 144.24 | $[Xe]4f^46s^2$ | 7.00 | HCP |
| Promethium | Pm | 61 | (145) | $[Xe]4f^56s^2$ | 7.26 | Hexagonal |
| Samarium | Sm | 62 | 150.4 | $[Xe]4f^66s^2$ | 7.5 | Rhombohedral |
| Europium | Eu | 63 | 151.97 | $[Xe]4f^76s^2$ | 5.24 | BCC |
| Gadolinium | Gd | 64 | 157.25 | $[Xe]4f^75d^16s^2$ | 7.90 | HCP |
| Terbium | Tb | 65 | 158.92 | $[Xe]4f^96s^2$ | 8.22 | HCP |
| Dysprosium | Dy | 66 | 162.50 | $[Xe]4f^{10}6s^2$ | 8.55 | HCP |
| Holmium | Ho | 67 | 164.93 | $[Xe]4f^{11}6s^2$ | 8.80 | HCP |
| Erbium | Er | 68 | 167.26 | $[Xe]4f^{12}6s^2$ | 9.06 | HCP |
| Thulium | Tm | 69 | 168.93 | $[Xe]4f^{13}6s^2$ | 9.32 | HCP |
| Ytterbium | Yb | 70 | 173.04 | $[Xe]4f^{14}6s^2$ | 6.90 | FCC |
| Lutetium | Lu | 71 | 174.97 | $[Xe]4f^{14}5d^16s^2$ | 9.84 | HCP |
| Hafnium | Hf | 72 | 178.49 | $[Xe]4f^{14}5d^26s^2$ | 13.3 | HCP |
| Tantalum | Ta | 73 | 180.95 | $[Xe]4f^{14}5d^36s^2$ | 16.4 | BCC |
| Tungsten | W | 74 | 183.84 | $[Xe]4f^{14}5d^46s^2$ | 19.3 | BCC |
| Rhenium | Re | 75 | 186.21 | $[Xe]4f^{14}5d^56s^2$ | 21.0 | HCP |

| Element | Symbol | Z | Atomic mass (g mol$^{-1}$) | Electronic Structure | Density (g cm$^{-3}$) (*at 0 °C, 1 atm) | Crystal in Solid State |
|---|---|---|---|---|---|---|
| Osmium | Os | 76 | 190.2 | [Xe]$4f^{14}5d^6 6s^2$ | 22.6 | HCP |
| Iridium | Ir | 77 | 192.22 | [Xe]$4f^{14}5d^7 6s^2$ | 22.5 | FCC |
| Platinum | Pt | 78 | 195.08 | [Xe]$4f^{14}5d^9 6s^1$ | 21.4 | FCC |
| Gold | Au | 79 | 196.97 | [Xe]$4f^{14}5d^{10}6s^1$ | 19.3 | FCC |
| Mercury | Hg | 80 | 200.59 | [Xe]$4f^{14}5d^{10}6s^2$ | 13.55 | Rhombohedral |
| Thallium | Tl | 81 | 204.38 | [Xe]$4f^{14}5d^{10}6s^2p^1$ | 11.8 | HCP |
| Lead | Pb | 82 | 207.2 | [Xe]$4f^{14}5d^{10}6s^2p^2$ | 11.34 | FCC |
| Bismuth | Bi | 83 | 208.98 | [Xe]$4f^{14}5d^{10}6s^2p^3$ | 9.8 | Rhombohedral |
| Polonium | Po | 84 | (209) | [Xe]$4f^{14}5d^{10}6s^2p^4$ | 9.2 | SC |
| Astatine | At | 85 | (210) | [Xe]$4f^{14}5d^{10}6s^2p^5$ | — | — |
| Radon | Rn | 86 | (222) | [Xe]$4f^{14}5d^{10}6s^2p^6$ | 0.0099* | Rhombohedral |
| Francium | Fr | 87 | (223) | [Rn]$7s^1$ | — | — |
| Radium | Ra | 88 | 226.02 | [Rn]$7s^2$ | 5 | BCC |
| Actinium | Ac | 89 | 227.02 | [Rn]$6d^1 7s^2$ | 10.0 | FCC |
| Thorium | Th | 90 | 232.04 | [Rn]$6d^2 7s^2$ | 11.7 | FCC |
| Protactinium | Pa | 91 | (231.03) | [Rn]$5f^2 6d^1 7s^2$ | 15.4 | BCT |
| Uranium | U | 92 | (238.05) | [Rn]$5f^3 6d^1 7s^2$ | 19.07 | Orthorhombic |

"I don't really start until I get my proofs back from the printers. Then I can begin serious writing."

John Maynard Keynes (1883–1946)

appendix

# E

# Constants and Useful Information

## Physical Constants

| | | |
|---|---|---|
| Atomic mass unit | amu | $1.66054 \times 10^{-27}$ kg |
| Avogadro's number | $N_A$ | $6.02214 \times 10^{23}$ mol$^{-1}$ |
| Bohr magneton | $\beta$ | $9.2740 \times 10^{-24}$ J T$^{-1}$ |
| Boltzmann constant | $k$ | $1.3807 \times 10^{-23}$ J K$^{-1}$ $= 8.6174 \times 10^{-5}$ eV K$^{-1}$ |
| Electron mass in free space | $m_e$ | $9.10939 \times 10^{-31}$ kg |
| Electron charge | $e$ | $1.60218 \times 10^{-19}$ C |
| Gas constant | $R$ | 8.3145 J K$^{-1}$ mol$^{-1}$ or m$^3$ Pa K$^{-1}$ mol$^{-1}$ |
| Gravitational constant | $G$ | $6.6742 \times 10^{-11}$ N m$^2$ kg$^{-2}$ |
| Permeability of vacuum or absolute permeability | $\mu_o$ | $4\pi \times 10^{-7}$ H m$^{-1}$ (or Wb A$^{-1}$ m$^{-1}$) |
| Permittivity of vacuum or absolute permittivity | $\varepsilon_o$ | $8.8542 \times 10^{-12}$ F m$^{-1}$ |
| Planck's constant | $h$ | $6.626 \times 10^{-34}$ J s $= 4.136 \times 10^{-15}$ eV s |
| Planck's constant/$2\pi$ | $\hbar$ | $1.055 \times 10^{-34}$ J s $= 6.582 \times 10^{-16}$ eV s |
| Proton rest mass | $m_p$ | $1.67262 \times 10^{-27}$ kg |
| Rydberg constant | $R_\infty$ | $1.0974 \times 10^{7}$ m$^{-1}$ |
| Speed of light | $c$ | $2.9979 \times 10^{8}$ m s$^{-1}$ |
| Stefan's constant | $\sigma_s$ | $5.6704 \times 10^{-8}$ W m$^{-2}$ K$^{-4}$ |

## Useful Information

| | | |
|---|---|---|
| Acceleration due to gravity at 45° latitude | $g$ | 9.81 m s$^{-2}$ |
| $kT$ at $T = 293$ K (20 °C) | $kT$ | 0.02525 eV |
| $kT$ at $T = 300$ K (27 °C) | $kT$ | 0.02585 eV |
| Bohr radius | $a_o$ | 0.0529 nm |
| 1 angstrom | Å | $10^{-10}$ m |
| 1 micron | μm | $10^{-6}$ m |

1 eV $= 1.6022 \times 10^{-19}$ J

1 kJ mol$^{-1}$ $= 0.010364$ eV atom$^{-1}$

1 atmosphere (pressure) $= 1.013 \times 10^{5}$ Pa

## LED Colors

The table gives the wavelength ranges and colors as usually specified for LEDs.

| Color | Blue | Emerald green | Green | Yellow | Amber | Orange | Red orange | Red | Deep red | Infrared |
|---|---|---|---|---|---|---|---|---|---|---|
| $\lambda$ (nm) | $\lambda < 500$ | 530–564 | 565–579 | 580–587 | 588–594 | 595–606 | 607–615 | 616–632 | 633–700 | $\lambda > 700$ |

## Visible Spectrum

The table gives the typical wavelength ranges and color perception by an average person.

| Color | Violet | Blue | Green | Yellow | Orange | Red |
|---|---|---|---|---|---|---|
| $\lambda$ (nm) | 390–455 | 455–492 | 492–577 | 577–597 | 597–622 | 622–780 |

## Complex Numbers

$j = (-1)^{1/2} \qquad j^2 = -1$

$\exp(j\theta) = \cos\theta + j\sin\theta$

$Z = a + jb = re^{j\theta} \qquad r = (a^2 + b^2)^{1/2} \qquad \tan\theta = \dfrac{b}{a}$

$Z^* = a - jb = re^{-j\theta} \qquad \mathrm{Re}(Z) = a \qquad \mathrm{Im}(Z) = b$

$\text{Magnitude}^2 = |Z|^2 = ZZ^* = a^2 + b^2 \qquad \text{Argument} = \theta = \arctan\left(\dfrac{b}{a}\right)$

$\cos\theta = \dfrac{1}{2}(e^{j\theta} + e^{-j\theta}) \qquad \sin\theta = \dfrac{1}{2j}(e^{j\theta} - e^{-j\theta})$

## Expansions

$$e^x = 1 + x + \frac{1}{2!}x^2 + \frac{1}{3!}x^3 + \cdots$$

$$(1+x)^n = 1 + nx + \frac{n(n-1)}{2!}x^2 + \frac{n(n-1)(n-2)}{3!}x^3 + \cdots$$

Small $x$: $\quad (1+x)^n \approx 1 + nx \qquad \sin x \approx x \qquad \tan x \approx x \qquad \cos x \approx 1$

Small $\Delta x$ in $x = x_o + \Delta x$: $\quad f(x) \approx f(x_o) + \Delta x \left(\dfrac{df}{dx}\right)_{x_o}$

# index

"We have a habit in writing articles published in scientific journals to make the work as finished as possible, to cover up all the tracks, to not worry about the blind alleys or describe how you had the wrong idea first, and so on. So there isn't any place to publish, in a dignified manner, what you actually did in order to get to do the work."

*Richard P. Feynman*
*Nobel Lecture, 1966*